THE SYNTHESIS, ASSEMBLY AND TURNOVER OF CELL SURFACE COMPONENTS

4

CELL SURFACE REVIEWS
VOLUME 4

NORTH-HOLLAND PUBLISHING COMPANY
AMSTERDAM · NEW YORK · OXFORD

THE SYNTHESIS, ASSEMBLY AND TURNOVER OF CELL SURFACE COMPONENTS

Edited by

GEORGE POSTE

*Department of Experimental Pathology,
Roswell Park Memorial Institute, Buffalo, N.Y.*

and

GARTH L. NICOLSON

*Department of Developmental and Cell Biology,
University of California, Irvine, CA*

1977

NORTH-HOLLAND PUBLISHING COMPANY
AMSTERDAM · NEW YORK · OXFORD

PUBLISHED BY:
Elsevier/North-Holland Biomedical Press
335 Jan Van Galenstraat, P.O. Box 211
Amsterdam, The Netherlands

SOLE DISTRIBUTORS FOR THE U.S.A. AND CANADA:
Elsevier North-Holland, Inc.
52 Vanderbilt Avenue
New York, New York 10017

Library of Congress Cataloging in Publication Data

Main entry under title:

The Synthesis, assembly, and turnover of cell sur-
    face components.

    (Cell surface reviews; v. 4)
    Bibliography: p.
    Includes index.
    1.   Plasma membranes.   2.   Cytochemistry.
3.   Biosynthesis.   I.   Poste, George.   II.   Nicolson,
Garth L. [DNLM: 1. Cell membrane—Metabolism.
2. Cell wall—Metabolism. W1 CE1283 v. 4 / QH601
P857s]
QH601.S95   574.8'75   77-2016
ISBN 0-444-00232-4

General preface

Research on membranes and cell surfaces today occupies center stage in many areas of biology and medicine. This dominant position reflects the growing awareness that many important biological processes in animal and plant cells and in microorganisms are mediated by these structures. The extraordinary and unprecedented expansion of knowledge in molecular biology, genetics, biochemistry, cell biology, microbiology and immunology over the last fifteen years has resulted in dramatic advances in our understanding of the properties of the cell surface and heightened our appreciation of the subtle, yet complex, nature of cell surface organization.

The rapid growth of interest in all facets of research on cell membranes and surfaces owes much to the convergence of ideas and results from seemingly disparate disciplines. This, together with the recognition of common patterns of biological organization in membranes from highly different forms of life, has led to a situation in which the sharp boundaries between the classical biological disciplines are rapidly disappearing. The investigator interested in cell surfaces must be at home in many fields, ranging from the detailed biochemical and biophysical properties of the molecules and macromolecules found in membranes to morphological and phenomenological descriptions of cellular structure and cell-to-cell interactions. Given the broad front on which research on cell surfaces is being pursued, it is not surprising that the relevant literature is scattered in a diverse range of journals and books, making it increasingly difficult for the active investigator to collate material from several areas of research. Thus, while scientists are becoming increasingly specialized in their techniques, and in the nature of the problems they study, they must interpret their results against an intellectual and conceptual background of rapidly expanding dimension. It is with these conflicting demands and needs in mind that this series, to be known under the collective title of CELL SURFACE REVIEWS, was conceived.

CELL SURFACE REVIEWS will present up-to-date surveys of recent advances in our understanding of membranes and cell surfaces. Each volume will contain authoritative and topical reviews by investigators who have contributed to progress in their respective research fields. While individual reviews will provide comprehensive coverage of specialized topics, all of the reviews published within each volume will be related to an overall common theme. This format represents a departure from that adopted by most of the existing series of "review" publications which usually provide heterogeneous collections of reviews on unrelated topics. While this latter format is considerably more convenient from an editorial standpoint, we feel that publication together of a number of related reviews will better serve the stated aims of this series—to bridge the information and specialization "gap" among investigators in related areas. Each volume will therefore present a fairly complete and critical survey of the more important and recent advances in well defined topics in biology and medicine. The level will be advanced, directed primarily to the needs of the research worker and graduate students.

Editorial policy will be to impose as few restrictions as possible on contributors. This is appropriate since the volumes published in this series will represent collections of review articles and will not be definitive monographs dealing with all aspects of the selected subject. Contributors will be encouraged, however, to provide comprehensive, critical reviews that attempt to integrate the available data into a broad conceptual framework. Emphasis will also be given to identification of major problems demanding further study and the possible avenues by which these might be investigated. Scope will also be offered for the presentation of new and challenging ideas and hypotheses for which complete evidence is still lacking.

The first four volumes of this series will be published within one year, after which volumes will appear at approximately one year intervals.

George Poste
Garth L. Nicolson
Editors

Contents of previous volumes

Contents of Forthcoming Volumes

Preface

The concept that the various structural elements of the living cell are subject to constant turnover and renewal is now sufficiently well established that departures from this scheme are viewed as exceptions rather than the rule. Both eukaryotic and prokaryotic cells must constantly resynthesize, assemble, and organize their surfaces in response to a myriad of changing environmental stimuli and genetically programmed routines. Such processes not only play a vital role in maintaining the specificity of cell surface organization but also contribute to dynamic changes in cell surface organization and to the expression of new surface properties. Often the biosynthesis of structural precursors for the plasma membrane and associated structures such as cell walls takes place some distance from their final point of assembly at the cell surface, and cells have evolved varied and complex systems for transporting precursors between these sites. Once assembled, surface structures display an extraordinary degree of functional diversity, ranging from the tremendously strong supporting walls of plant cells and bacteria to the highly specialized plasma membranes found in the different cell populations in multicellular organisms. The mechanisms by which plasma membranes and other surface structures are synthesized and assembled are fundamental to the genesis of this diversity—each cell type having evolved its own characteristic sequence of synthesis and assembly. The complexity and diversity of these synthetic and assembly processes are matched, however, by the equally wide range of mechanisms employed in the turnover, reutilization and degradation of surface components.

These aspects of cell surface organization have been a major area of research interest over the last twenty years. The ubiquity of synthetic and turnover processes at the surface of living cells has also dictated that information relevant to these important phenomena has been obtained from studies on widely differing cell types spanning the entire taxonomic spectrum. This, together with the expansion of the literature on this subject and the wide range of journals and books in which information is now scattered, has made it increasingly difficult for the active investigator to keep informed of developments even in his own specialization let alone of progress in other fields that might also be pertinent. These considerations were uppermost in our minds when selecting the broad ranging group of articles on membrane synthesis, assembly and turnover for publication in the present volume, the fourth in the *Cell Surface Reviews* series. The thirteen chapters, written by leading authorities, provide a comprehensive survey of the metabolism of cell surface components in both eukaryotic and prokaryotic cells. The contributors have been encouraged to point out possible similarities as well as dissimilarities in the mechanisms utilized by different cell types in the renewal and turnover of their surfaces. Emphasis has also been given throughout the volume to those aspects of membrane assembly and turnover that may be of importance in influencing the dynamic properties of membranes and associated structures. This perspective makes the present volume a useful companion publication to the third volume of *Cell Surface Reviews* published recently which reviewed membrane dynamics from a structural standpoint. Together these two volumes provide one of the most complete treatments of membrane organization and dynamics available to date and they should serve as valuable sources of ideas and information.

We thank the contributors to this volume and express our appreciation of their willingness to accept editorial suggestions. We would again like to thank Adele Brodginski, Judy Kaiser, Molly Terhaar, and Shirley Guagliardi for their untiring help in editing the manuscripts, and Ms. Margaret M. Quinlin of Elsevier/North-Holland for her assistance throughout the production stages.

August, 1976

G. Poste
Buffalo, New York

G.L. Nicolson
Irvine, California

List of contributors

The number in parentheses indicates the page on which the author's contribution begins.

John G. BLUEMINK (403), Hubrecht Laboratory, International Embryological Institute, Uppsalalaan 8, Utrecht, The Netherlands.

Geoffrey M.W. COOK (85), Strangeways Research Laboratory, Worts Causeway, Cambridge CB1 4RN, England.

A.J. CRANG (247), Department of Biology, University of York, Heslington, York YO1 5DD, England.

L. DANEO-MOORE (597), Department of Microbiology and Immunology, Temple University School of Medicine, Philadelphia, Pennsylvania 19140, U.S.A.

Siegfried W. DE LAAT (403), Hubrecht Laboratory, International Embryological Institute, Uppsalalaan 8, Utrecht, The Netherlands.

Darrell DOYLE (137), Department of Molecular Biology, Roswell Park Memorial Institute, Buffalo, New York 14263, U.S.A.

Peter ELSBACH (363), Department of Medicine, New York University School of Medicine, New York, New York 10016, U.S.A.

Judith EVANS (165), Department of Biological Sciences, Columbia University, New York, New York 10027, U.S.A.

Richard M. FRANKLIN (803), Abteilung Strukturbiologie, Biozentrum der Universitat Basel, Klingelbergstrasse 70, CH-4056 Basel, Switzerland.

Jean-Marie GHUYSEN (463), Service de Microbiologie, Institut de Botanique, Universite de Liege, Sart-Tilman - 4000, Liege, Belgium.

Eric HOLTZMAN (165), Department of Biological Sciences, Columbia University, New York, New York 10027, U.S.A.

Leevi KÄÄRIÄINEN (741), Department of Virology, University of Helsinki, Haartmaninkatu 3, SF 00290 Helsinki 29, Finland.

D. James MORRÉ (1), Department of Medicinal Chemistry and Pharmacognosy and Department of Biological Sciences, Purdue University, West Lafayette, Indiana 47907, U.S.A.

D.H. NORTHCOTE (717), Department of Biochemistry, University of Cambridge, Cambridge CB2 1QW, England.

Ossi RENKONEN (741), Department of Biochemistry, University of Helsinki, Haartmaninkatu 3, SF 00290 Helsinki 29, Finland.

M.G. RUMSBY (247), Department of Biology, University of York, Heslington, York, YO1 5DD, England.

Samuel SCHACHER (165), Department of Biological Sciences, Columbia University, New York, New York 10027, U.S.A.

G.D. SHOCKMAN (597), Department of Microbiology and Immunology, Temple University School of Medicine, Philadelphia, Pennsylvania 19140, U.S.A.

Tracy M. SONNEBORN (829), Department of Zoology, Indiana University, Bloomington, Indiana 47401, U.S.A.

Saul TEICHBERG (165), Department of Pediatrics and Laboratories (and Cornell University), North Shore University Hospital, Manhassett, New York 12030, U.S.A.

John TWETO (137), Department of Molecular Biology, Roswell Park Memorial Institute, Buffalo, New York 14263, U.S.A.

Contents

1   The Golgi apparatus and membrane biogenesis, by D.J. Morré

2 Biosynthesis of plasma membrane glycoproteins, by G.M.W. Cook

3 Turnover of proteins of the eukaryotic cell surface, by J. Tweto and D. Doyle

4   *Origin and fate of the membranes of secretion granules and synaptic vesicles: membrane circulation in neurons, gland cells and retinal photoreceptors, by E. Holtzman, S. Schacher, J. Evans, and S. Teichberg*

5   *The myelin sheath — a structural examination, by M.G. Rumsby and A.J. Crang*

6   *Cell surface changes in phagocytosis, by P. Elsbach*

9   *The bacterial cell surface in growth and division, by L. Daneo-Moore
    and G.D. Shockman*

*10 The synthesis and assembly of plant cell walls: possible control mechanisms,
    by D.H. Northcote*

*11 Envelopes of lipid-containing viruses as models for membrane assembly
    by L. Kääriäinen and O. Renkonen*

12 In vitro and in vivo assembly of bacteriophage PM2: a model for protein-lipid interactions, by R.M. Franklin

13 Local differentiation of the cell surface of Ciliates: their determination, effects, and genetics, by T.M. Sonneborn

The Golgi apparatus and membrane biogenesis

1

D. James MORRÉ

1. Introduction

In current concepts of membrane biogenesis some variation of a directed assembly mechanism is usually envisioned (Cook and Stoddart, 1973). In a "single step" mechanism all components of the membrane—protein, lipid, and carbohydrate—are assembled simultaneously. Alternatively, a "multistep" mechanism involves primary synthesis of a basic membrane of lipid and intrinsic protein to which additional constituents such as enzymes, sugar moieties, and specific lipids are added sequentially.

Although the evidence is largely indirect, the multistep mechanism is indicated for plasma membranes (Cook and Stoddart, 1973; Morré and VanDer-Woude, 1974), Golgi apparatus (Morré et al., 1971d), and perhaps other cytoplasmic membranes as well (Morré, 1975). The biochemical pathways and mechanisms for synthesis of the lipids and proteins, and lesser amounts of asymmetrically distributed carbohydrates and other prosthetic groups from which membranes are derived, are reasonably well established (cf. Morré, 1975). Most are synthesized in a stepwise manner with incorporation or insertion into the membrane occurring nearly simultaneously with synthesis. A major unanswered question is where are the constituents common to all cellular membranes synthesized and assembled? If membrane specialization were consistently preceded by synthesis of basic constituents of the membrane core, then sites and mechanisms of membrane modification would be more clearly implicated. If, for example, certain intrinsic membrane proteins and lipids of the Golgi apparatus are synthesized exclusively in endoplasmic reticulum and/or nuclear envelope (these two membrane structures being more or less equivalent in their potential membrane biogenetic capabilities), then the Golgi apparatus role in membrane biogenesis would be largely one of membrane modification (i.e., membrane differentiation).

Morphological investigations (see Morré et al., 1971d for references), make it possible to conclude that a principal function of the Golgi apparatus is to pro-

G. Poste and G.L. Nicolson (eds.)   The Synthesis, Assembly and Turnover of Cell Surface Components
© *Elsevier North-Holland Biomedical Press, 1977.*

duce secretory vesicles whose membranes are plasma membrane-like and capable of fusing with (adding new material to) the plasma membrane (Mollenhauer and Morré, 1966). These vesicles are encountered at the peripheries of cisternae at a maturing or distal face of the apparatus. In favorable examples (e.g., Schnepf and Koch, 1966a,b; Brown, 1969), their formation, migration to the cell surface, and fusion with the plasma membrane have been observed in living cells with the aid of the light microscope. Yet one cannot conclude that the Golgi apparatus is the synthesis site of the membrane constituents of the secretory vesicles that are destined to add new plasma membrane to the cell surface. Actually, evidence favors synthesis within the endoplasmic reticulum.

At the opposite (proximal or forming) face of the Golgi apparatus, collections of small, presumably endoplasmic reticulum-derived, vesicles are aligned in the space between the endoplasmic reticulum and the forming cisterna of the Golgi apparatus proximal to the endoplasmic reticulum. A similar disposition of presumably nuclear envelope-derived vesicles is encountered with perinuclear Golgi apparatus. Although these vesicles are thought to coalesce to form new Golgi apparatus cisternae (see Morré et al., 1971d for discussion), this view has been challenged recently (Whaley et al., 1975)[1]

Membrane differentiation within the Golgi apparatus is supported by observations of morphological differences in thickness and staining intensity of membranes. The dimensions and appearance of the membranes of Golgi apparatus cisternae are intermediate between those of endoplasmic reticulum and plasma membrane (see Morré and Mollenhauer, 1974, for references). In certain organisms, such as the fungus, *Pythium ultimum* (Grove et al., 1968), it has been possible to demonstrate a transition in membrane staining and thickness among individual cisternae within the separate stacks of cisternae (dictyosomes) of the Golgi apparatus (Fig. 1). Membranes of the cisterna or cisternae at the face proximal to endoplasmic reticulum or nuclear envelope (forming face) have a morphology that is like endoplasmic reticulum; cisternal membranes of the opposite or maturing face (as well as those of the secretory vesicles) are similar to

---

[1]The major argument against this concept seems to be that secretory products usually appear first toward the distal face (e.g., Whaley, 1975), while the transition vesicles are suggested to coalesce at the opposite or forming face (Morré et al., 1971d). This argument, however, is weakened by the realization that certain carbohydrate secretion products actually may be synthesized progressively within the cisternae and vesicles of the mature face of the Golgi apparatus (see, for example, Morré and VanDerWoude, 1974), while in liver, at least, the entry of the lipoprotein particles into the preformed secretory vesicles clearly does not involve the transition vesicles of the forming face but is rather via the peripheral elements ("boulevard périphérique") of the Golgi apparatus (Morré et al., 1971b, 1974a; Ovtracht et al., 1973). In either direct synthesis or peripheral input the formation and differentiation of membranes and the elaboration or acquisition of secretory products may be separated for a time both temporally and spatially (see, Morré et al., 1974a for kinetic evidence for independent migration of membrane proteins and proteins of the secretory product in liver). This hardly constitutes a serious argument against the participation of transition vesicles in the flow of membrane constituents between endoplasmic reticulum and the forming face of the Golgi apparatus.

plasma membrane. This differential staining pattern suggests that the *Golgi apparatus is a site of endomembrane differentiation* within the cell from endoplasmic reticulum-like to plasma membrane-like. But, equally important from the standpoint of this review, these observations also clearly imply that the *endoplasmic reticulum is a major site of biogenesis* of membrane constituents of the Golgi apparatus and, through fusion of Golgi apparatus-derived secretory vesicles, of plasma membrane constituents. As was emphasized previously (Morré et al., 1971d; Morré, 1975), perhaps no membranous cell component is completely independent of the endoplasmic reticulum in its biogenesis.

In in vivo experiments with rats, isotopically labeled amino acids are rapidly incorporated without discernible lag into membranes of rough endoplasmic reticulum with delayed or subsequent incorporation into other membrane fractions (Franke et al., 1971c; Taylor et al., 1973; Morré et al., 1974a). Major components of membrane lipids are synthesized by enzyme systems that terminate more or less exclusively in endoplasmic reticulum (Wilgram and Kennedy, 1963; Wirtz and Zilversmit, 1968; McMurray and McGee, 1972; Dawson, 1973; Van Den Bosch, 1974). Glycoproteins and glycolipids are synthesized in a stepwise manner (Roseman, 1970) but, for glycoproteins at least, evidence indicates that polypeptides of protein portions synthesized on polysomes migrate to or are inserted into new membranes such as those of the endoplasmic reticulum where initial glycosylation takes place. Additional sugar residues are added successively to the growing oligosaccharide prosthetic group in different compartments of the transitional endomembranes, especially the Golgi apparatus (Schachter, 1974a,b). Also, the Golgi apparatus may influence the organization of the cell surface through the synthesis of carbohydrate moieties for the cell coat (Dauwalder et al., 1972; Whaley et al., 1972). However, in spite of certain important membrane modifications that may occur in the Golgi apparatus, the endoplasmic reticulum contains the most complete array of lipid, protein, and carbohydrate biosynthetic machinery and is most clearly implicated as a major site of membrane biogenesis. In contrast, plasma membranes are potentially the least autonomous of the cellular endomembranes in terms of their biogenesis.

Thus, in organizing this review, it has been assumed that the bulk of Golgi apparatus membranes are synthesized in the rough endoplasmic reticulum. Membrane constituents are thought to separate from rough endoplasmic reticulum, transform into smooth membranes, and give rise to Golgi apparatus cisternae either by fusion of small endoplasmic reticulum-derived vesicles or through direct continuities with smooth endoplasmic reticulum. As the endoplasmic reticulum-contributed materials fuse to form the flattened cisternae at the forming face of the Golgi apparatus, a balance is maintained by simultaneous transformation of other membranes of the Golgi apparatus into membranes of secretory vesicles at the opposite or maturing face. In discussing the role of the Golgi apparatus in membrane biogenesis, emphasis is placed on membrane modifications with restriction of exclusive biosynthetic activities to those constituents unique to Golgi apparatus and/or the plasma membrane.

4

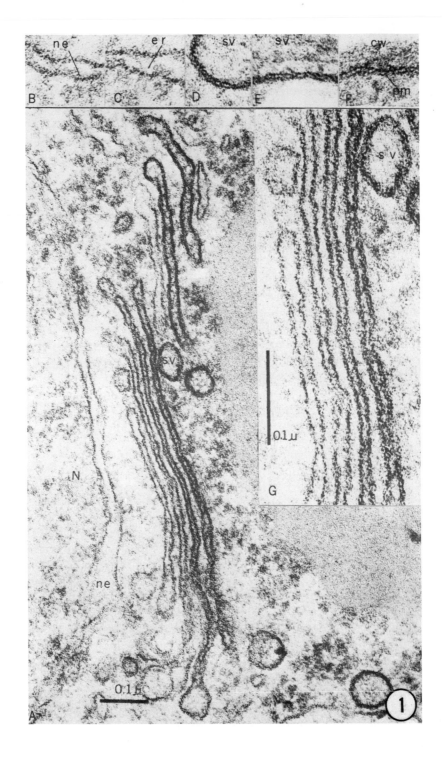

2. Golgi apparatus functioning as part of an integrated endomembrane system

In recognition of the interdependence of intercellular membranes in membrane biogenesis and other functional activities, the concept of an endomembrane system was proposed (Morré et al., 1971c; Morré and Mollenhauer, 1974). The system is viewed as a functional continuum of membrane types in which processes of membrane biogenesis, membrane differentiation, and membrane flow are combined to account for Golgi apparatus origins and function in secretion as well as the biogenesis and transformation of other internal membranes of the cell.

The *endomembrane system* is the term for the structural and developmental continuum of internal membranes that characterizes the cytoplasm of eukaryotic cells. Included within this system are the nuclear envelope, rough and smooth endoplasmic reticulum, Golgi apparatus, and various cytoplasmic vesicles. Plasma membranes, vacuole membranes, and lysosomes are regarded as end-products of the system. Endomembrane components are distinguished biochemically from the semiautonomous organelles (with definite inner and outer membrane systems), such as chloroplasts and mitochondria, by the absence of DNA and cytochrome oxidase (see Jarasch and Franke, 1974 for detailed studies of the nuclear envelope); the inability to generate ATP through respiratory or photosynthetic chain-linked phosphorylation; and a high degree of functional interdependence.

The major parts of the system are illustrated diagramatically in Figure 2 and have, or are expected to have, properties in common. Yet, since each part differs from the others structurally, positionally, and chemically, there are also properties that are unique to each part. Most important, functional activities such as those of the Golgi apparatus in glycosylation of glycolipids and glycoproteins, and the very existence of the endomembrane system per se, may depend upon the integrated activities of its component parts.

3. Biosynthetic capabilities of Golgi apparatus relevant to membrane biogenesis

Certain glycosyltransferases (Fleischer et al., 1969; Morré et al., 1969; Schachter et al., 1970; Table 4; and see Schachter, 1974a,b for review) are concentrated in

---

Fig. 1. Endomembrane differentiation in the fungus *Pythium ultimum* as revealed by a progressive increase in membrane thickness and staining intensity across the stacks of dictyosome cisternae (from endoplasmic reticulum or nuclear envelope-like at one pole to plasma membrane-like at the opposite pole). (a) A dictyosome adjacent to the nucleus (N) and associated secretory vesicles (sv). The membrane of the dictyosome cisterna nearest the nuclear envelope (ne) is similar to nuclear envelope. The membranes of each successive cisterna stain more intensely and appear thicker toward the opposite pole. (b) Membranes of nuclear envelope (ne). (c) Membranes of endoplasmic reticulum (er). (d, e) Membranes of secretory vesicles (sv). (f) Plasma membrane (pm) adjacent to the cell wall (cw). (g) Enlargement of a portion of the dictyosome of (a) (Courtesy of Grove et al. [1968].)

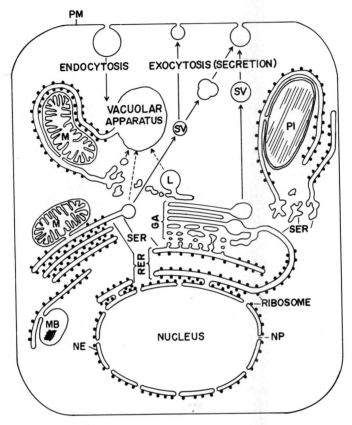

Fig. 2. Diagramatic representation of the endomembrane system as an interassociated complex of interacting components. GA = Golgi apparatus; RER = rough endoplasmic reticulum; SER = smooth endoplasmic reticulum; PM = plasma membrane; SV = secretory vesicle; L = lysosome; MB = microbody; NE = nuclear envelope; NP = nuclear pore; M = mitochondrion; Pl = plastid. (Adapted from Morré et al. [1974b].)

the Golgi apparatus in comparison with other fractions. Inosine diphosphatase may be concentrated in Golgi apparatus of plants (Ray et al., 1969; see Dauwalder et al., 1969 for review of cytochemical evidence), while thiamine pyrophosphatase provides a cytochemical marker for the Golgi apparatus of mammalian cells (Novikoff et al., 1962; 1971). Sulfotransferases have been localized in Golgi apparatus fractions from rat testis (Knap et al., 1973) and from kidney (Fleischer and Zambrano, 1973; Fleischer et al., 1974). Other membrane biosynthetic activities are either absent from Golgi apparatus or are shared with endoplasmic reticulum and/or nuclear envelope.

3.1 Biosynthesis of membrane lipids

The same five major phospholipids are present in all major classes of membranes within the endomembrane system (see Keenan and Morré, 1970 for data from rat liver). The phospholipid to protein ratios vary among endomembrane types as does the relative amount of individual phospholipids. The phospholipid content is greatest in plasma membrane and Golgi apparatus and least in endoplasmic reticulum and nuclear envelope. The major differences in lipid distribution are an increasing proportion of sphingomyelin (nuclear envelope = endoplasmic reticulum < Golgi apparatus < plasma membrane) and a compensating decrease in phosphatidylcholine. The contents of phosphatidylethanolamine, phosphatidylserine, and phosphatidylinositol are essentially constant. With cell fractions from mammary gland of the rat and bovine, the patterns of lipid distribution are similar to those of rat liver, except that with fractions from mammary gland of the rat, in addition, an increasing proportion of phosphatidylethanolamine and a decreasing proportion of phosphatidylinositol were observed when endoplasmic reticulum, Golgi apparatus, and plasma membrane were compared (Keenan et al., 1974c). Determinations for endomembranes of pancreas (Meldolesi et al., 1971) show trends similar to those of liver (Keenan and Morré, 1970; see also Zambrano et al., 1975 for a recent confirmation of the findings for liver).

Translated on a protein basis, the phosphatidylcholine content of endomembranes remains essentially constant, whereas there is a net increase in the content of the serine, inositol, and ethanolamine phosphatides going from endoplasmic reticulum to Golgi apparatus to plasma membrane. There is also a very large net increase in sphingomyelin content and a parallel increase in cholesterol content (Yunghans et al., 1970). Thus, a constant ratio between cholesterol and sphingomyelin is maintained (Patton, 1970). The fatty acid composition of the various phospholipids also shows a pattern of change, with those from the Golgi apparatus frequently being intermediate between those from endoplasmic reticulum and plasma membrane. For example, the proportion of unsaturated fatty acids increases (endoplasmic reticulum < Golgi apparatus < plasma membrane) while the contents of 16:0 fatty acids of phosphatidylcholine decrease (endoplasmic reticulum > Golgi apparatus > plasma membrane). These findings demonstrate that different endomembranes have characteristic fatty acid compositions and also indicate that selective incorporation and/or exchange of these phospholipids or their acyl moieties may take place within the Golgi apparatus (Keenan and Morré, 1970).

There are several mechanisms by which the accretion of lipids into membranes may occur. Synthesis of phospholipids, especially phosphatidylcholine, are characteristic reactions of the endoplasmic reticulum (Higgins and Barrnett, 1971; Stein and Stein, 1971; Fig. 3; McMurray and McGee, 1972). Golgi apparatus also have a modest capacity to synthesize phospholipids (Jelsema and Morré, in preparation; Table 1; but see Van Golde et al., 1974). To account for the characteristic lipid composition of the Golgi apparatus, regions of endoplasmic reticulum enriched in the required lipids might be transferred by a flow mechanism. However, synthesis by Golgi apparatus and/or transfer of specific lipids by exchange (Wirtz and Zilversmit, 1968, 1969; Akiyama and Sakagami, 1969; McMurray and Dawson, 1969; Kamath and Rubin, 1973; see Wirtz, 1974, for a review) involving carrier proteins of the soluble cytoplasm (Wirtz and

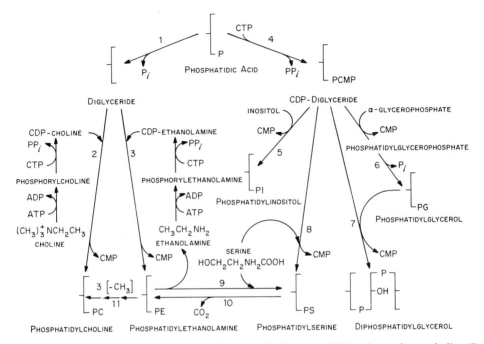

Fig. 3. Summary of biosynthetic pathways for glycerolipids. 1–3 = CDP-amine pathway: choline (2) and ethanolamine (3) phosphotransferase steps and the formation of diglycerides from phosphatidic acid catalyzed by phosphatidate phosphatase (1). 4–8 = CDP-diglyceride pathway: formation of CDP-diglyceride (4), phosphatidylinositol (5), phosphatidylglycerol (6), diphosphatidylglycerol (7) and phosphatidylserine (8) by synthesis. 9 = Formation of phosphatidylserine by exchange (the major pathway of phosphatidylserine formation in liver). 10 = Formation of phosphatidylethanolamine by decarboxylation of phosphatidylserine. 11 = Biosynthesis of phosphatidylcholine by direct methylation of phosphatidylethanolamine.

Zilversmit, 1969; Kamath and Rubin, 1973), provide alternative mechanisms.

According to Van Golde et al. (1971, 1974), the only phospholipid biosynthetic activity found associated with the Golgi apparatus of rat liver is α-glycerophosphotransferase. These findings are based on analyses of preparations estimated to contain 80% Golgi "vesicles" (Fleischer et al., 1969). In contrast, analyses of intact Golgi apparatus stacks (dictyosomes) show that Golgi apparatus from rat liver (Table 1) have a nearly full complement of glycerophospholipid biosynthetic enzymes as suggested earlier from studies with plants (Morré, 1970; Morré et al., 1970). For synthesis of phosphatidylcholine, phosphatidylethanolamine, and phosphatidylinositol, the endoplasmic reticulum is the primary site for the terminal enzymes. Of the total activity, 80% to 90% is accounted for by endoplasmic reticulum, but most of the remainder may be due to Golgi apparatus (Williamson and Morré, 1976; Jelsema and Morré, in preparation; Table 1). The specific activities of each of these enzymes in the Golgi apparatus are approximately 40% that of the endoplasmic reticulum, a

Table 1
Distribution of the terminal enzymes of biosynthesis of glycerophosphatides among purified cell fractions of rat liver

Cell fraction	Phosphoglyceride biosynthetic enzymes [a]							Marker enzymes [b]			
	PC	PE	CDP-DG	PI	PI [c]	PS [c]	PG	G-6-Pase	Gal—Tase	AMPase	Succ—INT redase
Total homogenate	56	34	51	180	250	79	185	2.8	1.4	2.0	2.0
Endoplasmic reticulum	232	142	164	820	1020	151	472	31.0	4.5	3.2	1.4
Golgi apparatus	86	58	140	230	410	40	322	2.8	90.1	3.6	0.15
Plasma membrane	2	3	36	6	4	53	88	0.8	1.5	77.6	1.85
Supernatant	3	2	4	17	1	42	3	0.35	0.75	0.01	0.01
Mitochondria											26.9

Source: Morré and Jelsema (in preparation).

[a] Units of specific activity are nmoles product formed/hr./mg. protein. The identity of products was verified by thin-layer chromatography. Cell fractions were prepared according to Morré (1973). The soluble supernatant is the microsome-free supernatant after centrifugation for 2.5 hr. at 100,000 × g. Assays for terminal glycerophosphatide biosynthesizing enzymes were as follow: PC (phosphatidylcholine), CDP-choline 1,2-diglyceride cholinephosphotransferase; PE (phosphatidylethanolamine), CDP-ethanolamine 1,2-diglyceride ethanolaminephosphotransferase; CDP-DG (CDP-diglyceride), cytidyl transferase; PI (phosphatidylinositol), complete synthesis, Williamson and Morré (1976); PS (phosphatidylserine), exchange reaction only; PG (phosphatidylglycerol), α-glycerophosphotransferase.

[b] Units of specific activity are μmoles inorganic phosphorous formed/hr./mg protein for glucose-6-phosphatase (G-6-Pase) a marker for endoplasmic reticulum and 5′-nucleotidase, with 5′-AMP as substrate (AMPase) a marker for plasma membrane; μmoles INT reduced/hr./mg. protein for succinate-INT reductase (Succ-INT redase) a marker for mitochondria; and nmoles acceptor-dependent disappearance of UDP-galactose/hr./mg. protein for galactosyl transferase (Gal-Tase) a marker for Golgi apparatus. Quantitative electron microscope morphometry carried out in parallel verified that the fraction purity exceeded 85% for all fractions.

[c] By exchange.

level much too high to be explained as contamination of the fraction by fragments of endoplasmic reticulum.

Evidence from in vivo labeling studies suggested that the Golgi apparatus should synthesize both phosphatidylinositol and sphingomyelin (Morré et al., 1974a). Golgi apparatus carry out the complete synthesis of phosphatidylinositol from α-glycerophosphate, coenzyme A, oleic acid, ATP, CTP, and myo-inositol in the presence of soluble enzymes (Williamson and Morré, 1976). In the absence of membranes no phosphatidylinositol is synthesized. Thus, Golgi apparatus must have at least the terminal enzymes for phosphatidylinositol synthesis.

Previous studies (Morré et al., 1970, 1971b) demonstrated the presence of biosynthetic enzymes of phosphatidylcholine in Golgi apparatus of rat liver, in contrast to the findings of Van Golde et al. (1971, 1974). Golgi apparatus of rat liver contain a phosphoryl-choline cytidyltransferase of high specific activity (Morré et al., 1970, 1971b). Although primarily a soluble enzyme in rat liver, choline kinase activity is also exhibited by carefully isolated Golgi apparatus from this tissue (Morré et al., 1971b). Both enzymes are potentially important to either sphingomyelin or phosphatidylcholine biosynthesis, or both (Kennedy, 1961). In total activity, the terminal choline and ethanolamine phosphotransferases of endoplasmic reticulum accounted for nearly 90% while comparable activities in the Golgi apparatus accounted for approximately 1%. In specific activity these enzymes of Golgi apparatus were 30 to 40% those of endoplasmic reticulum (Table 1).

Enzymes of CDP-glyceride and phosphatidylglycerol biosynthesis are present in endoplasmic reticulum (about 65% of the total for cytidyltransferase). However, the specific activities of the Golgi apparatus approached 70 to 80% those of endoplasmic reticulum (Table 1). Phosphatidate phosphatase, important in the formation of diglycerides from phosphatidic acid, was reported previously for Golgi apparatus of rat liver (Morré et al., 1971b).

The presence of a CDP-glyceride serine phosphotransferase, even in endoplasmic reticulum of rat liver, remains to be established (Bjerve, 1973). However, the enzymes catalyzing the formation of phosphatidylserine by exchange are present in both the endoplasmic reticulum and the Golgi apparatus fractions, the specific activity of the Golgi apparatus being about 25% that of the endoplasmic reticulum (Table 1).

Thus Golgi apparatus membranes appear to be potentially independent in the synthesis of at least those phospholipids required for the transformation of membranes from endoplasmic reticulum-like to plasma membrane-like. In general, our findings agree with those of Higgins and Barrnett (1972) and support their suggestion of an active role for the Golgi apparatus in phospholipid biosynthesis.

In contrast to Golgi apparatus, the plasma membrane seems entirely dependent on other membranes or cell components for phosphatidylcholine, phosphatidylethanolamine, and phosphatidylinositol, but appears capable of synthesizing CDP-diglyceride, diglyceride, phosphatidylglycerol, and phosphatidylserine (the latter by exchange) (Van Golde et al., 1971, 1974; Jelsema and Morré, in preparation; Table 1). The high phosphatidylserine and phosphatidylglycerol contents of rat liver plasma membrane may reflect this biosynthetic activity (McMurray and McGee, 1972).

The possibility of extensive redistribution of intact phospholipids among cellular membranes is supported by demonstrations of exchange of intact lipids between plasma lipoprotein and erythrocyte membranes (Van Deenan and de Grier, 1964; Sakagami et al., 1965; Helmkamp et al., 1974) and between microsomes and mitochondria of rat liver in vitro (Wirtz and Zilversmit, 1968; McMurray and Dawson, 1969). The exchange appears to be mediated by an exchange protein (McMurray and McGee, 1972) that facilitates the exchange of phosphatidylcholine (McMurray and McGee, 1972) and phosphatidylinositol (McMurray and Dawson, 1969; Helmkamp et al., 1974) more readily than phosphatidylethanolamine. Jungalwala and Dawson (1970) provide evidence that the exchange of intact phospholipids occurs in vivo. Transfer of intact phospholipids from endoplasmic reticulum to Golgi apparatus via exchange proteins of the soluble cytoplasm has not been investigated.

In studies in which ^{14}C-choline was supplied to plant stems (Morré, 1970), endoplasmic reticulum (microsomes) and Golgi apparatus membranes incorporated ^{14}C-choline into their membranes at nearly the same rate on a protein basis and with no detectable lag. These findings are most consistent with incorporation by net synthesis of lecithin or by exchange at the Golgi apparatus rather than by transfer from endoplasmic reticulum either by a flow mechanism or by cytoplasmic exchange proteins. In contrast, the time sequence of incorporation and short-term turnover of ^{3}H-glycerol administered to cells of *Acanthamoeba palestinensis* into endoplasmic reticulum-, Golgi apparatus-, and plasma membrane-rich fractions was compatible with transfer of membrane lipids from endoplasmic reticulum to Golgi apparatus to plasma membrane (Chlapowski, 1969; Chlapowski and Band, 1971a,b). Because of the many factors involved, kinetic analyses of synthesis and transfer of labeled phospholipids are expected to be relatively more complex than those of proteins or carbohydrate moieties of glycoproteins or glycolipids.

Considerably less is known to account for an increasing sterol content. Sterol biosynthesis is at least partly a function of internal cytomembranes (see Gaylor, 1972 for review) and the biosynthetic enzymes, which are membrane-associated, enter the microsome fraction (Gaylor, 1972; Tamaoki, 1973; Scallen et al., 1974). The specific cellular location is not known except that the enzymes are consistently associated with smooth membranes (Goldfarb, 1972), i.e., smooth endoplasmic reticulum and/or portions of the Golgi apparatus.

Enzymes of most membrane types, including Golgi apparatus (Keenan and Morré, unpublished observations) and plasma membrane (Mulder and Van-Deenen, 1965; Stabl and Trams, 1968; Stein et al., 1968; Lee et al., 1973; Stadler and Franke, 1973), catalyze exchange reactions (McMurray and Magee, 1972;

Van Den Bosch et al., 1972; Kamath and Rubin, 1973; Lee and Snyder, 1973; Van Den Bosch, 1974) including the incorporation of free fatty acids into phospholipids (Stadler and Franke, 1973).

3.2 Biosynthesis of membrane proteins

Membrane proteins are generally divided into two broad categories based on composition and location in the membrane (Bretscher, 1973; Singer, 1974; Nicolson et al., 1977). Extrinsic or peripheral membrane proteins are associated with the surfaces of the membrane by ionic interactions and can be extracted by salt solutions or by metal chelating agents (Bretscher, 1973). One example is cytochrome c (Ernster and Kuylenstierna, 1970), which is readily dissociated from the mitochondrial membrane.

Intrinsic, or integral proteins, on the other hand, are associated by hydrophobic regions and with lipids. They can be removed only by chaotropic and other agents that disrupt hydrophobic interactions. Generally, intrinsic proteins are isolated in association with lipids and some, such as glucose-6-phosphatase (Carvo et al., 1974), require an association with lipids for enzymatic activity.

3.2.1 Comparison of protein composition and distribution among endomembranes

Except for the early studies of Yunghans et al. (1970), most comparisons of protein compositions of cellular endomembranes by electrophoretic analysis have monitored only total proteins. It is not known whether differences observed were due to extrinsic, and hence loosely associated, surface proteins, or to differences in the intrinsic proteins of the membrane's interior. The protein bands observed are for the most part unidentified, uncharacterized, and probably incompletely resolved. As emphasized by various investigators (e.g., Meldolesi, 1974a), even with partially purified preparations, similar bands can contain completely different proteins, and the same peptide chains, found in more than one endomembrane component, may have different rates of migration due to molecular changes that occur after the termination of synthesis of the peptide chain (e.g., proteolytic cleavage, glycosylation). Certainly, endoplasmic reticulum, Golgi apparatus, and plasma membranes would be expected to contain more proteins than the 30 to 40 electrophoretic bands usually resolved by electrophoretic methods.

Despite these limitations, findings from acrylamide gel electrophoresis confirm the uniqueness of endomembrane components with respect to protein composition as predicted from enzyme distributions (Evans, 1970; Yunghans et al., 1970; Zahler et al., 1970; Meldolesi and Cova, 1971, 1972a; Helgeland et al., 1972; Borens and Kasper, 1973; Bailey et al., 1974; Kartenbeck, 1974; Meldolesi, 1974b; Franke and Kartenbeck, 1976; Elder and Morré, 1976a), but still show the ordered progression of change indicated by morphological studies (Evans, 1970; Hinman and Philips, 1970; Yunghans et al., 1970; Bailey et al., 1974; Hodson and Brenchley, 1976; Elder and Morré, 1976a,b). This is especially evident in comparisons where the ribosomal, extrinsic, and secretory proteins are first removed (e.g., the stripping procedure of Franke et al., 1971c), leaving only a fraction of the more intrinsic proteins of the membrane core for analysis (Yunghans et al., 1970; Bailey et al., 1974; Elder and Morré, 1976a; and Fig. 2). Clearly, much needed clarification of these apparent similarities and differences among proteins of different endomembranes must come from studies of purified and defined intrinsic membrane proteins.

Although a few intrinsic membrane proteins have now been isolated and characterized (e.g., cytochrome b_5 of endoplasmic reticulum, Stritmatter et al., 1972; and the major erythrocyte glycoprotein, glycophorin, Winzler, 1969; Tomita and Marchesi, 1975), comparisons of proteins of known function and/or composition among different endomembranes have been limited to a single enzymatic complex, NADH-oxidoreductase or NADH-cytochrome c oxidoreductase. This activity is found in all endomembrane components of rat liver (Morré et al., 1971b, 1974a; Bergeron et al., 1973a; Table 2; see, however, Fleischer and Fleischer, 1970), an observation verified by cytochemistry (Morré et al., 1974b). With either cytochrome c or ferricyanide as electron acceptor, the activities in endoplasmic reticulum, Golgi apparatus, and plasma membrane are kinetically indistinguishable (Frantz, 1973; Morré et al., 1974a). Microsomal cytochrome b_5, the normal electron acceptor for this oxidoreductase is also present in endoplasmic reticulum and Golgi apparatus from rat liver (Ichikawa and Yamamoto, 1970; Morré et al., 1971a; Fleischer et al., 1971) and has recently been demonstrated in milk fat globule membranes of several species (Jarasch et al., in press). The membranes that surround the fat globules of milk are plasma membrane derivatives (Keenan et al., 1970; Patton and Keenan,

Table 2
Summary of microsomal electron transport activities comparing endoplasmic reticulum (ER) and Golgi apparatus (GA) fractions from rat liver

Constituent	Units	Cell Component	Specific Activity or Amount/mg. Protein[a]						
			I	II	III	IV	V	VI	VII
NADH-cyt c reductase	μmoles/hr.	ER	48.5	75.6	84.0	16.8		78.6	
		GA	6.3	18.0	5.3	2.4			
NADH-ferricyanide reductase	μmoles/hr.	ER		208.2				181.2	
		GA		86.4				63.6	
NADPH-cyt c reductase	μmoles/hr.	ER	5.0	6.0	3.8	1.4			
		GA	0.6	0.8	0.7	0.2			
Cytochrome b5	mμmoles	ER	0.6				1.3		0.8
		GA	0.3				1.0		0.22
Cytochrome P450	mμmoles	ER	0.6				0.4		0.7
		GA	0.2				0.16		0.22
Total iron	mμatoms	ER	35.3						
		GA	36.4						

[a] I, Morré et al. (1971a); II, Morré (1974a); III, Fleischer and Fleischer (1970); IV, Bergeron et al. (1963) GF2; V, Fleischer et al. (1971); VI, Frantz (1973), VII, E. D. Jarasch, unpublished.

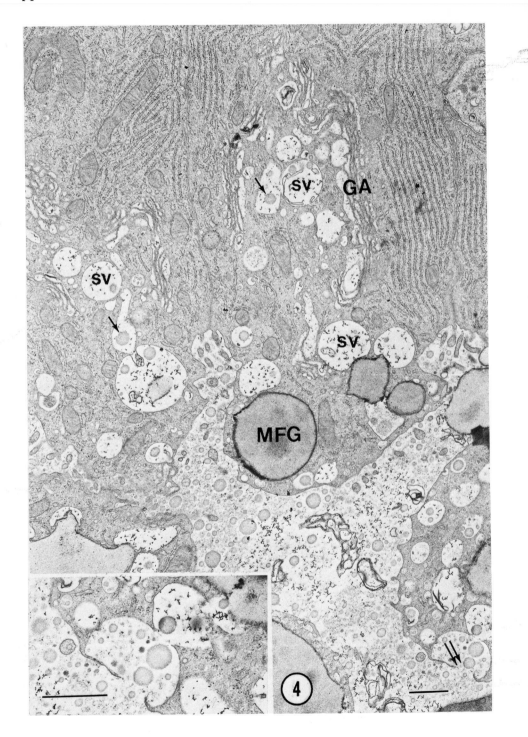

1975) to which considerable material is apparently contributed directly by secretory vesicles from the Golgi apparatus (Wooding, 1973; Fig. 4).

3.2.2. Synthesis of membrane proteins
by free versus membrane-bound polyribosomes

Initially, it was suggested that membrane proteins would be synthesized on bound polyribosomes (Palade, 1956). This suggestion was especially attractive since it would allow direct insertion of the relatively hydrophobic membrane proteins into the membrane, obviating the need for them to migrate through the hydrophilic cytoplasm. In vivo studies that monitored incorporation of radioactive amino acids into various cell fractions (Franke et al., 1971c; Taylor et al., 1973; Morré et al., 1974a) supported this hypothesis in that early incorporation into membrane proteins was into endoplasmic reticulum. These studies, however, only implicated rough endoplasmic reticulum as an early site of membrane assembly and did not exclude the possibility that synthesis occurred elsewhere.

Some membrane proteins seem to be synthesized exclusively or predominantly on free polyribosomes, e.g., NADPH cytochrome c reductase (Ragnotti et al., 1969; Lowe and Hallinan, 1973; see also Hughes, 1974) while others, e.g., 5'-nucleotidase, appear to be synthesized predominantly on membrane-bound polyribosomes (Bergeron et al., 1975). Additionally, Lodisch (1973) reported that free polyribosomes isolated from phenylhydrazine-treated rabbits synthesize two predominant species of surface membrane protein of mature erythrocytes. Studies by González-Cadavid and de Córdova (1974) demonstrated that both free and membrane-bound polyribosomes of rat liver carried out the synthesis of cytochrome c but concluded that, in vivo, cytochrome c was probably synthesized predominantly by rough endoplasmic reticulum since this fraction accounted for more than 70% of the total polyribosomes of rat liver.

In a more recent study, Elder and Morré (1976a) compared the abilities of total rough endoplasmic reticulum, polyribosomes released from rough endoplasmic reticulum, and free polyribosomes from rat liver to incorporate amino acids into intrinsic proteins using in vitro assay procedures. Polyribosomes bound to endoplasmic reticulum were shown to have the greatest capacity to synthesize radioactive products that either copurified with intrinsic proteins of endoplasmic reticulum or were precipitated by antisera raised against intrinsic

---

Fig. 4.   Electron micrograph of a portion of the epithelium of a lactating rat mammary gland. Large secretory vesicles (sv) containing structures morphologically recognizable as casein micelles (arrows) are formed from the Golgi apparatus (GA) and migrate to and fuse with (double arrows, lower right and inset) the apical plasma membrane. Simultaneously, apical plasma membrane is expended as globules of milk fat (MFG) are released into the gland lumen. The membrane that envelops the milk fat globule is derived from and biochemically similar to the apical plasma membrane (see Figs. 17, 18). During the period of maximum lactation, it has been estimated that mammary epithelial cells must replace plasma membrane equivalent to their entire apical surface within 8–10 hr. (Franke et al., 1976) so that a significant contribution of the Golgi apparatus to plasma membrane biosynthesis is expected in these cells. Glutaraldehyde-osmium tetroxide fixation. Scale line = 1 μ (From a study with Prof. T. W. Keenan, Purdue University, Lafayette, Indiana.)

proteins of endoplasmic reticulum. Although most of the radioactive products synthesized by bound polyribosomes were distinct from those synthesized by free polyribosomes, a few radioactive products synthesized by free polyribosomes also copurified with intrinsic membrane proteins; at least three components having the same apparent electrophoretic mobilities were among the synthesized products of both free and bound polyribosomes. These results with rat liver show no absolute segregation between free and bound polyribosomes in the synthesis of intrinsic membrane proteins of endoplasmic reticulum. The majority of these proteins, however, appear to be synthesized by polyribosomes bound to the endoplasmic reticulum.

3.2.3. Synthesis of membrane proteins
by Golgi apparatus-associated polyribosomes

Franke et al. (1972) and Franke and Scheer (1972) (see also Mollenhauer and Morré, 1974) reported that dictyosomes of plant Golgi apparatus had as a consistent feature of their morphology a single polyribosome or a group of several polyribosomes associated with the immature or forming faces of the cisternal stacks. In *Euglena gracilis*, the association of polyribosomes with dictyosomes was particularly striking: for each dictyosome, a single row of polyribosomes occupied a clearly defined zone limited on one side by rough endoplasmic reticulum and on the other by the immature dictyosome cisternae (Mollenhauer and Morré, 1974). Golgi apparatus-associated polyribosomes have been since reported for rat liver (Elder and Morré, 1976a; Fig. 5).

Golgi apparatus-associated polyribosomes are free from the membranes in a conventional sense. They are localized in a specialized region of the cytoplasm that has been termed a zone of exclusion (Morré et al., 1971d). These zones of exclusion surround the Golgi apparatus of all cell types examined including hepatocytes. Originally, it was suggested that these polyribosomes might be responsible for the synthesis of a few proteins or enzymes specifically localized in Golgi apparatus, such as one or more glycosyltransferases (Mollenhauer and Morré, 1974).

The Golgi apparatus polyribosomes from rat liver are active in an in vitro protein-synthesizing system (Table 3). On an RNA basis, Golgi apparatus fraction were found to be as active as rough endoplasmic reticulum in the incorporation of ^{14}C-leucine into high salt plus deoxycholate-insoluble proteins. Also, the incorporation was not due simply to the presence of membranes since isolated plasma membranes were incapable of synthesis of proteins in vitro (Elder and Morré, 1976a; see also Table 3).

Comparisons of the synthesized products of the Golgi apparatus fraction with the products of free polyribosomes and of rough endoplasmic reticulum showed that the majority of the radioactive incorporation products had distinct patterns among the three systems; the results were not due to indiscriminate cross contamination of Golgi apparatus by either free polyribosomes or rough endoplasmic reticulum. Instead, it appeared that the Golgi apparatus fractions contained a unique population of "free" polyribosomes. When analyzed by extraction with

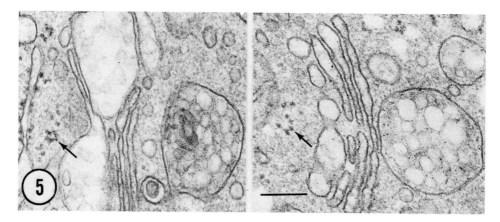

Fig. 5.   Portions of rat hepatoma cells showing polyribosomes (arrows) adjacent to dictyosomes. The micrographs are oriented to depict maturation of secretory vesicles from left to right. Dictysomes are surrounded in the cell by a differentiated region of cytoplasm or zone of exclusion in which the several Golgi apparatus-associated polysomes are embedded, usually in the near proximity of the forming face of the Golgi apparatus. The polysomes are not directly associated with membranes and represent a class of free polysomes. Such Golgi apparatus-associated polyribosomes have been shown to support in vitro synthesis of proteins (Elder and Morré, 1976a) and add a new dimension, protein synthesis, to the biosynthetic capacity of the Golgi apparatus complex.

1.5 M KCl and 0.1% deoxycholate, several proteins synthesized by the Golgi apparatus-associated polyribosomes remained insoluble. The majority of these proteins had electrophoretic mobilities similar to proteins in plasma membranes, which were absent from endoplasmic reticulum and were relatively minor components of Golgi apparatus.

A capacity for synthesis of intrinsic membrane proteins at or near the Golgi apparatus is consistent with in vivo studies of the kinetics of incorporation of labeled amino acids into membrane proteins (Ray et al., 1968; Franke et al., 1971c; Morré et al., 1974a). Pulse-chase experiments with H^3-leucine (Morré et

Table 3
Incorporation in vitro of ^{14}C-leucine into protein insoluble in 1.5 M KCl and 0.1% deoxycholate by cell fractions from rat liver

Fraction	Protein/RNA	Radioactivity	
		cpm/mg protein	cpm/mg RNA
Rough endoplasmic reticulum	2.5	810	2025
Golgi apparatus	8.2	340	2780
Plasma membrane	a	0	—

Source: Elder and Morré (1976a).
[a] No RNA was detected.

al., 1974a) showed that incorporation into the Golgi apparatus fraction was biphasic (see also Franke et al., 1971c) for similar results with [14]C-*guanido*-arginine). Maximum labeling of the intrinsic proteins of Golgi apparatus membranes occurred at 20 minutes and correlated with loss of radioactivity from the endoplasmic reticulum fraction. However, incorporation into intrinsic membrane proteins also occurred immediately after injection of label at nearly the same rate on a specific activity basis as that into intrinsic proteins of the rough endoplasmic reticulum. This first kinetic phase could be due to insertion of membrane proteins directly at the Golgi apparatus without synthesis or passage from rough endoplasmic reticulum (Elder and Morré, 1976a) or, as suggested previously (Franke et al., 1971c; Morré et al., 1974a), incorporation into endoplasmic reticulum occurs very near the Golgi apparatus, or both.

Similar studies with [14]C-*guanido*-arginine (Franke et al., 1971c) or ^{14}C-leucine (Ray et al., 1968) as label show that incorporation into plasma membranes is also biphasic. Although one phase of incorporation occurs between 30 and 60 minutes, simultaneously with loss of radioactivity from the Golgi apparatus (Franke et al., 1971c; Morré et al., 1974a), radioactivity also appears in the plasma membrane 15 to 20 minutes after administration of label during the first phase of Golgi apparatus labeling. This incorporation may represent, in part, the early appearance in plasma membranes of intrinsic membrane proteins synthesized by Golgi apparatus-associated polyribosomes and incorporated via secretory vesicles (Elder and Morré, 1976a). A previous explanation suggested direct flow from endoplasmic reticulum to the plasma membrane; early incorporation into plasma membrane also correlates with loss of label from both rough and smooth endoplasmic reticulum (Franke et al., 1971c; Morré et al., 1974a).

A puzzling observation encountered in experiments with both rat liver (Franke et al., 1971; Morré et al., 1974a) and plant stems (Morré and Van Der-Woude, 1974) now explained by these recent findings is that the peak specific activity of intrinsic membrane proteins of the Golgi apparatus was always higher than that of rough endoplasmic reticulum. Input of new membrane proteins from two sources (flow from endoplasmic reticulum and direct insertion at the Golgi apparatus) provides a mechanism whereby the total membrane proteins transferred to Golgi apparatus may have a specific activity which exceeds that of the total endoplasmic reticulum (Elder and Morré, 1976a).

In addition, the phenomenon of Golgi apparatus-associated polyribosomes offers a new mechanism whereby the protein composition of membranes might be altered to facilitate membrane differentiation. To what extent participation of other "membrane-associated" classes of free polyribosomes participate in the synthesis of membrane proteins, as for example direct insertion into plasma membrane, remains to be investigated.

Input of certain membrane proteins at or near the Golgi apparatus is required to account adequately for the functional polarity of the Golgi apparatus and to explain the presence of protein bands or enzymatic activities in Golgi apparatus that are absent from endoplasmic reticulum (Morré and Mollenhauer, 1974). Yet the relative amounts and limited numbers of proteins synthesized by the

Golgi apparatus complex limits this biosynthetic role to one of modification. Many of the intrinsic proteins of cellular endomembranes, including those of the Golgi apparatus, still appear to be synthesized at or near the rough endoplasmic reticulum (Elder and Morré, 1976a).

3.3. Glycosylation of membrane glycoproteins

Generally, plasma membranes and membranes of the Golgi apparatus are thought to be asymmetrically organized with carbohydrate prosthetic groups. With plasma membranes, they are concentrated on the external surface (Benedetti and Emmelot, 1967; Martinez-Palomo, 1970; Winzler, 1970; Rambourg, 1971; Hirano et al., 1972; Roland, 1973; Leblond and Bennett, 1974; Nicolson and Singer, 1974; Wood et al., 1974). Their precise locations can be visualized with the electron microscope by means of carbohydrate complexing reagents combined with heavy metals, i.e., lectins coupled with ferritin (Hirano et al., 1972), ferritin-complexed antibodies (Nii et al., 1968), and so on. With cytoplasmic membranes such as the Golgi apparatus, the inner surfaces of the cisterna are equivalent to the external surface of the plasma membrane. When vesicles derived from Golgi apparatus cisternae fuse with the plasma membrane, a correct final orientation of carbohydrate residues is already established.

Glycoproteins, although enriched in the plasma membrane of eukaryotes (Emmelot, 1973), are ubiquitous as constituents of intracellular membranes. Some are present as constituents of secretory products within the membrane lumina. Others exist as glycoproteins intrinsic to the membrane. Comparing only membrane glycoproteins that remain after 1.5 M KCl-0.1% deoxycholate treatment of the membranes, a pattern of increasing complexity and apparent molecular weight is seen comparing nuclear envelope, endoplasmic reticulum, Golgi apparatus, and plasma membranes of rat liver (Elder and Morré, 1976b). These findings agree with determinations of the carbohydrate composition of the various membranes, which ranges from 2% (Emmelot et al., 1974) to 7% (Dod and Gray, 1968) of the dry weight for plasma membrane (average 4%, Franke and Kartenbeck, 1976) to 0.5% for mitochondria (Martin and Bosman, 1971), about 1% for endoplasmic reticulum, and 2% for Golgi apparatus (Franke and Kartenbeck, 1976). Of this, about 10% is glycolipid and the remainder is assumed to be largely glycoprotein.

Nearly all (Kasper, 1974; Franke and Kartenbeck, 1976) of the nuclear membrane sugars are neutral or amino sugars, while in plasma membrane about 25% of the total carbohydrates are sialic acid (Franke and Kartenbeck, 1976). Sialic acid contents of endoplasmic reticulum and Golgi apparatus are intermediate (Golgi apparatus > endoplasmic reticulum) between these two extremes (Franke and Kartenbeck, 1976; Keenan et al., 1972b; unpublished results). The relative proportions of amino sugars are relatively constant among the various endomembranes, while certain neutral sugars such as galactose and fucose are concentrated in plasma membrane and Golgi apparatus of rat liver.

The progressive elaboration of membrane coat materials is especially evident at the Golgi apparatus where a marked polarity is shown by the increase in reactivity with carbohydrate complexing reagents from the forming to the maturing pole of the stacked cisternae (e.g., Rambourg et al. 1969; Rambourg, 1971). The combined activities of endoplasmic reticulum, Golgi apparatus, secretory vesicles, and perhaps even the plasma membrane per se appear to be involved in coat formation as the membranes migrate and differentiate within a multistep mechanism of membrane assembly. A time dimension is provided when cytochemistry is combined with autoradiographic data (e.g., Leblond and Bennett, 1974; also see discussion of this method in chapter 2 of this volume). Based on electron microscopic analyses, certain acidic groups appear to be acquired primarily in the last 1 to 3, most mature, Golgi apparatus cisternae (Berlin, 1968; Stockem, 1969), while other groups may be added to membranes of secretory vesicles (see, Section 3.4, this chapter).

Because of the assembly line nature of glycoprotein biosynthesis (Roseman, 1970; Schachter, 1974a,b), certain sugar groups appear to be added preferentially at the Golgi apparatus, e.g., galactose, fucose, N-acetyl neuraminic acid and, possibly, distal N-acetylglucosamines (Morré et al., 1969; Whur et al., 1969; Bouchilloux et al., 1970; Haddad et al., 1971; Kirby and Roberts, 1971; Wagner and Cynkin, 1971; Sturgess et al., 1972, 1973; Schachter, 1974a,b; see Leblond and Bennett, 1974 for review; and Table 4) while others, such as mannose and the more internal glucosamines (Whur et al., 1969; Bouchilloux et al., 1970; Uhr, 1970; Zagury et al., 1970; Haddad et al., 1971), are incorporated earlier at the endoplasmic reticulum soon after peptide synthesis. Glycosylation of the aglycoproteins may begin even while the polypetide chain is still attached to the ribosome (Molnar et al., 1965; Lawford and Schachter, 1966; Molnar and Sy, 1967; Hallinan et al., 1968; Schenkein and Uhr, 1970; Sherr and Uhr, 1970; Redman and Cherian, 1972). Additionally, the Golgi apparatus has been suggested to be the main subcellular site for the modification of N-acetyl neuraminic acid, which occurs frequently during the polymerization of oligosaccharide chains of glycoproteins (Schauer et al., 1974).

Thus, glycosylation of glycoproteins may occur as the result of the interaction of several different endomembrane components including perhaps even the nuclear envelope (Richard et al., 1975). It is stepwise and begins shortly after peptide synthesis, continuing in the Golgi apparatus and other smooth membrane transition elements. As a result, certain classes of incomplete glycoprotein chains may be characteristic of Golgi apparatus membranes (Elder and Morré, 1976b). The sugars added and the glycosyltransferases involved are different in different compartments. Certain glycosyltransferases characterize the Golgi apparatus (Table 4). Although they are thought to be restricted to internal endomembranes (Golgi apparatus or Golgi apparatus + endoplasmic reticulum), Roth et al. (1971) have reported glycosyltransferases to be present at the cell surface. Currently, however, the occurrence of cell surface glycosyltransferases is much debated (Deppert et al., 1974; Harwood et al., 1975; Keenan and Morré, 1975; Shur and Roth, 1975).

Based on cell-free synthesis and mixing experiments, it has been suggested

Table 4
Concentration of glycoprotein glycosyltransferases in Golgi apparatus-rich fractions from rat liver[a]

Transferase	Acceptor[b]	Relative Specific Activity[c]	Reference
Sialyltransferase	AGP (-SA)	92	Schachter et al. (1970)
		41	Schachter (1974a)
		41	Schachter (1974a)
	ApoVLDL (-SA)	42	Schachter et al. (1970)
Fucosyltransferase	AGP (-SA, -Gal)	39	Schachter (1974a)
N-acetylglucosaminyltransferase	AGP (-SA, -Gal, -GlcNAc) endogenous	50	Schachter et al. (1970)
		54	Schachter (1974a)
		100	Morré et al. (1969)
Galactosyltransferase		51	Schachter et al. (1970)

[a] While most biochemical data on the subcellular distribution of glycosyltransferases has derived from studies with rat liver or plants, preliminary analyses suggest a similar distribution of glycoprotein glycosyltransferases in Golgi apparatus of rat testes (Schachter, 1975a).
[b] AGP, Alpha acidglycoprotein treated enzymatically to remove specific sugars; SA, sialic acid; Gal, galactose; GlcNac, N-acetylglucosamine.
[c] Specific activity relative to the total homogenate.

that glycoproteins with completed oligosaccharide chains are released from the Golgi apparatus to the cytosol and are subsequently transferred to microsomes as constitutive membrane components (Autuori et al., 1975a,b; Elhammer et al., 1975). However, these glycoproteins are presumably associated to a large extent with the outer membrane surface (Elhammer et al., 1975), while most native glycoproteins of endoplasmic reticulum are considered to be transmembrane proteins with the oligosaccharide residue directed toward the interior, or luminal surface. Synthesis of endoplasmic reticulum constituents, even surface-associated ones, by membranes of Golgi apparatus is a relatively new concept. Complications due to the proposed involvement of cytoplasmic pools and the various alternative flow and biosynthetic routes will make direct in vivo verification of this notion especially difficult.

Thus, a central role is indicated for the Golgi apparatus in the biosynthesis of membrane glycoproteins. The degree of participation, however, has been suggested to vary from complete (to include glycoproteins of endoplasmic reticulum, Autuori et al., 1975a,b) to minor (predominance of surface transferases, Shur and Roth, 1975). The bulk of the evidence favors a major participation of the Golgi apparatus in the ordered sequential addition of sugars, with certain contributions from glycosyltransferases associated with both endoplasmic reticulum and (although direct evidence is unavailable for mammalian cells) the plasma membrane.

3.4. Glycosylation of membrane glycolipids

Glycolipids, particularly glycosphingolipids such as gangliosides, are also ubiquitous among mammalian membranes (Svennerholm, 1970). They are not restricted to neural tissues. Gangliosides are enriched in plasma membranes (Dod and Gray, 1968; Klenk and Choppin, 1970; Renkonen et al., 1970; Weinstein et al., 1970; Gahmberg, 1971; Keenan et al., 1972a,b; Yogeeswaran et al., 1972) but are also found associated with intercellular membranes (Keenan et al., 1972; Critchley et al., 1973), especially the Golgi apparatus (Keenan et al., 1972b, 1974c).

The functions of glycolipids in cellular membranes are just beginning to be identified. Originally speculated to be important to membrane structure, these molecules now appear to be involved in diverse aspects of cellular physiology (Critchley and Vicker, 1977). An important regulatory role, in addition to the presumed structural role, in now indicated. Several human blood group antigens have been characterized as glycophingolipids. These include the antigens A, B, H, Le[a], Le[b], P and p[k] (e.g., Rapport and Graf, 1969; Hakomori, 1970; Wiegandt, 1971; Stoffel, 1971; Naiki and Marcus, 1974). Lactosylceramide is a hapten (cytolipin H) found in most tissues including human tumors (Rapport et al., 1958), as are isoglotoside (Rapport and Graf, 1969) and galactosylceramide (Rapport and Graf, 1969). "Paragloboside" has been implicated as a tumor-associated antigen of hamster NIL-Py tumor (Sundsmo and Hakomori, 1976).

Because of their association with the exterior surface of the plasma membrane of the cell, the specific oligosaccharide sequences of glycolipids become important candidates as receptors for toxins, drugs, and transmitter substances. Considerable evidence now exists that receptors for serotonin and acetylcholine as well as for toxins of cholera, botulism, and tetanus are gangliosides (Simpson and Rapport, 1971; King and van Heyningen, 1973; Cuatrecacas, 1974; Staerk et al., 1974; van Heyningen, 1974). Glycolipids from brain also bind viruses (Haywood, 1974) and certain pharmacological agents interact stereospecifically with glycolipids (Loh et al., 1974), observations which suggest additional functions of glycolipids as surface receptors.

Considerable evidence exists for altered glycolipids associated with neoplastic cell transformations and malignancy (see Brady and Fishman, 1974; Hakomori, 1975; Richardson et al., 1975; and Critchley and Vicker, 1977 for reviews). Since the Golgi apparatus is an important site of carbohydrate addition to glycosphingolipids, those cell surface functions mediated by gangliosides must be regulated by Golgi apparatus function in some way (Richardson et al., 1975).

Elucidation of ganglioside biosynthesis pathways has come primarily from studies with brain (Svennerholm, 1970). Similar gangliosides are found associated with rat liver membranes; these glycolipids and the biosynthetic pathways for rat liver are summarized in Figure 6. The enzymes of the pathway LacCer $\rightarrow$ G_{M3} = G_{M2} $\rightarrow$ G_{M1} $\rightarrow$ G_{D1a} are concentrated in Golgi apparatus fractions from rat liver, present at much lower levels in rough endoplasmic reticulum, and could not be demonstrated for plasma membrane fractions (Keenan et al., 1974b). Because sialic acid was not incorporated into exogenous G_{D1a} from CMP-NAN, G_{D1a} appears to be an endproduct of this monosialoganglioside pathway (Richardson et al., in press). Approximately 85% of the total homogenate activity for each of these enzymes was accounted for, based on recovery of Golgi apparatus and endoplasmic reticulum markers. This pathway is the first direct demonstration for the completion of a major series of plasma membrane constituents exclusively by the Golgi apparatus.

The possibility of a second pathway of ganglioside biosynthesis by rat liver Golgi apparatus (LacCer $\rightarrow$ G_{M3} $\rightarrow$ G_{D3} $\rightarrow$ G_{D2} $\rightarrow$ G_{D1b}) was provided by demonstration of the formation of G_{D3} by a CMP-NAN:G_{M3} sialyltransferase (Keenan et al. 1974b). Particulate enzymes for conversion of G_{D3} $\rightarrow$ G_{D2}, and G_{D2} $\rightarrow$ G_{D1b} are present in rat liver as well as a sialyltransferase of low specific activity that transfer a third sialic acid from CMP-NAN with G_{D1b} as acceptor. Catalysis of the addition of a third moiety of NAN from CMP-NAN to exogenous G_{D3}, also a Golgi apparatus-associated activity, indicates a branch of the disialoganglioside pathway that may give rise to the trisialoganglioside G_{T3} (Richardson et al., in press). Thus, three interrelated pathways for the synthesis of complex gangliosides are potentially operative in rat liver (Fig. 6). The disialoganglioside pathway with its branch emerges as a major source of trisialoganglioside. The relevant sialyltransferases for each of the three pathways are present in the Golgi apparatus.

As with glycoproteins, certain classes of incomplete glycolipid chains may be characteristic of Golgi apparatus membranes. For example, the ratio of mono- to

24

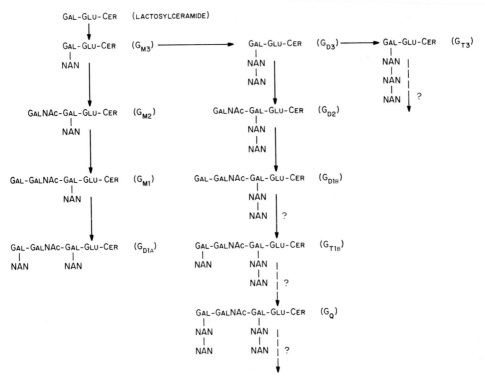

Fig. 6.   The ganglioside biosynthetic pathways of rat liver Golgi apparatus. Sugars are added sequentially to lactosylceramide in reactions catalyzed by specific glycosyltransferases. The product of one glycosyltransferase reaction becomes the acceptor substrate for the subsequent reaction with another nucleotide sugar donor. $G_{M3} \rightarrow G_{Dla}$ = monosialoganglioside pathway. $G_{D3} \rightarrow G_{Dlb}$ = disialoganglioside pathway. Cer = ceramide (N-acylsphingosine). Gal = galactose. GalNAc = N-acetylgalactosamine. NAN = N-acetylneuraminic acid (sialic acid). (Studies of Keenan et al. [1974b], and W. D. Merritt, C. L. Richardson, T. W. Keenan, and D. J. Morré [in preparation].)

digalactosylcerebrosides is > 1 for Golgi apparatus and ≈ 1 for plasma membrane of bovine mammary gland (Keenan et al., 1974c). Again, the sugars added and the glycosyltransferases involved vary in different compartments. Efforts to demonstrate the enzymes for formation of lactosylceramide, the immediate precursor to the gangliosides, by rat liver Golgi apparatus has been unsuccessful (Richardson, Keenan and Morré, unpublished observations), although the bulk of the recovered activity is microsomal (Morré et al., 1974b). Compartmentalization may even extend to Golgi apparatus subfractions where the glycosyl transferases that add the terminal sialic acid to the disialogangliosides are most concentrated in fractions of secretory vesicles (Table 5).

With plant cells, a phosphotungstic acid stain at low pH has been used as a selective stain for plasma membranes (Roland, 1969; Roland et al., 1972). The staining component, although not yet characterized chemically, is extracted with

Table 5

Glycolipid-glycosyltransferase activities of Golgi apparatus subfractions from rat liver comparing assays plus or minus added acceptors (monosialogangliosides, G_{M2} or G_{M1}) with UDP-galactose (for G_{M2}) or CMP-N-acetylneuraminic acid (for G_{M1}) as donors[a]

	Specific Activity[b]							
	UDP-Galactose: G_{M2} galactosyl transferase				CMP-NAN: G_{M1} Sialyltransferase			
Fraction[c]	A. (+ Acceptor, G_{M2})	B. (- G_{M2})	A. - B.	Ratio[d]	A. (+ Acceptor, G_{M1})	B. (- G_{M1})	A. - B.	Ratio[d]
Total homogenate	0.08	0.01	0.07	1	0.08	0.01	0.07	1
Intact Golgi apparatus	5.4	0.2	5.2	74	1.9	0.1	1.8	26
Purified cisternae	8.1	0.3	7.8	111	3.4	0.2	3.2	45
Secretory vesicles	12.3	0.3	12.0	171	6.1	0.2	5.9	84

[a] The products of the reaction are a monosialoganglioside (G_{M1}) with G_{M2} as acceptor and a disialoganglioside (G_{D1a}) with G_{M1} as acceptor.

[b] Units of specific activity are μmoles/hr./mg. protein. Incubations were for 2 hr. at 37°. The reaction mixture contained in a total volume of 0.1 ml: for galactosyl transferase, G_{M2}, 0.05 μmole; UDP-gal, 0.05 μmole (7 × 10⁵ cpm/μmole); Tween 80-Triton CF-54 (1:2, w/v), 0.6 mg; cacodylate-HCl buffer, pH 6.0, 15 μmoles; MnCl₂, 2.5 μmoles and for sialyltransferase, G_{M1}, 0.1 μmole, CMP-NAN, 0.05 μmole (5 × 10⁵ cpm/μmole); Tween 80-Triton CF-54 (1:2, w/v), 0.6 mg.; cacodylate-HCl, pH 6.3, 15 μmoles, MgCl₂, 0.1 μmole. Unpublished data of W. D. Merritt (Morré et al., in press).

[c] Procedures for preparation of fractions followed those referenced for Golgi apparatus (Morré et al., 1969), secretory vesicles (Merritt and Morré, 1973) and purified cisternae (Ovtracht et al., 1973).

[d] Ratio to total homogenate.

organic solvents and has chromatographic properties consistent with glycolipid rather than glycoprotein (Yunghans et al., in preparation). By use of this stain, membranes of secretory vesicles derived from the Golgi apparatus have been shown to acquire progressively the cytochemical characteristics of plasma membranes (Vian and Roland, 1972). The staining procedure has been especially useful in establishing (1) that membranes of secretory vesicles acquire at least one characteristic of plasma membranes in advance of their fusion with the plasma membrane (Roland, 1969), and (2) the progressive nature of the transformation from one pole of the dictyosome to the other (Frantz et al., 1973). The transformation begins in patches and coincides with the accumulation of fibrillar material in the vesicle interiors of the vesicles (Roland, 1969).

This aspect of membrane differentiation, the acquisition by Golgi apparatus of the component that stains with phosphotungstic acid at low pH, has been achieved with isolated Golgi apparatus in vitro. Frantz et al. (1973) showed that dictyosomes of soybean (*Glycine max*) hypocotyls did not stain (except for mature secretory vesicles) with the procedure, either in situ or when freshly isolated. However, after incubation of isolated dictyosome pellets for either 2 hours at 25°C or for 8 hours at 0°C, reactivity was acquired. Acquisition of staining capacity began at the mature pole or face of the dictyosome and progressed toward the forming pole or face.

3.5. Formation of sugar nucleotides and other activated intermediates of glycosylation reactions

The glycosyltransferases of Golgi apparatus that function in glycosylation of membrane glycoproteins and glycolipids transfer glycoses from appropriate nucleotide sugars to acceptors, which are usually incomplete carbohydrate side chains of glycoproteins or glycolipids, according to the following overall scheme: nucleotide-glycose + acceptor → glycose-acceptor + nucleotide. The enzymes are usually specific for both acceptor and donor. In some situations the acceptor may be a lipid intermediate analogous to the oligosaccharide-polyisoprenol derivatives that function in the biosynthesis of polymers of the bacterial envelope (Osborn, 1969; Lennarz and Scher, 1972 and Ghuysen, this volume, p. 463).

Within the Golgi apparatus, the glycosyltransferases are oriented at the inner or environmental surface of the vesicle membranes so that disruption of the vesicles with detergents or other means facilitates access to the substrate (Wilkinson et al., 1976). Significantly, this coincides with the asymmetric substitution of the membranes with carbohydrate groups of glycolipids and glycoproteins (Benedetti and Emmelot, 1968; Cook and Stoddart, 1973; see section 3.3). That these carbohydrate groups are exposed on the inner face of internal membranes has been confirmed by the inaccessibility of intact vesicles to lactoperoxidase labeling (Kreibich and Sabatini, 1973; Kreibich et al., 1974) and concanavalin A binding (Keenan et al., 1974a). Thus the conclusion is reached that glycosyltransferase enzymes of the Golgi apparatus function by the addition of sugars within that body's lumen, rather than at the Golgi apparatus-cytoplasm inter-face.

This spatial arrangement of glycosyltransferases poses a problem in logistics. How do charged and hydrophilic substrates such as sugar nucleotides traverse the Golgi apparatus membrane in situ when they are apparently excluded from doing so with isolated membrane preparations? If they do cross the membrane, a facilitated transport mechanism may be involved (Kuhn and White, 1976).

At least two possibilities may explain the apparent inability of sugar nuc-leotides to cross Golgi apparatus membranes. Berthillier and Got (1974) have provided evidence that all enzymes and all intermediary metabolites of the pathway for synthesis of UDP-glucose are present in the microsomal membranes of rat liver. The glucokinase activity was subsequently localized in a Golgi apparatus-rich fraction (Berthillier et al., 1973, 1976) as was a portion of the UTP-glucose 1-phosphate uridyltransferase and phosphoglucomutase (Berth-illier et al., 1976). By employing detergent treatment, freezing and thawing, and sonication, the glucokinase activity was shown to be either soluble in the Golgi apparatus lumen or loosely associated with the inside of the Golgi apparatus cisternae or vesicles. The Golgi apparatus glucokinase differs from the cyto-plasmic enzyme and the two do not cross-react immunologically (Berthillier et al., 1976). Enzymes of UDP-glucose formation are also present in Golgi apparatus fractions from onion (*Allium cepa*) stem (Morré and Mollenhauer, 1976) where they might function in concert with a glucan synthetase of polysaccharide forma-tion (VanDerWoude et al., 1974). The need for a pathway to form UDP-glucose by rat liver Golgi apparatus is less clear. More important, these findings open the way for additional studies to determine if other sugar nucleotides relevant to glycolipid and glycoprotein biosynthesis are generated within the lumens of Golgi apparatus cisternae.

The second possibility involves isoprenoid carrier lipids (Lennarz and Scher, 1972). Lipid-linked intermediates of bacterial antigen formation were purified in 1966 (Dankert et al., 1966) and subsequently identified as sugar derivatives of polyisoprenoids (see Osborn, 1969; Lennarz and Scher, 1972, and chapter 8, this volume for reviews). Specifically, undecaprenol functions in the synthesis of bacterial cell wall peptidoglycans, capsular polysaccharides, and mannans. Dolicol, another isoprenoid, has been shown to accept mannose in yeast (Tanner et al., 1971) and has been identified as an intermediate in microsomal prepara-tions of rabbit and pig liver (Behrens and LeLoir, 1970; Behrens et al., 1971a,b; Molnar et al., 1971; Richards and Hemming, 1972; Behrens et al., 1973; Evans and Hemming, 1973; LeLoir et al., 1973; Oliver et al., 1975). A dolichol inter-mediate has also been identified or implicated in glycoprotein synthesis in hen oviduct, bovine thyroid, calf pancreas, mouse myeloma, and human lymphocytes (Caccam et al., 1969; Baynes et al., 1973; Waechter et al., 1973; Hsu et al., 1974; Tkacz et al., 1974; Wedgewood et al., 1974). The α-D-mannopyranosyl dolichol phosphate intermediate was chemically characterized (Warren and Jeanloz, 1973, 1975); dolichol consists of twenty isoprene units with the terminal unit saturated (Richards and Hemming, 1972).

In addition to dolichol, the tetraisoprenoid retinol has also been implicated in glycosyl transfers (DeLuca et al., 1969a; Levinson and Wolf, 1972; Lucas et al., 1975). DeLuca and co-workers provide evidence that the intermediate is a phosphorylated retinol derivative; double label experiments showed that glucose and glucuronic acid formed retinyl-phosphate-sugar compounds, whereas xylose and glucosamine did not (DeLuca et al., 1973). In vitamin A deficiency, the absence of these intermediates has been suggested to be responsible for the depletion of mucous and changes in the immunologic properties of the cell surface (DeLuca et al., 1972, 1975). These and other experiments (Wolf and DeLuca, 1970; Helting and Peterson, 1972; Rosso et al., 1975) support at least a limited function of retinol as a lipid intermediate of sugar glycosylation.

Although they are not localized there, Golgi apparatus contains high concentrations of retinyl palmitate (Nyquist et al., 1971a; see, however, Fleischer et al., 1973; Zambrano et al., 1975) in addition to ubiquinones-9 and -10 (Nyquist et al., 1970) and vitamin K (Nyquist et al., 1971b). Preliminary evidence for multiple enzymes in the dolichol-utilizing pathway of glycoprotein biosynthesis in Golgi apparatus of rat liver has recently appeared (Vessey et al., 1976). Rat liver Golgi apparatus appear to contain free dolichol (Richardson, C. L. and Prokopeak, A., personal communication) but sugar transfers involving dolichol or other lipid intermediates remain to be established for Golgi apparatus.

3.6. Sulfation reactions

Sulfotransferase activities have been localized in Golgi apparatus fractions from kidney (Fleischer and Zambrano, 1973; Fleischer et al., 1974), and rat testis (Knap et al., 1973), and for smooth microsomes of chondrocytes (Horwitz and Dorfman, 1968). These findings confirm cytochemical studies that first suggested Golgi apparatus were important sites of polysaccharide sulfation. In different mammalian cell types sulfate was detected initially in Golgi apparatus, using electron microscope autoradiography to localize radioactivity in cells exposed for very short periods to ^{35}S-sulfate (Godman and Lane, 1964; Lane et al., 1964; Berg and Young, 1971; Young, 1973).

3.7. Summary of membrane biosynthetic capabilities of the Golgi apparatus complex

Being far from biochemically inert, the Golgi apparatus complex (cisternal stacks plus associated tubules, secretory vesicles, and surrounding zone of exclusion) possess the terminal enzymes of glycerolipid biosynthesis, nearly complete glycoprotein and glycolipid glycosylation machinery, and at least a limited capacity for synthesis and insertion of membrane proteins via a system of Golgi apparatus-associated polyribosomes within the zone of exclusion. When intact Golgi apparatus are isolated without loss of important components, all of these biosynthetic activities are demonstrable in cell-free systems.

However, Golgi apparatus membranes are not structurally homogeneous and are not expected to be homogeneous in their biosynthetic capacities. Each cisterna within the stack is different from the others. Secretory vesicles are enriched in certain activities and doubtless lack other activities. The system of connecting tubules (Ovtracht et al., 1973) that extends for several microns from the center of each cisternal plate may add to the biosynthetic capabilities of the complex to provide, for example, a potential site of steroid and phospholipid synthesizing enzymes. This extreme heterogeneity of structure is not always appreciated by biochemists and may explain why certain enzymatic activities are demonstrable in some Golgi apparatus preparations and not in others (e.g., compare Table 1 and results of Van Golde et al., 1971, 1974). Additionally, in assessing biosynthetic capabilities of total Golgi apparatus, isolated fractions must be representative of the state of the organelle in situ, including the preservation of those parts that normally are directly connected or function in concert (e.g., Morré et al., 1970; 1971b). Isolation procedures that deliberately disrupt Golgi apparatus structure (Bergeron et al., 1973a; Ehrenreich et al., 1973; Merritt and Morré, 1973; Ovtracht et al., 1973) and yield Golgi apparatus subfractions lead to biochemical data applicable to whatever part of the Golgi apparatus is isolated but should not be regarded as representative of the Golgi apparatus complex per se. The use of the terms "Golgi vesicle" and "Golgi membrane" to denote some representative fragment of the total structure are misleading since (as emphasized in the next section) a major function of the Golgi apparatus involves the differentiation of membranes from endoplasmic reticulum-like to plasma membrane-like across as well as within the stacked cisternae.

Although the potential for considerable biogenesis of membrane constituents may exist within the Golgi apparatus complex, the exercise of this potential may still be in terms of membrane modification rather than extensive de novo biogenesis of totally new membrane. As emphasized in the introduction, it is the prejudice of the author that many membrane materials (especially glycerolipids and proteins) which arrive at the Golgi apparatus are the products of combined synthesis by endoplasmic reticulum (or, less frequently, by nuclear envelope) and the peripheral elements of the Golgi apparatus. To the extent that Golgi apparatus-derived membranes contribute to plasma membrane, it is the function of the Golgi apparatus to appropriately modify these endoplasmic reticulum- (or nuclear envelope-) derived membranes and/or membrane precursors to ensure a compatible insertion into the plasma membrane.

4. The process of membrane differentiation: A major Golgi apparatus function

The Golgi apparatus is a component of the endomembrane system of eukaryotic cells (Morré et al., 1971d; Morré and Mollenhauer, 1974) with structural and biochemical properties intermediate between those of the endoplasmic reticulum and those of the plasma membrane. The Golgi apparatus is involved in the synthesis and/or transport of secretory products destined for passage out of the cell (Beams and Kessel, 1968; Mollenhauer and Morré, 1966a; Whaley, 1975) or to intracellular components such as lysosomes (Novikoff et al., 1971). One of the most common functions of the Golgi apparatus, however, may be its ability to transform membranes (Grove et al., 1968; Morré et al., 1971d; Morré and Mollenhauer, 1974; Franke and Kartenbeck, 1976) from those resembling endoplasmic reticulum to those that are similar to plasma membrane.

The membrane constituents, including both lipids and proteins, that are transformed by Golgi apparatus and utilized during the production of secretory vesicles must somehow be replaced. This replacement process implicates other components of the endomembrane system, chiefly endoplasmic reticulum and nuclear envelope, as important to the maintenance of Golgi apparatus structure and function. The conversion of one type of membrane to another or the structural and chemical modification of existing membranes is *membrane differentiation*, i.e., "the change in composition or organization which accompanies the conversion of one type of membrane to another" (Morré et al., 1974a). The physical transfer of bulk membrane or of parts of a membrane from one cell component to another is *membrane flow* (Franke et al., 1971c). These two cellular processes, in conjunction with established processes of biosynthesis of membrane constituents and secretion, constitute the basis for the functional description of the Golgi apparatus role in membrane biogenesis discussed in the remainder of this review.

4.1. Morphological manifestations of membrane differentiation

The existence of different membrane types among various cytomembranes was first emphasized in the reports of Sjöstrand (1956, 1963, 1968). Ledbetter (1962), Ueda (1966), and Yamamoto (1963). In general, membranes of endoplasmic reticulum were thinner, with a less pronounced dark-light-dark pattern than plasma membrane, whereas membranes of the Golgi apparatus were generally recognized to be intermediate in appearance.

4.1.1. Changes in thickness and staining intensity of membranes
In the fungus *Pythium ultimum,* dictyosome membranes were seen to be differentiated across the stack of cisternae so that those at the proximal pole appeared similar to endoplasmic reticulum and nuclear envelope, whereas those at the distal pole (including vesicle membranes) were similar to plasma membrane (Grove et al., 1968; Fig. 1). The differences in this study were demonstrated by changes in membrane thickness and in staining intensity so that differences were maximally enhanced. Intercalary cisternae of Golgi apparatus were morphologically intermediate, each successive cisterna progressing toward the distal pole becoming more like plasma membrane (denser, thicker, and showing the dark-light-dark pattern more clearly). The occurrence of dissimilar membranes in dictyosomes had been observed in other organisms (Ledbetter, 1962; Yamamoto, 1963, Hicks, 1966; Ueda, 1966; Whaley et al., 1966; Porter et al., 1967; Sakai and Shigenaka, 1967). However, the pattern of membrane differen-

tiation in *Pythium* (since observed in a variety of plant and animal cell types) was sufficiently striking to lead to a concept of Golgi apparatus function whereby membrane was received from endoplasmic reticulum or nuclear envelope (the generating elements) and sequentially transformed into plasma membranes (an end product).

In general, membrane thickness progresses from thin to thick along the nuclear envelope-endoplasmic reticulum-secretory vesicle-plasma membrane export route. Staining intensity, however, can proceed either from lightly stained for endoplasmic reticulum to darkly stained for plasma membrane, which is the usual observation (cf. Fig. 1), or vice versa (dark to light), depending on the tissue, the conditions of fixation, and the procedure for poststaining of the sections (see Morré and Mollenhauer, 1974, Fig. 3.13). Yet dictyosome membranes of the proximal pole (forming face) always resemble endoplasmic reticulum, while those at the distal pole (maturing face) or membranes of secretory vesicles always resemble plasma membrane independent of the staining pattern. I am not aware of any exceptions to this observation. Thus, membrane staining and/or morphology seems to reflect intrinsic properties of the membrane and to provide one reliable criterion for membrane differentiation.

4.1.2. Changes in the organization of membrane constituents

The transitional nature of the Golgi apparatus is also evident in the organization of membrane constituents for endomembranes of rat liver (Morré et al., 1974; Table 6; Figs. 7,8) and plants (Staehelin and Kiermayer, 1970; Vian, 1972, 1974) Morré and Mollenhauer, 1976). Independent confirmation of the electron microscope measurements of membrane thickness from thin to thick across the stacked cisternae has come from low-angle x-ray diffraction analysis (Table 6). The center to center spacing between the phospholipid polar groups in the bilayer was determined by Blaurock, Wilkins, and their colleagues to be in the range of 45 to 52 Å for plasma membranes from *Mycoplasma laidlawii*, erythrocyte ghosts, and nerve endings (Wilkins et al., 1971; Blaurock, 1972). Using this method, endoplasmic reticulum fragments from rat liver had a center to center

Table 6
Differentiation of endomembranes of rat liver

Parameter	Endoplasmic Reticulum	Golgi Apparatus	Plasma Membrane
Membrane thickness (OsO₄-fixation)[a]	65 Å	65–85 Å	85 Å
Spacing between phosphate head groups in lipid bilayer[b]	40 Å	45 Å[d]	50 Å
Lipid content[c]	30%	35%	40%
Protein content[c]	70%	65%	60%

[a] Morré et al. (1971b); Morré, unpublished results.
[b] Morré et al. (1974b).
[c] Keenan and Morré (1970); Yunghans et al. (1970).
[d] Average of all cisternae in the stack.

32

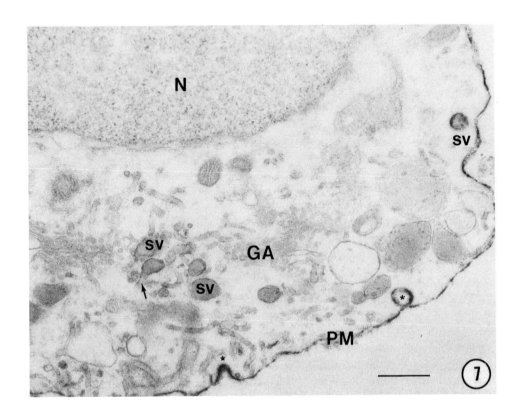

spacing between phospholipid polar groups of 40 Å, whereas distributions of Golgi apparatus showed a spacing of 45 Å; values for plasma membrane were about 50 Å (Morré et al., 1974a). These results confirm measurements of membrane thickness from electron micrographs.

Counts of the number and distribution of intramembranous particles revealed from freeze-fracture-etch analysis by Vian (1974) showed a progressive increase in numbers of particles comparing endoplasmic reticulum, Golgi apparatus, secretory vesicles, and plasma membrane of root tips of pea (*Pisum sativum*). Increases were apparent on both fracture faces and were reflected in distribution patterns. The final density and arrangement of characteristic plasma membrane particles was achieved in secretory vesicles prior to their fusion with the plasma membrane (Vian, 1974). Staehelin and Kiermayer (1970) obtained similar results for dictyosomes of the alga *Micrasterias denticulata*. An increase in particle density was observed from the forming to the maturing face of the Golgi apparatus, but particle density was greatly reduced at sites of attachment of secretory vesicles to the dictyosomes. Opposite findings were obtained by Pfen-

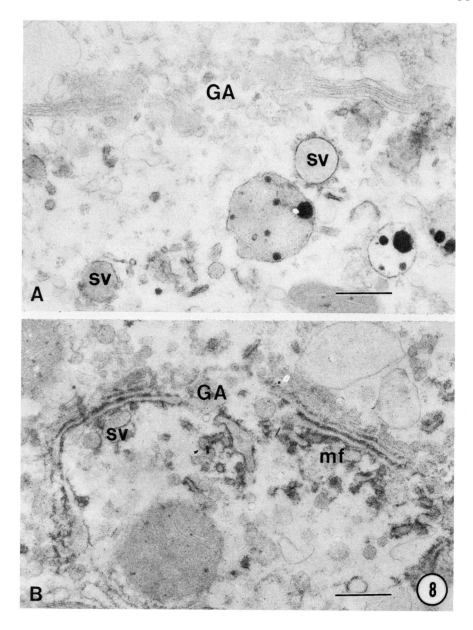

Figs. 7 and 8. Endomembrane differentiation in rat hepatocytes as visualized by cytochemical localization of an NADH-ferricyanide oxidoreductase (Morré et al., 1974b). Liver was fixed for 30 min. with 0.1% glutaraldehyde and incubated with a ferricyanide concentration of 5 mM to intensify reactivity of plasma membrane (PM). These reaction conditions appear to result in localization of an NADH oxidase activity restricted to the plasma membrane, presumptive endocytotic vesicles (*) (Fig. 7), membranes of mature secretory vesicles (sv) of the Golgi apparatus (GA) (Figs. 7, 8) and, sometimes, cisternal membranes or tubules (arrow, Fig. 7) of the mature face (mf) of the Golgi apparatus (Fig. 8b). N = nucleus. Scale line = 0.5μ. (From a study with Prof. E. L. Vigil, Marquette University, Milwaukee, Wisconsin.)

ninger and Bunge (1974) for nerve fibers. Here, the plasma membranes of the growing tips (growth cones) were characterized by an unusually small number of particles. The cytoplasmic vesicles interpreted to be in the process of fusing with the plasma membrane during tip growth of the fiber also were without intramembranous particles although they were presumably derived from cytoplasmic transition elements that included Golgi apparatus. While they support an intracytoplasmic origin of plasma membrane in this tip-growing system, the results strongly suggest that newly formed plasma membranes are particle-free, and that only at a later stage are particles added to the membrane "matrix" (Pfenninger and Bunge, 1974; Virtanen and Wartiovaara, 1974; DeCamilli et al., 1976; chapter 7, this volume, p. 403).

The arrangement of intramembrane particles, however, must be viewed as dynamic and subject to rapid change in response to environmental and developmental signals (Allen, 1975; Satir, 1974a). Of particular interest are internal arrays of particles called fusion rosettes, found in plasma and vesicle membranes of secretory cells (Vian, 1972) and thought to facilitate membrane recognition and fusion (Satir, 1974a). It must be emphasized that although the 75 Å intramembranous particles observed by electron microscopy have been suggested to be proteins or glycoproteins inserted into the lipid layer (Pinto Da Silva et al., 1971; Tillack et al., 1972), their identification with enzymes or antigens known to be associated with different membranes is generally inconclusive (Tillack et al., 1974; Franke and Kartenbeck, 1976).

4.1.3. Changes shown by cytochemistry

Cytochemistry has provided many important contributions to the concept of membrane differentiation. Much relevant work relating to the visualization of carbohydrate groups by electron microscopy is reviewed in sections 3.3 and 3.4 dealing with glycosylation of membrane glycoproteins and glycolipids and will not be repeated here. The detection of enzymes by cytochemical procedures (enzyme cytochemistry) affords a degree of precision in enzyme localization comparable to that for carbohydrates. Significantly, enzyme cytochemistry permits visualization of enzymatic gradients within cell components or systems of cell components and a means of assessing functional heterogeneity of seemingly homogenous structures.

Supporting evidence for membrane differentiation within the Golgi apparatus has come from enzyme cytochemistry. As examples, plasma membrane (Mg^{2+}-ATPase: Benedetti et al., 1973; Scheer and Franke, 1969; alkaline phosphatase: Bainton and Farquhar, 1968; Hugon et al., 1973; Ryder and Bowen, 1974; adenylate cyclase: Yunghans and Morré, in preparation; 5'-nucleotidase: Farquhar et al., 1974; Little and Widnell, 1975), endoplasmic reticulum (glucose-6-phosphatase: Leskes and Siekevitz, 1969; Leskes et al., 1971; Hugon et al., 1973; acetylcholine esterase: Kreutzberg et al., 1975; NADH-ferricyanide oxidoreductase: Morré et al., 1974b; acetyl coenzyme A carboxylase: Higgins and Yates, 1974), Golgi apparatus (thiamine pyrophosphatase: Novikoff et al., 1962, 1971; Cheetham et al., 1971; Sanders and Singal, 1975; nucleoside diphosphatase: Novikoff et al., 1962; Dauwalder et al., 1969; acyl transferase: Higgins and Barrnett, 1971, 1972; Benes et al., 1972; Levine et al., 1972) and lysosomal (acid phosphatase: Novikoff et al., 1962; Brandes, 1965; Moe et al., 1965; Boutry and Novikoff, 1975; Goldfischer, 1965) markers have been analyzed.

Enzyme cytochemistry also permits the distinction among enzyme activities associated with the cisternal lumina (e.g., acid phosphatase) and enzyme activities associated with the cisternal membranes (e.g., NADH oxidoreductases). Most activities show a distinct polarity across the stacked cisternae, with endoplasmic reticulum activities concentrated at the immature face of the apparatus (Novikoff et al., 1962; Bainton and Farquhar, 1968; Hugon et al., 1972; Pelletier and Novikoff, 1972; Higgins and Yates, 1974) and plasma membrane activities concentrated at the mature face of the apparatus (Friend and Farquhar, 1967; Pelletier and Novikoff, 1972; Hugon et al., 1973; Farquhar et al., 1974; Ryder and Bowen, 1974; Figs. 7 and 8) or absent. Exceptions include alkaline phosphatase in immature specific granules and Golgi apparatus cisternae in the "outer" face of the Golgi apparatus of polymorphonuclear leukocytes (Bainton and Farquhar, 1968) and certain activities (e.g., Dauwalder et al., 1969; Kreutzburg et al., 1975) that appear at both faces. Other activities appear in secretory vesicles (Farquhar et al., 1974), granules (Eppig and Dumont, 1972), lysosomes (Novikoff et al., 1962; Goldfischer, 1965), or in the peripheral Golgi apparatus tubules (Leskes and Siekevitz, 1969; Kartenbeck, 1974).

Osmication and related impregnation techniques have also been employed to visualize the marked polarity of the stacked cisternae of dictyosomes (Friend and Murray, 1965; Poux, 1973; Dauwalder and Whaley, 1974b). These techniques also demonstrate a relatedness of endoplasmic reticulum cisternae and the proximal cisternae of the Golgi apparatus (cf. Morré and Mollenhauer, 1974).

4.2. Biochemical manifestations of membrane differentiation

The biochemical basis for membrane differentiation has been sought through studies that compare endoplasmic reticulum, Golgi apparatus, and plasma membrane fractions isolated from rat liver (Morré et al., 1971,a,b; and references cited in Morré et al., 1971a; Morré et al., 1974a; Morré and Mollenhauer, 1974). If Golgi apparatus function in the conversion of endoplasmic reticulum membranes to plasma membranes, the composition of Golgi apparatus membranes should reflect this transformation (Keenan and Morré, 1970). Similarly, a comparison of endoplasmic reticulum membranes (or nuclear envelope, see Franke 1974a,b; Franke and Scheer, 1974) and plasma membranes will indicate the biochemical changes required to effect the transformation.

The transitional nature of Golgi apparatus membranes, revealed first in morphological studies (section 4.1), is reflected in the lipid and protein composition of the membranes for liver (Keenan and Morré, 1970; Yunghans et al., 1970; Franke and Kartenbeck, 1976; Elder and Morré, 1976a,b). Phospholipid and fatty acids of the major lipid classes found in Golgi apparatus are intermediate between those of the endoplasmic reticulum (or nuclear envelope) and plasma membrane (section 3.1; Keenan and Morré, 1970; Franke, 1974a,b). At present levels of resolution, all endomembrane fractions (rough endoplasmic reticulum, smooth endoplasmic reticulum, Golgi apparatus, and plasma membranes) have major protein bands in common, based on analyses by polyacrylamide disc-gel electrophoresis comparing apparent molecular weights (Section 3.2; Yunghans et al., 1970; Morré, 1974a; Franke and Kartenbeck, 1976; Hodson and Brenchley, 1976; Elder and Morré, 1976a). As discussed in the sections that follow, enzymatic activities characteristic of plasma membranes, i.e., plasma membrane marker enzymes, appear to be acquired either at the Golgi apparatus or by an insertion/activation mechanism near the plasma membrane, whereas enzyme activities characteristic of endoplasmic reticulum membranes appear to be lost. For a discussion of the opposite view involving limited intermixing of membranes, see Fleischer and Fleischer (1971), Meldolesi and Cova (1972b), Bergeron et al. (1973a), and Dauwalder and Whaley (1974b).

4.2.1. Analyses of plasma membrane markers

Enzymatic activities characteristic of plasma membranes of rat liver, for example, include 5′-nucleotidase, Mg^{2+}-ATPase, adenylate cyclase, alkaline phosphatase, and the ion-stimulated ATPases (i.e., Na^+-K^+-stimulated Mg^{2+} ATPases). Although they are concentrated in the plasma membrane, most exist also in endoplasmic reticulum and Golgi apparatus (Morré et al., 1971a; Fig. 9) where their localization has since been visualized and confirmed by cytochemistry. With 5′-nucleotidase, the activity in Golgi apparatus seems to be concentrated in membranes of mature secretory vesicles (Farquhar et al., 1974). As is characteristic of many plasma membrane markers, 5′-nucleotidase increases gradually from nuclear envelope/endoplasmic reticulum to Golgi apparatus and then rises sharply at the plasma membrane (Morré et al., 1971a; Fig. 9, see also Bergeron et al., 1975). With other enzymes (e.g., certain nucleoside diphosphatases) and chemical constituents (e.g., spingomyelin, cholesterol, sialic acid, neutral sugars) the relative specific activity (or amount per unit mass of membrane) in the Golgi apparatus fractions is at a level more nearly intermediate between those of the two reference fractions (endoplasmic reticulum < Golgi apparatus < plasma membrane).

The plasma membrane is a principal site of hormone and virus receptors and of other informational molecules that mediate cell-cell interactions, cell-environment interactions, and cell, organ and tissue specificity (see, for example, Cook and Stoddart, 1973; Clarkson and Baserga, 1974; Cuatrecasas, 1974). Many of these are proteins or glycoproteins, but glycolipids have also been described as membrane receptors (see section 3.4). Occasionally, such receptors may already exist in the Golgi apparatus. The endoplasmic reticulum and, to a lesser extent, the Golgi apparatus from rat liver exhibit glucagon-stimulated adenylate cyclase activities (Morré et al., 1974c; see McKeel and Jarett, 1974 for data from porcine adenohypophysis fractions). Also, the Golgi apparatus from

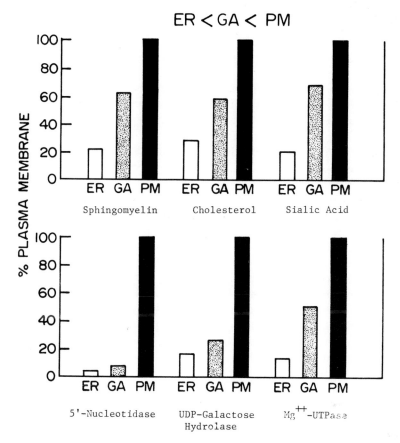

Fig. 9. Biochemical evidence for endomembrane differentiation. Relative specific activities or amounts of constituents concentrated in plasma membranes (PM) comparing endoplasmic reticulum (ER), Golgi apparatus (GA) and plasma membrane (PM) of rat liver. Other constituents showing a similar trend (ER < GA < PM) include Mg^{2+} -nucleoside triphosphatases (ATP, UTP, CTP, ITP or TTP as substrate), alkaline phosphatase nucleoside diphosphatase (ADP, CDP or TDP as substrate), cerebrosides, higher gangliosides, total glycoprotein carbohydrate, and saturated fatty acids. (Studies of Cheetham et al. [1970]; Keenan and Morré [1970]; Keenan et al. [1972b]; Morré et al. [1971b, 1974b] and Yunghans et al. [1970].)

rat liver is reported to bind insulin (Bergeron et al., 1973b). Functionally, such activities may be viewed as in transit to their site of primary action at the sinusoidal surface (Evans et al., 1973) or contributing in some unknown manner directly to intracellular recognition and/or regulatory phenomena.

The situation with regard to ion-stimulated ATPases, especially the classic Na^+-K^+-ATPase, is clearly an exception. This activity has not been observed in Golgi apparatus (Morré et al., 1974b; Elder and Morré, 1976a) and, although present in sarcoplasmic reticulum of muscle (Stewart and MacLennon, 1974), may be unique to plasma membranes of other cell types such as rat liver. Prostaglandin E 1 receptors were localized in plasma membranes of rat liver but were not found in rough microsomes, Golgi apparatus, or nuclei (Smigel and Fleischer, 1974).

4.2.2. Endoplasmic reticulum and/or nuclear envelope markers

Enzymatic activities characteristic of endoplasmic reticulum of rat liver include glucose-6-phosphatase, certain nucleoside diphosphatases, and an electron transport system unique to endoplasmic reticulum (Table 2; Fig. 10). Generally these enzymes are present in the Golgi apparatus at levels 15 to 30% of those found in the endoplasmic reticulum. Their activities are even lower or absent from the plasma membrane.

With few exceptions (e.g., microsomal oxidoreductases of nuclear envelope do not increase following administration of drugs such as phenobarbital [Kasper, 1971]), the enzymatic activities of nuclear envelope and of rough endoplasmic reticulum are at least qualitatively similar if not identical (Franke, 1974a,b; Franke and Scheer, 1974). The functional equivalence of nuclear envelope and rough endoplasmic reticulum is supported by the observations of Flickinger (1970; 1974a), where rough endoplasmic reticulum was shown to connect with nuclear envelope during the repair of experimentally damaged nuclei in ameba.

For glucose-6-phosphatase, the generally accepted marker enzyme for endoplasmic reticulum of rat liver, a controversy exists as to whether the low activity in the Golgi apparatus fraction (Morré et al., 1971a) is indigenous to this cell component or due to contamination of the fraction by endoplasmic reticulum or lysosomes (Fleischer and Fleischer, 1970; Bergeron et al., 1973a; Morré et al., 1974b). Cheetham et al. (1970) and Farquhar et al. (1974) were unable to demonstrate glucose-6-phosphatase in Golgi apparatus either in situ or with cell fractions by cytochemistry (see, however, below). Findings of Morré et al. (1974b) show a low level of activity, approximately 10% that of endoplasmic reticulum, which could not be attributed to either an unspecific lysosomal acid hydrolase or to contamination by endoplasmic reticulum. This low level of activity, even if concentrated in cisternae of the forming face (the specific activity of glucose-6-phosphatase in peripheral tubules and cisternal plates is nearly equal [Morré et al., 1971a; Ovtracht et al., 1973]), might elude cytochemical detection (e.g., Cheetham et al., 1970; Farquhar et al., 1974) and is evident in cytochemical preparations of others (e.g., Leskes et al., 1971; Hugon et al., 1973; Kartenbeck, 1974).

Electron transport is ubiquitous in biological systems and the endomembrane system is no exception. In parallel with mitochondrial electron transport, the electron transport system of endomembranes (microsomal electron transport) possesses flavoproteins in asociation with nonheme iron. These flavoproteins oxidize nicotinamide nucleotides (NADH and NADPH) at the low-potential end of the chain. Chains or arrays of cytochromes at intermediate potentials pass electrons to high-potential acceptors and finally to oxygen. However, the components and functions of the mitochondrial and microsomal electron transport systems are distinct. None of the NADH or NADPH-oxidizing flavoproteins in

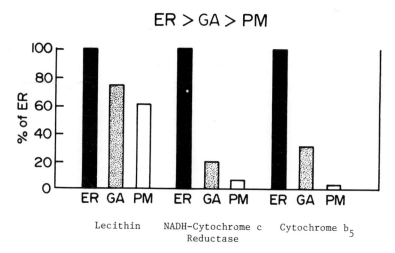

Fig. 10. Biochemical evidence for endomembrane differentiation. Relative specific activities or amounts of constituents concentrated in membranes of endoplasmic reticulum (ER) comparing endoplasmic reticulum, Golgi apparatus (GA), and plasma membrane (PM) of rat liver. Other constituents showing a similar trend (ER > GA > PM) include NADPH-cytochrome c oxidoreductase, UDP-glucuronyl transferases, arylsulfatase c, glucose-6-phosphatase, nucleoside diphosphatase (UDP, GDP or IDP as substrate), cytochrome P_{450}, and unsaturated fatty acids. (Studies of Keenan and Morré [1970]; Morré et al. [1971a,b; 1974 a,b] and Yunghans et al. [1970].)

microsomes are identical with the corresponding mitochondrial enzymes. Cytochrome b_5 and cytochrome P_{450} (a b-type cytochrome) are readily distinguishable from mitochondrial b-type cytochromes (Omura et al., 1965). Yet very little is known about the high-potential end of the electron transport chain of the endomembrane system. It seems to function solely in reactions involving transfer of reducing equivalents, as in mixed function oxidations of drug detoxification and carcinogen activation or in fatty acid desaturations, rather than in ATP generation (Siekevitz, 1965).

Generally, the ratio of P_{450} cytochrome b_5 of Golgi apparatus is somewhat less than that for endoplasmic reticulum (Ichikawa and Yamamoto, 1970; Morré et al., 1971a; Table 2). However, Fleischer et al. (1971) report little or no P_{450} in Golgi apparatus fractions from bovine liver. Cytochrome P_{450} is present in Golgi apparatus carefully isolated from rat liver (Morré et al., 1971a; Jarasch E. D., unpublished observations) at a level about one-third that of the endoplasmic reticulum. The total iron contents of Golgi apparatus and of endoplasmic reticulum are nearly identical (Table 2) due to the presence of considerable nonheme iron in Golgi apparatus.

Golgi apparatus from rat liver prepared by Fleischer and Fleischer (1970) contained little or no rotenone-insensitive NADH- or NADPH- cytochrome c reductase activity, whereas Golgi apparatus fractions of bovine liver had substantial levels of these activities (Fleischer et al., 1969). These findings contrast those of Morré and co-workers (Morré et al., 1971a, 1974a,b; Frantz, 1973) and

Bergeron et al. (1973a), who report these activities in Golgi apparatus fractions of greater than 90% purity at 0.12 (Morré et al., 1971a) to 0.25 (Bergeron et al., 1973a) the levels in endoplasmic reticulum (Table 2). By cytochemical methods, NADH-oxidoreductases are found concentrated in membranes of the immature face of the Golgi apparatus plus peripheral tubules (Morré et al., 1974b; Vigil E. L., and Morré, D. J., unpublished observations). In this regard the NADH and NADPH oxidoreductase activities were lowest in the lightest Golgi apparatus fractions of Bergeron et al. (1973a), i.e., those presumably derived from the lipid-filled secretory vesicles released during subfractionation (Merritt and Morré, 1973). These findings, along with evidence concerning plasma membrane markers and cytochemical findings already summarized suggest a functional polarity of the Golgi apparatus in which endoplasmic reticulum constituents are concentrated at the forming face and are gradually lost or replaced by plasma membrane constituents toward the maturing or secretory face of the stacked cisternae.

Progressive loss of enzyme activities and constituents of the electron transport system of endoplasmic reticulum (first the enzyme activities, then the prosthetic groups [the cytochromes]) within the Golgi apparatus has been suggested. This possibility was based on the ratios of these constituents in Golgi apparatus relative to endoplasmic reticulum plus the observation that the total iron (heme + nonheme) contents of endoplasmic reticulum and Golgi apparatus of rat liver were similar (Morré et al., 1971a). Much of the iron of liver Golgi apparatus fractions may be present as hemosiderin granules in the secretory vesicles and associated with lysosomes, an observation consistent with the findings of Schmid et al. (1966) that microsomal P_{450} is an important precursor of bile pigment.

More recent findings show that flavoprotein enzymes are present in Golgi apparatus (flavin contents of Golgi apparatus may exceed that of endoplasmic reticulum) where they transfer electrons from NADH to ferricyanide (thus bypassing both cytochrome b_5 and P_{450}) at rates more nearly equal to those of endoplasmic reticulum (Morré et al., 1974a). Thus, even though cytochromes of the b-type can be detected spectrally at levels 30% those of endoplasmic reticulum, their inability to accept electrons from the flavoprotein reductases may account for the much lower electron transfer levels to exogenous cytochrome acceptors (e.g., cytochrome c) rather than an exclusion of the flavoprotein enzymes from the membrane. Flavoprotein NADH reductase activity is also indigenous to plasma membranes (Frantz, 1973; Vigil et al., 1973; Emmelot et al., 1974; Jarasch et al., in press). This activity may function at the plasma membrane in some species as part of the more familiar xanthine oxidase complex (E. D. Jarasch and W. W. Franke, personal communication).

4.2.3. Analysis of Golgi apparatus markers
and mechanisms of membrane differentiation
Most of the enzymes studied in cell fractions, where comparisons with corresponding fractions of endoplasmic reticulum and plasma membrane are possible (Figs. 9 and 10), have specific activities intermediate between those of the endo-

plasmic reticulum and plasma membrane. Only thiamine pyrophosphatase (e.g., Cheetham et al., 1973) and certain glycosyl- and sulfotransferase activities (section 3.3–3.6) are concentrated in the Golgi apparatus in comparison with the other fractions (Fig. 11).

The enzymes concentrated in the Golgi apparatus for elaboration of glycoproteins and glycolipids, as well as the lipid and protein biosynthetic activities discussed in section 3, are sufficient to account for glycerophospholipid, glycoprotein, and glycolipid changes attendant to membrane differentiation. The very large net increase in the content of sphingomyelin and cholesterol (Keenan and

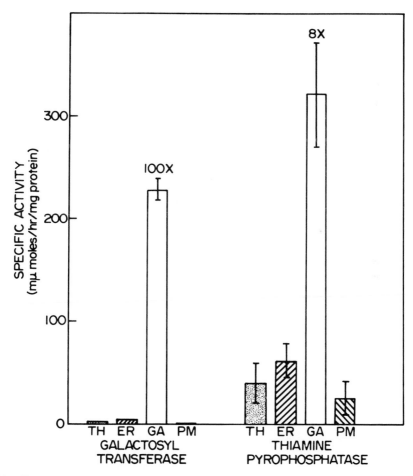

Fig. 11. Enzymes concentrated in Golgi apparatus. Specific activities of galactosyltransferase and thiamine pyrophosphatase comparing total homogenate (TH), endoplasmic reticulum (ER), Golgi apparatus (GA), and plasma membrane (PM) fractions from rat liver. Whereas Golgi apparatus fractions show a 100-fold enrichment relative to the total homogenate for galactosyltransferase, the comparable enrichment for thiamine pyrophosphatase in only 8-fold. (Studies of Morré et al. [1969] and Cheetham et al. [1971].)

Morré, 1970) may also be due to biosynthetic activities at or near the Golgi apparatus, but information on the subcellular location of sphingomyelin and sterol biosynthetic enzymes within the microsomal fraction is incomplete. Synthesis occurs elsewhere and transfer is to either Golgi apparatus or plasma membrane via a flow mechanism or cytoplasmic carrier proteins (see Wirtz, 1974 for sphingomyelin, and Dennick, 1972 and Scallen et al., 1974 for cholesterol).

The nature and mechanism of changes in enzyme activities during membrane differentiation are unknown. On gel electrophoresis, membranes of rough endoplasmic reticulum exhibit a greater number of protein bands than either Golgi apparatus or plasma membrane (Fleischer and Fleischer, 1970; Yunghans et al., 1970; Zahler et al., 1970; Meldolesi and Cova, 1972a; Franke and Kartenbeck, 1976; Elder and Morré, 1976a). This suggests exclusion of certain membrane proteins during membrane differentiation. An exclusion phenomenon may help to explain results of analyses of certain types of secretory granules such as chromaffin granules of adrenal cortex (Winkler and Hörtnagl, 1973) and zymogen granules from pancreas (McDonald and Ronzio, 1972; Meldolesi and Cova, 1972a) or parotid gland (Castle et al., 1975; Wallach et al., 1975). Here, fewer protein bands remain in the vesicle membranes than the 30 to 40 electrophoretic bands for endoplasmic reticulum (microsomes). Exclusion of proteins is also necessary to account for the formation of viral envelopes, where host lipids but no host proteins are thought to be retained (Rifkin and Quigley, 1974).

Some protein bands (not all of which are glycoproteins) and enzymatic activities (e.g., glycosyl transferases) appear in Golgi apparatus but are not found to any great extent in either endoplasmic reticulum or nuclear envelope. Other activities, such as ion-stimulated ATPases, are found only in plasma membranes (Morré et al., 1974b; Elder and Morré, 1976a). For these, an insertion mechanism might be operative; the Golgi apparatus-associated polyribosomes (Section 3.2) provide one potential source of such new proteins specifically destined for insertion into Golgi apparatus membranes. An insertion mechanism is also favored by the findings of Heine et al. (1972) for virus envelope formation, where a complete cessation of insertion of new host proteins into the membranes accompanies the appearance of viral-induced proteins and glycoproteins within an apparently existing membrane.

5. Membrane flow

The concept of membrane flow (Franke et al., 1971) is predicated on the assumption that biogenesis of cellular endomembranes is facilitated by physical transfer of membrane material from one cell component to another. Applied originally to endocytosis (Bennett, 1956), membrane flow provides an equally attractive mechanism to explain exocytosis as well as transfer processes associated with membrane biogenesis and differentiation (Morré and Mollenhauer, 1974). Without some sort of flow mechanism to account for transport of membrane constituents from one region of the cell to another, processes of membrane differentiation in the Golgi apparatus would explain the origins of no new membrane per se and would represent only a static manifestation of functional polarity.

5.1 Evidence

As emphasized by Franke and Kartenbeck (1976), membrane flow is not necessarily random but may be highly selective for specific membrane components or constituents. It is now nearly certain that some components or constituents are transferred while others are excluded. Thus, evidence for bulk flow of membranes is insufficient and indications for flow of membrane constituents, either singly or as discrete multicomponent complexes, must also be sought.

5.1.1. Morphological manifestations

Golgi apparatus of both plant and animal cells (Morré et al., 1971c) are organized into stacks of cisternae (dictyosomes). Dictyosomes are polarized structures, and polarity is expressed by differences in the form and composition of successive cisternae and by the nature of dictyosome association with other endomembrane components. The forming face, or proximal pole of the dictyosome is frequently associated with either endoplasmic reticulum or nuclear envelope while the maturing face, or distal pole is involved with the release of mature secretory vesicles.

The relationship between Golgi apparatus and endoplasmic reticulum is most obvious as a single cisterna of endoplasmic reticulum aligned along the forming face of a single dictyosome or group of dictyosomes (Friend, 1965; also see Morré et al., 1971d for additional references). A similar relationship occurs between the Golgi apparatus and the nuclear envelope when dictyosomes lie adjacent to the nuclear envelope. There are small evaginations, or blebs from the cisternal walls of these parts of associated endoplasmic reticulum or nuclear envelope, and small vesicular profiles (coated or uncoated) are observed between the endoplasmic reticulum (or nuclear envelope) and the nearest cisterna of the dictyosome. These relationships are illustrated in Figures 12 a,b and diagramatically in Figure 2. Direct connections between endoplasmic reticulum and Golgi apparatus cisternae and between nuclear envelope and Golgi apparatus cisternae in this region have been reported for fungi (Bracker et al., 1971). Electron microscope observations by numerous investigators (e.g., Kessel, 1971; and see Morré et al., 1971d for literature) support the early proposals of Dalton (1961), Zeigel and Dalton (1962), and Novikoff et al. (1962) that these vesicles function in the exchange of membrane precursors and possibly of other materials between contiguous endomembrane components such as endoplasmic reticulum and Golgi apparatus.

In other studies, relationships between endoplasmic reticulum and Golgi apparatus have been shown to involve direct connections between the two structures at the dictyosome peripheries (Mollenhauer et al., 1975; Franke and Kartenbeck, 1976; Mollenhauer and Morré, 1976b) and directly between endoplas-

44

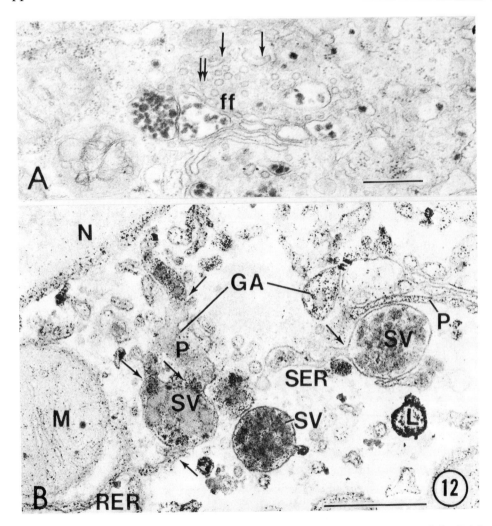

Fig. 12. Regions of endoplasmic reticulum-Golgi apparatus continuity. (a) Portion of the Golgi apparatus of rat liver showing numerous transition vesicles at the forming face (ff) of the dictyosome. The origin of these vesicles is thought to be endoplasmic reticulum (arrows). From there, the vesicles are presumed to migrate and fuse to form new dictyosome cisternae (double arrow). (b) Transport of lipoprotein particles from smooth endoplasmic reticulum (SER) to secretory vesicles (SV) of the Golgi apparatus (GA) of rat liver seems to occur via direct tubular connections (single arrows). To increase the visibility of the tubules, the tissue was stained for the cytochemical localization of arylsufatases. P = platelike portion of Golgi apparatus cisterna. RER = rough endoplasmic reticulum. L = lysosome. M = mitochrondrion. N = nucleus. Scale marker = 0.5μm. (Courtesy of Morré et al. [1974a].)

mic reticulum and forming secretory vesicles (Morré et al., 1961b, 1974a; Ovtracht et al., 1973; Fig. 12 c, Figs. 15 and 16). The extent to which such direct connections play a part in dynamic endoplasmic reticulum-Golgi apparatus associations is unknown. More common, perhaps, is the involvement of small primary vesicles that provide transient connections, or of intervening tubules or vesicles that bridge adjoining cell components (Figs. 12 a,b; and see Morré and Mollenhauer, 1974).

The secretory vesicle is the most extensively documented example of a cell component derived from one membrane structure (the Golgi apparatus or other transition element), which is released from that membrane structure to migrate to and coalesce with another membrane structure (the plasma membrane) to effect the physical transfer of membrane (membrane flow). Recently, an origin of secretory vesicles at the Golgi apparatus has been visualized for plant roots in experiments where cytochalasin B was used to block vesicle migration (Mollenhauer and Morré, 1976a) so that vesicles accumulate in large numbers at the mature face around each dictyosome. Secretory vesicle migration has been monitored in certain algae using the light microscope and cinematography (Schnepf and Kock, 1966a,b; Brown, 1969). Here, both the origin of the vesicle from the Golgi apparatus and its final fusion with the plasma membrane may be viewed in living cells.

5.1.2. Other evidence

Experimental evidence for the continuous renewal of dictyosome cisternae was first reported by Grimstone (1959) from experiments with starved and refed *Trichonympha*. Other examples exist where the numbers and size of dictyosome cisternae increase or decrease with changes in secretory activity, or under conditions where secretory vesicle production is stopped or stimulated experimentally (see Mollenhauer et al., 1975 for references). Since maturing cisternae are observed to separate ultimately from the dictyosome in at least those systems where markers are available (Brown, 1969; Falk, 1969; Mollenhauer, 1971; Dobberstein and Kiermayer, 1972; Kiermayer and Dobberstein, 1973), maintenance of a constant number of dictyosome cisternae must depend upon some form of membrane renewal.

Progressive appearance of membrane proteins labeled with amino acids (Ray et al., 1968; Franke et al., 1971c; Morré et al., 1974a; and section 3.2), of labeled membrane lipids (Chlapowsky and Band, 1971; Morré et al., 1974a), or of drug-induced NADPH-oxidoreductase activity (Morré et al., 1974) in rough endoplasmic reticulum, smooth endoplasmic reticulum, and Golgi apparatus are consistent with a flow mechanism. During regeneration of flagella in the green alga *Ochromonas*, extracellular appendages called mastigonemes provide a visual marker to monitor the flow of membrane from the nuclear envelope where the appendages first appear, via the endoplasmic reticulum and Golgi apparatus, to their final destination on the flagellar plasma membrane (Bouck, 1969, 1971; Hill and Outka, 1974; Morré and Mollenhauer, 1976 for additional references).

5.2. Kinetics

From autoradiographic analyses of the movement of secretory products (Neutra and LeBlond, 1966a,b), direct observations of algal cells (Schnepf and Koch, 1966a,b; Brown, 1969; Williams, 1974) and of surface contributions from membranes of secretory vesicles to the plasma membrane (Schnepf, 1961, 1969; Heinrich, 1973; Morré and Mollenhauer, 1974), it has been estimated that each dictyosome turns over within 8 to 40 minutes average (20 min) and that mature cisternae are released at the rate of one every 1 to 4 minutes. These findings are consistent with kinetics of accumulation and disappearance of both bulk (Franke et al., 1971c) or individual (Morré et al., 1974a, 1976; Kartenbeck, 1974; Franke and Kartenbeck, 1976) membrane proteins through the Golgi apparatus. Labeled membrane proteins first appear at the Golgi apparatus within 1 minute after administration of label, reach a plateau between 10 and 20 minutes and then drop sharply, so that after 30 minutes further losses of radioactivity follow an exponential decay curve from which an average half-life of 37.5 hours for the remaining membrane proteins has been determined (Franke et al., 1971c). These kinetic fluctuations in Golgi apparatus correlate with corresponding changes in endoplasmic reticulum and plasma membrane consistent with a dynamic model for Golgi apparatus functioning. Apparently, transfer of membrane from endoplasmic reticulum and the addition of newly synthesized proteins at the Golgi apparatus are completed between 10 and 15 minutes after isotope injection. Membrane differentiation and/or maturation then continues for another 10 to 20 minutes, so that by 20 to 30 minutes the membrane components exit from the Golgi apparatus as membranes surrounding secretory vesicles or as released cisternae or cisternal fragments. Migration of vesicles is rapid, the T_4 of secretory vesicles in the cytoplasm may be less than 1 minute (Morré et al., 1967; Morré and Mollenhauer, 1974; Schnepf, 1961; Bowles and Northcote, 1974) in plant cells. As the vesicle membranes fuse with the plasma membrane, they are incorporated into the plasma membrane, a process that probably occurs within milliseconds (Satir, 1974a). Thus, the labeling patterns of membrane proteins of the Golgi apparatus are consistent with a total elapsed time of about 20 minutes for cisternal formation and breakdown, as deduced previously from autoradiographic studies (Neutra and Leblond, 1966a,b).

5.3. Vectorial flow differentiation of membranes

One source of misunderstanding and confusion in the interpretation of findings relevant to membrane flow and differentiation is the incorrect assumption that the two processes are mutually exclusive. Beginning with the early morphological studies (e.g., Grove et al., 1968), it was apparent that for membrane flow to adequately explain the origin of Golgi apparatus membranes from endoplasmic reticulum (or of plasma membrane from either Golgi apparatus or endoplasmic reticulum) considerable modification and/or reorganization of membrane composition was required (e.g., Morré and Mollenhauer, 1974; Fig. 3.19, Franke and Kartenbeck, 1976). Proteins, especially, must be both added and deleted (enzymes perhaps activated and inactivated, in addition) to achieve the final composition of the fully differentiated membrane. Similarly, considerable remodeling of the lipid and carbohydrate composition of the membrane is also indicated. The process may not be as simple as lateral fluidity of surface membranes (Frye and Edidin, 1970; reviewed by Singer, 1974), although even here fluid components with relatively high mobility and a tendency to redistribute randomly are found along with fixed components that are nonrandom in distribution and possibly constrained in some manner (Edelman et al., 1973; Nicolson, 1974; Pinto Da Silva and Martínez-Palomo, 1975; Nicolson et al., 1977). It is perhaps best to regard the process as *flow differentiation* where the two processes, membrane flow and membrane differentiation, operate in concert to ensure that membrane flow is not random but highly selective so that some components are transferred while others are excluded (see Nicolson et al., 1977).

Excess membrane added to the cell surface of nonexpanding cells by secretory or other types of vesicles must be returned to the cell interior to maintain a relatively constant amount of plasma membrane. Some membranes from the cell surface return as exocytotic vesicles (see review chap. 4, this volume). As perhaps an extreme example, macrophages and L cells are reported to form in the order of 125 pinocytotic vesicles per minute and to interiorize the equivalent of their cell surfaces every 33 to 125 minutes (Steinman et al., 1976). Membrane degradation at the molecular level (Fawcett, 1962) is also indicated as a mechanism of membrane retrieval especially in slime-secreting cells where internalization of intact membranes is not observed (Morré and Mollenhauer, 1974; Schnepf and Busch, 1976).

5.4. Cyclic membrane flow

Some evidence suggests the synaptic vesicles originate in Golgi apparatus or other transitional elements (Gray, 1970; Brunngraber, 1972; Griffith and Bondareff, 1973). Uncertainty still exists and related questions involve the possible recycling of membrane-bounded compartments within the region of the synapse (Holtzman et al., 1971; Douglas et al., 1971; Ceccarelli et al., 1973; Heuser and Reese, 1973; Chap. 4, this volume). Secretion of catecholamines (Winkler, 1971) and zymogens (see section 6.1.4) have been considered in relation to cyclic processes in which the granule membranes originate in the Golgi apparatus, are exocytosed, and their membranes then returned to the cytoplasm. The fate of the interiorized membrane is unclear, and there is insufficient evidence to decide if the membranes "retrieved" by endocytosis or degradation are the same as the original membrane of the secretory granule that fused with the plasma membrane. A simple shuttle mechanism is considered unlikely because of the implications inherent in all perpetual motion machines. A major tenet of biochemistry is that synthetic pathways are rarely if ever the reverse of degradative pathways, and vice versa.

6. Specific examples of Golgi apparatus participation in membrane biogenesis

Several examples of well-documented contributions of Golgi apparatus membranes to plasma membrane have already been discussed. Although plant cells

48

provide the clearest examples, significant contributions of Golgi apparatus or Golgi apparatus-derived secretion vesicles to plasma membranes of mammalian cells are also indicated.

6.1. Role of the Golgi apparatus
in biogenesis of plasma membranes

Generally, a role of the Golgi apparatus in biogenesis of plasma membranes is deduced from morphological observations, i.e., elaboration of secretory vesicles and their subsequent coalescence with the plasma membrane. Thousands of publications over the past three decades have supported this general observation. Since the selection of examples must be somewhat arbitrary, many excellent studies cited previously are not mentioned here. For additional examples, the reviews by Mollenhauer and Morré (1966a), Schramm (1967), Beams and Kessel (1968), Dauwalder et al. (1969), Schnepf (1969), Morré et al. (1971d), Dauwalder et al. (1972), Cook (1973), and Morré and Mollenhauer (1976b), and the recent book by Whaley (1975) may be consulted.

6.1.1. Algae and epithelial cells of the urinary bladder
Examples from algae include membrane contributions associated with the stepwise assembly and secretion of complex cell wall subunits and secretion of surface scales of Chrysophytes as detailed by Brown, Franke, Herth, and co-workers (Brown et al., 1969, 1970, 1973; Herth et al., 1972; Williams, 1974), and in Prasinophytes (Moestrup and Thompsen, 1974), certain fungi (Darley et al., 1973) and Labyrinthula-like organisms (survey in Dykstra, 1976); in water expulsion (Schnepf and Koch, 1966a,b), flagellar regeneration (Bouck, 1969, 1971; Hill and Outka, 1974), and primary (Kiermayer, 1973, and Ueda and Noguchi, 1976 for literature) and secondary (Dobberstein and Kiermayer, 1972; Kiermayer and Dobberstein, 1973; Ueda and Noguchi, 1976) wall formation in desmids. Other examples are summarized in the excellent book by Dodge (1973). Even if only the membrane contribution to surface materials is considered, the Golgi apparatus-mediated contribution is substantial. In *Vacuolaria* and *Glaucocystis,* Schnepf and Koch (1966a,b) estimate that an amount of Golgi apparatus membrane equivalent to the total surface of the cell is secreted every 10 minutes. Thus, during active water secretion, the plasma membrane must turn over six times every hour; the immediate source of new plasma membrane is the Golgi apparatus.

In other examples, Golgi apparatus of algal cells for special cisternae are the source of highly differentiated plasma membrane regions during specific developmental stages (Falk, 1969; Dobberstein and Kiermayer, 1972; Kiermayer and Dobberstein, 1973; Ueda and Noguchi, 1976). Unusual disclike vesicles are formed with very thick membranes (160–200 Å) and 200 Å globular particles on the inner surface. In some ways they seem structurally analogous to the specialized plasma membrane surface of the bladder epithelium that also receives substantial contributions from similarly differentiated Golgi apparatus cisternae (Hicks, 1966). These contributions by Golgi apparatus to plasma membranes are especially important since they seem to involve no special secretory products, only the delivery of preformed plasma membrane units.

6.1.2. Tip-growth

Cells that elongate by tip growth, such as elongating nerve fibers (Pfenninger and Bunge, 1974), pollen tubes, rhizoids, fungal hyphae, and plant hairs (Roelofsen, 1959), obtain new plasma membrane predominantly, if not exclusively, via secretory vesicles from the Golgi apparatus (Fig. 13; for literature, see Grove et al., 1970). For pollen tubes of Easter lily *(Lilium longiflorum)*, Morré and VanDerWoude (1974) calculated that the Golgi apparatus in these cells produce and export more than 1000 secretory vesicles per minute to generate 300 μm^2 of new plasma membrane per minute during steady-state growth.

6.1.3. Mucous and slime secretion

In a classic series of investigations, Neutra and Leblond (1966a,b; Peterson and Leblond, 1964a,b) utilized autoradiography in combination with electron microscopy to show that mucin secretion in goblet cells of the colon is continuous. The mucous is delivered to the surface of the intestine in the form of secretory vesicles (mucigen granules). They calculated that a distal or mature Golgi apparatus cisterna is converted into mucigen granules every 2 to 4 minutes to account for the kinetics of labeling and the number of vesicles produced. The membranes of the mucigen granules are presumably incorporated into the plasma membrane. Progressive acquisition of carbohydrate products reacting as cell coat materials by Golgi apparatus and secretion granules have been shown by electron microscope cytochemistry for a variety of intestinal cells including those of enterocytes which do not elaborate a mucous secretion (L. Ovtracht and J.-P. Thiéry, personal communication).

With sugars, the assumption is usually made that either secretory products or coat materials are preferentially labeled (Leblond and Bennett, 1974). However, the serine-N-acetylgalactosamine class of mucins contains no mannose (Schachter, 1974a). Thus in mucous glands such as the intestinal goblet cells, the mannose label for autoradiography might be a specific membrane label (H. Schachter, personal communication).

Nearly parallel results have been obtained with mucin or slime-secreting cells of plants (see Schnepf, 1969; Morré and Mollenhauer, 1974, 1976b, for literature). Electron microscopy shows progressive changes in membrane staining and dimensions from one face of a dictyosome to the other in slime-secreting cells of the maize root cap (Morré and Mollenhauer, 1974), from a membrane type that resembles endoplasmic reticulum to one that resembles plasma membrane. Roland et al. (Roland, 1969; Roland and Vian, 1971; Frantz, 1973) used phosphotungstic acid of low pH to show that mucin-containing secretory vesicles acquired plasma membrane characteristics in advance of their fusion with the plasma membrane. Recently, the origin of these vesicles from the Golgi apparatus has been dramatically visualized for outer cap cells of the maize root in

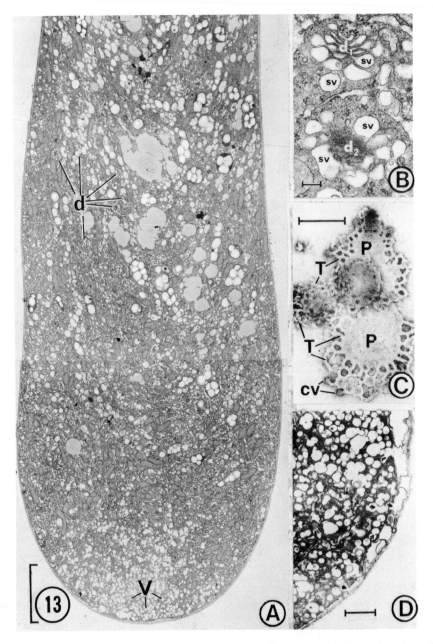

Fig. 13.   Apical regions of pollen tubes of Easter lily *(Lilium longiflorum)*. (a) Approximately median
longitudinal section showing secretory vesicles (sv) concentrated at the apex of the tube and the
numerous dictyosomes (d) of the Golgi apparatus located distal to the apex, from which the vesicles
arise. Montage of three electron micrographs. Glutaraldehyde-acrolein-osmium tetroxide fixation.
Scale marker = 5 μm. (b) Two dictyosomes (d). One is sectioned tangentially (d $_2$) to show the secretory
vesicles (sv) attached to the central platelike portion of the cisternae via the system of peripheral

experiments where cytochalasin B was used to block vesicle migration to the cell surface (Mollenhauer and Morré, 1976a). The production of vesicles continues, so that vesicles accumulate in great numbers at the mature face in the vicinity of each dictyosome.

As with mucous-secreting cells of the intestine, membrane turnover in slime-secreting cells in plants must be considerable. From morphometric measurements, Schnepf and Busch (1976) calculated that membranes of secretory vesicles replace 3% of the plasma membrane each minute in mucous gland of *Mimulus tilingii*. This rate is by no means an extreme. It is similar to that for outer root cap cells of maize and lower than that for algae and special slime glands of insectivorous plants (Schnepf, 1961).

In contrast to dividing and tip-growing cells, mucous and slime-secreting cells normally are not growing; neither does the plasma membrane increase in surface area or thickness. Thus, the plasma membrane must turn over at a rate commensurate with the rate of addition of new membrane, i.e., 3% per minute during steady-state secretion. In spite of secretion times of 20 hours or longer and rates of more than 500 vesicles per minute, there are no morphological indications that membrane flow from Golgi apparatus to plasma membrane in these cells is compensated for by backflow in the form of vesicles or other organized membrane structures (Morré and Mollenhauer, 1974; Schnepf and Busch, 1976). Rather, reversed flow of membrane materials must take place in micelles or molecules, as recently demonstrated for membranes of transplanted nuclei of *Amoeba proteus* (Maruta and Goldstein, 1975).

Here, and with protein and lipoprotein secreting cells discussed in the next section, the contribution of secretory vesicles to the plasma membrane is of questionable functional significance in contrast to the net plasma membrane biogenesis of, for example, tip-growing cells. In the latter a fully functional plasma membrane appears to be formed, while in nongrowing secretory cells such a condition need not be fulfilled. Only some type of "minimal" surface membrane that maintains permeability and ensures the efficient continued operation of the secretory cycle need be delivered.

6.1.4. Protein and lipoprotein secretion

Secretion and the role of the Golgi apparatus in that process has been most thoroughly studied in hepatocytes active in secretion of albumin and circulating

---

Fig. 13 *(continued)*
tubules. Scale line = 0.2 μm. (c) A dictyosome, isolated and negatively stained with potassium phosphotungstate, to show the central platelike region (P) and the system of peripheral tubules (T). The small cisterna from near the forming face (top) is almost entirely tubular while the cisternae nearer the maturing face have extensive platelike regions. Coated vesicles (cv) attached to the cisternal tubules are a consistent feature of all dictyosome cisternae. Scale marker = 0.5μm. (d) Enlargement of the pollen tube apex. Images of vesicle fusion (small arrows) are common in this region. Scale line = 1 μm. (Courtesy of Morré and VanDerWoude [1974].) In this and other tip-growing systems, such as neurons, additions of Golgi apparatus-derived vesicles provide an attractive mechanism for surface growth.

lipoproteins, chromaffin cells of the adrenal medulla, and especially in zymogen-secreting cells of pancreas and parotid gland. Although emphasis is frequently placed on secretion and the processing of products destined for secretion, these studies have also yielded information relevant to the role of the Golgi apparatus in membrane biogenesis.

Secretion of zymogens by the pancreas and parotid gland and secretion of catecholamines by chromaffin cells of the adrenal medulla exhibit many parallels. In contrast to liver, secretory vesicles appear to be formed at or near the Golgi apparatus and then function as condensing vacuoles, i.e., they are filled with secretory products over periods of several hours. During the filling period, the vesicles do not increase in diameter (Jamieson and Palade, 1967b) but instead accumulate secretory products gradually increasing in electron opacity. The filling of the condensing vacuoles may occur several microns distant from the Golgi apparatus. It is not known whether the condensing vacuoles or secretory granules are even attached to the Golgi apparatus during much of the filling period.

Clearly, the zymogen granule proteins are synthesized at the rough endoplasmic reticulum and initially sequestered within the lumen of the endoplasmic reticulum (Morris and Dickman, 1960; Siekevitz and Palade, 1958, 1960; see Schram, 1967, and Palade, 1975 for reviews). However, in pancreas, the Golgi apparatus cisternae seem to play only a minor role in mediating the entry of these proteins into the zymogen granules. The proteins appear to enter via a peripheral route (Jamieson and Palade, 1967a).

Several mechanisms have been proposed to account for transfer from endoplasmic reticulum to condensing vacuoles or secretory granules. In the first mechanism, small primary vesicles are suggested to bleb off endoplasmic reticulum, migrate to the zymogen granule, and fuse with it to deliver small "quanta" of both secretory proteins and membranes. Since this type of delivery mechanism must lead to a significant accumulation of excess membrane at the zymogen granule, a shuttle mechanism has been invoked (Meldolesi, 1974a,b; Palade, 1975) whereby empty vesicles return to the endoplasmic reticulum to recycle membrane and permit accumulation of secretory products in the condensing vacuoles.

An alternative mechanism is the entry of secretory products via direct tubular connections, as has been demonstrated for secretion of lipoproteins in rat liver (Morré et al., 1971b, 1974a; Ovtracht et al., 1973). Vesicles are formed from Golgi apparatus and products enter via the system of peripheral smooth tubules independently of further Golgi apparatus activity and without any need to invoke a shuttle mechanism.

In rat liver, the lipoprotein-containing vesicles are formed and filled while attached to the Golgi apparatus (Claude, 1970; Glaumann et al., 1975). Once released, they migrate rapidly to the plasma membrane and do not accumulate in the cytoplasm unless the process is inhibited, as for example by colchicine (Stein et al., 1974). In contrast, mature secretory ganules of pancreas, parotid gland, and chromaffin cells accumulate in the cytoplasm. In exocrine cells of pancreas they occupy about 7% of the cytoplasmic volume (Bolender, 1974) and up to as much as half of the cellular volume in parotid gland (Castle et al., 1975). Fusion is induced by secretagogue drugs or endogenous stimuli. The large accumulations of mature granules that may occur in the absence of secretagogue stimuli offer an experimental advantage in that mature granules can be isolated in sufficient quantity to permit biochemical analyses. The membranes of mature zymogen granules are composed of relatively few proteins that seem in part to be unique to this membrane for both parotid gland (Castle et al., 1975, Wallach et al., 1975); and pancreas (McDonald and Ronzio, 1972; Meldolesi and Cova, 1972a,b); as well as for chromaffin granules (Winkler and Hörtnagl, 1973). The published gel patterns of both zymogen granules and chromaffin granule membranes rule out direct transfer of bulk endoplasmic reticulum membranes. A lack of correspondence of granule membrane proteins with proteins of the plasma membrane remains to be demonstrated.

In spite of earlier results to the contrary (Amsterdam et al., 1971), it now appears that relative to secretory proteins of the granule contents, membrane proteins of zymogen granules of the parotid are poorly labeled, if at all, by exogenously supplied amino acids under conditions where the secretory proteins are heavily labeled (Wallach et al., 1975). A similar situation was reported with the membrane proteins of the zymogen granules of the pancreas (McDonald and Ronzio, 1972; Meldolesi, 1974a,c) and with chromaffin granules of the bovine adrenal medulla (Winkler et al., 1970, 1972). A proline-rich fraction selectively extracted from the zymogen granule membranes of parotid gland by 0.15 M NaCl or dilute buffer at pH 4.5 was labeled with ^{3}H-proline; however, much protein was unlabeled (Wallach et al., 1975).

As emphasized by Wallach et al. (1975), autoradiographic studies show that the labeled secretory proteins are concentrated in new secretory granules rather than being introduced randomly into mature old granules (Jamieson and Palade, 1967b; Winkler et al., 1972). Thus proteins of the granule membrane must be ready in the cell in some intermediate pool awaiting the arrival of the newly synthesized export proteins. Probably this pool is the secretory vesicles or "immature" zymogen granules of the Golgi apparatus (Wallach et al., 1975). That the granule membranes are significantly older than the proteins of the granule content in both zymogen and chromaffin cells may be an important factor to account for low specific activities of the membrane proteins.

Electron micrographs show that the secretory process in the parotid gland involves fusion of the secretory granule membrane with the cell membrane (Amsterdam et al., 1969; DeCamilli et al., 1976). Since both the zymogen and the chromaffin cells are nongrowing, the large amounts of new plasma membrane delivered in each secretory cycle must somehow be removed. A process of resorption is indicated from morphological studies (Amsterdam et al., 1969), but the fate of the resorbed membrane is unclear (Wallach et al., 1975). One suggestion is that after the membrane is resorbed its constituents are reutilized directly in the formation of new secretory granules (Palade, 1958; Hokin, 1968) via a second shuttle (Palade, 1975). Another possibility is that the secretory granule

membranes are degraded and the products used in redirected synthesis according to the concepts outlined by deDuve and collaborators (deDuve and Wattiaux,
1966; deDuve, 1969). In the study of Geuze and Poort (1973), ferritin markers
were used to establish that excess membrane was internalized in the form of
endocytotic vesicles in cells of the exocrine pancreas. After 1 hour, ferritin was
also present in lysosomes and digestive vacuoles along with whorls of membranelike material. In no instance, however, was ferritin observed within condensing vacuoles. In other studies, internalized plasma membrane receptors for
anti-immunoglobulin antibodies (Antoine et al., 1974) and ricin receptors
(Gonatas et al., 1975) appear associated with smooth vesicles in the area of the
Golgi apparatus equivalent to Novikoff's [Golgi apparatus-endoplasmic reticulum-lysosome (GERL)]. The ricin-horseradish peroxidase label appears within cisternae of the GERL complex (Gonatas et al., 1975). In the study of Cope
and Williams (1976), morphometric methods were used to monitor quantitatively the membrane content of various cell components after isoproterenol-
induced secretion in the rabbit parotid gland. Upon greater than 95% depletion
of granules, about 1340 μm^2 of granule membrane fused with the plasma membrane. Two hours later, 1150 μm^2 of this had been eliminated. However, only a
small increase in intracellular smooth membranes was recorded at any time
following degranulation and the authors could find "no evidence that the
zymogen granule membrane is stored as smooth membrane fragments either in
the region of the Golgi apparatus or elsewhere in the cytoplasm." Such findings
might argue against a shuttle mechanism of membrane recycling and favor
membrane recycling via breakdown and resynthesis.

6.1.5. Cell surface regeneration in amoebae

Amoebae provide a unique opportunity for the study of cell coat formation and
plasma membrane removal (Nachmias, 1966). A high rate of endocytotic turnover demands replacement of 0.14 to 0.21% of the membrane surface area per
minute (Stockem, 1972; Komnick et al., 1973). A role for the Golgi apparatus in
this process was suggested by Stockem (1969), who concluded from cytological
studies that large vesicles derived from Golgi apparatus were the source of new
plasma membrane as well as of the mucous layer (see also Wise and Flickinger,
1970).

More definitive evidence has come from investigations of Flickinger (1975) in
which amoebae were injected individually with a solution of ^{3}H-mannose and
examined by radioautography at intervals between 10 minutes and 24 hours
after injection. The Golgi apparatus was heavily labeled 30 minutes after injection, followed by groups of small cytoplasmic vesicles 1 to 2 hours after injection.
Labeling of the cell surface was low initially but rose rapidly after 2 hours to a
peak at 12 hours. The pattern of labeling implicates aggregates of small vesicles
and tubules (35 to 50 nm in diameter) in the movement from Golgi apparatus to
cell surface. Other types of vesicles, including the large, so-called "fringed vacuoles", are more clearly endocytotic in origin since they became heavily labeled
only after peak labeling of the cell surface. Similar conclusions were reached by
Sanders (1970) from studies of stratified and bisected amoebae.

The assumption that coat and membrane move as a unit in amoebae may be unwarranted. However, as summarized by Flickinger (1975), the carbohydrate-rich coat is firmly attached to the underlying membrane, the coat and membrane are isolated as a unit, and the coat of amoebae should be regarded as an integral part of the surface membrane.

6.1.6. Cell division

Most of the evidence for a role of the Golgi apparatus in plasma membrane biogenesis in dividing cells comes from studies with plants; far less is known about mammalian cells (see Morré et al., 1971c; Sanders and Singal, 1975 for references).

As emphasized by Whaley (1975), "that most plant cells divide by cell plate formation and not by cleavage or furrowing, presents an unusual opportunity to determine the actual contribution of the Golgi apparatus." Vesicles of en-domembrane origin, at least some of which are derived from the Golgi apparatus (Mollenhauer and Morré, 1966a for early literature), form the new wall (cell plate) which, in plants, separates daughter cells following division. The membranes of the vesicles are a source of new plasma membrane, while the contents of the vesicles contribute wall materials or enzymes (Whaley and Mollenhauer, 1963; Frey-Wyssling et al., 1964; Whaley et al., 1966; Cronshaw and Esau, 1968; Hepler and Jackson, 1968; Pickett-Heaps, 1967; 1968; Roberts and Northcote, 1970; Dauwalder and Whaley, 1974a). A role for coated vesicles in this process, possibly to remove excess membrane, has also been shown (Franke and Hearth, 1974).

Since cell plate formation begins in the midregion of the cell and extends bidirectionally from this point, formation of new plasma membrane by extension of an existing membrane is eliminated as a mechanism. Thus, there seems to be little doubt that, in plant cells, the new plasma membranes delimiting the cell plate are contributed by the Golgi apparatus and or other vesicles of endomembrane origin.

6.1.7. Ciliated protozoans

Membranes of mucocyst, trichocyst, contractile, and food vacuoles constitute additional examples of potentially important contributions of the Golgi apparatus to membrane biogenesis (McKanna, 1973; Plattner, 1974; Satir, 1974a,b). With trichocysts of *Paramecium*, Hausman and Allen (1976) concluded that after the trichocyst membrane fuses with the plasma membrane, the majority of the membrane is released back into the cytoplasm, i.e., no more than a transient association with the plasma membrane. This interpretation differs from the situation in *Tetrahymena* where membranes of exocytosed mucocysts are thought to be incorporated into the plasma membrane at least for a time (Satir, 1974a,b).

6.1.8. Mammary gland and milk secretion

The question of membrane flow and the potential contribution of Golgi apparatus to the formation of the plasma membrane can be appreciated best in the

56

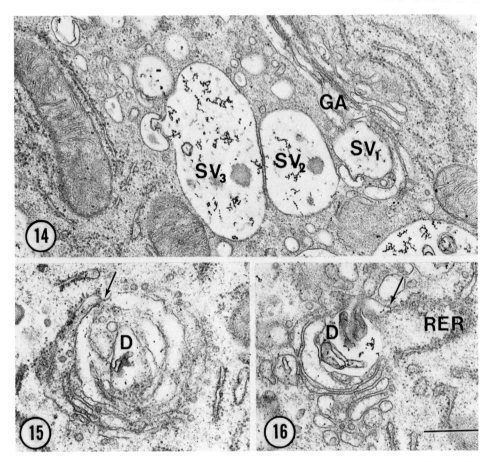

Figs. 14–16. The endoplasmic reticulum-Golgi apparatus-secretory vesicle complex of epithelial cells of the lactating rat mammary gland.

Fig. 14. Formation of secretory vesicles (sv) containing casein micelles (arrows) from the Golgi apparatus (GA). These vesicles migrate to and fuse with the apical plasma membrane to replenish the apical plasma membrane expended by envelopment of milk fat globules during milk secretion (see Fig. 17).

Figs. 15–16. Dictyosomes (D) of the mammary gland Golgi apparatus seen in face view to illustrate direct membrane continuity between rough endoplasmic reticulum (RER) and the smooth membrane tubules that connect directly to the cisternal membranes (arrows). Numerous transition vesicles are also seen at the dictyosome periphery. Glutaraldehyde-osmium tetroxide fixation. Scale line = 0.5μm. (From a study with Prof. T. W. Keenan, Purdue University, Lafayette, Indiana.)

lactating mammary gland, where morphological (Bargman and Knop, 1959; Helminen and Ericsson, 1968; Hollman, 1959; Wellings, 1969; Linzell and Peaker, 1971; Wooding, 1971a,b; Monis et al., 1975; Franke et al., 1976; Figs. 4 and 14 to 17) and biochemical (Dowben et al., 1967; Keenan et al., 1970) investigations suggest an extensive if not complete derivation of plasma membrane from secretory vesicles of the Golgi apparatus during milk secretion.

Figure 4 gives an overview of an epithelial cell of rat lactating gland. Golgi apparatus and secretory vesicles occupy a large proportion of the basal cytoplasm. From there the vesicles migrate to the luminal surface of the cell where they release their contents, as shown by the appearance of casein micelles in the gland lumen.

Milk fat is also secreted. Droplets without membranes found in the cytoplasm acquire a membrane as they pass through the plasma membrane and enter the gland lumen (Bargman and Knop, 1959; Kurosumi et al., 1971; Patton and Fowkes, 1967; and Fig. 17). Thus, mammary gland offers an experimental advantage for study of membrane flow because the extruded milk fat carries with it a portion of the plasma membrane as it leaves the cell to enter the lumen. In effect, the milk fat droplets specifically isolate the apical or luminal surface of the plasma membrane, which can then be recovered from the milk and analyzed. As summarized by Patton and Keenan (1975), the composition of milk fat globule

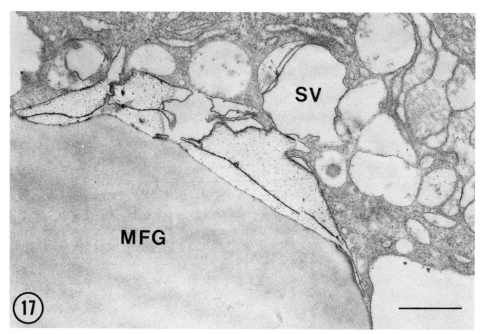

Fig. 17.   Same as Figure 4, showing a late stage in the discharge of a milk fat globule (MFG) by an epithelial cell of the lactating rat mammary gland. The extruded globule of milk fat is enveloped by a portion of the apical plasma membrane. A direct contribution from membranes derived from secretory vesicles (SV) of the Golgi apparatus occurs at the base of the fat globule to completely surround the globule with plasma membrane and ensure immediate repair of the ruptured surface (shown diagramatically in Fig. 18). Note a more intense staining of the membranes of the mature secretory vesicles that is indicative of membrane differentiation. Scale line = 1 μm. (From a study with Prof. T. W. Keenan, Purdue University, Lafayette, Indiana.)

membranes is very similar in terms of phospholipid, fatty acid, neutral lipid, glycolipid, and overall protein composition to plasma membranes. There are also morphological similarities (Monis et al., 1975). The two membranes differ primarily in the absence from the milk fat globule membrane of a Na^+-K^+-stimulated-Mg^{2+}-ATPase, adenylate cyclase activity, and prolactin binding sites. Based on studies with rat liver, these activities are also expected to be absent from membranes of the Golgi apparatus.

Details of the origin of the membrane surrounding the milk fat globule are provided in Figures 14 and 17. As it approaches the cell surface, each fat globule is frequently surrounded by secretory vesicles from the Golgi apparatus (Fig. 17). The secretory vesicles are identified by the thick, darkly-staining membranes (Helminen and Ericsson, 1968) and by the casein micelles of their content (Bucheim and Welsch, 1973). Bristle coats may partially or totally cover the membrane surface (Franke et al., 1976). In certain pathological situations, the vesicles may fuse to surround the milk fat with plasma membrane deep within the cytoplasm (Wooding, 1973). During normal fat extrusion, the apical portion of the droplet acquires membrane from existing plasma membrane but the basal portion of the globule may be derived directly from fusion of the basal secretory vesicles. As diagramed in Figure 18, half of the membrane contributed by these vesicles goes with the existing milk fat globule and half remains to reform the ruptured plasma membrane. Sometimes these vesicles fuse "prematurely" and trap a cytoplasmic crescent in the space between the milk fat globule and the plasma membrane (Helminen and Ericsson, 1968; Kurosumi et al., 1968).

In retrospect, the analysis of milk fat globule membranes offers an opportunity, unique among current experimental systems, of evaluating the biochemical characteristics of plasma membrane derived perhaps exclusively from membranes of Golgi apparatus-derived secretory vesicles. The findings show that in major phospholipids and membrane proteins, the composition of the vesicle membranes is similar to that of plasma membranes derived principally from lateral and apical surfaces of the glandular epithelium. In this regard, results from mammary gland may differ from those from pancreas, parotid gland, and chromaffin cells where a unique composition of the granule membrane has been indicated. Also, in mammary gland, there is no need to invoke a shuttle or recycling mechanism since input of new plasma membrane from secretory vesicles may be compensated for by release to the milk in the form of membrane surrounding milk fat globules.

6.2. Golgi apparatus and acrosome formation

The Golgi apparatus, or membranes derived from the Golgi apparatus, is the immediate precursor of the acrosome and acrosomal membrane during spermiogenesis (Fig. 19). Golgi apparatus of mammalian germ cells are mobilized early in spermiogenesis to synthesize or sequester constituents of the proacrosomal granules from which the acrosome is derived (Burgos and Fawcett, 1955; Fawcett and Burgos, 1956; Burgos et al., 1970; Sandoz, 1970a,b; Susi et

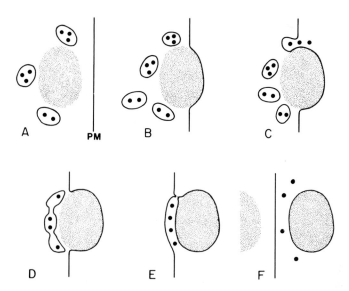

Fig. 18. Diagramatic representation of milk fat globule extrusion as facilitated by secretory vesicles derived from the Golgi apparatus. (a) A globule of milk fat (stippled area) approaches the plasma membrane (PM), already partially surrounded by casein-containing secretory vesicles (the smaller ellipsoids: the black dots represent casein micelles). (b) The fat globule interfaces with the apical plasma membrane and extrusion begins. (c) During fat extrusion, some of the secretory vesicles fuse with the plasma membrane to replenish the apical plasma membrane. (d) The remaining vesicles fuse at the base of the fat globule to form a large subsurface cisterna. (e) Fusion of this subsurface cisterna with the apical plasma membrane completes the process of milk fat extrusion and simultaneously serves to repair the apical surface at the point of extrusion. (f) The globule of milk fat now enters the milk, completely enveloped by a membrane derived from the epithelial surface. The coordinated coalescence of secretory vesicles from the Golgi apparatus allows for the simultaneous discharge of casein, lactose, ions, and other constituents carried by the secretory vesicles, and for the steady-state maintenance of the apical plasma membrane.

al., 1971). These constituents include the contents as well as the membranes of the acrosomes and must provide for the various functional requirements of the acrosomes during fertilization (Bedford, 1970; Franklin et al., 1970; Bedford and Nicander, 1971; Zamboni, 1971; Yanagimachi, 1973).

Little direct information is available regarding flow of membrane through the Golgi apparatus of spermatids. The directionality of the process has been inferred from morphological observations.

Golgi apparatus are thought to receive products from the. endoplasmic reticulum and to transfer them to the forming acrosomes (Sandoz, 1970a,b; Susi et al., 1971). The materials may be transferred by vesicles, first from the endoplasmic reticulum to the forming faces of the dictyosomes, and then from the mature dictyosome cisternae to the acrosome. The presence of coated vesicles in zones at both faces of the Golgi apparatus suggest a role for coated vesicles in the transport process (Fig. 19; Mollenhauer et al., 1976).

60

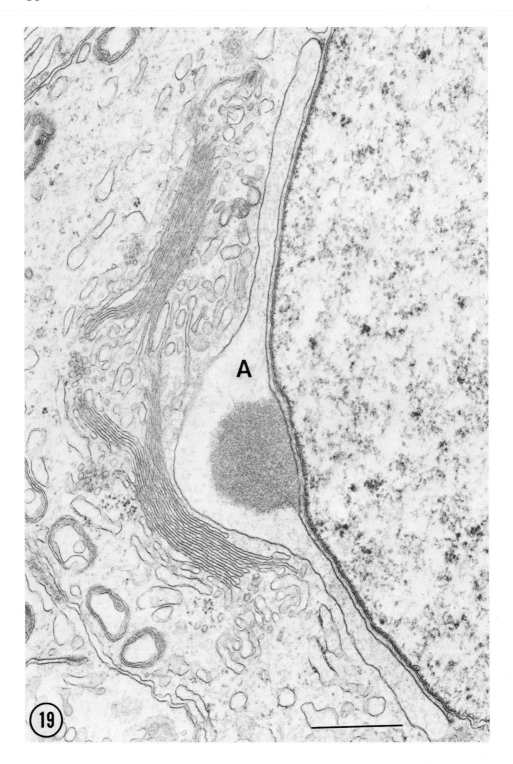

A

Thick cisternae of spermatocyte Golgi apparatus appear during the proacrosomal stages of spermiogenesis and remain throughout acrosome development. Similarities among the membranes of the thick cisternae after tannic acid-aldehyde fixation and of the acrosome suggest that a function of the thick cisternae is to transform membranes destined for incorporation in the acrosome. These thick cisternae are positionally equivalent to the GERL complex of Novikoff (Essner and Novikoff, 1962) involved in the formation of lysosomes. Acrosomes are considered to be a specialized form of lysosome (Dott, 1973) and contain enzymes that are released during fertilization to digest the outer investments of the egg and allow entry of the sperm (Yanagimachi, 1973). The thick cisternae appear to originate from conventionally appearing membranes of the Golgi apparatus (Mollenhauer et al., 1976).

6.3. Golgi apparatus and biogenesis
of lysosomes and plant vacuoles

Although details are lacking, a central role of the Golgi apparatus is indicated in the formation of primary lysosomes and in the packaging of the hydrolytic enzymes contained therein (deDuve and Wattiaux, 1965; Cook, 1973). Primary lysosomes originate within a system of transition elements that includes smooth endoplasmic reticulum, Golgi apparatus, or both (Essner and Novikoff, 1962; Novikoff et al., 1962; Novkioff and Shin, 1964; Brandes, 1965; Moe et al., 1965; Novikoff et al., 1971) the GERL (Golgi apparatus-endoplasmic reticulum-lysosome complex) in the terminology of Novikoff.

The contents of cholesterol, sphingomyelin, and free fatty acids of lysosomal membranes from rat liver are intermediate between those of the Golgi apparatus and the plasma membrane (Henning and Heidrich, 1974). The findings support morphological evidence for formation of primary lysosomes via the Golgi apparatus.

An origin of plant vacuoles from vesicles or cisternae derived from Golgi apparatus has also been suggested (Marinos, 1963; Ueda, 1966; Matile and Moor, 1968; Coulomb et al., 1972; McBride and Cole, 1972). However, most authors favor the endoplasmic reticulum or an endoplasmic reticulum-derived provacuolar structure as the ultimate source of the tonoplast (See Morré, 1975; Morré and Mollenhauer, 1976b for literature).

6.4. Origin and continuity of Golgi apparatus

Although frequently overlooked, an important consideration of the Golgi apparatus in membrane biogenesis derives from a potential role in its own replica-

---

Fig. 19. Electron micrograph of a rat spermatid showing a portion of a Golgi apparatus adjacent to a forming acrosome (A). The acrosome and acrosomal membrane are generated, at least in part, by the Golgi apparatus. Acrosomes may represent a form of lysosome, and contain enzymes that are released during fertilization to digest the outer investments of the egg and allow entry of the sperm. Scale line = 0.5 μm. (From a study with Dr. Hilton H. Mollenhauer, USDA, College Station, Texas.)

tion (Morré et al., 1971d). In one example, during dictyosome multiplication in the fungus *Pythium aphanidermatum*, dictyosome doubling is accompanied by an extension of the forming face regions where dictyosomes interface with a region of endoplasmic reticulum lacking ribosomes. Cisternae having twice normal diameter are observed in stages that correspond to their derivation from the caolescence of small evaginations or transition vesicles derived from an extended forming face region of the endoplasmic reticulum. Complete dictyosomes with all cisternae of twice normal diameter are found in addition to "hybrid" dictyosomes with two stacks of cisternae having normal dimensions located side by side directly underneath forming face regions of normal dimensions and on top of a single stack of cisternae with twice the normal dimensions. The hybrid structures are assumed to originate by division of an extended forming face region so that two dictyosomes are formed where only one existed previously. Thus, the coalescence of endoplasmic reticulum- or nuclear envelope-derived transition vesicles could account for membrane input during cisterna formation and dictyosome multiplication as well as the de nova origin of so-called dictyosomal prestages (Morré et al., 1971d). Again, the role of the Golgi apparatus per se would be one of modification and regulation rather than of strict de novo membrane biogenesis.

7. *Summary and concluding comments*

Membrane biogenesis results from a complex set of interactions among the cytoplasm (cytosol plus guide elements), endomembranes (nuclear envelope, endoplasmic reticulum, Golgi apparatus, and other transition elements and organelles) and surface membranes (plasma membrane plus vacuolar apparatus). Aspects include biosynthesis and transport of low molecular weight precursor molecules and activated intermediates, the assembly and migration of macromolecules and macromolecular complexes, and the formation and flow of preformed membrane materials. Although evidence favors a role for the Golgi apparatus in membrane modification, important biosynthetic contributions by the total Golgi complex are indicated as well. These include glycoprotein and glycolipid glycosylations, biosynthesis of both phospholipids and neutral lipids, and the synthesis and insertion of a limited number of membrane proteins.

The role of the Golgi apparatus in membrane modification may be quite complex. Since the Golgi apparatus receives preformed membranes from endoplasmic reticulum or nuclear envelope via transition vesicles or direct tubular connections, major modifications in membrane composition are indicated to adequately explain membrane differentiation across the stacked cisternae. Constituents must be removed as well as added. Enzymes must be inactivated as well as activated. The final result is a membrane that usually resembles plasma membrane both morphologically and biochemically.

At the mature face of the Golgi apparatus, several membrane types derive. These include various types of secretory vesicles destined to make permanent contributions to the plasma membrane, condensing vacuoles that give rise to mature secretion granules, primary lysosomes for internal digestive events, and the proacrosomal granules that form the acrosome during spermiogenesis. In some secretory vesicles, where permanent contributions are made to the plasma membrane, the result is growth of the cell surface. With condensing vacuoles destined to become secretion granules, and certain other specialized types of exocytotic vacuoles, a permanent contribution to plasma membrane is less clear. Manifestations of differentiation that involve replacement or modification of cell surfaces might result along with a continual replacement of surface material to balance degradation during turnover. Certainly, in nongrowing secretory cells evidence supports a mechanism whereby a balance is achieved between exocytotic discharge and endocytotic uptake of surface membranes. In either situation, the secretory vesicle or granule provides a clear example of bulk flow of membrane from a site of formation (the Golgi apparatus) to a site of utilization (the plasma membrane). Among the best documented examples of the latter are secretory epithelia of the mammary gland, where excess plasma membranes derived from Golgi apparatus are discharged into the milk as membranes that surround the milk fat globules.

Although individual enzymatic steps are rapidly being elucidated for biosynthesis of most membrane constituents, much less is known about the biochemistry and regulation of membrane differentiation. The latter is an important aspect of the total events of membrane biogenesis especially concerning the Golgi apparatus. The transducing mechanisms by which chemical potential is converted into a regulated vectorial (or cyclic) lateral displacement of membrane constituents is unknown.

The Golgi apparatus is perhaps unique among prevailing experimental systems for the study of membrane biogenesis because it may be viewed as undergoing rapid turnover in vivo so that the formation and release of new cisternae may occur every 1 to 4 minutes. Such a dynamic system is not readily amenable to correlative biochemical and ultrastructural investigation, especially in cell-free systems. This, coupled with the realization that flow differentiation of membranes is selective, i.e., certain constituents are excluded and others are inserted, places Golgi apparatus function in membrane biogenesis and its regulation among the more challenging extant problems of contemporary membrane biology.

References

Akiyama, M. and Sakagami, T. (1969) Exchange of mitochondrial lecithin and cephalin with those in rat liver microsomes. Biochim. Biophys. Acta 187, 105–112.

Allen, R. D. (1975) Intramembrane changes accompanying membrane recycling in Paramecium. J. Cell Biol. 67, 7a.

Amsterdam, A., Ohad, I. and Schramm, M. (1969) Dynamic changes in the ultrastructure of the acinar cell of the rat parotid gland during the secretory cycle. J. Cell Biol. 41, 753–773.

Amsterdam, A., Schramm, M., Ohad, I., Salamon, Y. and Selinger, Z. (1971) Concomitant synthesis of membrane proteins of the secretory granules in rat parotid gland. J. Cell Biol. 50, 187–200.

Antoine, J. C., Avrameas, S., Gonatas, N. K., Stieber, A. and Gonatas, J. O. (1974) Plasma membrane and internalized immunoglobulins of lymph node cells studies with conjugates of antibody of its Fab fragments with horseradish peroxidase. J. Cell Biol. 63, 12–23.

Autuori, F., Svensson, H. and Dallner, G. (1975a) Biogenesis of microsomal membrane glycoproteins in rat liver. I. Presence of glycoproteins in microsomes and cytosol. J. Cell Biol. 67, 687–699.

Autuori, F., Svensson, H. and Dallner, G. (1975b) Biogenesis of microsomal membrane glycoproteins in rat liver. II. Purification of soluble glycoproteins and their incorporation into microsomal membranes. J. Cell Biol. 67, 700–714.

Bailey, D. J., Murray, R. K. and Rolleston, F. S. (1974) Electrophoretic studies of the proteins of rat liver endoplasmic reticulum. Can. J. Biochem. 52, 1003–1012.

Bainton, D. F. and Farquhar, M. G. (1968) Difference in enzyme content of azurophil and specific granules of polymorphonuclear leukocytes. II. Cytochemistry and electron microscopy of bone marrow cells. J. Cell Biol. 39, 299–317.

Bargmann, W. and Knoop, A. (1959) Über die Morphologie der Milchsekretion. Licht-und elektronenmikroskopische Studien an der Milchdruse der Ratte. Z. Zellforsch. Mikrosk. Anat. 49, 344–388.

Baynes, J. W., Hsu, A.-F. and Heath, E. C. (1973) The role of mannosyl-phosphoryl-dihydropolyisoprenol in the synthesis of mammalian glycoproteins. J. Biol. Chem. 248, 5693–5704.

Beams, H. W. and Kessel, R. G. (1968) The Golgi apparatus: structure and function. Int. Rev. Cytol. 23, 209–276.

Bedford, J. M. (1970) Sperm capacitation and fertilization in mammals. Biol. Reprod. 2 (Suppl), 128–158.

Bedford, J. M. and Nicander, L. (1971) Ultrastructural changes in the acrosome and sperm membrane during maturation of spermatozoa in the testis and epididymis of the rabbit and monkey. J. Anat. 108, 527–543.

Behrens, N. H. and Leloir, L. F. (1970) Dolichol monophosphate glucose: an intermediate in glucose transfer in liver. Proc. Nat. Acad. Sci, U.S.A. 66, 153–159.

Behrens, N. H., Parodi, A. J., Leloir, L. F. and Krisman, C.R. (1971a) The role of dolichol monophosphate in sugar transfer. Arch. Biochem. Biophys. 143, 375–383.

Behrens, N. H., Parodi, A. J. and Leloir, L. F. (1971b) Glucose transfer from dolichol monophosphate glucose: the product formed with endogenous microsomal acceptor. Proc. Nat. Acad. Sci. U.S.A. 68, 2857–2860.

Behrens, N. H., Carminatti, H., Stanleoni, R. J., Leloir, L. F. and Cantarella, A. I. (1973) Formation of lipid-bound oligosaccharide containing mannose—their role in glycoprotein synthesis. Proc. Nat. Acad. Sci. U.S.A. 70, 3390–3394.

Benedetti, E. L. and Emmelot, P. (1967) Studies on plasma membranes. IV. The ultrastructural localization and content of sialic acid in plasma membranes isolated from rat liver and hepatoma. J. Cell Sci. 2, 499–512.

Benedetti, E.L. and Emmelot, P. (1968) Structure and function of plasma membranes isolated from liver. In: Ultrastructure in Biological Systems (Dalton, A. and Haguenau, F., eds.), pp. 33–120, Academic Press, New York and London.

Benedetti, E. L., Dumia, I. and Diawara, A. (1973) The organization of the plasma membrane in mammalian cells. Eur. J. Cancer 9, 263–272.

Benes, F. M., Higgins, J. A. and Barrnett, R. J. (1972) Fine structural localization of acyl transferase activity in rat hepatocytes. J. Histochem. Cytochem. 20, 1031–1040.

Bennett, H. S. (1956) The concepts of membrane flow and membrane vesiculation as mechanisms for active transport and ion pumping. J. Biophys. Biochem. Cytol. 2 (Suppl.), 99–103.

Berg, N. B. and Young, R. W. (1971) Sulfate metabolism in pancreatic acinar cells. J. Cell Biol. 50, 469–483.

Bergeron, J. J. M., Berridge, M. V. and Evans, W. H. (1975) Biogenesis of plasmalemmal glycoproteins. Intracellular site of synthesis of mouse liver plasmalemmal 5'-nucleotidase as determined by the sub-cellular location of messenger RNA coding for 5'-nucleotidase. Biochim. Biophys. Acta 407, 325–337

Bergeron, J. J. M., Ehrenreich, J. H., Siekevitz, P. and Palade, G. E. (1973a) Golgi fractions prepared from rat liver homogenates. II. Biochemical characterization. J. Cell Biol. 59, 73–88.

Bergeron, J. J. M., Evans, W. H. and Geschwind, I. I. (1973b) Insulin binding to rat liver Golgi fractions. J. Cell Biol. 59, 771–776.

Berlin, J. D. (1968) The ultrastructural localization of acid mucopolysaccharides in the intestine after irradiation. Rad. Res. 34, 347–356.

Berthillier, G. and Got, R. (1974) Biosynthèse de l'UDPglucose par les microsomes des hépatocytes de rat. Biochim. Biophys. Acta 362, 390–402.

Berthillier, G., Dubois, P. and Got, R. (1973) Glucokinase microsomique du foie de rat. Localisation dans une fraction riche en appareil de Golgi. Biochim. Biophys. Acta 293, 370–378.

Berthillier, G., Coleman, R. and Walker, D. G. (1976) The topographical location and unique nature of a glucokinase associated with the Golgi apparatus of rat liver. Biochem. J. 154, 193–201.

Bjerve, K. S. (1973) The Ca^{2+}-dependent biosynthesis of lecithin, phosphatidylethanolamine and phosphatidylserine in rat liver subcellular particles. Biochim. Biophys. Acta 296, 549–562.

Blaurock, A. E. (1972) Locating protein in membranes. Chem. Phys. Lipids 8, 285–291.

Bolender, R. P. (1974) Sterological analysis of the guinea pig pancreas. I. Analytical model and quantitative description of nonstimulated pancreatic exocrine cells. J. Cell Biol. 61, 269–287.

Bornens, M. and Kaspar, C. B. (1973) Fractionation and partial characterization of proteins of the bileaflet nuclear membrane from rat liver. J. Biol. Chem. 248, 571–579.

Bouchilloux, S., Chaubaud, O., Michel-Béchte, M., Ferrand, M. and Athouël-Haon (1970) Differential localization in thyroid microsomal subfractions of a mannosyltransferase, two N-acetylglucosaminyltransferases and a galactosyltransferase. Biochem. Biophys. Res. Commun. 40, 314–320.

Bouck, G. B. (1969) Extracellular microtubules. The origin, structure and attachment of flagellar hairs in Fucus and Ascophyllum antherozoids. J. Cell Biol. 40, 446–460.

Bouck, G. B. (1971) The structure, origin, isolation and composition of the tubular mastigonemes of the Ochromonas flagellum. J. Cell Biol. 50, 362–384.

Boutry, J.-M. and Novikoff, A. B.: (1975) Cytochemical studies on Golgi apparatus, GERL, and lysosomes in neurons of dorsal root ganglia in mice. Proc. Nat. Acad. Sci. U.S.A. 72, 508–512.

Bowles, D. J. and Northcote, D. H. (1974) The amounts and rates of export of extracellular polysaccharides in the root tissues of maize. Biochem. J. 130, 1133–1145.

Bracker, C. E., Grove, S. N., Heintz, C. E. and Morré, D. J. (1971) Continuity between endomembrane components in hyphae of Pythium spp. Cytobiologie 4, 1–8.

Brady, R. O. and Fishman, P. H. (1974) Biosynthesis of glycolipids in virus-transformed cells. Biochim. Biophys. Acta 355, 121–148.

Brandes, D. (1965) Observations on the apparent mode of formation of "pure" lysosomes. J. Ultrastruct. Res. 12, 63–80.

Bretscher, M. S. (1973) Membrane structure: Some general principles. Science 181, 622–628.

Brown, R. M. (1969) Observations on the relationship of the Golgi apparatus to wall formation in the marine chrysophycean alga, Pleurochrysis scherffelii Pringsheim. J. Cell Biol. 41, 109–123.

Brown, R. M., Franke, W. W., Kleinig, H., Falk, H. and Sitte, P. (1969) A cellulosic wall component produced by the Golgi apparatus of Pleurochrysis scherffelii. Science 166, 894–896.

Brown, R. M., Franke, W. W., Kleinig, H., Falk, H. and Sitte, P. (1970) Scale formation in chrysophycean algae. I. Cellulosic and noncellulosic wall components made by the Golgi apparatus. J. Cell Biol. 45, 246–271.

Brown, R. M., Herth, W., Franke, W. and Ramanovicz, D. (1973) The role of the Golgi apparatus in the biogenesis and secretion of a cellulosic glycoprotein in Pleurochrysis: A model system for the synthesis of structural polysaccharides. In: Biogenesis of Plant Cell Wall Polysaccharides (Loewus, F., ed.), pp. 207–258, Academic Press, New York.

Brunngraber, E. G. (1972) Biochemistry, function and neuropathology of the glycoproteins in brain tissue. In: Functional and Structural Proteins of the Nervous System (Davidson, A. N., Mandel, P. and Morgan, I. G., eds.), pp. 109–133, Plenum Press, New York.

Buchheim, W. and Welsch, U. (1973) Evidence for the submicellar composition of casein micelles on

88888888888888

the basis of electron microscopical studies. Neth. Milk Dairy J. 27, 163–180.

Burgos, M. H. and Fawcett, D. W. (1955) Studies on the fine structure of the mammalian testis. I. Differentiation of spermatids in the cat (*Felis domestica*) J. Biophys. Biochem. Cytol. 1, 287–300.

Burgos, M. H., Vitale-Calpe, R. and Aoki, A. (1970) Fine structure of the testis and its functional significance. In: The Testis (Johnson, A. D., Gomes, W. R. and VanDemark, N. L., eds.), vol. 1, pp. 551–577, Academic Press, New York.

Caccam, J. F., Jackson, J. J. and Eylar, E. H. (1969) The biosynthesis of mannose-containing glyco-proteins. A possible lipid intermediate. Biochem. Biophys. Res. Commun. 35, 505–511.

Carvo, M., Maddaiah, V. T., Collipp, P. J. and Chen, S. Y. (1974) Hepatic microsomal membrane: Activation of glucose-6-phosphatase. FEBS Lett. 39, 102–104.

Castle, J. D., Jamieson, J. D. and Palade, G. E. (1975) Secretion granules of the rabbit parotid gland. Isolation, subfractionation, and characterization of membrane and content subfractions. J. Cell Biol. 64, 182–210.

Cheetham, R. D., Morré, D. J. and Yunghans, W. (1970) Isolation of a Golgi apparatus-rich fraction from rat liver. II. Enzymatic characterization and comparison with other cell fractions. J. Cell Biol. 44, 492–500.

Cheetham, R. D., Morré, D. J., Pannek, C. and Friend, D. S. (1971) Isolation of a Golgi apparatus-rich fraction from rat liver. IV. Thiamine pyrophosphatase. J. Cell Biol. 49, 899–905.

Ceccarelli, B., Hurlbut, W. P. and Mauro, A. (1973) Turnover of transmitter and synaptic vesicles at the frog neuromuscular junction. J. Cell Biol. 57, 499–524.

Clarkson, B. and Baserga, R., (1974) Eds. Control of Proliferation in Animal Cells. Cold Spring Harbor Lab., Cold Spring Harbor, New York.

Claude, A. (1970) Growth and differentiation of cytoplasmic membranes in the course of lipoprotein granule synthesis in the hepatic cell. I. Elaboration of elements of the Golgi complex. J. Cell Biol. 47, 745–766.

Cook, G. M. W. (1973) The Golgi apparatus: form and function. In: Lysosomes in Biology and Pathology (Dingle, J. T., ed), vol. 3, pp. 237–277, Elsevier/North-Holland, Amsterdam.

Cook, G. M. W. and Stoddart, R. W. (1973) Surface Carbohydrates of the Eukaryotic Cell, Academic Press, London.

Cope, G. H. and Williams, M. A. (1976) Quantitative analyses of the constituent membranes of parotid acinar cells and of the changes evident after induced exocytosis. Z. Zellforsch. 145, 311–330.

Critchley, D. R. and Vicker, M. G. (1977) Glycolipids as membrane receptors important in growth regulation and cell-cell interactions. In: Dynamic Aspects of Cell Surface Organization (Poste, G. and Nicolson, G. L., eds.), Cell Surface Reviews, vol. 3. Elsevier/North-Holland, Amsterdam.

Critchley, D. R., Graham, J. M. and MacPherson, I. (1973) Subcellular distribution of glycolipids in a hamster cell line. FEBS Lett. 32, 37–40.

Cronshaw, J. and Esau, K. (1968) Cell division in leaves of *Nicotiana*. Protoplasma 65, 1–24.

Cuatrecasas, P. (1974) Membrane receptors. Ann. Rev. Biochem. 43, 169–214.

Dalton, A. J. (1961) Golgi apparatus and secretion granules. In: The Cell (Brachet, J. and Mirsky, A. E., eds.), Vol. 2, pp. 603–617, Academic Press, New York.

Dankert, M., Wright, A., Kelley, W. S. and Robbins, P. W. (1966) Isolation, purification and properties of the lipid-linked intermediates of O-antigen biosynthesis. Arch. Biochem. Biophys. 116, 425–435.

Darley, W. M., Porter, D. and Fuller, M. S. (1973) Cell wall composition and synthesis via Golgi-directed scale formation in the marine eukaryote, *Schizochytrium aggregatum* with a note on *Thraustochytrium* sp. Arch. Mikrobiol. 90, 89–106.

Dauwalder, M. and Whaley, W. G. (1974a) Patterns of incorporation of [^{3}H]-galatose by cells of *Zea mays* root tips. J. Cell Sci. 14, 11–27.

Dauwalder, M. and Whaley, W. G. (1974b) Staining of cells of *Zea mays* root apices with the osmium-zinc iodide and osmium impregnation techniques. J. Ultrastruct. Res. 45, 279–296.

Dauwalder, M., Whaley, W. G. and Kephart, J. E. (1969) Phosphatases and differentiation of the Golgi apparatus. J. Cell Sci. 4, 455–497.

Dauwalder, M., Whaley, W. G. and Kephart, J. E. (1972) Functional aspects of the Golgi apparatus. Sub-Cell. Biochem. 1, 225–276.

Dawson, R. M. C. (1973) The exchange of phospholipids between cell membranes. Sub-Cell. Biochem. 2, 69–89.

DeCamilli, P., Peluchetti, D. and Meldolesi, J. (1976) Dynamic changes of the luminal plasmalemma in stimulated parotid acinar cells. A freeze-fracture study. J. Cell Biol. 70, 59–74.

DeDuve, C. (1969) The lysosome in retrospect. In: Lysosomes in Biology and Pathology (Dingle, J. T. and Fell, H. B., eds.), vol. 1, pp. 3–40, Elsevier/North-Holland, New York.

DeDuve, C. and Wattiaux, R. (1966) Functions of lysosomes. Ann. Rev. Physiol. 28, 435–492.

DeLuca, L., Schumacher, M. and Wolf, G. (1969) Biosynthesis of a fucose-containing glycopeptide from rat small intestine in normal and vitamin A-deficient conditions. J. Biol. Chem. 245, 4551–4558.

DeLuca, L., Maestri, N., Bonanni, F. and Nelson, D. (1972) Maintenance of epithelial cell differentiation. Mode of action of vitamin A. Cancer 30, 1326–1331.

DeLuca, L., Maestri, N., Rosso, G. and Wolf, G. (1973) Retinol glycolipids. J. Biol. Chem. 248, 641–648.

DeLuca, L., Silverman-Jones, C. S. and Barr, R. M. (1975) Biosynthetic studies on mannolipids and mannoproteins of normal and vitamin A-depleted hamster livers. Biochim. Biophys. Acta 409, 342–359.

Dennick, R. G. (1972) The intracellular organization of cholesterol biosynthesis. Steroids Lipids Res. 3, 236–256.

Deppert, W., Werchau, H. and Walter, G. (1974) Differentiation between intracellular and cell surface glycosyl transferase: Galactosyl transferase activity in intact cells and cell homogenates. Proc. Nat. Acad. Sci. U.S.A. 71, 3068–3072.

Dobberstein, B. and Kiermayer, O. (1972) Das Auftreten eines besonderen Typs von Golgivesikeln während der Sekundärwandbildung von *Micrasterias denticulata* Breb. Protoplasma 75, 185–194.

Dod, B. J. and Gray, G. M. (1968) The lipid composition of rat-liver plasma membranes. Biochim. Biophys. Acta 150, 397–404.

Dodge, J. D. (1973) The Fine Structure of Algal Cells, Academic Press, New York.

Dott, H. M. (1973) Lysosomes and lysosomal enzymes in reproduction. In: Advances in Reproductive Physiology (Bishop, M. W. H., ed.), vol. 6, pp. 231–277, Academic Press, New York.

Douglas, W. W., Nagasawa, J. and Schulz, R. (1971) Electron microscopic studies on the mechanism of secretion of pituitary hormones and significance of microvesicles ("synaptic vesicles"): evidence of secretion by exocytosis and formation of microvesicles as a byproduct of this process. Mem. Soc. Endocrinol. 19, 353–378.

Dowben, R. M., Brunner, J. R. and Philpot, D. E. (1967) Studies on milk fat globule membranes. Biochim. Biophys. Acta 135, 1–10.

Dykstra, M. J. (1976) Wall and membrane biogenesis in the unusual Labyrinthulidlike organism *Sorodiplophrys stercorea*. Protoplasma 87, 329–346.

Edelman, G. M., Yahara, I. and Wang, J. L. (1973) Receptor mobility and receptor-cytoplasmic interactions in lymphocytes. Proc. Nat. Acad. Sci. U.S.A. 70, 1442–1446.

Ehrenreich, J. H., Bergeron, J. J. M., Siekevitz, P. and Palade, G. E. (1973) Golgi fractions prepared from rat liver homogenates. I. Isolation procedure and morphological characterization. J. Cell Biol. 59, 45–72.

Elder, J. H. and Morré, D. J. (1976a) Synthesis *in vitro* of intrinsic membrane proteins by free, membrane-bound, and Golgi apparatus-associated polyribosomes from rat liver. J. Biol. Chem. 251, 5054–5068.

Elder, J. H. and Morré, D. J. (1976b) Distribution of glycoproteins among subcellular fractions from rat liver. Cytobiologie 13, 279–284.

Elhammer, A., Svennsson, H., Auturoi, F. and Dallner, G. (1975) Biogenesis of microsomal membrane glycoproteins in rat liver. III. Release of glycoproteins from the Golgi fraction and their transfer to microsomal membranes. J. Cell Biol. 67, 715–724.

Emmelot, P. (1973) Biochemical properties of normal and neoplastic cell surfaces: a review. Eur. J. Cancer 9, 319–333.

Emmelot, P., Bos, C. J., Van Hoeven, R. P. and van Blitterswizk, W. J. (1974) Isolation of plasma membranes of rat and mouse livers and hepatomas. Meth. Enzymol. 31A, 75–90.

Ernster, L. and Kuylenstierna, B. (1970) Outer membranes of mitochondria. In: Membranes of Mitochondria and Chloroplasts (Racker, E., ed.), pp. 172–212, Van Nostrand, New York.

Essner, E. and Novikoff, A. (1962) Cytological studies on two functional hepatomas: interrelations of endoplasmic reticulum, Golgi apparatus and lysosomes. J. Cell Biol. 15, 289–312.

Evans, P. J. and Hemming, F. W. (1973) The unambiguous characterization of dolichol phosphate mannose as a product of mannosyl transferase in pig liver endoplasmic reticulum. FEBS Lett. 31, 335–338.

Evans, W. H. (1970) Glycoproteins of mouse liver smooth microsomal and plasma membrane fractions. Biochim. Biophys. Acta 211, 578–581.

Evans, W. H., Bergeron, J. J. M. and Geschwind, I. I. (1973) Distribution of insulin receptor sites among liver plasma membrane subfractions. FEBS Lett. 34, 259–262.

Falk, H. (1969) Fusiform vesicles in plant cells. J. Cell Biol. 43, 167–174.

Farquhar, M. G., Bergeron, J. J. M. and Palade, G. E. (1974) Cytochemistry of Golgi apparatus fractions prepared from rat liver. J. Cell Biol. 60, 8–25.

Fawcett, D. W. (1962) Physiologically significant specializations of the cell surface. Circulation 26, 1105–1125.

Fawcett, D. W. and Burgos, M. H. (1956) Observations on the cytomorphosis of the germinal and interstitial cells of the human testis. Ciba Found. Colloq. Ageing 2, 86–99.

Fleischer, B. and Fleischer, S. (1970) Preparation and characterization of Golgi membranes from rat liver. Biochim. Biophys. Acta 219, 301–319.

Fleischer, B. and Fleischer, S. (1971) Comparison of cellular membranes of liver with emphasis on the Golgi complex as a discrete organelle. Biomembranes 2, 75–94.

Fleischer, B. and Zambrano, F. (1973) Localization of cerebroside-sulfotransferase activity in the Golgi apparatus of rat kidney. Biochem. Biophys. Res. Commun. 52, 951–958.

Fleischer, B. and Zambrano, F. (1974) Golgi apparatus of rat kidney. Preparation and role in sulfatide formation. J. Biol. Chem. 249, 5995–6003.

Fleischer, B., Fleischer, S., and Ozawa, H. (1969) Isolation and characterization of Golgi membranes from bovine liver. J. Cell Biol. 43, 59–79.

Flesicher, B., Zambrano, F. and Fleischer, S. (1974) Biochemical characterization of the Golgi complex of mammalian cells. J. Supramol. Struct. 2, 737–750.

Fleischer, S., Fleischer, B., Assi, A. and Chance, B. (1971) Cytochrome b_5 and P-$_{450}$ in liver cell fractions. Biochim. Biophys. Acta 222, 194–200.

Flickinger, C. J. (1969) The development of Golgi complexes and their dependence upon the nucleus in Amebae. J. Cell Biol. 43, 250–262.

Flickinger, C. J. (1970) The fine structure of the nuclear envelope in Amebae: Alterations following nuclear transplantation. Exp. Cell Res. 60, 225–236.

Flickinger, C. J. (1974a) The role of endoplasmic reticulum in the repair of Ameba nuclear envelope damaged microsurgically. J. Cell Sci. 14, 421–437.

Flickinger, C. J. (1974b) Radioactive labeling of the Golgi apparatus by microinjection of individual amebae. Exp. Cell Res. 88, 415–418.

Flickinger, C. J. (1975) The relation between the Golgi apparatus, cell surface, and cytoplasmic vesicles in Amoeba studied by electron microscope radioautography. Exp. Cell Res. 96, 189–201.

Franke, W. W. (1974a) Nuclear envelopes. Structure and biochemistry of the nuclear envelope. Phil. Trans. Roy. Soc. Lond. (Ser. B) 268, 67–93.

Franke, W. W. (1974b) Structure, biochemistry, and functions of the nuclear envelope. Int. Rev. Cytol., Suppl. 4, 71–236.

Franke, W. W. and Eckert, W. A. (1971) Cytomembrane differentiation in a ciliate, *Tetrahymena pyriformis*. II. Bifacial cisternae and tubular formations. Z. Zellforsch. 122, 244–253.

Franke, W. W. and Herth, W. (1974) Morphological evidence for de novo formation of plasma membrane from coated vesicles in exponentially growing cultured plant cells. Exp. Cell Res. 89, 447–451.

Franke, W. W. and Kartenbeck, J. (1976) Some principles of membrane differentiation. In: Progress in Differentiation Research (Müller-Bérat, N., ed.) pp. 213–243, Elsevier/North-Holland, New York.

Franke, W. W. and Scheer, U. (1972) Structural details of dictyosomal pores. J. Ultrastruct. Res. 40, 132–144.

Franke, W. W. and Scheer, U. (1974) Structures and functions of the nuclear envelope. In: The Cell Nucleus (H. Busch, ed.), vol. 1, pp. 219–347, Academic Press, New York.

Franke, W. W., Eckert, W. A. and Krien, S. (1971a) Cytomembrane differentiation in a ciliate, *Tetrahymena pyriformis*. I. Endoplasmic reticulum and dictyosomal equivalents. Z. Zellforsch. 119, 577–604.

Franke, W. W., Kartenbeck, J., Zentgraf, H., Scheer, U. and Falk, H. (1971b) Membrane to membrane cross-bridges. A means to orientation and interaction of membrane faces. J. Cell Biol. 51, 881–888.

Franke, W. W., Morré, D. J., Deumling, B., Cheetham, R. D., Kartenbeck, J., Jarasch, E. D., and Zentgraf, H. W. (1971c) Synthesis and turnover of membrane proteins in rat liver: An examination of the membrane flow hypothesis. Z. Naturforsch. 26b, 1031–1039.

Franke, W. W., Herth, W., VanDerWoude, W. J. and Morré, D. J. (1972a) Tubular and filamentous structures in pollen tubes: Possible involvement as guide elements in protoplasmic streaming and vectorial migration of secretory vesicles. Planta 105, 317–341.

Franke, W. W., Kartenbeck, J., Krien, S. VanDerWoude, W. J., Scheer, U. and Morré, D. J. (1972b) Inter- and intracisternal elements of the Golgi apparatus: A system of membrane-to-membrane cross-links. Z. Zellforsch. 132, 365–380.

Franke, W. W., Lüder, M. R., Kartenbeck, J., Zerban, H. and Keenan, T. W. (1976) Involvement of vesicle coat material in casein secretion and surface regeneration. J. Cell Biol. 69, 173–195.

Franklin, L. E., Barros, C. and Fussell, E. N. (1970) The acrosomal region and the acrosome reaction in sperm of the golden hamster. Biol. Reprod. 3, 180–200.

Frantz, C. E. (1973) NADH: Ferricyanide oxidoreductase in rat liver plasma membrane. M.S. Thesis, Purdue University.

Frantz, C., Roland, J. C., Williamson, F. A., and Morré, D. J. (1973) Différenciation *in vitro* des membranes des dictyosomes. C. R. Acad. Sci. Paris 277, 1471–1474.

Frey-Wyssling, A., López-Sáez, J. F. and Mühlethaler, K. (1964) Formation and development of the cell plate. J. Ultrastruct. Res. 10, 422–432.

Friend, D. S. (1965) The fine structure of Brunner's gland in the mouse. J. Cell Biol. 25, 563–576.

Friend, D. S. and Murray, M. J. (1965) Osmium impregnation of the Golgi apparatus. Am. J. Anat. 117, 135–149.

Friend, D. S. and Farquhar, M. G. (1967) Functions of coated vesicles during protein absorption in the rat vas deferens. J. Cell Biol. 35, 357–376.

Frye, L. D. and Edidin, M. (1970) The rapid intermixing of cell surface antigens after formation of mouse-human heterokaryons. J. Cell Sci. 7, 319–335.

Gahmberg, C. G. (1971) Protein and glycoprotein of hamster kidney fibroblast (BHK 21/13) plasma membranes and endoplasmic reticulum. Biochim. Biophys. Acta 249, 81–95.

Gaylor, J. L. (1972) Microsomal enzymes of sterol biosynthesis. Adv. Lipid Res. 10, 89–141.

Geuze, J. J. and Poort, C. (1973) Cell membrane resorption in the rat exocrine pancreas cell after *in vivo* stimulation of the secretion as studied by *in vitro* incubation with extracellular space markers. J. Cell Biol. 57, 159–174.

Glaumann, H., Bergstrand, A. and Ericsson, J. L. E. (1975) Studies on the synthesis and intracellular transport of lipoprotein particles in rat liver. J. Cell Biol. 64, 356–377.

Godman, G. C. and Lane, N. (1964) On the site of sulfation in the chondrocyte. J. Cell Biol. 21, 353–366.

Goldfarb, S. (1972) Submicrosomal localization of hepatic 3-hydroxy-3-methylglutaryl coenzyme A (HMG-CoA) reductase. FEBS Lett. 24, 153–155.

Goldfischer, S. (1965) The cytochemical demonstration of arylsulfatase activity by light and electron microscopy. J. Histochem. Cytochem. 13, 520–523.

Gonatas, N. K., Steiber, A., Kim, S. U., Graham, D. I. and Avrameas, S. (1975) Internalization of neuronal plasma membrane ricin receptors into the Golgi apparatus. Exp. Cell Res. 94, 426–431.

González-Cadavid, N. F. and Sáez de Córdova, C. (1974) Role of membrane-bound and free polyribosomes in the synthesis of cytochrome c in rat liver. Biochem. J. 140, 157–167.

Gray, E. G. (1970) The question of relationship between Golgi vesicles and synaptic vesicles in *Octapus* neurons. J. Cell Sci. 7, 189–201.

Griffith, D. L. and Bondareff, W. (1973) Localization of thiamine pyrophosphatase in synaptic vesicles. Am. J. Anat. 136, 549–556.

Grimstone, A. V. (1959) Cytoplasmic membranes and the nuclear membrane in the flagellate *Trichonympha*. J. Biophys. Biochem. Cytol. 6, 369–378.

Grove, S. N., Bracker, C. E. and Morré, D. J. (1968) Cytomembrane differentiation in the endoplasmic reticulum-Golgi apparatus-vesicle complex. Science 161, 171–173.

Grove, S. N., Bracker, C. E. and Morré, D. J. (1970) An ultrastructural basis for hyphal tip growth in *Pythium ultimum*. Am. J. Bot. 57, 245–266.

Haddad, A. M., Smith, D., Herscovics, A., Nadler, N. J. and Leblond, C. P. (1971) Radioautographic study of *in vivo* and *in vitro* incorporation of fucose-^{3}H into thyroglobulin by rat thyroid follicular cells. J. Cell Biol. 49, 856–882.

Hakomori, S. (1970) Glycosphingolipids having blood-group ABH and Lewis specificities. Chem. Phys. Lipids 5, 96–115.

Hakomori, S. (1975) Structure and organization of cell surface glycolipids, dependency on cell growth and malignant transformation. Biochim. Biophys. Acta 417, 55–89.

Hallinan, T., Murty, C. N. and Grant, J. H. (1968) Early labeling with glucosamine-^{14}C of granular and agranular endoplasmic reticulum and free ribosomes from rat liver. Arch. Biochem. Biophys. 125, 715–720.

Harwood, R., Grant, M. E. and Jackson, D. S. (1975) Studies on the glycosylation of hydroxylysine residues during collagen biosynthesis and the subcellular location of collagen: galactosyltransferase and collagen glucosyltransferase in tendon and cartilage cells. Biochem. J. 152, 291–302.

Hausmann, K. and Allen, R. D. (1976) Membrane behavior of exocytic vesicles. II. Fate of the trichocyst membranes in *Paramecium* after induced trichocyst discharge. J. Cell Biol. 69, 313–326.

Haywood, A. M. (1974) Characteristics of Sendai virus receptors in a model membrane. J. Mol. Biol. 83, 427–436.

Heine, J. W., Spear, P. G. and Roizman, B. (1972) Proteins specified by herpes simplex virus. VI. Viral proteins in the plasma membrane. J. Virol. 9, 431–439.

Heinrich, G. (1973) Die Feinstruktur der Tichom-Hydathoden von *Monarda fistulosa*. Protoplasma 77, 271–278.

Helgeland, L., Christensen, T. B. and Janson, T. L. (1972) The distribution of protein-bound carbohydrates in submicrosomal fractions from rat liver. Biochim. Biophys. Acta 286, 62–71.

Helminen, H. J. and Ericsson, J. L. E. (1968) Studies on mammary gland involution. I. On the ultrastructure of the lactating mammary gland. J. Ultrastruct. Res. 25, 193–213.

Helmkamp, G. M., Harvey, M. S., Wirtz, K. W. A. and Van Deenen, L. L. M. (1974) Phospholipid exchange between membranes. Purification of bovine brain proteins that preferentially catalyze the transfer of phosphatidyl inositol. J. Biol. Chem. 249, 6382–6389.

Helting, T. and Peterson, P. A. (1972) Galactosyltransfer in mouse mastocytoma: synthesis of a galactose-containing polar metabolite of retinol. Biochem. Biophys. Res. Commun. 46, 429–436.

Henning, R. and H.-G. Heidrich (1974) Membrane lipids of rat liver lysosomes prepared by free-flow electrophoresis. Biochim. Biophys. Acta 345, 326–335.

Hepler, P. K. and Jackson, W. T. (1968) Microtubules and early stages of cell-plate formation in the endosperm of *Haemanthus katherinae* Baker. J. Cell Biol. 38, 437–466.

Herth, W., Franke, W. W. and VanDerWoude, W. J. (1972a) Cytochalasin stops tip growth in plants. Naturwissenschaften 59, 38–39.

Herth, W., Franke, W. W., Stadler, J., Bittiger, H., Keilich, G., and Brown, R. M. (1972b) Further characterization of the alkali-stable material from the scales of *Pleurochrysis scherffelii:* A cellulosic glycoprotein. Planta 105, 79–92.

Heuser, J. E. and Reese, T. E. (1973) Evidence for recycling of synaptic vesicle membranes during transmitter release at the frog neuromuscular junction. J. Cell Biol. 57, 315–344.

Hicks, R. M. (1966) The function of the Golgi complex in transitional epithelium. Synthesis of the thick cell membrane. J. Cell Biol. 30, 623–643.

Higgins, J. A. and Barrnett, R. J. (1971) Fine structural localization of acyltransferases: The monoglyceride and α-glycerophosphate pathways in intestinal absorptive cells. J. Cell Biol. 50, 102–120.

Higgins, J. A. and Barrnett, R. J. (1972) Studies of the biogenesis of smooth endoplasmic reticulum membranes in livers of phenobarbital-treated rats. I. The site of activity of acyltransferases involved in the synthesis of the membrane phospholipid. J. Cell Biol. 55, 282–298.

Higgins, J. A. and Yates, R. D. (1974) Acetyl coenzyme A carboxylase. In: Electron Microscopy of Enzymes (Hyat, M. A., ed), vol. 3, pp. 153–165, Van Nostrand Reinhold, New York.

Hill, F. G. and Outka, D. E. (1974) The structure and origin of mastigonemes in *Ochromonas minute* and *Monas* sp. J. Protozool. 21, 299–312.

Hinman, N. D. and Philips, A. H. (1970) Similarity and limited multiplicity of membrane proteins from rough and smooth endoplasmic reticulum. Science 170, 1222–1223.

Hirano, H., Parkhouse, B., Nicolson, G. L., Lennox, E. S. and Singer, S. J. (1972) Distribution of saccharide residues on membrane fragments from a myeloma-cell homogenate; its implications for membrane biogenesis. Proc. Nat. Acad. Sci. U.S.A. 69, 2945–2949.

Hodson, S. and Brenchley, G. (1976) Similarities of the Golgi apparatus membrane and the plasma membrane in rat liver cells. J. Cell Sci. 20, 167–182.

Hokin, L. E. (1968) Dynamic aspects of phospholipids during secretion. Int. Rev. Cytol. 23, 187–208.

Hollman, K.-H. (1959) L'ultrastructure de la glande mamaire normale de la souris en lactation. Etude au microscope électronique. J. Ultrastruct. Res. 2, 423–443.

Holtzman, E., Freeman, A. R. and Kashner, L. A. (1971) Stimulation dependent alterations in peroxidase uptake by lobster neuromuscular junctions. Science 173, 733–736.

Horwitz, A. L. and Dorfman, A. (1968) Subcellular sites for synthesis of chondromucoprotein of cartilage. J. Cell Biol. 38, 358–368.

Hsu, A.-F., Baynes, J. W. and Heath, E. C. (1974) The role of dolichol-oligosaccharide as an intermediate in glycoprotein biosynthesis. Proc. Nat. Acad. Sci. U.S.A, 61, 2391–2395.

Hughes, R. C. (1974) Biogenesis of surface membranes. Nature 249, 414.

Hugon, J. S., Maestracci, D. and Ménard, D. (1973) Stimulation of glucose-6-phosphatase in the mucosal cell of the mouse intestine. J. Histochem. Cytochem. 21, 426–440.

Jamieson. J. D. and Palade, G. E. (1967a) Intracellular transport of secretory proteins in the pancreatic exocrine cell. I. Role of the peripheral elements of the Golgi complex. J. Cell Biol. 34, 577–596.

Jamieson, J. D. and Palade, G. E. (1967b) Intracellular transport of secretory proteins in the pancreatic exocrine cell. II. Transport to condensing vacuoles and zymogen granules. J. Cell Biol. 34, 597–615.

Jamieson, J. D. and Palade, G. E. (1968) Intracellular transport of secretory proteins in the pancreatic exocrine cell. III. Dissociation of intracellular transport from protein synthesis. J. Cell Biol. 39, 580–588.

Jarasch, E.-D. and Franke, W. W. (1974) Is cytochrome oxidase a constituent of nuclear membranes? J. Biol. Chem. 249, 7245–7254.

Jarasch, E. D., Bruder, G., Keenan, T. W. and Franke, W. W. (1977) Redox components in milk fat globule membranes and rough endoplasmic reticulum from lactating mammary gland. J. Cell Biol., in preparation.

Jelsema, C. L. and Morré, D. J. (1977) Distribution of phospholipid biosynthetic enzymes among cell components of rat liver. In preparation.

Jungalwala, F. B. and Dawson, R. M. C. (1970) Phospholipid synthesis and exchange in isolated liver cells. Biochem. J. 117, 481–490.

Kamath, S. A. and Rubin, E. (1973) The exchange of phospholipids between subcellular organelles of the liver. Arch. Biochem. Biophys. 158, 312–322.

Kartenbeck, J. (1974) Kontinuierliche und diskontinuierliche Membranbezeihungen in der Rattenhepatocyte und in Hepatom-Zellen. Doctoral thesis. University of Freiburg, Germany, pp. 1–233.

Kasper, C. B. (1971) Biochemical distinctions between the nuclear and microsomal membranes from rat hepatocytes. The effect of phenobarbital administration. J. Biol. Chem. 246, 577–581.

Kasper, C. B. (1974) Chemical and biochemical properties of the nuclear envelope. In: The Cell Nucleus (Busch, H., ed.), vol. 1, pp. 349–384, Academic Press, New York.

Keenan, T. W. and Morré, D. J. (1970) Phospholipid class and fatty acid composition of Golgi apparatus isolated from rat liver and comparison with other cell fractions. Biochemistry 9, 19–25.

Keenan, T. W. and Morré, D. J. (1975) Glycosyltransferases: Do they exist on the surface membrane of mammalian cells? FEBS Lett. 55, 8–13.

Keenan, T. W., Morré, D. J., Olson, D. E., Yunghans, W. N. and Patton, S. (1970) A biochemical and morphological comparison of plasma membrane and milk fat globule membrane from bovine mammary gland. J. Cell Biol. 44, 80–93.

Keenan, T. W., Huang, C. M. and Morré, D. J. (1972a) Gangliosides: Nonspecific localization in the surface membranes of bovine mammary gland and rat liver. Biochem. Biophys. Res. Commun. 47, 1277–1283.

Keenan, T. W., Morré, D. J. and Huang, C. M. (1972b) Distribution of gangliosides among subcellular fractions from rat liver and bovine mammary gland. FEBS Lett, 24, 204–208.

Keenan, T. W., Franke, W. W. and Kartenbeck, J. (1974a) Concanavalin A binding by isolated plasma membranes and endomembranes from liver and mammary gland. FEBS Lett. 44, 274–278.

Keenan, T. W., Morré, D. J. and Basu, S. (1974b) Ganglioside biosynthesis: Concentration of glycophingolipid glycosyltransferases in Golgi apparatus from rat liver. J. Biol. Chem. 249, 310–315.

Keenan, T. W., Morré, D. J. and Huang, C. M. (1974c) Membranes of the mammary gland. In: Lactation: A Comprehensive Treatise (Larson, B. L. and Smith, V. R., eds.), pp. 191–233, Academic Press, New York.

Kennedy, E. P. (1961) Biosynthesis of complex lipids. Fed. Proc. 20, 934–940.

Kessel, R. G. (1971) Origin of the Golgi apparatus in embryonic cells of the grasshopper. J. Ultrastruct. Res. 34, 260–275.

Kiermayer, O. (1973) Feinstrukturelle Grundlagen der Cytomorphogenese. Ber. Deutsch. Bot. Ges. 86, 287–291.

Kiermayer, O. and Dobberstein, B. (1973) Membrankomplexe dictyosomaler Herkunft als Matrizen für die extraplasmatische Synthese und Orientierung von Mikofibrillen. Protoplasma 77, 437–451.

King, C. A. and Van Heyningen, W. E. (1973) Deactivation of cholera toxin by sialidase resistant monosialoganglioside. J. Infect. Dis. 127, 639–647.

Kirby, E. G. and Roberts, R. M. (1971) The localized incorporation of ^{3}H-L-fucose into cell-wall polysaccharides of the cap and epidermis of corn roots. Autoradiographic and biosynthetic studies. Planta 99, 211–221.

Klenk, H.-D. and Choppin, P. W. (1970) Plasma membrane lipids and parainfluenza virus assembly. Virology 40, 939–947.

Knap, A., Kornblatt, M. J., Schachter, H. and Murray, R. K. (1973) Studies on the biosynthesis of testicular sulfoglycerogalactolipid: Demonstration of a Golgi-associated sulfotransferase activity. Biochem. Biophys. Res. Commun. 55, 179–186.

Komnick, H., Stockem, W. and Wohlfarth-Botterman, K. E. (1973) Cell motility: Mechanisms in protoplasmic streaming and ameboid movement. Int. Rev. Cytol. 34, 169–249.

Kreibich, G. and Sabatini, D. D. (1973) Microsomal membranes and the translational apparatus of eukaryotic cells. Fed. Proc. 32, 2133–2138.

Kreibich, G., Hubbard, A. L. and Sabatini, D. D. (1974) On the spatial arrangement of proteins in microsomal membranes from rat liver. J. Cell Biol. 60, 616–627.

Kreutzburg, G. W. Tóth, L. and Kaiya, H. (1975) Acetylcholine-esterase as a marker for dendritic transport and dendritic secretion. Adv. Neurol. 12, 269–281.

Kuhn, N. J. and White, A. (1976) Evidence for specific transport of uridine diphosphate galactose across the Golgi membrane of rat mammary gland. Biochem. J. 154, 243–244.

Kurosumi, K., Kobayashi, Y. and Baba, N. (1968) The fine structure of mammary glands of lactating rats, with special reference to the apocrine secretion. Exp. Cell Res. 50, 177–192.

Lane, N. L., Caro, L., Otero-Vilardebó, L. R. and Godman, G. C. (1964) On the site of sulfation in colonic goblet cells. J. Cell Biol. 21, 339–352.

Lawford, G. R. and Schachter, H. (1966) Biosynthesis of glycoprotein by liver. The incorporation *in vivo* of ^{14}C-glucosamine into protein-bound hexosamine and sialic acid of rat liver subcellular fractions. J. Biol. Chem. 241, 5408–5418.

Leblond, C. P. and Bennett, G. (1974) Elaboration and turnover of cell coat glycoproteins. In: The Cell Surface in Development (Moscona, A. A., ed.), pp. 29–49, John Wiley, New York.

Ledbetter, M. C. (1962) Observations on membranes in plant cells fixed with OsO$_4$. In: Proceedings 5th International Congress of Electron Microscopy. (Breese, S. S., ed.), pp. 1–10, Academic Press, New York.

Lee, T. C. and Snyder, F. (1973) Phospholipid metabolism in rat liver endoplasmic reticulum. Structural analyses, turnover studies and enzymatic activities. Biochim. Biophys. Acta 291, 71–82.

Lee, T. C., Stephens, N., Moehl, A. and Snyder, F. (1973) Turnover of rat liver plasma membrane phospholipids. Comparison with microsomal membranes. Biochim. Biophys. Acta 291, 86–92.

Leloir, L. F., Stanleoni, R. J., Carminatti, H. and Behrens, H. H. (1973) The biosynthesis of a N-N' diacetyl chitobiose containing lipid by liver microsomes: a probable dolichol pyrophosphate derivative. Biochem. Biophys. Res. Commun. 52, 1285–1292.

Lennarz, W. J. and Scher, M. G. (1972) Metabolism and function of polyisoprenol sugar intermediates in membrane-associated reactions. Biochim. Biophys. Acta 265, 417–441.

Leskes, A., Siekevitz, P. and Palade, G. E. (1971) Differentiation of endoplasmic reticulum in hepatocytes. I. Glucose-6-phosphatase distribution *in situ*. J. Cell Biol. 49, 264–287.

Levine, A. M., Higgins, J. A. and Barrnett, R. J. (1972) Biogenesis of plasma membranes in salt glands of salt-stressed domestic ducklings: Localization of acyl transferase activity. J. Cell Sci. 11, 855–873.

Levinson, S. S. and Wolf, G. (1972) The effect of vitamin A acid on glycoprotein synthesis in skin tumors (keratoacanthomas). Cancer Res. 32, 2248–2252.

Lin, C.-T. and Chang, J. P. (1975) Electron microscopy of albumin synthesis. Science 190, 465–467.

Linzell, J. L. and Peaker, M. (1971) Mechanism of milk secretion. Physiol. Rev. 51, 564–597.

Little, J. S. and Widnell, C. C. (1975) Evidence for the translocation of 5'-nucleotidase across hepatic membranes *in vivo*. Proc. Nat. Acad. Sci. U.S.A. 72, 4013–4017.

Lodish, H. F. (1973) Biosynthesis of reticulocyte membrane proteins by membrane-free polyribosomes. Proc. Nat. Acad. Sci. U.S.A. 70, 1526–1530.

Loh, H. H., Cho, T. M., Wu, Y.-C. and Way, E. L. (1974) Stereospecific binding of narcotics to brain cerebrosides. Life Sci. 14, 2231–2245.

Lowe, D. and Hallinan, T. (1973) Preferential synthesis of a membrane-associated protein by free polyribosomes. Biochem. J. 136, 825–828.

Lucas, J. J., Waechter, C. J. and Lennarz, W. J. (1975) The participation of lipid-linked oligosaccharide in synthesis of membrane glycoproteins. J. Biol. Chem. 250, 1992–2002.

Marinos, N. G. (1963) Vacuolation in plant cells. J. Ultrastruct. Res. 9, 177–185.

Martin, S. S. and Bosman, H. B. (1971) Glycoprotein nature of mitochondrial structural protein and neutral sugar content of mitochondrial proteins and structural proteins. Exp. Cell Res. 66, 59–64.

Martínez-Palomo, A. (1970) The surface coats of animal cells. Int. Rev. Cytol. 29, 29–75.

Maruta, H. and Goldstein, L. (1975) The fate and origin of the nuclear envelope during and after mitosis in *Amoeba proteus*. I. Synthesis and behavior of phospholipids of the nuclear envelope during the cell life cycle. J. Cell Biol. 65, 631–645.

Mata, L. R. (1976) Dynamics of HRPase absorption in the epithelial cells of the hamster seminal vesicles. J. Microscop. Biol. Cell. 25, 127–132.

Matile, P. and Moor, H. (1968) Vacuolation: Origin and development of the lysosomal apparatus in root-tip cells. Planta, 80, 159–175.

McBride, D. L. and Cole, K. (1972) Ultrastructural observations on germinating monospores in *Smithora naiadum* (Rhodophyceae, Bangiophycidae). Phycologia 11, 181–191.

McDonald, R. J. and Ronzio, R. A. (1972) Comparative analysis of zymogen granule membrane polypeptides. Biochem. Biophys. Res. Commun. 49, 377–382.

McKanna, J. A. (1973) Membrane recycling: vesiculation of the *Amoeba* contractile vacuole at systole. Science 179, 88–92.

McKeel, D. W. and L. Jarett (1974) The enrichment of adenylate cyclase in the plasma membrane and Golgi subcellular fractions of porcine adenohypophysis. J. Cell Biol. 62, 231–236.

McMurray, W. C. and Dawson, R. M. C. (1969) Phospholipid exchange reactions within the liver cell. Biochem. J. 112, 91–108.

McMurray, W. C. and Magee, W. L. (1972) Phospholipid metabolism. Ann. Rev. Biochem. 41, 129–160.

Meldolesi, J. (1974a) Secretory mechanisms in pancreatic acinar cells. Role of the cytoplasmic membranes. In: Advances in Cytopharmacology (Ceccarelli, B., Clementi, F. and Meldolesi, J., eds.), vol. 2 pp. 71–85, Raven Press, New York.

Meldolesi, J. (1974b) Membranes and membrane surfaces. Dynamics of cytoplasmic membranes in pancreatic acinar cells. Phil. Trans. R. Soc. Lond. Ser. B 268, 39–53.

Meldolesi, J. (1974c) Dynamics of cytoplasmic membranes in guinea pig pancreatic acinar cells. I. Synthesis and turnover of membrane proteins. J. Cell Biol. 61, 1–13.

Meldolesi, J. and Cova, D. (1971) Synthesis and interactions of cytoplasmic membranes in the pancreatic exocrine cells. Biochem. Biophys. Res. Commun. 44, 139–143.

Meldolsei, J. and Cova, D. (1972a) Composition of cellular membranes in the pancreas of the guinea pig. IV. Polyacrylamide gel electrophoresis and amino acid composition of membrane proteins. J. Cell Biol. 55, 1–18.

Meldolesi, J. and Cova, D. (1972b) Synthesis and interactions of cytoplasmic membranes in the acinar cells of the guinea pig pancreas. In: Role of Membranes in Secretory Processes (Bolis, L., ed.), pp. 62–71. Elsevier/North-Holland, Amsterdam.

Meldolesi, J., Jamieson, J. D. and Palade, G. E. (1971) Composition of cellular membranes in the pancreas of the guinea pig. II. Lipids J. Cell Biol. 49, 130–149.

Merritt, W. D. and Morré, D. J. (1973) A glycosyl transferase of high specific activity in secretory vesicles derived from Golgi apparatus of rat liver. Biochim. Biophys. Acta 304, 397–407.

Moe, H., Rostgaard, J. and Behnke, O. (1965) On the morphology and origin of virgin lysosomes in the intestinal epithelium of the rat. J. Ultrastruct. Res. 12, 396–403.

Moestrup, Ø. and Thomsen, H. A. (1974) An ultrastructural study of the flagellate *Pyraminonas orientalis* with particular emphasis on Golgi apparatus activity and the flagellar apparatus. Protoplasma 81, 247–269.

Mollenhauer, H. H. (1971) Fragmentation of mature dictyosome cisternae. J. Cell Biol. 49, 212–214.

Mollenhauer, H. H. and Morré, D. J. (1966) Golgi apparatus and plant secretion. Ann. Rev. Plant Physiol. 17, 27–46.

Mollenhauer, H. H. and Morré, D. J. (1974) Polyribosomes associated with the Golgi apparatus. Protoplasma 79, 333–336.

Mollenhauer, H. H. and Morré, D. J. (1976a) Cytochalasin B, but not colchicine, inhibits migration of secretory vesicles in root tips of maize. Protoplasma 87, 39–48.

Mollenhauer, H. H. and Morré, D. J. (1976b) Transition elements between endoplasmic reticulum and Golgi apparatus in plant cells. Cytobiologie 13, 297–306.

Mollenhauer, H. H., Morré, D. J. and VanDerWoude, W. J. (1975) Endoplasmic reticulum-Golgi apparatus associations in maize root tips. Mikroskopie 31, 257–272.

Mollenhauer, H. H., Hass, B. S. and Morré D. J. (1976) Membrane transformations in Golgi apparatus of rat spermatids: A role for thick cisternae and two classes of coated vesicles in acrosome formation. J. Microscop. Biol. Cell. 27, 33–36.

Molnar, J. and Sy, D. (1967) Attachment of glucosamine to protein at the ribosomal site of rat liver. Biochemistry 6, 1941–1947.

Molnar, J., Robinson, G. B. and Winzler, J. (1965) Biosynthesis of glycoproteins. 4. The subcellular sites of incorporation of glucosamine-1-^{14}C into glycoprotein of rat liver. J. Biol. Chem. 240, 1882–1888.

Molnar, J., Chao, H. and Ikehara, Y. (1971) Phosphoryl-N-acetylglucosamine transfer to a lipid acceptor of liver microsomal preparations. Biochim. Biophys. Acta 239, 401–410.

Monis, B., Rovasio, R. A. and Valentich, M. A. (1975) Ultrastructural characterization by ruthenium red of the surface of the fat globule membrane of human and rat milk with data on carbohydrates of fractions of rat milk. Cell Tissue Res. 157, 17–24.

Morré, D. J. (1970) *In vivo* incorporation of radioactive metabolites by Golgi apparatus and other cell fractions of onion stem. Plant Physiol. 45, 791–799.

Morré, D. J. (1973) Isolation and purification of organelles and endomembrane components from rat liver. In: Molecular Techniques and Approaches in Developmental Biology. (Chrispeels, M. J., ed.), pp. 1–27, John Wiley, New York.

Morré, D. J. (1975) Membrane biogenesis. Ann. Rev. Plant Physiol. 26, 441–481.

Morré, D. J. and Mollenhauer, H. H. (1974) The endomembrane concept: a functional integration of endoplasmic reticulum and Golgi apparatus. In: Dynamic Aspects of Plant Ultrastructure (Robards, A. W., ed), pp. 84–137, McGraw-Hill, New York and London.

Morré, D. J. and Mollenhauer, H. H. (1976) Interactions among cytoplasm, endomembranes, and the cell surface. In: Transport in Plants. III. Intracellular Interactions and Transport Processes. (Stocking, C. R. and Heber, U., eds.), pp. 288–344, Encyclopedia of Plant Physiology, New Series, Springer-Verlag, Berlin, Heidelberg and New York.

Morré, D. J. and VanDerWoude, W. J. (1974) Origin and growth of cell surface components. In: Macromolecules Regulating Growth and Development (Hay, E. D., King, T. J. and Papaconstantinou, J., eds.), pp. 81–111, Academic Press, New York.

Morré, D. J., Jones, D. D. and Mollenahuer, H. H. (1967) Golgi apparatus mediated polysaccharide secretion by outer root cap cells of Zea mays. I. Kinetics and secretory pathway. Planta 74, 286–301.

Morré, D. J., Merlin, L. M. and Keenan, T. W. (1969) Localization of glycosyl transferase activities in a Golgi apparatus-rich fraction isolated from rat liver. Biochem. Biophys. Res. Commun. 37, 813–819.

Morré, D. J., Nyquist, S. and Rivera, E. (1970) Lecithin biosynthetic enzymes of onion stem and the distribution of phosphoryl choline-cytidyl transferase among cell fractions. Plant Physiol. 45, 800–804.

Morré, D. J., Franke, W. W., Deumling, B., Nyquist, S. E. and Ovtracht, L. (1971a) Golgi apparatus function in membrane flow and differentiation: origin of plasma membranes from endoplasmic reticulum. Biomembranes 2, 95–104.

Morré, D. J., Keenan, T. W. and Mollenhauer, H. H. (1971b) Golgi apparatus function in membrane transformations and product compartmentalization: Studies with cell fractions isolated from rat liver. In: Advances in Cytopharmacology (Clementi, F. and Ceccarelli, B., eds.), vol. 1, pp. 159–182, Raven Press, New York.

Morré, D. J., Merritt, W. D. and Lembi, C. A. (1971c) Connections between mitochondria and endoplasmic reticulum in rat liver and onion stem. Protoplasma 73, 43–49.

Morré, D. J., Mollenhauer, H. H. and Bracker, C. E. (1971d) The origin and continuity of Golgi apparatus. In: Results and Problems in Cell Differentiation. II. Origin and Continuity of Cell Organelles (Reinert, T. and Ursprung, H., eds.), pp. 82–126, Springer-Verlag, Berlin.

Morré, D. J., Keenan, T. W. and Huang, C. M. (1974a) Membrane flow and differentiation: Origin of Golgi apparatus membranes from endoplasmic reticulum. In: Advances in Cytopharmacology (Ceccarelli, B., Clementi, F. and Meldolesi, J., eds.) vol. 2, pp. 107–125, Raven Press, New York.

Morré, D. J., Yunghans, W. N., Vigil, E. L. and Keenan, T. W. (1974b) Isolation of organelles and endomembrane components from rat liver: Biochemical markers and quantitative morphometry. In: Methodological Developments in Biochemistry (Reid, E., ed), vol. 4, pp. 195–236, Longmans, London.

Morris, A. J. and Dickman, S. R. (1960) Biosynthesis of ribonuclease in mouse pancreas. J. Biol. Chem. 235, 1404–1408.

Mulder, E. and Van Deenen, L. L. M. (1965) Metabolism of red-cell lipids. I. Incorporation in vitro of fatty acids into phospholipids from mature erythrocytes. Biochim. Biophys. Acta 106, 106–117.

Nachmias, V. T. (1966) A study by electron microscopy of the formation of new surface by Chaos chaos. Exp. Cell Res. 43, 483–601.

Naiki, M. and Marcus, D. M. (1974) Human erythrocytes P and P^k blood group antigens: Identification as glycosphingolipids. Biochem. Biophys. Res. Commun. 60, 1105–1111.

Neutra, M. and Leblond, C. P. (1966a) Synthesis of the carbohydrate of mucus in the Golgi complex as shown by electron microscope radioautography of goblet cells from rats injected with glucose-H^3. J. Cell Biol. 30, 119–136.

Neutra, M. and Leblond, C. P. (1966b) Radioautographic comparison of the uptake of galactose-H^3 and glucose-H^3 in the Golgi region of various cells secreting glycoproteins or mucopolysaccharides. J. Cell Biol. 30, 137–150.

76

Nève, P., Williams, C. and Dumont, J. E. (1970) Involvement of the microtubule-microfilament system in thyroid secretion. Exp. Cell Res. 63, 457–460.

Nicolson, G. L. (1974) The interactions of lectins with animal cell surfaces. Int. Rev. Cytol. 39, 89–190.

Nicolson, G. L., Poste, G. and Ji, T. H. (1977) The dynamics of cell membrane organization. In Dynamic Aspects of Cell Surface Organization (Poste, G. and Nicolson, G. L., eds.), Cell Surface Reviews, vol. 3, pp. 1–75, Elsevier/North-Holland, Amsterdam.

Nicolson, G. L. and Singer, S. J. (1974) The distribution and asymmetry of mammalian cell surface saccharides utilizing ferritin conjugated plant agglutinins as specific saccharide stains. J. Cell Biol. 60, 236–248.

Nii, S., Morgan, C. Rose, H. M. and Hsu, K. C. (1968) Electron microscopy of herpes simplex virus. IV. Studies with ferritin-conjugated antibodies. J. Virol. 2, 1172–1184.

Novikoff, A. B. and Shin, W. Y. (1964) The endoplasmic reticulum in the Golgi zone and its relations to microbodies, Golgi apparatus and autophagic vacuoles in rat liver cells. J. Microscop. 3, 187–206.

Novikoff, A. B., Essner, E., Goldfischer, S. and Heus, M. (1962) Nucleoside-diphosphatase activities of cytomembranes. In: The Interpretation of Ultrastructure (Harris, R. J. C., ed.), pp. 149–192, Academic Press, New York and London.

Novikoff, P. M., Novikoff, A. B., Quintana, N. and Hauw, J.-J. (1971) Golgi apparatus, GERL, and lysosomes of neurons in dorsal root ganglia, studied by thick section and thin section cytochemistry. J. Cell Biol. 50, 859–886.

Nyquist, S. E., Barr, R. and Morré, D. J. (1970) Ubiquinone from rat liver Golgi apparatus fractions. Biochim. Biophys. Acta 208, 532–539.

Nyquist, S. E., Crane, F. L. and Morré, D. J. (1971a) Vitamin A: Concentration in the rat liver Golgi apparatus. Science 173, 939–941.

Nyquist, S. E., Matschiner, T. T. and Morré, D. J. (1971b) Distribution of vitamin K among rat liver cell fractions. Biochim. Biophys. Acta 244, 645–649.

Oliver, G. J., Harrison, J. and Hemming, F. W. (1975) The mannosylation of dolichol-diphosphate oligosaccharides in relation to the formation of oligosaccharides and glycoproteins in pig liver endoplasmic reticulum. Eur. J. Biochem. 58, 223–229.

Omura, T., Sato, R., Cooper, D. Y., Rosenthal, O. and Estabrook, R. W. (1965) Function of cytochrome P-450 of microsomes. Fed. Proc. 24, 1181–1189.

Orei, L., Malaisse-Lagae, F., Ravazzola, M. and Amherdt, M. (1973) Exocytosis-endocytosis coupling in the pancreatic beta cell. Science 181, 561–562.

Osborn, M. J. (1969) Structure and biosynthesis of the bacterial cell wall. Ann. Rev. Biochem. 38, 501–533.

Ovtracht, L., Morré, D. J., Cheetham, R. D. and Mollenhauer, H. H. (1973) Subfractionation of Golgi apparatus from rat liver: Method and morphology. J. Microscopie 18, 87–102.

Palade, G. E. (1956) The endoplasmic reticulum. J. Biophys. Biochem. Cytol. 2 (Suppl.), 85–97.

Palade, G. E. (1959) Functional changes in the structure of cell components, In: Subcellular Particles (Hayashi, T., ed.), pp. 64–83, Ronald Press, New York.

Palade, G. E. (1975) Intracellular aspects of the process of protein secretion. Science 189, 347–358.

Patton, S. (1970) Correlative relationship of cholesterol and sphingomyelin in cell membranes. J. Theoret. Biol. 29, 489–491.

Patton, S. and Fowkes, F. M. (1967) The role of plasma membrane in the secretion of milk fat. J. Theoret. Biol. 15, 274–281.

Patton, S. and Keenan, T. W. (1975) The milk fat globule membrane. Biochim. Biophys. Acta 415, 273–309.

Pelletier, G. and Novikoff, A. B. (1972) Localization of phosphatase activities in the rat anterior pituitary gland. J. Histochem. Cytochem. 20, 1–12.

Peterson, M. and Leblond, C. P. (1964a) Synthesis of complex carbohydrates in the Golgi region, as shown by radioautography after injection of glucose. J. Cell Biol. 21, 143–148.

Peterson, M. and Leblond, C. P. (1964b) Uptake by the Golgi region of glucose labelled with tritium in the 1 or 6 position, as an indicator of synthesis of complex carbohydrates. Exp. Cell Res. 34, 420–423.

Pfenninger, K. H. and Bunge, R. P. (1974) Freeze-fracturing of nerve growth cones and young fibers. A study of developing plasma membrane. J. Cell Biol. 63, 180–196.

Pickett-Heaps, J. D. (1967) Further observations on the Golgi apparatus and its functions in cells of the wheat seedling. J. Ultrastruct. Res. 18, 287–303.

Pickett-Heaps, J. D. (1968) Further ultrastructural observations on polysaccharide localization in plant cells. J. Cell Sci. 3, 55–64.

Pinto Da Silva, P. and Martínez-Palomo, A. (1975) Distribution of membrane particles and gap junctions in normal and transformed 3T3 cells studied *in situ* in suspension, and treated with concanavalin A. Proc. Nat. Acad. Sci. U.S.A. 72, 572–576.

Pinto Da Silva, P., Douglas, S. D. and Branton, D. (1971) Localization of A antigen sites on human erythrocyte ghosts. Nature 232, 194–196.

Plattner, H. (1974) Intramembranous changes on cation uptake triggered exocytosis in *Paramecium*. Nature 252, 722–724.

Porter, K. R., Kenyon, K. and Badenhausen, S. (1967) Specialization of the unit membrane. Protoplasma 63, 262–274.

Poux, N. (1973) Observation en microscopie électronique de cellules végétales imprégnées par l'osmium. C. R. Acad. Sci. Paris Ser. D 276, 2163–2166.

Ragnotti, G., Lawford, G. R. and Campbell, P. N. (1969) Biosynthesis of microsomal nicotinamide-adenine dinucleotide phosphate-cytochrome c reductase by membrane-bound and free polysomes from rat liver. Biochem. J. 112, 139–147.

Rambourg, A. (1971) Morphological and histochemical aspects of glycoproteins at the surface of animal cells. Int. Rev. Cytol. 31, 57–114.

Rambourg, A., Hernandez, W. and Leblond, C. P. (1969) Detection of complex carbohydrates in the Golgi apparatus of rat cells. J. Cell Biol. 40, 395–414.

Rapport, M. M. and Graf, L. (1969) Immunochemical reactions of lipids. Progr. Allergy 13, 273–331.

Rapport, M. M., Graf, L., Skipski, V. P. and Alonzo, N. F. (1958) Cytolipin H, a pure lipid hapten isolated from human carcinoma. Nature 181, 1803–1804.

Ray, P. M., Shininger, T. L. and Ray, M. M. (1969) Isolation of β-glucan synthetase particles from plant cells and identification with Golgi membranes. Proc. Nat. Acad. Sci. U.S.A. 64, 605–612.

Ray, T. K., Lieberman, I. and Lansing, A. E. (1968) Synthesis of the plasma membrane of the liver cell. Biochem. Biophys. Res. Commun. 31, 54–58.

Redman, C. M. and Cherian, M. G. (1972) The secretory pathways of rat serum glycoproteins and albumin. Localization of newly formed protein within the endoplasmic reticulum. J. Cell Biol. 52, 231–245.

Renkonen, G., Gahmberg, C. G., Simons, I. and Kääriäinen, L. (1970) Enrichment of gangliosides in plasma membranes of hamster kidney fibroblasts. Acta Chem. Scand. 24, 733–735.

Richard, M., Martin, A. and Louisot, P. (1975) Evidence for glycosyltransferases in rat liver nuclei. Biochem. Biophys. Res. Commun. 64, 108–114.

Richards, J. B. and F. W. Hemming (1972) The transfer of mannose from guanosine diphosphate mannose to dolichol phosphate and protein by pig liver endoplasmic reticulum. Biochem. J. 130, 77–93.

Richardson, C. L., Baker, S. R., Morré, D. J. and Keenan, T. W. (1975) Glycophingolipid synthesis and tumorigenesis. A role for the Golgi apparatus in the origin of specific receptor molecules of the mammalian cell surface. Biochim. Biophys. Acta 417, 175–186.

Rifkin, D. B. and Quigley, J. P. (1974) Virus-induced modification of cellular membranes related to viral structure. Ann. Rev. Microbiol. 28, 325–351.

Roberts, K. and Northcote, D. H. (1970) The structure of sycamore cells during division in a partially synchronized suspension culture. J. Cell Sci. 6, 299–321.

Roelofsen, P. A. (1959) The Plant Cell-Wall. Hand. der Pflanzenanatomie 2 (Zimmerman, W. and Ozenda, P. G., eds.), 325 pp., Gebr. Borntraeger, Berlin-Nikolassee.

Roland, J.-C. (1969) Mise en évidence sur coupes ultrafines de formations polysaccharidiques directement associées au plasmalemme. C. R. Acad. Sci. Paris 269, 939–942.

Roland, J.-C. (1973) The relationship between the plasmalemma and plant cell wall. Int. Rev. Cytol. 36, 45–92.

Roland, J.-C. and Vian, B. (1971) Réactivité du plasmalemme végétal. Etude cytochimique. Protoplasma 73, 121–137.

Roland, J.-C., Lembi, C. A. and Morré, D. J. (1972) Phosphotungstic acid-chromic acid as a selective electron-dense stain for plasma membranes of plant cells. Stain Technol. 47, 195–200.

Roseman, S. (1970) The synthesis of complex carbohydrates by multiglycosyltransferase systems and their potential function in intercellular adhesion. Chem. Phys. Lipids 5, 270–297.

Rosso, G., DeLuca, L., Warren, C. D. and Wolf, G. (1975) Enzymatic synthesis of mannosyl retinyl phosphate from retinyl phosphate and guanosine diphosphate mannose. J. Lipid Res. 16, 235–242.

Roth, S., McGuire, E. J. and Roseman, S. (1971) Evidence for cell-surface glycosyl transferases. Their potential role in cellular recognition. J. Cell Biol. 51, 536–547.

Ryder, T. A. and Bowen, I. D. (1974) The fine structural localization of alkaline phosphatase in *Polycelis tenuis* (Ijima). Protoplasma 79, 19–29.

Sakagami, T., Minari, O. and Orii, T. (1965) Behavior of plasma lipoproteins during exchange of phospholipids between plasma and erythrocytes. Biochim. Biophys. Acta 98, 111–116.

Sakai, A. and Shigenaka, M. (1967) Behavior of cytoplasmic membranous structures in the spermatogenesis of the grasshopper *Atractomorpha bedeli*, Bolivar Cytologia (Tokyo) 32, 72–86.

Sanders, E. J. (1970) Pinocytosis in centrifuged and bisected amoebae. Exp. Cell Res. 61, 461–465.

Sanders, E. J. and Singal, P. K. (1975) Furrow formation in *Xenopus* embryos. Involvement of the Golgi body as revealed by ultrastructural localization of thiamine pyrophosphatase activity. Exp. Cell Res. 93, 219–224.

Sandoz, D. (1970a) Étude cytochemique des polysaccharides au cours de la spermatogenése d'un amphibien anoure: Le discoglosse *Discoglossus pictus* (Otth.). J. Microscop. Paris 9, 243–262.

Sandoz, D. (1970b) Évolution des ultrastructures au cours de la formation de l'acrosome du spermatozoide chez la souris. J. Microscop. Paris 9, 535–558.

Satir, B. (1974a) Ultrastructural aspects of membrane fusion. J. Supramol. Struc. 2, 529–537.

Satir, B. (1974b) Membrane events during the secretory process. Soc. Exp. Biol. Symp. 28, 399–418.

Scallen, T. J., Srikantaiah, M. V., Seetharam, B., Hansbury, E. and Gavay, K. L. (1974) Sterol carrier protein hypothesis. Fed. Proc. 33, 1733–1746.

Schachter, H. (1974a) The subcellular sites of glycosylation. Biochem. Soc. Symp. 40, 57–71.

Schachter, H. (1974b) Glycosylation of glycoproteins during intracellular transport of secretory products. In: Advances in Cytopharmacology (Ceccarelli, B., Clementi, F. and Meldolesi, J., eds.), vol. 2, pp. 207–218, Raven Press, New York.

Schachter, H., Jabbal, I., Hudgin, R. L., Pinteric, L., McGuire, E. J. and Roseman, S. (1970) Intracellular localization of liver sugar nucleotide glycoprotein glycosyl-transferases in a Golgi-rich fraction. J. Biol. Chem. 245, 1090–1100.

Schauer, R., Buscher, H.-P. and Casals-Stenzel, J. (1974) Sialic acids: Their analysis and enzymatic modification in relation to the synthesis of submandibular-gland glycoproteins. Biochem. Soc. Symp. 40, 87–116.

Scheer, U. and Franke, W. W. (1969) Negative staining and adenosine triphosphatase activity of annulate lamellae of newt oocytes. J. Cell Biol. 42, 519–533.

Schenkein, I. and Uhr, J. W. (1970) Immunoglobulin synthesis and secretion. I. Biosynthetic studies of the addition of the carbohydrate moieties. J. Cell Biol. 46, 42–51.

Schmid, R., Marver, H. S. and Hammaker, L. (1966) Enhanced formation of rapidly labeled bilirubin by phenobarbital: Hepatic microsomal cytochromes as a possible source. Biochem. Biophys. Res. Commun. 24, 319–328.

Schnepf, E. (1961) Quantitative zusammenhänge zwischen der Sekretion des Fangschleimes und den Golgi-Strukturen bei *Drosophyllum lusitanicum*. Z. Naturforsch. 16b, 605–610.

Schnepf, E. (1969a) Sekretion und Exkretion bei Pflanzen. Protoposmatologia Handbuch der Protoplasmaforschung 8, 1–181.

Schnepf, E. (1969b) Membranfluss und membrantransformation. Ber. Deutsch. Bot. Ges. 82, 407–413.

Schnepf, E. and Busch, J. (1976) Morphology and kinetics of slime secretion in glands of *Mimulus tilingii* Z. Pflanzenphysiol 69, 62–71.

Schnepf, E. and Koch, W. (1966a) Golgi-Apparat und Wasserausscheidung bei *Glaucocystis*. Z. Pflanzenphysiol. 55, 97–109

Schnepf, E., and Koch, W. (1966b) Über die Entstehung der pulsierenden Vacuolen von *Vacuolaria virescens* (Choromonadophyceae) aus dem Golgi-Apparat. Arch. Mikrobiol. 54, 229–236.

Schramm, M. (1967) Secretion of enzymes and other macromolecules. Ann. Rev. Biochem. 36, 307–320.

Sherr, C. J. and Uhr, J. W. (1970) Immunoglobulin synthesis and secretion. V. Incorporation of leucine and glucosamine into immunoglobulin on free and bound polyribosomes. Proc. Nat. Acad. Sci. U.S.A. 66, 1183–1189.

Shur, L. D. and Roth, S. (1975) Cell surface glycosyl transferases. Biochim. Biophys. Acta 415, 473–512.

Siekevitz, P. (1965) Origin and functional nature of microsomes. Fed. Proc. 24, 1153–1155.

Siekevitz, P. and Palade, G. E. (1958) A cytochemical study on the pancreas of the guinea pig. I. Isolation and enzymatic activities of cell fractions. J. Biophys. Biochem. Cytol. 4, 203–217.

Siekevitz, P. and Palade, G. E. (1960) A cytochemical study on the pancreas of the guinea pig. V. *In vivo* incorporation of leucine-1-C^{14} into the chymotrypsinogen of various cell fractions. J. Biophys. Biochem. Cytol. 7, 619–630.

Simpson, L. L. and Rapport, M. M. (1971) Ganglioside inactivation of botulism toxin. J. Neurochem. 18, 1341–1343.

Singer, S. J. (1974) The molecular organization of membranes. Ann. Rev. Biochem. 43, 805–833.

Sjöstrand, F. S. (1956) The ultrastructure of cells as revealed by the electron microscope. Int. Rev. Cytol. 5, 455–533.

Sjöstrand, F. S. (1963) A comparison of plasma membrane, cytomembranes, and mitochondrial membranes with respect to ultrastructural features. J. Ultrastruct. Res. 9, 561–580.

Sjöstrand, F. S. (1968) Ultrastructure and function of cellular membranes in ultrastructure in biological systems. In: The Membranes (Dalton, A. J. and Haguenau, F., eds.), pp. 151–210, Academic Press, New York.

Smigel, M. and Fleischer, S. (1974) Characterization and localization of prostaglandin E_1 receptors in rat liver plasma membranes. Biochim. Biophys. Acta 332, 358–373.

Stadler, J. and Franke, W. W. (1973) Nuclear membranes and plasma membranes from hen erythrocytes. III. Localization of activities incorporating fatty acids into phospholipids. Biochim. Biophys. Acta 311, 205–213.

Staerk, J., Ronnenberger, H. J., Weigandt, H. and Ziegler, W. (1974) Interaction of ganglioside C Gtetl and its derivatives with choleragen. Eur. J. Biochem. 48, 103–110.

Stahelin, L. A. and Kiermayer, O. (1970) Membrane differentiation in the Golgi complex of *Micrasterias denticulata* Bréb. visualized by freeze-etching. J. Cell Sci. 7, 787–792.

Stahl, W. L. and Trams, E. G. (1968) Synthesis of lipids by liver plasma membranes. Incorporation of acyl-coenzyme A derivatives into membrane lipids *in vitro*. Biochim. Biophys. Acta 163, 459–471.

Stein, O. and Stein, Y. (1971) Light and electron microscopic radioautography of lipids: Techniques and biological applications. Adv. Lipid Res. 9, 1–72.

Stein, O., Sanger, L. and Stein, Y. (1974) Colchicine-induced inhibition of lipoprotein and protein secretion into the serum and lack of interference with secretion of biliary phospholipids and cholesterol by rat liver *in vivo*. J. Cell Biol. 62, 90–103.

Stein, Y., Widnell, C. and Stein, O. (1968) Acylation of lysophosphatides by plasma membrane fractions of rat liver. J. Cell Biol. 39, 185–198.

Steinman, R. M., Brodie, S. E. and Cohn, Z. A. (1976) Membrane flow during pinocytosis. A stereological analysis. J. Cell Biol. 68, 665–687.

Stewart, P. S. and MacLennon, D. H. (1974) Surface particles of sarcoplasmic reticulum membranes. Structural features of the adenosine triphosphatase. J. Biol. Chem. 249, 985–993.

Stockem, W. (1969) Pinocytose und Bewegung von Amoben. III. Die Funktion des Golgiapparates von *Amoeba proteus* und *Chaos chaos*. Histochemie 18, 217–240.

Stockem, W. (1972) Membrane turnover during locomotion of *Amoeba proteus*. Acta Protozool. 11, 88–93.

Stoffel, W. (1971) Sphingolipids. Ann. Rev. Biochem. 40, 57–82.

Strittmatter, P., Rogers, M. and Spatz, L. (1972) The binding of cytochrome b_5 to liver microsomes. J. Biol. Chem. 247, 7188–7194.

Sturgess, J. M., De La Iglesia, F. A., Minaker, E., Mitranic, M. and Moscarello, M. A. (1974) The

Golgi complex. II. The effects of aminonucleosidase on ultrastructure and glycoprotein biosynthesis. Lab. Invest. 31, 6–14.

Sturgess, J. M., Minaker, E., Mitranic, M. M. and Moscarello, M. A. (1973) The incorporation of L-fucose into glycoproteins in the Golgi apparatus of rat liver and in serum. Biochim. Biophys. Acta 320, 123–132.

Sturgess, J. M., Mitranic, M. and Moscarello, M. A. (1972) The incorporation of D-glucosamine ³H into the Golgi complex from rat liver and into serum glycoproteins. Biochem. Biophys. Res. Commun. 46, 1270–1277.

Sundsmo, J. S. and Hakomori, S.-I. (1976) Lacto-N-neolactosylceramide ("paragloboside") as a possible tumor associated surface antigen of hamster NILpy tumor. Biochem. Biophys. Res. Commun. 68, 799–806.

Susi, F. R., Leblond, C. P. and Clermont, Y. (1971) Changes in the Golgi apparatus during spermiogenesis in the rat. Am. J. Anat. 130, 251–268.

Svennerholm, L. (1970) Ganglioside metabolism. In: Comprehensive Biochemistry (Florkin, M. and Stotz, E. A., eds), vol. 18, pp. 201–227, Elsevier, Amsterdam.

Tamaoki, B. (1973) Steroidognesis and cell structure. Biochemical pursuit of sites of steroid biosynthesis. J. Steroid Biochem. 4, 89–118.

Tanner, W., Jung, P. and Behrens, N. H. (1971) Dolichol monophosphates: mannosyl acceptors in a particulate in vitro system of S. cerevisiae. FEBS Lett. 16, 245–248.

Taylor, J. M., Dehlinger, P. J., Dice, J. F. and Schimke, R. T. (1973) The synthesis and degradation of membrane proteins. Drug Metab. Dispos. 1, 84–91.

Thiéry, J.-P. (1969) Role de l'appareil de Golgi dans la synthese des mucopolysaccharides étude cytochimique. I. Mise en évidence de mucopolysaccharides dans les vésicules de transition entre l'ergastoplasme et l'appareil de Golgi. J. Microscop. 8, 689–708.

Tillack, T. W., Scott, R. E. and Marchesi, V. T. (1972) The structure of erythrocyte membranes studied by freeze-etching. II. Localization of receptors for phytohemagglutinin and influenza virus to the intramembranous particles. J. Exp. Med. 135, 1209–1227.

Tillack, T. W., Boland, R. and Martonosi, A. (1974) The ultrastructure of developing sacroplasmic reticulum. J. Biol. Chem. 249, 624–633.

Tixier-Vidal, A., Picart, R. and Moreau, M.-F. (1976) Endocytose et sécrétion dans les cellules antéhypophysaires en culture. Action des hormones hypothalamiques. J. Microscop. Biol. Cell. 25, 159–172.

Tkacz, J. S., Herscovics, A., Warren, C. D. and Jeanloz, R. W. (1974) Mannosyltransferase activity in calf pancreas microsomes. Formation from guanosine diphosphate-D-¹⁴C-mannose of a ¹⁴C-labelled mannolipid with properties of dolichylmannopyranosylphosphate. J. Biol. Chem. 249, 6372–6381.

Tomita, M. and Marchesi, V. T. (1975) Amino-acid sequence and oligosaccharide attachment sites of human erythrocyte glycophorin. Proc. Nat. Acad. Sci. U.S.A. 72, 2964–2968.

Ueda, K. (1966) Fine structure of Chlorogonium elongatum with special reference to vacuole development. Cytologia Tokyo 31, 461–472.

Ueda, K. and Noguchi, T. (1976) Transformation of the Golgi apparatus in the cell cycle of a green alga, Micrasterias americana. Protoplasma 87, 145–162.

Uhr, J. W. (1970) Intracellular events underlying synthesis and secretion of immunoglobulin. Cell. Immunol. 1, 228–244.

Van Deenen, L. L. M. and de Grier, J. (1964) Chemical composition and metabolism of lipids in red cells of various animal species. In: The Red Blood Cell (Bishop, C. and Surgenor, D. M., eds.), pp. 243–302, Academic Press, New York.

Van Den Bosch, H. (1974) Phosphoglyceride metabolism. Ann. Rev. Biochem. 43, 243–277.

Van Den Bosch, H., Van Golde, L. M. G. and Van Deenen, L. L. M. (1972) Dynamics of phosphoglycerides. Ergeb. Physiol. Biol. Chem. Exp. Pharmakol. 66, 13–145.

Van Der Woude, W. J., Lembi, C. A., Morré, D. J., Kindinger, J. I. and Ordin, L. (1974) β-glucan synthetases of plasma membrane and Golgi apparatus from onion stem. Plant Physiol. 54, 333–340.

Van Golde, L. M. G., Fleischer, B. and Fleischer, S. (1971) Some studies on the metabolism of phospholipids in Golgi complex from bovine and rat liver in comparison to other subcellular fractions. Biochim. Biophys. Acta 249, 318–330.

Van Golde, L. M. G., Raben, J., Batenburg, J. J., Fleischer, B., Zambrano, F. and Fleischer, S. (1974) Biosynthesis of lipids in Golgi complex and other subcellular fractions from rat liver. Biochim. Biophys. Acta 360, 179–192.

Van Heyningen, W. E. (1974) Gangliosides as membrane receptors for tetanus toxin, cholera toxin and serotonin. Nature 249, 415–417.

Vessey, D. A., Lysenko, N. and Zakim, D. (1975) Evidence for multiple enzymes in the dolichol utilizing pathway of glycoprotein biosynthesis. Biochim. Biophys. Acta 428, 138–145.

Vian, B. (1972) Aspects, en cryodécapage, de la fusion des membranes des vésicules cytoplasmiques et du plasmalemme lars des phénomenes de sécrétion végétale. C. R. Acad. Sci. Paris Ser. D. 275, 2471–2474

Vian, B. (1974) Précisions fournies par le cryodécapage sur la restructuration et l'assimilation au plasmalemme des membranes des dérivés golgiens. C. R. Acad. Sci. Paris Ser. D 278, 1483–1486.

Vian, B. and Roland, J. C. (1972) Différenciation des cytomembranes et renouvelement du plasmalemme dans les phénomenes de sécrétions végétales. J. Microscop. 13, 119–136.

Vigil, E. L., Morré, D. J., Frantz, C. and Huang, C. M. (1973) A NADH-ferricyanide oxido-reductase from plasma membranes of rat liver. J. Cell Biol. 59, 353a.

Virtanen, I. and Wartiovaara, J. (1974) Virus-induced cytoplasmic membrane structures associated with Semliki Forest virus infection studied by the freeze-etching method. J. Virol. 13, 222–225.

Waechter, C. J., Lucas, J. J., and Lennarz, W. J. (1973) Membrane glycoproteins. I. Enzymatic synthesis of mannosyl phosphoryl-polyisoprenol and its role as a mannosyl donor in glycoprotein synthesis. J. Biol. Chem. 248, 7570–7579.

Wagner, R. R. and Cynkin, M. A. (1971) Glycoprotein biosynthesis. Incorporation of glycosyl groups into endomembrane acceptors in a Golgi apparatus-rich fraction of liver. J. Biol. Chem. 246, 143–151.

Wallach, D., Kirshner, N. and Schramm, M. (1975) Non-parallel transport of membrane proteins and content proteins during assembly of the secretory granule in rat parotid gland. Biochim. Biophys. Acta 375, 87–105.

Warren, C. D. and Jeanloz, R. W. (1973) The characterization of glycolipids derived from long chain polyprenols: chemical synthesis of α-D-mannopyranosyl dolichol phosphate. FEBS Lett. 31, 332–334.

Warren, C. D. and Jeanloz, R. W. (1975) Synthesis of P'-dolichol P^{32}-α-D-mannopyranosyl pyrophosphate. The acid and alkaline hydrolysis of polyisoprenol α-D-mannopyranosyl mono- and pyrophosphate diesters. Biochemistry 14, 412–419.

Wedgewood, J. F., Strominger, J. L., and Warren, C. D. (1974) Transfer of sugar compounds to endogenous and synthetic dolichyl phosphate in human lymphocytes. J. Biol. Chem. 249, 6316–6324.

Weinstein, D. B., Marsh, J. B., Glick, M. C., and Warren, L. (1970) Membranes of animal cells. VI. The glycolipids of the L cell and its surface membrane. J. Biol. Chem. 245, 3928–3937.

Wellings, S. R. (1969) Ultrastructural basis of lactogenesis. In: Lactogenesis (Reynolds, M. and Folley, S. J., eds.), pp. 5–25, University of Pennsylvania Press, Philadelphia.

Whaley, W. G. (1975) The Golgi Apparatus. Cell Biology Monographs 2, 1–190.

Whaley, W. G. and Mollenhauer, H. H. (1963) The Golgi apparatus and cell plate formation—a postulate. J. Cell Biol. 17, 216–221.

Whaley, W. G., Dauwalder, M. and Kephart, J. E. (1966) The Golgi apparatus and an early stage of cell plate formation. J. Ultrastruct. Res. 15, 169–180.

Whaley, W. G., Dauwalder, M. and Kephart, J. E. (1972) Golgi apparatus: Influence on cell surfaces. Science 175, 596–599.

Whaley, W. G., Dauwalder, M. and Leffingwell, T. P. (1975) Differentiation of the Golgi apparatus in the genetic control of development. In: Current Topics in Developmental Biology (Moscona, A. A. and Monroy, A., eds.), vol. 10, pp. 161–186, Academic Press, New York.

Whur, F., Herscovics, A. and Leblond, C. P. (1969) Radioautographic visualization of the incorporation of galactose-^{3}H and mannose-^{3}H by rat thyroids in vitro in relation to the stages of thyroglobulin synthesis. J. Cell Biol. 43, 289–311.

Wiegandt, H. (1973) Gangliosides of extraneural origin. Z. Physiol. Chem. Hoppe-Seyler's 354, 1049–1056.

Wilgram, G. J. and Kennedy, E. P. (1963) Intracellular distribution of some enzymes catalyzing reactions in the biosynthesis of complex lipids. J. Biol. Chem. 238, 2615–2619.

Wilkins, M. H. F., Blaurock, A. E. and Engelman, D. M. (1971) Bilayer structure in membranes. Nature New Biol. 230, 72–76.

Wilkinson, F. E., Morré, D. J. and Keenan, T. W. (1976) Ganglioside biosynthesis. Characterization of uridine diphosphate galactose: GM$_2$ galactosyltransferase in Golgi apparatus from rat liver. J. Lipid Res. 17, 146–153.

Williams, D. C. (1974) Studies of protistan mineralization. I. Kinetics of coccolith secretion in *Hymenomonas carterae*. Calcif. Tissue Res. 16, 227–237.

Williamson, F. A. and Morré, D. J. (1976) Distribution of phosphatidylinositol biosynthetic activities among cell fractions from rat liver. Biochem. Biophys. Res. Commun. 68, 1201–1205.

Winkler, H. (1971) The membrane of the chromaffin granule. Phil. Trans. Roy. Soc. Lond. Ser. B 261, 293–303.

Winkler, H. and Hörtnagl, H. (1973) Composition and molecular organization of chromaffin granules. In: Frontiers in Catecholamine Research (Usdin, C. and Snyder, S. H., eds.), pp. 415–421, Pergamon Press, New York.

Winkler, H., Hörtnagl, H. and Smith, A. D. (1970) Membrane of the adrenal medulla. Behavior of insoluble proteins of chromaffin granules on gel electrophoresis. Biochem. J. 118, 303–310.

Winkler, H., Schöpf, J. A. L., Hörtnagl, H., and Hörtnagl, H. (1972) Bovine adrenal medulla: Subcellular distribution of newly synthesized catecholamines, nucleotides and chromogranins. Naunyn-Schmiedeberg's Arch. Pharmacol. 273, 43–61.

Winzler, R. J. (1969) A glycoprotein in human erythrocyte membranes. In: The Red Cell Membrane (Jamieson, G. A. and Greenawalt, T. J., eds.), pp. 157–171, Lippincott, Philadelphia.

Winzler, R. J. (1970) Carbohydrates in cell surfaces. Int. Rev. Cytol. 29, 77–125.

Wirtz, K. W. A. (1974) Transfer of phospholipids between membranes. Biochim. Biophys. Acta 344, 95–117.

Wirtz, K. W. A. and Zilversmit, D. B. (1968) Exchange of phospholipids between liver mitochondria and microsomes *in vitro*. J. Biol. Chem. 243, 3596–3602.

Wirtz, K. W. A. and Zilversmit, D. B. (1969) Participation of soluble liver proteins in the exchange of membrane phospholipids. Biochim. Biophys. Acta 193, 105–116.

Wise, G. E. and Flickinger, C. J. (1970) Relation of the Golgi apparatus to the cell coat in amebae. Exp. Cell Res. 61, 13–23.

Wolf, G. and LeLuca, L. (1970) Incorporation of radioactivity from retinol-^{14}C into the lipid moiety of a mannose lipid compound by rat liver. Fed. Proc. 29, 1771.

Wood, J. G., McLaughlin, B. J. and Barber, R. P. (1974) The visualization of concanavalin A binding sites in Purkinje cell somata and dendrites of rat cerebellum. J. Cell Biol. 63, 541–549.

Wooding, F. B. P. (1971a) The structure of the milk fat globule membrane. J. Ultrastruct. Res. 37, 388–400.

Wooding, F. B. P. (1971b) The mechanism of secretion of the milk fat globule. J. Cell Sci. 9, 805–821.

Wooding, F. B. P. (1973) Formation of the milk fat globule membrane without participation of the plasmalemma. J. Cell Sci. 13, 221–235.

Yamamoto, T. (1963) On the thickness of the unit membrane. J. Cell Biol. 17, 413–422.

Yanagimachi, R. (1973) Behavior and functions of the structural elements of the mammalian sperm head in fertilization. In: The Regulation of Mammalian Reproduction (Segal, S. J., Crozier, R., Corfman, P. A. and Condliffe, P. G., eds.), pp. 215–230, Charles C Thomas, Springfield, Illinois.

Yogeeswaran, G., Sheinen, R., Wherrett, J. R. and Murray, R. K. (1972) Studies on the glycophingolipids of normal and virally transformed 3T3 mouse fibroblasts. J. Biol. Chem. 247, 5146–5158.

Young, R. W. (1973) The role of the Golgi complex in sulfate metabolism. J. Cell Biol. 57, 175–189.

Yunghans, W. N. and Morré, D. J. (1977) Distribution of adenylate cyclase among membrane fractions of rat liver. J. Cell Biol. To be published.

Yunghans, W. N., Keenan, T. W. and Morré, D. J. (1970) Isolation of a Golgi apparatus-rich fraction from rat liver. III. Lipid and protein composition. Exp. Mol. Path. 12, 36–45.

Zagury, D., Uhr, J. W., Jamieson, J. D. and Palade, G. E. (1970) Immunoglobulin synthesis and secretion. I. Radioautographic studies of site of addition of carbohydrate moieties and intracellular transport. J. Cell Biol. 46, 52–63.

Zahler, W. L., Fleischer, B. and Fleischer, S. (1970) Gel electrophoresis patterns of the proteins of organelles isolated from bovine liver. Biochim. Biophys. Acta 203, 283–290.

Zambrano, F., Fleischer, S. and Fleischer, B. (1975) Lipid composition of the Golgi apparatus of rat kidney and liver in comparison with other subcellular organelles. Biochim. Biophys. Acta 380, 357–369.

Zeigel, R. F. and Dalton, A. J. (1962) Speculations based on the morphology of the Golgi system in several types of protein secreting cells. J. Cell Biol. 15, 45–54.

Biosynthesis of plasma membrane glycoproteins

2

G. M. W. COOK

1. Introduction

About fifteen years ago important advances were made in understanding the chemistry of the cell surface by histochemical and electrokinetic techniques that established the presence of complex carbohydrates[1] at the cell periphery (see Cook and Stoddart, 1973 for a review). This undoubtedly laid the foundation for the intense interest in cell surfaces that has dominated cell biology in recent years. It is now widely accepted that complex carbohydrates are an important feature of the cell surface and much evidence exists to implicate glycoproteins in a wide range of cell interaction phenomena. With the growing realization of the importance of glycoproteins in the plasma membrane work on the biosynthesis of these compounds has increased. This review covers the progress that has been made in the biosynthesis of membrane glycoproteins and also the wider problems of membrane turnover and assembly demanded by the recently realized dynamic nature of the cell surface (Poste and Nicolson, 1977).

In addition to the actual mechanism of glycoprotein biosynthesis there is the question of identifying the site or sites of synthesis within the cell as well as elucidating the means by which the cell controls this synthesis. As this chapter shows, a considerable amount of progress has been made with the first two aspects of the subject, though the question of control is one on which there is still much work to be done.

Before turning to the various aspects of plasma membrane glycoprotein biosynthesis the structural features of these macromolecules will be dealt with, since a knowledge of glycoprotein structure is a prerequisite for tackling the problem.

---

[1]The term complex carbohydrates has been used to include glycoproteins, mucopolysaccharides, glycogen, and glycolipids. When this term is used in a histochemical context it need not include glycolipids, since this group of compounds may be removed by the organic solvents used in the histochemical processing (Neutra and Leblond, 1966).

G. Poste and G.L. Nicolson (eds.)   The Synthesis, Assembly and Turnover of Cell Surface Components
© Elsevier/North-Holland Biomedical Press, 1977.

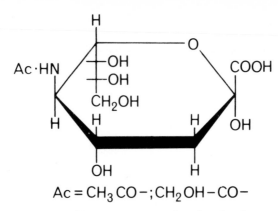

Ac = CH₃CO−; CH₂OH−CO−

$Ac = CH_3 CO-; CH_2 OH-CO-$

Fig. 1. The sialic acids. Sialic acid is the group name for all acylated neuraminic acids; the term neuraminic acid is reserved for the unsubstituted structure $C_9H_{17}O_8N$ common to all. This figure depicts two commonly occurring sialic acids. N-acetylneuraminic acid ($A_c = CH_3CO^-$ and N-glycolylneuraminic acid ($A_c = CH_2OH\text{-}CO$). The molecule can, in addition, be acylated at other positions (e.g., C-4,C-7) in N-acetyl-O-diacetylneuraminic acid and (at C-4) in N,O-diacetylneuraminic acid. In the glycoproteins the sialic acid residue is linked α-ketosidically at C-2, usually to galactose or N-acetylgalactosamine; this is the bond that is sensitive to neuraminidase (EC 3.2.1.18). (Reproduced with permission from Kemp et al. [1973].)

2. Structural chemistry of membrane glycoproteins

Generally, glycoproteins may be described as consisting of a polypeptide back-bone to which heteropolymerized carbohydrate units are covalently attached. This term covers an enormous variety of different biopolymers, ranging from those conjugated proteins that possess one prosthetic group (e.g., bovine ribonuclease B) to those which contain several hundred (e.g., ovine submaxillary gland glycoprotein with 800 units per molecule). The prosthetic groups in turn can range in size from one sugar residue (e.g., tropocollagen) to some seventeen residues, as in the case of fetuin. Naturally this variation in structure has an influence on the properties of the molecule, ranging from molecules that are predominantly carbohydrate to those that are largely protein.

Characteristically, the prosthetic glycoprotein groups contain the neutral sugars D-galactose, D-mannose, L-fucose, and occasionally D-glucose, along with the amino sugars D-glucosamine and D-galactosamine. These latter sugars are invariably present as the N-acetyl derivative. In addition to these monosac-charides, members of an important family of nine carbon sugars, the nonulosaminic acids, more commonly referred to as sialic acids, are often pres-ent (Fig. 1). Sialic acid is an agreed (Blix et al., 1957) generic name for all the acylated neuraminic acids of which N-acetylneuraminic acid and the N-glycolylneuraminic acids are common examples. Neuraminic acid, on which this group of compounds is based, is a polyhydroxy 5-(or δ) amino acid that may be regarded as a monosaccharide because the primary alcohol group of ketononose (nonulose) has been oxidized to a carboxyl group, and the hydroxyl groups on C_3 and C_5 have been replaced by a hydrogen and an amino group respectively.

From the foregoing it could be argued that a number of other protein-carbohydrate complexes come within the acceptable definition of a glycoprotein; for example, proteoglycans consist of glycosaminoglycan (or mucopolysaccharide) side chains attached to a protein "backbone." A distinction between the two classes of compounds can, however, be drawn by reference to the actual nature of the carbohydrate groups. Apart from the scarcity of sulfate and a lack of uronic acids in their prosthetic groups, the glycoproteins possess relatively shorter carbohydrate side chains than the proteoglycans. The side chains of the proteoglycans are not only much longer but are unbranched and composed of repeating disaccharide units. While the proteoglycans are not within the scope of this chapter, it is perhaps wise initially to define as clearly as possible the type of molecule being discussed here, although it will be obvious from the broad resemblances existing between glycoproteins and proteoglycans that several biosynthetic features are similar for both types of compounds.

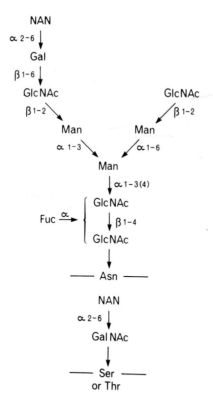

Fig. 2. Diagram illustrating the carbohydrate prosthetic group of a typical "serum type" glycoprotein (carbohydrate group of human IgG immunoglobulin) *above* and the carbohydrate group characteristic of a "mucin type" glycoprotein (principal carbohydrate moiety of bovine submaxillary mucin) illustrated *below*. In the serum type glycoproteins, the carbohydrate to protein linkage is via an N-glycosidic bond involving the amide nitrogen of asparagine and the Cl of the amino sugar residue, while in the mucin type O-glycosidic bonds to serine or threonine residues are found.

Reference has already been made to the wide variety of molecules defined as glycoproteins, and in an effort to further classify this rather diverse group of compounds workers have used the type of carbohydrate protein linkage present in a particular glycoprotein. A common type of carbohydrate protein linkage that occurs in the plasma proteins is an alkali-stable N-glycosidic bond involving the amide nitrogen of asparagine and the C_1 of a β-N-acetyl glucosamino pyranosyl residue, hence glycoproteins containing this linkage have been classified (Kraemer, 1970) as "plasma type" or "serum type" (Fig. 2) (Roseman, 1970). In the mucins, the carbohydrate groups are linked to the polypeptide by means of an alkali-labile O-glycosidic bond to serine and threonine residues —the sugar residue at the linkage point being invariably N-acetylgalactosamine. Again, by example, those glycoproteins containing O-glycosidic bonds have been referred to as "mucin type" (Kraemer, 1970; Roseman, 1970). Two other types of carbohydrate protein linkage occur in glycoproteins: the binding of sugar O-β-glycosidically to hydroxylysine as in collagen, and the S-glycosidic linkage between cysteine and carbohydrate in the erythrocyte membrane. Thus glycoproteins bearing the former linkage are referred to as "collagen type." Kraemer (1970) points out that such a classification, based on the name of the best-known example, is not very satisfactory and illustrates this contention in the case of the soluble glycoproteins by reference to ovalbumin, an estrogen-induced secretion product of the hen oviduct mucosa which, on the basis of carbohydrate-protein linkage, is classified as a "plasma" or "serum type" glycoprotein. In addition to drawing attention to the unsatisfactory nature of "plasma type" as a descriptive term, Kraemer (1970) points out that "mucin type" encompasses not only submaxillary mucin glycoproteins but also the M and N blood group active materials. These antigens can be removed from the surface of the human erythrocyte as glycopeptides by proteolytic enzyme digestion (Cook et al., 1960; Mäkelä et al., 1960; Uhlenbruck, 1961; Cook and Eylar, 1965), and Kraemer (1970) rightly states that these blood group substances "hardly satisfy anybody's definition of a mucin." Lacking a detailed knowledge of membrane glycoprotein structure and the unsatisfactory nature of classifying soluble glycoproteins, how justifiable is it to apply the serum type or mucin type descriptions to these membrane constituents?

Although considerable progress has been made with isolation of purified plasma membrane fractions as a prelude to isolating constituent proteins in a soluble form, the problems of elucidating the detailed chemical structure of such small quantities of material are still formidable. At present our knowledge of the detailed structure of membrane glycoproteins is confined to very few examples, the principal one being the major sialoglycoprotein of the human erythrocyte. Consequently, our ideas on the structure of these materials are largely influenced by the findings of studies on the soluble glycoproteins. There are, however, an increasing number of examples of membrane glycoproteins that have been obtained in sufficient quantity to enable a study of their primary structure. Reference was made earlier to the M and N blood group active substances of the human erythrocyte membrane. An analysis of M and N active glycopeptides enriched in the amino acids surrounding the carbohydrate linkage point, which are released from red cells by pronase, showed appreciable quantities of serine are present and suggested that the predominant type of carbohydrate-protein linkage in these antigens is O-glycosidic (Cook and Eylar, 1965). This suggestion was subsequently confirmed by Winzler et al. (1967) by means of alkaline cleavage. In addition, Eylar (1965) observed that the much higher ratio of galactosamine to glucosamine would also distinguish this glycoprotein from the plasma proteins. Subsequently it has been shown that in addition to bearing the M and N blood group determinants the major sialoglycoprotein of the human erythrocyte also possesses PHA (phytohemagglutinin of *Phaseolus vulgaris*) receptors (Marchesi et al. 1972). The PHA-binding sites have also been isolated as glycopeptides (Kornfeld and Kornfeld, 1969), though by contrast with the M and N blood group active sites it appears that the oligosaccharide unit in the PHA receptor is linked to the polypeptide by an N-glycosidic linkage (Kornfeld and Kornfeld, 1970). On the basis of this information it can be seen that though this membrane glycoprotein has structural features consistent with mucin type, some serum character is present. Jackson et al. (1973) suggest that the complete molecule may have at least eighteen alkali-labile carbohydrate-protein linkages and two or three asparagine-carbohydrate linkages. More recently, the same group (Tomita and Marchesi, 1975) has revised this suggestion to fifteen oligosaccharide groups linked via alkali labile O-glycosidic bonds to polypeptide with one more complex carbohydrate group linked to asparagine. Other examples of membrane glycoproteins possessing N-glycosidic linkages are appearing in the literature (Pepper and Jamieson, 1969) and it would clearly appear that to assign plasma membrane glycoproteins to the mucin or serum category is unsatisfactory. Marchesi et al. (1972) have made the suggestion that membrane glycoproteins be called glycophorins (Greek $\phi o \rho \epsilon \omega$; to bear), so that they may be regarded as a special class of glycoproteins whose role is to bear the oligosaccharides that act as receptors or recognition sites at the cell surface. These workers suggest that as membrane glycoproteins are characterized they could be distinguished by cell type or alphabetically.

While the classification of membrane glycoproteins as either serum or mucin type may be unsatisfactory, the importance of the type of carbohydrate protein linkages typified by this nomenclature in predicting oligosaccharide structure should not be underestimated. While pointing out the weaknesses of the above classification, Kraemer (1970) argues that a knowledge of the protein-carbohydrate linkage in an unknown glycoprotein structure provides considerable predictive power about the structural patterns that may be expected to be found in the carbohydrate groups. For example, a common feature of plasma type glycoproteins would appear to be a "core" structure composed of mannose and N-acetylglucosamine, some of the latter sugar being present as an N-acetylchitobiose structure linked to the asparagine residues of the polypeptide. This core structure may be linked with "peripheral" units consisting of either sialic acid $\xrightarrow{\alpha 2 - 3}$ Gal^2 or Fuc. $\xrightarrow{\alpha 1 - 3}$ Gal, and these units are linked via N-acetylglucosamine residues in the core structure. In contrast, the carbohydrate groups of the mucin type glycoproteins lack mannose and the core structure referred to above. The sugar residue forming the alkali-labile O-glycosidic linkage to either serine and threonine is invariably N-acetylgalactosamine. Equivalent peripheral sequences for this type of glycoprotein include sialic acid $\xrightarrow{2-6}$ Gal NAc, Fuc $\xrightarrow{1-2}$ Glc NAc. In both types of glycoprotein, fucose and the sialic acids always occur in the nonreducing terminal position. For the reasons discussed earlier these structural patterns have been deduced largely from the soluble glycoproteins of the body fluids, so it is pertinent to see how they apply to membrane glycoproteins. In the major sialoglycoprotein of human erythrocytes, Winzler and his colleagues (Thomas and Winzler, 1969a, b; Adamany and Kathan, 1969) have shown that a tetrasaccharide N-acetyl-neuraminyl-(2→3)-β-O-galactopyranosyl-(1→3) [N-acetyl-N-acetylneuraminyl-(2→6)] -D-N-acetyl-galactosamine linked to protein via an labile O-glycosidic linkage is associated with the M and N blood group activity; this carbohydrate structure possesses the features predicted for a mucin type glycoprotein. The same molecule also possesses receptors for wheat germ agglutinin and PHA. In the case of the latter lectin Kornfeld and Kornfeld (1970), working with a tryptic glycopeptide derived from human erythrocyte stroma (which would inhibit hemagglutination and the mitogenic effect of this hemagglutinin), demonstrated that the carbohydrate unit joined by an N-acetyl-glucosaminyl-asparagine linkage possessed many of the structural features expected for a serum type glycoprotein. Wheat germ agglutinin binds to a number of cell surfaces causing cell agglutination, and Burger and Goldberg (1967), using a range of hapten inhibitors, presented evidence consistent with N-acetylglucosamine being an important constituent of the receptor site. Using L1210 cells, it was found that their agglutination by the wheat germ lectin could be inhibited by N-acetylglucosamine. All of the other sugars commonly found in glycoproteins were ineffective and even the free hexosamine failed to prevent agglutination. Interestingly, from the point of view of core structure, Burger and Goldberg (1967) found that N-acetylchitobiose was even more effective than N-acetylglucosamine in inhibiting wheat germ-mediated cell agglutination. As mentioned earlier, a N-acetylchitobiose residue linked to asparagine appears to be a distinctive feature of the core structure of serum type glycoproteins. In this context the experiments of Shier (1971), using a synthetic antigen-comprising poly (L-aspartic acid) having some of its free carboxyl groups substituted by N-acetylchitobiose, are of particular relevance. This antigen elicits an immune response in mice, cross-reacting with wheat germ receptors on tumor cell surfaces, and such mice have a greater potential for rejecting methylcholanthrene-induced tumors than do nonimmunized animals. These experiments appear to confirm that a structural feature of the soluble serum type glycoproteins can also be expected to be an important feature of cell surfaces. In addition, Toyoshima et al. (1972) found, by examining the effect of sequentially glycosidase-degraded glycopeptides from porcine thyroglobulin on the stimulation of incorporation of [^{3}H]-thymidine into human lymphocytes by a number of phytomitogens, that the structure (Man)$_3$ GlcNAc-GlcNAc probably represents a common determinant sugar sequence for the binding of these materials in lymphocyte transformation. This structure is conceivably the receptor site on the lymphocyte cell surface. If so, this work, and the earlier evidence for N-acetylchitobiose units linked to asparagine being a feature of cell surfaces, greatly strengthens the contention that a knowledge of the carbohydrate-protein linkage allows some prediction about oligosaccharide structure and also holds for membranes. More recently Muramatsu et al. (1975), subjecting mannose-labeled glycopeptides from SV-40 transformed fibroblasts to enzymatic degradation, using endo β-N-acetyl-glucosaminidases D and H of *Diplococcus pneumoniae* and *Streptomyces griseus*, respectively, have provided evidence of structural homology between oligomannosyl cores of cellular and nonmembrane glycoproteins.

From a discussion of the structure of glycoproteins it can be seen that a study of their biosynthesis will be directed toward answering such questions as where in the cell does initial glycosylation of protein take place, and what determines whether a particular amino acid residue will be glycosylated. Does the formation of N-glycosidic bonds and O-glycosidic bonds occur at the same subcellular site? Is there a special mechanism responsible for the biosynthesis of the core region in the serum type glycoproteins and, if so, where in the cell does this take place? Where are the more peripheral sugars added? How is glycoprotein synthesis controlled?

The cell elaborates glycoproteins both as soluble secreted materials and as membrane components. Although our knowledge of the structure of membrane glycoproteins is presently limited, there appears to be good evidence that some of the structural features of membrane glycoproteins are present in soluble secreted forms. It is possible, therefore, that certain similarities exist in the mechanism of their biosynthesis, though this begs the question of how the cell is able to distinguish between those glycoproteins destined for export and those that are to be retained by the cell membrane.

Problems such as these have been investigated using whole cell and cell-free experiments, and these approaches are discussed in the following sections.

3. Intact cell studies

Due to increased knowledge about the chemical structure of secreted soluble glycoproteins, attention focused on the means by which the cell assembled glycoproteins. Much of our understanding of membrane glycoprotein structure has been inferred from the knowledge obtained about the structure of soluble glycoproteins. Consequently, in designing experiments to examine the biosynthesis of glycoproteins of the plasma membrane, the type of structure synthesized has been assumed to bear chemical similarities to the secreted glycoproteins and so far there is no reason to suppose that this approach is not valid. Even today the detailed structural knowledge of membrane glycoproteins is confined

---

[2]Some investigators consider the sialic acid →Gal →GlcNAc sequence to represent a peripheral unit.

to very few examples, the principal ones being the major glycoprotein of the human erythrocyte and of some lectin receptors isolated from various tumor cells. Nevertheless, by ignoring this lack of detailed structural information and working on the basis of inferred structure, a considerable body of experimental findings was obtained that outlines the main features of membrane glycoprotein construction. This approach should not be allowed to obscure the difficulties still facing the investigator in this area. In the case of a soluble glycoprotein it is possible to obtain appreciable quantities of highly purified material, to which antibodies may be raised or from which characterized fragments obtained by controlled enzymatic or chemical means can be isolated for use in characterizing labeled products produced in an appropriate biosynthetic system. In plasma membrane glycoprotein biosynthesis the investigator rarely has such tools available. Often, only small quantities of biological material are available and although it may be possible to isolate a plasma membrane fraction from the system under study, the difficulty of establishing that any radioactive product present is related to any one particular membrane component is enormous. Certainly the methods available for isolating and characterizing highly purified plasma membranes are increasing and this can only aid the elucidation of the synthesis of their constituent glycoproteins. In many of the early studies reported in the literature, detailed characterization of the isolated membrane fractions is often lacking or incomplete because the methods for characterizing such material were not then available. Hence some reinterpretation of earlier biosynthetic work may be necessary in light of more recent findings on membrane isolation and characterization.

In intact cell studies two main experimental approaches have been followed to establish the subcellular sites of glycosylation, namely subcellular fractionation and autoradiography. In the former method incubation of the cell with an appropriate radiolabeled intermediate (e.g., ^{14}C-glucosamine) is followed by subcellular fractionation and an examination of label incorporation into macromolecular species (usually measured as a delipidated trichloroacetic acid-insoluble product) within the various compartments of the cell. The aim of such kinetic studies is to establish which organelles participate in the biosynthetic process and the sequence in which they are involved. Using this approach, it is possible to investigate by hydrolyzing the isolated cell fraction under appropriate conditions whether the labeled intermediate has been metabolized into other pathways. If there is any appreciable conversion of the applied label into other metabolites this problem can be overcome by isolating purified sugar fractions for subsequent estimation of radioactivity. As mentioned above, the problem of assigning such data to any one particular glycoprotein, though theoretically possible, is fraught with various practical difficulties.

A second method for the study of membrane glycoprotein biogenesis with intact cells follows a similar kinetic approach to that described earlier, but the cell integrity is maintained and the incorporated label (usually tritium) is followed by autoradiography. This technique has the advantage that the sequence of biochemical events can be more easily followed than is possible using a scheme of cell fractionation with its danger of attendant artifacts. There are, however, two major problems associated with the autoradiographic approach to the problem: unless the cell is making primarily a single glycoprotein it is impossible to distinguish between various glycoproteins, and it is essential in the autoradiographic technique that a metabolically stable precursor be used.

The kinetics of incorporation of radioactive precursors into the glycoproteins of subcellular organelles probably follows a number of different patterns. Initial incorporation of label might take place in the rough endoplasmic reticulum of the cell with subsequent transfer of partially synthesized glycoprotein to the smooth endoplasmic reticulum and Golgi apparatus for further glycosylation. Finally, the terminal product, is either secreted or, for surface glycoproteins, incorporated into the plasma membrane. Alternatively the label might not be incorporated at the rough endoplasmic reticulum but only at the level of the Golgi apparatus, in which case the labeling pattern would be quite different. However, it is possible to differentiate between these two mechanisms by stop-flow methods where the location of radioactive products at varying degrees of completeness are examined at different time intervals after the administration of radioactive precursors to whole animals. In the first case one would expect to find initially that the protein of the rough endoplasmic reticulum became increasingly labeled and only when a substantial level of incorporation had been achieved would any appreciable labeling of the Golgi apparatus be anticipated. Further, it would be expected that once the peak of activity in the rough endoplasmic reticulum had passed, it would be followed by a peak of activity in the Golgi apparatus followed by subsequent labeling of the plasma membrane. Alternatively, if no incorporation of label is detected in the rough endoplasmic reticulum, but a substantial amount of incorporation occurs in the Golgi apparatus followed by subsequent labeling of the plasma membrane, then the second mechanism might be operative. Incorporation of the same sugar at both the level of the rough endoplasmic reticulum and the Golgi apparatus would be characterized by coincident labeling of these organelles together with incorporation of activity into the plasma membrane as elements of the endoplasmic reticulum lose their radioactivity. By using different sugars it should be possible to elucidate the pattern of incorporation for a particular glycose residue, although it will be seen from the above discussion that if the administered sugar is metabolized to any appreciable extent to yield other labeled saccharides then the interpretation of the experimental results becomes extremely difficult, especially if incorporation is being monitored by autoradiography.

3.1. Use of radioactive precursors and subcellular fractionation

The main objective of this type of study is to understand the process and site of glycosylation, assuming in the majority of cases that a common mechanism exists for the synthesis of the polypeptide portion of glycoproteins and nonglycosylated proteins.

Most of the early work on membrane glycoprotein biosynthesis was per-

formed with tumor cells, which reflects the attempts to relate the invasive behavior of malignant cells to changes in molecular architecture at the cell
periphery. With various tumors, detailed electrokinetic examination of the cells,
coupled with electron microscopic studies of the cell periphery, have established
that sialoglycoproteins are a significant molecular species at the outer surface of
the plasma membrane. Using ascites tumors in particular, which yield free-living
single cells, it was possible (without resorting to various chemical or enzymatic
treatments that might reasonably be considered damaging to the cell surface) to
obtain sizable quantities of cells with which to follow the incorporation of appropriate intermediates into various organelles. One disadvantage of such
tumor systems is the difficulty of finding a suitable normal counterpart for
comparative purposes.

The Ehrlich ascites carcinoma cell was the first tumor cell to be subjected to
detailed electrokinetic examination (Cook et al., 1962), and sialic acids were
shown (Wallach and Eylar, 1961; Cook et al., 1962) to be an important ionogenic
species at the surface of this cell. Thus, this cell has been used by various groups
interested in studying the biosynthesis of membrane glycoproteins. Using the in
vitro incorporation of ^{14}C-glucosamine as well as ^{14}C-serine or ^{14}C-leucine into
these cells as a measure of glycoprotein and protein biosynthesis, respectively,
Cook et al. (1965) showed that these precursors were rapidly incorporated into
the membranes of the cell. Of particular interest was the finding that when the in
vitro incorporation of these precursors by intact cells was studied in the presence
of protein synthesis inhibitors, puromycin and tenuazonic acid, though the incorporation of the labeled amino acids into cellular protein was almost completely abolished, their effect on glucosamine incorporation was less marked. It
was found (Cook et al., 1965) that glycoprotein synthesis could continue for
periods of up to 4 hours even though protein synthesis was completely inhibited.
Subsequently, Molnar et al. (1965) reported a similar effect of puromycin on the
biosynthesis of glycoproteins by the Ehrlich ascites carcinoma cell. In their experiments the cells were incubated with the drug for various periods of time
before the addition of glucosamine. At concentrations of puromycin that inhibited amino acid incorporation by more than 90% the effect on glucosamine
uptake was only 14% (Molnar et al., 1965a), values which are in good agreement
with those found when the drug is added at the same time as the precursor
(Cook et al., 1965). There are many explanations for the fact that glucosamine
incorporation is resistant to puromycin, such as the possibility that a large pool of
unglycosylated membrane protein already exists in the cell, that the carbohydrate components turn over relatively slowly, or the less likely explanation that
exchange takes place with the oligosaccharide portion of the glycoprotein. The
first of these explanations is particularly relevant to the question of whether
some glycosylation takes place while the protein is associated with the polysomal
complex. Certainly the finding (Cook et al., 1965) that puromycin and
tenuazonic acid (both of which inhibit protein synthesis but by different
mechanisms) have initially only a small effect on glycoprotein synthesis suggests
that glycosylation occurs after the protein has left the polysomes. To test this

possibility, Cook et al., (1965) performed sucrose density gradient analysis of deoxycholate-dissociated rough-surfaced membranes isolated from intact cells that had been incubated for various periods of time with labeled glucosamine and amino acids. While appreciable quantities of labeled amino acids were associated with polysomal and ribosomal RNA, at no time was it possible to demonstrate any association of labeled sugar with polysomes or ribosomes. In addition to suggesting that glycoprotein biosynthesis as assessed by glucosamine incorporation is a postribosomal event, these studies indicated that the attachment of carbohydrate to polypeptide is unlikely to be controlled at the RNA level by means of a specific RNA-carrying glucosamine-amino acid intermediate. Sinohara and Sky-Peck (1965), studying the specific activity of subcellular fractions of mouse liver at varying intervals following the intraperitoneal injection of ^{14}C-glucosamine, found that the deoxycholate-soluble portion of microsomal particulates was the most active. As part of this study these authors specifically sought a soluble RNA amino acid-hexosamine complex without success and were forced to conclude that since glucosamine is incorporated mainly into a deoxycholate-soluble portion of microsomes, the carbohydrate moiety of the glycoprotein was probably "attached after completion of the polypeptide chain but before its release into cytoplasm." In a similar study using ^{14}C-glucosamine, but performed on rat liver, Helgeland (1965) found a small but significant amount of radioactivity associated with the ribosomes even after deoxychloate extraction. Helgeland believed that insufficient extraction was not a suitable explanation for this result because no radioactivity was found in a second deoxycholate extract, although adsorption of labeled material could not be excluded. While agreeing that it would be tempting to suggest that the extractable proteins of the membranes were the site of carbohydrate attachment to the protein, Helgeland (1965) was cautious of such an explanation in view of the radioactivity of the ribosomes and suggested that further time-course studies should help resolve the problem.

In the case of the Ehrlich ascites tumor cell, Molnar et al. (1969) have carried out pulse-lable experiments under in vitro and in vivo conditions with ^{14}C-glucosamine. They found that the label was taken up very quickly and was present in the cells in a trichloroacetic acid-soluble form at first (a point that will be discussed later) and was subsequently taken up into the glycoproteins of the cell mainly as sialic acid and N-acetylglucosamine residues. Under in vivo labeling conditions, the isolated plasma membrane fraction of these cells had higher specific activities for both sialic acid and N-acetylglucosamine than for isolated fragments of the endoplasmic reticulum. These activities reached a plateau after 3 hours, by which time the acid-soluble radioactivity had disappeared. A similar picture was found with in vitro labeling, except that the specific activity of the endoplasmic reticulum fractions never approached that of the isolated plasma membrane material. If a single site mechanism is anticipated for the biosynthesis of membrane glycoprotein, one would expect the fraction representing this site to be labeled first and its radioactivity to go through a maximum before the other fractions (presumably the plasma membrane in the case of cell surface glycoproteins) representing the terminal pool became labeled. However, Molnar et al. (1969) were unable to find a precursor-product relationship consistent with a single site hypothesis under either in vivo or in vitro conditions. Their observations would, however, favor the view that each fraction can catalyze the incorporation of carbohydrate into endogenous acceptors, a view they support by demonstrating glycosyltransferase activity in their various subcellular fractions (see section 7.1). Molnar et al. (1969) suggest that their results indicate glycosyltransferases are present in both the endoplasmic reticulum and the plasma membrane. In this latter site they suggest these enzymes would be able to repair the surface when carbohydrates are cleaved off by exogenous enzymes (see also section 7.1.). Alternatively, these investigators believe that these activities permit the completion of oligosaccharide units at the final location of membrane-bound glycoproteins (see section 6). Such deductions rely heavily on the composition of the fractions under study being representative of the membrane system from which they are derived.

It is quite possible that different mechanisms are operative, depending on the cell type. Puromycin markedly inhibited the incorporation of glucosamine into hepatic and plasma glycoproteins of the rat (Molnar et al., 1964), though liver, as the main site of plasma glycoprotein synthesis, may not be an ideal material for studying membrane glycoprotein synthesis. As Eylar (1965) has pointed out, the mechanisms of assembly of membrane glycoproteins and secretory glycoproteins may be different. This is an aspect of complex carbohydrate biochemistry about which we still lack definitive information.

Thus it is relevant to consider whether the Ehrlich ascites carcinoma system is complicated by the presence of an appreciable amount of secreted glycoprotein synthesis. In a study specifically directed toward examining the production of extracellular radioactive macromolecules by Ehrlich ascites carcinoma cells during incubation with ^{14}C-glucosamine, Molnar et al. (1965b) showed that a considerable amount of high molecular weight radioactive material appears outside the cells in the ascites fluid in vivo and in the suspending media in vitro. In vivo, most of the injected radioactivity was taken up by the tumor cells and not into the plasma or liver proteins. The subsequent label found in the ascites fluid is likely therefore to be produced by the carcinoma cells. These extracellular materials are digested by pronase to yield smaller molecular entities, suggesting that the original materials are actually glycoproteins. In these extracellular materials, Molnar et al., (1965b) found that very high amounts of the incorporated radioactivity is present as galactosamine while in the whole cells the label is principally glucosamine. Cook et al. (1965) found that over 50% of the ^{14}C-glucosamine incorporated was present as galactosamine, and Langley and Ambrose (1964) isolated a glycopeptide from the Ehrlich ascites tumor cell that contained nearly equimolar quantities of galactosamine and sialic acid. The galactosamine detected by Cook et al. (1965) might represent material to be exported from the cells that is associated with the membrane fractions at the time of their isolation. In the studies of Langley and Ambrose (1964, 1967) the galactosamine was found in a glycopeptide fraction, which might be unrepresentative of the majority of carbohydrate groups found in the membrane glycoproteins of this cell. In addition, Kraemer (1967) has pointed out that Langley and Ambrose did not demonstrate that during the incubation of the tumor cells with enzyme the cell count and viability remained constant. Thus it is possible that the isolated glycopeptide was a degradation product from damaged cells, and it would be extremely difficult to determine whether it was derived from a membrane glycoprotein or from a macromolecule destined to be "secreted." Molnar et al. (1965b) suggest that the extracellular glycoprotein found in their experiments is unlikely to be derived from generalized destruction of the cells, because if this were the case one would expect the ratios of radioactive glucosamine and galactosamine in the extracellular material to be similar to those found in the whole cells. Molnar et al. (1965b) prefer to interpret their data on the basis of a differential leakage of certain macromolecules from the cells both in vivo and in vitro. From what is now known of the structure of membrane glycoproteins, it would be surprising to find a molecule that contained exclusively glucosamine (presumably N-acetyl derivative) in its constituent carbohydrate groups and no galactosamine. Whether this result indicates that there is a preferential turnover of galactosamine-enriched membrane glycoproteins or that there is dual assembly by the cell of glycoprotein (enriched in galactosamine) destined for export, as opposed to cellular glycoproteins relatively enriched in glucosamine destined for the plasma membrane, is still unclear. Certainly there is a need to distinguish between "membrane" and "serum" glycoprotein metabolism in biosynthetic studies.

In view of the possible complications that may arise when studying membrane glycoprotein synthesis in an intact cell displaying an appreciable degree of glycoprotein secretion, the extension of work on intact Ehrlich ascites carcinoma cells to HeLa cells by Bosmann et al. (1969) is of particular interest. The HeLa cell, like the Ehrlich ascites carcinoma cell, was one of the first (tumor-derived) cell lines to undergo relatively extensive physiochemical investigation of its cell surface. Detailed quantitative studies by Shen and Ginsburg (1968) on the sugar content of HeLa cells have shown that the neutral sugars galactose, mannose, and fucose are present, together with sialic acids and the amino sugars galactosamine and glucosamine. Trypsin treatment removed from one to two thirds of the total of each sugar in cells grown in suspension and monolayer cultures, respectively. By contrast, Marcus and Schwartz (1968) found that only 10% of the sialic acid at most was removed by trypsin treatment of either type of HeLa cell culture. Nonetheless, the combined physical and biochemical evidence would certainly indicate that sialic acid containing materials almost certainly linked to protein are present at the cell surface. Apart from the knowledge that sialic acid containing material is present at the periphery of this particular cell, studies with HeLa cells have the special advantage that only a very low level of glycoprotein secretion appears to be taking place. In the case of ^{14}C-glucosamine, Bosmann et al. (1969) were able to show that less than 1% of the protein-bound hexosamine was found in secreted glycoprotein or glycolipid, and although this figure was slightly higher (approximately 5%) when ^{14}C-fucose of ^{3}H-leucine were used, these authors concluded that under their conditions of labeling most of the HeLa cell glycoprotein is membrane bound or intracellular. An important aspect of the work by Bosmann et al. (1969) on HeLa cells is their finding regarding the relative distribution of radioactivity in the smooth and plasma membranes of this cell as a function of time. In the early periods of the experiment (up to half an hour), ^{14}C-glucosamine was found to predominate in the smooth internal membranes of the cell. However, when chased with ^{12}C-glucosamine the distribution of this label changed and after 3½ hours the major proportion of the activity was found in the plasma membranes. On the basis of this result, and the finding that over short periods of time ^{14}C-glucosamine and ^{14}C-fucose were incorporated into the protein of the smooth internal membranes and plasma membrane while ^{3}H-leucine was found in the rough endoplasmic reticulum membranes or soluble protein, Bosmann et al. (1969) concluded that "naked" polypeptide chains synthesized at the polysomes migrate to the smooth internal membranes of the cell where glycosylation takes place, to be followed by migration to the plasma membrane. They also concluded that in the HeLa cell glucosamine is not attached to polypeptide at the ribosomal level. Pointing out that the separation of ribosomes or rough membranes from contaminating smooth membrane containing ^{14}C-glucosamine is difficult, it was nevertheless suggested (Bosmann et al., 1969) that possibly some secreted glycoproteins "which have an asparaginyl-glucosylamine linkage the initial sugar-amino acid linkage forms at the ribosomes." Bosmann et al. (1969) proposed that the smooth internal membranes might be identical with the Golgi apparatus. The authors noted that the smooth internal membrane fraction had a relatively high UDPase activity, an enzyme marker which they regard as characteristic of this organelle.

The use of glucosamine in the above studies as a marker for glycoprotein is particularly apt in view of the earlier work by Kornfeld and Ginsburg (1966) who showed, using radioactively labeled glucosamine, that this hexosamine is initially incorporated almost exclusively into N-acetyl hexosamine and N-acetyl-neuraminic acid. While showing that the labeling of HeLa cells in culture with ^{14}C-glucosamine provides a sensitive means for studying the components of this cell that contain amino sugars, Kornfeld and Ginsburg (1966) also demonstrated that eventually two thirds of the label is secreted into the medium as high molecular-weight compounds containing amino sugar and that the label remaining with the cells is gradually metabolized. Bosmann et al. (1969) regard the HeLa cell as particularly useful for studying plasma membrane glycoproteins because nearly all the glycosubstances synthesized are membrane bound or intracellular, a finding that appears to differ from the results of Kornfeld and Ginsburg (1966). The apparent discrepancy between these two results may be reconciled by an examination of the time scales used by the two groups. In the work of Bosmann et al. (1969) the incorporation of both glucosamine and fucose into cellular and secreted material was studied for up to 1 hour, and in their pulse-chase experiments with glucosamine the procedure was completed within a total of 4 hours. Kornfeld and Ginsburg (1966) found that appreciable amounts of activity were only present in the media after some 60 to 80 hours following 4 and 36 hour pulses with ^{14}C-glucosamine, respectively. Thus the time scales used by the two groups of investigators might well explain the apparently differing results with regard to "secreted" macromolecules. Bosmann et al. (1969) were careful to use HeLa cells that were in the log phase of growth, when it is likely (see section 7.1) that label incorporated into the plasma membrane will be retained within this organelle as opposed to nondividing cell populations in which material will probably be "secreted" into the suspending medium. The other label used by Bosmann et al. (1969) is fucose, and in this respect the later studies of Atkinson and Summers (1971) on incorporation of ^{3}H-fucose into HeLa cells are especially interesting. Atkinson and Summers (1971) concluded, on the basis of cell fractionation studies, that most of the L-fucose incorporated by the cells over a 4-hour growing period is located in the plasma membrane—a finding in full agreement with the results of Bosmann et al. (1969). L-fucose is a particularly appropriate label for glycoprotein studies because, as Kaufman and Ginsburg (1968) showed, fucose is incorporated into glycoproteins without being converted to other sugars. In addition, there is little conversion of this sugar into glycogen or amino acids (Coffey et al., 1964). These properties also make the sugar particularly useful for autoradiographic studies.

3.2. Autoradiography

In addition to following the incorporation of glucosamine into HeLa cells by cell fractionation studies, this cell has also been the subject of an autoradiographic study by Reith et al. (1970). The autoradiographic technique is discussed in detail elsewhere in this volume, and the remarks in this chapter will be confined to those examples which directly complement the studies outlined in the preceding section. As pointed out by Reith et al. (1970), radioautography has been used to provide information on the route of glycoprotein synthesis in various specialized cells, and the investigators chose to examine the HeLa cell to see if they could obtain information on the biosynthesis of surface carbohydrates in a nonspecialized cell. Reith et al. (1970) also make the point that by using a tissue culture cell it is possible to obtain a high specific activity in vitro.

Both light and electronmicroscopic autoradiography were used to examine HeLa cells after incubation in monolayer culture with ^{3}H-glucosamine (Reith et al., 1970). Using the former procedure, moderate radioactivity was shown throughout the cell after incubation with labeled hexosamine for only 5 minutes, although considerable amounts of activity were present after 5 hours in the perinuclear region of the cells. When cells incubated under the latter conditions were examined by electron microscopic autoradiography, considerable reaction was found over the Golgi apparatus zone with all elements of this complex being labeled. In addition to labeling the Golgi apparatus, dense organelles, probably lysosomal in origin, also gave an autoradiographic reaction and numerous grains were found over the cell periphery. These results greatly strengthen the earlier cell fractionation data of Bosmann et al. (1969) regarding the probable importance of the Golgi apparatus in the synthesis of plasma membrane glycoproteins.

Autoradiographic studies have been used for studying the biosynthesis of a number of major glycoproteins, and in all these studies the Golgi apparatus has been demonstrated to play a central role in the glycosylation of protein (reviews, Whaley et al., 1972; Cook, 1973; Whaley, 1975; and Chapter 1, this volume). It is appropriate to consider, in the context of this chapter, whether all the glycosylation of membrane protein takes place in the Golgi apparatus. When considered in conjunction with the cell fractionation studies, the work of Reith et al. (1970) with ^{3}H-glucosamine would certainly indicate that the membranes of the Golgi apparatus are a site for the incorporation of hexosamine in protein(s) destined for the plasma membrane. Such studies, however, do not show whether the label represents glucosamine residues involved in the initial glycosylation of aspargine residues present in the polypeptide or if the autoradiographic reaction is indicative of the site of incorporation of more peripheral hexosamine residues. As there is a possibility of lipid intermediate involvement in the formation of N-glycosides (section 5), special care may be needed to detect such intermediates in autoradiography. The important point has already been made (Neutra and Leblond, 1966) that because of the solubility of glycolipids in a number of the organic solvents used in histological processing glycolipids may possibly be excluded from such preparations. With regard to the use of the autoradiographic technique for studying the glycosylation of nascent protein, the studies of Whur et al. (1969) on the incorporation of ^{3}H-galactose, ^{3}H-mannose and ^{3}H-leucine by rat thyroid in vitro in a study of the synthesis of thyroglobulin, albeit a secreted glycoprotein, are of special relevance. In this study the incorporation of leucine and mannose, a sugar found in the inner core of serum type carbohydrate groups of glycoproteins, occurred in the rough endoplasmic reticulum, the sugar being added as soon as the polypeptide was released into the cisternae of this part of the ergostoplasm. As might be expected, puromycin completely inhibited the incorporation of leucine and the core sugar mannose but had no effect on galactose uptake during the first hour. Galactose is an example of a sugar occupying a more peripheral position in the completed glycoprotein. As regards the completion of this glycoprotein, Haddad et al. (1971) found that the terminal sugar ^{3}H-fucose was added within the Golgi apparatus.

For comparable autoradiographic information on where the oligosaccharide chains of membrane glycoproteins are completed we must turn to the work of Bennett et al. (1974), who used light and electronmicroscope autoradiography to study the incorporation of ^{3}H-fucose into over 50 cell types of the rat. The main conclusion was that the carbohydrate groups of surface glycoproteins are to a large extent completed in the Golgi apparatus. Earlier work by Ito (1969) on electron microscopic autoradiography of mannose and galactose incorporation by cat intestine in vitro and Bennett (1970) on incorporation of galactose by the columnar cells of the duodenal epithelium of the rat in vivo indicated the importance of the Golgi apparatus as a major site of glycosylation of proteins destined for the cell periphery. In a later study, Bennett et al. (1974) extended observations on the epithelium of the small intestine of the rat to other cell types. These investigators administered ^{3}H-fucose intravenously to young rats, and then sacrificed the animals at various times between 2 minutes and thirty hours later. The fate of the label was followed initially in various tissues by light microscope autoradiography and later by an electron microscope investigation of the epithelium of the small intestine, an example of cells undergoing continuous renewal, as well as the kidney in which the epithelial cells of the proximal and distal tubules acted as models for cells which do not undergo renewal. Light microscope autoradiography showed that in nearly all cell types a discrete paranuclear clump of silver grains is formed within 2 to 10 minutes of injection of the label, the location of this clump corresponding to the Golgi apparatus of the cell. The intensity of the reaction varied from very high in colonic epithelium to weak in lymphocytes. When electron microscope autoradiographs were made at early times in the duodenal villus columnar cells and the kidney proximal and distal tubule cells, it was found that the reaction was confined to the saccules of the Golgi apparatus. After various periods of time the reaction in the region of the Golgi apparatus migrated to the cell surface although the intensity of the reaction varied among the different cells examined. Reactions were also observed with time in other parts of some cells, for example over lysosomes and secretory material, the latter being especially evident in mucous-secreting cells. The observations of Bennett et al. (1974) are important for a number of reasons: not only do they strengthen the view that the Golgi apparatus plays an important role in the biosynthesis of plasma membrane glycoproteins, but their results also indicate that the carbohydrate groups of these macromolecules are largely completed in this organelle. As these authors observed, the incorporation of fucose into glycoproteins occurs almost invariably in the Golgi apparatus, and since fucose is known to occupy a terminal position in the sugar chains of glycoproteins, this organelle would be the completion site of the carbohydrate side chains ending in fucose. An added point concerns the type of carbohydrate moiety with which fucose is associated. As Bennett et al. (1974) remark, this sugar usually occurs together with subterminal galactose residues as well as N-acetyl-hexosamines and an inner core of mannose residues in a serum type glycoprotein. While the actual site of glycosylation of asparagine residues may be uncertain, if the reaction takes place while nascent polypeptide is associated with the polysomes or released into the cisternae of the rough endoplasmic reticulum, it appears that autoradiographic evidence is clear-cut about the cell site where the completion of carbohydrate groups linked N-glycosidically to protein occurs. As components of both the cell surface and tissue secretions, it seems that both are completed before leaving the Golgi apparatus (a point that will be discussed later in section 6). It should be added that autoradiographic studies have not been able so far to define the subcellular site at which sialosyl residues, the other terminally occurring saccharides, are added to glycoproteins.

In addition to providing information on the subcellular assembly sites of plasma membrane glycoproteins, the study of Bennett et al. (1974) also gives some idea of the considerable variation in turnover rates of cell peripheries (as implied by the differences in reactivity in the autoradiographic method) among different cell types and their surfaces. That turnover is actually taking place is evident from results obtained with the kidney tubule cells, though in a rapidly renewing population of cells, i.e., gastrointestinal epithelia, it is apparent that glycoproteins are involved in growth as well as turnover processes.

4. The use of cell-free systems for studying membrane glycoproteins

It is now widely recognized that polysaccharides are synthesized from their respective glycose residues by a sequence of reactions involving the conversion of glucose to various monosaccharides and their derivatives and "activation" of these monosaccharides to form nucleoside phosphate sugar donors (Fig. 3), whose glycose residues are then transferred to the appropriate aglycone or growing carbohydrate chain by means of a group of enzymes, the glycosyltransferases. The intermediary metabolism of these sugars, especially the metabolism of amino sugars and sialic acids and the importance of glucosamine in this area, has been reviewed in detail by Warren (1966), while the role of sugar nucleotides as activated donors in the biosynthesis of polysaccharides has been surveyed by Leloir (1964) and by Schachter and Roden (1973).

Work with cell-free systems has not only enlarged our insight into the actual process of glycosylation and the subcellular sites involved but is an essential adjunct for purifying glycosyltransferases from cell fractions shown to be enriched in the relevant activity. The specificity of these transferases, directed toward both the sugar nucleotide and the acceptor, permits biosynthesis of the carbohydrate groups with their structure defined in terms of sequence, position, and anomeric configuration, as well as various branch points. It is therefore important to study these enzymes in as much detail as possible.

The development of cell-free systems for studying glycoprotein synthesis followed closely from the work performed with intact cells and radioactively labeled precursors. Toward the end of 1965 a cell-free system, derived from the Ehrlich ascites carcinoma cell, was first described for the synthesis of glycoproteins (Eylar and Cook, 1965). This work, which stemmed from earlier work (Cook et al., 1965) on intact Ehrlich ascites tumor cells, was an attempt to increase knowledge of the synthesis site of the oligosaccharide units of membrane glycoproteins, as well as the means by which the cell regulates the construction of its surface membrane. These authors demonstrated that the enzymes which mediated the incorporation of glucosamine and galactose (this latter sugar was shown to be incorporated from the appropriate sugar nucleotide, UDP galactose) in membrane as well as soluble glycoprotein were located in a postmicrosomal particulate fraction. At the time Eylar and Cook (1965) suggested that this fraction, which was inactive in protein synthesis, represented either (1) fragments of the surface membrane, (2) the Golgi apparatus, or (3) a new, undefined structure. The Golgi apparatus was thought to be the most likely origin of the postmic-

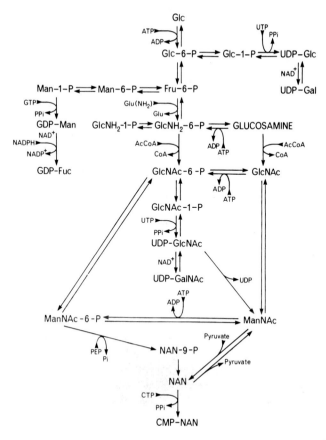

Fig. 3. An outline of the enzymatic reactions undergone by sugars to form appropriate sugar nucleotides utilized by the cell in the synthesis of the carbohydrate prosthetic groups of glycoproteins. (see also section 4 in text). This figure also summarizes the reactions by which hexosamines and their derivatives are ultimately derived from glucose as well as the reactions involved in the synthesis of the appropriate sugar nucleotides of the neutral sugar residues found in glycoproteins, galactose, and mannose. For clarity glucosamine, a widely used intermediate for the study of glycoprotein biosynthesis, is indicated in capital letters.

rosomal fraction in view of preliminary cytochemical evidence (Gasic and Gasic, 1963) that this organelle represents a site of surface glycoprotein synthesis. Using this cell-free system it was possible to show that the membranes bearing ribosomes, though completely inactive in the synthesis of oligosaccharide, were able to support protein synthesis. It was therefore suggested that oligosaccharides were synthesized in the smooth membranes of the cell, probably the Golgi apparatus, while the polypeptide acceptor was synthesized on the rough membranes. When the original work was performed, it was thought that the carbohydrate units might migrate to the rough membranes where oligosaccharide would be attached to protein; from current knowledge, however, the protein is believed to be glycosylated as it moves through the channels of the endoplasmic reticulum. The rough membranes contained appreciable quantities of polypeptide that stimulated the uptake of sugar by the postmicrosomal particulate fraction and, when in the intact cell, would be transferred subsequently to the Golgi apparatus for glycosylation. The postulate in this work (Eylar and Cook, 1965) that the smooth membranes of the cell represent the major site of intracellular glycosylation is, however, entirely in accord with present ideas on glycoprotein synthesis. Following this study, Caccam and Eylar (1970) reexamined the Ehrlich ascites carcinoma system paying special attention to the UDP galactose glycoprotein: galactosyltransferase activity reported by Eylar and Cook (1965). Caccam and Eylar (1970) were able to purify the latter activity some 46-fold and to show that it was intimately bound to smooth internal cell membranes. This confirmed the earlier suggestion of Eylar and Cook (1965) that the postmicrosomal particulate fraction represented the Golgi apparatus. Caccam and Eylar (1970) found that the enzyme could be solubilized from the membranes by treatment with nonionic detergents such as Triton X-100, or by the use of enzymes such as phospholipase A. Information on the solubilization of these activities is especially important when devising methods for purifying these membrane-bound activities.

Membrane-bound glycosyltransferase activities have also been examined in the HeLa cell (Bosmann et al., 1968a,b; Hagopian et al., 1968). These cells were fractionated (Bosmann et al., 1968a) into seven membrane fractions, two of which were characterized as "smooth internal membranes" and plasma membranes respectively. Hagopian et al. (1968) showed that polypeptidyl: N-acetylgalactosaminyl and glycoprotein: galactosyl transferases were present in the smooth internal membranes of the cell while collagen: glucosyl transferase activity is associated with the plasma membrane. The polypeptidyl: N-acetylgalactosaminyl transferase activity, which is purified 45- to 50-fold in the smooth internal membrane fraction and can be solubilized from the membranes by Triton X-100, is of special interest. This enzyme initiates assembly of the carbohydrate groups by the forming the protein-carbohydrate linkage between N-acetylgalactosamine and certain serine and threonine residues present in the polypeptide acceptor. In addition to the study of these transferases, Bosmann et al. (1968b) investigated two glycoprotein:fucosyl transferases in the HeLa cell: one appears to be involved in the formation of fucosyl-$(1\rightarrow4)$-N-acetyl-glucosamine linkage and the other is probably responsible for the formation of α-L-fucosyl-$(1\rightarrow2)$-D-galactose. Again, the smooth internal membranes of the cell were shown to be the subcellular localization of these activities. Hagopian et al. (1968), pointing out that work with intact HeLa cells showed that most of the glycoprotein produced by this cell is incorporated into membranes (Bosmann et al., 1969), suggested that their findings from cell-free studies on distribution of glycosyl transferases indicate that smooth internal membranes are the subcellular site of synthesis of those membrane glycoproteins destined for the cell surface. It appears that not only is the initial carbohydrate-protein (for "mucin" type carbohydrate groups) linkage formed at the smooth internal membranes but the oligosaccharide chains are completed at the same subcellular site. This latter deduction was made on the basis of cell-free work on the location of fucosyl transferases (Bosmann et al., 1968b). The particular importance of these latter activities for membrane biogenesis is that one of the fucosyl transferases appears to be synthesizing an α-L-fucosyl- $(1\rightarrow2)$-O-D-galactose sequence that is part of H substance, and from serological evidence is known to be present on the surface of HeLa cells (Kelus et al., 1959). This latter enzyme apparently has highly specific macromolecule receptor requirements, since it actively recognizes terminal galactosyl residues in desialyzed afucosyl porcine submaxillary glycoprotein but will not transfer fucosyl residues from GDP-fucose to asialofetuin or asialo α_1 acid glycoprotein. Apart from helping localize the subcellular site of glycosylation of membrane protein, this cell-free work supports the concept of a "one enzyme-one linkage" mechanism for the formation of glycoproteins in which a specific transferase exists in the cell for the synthesis of each linkage found in the carbohydrate groups.

That the "smooth internal membrane" fractions represent membranes of the Golgi apparatus was based upon their content of UDPase (Bosmann et al., 1968a), an activity that is considered to be a component of this organelle on the basis of cytochemical evidence (Novikoff and Goldfischer, 1961). As Hagopian et al. (1968) point out, the reaction in which galactose and hexosamine residues are transferred from their corresponding donor UDP-sugar with the formation of UDP would be driven toward synthesis following hydrolysis of the latter compound by the appropriate nucleoside diphosphatase. Schachter and Roden (1973) have criticized the characterization by Bosmann et al. (1968a) of their smooth-surfaced membrane fraction on the basis that UDPase may not be a reliable marker for the Golgi apparatus, and suggest that the presence of plasma membrane in this fraction has not been ruled out. Certainly, as Hagopian et al. (1968) stressed, it is a pity that the classical structure of the Golgi apparatus is not maintained during their purification process, although in view of the subsequent autoradiographic work of Bennett et al. (1974) that clearly demonstrates fucose is added within the Golgi apparatus, and as fucosyl transferases are enriched in their smooth membrane material, it would seem clear that Hagopian et al. (1968) are dealing with membranes derived from this organelle. That Bennett et al. (1974) find no significant autoradiographic reaction over the cell surface at early time intervals after labeling would indicate "a lack of significant incorporation of fucose into glycoproteins at the cell surface" and therefore any contamination of the "smooth internal membrane" fraction with fragments of the plasma membrane is unlikely to alter the interpretation of Hagopian et al. (1968) on their results.

The isolation of Golgi apparatus-enriched fractions in which the classical morphology of the dictysomes is preserved presents special problems, especially when using tumors as starting material (Hudgin et al., 1971; Cook, 1973; Ovtracht et al., 1973). It appears that intact dictyosomes rapidly become unstacked in tumor homogenates, due perhaps to increased levels of lysosomal enzymes (Ovtracht et al., 1973). Tumor cells have provided the biochemist with sizable quantities of readily accessible material for studying glycoproteins and, even though the isolated fractions are often poorly characterized morphologically, there can be no doubt that when the data obtained from cell-free and intact cell studies are combined the case for the involvement of the Golgi apparatus in membrane glycoprotein assembly is strong. The role of the Golgi apparatus in membrane biosynthesis is discussed in Chapter 1 in this volume by D. J. Morré.

5. Lipid intermediates in glycoprotein biosynthesis

In the work discussed in the earlier sections of this chapter the role of sugar nucleotides as donors of glycose residues has been emphasized. More recently, attention has focused increasingly on the role of lipid-linked sugars in the glycosylation of proteins. Previously a number of studies on the synthesis of the lipopolysaccharide of *Salmonella typhimurium* (Osborn and Weiner, 1967, 1968) and *Salmonella newington* (Wright et al., 1967) and the peptidoglycan of *Micrococcus lysodeikticus* (Higashi et al., 1967) had clearly demonstrated that lipid pyrophosphate-linked sugars, formed by the transfer of glycose-1-phosphate residues from sugar nucleotide to carrier lipid phosphate or glycose residues to sugar linked to lipid, are important intermediates in the biosynthesis of bacterial oligosaccharides. Both the latter papers showed that the lipid involved was a polyisoprenoid compound. With the interest in lipid-linked intermediates in the synthesis of bacterial glycans, it is not surprising that biochemists should turn their attention to eukaryotic cells to see whether similar components are involved in the glycoprotein biosynthesis of these cells.

In the case of animal cells, Caccam et al. (1969) were the first to present evidence of an enzyme that catalyzed the transfer of mannose from GDP-mannose to lipid. In considering the general applicability of the one linkage-one enzyme mechanism to the biosynthesis of glycoproteins, these authors investigated glycoproteins secreted into blood plasma and egg white and paid particular attention to the incorporation of mannose into these compounds. This latter sugar is a component of soluble secreted glycoproteins but is rarely found in mucins. Using GDP-mannose-^{14}C these authors investigated the ability of a number of tissues to incorporate mannose into lipid and protein fractions including hen oviduct, rabbit liver, myeloma cells, and HeLa cells. Tissues such as liver and oviduct were shown to contain enzymes for the transfer of mannose to lipid and protein and these tissues secrete quantities of plasma and egg white glycoproteins. Of particular interest was the finding that HeLa cells, which apart from collagen do not secrete any appreciable amounts of glycoprotein, did not incorporate any significant amounts of mannose. By contrast, in myeloma cells a very high level of mannose incorporation was achieved with a ratio of 6–10:1 of mannose incorporated in their assay into lipid relative to protein. Actually the lipid:protein ratio exceeds that found for liver. The myeloma cells secrete an IgG immunoglobulin, and Caccam et al. (1969) concluded from these results that mannose is transferred from the sugar nucleotide to both lipid and proteins by cells that are actively engaged in the secretion of mannose containing glycoproteins. In addition, the cellular site of mannose incorporation was studied using rabbit liver. A postmitochondrial supernatant fluid derived from homogenized rabbit liver was fractionated into rough and smooth endoplasmic reticulum using conventional sucrose/cesium chloride gradient procedures (Dallner et al., 1966) and the fractions were tested for transferase activity. Interestingly, most of the enzymes appeared to be concentrated in the smooth membranes, though other sites were not excluded by Caccam et al. (1969) as they found over 43% of the activity in a 20,000 × g pellet. Since no increase in specific activity was discovered, this result is believed to be due to adsorbed smooth membranes. With regard to the role of mannolipid, pulse-chase experiments, in which cold GDP mannose is added after 10 minutes of labeling, showed a fall in the radioactivity of the lipid accompanied by a gradual rise in the level of radioactivity in the protein, suggesting that the lipid is an intermediate in glycoprotein synthesis. These results are important because, as Caccam et al. (1969) point out, they exemplify a mechanism for the synthesis of glycoprotein more complicated than the one enzyme-one linkage type. Evidently, since it involves a lipid intermediate, at least two transferases would be involved: one mediating the synthesis of the mannolipid and the other the sugar transfer from the intermediate to glycoprotein. The authors (Caccam et al., 1969) appear to favor the view that this more complicated mechanism is applicable to secreted rather than membrane glycoproteins. As will be apparent from work discussed later in this section, however, various investigators consider these lipid intermediates to be important in the biosynthesis of membrane glycoproteins.

5.1 Nature of acceptor lipid-dolichol phosphate

5.1.1 Formation of glucose and
glucose-containing oligosaccharide derivatives
In the work already discussed, the mannolipid was subjected to some preliminary characterization and its properties were found to correspond closely to those of mannosyl-1-phosphoryl-polyisoprenol. Caccam et al. (1969) observed that such a compound had been shown earlier to be an intermediate in the synthesis of mannans in bacteria by Scher et al. (1968) and is an obvious example of an instructive analogy between a bacterial system and a mechanism present in a eukaryotic cell.

In 1970 Behrens and Leloir working with rat and pig livers produced further evidence for the formation of polyprenol-linked sugar in eukaryotic cells. They showed that a rat liver microsomal fraction (containing both rough and smooth microsomes) was able to mediate the transfer of glucose from UDP glucose to a lipid acceptor. Lipid acceptor was prepared from pig liver and the amount of glucosylation was proportional to the amount of acceptor added. This acceptor had the properties of a phospholipid and its infrared spectra had some similarities to that of dolichol (Fig. 4). In addition it was found that the unsaponifiable acceptor fraction of liver on phosphorylation was glucosylated as was dolichol following chemical phosphorylation. On the basis of this finding,

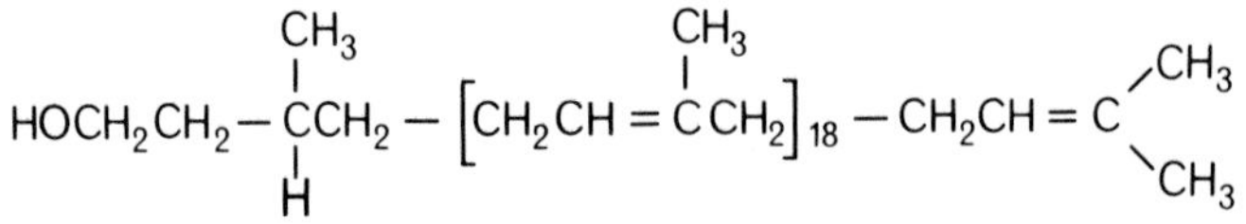

$$HOCH_2CH_2 - \overset{\overset{\displaystyle CH_3}{|}}{\underset{\underset{\displaystyle H}{|}}{C}}CH_2 - \left[CH_2CH = \overset{\overset{\displaystyle CH_3}{|}}{C}CH_2\right]_{18} - CH_2CH = C \overset{\diagup CH_3}{\diagdown CH_3}$$

Fig. 4. Structure of dolichol. Dolichols were first discovered in mammalian tissues by Burgos et al. (1963) and consist of a group of polyisoprenols of considerable chain length, about C_{100}, hence their name dolichol (Greek *dolichos*, long). In the above structure the terminal isoprene unit is shown to be saturated, the hydroxyl group being involved in forming a linkage with phosphate to sugar in the various intermediates described in the text.

and the identical chromatographic behavior of the acceptor and chemically phosphorylated dolichol, it was deduced that the neutral acceptor lipid is dolichol monophosphate and that in this study dolichol monophosphate glucose had been formed. Behrens and Leloir showed that glucosylated lipid (dolichol monophosphate glucose) could, in the presence of microsomes, act as a donor of glucosyl residues to a material that precipitated at the organic solvent water interface although the counts then gradually became water-soluble. Though the insoluble product was not characterized, it appeared to be a protein while the soluble material was mainly glucose presumably released by a glucosidase. As Behrens and Leloir (1970) point out, one of the few glycoproteins containing glucose is collagen. However, no evidence for the formation of the carbohydrate residues of this glycoprotein was found.

Following this work, Behrens et al. (1971) examined the glucosylated product in greater detail, although its properties had to be followed by the radioactivity of the compound because only small quantities of material were available. In their earlier study, Behrens and Leloir (1970) had suggested that the compound formed by the transfer of glucose from dolichol monophosphate glucose under the influence of the microsomal enzymes was a glycoprotein. This implication was prompted largely by the finding that this material was insoluble in lipid solvents and was precipitated by trichloroacetic acid. Further work (Behrens et al., 1971) showed that this radioactive product, referred to as GEA (glucosylated endogenous acceptor), became soluble in chloroform-methanol mixtures with a high water content, although it is insoluble in water unless detergent is present. On treatment with 10% aqueous ammonia, GEA gave rise to a water-soluble, negatively charged compound, whose charge could be removed by treatment with purified *Escherchia coli* alkaline phosphatase to produce a material that had no net charge. Acetolysis gave rise to a series of products which, on the premise of chromatographic evidence, behaved as oligosaccharides. On the basis of this type of experimentation, Behrens et al. (1971) suggested that GEA also contains a dolichol residue that is joined through a phosphate or pyrophosphate linkage to an oligosaccharide containing about 20 monosaccharide residues. Evidence for the lipid being dolichol was based largely on gel chromatography in the presence of deoxycholate. The principle of this procedure is that the deoxycholate forms an inclusion compound with lipids and the number of deoxycholate molecules that bind with lipid depends on the chain length of the fatty acid.

Using this technique, the molecular weight of dolichol monophosphate was found to be 11,300 and assuming that dolichol has a molecular weight of 1500 this would indicate that 24 molecules of detergent are associated with the dolichol. For GEA a molecular weight value of 14,300 was obtained by this method. It was noted that the difference (3000) in molecular weight between dolichol monophosphate and GEA would correspond to the hydrophilic moieties associated with the latter material. Further studies by Parodi et al. (1972) on endogenous acceptor provided evidence for the lipid moiety being dolichol and hence that this lipid is the hydrophobic moiety of GEA. This paper also noted that mild acid treatment of GEA liberated two substances: one was dolichol monophosphate and the other could generate dolichol monophosphate after stronger acid treatment, presumably as a result of the conversion of the pyrophosphate to the monophosphate derivative. It was suggested that this was evidence for the presence of a pyrophosphate bridge joining the hydrophilic oligosaccharide chain to the lipophilic moiety. Having established something of the chemical nature of the latter moiety, Parodi et al. (1973) turned their attention to the large oligosaccharide of GEA which, on the basis of paper electrophoresis, is neutral. In addition to containing glucose residues, the oligosaccharide, which can be obtained from GEA by methanolysis, was found to contain two hexosamine residues. The presence of hexosamine was deduced from the result of alkaline treatment, which generated two positively charged derivatives thought to be due to de-N-acetylation, especially as the charge could be neutralized by N-acetylation. The importance of this oligosaccharide-linked lipid (GEA) in glycoprotein synthesis will be discussed later. However, the formation of GEA is not limited to rat liver. Parodi et al. (1973) provided evidence for its formation in such tissues as rat brain and kidney, pig thyroid, and human lymphocytes. Of particular interest from the point of view of the plasma membrane is that while the enzymes responsible for the synthesis of dolichol monophosphate glucose from UDP glucose are absent from plasma membranes there is evidence of some activity associated with these membranes that mediates the glucosylation of another dolichol derivative from dolichol monophosphate glucose to form the endogenous acceptor (Dallner et al., 1972).

5.1.2. Formation of mannose-containing oligosaccharide derivatives of dolichol phosphate

In the introduction to section 5 attention was drawn to the formation of a mannolipid described by Caccam et al. (1969). Since then many groups have provided information on the occurrence of lipid-bound mannose in other systems. Working with pig liver microsomes, Richards and Hemming (1972) showed that when this fraction was incubated with GDP ^{14}C-mannose 10 to 40% of the label was transferred to mannolipid and 1 to 3% to mannoprotein. The mannolipid was found from its chromatographic behavior on silicic acid and DEAE-cellulose acetate to resemble undecaprenol phosphate mannose. While the mannolipid was resistant to dilute alkali, the mannose was bound to the lipid moiety by a very acid-labile linkage. These properties suggest that the mannose was linked through phosphate to a polyisoprenoid alcohol. Unfortunately, it was not possible at the time to isolate sufficient purified material to subject the lipid to physiocochemical characterization. Consequently, Richards and Hemming (1972) had to identify the lipid present by indirect methods. The problem of only small quantities of such lipid intermediates being available for direct characterization has been encountered by others (Behrens et al., 1971) and was referred to earlier in section 5.1.1. Dolichol phosphate stimulated the incorporation of labeled mannose into mannolipid as well as protein. Other exogenous polyprenol phosphates, including betulaprenol and solanesol phosphates, could also act as acceptors although they were less efficient. When exogenous ^{3}H-dolichol prepared from *Phytophthora cactorum* was used, a double-labeled lipid was formed from which ^{14}C-mannose and ^{3}H-dolichol phosphate could be recovered following acid hydrolysis. On the basis of the similarity of chromatographic properties of material from pig liver, which stimulates the transfer of ^{14}C-mannose from sugar nucleotide to mannolipid to dolichol phosphate and dolichol phosphate mannose, further evidence of the importance of this lipid was obtained. While the relationship of GDP-mannose to mannolipid was quite apparent, with the formation of mannolipid being readily reversible by an excess of GDP, the relationship of mannolipid to mannoprotein was considered by Richards and Hemming (1972) to be less certain. They concluded that their evidence favored the mannolipid acting as an intermediate between GDP-mannose and glycoprotein, though no evidence was found at that time for an oligosaccharide complex. Characterization of the mannolipid as dolichol phosphate mannose was strengthened when Evans and Hemming (1973) working with a bulk preparation of pig liver microsomes were able to characterize the molecule by performing infrared and nuclear magnetic resonance spectroscopy on the isolated mannolipid and to compare its properties with those of synthetic dolichol phosphate mannose. More recently, Hemming and his colleagues (Oliver and Hemming, 1975; Oliver et al., 1975) have obtained evidence for the transfer by pig liver microsomal preparations of mannose from GDP-mannose to lipid-linked oligosaccharides. In addition to describing the properties of their oligosaccharide portion (Oliver and Hemming, 1975), these researchers have evidence for the view that dolichol diphosphate oligosaccharide is directly related to the formation of glycoprotein in the pig liver system. In addition to data obtained from the work with pig liver, Martin and Thorne (1974a,b) were able to show that endogenous dolichol in the rat liver was involved in the formation of lipid-linked precursors of glycoprotein. While it is quite clear from several of the papers already discussed that microsomes are able to catalyze the formation of mannolipids from GDP-mannose and exogenous dolichol phosphate, Martin and Thorne (1974b) rightly pointed out that the question remains of whether dolichol phosphate is the natural acceptor, especially as retinol phosphate has also been cited as an acceptor (see section 5.2). In an accompanying paper (Martin and Thorne, 1974a), it was demonstrated that it was possible to label the dolichol of rat liver by injecting partially hepatectomized animals with 4S-^{3}H-mevalonate, the precursor of *cis*-isoprene residues. Optimal conditions for labeling the rat liver dolichol were determined, and Martin and Thorne (1974a) found that maximal incorporation of label was obtained by injecting the animals with labeled mevalonate 48 hours after hepatectomy and waiting 12 hours before killing the animals. Though the microsomal fraction contained label, the highest concentration of radioactive dolichol was found in the mitochondrial and nuclear-debris fractions (Martin and Thorne, 1974a). Using rat liver microsomes labeled by this method, Martin and Thorne (1974b) were able to show that following incubation with GDP ^{14}C-mannose a double-labeled lipid, equivalent to about one third of the total ^{14}C-mannolipid, could be isolated. The mannolipid obtained in this experiment was identical in chromatographic and stability properties to that obtained when exogenous dolichol phosphate was used. The findings of Martin and Thorne (1974b) are particularly important as they not only showed that dolichol phosphate acts as an acceptor of mannose but also that it is an endogenous acceptor, which had not been demonstrated unequivocably until then. Martin and Thorne (1974b) were also able to use the regenerating rat liver system to test whether retinol was involved in the formation of mannolipids. Using vitamin A-deficient rats, the administration of ^{14}C-retinol and ^{3}H-mevalonate enabled these investigators to show that dolichol phosphate is not the only natural acceptor of mannose but that retinol is also present in the mannolipid fraction in similar amounts.

In addition to observations on the liver, further work on mannose lipid intermediates in glycoprotein biosynthesis has also been done with hen oviduct membranes (Waechter et al., 1973), the system studied originally by Caccam et al.(1969). As with the liver studies, membranous particulate preparations of hen oviduct catalyze the transfer of mannose from GDP-mannose to endogenous acceptors present in the membranes. One of the mannosylated acceptors was characterized as mannosyl phosphoryl polyisoprenol and, on the basis of chromatographic evidence, is identical to mannosyl phosphoryl dolichol. Interestingly, as with liver systems, exogenous dolichol stimulates the synthesis of this mannosyl lipid. Ficaprenol phosphate also possesses acceptor activity although on an equimolar basis it is less efficient than dolichol phosphate. This result, apart from providing information on the probable nature of the lipid, has the additional importance of suggesting that the exogenous dolichol phosphate is not stimulating activity by virtue of acting as a detergent. The mannosyl polyisoprenol synthetase reaction is dependent on metal ions, of which Mn^{2+} is the most efficacious, a property that is useful when studying the role of this material as an intermediate in glycoprotein synthesis (section 5.3). In addition to mannosyl phosphoryl polyisoprenol, Waechter et al. (1973) demonstrated the formation of at least two other components, termed soluble mannosylated endogenous acceptor (mannosyl s-acceptor) and residual mannosylated endogenous acceptor (mannosyl r-acceptor). The former material is insoluble in chloroform:methanol (2:1 v/v) as well as water although it is soluble in a mixture of chloroform:methanol:water (1:1:0.3 by volume), a property reminiscent of the endogenous acceptor GEA discussed previously. Waechter et al. (1973) suggested that mannosyl s-acceptor represents an oligosaccharide containing com-

ponent. In a subsequent paper Lucas et al. (1975) examined and partially characterized this latter component. On the basis of the behavior of the oligosaccharide
released by mild acid hydrolysis from the acceptor on paper chromatography
(distance migrated on the chromatogram compared with malto-oligosaccharides
of known chain length), it was considered to consist of seven to nine glycose
residues. Apparently, N-acetylglucosamine is situated at the reducing end of the
oligosaccharide as judged by the identification of tritium-labeled
N-acetylglucosaminitol in N-acetylated acid hydrolysates of oligosaccharide
which had been treated with sodium borotritide. Mannose is situated at the
nonreducing terminus. Of special interest is the finding that when UDP-N-acetyl
^{14}C-glucosamine is incubated in the absence of GDP-mannose with hen oviduct
membranes a ^{14}C-chitobiosyl lipid is formed that is believed to consist of
chitobiose linked by a mono- or pyrophosphoryl link to lipid; little s-acceptor is
formed. However, the latter material containing an N-acetyl ^{14}C-glucosaminyl
residue was produced when GDP-mannose was added concomitantly with the
labeled UDP-N-acetyl glucosamine. Treatment of this material with
α-mannosidase[3] released over 80% of the label as chitobiose. The label was
incorporated into the second glycose residue because the terminal reducing end
of the chain was found to be unlabeled, suggesting that the particulate enzyme
preparation contains an endogenous lipid-linked N-acetylglucosamine compound. In addition to providing the above information on the nature of the
oligosaccharide chain of mannosyl s-acceptor, Lucas et al. (1975) concluded, on
the basis of experiments with bacterial alkaline phosphatase, that this chain is
linked to lipid via a phosphate or pyrophosphate bond attached to the chloride
position of the terminal N-acetyl glucosamine residue. The role of this mannose
containing oligosaccharide-lipid in glycoprotein biosynthesis and its relationship
to the other endogenous acceptor (mannosyl r-acceptor) will be discussed in a
later section.

Before turning to the question of whether the various mannosylated lipids
found in liver and hen oviduct are true intermediates in the formation of
glycoproteins, there is one other system of particular relevance to this topic: the
synthesis of glycoproteins by the mouse myeloma tumor MOPC-46B (Baynes et
al., 1973). This tumor is an especially useful model for studying the incorporation of mannose into glycoproteins because the cell synthesizes a K-type immunoglobulin light chain (K46), which is a glycoprotein with a molecular weight
of approximately 24,000 with a single serum-type carbohydrate group. This
carbohydrate group contains four mannose residues in addition to eleven other
glycose units, and is attached to asparagine residue 34 in the polypeptide chain.
Apart from producing a well-characterized product against which antibodies can
be used for immunoprecipitation procedures, it appears that the cell is highly
directed toward synthesizing this particular glycoprotein. Labeling studies show

---

[3]Lucas et al. (1975) stress that because all the mannose residues were released by α-mannosidase does
not necessarily prove that all these units are linked by α-glycosidic bonds, as the enzyme preparation
may be contaminated with low levels of β-mannosidase.

that at least 50% of the cell's glycoprotein synthesis is represented by K-46 which, in turn, accounts for between 35 and 60% of the total protein synthesized by these cells. Baynes et al. (1973) have provided evidence for the formation of lipid intermediates in this cell. These workers concluded that microsomes prepared from the MOPC-46B tumor contain enzymes that are capable of transferring mannosyl residues from GDP-mannose to endogenous lipid and protein acceptors. Evidence was provided that a mannolipid was formed by the reversible transfer of mannose from the sugar nucleotide to endogenous phospholipid. On the basis of chemical and mass spectrometric data the mannolipid synthesized was considered to be mannosyl-monophosphoryl-dihydropolyisoprenol. In common with some of the earlier liver work (Richards and Hemming, 1972), Baynes et al. (1973) were unable to find any evidence for a lipid that contained more than one mannose residue or both mannose and N-acetylglucosamine residues. The latter authors, however, pointed out that a low molecular weight compound (approximately 2200) was formed as a major product when mannolipid was incubated with tumor cell microsomes. Baynes et al. (1973) speculated that this compound, which contained mannose and phosphate, might be a degradation product of an oligosaccharide-lipid and therefore the described mannosyl-monophosphoryl-dihydropolyisoprenol may not be the sole lipid intermediate involved in the glycosylation of the protein of these cells. The likelihood of degradation in these systems is obviously a problem. For example, Vessey and Zakim (1975) have drawn attention to the problems raised by a membrane-bound pyrophosphatase that is present in sizable amounts in a number of different membrane preparations and for which GDP-mannose is a substrate. These authors make the point that the Mn^{2+}-dependent transfer reaction between GDP-mannose and dolichol phosphate in liver systems cannot be followed accurately for any length of time, as the incorporation of mannose into endogenous acceptors is limited due to the hydrolysis of the donor sugar nucleotide. Apparently AMP is an effective inhibitor of pyrophosphatase and may be used to prevent the diversion of GDP-mannose into other metabolic pathways. The relevance of such inhibitors to the formation of lipid intermediates will be returned to later in the next section.

In 1974 Hsu et al. reported further details on the water-soluble 2200 (1800–2000 quoted in the 1974 paper) molecular weight substance described by this group in the previous year (Baynes et al., 1973). On the basis of various degradation studies, it was concluded that this low molecular weight compound is an oligosaccharide phosphate containing five mannose and two N-acetylglucosamine residues. The potential reducing terminus is a N-acetylglucosamine residue that is substituted with a phosphomonoester residue. These workers also found that if incubation mixtures of MOPC-46B microsomes and GDP ^{14}C-mannose were solubilized in 2% w/v Triton X-100, it was possible to separate lipids from small molecules by gel filtration on Sephadex G75. The lipid-containing fractions could then be quantitatively separated by affinity chromatography on insolubilized concanavalin A (concanavalin A-Sepharose) into two fractions: ^{14}C-mannose phosphosyl dolichol (which passes through the bed of insolubilized lectin) and ^{14}C-mannose-oligosaccharide pyrophosphoryl dolichol (which binds to the affinity column). The retained compound can, however, be eluted with α-methylmannopyranoside. Thus it appears that the water-soluble oligosaccharide phosphate described above is obtained as a result of the degradation of the ^{14}C-mannose-oligosaccharide pyrophosphoryl dolichol which can be retained on the insolubilized lectin.

5.2. Retinol glycolipids

Reference has been made to the possibility that other lipids, particularly retinol (vitamin A), may act as an acceptor of sugars. Using a rat liver "membrane fraction," probably consisting of rough and smooth endoplasmic reticulum as well as fragments of the Golgi apparatus and plasma membrane, De Luca et al. 1973) presented evidence for the formation of retinol mannolipid. The synthesis of this glycolipid from GDP-mannose was dependent on ATP and Mn^{2+}, and although material derived from vitamin A-deficient rats possessed some acceptor activity (probably due to some residual retinol [or dolichol]), the addition of exogenous retinol markedly stimulated the formation of mannolipid. Normal rat livers, which contain an inherently high level of retinol, gave a maximal rate of formation of mannolipid with or without the addition of retinol. Intravenous administration of GDP ^{14}C-mannose to vitamin A-deficient rats fed on ^{3}H-retinol resulted in the in vivo formation of ^{14}C and ^{3}H-mannolipid. Double-labeled glycolipid could also be produced enzymatically in vitro and the molar ratio of retinol:mannose was found to be 1:1 from such material. De Luca et al. (1973) also noted that when GDP ^{14}C-mannose, ^{3}H-retinol, and β[32]P-GDP-mannose were incubated together in their system, double-labeled ^{14}C, ^{3}H-mannolipid could be isolated. This material was devoid of any trace of ^{32}P, indicating that the mannolipid must lack any pyrophosphate bonds. When ^{3}H-retinol, γ[32]P-ATP and GDP ^{14}C-mannose were tested together, a triple-labeled ^{14}C,[3]H,[32]P-mannolipid was produced, presumably as the result of the formation of retinol phosphate. Indeed, using chemically prepared retinyl phosphate as substrate, Rosso et al. (1975) have presented evidence for the enzymatic synthesis of mannosyl retinyl phosphate in rat liver.

While demonstrating the formation of retinol-containing mannolipid, De Luca et al. (1973) do not exclude the formation of dolichol glycolipids and actually show that in their system two mannolipids are formed, one of which almost certainly contains a more stable isoprenol that may possibly be dolichol. These authors make the point that the incorporation of mannose from the sugar nucleotide into lipid is much less in the absence of ATP than in the absence of retinol. If the reaction in the absence of retinol were due entirely to endogenous dolichol, which they point out does not require ATP, then the two values should be the same; therefore they suspect other isoprenols may be important. In light of the later work of Vessey and Zakim (1975) on the problem associated with endogenous pyrophosphatases, discussed earlier in this section, it seems more likely that the ATP is preventing the breakdown of GDP-mannose.

Vessey and Zakim (1975) found that a number of nucleotide mono-, di-, and triphosphates are inhibitors of microsomal pyrophosphatase of which ATP is the most effective.

Mannose is not unique among sugars being incorporated into retinol glycolipids since other monosaccharide residues may also be transferred to this lipid. De Luca et al. (1973) showed that when UDP ^{14}C-glucose was incubated with ^{3}H-retinol in their rat liver membrane system a double-labeled glucolipid was formed. However, UDP-N-acetyl ^{14}C-glucosamine, UDP-N-acetyl ^{14}C-galactosamine and UDP ^{14}C-xylose were not incorporated into glycolipid in their system, a finding that De Luca et al. (1973) suggest makes it unlikely that ^{14}C-mannolipid is simply contaminated with ^{3}H-retinol.

At about the same time, Helting and Peterson (1972) reported that a microsomal fraction prepared from mouse mastocytoma catalyzed the transfer of a galactosyl residue from UDP ^{14}C-galactose to endogenous lipid which, upon extraction with chloroform-methanol, behaved as a single radioactive peak on DEAE Sephadex. When UDP ^{3}H-galactose and ^{14}C-retinol were used together in their system, Helting and Peterson found that material behaving as a single peak on DEAE Sephadex was obtained. On silica gel this material showed a compound containing both radioactive isotopes which analysis revealed to be retinol phosphate galactose.

5.3. Evidence for the participation of lipid-linked oligosaccharides in the synthesis of glycoproteins: particular relevance to the plasma membrane

In the previous sections, evidence for the formation of lipid-linked oligosaccharides has been described in detail for three separate systems. From the viewpoint of this chapter the relevant question is: What role do these glycolipids play in the biosynthesis of glycoproteins? Is there any definitive evidence for their being intermediates in the formation of these macromolecules and, if so, are such compounds of special relevance to either secreted or membrane glycoproteins, or both?

From what has been discussed regarding the chemical structure of the lipid-linked oligosaccharides it will be obvious that the carbohydrate groups of these compounds bear a close resemblance to the core structure found in the carbohydrate groups of serum-type glycoproteins. Thus it is tempting to suggest that the cell synthesizes a complete core structure on a lipid carrier and that the oligosaccharide is then transferred to the appropriate protein acceptor.

Behrens et al. (1973) have provided evidence for the role of oligosaccharide-dolichol derivatives in glycoprotein synthesis in the liver. Further to their earlier work on glucosylated endogenous acceptor (GEA) (section 5.1.1), Behrens et al. (1973) showed that by incubating liver microsomes with GDP-mannose in addition to dolichol monophosphate mannose, oligosaccharide-containing materials were formed similar to those synthesized when UDP-glucose was used as a donor. In both cases alkaline treatment produced evidence for the presence of two acetyl-hexosamine residues in the respective oligosaccharide chains. What

116

then is the relationship of this mannosylated endogenous acceptor to GEA? Behrens et al. (1973) point out that a connection between the two has not been proved experimentally. It is likely, however, that mannosyl residues are added, from either sugar nucleotide or (more probably) dolichol monophosphate mannose, to a dolichol pyrophosphate N.N' diacetyl chitobiose (Fig. 5). Evidence for this latter material having been presented earlier (Leloir et al., 1973), an intermediate with this structure was in keeping with the inferred presence of two N-acetylhexosamine residues found in the various lipid-linked oligosaccharides. Working with rabbit liver, Molnar et al. (1971) have presented evidence using β^{32}P-UDP-N-acetyl ^{14}C-glucosamine that a lipid material was formed containing ^{32}P and ^{14}C while only ^{14}C was found in protein products. Clearly, this shows that both phosphate and sugar residues are transferred from this sugar nucleotide to lipid, which then acts as an intermediate resulting in the transfer of labeled sugar

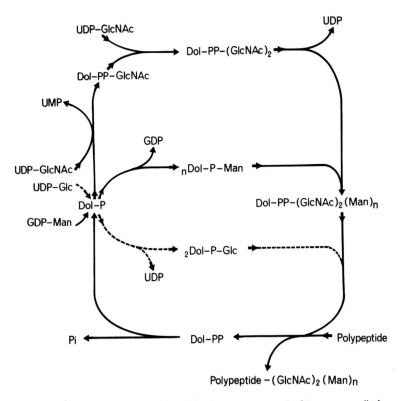

Fig. 5.   Possible pathways for the assembly of the "core structures" of "serum type" glycoproteins using dolichol phosphate as a coenzyme. The dotted pathway shows the possible relationship between mannosylated endogenous acceptor and the synthesis of glucosylated endogenous acceptor. In the diagram the subscripts n and 2, preceding Dol-P-Man and Dol-P-Glc respectively, indicated that n mannose residues and two glucose residues can be transferred to the endogenous acceptor by this reaction. The diagram is not intended to indicate that n and 2 dolichol phosphate sugar molecules are utilized in one step.

only to protein. It was shown by paper chromatographic separation that the oligosaccharide of the mannosylated endogenous acceptor contained up to sixteen residues. When considering that the oligosaccharide of GEA consists of some twenty glycose residues having a molecule weight of 3500, it is tempting to speculate that this latter material is formed by the addition of two glucose residues to the mannosylated endogenous acceptor. Although this is a reasonable assumption it needs experimental verification. However, what is particularly interesting is that these lipid-linked oligosaccharides appear to be important as intermediates in the synthesis of glycoproteins; the oligosaccharide moiety of these materials is transferred by microsomal enzymes to a trichloroacetic acid-insoluble product. As radioactivity was lost from the lipid fraction, increasing amounts of activity were found in the latter material. Again, using labeled material demonstrates that this latter product is solubilized following treatment with either of the proteolytic enzymes, pronase or trypsin. This finding suggests that the oligosaccharide is being transferred to protein, though it is interesting that in the case of the insoluble radioactive product formed from mannosylated endogenous acceptor the amounts solubilized by trypsin are greater than those solubilized by pronase (Behrens et al., 1973). As pronase has a wider specificity than trypsin the reverse might have been expected, though Behrens et al. (1973) point out that the amounts solubilized are not corrected for self-absorption, which may explain the experimental findings. In the case of GEA it may be shown that following the transfer of oligosaccharide to protein, pronase digestion of the product yields glycopeptides that bind to concanavalin A. In addition to the cited work of Leloir and his colleagues (1973), Hemming and his group (Oliver et al. 1975) have been able to show that the transfer of mannose to glycoprotein in pig liver is via dolichol diphosphate oligosaccharide. The transfer of label to "insoluble polymer" over a 1-hour period was approximately 13% using the latter intermediate as opposed to 3% for GDP-mannose and dolichol phosphate mannose. Insoluble polymer contained both glycoprotein and oligosaccharide. The release of oligosaccharide by hydrolysis in vitro competes with the transfer to protein, an effect that is accentuated by using excess EDTA.

If, as seems likely, these lipid acceptors are intermediates in the biosynthesis of glycoprotein, it would appear that a whole core structure, ready for transfer to an appropriate asparagine residue in a polypeptide acceptor, is being synthesized on the lipid. The number of mannose residues attached to the chitobiose-like residue would appear to be much greater than that normally found in the serum type glycoproteins. It is possible that by using these isolated microsomal systems an extended mannan is being formed artificially since the controlling glycosyl transferases necessary for the completion of the saccharide chains are absent. If this is the case, this would suggest that these enzymes are located in another subcellular site (section 3.2). While the evidence for the role of mannosylated and glucosylated endogenous acceptors in glycoprotein biosynthesis in liver is persuasive, it is difficult to establish from this data whether they are involved in the production of both secreted and membrane glycoproteins or only in the former. The finding of Parodi et al. (1973) that GEA is obtained in variable yields when rat brain or kidney as well as human lymphocytes or pig thyroid are used as a source of enzyme would be in accord with the view that this type of intermediate is not required solely for the synthesis of plasma glycoproteins. The presence of glucose in GEA is of particular interest because (as observed in the introduction to this chapter) the presence of glucose in glycoproteins is rather limited, being principally associated with the collagens. In addition, Behrens and Leloir (1970) could find no evidence for the type of glycoprotein being formed in the systems they studied. Again, it is possible that glucosylated oligosaccharide is an artifact of the system perhaps as the result of a glycogen synthetase type of reaction.

The work of Leloir and his colleagues (Behrens et al., 1973) is not alone in suggesting that these lipid-linked oligosaccharides represent a "preassembled sugar core" that is transferred intact to a protein acceptor. Again, in the mouse myeloma system, when radioactively labeled mannose contained oligosaccharide pyrophosphoryl dolichol is incubated with microsomes, radioactivity is incorporated into protein (Hsu et al. 1974). Digestion with pronase, subtilisin, trypsin, or pepsin produced low molecular weight products which, on the basis of their behavior on electrophoresis in acid (pH 1.9) and alkaline (pH 7.7) buffer, were shown to be zwitterionic, a finding consistent with their being glycopeptides (Baynes et al., 1973). In addition, the product can be solubilized in various detergents and following such solubilization appears to behave on gel filtration as a compound of molecular weight 20,000 to 50,000. Of special significance regarding the characterization of this product is the finding that 10 to 20% of the radioactive protein is immunoprecipitable with specific anti-κ 46B antibody (Hsu et al., 1974). Evidence for mannolipid being an intermediate in the formation of glycoprotein was provided by Baynes et al. (1973). These workers listed three main arguments in support of this view. First, incorporation of ^{14}C-mannose into protein was observed even after the destruction of the initial substrate GDP-mannose by endogenous sugar hydrolases, though an increase in radioactivity in protein was associated with a loss of radioactivity from the mannolipid fraction. Second, addition of EDTA to the system after mannolipid had been formed resulted in the continued incorporation of label into protein to the extent of that present in the mannolipid. The transfer of mannose from GDP-mannose to lipid, as well as protein, is sensitive to EDTA but the transfer of counts from exogenously added mannolipid is not inhibited, thus ruling out the possibility that the sugar lipid is acting as a source of activated mannosyl residues for the generation of GDP-mannose for the subsequent direct glycosylation of protein. The third argument in support of mannolipid acting as a direct donor of mannosyl residues to protein is that when microsomes are pulsed briefly with ^{14}C-mannose followed by a chase with large excess of unlabeled sugar nucleotide, incorporation of label into lipid ceases immediately. However, incorporation into protein continues again to an extent proportional to the amount of mannolipid formed before adding the chase of unlabeled sugar nucleotide. Though the above arguments (Baynes et al., 1973) were advocated at a time when these workers were unable to detect any species of lipid containing sugar other than a single mannose residue, their points would appear to rule out the possibility of GDP-mannose acting as a direct donor of sugar to protein. When combined with later evidence (Hsu et al., 1974) for the rapid transfer of mannose from monomannosyl lipid to oligosaccharide lipid a strong case can be made for lipid intermediates in glycoprotein synthesis, especially since by adding exogenous mannose oligosaccharide pyrophosphoryl lipid to microsomes it can be demonstrated that this compound appears to be a primary donor of sugar in protein glycosylation (Hsu et al., 1974).

In the hen oviduct system studied by Waechter et al. (1973), similar arguments to those raised above have been put forward to show that mannosylated lipids are obligatory intermediates in glycoprotein synthesis.[4] As with the myeloma system discussed previously, the effect of chasing with unlabeled GDP-mannose was an immediate cessation of the synthesis of mannosyl phosphoryl polyisoprenol but a continued synthesis of oligosaccharide lipid (mannosyl s-acceptor) and mannosylated proteins (mannosyl r-acceptor). Addition of exogenous mannosyl phosphoryl polyisoprenol serves as a donor for both the above types of receptors (Waechter et al., 1973). Further evidence for the participation of oligosaccharide-lipid (mannosyl s-acceptor) in glycoprotein synthesis has been provided more recently by Lucas et al. (1975). These authors used oligosaccharide-lipid that had been doubly labeled with N-acetyl ^{3}H-glucosamine and ^{14}C-mannose and showed that doubly labeled oligosaccharide was transferred to protein. Digestion of the product with proteolytic enzymes resulted in the production of glycopeptides containing both labels. As Lucas and his colleagues point out, this is the most direct evidence to date for the enzymatic transfer en bloc of an oligosaccharide chain from lipid to protein.

The earlier paper by Waechter et al. (1973) drew attention to the finding that the mannosylated protein is tightly associated with the membranes of the enzyme preparation and suggested it was possible that these lipid intermediates play an important role in the assembly of membrane glycoproteins. Using SDS-polyacrylamide gel electrophoresis, Lucas et al. (1975) showed that at least six polypeptide chains were labeled in the hen oviduct system. However, ovalbumin does not correspond to any of these components. Using an anti-ovalbumin less than 10% of the labeled mannose containing protein was precipitated in this system. On the basis of this information, Lucas et al. (1975) go so far as to suggest that perhaps the lipid-mediated assembly of glycoprotein operates only for the membrane glycoproteins under the conditions studied and is not operative for soluble, secretory glycoproteins—a theme that Lennarz (1975) adopts in a recent review. Lennarz states that two distinct types of reactions appear to be involved in the biosynthesis of glycoproteins: the process (see earlier discussion, section 4) by which single sugars are sequentially transferred from the appropriate sugar nucleotide derivatives to the distal portion of incomplete oligosaccharide chains

---

[4]In this system the component formed is termed residual mannosylated acceptor (mannosyl r-acceptor) and characterized by gel filtration and electrophoresis before and after proteolytic enzyme digestion as consisting of a group of mannosylated proteins.

of glycoproteins, as well as the more complicated mechanism involving the assembly of the core oligosaccharide chains on a lipid carrier. Lest it be assumed from this statement that it is not possible for a sugar residue to be transferred directly from a sugar nucleotide to an appropriate amino acid residue present in a polypeptide, the example of UDP-N-acetylgalactosamine polypeptidyl N-acetyl-galactosaminyltransferase is a case where the first mechanism is operative for the formation of a carbohydrate protein linkage. As Lennarz (1975) points out, a number of glycosyltransferases have been highly purified and, when used to catalyze the transfer of single sugars to purified acceptors, there is little possibility that lipid intermediates are involved. That the two mechanisms relate solely to "distal" and "core sugars" respectively (Lennarz, 1975) is perhaps a convenient division, though an oversimplification in that the O-glycosidic linkage of sugar to serine and theonine residues in a polypeptide can take place by a mechanism that is more characteristic of the assembly of the distal portions. Perhaps it is more appropriate to regard lipid intermediates as being involved in the biosynthesis of the carbohydrate groups of serum-type glycoproteins, although once the core has been assembled the addition of further glycose residues by a one enzyme-one linkage mechanism may be operative.

What is the advantage of having a lipid-mediated assembly process for the formation of a glycoprotein? One working hypothesis (Lennarz, 1975) is that this mechanism allows for the conversion of nascent proteins already associated with a membrane to be converted to a glycoprotein by the addition of a core sugar. Lennarz (1975), noting that glycoproteins are asymmetrically arranged in membranes with the oligosaccharide chains beyond the hydrophobic region of the membrane at the periphery of the structure, suggests that perhaps the potential membrane glycoproteins are introduced into the membrane without their carbohydrate groups and that the lipid intermediates could be responsible for glycosylating amino acid residues located in the interior of the membrane (Fig. 6). Once the hydrophilic sugar residue is linked to the protein, one might expect conformational rearrangement of the protein to occur with the sugar residues being oriented to the surface of the membrane. Alternatively, an oligosaccharide linked to lipid might be a convenient way of glycosylating those amino acids that are close to the surface of the membrane, the lipid being associated with the hydrophobic environment of the membrane and the oligosaccharide oriented toward the hydrophilic exterior. Although the proposed mechanism could take place in a number of different membrane systems, Lennarz (1975) has described it as occurring at the plasma membrane, citing as evidence the recent finding that using GDP-mannose as the glycosyl donor mannosylated intermediates probably linked to lipid are formed by several cell types, including chick embryo liver cells (Arnold et al., 1973), normal and transformed mouse fibroblasts (Patt and Grimes, 1974), and intact oviduct cells (Struck and Lennarz, 1975). If GDP-mannose cannot penetrate the cell membrane, then it is argued (Lennarz, 1975) that the lipid-mediated synthesis of oligosaccharide must take place at the cell surface. In the studies outlined earlier, microsomal preparations have been used as a source of enzyme which, as admitted by some authors (De Luca et al.,

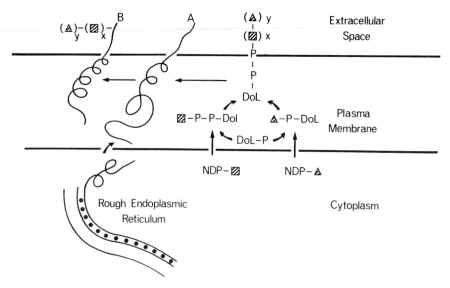

Fig. 6. A model proposed by Lennarz (1975) for the glycosylation of plasma membrane glycoprotein via the lipid intermediate pathway. This figure has been redrawn following the scheme proposed by Lennarz; the shaded square and triangular symbols represent two different glycosyl residues. Such a glycosylation scheme need not necessarily be confined to the plasma membrane. A = unglycosylated protein. B = complete glycoprotein.

1973), is likely to contain fragments of membrane derived from a number of different sources. Lennarz (1975) points out that in no case has a thorough study been made of the purity of the subcellular fractions, a serious omission if one is attempting to localize the subcellular site of the assembly of a glycoprotein. Presumably, if a core structure is being added to protein at the cell periphery, it would also be necessary to add the more distal sugars of the carbohydrate groups of the complete glycoprotein at the same site.

The search for glycosyltransferases at the cell surface has received considerable impetus from the suggestion of Roseman (1970) that the presence of these enzymes at the cell periphery is responsible for the recognition of appropriate sugar residues on apposing cell surfaces. The formation of such an enzyme substrate complex is a possible mechanism for intercellular adhesion. It is suggested that should *trans*-glycosylation occur, the adhesion would be broken. While providing a novel explanation for the formation of mutable adhesions between cells, a difficulty associated with this theory involves the mechanism by which the appropriate endogenous sugar nucleotides pass through the cell membrane to become effective at the cell periphery. If the appropriate sugar donor is linked to lipid this problem can be overcome. In a chapter devoted to the biosynthesis of membrane glycoproteins it would probably not be appropriate to go into lengthy detail about the arguments for or against the view that glycosyltransferases present at the cell surface mediate intercellular adhesion, although the finding that glycolipids affixed covalently to glass were glycosylated upon contact by cells using a reaction that proceeds through a retinol phosphate glycoside are of particular interest to this section (Yogeeswaran et al., 1975). In the next section evidence that ectoglycosyltransferases—membrane-bound enzymes whose active site is accessible from the exterior of the cell—are involved in the synthesis of complex carbohydrates will be discussed briefly.

6. The role of the cell surface in glycoprotein biosynthesis

The evidence for glycosyltransferases being associated with intracellular membranes has already been discussed, though as pointed out in the preceding section there is now increasing evidence that the cell also possesses glycosyltransferases as ectoenzymes. This latter term is used to denote enzymes that are present in or on the external surface of the plasma membrane. Evidence to support such a location has been obtained in several studies (Roth et al., 1971; Bosmann, 1972; Roth and White, 1972; Bosmann et al., 1974; Webb and Roth, 1974; Lloyd and Cook, 1974; Patt and Grimes, 1974; McLean and Bosmann, 1975; Patt and Grimes, 1975). While the suggestions of Roseman (1970) on ectoglycosyltransferases and their possible role in intercellular adhesion are outside the scope of this chapter; the possibility that they have a synthetic role to play, either by completing oligosaccharide chains or repairing sublethal "lesions" at the cell surface is relevant to the topic of plasma membrane glycoprotein biosynthesis and turnover and will therefore be considered. It should be pointed out that the presence of glycosyltransferases at the cell surface is by no means universally accepted. Recently, Keenan and Morré (1975) examined the evidence for the existence of glycosyltransferases on the surface membrane of mammalian cells and concluded that the assignment to glycosyltransferases of specific roles in phenomena such as cell recognition and adhesion is premature. It is also argued that there appears to be no valid basis for the conclusion that ectoglycosyltransferases of membrane biogenesis are localized on the cell surfaces. Keenan and Morré (1975) suggest from fractionation studies on isolated Golgi-enriched material that this particular organelle, in combination with endoplasmic reticulum, accounts for a large proportion of the total glycosyltransferase activity present in the cell (at least in the liver) and that only a small proportion of such activity remains for distribution among the other membranes of the cell. Furthermore, they suggest that isolated plasma membranes in which the criteria for a lack of contamination by Golgi apparatus and other intracellular membranes is well defined lack glycosyltransferase activity. The evidence that glycosyltransferases are accessible from the exterior of the cell is based on experiments in which suspensions or monolayers of intact cells are incubated with radioactively labeled sugar nucleotides (it is supposed that sugar nucleotides are not transported through the plasma membrane) and from which transferase activity is subsequently determined on the basis of precipitable counts. Sometimes exogeneous glycoproteins have been tested for acceptor activity in those systems (Bosmann et al., 1974; Lloyd and Cook, 1974; Patt and Grimes, 1975). Keenan and Morré (1975) argue, however, that such evidence is open to criticism on two counts: while the cell may be impermeable to the labeled sugar nucleotide, the presence of plasma membrane nucleotide pyrophosphatase (Evans et al., 1973) located on the external surface of the cell (Evans, 1974) would generate free labeled sugar that could then be incorporated into the cell. Also, the incorporation of label could be a result of intracellular glycosyltransferases being released into the incubation medium as a result of cell lysis or secretion. Certainly these two arguments must be given serious consideration before assigning transferase activity to the cell periphery on the basis of incorporation of labeled sugar from the appropriate sugar nucleotide by intact cells. Since the work on nucleotide pyrophosphatase (Evans et al., 1973; Evans, 1974) was performed on liver material, it is possible that such activity is not universally present at cell peripheries although the work of Deppert et al. (1974) has demonstrated the hydrolysis of UDP-galactose to free galctose by intact baby hamster kidney (BHK) cells. Deppert et al. (1974) suggest that this data, together with the absence of detectable cell surface galactosyltransferase activity in their experiments, rules out the possibility that this enzyme plays a general role in cell surface phenomena, though they do indicate their results do not exclude the possibility that other glycocyltransferases are located on the cell surface. Porter and Bernacki (1975a,b) have shown by electron microscope autoradiography that while UDP-^{3}H-galactose is probably enzymatically cleaved extracellularly by L-1210 murine leukemia cells; the same cell can transfer N-acetylneuraminic acid from its nucleotide sugar to endogenous glycoprotein by what appears to be an ectoglycosyltransferase. Incidentally, the role of ectoglycosyltransferases in mediating intercellular adhesion (Roseman, 1970) would not necessarily be impaired by the presence of surface nucleotide pyrophosphatases as the initial recognition and adhesion between cells formed as a result of enzyme-substrate binding between glycosyltransferase and nonreducing terminal sugar residues would not be sensitive to this hydrolase. Admittedly, the subsequent transglycosylation reaction (which, unlike enzyme/substrate binding, does involve sugar nucleotide) postulated by Roseman (1970) to explain the de-adhesion of cells would be affected. Patt and Grimes (1975) and Yogeeswaran et al. (1974), have observed that intracellular sugar can be transferred to exogeneous glycoprotein acceptor and they suggest that "the most likely role for ectoglycosyltransferases is the process of cell surface and extracellular complex polysaccharide synthesis using intracellular substances." Presumably, sugar lipid intermediates would offer a particularly suitable way of allowing donor sugars to pass through the cell's permeability barrier (Lennarz, 1975), thus allowing ectoenzymes operating at the cell periphery to utilize intracellular intermediates. This mechanism would obviate the problem raised by the finding of surface nucleotide pyrophosphatase activity. The latter enzyme could also serve to exclude foreign messenger RNA from the cytoplasmic environment of the cell (Aronson, 1975). As Aronson (1975) points out, the presence of 5'-nucleotidase, an enzyme that occurs on the exterior surface of some cells, would be ideally situated for hydrolysis of the 5'-nucleotide products of nucleotide pyrophosphatase to nucleoside and inorganic phosphate.

The second objection raised by Keenan and Morré (1975) to the use of intact cell studies for assigning glycosyltransferases to the cell surface is the problem of discerning whether the incorporation of radioactivity into appropriate acceptors is due to "solubilized" glycosyltransferases, which have been released into the suspending medium, as opposed to true plasma membrane-bound ectoenzymes. While glycosyltransferases are membrane-bound within cells, these enzymes could be freed by cell lysis and the authors cite the finding of soluble glycosyltransferases in serum and other body fluids as an example of the release of these enzymes by some cells. Keenan and Morré (1975) rightly suggest that the presence of intracellular enzyme activity escaping as a result of cell lysis could be "tested by removal of intact cells and assay of the resultant supernatant for glycosyltransferase activities" but state incorrectly that this type of experiment has apparently never been tested. For example, in the study by Lloyd and Cook (1974) on the possible role of ectoglycosyltransferases in the aggregation of 16C malignant rat dermal fibroblasts, cell-free supernatant prepared by incubating an identical batch of cells in suspending buffer simultaneously with the ectoglycosyltransferase assay was subsequently examined for glycosyltransferase activity (glycoprotein:sialyltransferase and glycoprotein:galactosyltransferase, together with appropriate endogenous controls). It was found (Lloyd and Cook, 1974) that with the Tris buffered medium used by Roth et al. (1971) significant quantities of glycoprotein:sialyltransferase activity could be found in the supernatant fluid and, in addition, high levels of glycoprotein:galactosyltransferase were released. However, when a phosphate-buffered saline was used, no leakage of glycoprotein:sialyltransferase into the supernatant medium could be detected although the intact cells still displayed endogenous and exogenous ectosialyltransferase activity in this buffer. In the latter experiment it was not possible to test for glycoprotein:galactosyltransferase activity because of the dependence of this enzyme on manganous ions that are absent from the phosphate-buffered medium—the purpose of this study not being to derive optimal enzyme activity but to determine activity under conditions in which the cells are known to adhere. The finding (Lloyd and Cook, 1974) that suspensions of intact cells, in which it can be demonstrated that there is no leakage of transferase, can transfer labeled sugar from exogenous sugar nucleotide to exogenous glycoprotein acceptors would certainly support the ectoglycosyltransferase concept. As Keenan and Morré (1975) point out, exogenous acceptors "can reasonably be expected not to enter intact cells."

In the system studied by Lloyd and Cook (1974) it is possible that though ectoglycosyltransferases are present and can be demonstrated in vitro by the use of neuraminidase-treated cells or asialoglycoprotein acceptors, in vivo they are inactive. Indeed, in the case of L1210 leukemic cells, Porter and Bernacki (1975b) have exposed additional endogenous acceptor sites for the sialyl ectoglycosyltransferase of this cell by treatment with exogenous neuraminidase. The possibility that an endogenous neuraminidase (Bosmann 1972; Visser and Emmelot, 1973) acting as an ectoenzyme generates effective acceptors for a surface glycosyltransferase should, however, be considered.

The results obtained in autoradiographic studies strongly suggest that the majority of glycoproteins are completed in the Golgi apparatus whether they are of the mucin or serum type. In view of the results obtained in autoradiographs with fucose (a terminal sugar in serum type carbohydrate groups) it is difficult to see how the plasma membrane can represent a major site (cf., Lennarz, 1975) for the addition, from lipid intermediates, of core sugars to asparagine residues. It is possible, as Schacter (1974) suggests, that glycosyltransferases occur at the cell periphery as the result of membrane flow (see section 7.2). As these enzymes are constituents of the Golgi apparatus and are considered to be precursors of plasma membrane, it would not be surprising to find that such activities are carried to the cell surface. Whether, once in the cell periphery, they perform a special function such as intercellular adhesion in which a sugar-retinol intermediate is utilized is an intriguing question that needs further investigation. Certainly, as noted by Schachter (1974), not particularly active cell surface glycosyltransferases may nevertheless perform some special function. On the basis of present evidence, however, the contention (Patt and Grimes, 1975) that a "significant part of cellular complex polysaccharide synthesis occurs at the cell surface" should be treated with due caution.

7. Biosynthesis of membranes

The study of plasma membrane glycoprotein biosynthesis has dealt with such topics as details of the chemical reactions responsible for the polymerization of the various monosaccharide residues and the location in the cell of the various stages in macromolecule assembly. Other areas of interest concern the mechanism by which the cell distinguishes between those glycoproteins destined for export as opposed to those that form components of the various membrane systems of the cell. At present this question is unresolved. A considerable body of information is available, however, on the biochemical reactions involved in glycosylation and on the subcellular location of the various enzymes involved in the process. While one can speculate on the factors that control glycoprotein biosynthesis, much experimental work is required in this area. The assembly of polypeptides is controlled by an accurate template mechanism, although the oligosaccharide groups are secondary gene products—the glycosyltransferases being the primary gene product. It appears that the initiation of glycosylation and the subsequent elongation and termination of the oligosaccharide groups are controlled by the substrate specificity of the particular transferase.

If each carbohydrate group were synthesized by a separate group of glycosyltransferases, it would be necessary for the cell to possess an enormous number of different enzymes, since not only do most cells synthesize more than one glycoprotein but each glycoprotein may have more than one type of carbohydrate group. Alternatively, the number of glycosyltransferases required by the cell would be reduced if each glycoprotein being synthesized passed through different "assembly lines" although whether such lines exist in the cell awaits confirmation. The work of Ovtracht et al. (1973) on the subfractionation of the Golgi apparatus is particularly relevant to this type of study. Although much remains to be learned about the biosynthesis of membrane glycoproteins, with present knowledge it is at least possible to make some generalizations regarding the incorporation of such macromolecules into the final plasma membrane.

7.1. Membrane turnover

Before discussing membrane assembly in any detail, it is essential to distinguish between those experimental models in which there is de novo synthesis of membrane, as opposed to those systems in which one is examining turnover of the plasma membrane. Some of the most detailed information on membrane turnover has been provided by Warren and Glick (1968) who studied the fate of such precursors as ^{14}C-labeled D-glucose, L-leucine; L-valine and D-glucosamine in the surface membranes of mouse L-strain fibroblasts. Using a general precursor such as D-glucose, the rate of increase in the specific activity of plasma membranes of both dividing and nondividing cells was approximately the same. For the nondividing cell such a result is only compatible with membrane turnover, if the assumption is made that the mass of membrane in such cells remains constant. In dividing cells, where additional membrane is being synthesized, an increase in specific activity is not surprising although the question of membrane turnover cannot be entirely excluded.

These results are not confined to the use of glucose, similar results being reported with precursors of protein and carbohydrate (Warren and Glick, 1968). In an effort to measure the extent of turnover in the dividing cell, Warren and Glick (1968) measured the removal of ^{14}C-labeled material from cells grown in cold medium following heavy labeling. The authors found that the rate of fall in specific activity was slightly lower than that predicted by taking into account dilution caused by an increase in cell number. The overall conclusion from this work must be that nondividing and dividing cells are synthesizing similar amounts of membrane that is incorporated into the dividing cell, where it remains at the surface, while in the nondividing cell membrane turnover is taking place with the expulsion of some of the material and incorporation of fresh components. This appears to be a general feature for all membranes and not just for the plasma membrane. For example, Warren and Glick (1968) found a similar labeling pattern for a particulate fraction that they considered consisted largely of the internal membranes of the cell. A general note of caution (Cook, 1971) has been sounded, however, in accepting the L cell as a representative model in view of the observations of Kindig and Kirsten (1967) that virus-like particles have been found in the cells of a number of L-cell cultures. These particles are associated with intracellular vesicles and also bud from the surface of the cell. If there were an appreciable release of virus-like particles from the cells under study, an apparently high rate of membrane turnover in the nondividing cell could measure the synthesis and release of such particles, each particle on its release acquiring a membrane from the surface membrane of the host cell (see chapter 11, this volume). It is interesting to note from the work of Rieber et al. (1975) on the turnover of cell surface proteins in normal rat kidney cells transformed by a temperature-sensitive mutant of Rous sarcoma virus (B77) that high-molecular-weight surface components are normally subject to a significant turnover which may be accelerated during growth or transformation. Certainly the presence of virus either as a result of shedding mechanism or transformation per se appears to exert a significant effect on the rate of membrane turnover.

An important feature of membrane turnover is that such a process provides the cell with a means of repairing sublethal autolytic damage done to the cell periphery. Warren (1969) has shown that when L cells are treated with the proteolytic enzyme trypsin, there is no specific repair process operating; instead, the plasma membrane is replaced by the normal process of membrane turnover. This contrasts with the suggestion made by Molnar et al. (1969) who, on the basis of their work on glycosyltransferase activities of isolated fractions from Ehrlich ascites cells, suggested that these enzymes situated in the plasma membranes may carry out repair processes when sugars are cleaved from the cell membrane by exogenous enzymes. Schachter and Roden (1973) noted that "Molnar et al. (1969) found this so-called plasma membrane fraction to be significantly enriched in N-acetylhexosaminyl and galactosyltransferase activities," but they believe that in the plasma membrane fraction of Molnar et al. (1969) elements of both Golgi and plasma membrane are present and consider that the identification of this fraction is not conclusive. As stressed earlier, the need to obtain well-characterized fractions in this type of study is essential. Is it possible that the cell's response to a lesion caused by an exogenous glycosidase (as opposed to a protease) might be different and thus provide a rationale for the findings of Molnar et al. (1969)? In this respect the work of Hughes et al. (1972) on the regeneration of surface glycoproteins of TA3 cells (a transplantable spontaneous mammary adenocarcinoma of the A strain mice) following treatment with neuraminidase is particularly pertinent. These investigators showed that the turnover rate of membrane glycoproteins lacking sialic acid residues is the same as that occurring in nondividing cells not exposed to the glycosidase but cultured under identical conditions. This result not only substantiates for the TA3 cell the earlier results obtained with L cells by Warren et al. (1968), but again suggests that a specific repair mechanism is not operative in the plasma membrane following glycosidase treatment. In agreement with the findings of Warren et al. (1968), it appears that damaged oligosaccharide chains are replaced in the course of membrane turnover by de novo synthesis of completed carbohydrate groups. Hughes et al. (1972) found that during turnover high-molecular-weight surface glycoproteins were released into the extracellular medium of cultured TA3 cells, However, in view of the more recent work of Harms and Reutter (1974) on the half-life of N-acetylneuraminic acid in liver and hepatoma membranes and the finding (Gurd and Evans, 1973) that sialoproteins and asialoproteins in membranes are degraded at different rates, it appears that plasma membranes may not be degraded as a single entity. The possibility that different constituents of the membrane are turning over at different rates is interesting and does not invalidate the findings discussed earlier that there are no specific repair mechanisms operative. Rather, it suggests that the restoration of a particular component at the cell surface will depend on the turnover of that particular constituent rather than on the membrane as a whole. Naturally these findings will relate to the question of membrane biogenesis.

7.2. Mechanisms of membrane biogenesis

Having discussed the biosynthesis of glycoproteins per se, it is relevant to consider how these macromolecules are assembled into the completed membrane. For the purposes of this chapter a bilayer, as opposed to a subunit model for the plasma membrane, will be assumed; the relevance of a subunit model in light of biosynthetic studies has been discussed elsewhere (Cook, 1971). Membranes can be synthesized either by a single- or a multistep mechanism. In the single-step process it is implied that the various components of the membrane are assembled simultaneously, while a multistep mechanism envisages the synthesis of a relatively stable, primary membrane composed of lipid and some protein to which various constituents are added.

Dallner et al. (1966a) have reported findings of direct relevance to membrane biogenesis using embryonic and neonatal hepatocytes of the rat. Tracer experiments using ^{14}C-leucine and ^{14}C-glycerol coupled with cell fractionation suggest that new membrane is synthesized in the rough endoplasmic reticulum and is then transferred to the smooth endoplasmic reticulum. This synchronous assembly of lipid and protein is in accord with the single-step hypothesis for membrane biogenesis. However, in an accompanying paper it was shown that microsomal enzymes were synthesized at different times of rat liver development, which is more in keeping with a multistep process (Dallner et al., 1966b). The subsequent findings of this group (Omura et al., 1967) that the half-lives of protein and lipid in these membranes are different would also argue in favor of a multistep assembly.

In this chapter it is necessary to consider how the system described above is concerned with the biogenesis of intracellular membranes, applies to the plasma membrane and, in particular, to its constituent glycoproteins. Using the finding of Warren and Glick (1968) that the labeling pattern of plasma membranes and particulate matter is similar, and assuming that the latter fraction is representative of internal membranes, a direct relationship between the synthesis of intracellular membranes and the plasma membrane is evident. That membrane glycoproteins and glycolipids are synthesized at different rates (Evans and Gurd, 1971) and lipid and glycolipid are synthesized late in the cell cycle (Bosmann and Winston, 1970) would be difficult to explain on the basis of a one-step assembly involving the combination of a homogeneous group of membrane subunits.

In membrane glycoprotein synthesis, it is apparent that protein synthesized at the rough endoplasmic reticulum is then glycosylated in the smooth membranes of the cell. Thus, although the bulk of the sugar may be added at one site in the cell, such as the Golgi apparatus, and though (as suggested above) a completed internal oligosaccharide core may be added to protein in one step (in N-glycosidically-linked sugar) via a lipid intermediate, protein clearly undergoes a migration from rough endoplasmic reticulum to smooth endoplasmic reticulum and, ultimately, to the cell surface. This process is in agreement with the assembly line mechanism discussed in this section. This mechanism has been invoked by Hirano et al. (1972) to explain the exclusive localization of carbohydrate on the outer surfaces of the plasma membrane. The exclusive localization of sialosyl residues at the extracellular surface of the cell membrane was demonstrated originally by Benedetti and Emmelot (1967) in isolated rat-liver plasma membranes, using colloidal ion hydroxide as a marker for the sialic acid residues and the junctional complexes as a morphological marker for the extracellular surface of the membrane. Further, using ferritin-conjugated lectins, Nicolson and Singer (1971) found by electron microscopy an asymmetry of lectin-binding sites on the two membrane surfaces. These authors concluded, on the basis of this information and earlier data (Benedetti and Emmelot, 1967), that an asymmetric distribution of oligosaccharide is a general feature of the plasma membrane of eukaryotic cells. Further studies by Hirano et al. (1972) demonstrated that an asymmetric distribution of sugar residues was also a feature of internal cell membranes of P3K murine plasmacytoma cells which, in addition to producing membrane glycoproteins, also synthesize an IgG1 myeloma protein with covalently attached oligosaccharides. Hirano et al. (1972) also made the important finding that concanavalin A, presumably interacting with "core" sugars, binds to the cisternal side of the rough endoplasmic reticulum while *Ricinus communis* lectin, which is specific for galactosyl residues, does not stain these structures. Smooth membranes were stained by this latter lectin, consistent with the view that they represent the addition site of the more peripheral sugars to the growing oligosaccharide chain. That staining was asymmetric in all cases enabled Hirano et al. (1972) to postulate that the biogenesis of membranes and their glycoproteins involves an assembly line mechanism as shown in Figure 7.

8. Conclusions

Great strides have been made in understanding membrane glycoprotein biosynthesis. Although a number of questions regarding the control of synthesis and the relation of amino acid sequence to the site of glycosylation on the polypeptide chain remain to be investigated further, some fundamental principles governing the synthesis of these important macromolecules are now clear. When considered in relation to membrane assembly and turnover, these are complex mechanisms that serve to emphasize the dynamic properties of the plasma membrane.

130

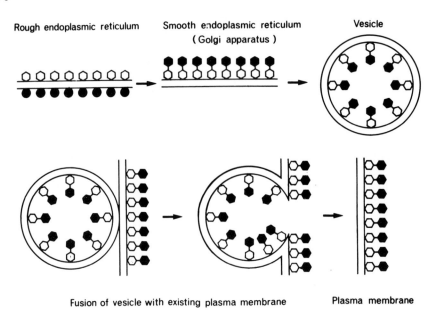

Rough endoplasmic reticulum   Smooth endoplasmic reticulum   Vesicle
(Golgi apparatus)

Fusion of vesicle with existing plasma membrane   Plasma membrane

Fig. 7.   Schematic representation of the mechanism for the biogenesis of plasma membranes of
eukaryotic cells after the scheme proposed by Hirano et al. (1972). In their scheme an intracellular
membrane "assembly line" mechanism is envisaged, leading penultimately to the formation of vesicles, the new surface membrane being generated by the fusion of these vesicles with the existing
plasma membrane. The large, filled circles represent ribosomes. At different stages along the assembly line it is suggested that saccharide units (represented by open and filled hexagonal symbols) are
added successively to growing carbohydrate groups on membrane-bound glycoproteins and
glycolipids. In this scheme carbohydrate was envisaged as being added at the rough endoplasmic
reticulum, though on present evidence (see text) the glycosylation of membrane protein, as opposed
to secreted glycoproteins, takes place predominantly in the Golgi apparatus. As Hirano et al. (1972)
showed diagramatically, the carbohydrate exists on the cell surface as a trisaccharide and in the
vesicle as a disaccharide. In the present figure, the carbohydrate in the vesicle and on the cell surface
is shown diagramatically as a disaccharide. The portrayal of the same carbohydrate moiety in the
vesicle and on the cell surface is for clarity and should not be taken as an indication that the synthesis
of surface oligosaccharide moieties are necessarily completed at the vesicular stage. The above
scheme explains the observations of various workers (Benedetti and Emmelot, 1967; Nicolson
and Singer, 1971) that cell membrane oligosaccharides are located exclusively on the outer surface
of the plasma membrane.

Acknowledgments

The author wishes to thank Mr. Michael Abercrombie, F. R. S., for his interest in
this work. The author is also grateful to his colleagues, in particular to Dr.
Kareen Thorne and Dr. C. W. Lloyd, who have read and commented on various
drafts of the manuscript. Special thanks are due to Mrs. E. Broad for painstaking
secretarial help and also to Miss J. Lander, Miss C. Reid and Mrs. Sheila Keep for
bibliographic assistance. The author is a Member of the External Scientific Staff
of the Medical Research Council, London.

References

Adamany, A. M. and Kathan, R. H. (1969) Isolation of a tetrasaccharide common to MM, NN and MN antigens. Biochem. Biophys. Res. Commun. 37, 171–178.

Arnold, D., Hommel, E. and Risse, H.-J. (1973) Glycosyl transfer activities in intact liver cells of embryonic chick. Biochem. Biophys. Res. Commun. 54, 100–107.

Aronson, N. N. (1975) Hepatocyte sialoglycoprotein. Nature 253, 380.

Atkinson, P. H. and Summers, D. F. (1971) Purification and properties of HeLa cell plasma membranes. J. Biol. Chem. 246, 5162–5175.

Baynes, J. W., Hsu, A.-F. and Heath, E. C. (1973) The role of mannosyl-phosphoryl-dihydropolyisoprenol in the synthesis of mammalian glycoproteins. J. Biol. Chem. 248, 5693–5704.

Behrens, N. H. and Leloir, L. F. (1970) Dolichol monophosphate glucose: An intermediate in glucose transfer in liver. Proc. Nat. Acad. Sci. U.S.A. 66, 153–159.

Behrens, N. H., Parodi, A. J. and Leloir, L. F. (1971) Glucose transfer from dolichol monophosphate glucose: The product formed with endogenous microsomal acceptor. Proc. Nat. Acad. Sci. U.S.A. 68, 2857–2860.

Behrens, N. H., Carminatti, H., Staneloni, R. J., Leloir, L. F. and Cantarella, A. I. (1973) Formation of lipid-bound oligosaccharides containing mannose. Their role in glycoprotein synthesis. Proc. Nat. Acad. Sci. U.S.A. 70, 3390–3394.

Benedetti, E. L. and Emmelot, P. (1967) Studies on plasma membranes, IV. The ultrastructural localization and content of sialic acid in plasma membranes isolated from rat liver and hepatoma. J. Cell Sci. 2, 499–512.

Bennett, G. (1970) Migration of glycoprotein from Golgi apparatus to cell coat in the columnar cells of the duodenal epithelium. J. Cell Biol. 45, 668–673.

Bennett, G., Leblond, C. P. and Haddad, A. (1974) Migration of glycoprotein from the Golgi apparatus to the surface of various cell types as shown by radioautography after labeled fucose injection into rats. J. Cell Biol. 60, 258–284.

Blix, G., Gottschalk, A. and Klenk, E. (1957) Proposed nomenclature in the field of neuraminic and sialic acids. Nature 179, 1088.

Bosmann, H. B. (1972) Platelet adhesiveness and aggregation 11. Surface sialic acid, glycoprotein: N-acetylneuraminic acid transferase, and neuraminidase of human blood platelets. Biochim. Biophys. Acta 279, 456–474.

Bosmann, H. B. and Winston, R. A. (1970) Synthesis of glycoprotein, glycolipid, protein, and lipid in synchronized L5178Y cells. J. Cell Biol. 45, 23–33.

Bosmann, H. B., Hagopian, A. and Eylar, E. H. (1968a) Cellular membranes: The isolation and characterization of the plasma and smooth membranes of HeLa cells. Arch. Biochem. Biophys. 128, 51–69.

Bosmann, H. B., Hagopian, A. and Eylar, E. H. (1968b) Glycoprotein biosynthesis: The characterization of two glycoprotein: fucosyl transferases in HeLa cells. Arch. Biochem. Biophys. 128, 470–481.

Bosmann, H. B., Hagopian, A. and Eylar, E. H. (1969) Cellular membranes: The biosynthesis of glycoprotein and glycolipid in HeLa cell membranes. Arch. Biochem. Biophys. 130, 573–583.

Bosmann, H. B., Case, K. R. and Morgan, H. R. (1974) Surface biochemical changes accompanying primary infection with Rous sarcoma virus I. Electrokinetic properties of cells and cell surface glycoprotein: glycosyl transferase activities. Exp. Cell Res. 83, 15–24.

Burger, M. M. and Goldberg, A. R. (1967) Identification of a tumor-specific determinant on neoplastic cell surfaces. Proc. Nat. Acad. Sci. U.S.A. 57, 359–366.

Burgos, J., Hemming, F. W., Pennock, J. F. and Morton, R. A. (1963) Dolichol: a naturally-occurring C 100 isoprenoid alcohol. Biochem. J. 88, 470–482.

Caccam, J. F., and Eylar, E. H. (1970) Glycoprotein biosynthesis: purification and characterisation of a glycoprotein: galactosyl transferase from Ehrlich ascites tumor cell membranes. Arch. Biochem. Biophys. 137, 315–324.

132

Caccam, J. F., Jackson, J. J. and Eylar, E. H. (1969) The biosynthesis of mannose-containing glycoproteins: A possible lipid intermediate Biochem. Biophys. Res. Commun. 4, 505–511.

Coffey, J. W., Miller, O. N. and Sellinger, O. Z. (1964) The metabolism of L-fucose in the rat. J. Biol. Chem. 239, 4011–4017.

Cook, G. M. W. (1971) Membrane structure and function. Ann. Rev. Plant Physiol. 22, 97–120.

Cook, G. M. W. (1973) The Golgi apparatus: Form and function. In: Lysosomes in Biology and Pathology (Dingle, J. T., ed.), pp. 237–277, Elsevier North-Holland, Amsterdam.

Cook, G. M. W. and Eylar, E. H. (1965) Separation of the M and N blood-group antigens of the human erythrocyte. Biochim. Biophys. Acta 101, 57–66.

Cook, G. M. W. and Stoddart, R. W. (1973) Surface Carbohydrates of the Eukaryotic Cell, 346 pp., Academic Press, London.

Cook, G. M. W., Heard, D. H. and Seaman, G. V. F. (1960) A sialomucopeptide liberated by trypsin from the human erythrocyte. Nature 188, 1011–1012.

Cook, G. M. W., Heard, D. H. and Seaman, G. V. F. (1962) The electrokinetic characterization of the Ehrlich ascites carcinoma cell. Exp. Cell Res. 28, 27–38.

Cook, G. M. W., Laico, M. T. and Eylar, E. H. (1965) Biosynthesis of glycoproteins of the Ehrlich ascites carcinoma cell membranes. Proc. Nat. Acad. Sci. U.S.A. 54, 247–252.

Dallner, G., Siekevitz, P. and Palade, G. E. (1966a) Biogenesis of endoplasmic reticulum membranes. I. Structural and chemical differentiation in developing rat hepatocyte. J. Cell Biol. 30, 73–96.

Dallner, G., Siekevitz, P. and Palade, G. E. (1966b). Biogenesis of endoplasmic reticulum membranes. II. Synthesis of constitutive microsomal enzymes in developing rat hepatocyte. J. Cell Biol. 30, 97–117.

Dallner, G., Behrens, N. H., Parodi, A. J. and Leloir, L. F. (1972) Subcellular distribution of dolichol phosphate. FEBS Lett. 24, 315–317.

DeLuca, L., Maestri, N., Rosso, G. and Wolf, G. (1973) Retinol glycolipids. J. Biol. Chem. 248, 641–648.

Deppert, W., Werchau, H. and Walter, G. (1974) Differentiation between intracellular and cell surface glycosyl transferases: Galactosyl transferase activity in intact cells and in cell homogenate. Proc. Nat. Acad. Sci. U.S.A. 71, 3068–3072.

Evans, P. J. and Hemming, F. W. (1973) The unambiguous characterization of dolichol phosphate mannose as a product of mannosyl transferase in pig liver endoplasmic reticulum. FEBS Lett. 31, 335–338.

Evans, W. H. (1974) Nucleotide pyrophosphatase, a sialoglycoprotein located on the hepatocyte surface. Nature 250, 391–394.

Evans, W. H. and Gurd, J. W. (1971) Biosynthesis of liver membranes. Incorporation of [³H] leucine into proteins and of [¹⁴C] glucosamine into proteins and lipids of liver microsomal and plasma-membrane fractions. Biochem. J. 125, 615–624.

Evans, W. H., Hood, D. O. and Gurd, J. W. (1973) Purification and properties of a mouse liver plasma-membrane glycoprotein hydrolysing nucleotide pyrophosphate and phosphodiester bonds. Biochem. J. 135, 819–826.

Eylar, E. H. (1965) On the biological role of glycoproteins. J. Theoret. Biol. 10, 89–113.

Eylar, E. H. and Cook, G. M. W. (1965) The cell-free biosynthesis of the glycoprotein of membranes from Ehrlich ascites carcinoma cells. Proc. Nat. Acad. Sci. U.S.A. 54, 1678–1685.

Gasic, G. and Gasic, T. (1963) Origin of the surface sialomucins in free tumor cells. Proc. Am. Assoc. Cancer Res. 4, 22.

Gurd, J. W. and Evans, W. H. (1973) Relative rates of degradation of mouse-liver surface-membrane proteins. Eur. J. Biochem. 36, 273–279.

Haddad, A., Smith, M. D., Herscovics, A., Nadler, N. J. and Leblond, C. P. (1971) Radioautographic study of in vivo and in vitro incorporation of fucose-³H into thyroglobulin by rat thyroid follicular cells. J. Cell Biol. 49, 856–878.

Hagopian, A., Bosmann, H. B. and Eylar, E. H. (1968) Glycoprotein biosynthesis: The localization of polypeptidyl:N-acetylgalactosaminyl, collagen:glucosyl and glycoprotein:galactosyl transferases in HeLa cell membrane fractions. Arch. Biochem. Biophys. 128, 387–396.

Harms, E. and Reutter, W. (1974) Half-life of N-acetylneuraminic acid in plasma membranes of rat liver and Morris hepatoma 7777. Cancer Res. 34, 3165–3172.

Helgeland, L. (1965) Incorporation of radioactive glucosamine into submicrosomal fractions isolated from rat liver. Biochim. Biophys. Acta 101, 106–112.

Helting, T. and Peterson, P. A. (1972) Galactosyltransfer in mouse mastocytoma: Synthesis of a galactose-containing polar metabolite of retinol. Biochem. Biophys. Res. Commun. 46, 429–436.

Higashi, Y., Strominger, J. L. and Sweeley, C. C. (1967) Structure of a lipid intermediate in cell wall peptidoglycan synthesis: A derivative of a C_{55} isoprenoid alcohol. Proc. Nat. Acad. Sci. U.S.A. 57, 1878–1884.

Hirano, H., Parkhouse, B., Nicolson, G. L., Lennox, E. S., and Singer, S. J. (1972) Distribution of saccharide residues on membrane fragments from a myeloma-cell homogenate: Its implications for membrane biogenesis. Proc. Nat. Acad. Sci. U.S.A. 69, 2945–2949.

Hsu, A.-F., Baynes, J. W. and Heath, E. C. (1974) The role of a dolichol-oligosaccharide as an intermediate in glycoprotein biosynthesis. Proc. Nat. Acad. Sci. U.S.A. 71, 2391–2395.

Hudgin R. L., Murray, R. K., Pinteric, L., Morris, H. P. and Schachter, H. (1971) The use of nucleotide-sugar:glycoprotein glycosyltransferase to assess Golgi apparatus function in Morris hepatomas. Can. J. Biochem. 49, 61–70.

Hughes, R. C., Sanford, B. and Jeanloz, R. W. (1972) Regeneration of the surface glycoproteins of a transplantable mouse tumor cell after treatment with neuraminidase. Proc. Nat. Acad. Sci. U.S.A. 69, 942–945.

Ito, S. (1969) Structure and function of the glycocalyx. Fed. Proc. 28, 12–25.

Jackson, R. L., Segrest, J. P., Kahane, I. and Marchesi, V. T. (1973) Studies on the major sialoglycoprotein of the human red cell membrane. Isolation and characterization of tryptic glycopeptides. Biochemistry 12, 3131–3138.

Kaufman, R. L. and Ginsburg, V. (1968) The metabolism of L-fucose by HeLa cells. Exp. Cell Res. 50, 127–132.

Keenan, T. W. and Morré, D. J. (1975) Glycosyltransferases: Do they exist on the surface membrane of mammalian cells? FEBS Lett. 55, 8–13.

Kelus, A., Gurner, B. W. and Coombs, R. R. A. (1959) Blood group antigens on HeLa cells shown by mixed agglutination. Immunology 2, 262–267.

Kemp, R. B., Lloyd, C. W. and Cook, G. M. W. (1973) Glycoproteins in cell adhesion. In: Progress in Surface and Membrane Science (Danielli, J. F., Rosenberg, M.D. and Cadenhead, D. A., eds.), pp. 271–318, Academic Press, New York.

Kindig, D. A. and Kirsten, W. H. (1967) Virus-like particles in established murine cell lines: Electron-microscopic observations. Science 155, 1543–1545.

Kornfeld, R. and Kornfeld, S. (1970) The structure of a phytohemagglutinn receptor site from human erythrocytes. J. Biol. Chem. 245, 2536–2545.

Kornfeld, S. and Ginsberg, V. (1966) The metabolism of glucosamine by tissue culture cells. Exp. Cell Res. 41, 592–600.

Kornfeld, S. and Kornfeld, R. (1969) Solubilization and partial characterisation of a phytohemagglutinin receptor site from human erythrocytes. Proc. Nat. Acad. Sci. U.S.A. 63, 1439–1446.

Kraemer, P. M. (1967) Regeneration of sialic acid on the surface of Chinese hamster cells in culture. II. Incorporation of radioactivity from glucosamine-1-^{14}C. J. Cell Physiol. 69, 23–34.

Kraemer, P. M. (1970) Complex carbohydrates of animal cells: Biochemistry and physiology of the cell periphery. In: Biomembranes (Manson, L. A., ed.), pp. 67–190, Plenum Press, New York.

Langley, O. K. and Ambrose, E. J. (1964) Isolation of a mucopeptide from the surface of Ehrlich ascites tumour cells. Nature 204, 53–54.

Langley, O. K. and Ambrose, E. J. (1967) The linkage of sialic acid in the Ehrlich ascites-carcinoma cell surface membrane. Biochem. J. 102, 367–372.

Leloir, L. F. (1964) The biosynthesis of polysaccharides. In: Proceedings of the Plenary Sessions 6th International Congress of Biochemistry, pp. 15–29, Federation of American Societies for Experimental Biology, New York.

Leloir, L. F., Staneloni, R. J., Carminatti, H. and Behrens, N. H. (1973) The biosynthesis of a N,N'-diacetylchitobiose containing lipid by liver microsomes. A probable dolichol pyrophosphate derivative. Biochem. Biophys. Res. Commun. 52, 1285–1292.

Lennarz, W. J. (1975) Lipid linked sugars in glycoprotein synthesis. Science 188, 986–991.

Lloyd, C. W. and Cook, G. M. W. (1974) On the mechanism of the increased aggregation by neuraminidase of 16C malignant rat dermal fibroblasts *in vitro*. J. Cell Sci. 15, 575–590.

Lucas, J. J., Waechter, C. J. and Lennarz, W. J. (1975) The participation of lipid-linked oligosaccharide in synthesis of membrane glycoproteins. J. Biol. Chem. 250, 1992–2002.

Mäkelä, O., Miettinen, T. and Presola, R. (1960) Release of sialic acid and carbohydrates from human red cells by trypsin treatment. Vox Sang. 5, 492–496.

Marchesi, V. T., Tillack, T. W., Jackson, R. L., Segrest, J. P. and Scott, R. E. (1972) Chemical characterization and surface orientation of the major glycoprotein of the human erythrocyte membrane. Proc. Nat. Acad. Sci. U.S.A. 69, 1445–1449.

Marcus, P. I. and Schwartz, V. G. (1968) Monitoring molecules of the plasma membrane: Renewal of sialic acid-terminating receptors. In: Biological Properties of the Mammalian Surface Membrane (Manson, L. A., ed.), pp. 143–147, Wistar Institute Press, Philadelphia.

Martin, H. G. and Thoner, K. J. I. (1974a) Synthesis of radioactive dolichol from [4S -³H] mevalonate in the regenerating rat liver. Biochem. J. 138, 277–280.

Martin, H. G. and Thorne, K. J. I. (1974b) The involvement of endogenous dolichol in the formation of lipid-linked precursors of glycoprotein in rat liver. Biochem. J. 138, 281–289.

McLean, R. J. and Bosmann, H. B. (1975) Cell-cell interactions: Enhancement of glycosyl transferase ectoenzyme systems during *Chlamydamonas* gametic contact. Proc. Nat. Acad. Sci. U.S.A. 72, 310–313.

Molnar, J., Robinson, G. B. and Winzler, R. J. (1964) The biosynthesis of glycoproteins III. Glucosamine intermediates in plasma glycoprotein synthesis in livers of puromycin-treated rats. J. Biol. Chem. 239, 3157–3162.

Molnar, J., Lutes, R. A. and Winzler, R. J. (1965a) The biosynthesis of glycoproteins. V. incorporation of glucosamine-1-¹⁴C into macromolequles by Ehrlich ascites carcinoma cells. Cancer Res. 25, 1438–1445.

Molnar, J., Teegarden, D. W. and Winzler, R. J. (1965b) The biosynthesis of glycoproteins. VI. production of extracellular radioactive macromolecules by Ehrlich ascites carcinoma cells during incubation with glucosamine-¹⁴C. Cancer Res. 25, 1860–1866.

Molnar, J., Chao, H. and Markovic, G. (1969) Subcellular site of structural glycoprotein synthesis in Ehrlich ascites tumor. Arch. Biochem. Biophys. 134, 533–538.

Molnar, J., Chao, H. and Ikehara, Y. (1971) Phosphoryl-N-acetylglucosamine transfer to a lipid acceptor of liver microsomal preparations. Biochim. Biophys. Acta. 239, 401–410.

Muramatsu, T., Koide, N. and Ogata-Arakawa, M. (1975) Analysis of oligomannosyl cores of cellular glycopeptides by digestion with endo-β-N-acetylglucosaminidases. Biochem. Biophys. Res. Commun. 66, 881–888.

Neutra, M. and Leblond, C. P. (1966) Synthesis of the carbohydrate of mucus in the Golgi complex as shown by electron microscope radioautography of goblet cells from rats injected with glucose-H³. J. Cell Biol. 30, 119–136.

Nicolson, G. L. and Singer, S. J. (1971) Ferritin-conjugated plant agglutinins as specific saccharide stains for electron microscopy: Application to saccharides bound to cell membranes. Proc. Nat. Acad. Sci. U.S.A. 68, 942–945.

Novikoff, A. B. and Goldfischer, S. (1961) Nucleosidediphosphatase activity in the Golgi apparatus and its usefulness for cytological studies. Proc. Nat. Acad. Sci. U.S.A. 47, 802–810.

Oliver, G. J. A. and Hemming, F. W. (1975) The transfer of mannose to dolichol diphosphate oligosaccharides in pig liver endoplasmic reticulum. Biochem. J. 152, 191–199.

Oliver, G. J. A., Harrison, J. and Hemming, F. W. (1975) The mannosylation of dolichol-diphosphate oligosaccharides in relation to the formation of oligosaccharides and glycoproteins in pig-liver endoplasmic reticulum. Eur. J. Biochem. 58, 223–229.

Omura, T., Siekevitz, P. and Palade, G. E. (1967) Turnover of constituents of the endoplasmic reticulum membranes of rat hepatocytes. J. Biol. Chem. 242, 2389–2396.

Osborn, M. J. and Weiner, I. M. (1967) Mechanism of biosynthesis of the lipopolysaccharide of *Salmonella*. Fed. Proc. 26, 70–76.

Osborn, M. J. and Weiner, I. M. (1968) Biosynthesis of a bacterial lipopolysaccharide. VI. mechanism of incorporation of abequose into the O-antigen of *Salmonella typhimurium*. J. Biol. Chem. 243, 2631–2639.

Ovtracht, L., Morré, D. J., Cheetham, R. D. and Mollenhauer, H. H. (1973) Subfractionation of Golgi apparatus from rat liver: Method and morphology. J. Microscop. 18, 87–102.

Parodi, A. J., Behrens, N. H., Leloir, L. F. and Dankert, M. (1972) Glucose transfer from dolichol monophosphate glucose the lipid moiety of the endogenous microsomal acceptor. Biochim. Biophys. Acta. 270, 529–536.

Parodi, A. J., Staneloni, R., Cantarella, A. I., Leloir, L. F., Behrens, N. H., Carminatti, H. and Levy, J. A. (1973) Further studies on a glycolipid formed from dolichyl-D-glucose monophosphate. Carbohydrate Res. 26, 393–400.

Patt, L. M. and Grimes, W. J. (1974) Cell surface glycolipid and glycoprotein glycosyltransferases of normal and transformed cells. J. Biol. Chem. 249, 4157–4165.

Patt, L. M. and Grimes, W. J. (1975) Ectoglycosyltransferase activity in suspensions and monolayers of cultured fibroblasts. Biochem. Biophys. Res. Commun. 67, 483–490.

Pepper, D. S. and Jamieson, G. A. (1969) Studies on glycoproteins. III. Isolation of sialoglycopeptides from human platelet membranes. Biochemistry 8, 3362–3369.

Porter, C. W. and Bernacki, R. J. (1975a) Ultrastructural evidence for ectoglycosyltransferase systems. Nature 256, 648–650.

Porter, C. W. and Bernacki, R. J. (1975b) EM autoradiographic evidence for plasma membrane glycosyltransferase activity. J. Cell Biol. 67, 341a.

Poste, G. and Nicolson, G. L., eds. (1977) Dynamic Aspects of Cell Surface Organization, Cell surface Reviews, vol. 3. Elsevier/North-Holland, Amsterdam.

Reith, A., Oftebro, R. and Seljelid, R. (1970). Incorporation of ^{3}H-glucosamine in HeLa cells as revealed by light and electron microscopic autoradiography. Exp. Cell Res. 59, 167–170.

Richards, J. B. and Hemming, F. W. (1972) The transfer of mannose from guanosine diphosphate mannose to dolichol phosphate and protein by pig liver endoplasmic reticulum. Biochem. J. 130, 77–93.

Rieber, M., Bacaloa, J. and Alonso, G. (1975) Turnover of high-molecular-weight cell surface proteins during growth and expression of malignant transformation. Cancer Res. 35, 2104–2108.

Roseman, S. (1970) The synthesis of complex carbohydrates by multiglycosyltransferase systems and their potential function in intercellular adhesion. Chem. Phys. Lipids. 5, 270–297.

Rosso, G. C., DeLuca, L., Warren, C. D. and Wolf, G. (1975) Enzymatic synthesis of mannosyl retinyl phosphate from retinyl phosphate and guanosine diphosphate mannose. J. Lipid Res. 16, 235–243.

Roth, S. and White, D. (1972) Intercellular contact and cell-surface galactosyl transferase activity. Proc. Nat. Acad. Sci. U.S.A. 69, 485–489.

Roth, S., McGuire, E. J. and Roseman, S. (1971) Evidence for cell-surface glycosyltransferases. Their potential role in cellular recognition. J. Cell Biol. 51, 536–547.

Schachter, H. (1974) The subcellular sites of glycosylation. Biochem. Soc. Symp. 40, 50–71.

Schachter, H. and Roden, L. (1973) The biosynthesis of animal glycoproteins. In: Metabolic Conjugation and Metabolic Hydrolysis (Fishman, W. H., ed.), vol. 3, pp. 1–149. Academic Press, New York.

Scher, M., Lennarz, W. J. and Sweeley, C. C. (1968) The biosynthesis of mannosyl-l-phosphoryl-polyisoprenol in *Micrococcus lysodeikticus* and its role in mannan synthesis. Nat. Acad. Sci. U.S.A. 59, 1313–1320.

Shen, L. and Ginsburg, V.(1968) Release of sugars from HeLa cells by trypsin. In: Biological Properties of the Mammalian Surface Membrane (Manson, L. A., ed.), pp. 67–71, Wistar Institute Press, Philadelphia.

Shier, W. T. (1971) Preparation of a "chemical vaccine" against tumor progression. Proc. Nat. Acad. Sci. U.S.A. 68, 2078–2082.

Shur, B. D. and Roth, S. (1975) Cell surface glycosyltransferases. Biochim. Biophys. Acta 415, 473–512.

Sinohara, H. and Sky-Peck, H. H. (1965) Soluble ribonucleic acid and glycoprotein biosynthesis in the mouse liver. Biochim. Biophys. Acta. 101, 90–96.

Struck, D. and Lennarz, W. J. (1975) Synthesis of lipid-linked sugars and glycoproteins in oviduct cell preparations. Fed. Proc. 34, 678.

Thomas, D. B. and Winzler, R. J. (1969a) Oligosaccharides from M-active sialoglycopeptides of human erythrocytes. Biochem. Biophys. Res. Commun. 35, 811–818.

Thomas, D. B. and Winzler, R. J. (1969b) Structural studies on human erythrocyte glycoproteins. Alkali-labile oligosaccharides. J. Biol. Chem. 244, 5943–5946.

Tomita, M. and Marchesi, V. T. (1975) Amino-acid sequence and oligosaccharide attachment sites of human erythrocyte glycophorin. Proc. Nat. Acad. Sci. U.S.A. 72, 2964–2968.

Toyoshima, S., Fukada, M. and Osawa, T. (1972) Chemical nature of the receptor site for various phytomitogens. Biochemistry 11, 4000–4005.

Uhlenbruck, G. (1961) Action of proteolytic enzymes on the human erythrocyte surface. Nature 190, 181.

Vessey, D. A. and Zakin, D. (1975) Characterization of the reaction of GDP-mannose with dolichol phosphate in liver membranes. Eur. J. Biochem. 53, 499–504.

Visser, A. and Emmelot, P. (1973) Studies on plasma membranes XX. Sialidase in hepatic plasma membranes. J. Membrane Biol. 14, 73–84.

Waechter, C. J., Lucas, J. J. and Lennarz, W. J. (1973) Membrane glycoproteins. 1. Enzymatic synthesis of mannosyl phosphoryl polyisoprenol and its role as a mannosyl donor in glycoprotein synthesis. J. Biol. Chem. 248, 7570–7579.

Wallach, D. F. H. and Eylar, E. H. (1961) Sialic acid in the cellular membranes of Ehrlich ascites-carcinoma cells. Biochim. Biophys. Acta. 52, 594–596.

Warren, L. (1966) The biosynthesis and metabolism of amino sugars and amino sugar-containing compounds. In: Glycoproteins, Their Composition, Structure and Function (Gottschalk, A., ed.), B.B.A. Library, vol. 5, pp. 570–593, Elsevier/North-Holland, Amsterdam.

Warren, L. (1969) The biological significance of turnover of the surface membrane of animal cells. In: Current Topics in Developmental Biology (Moscona, A. A. and Monroy, A., eds.), vol. 6, pp. 197–222, Academic Press, New York.

Warren, L. and Glick, M. C. (1968) Membranes of animal cells. II. The metabolism and turnover of the surface membrane. J. Cell Biol. 37, 729–746.

Webb, G. C. and Roth, S. (1974) Cell contact dependence of surface galactosyltransferase activity as a function of the cell cycle. J. Cell Biol. 63, 796–705.

Whaley, W. G. (1975) The Golgi apparatus. Cell Biology Monographs: Continuation of Protoplasmatologia (Alfert, M., Beerman, W., Rudkin, G., Sandritter, W. and Sitte, P., eds.), vol. 2, 190 pp. Springer-Verlag, Vienna and New York.

Whaley, W. G., Dauwalder, M. and Kephart, J. E. (1972) Golgi apparatus: Influence on cell surfaces. A role in the assembly of macromolecules makes the organelle a determinant of cell function. Science 175, 596–599.

Whur, P., Herscovics, A. and Leblond, C. P. (1969) Radioautographic visualization of the incorporation of galactose-^{3}H and mannose -^{3}H by rat thyroids in vitro in relation to the stages of thyroglobulin synthesis. J. Cell Biol. 43, 289–311.

Winzler, R. J., Harris, E. D., Pekas, D. J., Johnson, C. A. and Weber, P. (1967) Studies on glycopeptides released by trypsin from intact human erythrocytes. Biochemistry 6, 2195–2202.

Wright, A., Dankert, M., Fennessey, P. and Robbins, P. W. (1967) Characterization of a polyisoprenoid compound functional in O-antigen biosynthesis. Proc. Nat. Acad. Sci. U.S.A. 57, 1798–1803.

Yogeeswaran, G., Laine, R. A. and Hakomori, S. (1974) Mechanism of cell contact-dependent glycolipid synthesis: Further studies with glycolipid-glass complex. Biochem. Biophys. Res. Commun. 59, 591–599.

Yogeeswaran, G., Laine, R. A. and Hakomori, S. (1975) Participation of retinol phosphate sugar glycolipid in glycosylation of glycosphingolipid-glass complex on cell contact. Fed. Proc. 34, 645.

Turnover of proteins of the eukaryotic cell surface

3

John TWETO and Darrell DOYLE

1. Introduction

Several recent reviews have covered much of the general literature on the degradation of cellular components (Goldberg and Dice, 1974; Schimke and Doyle, 1970; Schimke, 1970; Schimke, 1973). In addition, other reviews have dealt specifically with degradation of membrane proteins (Siekevitz, 1972; Taylor et al., 1973; Schimke, 1975). Therefore, we have chosen to restrict this review to those papers that deal with aspects of biogenesis and degradation of the plasma membrane proteins of eukaryotic cells. Such a limited discussion is appropriate in view of the considerable amount of recent work involving various aspects of plasma membrane protein turnover, much of which is often contradictory. We will attempt to resolve these contradictions in favor of a model for turnover of the plasma membrane which proposes that the proteins are degraded as structural units. Although this model implies that the nonprotein constituents of the membrane are similarly degraded, we will not attempt to evaluate the data dealing with these constituents.

The importance of protein degradation in the control of protein levels has been documented extensively (Schimke and Doyle, 1970; Schimke, 1970; Schimke, 1973) and the theoretical (Buchanan, 1961; Reiner, 1953; Koch, 1962) and practical (Schimke, 1970, 1973; Schimke and Doyle, 1970; Goldstein and Brown, 1974; Doyle and Tweto, 1975) aspects of the measurement of degradation have been described. Much of this same background information can be applied to the study of membrane protein turnover. We will therefore not discuss methodology except when necessary to interpret various experiments. We believe that an understanding of the general features of membrane degradation will provide important insight into the complex process of membrane biogenesis.

2. Problems involved in the study of membrane proteins

Before proceeding with a critique of the work on membrane turnover, it will be useful to mention some of the properties of membranes that make their study

G. Poste and G.L. Nicolson (eds.) *The Synthesis, Assembly and Turnover of Cell Surface Components*
© Elsevier/North-Holland Biomedical Press, 1977.

especially difficult. One obvious difference between membrane proteins and cytoplasmic proteins is solubility. Membrane proteins exist within or partially within a hydrophobic matrix, some apparently moving about quite freely within the plane of the membrane but presumably restricted from the cytosol (Singer and Nicolson, 1972; Singer, 1974a; Singer, 1974b; Nicolson, 1976; Nicolson et al., 1977). The hydrophobic nature of these proteins makes it particularly difficult to imagine how they could exist free for any length of time in the cytosol, such as during their synthesis or degradation (see section 3).

The hydrophobic nature of membrane proteins makes them particularly difficult to purify. Consequently, studies on the degradation of membrane proteins have been done with complex mixtures of proteins partially resolved on SDS acrylamide gels or by indirect methods. These types of analyses are less satisfactory than studies done with proteins purified to homogeneity.

A second property of membrane proteins not observed with soluble proteins is that they have a definite organization within the lipid bilayer. Membranes exhibit a sidedness, with some of their protein constituents partially exposed only on one side of the bilayer, some on the other, and others exposed on both sides (Hirano et al., 1972; Kreibich et al., 1974; Nicolson and Singer, 1974; Steck, 1974; Hunt and Brown, 1975). This property has proved to be extremely helpful in the study of the turnover of some plasma membrane proteins since it allows identification of only those proteins exposed on the external surface of the cell; e.g., by specific labeling using vectorial reagents such as lactoperoxidase-catalyzed iodination. For practical reasons this approach to the study of membrane protein turnover has been limited to cultured cells. The topological organization of membrane proteins places restrictions on models of biogenesis and presumably degradation, since proteins present on one side of the bilayer do not appear to be able to switch sides at any appreciable rate (Edidin, 1974; Nicolson and Singer, 1974; Cherry, 1975). Any model describing the process of membrane biogenesis must satisfy the topological requirements of membrane proteins.

A third property to remember in attempting to understand studies involving membrane proteins is that the various types of membranes are defined most clearly by morphological criteria. Thus, the plasma membrane is best described as that membrane which defines the cell periphery. When the plasma membrane is ruptured by homogenization, many of the characteristics of this membrane that distinguish it from other cell membranes are lost. Consequently, it is difficult to determine whether a preparation of plasma membrane is homogeneous. One must presume that studies involving the degradation of plasma membrane proteins are all complicated to a greater or lesser extent by the presence in the plasma membrane fraction of Golgi, lysosomal, and endoplasmic reticulum membranes. Soluble proteins may also be present if there is extensive vesicle formation during membrane preparation, since soluble proteins may be trapped within the vesicles. Use of cultured cells may avoid these problems to some extent, since it is possible to isolate the plasma membranes as intact ghosts. Thus, studies using this type of membrane preparation may be relatively more reliable.

The problem of purity of the plasma membrane fraction is further complicated because there is no clear criterion for what constitutes a membrane protein. A protein may be peripherally associated with the membrane and copurify with it, but may not be characteristic of integral membrane proteins in terms of biogenesis or degradation. For the purposes of this review, membrane-associated cytoskeletal elements such as actin or tubulin, (Poste et al., 1975; Yahara and Edelman, 1975; Nicolson et al., 1977) and secretory proteins such as various glycoproteins involved in cell-substrate adherence of monolayer cultures (Revel and Wolker, 1973) will be considered essentially nonmembrane. If a protein in the process of being secreted can become associated with the membrane, it is obvious that measurements of overall membrane degradation will be inaccurate to the extent that label in this secretory protein contributes to the total label of the plasma membrane fraction. The presence of secretory proteins in the plasma membrane fraction can be a significant problem when membrane glycoprotein turnover is measured, since it has often been observed that liver and cultured cells secrete glycoproteins into the medium (Hughes et al., 1972; Vitetta an Uhr, 1972; Wilson et al., 1972; Riodan et al., 1974; Schmidt-Ullrich et al., 1974; Yu and Cohen, 1974; Kaplan and Moskowitz, 1975b).

One of the most interesting properties of the plasma membrane is that it is a dynamic organelle. Not only are the membrane proteins and lipids capable of rapid movement within the plane of the membrane (Singer and Nicolson, 1972; Edidin, 1974; Singer, 1974a; Singer, 1974b; Cherry, 1975; Nicolson, 1976; Nicolson et al., 1977), but more important for the present discussion, they appear to be undergoing removal from the surface at prodigious rates. Recent studies by Steinman et al. (1976) suggest that the process of pinocytosis in macrophages and mouse L cells results in the interiorization of an amount of membrane equal to the entire cell surface within 2 hours. In experiments with normal growing cells, they have shown that the internalized pinocytotic vesicles rapidly fuse with lysosomes. The total surface area of the lysosomes and the vesicles, however, remains at a steady-state level which is much smaller than that of the surface area ingested. The fact that interiorization of membrane is much faster than membrane degradation (see page 153) implies that this membrane is recycled. Thus, studies directed at the surface of the cell may be examining only a portion of the life history of the plasma membrane. For instance, plasma membrane degradation may occur by some side process associated with membrane recycling. Before this subject can be discussed, however, it is necessary to describe the two conflicting models of membrane degradation and the data supporting them.

3. Models for membrane degradation

Work on the degradation of plasma membrane proteins has been done primarily with rodent liver in vivo, and with a large variety of cultured cells both in suspension and in monolayer. Although similar techniques have been employed in all studies, the data have been interpreted in two conflicting ways. Schimke and coworkers (Taylor et al., 1973; Schimke, 1975) have concluded from turnover studies of rat liver membranes that proteins of the endoplasmic reticulum and the plasma membrane are degraded with different half-lives and that a correlation exists between the half-life of a membrane protein and its subunit molecular weight, with larger molecules being degraded faster. The importance of this observation is that it implies that the individual protein components of the membrane must dissociate and associate freely with the membrane matrix by some mechanism. While free in the cytoplasm, the proteins are subject to degradation. The surviving molecules in the cytoplasmic pool mix with the newly synthesized molecules, maintaining a constant pool size and can then be inserted into the membrane. In this model, dissociation from the membrane cannot be the rate-limiting step in membrane protein degradation because both "old" and newly synthesized membrane proteins must be degraded from the same pool to maintain first order kinetics. Given our present understanding of membrane structure, the greatest difficulty with this model is in conceiving how hydrophobic membrane proteins can exist free for any length of time in the cytoplasm without being irreversibly denatured, and how the hydrophilic structure exposed on the external surface of the cell can be moved or rotated through the lipid bilayer. This latter process is especially difficult to understand in membrane glycoproteins.

An alternative mechanism for membrane protein turnover is suggested by recent data from the authors' laboratory (Tweto and Doyle, 1976) as well as that of Cohn (Hubbard and Cohn, 1975b) and Roberts (1974, 1975b). These authors have concluded that the half-lives of a large number of different plasma membrane proteins are very similar or the same, and therefore that the membrane may be degraded as a unit. This model suggests that the proteins of the membrane must be inserted into the membrane in coordinate fashion similar to the membrane flow hypothesis of Palade (1959). The chief difficulty with this model is to reconcile it with the data supporting the heterogeneous turnover model.

3.1. Experimental evidence for heterogeneous degradation

3.1.1. Direct measurements
The double isotope technique devised by Arias et al. (1969) is commonly used to study the turnover of various membrane fractions. This technique, as originally applied, involves administration of a labeled amino acid, [^{14}C]leucine, to animals followed after 3 to 4 days or longer by a second administration of [^{3}H]-labeled amino acid. After a brief interval (3–4 hr.), the animals are killed, the tissue fractionated, and the proteins separated, usually by sodium dodecylsulfate (SDS) polyacrylamide gel electrophoresis. The ratio of ^{3}H/^{14}C is then determined in the separated fractions. The amount of ^{14}C remaining in any cell fraction or individual protein is a reflection of the effects of both synthesis and degradation occurring in the period between the two isotope administrations. The amount of ^{3}H present in the sample is a reflection primarily of synthesis because of the

short period between injection and death. This technique allows the determination of two points on a decay curve. ^{3}H is a measure of radioactivity incorporated into each fraction at t=0, and ^{14}C is the amount of radioactivity remaining after 3 to 4 days or some other interval. A high value of the ratio indicates relatively rapid degradation, and conversely, a low ratio shows relatively slow degradation.

Using the double isotope method Arias et al. (1969) found that rat liver microsomal proteins resolved by DEAE cellulose chromatography exhibited quite different ratios, indicating different half-lives for the separated membrane proteins of this organelle. Dehlinger and Schimke (1971) confirmed and extended this result by looking at the ratios of microsomal, soluble, and plasma membrane proteins separated by SDS acrylamide gel electrophoresis. After administration of isotopes as described above and separation of the proteins by electrophoresis, the gels were serially sliced into many sections and the isotope ratio of each was determined. Heterogeneity was noted, manifested primarily as a decline in isotope ratio from the top of the gel to the bottom. Since migration distance in SDS acrylamide gels is a measure of polypeptide molecular weight (Weber and Osborn, 1969; Dunker and Rueckert, 1969), these results indicated that proteins of the different cell fractions have different half-lives and, interestingly, that large polypeptides are generally degraded faster than small ones. The trend of decreasing ratio with increasing mobility, while present, was noticeably less conspicuous for the plasma membrane fraction as compared with the cytoplasmic and endoplasmic reticulum fractions.

Other in vivo double labeling studies have led various authors to similar conclusions. Gurd and Evans (1973) examined the relative degradation rates of polypeptides from mouse liver plasma membranes separated by Sephadex chromatography in the presence of SDS. The heterogeneity observed in the isotope ratios was limited solely to a trend of increasing ratio with increased retention volume or smaller size, since the isotope ratio in these experiments was reversed. This trend was slight compared with the data of Dehlinger and Schimke, which may reflect the longer time (8 hr.) between the second isotope administration and sacrifice. Gurd and Evans also attempted to subfractionate the plasma membrane into various solubility classes. Upon separation of the proteins of these classes by Sephadex chromatography, little heterogeneity was observed except for a class enriched for glycoproteins. While the heterogeneity found was not pronounced, these authors concluded that there was a correlation between subunit size and degradation rate. In another study, the turnover of rat liver plasma membrane glycoproteins was examined by double label methods using either glucosamine or fucose, and was compared to that of leucine-labeled membrane proteins (Landry and Marceau, 1975). After separating the glycoproteins on SDS acrylamide gels, a marked heterogeneity in isotope ratio was found with each precursor. This heterogeneity occurred primarily in nearby or adjacent gel slices, and any correlation of isotope ratio with migration distance was ambiguous. However, this study concluded that glycoproteins are not a special class with regard to their turnover, and that they are degraded in a way similar to that found by Dehlinger and Schimke for leucine-labeled membrane proteins. It is not clear from these studies, however, whether the protein backbone of glycoproteins is degraded at the same time as the labeled sugars.

In studies on the turnover of rat intestinal brush border membrane proteins, Alpers (1972) has found heterogeneity of isotope ratios and of the correlation between size and degradation rate constant described for the polypeptides of liver plasma membrane. However, the complexity of this particular system makes the results of the turnover study difficult to interpret. Not only are the isotope ratios in rat intestine affected by cell migration (Bennett et al., 1974) and extracellular proteases (Alpers and Tedesco, 1975), but the complex manner in which the experiments were performed makes it difficult to interpret the value of the ratios. Since the time between the first and second administrations of isotope was short (10 hr.) compared to the duration of the second pulse (8 hr.), very little degradation would have been expected, and very little was actually found. The control mean ratio 8 hours after simultaneous injection of both isotopes was similar to the experimental ratios of the separated proteins when the interval between injections was 10 hours.

Kaplan and Moskowitz (1975b) used the double label technique to examine the relative degradation rates of plasma membrane proteins from a cultured cell line of Rhesus monkey kidney cells (LLC-MK$_2$). Actively dividing cells were grown for 6 hours in the presence of [^{14}C]leucine, washed, and then grown for 60 hours in unlabeled medium. After this period the cells were grown for 6 hours in the presence of [^{3}H]leucine, harvested, and the isolated plasma membrane proteins separated on SDS polyacrylamide gels. They found a dramatic heterogeneity in the isotope ratios of the gel slices. Generally, the slower migrating polypeptides had higher ratios of ^{3}H to ^{14}C, suggesting shorter half-lives. Similar results were obtained when isotopes of D-glucosamine were used to label the membranes. The authors therefore concluded that in exponentially growing cultures plasma membrane proteins are degraded with heterogeneous rates. However, one of the basic assumptions of the double label technique is that the rate of synthesis for all the proteins examined is the same at the time both labels are administered (Arias et al., 1969). In the experiments of Kaplan and Moskowitz, the rates of synthesis may not have been the same at the time of the first administration and 60 hours later at the second administration since exponentially growing cultures are composed of cells in all stages of the cell cycle. The protein composition (Johnsen et al., 1975) and properties (Cikes and Friberg, 1971; Fox et al., 1971; Pasternak et al., 1971; Shoman and Sachs, 1972; Noonan et al., 1973; Revoltella et al., 1974; Johnsen et al., 1975) of the plasma membrane of various cultured cells may differ during the cell cycle. Cells in sparsely seeded cultures (first administration of precursor) may exhibit different growth characteristics than those in nearly confluent cultures (second administration of precursor). The data of Kaplan and Moskowitz may therefore reflect differences in membrane protein synthetic rate rather than degradation rate.

3.1.1.1. Alternative interpretation of direct measurement data   Other explanations of the data suggesting heterogeneity of turnover rates among rat liver plasma membrane proteins are possible. For example, cells other than hepatocytes, i.e.,

Kupffer cells, blood cells, connective tissue cells, and so on could contribute constituents to the plasma membrane fraction of liver. Also, preparations of plasma membrane from various tissues probably do not represent a homogeneous population of structures. Areas of the plasma membrane with apparently distinct functional capacities have been distinguished morphologically in liver. It might be expected that specialized regions such as the sinusoidal surface, bile cannicula, and the nexus would not behave similarly with regard to their turnover. For example, since exchange between blood and liver cells presumably occurs at the sinusoidal surface (Keik et al., 1970; Novikoff and Essner, 1960), it might be expected that the membrane proteins at this surface would be less metabolically stable than those of the nexus region (Jamieson and Palade, 1967a; Jamieson and Palade, 1967b; Jamieson and Palade, 1968a; Jamieson and Palade, 1968b; Glaumann and Ericsson, 1970; Amsterdam et al., 1971; Meldolesi, 1974).

Where subfractionation of the plasma membrane has been attempted (Evans and Gurd, 1972; Toda et al., 1975), an uneven distribution of membrane-bound enzymes has been observed. Thus the heterogeneity of degradation rate constants among proteins in preparations of liver plasma membranes may reflect homogeneous degradation of different specialized regions of the membrane. Some evidence exists to support this possibility. Simon et al. (1970) fractionated plasma membrane preparations from liver based on differential solubility in aqueous pyridine or K_2CO_3 solutions. After double labeling the animals they observed that the soluble fractions in both solvents had similar rates of turnover that were lower than those of the insoluble fraction and the unfractionated membrane. The insoluble fractions had similar isotope ratios that were higher than those of the total plasma membrane. The protein compositions of all fractions were similar as judged by gel electrophoresis. However, the isotope ratios of the separated proteins from each fraction were not measured to determine whether individual proteins were being degraded heterogeneously. Evans and Gurd (1972) have subfractionated mouse liver plasma membrane by similar methods, except that in addition to the pyridine extraction technique they employed a sarcosyl-tris extraction. After double label isotope administration, the various solubility classes of plasma membrane proteins were counted. Each class had distinct isotope ratios. Interestingly, a fraction enriched for tight junctions appeared to be turning over at a markedly slower rate than the total membrane or any other membrane subfraction including one that contained "vesicies," possibly corresponding to sinusoidal or cannilicular surfaces. These authors attempted to compare protein ratios of several of the solubility classes by Sephadex chromatography in the presence of SDS. Only one class showed significant heterogeneity of ratio and the subunit size-degradation correlation as found for total plasma membrane (see page 145). Since this class was composed of "vesicles," the heterogeneity could be explained by the presence of secretory proteins and/or exocytotic vesicle membranes of differing origins.

Contamination of the plasma membrane preparation with other cell fractions can yield artifactual heterogeneity of isotope ratios. For example, Dehlinger and Schimke (1971) found that their plasma membrane fraction contained 5.4% of the glucose-6-phosphatase activity found in the rough endoplasmic reticulum and 6.8% of that found in the smooth endoplasmic reticulum. The plasma membrane fraction also had 10% of the cytochrome oxidase activity found in the mitochondrial fraction. Several studies using both double isotope techniques (Omura et al., 1967; Arias et al., 1969; Dehlinger and Schimke, 1971; Dehlinger and Schimke, 1972) and measurement of degradation rates of individual enzymes (Omura et al., 1967; Bock et al., 1971; Kuriyama and Omura, 1971; Omura and Kuriyama, 1971; Edwards and Gould, 1972) have shown that proteins of the endoplasmic reticulum are degraded with different half-lives. Thus, contamination of the plasma membrane preparation with these fractions could result in an apparent heterogeneity of degradation rates of plasma membrane proteins especially if the half-life of all plasma membrane proteins is the same and long relative to that of contaminating heterogeneously degraded proteins.

Contamination of the plasma membrane preparations with soluble proteins could produce the appearance of heterogeneity in degradation rates and, more importantly, could explain the correlation of isotope ratio with apparent molecular weight. Cytoplasmic proteins show subunit size-degradation rate correlation (Dehlinger and Schimke, 1971; Glass and Doyle, 1972); thus it is important to determine whether the membrane preparation is contaminated with these proteins since the size-degradation rate correlation is the main evidence for the heterogeneous turnover model (Taylor et al., 1973; Schimke, 1975). The only reasonable way that a size-degradation rate correlation could be explained is that the proteins leave the membrane matrix before being degraded.

The conclusion that membrane protein degradation is related to subunit molecular weight is based entirely upon the results of SDS polyacrylamide gel electrophoresis or Sephadex chromatography in the presence of SDS. While it is clear that soluble proteins migrate according to molecular weight in such systems (Dunker and Rueckert, 1969; Weber and Osborn, 1969) this relationship has been questioned for membrane proteins. There is reason to suspect that membrane proteins, especially glycoproteins, may migrate anomalously on SDS acrylamide gels. Two membrane proteins that have been carefully examined, glycophorin (Grefrath and Reynolds, 1974) and rhodopsin (Frank and Rodbard, 1975), exhibit different molecular weights depending on the techniques used to measure them. Sedimentation equilibrium studies and amino acid analysis of glycophorin have indicated that this protein has a molecular weight of 31,000 (Jackson et al., 1973; Grefrath and Reynolds, 1974; Tomita and Marchesi, 1975). On SDS gels the apparent molecular weight of glycophorin dimers is 100,000 (Steck, 1974). Rhodopsin has an apparent molecular weight by SDS gel electrophoresis of 35–40,000, but analysis of its retardation coefficient yields the more precise estimate of 30–32,000. The discrepancy in molecular weight measurements of membrane proteins by SDS gel electrophoresis can be attributed to anomalous binding of the detergent and formation of aggregates (Grefrath and Reynolds, 1974; Clarke, 1975; Robinson and Tanford, 1975). The amount of detergent bound affects the Stoke's radius and charge density. Thus, the migration distance after electrophoresis of various membrane proteins may appear greater or smaller depending upon their individual affinities for detergent. Although this problem has not been shown to occur extensively, the suggested subunit correlation should be viewed as tenuous especially with respect to glycoproteins (Kaplan and Moskowitz, 1975b).

3.1.2. Indirect evidence

In the previous discussion, general labeling methods using amino acid or sugars as precursor were employed to examine the turnover of a large number of membrane polypeptides. Different methods have been used to study the disappearance or reappearance of several specific surface components, primarily "receptors," of various cultured cells. In these studies it is not clear how such changes in the turnover of these individual surface components reflect the turnover of the plasma membrane as a whole, since little or no effort was made to measure directly the degradation of the other proteins in the plasma membrane. These changes in "receptor" concentration are of interest because they are characterized by very rapid kinetics, total disappearance or reappearance occurring within a few hours after experimental perturbation. This time span is an order of magnitude faster than the overall degradation of the plasma membrane. For example, the plasma membrane proteins in rat liver show an average half-life of 2 to 3 days (Arias et al., 1969; Franke et al., 1971), in rat liver hepatoma cells (HTC cells) the value is 100 hours (Tweto and Doyle, 1976), in mouse L cells (Hubbard and Cohn, 1975) and HeLa cells (Proctor, 1974) it is 20 to 30 hours, in BHK cells more than 16 hours (Pearlstein and Waterfield, 1974), and in Mk-2 cells (Kaplan and Moskowitz, 1975a) 80 to 100 hours. The plasma membrane of the macrophage may have a relatively fast degradation rate of 7 to 12 hours (Nachman et al., 1971). The great difference in the degradation kinetics of individual surface components compared with the whole plasma membrane makes the data described in this section supportive of the heterogeneous turnover model.

Two clear examples exist of a rapid turnover in plasma membrane components. Turner et al. (1972) examined the reappearance of the histocompatibility antigen HLA-2 after its removal by papain treatment of lymphocytes and found complete regeneration of the antigen within 6 hours. Since the time course of reappearance is determined by k_d (the rate constant of degradation) (Berlin and Schimke, 1965), the result indicates a half-life of less than 6 hours for this surface antigen. Reappearance is inhibited completely by puromycin and partially by actinomycin D, suggesting that new synthesis rather than internal pools is responsible for the rapid regeneration of the antigen. Cook and his coworkers (Vaughan and Cook, 1972; Cook et al., 1975; Cook et al., 1976) have observed similar kinetics in studies of the Na^+, K^+ ATPase (adenosine triphosphatase) of HeLa cells. These authors found that after partial blockade of the cation transport system with ouabain, washed cells recovered alkali-dependent cation transport capacity within 3 to 6 hours. This recovery was puromycin-sensitive.

Treatment of TA3 cells (cultured after growth in mouse ascitic fluids) with neuraminidase (Hughes et al., 1972) increased their sensitivity to a cytotoxic factor in guinea pig serum. Resistance to this cytotoxic factor returned to normal with a half-life of about 10 hours. If this return is due to de novo biosynthesis of a surface protein and not simply to replacement of the sugar moiety, then some selective mechanism must exist for its replacement in the membrane. This may be the case, because measurements of the degradation of total glucosamine-labeled proteins (Hughes et al., 1972) in these cells indicate a half-life of 20 to 30 hours. This latter value may, of course, be high due to the effects of precursor reutilization.

Hynes (1973) has reported the presence of a readily iodinated, high-molecular-weight, transformation-sensitive glycoprotein on the surface of hamster fibroblasts (LETS protein). The turnover of a protein with these characteristics has been studied in three different cell lines. Using HT_3-KR cells transformed with a temperature-sensitive B77 virus, Rieber et al. (1975) have observed that a protein analogous to the LETS protein disappeared within 16 hours after culture at the permissive temperature. In a more quantitative study using BHK cells, Pearlstein and Waterfield (1974) found that in untransformed cells the half-life of this protein after iodination is approximately 16 hours. Their kinetic data indicate, however, that the degradation is biphasic with a rapidly turning over component ($t_{1/2} = 16$ hr.) and a slow component ($t_{1/2} >> 16$ hr.). The rapid loss of label following iodination is reminiscent of similar observations by Hubbard and Cohn (1975) and Tweto and Doyle (1976), and was attributed to handling during iodination. The possibility that the half-life of the large external glycoprotein is relatively long is further indicated by results of Yamada and Weston (1975) with chick embryo fibroblasts, who found a half-life of 45 hours for the LETS protein. Interpretation of the LETS protein degradation measurements is complicated because its presence depends upon the phase of the cell cycle (Hynes and Bye, 1974). In hamster fibroblasts, the protein appears to be at its highest level in G_1 and disappears in M, as detected by iodination. In these studies, it is not clear whether the protein actually disappears or becomes less accessible to the lactoperoxidase probe. In addition, this protein may be a peripheral membrane protein or associated with a special structure, and therefore its metabolism may be atypical compared to other surface integral proteins (Graham et al., 1975).

Degradation of low density lipoprotein (LDL) receptors on the surface of cultured human fibroblasts was examined by following the release into the medium of acid-soluble radioactivity from ^{125}I-LDL (Goldstein and Brown, 1974). While exact measurement of the half-life of the receptor was not attempted, the label was apparently degraded rapidly, only a few hours being required for the amount of label in the medium to double. It cannot be determined from these data whether the degradation of ^{125}I-LDL reflects degradation of the receptor, as opposed to recycling after internalization. However, Werb and Cohn (1971) reported that after removal of lipoprotein receptors from the macrophage membrane by trypsin, regeneration of the binding sites was complete after 7 hours in culture.

The possibility that cells can modulate the amount of certain surface receptors is also of interest to the present discussion. Gavin et al. (1974) have reported that the concentration of insulin receptors on the surface of lymphocytes decreases after chronic exposure to low levels of insulin. The effect is apparently rapid and depends upon the concentration of insulin used. After preincubation of the cells with 10^{-6}M insulin, the number of binding sites decreases by half in about 1 hour; with 10^{-8}M insulin, the time is 8 hours. While the disappearance of binding sites could be due to changes in both degradation and synthesis, the rapid loss from the surface could also occur by endocytosis without concomitant degradation. These authors also showed that complete regeneration to the control level of binding sites occurred within 16 hours after returning the cells to insulin-free medium. A similar observation has been made with the TL antigens of mouse ASL-1 leukemia cells. Yu and Cohen (1974) measured the amount of fucose-labeled antigen in cells incubated in the presence of anti-TL antisera or normal control mouse serum for 10 hours. After extraction of the antigen with detergent they found that cells incubated with anti-TL sera had only 20% of the amount of antigen found in cells treated with control serum. They also observed that the rate of disappearance of TL antigen labeled with fucose or ^{125}I was faster in cells incubated with anti-TL sera (modulated) versus cells incubated with normal mouse serum. The latter observations are complicated by the facts that the anti-TL antiserum was not monospecific and antigen was shed into the medium. Therefore, changes in concentration of the surface antigen may actually reflect shedding rather than internalization and degradation. Since the apparent degradation rate of other antigens was not changed, it is possible that overall membrane degradation is not altered during modulation.

Thyrotropin-releasing hormone apparently has a regulatory effect on the concentration of its receptor in rat pituitary GH_3 cells. Hinkle and Tashjian (1975) have shown that the level of hormone receptor decreases when cells are grown in the presence of low hormone levels. The rate of receptor loss from cells incubated with hormone and the rate of reappearance of receptors after removal of hormone are consistent with a receptor half-life of about 20 hours. It is not clear whether this example of modulation is due to changes in the synthetic rate of the receptor or to its degradation rate, since the rate of disappearance of receptors was decreased by incubation in the presence of puromycin. However, the long half-life and the similarity of half-lives in both modulated and unmodulated cells indicate that the results may be explained by alteration of synthetic rate of the receptor, with the rate of disappearance reflecting general membrane degradation.

The relatively long half-lives reported for two other cell surface components, acetylcholinesterase (ACh) of rat myoblasts (Derreotes and Fabrough, 1975; Fambrough and Rash, 1971) and IgM in bone marrow lymphocytes (Andersson et al., 1974), are consistent with degradation of the whole membrane. The binding of labeled α-bungarotoxin to ACh has been shown to be very strong and specific (Fambrough and Rash, 1971; Derreotes and Fambrough, 1975). This has allowed measurement of ACh turnover on the surface and in an intracellular pool of myoblasts. Both pools have an apparent half-life of 22 to 24 hours. The communication of the intracellular pool with the surface pool will be discussed further below.

Andersson et al. (1974) separated bone marrow lymphocytes into two classes based on size. They reported that large lymphocytes turn over IgM with a half-life of 4 hours. The IgM of these cells is present both within the cell and on the surface, and the rapid turnover probably reflects secretion of IgM into the medium. The small cells, however, have 90% of their IgM exposed externally on their surface membrane, and the half-life of the IgM molecules from these cells is 20 to 80 hours. The synthesis of IgM in the small B cells is very sensitive to actinomycin D, while that in the large B cells is less so. However, after stimulation of the B cells with mitogen, the synthesis of IgM becomes insensitive to actinomycin D, the amount of IgM on the surface and in the medium increases, and the half-life drops to 4 hours (Melchers and Andersson, 1974). The increase in surface-bound IgM after mitogenic stimulation is interesting because it appears to come from surface membrane fusion of exocytotic vesicles containing secretory IgM. The presence of IgM on the surface in this case is transitory, and its turnover is not characteristic of the plasma membrane as a whole. Measurements of the synthesis rates of membrane-bound and secreted immunoglobulins in chicken lymphoid cells (Choi, 1976) and direct assay after puromycin block in diploid human lymphocytes (Lerner et al., 1972) have indicated that the two forms may arise from different metabolic pathways, in agreement with the observations of Andersson et al. (1974).

Clearly, from the preceding discussion, care must be taken in interpreting degradation measurements made in specialized secretory cells in terms of a general model for the turnover of membrane proteins. The plasma membranes of lymphoid cells (Vitetta and Uhr, 1972; Andersson et al., 1974; Melchers and Andersson, 1974), parotid gland cells (Amsterdam et al., 1971), and pancreatic acinar cells (Jamieson and Palade, 1967a, 1967b; Jamieson and Palade, 1968a, 1968b; Meldolesi, 1974) may be composed of specialized areas where fusion of secretory granules contributes atypical proteins to the plasma membrane. The turnover of the proteins in these areas may be different from the plasma membrane as a whole. Meldolesi (1974) has concluded that secretory granule membrane is recycled in acinar cells, while Amsterdam et al. (1971) believe that in parotid gland this membrane is degraded after secretion of its contents. The latter observation implies relatively rapid degradation of a portion of the plasma membrane. Additional evidence that the surface of some cells may contain specialized areas with different turnover rates than the rest of the membrane comes from studies of mouse melanoma cells by Varga et al. (1976). They found that the binding sites for β-melanotropin are localized in the region of the cell surface overlying the Golgi complex, and concluded that the mechanism of action of this hormone involves internalization of the hormone (complexed with its receptor) in vesicles that fuse with the Golgi. After release of the hormone, these authors propose that the receptors are recycled.

Tsan and Berlin (1971) have shown that specialized areas in which phagocytosis can be induced exist in the macrophage and polymorphonuclear leukocyte plasma membranes. During active phagocytosis of oil droplets, these cells internalize a large proportion of their plasma membrane without altering the levels of several membrane transport systems. Oliver et al. (1974) have also shown that removal of certain areas of the plasma membrane can be induced by experimental manipulation. The binding sites for wheat germ agglutinin and concanavalin A were selectively removed from the surface of polymorphonuclear leukocytes after lectin binding. To some extent the removal was specific to the lectin used and was presumably related to aggregation of the lectin-bound sites since vinblastine and colchicine inhibited the process. This selective process may be generalized to the removal of certain antigens from the surface of lymphocytes after binding of specific antibodies. Capping is induced by such binding and is followed by internalization of the antibody-antigen complexes (Loor et al., 1972; Karnovsky et al., 1972). Selective removal of immunoglobulins from the cell after antibody binding is indicated by the observation that H-2 antigens are not similarly removed (Wilson et al., 1972; Vitetta and Uhr, 1972). However, the mechanism for the selective loss of the surface-bound antibody may not be via endocytosis. Wilson et al. (1972) have suggested instead that the 2 hour half-life of surface immunoglobulins of thymus lymphocytes may result from active shedding. It is not clear whether these authors were examining the membrane-bound transient secretory immunoglobulin mentioned above although the short half-life would suggest this possibility.

In conclusion, while some of the data on the turnover of specific plasma membrane constituents is incompatible with the concept of unit degradation, it is clear from the above discussion that many of these studies are complicated by problems of surface specialization and active shedding. A discussion of studies that suggest the integral proteins of the plasma membrane are degraded as a unit follows.

3.2 Evidence supporting unit degradation

3.2.1. Indirect support
Unit degradation of the plasma membrane is consistent with the concept of membrane flow described by Palade (1959). According to this hypothesis, the biogenesis of membranes involves transfer of membrane material from the site of synthesis, probably associated with the rough endoplasmic reticulum, through the smooth endoplasmic reticulum and Golgi to the plasma membrane. Differences in protein composition of various membranes result from assembly-line type modification during biogenesis. The plasma membrane is then formed by fusion with the existing membrane of vesicles analogous to secretion vesicles of the pancreas (Jamieson and Palade, 1967a, 1967b; Jamieson and Palade 1968a, 1968b). Abundant evidence supporting membrane flow has been presented (Caro and Palade, 1964; Jamieson and Palade, 1967a, 1967b; Jamieson and Palade, 1968a, 1968b; Ray et al., 1968; Ashley and Peters, 1969; Franke, 1971; Kawasaki and Yamashira, 1971; Atkinson, 1973; Bennett et al., 1974; Riordan et al., 1974; Atkinson, 1975) and reviewed (Franke et al., 1971) and will not be

described here, except to present the general findings. Pulse-labeling experiments with amino acids show that the specific activities of various liver membrane fractions reach maximums at different times (Ray et al., 1968; Franke et al., 1971). The order of these maximums suggests a precursor-product relationship with the rough and smooth microsomes being labeled most rapidly, followed by the Golgi and finally by the plasma membrane. Studies utilizing inhibitors of protein synthesis suggest that membrane processing occurs after initial synthesis of the protein components. Studies using labeled fucose or glucosamine have shown that the addition site of these sugars might be the Golgi (Kawasaki and Yamashira, 1971; Atkinson, 1973; Bennett et al., 1974; Riordan et al., 1974; Atkinson, 1975). The membrane-flow hypothesis implies that the plasma membrane is prefabricated in the cytoplasm and is assembled by fusion of the prefabricated units with the existing plasma membrane. If this hypothesis is correct, it would be surprising to find that disassembly of the plasma membrane involves independent removal of proteins at varying rates. Such removal implies that the protein composition of the membrane fluctuates continuously, since the components of the membrane are inserted coordinately but removed at different rates.

Evidence of a different sort has been interpreted in terms of the membrane-flow hypothesis. Hirano et al. (1972) and Nicolson and Singer (1974) have examined the distribution of lectin receptors on the surface membrane of a variety of mammalian cell types using ferritin-conjugated lectins. These authors found that the lectin receptors are uniformly confined to the external surface of the cell in a random pattern. They concluded that this asymmetric distribution could result if the polysaccharide side chains were added to the membrane proteins by enzymes confined to the lumen of precursor intracellular vesicles. Subsequent fusion of the vesicles with the plasma membrane would then lead to the localization of saccharide residues on the outside of the cell. Their results also suggest that the glycoproteins do not rotate appreciably in the membrane. These findings do not support the heterogeneous turnover model, since if proteins were individually removed and reinserted into the membrane some saccharide residues might be expected to be exposed on the internal face of the plasma membrane.

3.2.2. Direct turnover measurements

Direct measurements of plasma membrane protein turnover have lead a number of authors to conclude that these proteins are degraded at the same rates. Some of this data, however, is equivocal, being based upon comparisons of SDS polyacrylamide gel labeling patterns of membrane fractions isolated at various times after administration of labeled precursors. Quirk et al. (1973) measured the relative degradation rates of kidney brush border proteins labeled with glucosamine. Little difference was found in the relative rates of loss of label from several polypeptide bands resolved electrophoretically. Similar experiments with thymocyte plasma membranes labeled with amino acids (Schmidt-Ullrich et al., 1974) have indicated little difference in the half-life of proteins separated on gels by electrophoresis, except for the apparent rapid turnover of excreted components. Amino acid-labeling of the macrophage membrane (Nachman et al., 1971), followed after one half-life by separation of the labeled proteins, showed no difference in the relative labeling patterns of the proteins compared to that found at time zero. These studies, particularly the last, are subject to the criticism that the technique is not very quantitative since it depends upon identical yields of various proteins in separate preparations. Also, it is unclear how sensitive the technique is to small but significant differences in turnover rates among the separated proteins. Kaplan and Moskowitz used the double labeling procedure described in sections 3.1.1. and 3.1.2. to examine the relative degradation rates of membrane proteins in nongrowing MK_2 monkey kidney cells (Kaplan and Moskowitz, 1975b). Employing this more sensitive technique, these investigators found little heterogeneity in degradation rate except for a secreted component. Similar results were obtained with labeled glucosamine (Kaplan and Moskowitz, 1975b).

Measurements of the degradation of plasma membrane proteins labeled with amino acids or sugars are subject to misinterpretation due to possible contamination of the plasma membrane with other cell fractions, as discussed earlier. This problem can be circumvented with cultured cells by selective labeling of the surface proteins with membrane impermeable reagents. Labeled acetic anhydride has been employed on the surface proteins of CHO cells and skin fibroblasts (Roberts and Yuan, 1974; Roberts and Yuan, 1975b). This reagent has the potential of yielding valid estimates of the half-life of labeled proteins, since acetylation does not appear to affect the degradation of proteins and the label is not reutilized (Roberts and Yuan, 1975a). Unfortunately, since the plasma membrane is not totally impermeable to acetic anhydride, some interior proteins may become labeled (Roberts and Yuan, 1975a) and purification of the plasma membrane is necessary. Despite this disadvantage, Roberts and Yuan (1975b) were able to measure the relative degradation rates of CHO cell and skin fibroblast plasma membrane proteins after separation by SDS acrylamide gel electrophoresis. They employed a variation of the double isotope technique that involved labeling nongrowing cells with [^{3}H] acetic anhydride followed after a culture period of 2 or 4 days by a second acetylation with [^{14}C] acetic anhydride. During this period, extensive degradation of the ^{3}H label was found, but after separation of the labeled proteins no pronounced variation in isotope ratio was observed with either cell type. Thus, it was concluded that all labeled proteins of the plasma membrane are degraded with similar half-lives.

Specific labeling of the proteins exposed on the surface of cultured cells can be achieved by iodination catalyzed by lactoperoxidase. The large molecular size of lactoperoxidase prevents its entry into intact cells. In addition, careful experiments have demonstrated that the label is confined to protein tyrosine, does not alter cell physiology in any significant way (including the turnover of labeled protein), is not reutilized, and is associated with many and possibly most of the proteins present in the plasma membrane (Ryser, 1963; Hubbard and Cohn, 1972; Nachman et al., 1973; Hubbard and Cohn, 1975; Tweto et al., 1976).

Tweto and Doyle (1976) employed the lactoperoxidase labeling method to measure the relative degradation rates of the plasma membrane proteins to rat hepatoma cells (HTC cells). These authors used a double isotope technique that involved labeling one culture of cells with ^{125}I and a second culture with ^{131}I. The culture labeled with ^{131}I was frozen to avoid protein degradation, while the ^{125}I-labeled culture was allowed to metabolize label at 37°C for various periods up to 96 hours. This method of labeling avoids the criticism that the cells are not equivalent during the first and second labeling periods. After a length of time in culture, aliquots of each labeled culture were mixed and the plasma membrane proteins separated by SDS polyacrylamide gel electrophoresis. The isotope ratio of the separated membrane proteins was determined after slicing the gel and was found to be invariant over the length of the gel. Since as much as half of the ^{125}I label was degraded over the period of the experiment, this result indicates that there is no heterogeneity in degradation rates among the many different plasma membrane proteins. Growing and nongrowing cells behaved the same. Identical results were obtained by double labeling experiments using leucine, followed by careful plasma membrane isolation. However, fucose-labeled plasma membrane proteins showed some heterogeneity in the lose of incorporated fucose. Hubbard and Cohn (1975) also observed no difference in the half-lives of iodinated plasma membrane proteins of mouse L cells. In these experiments phagocytosis was induced by culturing cells in the presence of latex beads, and sufficient time elapsed for about half of the iodine-labeled protein to be degraded. Mastro et al. (1974) also concluded that the surface proteins still associated with BHK cells after trypsin treatment are degraded as a unit.

4. Discussion and conclusion

While the measurement of protein degradation appears to be a simple process, the abundant literature dealing with the theoretical (Reiner, 1953; Buchanan, 1961; Kock, 1962) and practical (Schimke, 1970; Schimke and Doyle, 1970; Schimke, 1973; Goldberg and Dice, 1974; Doyle and Tweto, 1975) aspects of such measurements indicates that this is not the case. Degradation measurement of a single homogeneous soluble protein is much simpler than attempting to characterize the degradation of a complex mixture of membrane proteins that can only be separated by their mobility upon electrophoresis in SDS polyacrylamide gels. Often additional complications are introduced by the idiosyncrasies of each system in which membrane turnover is studied. Further, the different methods used to measure protein turnover require certain assumptions, some of which have been mentioned earlier. The validity of these assumptions is largely unproven. For all of the above reasons we have not attempted a detailed critique of each study on plasma membrane turnover. Rather, we have sought to present (albeit with a particular bias) the views of a number of workers, emphasizing the differences in their conclusions. The result has been a division of the current ideas about degradation of plasma membrane proteins into two positions, exemplified by the heterogeneous turnover model and the unit turnover model. Data exist to support both models, and acceptance of one or the other model requires that some data be ignored. We propose the following model as one that best accommodates most of the results on plasma membrane biogenesis and turnover:

1. The proteins of the plasma membrane are degraded by a process that involves internalization of large areas (in molecular terms) of the cell surface.
2. The entire cell surface is not of uniform composition in most tissues, and functionally different areas may have distinct rates of internalization and/or degradation. In cultured cells that have lost their differentiated functions to some degree, the surface is correspondingly more uniform, and therefore degradation rates of surface proteins are more uniform.
3. In some cells, various membrane proteins are not removed by internalization but rather by shedding into the medium.

Except for point (3), this model does not attribute heterogeneity of turnover to removal (internalization) of individual proteins from the membrane at different rates. Such a mechanism for membrane turnover is difficult to accept on thermodynamic grounds, and actually no absolute evidence exists to support it. Unequivocal evidence might be the clear demonstration of plasma membrane proteins free in the soluble compartment of the cell. In regard to the latter, Lodish (1973; Lodish and Small, 1975) has found that two proteins associated with the rabbit reticulocyte plasma membrane can be synthesized by free polysomes in cell lysates. If these are authentic membrane proteins and if they are synthesized on free polysomes, it is likely that they do exist free for some limited interval in the cytoplasm. It is not known whether bound polysomes also synthesize these proteins, although it appears that the major product of bound polysomes in reticulocytes is globin (Woodward et al., 1973; Lodish and Desalu, 1973). Furthermore, some erythrocyte "membrane" proteins such as glyceraldehyde phosphate dehydrogenase or actin (Steck, 1975) might be expected to be synthesized on free polysomes. These two proteins are present in other "nonmembrane" compartments of the cell and may not penetrate much, if at all, into the lipid bilayer. The real issue is whether integral transmembrane proteins such as glycophorin are made on free or membrane-bound polysomes. We suggest that any protein that spans the membrane is synthesized on membrane-bound polysomes. The site of synthesis of membrane proteins is relevant to the model presented above. If large areas of membrane are removed by the process of interiorization and degradation, the simplest way for the cell to maintain a constant membrane protein composition would be to replace these areas with similarly composed, newly synthesized areas. We agree with the suggestion of Blobel and Dobberstein (1975a) that some membrane proteins are synthesized by bound ribosomes on preformed intracellular membranes by some variation of the process described for secretory proteins (Blobel and Dobberstein, 1975a,b), and that they are not free in the cytoplasm at any time. Aside from the studies of Lodish mentioned above, no information exists on the site of membrane protein synthesis (Rolleston, 1974). However, it should be pointed out that two plasma membrane proteins, glycophorin (Marchesi et al., 1972) and H-2 antigen (Henning et al., 1976), have their amino terminals exposed on the outside of the cell. This external location could be expected if membrane proteins were synthesized on membrane-bound ribosomes as proposed here. Further research on these subjects could provide a test for the above model and is being pursued in the author's laboratory.

The model has other implications with regard to plasma membrane protein synthesis. In the steady state, the equation describing the relationship between the rate of synthesis and the rate of degradation is $dP/dT = 0 = k_s - k_d P$ (Schimke, 1970, 1973; Schimke and Doyle, 1970; Goldberg and Dice, 1974), where P is the concentration of a protein, k_s the zero order rate constant of its synthesis, and k_d the first order constant of its degradation. If k_d is the same for all proteins within an area of the cell surface, as suggested by the model, then the relative levels of various proteins in that area are determined solely by their synthetic rates. For the protein composition of an area to remain constant some means of coordinate control of synthesis must be maintained. If the function of a particular membrane protein has anything to do with its level or concentration in the membrane, some feedback mechanism may exist to "inform" the synthetic machinery of this level. Impairment of function by treatment of the cell with hydrolytic enzymes (Hughes et al., 1972; Turner et al., 1972; Hynes, 1973; Mastro et al., 1974; Pearlstein, 1974) or by inhibition with specific agents such as α-bungarotoxin (Fambrough and Rash, 1971; Derreotes and Fambrough, 1975) or ouabain (Vaughan and Cook, 1972; Cook et al., 1975, 1976) might result in increased synthetic rates of affected proteins. Increased synthesis may result in a rapid return and/or a temporary elevation of level for affected proteins, the nonfunctional proteins remaining until the slow process of degradation eliminates them. Some attempt has been made to measure changes in the synthetic rate of the ouabain receptor after partial blockade. Negative results led Cook et al. (1976) to conclude that the rapid return of Na^+, K^+ ATPase function after partial blockade is due to increased degradation of the enzyme. However, the result may reflect a lack of sensitivity in the measurement of synthesis or a special degradation mechanism for this vital enzyme. The modulation phenomenon of antigen and insulin receptors may also be related to alterations of synthetic rates.

One problem with invoking feedback mechanisms is that it is difficult to understand how a protein whose function is on the external cell surface, i.e., immunoglobulin on lymphocytes, could communicate rapidly with the machinery for its synthesis. However, recent studies from Cohn's laboratory (Steinman et al., 1976) suggest that the entire plasma membrane is internalized every few hours and, by implication, returned to the cell surface at the same rate. The details of this process are unknown, although the process allows a means by which an accounting of the levels of plasma membrane proteins could be made intracellularly. Insufficient data exists to permit speculation about how the biogenesis of the plasma membrane is related to the recycling phenomenon, but it seems plausible to hypothesize that degradation of this organelle is linked to recycling. Since the rate of endocytosis is much greater than the half-life of plasma membrane proteins, it is clear that endocytosis need not necessarily result in degradation of the plasma membrane. Rather, it seems that degradation may be a result of some side path occurring during or after fusion of the pinocytotic vesicles with lysosomes. Possibly, a small percentage of the internalized membrane is altered sufficiently by lysosomal enzyme action to make the proposed (Steinman et al., 1976) pinching off of small vesicles unfavorable. These altered areas might remain associated with the lysosome for a sufficient time to be degraded. It is also possible that some lysosomes are specially designed to degrade membrane-bound proteins. Chance fusion of pinocytotic vesicles with these special lysosomes would then result in their degradation. Whatever the mechanism, it is possible that the degradation of some proportion of the incoming pinocytotic vesicles is simply a result of chance, occurring as a result of metabolism peripheral to the process of membrane recycling. The probability that vesicles will be degraded may be different for different cell types or under different physiological conditions. Several measurements of the half-life of plasma membrane proteins (Franke et al., 1971; Nachman et al., 1971; Pearlstein and Waterfield, 1974; Proctor, 1974; Hubbard and Cohn, 1975; Kaplan and Moskowitz, 1975b, Tweto and Doyle, 1976) have indicated that this rate varies appreciably among different cultured cells.

Measurement of the pinocytosis rates mentioned above were done using rapidly growing L cells and in nongrowing macrophages. The results indicate that the rate of pinocytosis is rapid in both cell types. If pinocytosis is independent of growth rate and is the initial step in membrane turnover, then the rate of membrane degradation should also be independent of growth rate. However, several authors have concluded that rapidly growing cells are not degrading membrane while stationary cells are. Warren and Glick (1968) measured the decrease in specific activity of valine-labeled membrane proteins in growing L cells and found no dilution of label that could not be accounted for by cell growth. Roberts and Yuan (1974) concluded from experiments similar to those of Warren and Glick (1968) that the plasma membrane of CHO cells is not being degraded during active growth. Hughes et al. (1972) reached a similar conclusion with regard to membrane glycoproteins of mouse ascites cells. The opposite conclusion, i.e., membrane turnover is not growth related, was reached by Kaplan and Moskowitz (1975a) with cultured monkey kidney cells, Melchers and Andersson (1974) with mitogen-stimulated mouse β-lymphocytes, Derreotes and Fambrough (1975) with developing muscle myoblasts, Hubbard and Cohn (1975) with iodinated mouse L cells, and Tweto and Doyle (1976) with iodinated HTC-cell membrane proteins. The probable explanation for the discrepancy in these conclusions is that no growth-related difference in turnover of plasma membrane proteins exists but that measurement by specific activity dilution is too insensitive to detect the turnover. Alternatively, increased reutilization of the labeled precursor during growth could account for the result.

The degradation of plasma membrane glycoproteins labeled with fucose or glucosamine has been reported to be heterogeneous (Kawasaki and Yamashira, 1971; Hubbard and Cohn, 1975; Landry and Marceau, 1975; Kaplan and Moskowitz, 1975b) with two exceptions (Quirk et al., 1973; Kaplan and Moskowitz, 1975b). This heterogeneity could be due to the activity of exoglycosidases (Roseman, 1970; Harms and Reutter, 1974; Patt and Grimes, 1974; Shur and Roth, 1975). Alternatively, the selective turnover of the saccharide portion of glycoproteins could take place during membrane recycling or represent a special mechanism for the turnover of glycoproteins as a whole (possibly by shedding). Insufficient data exists for speculation about which, if any, of these mechanisms apply.

Membrane recycling offers a possible explanation for the apparent disagreement of the model proposed here and the concept of membrane fluidity. We suggest that one reason for heterogeneity in degradation rates of plasma membrane proteins might be that the plasma membrane is composed of distinct functional areas that turn over at different rates. The concept of membrane fluidity suggests that diffusion of proteins embedded in the lipid bilayer results in their free movement over the cell surface. The proposed specialization of certain areas of the cell surface could be explained by invoking barriers to diffusion as a result of interaction of surface proteins with structural elements in the underlying cytoplasm as suggested for the red cell membrane (Steck, 1974) and other cells (Nicolson, 1976) or by specific interaction of the membrane proteins with each other in certain areas (Singer and Nicolson, 1972; Singer, 1974; Nicolson, 1976). However, redistribution of membrane proteins might also result from the indirect movement of plasma membrane proteins through the cytoplasm and reappearance at new topographical positions on the membrane, rather than direct diffusion in the plane of the membrane to the new positions.

Independent support of membrane recycling and the turnover model proposed here could be obtained if plasma membrane could be demonstrated to exist intracellularly. One property of an intracellular pool of plasma membrane derived via pinocytosis would be that proteins exposed on the surface of the cell should be hidden on the inside of internal vesicles. Fambrough and Derreotes (1971) have found an internal pool of acetylcholinesterase (ACh) that is degraded at the same rate as the surface pool. The size of this pool is about that of the internal pool expected if the membrane is recycled. In addition, the ACh of the internal pool could not be detected by α-bungarotoxin binding in the absence of detergent. This result suggests that the α-bungarotoxin receptor in this case is inside vesicles. In another study, Cook et al. (1976) have found a large internal pool of ouabain-binding sites but it does not appear to be recycled.

Clearly, from the present discussion, much more work is required for an understanding of the mechanism of plasma membrane degradation. We have presented a model for the degradation of the membrane based on unit turnover of the proteins in the membrane. This model was derived primarily from experiments done in our laboratory and that of Cohn and his collaborators. Our hope is that the model will provide a framework for further research on membrane turnover, but only time will tell if our bias in choice of model is correct.

NOTE ADDED IN PROOF   Since the data concerning the turnover of plasma membrane proteins in rodent liver were obtained under the assumption that the predominant cell type of that tissue is hepatocyte, a recent publication supports our explanation for heterogeneous degradation. A. Blowin, R. P. Bolender, and E. R. Weibel (1977: J. Cell Biol. 72, 441–455) in a paper entitled "Distribution of organelles and membranes between hepatocytes and nonhepatocytes in the rat liver parenchyma" have reported that hepatocytes comprise 78% of the parenchymal volume and nonhepatocytes 6.3%. However, nonhepatocytes contain 26.5% of the total plasma membrane. It is apparent that heterogeneity in the degradation of plasma membrane proteins could be explained by the presence of membranes from numerous cell types.

Acknowledgments

Support for work in the authors' laboratory, which has led to the conclusions mentioned in this review, was obtained from United States Public Health Service grants CA17149, HD08410, GM19521, and F22 AM03382.

References

Alpers, D. H. (1972) The relation of size to the relative rates of degradation of intestinal brush border proteins. J. Clin. Invest. 51, 2621–2630.

Alpers, D. H. and Tedesco, F. J. (1975) The possible role of pancreatic proteases in the turnover of intestinal brush border proteins. Biochim. Biophys. Acta 401, 28–40.

Amsterdam, A., Schramm, M., Ohad, I., Salomon, Y. and Selinger, A. (1971) Concomitant synthesis of membrane protein of the secretory granule in rat parotid gland. J. Cell Biol. 61, 1–13.

Andersson, J., Lafleur, L. and Melchers, F. (1974) IgM in bone marrow-derived lymphocytes. Synthesis, surface deposition, turnover, and carbohydrate composition in unstimulated mouse B cells. Eur. J. Immunol. 4, 170–180.

Arias, I. M., Doyle, D. and Schimke, R. T. (1969) Studies on the synthesis and degradation of proteins of the endoplasmic reticulum of rat liver. J. Biol. Chem. 244, 3303–3315.

Ashley, C. A. and Peters, T., Jr. (1969) Electron microscopic radioautographic detection of sites of protein synthesis and migration in liver. J. Cell Biol. 43, 237–249.

Atkinson, P. H. (1973) HeLa cell plasma membranes. In: Methods in Cell Biology (Prescott, D., ed.), vol. 8, pp. 158–188, Academic Press, New York.

Atkinson, P. H. (1975) Synthesis and assembly of HeLa cell plasma membrane glycoproteins and proteins. J. Biol. Chem. 250, 2123–2134.

Banker, G. A. and Cotman, C. W. (1972) Measurement of free electrophoretic mobility and retardation coefficient of protein-sodium dodecyl sulfate complexes by gel electrophoresis. A method to validate molecular weight estimates. J. Biol. Chem. 247, 5856–5861.

158

Bennett, G., Leblond, C. P., and Haddad, A. (1974) Migration of glycoproteins from the Golgi apparatus to the surface of various cell types as shown by radioautography after labeled fucose injection into rats. J. Cell Biol. 60, 258–284.

Berlin, C. M. and Schimke, R. T. (1965) Influence of turnover rates on the responses of enzymes to cortisone. Mol. Pharmacol. 1, 149–156.

Blobel, G. and Dobberstein, B. (1975a) Transfer of proteins across membranes. I. Presence of proteolytically processed and unprocessed nascent immunoglobulin light chains on membrane-bound ribosomes of murine myeloma. J. Cell Biol. 67, 835–851.

Blobel, G. and Dobberstein, B. (1975b) Transfer of proteins across membranes. II. Reconstitution of functional rough microsomes from heterologous components. J. Cell Biol. 67, 852–862.

Bock, K. W., Siekevitz, and Paladè, G. E. (1971) Localization and turnover studies of membrane nicotinamide adenine dinucleotide glycohydrolase in rat liver. J. Biol. Chem. 246, 188–195.

Buchanan, D. L. (1961) Analysis of continuous dosage isotope experiments. Arch. Biochem. Biophys. 94, 489–499.

Caro, L. G. and Palade, G. E. (1964) Protein synthesis storage and discharge in the pancreatic exocrine cell. An autoradiographic study. J. Cell Biol. 20, 473–495.

Cikes, M. and Friberg, S., Jr. (1971) Expression of H-2 and Moloney leukemia virus-determined cell-surface antigens in synchronized cultures of a mouse cell line. Proc. Nat. Acad. Sci. U.S.A. 68, 566–569.

Cherry, R. J. (1975) Protein mobility in membranes. FEBS Lett. 55, 1–7.

Choi, Y. S. (1976) Biosynthesis of membrane bound Ig and secretion of Ig by chicken lymphoid cells. Biochemistry 15, 1037–1042.

Clarke, S. (1975) The size and detergent binding of membrane proteins. J. Biol. Chem. 250, 5459–5469.

Cook, J. S., Vaughan, G. L., Proctor, W. R. and Brake, E. T. (1975) Interaction of two mechanisms of regulating alkali cations in HeLa cells. J. Cell Physiol. 86, 59–70.

Cook, J. S., Will, P. C., Proctor, W. R. and Brake, E. T. (1976) Turnover of Ouabain-binding sites and plasma membrane proteins in HeLa cells. In: Biogenesis and Turnover of Membrane macromolecules (Cook, J. S., ed.), pp. 114–137, Raven Press, New York.

Dehlinger, P. J. and Schimke, R. T. (1971) Size distribution of membrane proteins of rat liver and their relative rates of degradation. J. Biol. Chem. 246, 2574–2583.

Dehlinger, P. J. and Schimke, R. T. (1972) Effects of phenobarbital 3-methylcholanthrene and hematin on the synthesis of protein components of rat liver microsomal membranes. J. Biol. Chem. 247, 1257–1264.

Derreotes, P. W. and Fambrough, D. M. (1975) Acetylcholine receptor turnover in membranes of developing muscle fibers. J. Cell. Biol. 65, 335–358.

Doyle, D. and Tweto, J. (1975) Measurement of protein turnover in animal cells. In: Methods in Cell Biology (Prescott, D.M., ed.), vol. 10, pp. 235–260, Academic Press, New York.

Dunker, A. K. and Rueckert, R. R. (1969) Observations on molecular weight determinations on polyacrylamide gel. J. Biol. Chem. 244, 5074–5080.

Edidin, M. (1974) Rotational and translational diffusion in membranes. Ann. Rev. Biophys. Bioeng. 3, 179–202.

Edwards, P. A. and Gould, R. G. (1972) Turnover rate of hepatic 3-hydroxy-3-methylglutanyl coenzyme A reductase as determined by use of cycloheximide. J. Biol. Chem. 247, 1520–1524.

Evans, W. H. and Gurd, J. W. (1972) Preparation and properties of nexuses and lipid-enriched vesicles from mouse liver plasma membrane. Biochem. J. 128, 691–700.

Fambrough, D. and Rash, J. E. (1971) Development of acetylcholine sensitivity during myogenesis. Develop. Biol. 26, 55–68.

Fox, T. O., Sheppard, J. R. and Burger, M. M. (1971) Cyclic membrane changes in animal cells: transformed cells permanently display a surface architecture detected in normal cells only during mitosis. Proc. Nat. Acad. Sci. U.S.A. 68, 244–247.

Frank, R. W. and Rodbard, D. (1975) Precision of sodium dodecyl sulfate-polyacrylamide gel electrophoresis for the molecular weight estimation of a membrane glycoprotein: studies on bovine rhodopsin. Arch. Biochem. Biophys. 171, 1–13.

Franke, W. W., Mooré, D. J., Deumling, B., Cheetham, R. D., Kartenbeck, J., Jarasch, E., and Zentgraf, H. (1971) Synthesis and turnover of membrane proteins in rat liver: an examination of the membrane flow hypothesis. Z. Naturforsch. 26b, 1031–1039.

Gavin, J. T., III, Roth, J., Neville, D. M., Jr., deMeytes, P. and Buell, D. M. (1974) Insulin-dependent regulation of insulin receptor concentrations: a direct demonstration in cell culture. Proc. Nat. Acad. Sci. U.S.A. 71, 84–88.

Glass, R. D., and Doyle, D. (1972) On the measurement of protein turnover in animal cells. J. Biol. Chem. 247, 5234–5242.

Glaumann, H. and Ericsson, J. L. E. (1970) Evidence for the participation of the Golgi apparatus in the intracellular transport of nascent albumin in the liver cell. J. Cell Biol. 47, 555–567.

Goldberg, A. L. and Dice, J. F. (1974) Intracellular protein degradation in mammalian and bacterial cells. Ann. Rev. Biochem. 43, 835–869.

Goldstein, J. L. and Brown, M. S. (1974) Binding and degradation of low density lipoproteins by cultured human fibroblasts. J. Biol. Chem. 249, 5153–5162.

Graham, J. M., Hynes, R. O., Davidson, E. A. and Bainton, D. F. (1975) The location of proteins labeled by the ^{125}I-lactoperoxidase system in the NIL 8 hamster fibroblast. Cell 4, 353–365.

Grefrath, S. P. and Reynolds, J. A. (1974) The molecular weight of the major glycoprotein from the human erythrocyte membrane. Proc. Nat. Acad. Sci. U.S.A. 71, 3913–3916.

Gurd, J. W. and Evans, W. H. (1973) Relative rates of degradation of mouse liver surface-membrane proteins. Eur. J. Biochem. 36, 273–279.

Harms, E. and Reutter, W. (1974) Half-life of N-acetylneuraminic acid in plasma membranes of rat liver and Morris hepatoma 7777. Cancer Res. 34, 3165–3172.

Henning, R., Milner, R. J., Reske, K., Cunningham, B. A., and Edelman, G. (1976) Subunit structure, cell surface orientation, and partial amino acid sequence of murine histocompatibility antigens. Proc. Nat. Acad. Sci. U.S.A. 73, 118–122.

Hinkle, P. M. and Tashnian, A. H., Jr., (1975) Thyrotropin-releasing hormone regulates the number of its own receptors in the GH$_3$ strain of pituitary cells in culture. Biochemistry 14, 3845–3851.

Hirano, H., Parkhouse, B., Nicolson, G. L., Lennox, E. S., and Singer, S. J. (1972) Distribution of saccharide residues on membrane fragments from a myeloma-cell homogenate: its implications for membrane biogenesis. Proc. Nat. Acad. Sci. U.S.A. 69, 2945–2949.

Hubbard, A. L. and Cohn, Z. A. (1972) The enzymatic iodination of the red cell membrane. J. Cell Biol. 55, 390–405.

Hubbard, A. L. and Cohn, Z. A. (1975a) Externally disposed plasma membrane proteins. I. Enzymatic iodination of mouse L cells. J. Cell Biol. 64, 438–460.

Hubbard, A. L. and Cohn, Z. A. (1975b) Externally disposed plasma membrane protein. II. Metabolic fate of iodinated polypeptides of mouse L cells. J. Cell Biol. 64, 461–479.

Hughes, R. C., Sanford, B., and Jeanloz, R. W. (1972) Regeneration of surface glycoproteins of a transplantable mouse tumor cell after treatment with neutraminidase. Proc. Nat. Acad. Sci. U.S.A. 69, 942–945.

Hunt, R. C. and Brown, J. C. (1975) Identification of a high molecular weight trans-membrane protein in mouse L cells. J. Mol. Biol. 97, 413–422.

Hynes, R. O. (1973) Alteration of cell-surface proteins by viral transformation and by proteolysis. Proc. Nat. Acad. Sci. U.S.A. 70, 3170–3174.

Hynes, R. O. and Bye, J. M. (1974) Density and cell cycle dependence of cell surface proteins in hamster fibroblasts. Cell 3, 113–120.

Jackson, R. L., Segrest, J. P., Kahane, I. and Marchesi, V. T. (1973) Studies on the major sialoglycoprotein of the human red cell membrane. Isolation and characterization of tryptic glycopeptides. Biochemistry 12, 3131–3138.

Jamieson, J. D. and Palade, G. E. (1967a) Intracellular transport of secretory proteins in the pancreatic exocrine cell. I. Role of the peripheral elements of the Golgi complex. J. Cell Biol. 34, 577–596.

Jamieson, J. D. and Palade, G. E. (1967b) Intracellular transport of secretory proteins in the pancreatic exocrine cell. II. Transport of condensing vacuoles and zymogen granules. J. Cell Biol. 34, 597–615.

Jamieson, J. D. and Palade, G. E. (1968a) Intracellular transport of secretory proteins in the pancreatic exocrine cell. III. Dissociation of intracellular transport from protein synthesis. J. Cell Biol. 39, 580–588.

Jamieson, J. D. and Palade, G. E. (1968b) Intracellular transport of secretory proteins in the pancreatic exocrine cell. IV. Metabolic requirements. J. Cell Biol. 39, 589–603.

Johnsen, S., Stokke, T. and Prydz, H. (1975) HeLa cell plasma membranes: Changes in membrane protein composition during the cell cycle. Exp. Cell Res. 93, 245–251.

Kaplan, J. and Moskowitz, M. (1975a) Studies on the turnover of plasma membranes in cultured mammalian cells. I. Rates of synthesis and degradation of plasma membrane proteins and carbohydrates. Biochim. Biophys. Acta 389, 290–305.

Kaplan, J. and Moskowitz, M. (1975b) Studies on the turnover of plasma membranes in cultured mammalian cells. II. Demonstration of heterogeneous rates of turnover for plasma membrane proteins and glycoproteins. Biochim. Biophys. Acta 389, 306–313.

Karnovsky, M. J., Unanue, E. R. and Leventhal, M. (1972) Ligand-induced movement of lymphocyte membrane macromolecules. II. Mapping of surface moieties. J. Exp. Med. 136, 907–930.

Kawasaki, T. and Yamashina, I. (1971) Metabolic studies of rat liver plasma membrane using D-[1-^{14}C] glucosamine. Biochim. Biophys. Acta 225, 234–238.

Koch, A. L. (1962) The evaluation of the rates of biological processes from tracer kinetic data. I. The influence of labile metabolic pools. J. Theoret. Biol. 3, 283–303.

Kreibich, G., Hubbard, A. L. and Sabatini, D. D. (1974) On the spatial arrangement of proteins in microsomal membranes from rat liver. J. Cell Biol. 60, 616–627.

Kuriyama, Y. and Omura, T. (1971) Different turnover behavior of phenobarbital-induced and normal NADPH-cytochrome c reductases in rat liver microsomes. J. Biochem. 69, 659–669.

Landry, J. and Marceau, N. (1975) The relative rates of degradation of the plasma membrane glycoproteins from normal rat liver. Biochim. Biophys. Acta 389, 154–161.

Lerner, R. A., McConahey, P. J., Jansen, I., and Dixon, F. K. (1972) Synthesis of plasma membrane-associated and secretory immunoglobulin in diploid lymphocytes. J. Exp. Med. 134, 1016–1035.

Lodish, H. F. (1973) Biosynthesis of reticulocyte membrane proteins by membrane-free polyribosomes. Proc. Nat. Acad. Sci. U.S.A. 70, 1526–1530.

Lodish, H. F. and Desalu, O. (1973) Regulation of synthesis of non-globin proteins in cell-free extracts of rabbit reticulocytes. J. Biol. Chem. 248, 3520–3527.

Lodish, H. F. and Small, B. (1975) Membrane proteins synthesized by rabbit reticulocytes. J. Cell Biol. 65, 51–64.

Loor, F., Forni, L. and Pernis, B. (1972) The dynamic state of the lymphocyte membrane. Factors affecting the distribution and turnover of surface immunoglobulins. Eur. J. Immunol. 2, 203–212.

Marchesi, V. T., Tillack, T. W., Jackson, R. L., Segrest, J. P. and Scott, R. E. (1972) Chemical characterization and surface orientation of the major glycoprotein of the human erythrocyte membrane. Proc. Nat. Acad. Sci. U.S.A. 69, 1445–1449.

Mastro, A. M., Beer, C. T. and Mueller, G. C. (1974) Iodination of plasma membrane proteins of BHK cells in different growth states. Biochim. Biophys. Acta 352, 38–51.

Melchers, F. and Andersson, J. (1974) IgM in bone marrow derived lymphocytes. Changes in synthesis, turnover, and in numbers of molecules on the surface of B cells after mitogenic stimulation. Eur. J. Immunol. 4, 181–188.

Meldolesi, J. (1974) Dynamics of the cytoplasmic membranes in guinea pig pancreatic acinar cells. I. Synthesis and turnover of membrane proteins. J. Cell Biol. 34, 577–615.

Nachman, R. L., Ferris, G., and Hirsch, J. G. (1971) Macrophage plasma membrane. II. Studies on synthesis and turnover of protein constituents. J. Exp. Med. 133, 807–820.

Nachman, R. L., Hubbard, A. and Ferris, B. (1973) Iodination of the human platelet membrane. Studies of the major surface glycoprotein. J. Biol. Chem. 248, 2929–2936.

Nicolson, G. L. (1976) Transmembrane control of the receptor on normal and tumor cells. I. Cytoplasmic influence of cell surface components. Biochim. Biophys. Acta 457, 57–108.

Nicolson, G. L. and Singer, S. J. (1974) The distribution and asymmetry of mammalian cell surface saccharides utilizing ferritin-conjugated plant agglutinins as specific saccharide stains. J. Cell Biol. 60, 236–248.

Nicolson, G. L., Poste, G. and Ji, T. H. (1977) The dynamics of cell membrane organization. In:

Dynamic Aspects of Cell Surface Organization, vol. 3, Cell Surface Reviews (Poste, G. and Nicolson, G. L., eds.), pp. 1–73, Elsevier/North-Holland, Amsterdam.

Noonan, K. D., Levine, A. J. and Burger, M. M. (1974) Cell cycle dependent changes in the surface membrane as detected with ^{3}[H] concanavalin A. J. Cell Biol. 58, 491–497.

Novikoff, A. G. and Essner, E. (1960) The liver cell. Some new approaches to its study. Am. J. Med. 29, 102–131.

Oliver, J. M., Ukena, T. E., and Berlin, R. D. (1974) Effect of phagocytosis and colchicine on the distribution of lectin-binding sites on cell surfaces. Proc. Nat. Acad. Sci. U.S.A. 71, 394–398.

Omura, T. and Kuriyama, Y. (1971) Role of rough and smooth microsomes in the biosynthesis of microsomal membranes. J. Biochem. 69, 651–658.

Omura, T., Siekevitz, P. and Palade, G. (1967) Turnover of constituents of the endoplasmic reticulum membranes of rat hepatocytes. J. Biol. Chem. 242, 2389–2396.

Palade, G. E. (1959) Functional changes in structure of cell components. In: Subcellular Particles. (Hayashi, T., ed.), pp. 64–83, Ronald Press, New York.

Pasternak, C. A., Warmsley, A. M. H. and Thomas, D. B. (1971) Structural alterations in the surface membrane during the cell cycle. J. Cell Biol. 50, 562–564.

Patt, L. M. and Grimes, W. J. (1974) Cell surface glycolipid and glycoprotein glycosyl-transferases of normal and transformed cells. J. Biol. Chem. 249, 4257–4265.

Pearlstein, E. and Waterfield, M. D. (1974) Metabolic studies on ^{125}I- labeled baby hamster kidney cell plasma membranes. Biochim. Biophys. Acta 362, 1–12.

Poste, G., Papahadjopoulos, D. and Nicolson, G. L. (1975) Local anesthetics affect transmembrane cytoskeletal control of mobility and distribution of cell surface receptors. Proc. Nat. Acad. Sci. U.S.A. 72, 4430–4434.

Proctor, W. R. (1974) Membrane protein turnover and specific membrane enzyme activities during the HeLa cell cycle. Ph. D. thesis, University of Tennessee.

Quirk, S. J., Byrne, J. and Robinson, G. R. (1973) Studies on the turnover of protein and glycoprotein components in rabbit kidney brush borders. Biochem. J. 132, 501–508.

Ray, T. K., Lieberman, I. and Lansing, A. I. (1968) Synthesis of the plasma membrane of the liver cell. Biochem. Biophys. Res. Commun. 31, 54–58.

Reik, L., Petzold, G. L., Higgins, J. A., Greengard, P., Barrnett, R. J. (1970) Hormone-sensitive adenyl cyclase: cytochemical localization in rat liver. Science 168, 382–384.

Reiner, J. M. (1953) The study of metabolic turnover rates by means of isotopic tracers. I. Fundamental relations. Arch. Biochem. Biophys. 46, 53–79.

Revel, J. P. and Wolker, K. (1973) Electron microscope investigations of the underside of cells in culture. Exp. Cell Res. 78, 1–14.

Revoltella, R., Bertolini, L. and Pediconi, M. (1974) Unmasking of nerve growth factor membrane-specific binding sites in synchronized murine CL300 neuroblastoma cells. Exp. Cell Res. 85, 89–94.

Rieber, M., Bacalao, J. and Alonso, G. (1975) Turnover of high-molecular-weight cell surface protein during growth and expression of malignant transformation. Cancer Res. 35, 2104–2108.

Riordan, J. R., Mitranic, M., Slavik, M. and Moscarello, M. A. (1974) The incorporation of 1-[^{14}C] fucose into glycoprotein fractions of liver plasma membranes. FEBS Lett. 47, 248–251.

Roberts, R. M. and Yuan, B. O. (1974) Chemical modification of the plasma membrane polypeptides of cultured mammalian cells as an aid to studying protein turnover. Biochemistry 13, 4846–4855.

Roberts, R. M. and Yuan, B. O. (1975a) Radiolabeling of mammalian cells in tissue culture by use of acetic anhydride. Arch. Biochem. Biophys. 171, 226–233.

Roberts, R. M. and Yuan, B. O. (1975b) Turnover of plasma membrane polypeptides in non-proliferating cultures of Chinese hamster ovary cells and human skin fibroblasts. Arch. Biochem. Biophys. 171, 234–244.

Robinson, N. C. and Tanford, C. (1975) The binding of deoxycholate, Triton X-100, sodium dodecyl sulfate, and phosphatidylcholine vesicles to cytochrome b5. Biochemistry 14, 369–378.

Rolleston, F. S. (1974) Membrane-bound and free ribosomes. Sub-Cell. Biochem. 3, 91–117.

Roseman, S. (1970) The synthesis of complex carbohydrates by multiglycosyl-transferase systems and their potential function in intercellular adhesion. Chem. Phys. Lipids. 5, 270–299.

162

Ryser, H. J.-P. (1963) Comparison of the incorporation of tyrosine and its iodinated analogs into the proteins of Ehrlich ascites tumor cells and rat liver slices. Biochim. Biophys. Acta 78, 759–762.

Schimke, R. T. (1970) Regulation of protein degradation in mammalian tissues. In: Mammalian Protein Metabolism, (Munro, H.N., ed.), vol. 4, pp. 177–228, Academic Press, New York.

Schimke, R. T. and Doyle, D. (1970) Control of enzyme levels in animal tissues. Ann. Rev. Biochem. 39, 929–976.

Schimke, R. T. (1973) Control of enzyme levels in mammalian tissues. Adv. Enzymol. 37, 135–187.

Schimke, R. T. (1975) Turnover of membrane proteins in animal cells. Meth. Membr. Biol. 3, 201–236.

Schmidt-Ullrich, R., Wallach, D. F. H., and Ferber, E. (1974) Concanavalin A augments the turnover of electrophoretically defined thymocyte plasma membrane proteins. Biochim. Biophys. Acta 356, 288–299.

Shoham, J. and Sachs, L. (1972) Differences in the binding of fluorescent concanavalin A to the surface membrane of normal and transformed cells. Proc. Nat. Acad. Sci. U.S.A. 69, 2479–2482.

Shur, B. D. and Roth, S. (1975) Cell surface glycosyltransferases. Biochim. Biophys. Acta 415, 473–512.

Siekevitz, P. (1972) Biological membranes: The dynamics of their organization. Ann. Rev. Microbiol. 34, 117–140.

Simon, F. R., Blumenfeld, O. O. and Arias, I. M. (1970) Two protein fractions obtained from hepatic plasma membranes. Studies of their composition and differential turnover. Biochim. Biophys. Acta 219, 349–360.

Singer, S. J. and Nicolson, G. L. (1972) The fluid mosaic model of the structure of cell membranes. Science 175, 720–731.

Singer, S. J. (1974a) Molecular biology of cellular membranes with applications to immunology. Adv. Immunol. 19, 1–66.

Singer, S. J. (1974b) The molecular organization of membranes. Ann. Rev. Biochem. 43, 805–833.

Steck, T. L. (1974) The organization of proteins in the human red blood cell membrane: A review. J. Cell Biol. 74, 1–19.

Steinman, R. M., Silver, J. M., and Cohn, Z. A. (1974) Pinocytosis in fibroblasts. Quantitative studies in vitro. J. Cell Biol. 63, 949–969.

Steinman, R. M., Brodie, S. E. and Cohn, Z. A. (1976) Membrane flow during pinocytosis: a stereologic analysis. J. Cell Biol. 68, 665–687.

Taylor, J. M., Dehlinger, P. J., Dice, J. F. and Schimke, R. T. (1973) The synthesis and degradation of membrane proteins. Drug Metab. Disp. 1, 84–91.

Toda, G., Oka, H., Oda, T. and Ikeda, Y. (1975) Subfractionation of rat liver plasma membrane. Uneven distribution of plasma membrane bound enzymes of the liver cell surface. Biochim. Biophys. Acta 413, 52–64.

Tomita, M. and Marchesi, V. T. (1975) Amino-acid sequence and oligosaccharide attachment sites of human erythrocyte glycophorin. Proc. Nat. Acad. Sci. U.S.A. 72, 2964–2968.

Tsan, M. and Berlin, R. D. (1971) Effect of phagocytosis on membrane transport of nonelectrolytes. J. Exp. Med. 134, 1016–1035.

Turner, M. J., Strominger, J. L. and Sanderson, A. R. (1972) Enzymic removal and re-expression of a histocompatibility antigen, HL-A2, at the surface of human peripheral lymphocytes. Proc. Nat. Acad. Sci. U.S.A. 69, 200–202.

Tweto, J. and Doyle, D. (1976) Turnover of the plasma membrane proteins of hepatoma tissue culture cells. J. Biol. Chem. 251, 872–882.

Tweto, J., Friedman, E. and Doyle, D. (1976) Proteins of the hepatoma tissue culture cell plasma membrane. J. Supramol. Struct. 4, 141–159.

Varga, J. M., Moellman, G., Fritsch, P., Godawska, E., Lerner, A. B. (1976) Association of cell surface receptors for melanotropin with the Golgi region in mouse melanoma cells. Proc. Nat. Acad. Sci. U.S.A. 73, 559–562.

Vaughan, G. L. and Cook, J. S. (1972) Regeneration of cation-transport capacity in HeLa cell membranes after specific blockade by ouabain. Proc. Nat. Acad. Sci. U.S.A. 69, 2627–2631.

Vitetta, E. S. and Uhr, J. W. (1972) Cell surface immunoglobulin V. Release from murine splenic lymphocytes. J. Exp. Med. 136, 676–696.

Warren, L. and Glick, M. C. (1968) Membranes of animal cells. II. The metabolism and turnover of the surface membrane. J. Cell Biol. 37, 729–746.

Weber, K. and Osborn, M. (1969) The reliability of molecular weight determinations by dodecyl sulfate-polyacrylamide gel electrophoresis. J. Biol. Chem. 244, 4406–4412.

Werb, Z. and Cohn, Z. A. (1971) Cholesterol metabolism in the macrophage. II. Alteration of subcellular exchangeable cholesterol components and exchange in other cell types. J. Exp. Med. 134, 1570–1590.

Werb, Z. and Cohn, Z. A. (1972) Plasma membrane synthesis in the macrophage following phagocytosis of polystyrene latex particles. J. Biol. Chem. 247, 2439–2446.

Wilson, J. P., Nossal, G. J. V. and Lewis, H. (1972) Metabolic characteristics of lymphocyte surface immunoglobulins. Eur. J. Immunol. 2, 225–232.

Woodward, W. R., Adamson, S. D., McQueen, A. M., Larson, J. W., Estvanik, S. M., Wilairat, P. and Herbert, E. (1973) Globin synthesis on reticulocyte membrane-bound ribosomes. J. Biol. Chem. 248, 1556–1561.

Yahara, I. and Edelman, G. M. (1975) Electron microscopic analysis of the modulation of lymphocyte receptor mobility. Exp. Cell Res. 91, 125–142.

Yamada, K. M. and Weston, J. A. (1975) The synthesis, turnover, and artificial restoration of a major cell surface glycoprotein. Cell 5, 75–81.

Yu, A. and Cohen, E. P. (1974) Studies on the effect of specific antisera on the metabolism of cellular antigens. II. The synthesis and degradation of TL antigens of mouse cells in the presence of TL antiserum. J. Immunol. 112, 1296–1307.

Origin and fate of the membranes of secretion granules and synaptic vesicles: membrane circulation in neurons, gland cells and retinal photoreceptors

4

Eric HOLTZMAN, Samuel SCHACHER,
Judith EVANS, and Saul TEICHBERG

1. Introduction and background

This chapter is concerned with the formation, intracellular "circulation," and eventual degradation of the membranes surrounding secretory "packages"—secretion granules and comparable structures of gland cells, and the vesicles that abound at the transmitting terminals and varicosities of axons. The pertinent experimental tools and approaches currently available do not permit direct attacks upon several crucial areas of uncertainty; we lack reliable means to follow the membranes per se through all the stages of their life cycle. Thus, the sparse and fragmentary direct biochemical information about the membranes must be extensively supplemented with inferences from morphological studies and from indirect evidence, particularly from considerations of the behavior of the *contents* of the bodies delimited by the membranes. For example, for the gland cells with which we will be concerned, it is clear that much, if not all of the release of secretions depends upon *exocytosis*, the fusion of the membrane delimiting a secretory body with the plasma membrane. Therefore it is common to utilize the release of secretions under various experimental conditions (presence or absence of divalent cations, alterations in levels of energy metabolism, etc.) to monitor the effects of these conditions on relevant membrane behavior. The limitations of such approaches, especially for conclusions about the subtler aspects of membrane turnover, must be kept clearly in mind for all our discussions. Similarly, a key process in the events with which we will deal is *endocytosis* (DeDuve and Wattiaux, 1966), the formation of a cytoplasmic, membrane-

G. Poste and G.L. Nicolson (eds.)   *The Synthesis, Assembly and Turnover of Cell Surface Components*
© *Elsevier/North-Holland Biomedical Press, 1977.*

delimited vesicle or vacuole by pinocytic or phagocytic "budding" from the cell surface. To follow the behavior of endocytic structures, recourse is often had to extracellularly administered "tracers" that label the structures as they form and permit their identification within the cell. Microscopically detectable macromolecules such as ferritin or horseradish peroxidase (HRP; Graham and Karnovsky, 1966; Straus, 1967) are most frequently used for this but related work is also being done with antibodies, lectins, or toxins that bind to specific cell surface macromolecules (DePetris and Raff, 1973; Berg and Hall, 1975; Devreotes and Fambrough, 1975), with radioactively labeled exogenous proteins (Davidson, 1975) and with approaches such as the peroxidase-mediated iodination of cell surface proteins (Hubbard and Cohn, 1975). Some of these tracers may strongly influence the intracellular behavior of the structures in which they come to be contained. For example, lectins can inhibit fusion of endocytic vesicles with lysosomes (Edelson, 1974; Goldman, 1974). While such dramatic effects are generally not observed with HRP or with cell-surface iodination, the possibilities of less evident tracer-induced perturbations are difficult to evaluate adequately.

The conceptual context for our discussion is provided by several chapters in this volume and by other recent reviews (Siekevitz, 1972; Singer and Rothfeld, 1973; Taylor et al., 1973; Schimke and Katunuma, 1975; Whaley, 1975; Holtzman, 1976). Briefly stated, present general notions of membrane assembly and turnover involve two disparate bodies of information whose different emphases have yet to be adequately integrated. On the one hand, many biochemists stress the heterogeneity of turnover rates of the different lipids and proteins of a given type of membrane (Arias et al., 1969; Kuriyama et al., 1969; Siekevitz, 1972; Schimke, 1974; Kaplan, 1975; Schimke and Katunuma, 1975). This has led them to views centered around molecule-by-molecule exchange and growth mechanisms in which a membrane supposedly can remain morphologically intact while lipids and proteins enter from, or depart to, soluble pools, or are inserted by "special devices," such as the ribosomes that are bound to the rough endoplasmic reticulum and perhaps to certain other membranes (e.g., the studies on chloroplasts by DePetrocellis et al., 1970; Chua et al., 1973; Eytan et al., 1974; Philippovich et al., 1974). Increases in membrane mass may occur by widespread addition of new molecules throughout the expanse of a preexisting membrane. There also has been speculation about more localized addition of macromolecules to a restricted zone of a growing membrane, followed perhaps by lateral diffusion of the molecules so that "older" and "newer" molecules become intermixed (Singer and Rothfeld, 1973), though it is recognized that congregating or stabilizing mechanisms must exist which permit membranes to maintain such chemically and structurally stabilized regions as intercellular junctions (Staehelin, 1974; Gilula, 1975; Montessano et al., 1975). The possibility is also being tested that some membranes may grow in part by early incorporation of lipid-synthesizing enzymes leading to local synthesis of membrane lipids, the proteins being "imported" from elsewhere in the cell (Benes et al., 1974; Higgins, 1974).

Morphologists, on the other hand, for obvious technical reasons, have focused on processes of "bulk" membrane "flow" (Bennett, 1969) or transport processes such as exocytosis and endocytosis in which membranes move as structurally intact units from one part of the cell to another. Thus, for several cell types, a circumstantial case can be made for assembly of portions of the plasma membrane in the Golgi apparatus, and insertion of the membrane into the cell surface by exocytic-like fusion of Golgi-derived vacuoles. The major evidence for this is morphological in that distinctive looking membranes are found in various cell types in appropriate configurations (cf. Hicks, 1974; Morré et al., 1974; Porter et al., 1967) and surface coat and cell wall materials also seem to follow such a route (Bennett and Haddad, 1974; Masur et al., 1972; Vian and Roland, 1972; Wise and Flickenger, 1970). In addition, there are special cases, such as the formation and functioning of the acrosomes of sperm (reviewed in Whaley, 1975), in which Golgi-derived membranes appear to engage in unique activities at the cell surface. Some biochemical data on the kinetics of accumulation of radioactive labels (e.g., Chlapowski et al., 1971) and on comparisons of Golgi or endoplasmic reticulum membranes with plasma membranes (e.g., Hirano et al., 1972; Hodson and Brenchley, 1972; Bergeron et al., 1973; Morré et al., 1974; Little and Widnell, 1975) may also point in the same direction.

The microscopically visible aspects of membrane degradation involve incorporation of structures within lysosomes (reviewed in Holtzman, 1976). Various cytoplasmic organelles are observed to undergo destruction within autophagic vacuoles, and membrane internalized from the cell surface by endocytosis often appears ultimately to be incorporated with multivesicular bodies (Fig. 1). (As reviewed below, endocytized membrane may sometimes be returned to the cell surface, rather than being degraded.)

While the simplest concepts of bulk transport and degradation seem to imply homogeneous behavior of all the molecular components of a given structure, there is no necessary contradiction between the biochemists' molecule-by-molecule processes and the microscopists' "bulk" processes. Clearly, both could occur simultaneously. What is needed now is a systematic study of the relative importance of different mechanisms under varying conditions, clarification of the events by which structures and individual molecules come to be incorporated within lysosomes (see Holtzman, 1976 for discussion and references), and detailed work on the turnover of specific molecules. For example, a priori, one might expect some lipids to exchange between membranes and soluble pools more readily than "integral" membrane proteins, and perhaps there are corresponding differences in turnover. Some data suggest turnover coordination of certain proteins of a given membrane (reviewed in Holtzman, 1976), including the plasma membrane (Kaplan, 1975; Tweto and Doyle, 1976, and chapter 3 in this volume). There are other interesting considerations that may prove quite useful when factored into "reconciliations" of the biochemical and morphological data. For instance, it appears that under various experimental circumstances endocytosis selectively internalizes certain plasma membrane proteins, leaving others behind. This may not always depend upon "permanent" membrane mosaicism; rather, it may sometimes involve movement of proteins within the

168

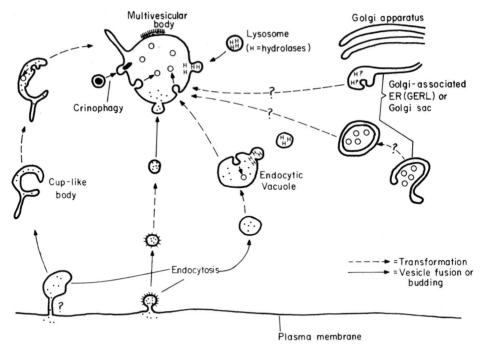

Fig. 1. Formation and functioning of multivesicular bodies (see Holtzman, 1976 for details and references). It is reasonably well established that the internal vesicles of multivesicular bodies (MVBs) can form by the invagination and budding of the surface membrane of the MVB and that membrane can be added to the MVB surface from endocytic structures. Morphological evidence tentatively suggests that some MVB precursors, such as cuplike bodies, may form directly from the cell surface while others may form from Golgi-associated membrane systems. There are also indications that MVBs can arise by the autophagiclike sequestration of vesicle-rich regions of cytoplasm (or of cisternae that break up into vesicles) within a sac derived from the ER or Golgi apparatus. Lysosomal hydrolases seem to enter MVBs through fusions with preexisting lysosomes, but some of the hydrolases may also be present in MVB precursor structures arising near the Golgi apparatus. Crinophagy refers to the fusion of secretion granules with MVBs or other lysosomes.

plane of the membrane so that an initially "homogeneous" region is converted into one in which membrane portions destined for endocytosis are selectively enriched in particular components (DePetris and Raff, 1973; Berlin et al., 1974; Ryan et al., 1974; Raff, 1976). (Microtubules, microfilaments, and other "structural" or "contractile" components in the cytoplasm subjacent to the plasma membrane are the prime candidates as governing agents for such behavior.)

For reviews of the basic biochemistry of the membranes of secretion granules and synaptic vesicles see DeRobertis, 1967; Ceccarelli et al., 1974; Hall et al., 1974; Winkler et al., 1974; Castle et al., 1975; Hortnagel, 1976. The membranes are composed of the usual variety of lipids and a few cell-type specific proteins, including some glycoproteins. In terms directly applicable to our discussions, few remarkable features have been reported thus far; some specific ones will be mentioned below in their appropriate contexts.

2. Membrane origins

In this section we will deal with the sites at which the membranes delimiting secretion granules and those bounding synaptic vesicles seem to be originally assembled. The question of reuse of "old" membranes to form new secretory structures will be considered in later sections.

2.1 Origin of the macromolecular contents of secretory bodies

Proteins or polypeptides are included in the contents of the secretory structures formed by many gland cells, by neurosecretory cells, and by some types of neurons. The proteins may constitute the characteristic secreted cell product (e.g., the zymogens of the exocrine pancreas). Alternatively, when the characteristic product is a low-molecular-weight hormone or transmitter (e.g., the catecholamines) proteins serve as enzymes or storage factors that participate in the synthesis or accumulation of the product within the secretion granule or synaptic vesicle (e.g., Kirschner, 1972; Whittaker, 1974; Winkler and Smith, 1974). It is generally assumed that the proteins and polypeptides destined for incorporation into secretory structures are synthesized on the ribosomes of the rough endoplasmic reticulum (ER) and enter the cisternae of the ER for transport to the sites at which they are packaged into granules or vesicles. This seems an appropriate assumption in most cases, though reasonably direct proof is available only for a relatively few cell types (see discussion in the reviews by Castle et al., 1975; Jamieson, 1975; Olsen et al., 1975; Palade, 1975; Whaley, 1975; Novikoff and Holtzman, 1976).

Polysaccharides or related peptidoglycans are also included in many secretory structures. In some tissues, secreted polysaccharides play major extracellular roles (e.g., as constituents of connective tissue matrices). In others, their roles may be at least partially intracellular—their interaction with proteins in maturing secretion granules may, for example, be important in the condensation processes leading to the establishment of an electron-dense granule with highly concentrated content. (See Jamieson, 1975 and Palade, 1975 for the proposal that condensation depends on the aggregation of proteins and polyanions producing an osmotic efflux of water.) Polysaccharides of secretion granules are thought to be synthesized by the Golgi apparatus (see Berg and Austin, 1976; and review by Whaley, 1975).

Both the ER and the Golgi apparatus possess enzymes that can glycosylate proteins. Evidently, the saccharide side chains of secretory glycoproteins are initiated in the ER and, in some cases, completed in the Golgi apparatus (reviewed in Whaley, 1975 and by Morré and Cook in this volume). The ER is also thought to carry out such modifications of secretory proteins as hydroxylation of the prolines in procollagen (Olsen et al., 1975). Precisely where other modifications take place is not certain. For example, the limited proteolysis that converts certain prohormones into their final active forms is thought to take place at about the time they are packaged into secretion granules, but whether this occurs in the ER, the Golgi apparatus, or the secretion granules is still in dispute (Steiner et al., 1974; Novikoff et al., 1975; Palade, 1975; Gainer et al., 1977).

Apparently, the synthesis of secretory lipoprotein components is carried out by the ribosomes of the rough ER and enzymes of the rough and smooth endoplasmic reticulum (Claude, 1970; Glaumann et al., 1975; Novikoff and Holtzman, 1976). The Golgi apparatus may make a contribution as well, though this is not adequately understood.

In the discussion below we will use information on the formation or behavior of lysosomes and peroxisomes. Although these organelles resemble secretory bodies because they are membrane-delimited collections of specific macromolecules, and the lysosomes are able to undergo fusion with vesicles or vacuoles derived from the plasma membrane, the similarities are only partial.

2.2. The formation of membrane-delimited "packages" in gland cells

2.2.1. Where does it occur?

The general consensus is that during or immediately after their synthesis, secretory macromolecules become sequestered within membrane-delimited channels or compartments such as the ER and that they remain within such compartments until they are released from the cell (see DeDuve et al., [1974] for relevant possible "deviations" in the case of peroxisomes, where some enzymes conceivably enter packages from soluble pools. Also see Rothman [1975] for a dissenting view on the pancreas, in an article valuable for stressing several points that still require clarification.).

In most of the familiar gland cells, the actual packaging of secretions into membrane-delimited granules (or, in some cell types, into comparable structures that may lack a notably electron-dense core) occurs mainly in the vicinity of the Golgi apparatus (e.g., Palade, 1959, 1975; Holtzman and Dominitz, 1968; Smith, 1972; Abrahams and Holtzman, 1973; Farquhar et al., 1975; Jamieson, 1975; Novikoff et al., 1975; Whaley, 1975). Generally the packages appear to form by the budding of dilated regions of membrane-delimited sacs or tubules. In the exocrine pancreas, however, the immediate precursors of secretion granules are "condensing" vacuoles found close to the Golgi apparatus (Jamieson, 1975; Palade, 1975). These vacuoles may actually arise as dilated portions of sacs, but the relations of condensing vacuoles, Golgi apparatus, and Golgi-associated ER are still to be finally delineated and may differ significantly under altered conditions and in different species (cf. Palade, 1975; Novikoff et al., 1976; Slot et al., 1976). The routes and mechanisms by which proteins destined for secretion reach the appropriate packages seem to vary in different cell types, and perhaps even for different products of a given cell type (note, for example, the apparent use of different portions of the Golgi apparatus for packaging of different materials in developing polymorphonuclear leukocytes [Bainton and Farquhar, 1966]). From extensive studies on the exocrine pancreas of guinea pig, and less detailed work on a variety of other tissues, it is believed that secretory proteins are often carried from the ER by vesicles that bud from the reticulum and fuse with sacs of the Golgi apparatus or with condensing vacuoles. In the pancreas, energy seems to be required for this transport (Jamieson, 1975; Palade, 1975). Limited work on other gland cells suggests this may be a widespread requirement (Howell and Whitfield, 1973). In the adrenal medulla (Holtzman et al., 1973), endocrine pancreas (Lazarus et al., 1966; Novikoff et al., 1975), and thyroid (Novikoff et al., 1974) evidence from morphological and cytochemical studies suggests that a different route applies in that some of the secretion bodies seem to arise as buds from smooth endoplasmic reticulum that may be located close to the Golgi apparatus but is in direct continuity with rough endoplasmic reticulum. (This more or less "direct" route has been most carefully detailed for transport of lysosomal hydrolases; the Golgi-associated endoplasmic reticulum from which lysosomes seem to form has been called GERL by Novikoff et al., 1971.) In other cases, morphological evidence suggests that ER to Golgi transport involves continuities between endoplasmic reticulum and Golgi sacs, or that cisternae of ER may transform into Golgi sacs (Flickinger, 1969; Claude, 1970; and see Holtzman et al., 1973 and Holtzman, 1976, 1977 for additional references and discussion of the often equivocal evidence concerning the various sorts of ER-Golgi relations outlined in this paragraph).

2.2.2. Where do the membranes come from?

Presumably the close structural and dynamic relations between the ER and the Golgi apparatus evident in the cells described above facilitate metabolic interactions between the two systems and accomplish such processes as the admixing of ER and Golgi products or the movement of a partially completed glycoprotein from the ER into a Golgi-associated compartment in which terminal glycosylations can occur. It remains for future studies to document such events more adequately and to make sense of the observed variations in terms of the ways in which different molecules are packaged appropriately into separate or common compartments.

For our purposes, the central point is that the same mechanisms which transport proteins from the ER to the Golgi apparatus also afford potential routes for the provision of membrane for the secretory packages. One could, for example, conclude that the membranes of the vesicles that bud from the ER are a source of the membranes that come to delimit mature secretion granules, or that the continuities and transformations alluded to above permit bulk movement or "flow" of membrane from the rough ER to the Golgi apparatus, or to Golgi-associated smooth ER, and then to secretory bodies. Since the ER is known to synthesize both lipids and proteins, it is logical to consider a possible role as a

172

source of membranes. The bound ribosomes offer obvious advantages for insertion of membrane proteins with extensive hydrophobic regions. For example, "integral" proteins might enter the membranes as they are synthesized through events analogous to those that permit secretory proteins to pass across the ER membrane (for discussion and references see Singer and Rothfield, 1973; Blobel and Dobberstein, 1975; Palade, 1975). There are some biochemical findings consistent with the origin of various membrane components in the ER (e.g., Hirano et al., 1972; Morré et al., 1974; and chapter 1 in this volume) including work that some interpret as reflecting membrane movement from the ER to the Golgi apparatus and then to the cell surface (Chlapowski et al., 1971; Morré et al., 1974; but see also Atkinson, 1975; Lodish and Small, 1975; Pfenninger and Bunge, 1974 for evidence suggesting that not all plasma membrane proteins follow this route; and see Autori et al., 1975, and Elhammer et al., 1975 for findings interpreted as indicating passage of some glycoproteins from the Golgi apparatus back to the ER.)

The simplest views of membrane flow cannot be sustained. Among other difficulties, it seems fairly clear that many of the secretory proteins that reach the Golgi region arrive in dilute forms and undergo the substantial condensation mentioned earlier, either in Golgi sacs or in forming secretory bodies. Especially for vesicle-mediated transport, this seems to imply that more membrane is required to bring the proteins from the ER than can be used to delimit the final packages that generally are roughly spherical and thus have "minimal" surface areas. This problem may actually be more apparent that real, if, as some believe, the transport vehicles do not fuse permanently with Golgi-associated structures but shuttle back and forth repeatedly between the ER and forming secretory bodies. Alternatively, perhaps some of the "surplus" membrane is degraded. Multivesicular bodies and other lysosomes that can participate in autophagy are quite common in the Golgi region of many cell types and might function in this capacity (Fig. 1; and see review by Holtzman, 1976). One might also argue, for example, that some of the "coated" vesicles seen near, or fused with, mature secretion granules (cf., Benedeczy and Smith, 1972; Novikoff et al., 1975) actually are removing excess membrane rather than contributing Golgi or ER products to the granule contents as is usually suggested (see also Kramer and Geuze, 1974).

2.2.3. Biochemical perspectives related to "membrane flow"
One of the first detailed biochemical studies on the turnover of secretion granule membranes (in parotid gland; Amsterdam et al., 1971) appeared to show that the membrane proteins were synthesized at the same time as the contents. This would be strong support for a membrane flow scheme. However, problems with adequate separation of the membranes from the contents frequently plague work on secretion granules (Castle et al., 1975) and may have contributed to these results. Studies using radioactive labeling approaches on adrenal medulla (Winkler et al., 1972) and exocrine pancreas (Meldolesi, 1974) indicate that the apparent half-lives of secretion granule membrane molecules are much longer

than those of the secretory contents. Of potential interest in light of possible membrane relations is the fact that in the exocrine pancreas the average half-life of the granule membrane (5 days) is similar to that of ER membrane proteins (4.5 days), though the heterogeneous patterns underlying these averages differ for the two compartments. These findings are thought by some to be significant in analysis of the overall membrane life cycle. This will be discussed in detail further on. For the problem at hand, that of membrane origin, they merely suggest that the membrane in which a secretion granule is clad is not made of molecules synthesized simultaneously with the contents. For membrane-flow hypotheses, one might maintain that this reflects the existence of appreciable pools from which secretion granule membranes are drawn. The great disparity in turnover rates between contents and membranes may also tentatively support the widely accepted concept that transport in the ER actually involves movement of material within the membrane-delimited channels rather than simply the movement of entire elements of the reticulum, membrane plus contents. The repeated "shuttling" of vesicles between the ER and Golgi-associated structures, if this does occur, might also be a factor in this disparity; for example, a vesicle might make several round trips before fusing permanently with a Golgi sac, or vacuole.

A more serious challenge to membrane-flow views comes from studies on the composition of the pertinent membranes. In systems such as the adrenal medulla (Winkler et al., 1974; Hortnagl, 1976) and pancreas (Meldolesi, 1974), secretion granule membranes seem distinctly different in their lipids and proteins from the membranes of rough or smooth microsomes (i.e., cell fractions including, to various extents, rough ER, smooth ER, plasma membrane, or Golgi apparatus). These differences may relate directly to biological roles. Thus, the ER membrane is a relatively permeable one, which may tie in with the extensive transfers of molecules occurring across it. In contrast, features such as the high content of sphingomyelin and cholesterol reported for some secretion granules, may produce a low permeability membrane whose incorporation in the cell surface during exocytosis minimally perturbs the cell's ionic and water balances (see Palade, 1975). One might also conclude from the occurrence of such differences that the extensive mixing of membranes of different compartments seemingly implied by straightforward membrane flow schemes does not take place. It could follow that secretion granule membranes originate not from the ER but rather from a presently unknown source, presumably in the Golgi region, that has independent growth and assembly capacities which permit it to sustain itself in the face of use of membrane for granule formation. Molecule-by-molecule replenishment that compensates for membrane withdrawal could be invoked. This does not violate the widely held concept that membranes form from preexisting membranes. For the processes discussed previously it also raises a number of reasonable corollaries. For example, it fits well with the proposal that the ER-derived vesicles do in fact shuttle back and forth rather than fuse permanently with condensing vacuoles or Golgi sacs. Also, it suggests that the morphological configurations interpreted as reflecting transformation of ER into Golgi apparatus are misleading, or only of secondary importance quantitatively, or that they relate to processes other than formation of secretion granule membrane.

2.2.4. Membrane transformation

Membrane-flow enthusiasts can construct ad hoc arguments to bypass these difficulties. For example, it could be that nascent secretion granule membrane represents some specialized portion of the ER that is not detectable in microsome preparations because it is not distinctive enough in pertinent characteristics to be separable by centrifugation or recognizable microscopically. In general, aside from the distinctions between rough and smooth ER, demonstration of heterogeneity in the ER in situ (Leskes et al., 1971) or among membranes of ER-derived microsomes has proved quite difficult, and the results are still cloudy (e.g., even for relatively homogeneous sources of microsomes, like liver, it is very hard to distinguish cell-to-cell variations from intracellular ones). Some heterogeneity in purified ER microsomes has been reported (Beaufay et al., 1974; De Pierre and Dallner, 1975), however, raising the possibility that different ER regions vary in their membrane composition. And, in morphological terms, formation of lysosomes and peroxisomes seems to involve special ER regions (Novikoff et al., 1971; Novikoff and Novikoff, 1973; Reddy and Svoboda, 1973), though it cannot yet be concluded that the actual synthesis of the macromolecules for these organelles is similarly localized to discrete ER zones. Related to this are hints that certain ER-derived "packages," such as peroxisomes, may be delimited by membranes that are "special" or unusual in composition, permeability, or other features when contrasted with the more traditionally studied "packages" such as lysosomes or secretion granules (Leighton et al., 1975). By comparison with the ER, relatively little membrane is invested in secretion granules (for the exocrine pancreas, morphometric studies suggest that the total surface area of the zymogen granule population equals only 5% of the area of the ER (see Bolender, 1975). Thus, a special ER source of granule membranes may be almost impossible to find by use of present biochemical procedures. (See also section 2.3.4 below for possible specialization of different regions of smooth endoplasmic reticulum in photoreceptors.)

Compromise is also possible. Hypotheses for the origin or interrelations of various cellular membranes often involve extensive membrane transformations through gain and loss of lipid and protein molecules (Siekevitz, 1972; Eytan et al., 1974; Novikoff and Holtzman, 1976). ER membrane might be modified into the very different membrane that eventually turns up at the surface of a secretion granule. Earlier we mentioned some of the post-translational changes occurring in secretory proteins produced by enzymes of the ER or Golgi apparatus. Membrane proteins might well be treated similarly. We also mentioned the possibility that vesicles forming from a membrane might be selectively enriched or depleted of specific membrane components. Somewhat more elaborate modifications could occur in the Golgi apparatus—progressive alteration in membrane thickness among sacs at different levels of the Golgi stacks has been noted in several cases and understood to reflect corresponding molecular remodeling by which a thin ER membrane is converted to a thicker plasma membrane that eventually is exocytically inserted into the cell surface (Morré et al., 1974; Whaley, 1975; and chapter 1 in this volume). (See also Little and Widnell, 1975 for the suggestion that proteins and other macromolecules may undergo major changes in conformation within an ER or Golgi membrane as it transforms into plasma membrane.) Molecule-by-molecule modification could also occur after the membrane has been emplaced at the secretion granule surface. For example, there are data that strongly suggest that individual proteins can be added to preexisting plasma membranes (Singer and Rothfield, 1973; Lodish and Small, 1975). Application of a similar mechanism to the secretion granule membrane might generate the working hypothesis that the relevant Golgi-associated sites do receive membrane from the ER but that this membrane is locally modified and perhaps even expanded through molecule-by-molecule replacement and addition. The ER membrane contribution might occur along the same routes used for secretory protein transport, but perhaps there are separate routes and mechanisms for handling membranes and contents (perhaps the stacked sacs of the Golgi apparatus represent stages in membrane elaboration or transformation, and the various transport mechanisms outlined earlier reflect ways in which different materials are brought into association with the proper membrane [see, for example, Claude, 1970; Holtzman, 1971; Novikoff et al., 1971; Holtzman et al., 1973; Morré et al., 1974; Palade, 1975; Whaley, 1975]).

Clearly, more information is needed to sort out the spectrum of possibilities. Work now underway on comparing the biochemical, cytochemical, and morphological properties of different structures of the Golgi region (Novikoff et al., 1971; Farquhar et al., 1974) should clarify the relations among the sacs of the Golgi stack, elucidate the differences that may exist among transport routes and mechanisms for different components of a given cell, and provide added insight into such features of the structures as the elaborate fenestration noted in some of the sacs present in or close to the Golgi stack (for example, do the fenestrations provide a large surface area for metabolic interactions with adjacent structures [cf. Novikoff et al., 1971], or do they reflect membrane alterations related to the budding and fusing of vesicles and vacuoles?). At present, however, we do not have even basic data on such questions as the nature of the devices that control organelle movements within the Golgi region: Do the vesicles that move from the ER to Golgi structures move at random and make the appropriate fusions largely because of propinquities between the "donor" and "recipient" organelles; or are the vesicles somehow guided, and the fusions controlled by membrane recognition mechanisms? Is the fact that some of the vesicles are "coated" important in the regulation of these processes (see below)? Are the numerous microtubules often seen near the Golgi apparatus important in controlling local traffic? (the microtubules of the Golgi area are often thought to be implicated in longer range transport, such as migration of endocytic structures, from the cell surface to the Golgi region or the movement of exocytic structure in the other direction). And, finally, what maintains the geometry of the Golgi stack—is some crucial clue to be found in the dense material present between the sacs in some cell types (see Whaley, 1975, for review and references), or in the disappearance or dispersal of the apparatus that may follow enucleation (Flickinger, 1969) or colchicine treatment (Maskalewski et al., 1976)?

Paigen and coworkers (e.g., Swank and Paigen, 1973) have shown that the localization of Beta-glucuronidase in different cell compartments (ER vs. lysosomes) depends on its interactions with other proteins that might help anchor the enzyme in membranes. Such proposals, and especially the "signal hypothesis" advanced recently to explain the interactions of ribosomes and the ER (Blobel and Dobberstein, 1975; Ojakian et al., 1977) may eventually help resolve some of the dilemmas outlined above. Does the cell use different "signals" to bind polysomes specifically to different membrane regions? Would this permit "specialization" of ER regions (see e.g., the comments above, and those on agranular reticulum in photoreceptors in section 2.3.4. below) or allow certain proteins to be inserted directly from polysomes into membranes other than the ER, such as the plasma membrane?

2.3. The formation of synaptic vesicles

2.3.1. The role of perikarya

Most, perhaps all, of the proteins and other macromolecules important for synaptic function are synthesized in the perikaryon and transported down the axon to the presynaptic terminal (Lasek, 1970; Dahlstrom, 1973; Barondes, 1974; Ochs, 1974; Hall et al., 1974; Grafstein, 1975; Jones, 1975; Lubinska, 1975). There have been reports of local protein synthesis in the terminals but their reliability and significance are yet to be established (see Mahler et al., 1975 for a critical review). Complicating factors in the interpretation of experimental data on terminals range from contamination of the cell fractions studied to the possibility that the synthesis is exclusively intramitochondrial. Exchange reactions involving portions of lipid molecules may occur locally in terminals, and there is a claim that acyl transferases may be present along axons (Benes et al., 1974). At present, however, there is no strong reason to conclude that extensive new synthesis of lipids occurs in terminals.

On the other hand, local synthesis of neurotransmitters and the extensive uptake of experimentally administered exogenous transmitters clearly take place in axon terminals (Hebb, 1972; Potter, 1972; Taxi and Sotelo, 1973; Hall et al., 1974; Jones, 1975). Certain central enzymes of transmitter synthesis, such as choline-acetyl-transferase, and perhaps tyrosine hydroxylase, are thought to be located largely in the "soluble" axoplasm outside the synaptic vesicles.[1] Others, such as dopamine-β-hydroxylase (DBH) appear to be components of the vesicle membrane or contents. When nerve terminals are isolated (as "synaptosomes"; cf. De Robertis, 1967; Whittaker, paper in Pappas and Purpura, 1972; Jones,

---

[1]There remain, however, adherents to the view that some of the choline-acetyl-transferase is associated with synaptic vesicles, at least in certain species (DeRobertis et al., 1975), and the same may be true of other seemingly soluble enzymes.

1975) and then ruptured, one commonly finds at least two "pools" of transmitter, a soluble pool and a vesicle-associated one. There are many problems in studying these pools: the possibilities of artifactual disruption of labile structures are difficult to evaluate; the terminal and vesicle populations studied may be heterogeneous to unknown extents; and relatively minor changes in ionic or other environmental conditions may affect the distribution of the transmitters. Thus there are many unresolved disputes.

At least for synapses of the cholinergic and adrenergic type it seems likely that the existence of multiple transmitter pools reflects, in some sense, the cooperation between vesicles and soluble enzymes in transmitter synthesis and storage and also the ability of terminals to reaccumulate components such as catecholamines released during transmission or choline produced extracellularly from released acetylcholine. The implications of available information about transmitter storage for the nature and properties of the vesicle membranes are still largely unexplored. A popular, though not unchallenged, view for catecholamine storage in the adrenal medulla is that a special energy-dependent membrane-associated transport system (perhaps an Mg^{2+}-ATPase) is involved in the passage of catecholamines into the secretory granules (Kirschner, 1972; Winkler and Smith, 1974), but the extent to which similar enzymes are required for the various types of synaptic vesicles is unclear. Within the vesicles, the transmitters may be stored as complexes with ATP, divalent cations, and proteins (Smith et al., 1970; Weinshilbaum et al., 1971; Pappas and Purpura, 1972; Rajan et al., 1974). A number of the known transmitters, such as acetylcholine, are electrically charged and the storage complexes presumably involve ionic bonds that neutralize the charges. For the adrenal medulla it is suggested that such complexes permit the maintenance of high concentrations of catecholamines within secretion granules isolated from the cell and incubated under conditions in which their membranes seem freely permeable to the amines (Kirschner, 1972). The native permeabilities of synaptic vesicle membranes to transmitters might thus be appreciable and the vesicles still be able to concentrate transmitters by forming intravesicular complexes, perhaps without requiring special transmitter transport across vesicle membranes. (See Holtzman, 1977 for references and further discussion, and Gainer et al., 1977 for their report on the enzymatic processing of neurosecretory materials that may occur within neurosecretory granules during their transport in axons.)

Insufficient data exists on the membranes of the various intracellular compartments of neurons to permit definitive or even detailed comparisons of synaptic vesicles with the ER or Golgi apparatus. However, from work with synaptosomes, it appears that the membranes of the vesicles in a given type of terminal differ in composition from the corresponding terminal plasma membrane (DeRobertis, 1967; Whittaker, 1974; Jones, 1975). Similarly, as discussed below, the available data on biochemical turnover of synaptic vesicle membranes cannot be interpreted unambiguously. Thus, virtually all that we know about the formation of synaptic vesicles comes from microscopy and allied cytochemical and autoradiographic methods.

In neurosecretory neurons the materials to be secreted include such components as peptides, which function as hormones or hormone-releasing factors. These are associated with "carrier" proteins within fairly large secretory vesicles (Walter, 1975). The secretory vesicles usually have a distinct electron-dense content. Vesicles of similar appearance seem to form from sacs in the Golgi region, and thus it is widely assumed that neurosecretory vesicles generally form in the Golgi region and are transported to terminals (Scharrer and Brown, 1961; Bargman, 1966; Howes et al., 1974; Osinchak, 1974). A similar assumption is often made for "conventional" neurons. Vesicles of various types, including some in the size range of synaptic vesicles (30–100 nm for the familiar vertebrate systems), abound near the perikaryal Golgi apparatus (Gray, 1970; Peters et al., 1970). Among these are structures resembling, at least morphologically, the distinctive-looking dense-cored vesicles found in various types of presynaptic terminals (Hokfelt, 1969; Taxi and Sotelo, 1973; Teichberg and Holtzman, 1973; Richards and Tranzer, 1975). This is most readily observed for the so-called large dense-cored vesicles (diameters of roughly 75–100 nm) that occur in terminals of several neuron classes as a minority population along with more numerous, smaller vesicles of types specific for given neurons. The roles of these large dense-cored vesicles are not understood. In adrenergic neurons both large and small dense-cored vesicles may function in catecholamine storage and release (Eranko, 1971; Geffen and Livett, 1971; Lagerkrantz, 1971; Dahlstrom, 1973; Hokfelt, 1974; Klein and Klein, 1974; but see also Taxi and Sotelo, 1973), but for other neurons it is not even universally agreed that the large dense-cored vesicles are involved in neurotransmission (see below). Other types of recognizable synaptic vesicles also have been noted in perikarya and axons (e.g., Thompson et al., 1973; Goldberg et al., 1976), and there have been observations of extensive arrays of synaptic vesicles arranged along axonal microtubules (Smith, 1971). This has been plausibly interpreted as reflecting microtubule-mediated transport of the vesicles from the perikarya, although, of course, the direction of motion is not evident from electron micrographs.

(Pickel et al., 1977, very recently reported the presence of substance P, demonstrable immunocytochemically, in large dense-cored vesicles, raising the hope that some of the ambiguities just alluded to may soon be resolved. See also Holtzman, 1977 for additional review of recent literature on synaptic vesicles.)

2.3.2. The axonal agranular reticulum
Taken as a whole, the evidence for vesicle formation in perikarya, though disturbingly sketchy, suggests strongly that neurons may resemble gland cells in their employment of Golgi-associated membrane systems for packaging processes. This may not be the whole story, however, or even necessarily the most important phase of it. In general, structures resembling synaptic vesicles are not prominent in normal axons and in adrenergic neurons, the larger dense-cored vesicles predominate in the axons and perikarya whereas the smaller ones are more abundant at terminals (e.g., Taxi and Sotelo, 1973). To some extent the apparent low frequency of vesicles in axons may simply reflect their temporary

dispersal in the large volume of "flowing" axoplasm during transport. This assumption has been used to explain the fact that ligated or otherwise interrupted axons may show striking accumulations of vesicles immediately proximal to interruptions, as if the vesicles are piling up there (for discussion and references, see Pelligrino-de-Iraldi and DeRobertis, 1970; Banks and Tomlinson, 1971; Geffen and Livett, 1971; Dahlstrom, 1973; Holtzman et al., 1973). Presumably, in normal neurons the terminals would represent similar "accumulation" sites for vesicles coming from the perikarya. However, evidence is growing that synaptic vesicles may also form locally in axons. Later, we will discuss how this may occur by vesicle recycling phenomena accompanying neurotransmission. Here we restrict our discussion to a review of the findings implicating an axonal system of membrane-delimited sacs and tubules that may transport materials from the perikaryon and give rise to synaptic vesicles. This system includes elements of axonal endoplasmic reticulum referred to as agranular since they lack ribosomes. One of the earliest indications that it might participate in vesicle formation came from studies of the zones in interrupted axons where vesicles accumulate. At these zones one often encounters morphological configurations, suggesting vesicle formation by budding from tubules (see, for example, Pelligrino-de-Iraldi and DeRobertis, 1970; Rodriguez-Echandia, 1970). This offers an explanation for vesicle accumulation alternative to or additional to the notions of damming up of transport of preformed structures.

Before proceeding to a detailed discussion of the axonal reticulum, two cautionary notes should be sounded. First, it is premature to argue that the sacs and tubules of axons constitute a single channel system or even that all of them are legitimately regarded as endoplasmic reticulum. As will be discussed subsequently in this chapter, evidence exists for heterogeneity or complexity of the structures at least in terms of their roles. Correspondingly, not only neurotransmission but also the movement of "trophic" substances, the maintenance or growth of the axon surface and other metabolic processes must be served by axonal transport, and movement of the materials necessary for these processes might also involve the agranular reticulum (Yamada, 1971; Bunge, 1973; Pfenniger and Bunge, 1974). Second, the evidence at hand concerning the functioning of the agranular reticulum is almost entirely circumstantial. We do not yet have a convincing picture of the details of transport of any specific synaptic macromolecule—perhaps the currently promising immunohistochemical work on components such as dopamine-β-hydroxylase (see Eranko, 1971 for background and Pickel, 1976 for progress report) will eventually provide the needed links between morphology and chemistry, but this has not been accomplished to date.

Serial section studies (Teichberg and Holtzman, 1973; Ducros, 1974) and, more recently, high-voltage electron microscopy of relatively thick preparations (Droz, 1975) have clearly demonstrated that many of the agranular sacs and tubules in axons are interconnected and thus form a continuous channel system through long stretches of the axon (Figs. 2,3). Although there may be a structurally continuous system connecting the perikaryon with the terminal, this has yet

180

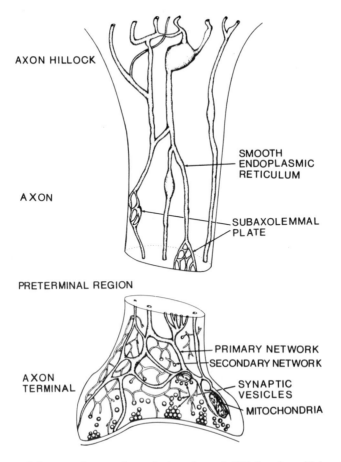

AXON HILLOCK

SMOOTH
ENDOPLASMIC
RETICULUM

AXON

SUBAXOLEMMAL
PLATE

PRETERMINAL REGION

PRIMARY NETWORK
SECONDARY NETWORK

AXON
TERMINAL

SYNAPTIC
VESICLES

MITOCHONDRIA

Fig. 2. Diagram of the axonal agranular reticulum (smooth ER) based on high-voltage electron microscope studies of thick sections (Fig. 3). (Courtesy of B. Droz; reproduced from Droz et al. [1975] Brain Res. 93, 1.)

to be demonstrated definitively. That the system is a form of endoplasmic reticulum is suggested (at least for the few cases studied so far) by its similarities to the perikaryal rough ER. For example, in cultured chick sympathetic ganglia we (Teichberg and Holtzman, 1973) have demonstrated the presence of two cytochemical ER "marker" enzymes, glucose-6-phosphatase and inosine diphosphatase in the axonal reticulum. These enzymes are also present in the perikaryal Nissl substance (and in the ER of other cell types). Furthermore, the axonal system shares with the perikaryal system a characteristic "thinness" in its delimiting membrane. The membrane is roughly 50Å in width (in contrast to the thicker (70 Å or more) membranes present at the cell surface and in portions of the Golgi apparatus). In perikarya continuities are usually encountered between rough and agranular endoplasmic reticulum, and in the neurons we have studied elements of the agranular reticulum are found in the perikaryal regions

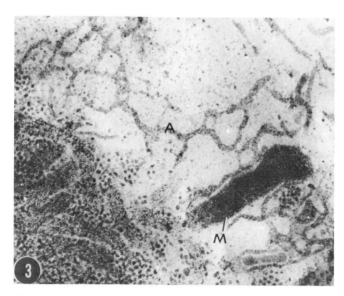

Fig. 3. Portion of a presynaptic terminal in a 1 μm thick section of chick ciliary ganglion stained with uranyl and lead. A, agranular reticulum; M, a mitochondrion. (Courtesy of B. Droz; reproduced from Droz et al. [1975] Brain Res. 93, 1.)

where the axon originates. However, it is not yet established whether the axonal reticulum is directly continuous with the perikaryal ER.

In terminals and in comparable axon regions, such as the varicosities of catecholaminergic axons, the axonal reticulum branches into a network of interconnected tubules. This has been demonstrated directly for a few cases, such as neuromuscular junctions (VonDuring, 1967) and chick ganglia (Teichberg and Holtzman, 1973; Droz, 1975), using serial section reconstruction and high voltage microscopy. In various other cases, membrane-delimited networks have been detected in terminals but a full description of their relations to the reticulum along axons has yet to be completed. This is true for example of vertebrate retinal photoreceptors, a cell type useful for our later discussions. In several animal species, elaborate branched networks have been described in the photoreceptor terminals (Yamada, 1965; Samorajski et al., 1966; Lovas, 1971; Pelligrino-de-Iraldi and DeRobertis, 1971). Similar networks have been noted in a few instances in the axonlike processes of the cells (Lovas, 1971), but usually the axons contain more conventional agranular sacs or tubules that can be shown in serial section to extend for considerable distances along the axon (Evans and Holtzman, unpublished observations). The axonal sacs and tubules and the terminal networks share cytochemical features, such as their ability to bind lead (see below and Fig. 5), but the extent of their continuity with each other has yet to be ascertained. (It is also quite important to bear in mind for our subsequent discussion that not all tubular networks in terminals are necessarily part of the ER. Some may be continuous with the cell surface; see Holtzmann, 1977 for additional discussion and references, and Bunge, 1973.)

Few biochemists studying axonal transport have paid much attention to the agranular reticulum, probably because adequate means for its isolation and analysis are presently unavailable. Cytochemically, however, acetylcholinesterase activity is commonly found in axonal sacs and tubules as well as in the perikaryal ER (see Eranko [1971] for reviews and other references to the extensive literature). It has been assumed that the enzyme detected in the axons is in transit perhaps partly to the plasma membrane or for other "extracellular" use (cf. Tennyson and Brzin, 1970; Chubb and Smith, 1975; Somogyi et al., 1975), though this is still uncertain. In injured neurons, acid hydrolase activities are demonstrable in axonal agranular sacs and tubules (Fig. 22; and Holtzman and Novikoff, 1965; Whitaker and LaBella, 1972; Holtzman, 1976) and these seem destined for inclusion in lysosomes or for related digestive roles. Hence the reticulum can seemingly transport enzymes. More recently, autoradiographic studies have demonstrated directly that the agranular reticulum is a route for relatively rapid transport of proteins (Byers, 1974; Droz, 1975).

2.3.3. Axonal reticulum and synaptic vesicles

Does the reticulum-mediated transport involve synaptic vesicle components (cf. Palay, 1958)? It seems quite likely that it does. Morphological signs of vesicles budding from axonal sacs or tubules have been encountered in various types of neurons (Fillenz, 1970; Machado, 1971; Korneliussen, 1972; Teichberg and Holtzman, 1973; see also Fig. 4). In the adrenergic neurons of cultured chick sympathetic ganglia, some of the vesicles that appear to be forming from the agranular reticulum are in the size range of the small-dense cored vesicles mentioned above (Teichberg and Holtzman, 1973); that these vesicles are able to accumulate catecholamines can be shown through use of the "false" transmitter, 5-hydroxy-dopamine, whose presence in the vesicles can be detected microscopically. In rat adrenergic neurons, a cytochemical chromaffin reaction used to detect amine storage sites produces electron-dense reaction products in both dense-cored vesicles and in networks of agranular tubules present in perikarya and in terminals (Richards and Tranzer, 1975).

In several types of neurons both the agranular reticulum and synaptic vesicles "stain" with the zinc-iodide-osmic (ZIO) method. (Akert et al., 1971; Barlow and Martin, 1971; Pelligrino-de-Iraldi and DeRobertis, 1971; Stelzner, 1971). The chemical basis of this procedure is not understood, but the results suggest a relationship between the reticulum and the vesicles as do some of our recent findings on frog photoreceptors. In both rods and cones, virtually all the synaptic vesicles show electron-dense deposits (Fig. 5) in preparations incubated in lead-containing cytochemical media, designed originally for the demonstration of certain phosphatases, but used without inclusion of the substrate (glucose-6-phosphatase, thiamine-pyrophosphatase, nucleoside diphosphatase and acid phosphatase media were used successfully; for references giving their composition, see Teichberg and Holtzman, 1973). Preincubation of the tissue in buffer lacking both substrate and lead does not affect subsequent production of the deposits, suggesting that endogenous substrates are not responsible. Thus it appears likely that we are seeing nonenzymatic binding of the lead present in the media. This lead binding may relate to the presence of transmitter storage complex components mentioned earlier or to other divalent-cation binding sites discussed by others in various contexts (Boyne et al., 1975; Pappas and Rosen, 1976). However, so little is known of the transmitters in photoreceptors that this idea cannot now be evaluated. For our purposes the phenomenon is interesting chiefly because the other structures that "stain" most consistently to an appreciable extent (Figs. 6,7) in addition to the vesicles are the tubular networks in terminals and the agranular reticulum present in the rod axons and in the perinuclear cytoplasm and adjoining portions of the myoid regions of both rods and cones (the myoids are the photoreceptor regions comparable to the perikarya of conventional neurons).[2] Configurations suggesting formation of vesicles of the size and appearance of synaptic vesicles from agranular reticulum are commonly encountered in the photoreceptors (Fig. 4). Several other authors have also noted electron-dense deposits of uncertain provenance in synaptic vesicles and agranular reticulum of other neurons (Kokko and Barrnett, 1971; Gray and Paula-Barbassa, 1974).

In the photoreceptors, lead-binding of variable intensity and frequency is seen in the structures of the Golgi regions. Usually some vesicles near the Golgi apparatus show the electron-dense deposits, and frequently a single sac near one face of the Golgi stack also "stains" (Fig. 7). Occasionally the deposits are present in several or all of the Golgi sacs. Comparable findings for the Golgi region emerge from ZIO studies (see, for example, Stelzner, 1971).

The autoradiographic investigations alluded to earlier show that some of the proteins transported relatively rapidly along axons are destined for inclusion in synaptic vesicles (see Koenig et al., 1973; Schonbach et al., 1973; and Droz, 1975; for retinal photoreceptors, see Young and Droz, 1968). From the observed kinetics it seems reasonable to suggest that they get there directly from the agranular reticulum. This is of interest in light of the biochemical data suggesting that at least some synaptic vesicle components, such as DBH, are transported relatively

---

[2]In frogs, the rods and cones differ notably in organization, and this may be relevant to the origin and transport of synaptic vesicles (see Figs. 4–7). In cones, the nucleus is located immediately adjacent to the terminal, and vesicles resembling those present in the terminals in size range (30–60 nm diameter) and lead binding are numerous lateral to the nucleus and in the perinuclear cytoplasm on the myoid side of the nucleus. Agranular reticulum is relatively abundant in the perinuclear cytoplasm and in certain other locations, such as the regions rich in longitudinally oriented microtubules and agranular reticulum that are part of the cytoplasm "connecting" the nucleus to the outer segment. (Such regions are seen in both rods and cones; Fig. 6.) In rods, fewer vesicles resembling those of terminals are seen in the myoid region. Those that are present tend to be found near portions of the Golgi apparatus in the perinuclear cytoplasm. Agranular reticulum is located in the latter region as well as in the axon that connects the nuclear region to the terminal (cones do not possess axons). Detailed comparison of synaptic vesicle formation in these two cell types may prove very useful in delineating the kinds of variations to be anticipated in conventional neurons.

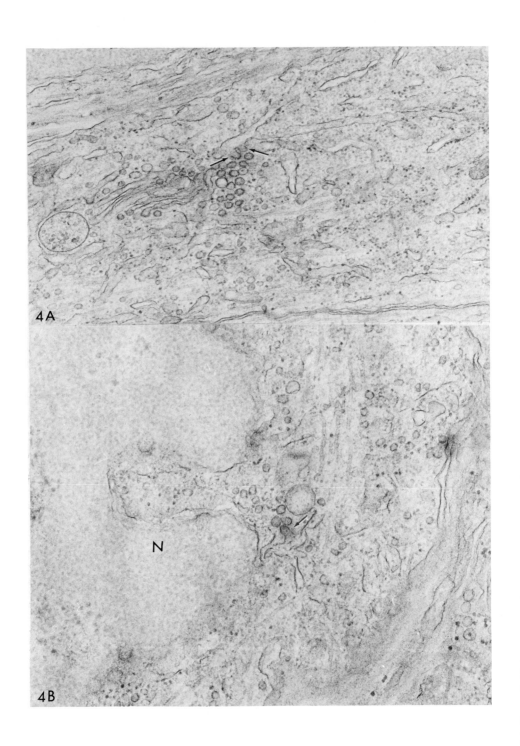

rapidly (Brimijoin, 1972; Thoenen et al., 1973; Wooten, 1973). Some axonal membrane components are also transported rapidly (Krusier-Brevart et al., 1974; Abe et al., 1975).

On the whole these findings comprise a defensible case for direct participation of the axonal reticulum in the formation of synaptic vesicles. Yet some ambiguities persist. In chick sympathetic ganglia, for example, few synaptic vesicles show the cytochemical phosphatase activities demonstrable in the ER (Teichberg and Holtzman, 1973). In rodent retinas, the tubular networks of the photoreceptor terminals are capable of oxidizing the cytochemical reagent diaminobenzidine at high pH, but neither synaptic vesicles nor the cisternae of the ER show such activity (Leuenberger and Novikoff, 1975). The rough ER of some neurons shares ZIO reactivity with the smooth ER (e.g., Joo et al., 1973), but in photoreceptors the rough ER usually shows little or no lead binding (Fig. 6), even though many direct continuities with portions of the smooth ER are observed. There are "precedents" in secretory cells for such observations—secretory granules forming from phosphatase-containing Golgi-associated sacs may lack demonstrable phosphatase activity or show it only when immature (see Osinchak, 1964; Smith and Farquhar, 1966; Holtzman and Dominitz, 1968; Novikoff et al., 1975). A detailed explanation of membrane relationships or transformation or of enzyme inactivation must, of course, be developed eventually to account for such observations.

There have been two reports that synaptic vesicles and axonal agranular reticulum both show thiamine pyrophosphatase (TPPase) activity, an activity that is most commonly detected in the Golgi apparatus (cf. Novikoff et al., 1971); the authors indicate that they examined substrate-free controls, which diminishes the possibility that they are actually demonstrating the nonenzymatic lead binding described above (Griffith and Bondareff, 1973; Csillik et al., 1974; also see Kadota and Kadota for their report that nucleoside diphosphatase is demonstrable biochemically in vesicles isolated from synapses). Generally, we have not found TPPase activity in synaptic vesicles of neurons or in the agranular reticulum in axons, although it is easily demonstrated in the Golgi apparatus of all neurons we have investigated. In photoreceptors incubated for TPPase activity, reaction product not attributable to nonenzymatic lead binding is visible in 5% of rod terminal vesicles and in occasional small tubules in the same terminals. No such product is seen in cones. We will discuss other ambiguities growing out of the observation of endocytized tracers in axonal sacs later on.

Fig. 4. Portions of cone inner segments from frog retinas. (A) The myoid region just above the nucleus (B) A portion of the cell lateral to the nucleus (N). These micrographs illustrate the apparent origin of vesicles from the agranular reticulum (arrows). Such configurations are not uncommon throughout the regions of agranular reticulum in the cell, and are not confined to the areas adjacent to the Golgi apparatus (cf. Fig. 6). The vesicles in question resemble the photoreceptor synaptic vesicles in size (diameters 30–60 nm) and appearance and evidently share with the synaptic vesicles the ability to bind lead from suitable incubation media (Figs. 5–7). In frogs the cone nucleus abuts directly on the synaptic region (Fig. 16), there being no intervening axon (cf. Rodiek, 1973) one might surmise that vesicles can move readily from the agranular reticulum to the terminals. However, the evidence is circumstantial. A, 39,000×; B, 45,000×.

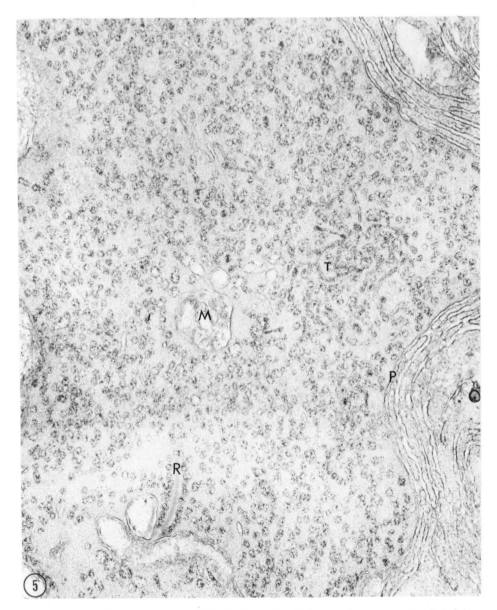

Fig. 5. Portion of a presynaptic terminal of a rod cell from a frog retina incubated in a glucose-6-phosphatase medium (Leskes et al., 1971) from which the glucose-6-phosphate was omitted. R, a synaptic ribbon; P, the terminal's plasma membrane. Electron-dense deposits are seen in virtually all the synaptic vesicles and in a network of membrane-delimited tubules (T). The deposits are far less evident or absent at other sites, such as within the vesicles of a multivesicular body (M). 45,000×.

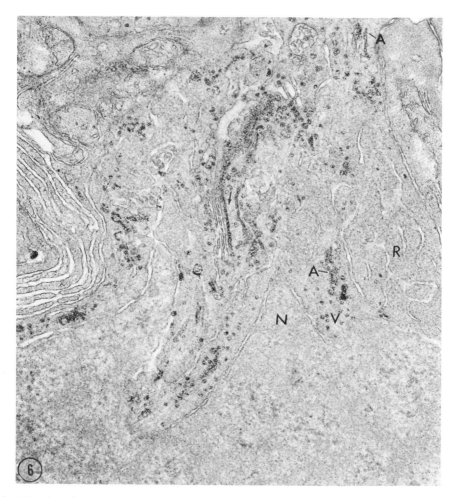

Fig. 6. Portion of a cone from a frog retinal preparation incubated as in Fig. 5, showing the cytoplasm near the nucleus (N). Electron-dense deposits are seen in numerous small vesicles (V), in agranular reticulum (A), and in a sac associated with the Golgi apparatus (arrow). None is present in the rough ER (R). 32,000×.

For the most part, students of axonal transport have attributed central roles as motive and orienting elements, to the microtubules and axonal filaments (Lasek, 1970; Banks et al., 1971; Geffen and Livett, 1971; Dahlstrom, 1973; Bunt and Lund, 1974; Lubinska, 1975; Yen et al., 1976). This has been based largely on inference, sometimes quite ingenious, from rather indirect evidence which, unfortunately, can rarely be interpreted unambiguously. For example, although inhibitor effects of colchicine are interpreted conventionally as reflecting involvement of microtubules, colchicine may affect membranes as well as microtubules (for discussion of this question and references see Douglas and

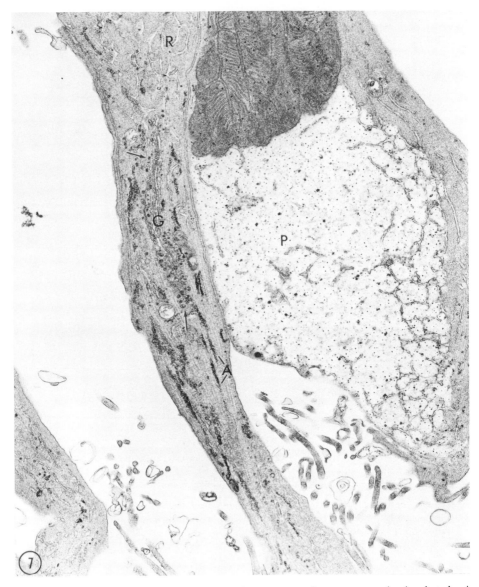

Fig. 7. Portions of the inner segments of retinal photoreceptors from a preparation incubated as in Fig. 5. P, the paraboloid region of a double cone (accessory member); the glycogen has been extracted during tissue preparation so the network of endoplasmic reticulum shows well. The cell at the left is probably the other member (principal member) of the double cone pair, although we could not observe the outer segment due to an inappropriate plane of section. This cell illustrates unusually well the kind of region rich in longitudinally oriented microtubules and agranular reticulum we have observed to be present (with varying prominence) in both rods and cones. Note the electron-dense deposits in vesicles (arrows) and agranular reticulum (A), and the absence of such deposits in the Golgi apparatus (G) and rough ER (R). 18,000×.

Sorimachi, 1972; Allison and Davies, 1974; Lubinska, 1975; Redman et al., 1975). Even if its influence on axonal transport depends upon microtubule disruption, this need not indicate exclusively a direct role of the tubules in movement of other materials; changes in the axon resulting from alteration of the structural supporting system represented by the tubules may well have a variety of indirect effects. We do not know precisely what is involved in transport mediated by the axonal agranular reticulum. Does the overall energy dependency of rapid axonal transport (Ochs, 1974) extend to transport in the reticulum? What is the relative rate of movement of the membrane of the axonal reticulum as compared to its contents? Could microtubules or filaments contribute to these movements? Lacking such information, it is difficult to integrate the views developed in the present section with the more frequently expressed proposals concerning axonal transport. Clearly, this is an important matter for future research. The observations of close relations, or apparent interaction of microtubules and neuronal agranular reticulum strongly merit follow-up (see e.g., Lieberman, 1971 and also Byers et al., 1973).

2.3.4. The Golgi apparatus

Our arguments are not intended to exclude roles of the Golgi apparatus or systems associated with the Golgi apparatus in the formation of synaptic vesicles. As in gland cells, in the various neurons we have investigated, several types of close interrelation between Golgi sacs and endoplasmic reticulum are readily evident. For example, as initially demonstrated for the Golgi-associated ER involved in lysosome formation (GERL; Novikoff and Novikoff, 1971), the smooth-surfaced sacs and tubules located at one face of the neuronal Golgi apparatus may include structures in direct continuity with the rough ER. Without careful study it is often impossible to decide whether a particular Golgi-associated sac or tubule is best thought of as part of the Golgi stack or as Golgi-associated ER. In fact, the actual usefulness of such distinctions must await elucidation of the metabolic interactions of ER and Golgi apparatus. From such considerations and the various lines of cytochemical and morphological evidence outlined earlier we tend at present to suspect that synaptic vesicles may originate both near the Golgi apparatus and in axons, and that this apparent dual origin of synaptic vesicles may actually represent geometric variation on a common theme. The same types of membranes may be present near the Golgi apparatus and in the axon and the same kinds of processes may occur in both locales. (A similar "dual" system may operate for lysosome formation in injured neurons —see review by Holtzman, 1976.) Unfortunately, there is no quantitative information on the relative importance of the two routes in a given neuron or in different types of neurons. Lacking constraining facts, one might speculate that in some way the neurosecretory cell represents or resembles an evolutionary forerunner that relies on the Golgi-associated membrane systems in similar fashion to its glandlike relatives. Conventional neurons, with their interplay of vesicles and soluble enzymes and pools, may have evolved to a greater utilization of the axonal route, which might facilitate the rapid and simple supply of vesicle

190

macromolecules. This, together with membrane recycling discussed later, may be important in permitting terminals to sustain prolonged periods of intensive transmission. From an evolutionary standpoint the suggested ER origin of "synaptic" vesicles of the primitive organism, *Ctenophores*, may prove to be particularly telling (Fig. 8; Hernandez-Nicaise, 1973). It could also be significant that many of the studies that suggest origin of synaptic vesicles from the axonal agranular reticulum were done on adrenergic neurons growing in situ, or in culture (e.g., Fillenz, 1970; Machado, 1971; Teichberg and Holtzman, 1973). While it appears that agranular reticulum is involved in vesicle formation in other neurons as well (Kornelieussen, 1972; Droz, 1975; Richards and Tranzer, 1975; and Fig. 6) the comparative importance of the reticulum and the Golgi apparatus could vary in different neuron types or during different periods of neuronal growth and maintenance (cf. Stelzner et al., 1971). Along these lines, it is interesting that the presynaptic terminals of certain developing neurons in the

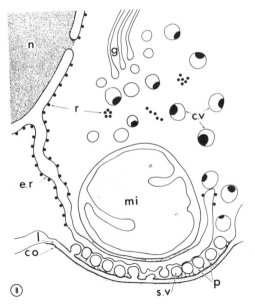

Fig. 8. Diagram illustrating the working hypotheses proposed by Hernandez-Nicaise, concerning the organization of a *Ctenophore* neurite at the region where it makes a synapticlike contact with an epithelial cell. Co, a cytoplasmic coat on the postsynaptic membrane; l, the synaptic cleft; p, densities in the presynaptic membrane; n, the nucleus of the presynaptic cell; r, ribosomes; mi, a mitochondrion. The diagram illustrates the observation that there are vesicles at the contacts (sv), agranular cisternae and mitochondria disposed as shown. Continuities of the cisternae with ribosome-studded endoplasmic reticulum (ER) are thought to occur on the basis of some suggestive micrographs, although full substantiation and evaluation of the extent of such continuities may require serial sections. The apparent origin of the "synaptic" vesicles from the agranular cisternae is inferred from configurations suggesting budding. Hernandez-Nicaise indicates that further work is needed to demonstrate that the "synaptic" vesicles release transmitters, to determine the significance of the exocyticlike configurations seen at the plasma membrane, and to evaluate the relations of the "synaptic" vesicles with some other vesicles (cv), apparently produced by the Golgi apparatus (g). (Courtesy of J. Neurocytology 2, 249, 1973; see also Couteaux, 1974).

visual system of *Daphnia* contain what seem to be stacked sacs of the Golgi apparatus (E. Macagno, unpublished).

It should be kept in mind that even if the Golgi apparatus per se does not generate all neurotransmitter vesicles, there may be other major tasks that help explain its extraordinary elaborateness in neurons. Apart from its likely cooperation with the ER in glycoprotein synthesis (and possibly in lysosome formation), the most obvious role would be in the maintenance of neuronal surface organization involving both the plasma membrane and the extracellular coats (cf. Elam and Peterson, 1976). Golgi vesicles have recently been suggested as a source of plasma membrane in perikarya (e.g., Rees et al., 1976), and this might be extended to axons as well. As implied earlier, some investigators consider that the axonal agranular reticulum may also be involved in surface maintenance and expansion, at least in growing axons (Bunge, 1973). Even if this is so, it would not exclude contributions by axonally-transported Golgi-derived vesicles (cf. Pfenninger and Bunge, 1974). For photoreceptors, the Golgi apparatus seems to participate in maintenance of the outer segments (Young and Droz, 1968), which presumably includes involvement in some aspects of membrane elaboration. The actual formation of the outer segment discs is based on invagination of the plasma membrane, but the relevant transport from the Golgi apparatus through the ellipsoid and cilium to the outer segment is not understood. Various potential participants can be identified, such as the groups of smooth-surfaced membrane-delimited tubules distinct from the Golgi apparatus and located at the base of the ellipsoid region (Fig. 9). These are most evident in rods. They are continuous with the rough ER, but differ from the other regions of the agranular reticulum discussed above in that they do not bind lead (A. Mercurio and Holtzman, unpublished). Given their location and the roles of the ER in lipid synthesis and in the transport and modification of proteins, one might anticipate that these systems will turn out to contribute to the maintenance of the outer segment. Interestingly, vesicles and a few tubules or sacs with contents reminiscent of the tubules in Fig. 9 are seen in the cytoplasm of the inner segment immediately adjacent to the outer segment (see also Young, 1968). In this region we also see occasional configurations suggestive of exocytosis and have demonstrated apparent uptake of horseradish peroxidase through endocytosis, suggesting that these processes may play some pertinent roles in membrane turnover, as they do in other cells discussed in this chapter. (For references and discussion on the migration of membrane through the outer segment and its eventual degradation by the pigment epithelium, see Holtzman, 1976.)

3. Membrane cycling during secretion or transmitter release

3.1. Some general considerations and background

As in axonal transport discussed earlier, the intracellular transport of secretion granules or synaptic vesicles to the sites where they release their contents has

192

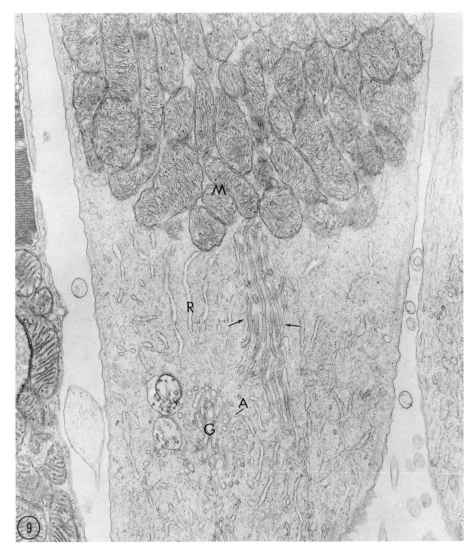

Fig. 9. Portion of a rod inner segment of a frog, showing the cytoplasm just below the mitochondria-rich ellipsoid (M). Golgi apparatus is seen at G and rough ER at R; note the Golgi-associated ER at A. Arrows indicate a group of membrane-delimited structures with modestly electron-dense contents and no ribosomes on their outer surfaces. Such structures are often seen in this region of the cell. Cross-sectional views suggest that many of them are tubules. However, many clearly are flat sacs and serial sections demonstrate continuities with the rough ER (Mercurio and Holtzman, unpublished). 19,000×.

often been suggested as involving cytoplasmic microtubules and/or the microfilament systems of the type that have aroused much recent interest among investigators of cellular motility (Goldman et al., 1973; Lazarides and Weber, 1974; Pollard and Weihung, 1974). These views derive from general considerations of the possible roles of microtubules (Soifer, 1975) and microfilaments (Allison and Davies, 1974; Pollard and Weihung, 1974; Berl, 1975) as "guiding" and "contractile" elements. These views also arise in part from the observed effects of colchicine, cytochalasin B, and other related agents in modifying the release of various secretion products and neurotransmitter substances (cf. Soifer, 1975; Rubin et al., 1976) and the release of enzymes from leukocytes (reviewed in Soifer, 1975 and Holtzman, 1976). Limited support for the involvement of cytoskeletal elements in vesicle transport has also been provided by direct microscopic evaluation of the association of microtubule or filament populations with secretion granules or vesicles under various circumstances and by direct biochemical studies on the state of tubulins or contractile proteins in cells of interest (see Pipeleers et al., 1976; and also Burridge and Phillips, 1975 for an effort to demonstrate that actin and myosin can interact directly with the membranes of adrenal medullary secretion granules). However, the pertinent data are complex and, as previously emphasized, the effects of cytochalasin B and colchicine may involve cellular elements other than the microfilaments and microtubules and are thus difficult to evaluate (for discussion and references, see Allison and Davies, 1974; Winkler et al., 1974; Lubinska, 1975; Palade, 1975; Redman et al., 1975; Mollenhaur and Morré, 1976). The validity of the various schemes involving microtubules and/or microfilaments that have been suggested cannot therefore be reliably evaluated at present. For cells such as those of the exocrine pancreas, passage of zymogen granules from the Golgi region to the cell surface neighborhood may be partly a random process that requires no elaborate special guidance beyond that provided by the overall distribution of ER, Golgi apparatus, and other organelles. Palade (1975) argues that newer and older secretion granules mix relatively freely in the cell and there is no preferential discharge of older granules; Sharoni et al. (1976) disagree. For neurons, there is no general agreement about the significance of data, suggesting that under some circumstances recently synthesized transmitter molecules are preferentially released or that at a given time terminals contain distinct subpopulations of vesicles with differing levels of transmitters or other components (Smith et al., 1970; Birks and Fitch, 1974; Hall et al., 1974; Heuser, 1976; Zimmerman and Whittaker, 1974; see also below). Given the disputes it is entirely unclear what might be implied in terms of selective vesicle use. Tracer studies summarized later on suggest that older and newer vesicles may be used for transmission substantially at random in some situations. Microtubules are not abundant in presynaptic terminals, although some are seen in large terminals such as those of retinal photoreceptors (for example, we regularly encounter a few microtubules among the synaptic vesicles in frog photoreceptors). Nor are readily identifiable filaments prominent, although there has been speculation about the presence of labile filamentous networks, and contractile proteins in terminals (Berl, 1975;

Gray, 1975). Gray (1976) reports that proper preparative techniques reveal microtubules to be more numerous in synapses such as those of retinal photoreceptors than is usually thought, and to be disposed in interesting patterns.

In the immediate vicinity of the sites where secretions or transmitters are released, morphological specializations exist that may participate in governing the interactions of secretory structures and the plasma membrane. In the exocrine pancreas and other gland cells (Allison and Davies, 1974; Palade, 1975; Franke et al., 1976), a meshwork of fine filaments directly underlying the cell surface may influence those granules that are immediately adjacent to the surface. In presynaptic terminals of neurons, special electron-dense material may be disposed near the plasma membrane in patterns to "control" access of vesicles to key plasma membrane sites. The synaptic ribbons of photoreceptors (Figs. 11,19) may represent one such configuration (Rodiek, 1973; Dowling, 1975; Pappas, 1975). In other types of neurons, less readily evident, apparently regular arrangements of electron-dense material are found along the cytoplasmic surface of the terminal membrane (Akert et al. paper in Pappas and Purpura, 1972; Pfenninger et al., 1972; Couteaux, 1974; Jones, 1975). Such components may interact in various ways with membrane macromolecules and thus "pull" or "guide" the granules or vesicles or suitably influence plasma membrane conformations; or, some of these cytoskeletal structures might provide steric "barriers" that prevent contact between the secretory vesicles and the plasma membrane and must be overcome by the mechanisms producing release of secretions (Allison and Davies, 1974; Pollard and Weihung, 1974; Burridge and Phillips, 1975; Stossel and Hartwig, 1976). The structures might also anchor specific membrane macromolecules or otherwise influence their distribution in the membrane.

As mentioned at the beginning of this chapter, the actual release of secretions from many gland cells occurs via exocytosis. Shortly, we will provide evidence that is probably true for neurons as well. It is frequently asserted that such release depends on a supply of metabolic energy, but direct evidence is available only in a few cases, involving certain gland cells (e.g., Palade, 1975; also see Allison and Davies, 1974; Soifer, 1975; Holtzman, 1976 for additional references). Precisely how the energy is "used" is not known. It could be required for membrane reorganizations related to exocytic fusions or for the initial events by which secretory structures are brought into close proximity to the plasma membrane. Cyclic nucleotides may also be important in controlling release, perhaps through influencing phosphorylations or other modifications of proteins of the membranes or transport systems involved (Allison and Davies, 1974; Palade, 1975; Weissmann and Claiborne, 1975; Haymovits and Scheele, 1976). In many cases the presence of calcium or certain other divalent cations has been shown to be required for secretion (see Ceccarelli et al., 1974; Pappas and Purpura, 1974; Quastel, 1974; Palade, 1975). In neurons and gland cells the factors triggering transmission or secretion are believed to produce a change in the ionic permeability of the plasma membrane, resulting in an influx of calcium ions (Bennett, 1974; Hall et al., 1974; Quastel, 1974). Some investigators consider that in the cell the calcium ions interact with contractile proteins or other structural or "mechanical" components that participate in release processes (see Allison and Davies, 1974 for review). Others suggest that the calcium ions interact directly with the plasma membrane or the membrane of the secretory structure, permitting the membranes to fuse or to approach more closely than would otherwise be possible (e.g., Poste and Allison, 1973; Heuser, 1976). Perhaps this involves the masking of negative charges in phospholipids, the establishment of ionic bridges between specific molecules in the two membranes, or conformational changes in particular membrane macromolecules. There is no a priori reason, however, why both of the above mechanisms should not be involved. (In some cases, the changes in intracellular calcium prior to exocytosis may reflect release of the ions from intracellular "stores" rather than changes in plasma membrane permeability.)

In morphological terms, the actual fusion of membranes during exocytosis seems to occur by a straightforward sequence of merger and elimination of the layers seen in membranes by conventional electron microscopy (Palade, 1975; Tandler and Poulsen, 1976). In protozoa, distinctive accumulations of the intramembrane particles visible with freeze-etch procedures are present at the points where secretory structures such as trichocysts and mucocysts attach to the plasma membrane (Satir and Satir, 1974; Plattner et al., 1975; Allen and Hausmann, 1976; Beisson et al., 1976). It is not yet clear whether these particles are important directly for exocytosis or whether they serve some related function, such as the maintenance of a close, relatively long-term relationship between the secretory organelle and the cell surface. One line of speculation as to their significance has centered on supposed roles in localized changes in membrane permeability (perhaps to Ca^{2+}). Or, they might serve as anchors for cytoplasmic materials, as direct, causal participants in exocytic (and endocytic?; cf. Orci and Perrelet, 1973) membrane reorganization or as recognition devices promoting specificities in fusions. In light of such possibilities, it is of obvious interest that extensive rowlike arrays of intramembrane particles (Fig. 10) have now been detected at the sites ("active zones") in the surfaces of neuromuscular junctions where transmitters appear to be preferentially released (Heuser et al., 1974; Ellisman et al., 1976; Heuser, 1976). The active zone also shows special arrangements of electron dense materials on the cytoplasmic surface of the nerve terminal membrane (Couteaux, 1974). (For an interesting recent detailed description of exocytosis based on freeze fracture techniques, see Pinto-da-Silva and Nogueira, 1977. These authors find that as the membranes fuse, intramembrane particles seem to be excluded from the diaphragm-like zone where the lipid layers of the two membranes have apparently merged prior to the actual opening of the secretory structure to the extracellular space.)

The chemistry of membrane fusion is incompletely understood. A fairly extensive literature exists on such topics as alterations in lipid turnover and exchange reactions in actively secreting cells (Hokin, 1968; 1969). This information has yet to be interpreted unequivocally in terms of detailed mechanisms, but it may relate to the rearrangements of membrane directly involved in exocytosis (and endocytosis). A special role of lysophospholipids in promoting exocytic

Fig. 10. Freeze-fracture view of a frog sartorius preparation showing a neuromuscular junction fractured along its zone of contact with the muscle. Arrows indicate rows of intramembrane particles present in the cytoplasmic face of the plasma membrane of the nerve terminal and concentrated along the "presynaptic ridges." This preparation was stimulated electrically during its fixation in a mixture of formaldehyde and glutaraldehyde. Consequently, dimplelike "vesicle sites" are fairly numerous near the rows of particles (s). These sites are thought to represent the interaction (fusion?) of synaptic vesicles with the plasma membrane. (Courtesy of J. Heuser; reproduced from Heuser et al. [1974] J. Neurocytology 3, 109.)

fusions has been postulated from effects of such lipids in model systems (Lucy, 1975) coupled with findings that in certain glands, notably the adrenal medulla, the secretion granule membranes are fairly rich in lysolecithin (Blaschko et al., 1967; Winkler and Smith, 1974). It is not always easy to ascertain that the apparent high concentrations of such components are not the artifactual result of phospholipase action during homogenization and centrifugation. Also, in nerve terminals the lysophosphilipid levels seem low (Baker et al., 1975). Thus generalization would seem premature. It also is too early to evaluate possible roles of membrane-associated enzymes, such as protein carboxymethylases (Diliberto et al., 1976) in promoting fusions.

One very important feature of membrane fusion that requires explanation is its specificity. For gland cells and neurons, secretion granules or synaptic vesicles fuse with only a limited portion of the cell surface—usually at a more or less well-defined "secretory pole" or "active zone." Fusion with structures other than the plasma membrane is generally not observed, although there are some exceptions. For example, in the phenomenon of crinophagy, secretion granules fuse with lysosomes (Fig. 1), and such fusions occur with enhanced frequency when surplus granules are being produced (Smith and Farquhar, 1966). In addition, during especially active secretion, cells may show release of secretory granules over a greater area of cell surface than is usully employed (e.g., Benedeczy and Somogyi, 1971). This may reflect specificity changes in the granules or in the plasma membrane or associated structures, or both. Other, simpler considerations may also be germane. For example, when exocytosis is rapid, secretion granule membranes may be extensively incorporated in the cell surface (see below); that such added membranes may contribute to broadening the area for potential release is suggested by the observation that in actively secreting cells, secretion granules can fuse with one another (Palade, 1975). The "recognition" devices underlying fusion specificities are unknown; presumably they depend on membrane macromolecules as well as on the mechanisms that bring the fusing structures into close juxtaposition. One difficulty in analyzing the problem is that we do not know how far the available body of data concerning plasma membrane composition and properties applies to localized regions of the cell surface such as those where secretion occurs. Presumably, most plasma membrane-rich cell fractions represent the entire cell surface, or unidentified areas thereof.

Recent progress reports on producing specific membrane fusions in vitro are encouraging and may foreshadow direct approaches to the questions raised here (Gratzl and Dahl, 1976; Oates and Touster, 1976).

The molecular mechanisms underlying endocytosis are less well understood than those of exocytosis. Even the widely assumed energy dependency is in dispute (reviewed in Poste and Allison, 1973; Allison and Davies, 1975, and Holtzman, 1976), and while there are numerous hints that both pinocytosis and phagocytosis involve interplay among membrane-associated receptors, other membrane molecules, cytoplasmic filaments, "contractile" proteins, microtubules, and so forth, few of the details are clear (see also the discussions in Berlin et al., 1974; DePetris and Raff, 1973; Ryan et al., 1974; Raff, 1976 concerning the requirements and controls governing movements of macromolecules within membranes that are destined for endocytosis).

3.2. Coupling of exocytosis and endocytosis in gland cells and protozoa

Among the immediate effects of secretory granule membrane fusion with the plasma membrane is an increase in the cell surface area. Since exocytic images are not difficult to observe under the microscope, especially with freeze-etch techniques (Orci et al., 1973a; Smith et al., 1973), it seems likely that the granule membrane remains fused with the plasma membrane for at least an appreciable period of time. However, even in very actively secreting cells, changes in the surface area are transient so there must be compensatory mechanisms that remove the secretion granule membranes from the surface or decrease the surface area by some other process. The only such mechanism that is currently well documented is the endocyticlike budding of vesicles, short tubules, and other structures from the surface membrane. This was suspected from general considerations (Bennett, 1969; Fawcett, 1962; Palade, 1975), and gained initial support from morphological studies on the parotid gland and other systems in which actively secreting cells were found to accumulate numerous "empty" vesicles in their apical cytoplasm (Amsterdam et al., 1969). There has been abundant confirmation of the existence of compensatory endocytosis from studies on several types of glands in which extracellular tracers were used to demonstrate enhanced endocytosis accompanying enhanced secretion (Abrahams and Holtzman, 1973; Orci et al., 1973b; Pelletier, 1973; Kramer and Geuze, 1974; Kalina and Robinowitz, 1975). For example, in work on the adrenal medulla (Abrahams and Holtzman, 1973), we detected a marked increase in uptake of horseradish peroxidase (HRP) during the first few hours after rats were stimulated to release their epinephrine (adrenalin) stores rapidly, through insulin induction of hypoglycemia (this condition sets off a neural mechanism that affects the secretory rate). The numerous tubules, vesicles, and other structures that incorporate the tracer accumulate whether or not HRP is present indicating that the endocytic events are not tracer-induced. Apparently, through endocytosis, the marked changes evident in cell surface topology when very active secretion has been induced, can be reversed within a few hours and the cell returned to "normal" (see Kramer and Geuze, 1974 for their estimates that it takes 1–3 hr. for such restoration in the exocrine pancreas).

At present, although no compelling evidence exists for surface-reduction mechanisms other than endocytosis, they cannot be ruled out. For example, the possibility that molecule-by-molecule dismantling of the surplus membrane occurs is difficult to test with present techniques but clearly merits further consideration.

In various cases, vesicles, with the cytoplasmic "coating" similar to that seen in pinocytic structures (Roth and Porter, 1964) have also been seen forming from those portions of the cell surface where exocytosis has just occurred (the sites are characterized by the indented "omega" configuration of the surface membrane, typical of exocytosis, and the presence of visible secretion granule contents in the adjacent extracellular space; Grynszpan-Winograd, 1971; Benedeczy and Smith, 1972). This is one of the few bits of available evidence directly involving the question of whether the membrane retrieved by endocytosis is actually the same membrane as that added to the surface by exocytosis. In these cases, apparently it is. However, endocytosis by actively secreting cells is by no means confined to the exact sites of exocytosis (see Kramer and Geuze, 1974), and there may be more complex patterns of membrane circulation than simple direct withdrawal. As will be discussed below, in the case of neurons a kind of membrane "flow" may occur along the cell surface, so retrieval can take place at sites distant from release. For gland cells, this also has plausible elements, though it remains to be explained how membrane could flow from the cell apex through the zones of intercellular junctions for subsequent retrieval at lateral or basal cell surfaces (cf. the suggestions of Kramer and Geuze, 1974), and the extent to which rates of endocytosis at these "nonsecretory" surfaces change during the events surrounding secretion has yet to be determined fully.

De Camilli et al. (1976) recently reported on freeze-fracture studies of the parotid gland. They observed that the plasma membrane and pertinent exocytic and endocytic structures differ in intramembrane particle distribution, and that endocytically-retrieved membrane resembles that of the secretion granules.

If the biochemical data indicating compositional differences between plasma membranes and secretion granule membranes are applicable to the specific regions of the plasma membrane where secretion occurs, then the implied absence of mixing could be construed as indicating that secretion granule membranes do not reside long at the cell surface. However, this in itself does not necessarily imply that the secretion granule is selectively retrieved from the cell surface since, in principle, net withdrawal of granule membranes could be accomplished even by relatively nonselective processes, if the plasma membrane itself undergoes reasonably rapid turnover (i.e., if the steady-state replacement of the cell surface proceeds at an appreciable pace, then the clearance of intermittently added material contributed by secretion granules and its replacement by "normal" cell surface could be accomplished as a matter of course, or simply by enhancement of the operation rates of the "ordinary" mechanisms). Few biochemical studies have been concerned directly with the membrane retrieval by gland cells. However, the fragmentary data available for systems such as the adrenal medulla suggests that the membrane retrieved shortly after secretion resembles secretion granule membrane in at least some respects (e.g., the presence of membrane-associated dopamine-β-hydroxylase; see Kirschner, 1972, 1974).

For protozoa, the morphological evidence on the circulation of exocytically added membrane seems somewhat clearer than for gland cells. It is possible that exocytically added mucocyst membrane remains "permanently" in the cell surface. Satir and Satir (1974) believe that mucocysts may serve to provide membrane for net surface growth. However, membranes of the trichocysts, contractile vacuoles, and residual bodies that fuse with the cell surface seem to be rapidly withdrawn through the direct formation of small vesicles and other endocytic structures from the exocytic configurations. For protozoa such as the ciliates, which have a complex specific surface architecture, such rapid withdrawal may help avoid unnecessary disruption of the architecture (Hausmann and Allen, 1976).

3.3. Exocytosis and endocytosis at nerve terminals

3.3.1. Exocytosis

Little doubt exists that extensive fusion of synaptic vesicles with the plasma membrane can occur at sites of transmitter release. Biochemically, release of transmitters is found to be accompanied by release of other components of the vesicle contests, most notably the nucleotides and proteins (Smith et al., 1970; Weinshilbaum et al., 1971; Musick and Hubbard, 1972; Boyne et al., 1975). This argues for an exocytic release mechanism that permits the "escape" of molecules too large to pass through membranes by other normal means. Direct morphological evidence for exocytosis came initially from work on neurosecretory systems in which the distinctive electron-dense contents of the secretory vesicle facilitated visualization of the process (Bargmann, 1966; Normann, 1969; Douglas et al., 1971). In conventional neurons, images suggestive of exocytosis are found with sectioned material and in freeze-etch preparations (Fig. 10). That these configurations are related to neurotransmission is supported by observations that they are especially frequent in preparations which are still actively releasing transmitters during the initial period of fixation for microscopy, and are "suppressed" by agents such as Mg^{2+} ions that inhibit neurotransmission (Heuser et al., 1974; Pfenninger and Rovainen, 1974). However, the exocytic images are not numerous under most conditions of specimen preparation or with special "quick-freezing" procedures designed to preserve the terminal in as "lifelike" a state as possible (Heuser, 1976). Evidently, except when fixation slows things down, the vesicle membrane remains in a recognizable exocytic configuration for a very short time (for vertebrate neuromuscular junctions estimates range down to a few milliseconds—Heuser, 1976). One is also given pause in evaluating the observations, by other findings, including the reported strong influence of technical details, such as the composition of fixatives, on the overall distribution and frequency of synaptic vesicles in terminals (Birks, 1974).

Firmer morphological evidence for exocytosis comes from studies in which induced, massive transmitter release is accompanied by disappearance of synaptic vesicles and corresponding increases in the cell surface area (or of the compartments participating in membrane retrieval discussed below). This has been observed for vertebrate neuromuscular junctions (Ceccarelli et al., 1973; Heuser and Reese, 1973) and for a few other systems (Atwood et al., 1972; Hamilton and Robinson, 1973; Zimmerman and Whittaker, 1974; Boyne et al., 1975; Model et al., 1975) in studies using agents such as black widow spider venom that deplete the vesicles in essentially irreversible fashion (Clark et al., 1972; Kawai et al., 1972; Holtzman et al., 1973; Hurlburt and Ceccarelli, 1974) or through employment of substantially reversible treatments such as prolonged repetitive electrical stimulation (Ceccarelli et al., 1972; Heuser and Reese, 1973; Pysh and Wiley, 1974; Model et al., 1975). In the latter cases, changes in terminal surface area are most convincingly seen when the temperature is kept low (10°C for frog preparations) or when stimulation is continued during the initial fixation period for microscopy. According to some reports the populations of vesicles that are depleted first are those located immediately adjacent to the terminal surface, as might be expected (Dickinson and Reese, cited in Heuser, 1976). Figure 11 illustrates changes in photoreceptor morphology that seem to signal the types of membrane rearrangements discussed in here occurring under relatively "physiological" conditions (Schaeffer and Raviola, 1976).

3.3.2. Endocytosis
The operation of endocytic membrane "retrieval" mechanisms at terminals is readily demonstrable through use of extracellular tracers such as horseradish peroxidase (HRP) or dextrans. Initially, this was documented for lobster neuromuscular junctions stimulated at high rates (Holtzman et al., 1971) and for neurosecretory cells (Bunt, 1969; Smith, 1971; Douglas et al., 1971; Nordmann et al., 1974; Pelletier et al., 1975). Subsequent extensive studies on electrically-stimulated frog neuromuscular preparations confirmed the original findings that stimulation of transmission also stimulates endocytosis and demonstrated directly the relationships between exocytic and endocytic processes (Ceccarelli et al., 1973; Heuser and Reese, 1973). Since similar transmission-related uptake of HRP has now been observed in the intact neurons of cultured mammalian spinal cord stimulated only to moderate levels of activity (Teichberg et al., 1975), the results probably do not depend either on nonphysiological rates of transmitter release, or on the fact that neuromuscular preparations have their axons cut (obviously, interruption of the axon might severely affect transport patterns or other pertinent processes). Other rodent central nervous tissue studies (Jorgensen and Mellerup, 1974) and our recent findings on HRP uptake by photoreceptors of the frog retina (Schacher et al., 1974; 1976) lead to similar conclusions. In the latter studies we have "reversed" the logic and used HRP uptake to monitor synaptic activity of terminals whose behavior is difficult to study by more direct means. Our findings on the effects of illumination (Figs. 14, 15, 16) correlate comfortably with predictions made from indirect physiological evidence, and suggest that tracer uptake may be broadly useful for such evaluations by both electron and light microscopy. We have also found that cobalt ions can mimic the effect of moderate light intensities in differentially inhibiting synaptic activity in rods with little effect on cones, a finding of interest in light of the supposed roles of divalent cations discussed above (Evans et al., 1977; we have confirmed these HRP observations with physiological measurements).

Tracers endocytosed by terminals accumulate rapidly within vesicles resembling the other synaptic vesicles. This is probably a major factor accounting for the repeated observation that it is difficult to engender marked depletion of synaptic vesicle numbers in terminals even by reasonably vigorous stimulation of transmis-

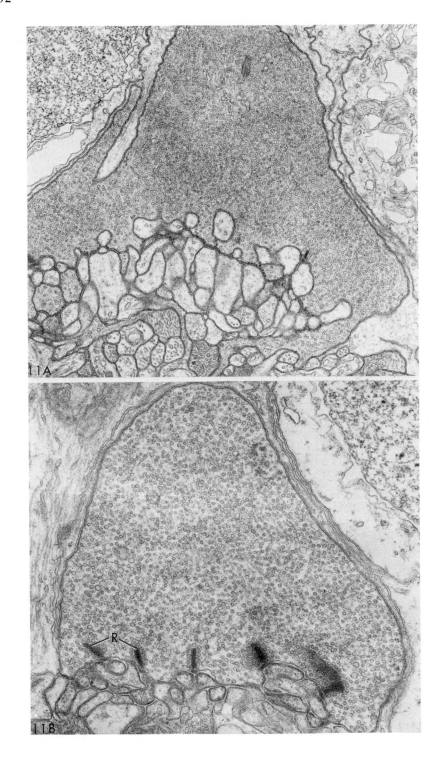

sion (see, for example, Bittner and Kennedy, 1970). Under ordinary circumstances, membrane retrieval can evidently "keep up" with membrane use. However, at very high stimulation rates or when the temperature is lowered, retrieval may lag behind, or the related "processing" of retrieved membranes to produce vesicles to be discussed shortly, may be slowed.[3] This apparent inability of endocytosis to keep up with exocytosis may explain findings such as our observation in the lobster neuromuscular system that the frequency of HRP-labeled synaptic vesicles was markedly increased if stimulated neurons were allowed to "rest" periodically by interrupting the stimulation for a few minutes. Interestingly, in work with HRP or other tracers by various groups, even with very extensive stimulation, the percentage of labeled vesicles in a given terminal rarely exceeds 30 to 50% and is generally lower. The significance of this is not known. Conceivably, only a subpopulation of the vesicles in a terminal engage in the processes leading to labeling. Or, perhaps all the vesicles can participate in such processes and the extent of labeling is simply a statistical matter reflecting the formation rates of new vesicles from the ER or other sources, the repeated reuse of a given vesicle for transmission (see below), unknown subtleties of vesicle distribution in the terminals, and so forth. A crucial problem in deciding among the alternatives is our ignorance about labeling efficiency—how many tracer molecules are needed for microscopic detection, and what is the probability of endocytic vesicles' acquiring the requisite number of tracer molecules under varying circumstances? Estimates could be based on the size of the vesicles and the concentration of tracer molecules in the media used. But lacking information about the time lapse between the initiation and the sealing off of a vesicle, as well as tracer binding to membranes, the figures obtained would be limited in value.

3.3.3. How are transmitters released?

Given that exocytosis and related compensatory endocytosis can occur at terminals, several major questions remain. First, does transmitter release depend completely on fusion of the vesicle membrane with the terminal membrane? Heuser and Reese's (1973) data suggest that during a 1 to 15-minute stimulation

---

[3]The effects of temperature, plus the fact that endocytic retrieval is severely altered or blocked with black widow spider venom, may provide opening wedges for experimental attacks on the metabolic events and membrane movements involved in retrieval.

Fig. 11. Presynaptic terminals of turtle cone cells from eye cup preparations. R. synaptic ribbons. (A) A retina maintained in the dark for one hour prior to fixation. (B) A preparation illuminated for 1 hour prior to fixation. The "dark" preparation shows elaborate infolding of its plasma membrane; the "light" preparation is far less extensively invaginated. These differences are thought to be due to the fact that neurotransmission is more extensive in the dark than the light (Figs. 14–16; and Dowling, 1975; Rodiek, 1973; Schacher et al., 1974, 1976). Apparently the addition of membrane to the cell surface is also more extensive in the dark, and this is interpreted as reflecting an exocytic mode of transmitter release (cf. also Schacher et al., 1974, 1976). (Courtesy of S.F. Schaeffer and E. Raviola, Cold Spring Harbor Symposium on the Synapse, 40, 521–528.)

enough vesicles disappear from neuromuscular junctions to explain the number of transmitter quanta thought to be released. This argues for a direct relationship between exocytosis and transmitter release. But the data must still be interpreted conservatively, especially since there are problems associated with transmitter release during fixation (Heuser, 1976) and difficulties in adequately assessing quantal numbers. Also, it is possible that some components thought to be released from vesicles may have additional sources at synapses (see, for example, the discussion relating to ATP by Meunier et al., 1975). Birks (1974) reported for cat sympathetic ganglia that depletion of acetylcholine stores does not directly parallel changes in the vesicle populations in the terminals and there is no notable expansion of the plasma membrane when vesicles are depleted. This has led him to advance the rather heterodox proposal that vesicles are at most indirectly related to transmitter release and although they may function in storage the actual release is directly from the "soluble" pools of transmitter present outside the vesicles in the terminals and depends on some type of voltage-dependent change in the permeability of the axon membrane to transmitter (Birks, 1974; Birks and Fitch, 1974). At present this proposal is valuable mainly for its emphasis on the gaps in current knowledge. It leaves unexplained many of the observations outlined above, and will become convincing only if supported by future analysis of the still murky relations between the different transmitter pools in terminals. However, at its boundaries this hypothesis merges with other interesting challenges to simple exocytosis models. For example, while it is true that vesicle proteins are detected in perfusates of actively transmitting preparations, the quantities of such proteins sometimes seem smaller relative to released transmitters than might be expected for complete exocytic release of vesicle contents (see, for example, Smith et al., 1970). Although there may be a simple explanation for this, such as the reincorporation of proteins into the terminals during endocytosis (see below) or their adsorption or uptake by non-neuronal cells present in the preparations (see also below for the possibility that only some of the vesicles in a terminal contain proteins), it may also mean that the release of transmitter does not require full or prolonged fusion of a vesicle with the plasma membrane. Perhaps only a very small aperture opens, or even only something resembling a membrane pore. Perhaps the fusion is so transient that an appreciable portion of the small molecules "leak" out but few of the large ones have the opportunity to do so (for discussion and references concerning such proposals, see Pfenninger et al., 1972; Smith, 1972; Hall et al., 1974; Heuser, 1976). The membrane cycling process discussed earlier may apply only to a percentage of the vesicles, namely, those that have fused too extensively with the plasma membrane "by accident," or those that have been depleted of their contents by repeated partial fusions, or some other special class (see Pfenninger and Rovainen, 1974). Since transmitters apparently can cross vesicle membranes during storage, one might even imagine that when vesicle membranes are incorporated into the cell surface the local surface permeability to transmitters is altered and molecules from the "soluble" pool can then pass out of the terminal. This would be congruent with hypotheses we have developed for permeability control in the toad bladder (Masur et al., 1971, 1972), and might also relate to the suggestion that a final equilibration of transmitters between a vesicle and the soluble transmitter pools occurs shortly before or at about the time of transmitter release[4] (Highstein and Bennett, 1975; Heuser, 1976). Each of these proposals requires at least as many special assumptions about the nature of the storage complexes, the heterogeneity of the vesicle population, the interpretation of microscopic images, and so on, as does the straightforward exocytosis model, and none is as well supported by the available evidence. Nonetheless, they all address themselves to important problems needing further study.

In summary, as discussed with additional details elsewhere (Holtzman, 1977) the available information does not permit an adequate decision among several alternative proposals as to the extent and significance of vesicle fusion with the surface during transmitter release.

3.3.4. Processes of membrane retrieval

Another area of controversy involves the timing and morphology of membrane retrieval. For example, if synaptic vesicles fuse only partially, or transiently, with the plasma membrane, then membrane retrieval would be virtually immediate and the "endocytic" vesicles would presumably be delimited by precisely the same bits of membrane as the "exocytic" vesicles. As with gland cells, this would explain the persistence of differences in composition between synaptic vesicles and the plasma membranes of terminals (DeRobertis, 1967; McBride et al., 1972; Whittaker, 1974). In neurosecretory systems, endocytic profiles seem to occur at sites where exocytosis has just taken place (Bunt, 1969; Douglas et al., 1971). For conventional neurons the situation is less certain. Occasionally extracellular tracer is seen within a vesicle continuous with the plasma membrane at the active site of a neuromuscular junction (Ceccarelli et al., 1972), but there is no indication that the vesicle in question was about to return to the cell's interior rather than being one that had just arrived at the surface and was about to complete its fusion and "disappear" into the surface. Heuser and Reese (1973) and Couteaux (1974) have stressed that even during the initial periods following stimulation, the regions near the active zones of neuromuscular junctions show no special concentration of vesicles identifiable as endocytic in origin either by the presence of a cytoplasmic coating or by their appropriate tracer content. Actually, most such vesicles seem to form at points distant from the active zone (e.g., where processes of Schwann cells are closely associated with the terminals). Such observations, plus viewing vesicle depletion and enlargement of the surface under suitable circumstances, indicate that membrane retrieval need not be

---

[4]If vesicle membranes are incorporated into the cell surface for any prolonged period, an explanation must be provided for the fact that this seemingly does not make the surface "permeable" to transmitter. Why should the vesicle membrane permit transmitters to enter the vesicles but, on insertion in the cell surface, not permit them to pass out of the terminal? (See Holtzman, 1977 for further discussion and references; there is some evidence that the surface may actually be "leaky"—Katz and Miledi, Proc. Roy. Soc. Lond. B 196, 59–172.)

immediate, and raise the probability that synaptic vesicle membrane can merge with the plasma membrane and perhaps "flow" to a "distant" site prior to retrieval. As further evidence for this hypothesis Heuser (1976) cites his observations on neuromuscular junctions, suggesting that during extensive release of transmitter intramembranous particles from the synaptic vesicles become widely dispersed in the plasma membrane surrounding the active zone.

Clearly, retrieval might occur by somewhat different mechanisms and at somewhat different rates under varying circumstances or in different types of terminals.

3.3.5. Coated vesicles and cisternae

There is an ongoing debate about whether most of the vesicles that retrieve membrane from the terminal surface are coated. These disputes stem from the extremely varied frequencies of coated vesicles encountered by different groups, even in the same types of preparations (see Zachs and Saito, 1969; Hurlburt and Ceccarelli, 1972; Heuser and Reese, 1973). Since coated vesicles seem able to "lose" their coats soon after forming (Roth and Porter, 1964; Friend and Farquhar, 1967; Douglas et al., 1971), the controversy may have a simple technical basis: differences in timing, details of fixation, or other relatively minor variations may be the culprits. (We, and others such as Gray, 1975, have noted variations in the apparent frequency of coated vesicles in the same types of terminals fixed for microscopy by varying procedures.)

The significance of the cytoplasmic coating of endocytic vesicles is still to be determined. A plausible assumption is that coating represents a special device involved in the pinching off of vesicles from larger surfaces (Roth and Porter, 1964; Kanesaki and Kadota, 1969), or that it reflects, as a fixation artifact, the attachment of local membrane regions to a more widely distributed cytoplasmic network of filaments (Gray, 1975). Certainly, it is reasonable to consider the coating as a potential site where any energy requirements for membrane reorganizations may originate (Model et al., 1975; Franke et al., 1976; Heuser, 1976). However, testing such ideas will have to await further extension of the recent progress in isolating and analyzing the vesicles (Kadota and Kadota, 1973; Pearse, 1976). The possibility that coated vesicles are selectively enriched in certain membrane proteins or lipids (cf., Heuser, 1976) is, of course, also germane.

Coating is widely observed on the vesicle surfaces forming from various intracellular surfaces such as certain sacs present in the Golgi region. For example, some of the lysosomal vesicles that originate from Golgi-associated membrane systems are coated (Friend and Farquhar, 1967; Holtzman et al., 1967; see also Locke and Huie, 1976). Thus, unless independent corroborative evidence exists, it is very dangerous to use coating as a membrane marker as is done occasionally. The fact that coating is present both on endocytic vesicles and on vesicles fused with nearby intracellular sacs or tubules does not necessarily mean that the endocytic structures are fusing with the intracellular ones.

This last point relates to the observation that various structures larger than either coated vesicles or synaptic vesicles exist among the intracellular sites of early accumulated endocytosed tracers in terminals. These structures include short tubules, larger sacs or vacuoles often referred to as cisternae, and cuplike bodies or other precursors of multivesicular bodies (Fig. 1). Many of them appear to participate in the degradative aspects of membrane turnover discussed in the next section. However, as originally formulated for neuromuscular junctions (Heuser and Reese, 1973), the most widely discussed, detailed scheme for membrane circulation from the cell surface to synaptic vesicles suggests that cisternae may have special significance as intermediates. Supposedly these form by fusion of endocytically-derived coated vesicles, then give rise to synaptic vesicles by budding. However, several problems exist with this as a potentially general model for recycling of synaptic vesicles. In systems such as photoreceptor terminals, tracer-containing vesicles are seen occasionally in configurations that suggest budding from small, endocytic tubules (Fig. 12). In these terminals, however, and in various other types, large cisternae are not prominent; and even in

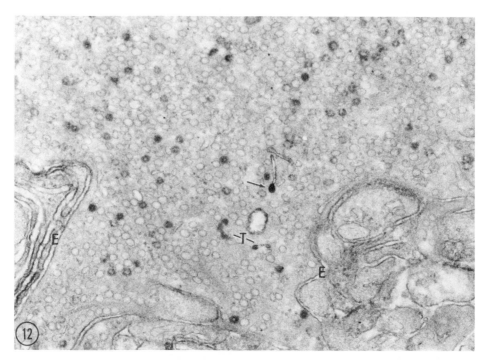

Fig. 12. Portion of a terminal of a rod cell from a frog retinal preparation maintained in the dark for 60 minutes with horseradish peroxidase present. Peroxidase reaction product is present in synaptic vesicles and in a few short tubules (T); one of the vesicles (arrow) is seen as if budding from a tubule. The paucity of reaction product in some of the tubules could reflect their continuity with the cell surface at points not included in the present thin-section; appropriate tubulelike invaginations of the plasma membrane are encountered in the terminals and the sparseness of the tracer may be due to the fact that we routinely rinse away much of the extracellular tracer prior to fixation (Schacher et al., 1976; note the absence of heavy deposits of reaction product in the extracellular space at E). 48,000×.

preparations where cisternae are found, they are most evident or persist longer under special circumstances (high rates of transmitter release and particular temperature levels—see Heuser and Reese, 1973; Model et al., 1975). As just outlined, the images suggesting coated-vesicle fusion are ambiguous. Furthermore, the participation of tubules and sacs in endocytosis has long been observed for many cell types where synaptic vesicle formation is not involved (e.g., De-Bruyn, 1975; Quatacker, 1975; Holtzman, 1976), and while such bodies may form through vesicle fusion, others may actually arise directly from the cell surface. Finally, freeze-etch studies of the cisternae have now shown that large portions of their surface lack the intramembranous particles characteristic of synaptic vesicles (Heuser, 1976). Thus the importance of cisternae is uncertain. Under certain conditions they may function to some extent in membrane recycling as initially envisaged, but their involvement in membrane degradation is also possible (cf. Heuser, 1976). At present, it seems likely that many synaptic vesicles of endocytic origin arise directly from the cell surface or from infoldings of the surface, perhaps by transformation of coated vesicles formed at these sites. However, it probably is too early to draw final conclusions about the relative importance of different recycling routes. The most prudent interpretation of available information probably is that the fusion of coated vesicles to form larger sacs which then break up into smaller vesicles has yet to be demonstrated decisively (see Heuser, 1976; Holtzman, 1977 for additional discussion).

4. The fate of retrieved membrane: reuse and degradation

4.1 Background and methodological problems

What happens to membrane once it has returned to intracellular compartments? Is it reused directly as membrane, is it disassembled into its constituent macromolecules, or is it broken down into small molecules which then reenter metabolism? (The alternatives were recognized early in the history of modern cell biology—see Palade, 1959; Fawcett, 1962; Bennett, 1969). In principle biochemical studies on membrane turnover could provide decisive evidence, and several studies have been cited as strong support for direct reuse of membranes or membrane macromolecules. For example, as mentioned earlier, in both pancreas and adrenal medulla the apparent protein turnover of secretion granule membranes is much slower than that of the corresponding granule contents. This has been used as the basis of an argument for the repeated reuse of the membranes (Meldolesi, 1974). Similarly, when one pulse-labels brain and then studies synaptosomes prepared at later intervals, the half-lives of the synaptic vesicle membranes appear to be on the order of several weeks (Rodriguez-de-Lores-Arnaiz et al., 1970; Van Hungen et al., 1970; De-Lores-Arnaiz et al., 1971; Marko and Cuenod, 1973). Based on reasonable estimates of the neurotransmission rates and the numbers of vesicles present at a given time, this would imply extraordinarily repetitive reuse of the vesicles—probably thousands to millions of times. There are, however, technical ambiguities that render interpretation uncertain. For example, potential problems with tracer reutilization and membrane purification have not always been adequately controlled; when labeled amino acids or other precursors can reenter synthetic pathways after having been released through degradation of the macromolecules in which they were originally incorporated, there is an artifactual lengthening of the apparent half-lives measured by conventional techniques (see Holtzman, 1976 for general review and references, and Mahler et al., 1975 for discussion of events in nervous tissue). Further, interpretation of radioactive-tracer data depends on assumptions about how precursor pools contribute to the structures being studied. For our present considerations, it may be misleading to treat the systems as if the secretion granule membranes or synaptic vesicles had direct access to the pools of low-molecular-weight precursors from which their macromolecules are formed. From our prior discussion one could conclude that in gland cells, the "pools" that contribute directly to granule formation may well be the membranes of the ER or the Golgi apparatus (hence the limited correlation between turnover of ER and secretion granule membranes mentioned earlier may prove important). In neurons it seems clear that the synaptic vesicles isolated from synaptosomes have a long "prior history" that must be considered. Their membranes are probably assembled at sites distant from the terminals, and are transported to the terminals by lengthy processes. Most such factors would produce an apparent slowing of turnover. The membranes would appear to last longer than they actually do, since even if a given bit of labeled membrane is completely degraded it can be replaced by other labeled membrane from the "reserves" in the ER, or from an axonal transport system. It is interesting that determination of the rates at which synaptic vesicle components such as dopamine-β-hydroxylase are transported along nerves suggest relatively rapid vesicle turnover. Insofar as ligation experiments provide accurate bases for judgment, sufficient quantities of such components seem to pass down axons to replace the supplies in terminals in one or a very few days (DePotter and Chubb, 1971; Brimijoin, 1972; Lauduron, 1974). If this is so, we would presume that vesicle half-lives are correspondingly short.

The extensive body of data indicating substantial lipid turnover in actively secreting cells also tends to suggest reasonably rapid rates of membrane replacement (Hokin, 1968; 1969). However, the full implications of the findings are difficult to evaluate because of the possibility of lipid exchange between membranes and soluble pools, and the uncertainty about what may transpire during the fusion and budding processes of endocytosis and exocytosis. Furthermore, there are some problematic findings such as the observation by Gerber et al. (1973) that the enhancement of phosphatidylinositol synthesis engendered in the pancreas by secretagogues is confined entirely to the microsome fraction; the effect does not produce enhanced labeling of secretion granules.

Before proceeding to a more detailed discussion of the fate of retrieved membrane in secretory cells and neurons, a brief note on other cell types may be

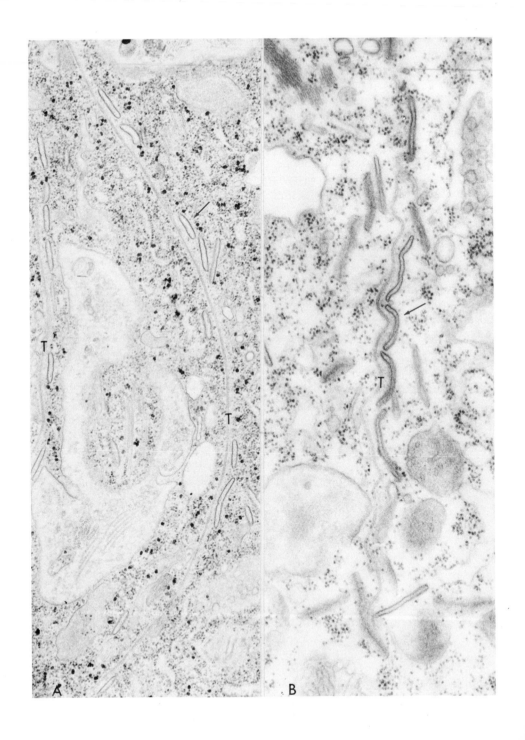

useful for the "precedents" established. In particular, findings in certain mammalian cells and protozoa seem to indicate that endocytosed membrane can either be reused or degraded depending on the cell and the circumstances. Thus, when fibroblasts engulf indigestible latex particles, the proteins of the internalized segment of plasma membrane that surround phagocytic vacuoles seem to undergo degradation within the cell (Hubbard and Cohn, 1975). On the other hand, circumstantial evidence for protozoa (and perhaps for fibroblasts pinocytosing horseradish peroxidase—Steinman et al., 1976) points toward reutilization of endocytosed membrane. Morphological studies on ciliates suggest that membrane taken into the cell as the delimiting membranes of food vacuoles can bud off maturing vacuoles and then be transported along microtubule-delineated routes (Fig. 13), back to appropriate portions of the cell surface into which they can be incorporated by fusion for possible reuse in endocytosis (see legend for Fig. 13).

Finally, if endocytosed membrane can be reused directly, then the question of the "fidelity" of retrieval may be a crucial one. If compensatory endocytosis sometimes occurs at sites distant from the corresponding exocytosis, as seems likely from the observations cited earlier, then to what extent and by what mechanisms does the cell channel or select the membrane destined for endocytosis to return usable membrane to the cytoplasm? The selectivity devices mentioned at the beginning of this chapter may provide an answer. Thus it is encouraging that preliminary results with freeze-etching (Heuser, 1976) hint at selective congregation of intramembranous particles at the endocytosis sites in neuromuscular junctions.

4.2. Gland cells

When adrenal medulla cells are stimulated to secrete most of their epinephrine stores rapidly by means of insulin-induced hypoglycemia, the membranes that return to the cell interior through the endocytic-like retrieval mechanisms (above) seem destined largely for degradation in lysosomes. Many large, multivesicular bodies (MVBs) accumulate, and studies with horseradish peroxidase (HRP) indicate that these are the primary sites to which the endocytic structures deliver their contents (Abrahams and Holtzman, 1973; see Koerker et al., 1973 for opposing views). The MVBs accumulate whether or not HRP is present. This

---

Fig. 13. Regions of cytoplasm from *Paramecium* illustrating the close association of distinctive-looking membrane-delimited flattened vesicles (arrows) with microtubules (T). When the microtubules are distorted during fixation (B) the vesicles seem to follow the new contours suggesting some type of linkage to the tubules. The vesicles shown are thought to be bounded by membranes pertaining to food vacuole formation. R. D. Allen (J. Cell Biol. 63, 904, 1974) has proposed that microtubule-mediated transport of the vesicles is part of a mechanism by which membrane can be added to the cytopharynx (by fusion), participate in endocytosis and subsequently be retrieved for recycling back to the cytopharynx by budding from the food vacuoles as digestion proceeds or during egestion of digestive residues. (Courtesy of Richard D. Allen; [1975] reproduced from J. Cell Biol 64, 503.)

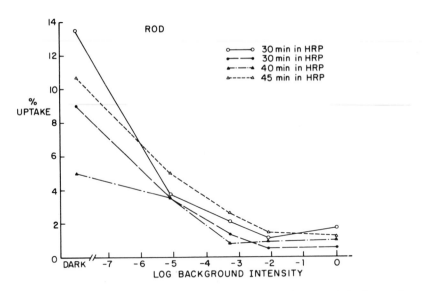

Fig. 14. Data illustrating the uptake of horseradish peroxidase by rod terminals in frog retinas maintained at different levels of illumination. (The illumination intensity increases from left to right as indicated; full intensity is 130 footcandles and the lower intensities are expressed as the log of the fraction of full intensity. (See Schacher et al., 1976 for details.) Percent uptake refers to the proportion of the synaptic vesicles that showed peroxidase. The four curves represent four replicate separate experiments, each involving the indicated exposures to peroxidase and each utilizing one retina per illumination point. Evidently, light induces a decline in the frequency of labeled vesicles, presumably reflecting a decline in the rate of neurotransmission (cf. Schacher et al., 1974, 1976; Dowling, 1975; Rodiek, 1973; Ripps et al., 1976).

central participation of MVBs is significant because the morphology of these lysosomes seems to reflect their involvement in membrane degradation; the internal vesicles arise by mechanisms that should be capable of transferring the membrane of endocytic or other structures from outside the lysosome to inside the body (Fig. 1; see also Hirsch et al., 1968; Arstila et al., 1971; Locke, 1975). As in many other tissues, the MVBs of the medulla evolve eventually into residual bodies as their contents undergo degradation. In the insulin-treated preparations, much lipid accumulates transiently in such bodies as in other circumstances where extensive lysosomal breakdown of membrane is occuring as, for example, in Schwann cells during myelin degradation (Holtzman and Novikoff, 1965). Probably the lysosomal lipases are overwhelmed initially by the quantities of material presented to them and can only degrade the lipids slowly (cf. Holtzman, 1976). Order-of-magnitude estimates suggest that the amounts of lipid which accumulate in the medulla cells are in line with the quantities expected for the amounts of membrane being degraded (Holtzman et al., 1973), but we hesitate to place much weight on this since little is known about the degradative rates in lysosomes (Holtzman, 1976). Furthermore, we are dealing with insulin-treated material and insulin has direct, probably nonlysosomal effects on lipid metabolism in some cell types. (See footnote 5, p. 213.)

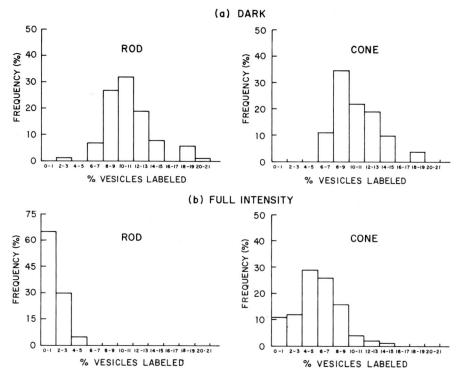

Fig. 15. Histograms indicating the effects of illumination on the uptake of horseradish peroxidase into rod and cone terminals maintained for 60 minutes in the dark or at an illumination intensity of 130 footcandles ("Full intensity"; see Schacher et al., 1976 for details). Each bar represents the average of values obtained in four separate experiments. In each experiment 20–30 rod terminals and a similar number of cone terminals were studied for the "light" ("full-intensity") condition and a comparable sized sample was used for the "dark" condition. Each terminal was scored as to the proportion of its synaptic vesicles containing peroxidase. The histograms show the relative frequencies of terminals with the indicated degree of labeling. As in Fig. 14, the data suggest that light induces a decline in synaptic activity (see also Figs. 11 and 16). As might be expected from physiological information, the effects of illumination are more dramatic with rods, than with cones.

---

[5]In principle, one should be able to obtain information about membrane degradation rates by counting the number of multivesicular bodies and determining the surface areas of their internal vesicles. (The surface area of each vesicle in an MVB is on the order of 0.0025–0.01 μm^2 and large MVBs may contain hundreds of vesicles. Synaptic vesicles fall in the same size range as MVB vesicles. Secretion granules 0.2–0.5 μm in diameter have surface areas on the order of 0.1–0.8 μm^2—see Holtzman, 1976). However, the underlying kinetics may be complex, since a given MVB may receive membrane by endocytic vesicle fusion over a period of many minutes to several hours and because the rates at which the vesicles are degraded and at which MVBs fuse with other lysosomes or transform into residual bodies are unknown. Such factors have not always been adequately considered by investigators who anticipate changes in lysosome number as a reflection of lysosome participation in membrane degradation.

Results similar to ours have been reported for the exocrine pancreas (Kramer and Geuze, 1974) and the parotid gland (Kalina and Robinowitch, 1975, but see also Cope and Williams, 1973). (Studies on systems other than gland cells, such as mammalian bladder (Hicks, 1974), toad bladder (Masur et al., 1971; 1972) and other cell types (Arstila et al., 1971), have also implicated multivesicular bodies and certain comparable lysosome types as major terminals for membrane circulation between the cell surface and intracellular compartments.) Thus one could conclude that much, if not most or all, of the membrane added to the surface of a gland cell is fated for degradation upon its return to the cytoplasm. New granules might be clad in membranes that are older than the granule contents (see above), but these could come from stores in the ER or Golgi apparatus rather than from direct recycling (in the stimulated adrenal medulla, replenishment of the granule population involves hypertrophy of the ER, perhaps reflecting in some degree the need to resupply the membrane stocks; Abrahams and Holtzman, 1973; Holtzman et al., 1973). However, it has been emphasized that several of the studies just cited involve abnormally rapid rates of secretion, and thus, the question has been raised as to whether the same routes and mechanisms operate under normal circumstances (Meldolesi, 1974). Perhaps the cell brings its lysosomal mechanisms into play only when there is a "surplus" of membrane. The phenomenon of crinophagy, mentioned earlier, could provide a precedent for this. It may be significant that MVBs are prominent among the lysosomes participating in crinophagy and that lipid droplets accumulate in MVB-derived erinophagic bodies, perhaps reflecting degradation of the secretion granule membranes. The extreme position in this argument seems untenable; even in gland cells that have not been specially stimulated to secrete, MVBs and related lysosomes exist, and the endocytosed tracers follow the same general routes as they do in stimulated cells (Holtzman and Dominitz, 1968; Nagasawa and Douglas, 1972). Overall, while involvement of lysosomes in normal turnover is very likely, it clearly would be a vast overstatement to assert that the lysosomal route has been shown consistently to be the predominant one. For example, it may still be argued that the lysosomes normally serve some sort of steady baseline membrane turnover (cf. Jamieson, 1975) coexisting with membrane reuse via a route that has yet to be identified.

This last point is given some added weight by observations made about pituitary gland cells (Pelletier, 1973; Farquhar et al., 1975). In several cases, HRP endocytosed by pituitary cells has been found to accumulate in sacs associated with the Golgi apparatus and in bodies the size of secretion granules. This could constitute an indication of direct reuse. However, uptake into Golgi-associated sacs seems quite limited by comparison with incorporation of the tracer into lysosomes of the same cells. It is also worth remembering that both secretion granules and lysosomes form from Golgi-associated sacs. Thus, although these observations obviously merit further study, they are, at best, only suggestive. We will return to this point later, when we discuss HRP uptake into axonal tubules.

It should also be kept in mind that tracer approaches of the sort we have used cannot rule out the movement of some membrane along routes different from those described above. For example, a proportion of endocytic structures might conceivably fuse with MVBs and empty their contents, and then bud back into the cytoplasm where, now unlabeled, they might participate in other events (see legend to Fig. 13). This would of course immensely complicate quantitative interpretation (e.g., of MVB labeling, as in footnote 5, or of the extent of labeling of the synaptic vesicle population, as in section 3.3.2).

4.3. Neurons

4.3.1. Degradation?

Some of the membrane internalized at nerve terminals seems destined for a lysosomal fate similar to that described for gland cells. In neurosecretory systems, where local refilling of membranes to form functional secretory granules seems unlikely, given the nature of the granule contents, endocytosed tracers are found to accumulate rapidly in structures larger than vesicles, many of which may be precursors of multivesicular bodies (Nordmann et al., 1974; Pelletier et al., 1975). Multivesicular bodies accessible to exogenous tracers are also found regularly in terminals of conventional neurons. We (Teichberg et al., 1975) have noted that in rodent spinal cords in culture, or in situ, roughly 5 to 10% of the terminals seen in a given thin section contain an MVB (or a readily recognizable, probable precursor structure—Rosenbluth and Wissig, 1964; Holtzman, 1976). For frog photoreceptor terminals the frequency is on the order of 5 to 25% or more (Schacher et al., 1976). Extrapolating to three dimensions, these figures indicate that MVBs are not rare in the terminals we have studied. Though MVB frequency in given terminals varies from preparation to preparation limiting confident comparisons, the frequency of MVBs does not seem to change notably when tracers such as HRP are introduced into the extracellular space, suggesting that the bodies do not accumulate in response to the presence of foreign material. Other types of lysosomes that might degrade membrane, such as autophagic vacuoles, are also found occasionally in normal terminals (Holtzman, 1969), at relatively low frequency. Neuromuscular junctions seem to contain few multivesicular bodies, or other recognizable lysosomes, although some MVBs have been reported (Von During, 1967).[6]

Generally, for terminals exposed to HRP, one can correlate the frequency of labeled MVBs with the rates of endocytic tracer uptake (see Figs. 16, 18; Teichberg et al., 1975; Schacher et al., 1976). But the correlations are not very strict. This may in part reflect the fact that a given MVB probably can receive tracers over an extended period and perhaps from more than one source (see footnote

---

[6]An interesting set of observations on the compound eyes of some invertebrates indicates that illumination leads to enhancement of endocyticlike internalization of the photoreceptor surface membranes and the consequent formation of numerous multivesicular bodies (Eguchi and Waterman, 1968; White, 1968). Presumably, at least in part, this accounts for the changes in membrane turnover rates associated with illumination (Krauhs et al., 1976). These observations could merely reflect a curious "morphological" convergence of very different phenomena, but they may also have some intriguing evolutionary implications that are only very dimly perceived at present.

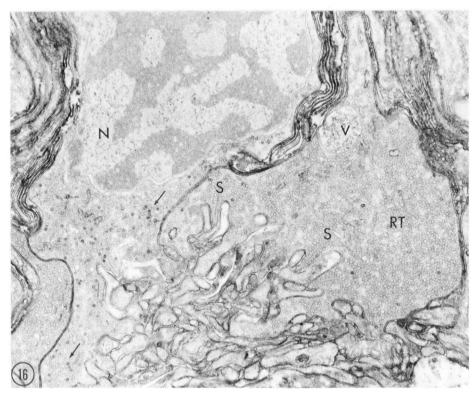

Fig. 16. Photoreceptor terminals from a frog retina maintained in the presence of horseradish peroxidase for 45 minutes at a light intensity expected, from indirect physiological evidence, to strongly affect rod synaptic activity but to have far less effect on the cones (−3.3 on the scale, Fig. 14; see Schacher et al., 1976 for details). The micrograph shows a cone terminal (note the nucleus at N) containing a moderate number of peroxidase-labeled vesicles (arrows) while the adjacent rod terminal (RT) shows far fewer such vesicles. V, a multivesicular body in the rod terminal; S, synaptic ribbons. 18,000×.

no. 5). For example, we suspect that MVB precursors can form directly from invaginations of the cell surface and subsequently acquire more tracer and more membrane through fusion with endocytic vesicles (the consequences of these two routes of delivery to MVBs in terms of membrane retrieval are the same, but their relative quantitative importance is yet to be determined). Hence even when endocytic rates are low, some labeling of MVBs may occur, although presumably the *amounts* of tracer involved would be smaller than when endocytosis is extensive. Since it is very difficult to evaluate quantities of HRP present in different structures from cytochemical preparations, we cannot actually judge the impact of such factors at present. Nor are we certain why the overall (total) frequencies of MVBs, labeled and unlabeled, often seem not to change dramatically under varying conditions (again, due to problems in enumerating small MVB and bodies at early stages in MVB formation, and to variability from preparation to preparation, this cannot be asserted with full confidence; for example, we some-times have the impression that MVBs are more frequent in active terminals [Teichberg et al., 1975]). Reliable interpretation of such relative "constancy" will depend on better understanding of MVB formation controls and features of the MVB life cycle, such as how long the bodies persist in recognizable form under varying conditions.

We have, however, appreciated another factor that may be of major significance for membrane turnover in terminals. During the last few years it has been well established that axons are capable of transport in retrograde fashion, that is in the terminal-to-perikaryon direction (Kristensson et al., 1971; LaVail and LaVail, 1974). This transport is interrupted by microtubule-disrupting agents (Bunt and Lund, 1974; LaVail and LaVail, 1974). Such transport might "make sense" biologically since few of the lysosome-related structures in normal terminals contain cytochemically demonstrable acid hydrolases; they could acquire such hydrolases from structures present in the region of the Golgi apparatus of the perikaryon (Novikoff et al., 1971; Holtzman, 1976) or photoreceptor myoid region (Schacher et al., 1973). Many neuron lysosomes are found near the Golgi apparatus and, as already mentioned, the Golgi-associated ER is a likely source of lysosomal hydrolases.

Tentative evidence in support of involvement of retrograde transport in membrane turnover in terminals comes from "chase" experiments in which neurons of cultured spinal cord have been initially exposed to HRP under varying conditions and then maintained ("chased") in tracer-free media (Holtzman et al., 1973; Teichberg et al., 1975). We have found, for example, that when the terminals are permitted to take up HRP and the chase is then carried out with media containing 10 mM Mg^{2+} (this "high Mg^{2+}" condition suppresses neuro-transmission), labeled vesicles disappear from the terminals sufficiently fast that after 8–16 hours of chasing only very few remain (Fig. 17). Concomitantly, during the chase there is a marked increase in the labeling of perikaryal structures, chiefly lysosomes and related bodies. To an unknown extent this increase may reflect transport of the HRP from dendrites, which are capable of endocytosis[7] (Pappas, cited in Pappas and Purpura, 1972). However, involvement of the terminals is suggested by the fact that the increase in perikaryal labeling during chases is far less evident when the initial exposure of the cultures

---

[7] Is this related to turnover of the postsynaptic membrane? Direct uptake of tracers by perikarya also seems to occur. For our work (Holtzman, 1969; Holtzman and Peterson, 1969), and that of Rosenbluth and Wissig (1964), such uptake is the simplest explanation for the configurations observed at the cell surface and for the fact that tracer appears intracellularly in the perikarya at the earliest times studied (within 35 minutes in neurons in the adrenal medulla (Holtzman and Peterson, 1969)). However, as stressed, perhaps overenthusiastically, by Rees et al. (1976), such uptake may actually be limited. In the work considered here it seems to have a minor impact.

Endocytosis by postsynaptic membranes might, of course, also involve uptake of materials released presynaptically. This could contribute to interaction of pre- and postsynaptic cells, although it should be borne in mind that most of the material endocytized by neurons is probably slated for incorporation in lysosomes. (As reviewed elsewhere (Holtzman, 1976; 1977) some biologically active materials actually require hydrolysis for their "activation" and others can "avoid" lysosomal degradation.)

218

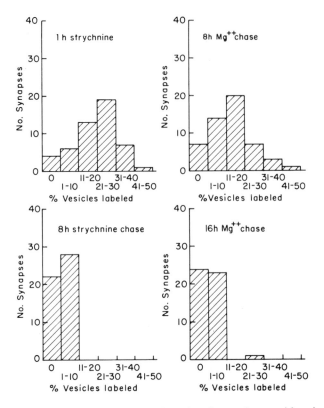

Fig. 17. Histograms summarizing the results of a "chase" experiment with cultured spinal cord explants (Teichberg et al., 1975 for details). Four cultures were exposed to horseradish peroxidase for 1 hour in the presence of strychnine, which enhances the rates of neurotransmission. One culture was now fixed ("1 hr strychnine") while the other three were rinsed and then maintained ("chased") in peroxidase-free medium either with strychnine or with 10 mM Mg^{2+} (which inhibits neurotransmission). After the indicated times the cultures were fixed and incubated to demonstrate peroxidase activity, and studied in the electron microscope. The first 50 synapses encountered in a representative area of each culture were scored as to the percent of their synaptic vesicles that contained the tracer. The data demonstrate a decline in the proportion of labeled vesicles during the Mg^{2+} chase and a faster decline with strychnine. (Courtesy Teichberg et al. [1975] J. Cell Biol. 67, 215.)

to HRP had been in the presence of high Mg^{2+}, so that few structures in the terminals acquire the tracer. As discussed further on, loss of tracers from terminals can apparently occur through exocytosis, but in the present experiments the use of high Mg^{2+} during the chases minimizes or abolishes this. Thus we have tentatively concluded that much of the disappearance of labeled vesicles in our high Mg^{2+} chases reflects transport to perikarya. Few labeled vesicles are seen in the axons of our preparations. However, in our observations (as in other studies of retrograde transport of exogenous macromolecules, that were not focused on the questions at issue here; e.g., LaVail and LaVail, 1974) MVBs seemed to predominate among the intraaxonal sites of HRP. Some labeled tubules that will

219

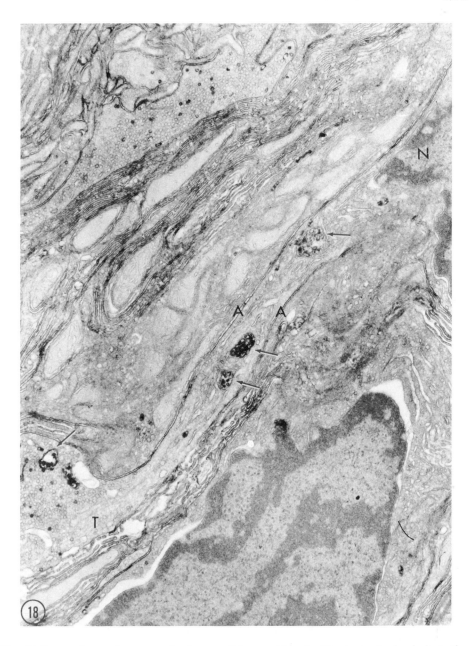

Fig. 18. From a frog retina exposed to horseradish peroxidase for 60 minutes in the dark. A, the borders of the axonlike process connecting a rod terminal (T) to the nuclear region (N) of the cell. Peroxidase is shown in several vesicles in the terminal and in multivesicular bodies in the terminal and axon (arrows). No tracer is present in the numerous longitudinally oriented axonal membrane-delimited sacs or tubules that presumably include agranular reticulum (a possible exception is present adjacent to the uppermost multivesicular body, but this probably is an infolding of the cell surface). 24,000×. (Courtesy of Schacher et al., J. Cell Biol. 70, 178–192).

be discussed later, and a few vacuoles or other bodies larger than synaptic vesicles also show tracer. Evidently, incorporation of vesicle contents and membranes into MVBs and other bodies occurs early in the transport process. In the axons and the perikarya these structures apparently evolve further; the tubules and vacuoles may contribute to MVBs, and fusions with hydrolase-containing perikaryal lysosomes probably occur, as implied above.

Other data supporting the general set of concepts developed in the preceding paragraph include those of Birks et al. (1972), who estimate that tracer-labeled multivesicular bodies persist in terminals of cultured sympathetic neurons for only an hour or two, and our own findings on the relative frequencies of HRP labeled vesicles and MVBs in terminals, axons, and myoids of photoreceptors at various times and under various illumination conditions (Fig. 18, and Schacher et al., 1976). It also has been reported recently that anti-dopamine-β-hydroxylase antibodies, when administered to living sympathetic neurons, are taken up at terminals and varicosities and rapidly transported back to perikarya (Jacobowitz et al., 1975; Ziegler et al., 1975). If, as the authors contend, this reflects retrograde transport of vesicles containing dopamine-β-hydroxylase (bound to their membranes?) it parallels the expectations of the arguments developed here. Finally, Litchy (1973) has demonstrated a stimulation-related increase in the rates of retrograde transport of HRP monitored biochemically, although he has not directly assessed the contributions of the terminals.

Overall, our studies suggest that labeled vesicles persist in quiescent terminals of the cultured spinal cord for average periods of 10 to 20 hours, which would be consistent with the lower ranges of estimates for vesicle turnover kinetics mentioned earlier.

4.3.2. Reuse

If vesicles formed from retrieved membrane persist in terminals for many hours after they have been formed, can they be used for transmission? Given the ability of systems such as neuromuscular junctions to sustain rapid rates of transmitter release for prolonged periods of time, even when they are no longer connected to their perikarya, reuse seems probable (Bittner and Kennedy, 1970; Heuser and Reese, 1973; Hurlburt and Ceccarelli, 1974). Replacing each vesicle used one time by a vesicle formed from new membrane would be a substantial metabolic task, and more economic alternatives provided by evolution may have been gratefully accepted. When examining the vesicle population of stimulated neurons in ordinary EM preparations, one does not encounter a distinctive subpopulation corresponding to the vesicles that would be labeled with tracers in appropriate experiments.[8] Of particular relevance perhaps is the fact that

---

[8]This may be an overstatement since, for example, in very active terminals some of the vesicles present seem larger than normal (Holtzman, unpublished), and there may be corresponding "large" quanta of transmitter (Heuser, 1976). But the ordinary-sized vesicles seemingly show no recognizable heterogeneity, even when direct comparisons are possible or when dextrans are used as the tracers (Ceccarelli et al., 1974). However, the possibility that retrieved vesicles may sometimes be

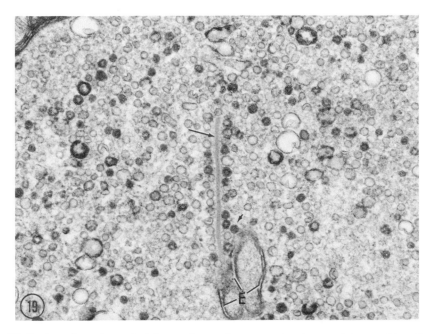

Fig. 19. Portion of a rod terminal from a frog retina exposed to horseradish peroxidase for 120 minutes in the dark. E, extracellular space. A number of the vesicles along the synaptic ribbon indicated by the long arrow contain peroxidase. Generally there is no notable difference in the frequency of such vesicles along the ribbon as compared with other regions of the terminal. The vesicle indicated by the short arrow shows the sort of relations to the cell surface that might be expected for an early stage of exocytosis; if so, this may reflect events occurring during the initial moments of fixation. 45,000×.

tracer-labeled vesicles are not segregated into special portions of the terminal. They intermingle, apparently at random, with the unlabeled vesicles. For example, in retinal photoreceptors, the population of vesicles present along synaptic ribbons becomes labeled to almost the same extent as does the overall vesicle population in the terminal (Fig. 19 and Schacher et al., 1976). This finding seems to indicate the absence of "special treatment" for previously used vesicles with respect to access to transmitter release sites.

Heuser and Reese (1973) showed that when neuromuscular junctions whose vesicles had previously been loaded with HRP were stimulated in tracer-free media, the HRP disappeared from the vesicle population. They could not demonstrate directly that this reflects the reuse of the vesicles for exocytosis with consequent release of the tracer protein to the extracellular space. But their findings strongly suggest such reuse, especially since recent work has demon-

---

smaller than "unused" ones has been raised by several studies (e.g., Smith, 1972, and more recent work by Zimmerman, DePotter and Chubb, and others reviewed in Holtzmann, 1977). The existence of such differences presumably would imply that at least some vesicle fusions with the cell surface involve more than the opening of a transient "pore" (see above).

strated that depletion of HRP from preloaded neuromuscular junctions can occur very rapidly (McMahan and Yee, 1975). Although one might cavil at the use of preparations with cut axons in such studies, our results with the intact cells of the spinal cord cultures offer some reassurance (Teichberg et al., 1975). We found that depletion of HRP-labeled vesicles from the terminals was notably faster when our chase experiments were done under conditions (presence of strychnine) that produce sustained synaptic activity than with the high Mg^{2+} medium used to quiet terminals (Fig. 17). Other problems are less easily dealt with, although far from fatal. For example, the depletion of HRP from preloaded terminals in Heuser and Reese's experiments seems to continue during an appreciable period after the depleting round of stimulation is terminated. Spontaneous release may be responsible, but this raises questions about the normalcy of the preparations (e.g., Miyamoto and Preveti, 1976). In addition, as mentioned earlier, reports from biochemical studies of frog sartorius and sciatic nerves indicate that stimulation enhances the rates of retrograde transport of HRP (Litchy, 1973). Heuser and Reese suggest that during the short periods they studied, retrograde transport could be ignored as a factor contributing to the disappearance of tracer, especially since no HRP appeared at ligations placed around the axons leading to the terminals with which they worked. This last point needs further support to be fully convincing since at present it is difficult to predict just how far retrograde transport can carry the tracers in ligated or cut axons over the time period used.

Overall it seems proper to regard the reuse of vesicles for exocytosis as at least provisionally demonstrated (see also Fig. 19). It may soon be possible to evaluate the relative quantitative importance of this as contrasted with the degradation and transport mechanisms outlined previously. It is tempting to advance relatively simple models such as those in which a vesicle moves at random within the terminal, undergoing degradation if it encounters an MVB, or being used for transmission if it reaches the special electron-dense materials found close to release sites. In such models the factors governing the relative rates of degradation and reuse presumably would be the frequency of MVBs and the details of their formation and fusion with endocytic structures, plus the rates of retrograde transport, the size of the vesicle population, the prevailing rates of transmitter release, the "efficiency" with which the electron-dense material along the presynaptic surface of the plasma membrane might "capture" a vesicle that approaches it, and so forth. (Obviously, if structures such as the endocytic tubules, discussed below, participate in membrane cycling and degradation, their involvement must also be factored into these considerations. We certainly do not mean to imply that MVBs are the only bodies important in retrograde transport.)

Does reuse of membrane for exocytosis necessarily imply reuse for transmitter release? Apparently not, since Ceccarelli and Hurlburt (1975) have shown that neuromuscular junctions depleted of acetylcholine by stimulation in the presence of hemicholinium, contain many vesicles but little acetylcholine and yet show substantial subsequent depletion of the vesicle population when exposed to black widow spider venom. This issue raises some specters, especially since it relates to a general problem: a notable stumbling block to acceptance of the idea that vesicle membrane can be reused repeatedly is the fact that synaptic vesicles may contain components such as proteins, for which no local synthetic system exists in the terminals. Although transmitters could easily reenter used vesicles, the reentry of proteins and nucleotides may be more difficult to explain. It has been demonstrated that some vesicle components may reappear in previously depleted terminals much more slowly than the vesicles themselves. This is true of the material responsible for the microscopically detectable binding of Ca^{2+} within vesicles of neuromuscular junctions (Boyne et al., 1975; Pappas and Rose, 1976). Zimmerman and Whittaker (1974) have reported that subsequent to depletion the terminals of *Torpedo* electric organ regain their acetylcholine content more rapidly than their adenine nucleotides. The significance of such considerations depends on the minimal requirements for vesicle filling and function. One proposal for adrenergic neurons maintains that only the "never-used" vesicles contain the full complement of catecholamines, proteins, and other components. The retrieved vesicles may lack some of these elements, but still possess enough of the crucial ones (e.g., dopamine-β-hydroxylase [DBH]) to function. It has been proposed that part of the DBH of the secretion granules in the adrenal gland is tightly bound to the delimiting membrane (cf. Kirschner, 1972). If this is true in adrenergic neuronal synaptic vesicles it may be important for the questions under consideration (cf. Smith et al., 1970; but see also Klein and Thureson-Klein for scepticism about membrane-bound DBH). According to one scheme, the larger dense-cored vesicles of adrenergic neurons might represent the "never-used" population and the smaller dense-cored vesicles, the products of retrieval (footnote 8). Cytochemical evidence of still uncertain interpretation has been adduced as support for the notion that both these vesicle populations store catecholamines but differ in their contents of other components, perhaps including proteins (Hokfelt, 1974; Till and Banks, 1976).

Alternative proposals suggest that the depletion of proteins and other non-transmitter vesicle contents occurs slowly, since only a little is lost each time a vesicle is used. This ties in with the proposals for partial rather than complete fusion of vesicles with the plasma membrane that was mentioned earlier. Perhaps the eventual passing of a threshold marks the vesicle for destruction rather than reuse. Yet a vesicle may also be able to recoup its losses through selective endocytosis—perhaps not only the membrane but also the soluble proteins are retrieved (cf. Smith et al., 1970; Heuser, 1976).

One is frustrated in choosing among such alternatives by inadequate available information about the heterogeneity of the vesicles in a given terminal and the details of transmitter storage and release. (For additional discussion and references, see Holtzman, 1977.)

4.3.3. Vesicle recycling
and the axonal agranular reticulum
In examining perikarya of neurons from HRP uptake experiments we have, on rare occasions, encountered peroxidase reaction product in sacs associated

224

closely with the Golgi apparatus (Holtzman et al., 1973; see also Turner and Harris, 1974). As with secretory cells, such configurations are of particular interest since they may reflect direct reuse of membrane. However, we have seen them so rarely under normal circumstances as to make interpretation virtually impossible.

Gonatas and coworkers (1975) have reported recently that when complexes of HRP and ricin are endocytosed by neurons they accumulate largely in Golgi-associated sacs (Fig. 20). The sacs may actually be those normally responsible for lysosome formation, but the perturbation is an interesting one. Does it indicate that membrane-associated receptors or other systems exist that switch the fate of an endocytic structure from one intracellular pathway to another, or otherwise control the timing or geometry of degradation (Raff, 1976)? The existence of such mechanisms would have strong implications for the issue considered here.[9] Fusion of plasma membrane-derived (endocytic) vacuoles with lysosomes has previously been shown to be inhibited by the presence of lectins (Edelson, 1974; Goldman, 1974) or certain microorganisms (review, Holtzman, 1976) within the vacuoles. Hypotheses to explain such effects and the more general phenomena of specificity of lysosomal fusions, have ranged from intervention of cyclic nucleotides (e.g., Lowrie et al., 1975; Oliver et al., 1976) or of microtubules and microfilaments (review, Holtzman, 1976) to the operation of special recognition devices, based perhaps on glycoproteins or other molecules in the lysosome surface (e.g., Feigensson et al., 1975). The microtubule-dependent surface membrane recycling system of protozoa discussed earlier (cf. Fig. 13) may be an apposite example. Interestingly freeze-fracture studies suggest that as it passes from one compartment to another the membrane of protozoan food vacuoles shows shifts in the intramembrane particle pattern that could conceivably influence the properties and fate of the structures it delimits (Allen, 1976).

Studies of tracer uptake by terminals, and of both retrograde and anterograde (orthograde) axonal transport of exogenous macromolecules such as HRP, have regularly found that some of the tracer enters elongate membrane-delimited tubules in the axons (Fig. 21 and Birks et al., 1972; Krishnan and Singer, 1973; Kaiserman-Abramof et al., 1974; LaVail and LaVail, 1974; Reperant, 1975; Weldon, 1975; Colman et al., 1976). This has led several authors to speculate about roles for the axonal agranular reticulum in retrograde transport (Kaiserman-Abramof et al., 1974; Reperant, 1974), and also concerns the ques-

---

[9]Other observations may also be explained along these lines. For example, different exogenous proteins seem to gain access to the retrograde transport system to very different extents (e.g., Bunt et al., 1976). Is this solely a matter of differential uptake, or does some sort of intracellular selection operate? Among the proteins that are transported, some are potentially active biologically. Nerve growth factor is one of these, and there has also been speculation that postsynaptic cells may send signals to perikarya of presynaptic cells via endocytosable molecules that undergo retrograde transport. While thoroughly definitive, direct demonstration of the biological activity of molecules following such routes has yet to be accomplished (see Paravicini et al., 1975 and Stoeckel and Thoenen, 1975 for progress and discussion), if they are active, they must presumably escape lysosomal degradation as mentioned in footnote 7.

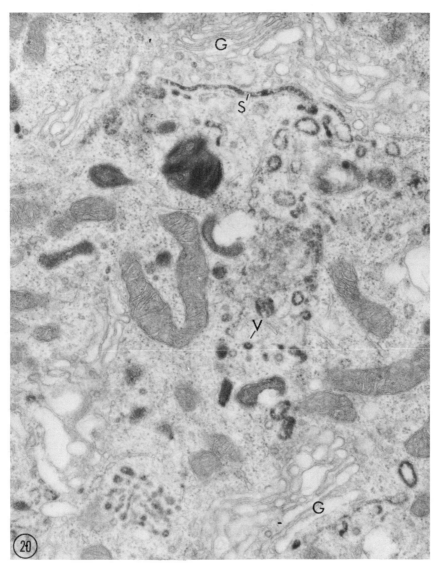

Fig. 20. Portion of the perikaryon of a cultured mouse cervical ganglion neuron that had been exposed to ricin-horseradish peroxidase conjugates for 1 hour (4°C) then rinsed and incubated in tracer-free medium at 36°C for an additional 3 hours prior to fixation and incubation, to demonstrate peroxidase activity. Reaction product is seen in sacs (S) and vesicles (V) near the Golgi apparatus (G). In the J. Cell Biol. paper (see bibliography), Gonatas and coworkers demonstrate that the Golgi associated structures that accumulate the exogenous lectins almost certainly correspond to acid-phosphatase-containing elements of GERL, the system from which lysosomes seem to form. (Courtesy of N. K. Gonatas; [1975] reproduced from Exp. Cell Res. 94, 426; Gonatas, Sterber, Kim, Graham and Avreamas).

226

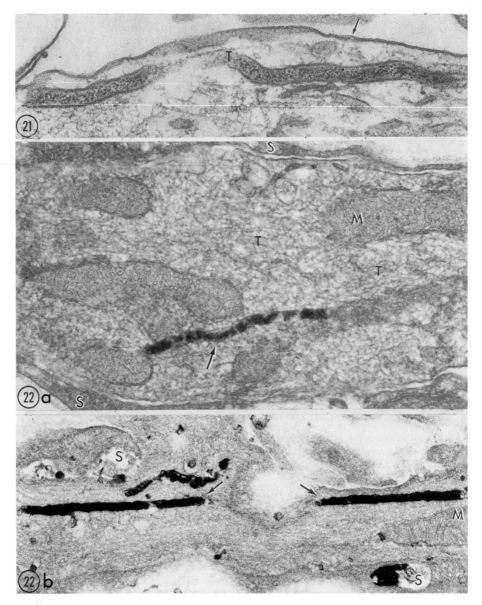

Fig. 21. Portion of a neurite from a cultured chick sympathetic ganglion exposed to ferritin for 120 minutes. Ferritin is seen in an elongate membrane-delimited tubule (T) with a moderately electron-dense content. The arrow indicates plasma membrane. (Courtesy of P. R. Weldon and R. I. Birks [1972] reproduced from J. Neurocytology 1, 311; Birks, Mackey and Weldon.) 75,000×. (approx).

Fig. 22. (A) Portion of an axon from a cultured mouse dorsal root ganglion fixed 2 days after x-irradiation and incubated to demonstrate acid phosphatase activity. Arrow indicates an acid phosphatase-containing tubule or sac in the axoplasm. Several other membrane-delimited sacs or tubules that lie nearby, lack reaction product (T). S, Schwann cells; M, a mitochondrion. 46,000×.

tions discussed here. If endocytosis delivers membrane to the axonal reticulum, then there is a simple explanation for the ways that vesicles might be recharged with proteins and other components. Obviously, problems will also exist that relate, for example, to the apparent use of the same channel system for (simultaneous?) transport in opposite directions.

The question is: Are the tracer-containing axonal tubules actually part of the axonal agranular reticulum? In most cell types, when endocytosed materials are seen in tubules, the tubules may be distinguished readily from the endoplasmic reticulum by features such as the absence of ribosomes. For the axons we do not have such simple criteria. It appears, however, that certain recognizable portions of the axonal reticulum are not readily accessible to exogenous tracers. In the central nervous system cultures and sympathetic ganglion cultures with which we have worked, we find little if any HRP in the networks of fine tubules present in terminals, and Teichberg and Bloom (1976) have recently demonstrated that this is the case even with prolonged (2–16 hr.) exposures to HRP (Fig. 23). In photoreceptors (Fig. 12) we see occasional small branched tubules with HRP, but these could be endocytic or still continuous with the plasma membrane, and on the whole the tubular networks do not acquire notable quantities of the tracer. Comparable findings have been reported for other systems (Heuser and Reese, 1974; Wessels et al., 1974; Weldon, 1975). Thus, while it would be foolish to insist that no exogenous tracers enter the agranular reticulum in terminals, the evidence thus far suggests that such entry is limited or occurs on a time scale such that we have missed most of it. Similar comments apply to the tracer-containing tubules along axons; in our material they are relatively rare, and most of axonal reticulum of the same axons shows no tracer. In other words, for the present, it seems best to think (tentatively) of the agranular reticulum and the tracer-containing endocytic tubules as distinct compartments. (See Holtzman, 1977 for additional references and discussion; perhaps this is one of the ways in which the axon channels transport with retrograde and orthograde transport involving different structures, and somewhat different mechanisms.)

However, this does not end the matter. The tracer-containing axonal tubules represent a potentially interesting set of structures and should be studied further. Many of them seem broader and more regular than the usual profiles of axonal reticulum and often appear to be more limited in length than the elements of the reticulum (although this is more an impression from individual thin sections and very limited work with serial sections rather than a fully documented fact). Clearly, some are elongate multivesicular bodies (Fig. 24) and others may simply be endocytic tubules. However, structures such as those in

---

Fig. 22 (*continued*)

(B) Axon from a similar irradiated preparation incubated for another lysosomal hydrolase, aryl sulfatase. Arrows indicate reaction product in elongate sacs or tubules. In their breadth and their relatively straight course the hydrolase-containing structures (see also A) are reminiscent of the axonal sacs or tubules that accumulate exogenous tracers in other preparations (cf. Fig. 21). The two sacs seen in this section are probably part of a single continuous structure. 27,000×. (Courtesy of Phil. Trans. Roy. Soc. London B 261,407.)

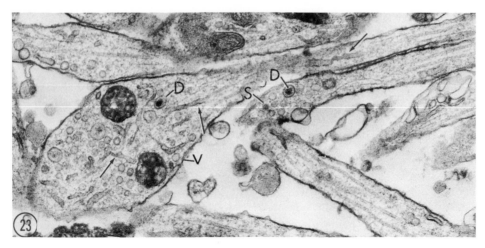

Fig. 23. Axons from a cultured chick sympathetic ganglion maintained in the presence of horse-radish peroxidase for 16 hours. Peroxidase is seen in two multivesicular bodies in a terminal or varicosity (V) and in a synaptic vesicle (S) in another axon, but is not evident in elements of the agranular reticulum (arrows; see Teichberg and Holtzman, 1973). The electron density of the dense-cored vesicles at D is largely intrinsic but conceivably some reflect the presence of exogenous peroxidase. (Courtesy of Teichberg and Bloom [1976]; Abstracts International Soc. for Cell Biology; J. Cell Biol. 70, 285A.)

Fig. 21 are not easily interpreted along such simple lines. In some morphological respects they are most reminiscent of the elongate lysosome-related sacs with electron-dense content or tubules that appear in injured neurons (Holtzman, 1971; Matthews, 1973; see also Fig. 22B). This raises the possibility (cf. Bunge, 1973; Wessels et al., 1974; Weldon, 1975) that some of the tracer-containing axonal tubules may actually be relatives of, or a special form of, agranular reticulum comparable to the Golgi-associated ER (GERL) and involved in lysosomal formation or functions. Since lysosomes are central participants in the disposal of endocytosed materials, this would be an aesthetically pleasing solution although its scientific validity certainly needs more thorough testing. It does tie in with the fact that axonal tubules and sacs seem to participate in acid hydrolase transport in injured neurons (Holtzman and Novikoff, 1965; Sotelo and Palay, 1971; Whitaker and LaBella, 1972; Holtzman, 1976) and in cultured sympathetic ganglia. From an examination of Fig. 22, one might conclude that hydrolase transport involves only some of the sacs or tubules present in an axon, and thus that more than one type of membrane-delimited agranular reticulum exist in axons. Perhaps the agranular reticulum discussed earlier as a transport system for synaptic vesicle components is accompanied by another system that is structurally discrete. In some injured axons, however, acid phosphatase is distributed extensively in axonal sacs or tubules (Holtzman and Novikoff, 1965; Holtzman, 1971; Holtzman et al., 1973) and negative results in enzyme cytochemistry are frail reeds at best. Thus other possibilities may not be excluded. For example, although there might be only one channel system it could be endowed with mechanisms permitting it to segregate and differentially direct its contents, and perhaps even to bud off both transmitter vesicles and lysosome-related tubules.

Related to these concerns is the observation that in the terminals, varicosities, and growth cones of some cultured neurons, both adrenergic and nonadrenergic, several of the vesicles with a dense-corelike content seem to acquire exogenous tracers (Fig. 25, and Birks et al., 1972; Weldon, 1975). Conceivably, this represents vesicle recycling of the type discussed earlier. Yet why are these vesicles often larger than typical synaptic vesicles, and what is the source of their core? The tracers used were the particulate ones, such as thorium dioxide, and one might have recourse to arguments that enable ignoring uncomfortable findings (e.g., when a few tracer particles are found in an unlikely spot, it is sometimes thought that they get there by displacement during fixation or during embedding and sectioning). But since the configurations under consideration are not very rare, such ad hoc explanations lack force. Some of the "dense cores" in the vesicles are themselves clearly membrane-delimited (Birks et al., 1972; Weldon, 1975; see also the "vesicle-within-a-vesicle," Fig. 25) and, even when this is not obvious, the possibility of tangential sectioning obscuring the presence of a membrane around the core must be entertained. However, while serial-section studies are needed to determine how many of the configurations actually are sections through elongate MVBs; it seems unlikely that all of them are. An interesting possibility is that the vesicles in question are not neurotransmitter vesicles but serve some other function in the terminals. Are they related to lysosomes (cf. Taxi and Sotelo, 1971; Matthews, 1973; Wessels et al., 1974)? One must be uneasy about such a suggestion, since the dense content is not an adequate marker (Holtzman, 1976) and few acid hydrolase-containing bodies are seen in normal axons and terminals. The cultured material used for the observations discussed here has yet to be adequately studied in this latter regard. Extending the speculation a little further, however, it might be suggested (H. Blaschko, personal communication) that some limited exocytic defecation of lysosomal residues occurs at terminals and that the vesicles we are discussing (and perhaps some of the axonal tubules considered earlier) participate in this action. No direct evidence exists to support these proposals but several facts lend them superficial plausibility. For example, when stimulated to secrete catecholamines, the adrenal medulla also seems to release a small proportion of its acid hydrolase content from a presently unidentified source in the cells (Schneider, 1970; Smith, 1972). In addition, images suggesting exocytic release of the contents of multivesicular bodies, and perhaps of the dense contents of some of the vesicles in question, have occasionally been encountered in the terminals of cultured neurons (Birks et al., 1972; Weldon, 1975).

The possible involvement of agranular reticulum and vesicles in presynaptic terminals in transport and release of cell surface components, or "trophic" materials should also be kept in mind.

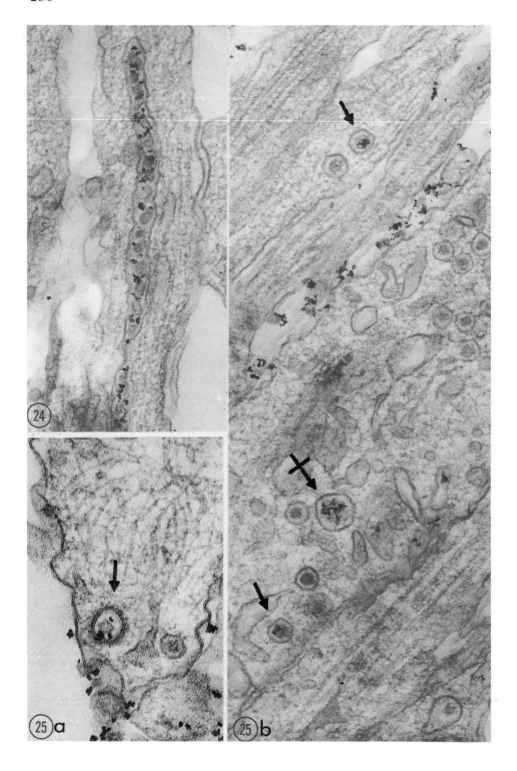

5. Closing Comments

In light of all the caveats and uncertainties involved it would be foolhardy to sum up all this discussion by advancing a general scheme for membrane turnover in gland cells and neurons. We have tried to present the evidence relating to the origin, circulation, and fate of the membranes and to bring forth, in explicit fashion, some of the perplexing implications and ambiguities of current views. To some extent we have overstressed the axonal agranular reticulum and the degradative phases of membrane cycling. This reflects our own research interests, coupled with the feeling that although it is presently somewhat neglected the reticulum will eventually be recognized as being of major importance in axonal transport. We also feel that the extraordinary abundance of lysosomes in neurons must signal some major role, in the intensive membrane turnover intrinsic to neuronal functioning. In some respects, the Golgi apparatus that is also a very prominent structure in neurons, remains the most mysterious of the cell compartments with which we have dealt (see also Palade, 1975). Detailed elucidation of its functioning must await further progress in analysis of the mechanisms of membrane assembly and intracellular transport.

Because of our prejudices as microscopists we have tended to argue our case in terms of visible phenomena or processes that can be reconstructed from micrographs, such as the movement of membrane from one compartment to another, the budding of vesicles from Golgi-associated sacs or the axonal reticulum, exocytosis and endocytosis, and the fusion of structures with lysosomes. The subtler phenomena we have been concerned with are more difficult to study—the passage of macromolecules into or out of membranes, the interaction of ions with surfaces, or the movement of transmitters between different pools. Nonetheless, it seems to us on reviewing the literature that impressive progress has been made in integrating the visible and the invisible to form a coherent picture of the cellular dynamics of secretion and neurotransmission, though formidable tasks obviously remain.

Acknowledgments

We are grateful to Drs. R.D. Allen, J. Heuser, J. Schwartz, and S. Schaeffer for sending us preprints of their manuscripts. Previously unpublished personal research described in the manuscript was supported by NIH grant NS 09475,

Fig. 24. Neurite of a cultured chick sensory ganglion exposed to thorium dioxide particles for 120 minutes. The tracer is seen in a tubular multivesicular body. (Courtesy of P. R. Weldon [1975]; reproduced from J. Neurocytol. 4, 341.)

Fig. 25. Neurites from cultures like those used for Fig. 24. (a) In 15 min exposure to thorium dioxide the tracer is seen within a coated vesicle (arrow) that also contains another membrane-delimited vesicle. (b) In 120 min. exposure the tracer is seen in dense-cored vesicles (arrows); the two smaller ones measure 850Å in diameter and thus qualify as "large" dense-cored vesicles. The larger vesicle falls outside the usual range for synaptic vesicles of mature neurons. (Courtesy of P. R. Weldon [1975]; reproduced from J. Neurocytol. 4, 341.)

NEUB (to E.H.) and NIH grant NS 12148 (to S.T.). S. Schacher was supported by a predoctoral training grant (NIH; GM 02087). Technical assistance was provided by Dr. Fe Reyes, and photographic materials were prepared by Ms. Irene Tarr and Ms. Tana Ross.

References

Abe, T., Haga, J. and Kurokawa, M. (1973) Rapid transport of phosphatidylcholine occurring simultaneously with protein transport in sciatic nerve. Bichem J. 136, 731–738.

Abrahams, S. and Holtzman, E. (1973) Secretion and endocytosis in insulin-stimulated rat adrenal medulla cells. J. Cell Biol. 56, 540–558.

Akert, K., Kawara, E. and Sandri, C. (1971) ZIO positive and ZIO negative vesicles in nerve terminals. Prog. Brain Res. 34, 305–317.

Allen, R. D. (1976) Freeze-fracture evidence for intramembrane changes accompanying membrane recycling in *Paramecium.* Cytobiologie, (in press).

Allen, R. D. and Hausmann, K. (1976) Membrane behavior of exocytic vesicles. 1. The ultrastructure of Paramecium trichocysts in freeze-fracture preparations. J. Ultrastruct. Res. 54, 224–234.

Allison, A. E. and Davies, P. (1974) Interactions of membranes, microfilaments and microtubules in endocytosis and exocytosis. In: Advances in Cytopharmacology (Ceccarelli, B., Clementi, F. and Meldolesi, J., eds.), vol. 2 pp. 237–248, Raven Press, New York.

Amsterdam, A., Ohad, I. and Schramm, M. (1969) Dynamic changes in the ultrastructure of the acinar cells of the rat parotid gland during the secretory cycle. J. Cell Biol. 40, 753–773.

Amsterdam, A., Ohad, I., Schramm, M., Salomon, Y. and Selinger, Z. (1971) Concomitant synthesis of membrane protein and exportable protein of the secretory granule in rat parotid gland. J. Cell Biol. 50, 187–200.

Arias, I. M., Doyle, D. and Schimke, R. T. (1969) Studies on the synthesis and degradation of proteins of the endoplasmic reticulum of rat liver. J. Biol. Chem. 244, 3303–3315.

Arstila, A., Jauregui, H., Chang, J. and Trump, B. (1971) Studies on cellular autophagocytosis. Lab. Invest. 24, 162–174.

Atkinson, P. H. (1975) Synthesis and assembly of Hela cell plasma membrane glycoproteins and proteins. J. Biol. Chem. 250, 2123–2134.

Atwood, H. L., Lang, F. and Morin, W. A. (1972) Synaptic vesicles: selective depletion in crayfish excitatory and inhibitory axons. Science 176, 1353–1355.

Autori, F., Svensson, M. and Dallner, G. (1975) Biogenesis of microsomal membrane glycoprotein in rat liver. J. Cell Biol. 67, 687–699, 700–714.

Bainton, D. F., and Farquhar, M. G. (1966) Origin of granules in polymorphonuclear leukocytes; two types derived from opposite faces of the Golgi apparatus. J. Cell Biol. 28, 277–301.

Baker, R. R., Dowdell, M. J. and Whittaker, V. P. (1975) The involvement of lysophosphoglycerides in neurotransmitter release: the composition and turnover of phospholipids of synaptic vesicles of guinea pig cerebral cortex and Torpedo electric organ and the effect of stimulation. Brain Res. 100, 629–644.

Banks, P., Mayor, D. and Tomlinson, D. R. (1971) Further evidence for the involvement of microtubules in the intra-axonal movement of noradrenaline storage granules. J. Physiol. 219, 755–761.

Bargmann, W. (1966) Neurosecretion. Int. Rev. Cytol. 19, 183–201.

Barlow, J. and Martin, R. (1971) Structural identification and distribution of synaptic profiles in the octupus brain using the zinc-iodine osmium method. Brain Res. 25, 241–253.

Barondes, S. H. (1974) Synaptic macromolecules: identification and metabolism. Ann. Rev. Biochem. 43, 147–168.

Beaufay, J., Amar-Costesec, A., Thines-Sempoux, D., Wibo, M., Robbi, M. and Berthet, J. (1974) Analytical studies of microsomes and isolated subcellular membranes from the rat liver III: subfractionation of the microsome fraction by isopycnic and differential centrifugation in density gradients. J. Cell Biol. 61, 213–231.

Beisson, J., Lefort-Tran, M., Pouphile, M., Rossignol, M. and Satir, B. (1976) Genetic analysis of membrane differentiation in Paramecium. Freeze-fracture study of the trichocyst cycle in wild-type and mutant strains. J. Cell Biol. 69, 1126–1143.

Benedeczky, I. and Smith, A. D. (1972) Ultrastructural studies of the adrenal medulla of golden hamster: origin and fate of secretory granules. Zeit. Zellforsch. 124, 367–386.

Benedeczky, I. and Somogyi, P. (1971) Ultrastructure studies of the adrenal medulla of normal and insulin treated hamsters. Cell Tissue Res. 162, 541–550.

Benes, F., Higgins, J. A. and Barrnett, R. J. (1974) Ultrastructural localization of phospholipid synthesis in rat trigeminal nerve during myelination. J. Cell Biol. 57, 613–629.

Bennett, G. and Haddad, A. (1974) Migration of glycoprotein from the Golgi apparatus to the surface of various cell types as shown by autoradiography after labelled fucose injection into rats. J. Cell Biol. 60, 258–284.

Bennett, H. S. (1969) Cell surface movements and recombinations. In: Handbook of Molecular Cytology (Lima da Faria, A., ed.), pp. 1294–1319. Elsevier/North-Holland, Amsterdam.

Bennett, M. V. L., ed. (1974) Synaptic Transmission and Neuronal Interaction, Soc. of General Physiologists Symposia, vol. 28, Raven Press, New York.

Berg, D. K. and Hall, Z. W. (1975) Loss of α-bungarotoxin from junctional and extrajunctional acetylcholine receptors in rat diaphragm muscle in vivo and in organ culture. J. Physiol. 252, 771–789.

Berg, N. B. and Austin, B. P. (1976) Intracellular transport of sulfated macromolecules in parotid acinar cells. Cell Tissue Res. 165, 215–226.

Bergeron, J. J. M., Evans, W. H. and Gerschwind, I. I. (1973) Insulin binding to rat liver Golgi fractions. J. Cell Biol. 59, 771–776.

Berl, S. (1975) The actomyosin-like system in nervous tissue. In: The Nervous System (Tower, D. B., ed.) vol. 1, pp. 565–574. Raven Press, New York.

Berlin, R. D., Oliver, J. M., Ukena, T. E. and Yin, H. H. (1974) Control of cell surface topography. Nature 247, 45–46.

Birks, R. I. (1974) The relationship of transmitter storage and release to fine structure in a sympathetic ganglion. J. Neurocytol. 3, 133–160.

Birks, R. I. and Fitch, S. J. G. (1974) Storage and release of acetylcholine in a sympathetic ganglion. J. Physiol. 240, 125–134.

Birks, R. I., Mackay, M. C. and Weldon, P. R. (1972) Organelle formation from pinocytotic elements in neurites of cultured sympathetic ganglia. J. Neurocytol. 1, 311–340.

Bittner, G. D. and Kennedy, D. (1970) Quantitative aspects of transmitter release. J. Cell Biol. 47, 585–592.

Blaschko, H., Firemark, H., Smith, A. D. and Winkler, H. (1967) Lipids of the adrenal medulla: lysolecithin, a characteristic constituent of chromaffin granules. Biochem. J. 104, 545–549.

Blobel, G. and Dobberstein, B. (1975) Transfer of proteins across membranes. J. Cell Biol. 67, 835–851, 852–862.

Bolender, R. P. (1974) Stereological study of pancreatic exocrine cells. In: Advances in Cytopharmacology (Ceccarelli, B., Clementi, F. and Meldolesi, J., eds.), vol. 2, pp. 99–106. Raven Press, New York.

Boyne, A. F., Bohan, T. P. and Williams, T. H. (1975) Changes in cholinergic synaptic vesicle populations and the ultrastructure of the nerve terminal membranes of narcine brasiliensis electric organ stimulated in vivo. J. Cell Biol. 67, 814–825.

Brimijoin, S. (1972) Transport and turnover of dopamine-β-hydroxylase in sympathetic nerves of the rat. J. Neurochem. 19, 2183–2193.

Brimijoin, S. (1974) Local changes in subcellular distribution of dopamine-beta-hydroxylase after blockade of axonal transport. J. Neurochem. 22, 347–358; 26, 35–40.

Bunge, M. B. (1973) Fine structure of nerve fibers and growth cones of isolated sympathetic neurons in culture. J. Cell Biol. 56, 713–740.

Bunt, A. H. (1969) Formations of coated and synaptic vesicles within neurosecretory axon terminals of the crustacean sinus gland. J. Ultrastruct. Res. 28, 411–421.

Bunt, A. H. and Lund, R. D. (1974) Vinblastine-induced blockage of orthograde and retrograde axonal transport of protein in retinal ganglion cells. Exp. Neurol. 45, 288–297.

Bunt, A. H., Hascke, R. H., Calkins, D. F. and Lund, R. D. (1976) Factors affecting retrograde transport of HRP in the visual system. Brain Res. 102, 152–155.

Burridge, K. and Phillips, J. H. (1975) Association of actin and myosin with secretory granule membranes. Nature 254, 526–529.

Byers, M. R. (1974) Structural correlates of rapid axonal transport: evidence that microtubules may not be involved. Brain Res. 75, 97–113.

Byers, M. R., Fink, B. R., Kennedy, R. D., Middaugh, M. E. and Hendrickson, A. G. (1973) Effects of lidocaine on axonal morphology, microtubules and rapid axonal transport in rabbit vagus nerve *in vitro*. J. Neurobiol. 4, 125–143.

Castle, J. D., Jamieson, J. D., and Palade, G. E. (1972) Radioautographic analysis of the secretory process in the parotid gland of the rabbit. J. Cell Biol. 53, 290–311.

Castle, J. D., Jamieson, J. D. and Palade, G. E. (1975) Secretion granules of the rabbit parotid gland. Isolation, subfractionation and characterization of the membrane and content subfractions. J. Cell Biol. 64, 182–210.

Ceccarelli, B. and Hurlburt, W. P. (1975) The effects of prolonged repetitive stimulation in hemicholinium on the frog neuromuscular junctions. J. Physiol. 247, 163–188.

Ceccarelli, B., Hurlburt, W. P. and Mauro, A. (1972) Depletion of vesicles from frog neuromuscular junctions during prolonged tetanic stimulation. J. Cell Biol. 54, 30–38.

Ceccarelli, B., Hurlburt, W. P. and Mauro, A. (1973) Turnover of transmitter and synaptic vesicles of the frog neuromuscular junction. J. Cell Biol. 57, 499–524.

Ceccarelli, B., Clementi, F. and Meldolesi, J., eds. (1974) Cytopharmacology of Secretion. Advances in Cytopharmacology, vol. 2, Raven Press, New York.

Chlapowski, F. J. and Band, R. N. (1971) Assembly of lipids into membranes in Acanthameba palestinesis. J. Cell Biol. 50, 634–651.

Chua, N. H., Blobel, G., Siekevitz, P. and Palade, G. E. (1973) Attachments of chloroplast polysomes to thylakoid membranes in Chlamydomonas reinhardti. Proc. Nat. Acad. Sci. U.S.A. 70, 1554–1558.

Chubb, I. W. and Smith, A. D. (1975) Release of acetylcholinesterase into the perfusate from the ox adrenal gland. Proc. Roy. Soc. Lond. B. 191, 263–269.

Clark, A. W., Hurlburt, W. P. and Mauro, A. (1972) Changes in the fine structure of the neuromuscular junction of the frog caused by black widow spider venom. J. Cell Biol. 52, 1–14.

Claude, A. (1970) Growth and differentiation of cytoplasmic membranes in the course of lipoprotein granule synthesis in the hepatic cell. J. Cell Biol. 47, 745–766.

Colman, P. R., Scalea, F. and Labrales, E. (1976) Light and electron observation of the anterograde transport of HRP in the optic system in mouse and rat. Brain Res. 102, 156–163.

Cope, G. H. and Williams, M. A. (1973) Quantitative analysis of the constituent membranes of parotid acinar cells and of the changes evident after exocytosis. Zeit Zellforsch. 145, 311–330.

Couteaux, R. (1974) Remarks on the organization of axon terminals in relation to secretory processes at synapses. In: Advances in Cytopharmacology (Ceccarelli, B., Clementi, F. and Meldolesi, J., eds.) vol. 2, pp. 369–379. Raven Press, New York.

Csillik, B., Knyihar, E., Laszlo, I. and Boncz, I. (1974) Electron histochemical evidence for the role of thiamine pyrophosphatase in synaptic transmission. Brain Res. 70, 179–183. See also Int. Rev. Neurobiol. 18, 69–140.

Dahlstrom, A. (1973) Aminergic transmission: Introduction and short review. Brain Res. 62, 441–460.

Davidson, S. J. (1975) Proteolytic activity within lysosomes and turnover of pinocytic vesicles. Biochim. Biophys. Acta 411, 282–290.

DeBruyn, P. D. H. (1975) Endocytosis, transfer tubules and lysosomal activity in myeloid sinusoidal endothelium. J. Ultrastruct. Res. 53, 133–141.

DeCamilli, P., Peluchetti, D. and Meldolesi, J. (1976) Dynamic changes of the luminal plasmalemma in stimulated parotid acinar cells; a freeze fracture study. J. Cell Biol. 70, 59–74.

DeDuve, C. and Wattiaux, R. (1966) Functions of lysosomes. Ann. Rev. Physiol. 28, 435–492.

DeDuve, C., Lazarow, P. B. and Poole, B. (1974) Biogenesis and turnover of rat liver peroxisomes. In: Advances in Cytopharmacology (Ceccarelli, B., Clementi, F. and Meldolesi, J., eds.), vol. 2, pp. 219–224. Raven Press, New York.

DeLores-Arnaiz, G. M., deCanal, M. A. and deRobertis, E. R. (1971) Turnover of proteins in subcellular fractions of rat cerebral cortex. Brain Res. 31, 179–184.

de Petris, S. and Raff, M. C. (1973) Normal distribution, patching and capping of lymphocyte surface immunoglobulins studied by electron microscopy. Nature New Biol. 241, 257–259.

de Pierre, J. W. and Dallner, G. (1975) Structural aspects of the membrane of the endoplasmic reticulum. Biochim. Biophys. Acta 415, 411–472.

de Potter, W. P. and Chubb, I. W. (1971) The turnover rate of noradrenergic vesicles. Biochem. J. 125, 375–376.

De Robertis, E. (1967) Ultrastructure and cytochemistry of the synaptic region. Science 156, 907–914.

De Robertis, E., Saez, F. A. and De Robertis, E. M. F. (1975) Cell Biology, fifth edit. W. B. Saunders, Philadelphia.

Devreotes, P. N. and Fambrough, D. M. (1975) Acetyl choline receptor turnover in membranes of developing muscle fibers. J. Cell Biol. 65, 353–358.

Diliberto, E., Viveros, O. H. and Axelrod, J. (1976) Localization of protein carboxymethylase and its endogenous substrates in the adrenal medulla—methylation of the chromaffin vesicle membrane. Fed. Proc. 35, 326a.

Douglas, W. W. and Sorimachi, M. (1972) Colchicine inhibits adrenal medullary secretion evoked by acetylcholine, without affecting that evoked by potassium. Br. J. Pharmacol. 45, 129–132.

Douglas, W. W., Nagasawa, J. and Schulz, R. A. (1971a) Coated microvesicles in neurosecretory terminals of posterior pituitary glands shed their coats to become smooth "synaptic" vesicles. Nature 232, 340–341.

Douglas, W. W., Nagasawa, J. and Schulz, R. (1971b) Electron microscopic studies on the mechanism of secretion of posterior pituitary hormones. Mem. Soc. Endocrinol. 19, 353–378.

Dowling, J. E. (1975) The vertebrate retina, In: The Nervous System (Tower, D. B., ed.), vol. 1, pp. 91–100. Raven Press, New York.

Droz, B. (1975) Synaptic machinery and axoplasmic transport: maintenance of neuronal connectivity, In: The Nervous System (Tower, D. B., ed.), vol. 1, pp. 111–127. Raven Press, New York. (See also Brain Res. 93:1–13.)

Ducros, C. (1974) Ultrastructural study of the organization of axonal agranular reticulum in *Octopus* nerve. J. Neurocytol. 3, 513–523.

Edelson, P. M. (1974) Effects of concanavalin A on mouse peritoneal macrophages. J. Exp. Med. 140, 1364–1368, 1378–1398.

Eguchi, E. and Waterman, T. (1968) Cellular basis for polarized light perception in the spider crab, Libinium. Zeit Zellforsch. 84, 87–101.

Elam, J. S. and Peterson, N. W. (1976) Axonal transport of sulfated glycoproteins and mucopolysaccharides in the garfish olfactory nerve. J. Neurochem. 26, 845–850.

Elhammer, A., Svenson, H., Autori, F. and Dallner, G. (1975) Biogenesis of microsomal membrane glycoprotein in rat liver. J. Cell Biol. 67, 715–724.

Ellisman, M. H., Rash, J. E., Staehelin, L. A. and Porter, K. R. (1976) Studies of excitable membranes. II. A comparison of specializations at neuromuscular junctional and non-junctional sarcolemmas of mammalian fast and slow twitch muscle fibers. J. Cell Biol. 68, 752–764.

Eranko, O. (1971) Histochemistry of Nervous Transmission, Prog. Brain Res. vol. 34, Elsevier/North-Holland, Amsterdam.

Evans, J., Hood, D. and Holtzman, E. (1977) Effects of aspartate and cobalt on horseradish peroxidase uptake in frog retina. Abst. Assoc. Res. Vis. Ophthalmol., p. 26. (Suppl. to Invest. Ophthalmol. April, 1977.)

Eytan, G., Jennings, R. C., Forti, G. and Ohad, I. (1974) Biogenesis of chloroplast membranes. J. Cell Biol. 249, 738–744.

Farquhar, M. G., Bergeron, J. J. M. and Palade, G. E. (1974) Cytochemistry of Golgi fractions prepared from rat liver. J. Cell Biol. 60, 8–25.

Farquhar, M. G., Skutelsky, E. H. and Hopkins, C. R. (1975) Structure and function of the anterior pituitary and dispersed pituitary cells: *in vitro* studies. In: The Anterior Pituitary (Tixier-Vidal, A. and Farquhar, M. G., eds.) pp. 84–135. Academic Press, New York.

Fawcett, D. W. (1962) Physiologically significant specializations of the cell surface. Circulation 26, 1105–1125.

Feigenson, M. E., Schnebli, H. P. and Baggiolini, M. (1975) Demonstration of ricin-binding sites on the outer face of azurophil and specific granules of rabbit polymorphonuclear leukocytes. J. Cell Biol. 66, 183–187.

Fillenz, M. (1970) The innervation of the cat spleen. Proc. Roy. Soc. Lond. B. 174, 459–468.

Flickinger, C. J. (1969) The development of Golgi complexes and their dependence on the nucleus in amoebae. J. Cell Biol. 43, 250–262.

Flickinger, C. J. (1975) The relationship between the Golgi apparatus, cell surface and cytoplasmic vesicles in amoebae studied by electron microscope autoradiography. Exp. Cell Res. 96, 189–201.

Franke, W. W., Luder, M. R., Kartenbeck, J., Zerban, H. and Keenan, T. W. (1976) Involvement of vesicle coat material in casein secretion and surface regeneration. J. Cell Biol. 69, 173–195.

Friend, D. S. and Farquhar, M. G. (1967) Functions of coated vesicles during protein absorption in the rat vas deferens. J. Cell Biol. 35, 357–376.

Gainer, H., Sarne, Y. and Brownstein, M. J. (1977) Biosynthesis and axonal transport of rat neurophypophyseal proteins and peptides. J. Cell Biol. 73, 366–381.

Geffen, L. B. and Livett, B. G. (1971) Synaptic vesicles in sympathetic neurons. Physiol. Rev. 51, 98–157.

Gerber, D., Davies, M. and Hokin, L. E. (1973) The effects of secretagogues on the incorporation of [2-^{3}H]-Myoinositol into lipid in cytological fractions in the pancreas of the guinea pig, in vivo. J. Cell Biol. 56, 736–745.

Gilula, N. B. (1975) Junctional membrane structure, In: The Nervous System (Tower, D. B., ed.), vol. 1. pp. 1–11. Raven Press, New York.

Glaumann, H., Bergstrom, A. and Ericsson, J. L. E. (1975) Studies on the synthesis and intracellular transport of lipoprotein particles in rat liver. J. Cell Biol. 64, 356–377.

Goldberg, D. J., Goldman, J. E. and Schwartz, J. H. (1976) Alterations in amounts and rates of serotonin transported in an axon of the giant cerebral neuron of Aphysia californica. J. Physiol (in press). (See also J. Cell Biol. 70, 304–318.)

Goldman, R. D. (1974) Effects of concanavalin A on phagocytosis by macrophages. FEBS Lett. 46, 290–213.

Goldman, R. D., Berg, G., Bushnell, A., Chang, C-M., Dickerman, L., Hopkins, N., Miller, M. L., Pollack, R. and Wang, E. (1973) Fibrillar systems in cell motility, In: Locomotion of Tissue Cells, Ciba Foundation Symposium 14, new series, pp. 83–107. Elsevier/North-Holland, Amsterdam.

Gonatas, N. K., Sterber, A., Kim, S. Y., Graham, D. I. and Avreamas, S. (1975) Internalization of neuronal plasma membrane ricin receptors into the Golgi apparatus. Exp. Cell Res. 94, 426–430; also J. Cell Biol. 73, 1–13.

Grafstein, B. (1975) The eyes have it: axonal transport and regeneration in the optic nerve, In: The Nervous System (Tower, D. B., ed.), vol. 1, pp. 147–151. Raven Press, New York.

Graham, R. C. and Karnovsky, M. J. (1966) The early stages of absorption of injected horseradish peroxidase in the proximal tubules of the mouse kidney: ultrastructural cytochemistry by a new technique. J. Histochem. Cytochem. 14, 291–302.

Gratzl, M. and Dahl, G. (1976) Ca^{2+} -induced fusion of Golgi-derived secretory vesicles isolated from rat liver. FEBS Lett. 62, 142–145.

Gray, E. G. (1970) The question of the relationship between Golgi vesicles and synaptic vesicles in octopus neurons. J. Cell Sci. 7, 189–201.

Gray, E. G. (1975) Synaptic fine structure and nuclear, cytoplasmic and extracellular networks. J. Neurocytol. 4, 315–339.

Gray, E. G., (1976) Microtubules in synapses of the retina. J. Neurocytol. 5, 361–370.

Gray, E. G. and Paula-Barbosa, M. (1974) Dense particles within synaptic vesicles fixed with acid and aldehyde. J. Neurocytol. 3, 487–496.

Griffith, D. L. and Bondareff, W. (1973) Localization of thiamine pyrophosphatase in synaptic vesicles. Am. J. Anat. 136, 549–556.

Grynszpan-Winograd, O. (1971) Morphological aspects of exocytosis in the adrenal medulla. Phil. Trans. Roy. Soc. Lond. B. 261, 291–298.

Hall, Z. W., Hildebrand, J. G. and Kravitz, E. A., eds. (1974) Chemistry of Synaptic Transmission, Chiron Press, Newton, Mass.

Hamilton, R. C. and Robinson, P. M. (1973) Disappearance of small vesicles from adrenergic nerve endings in the rat vas deferens caused by red back spider venom. J. Neurocytol. 2, 465–469.

Hausmann, K. and Allen, R. D. (1976) Membrane behavior of exocytic vesicles. II Fate of the trichocyst membranes in Paramecium after induced trichocyst discharge. J. Cell Biol., 69, 313.

Haymovits, A. and Scheele, G. (1976) Cellular cyclic nucleotides and enzyme secretion in the pancreatic acinar cell. Proc. Nat. Acad. Sci. U.S.A. 73, 156–160.

Hebb, C. (1972) Biosynthesis of acetylcholine in nervous tissue. Physiol. Rev. 52, 918–932.

Hendry, I. A., Stockel, K., Thoenen, M. and Iverson, L. L. (1974) The retrograde axonal transport of nerve growth factor. Brain Res. 68, 103–121.

Hernandez-Nicaise, M-L. (1973) The nervous system of ctenophores. III Ultrastructure of synapses. J. Neurocytol. 2, 249–263.

Heuser, J. (1976) Morphology of synaptic vesicle discharge and reformation at the frog neuromuscular junction. In: The Neuromuscular Junction (Thesloff, S., ed.), Academic Press, New York, (in press).

Heuser, J. E. and Reese, T. S. (1973) Evidence for recycling of synaptic vesicle membrane during transmitter release at the frog neuromuscular junction. J. Cell Biol. 57, 315–344.

Heuser, J. E. and Reese, T. S. (1974) Morphology of synaptic vesicle discharge and reformation at the frog neuromuscular junction, In: Synaptic Transmission and Neuronal Interaction (Bennett, M. V. L., ed.). Raven Press, New York.

Heuser, J. E., Reese, T. S. and Landis, D. M. D. (1974) Functional changes in frog neuromuscular junctions studies with freeze-fracture. J. Neurocytol. 3, 109–131.

Hicks, R. M. (1974) The ultrastructure and chemistry of the luminal plasma membrane of the mammalian urinary bladder: a structure with low permeability to water and ions. Phil. Trans. Roy. Soc. Lond. B. 268, 23–68.

Higgins, J. A. (1974) Studies on the biogenesis of smooth endoplasmic reticulum membranes in hepatocytes of phenobarbital-treated rats. J. Cell Biol. 62, 635–646.

Highstein, S. M. and Bennett, M. V. L. (1975) Fatigue and recovery of transmission at the Mauthner fiber-giant fiber synapse of the hatchetfish. Brain Res. 98, 229–242.

Hirano, M., Parkhouse, B., Nicolson, G. L., Lennox, E. S. and Singer, S. J. (1972) Distribution of saccharide residues on membrane fragments from myeloma cell homogeneates: its implications for membrane biogenesis. Proc. Nat. Acad. Sci. U.S.A. 69, 2945–2949.

Hirsch, J. G., Fedorko, M. E. and Cohn, Z. A. (1968) Vesicle fusion and formation at the surface of pinocytic vacuoles in macrophages. J. Cell Biol. 38, 629–632.

Hodson, S. and Brenchley, G. (1976) Similarities of the Golgi apparatus membrane and the plasma membrane in rat liver cells. J. Cell Sci. 20, 167–182.

Hokfelt, T. (1969) Distribution of noradrenaline-storing particles in peripheral adrenergic neurons as revealed by electron microscopy. Acta. Physiol. Scand. 76, 427–440.

Hokfelt, T. (1974) Morphological contributions to monoamine pharmacology. Fed. Proc. 33, 2177–2186.

Hokin, L. E. (1968) Dynamic aspects of phospholipids during protein secretion. Int. Rev. Cytol. 23, 187–208.

Hokin, L. E. (1969) Phospholipid metabolism and functional activity of nerve cells, In: The Structure and Function of Nervous Tissue (Bourne, G., ed.), vol. 3, pp. 161–184. Academic Press, New York.

Holtzman, E. (1969) Lysosomes in the physiology and pathology of neurons, In: Lysosomes in Biology and Pathology (Dingle, J. T. and Fell, H. B., eds.), vol. 1, pp. 192–216. Elsevier/North-Holland, Amsterdam.

Holtzman, E. (1971) Cytochemical studies of protein transport in the nervous system. Phil. Trans. Roy. Soc. Lond. B. 261, 407–421.

Holtzman, E. (1976) Lysosomes, A Survey. Cell Biology Monographs vol. 3. Springer-Verlag, Vienna.

Holtzman, E. (1977) Origin and fate of secretory packages, especially synaptic vesicles. Neuro-science—in press (vol. 2, June issue).

Holtzman, E. and Dominitz, R. (1968) Cytochemical studies of lysosomes, Golgi apparatus and endoplasmic reticulum in secretion and protein uptake by adrenal medulla cells of the rat. J. Histochem. Cytochem. 16, 320–336.

Holtzman, E. and Novikoff, A. B. (1965) Lysosomes in the rat sciatic nerve following crush. J. Cell Biol. 27, 651–669.

Holtzman, E. and Peterson, E. (1969) Protein uptake by mammalian neurons. J. Cell Biol. 40, 863–869.

Holtzman, E., Novikoff, A. B. and Villaverde, H. (1967) Lysosomes and GERL in normal and chromatolytic neurons of the rat ganglion nodosum. J. Cell Biol. 33, 419–436.

Holtzman, E., Freeman, A. R. and Kashner, L. A. (1971) Stimulation-dependent alterations in peroxidase uptake at lobster neuromuscular junctions. Science 173, 733–736.

Holtzman, E., Teichberg, S., Abrahams, S. J., Citkowitz, E., Crain, S. M., Kawai, N. and Peterson, E. R. (1973) Notes on synaptic vesicles and related structures, endoplasmic reticulum, lysosomes and peroxisomes in nervous tissue and the adrenal medulla. J. Histochem. Cytochem. 21, 349–385.

Hortnagel, H. (1976) Membrane of the adrenal medulla: a comparison of membranes of chromaffin granules with those of endoplasmic reticulum. Neuroscience 1, 9–18.

Howell, S. L. and Whitfield, M. (1973) Synthesis and secretion of growth hormone in the rat anterior pituitary. J. Cell Sci. 12, 1–21.

Howes, E. A., McLaughlin, B. J. and Heslop, J. P. (1974) The autoradiographic association of fast transported material with dense core vesicles in the central nervous system of Anodonta cygnea. Cell Tissue Res. 153, 545–558.

Hubbard, A. L. and Cohn, Z. A. (1975) Externally disposed plasma membrane proteins. II Metabolic fate of iodinated polypeptides of mouse L cells. J. Cell Biol. 64, 461–479.

Hurlburt, W. P. and Ceccarelli, B. (1974) Transmitter release and recycling of synaptic vesicle membrane at the neuromuscular junction. In: Advances in Cytopharmacology (Ceccarelli, B., Clementi, F. and Meldolesi, J., eds.), vol. 2, pp. 141–154. Raven Press, New York.

Jacobwitz, D. M., Ziegler, M. G. and Thomas, J. A. (19   ) In vivo uptake of antibody to dopamine-β-hydroxylase into sympathetic elements. Brain Res. 91, 165–170.

Jamieson, J. D. (1975) Membranes and secretion, In: Cell Membranes: Biochemistry, Cell Biology and Pathology (Weissmann, G. and Claiborne, R., eds.), pp. 143–152. HP Company, New York.

Jones, D. G. (1975) Synapses and synaptosomes, Chapman and Hall, London.

Joo, F., Halasz, N. and Parducz, A. (1973) Studies on the fine structural localization of zinc-iodide-osmium reaction in the brain. I. Some characteristics of localization in the perikarya of identified neurons. J. Neurocytol. 2, 393–405.

Jorgensen, O. S. and Mellerup, E. T. (1974) Endocytotic formation of rat brain synaptic vesicles. Nature 249, 770–771.

Kadota, K. and Kadota, K. (1973) A nucleoside diphosphatase present in a coated vesicle fraction from synaptosomes of guinea pig whole brain. Brain Res. 56, 371–376.

Kaiserman-Abramof, I. R., Haring, J., Nauta, W. and Lasek, R. (1974) Involvement of the agranular reticulum in the transport of horseradish peroxidase within neurons. J. Cell Biol. 63, 160A.

Kalina, M. and Rabinowitch, R. (1975) Exocytosis coupled to endocytosis of ferritin in parotid acinar cells from isoprenalin-stimulated rats. Cell Tissue Res. 163, 373–382.

Kanesaki, T. and Kadota, K. (1966) The vesicle in a basket; a morphological study of the coated vesicle isolated from the nerve endings of the guinea pig brain with special reference to the mechanism of membrane movements. J. Cell Biol. 42, 202–220.

Kaplan, J. (1975) Differential turnover of plasma membrane polypeptides. J. Cell Biol. 67, 199A.

Kawai, N., Mauro, A. and Grundfest, H. (1972) Effects of blackwidow spider venom on the lobster neuromuscular junction. J. Gen. Physiol. 60, 650–663. (See also J. Cell Biol. 68, 462–479.)

Kawana, E., Sandri, E. and Akert, K. (1971) Ultrastructure of growth cones in the cerebellar cortex of the neonatal rat and cat. Zeit. Zellforsch. 115, 284–298.

Kirschner, N. (1972) The adrenal medulla. In: The Structure and Function of Nervous Tissue (Bourne, G., ed.), vol. 5, pp. 164–204. Academic Press, New York. (See also p. 71 In: New Aspects of Storage and Release Mechanisms of Catecholamines [Schumann and Kroneberg, eds.], Springer-Verlag, Berlin.

Kirschner, N. (1974) Molecular organization of the chromaffin vesicles of adrenal medulla. In: Advances in Cytopharmacology (Ceccarelli, B., Clementi, F. and Meldolesi, J., eds.), vol. 2, pp. 265–272. Raven Press, New York.

Klein, R. and Thureson-Klein, A. K. (1974) Pharmacological aspects of large dense-cored adrenergic vesicles. Fed. Proc. 33, 2195–2206.

Koenig, H. L., di Giamberadino, L. and Bennett, G. (1973) Renewals of proteins and glycoproteins of synaptic constituents by means of axonal transport. Brain Res. 62, 413–417.

Koerker, R. L., Schneider, F. H. and Hahn, W. E. (1973) Quantitation of morphologic changes in rabbit adrenal medulla following catecholamine depletion. Fed. Proc. 32, 783(abstr.).

Kokko, A. and Barrnett, R. J. (1971) Dense content in synaptic vesicles produced by sequential cation-binding, alcohol treatment and osmium tetroxide fixation. Prog. Brain Res. 34, 319–327.

Korneliussen, M. (1972) Ultrastructure of normal and stimulated motor endplates with comments on the origin and fate of synaptic vesicles. Z. Zellforsch. 130, 28–57.

Kramer, M. F. and Geuze, J. J. (1974) Redundant cell-membrane regulation in the exocrine pancreas cells after pilocarpine stimulation of the secretion. In: Advances in Cytopharmacology (Ceccarelli, B., Clementi, F. and Meldolesi, J., eds.), vol. 2, pp. 87–98. Raven Press, New York.

Krauhs, J. M., Mahler, H. R., Minkler, G. and Moore, W. J. (1976) Synthesis and degradation of protein of visual receptor membrane in lateral eyes of Limulus. J. Neurochem. 26, 281–283.

Krishnan, N. and Singer, M. (1973) Penetration of peroxidase into peripheral nerve fibers. Am. J. Anat. 136, 1–14.

Kristensson, K., Olsson, Y., Sjöstrand, J. (1971) Axonal uptake and retrograde transport of exogenous proteins in the hypoglossal nerve. Brain Res. 32, 399–406.

Krugier-Brevart, V., Weiss, P. G., Mehl, E., Schubert, P. and Kreutzberg, G. W. (1974) Maintenance of synaptic membranes by the fast axonal flow. Brain Res. 77, 97–110.

Kuriyama, Y., Omura, T., Siekevitz, P. and Palade, G. E. (1969) Effects of phenobarbital on the synthesis and degradation of the protein component of rat microsomal membranes. J. Biol. Chem. 224, 2017–2026.

Lagerkrantz, H. (19) Isolation and characterization of sympathetic nerve trunk vesicles. Acta. Physiol. Scand. Suppl.#366.

Lasek, R. J. (1970) Protein transport in neurons. Int. Rev. Neurobiol. 13, 289–317.

Lauduron, P. (1974) Differential accumulation of enzymes in constricted nerves according to their subcellular distribution. In: Dynamics of Regeneration and Growth in Neurons (Fuxe, K., Olsson, K. and Zolterman, Y., eds.), pp. 245–256. Pergamon Press, New York.

La Vail, M. M. and La Vail, J. (1974) The retrograde intraaxonal transport of horseradish peroxidase in the chick visual system: a light and electron microscope study. J. Comp. Neurol. 157, 303–357.

Lazarides, E. and Weber, K. (1974) Actin antibody: the specific visualization of actin filaments in non-muscle. Proc. Nat. Acad. Sci. U.S.A. 71, 2268–2272.

Lazarus, S. S., Volk, B. W. and Barden, H. (1966) Localization of acid phosphatase activity and secretion mechanism in rabbit pancreatic β cells. J. Histochem. Cytochem. 14, 233–246.

Leighton, F., Colona, L. and Koenig, C. (1975) Structure, composition, physical properties and turnover of proliferated peroxisomes. J. Cell Biol. 67, 281–309.

Leskes, A., Siekevitz, P. and Palade, G. E. (1971) Differentiation of endoplasmic reticulum in hepatocytes. J. Cell Biol. 49, 264–287, 288–302.

Leuenberger, P. M. and Novikoff, A. B. (1975) Studies on microperoxisomes. VII:Pigment epithelial cells and other cell types of the retina of rodents. J. Cell Biol. 65, 324–334.

Lieberman, A. R. (1971) Microtubule-associated smooth endoplasmic reticulum in the frog brain. Z. Zellforsch. 116, 564–577.

Litchy, W. J. (1973) Uptake and retrograde transport of horseradish peroxidase in frog sartorius nerve in vitro. Brain Res. 56, 377–381.

Little, J. S. and Widnell, C. C. (1975) Evidence for the translocation of 5′-nucleotidase across hepatic membranes in vitro. Proc. Nat. Acad. Sci. U.S.A. 72, 4013–4017.

Locke, M. (1975) Organelle turnover, In: Pathological Changes in Cell Membranes (Trump, B. F., ed.). Academic Press, New York.

Locke, M. and Huie, P. (1975) Golgi-complex, Endoplasmic reticulum transition region has rings of beads. Science 188, 1219–1221.

Lodish, H. and Small, B. (1975) Membrane protein synthesized by rabbit reticulocytes. J. Cell Biol. 65, 51–64.

Lovas, B. (1971) Tubular networks in the terminal endings of the visual receptor cell in the human, the monkey, the cat and the dog. Z. Zellforsch. 121, 341–357.

Lowrie, D. B., Jackett, P. S. and Ratcliffe, N. A. (1975) Mycobacterium micoti may protect itself from intracellular destruction by release of cyclic AMP into phagosomes. Nature 254, 600–602.

Lubinska, L. (1975) On axoplasmic flow. Int. Rev. Neurobiol. 17, 241–296.

Lucy, J. A. (1975) The fusion of cell membranes, In: Cell Membranes, (Weissmann G. and Claiborne, R., eds.), pp. 75–83. HP Publishers, New York.

Machado, A. B. M. (1971) Electron microscopy of developing sympathetic fibers in the rat pineal body: the formation of granular vesicles. Prog. Brain Res. 34, 171–185.

Mahler, H. R., Gurd, J. W. and Wang, Y-J. (1975) Molecular topography in the synapse, In: The Nervous System (Tower, D. B., ed.), vol. 1, pp. 455–466. Raven Press, New York.

Marko, P. and Cuenod, M. (1973) Contributions of the nerve cell body to renewal of axonal and synaptic glycoproteins in the pigeon visual system. Brain Res. 62, 419–423.

Maskalewski, S., Thyberg, J. and Friberg, U. (1976) In vitro influence of colchicine on the Golgi complex in A and B cells of guinea pig islets. J. Ultrastruct. Res. 54, 304–317.

Masur, S. K., Holtzman, E., Schwartz, I. L. and Walter, R. (1971) Correlation between pinocytosis and hydroosmosis induced by neurohypophyseal hormones and adenosine 3',5'-cyclic nucleotide. J. Cell Biol. 49, 582–594.

Masur, S. K., Holtzman, E. and Walter, R. (1972) Hormone-stimulated exocytosis in the toad urinary bladder. J. Cell Biol. 52, 211–219.

Matthews, M. B. (1973) An ultrastructural study of axonal changes following constriction of post-ganglionic branches of the superior cervical ganglion in the rat. Phil. Trans. Roy. Soc. Lond. B. 264, 479–508.

McBride, W. J. and Van Tassel, J. (1972) Resolution of proteins from subfractions of nerve endings. Brain Res. 44, 177–187.

McMahan, U. J. and Yee, A. (1975) Rapid uptake and loss of HRP in motor nerve terminals of the snake. Proc. Soc. Neurosci., 5th Ann. Meet., 1975, p. 624.

Meldolesi, J. (1974) Secretory mechanisms in pancreatic acinar cells. Role of the cytoplasmic membranes. In: Advances in Cytopharmacology (Ceccarelli, B., Clementi, F. and Meldolesi, J., eds.), vol. 2, pp. 71–85. Raven Press, New York.

Meunier, F-M., Israel, M. and Lesbats, B. (1975) Release of ATP from stimulated nerve electroplaque junctions. Nature 257, 407–408.

Miyamoto, M. D. and Previti, M. A. (1976) Electrophysiological studies on the recycling of synaptic vesicles. Fed. Proc. 35, 696a.

Model, P. G., Highstein, S. M. and Bennett, M. V. L. (1975) Depletion of vesicles and fatigue of transmission at a vertebrate central synapse. Brain Res. 98, 209–228.

Mollenhauer, H. H. and Morré, D. J. (1976) Cytochalasin B, but not colchicine, inhibits migration of secretory vesicles in root tips of maize. Protoplasma 87, 39–48.

Montessano, R., Friend, D. S., Perrelet, A. and Orci, L. (1975) In vivo assembly of tight junctions in fetal rat liver. J. Cell Biol. 67, 310–319.

Morré, D. J., Keenan, T. W. and Huang, C. M. (1974) Membrane flow and differentiation: origin of Golgi apparatus membranes from endoplasmic reticulum. In: Advances in Cytopharmacology (Ceccarelli, B., Clementi, F. and Meldolesi, J., eds.), vol. 2, pp. 107–126. Raven Press, New York.

Musick, J. and Hubbard, J. (1972) Release of protein from mouse motor nerve terminals. Nature 237, 279–281.

Nagasawa, J. and Douglas, W. W. (1972) Thorium dioxide uptake into adrenal medullary cells and the problem of recapture of granule membranes following exocytosis. Brain Res. 37, 141–145.

Nordmann, J. J., Dreifuss, J. J., Baker, P. F., Ravazzolla, M., Malaisse-Lagae, F. and Orci, L. (1974)

Secretion dependent uptake of extracellular fluid by the rat neurohypophysis. Nature 250, 155–157.

Normann, T. C. (1969) Experimentally induced exocytosis of neuro-secretory granules. Exp. Cell Res. 55, 285–287.

Novikoff, A. B. and Holtzman, E. (1976) Cells and Organelles, 2nd ed. Holt, Reinhart and Winston, New York.

Novikoff, A. B. and Novikoff, P. M. (1973) Microperoxisomes. J. Histochem. Cytochem. 21, 963–966.

Novikoff, P. M., Novikoff, A. B., Quintana, N. and Hauw, J-J. (1971) Golgi apparatus, GERL and lysosomes of neurons in rat dorsal root ganglia studied by thick section and thin section cytochemistry. J. Cell Biol. 50, 859–886. (See also Proc. Nat. Acad. Sci., U.S.A. 73, 2781–2787.)

Novikoff, A. B., Novikoff, P. M., Ma, M., Shin, W. Y. and Quintana, N. (1974) Cytochemical studies of secretory and other granules associated with the endoplasmic reticulum in rat thyroid epithelial cells. In: Advances in Cytopharmacology (Ceccarelli, B., Clementi, F. and Meldolesi, J., eds.), vol. 2, pp. 349–368. Raven Press, New York.

Novikoff, A. B., Yam, A. and Novikoff, P. M. (1975) Cytochemical studies of secretory process in transplantable insulinoma of Syrian golden hamster. Proc. Nat. Acad. Sci. U.S.A. 72, 4501–4505.

Novikoff, A. B., Mori, M., Quintana, N. and Yam, A. (1976) Processing and packaging of secretory materials in the exocrine pancreas. J. Histochem. Cytochem, (in press).

Oates, P. J. and Touster, O. (1976) In vitro fusion of Acanthamoeba phagolysosomes. J. Cell Biol. 68, 319–331.

Ochs, S. (1974) Energy metabolism and supply of ≈P to the fast axonal transport mechanism in nerve. Fed. Proc. 33, 1049–1058.

Ojakian, G. K., Kreibich, G. and Sabatini, D. (1977) Mobility of ribosomes bound to microsomal membranes. J. Cell Biol. 72, 530–551.

Oliver, J. M., Krawiec, J. A. and Berlin, R. D. (1976) Carbamylcholine prevents giant granule formation in cultured fibroblasts from Beige (Chediak-Higashi) mice. J. Cell Biol. 69, 205–210.

Olsen, B. R., Berg, R. A., Kushida, Y. and Prockup, D. J. (1975) Further characterization of embryonic tendon fibroblasts and the use of immunofecritin techniques to study collagen biosynthesis. J. Cell Biol. 64, 340–355.

Orci, L. and Perrelet, A. (1973) Membrane associated particles: increase at sites of pinocytosis demonstrated by freeze-etching. Science 181, 868–869.

Orci, L., Amherdt, M., Malaisse-Lagae, F., Rouiller, C. and Renold, A. E. (1973a) Insulin release by emiocytosis: demonstration with freeze-etching technique. Science 179, 82–84.

Orci, L., Malaisse-Lagae, F., Ravazzola, M. and Amherdt, M. (1973b) Exocytosis-endocytosis coupling in the pancreatic beta cell. Science 181, 561–562.

Osinchak, J. (1964) Electron microscopic localization of acid phosphatase and thiamine pyrophosphatase activity in hypothalamic neurosecretory cells of the rat. J. Cell Biol. 21, 35–47.

Palade, G. E. (1959) Functional changes in the structure of cell components, In: Subcellular Particles (Hayashi, T., ed), pp. 64–83. Ronald Press, New York.

Palade, G. E. (1975) Intracellular aspects of the process of protein secretion. Science 189, 347–357.

Palay, S. L. (1958) The morphology of synapses in the central nervous system. Exp. Cell Res. Suppl. 5, 275–293.

Pappas, G. D. (1975) Ultrastructural basis of synaptic transmission, In: The Nervous System (Tower, D. B., ed.), vol. 1, pp. 19–30. Raven Press, New York.

Pappas, G. D. and Purpura, D. P. (1972) Structure and Function of Synapses. Raven Press, New York.

Pappas, G. D. and Rose, S. (1976) Localization of calcium deposits in the frog neuromuscular junction at rest and following stimulation. Brain Res. 103, 362–365.

Paravicini, U., Stoeckl, K. and Thoenen, H. (1975) Biological importance of retrograde axonal transport of nerve growth factor in adrenergic neurons. Brain Res. 84, 279–291.

Pearse, B. M. F. (1976) Clathrin, a unique protein associated with intracellular transport of membrane by coated vesicles. Proc. Nat. Acad. Sci., U.S.A. 73, 1255–1259.

Pelletier, G. (1973) Secretion and uptake of peroxidase by rat adenohypophyseal cells. J. Ultrastruct. Res. 43, 445–459.

Pelletier, G., Dupont, A. and Puvani, R. (1975) Ultrastructural study of the uptake of peroxidase by the rat median eminence. Cell Tissue Res. 156, 521–532.

Pelligrino-de-Iraldi, A. and De Robertis, E. (1970) Studies on the origin of the granulated and non-granulated vesicles, In: New Aspects of Storage and Release Mechanisms of Catecholamines. (Schumann, A. J. and Kroneberg, G., eds.), pp. 4–17. Springer-Verlag, New York.

Pelligrino-de-Iraldi, A. and Suburo, A. M. (1971) Presynaptic tubular structures in photoreceptor cells. Z. Zellforsch. 113, 39–43.

Peters, A., Palay, S. L. and de F. Webster, H. (1976) The Fine Structure of the Nervous System, 2nd ed. Hoeber, New York.

Pfenninger, K. H. and Bunge, R. P. (1974) Freeze-fracturing of nerve growth cones and young fibers: a study of developing plasma membranes. J. Cell Biol. 63, 180–196.

Pfenninger, K. H. and Rovainen, C. M. (1974) Stimulation and calcium dependence of vesicle attachment sites in the presynaptic membrane: a freeze cleave study on the lamprey spinal cord. Brain Res. 72, 1–23.

Pfenninger, K., Akert, K., Moor, H. and Sandri, S. (1972) The fine structure of freeze-fractured presynaptic membranes.

Philippovich, I. I., Bezmertnayer, I. N. and Oparin, A. I. (1973) On the localization of polyribosomes in the system of chloroplast lamellae. Exp. Cell Res. 79, 159–168.

Pickel, V. (1976) Immunocytochemical differentiation of noradrenergic, dopaminergic and serotonergic pathways in the central nervous system. J. Histochem. Cytochem, in press.

Pickel, V. M., Reis, D. J. and Leeman, S. E. (1977) Ultrastructural localization of substance P in neurons of rat spinal cord. Brain Res. 122, 534–540.

Pinto-da-Silva, P. and Nogueira, M. L. (1977) Membrane fusion during secretion: A hypothesis based on electron microscope observations of Phytophthora palmivora zoospores during encystment. J. Cell Biol. 73, 161–181.

Pipeleers, D. G., Pipeleers-Marichal, M. A. and Kipness, D. M. (1976) Microtubule assembly and the intracellular transport of secretory granules in pancreatic islets. Science 191, 88–89.

Plattner, H., Wolfram, D., Bachmann, L. and Wachter, E. (1975) Tracer and freeze-etching analysis of intracellular membrane junctions in Paramecium. Histochemistry 15, 1–25.

Pollard, T. B. and Weihung, R. R. (1974) Actin and myosin in cell movement. CRC Rev. Biochem. 2, 1–65.

Porter, K. R., Kenyon, K. and Badenhausen, S. (1967) Specializations of the unit membrane. Protoplasma 63, 262–274.

Poste, G. and Allison, A. C. (1973) Membrane fusion. Biochim. Biophys. Acta 300, 421–465.

Potter, C. T. (1972) Synthesis, storage and release of acetylcholine from nerve terminals, In: The Structure and Function of Nervous Tissue (Bourne, G., ed.), vol. 4, pp. 105–128. Academic Press, New York.

Pysh, J. J. and Wiley, G. G. (1974) Synaptic vesicle depletion and recovery in cat sympathetic ganglia stimulated in vivo. Evidence for transmitter secretion by exocytosis. J. Cell Biol. 60, 365–374.

Quastel, A. (1974) Excitation-secretion coupling at the mammalian neuromuscular junction, In: Synaptic Transmission and Neuronal Interaction (Bennett, M. V. L., ed.), Raven Press, New York.

Quatacker, J. (1975) Endocytosis and multivesicular bodies in rabbit luteal cells. Cell Tissue Res. 161, 541–554.

Raff, M. C. (1976) Self-regulation of membrane receptors. Nature 259, 265–266.

Rajan, K. S., Davis, J. M. and Colburn, R. W. (1974) Metal chelates in the storage and transport of neurotransmitters: Interaction of Cu^{2+} with ATP and biogenic amines. J. Neurochem. 22, 137–147.

Reddy, J. and Svoboda, D. (1973) Further evidence to suggest that microbodies do not exist as individual entities. Am. J. Pathol. 70, 421–438.

Redman, C. M., Banerjee, D., Howell, K. and Palade, G. E. (1975) Colchicine inhibition of plasma protein release from rat hepatocytes. J. Cell Biol. 66, 42–59.

Rees, R. P., Bunge, M. B. and Bunge, R. P. (1976) Morphological changes in the neurite growth cone and target neuron during synaptic junction development in culture. J. Cell Biol. 68, 240–263.

Reperant, J. (1975) The orthograde transport of horseradish peroxidase in the visual system. Brain Res. 85, 307–312.

Richards, J. G. and Tranzer, J. P. (1975) Localization of amine storage sites in the adrenergic cell body. J. Ultrastruct. Res. 53, 204–216.

Ripps, H., Shakib, M. and MacDonald, F. C. (1976) Peroxidase uptake by photoreceptor terminals of the skate retina. J. Cell Biol. 70, 86–96.

Rodiek, R. W. (1973) The Vertebrate Retina. Freeman, San Francisco.

Rodriquez-de Lores Arnaiz, L., Ziehen, M. and de Robertis, E. (1970) Neurochemical and structural studies on the mechanism of action of hemicholinium 3 at central cholinergic synapses. J. Neurochem. 17, 221–229.

Rodriguez-Echandia, E. L., Zamora, A. and Piezzi, R. S. (1970) Organelle transport in constricted nerve fibers of the toad, Bufo arenarium Hensel. Z. Zellforsch. 104, 409–428.

Rosenbluth, J. and Wissig, S. L. (1964) The distribution of exogenous ferritin in toad spinal ganglia and the mechanism of its uptake by neurons. J. Cell Biol. 23, 307–325.

Roth, T. F. and Porter, K. R. (1964) Yolk protein uptake in the oocyte of the mosquito *Aedes aegypti*. J. Cell Biol. 20, 313–332.

Rothman, S. S. (1975) Protein transport by the pancreas. Science 190, 747–753.

Rubin, L. L., Gorio, A. and Mauro, A. (1976) Effect of cytochalasin B on neuromuscular transmission in tissue culture. Brain Res. 104, 171–175.

Ryan, G. B., Borysenko, J. Z. and Karnovsky, M. J. (1974) Factors affecting the redistribution of surface-bound concanavalin A on human polymorphonuclear leukocytes. J. Cell Biol. 62, 351–363.

Samorajaski, T., Ordy, J. M. and Keefe, J. R. (1966) Structural organization of the retina in the tree shrew Tupara glis. J. Cell Biol. 28, 489–504.

Satir, P. and Satir, B. (1974) Design and function of site-specific particle arrays in the cell membrane. In: Control of Proliferation of Animal Cells (Baserga, R. and Clarkson, B., eds.), vol. 1, pp. 233–240 (Cold Spring Harbor Conf. on Cell Proliferation). Cold Spring Harbor Laboratory, New York.

Schacher, S., Holtzman, E. and Hood, D. C. (1973) Cytochemical studies of frog retinal photoreceptor cells. In: Expanded Abstracts, Proc. 3rd Ann. Meet. Soc. Neurosci., San Diego, p. 41, Brain Information Service, University of California, Los Angeles.

Schacher, S. M., Holtzman, E. and Hood, D. C. (1974) Uptake of horseradish peroxidase by frog photoreceptor synapses in the dark and the light. Nature 249, 261–263.

Schacher, S., Holtzman, E. and Hood, D. C. (1976) Synaptic activity of frog retinal photoreceptors: a peroxidase uptake study. J. Cell Biol. 70, 178–192.

Schaeffer, S. F. and Raviola, E. (1976) Ultrastructural analysis of functional changes in the synaptic endings of the turtle cone cells. Cold Spring Harbor Symp. Quant. Biol. 40, 521–528.

Scharrer, E. and Brown S. (1961) Neurosecretion XII. The formation of neurosecretory granules in the earthworm Lumbricus terrestris. Z. Zellforsch. 54, 530–540.

Schimke, R. T. (1974) The synthesis and degradation of membrane proteins. In: Advances in Cytopharmacology (Ceccarelli, B., Clementi, F. and Meldolesi, J., eds.), vol. 2, pp. 63–69. Raven Press, New York.

Schimke, R. T. and Katunuma, N., eds. (1975) Intracellular Protein Turnover. Academic Press, New York.

Schneider, F. M. (1970) Secretion from the bovine adrenal gland: release of lysosomal hydrolases. Biochem. Pharmacol. 19, 883–847.

Schonbach, J., Schonbach, C. H. and Cuenod, M. (1973) Distribution of transported proteins in the slow phase of axonal transport: an electron microscopic autoradiographic study. J. Comp. Neurol. 152, 1–16.

Sharoni, Y., Eimerl, S. and Schramm, M. (1976) Secretion of old versus new exportable protein in rat paratid slices. J. Cell Biol. 71, 107–122.

Siekevitz, P. (1972) Biological membranes: the dynamics of their organization. Ann. Rev. Physiol. 34, 117–140.

Singer, S. J. and Rothfeld, L. I. (1973) Synthesis and turnover of cell membranes. Neurosci. Res. Prog. Bull. vol. 11, No. 1.

Slot, J. W., Geuze, J. J. and Poort, C. (1976) Synthesis, and intracellular transport of proteins in the exocrine pancreas of the Frog (*Rana esculenta*) II. An *in vitro* study of the transport process and the influence of temperature. Cell Tissue Res. 167, 147–165.

Smith, A. D. (1972) Storage and secretion of hormones. Sci. Basis Med. Ann. Rev. pp. 74–102.

Smith, A. D., de Potter, W. P., Moerman, W. J. and de Schaepdryver, A. F. (1970) Release of dopamine-β-hydroxylase and chromagranin A upon stimulation of the splenic nerve. Tissue and Cell 2, 547–568.

Smith, D. S. (1971) On the significance of cross-bridges between microtubules and synaptic vesicles. Phil. Trans. Roy. Soc. Lond. B. 261, 395–404.

Smith, R. E. and Farquhar, M. G. (1966) Lysosome function in the regulation of the secretory process in cells of the anterior pituitary gland. J. Cell Biol. 31, 319–336.

Smith, U. (1971) Uptake of ferritin into neurosecretory terminals. Phil. Trans. Roy. Soc. Lond. B. 261, 391–394.

Smith, U., Smith, D. S., Winkler, H. and Ryan, J. W. (1973) Secretion from adrenal medulla studied by freeze etching. Science 179, 79–82.

Soifer, D., ed. (1975) The Biology of Cytoplasmic Microtubules. Ann. N.Y. Acad. Sci. 253, 1–848.

Somogyi, P., Chubb, I. and Smith, A. D. (1975) A possible structural basis for the extracellular release of acetyl cholinesterase. Proc. Roy. Soc. Lond. B. 191, 271–283.

Sotelo, C. and Palay, S. C. (1971) Altered axons and axon terminals in the lateral vestibular nucleus of the rat. Lab. Invest. 25, 653–671.

Staehelin, L. A. (1974) The structure and function of intracellular junctions. Int. Rev. Cytol. 39, 191–283.

Steiner, D. F., Kemmler, W., Tager, H. S. and Rubenstein, A. H. (1974) Molecular events taking place during intracellular transport of exportable proteins. The conversion of peptide hormone precursors. In: Advances in Cytopharmacology (Ceccarelli, B., Clementi, F. and Meldolesi, J., eds.), vol. 2, pp. 195–205. Raven Press, New York.

Steinman, R. M., Brodie, S. E. and Cohn, Z. A. (1976) Membrane flow during pinocytosis: a stereologic analysis. J. Cell Biol. 68, 665–687.

Stelzner, D. J. (1971) The relationship between synaptic vesicles, Golgi apparatus and smooth endoplasmic reticulum: a developmental study using the zinc-iodine-osmic technique. Z. Zellforsch. 120, 332–345.

Stoeckel, K. and Thoenen, H. (1975) Retrograde axonal transport of nerve growth factor: specificity and biological importance. Brain Res. 85, 337–341.

Stossel, J. P. and Hartwig, J. R. (1976) Interactions of actin myosin and a new actin-binding protein of rabbit pulmonary macrophages. J. Cell Biol. 68, 602–612.

Straus, W. (1967) Lysosomes, phagosomes and related particles, In: Enzyme Cytology (Roodyn, D. P., ed.), pp. 239–319. Academic Press, New York.

Swank, R. T. and Paigen, K. (1973) Genetic evidence for macromolecular β-glucuronidase complex in microsomal membranes. J. Molec. Biol. 77, 371–390.

Tandler, B. and Poulsen, J. H. (1976) Fusion of the envelope of mucous droplets with the luminal plasma membrane in acinar cells of the cat submandibular gland. J. Cell Biol. 68, 775–781.

Taxi, J. and Sotelo, C. (1973) Cytological aspects of the axonal migration of catecholamines and of their storage material. Brain Res. 62, 431–437.

Taylor, J. M., Dehlinger, P. J., Dice, J. F. and Schimke, R. T. (1973) The synthesis and degradation of membrane proteins. Drug Metabol. Dispos. 1, 84–91.

Teichberg, S. and Bloom, D. (1976) Uptake and fate of horseradish peroxidase in axons and terminals of sympathetic neurons. Abstracts 1st Int. Cong. J. Cell Biol. 70, 285A.

Teichberg, S. and Holtzman, E. (1973) Axonal agranular reticulum and synaptic vesicles in cultured embryonic chick sympathetic neurons. J. Cell Biol. 57, 88–108.

Teichberg, S., Holtzman, E., Crain, S. M. and Peterson, E. R. (1975) Circulation and turnover of synaptic vesicle membrane in cultured spinal cord neurons. J. Cell Biol. 67, 215–230.

Tennyson, V. M. and Brzin, M. (1970) The appearance of acetylcholinesterase in the dorsal root neuroblast of the rabbit embryo: a study by EM cytochemistry and microgasimetric analysis with the magnetic diver. J. Cell Biol. 46, 64–80.

Thoenen, M., Otten, U. and Oesch, F. (1973) Axoplasmic transport of enzymes involved in the synthesis of noradrenaline: relationship between the rate of transport and subcellular distribution. Brain Res. 62, 471–475.

Thompson, E. B., Kandel, E. R. and Schwartz, J. H. (1975) Axonal transport of vesicles: autoradiographic localization of ^{3}H-glycoproteins in identified neurons of Aplysia after intersomatic injection of ^{3}H-fucose. Proc. Soc. Neurosci., 5th Ann. Meet., p. 568.

Till, R. and Banks, P. (1976) Pharmacological and ultrastructural studies on the electron dense cores of the vesicles that accumulate in noradrenergic axons constricted in vitro. Neuroscience 1, 49–55.

Turner, P. T. and Harris, A. B. (1974) Ultrastructure of exogenous peroxidase in cerebral cortex. Brain Res. 74, 305–326.

Tweto, J. and Doyle, D. (1976) Turnover of the plasma membrane proteins of hepatoma tissue culture cells. J. Chem. Biol. 251, 872–882.

Vian, B. and Roland, J. C. (1972) Differentiation des cytomembranes et renouvellement du plasmalemme dans les phenomenes de secretions vegetales. J. Microscopie 13, 119–136.

Von During, M. (1967) Uber die Feinstruktur der motorischer endplatte von hoheren. Wirbeltieren. Z. Zellforsch. 80, 74–90.

Von Hungen, K., Mahler, H. R. and Moore, W. J. (1970) Turnover of protein and RNA in synaptic subcellular fractions from mouse brain. J. Biol. Chem. 243, 1415–1423.

Walter, R., ed. (1975) Neurophysins: Carriers of Peptide Hormones. Ann. N.Y. Acad. Sci. 248, 1–512.

Weinshilbaum, R. M., Thoa, N. B., Johnson, D. G., Kopin, I. J. and Axelrod, J. (1971) Proportional release of norepinephrine and dopamine-β-hydroxylase from sympathetic nerves. Science 174, 1349–1352.

Weldon, P. R. (1975) Pinocytotic uptake and intracellular distribution of colloidal thorium dioxide by cultured sensory neurites. J. Neurocytol. 4, 341–356.

Wessels, N. K., Leduena, M. A., Letourneau, P. C., Wrenn, J. J. and Spooner, B. S. (1974) Thorotrast uptake and transit in embryonic glia, heart fibroblasts and neurons in vitro. Tissue and Cell 6, 757–776.

Whaley, W. G. (1975) The Golgi Apparatus Cell Biology Monographs, vol. 2. Springer-Verlag, Vienna.

Whitaker, S. and LaBella, F. S. (1972) Ultrastructural localization of acid phosphate in the posterior pituitary of the dehydrated rat. Zeit. Zellforsch. 125, 1–15.

Whittaker, V. P. (1974) Molecular organization of the cholinergic vesicle. In: Advances in Cytopharmacology (Ceccarelli, B., Clementi, F. and Meldolesi, J., eds.), vol. 2, pp. 311–317. Raven Press, New York.

White, R. H. (1968) The effects of light and light deprivation upon the ultrastructure of the mosquito larval eye. III Multivesicular bodies and protein uptake. J. Exp. Zool. 169, 261–278.

Winkler, H. and Smith, A. D. (1974) The chromaffin granule and the storage of catecholamines. Handbook of Physiology, Endocrinology, VI, chap. 23, p. 321–339. American Society of Physiologists, Washington, D.C.

Winkler, H., Schopf, H., Hortnagl, H. and Hortnagl, H. (1972) Bovine adrenal medulla: subcellular distribution of newly synthesized catecholamines, nucleotides and chromagranins. Nauyn-Schmiedebergs Arch. Pharmakol. 273, 43–58.

Winkler, H., Schneider, F. H., Rufener, C., Nakane, P. K. and Hortnagel, H. (1974) Membranes of adrenal medulla: their role in exocytosis. In: Advances in Cytopharmacology (Ceccarelli, B., Clementi, F. and Meldolesi, J., eds.), vol. 2, pp. 127–140. Raven Press, New York.

Wise, G. E. and Flickenger, C. J. (1970) Relation of the Golgi apparatus to the cell coat in amoebae. Exp. Cell Res. 61, 13–23.

Wooten, G. F. (1973) Subcellular distribution and rapid axonal transport of dopamine-β-hydroxylase. Brain Res. 55, 491–494.

Yamada, E. (1965) Some observations on the membrane-limited structure within the retinal elements, In: Intracellular Membranous Structures (Sero, S. and Cowdry, E. V., eds.), pp. 49–63. Japan Soc. Cell Biol., Okayama, Japan.

Yamada, K., Spooner, B. S. and Wessels, N. K. (1971) Ultrastructure and function of growth cones and axons of cultured nerve cells. J. Cell Biol. 49, 614–635.

Yen, S. H., Dahl, D., Schachner, M. and Shelanski, M. L. (1976) Biochemistry of the filaments of brain. Proc. Nat. Acad. Sci. U.S.A. 73, 529–533.

Young, R. W. (1968) Passage of newly formed protein through the connecting cilium of retinal rods in the frog. J. Ultrastruct. Res. 23, 462–473.

Young, R. W. and Droz, B. (1968) The renewal of protein in retinal rods and cones. J. Cell Biol. 39, 169–184.

Zachs, S. I. and Saito, A. (1969) Uptake of exogenous horseradish peroxidase by coated vesicles in mouse neuromuscular junctions. J. Histochem. Cytochem. 17, 161–170.

Ziegler, M. G., Thomas, J. D. and Jacobowitz, D. M. (1976) Retrograde axonal transport of antibody to dopamine-β-hydroxylase. Brain Res. 104, 390–395.

Zimmerman, H. and Whittaker, V. P. (1974a) Effect of electrical stimulation on the yield and composition of synaptic vesicles from the cholinergic synapses of the electric organ of Torpedo: a combined biochemical, electrophysiological and morphological study. J. Neurochem. 22, 435–450.

Zimmerman, H. and Whittaker, V. P. (1974b) Different recovery rates of the electrophysiological, biochemical and morphological parameters in the cholinergic synapses of the Torpedo electric organ after stimulation. J. Neurochem. 22, 1109–1114.

The myelin sheath– a structural examination[1]

5

M. G. RUMSBY and A. J. CRANG

1. Introduction

Interest in the morphology and biochemistry of the myelin sheath in central and peripheral nerve tissue has been increasing steadily since the middle of the nineteenth century when myelin was first named. The history of the early research work on the myelin sheath has been well documented by Mokrasch (1971a). Two well known characteristics of myelin, its lipid-rich nature and its highly ordered structure, had been defined by the beginning of the twentieth century. Around the same period Thudichum was pioneering the isolation and characterization of a variety of lipid fractions from brain tissue, but the lack of suitable techniques precluded any direct study of the specific lipid components of the myelin sheath around nerve axons. In central nerve, tissue myelin accounts for about 50 to 60% of the dry weight of white matter and for more than 40% of the total lipid in brain (Norton and Autilio, 1966). Thus, much of the lipid examined by Thudichum would have been derived from myelin. In the last two or three decades the application of x-ray diffraction, electron microscopy, and other specialized physicochemical techniques to the problems of myelin structure, coupled with the development of sophisticated methods for the resolution and analysis of lipids and proteins, have produced a rapid increase in the volume of data available on all aspects of myelin biochemistry.

Reviews on myelin in recent years have dealt with the overall topic or have covered some narrower aspect of the subject. These have appeared in the form of complete monographs (Davison and Peters, 1970; Mokrasch et al., 1971; Morell, in preparation) or as chapters in broader treatments on the chemistry,

---

[1]Abbreviations: CN, central nerve; CNPH, 2′,3′-cyclic nucleotide-3′-phosphohydrolase; CNS, central nervous system; IPDL, intraperiod dense line; LPC, lysophosphatidylcholine; MDL, main dense line; PA, phosphatidic acid; PC, phosphatidylcholine; PE, phosphatidylethanolamine; PG, phosphatidylglycerol; PI, phosphatidylinositol; PL, proteolipid; PLP, proteolipid protein; PNS, peripheral nervous system; PS, phosphatidylserine; TPI, triphosphoinositide; SDS, sodium dodecyl sulfate;

G. Poste and G.L. Nicolson (eds.) The Synthesis, Assembly and Turnover of Cell Surface Components
© *Elsevier/North-Holland Biomedical Press, 1977.*

anatomy, development, or pathology of nerve tissue (Davison, 1968; Mokrasch, 1969, 1971b; O'Brien, 1970a,b; Norton, 1971, 1972, 1975; Hirano, 1972). Reviews dealing with specific aspects of myelin, such as lipids and proteins (Eichberg et al., 1969; Lebaron, 1969; O'Brien, 1970b; Shooter and Einstein, 1971; Davison et al., 1972; Einstein, 1972; Eylar, 1972, 1973; Folch-Pi, 1972, 1973; Carnegie and Dunkley, 1975), on the metabolism of myelin components (Smith, 1967; Davison, 1970; Lebaron, 1970; Davison et al., 1972; Ansell, 1973) and on the isolation of myelin from nerve tissue (Mokrasch, 1971b; Spohn and Davison, 1972; Norton, 1974, 1976) have also appeared.

The articles reveal much about the structure and form of the myelin sheath in normal, developing, and diseased systems. However, one area of myelin biochemistry that has not been summarized for some time concerns the molecular organization of the membrane system. Rapid advances have been made over the past few years in our understanding of the structure and dynamic state of lipids and proteins in biological membranes, and a review emphasising the chemical architecture and molecular organization of myelin as a membrane is timely. A molecular approach is also pertinent since over the last year or two a variety of techniques designed to investigate the surface molecular structure of membranes have been applied to myelin. This work has yielded results that clarify our ideas on the localization of specific proteins and lipids within the membrane structure. Similarly, the lipid components of myelin have been examined by specialized techniques that have provided valuable data on their structural and dynamic form in the membrane. The interaction of purified myelin proteins with lipids has also been studied in model systems in vitro. Collectively, the results from such work put previous ideas on structure in myelin on a more sound experimental basis.

The primary purpose of this review will be to examine data on the structure and dynamic form of the lipids and proteins in the compact myelin sheath. Results will be examined in light of current knowledge about the organization and structure of similar components in other natural membranes and in artificial model membrane systems. Our attention will be directed primarily to the myelin sheath in central nerve tissue as this system encompasses our main research interests. Also, more data is available on this system than on peripheral nerve myelin which is more difficult to isolate and purify. Significant points of difference and interest pertaining to peripheral nerve myelin will also be discussed. The references cited in this chapter will largely cover the period from 1970 to the beginning of 1976, but earlier work will be described where appropriate. References have been selected to fit the needs of this review and no attempt has been made to describe comprehensively all the literature in each area of the subject discussed.

2. Current concepts of membrane organization

Before attempting any consideration of molecular organization in the myelin sheath, it is necessary to review briefly the main points now known about the structure, form, and function of lipids and proteins in artificial lipid membrane systems and natural membrane structures. The findings can then be related to the myelin sheath to determine those that are applicable. Since the volume of data available on membrane structure is vast, only the main features will be covered. Wherever possible in this section reference will be made to review articles, as these summarize the results of the numerous research papers in the area.

General ideas and principles of membrane structure and function have been reviewed over the past few years by Chapman (1968), Korn (1969), Rothfield (1971), Singer (1971, 1974a, b, c), Vandenheuvel (1971), Bangham (1972), Finean (1972), Guidotti (1972), Phillips (1972), Singer and Nicolson (1972), Bretscher (1973, 1974), Lenaz (1973), Oseroff et al. (1973), Capaldi (1974), Lee et al. (1974), Marsh (1975) Nicolson et al. (1977).

2.1. Function of natural membranes

Membranes in living systems share two common properties. First, they act as barriers to the free diffusion of water, ions, and solutes and, in so doing, define the limits of a series of integrated reactions at the level of either the whole cell or organelles within the cell. This compartmentalization is necessary so that molecules may be brought together for reactions to proceed efficiently and to allow control systems to operate. Secondly, membranes provide a matrix for the localization of various enzyme systems.

Membranes differ markedly, however, in such features as level of enzyme activity, transport properties, receptor functions, and antigenicity. It is also clear that in addition to the two properties common to all membranes outlined above, other specialized functions and activities have evolved to suit the location and purpose of specific membrane systems. Thus the different membrane structures possess both common and specialized properties, and it is relevant to determine whether these are reflected in similarities and differences in chemical composition and structural organization.

2.2. Chemical composition of membranes

The development of reliable methods for subcellular fractionation (see Fleischer and Packer, 1974) has enabled biochemists to isolate and purify many different membrane structures from a variety of biological sources. Analysis of these preparations reveals that the two major components of all natural membrane systems are lipids and proteins. These occur in varying proportions in different membranes. In addition, low amounts of carbohydrate are present in certain membrane systems. Guidotti (1972) has summarized data on the lipid, protein, and carbohydrate content of a variety of purified membrane systems and, for the purpose of the present review, this can be supplemented by the recent data of Poduslo (1975b) on the plasma membrane of the oligodendrocyte which is reported to be 54% protein, 43% lipid, and 0.5% ganglioside on a dry weight basis.

Guidotti (1972) has divided membranes into three general classes based on their lipid-to-protein dry weight ratio. At one extreme are relatively "simple" membranes (our classification) such as myelin, which are characterized by a very low protein-to-lipid ratio of about 0.2:1. In the middle are membranes, such as the plasma membrane of animal cells, which have an approximately equal content of lipid and protein. Finally, there are structures, such as the inner membrane of mitochondria and the plasma membrane of bacteria, which have a high protein-to-lipid ratio of over 2:1. It is now clear that these differences in protein content reflect differences in the functional activity of membranes. The extremely low protein content of myelin relative to lipid indicates its low enzyme content and inert character. However, the high proportion of lipid is in keeping with the apparent role of the myelin sheath in central and peripheral nerve tissue as an effective barrier to the diffusion of ions and insulation around the axon (Bunge, 1968; Brazier, 1969; Norton, 1975). In summary, a lipid phase provides the basic permeability barrier in natural membranes. Where a membrane has other specialized activities, proteins and glycoproteins occur in the lipid matrix to provide the necessary metabolic, transport, and receptor functions.

Water, present in normal membranes as an essential structural component (Cerbon, 1974), accounts for about 20% or more of the wet weight. Membrane water is intimately associated with surface lipid and protein molecules. The vital role of water in maintaining membrane integrity has been shown clearly for myelin by Finean (1958) and Ladbrooke et al. (1968) among others. Estimates of the water content in myelin have been between 35 to 50% (Finean, 1957, 1960; Vandenheuvel, 1965; Gent et al., 1970; Blaurock, 1971; O'Brien, 1971). Ladbrooke et al. (1968) have indicated that 20% of the water associated with myelin is tightly bound to the polar groups of amphipathic lipids.

2.3. Structure and dynamic state of membrane lipids

The major membrane lipids have glycerol or 4-sphinganene as their parent alcohol; thus the main classes are the phosphoglycerides and the sphingolipids. Cholesterol accounts for a significant proportion of the total lipid of some membranes, notably the plasma membrane of animal cells, but this sterol is absent or is present in low concentrations in other membrane structures. The lipid and fatty acid composition of a variety of membranes has been reviewed by Rouser et al. (1968), Finean (1973), and phospholipid composition by White (1973). More detailed aspects of the composition and metabolism of lipids in nerve tissue and in membranes isolated from nerve tissue have been presented by Eichberg et al. (1969), Davison (1970), O'Brien (1970 a, b), Suzuki (1972), Veerkamp (1972) and Ansell (1973). The diversity of the lipid species found in different membranes and the range of their constituent fatty acids are revealed in these reviews.

2.3.1. Amphipathic lipids
The complex lipids occuring in membranes are amphipathic. The hydrocarbon chains of the constituent fatty acids comprise the nonpolar region of the lipid molecule, while the phosphate plus nitrogen base region of a phospholipid (or

the carbohydrate part of a glycolipid) constitutes the main polar groups. The amphipathic nature of complex lipids defines the way these molecules will behave in an aqueous environment. Nonpolar groups in lipid are sequestered from the aqueous phase, and polar groups interact with water so that a minimum energy level is reached. The structural behavior of amphipathic lipids in water at different levels of hydration and temperature has been studied in detail (reviews Luzzati, 1968; Abramson, 1971; Chapman, 1973a; and Shipley, 1973).

The main structural form adopted by most amphipathic lipids in the anhydrous state, and over a wide range of hydrated conditions, seems to be in the form of a bilayer (Levine and Wilkins, 1971; Levine, 1972; Shipley, 1973). This is a thermodynamically stable structure (Singer, 1971; Tien, 1971) in which hydrophobic association in the center of the bilayer and hydrophilic interactions at the surfaces in contact with the aqueous phase are maximized (Singer and Nicolson, 1972). Earlier ideas on the bilayer form of lipids (see Finean, 1972 for a history of the bilayer concept) that suggested a rather rigidly organized structure have been modified as more details on the physical and dynamic properties of lipid molecules in natural membranes and pure lipid systems have been derived. Thus a consideration of the mesomorphic phase properties of lipids has revealed that under appropriate conditions the fatty acid chains of complex lipids can have mobility in the middle of the bilayer due to rotation around the carbon-carbon single bonds of the hydrocarbon chains (Chapman, 1968). This procedures a low viscosity region in the center of the lipid bilayer.

Thermal and spectroscopic techniques have been used widely to study the phase transitions of complex lipids (e. g., Chapman et al., 1967; Chapman, 1968, 1973a, 1975; Phillips, 1972; Marsh, 1975). In the crystalline or gel state, amphipathic lipid molecules are rigidly arranged in bilayer form, with the fatty acyl chains extended in an all-*trans* structure. In the liquid-crystalline state, the bilayer characteristics are preserved through hydrophobic associations in the middle of the layer and through polar group interactions at the surface, but the hydrocarbon chains are melted and have considerable mobility. In this state individual lipid molecules can move in the bilayer plane because of the low viscosity fluid region in the center of the system (Marsh, 1975). The properties of a lipid molecule that influence the temperature at which the crystalline to liquid-crystalline phase transition occurs include the nature of the polar head group, the degree of hydration, the fatty acid chain length, the degree of unsaturation from *cis*-double bonds in fatty acids and the presence of cholesterol (e. g., Phillips, 1972; Traüble, 1972; Finean, 1973; Chapman, 1975; Marsh, 1975). Physical studies have revealed that in general the major proportion of lipid molecules in both hydrated bilayers composed of naturally occurring lipids and in natural membranes are in the liquid-crystalline state due to the large proportions of component *cis*-unsaturated fatty acids. The importance of the liquid-crystalline state for lipids in a bilayer in relation to the function and dynamic properties of natural membranes has been stressed by many writers (e. g., Singer, 1971, 1974a; Bretscher, 1973; Finean, 1973; Capaldi, 1974; Chapman, 1975).

Electron-spin resonance studies (review, Marsh, 1975) have defined the fluidity gradient that exists along fatty acyl chains in the liquid-crystalline state. The most fluid environment is at the terminal methyl end of the hydrocarbon chains at the center of the bilayer which can be in a very fluid condition having an apparent microviscosity of 2 poise (Cone, 1972; Cogan et al., 1973) that is comparable to that of a light mineral oil. The shape of the hydrophobic barrier in relation to the movement of water across amphipathic lipid bilayers has been discussed in detail by Griffith et al. (1974). Recently, the results of spectroscopic studies on the structure of lipids in artificial bilayers and natural membranes have been reviewed by Jost et al. (1971), Chapman (1972, 1973), Lee et al. (1974) and Marsh (1975). It is important to stress that, while the rate of diffusion for phospholipids in the bilayer plane for a pure phospholipid in the liquid-crystalline state is very fast (ca. 1 μm per sec.), movement of a complex lipid molecule from one side of the bilayer across to the other (flip-flop) is very much slower (Kornberg and McConnell, 1971; Marsh, 1975). Faster rates for the flip-flop movement of phospholipids in a bilayer have been observed in natural membranes where proteins may facilitate the process (Grant and McConnell, 1973; McNamee and McConnell, 1973) but the rate is still not as fast as lateral movement within the membrane plane.

Bilayers composed of mixed lipids generally show broader endothermic phase transitions compared with pure lipids. Broad phase transitions have been observed for mixtures of cerebroside and phosphatidycholine (Clowes et al., 1971) and phosphatidylcholine and phosphatidylethanolamine (quoted in Chapman, 1973b), and the results are interpreted as indicating that lipid molecules in a mixed bilayer system exist in distinctly defined regions of fluid and more crystallinelike structure (Chapman, 1973a, 1975; Marsh 1975). These separate phases coexist in equilibrium. Broad phase transitions are usually observed in thermal studies on natural membranes, and the term lateral phase separation has been applied to describe this condition where discrete regions of more crystalline and fluid lipid arise in the bilayer due to the presence of differing endothermic phase transitions of the various complex lipid types.

Shimshick and McConnell (1973a,b) have studied lateral phase separation in mixed phospholipid bilayers by electron-spin resonance techniques. They have speculated (Marsh, 1975) that considerable functional advantage may result from the structural state that exists in membranes exhibiting lateral phase separation because the high lateral compressibility available in the system may allow for insertion of proteins and permit changes in protein conformation to occur. Calcium, added to mixed phosphatidylserine-phosphatidylcholine and mixed phosphatidic acid-phosphatidylcholine bilayer systems, can induce lateral phase separation (Ohnishi and Ito, 1974; Papahadjopoulos et al., 1974; Ito and Ohnishi, 1974; Galla and Sackman, 1975). In these systems calcium interacts with negatively charged lipid species, bridging the molecules and causing solidification of phosphatidylserine or phosphatidic acid in discrete areas surrounded by more fluid phases of phosphatidylcholine. Magnesium is not as effective as calcium in inducing the same phase separation in these systems, and calcium has no effect on phosphatidylcholine bilayers alone. The pH of the aqueous system can also have a marked effect on the structure of negatively charged lipids (Traüble and Eibl, 1974; Verkleij et al., 1974). The conclusions reached from these studies about how external changes in pH and cation content can alter phospholipid form in bilayers have important implications for lipid structure and organization in natural membranes.

2.3.2. Cholesterol

The role of cholesterol in membranes is now understood in some detail (reviews, Phillips, 1972; Chapman, 1973a,b, 1975; Jain, 1975; Marsh, 1975). Cholesterol is believed to control the fluidity of fatty acyl chains in a bilayer of complex lipids so that a relatively constant and stable fluid environment is maintained in the center of the membrane. The size of the endothermic phase transition from a hydrated phospholipid is gradually decreased as increasing proportions of cholesterol are added to the system (Chapman, 1968). At a 1:1 molar ratio of cholesterol to phospholipid no endothermic transition peak can be detected by calorimetric techniques. This is thought to indicate that the fatty acyl chains of the phospholipid molecules are all in an intermediate fluid state. For phospholipids below the phase transition, cholesterol packs in between adjacent molecules up to a 1:1 molar ratio and in so doing allows acyl chains more freedom of movement. For phospholipids in the liquid-crystalline state, cholesterol can reduce excessive mobility in acyl chains because movement in the region of the hydrocarbon chains that are in contact with the rigid sterol molecule will be reduced. Spectroscopic studies have revealed that the effect of cholesterol is due to the way that the sterol molecule associates with a phospholipid molecule. The 3-β-hydroxyl group on the sterol nucleus is thought to associate with the phosphate of the phospholipid by hydrogen bonding (Darke et al., 1971, 1972). Molecular models indicate that the rest of the rigid cholesterol ring and side chain extend into the nonpolar region of the bilayer reaching to about the tenth carbon atom of fatty acyl chains. Interaction of this part of the fatty acid chains with cholesterol molecules restricts motion, while the mobility of methylene groups beyond the C10 position is increased in a 1:1 molar mixture of phospholipids with cholesterol because of the extra space available for movement (Darke et al., 1972). Similar conclusions have been reached in electron-spin resonance studies on cholesterol-phospholipid interactions in a bilayer (Marsh and Smith, 1972, 1973; Schreier-Mucillo et al., 1973). Cholesterol has the same effect on cerebroside at a 1:1 molar ratio (Oldfield and Chapman, 1972a). Thus cholesterol will tend to make lipids below their phase transition more fluid and will reduce the fluid nature of lipids above their transition temperature. At a constant temperature, such as body temperature, cholesterol will act to preserve a constant fluid environment in the middle of the hydrocarbon region of a membrane. In this way it can protect plasma membranes against possibly harmful changes in fluidity caused by alterations in fatty acid content created by variation in dietary lipids, for example. Cholesterol acts to stabilize membranes as dispersion force interactions in the hydrocarbon region increase. Papahadjopoulos et al. (1972) have discussed this role for cholesterol and state that "it is probably not coincidental that the most stable and metabolically inactive membrane (myelin) has the highest (1:1) molar ratio of cholesterol to phospholipid." At the same time, addition of cholesterol decreases the permeability of pure lipid bilayers (Papahadjopoulos et al., 1972). The maximum interaction between cholesterol and complex lipids seems to occur at a 1:1 molar ratio. Phillips and Finer (1974) have proposed that in bilayers containing lower ratios of cholesterol to phospholipid, lateral phase separation occurs in which discrete regions of 1:1 sterol to lipid complex separate with sterol-free phospholipid at the boundaries. The nonrandom distribution of cholesterol in phosphatidylcholine bilayers has also been considered by de Kruyff et al., (1974).

2.3.3. Water

The role of water in membrane structure is of great significance since hydrophobic bonding is a major feature determining orientation in the lipid bilayer. Details of the way the polar groups of lipids hydrate in bilayers have derived from the application of thermal and spectroscopic techniques (reviews Phillips, 1972; Chapman, 1968, 1973a,b,c, 1975). Under appropriate conditions, water molecules can pack around the head groups of phospholipids, causing an expansion of the bilayer structure. Hydration causes a slight lowering in the temperature at which the gel to liquid-crystalline transition occurs, since the rigid crystalline lattice structure expands on hydration and the motion of fatty acyl chains is thus increased (Williams and Chapman, 1970). Water molecules associated with complex lipids do not freeze at $0°C$ (Chapman et al., 1967), indicating that they are tightly bound to polar groups. For phosphatidylcholine the bound water forms a hydrate structure around the polar head group, and thermal data reveal that 10 moles of water per mole of phospholipid are bound at $0°C$. Spectroscopic data have shown further that 4 to 6 moles of water bind strongly to the polar part of the lipid and that another 5 to 6 moles then bind weakly (Salsbury et al., 1972; Gottlieb et al., 1973). Additional water, up to 40 wt %, will form a free layer in which water molecules have very free movement between the adjacent lipid bilayers.

The binding of water in this way is apparently independent of fatty acid composition, but Finer and Darke (1974) have shown that it varies considerably with different polar head groups. Phosphatidylcholine, phosphatidylethanolamine, and sodium phosphatidylserine show different hydration behaviors and water-binding energies. The same authors have proposed that while bound water molecules exchange rapidly with each other, they will exchange only slowly with bulk water. The binding of water in phospholipid:cholesterol bilayers has been studied by Inglefield et al. (1976). At a 1:1 molar ratio of lipid to sterol at room temperature some 12 water molecules correspond to complete hydration of the phosphatidylcholine-cholesterol unit. In the region of this hydration shell there is a fourfold decrease in water diffusion, probably due to the structured nature of the molecules in the hydration shell.

The formation of bound water molecules, or "icebergs", at a bilayer interface may be of considerable importance in interactions at the lipid surface. Tien and Ting (1968), investigating the water permeability of black lipid films, have suggested that the layer of ordered water covering the lipid surface is the rate-determining barrier to the penetration of the bilayer by water molecules. Solutions containing chaotropic ions that destroy water structure increase permeability through artificial lipid membranes (Tien, 1971). A considerable proportion of the water associated with a myelin-containing membrane preparation from nerve tissue shows "icelike" characteristics and is involved with membrane components by interactions stronger than those present in normal liquid water (Gent et al., 1970).

The interaction of cations and anions with the polar region of phospholipids in bilayers has been shown to affect a number of the physical lipid properties such as phase transition, surface potential and pressure, stability, and so on (Chapman, 1968, 1973a,b, 1975; Tien, 1971; Papahadjopoulos, 1973; Papahadjopoulos and Kimelberg, 1973; Traüble and Eibl, 1974).

2.3.4. Head group orientation

Considerable gaps exist in our knowledge of the way the polar head groups of complex lipids are oriented at bilayer surfaces. Indeed, the full details may be difficult to elucidate because polar head group orientation in natural membranes will be influenced by pH, by interaction with other lipids, proteins, ions and by various membrane-associated events. Currently, spectroscopic techniques are being used to examine this problem in artificial lipid and natural membrane systems. Recent results on head group orientation have been discussed by Horwitz (1972), Phillips (1972), Chapman (1973c) and Lee et al. (1974). Such studies reveal that a gradient of internal motion in a lipid molecule exists (Levine et al., 1972; reviewed by Lee et al., 1974). For bilayers of phosphatidylcholine, there is a large increase in motion going from the glycerol backbone along the fatty acids to the terminal methyl group and also the other way to the polar choline group. Hydration and the liquid-crystalline state greatly influence the mobility of the head group in lipid bilayers (Chapman, 1973c; Lee et al., 1974). In the rigid crystalline form the phosphorylcholine moiety of phosphatidylcholine may be curled up (Levine et al., 1968), but subsequent hydration of the bilayers causes an unfolding so that the head group adopts a more extended conformation. In this state it can undergo rapid motion (Phillips et al., 1972). For bilayers of phosphatidylethanolamine the phosphorylethanolamine head groups are either arranged approximately in the plane of the bilayer and folding back to give charge neutralization with an adjacent phosphatidylethanolamine molecule or the head groups interdigitate with opposing groups from the adjacent bilayer surface. Studies by Finer and Darke (1974) provide more data on the mobility of the polar head groups in these molecules.

While some information exists on the structural form of cerebrosides and cerebroside sulfates in the crystalline and water-dispersed form (review, Shipley, 1973) head group mobilities, orientation and levels of hydration around the galactose moiety have not been fully established.

2.4. Structure and state of membrane proteins

Current ideas on the structure and organization of protein molecules in membranes have been defined in several recent articles and reviews (Singer, 1971, 1974a,b,c; Guidotti, 1972; Singer and Nicolson, 1972; Bretscher, 1973, 1974; Coleman, 1973; Hughes, 1973, 1975; Oseroff et al., 1973; Vanderkooi, 1974). Experimentally, membrane proteins have been divided into two classes (Green, 1971; Capaldi and Vanderkooi, 1972; Singer and Nicolson, 1972). Extrinsic (peripheral) proteins are those that can be extracted from membranes by relatively mild conditions, such as treatment with salts, removal of divalent cations, and changes in pH. By contrast, intrinsic (integral) membrane proteins require organic solvents or detergents to remove them from their membrane location. Cytochrome c and spectrin are examples of extrinsic proteins and are thought to be associated with polar groups at the surface of the lipid phase in a membrane by charge-charge interactions. Cytochrome b_5 and glycophorin of the red cell membrane are more intimately associated with the membrane because part of the protein molecule penetrates into the nonpolar region of the lipid phase and is stabilized by hydrophobic interactions while ionic interactions with polar groups of lipids at the surface of the bilayer also occur. The depth to which such intrinsic proteins penetrate into the lipid phase of a membrane can vary. Evidence indicates that certain intrinsic proteins may span the lipid phase with hydrophilic polypeptide regions of the protein being exposed at both the interior and exterior membrane surfaces (e. g., Bretscher, 1973, 1974; Boxer et al., 1974; Steck, 1974). As might be expected where much of the polypeptide chain will be located in the hydrophobic region of the bilayer, amino acid analysis has revealed that many intrinsic proteins have a lower polarity than extrinsic ones (Capaldi and Vanderkooi, 1972). Some intrinsic proteins have their hydrophobic amino acid residues located close together in a distinct region of the polypeptide chain. This would be the part of the molecule that penetrates the hydrocarbon region of the membrane. Glycophorin and cytochrome b_5 are two intrinsic proteins in which the polypeptide chains have distinct regions of high hydrophobic and hydrophilic natures (reviews, Spatz and Strittmatter, 1971; Hughes, 1975).

Spectroscopic studies (Singer, 1971; Urry, 1972) have revealed that membrane proteins have little β-structure. Their overall conformation is similar to soluble proteins, which are globular, with much of their polypeptide chain as α-helix and random coil. Thus membrane proteins do not seem to cover the surfaces of membranes in monolayer form. The overall globular nature of proteins embedded in the lipid matrix of membranes has been revealed by the freeze-etch electron microscopy technique (Stolinski and Breathnach, 1975; McNutt, 1977).

Membrane proteins show rotational and lateral motion in the plane of the lipid layer (review, Marsh, 1975; and see Frye and Edidin, 1970; Cone, 1972; Poo and Cone, 1974). The movement is made possible by the fluid nature of the hydrophobic region in the lipid phase of the membrane. The diffusion of proteins in the plane of the membrane is 100 times slower than for lipid molecules, and this can be related to their larger molecular size, and the influence of boundary lipid and interactions at the aqueous surface. It is now believed that individual intrinsic proteins in membranes are surrounded by a closely associated layer of boundary lipid molecules. These are tightly associated with the protein by polar interaction at the surface and by hydrophobic interaction in the center of the bilayer so that fatty acyl chains in the boundary lipid molecules are immobilized (Jost et al., 1973a). The boundary lipid is essential for functional activity in the case of many membrane-bound enzymes (reviews, Coleman, 1973; Marsh, 1975; Kimelberg, 1977). As the lipid is tightly associated with the protein, it will diffuse with it in the plane of the membrane. It is also likely that the immobilized fatty acid chains of boundary lipid molecules will influence the mobility of acyl chains on free lipids in the immediate neighborhood of the protein in the membrane.

The effects of adding protein molecules to lipid bilayers and monolayers have provided information on the nature of lipid-protein interactions in membranes (e. g., Chapman, 1973, 1975; Finean, 1973; Papahadjopoulos, 1973; Papahadjopoulos and Kimelberg, 1973). Results of such studies have stressed the involvement of electrostatic and hydrophobic interactions in the association of a protein with a lipid bilayer. Most extrinsic proteins interact principally by charge-charge associations. For many intrinsic proteins both electrostatic and hydrophobic interactions are important, the former probably being involved in directing the initial contact between the protein and the lipid surface. Subsequently, the protein may rearrange so that some hydrophobic interaction occurs between part of the polypeptide chain and the hydrocarbon region of the lipid phase. This topic has been reviewed by Papahadjopoulos and Kimelberg (1973). More recently it has been suggested that some membrane proteins that are essentially extrinsic in character, such as cytochrome c and the myelin basic protein, also show a partial penetration and/or deformation of the lipid bilayer (Papahadjopoulos et al., 1975). The interaction of proteins with a lipid bilayer has been found to influence characteristics such as permeability, phase transition, and electrical properties. The influence of cholesterol on the interaction of a variety of proteins with phospholipid membranes has been studied by Papahadjopoulos et al. (1973). Generally the presence of the sterol in a membrane inhibits the ability of proteins to increase the permeability of phospholipid vesicles and to expand or penetrate lipid monolayers.

2.5. Membrane carbohydrate

In addition to lipid and protein some membrane systems, notably the plasma membranes of mammalian cells, contain carbohydrate-rich molecules that have been classified as glycoproteins and complex glycolipids. In most membranes the glycolipids are complex glycosphingolipids and gangliosides. Cerebrosides, which contain a single monosaccharide moiety, are a prominent component of myelin. The chemistry and structure of the membrane glycoproteins and glycolipids have been reviewed extensively (Hughes 1973, 1975; Glick 1974;

Gray 1974; Laine et al., 1974). These molecules contain short tri-, tetra-, and penta-saccharide units bound covalently to protein or to 4-sphingosene. These complex carbohydrate molecules in membranes have been implicated as antigens, as receptor sites for viruses, hormones and lectins and as important in cell-cell recognition and cell adhesion (reviews, Hughes, 1975; Critchley and Vicker, 1977). Such molecules may also be involved in the transfer of water and ions across lipid membranes (Hughes, 1975; Rothstein et al., 1976). The biochemistry of the complex glycosphingolipids and glycoproteins of nervous tissue has been reviewed by Brunngraber (1969), Svennerholm (1970a,b) and Bowen et al. (1974) and in relation to myelination by Brady and Quarles (1973).

2.6. Membrane structure

Current observations on the physical structure and state of lipids and proteins in membranes have been expressed in the fluid mosaic model for membrane structure, as shown in Fig. 1. This was first elaborated by Singer (1971) and Singer and Nicolson (1972). Recently, the molecular biology of the model has been discussed by Singer (1974a,c) and with an emphasis on immunology by Singer (1974b). The fluid mosaic model retains the concept of a thermodynamically-stable lipid bilayer as the basic structural matrix of the natural membrane, and incorporates findings on the dynamic and fluid nature of lipids in artificial and natural membranes. In the model, intrinsic proteins are inserted into the lipid phase and extrinsic proteins are located on the membrane surface to provide the functional activities of the system (see also Nicolson et al., 1977).

X-ray diffraction (review, Shipley, 1973), spectroscopic (reviews, Lee et al., 1974; Marsh, 1975), and calorimetric studies (review, Chapman, 1975) on a variety of natural membranes indicate that much of the lipid in the system is in

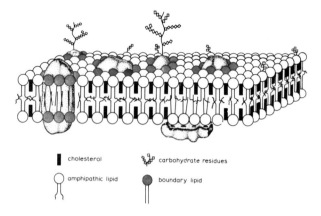

cholesterol   carbohydrate residues

amphipathic lipid   boundary lipid

Fig. 1. Diagramatic representation of a section through a biological membrane incorporating many of the current ideas on the structural organization of membrane lipids and proteins. The lipids are in a bilayer arrangement and form a matrix for the insertion of intrinsic proteins and the association of extrinsic proteins (cross-hatched). The fatty acyl chains of lipids in the bilayer are in an intermediate fluid state induced by the presence of cholesterol (solid rectangles). In this model extrinsic proteins at the membrane surface induce rigidification in fatty acyl chain mobility as does intrinsic protein. Boundary lipids associated with intrinsic proteins are shown with solid headgroups.

the bilayer form. A proportion of the lipid in the system may, however, adopt structurally different forms, the amount depending on both the type of membrane structure and on local interactions at the membrane surface. In a protein-rich membrane, much more lipid will be intimately associated with protein as boundary lipid and the lipid phase may then serve more as packing between closely spaced protein molecules. Current results also suggest that the lipid phase may not be homogeneous in its liquid-crystalline state due to lateral phase separation and protein effects. Evidence from the erythrocyte membrane also indicates that lipids are distributed asymmetrically between the two faces of the bilayer (Zwaal et al., 1973). Choline-containing phospholipids are on the outer face, and the more negatively charged lipids on the interior face. Convincing extrapolation of this finding to other plasma membranes awaits further experimentation. The asymmetric distribution of lipids in a membrane would, however, influence the properties of the two halves of the bilayer in overall charge, fluidity, diffusion of lipids and proteins in the plane of the bilayer and protein- and ion-binding capacity.

The asymmetric distribution of proteins in membranes has been established using chemical and enzymatic methods in studies with vesicles and by electron-histochemistry (Singer, 1974a,b,c). Glycoproteins and complex glycosphingolipids are distributed asymmetrically on the outer face of the plasma membrane (Parsons and Subjeck, 1972; Hughes, 1975). Proteins may be restricted to defined regions in a membrane by increasing the local viscosity of the immediate lipid phase in their vicinity or by being anchored at the cytoplasmic face of the membrane with other proteins or cytoplasmic structures (Nicolson et al., 1977).

From presently available data, a detailed picture of lipid and protein organization in certain membranes, such as the erythrocyte plasma membrane (Juliano, 1973; Steck, 1974; Rothstein et al., 1976), can be built up. Clearly, such models leave many unresolved questions about membrane structure and functioning, but, they also begin to reveal the precise nature of the structural interrelationships that exist between lipid and protein molecules to provide the many varied features of a natural membrane system.

The structural information available on the compact myelin sheath will now be reviewed. In the section above general findings on lipid and protein organization in artificial lipid and natural membranes have been briefly outlined, and such observations provide the guidelines and constraints that must be considered in thinking about molecular organization in the compact myelin sheath. In a later section the structural details available on compact myelin are assembled into a model of lipid and protein organization in this membrane system.

3. The myelin sheath

At the start of the Neurosciences Research Program (NRP) work session on myelin in 1971 there was apparently some initial discussion about whether myelin can be regarded as a typical membrane system (Mokrasch et al., 1971). In disagreement with the view of Green and Perdue (1966), who have stated that myelin "appears to have little in common with cellular membranes, either chemically or functionally," the present authors believe that myelin is typical of other membrane structures in at least two ways. First, myelin is composed of amphipathic lipids and proteins, as are all membranes. Second, like all individual membranes, myelin is a unique system. The chemical composition of a membrane, especially with regard to its protein components, is uniquely designed to provide for the requirements of the structure in its location in a cell. Myelin may be a "minimal membrane" (Norton, 1972) or a "simple membrane" (Guidotti, 1972) in that it has a low protein content relative to lipid. The protein and lipid composition of myelin is, however, unique among membranes and is designed specifically to suit its function in situ, as is the case for all membrane systems. Myelin is certainly typical of other membranes in these respects and, as we shall see, in others too.

3.1. Myelin in situ

The myelin sheath occurs as a sleeve around axons in peripheral and central nerve tissue. Axons over about 0.5μ in diameter in the peripheral nervous system (PNS) (Elfvin, 1968) are myelinated, while those under this figure are located within invaginations of the plasma membrane in the Schwann cell cytoplasm. Generally, in central nerve tissue all axons eventually acquire some degree of myelin covering (Matthews and Duncan, 1971), but the largest are the first to myelinate, becoming coated between 1.0 to $1.5\mu m$ in diameter. The myelin sheath is not continuous along the length of an axon. At regular intervals (the Nodes of Ranvier) the axolemma is exposed. Nodal length varies up to about 1 μm (Peters and Vaughn, 1970; Hildebrand, 1971). The structure of nerve fibers in this nodal region has been described in detail by Elfvin (1968), Peters (1968), Peters and Vaughn (1970), Peters et al. (1970), Hildebrand (1971), Livingstone et al. (1973), and Blank et al. (1974) among others. In both peripheral and central nerve tissue, compact internodal myelin is separated from the axolemma by a distance of not less than about 120Å. At the paranodal region, however, this distance is only 25 to 30Å in forming the specialized glial-axonal junction (Peters et al., 1970). It has been estimated that roughly 10% of the length of an axon is involved in intimate apposition with myelin sheath at glial-axonal junctions (Livingstone et al., 1973).

In peripheral nerves, each separate internodal myelin sheath along an axon is formed by an individual Schwann cell. In contrast, in central nerve tissue, the myelin-forming cell is generally agreed to be the oligodendrocyte (Bunge, 1968; Peters and Vaughn, 1970). This cell is responsible for the formation and maintenance of many internodal sheaths found mainly on closely adjacent axons. The body of the oligodendroglial cell is linked with each individual internodal myelin sheath by a cytoplasm-filled process that extends through the nerve tissue to the axon for up to $12\mu m$ in length (Bunge, 1968; Peters, 1968). Occasionally, a myelin sheath is closely adjacent to the body of the parent glial cell.

Peters and Vaughn (1970) have calculated that for rat optic nerve tissue there are 30 to 50 internodal myelin sheaths for every oligodendrocyte. These findings have been confirmed by Matthews and Duncan (1971) for fibers of different diameters in spinal cord. Here, individual oligodendrocytes produce from 18 to 60 internodal sheaths. The lower ratio of 18 nodes per oligodendrocyte was observed for axons with the largest diameters while the ratio of 60:1 was found in the tract with the smallest diameter fibers.

The length of an individual myelin sheath, the internodal distance, is correlated in a linear relationship with the diameter of the axon it surrounds and is often 1.5 to 2 orders of magnitude larger (Peters and Vaughn, 1970). It is not generally emphasized how long the internodal length of the myelin sheath is relative to the size of the myelin-forming cell (Fig. 2). Internodal lengths up to 1000 μm have been measured in peripheral nerve tissue for axons 15 μm in diameter. In the optic nerve, Peters and Proskauer (1969) have concluded that the average internodal length is about 126 μm. In spinal cord tracts, internodal lengths average 100 μm for fibers about 1 μm in diameter, while the average length is about 250 μm for mixed fibers from 0.5 to 5 μm in diameter (Matthews and Duncan, 1971).

It is relevant to details of myelination and myelin metabolism to note that internodal lengths in peripheral and central nerves of young animals are shorter than those on the same nerves in adults. The complement of Schwann cells or sheaths from oligodendrocytes gained at the start of myelination does not seem to increase as growth proceeds. Thus internodal length increases with axon growth during development (Peters and Vaughn, 1970; Peters et al., 1970).

The wide variation in internodal length observed in myelin sheaths in central and peripheral nerves arises because the results are taken from axons of various diameters. We have often wondered whether internodal myelin sheaths along the same axon are approximately constant in length. Answers to this question are not readily forthcoming for central nerve tissue, as the complexity of the tissue makes measurements along the length of single axons very difficult. For peripheral nerves a little more information is available and Lascelles and Thomas (1966) have commented that internodal lengths are uniformly short during myelination, nodes being 150 to 300 μm apart. Results of studies on the sural nerve in humans of different age groups (Lascelles and Thomas, 1966) indicate considerable variation in internodal distance in certain fibers from older subjects. Abnormally short internodes were found on fibers that also possessed internodes of normal length. On other fibers in the same subjects, internodes were uniformly short or long depending on axon diameter (see above). Fraher (personal communication) has indicated that internodal length on individual axons in the PNS varies by about 25%. This information is relevant to the question of the process that determines where nodes will occur along an axon during myelination.

Generally, the thickness of the compact myelin sheath is believed to be correlated linearly with axon caliber in the mature nerve fiber. This relationship holds for both peripheral and central nerve tissue but the correlation is lower in immature fibers (e. g., Samorajski and Friede, 1968; Webster, 1971; Williams and Wendell-Smith, 1971; Fraher, 1972; Friede, 1972; Boyd and Kalu, 1973 and

Sima, 1974). The same relationship has been extended by Friede (1972) in an attempt to correlate the rate of axonal growth with the rate of myelination by sheath cells. In immature peripheral nerve fibers the thickness of the developing myelin sheath and the axon circumference tend to vary along a given internode (Webster, 1971; Fraher, 1973). An earlier observation by Hildebrand (1972) that the periodicity of myelin lamellae in thick compact sheaths is narrower than in thin sheaths has been reevaluated. Using x-ray diffraction techniques, Hildebrand and Muller (1974) have shown that the periodicity is the same in thick and in thin sheaths and that the differences noted earlier arose artifactually during the preparation of samples for electron microscopy.

Some of the dimensions relating to the oligodendrocytes and compact myelin in central nerve tissue are shown in Fig. 2. Differences in size between the myelin-forming cell and the sheaths it produces and maintains are emphasized. The relatively small size of the oligodendrocyte cell body and process length compared with the large internodal length are contrasted in scale. Axon/sheath diameters and internodal lengths are roughly similar for myelin in peripheral nerve tissue, but it should be re-emphasized that each myelin sheath derives from a single Schwann cell wrapped around the axon. No process extending through the tissue is involved (Elfvin, 1968; Peters and Vaughn, 1970; Peters et al., 1970). On average, myelin sheaths in peripheral nerves, where axon diameters are larger, tend to have many lamellae; up to 95 layers have been recorded in one sheath (Friede and Samorajski, 1967). In central nerve tissue the number of lamellae is generally between 5 and 20 (Peters and Vaughn, 1970) as axon diameter is generally lower.

3.2. Structural studies on compact myelin in situ

In the previous section we attempted to put the myelin sheath into a kind of gross perspective in its situation in nerve tissue. In the following two sections, detailed structural results from x-ray diffraction and electron microscope studies on compact myelin in situ will be examined. These two techniques can be applied to derive structural details of the untreated normal myelin sheath in its location in normal nerve tissue and the results are of considerable importance. Most other data on myelin structure comes from work using treated or isolated preparations of the membrane system.

3.2.1. X-ray diffraction studies

The application of x-ray diffraction techniques to compact myelin has yielded structural data without the need to fix, stain, dehydrate, extract, or in any way treat the fresh, physiologically normal nerve tissue. Thus it is not surprising that this technique has been widely used to examine myelin for the dimensions of the system and to investigate details of molecular organization in the lamellae (e. g., Finean and Burge, 1963; Worthington and Blaurock, 1969a,b,; Blaurock, 1971, 1976; Caspar and Kirschner, 1971; Worthington and King, 1971; McIntosh and

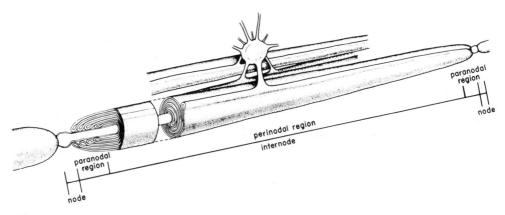

Fig. 2. Perspective scale drawing of the interrelationships between the oligodendroglial cell, its processes, and the myelin sheath in central nerve tissue. On average there are some 20–40 processes per oligodendrocyte (see text) that are up to 12 μ long. The length of the mature myelin sheath is usually at least an order of magnitude greater than the dimension of the oligodendroglial cell body.

Worthington, 1974a,b; Lalitha and Worthington, 1975). The results provide an electron density picture of structural detail averaged with respect to time and molecular organization, for it must be realized that even though sophisticated theoretical treatments may be applied to current diffraction results, small local variation in structure within the membrane and the movement of components during exposure time will not be detected in the averaged pattern recorded (Luzzati, 1974). Nevertheless, the dimensions of the repeating unit within compact myelin have been obtained (e. g., Blaurock and Worthington, 1969) and electron density profiles across myelin lamellae have given approximate dimensions for the lipid and nonlipid layers within the normal membrane structure in situ, as shown for central and peripheral nerve myelin in Fig. 3. The application of x-ray diffraction techniques to myelin has been discussed in detail by Finean (1966, 1969, 1973), Mokrasch et al., (1971) Worthington (1971, 1973), and Shipley (1973).

The interpretation of the x-ray data on myelin with regard to the construction of electron density profiles and strip models across the myelin repeating unit has been the subject of much debate (e.g., Akers and Parsons, 1972; Harker, 1972; Blaurock, 1973a, 1976; McIntosh and Worthington, 1973; Worthington, 1973). This has revolved largely around diffraction theory and a solution to the phase problem. Results up to 1971 have been discussed by Mokrasch et al. (1971). Here, it is acknowledged that the validity of the various electron density profiles and models presented for myelin up to that period (Finean and Burge, 1963; Finean, 1966, 1969; Worthington and Blaurock, 1969; Akers and Parsons, 1970; Blaurock, 1971; Caspar and Kirschner, 1971; Worthington and King, 1971) depend entirely on whether the correct phasing has been achieved for the diffraction orders used to obtain the profile. Confidence is expressed in profiles and models produced using the first five diffraction orders (low resolution 17 Å) because the majority of independent workers have reported the same sequence of phase signs, using different experimental approaches. More recently, this same phase sign sequence has been confirmed by McIntosh and Worthington (1974a) who used direct methods of structure analysis (Worthington et al., 1973) to obtain proof for the correctness of the phase choice. Thus we can be reasonably confident about the broad picture of molecular organization for myelin revealed by low resolution studies that have employed this agreed phase sequence.

Derivation of electron density profiles and strip models across the repeating unit of myelin are based on observations that in hypotonic media and in sucrose and glycerol solutions (e. g., Finean and Burge, 1963; Worthington and Blaurock, 1969b; Blaurock, 1971; McIntosh and Worthington, 1974a) the lipoprotein layers of compact myelin become separated by layers of water, while the molecular organization within the structural unit is not changed (for sucrose, see Worthington and Blaurock, 1969b; for glycerol over 20%, see McIntosh and Worthington, 1974a). Electron microscope studies have revealed that this separation of the layers in compact myelin occurs exclusively at the intraperiod dense line that corresponds to the external apposition region of the system (Finean and Burge, 1963; McIntosh and Robertson, 1976). Separation occurs at each external apposition site in peripheral nerve myelin, but only at alternate external appositions in central nerve myelin (Finean and Burge, 1963; McIntosh and Robertson, 1976).

Despite variation in small detail all electron density profiles and models for the repeating unit of myelin using the generally agreed-upon phase sequence from low resolution studies have been interpreted in lipid bilayer terms (Finean and Burge, 1963; Worthington and Blaurock, 1969a; Blaurock, 1971, 1976; Worthington and King, 1971; McIntosh and Worthington, 1974a; Lalitha and Worthington, 1975). As shown in Fig. 3, the repeating unit in myelin has two regions of low electron density, interpreted as locating the hydrocarbon chains of lipids. These are separated by peaks of high electron density that locate interaction between the polar groups of lipids and proteins and layers of intermembrane fluid. This interpretation for myelin is in general agreement with electron density profiles obtained for oriented lipid bilayers and the bilayer structure of lipids in natural membranes (Levine and Wilkins, 1971; Wilkins et al., 1971; Blaurock, 1973b, 1976; see also reviews, Levine, 1972; Shipley, 1973). The combined application of x-ray diffraction and electron microscope techniques has defined the origin of the electron-dense peaks on the basis of swelling studies; the high electron density peak is derived from the external apposition of membrane layers in the myelin sheath, while the narrower is from the cytoplasmic apposition region.

On an absolute electron density scale, minimum values for the center of the low electron density troughs are from 0.27 to 0.29 electrons/$Å^3$ (Worthington and Blaurock, 1969; Caspar and Kirschner, 1971; Blaurock, 1973b, 1976; McIntosh and Worthington, 1974a). These figures approximate well to the value of 0.27 electrons/$Å^3$ for hydrocarbon type molecules and to the data of Rand and Luz-

265

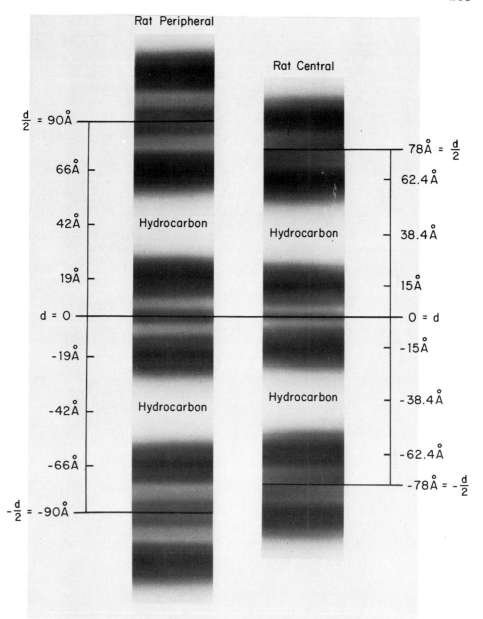

Fig. 3. Representation of the electron density profile of peripheral and central nerve myelin as revealed by x-ray diffraction. Light areas are regions of low electron density within the repeating unit (hydrocarbon chains); darkest areas are highest in electron density (polar groups of lipid and protein). The dimensions of the myelin repeating unit in peripheral and central systems are taken from the work of Caspar and Kirschner (1971).

zati (1968) for the low electron density region in a mixed cholesterol-phospholipid system from erythrocytes. Peaks in the profiles have absolute values from 0.37 to 0.40 electrons/$Å^3$. Such figures are consistent with the expected electron densities for hydrated polar groups in phospholipids, glycolipids, and proteins. The width of the lipid layer within the repeating unit of myelin as revealed by the peak to peak distance in electron density profiles, is from 45 to 50 Å for myelin in both peripheral and central nerve systems (Finean and Burge, 1963; Blaurock, 1971, 1976; Caspar and Kirschner, 1971; McIntosh and Worthington, 1974a; Lalitha and Worthington, 1975). Protein in the high electron density region may make a small contribution to the value, but the dimensions are in good agreement with similar data for pure, hydrated phospholipid bilayers and for mixed cholesterol-phospholipid bilayers in the liquid-crystalline state (Levine, 1972; Blaurock, 1973b, 1976; Shipley, 1973).

It is not possible to accommodate the long fatty acid chains of certain myelin lipids, notably the sphingolipids, in a rigid extended all-*trans* form within the 45 to 50 Å dimensions of the lipid region of the repeating unit for myelin. Fatty acyl chains may interdigitate on opposing sides of the membrane or lipids may be generally or partly in the liquid-crystalline state. Clowes et al. (1971) have noted that no interdigitation seems to occur in a mixture of cerebrosides with phosphatidylcholine. X-ray data has also revealed that the crystalline to liquid-crystalline phase transition in lipids is accompanied by a marked reduction in the width of the bilayer as acyl chains melt (Shipley, 1973). Thus, to be accommodated within the dimensions of the bilayer, it seems possible that at least the long chain fatty acids of myelin could be in a liquid-crystalline state. The width of the lipid layer in the repeating unit of myelin in central and peripheral nerve is similar in dimension. Fatty acid chain length, unsaturation, and a slight variation in lipid composition do not contribute significantly to the large difference in the width of the repeating unit observed for myelin from these two sources (Blaurock and Worthington, 1969).

Low resolution (17 Å) profiles of myelin show certain features of asymmetry within the repeating unit, but the validity of these details is debatable. Results from Blaurock (1971) and McIntosh and Worthington (1974a,b) indicate that the peak at the cytoplasmic apposition region has a higher absolute electron density value than the peak at the external apposition (but see Finean and Burge, 1963). This difference in peak heights has also been noted by Blaurock (1976). Such asymmetry may indicate a difference in chemical composition between the two regions. This is also suggested, for example, in the different staining behavior of the two regions following preparation for microscopy (section 3.2.2.). McIntosh and Worthington (1974b), confirming earlier observations, have noted that following rehydration of air-dried peripheral nerve, the height of the two peaks in the electron density profile is equalized. Freezing and thawing also significantly enhance the staining capacity of the external apposition region on electron microscopy (Finean and Burge, 1963). Another feature of asymmetry, especially in peripheral nerve myelin, is that the low density trough in electron density profiles is offset to the cytoplasmic apposition region within the repeating unit. This may indicate that fatty acids on one side of the membrane (presumably the external apposition region) are in a more extended form within the lipid bilayer. However, these asymmetry observations must await confirmation of the validity of the higher resolution profiles (Caspar and Kirschner, 1971; Worthington and McIntosh, 1974; and examination of these two sets of results by Blaurock, 1976) and results of other biochemical and physicochemical methods.

Low resolution studies provide a generally accepted indication that the lipids in the myelin sheath are for the most part arranged in bilayer form in situ. The possibility that a proportion of the lipid is organized in some other structural form within the bilayer structure cannot, however, be discounted. The low resolution data also indicate that protein within the membrane is usually located close to the hydrophilic surfaces of the lipid phase.

Results at 7 Å and 10 Å (moderate resolution) produce a more detailed picture of structure across the myelin repeating unit (Caspar and Kirschner, 1971; Worthington and King, 1971; Worthington, 1973; Worthington and McIntosh, 1974; and also see discussion by Blaurock, 1976) but uncertainty regarding the experimental approach and/or the allocation of phase signs means that the detail of the profiles cannot yet be completely accepted. Indeed, moderate resolution profiles and models reported to date, while supporting the bilayer form in myelin, all show significant differences in fine detail. The well-defined asymmetry of the hydrocarbon region in the profiles reported by Caspar and Kirschner (1971), which has been interpreted in terms of an uneven distribution of cholesterol (or a partial penetration by protein) within the halves of the bilayer, is not apparent in the more favored of the two profiles suggested by Worthington and McIntosh (1974) and recently by Blaurock (1976). A second profile by Worthington and McIntosh, derived by use of a slightly different phase sign allocation, shows asymmetry on the cytoplasmic side of the hydrocarbon region. This result should be contrasted with that of Caspar and Kirschner (1971), in which the shouldering in the hydrocarbon region is on the external apposition side. It is perhaps relevant to note that electron density profiles across both pure phospholipid bilayers at varying degrees of hydration and across mixed cholesterol:phospholipid bilayers show evidence of symmetrical shoulder features in the hydrocarbon regions (Rand and Luzzati, 1968; Levine and Wilkins, 1971; Levine, 1972; Blaurock, 1973b, 1976; Shipley, 1973; Ranck et al., 1974). An attempt has been made by Blaurock (1976) to reconcile the differences in the electron density profiles for myelin at moderate resolution as suggested by Caspar and Kirschner (1971) and Worthington and McIntosh (1974). The new and potentially powerful technique of neutron scattering (Mokrasch et al., 1971; Schoenborn, 1976) also provides strong support for the bilayer conclusion.

The relative ease with which solutions, especially nonionic ones such as sucrose and glycerol, can penetrate the compact myelin sheath at the extracellular apposition, causing the layers to separate by increasing hydration, indicates the hydrophilic nature of the region. Further, it shows that a tight association

268

through covalent or strong ionic forces is not involved between adjacent surfaces at the site. Interestingly, swelling in central nerve myelin seems to occur only at alternate external apposition regions (Finean and Burge, 1963; McIntosh and Robertson, 1976), but there appears to be no explanation for this since peripheral myelin swelling is at every external apposition (Finean and Burge, 1963). Fluid is almost certainly present at the external and cytoplasmic apposition sites in situ but the extent of structuring is not entirely clear. Both regions in low-resolution electron density profiles have absolute values around 0.33 to 0.35 electrons/Å³ (Blaurock, 1971, 1976; Caspar and Kirschner, 1971; McIntosh and Worthington, 1974a) that are similar to the value of 0.334 electrons/Å³ for water. The figures for these apposition regions are also significantly lower than the high density peaks in profiles. However, these electron density values cannot be taken as confirming the completely aqueous nature of the region because of the averaging nature of the method. Some protein and carbohydrate are probably located within these layers (Caspar and Kirschner, 1971). Neutron diffraction data also emphasizes the hydrated nature of the apposition regions (Mokrasch et al., 1971; Schoenborn, 1976). The inclusion of extra chemical components such as carbohydrate and/or an extra fluid layer accounts for the observation that the dimensions of the external apposition region in peripheral nerve myelin in situ are generally wider than the corresponding cytoplasmic apposition (Finean and Burge, 1963; Worthington, 1969; Caspar and Kirschner, 1971; McIntosh and Worthington, 1974a). The extra width of the external apposition in peripheral myelin explains the wider repeating unit compared with central nerve myelin (Blaurock and Worthington, 1969a). In central nerve myelin the external and cytoplasmic appositions are much more equal in width.

Conclusions. A bilayer interpretation for the structural state of lipid in compact myelin lamellae seems to fit the x-ray data best at both low and moderate resolution. Most experimenters agree on this interpretation from a variety of approaches. We therefore conclude that from the evidence to date, the lipid in myelin is mostly in bilayer form but the possibility that in some limited regions another structural state exists that is not detected in the diffraction work cannot be excluded. The width of the hydrocarbon region in the bilayer is too narrow to accommodate fatty acids of lipids in a fully extended all-*trans* form, especially for the sphingolipids. Acyl chains must either interdigitate or, for the most part, be in a melted liquid-crystalline state. More recent studies suggest the probability of this latter interpretation. Swelling studies indicate that strong interactions do not operate to hold the surfaces of the external apposition together. The cytoplasmic apposition region shows little tendency to separate under conditions when the external apposition layers will swell apart. There is no agreement from the x-ray data about an asymmetric distribution for cholesterol in the myelin lamellae. The possibility of limited penetration by protein into the lipid phase of the repeating unit (less than 10%) cannot be excluded.

3.2.2. Electron microscope studies
The electron microscope, through thin-section and freeze-fracture (etch) techniques, has provided information about the general structural organization of

compact myelin (reviews, Elfvin, 1968; Peters, 1968; Peters and Vaughn, 1970; Peters et al., 1970, 1976) and has also supplied hints from which finer details about the molecular architecture of the myelin lamellae may be gleaned. The reviews cited above generally summarize the electron microscopic evidence that compact myelin is in direct continuity with the plasma membrane of the Schwann cell in peripheral nerve tissue or the oligodendrocyte in central nerve tissue, and show how layers of plasma membrane appose to form the compact multilamellar myelin sheath. Clearly, apposition of adjacent cytoplasmic surfaces of the plasma membrane of the myelin-forming cell results in the appearance of the characteristic main dense line of compact myelin in peripheral and central nerve tissue (Fig. 4), while apposition of the external surfaces of the same membrane results in the less dense, intraperiod line. The electron microscope picture of compact myelin has provided an explanation of why the x-ray repeating unit of the system required interpretation in terms of a double bilayer structure and why symmetric planes occur at both the cytoplasmic and external apposition regions.

The dimensions of the repeating unit in compact myelin, as taken from electron micrographs (Fig. 4), are considerably narrower compared with x-ray results for the repeating unit of myelin in fresh nerve samples. This well-documented observation is due largely to the removal of water and some lipid from samples prepared for microscopic examination (Finean 1961; Hildebrand and Muller, 1974). The fundamental periodicity differences between peripheral and central nerve myelin samples are, however, preserved during preparation for electron microscopy (Karlsson, 1966).

Problems caused by fixation artifacts still arise to confuse structural issues. Thus, the earlier results of Hildebrand (1972), which had indicated that the periodicity of myelin in thin and in thick sheaths in central nerve tissue was different, have now been reappraised (Hildebrand and Muller, 1974). The later results show that thick and thin sheaths behave differently during preparation for microscopy to produce different periodicity measurements. Shrinkage on dehydration and embedding is more pronounced in thick sheaths than in thin ones. Of the explanations offered to explain this result, differences in water content and the involvement of nonuniform physical forces arising during preparative procedures seem to us a more likely explanation for the different shrinkage effects than known differences in the composition of compact myelin in brain and spinal cord (Mokrasch, 1969, 1971b; Morell et al., 1973; Lees and Paxman, 1974).

Lipids are lost from fixed myelin sheaths during dehydration and preparation for microscopy (Napolitano et al., 1967). To counter the need to use dehydrating solvents, Peterson and Pease (1972) have used a water-soluble glutaraldehyde-urea embedding medium for preservation of myelin structure. Micrographs of compact myelin embedded in this material have the normal periodicity of fresh nerve as revealed by x-ray diffraction, while the hydrocarbon regions have dimensions of 35 to 40 Å. This figure is considerably higher than that observed for the hydrocarbon region in micrographs of compact myelin fixed by conventional procedures, and is in good agreement with observed x-ray measurements on the

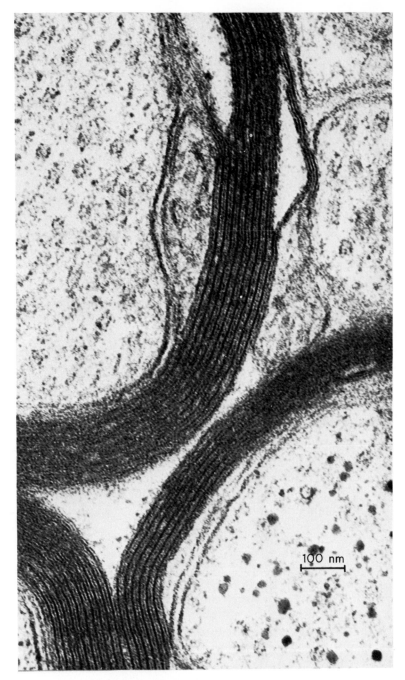

Fig. 4. Electron micrograph of central nerve myelin (rat optic nerve) showing main dense and intraperiod dense lines. Samples were fixed in glutaraldehyde and osmium tetroxide and embedded in medium Taab resin. Sections were poststained with uranyl acetate and lead citrate. The micrograph was kindly provided by Dr. N. A. Gregson.

width of the low electron density region of the myelin repeating unit (section 3.2.1.; Fig. 3). Despite this apparently excellent preservation of membrane structure, water-soluble embedding media do not appear to have achieved wide usage.

The modifying effects that fixation, staining, and other preparative procedures may have on compact myelin have not been fully defined at a molecular level. Thus a detailed elucidation of molecular organization and chemical architecture from a study of electron micrographs of myelin is not yet possible. The initial fixation of nerve tissue with glutaraldehyde does not introduce much contrast into the system (Peterson and Pease, 1972), and secondary fixation with osmium tetroxide and postsection overstaining with other reagents usually follows. Structurally, glutaraldehyde causes only a very small change in the periodicity of the myelin lamellae (Hildebrand and Muller, 1974), but it does serve to protect against the more serious changes that may occur when osmium tetroxide is used as a primary fixative (Finean, 1961) or against the disruptive effects of solvents (Napolitano et al., 1967) which modify or completely destroy the layered structure of unfixed, compact myelin (Rumsby and Finean, 1966a,b,c). Glutaraldehyde, by cross-linking proteins between primary amino groups (Richards and Knowles, 1968), induces a slight modification to the conformational structure of proteins (Lenard and Singer, 1968). Amino group-containing phospholipids can also interact with glutaraldehyde (Gigg and Payne, 1969) and may also be cross-linked to proteins by this fixative (Roozemond, 1969). In glutaraldehyde-fixed myelin, Wood (1973) has shown that proteins become too large to enter 10% SDS gels, and that while some cross-linking of phosphatidylethanolamine and phosphatidylserine to protein can be detected, the extractibility of cholesterol and remaining phospholipids from fixed myelin is not affected. The fate of cerebrosides in this system was not discussed by Wood and it would be interesting to know how this class of lipids, which predominates in myelin, reacts on fixation with glutaraldehyde. Polyhydroxy compounds have been reported to interact with the fixative (Hopwood, 1972).

Osmium tetroxide has been found to interact with the double bonds of fatty acids in lipid molecules (Korn, 1968, 1969). This may explain how osmium fixative acts on lipid bilayers to eliminate the motion of a spin-labeled probe and to alter the orientation of lipids perpendicular to the lipid plane (Jost et al., 1973). However, Dreher et al. (1967) have reported that osmium tetroxide does not easily penetrate to the double bonds of lipids in bilayer form and that the dark lines seen in osmium tetroxide-fixed samples outline the hydrophilic region of the system. Litman and Barrnett (1972) confirm this view reporting that osmium tetroxide hydrogen bonds to membrane proteins by an interaction that has been observed by Lenard and Singer (1968) to induce significant conformational changes in the molecular structure of proteins.

In interpreting the electron microscope image of membranes Finean (1973) is persuaded that "the highest densities are eventually to be found at the regions of interaction between lipid and protein components." In these terms the electron microscope appearance of myelin fits well with the bilayer model of structure indicated by x-ray electron density profiles across the repeating unit of the structure in fresh nerve (section 3.2.1.). The plasma membranes of both the Schwann cell and the oligodendrocyte show the dark-light-dark "tramline" appearance on glutaraldehyde and osmium tetroxide staining (Peters, 1968; Peters and Vaughn, 1970), as do the plasma membranes of other cells (Finean, 1969). Single bilayer, black lipid films that are known to be oriented with polar groups exposed outward to the aqueous medium (Tien, 1971) also show the same appearance on fixation with osmium tetroxide (Henn et al. 1967).

The intense and even staining of the main dense line in glutaraldehyde and osmium tetroxide-fixed samples of myelin (Fig. 4) contrasts with the appearance of the intraperiod dense line (IPDL), which is weaker in intensity and much more irregular in form. The difference in osmium tetroxide binding character cannot be due to inadequate penetration through the structure, since the IPDL defines the external apposition site, while the main dense line (MDL) is the cytoplasmic apposition region. These observations probably reflect differences in the chemistry of the two regions, the MDL being more intensely staining because of its greater osmiophilic nature or because more protein is concentrated there. X-ray studies have provided preliminary evidence to support this conclusion (section 3.2.1.). Since compact myelin is in direct continuity with the plasma membrane of the myelin-forming cell, this result, by extrapolation, could indicate that the external and cytoplasmic faces of this membrane are also asymmetric in stain-binding capacity and thus in chemical composition. A careful examination of published electron micrographs that show the plasma membrane of the myelin-forming cells (Figs. 1 to 12 in Peters and Vaughn, 1970) reveals that the external surface of the membrane is less densely stained and more irregular in appearance compared with the cytoplasmic side. The asymmetric distribution of proteins and of lipids in membranes is now a recognized feature of membrane organization (section 2.6.). However, we must be careful about relating structure in compact myelin to structure in the plasma membrane of the Schwann cell or the oligodendrocyte. Although these two systems are in direct continuity, in physical terms they have very differernt roles; a passive one for compact myelin in the sheath and an active role for the plasma membrane around the myelin-forming cells, which must presumably require a surface membrane showing the same range of activities that occur in the plasma membranes of other cells. Thus composition in terms of protein:lipid ratio and enzyme activity will vary widely in the two systems to match their functional properties even though they are in continuity. The biochemical evidence available at present on the composition of the oligodendrocyte plasma membrane supports this view (Poduslo, 1975). Thus at some point, probably close to the outer mesaxon, the chemical composition of the plasma membrane changes to that found in compact myelin.

In recent years interest has focused on the structural nature of the apposition sites in compact myelin in relation to how close opposing surfaces approach each other and to the chemical nature of the apposition region and whether tight chemical interactions hold the adjacent surfaces together across the apposition.

Finding answers to such questions will yield information on the organization of lipid and protein components in myelin. Although many conventional electron micrographs show the IPDL as a single irregular apposition region (e. g., Hall and Williams, 1971), peripheral nerve studies have found that the IPDL defining the external apposition region is resolved into two, thin, irregular dense lines separated by a light region 20 to 25Å in width (Napolitano and Scallen, 1969; Revel and Hamilton, 1969; Peters and Vaughn, 1970; Peterson, 1975). This same effect is also seen clearly in peripheral myelin embedded in glutaraldehyde-urea in which no dehydration steps have been used (Peterson and Pease, 1972). The double nature of the IPDL region is also apparent in some micrographs of central nerve myelin (see Figs. 1 to 28 in Peters and Vaughn, 1970; and comment in Reale et al., 1975) and in tissue fixed with a ferrocyanide-osmium mixture (Schnapp and Mugnaini, 1975). Substances such as colloidal lanthanum and lanthanum nitrate (but not ferritin, ruthenium red or horseradish peroxidase) can penetrate the IPDL region of compact peripheral nerve myelin (Revel and Hamilton, 1969; Klemm, 1970; Hall and Williams, 1971; Luft, 1971), forming a stained layer 50Å in width in the center of the external apposition region and thereby increasing the periodicity of the compact myelin (Hall and Williams, 1971). Soaking myelin in hypotonic media prior to treatment with lanthanum has been found to increase the uptake of the tracer at the external apposition region throughout the compact sheath (Hall and Williams, 1971). This is not surprising in view of the known effect of hypotonic solutions in causing adjacent myelin lamellae to separate at the external apposition region (Finean and Burge, 1963; McIntosh and Robertson, 1976). The ease with which the layers of the external apposition region swell apart in hypotonic media suggests that intersurface binding forces across the region cannot be very strong. Penetration of normal compact myelin by tracers may also indicate that the region has some, perhaps limited, continuity with the extracellular compartment around the cell. This is true for the paranodal region and for the periaxonal space between the axolemma and the compact sheath, which are also penetrated by lanthanum, ferritin, and microperoxidase presumably entering via the axon-glial junction region (e. g., Hall and Williams, 1971; Reese et al., 1971).

The question of whether lanthanum can penetrate the IPDL region of central nerve myelin seems less well resolved. It should be remembered that x-ray studies on fresh central nerve myelin indicate that adjacent surfaces at the IPDL region are more closely apposed compared with the same structure in peripheral myelin. In the CNS the dimensions of the IPDL are comparable with those of its cytoplasmic apposition. Further, swelling studies (McIntosh and Robertson, 1976) have revealed that central nerve myelin lamellae swell apart at alternate intraperiod dense lines and not at every one as in peripheral nerve myelin. Nevertheless, Hirano and Dembitzer (1969) have indicated that lanthanum can penetrate the IPDL region of compact myelin in central nerve tissue. The tracer was also observed to penetrate into the periaxonal space via the transverse bands of the paranodal region. In some sheaths, lanthanum, entering from the periaxonal space to the inner mesaxon, had penetrated the IPDL in inner lamellae of the compact sheath, as also observed by Revel and Hamilton (1969) for peripheral myelin. In contrast, Brightman and Reese (1969), using a perfusion system for introducing lanthanum and horseradish peroxidase, observed no penetration into the IPDL region of central nerve myelin by either tracer in mouse and fish brain. In some samples, lanthanum penetration into the periaxonal space was detected. As microperoxidase (molecular weight 2000 daltons can gain entry to the periaxonal space (Reese et al., 1971), it is probable that the inability of horseradish peroxidase (molecular weight 43,000 daltons) to penetrate is due to the larger size of the molecule. Although these studies tend to indicate that the axolemma-glial junction at the node in central and peripheral nerve tissue is not a completely tight seal and that ions, fluid, and tracer molecules can gain access to the internodal region (Blank et al., 1974), they do leave some doubt about whether the IPDL in central nerve myelin is accessible through the external mesaxon to the extracellular compartment. Schnapp and Mugnaini (1975) argue that lanthanum penetration of the IPDL region should not be used as definitive evidence that the pathway is open in vivo. They point out that in the study of Hirano and Dembitzer (1969) penetration by lanthanum in the extracellular fluid into the first few lamellae of compact sheaths was observed only at the periphery of tissue blocks, where some mechanical and necrotic damage may have occurred. Relevant to this whole question are recent freeze-fracture studies on the localization of zonulae occludentes in central myelin lamellae. These structures are specifically located immediately internal to the glial tongue (external mesaxon) and in corresponding areas of deeper lamellae (Tani et al., 1973; Dermietzel, 1974a; Reale et al., 1975; Schnapp and Mugnaini, 1975). However, it has yet to be determined whether the junction provided by these structures is leaky or tight (Claude and Goodenough, 1973). A tight junction in this location would probably constitute a barrier to the penetration of tracer molecules through the external mesaxon region to the IPDL in the outer lamellae. Movement of tracers through the periaxonal space (Brightman and Reese, 1969; Hirano and Dembitzer, 1969; Dermietzel, 1974a) would not be as impaired since the structural organization of the junction there is different (e. g., Livingstone et al., 1973; Schnapp and Mugnaini, 1975). Zonulae occludentes seem to be more sparsely located or more limited in number and extension in peripheral nerve myelin (Dermietzel, 1974a; Reale et al., 1975). If such is the case, these structures do not form a barrier to the penetration of lanthanum into the IPDL region (but see Schnapp and Mugnaini, 1975) as may occur in central nerve myelin.

In relation to molecular organization it would be of interest to determine how lanthanum penetrates the external apposition region in compact myelin and whether the localization of the tracer at the IPDL site defines a true space that exists in vivo or whether other factors, such as specific chemical binding, are involved. It has been suggested by Doggenweiler and Frenk (1965) that lanthanum acts as a "super" calcium molecule, staining the intercellular space by differential binding to the outer membrane layers. This has been discussed by

Hall and Williams (1971), who raise the possibility that lanthanum acts by binding specifically to surface components such as carbohydrates, proteins, and lipids. It is suggested that such binding could explain why the width of the lanthanum space in the external apposition region exceeds the normal 20 to 25 Å gap seen in micrographs of untreated nerves in which the IPDL is visualized as a double region (e.g., Revel and Hamilton, 1969; Peters and Vaughn, 1970; Peterson and Pease, 1972). Does the double nature of the IPDL represent the normal structural form at the external apposition region in situ? The answer is not entirely clear. The presence of an aqueous, and perhaps carbohydrate-rich, region separating adjacent surfaces at the external site would be in agreement with the x-ray data for peripheral nerve myelin, while the same gap would be narrower in central nerve myelin (section 3). A 20 to 25 Å fluid-filled space between adjacent membrane surfaces, maintained in vivo by inter-surface contacts between glycolipid and perhaps glycoprotein and/or protein, would be likely to collapse during solvent dehydration for microscopy unless strict precautions were taken. The fact that long-term staining of the IPDL with permanganate accentuates the intensity of the IPDL region to match that of the MDL suggests that interaction with carbohydrate-containing molecules at the site may be taking place. Such compounds will react with potassium permanganate to liberate manganese dioxide under relatively mild conditions. The labeling experiments of Poduslo and colleagues (section 8.2.) locate carbohydrate-rich molecules at the external surface of the membrane system.

The cytoplasmic apposition region defined by the MDL in electron micrographs (Fig. 4) is clearly different in character, as it shows little tendency to separate during swelling in hypotonic media (section 3.2.). Exposure of myelin to hypertonic solutions causes an increase in the width of the cytoplasmic apposition (Joy and Finean, 1963; Worthington and Blaurock, 1969a; and Blaurock, 1971), but this is far less than is experienced at the external apposition with hypotonic solutions. Such results provide evidence of the hydrophilic nature of this region, and suggest that the apposed surfaces must be much more tightly associated by different chemical interactions compared with those operating at the external junction region. Again, this is indicative of a different chemistry at the two sides of the membrane making up the compact myelin lamellae. It should be appreciated that the nature of the swelling processes initiated by hypo- and hypertonic solutions in their respective actions on myelin are different involving water layers with hypotonic solutions and penetration by salts with hypertonic solutions.

Separation of cytoplasmic apposition surfaces is observed at the Schmidt-Lanterman clefts in peripheral nerve myelin (and perhaps in central myelin), where cytoplasm and cytoplasmic features such as microtubules, membrane-bound dense bodies, and multivesicular bodies are located (see Hall and Williams, 1970). It is relevant to an understanding of the forces operating at the cytoplasmic junction that the cytoplasmic apposition on either side of the Schmidt-Lanterman cleft never splits beyond the original point of division under a variety of manipulative conditions. This observation by Hall and Williams (1970) contrasts markedly with the behavior of the external apposition layer and again serves to emphasize that these two apposition surfaces are chemically different. For a detailed discussion of the structure and possible function of the Schmidt-Lanterman clefts in compact myelin in peripheral and central nerve tissue, the reader is referred to papers by other experts (Friede and Samorajski, 1969; Williams and Hall, 1969; Peters and Vaughn, 1970; Peters et al., 1970; Hall and Williams, 1971; Hildebrand, 1971).

Electron microscope studies on compact myelin have provided details on the membrane system at a structural level, but molecular detail has not been forthcoming due to the limitations of the technique. Certain features of ultrastructure in the myelin lamellae have been confirmed and even extended by the application of freeze-fracture electron microscopy techniques to the system. The chief advantage of the freeze-fracture method is that the problems associated with dehydration and embedding, essential for thin-section work, are avoided. Freeze-fracture microscopy, however, has its own associated problems, such as difficulties in interpreting the image of the replica, angle of shadowing effects (Pinto da Silva and Miller, 1975), artifacts induced by vacuum incidents (Deamer et al., 1969; Stolinski and Breathnach, 1975), structural changes induced during freezing (Fluck et al., 1969; Zingsheim, 1972), and the effects that initial treatment with glutaraldehyde and glycerol may have on the organization of membrane-associated particles (Zingsheim, 1972; Pinto da Silva and Miller, 1975). There is general agreement that membranes fracture through their central hydrophobic core (Zingsheim, 1972; Branton, 1973; Stolinski and Breathnach, 1975). Thus the image observed in the replica is of the hydrophobic face of the membrane. The earlier freeze-fracture results of Bischoff and Moor (1967a,b, 1969) on compact myelin must therefore be reinterpreted in this light. It is also agreed that the 80 to 120Å diameter membrane-associated particles generally seen on the fracture faces of most membranes represent intrinsic proteins that penetrate the lipid layer of the membrane or which, in some cases, may span the bilayer (Stolinski and Breathnach, 1975; McNutt, 1977).

Before considering the results reported to date, it is worth thinking about what we may expect to see in freeze-fracture micrographs of compact myelin. Both x-ray diffraction and thin-section microscopy results have indicated a multilamellar structure for the compact sheath in both central and peripheral nerve tissue. The major proteins in central nerve myelin, the basic protein (BP) and the proteolipid protein (PLP), are both relatively small in molecular weight and are present in the compact membrane in approximately equimolar amounts, accounting for 80 to 85% of the total membrane protein (section 5.3.). Of these two major proteins the PLP has a distinctly lipophilic nature and is a good candidate to be an intrinsic protein. Thus it may be partially or even totally located in the lipid hydrocarbon phase of the membrane. The BP has characteristics like those of extrinsic proteins because it can be removed without the use of bilayer-disrupting agents such as solvents. This protein may show some penetration of the lipid phase but not to a depth that would be revealed in the hydrocarbon phase on freeze-fracturing. Our calculations (section 9.4.3.) indicate that a spherical proteolipid protein molecule with a molecular weight of 23,500 daltons

(Nussbaum et al., 1974a,b) would have a diameter of about 40 Å. A molecule of this size could just span the bilayer in compact myelin. These calculations also reveal that if this were the case the protein would occupy, at a maximum, about 20% of the hydrophobic surface area. If this is correct, freeze-fracture surfaces of compact myelin should show particles of protein occupying about 20% of the surface area depending on whether such particles cleave completely with one fracture surface or the other. Viewing this in perspective, an occupancy of a white surface by black circles at a density of 20% is shown for comparison in Fig. 5. We believe the particles of PLP should be smaller than those observed in other membrane systems. It should be noted that Vail et al. (1974) have obtained a figure of 85 to 90Å for the diameter of the N-2 proteolipid protein of human myelin using negative staining microscopy techniques. Substituting in our calculations, a protein of this diameter would have a molecular weight of over 275,000 daltons. Negative staining methods probably induce the aggregation of protein molecules during the process, which could account for this large figure. The result means that the observed diameter of the PLP could be much smaller than the 85 to 90 Å noted by Vail and coworkers. These authors also estimate that this diameter is significantly increased by the freeze-fracturing process especially the shadowing and coating procedures. Intrinsic proteins that span the lipid bilayer usually have molecular weights of over 100,000 daltons (e. g., Juliano, 1973) and thus are considerably larger than the figure we have taken for the proteolipid protein. In addition, some of the high molecular weight proteins of myelin will be intrinsic proteins and will penetrate the hydrophobic region of the membrane. Such components may also be visible on fracture surfaces of compact myelin and would presumably be more in agreement with the classic membrane-associated particles of other membrane systems. These proteins, however, would occupy only a very low density.

All of the freeze-fracture studies on compact myelin reported to date clearly show the multilamellar nature of myelin (Bischoff and Moor, 1967a,b, 1969;

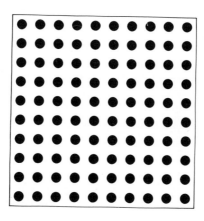

Fig. 5. Occupation of a white surface by dots at a level of 20% to help appreciate the calculated distribution of the proteolipid protein at the hydrocarbon fracture face of central nerve myelin.

Branton, 1967; Tani et al., 1973; Dermietzel, 1974; Pinto da Silva and Miller, 1975; Reale et al., 1975; Schnapp and Mugnaini, 1975; Stolinski and Breath-nach, 1975). The same has also been shown for myelin in isolated form (Wood et al., 1975; Rumsby and Lang, unpublished observations, and Fig. 6). These results further support the view that the lipid phase of myelin is generally organized in the bilayer form. The regular multilayered structure of the compact sheath revealed in the freeze-fracutre image is similar to results obtained with liposomes and lamellar lipid phases (Deamer et al., 1970; James and Branton, 1972). Lipid in other structural states such as the hexagonal form gives a very different image on freeze-fracturing (e.g., Olive, 1973). The fresh nerve myelin periodicity dimensions are well preserved in freeze-fracture images of compact myelin, and the characteristic periodicity difference between central and peripheral nerve myelin is also retained (Bischoff and Moor, 1967a; Dermietzel, 1974).

The question of whether compact myelin fracture faces show associated particles has been reconsidered but is not fully resolved. Initial studies on freeze-fractured myelin were interpreted as showing that the hydrophobic fracture planes were smooth and relatively free of the normal membrane-associated particles observed for all other membrane systems (Branton, 1967). It is relevant that this study was made with myelin in immature nerve tissue, and thus the results cannot necessarily be extrapolated to the adult system. There is little doubt, however, that the observed smooth fracture faces resemble closely those obtained by Deamer et al. (1970) with pure lipid systems. More recent freeze-fracture studies on myelin in the adult form in central and peripheral nerve tissue have also been interpreted as revealing the absence of significant particulate components on the fracture surfaces (Tani et al., 1973; Dermietzel, 1974; Reale et al., 1975; Stolinski and Breathnach, 1975). Fracture surfaces of isolated myelin were reported by Wood et al. (1975) to be essentially free of particles. Plaque-shaped structures observed in fracture planes of myelin by Bischoff and Moor (1967a,b, 1969) have since been interpreted as artifacts (Deamer et al., 1970; Schnapp and Mugnaini, 1975).

However, the results of two recent studies have been interpreted as showing the presence of membrane-associated particles on the fracture surfaces of compact myelin (Pinto da Silva and Miller, 1975; Schnapp and Mugnaini, 1975). Isolated central nerve myelin samples (Fig. 6) also show evidence of particulate components in the fracture plane (Rumsby and Lang, unpublished observations). Schnapp and Mugnaini (1975) noted that the particles in myelin are from 80 to 160Å in diameter but no density figures were quoted. The process of shadowing and coating increases the size of such particles significantly. No figure for particle size is quoted by Pinto da Silva and Miller (1975), but measurements from their electronmicrographs show the particles to be about 80 to 100Å in diameter. These authors also consider the reason other workers have not detected particles in myelin may be that examination has been confined to regions of high shadow angle in which particles have been obscured rather than resolved. This explanation is not entirely adequate to explain those studies in which fracture faces of compact myelin were found to lack particles and in

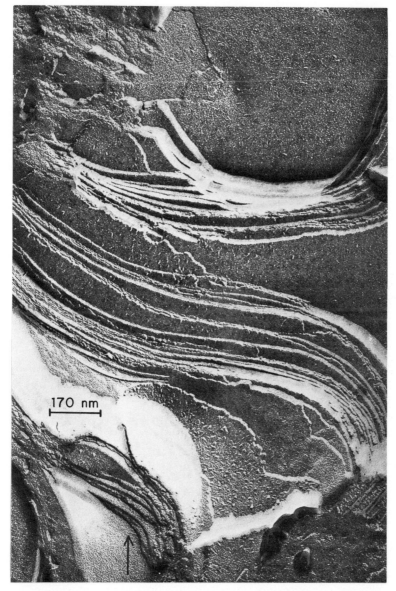

Fig. 6. Freeze-fracture electron micrograph of isolated central nerve myelin prepared by the method of Rumsby et al. (1970). The myelin sample was spray-frozen and examined using the method of Lang et al. (1976). Arrow denotes angle of shadow. (Courtesy of R. D. A. Lang.)

which ridges and edges of junctions of the zonulae occludentes in central nerve myelin were clearly seen (Tani et al., 1973; Dermietzel, 1974; Reale et al., 1975). We believe that the particles associated with the fracture faces of compact myelin have probably been missed because of their relatively small size and low density compared to the normal membrane-associated particles. It should be remembered that the limits of resolution with this technique are 30 to 40 Å. Thus particles of the PLP with a diameter of 40Å will be very difficult to resolve. Schnapp and Mugnaini (1975) have reported that in central nerve myelin the particles they see at the hydrophobic fracture face cleave predominantly with one fracture face. This may suggest only partial penetration of the lipid layer by the protein or an association of the particles with some other component of myelin. Unpublished results are cited by these authors to indicate that the particle-studded face is adherent to the cytoplasmic apposition region. Clearly, further high-resolution freeze-fracture observations are needed to resolve the disposition of protein particles in the hydrophobic fracture face of compact myelin.

Pinto da Silva and Miller (1975) reveal that glycerol as a cryoprotective agent in freeze-fracture work can result in particle aggregation and a lineup of particles across the lamellae in myelin. This is not seen in fresh myelin frozen immediately without any treatment. It is of interest that glutaraldehyde was found not to cause particle aggregation despite its known interaction with myelin proteins (Napolitano et al. 1967; Wood, 1973). In freeze-fracture studies on tight junctions, Staehelin (1973) reported that glutaraldehyde, but not glycerol, caused aggregation of particles into continuous ridges. Such anomalies will presumably be resolved by more work. The freeze-fracture studies on compact myelin have actually shown that closely related membrane structures, such as the axolemma in the perinodal region and the plasma membrane of the oligodendroglial cell, contain membrane-associated particles and are substantially different in appearance from images of compact myelin. These structures are more characteristic of classic membrane systems.

Two other important structural features of the compact myelin sheath have been examined by the freeze-fracture technique. One of these is the myelin-axolemma junction at the paranodal region in central and peripheral nerve tissue. Apparently, an extensive and unique union is established between the two closely adjacent membranes by particles attached to the axolemma that bridge across to the glial membrane. The reader is referred to other sources for further details (Peters and Vaughn, 1970; Livingstone et al., 1973; Dermietzel, 1974). The development of this junction region is of special interest in relation to the process of myelination.

The other important feature concerns the localization of tight junctions (zonulae occludentes) in central nerve myelin, especially at the interlamellar junction between the outermost lamella of myelin and the external tongue of the glial plasma membrane (outer mesaxon). This structure has been examined in reports by Tani et al. (1973), Dermietzel (1974), Reale et al. (1975) and Schnapp and Mugnaini (1975). The junctions appear as parallel ridges matched by furrows on the apposed fracture surface. The ridges appear to consist of rows of particles and may thus be more similar to endothelial cell junctions (Yee and Revel, 1975) than to tight junctions, especially since evidence exists that pretreatment of nerve tissue with glutaraldehyde or glycerol can change the appearance of linearly arranged particles into a continuous ridge form (Stolinski and Breathnach, 1975;

Dermietzel, personal communication). These ridges run parallel along the length of the sheath throughout its thickness at the outer tongue region (Dermietzel, 1974). Although most authors interpret the structure in terms of a junction, whether it is a tight or leakier junction in the terms of Claude and Goodenough (1973) has yet to be decided. The appearance of the structure in nerve tissue untreated with glycerol and glutaraldehyde will be of interest. The presence of similar zonulae occludentes in peripheral nerve myelin is not yet resolved (Dermietzel, 1974; Reale et al., 1975; Schnapp and Mugnaini, 1975).

The zonulae occludentes structure observed in fracture surfaces of compact central nerve myelin has been equated with the radial component which is apparent in thin sections of the same tissue fixed with potassium permanganate (e. g., Peters and Vaughn, 1970). The radial component appears as a series of thickenings in the IPDL of the sheath (Fig. 16 in Peters et al., 1968). Such thickenings are readily apparent through the sheath in the area underlying the external glial tongue, as is the zonulae occludentes structure in freeze-fracture images of compact myelin.

The thickenings of the IPDL in myelin, revealed as the ridges and depressions of the zonulae occludentes in freeze-fracture images, have been interpreted by Reale et al. (1975) as arising from the collapse of the gap region during dehydration for thin-section microscopy. In the particle area, where thickening is seen in thin sections, no collapse can occur. Although no definition of the type of molecule that could function here has been presented, perhaps the fact that permanganate is required to define the radial component in thin-section microscopy indicates the involvement of carbohydrate-containing molecules.

The function of the radial component (zonulae occludentes) in compact myelin may actually be to form a junction sealing the gap at the external mesaxon region and between cells and sheath in nerve tissue (Brightman and Reese, 1969). The significance of this structure in relation to the penetration of lanthanum and other tracers into the IPDL region of compact myelin has been discussed earlier in this section. The experiments of McIntosh and Robertson (1976) on central nerve myelin show that in the swollen state lamellae are kept in a compact form in certain areas by such structures.

Conclusions. The results of electron microscope studies on compact myelin provide further support for the x-ray diffraction view that structure in the myelin lamellae is based principally on a bilayer of lipid. The differential uptake of stain by the compact sheath suggests that there may be chemical differences between the external and the cytoplasmic membrane surfaces. Such differences are probably due to an asymmetry of protein (and lipid) within the myelin membrane. The weak and tight association of the external and cytoplasmic apposition surfaces respectively is confirmed. Little other detail at a molecular level is forthcoming. Freeze-fracture studies have provided more support for the bilayer concept for structure in the myelin sheath. The long accepted absence of particles in the hydrophobic region of the lamellae has been requestioned in recent reports. Calculations indicate that PLP particles may be difficult to resolve due to their small size.

4. Myelin in isolation

An appreciation of the molecular structure of a membrane can only be achieved if an understanding of its chemical composition is available. For myelin, the detailed knowledge that we have of the chemistry of the system (section 5) has been facilitated to a large extent by the fact that high purity membrane can be isolated in large amounts from nerve tissue. Central nerve white matter is 50 to 60% myelin on a dry weight basis (Norton and Autilio, 1966) and, using white matter as the starting material in a preparative procedure, a good yield of the isolated product can be expected even after rigorous purification. A distinct advantage in starting with white matter is that contamination of the final product with nuclei and nucleic acids is minimized (Rumsby et al., 1970). The feature of myelin that has made its isolation relatively easy is that compared with other systems (Guidotti, 1972) it has a very low buoyant density due to its lipid-rich nature (section 5). Myelin can be separated cleanly from other membranes that are more dense because of their higher protein content, using density gradient centrifugation techniques. Details of purification procedures for myelin have been reviewed by Mokrasch (1971b), Spohn and Davison (1972) and Norton (1974, 1976). References to specific research papers dealing with different aspects of myelin purification are given in these reviews. We would like, however, to emphasize our view that in working with isolated myelin a single procedure should be used routinely and that as far as possible animals of the same age should be used for the isolation. Thus a standardized sample of purified myelin is achieved, which can be characterized chemically and screened for contaminating membrane structures. This standardization ensures that isolated myelin preparations will be consistent in composition and form regardless of the type of experiment in progress. Unlike Mokrasch (1971b) we do not recommend that the method of myelin isolation be altered to suit the purpose of the experiment, except in certain cases, because the benefits of standardization will then be lost. Exceptions occur when myelin is to be isolated from peripheral nerve where high levels of collagen necessitate a different approach (summaries in Mokrasch, 1971b; Norton, 1974), or with immature nerve tissue where the proportion of membrane structures such as plasma membranes will be enriched relative to compact myelin. Here again, specific isolation procedures have been developed (Norton and Poduslo, 1973a). With regard to standardized myelin preparations it is worth noting that if one or two methods of isolation could be adopted at an international level, then at least one very important variable would be removed when results from different laboratories on myelin biochemistry are compared.

4.1. The form of isolated myelin

Since much of our data on myelin has come from studies on the isolated system, an interesting point for discussion concerns the structural nature of purified myelin and whether it bears any resemblance to its form in situ. Electron mic-

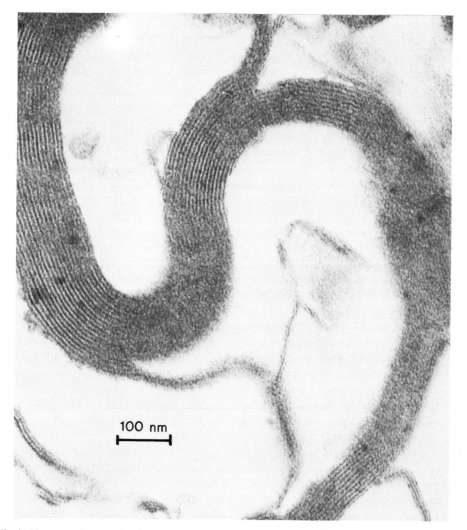

Fig. 7. Electron micrograph of isolated bovine cerebral myelin purified by our standard procedures and fixed in glutaraldehyde and osmium tetroxide. The sample was embedded in medium Taab resin, and sections were poststrained with uranyl acetate and lead citrate. The micrograph was provided by Dr. N. A. Gregson.

rographs of isolated myelin (Fig. 7) give the impression that preparations consist of sheets of membrane several lamellae in thickness and others of only single or few lamellae. Norton (1974, 1976) has suggested that when brain tissue is homogenized in sucrose, the myelin layers swell and peel away from the axon to reform as vesicles varying in diameter and of few or many lamellae in thickness. Probably some of the myelin in an isolated preparation is in a vesicle form, which will be seen in section in conventional thin-section micrographs. In our view much of a myelin preparation will be multilayered sheets of membrane, torn from the compact sheath by the shearing forces of homogenization but probably retaining some interlamellar interactions. The appearance in freeze-fracture micrographs (Fig. 6) suggests that the multilamellar sheets of isolated myelin exhibit some curvature. The presence of ions in the homogenization medium at the beginning of a myelin purification seems to interfere with the clean separation of the sheath from the axolemma (De Vries et al., 1972). This finding has been used as the basis for isolating an axolemma fraction from central nerve tissue (De Vries et al., 1975, 1976).

Density gradient centrifugation reveals that isolated myelin preparations are composed of fragments of myelin differing in size and thickness. If isolated myelin, purified initially by our routine procedure (Rumsby et al., 1970), is subsequently refractionated on a multistep, discontinuous sucrose density gradient (0.32 – 1.0M sucrose) and centrifuged at medium gravitational fields for short times (40,000g for 30 min), the originally homogeneous sample separates out into several discrete bands over the 0.32 to 0.656M density range (Torrie and Rumsby, unpublished observations). Subsequent equilibrium density gradient centrifugation for 24 hours at the same g forces results in all the bands sedimenting to the same position on the top of the 0.656M sucrose layer. This observation indicates that although the myelin fragments in the preparation are of the same density they vary in size. This conclusion has been confirmed by further analysis.

The multilamellar nature of isolated myelin preparations must be considered when the results of metabolic and enzymatic studies with myelin are interpreted. Substrates may not be freely accessible to enzymes localized in the inner regions of the multilamellar sheets of membrane. Certain enzymes associated with myelin, such as the cholesterol ester hydrolase (Eto and Suzuki, 1973) and the CNP hydrolase activity (Lees et al., 1974), are activated by detergents such as deoxycholate. In isolated myelin the detergent effect may disrupt multilayered sheets, allowing the exposure of more enzyme molecules and/or may extract the enzyme from the membrane. We have found this to be true for the nonspecific esterase activity, associated with isolated myelin samples (Rumsby et al., 1970, 1973), which is released from the lamellae by treatment with Triton X-100 and some other nonionic detergents (Rumsby et al., 1973; Torrie and Rumsby, 1975, and in preparation).

Contamination of myelin preparations by other intracellular membrane structures, and as a result of the artifactual redistribution of enzymes on homogenization when cell compartments become mixed, will continue to be a problem in work with isolated samples. Over the past years experimenters have become much more conscious of this problem, and currently electron microscope examination of purified preparations is not considered sufficient check for contamination. Assay of the preparation for specific enzyme markers that are representative of other membrane structures must be undertaken to gain an accurate picture of contamination. Such analyses can be supported by chemical analysis, although this is less effective because there are no obvious myelin-exclusive components, especially at the low levels where contamination would be present (Mokrasch, 1971b). Also, it may not be possible to remove all traces of certain enzymes thought to be derived from contamination by repeated purification. We have shown that during the purification of central nerve myelin preparations the activity of enzymes used as markers for mitochondria, soluble phase and lysosomes decrease steadily but then level off at a plateau which remains constant despite further purification and must be regarded as negligible or as naturally associated with the myelin (Rumsby et al., 1970; Riekkinen and Rumsby, 1972). The problem of contamination in myelin samples has been discussed fully elsewhere (Smith, 1967; Rumsby et al., 1970; Mokrasch, 1971b; Norton, 1974, 1976).

It can be assumed that isolated myelin preparations when purified by use of reasonable procedures bear a fair resemblance to compact myelin in situ and that little gross alteration of ultrastructural detail at a molecular level has occured during isolation, although there has been some splitting of lamellae and fragmentation of sheaths. A major point to be appreciated about compact myelin is that when the structure swells in hypotonic media during isolation, lamellae separate at the external apposition region. In peripheral myelin the lamellae separate at every external layer, while in central myelin this is apparent at alternate external layers (Finean and Burge, 1963; McIntosh and Robertson, 1976). Thus, in isolated myelin samples, the face of the multilayered membrane sheets that is exposed to the aqueous phase after purification will be the external apposition face. This will be discussed again in a later section on the labeling of myelin. The x-ray periodicity of isolated central nerve myelin samples is the same as compact myelin in situ, although the overall pattern is more diffuse and weak, probably due to the much less ordered nature of the isolated preparation (Rumsby, 1963; Wood et al., 1974; Glaisher, 1975).

We conclude this section on a note of caution for there are ways in which the structural and molecular form of the membrane preparation may certainly be altered during isolation. Hydrolytic enzymes, such as proteinases and phospholipases, can attack myelin, removing components such as the basic protein and degrading lipid molecules (McIlwain and Rapport, 1971; Sammeck and Brady, 1972; Roytta et al., 1974; Wood et al., 1974). Isolation of myelin from brain tissue long after death, delay in the isolation procedure after cell compartments have been mixed, the use of isolation temperatures above 4°C may all facilitate the action of lysosomal enzymes on myelin and lead to modification of the preparation. Recently Ansari et al. (1975) indicated that the action of such enzymes is accentuated by the use of fronzen and thawed brain tissue in isolation procedures. Control samples of nerve tissue chilled to 4°C after removal from the animal and used within 24 hours showed little sign of proteolytic action and loss of basic protein. Material frozen after removal and then thawed for use showed evidence of basic protein loss due to the release of hydrolytic enzymes from damaged lysosomes. However, many experimenters seem aware of these problems associated with preparative procedures and take appropriate precautions.

4.2. Lyophilized myelin

Concern must be expressed, however, about the use of lyophilized myelin samples in studies where conclusions are drawn about structure and organization in the myelin sheath (e.g., McIlwain and Rapport, 1971; Raghavan et al., 1973; Coles et al., 1974; Cohen et al., 1975). The effects of dehydration on the ultrastructure of the myelin sheath are summarized in section 7.1. These changes involve the formation of a dehydrated, residual lipoprotein structure and the separation from the original membrane system of lamellar lipid phases, probably of phospholipid and cholesterol (Finean, 1961). Such changes will occur in lyophilized myelin as will the modifications brought about in compact myelin by freezing and thawing (section 7.1.). When lyophilized myelin is rehydrated, evidence indicates that the phases of the dehydrated system do not reform to produce the same molecular organization found in fresh myelin in situ.

McIlwain (1973) found that nonuniform swelling processes occur when lyophilized myelin is rehydrated. This author commented that "it seems probable that changes in the organization of both myelin protein and lipid accompany swelling" and noted that when rehydrated lyophilized myelin is centrifuged at high gravitational forces (100,000g) protein-depleted lipid vesicles accounting for 10 to 15% of the original myelin phospholipid are left in the supernatant. Lees et al. (1974) reported that while freshly isolated myelin forms a fairly stable suspension, resuspended lyophilized myelin consists of particles of different sizes which do not remain suspended uniformly. These authors also show that the CNPH activity of fresh and resuspended lyophilized myelin samples differ and that the response of the two forms of preparation to various detergents is different. Electron micrographs of frozen and thawed myelin reveal that this treatment results in structural modifications in myelin structure that are apparent at the IPDL region and are seen as an increase in stain uptake, with IPDL becoming nearly equal to the MDL in staining intensity.

Thus, it cannot be assumed that the molecular organization of rehydrated lyophilized myelin is in any way the same as that of the intact, freshly isolated system or of the compact membrane in situ. Lyophilized myelin should not therefore be used for work from which conclusions about the molecular arrangement of lipid and protein in native myelin are to be drawn.

5. The chemical composition of myelin

5.1. Gross composition

One of the most striking features of the gross compositional data for myelin when compared with other isolated membrane systems is the high ratio of lipid to protein (Table 1). This gives the membrane a characteristic low density that enables experimenters to purify myelin free of other neural structures and subcellular fractions by sedimentation techniques. With the development of these relatively simple isolation procedures and appropriate analytical techniques, a wealth of compositional data for myelin isolated at different stages during development and from a number of species has appeared (e. g., Eng et al., 1968; Mokrasch, 1969, 1971b; Davison and Peters, 1970; Mokrasch et al., 1971; Norton, 1975, 1976). Such analyses have identified a number of lipid components, principally plasmalogens, cerebrosides, and cerebroside sulfate, which appear to be concentrated in myelin. The basic protein of myelin also appears to be specific for this membrane system.

5.2. Lipid components

Lipids are the most abundant organic molecules in myelin, comprising from 75 to 80% of the total dry weight; the remainder is protein. There are numerous reports on the lipid composition of myelin isolated from a number of vertebrate species (Cuzner et al., 1965) and particularly about myelin prepared from mammalian central nerve tissue (Nussbaum et al., 1963; Eichberg et al., 1964;

Table 1
Composition of central nerve myelin from different sources

	Man[a]	Ox[b]	Guinea Pig[c]	Rat[d]
Lipid percent dry weight	78.7	77.7	80	64
Protein percent dry weight	21.3	22.3	19	36
		moles per cent lipid		
Cholesterol	40.9	44.4	40.7	42.7
Phosphatidalethanolamine	11.6	11.5	10.0	14.2
Phosphatidylethanolamine	2.0	3.6	5.3	3.8
Phosphatidalserine	1.9	0.2 }	5.0	6.2
Phosphatidylserine	3.2	4.7 }		
Phosphatidylcholine	10.9	8.2	10.1	10.5
Sphingomyelin	4.7	5.1	4.9	3.4
Cerebroside	15.6	17.3 }	20.3	14.1
Cerebroside sulfate	4.1	2.5 }		2.8
Minor galactolipids		1.4		
Ceramide	0.1			
Inositol glycerophosphatides }				
Phosphatidic Acid }	4.5		0.6	1.0
Alkyl ether phosphatides }			1.7	

Data compiled from the following sources:
[a] O'Brien and Sampson, 1965.
[b] Norton and Autilio, 1965
[c] Eichberg et al., 1964
[d] Cuzner et al., 1965

Cuzner et al., 1965; Evans and Finean, 1965; Norton and Autilio, 1965, 1966; O'Brien and Sampson, 1965a,b; Thompson and Kies, 1965; Soto et al., 1966; Gerstl et al., 1967; Horrocks, 1967, 1968; Smith, 1967; Cuzner and Davison, 1968; Eng and Noble, 1968; Dalal and Einstein, 1969; MacBrinn and O'Brien, 1969; Singh et al., 1971; Gregson and Oxberry, 1972; Baumann et al., 1973; Norton and Poduslo, 1973b; Woelk and Borri, 1973; Fishman et al., 1975). Typical data for the lipid composition of central nerve myelin isolated from a number of mammalian species are presented in Table 1. Minor differences in composition between the different species studies are clearly apparent from the results, but it is not yet clear whether these represent true species variations or are due merely to differences in isolation procedures, purity, or analytical techniques (Norton, 1976). These methodological considerations preclude detailed examination of the composition differences reported by various workers. However, the general features of the lipid data that seem to be agreed upon will be considered in the context of their structural implications for the myelin membrane.

5.2.1. Phosphoglycerides

The major phosphoglycerides of compact myelin are ethanolamine phosphoglycerides, phosphatidylcholine, and phosphatidylserine. These different lipid species occur at an approximate molar ratio of 3:2:1 in human, ox, and rat brain myelin (Table 1). Although these lipids have certain structural similarities because they are all based on sn glycerol-3-phosphate, differences in polar head group chemistry such as configuration, charge, and hydration, and in fatty acid composition, means they may have very different structural functions in the membrane in situ.

Up to 80% of the ethanolamine phosphoglycerides of myelin are in the plasmalogenic (phosphatidal) form (O'Brien and Sampson, 1965a; Norton and Poduslo, 1973b; Fishman et al., 1975). In contrast, in grey matter, they are mainly in the diester form (O'Brien and Sampson, 1965a). The structural significance of the high proportions of these ethanolamine plasmalogens in myelin that differ from the diacyl form in the alk-1-enyl linkage only, and their proportionate increase relative to the diacyl form during myelination (Norton and Poduslo, 1973b; Fishman et al., 1975), is of considerable interest. Plasmalogenic forms of PC and PS are known, but the only other significant plasmalogenic lipid detected in myelin is phosphatidalserine. This accounts for only about 10% of the serine phosphoglycerides in the membrane (Mokrasch, 1969). The fatty acid and fatty aldehyde composition of phosphoglycerides isolated from human central nerve myelin are summarized in Fig. 8. The fatty aldehyde composition of serine and ethanolamine plasmalogens show a dominance of $C_{16:0}$, $C_{18:0}$ and $C_{18:1}$ aldehydes. The ester-linked fatty acids are of similar average chain length, but a significant proportion of longer chain length unsaturated fatty acids is found in PE and PS. Phosphatidylcholine has a high proportion of $C_{16:0}$ and $C_{18:1}$ fatty acids.

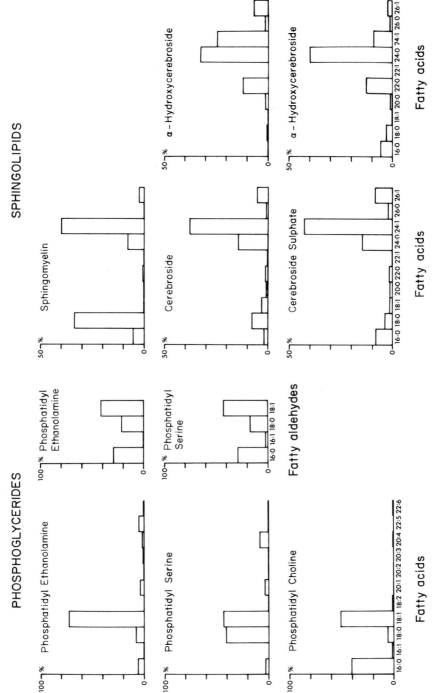

Fig. 8. Fatty acid and fatty aldehyde composition of myelin phosphoglycerides and sphingolipids. Data from O'Brien and Sampson (1965b).

5.2.2. Sphingolipids

The sphingolipids in general, and the cerebrosides in particular, are the most "myelin typical" lipids. The total amount of cerebroside in the brain has been shown to be directly proportional to the amount of myelin present (Norton and Poduslo, 1973b). The fatty acid composition data for myelin sphingolipids from adult human brain are presented in Fig. 8. The significant feature of these results, when compared with the fatty acid and fatty aldehyde composition of the phosphoglycerides in myelin, is the large proportion of long chain fatty acids (up to C26) in the myelin sphingolipids. Cerebroside and cerebroside sulfate lipid classes can be further subdivided based on whether they contain α hydroxy or unsubstituted fatty acids. Again, these are of long chain length. The sphingolipid components of grey matter contain fatty acids of shorter average chain length with smaller proportions of long chain fatty acids (O'Brien and Sampson, 1965b). A significant chain length difference between cerebrosides isolated from bovine central nerve myelin and axons has been reported (De Vries and Norton, 1974). It seems that cerebrosides from myelin have a longer average fatty acid chain length than the same lipid class purified from structurally related nerve axons. Myelin sphingolipids as a class appear peculiar to myelin in the high proportion of long chain fatty acids that they bring into the membrane. In this context it is significant to note that in developmental studies a decrease in PC and its partial replacement by sphingomyelin has been reported (Norton and Poduslo, 1973b; Fishman et al., 1975). The level of choline headgroups stays roughly the same in this exchange, but the main result of the decrease in PC and the elevation of sphingomyelin is to increase the proportion of long chain fatty acids in compact myelin. The functional significance of the high levels of long chain fatty acids in myelin sphingolipids in relation to structure in the membrane is not yet clear, though it must be presumed that they will have some influence on the overall state of membrane fluidity (see section 2).

5.2.3. Cholesterol

Cholesterol is the most abundant lipid molecule in myelin (Table 1). An empirical 1:1 molar relationship for the sum of ethanolamine phosphoglycerides plus sphingolipids to cholesterol has been proposed on the basis of metabolic data from rat brain myelin (Eng and Smith, 1966). This relationship is valid for various species and for myelin from the peripheral nervous system. (Horrocks, 1967; Eng and Noble, 1968) and also for myelin during development (Horrocks, 1968) and myelin isolated from the myelin-deficient *Quaking* mouse (Singh et al., 1971). A similar 1:1 phospholipid-to-cholesterol complex has also been proposed as a structural feature of myelin (Finean, 1953, 1961, 1962; Vandenheuvel, 1963, 1965). The question of whether these ratios of cholesterol to complex lipid are of structural significance in the compact myelin membrane is not resolved and must be interpreted in light of current understanding of membrane organization (section 2).

5.2.4. Minor lipid components

In addition to the major glycolipids of myelin, there are a number of minor carbohydrate-containing lipid species present in the system. Together these comprise 3% of the total glycolipids of myelin. Galactosyl diglyceride and the monoalkyl derivative of the same compound have been detected in brain and white matter lipid extracts (Norton and Brotz, 1963, 1967; Rumsby, 1967; Rumsby and Rossiter, 1968). The appearance of galactosyl diglycerides in brain tissue has been correlated with the period of active myelination (Pieringer et al., 1973), and the oligodendroglial cells have been shown to be the primary synthesis site (Deshmukh et al., 1974). Minor phospholipid components that have been identified as myelin constituents include the higher inositol-containing phospholipids, triphosphoinositide, and diphosphoinositide that account for between 3 and 6% and about 1% of the myelin lipid phosphorus respectively (Eichberg and Dawson, 1965; Gonzalez-Sastre et al., 1971; Hauser et al., 1971; Hauser and Eichberg, 1973), alkyl ether phospholipids and phosphatidic acid (Eichberg et al., 1964; Sheltawy and Dawson, 1968).

5.2.5. Gangliosides

Small amounts of gangliosides (about 0.15% by weight) have been identified as myelin components (Suzuki et al., 1967, 1968; Suzuki, 1970). In rats the monosialoganglioside, G4, comprises 80 to 90% of the total myelin ganglioside; similar patterns are found for beef brain myelin (Suzuki et al., 1968). In addition, human central nerve myelin has been shown to contain the sialogalactosyl ceramide, G7, as a major component (Leeden et al., 1973). This ganglioside is less abundant in other species and is unusual because it is the only ganglioside known derived from galactocerebroside that has a fatty acid and an α-hydroxy fatty acid composition similar to myelin cerebrosides (Leeden et al., 1973). The role of these complex glycolipid molecules and of the large proportions of glycolipid components in myelin is unclear.

5.2.6. Developmental changes

Changes in myelin lipid composition during development have been reported (Cuzner and Davison, 1968; Horrocks, 1968; Dalal and Einstein, 1969; Norton and Poduslo, 1973b; Fishman et al., 1975; Norton, 1976). The detail available from these studies is limited, however, by the necessary selection imposed by myelin isolation procedures and the spectrum of developmental stages of the myelin sheath during the myelination period. The general trend emerging from the reports is that immature myelin contains less cholesterol, galactolipid, phosphatidylethanolamine, plasmalogens, and sphingomyelin and a higher relative amount of phosphatidylcholine. A sequential relationship between the glial cell, immature myelin, and the mature myelin sheath is further supported by the first detailed report of the lipid composition of isolated calf oligodendroglial cells and a plasma membrane fraction from this source (Poduslo, 1975).

5.3. Myelin proteins

5.3.1. General details

Proteins in isolated myelin comprise from 20 to 30% of the total dry weight of the membrane (Table 1). Virtually all the protein components of myelin can be defined loosely as proteolipid proteins because of their solubility in chloroform:methanol mixtures (Mokrasch, 1969). However, the classic proteolipid protein of myelin remains in the organic phase when electrolytes are added to a chloroform:methanol solution of myelin to precipitate the water-soluble basic protein (Gonzalez-Sastre, 1970). The development of aqueous solvent systems that completely dissolve myelin, such as phenol:formic acid:water (Mehl and Wolfgram, 1969; Mehl, 1972) and sodium dodecylsulfate (Hulcher, 1963; Waehneldt and Mandel, 1972), has enabled experimenters to resolve myelin proteins by polyacrylamide gel electrophoresis. The resulting separations on both phenol:formic acid:water and SDS gels show quantitative similarities. With the exception of the rat and other rodents of the suborders *myomorpha* and *sciuromorpha,* which have two resolvable basic protein components (Martenson et al., 1970; Dunkley and Carnegie, 1974; Carnegie and Dunkley, 1975), the fastest migrating protein band on electrophoresis corresponds to the basic protein and the second slower moving major band to the proteolipid protein (cf. Mehl and Halaris, 1970 with Waehneldt and Mandel, 1972). A number of slower moving high-molecular-weight protein components are resolved behind the main proteolipid protein band. For central nerve myelin from several species a weight ratio for the proportion of proteolipid protein:basic protein:remaining myelin proteins of 5:3.5:2 has been obtained by electrophoretic separation (Mehl and Halaris, 1970). A similar ratio for human myelin has been reported using a Triton X-100-salt fractionation technique to resolve the proteins (Eng et al., 1968). Differences in the protein composition of myelin isolated from different regions of the central nervous system have been reported (Wolfgram and Kotorii, 1968; Mehl and Wolfgram, 1969; Einstein et al., 1970; Morell et al., 1973; Lees and Paxman, 1974). Generally spinal cord myelin contains less proteolipid protein than cerebral myelin. However, studies to examine the protein composition of myelin isolated from various anatomical regions of the brain have failed to detect any significant differences in the proportions of the major myelin proteins (Lees and Paxman, 1974).

5.3.2. The proteolipid protein

The general term proteolipid protein (PLP) was coined to describe the water-insoluble protein-lipid material that could be extracted from brain tissue with chloroform:methanol mixtures (Folch and Lees, 1951). Myelin proteolipid protein can be prepared free of bound lipid to yield the apo-protein. The proteolipid apo-protein is soluble in both water and chloroform:methanol mixtures, is resistant to digestion by trypsin and pepsin, and has a low content of acidic and basic amino acid residues but a high proportion of neutral and aromatic residues (see Folch-Pi, 1972, 1973 for a recent review of brain proteolipids). Using chromatographic techniques, Gagnon et al. (1971) have consistently demonstrated that highly purified myelin proteolipid apo-protein contains 2% fatty acids that cannot be removed by prolonged dialysis and are thus considered to be covalently bound (Stoffyn and Folch-Pi, 1971). The proteolipid apo-protein necessarily exists in two forms: a lipophilic form that is soluble in organic media, and a hydrophilic one soluble in aqueous media. Conformational changes in structure accompany the interchange between these two forms (review, Folch-Pi, 1973).

The molecular weight of the PLP of myelin as reported in the literature ranges from at least 150,000 daltons (Eng et al., 1968) to a value of about 25,000 daltons (Agrawal et al., 1972). This disparity is due no doubt to the marked tendency of myelin PLP to aggregate in denaturant systems (Folch-Pi, 1973; Lees and Paxman, 1974), and especially when boiled with SDS in the presence of β-mercaptoethanol, a technique frequently used to achieve the total solubilization of myelin proteins (Morell et al., 1975). The amino acid sequence of the N-terminal 20 residues of the proteolipid apo-protein from human and rat brain myelin have been found to be identical (Nussbaum et al., 1974a,b) and these studies give molecular weight estimates of 23,500 and 23,623 daltons for the human and rat myelin proteolipid apo-protein respectively. Furthermore, these authors suggest that from the first 20 residues sequenced and from the overall amino acid composition, the PLP appears to have the necessary character for an intrinsic (bimodal) protein (Green, 1971). This suggestion is consistent with the speculation of Eng and colleages (1968) who concluded that hydrophobic bonding is chiefly involved in the maintenance of PLP in its membrane situation. The possibility of ionic interactions are not excluded, however. The physical and chemical studies on proteolipids suggest that their general role in membrane structure is to provide a protein-lipid core capable of sequestering a number of membrane lipid (or protein) components (Folch-Pi, 1973). Thus proteolipids can be considered as structural proteins in membranes. The role of the PLP in compact myelin and a suggestion as to its structural location in the myelin lamellae is discussed in section 9. It is significant to note, however, that peripheral nerve contains little chloroform:methanol-soluble proteolipid protein (Folch and Lees, 1951). The absence of a protein component in peripheral nerve myelin, which has a mobility similar to the major PLP of central nerve myelin, has also been demonstrated by electrophoretic techniques (Mehl and Wolfgram, 1969; Greenfield et al., 1973; Singh and Spritz, 1974; Brostoff et al., 1975).

5.3.3. The basic protein (BP)

The basic protein has attracted more research interest than all the myelin membrane system components. This is almost certainly due to the demonstration that myelin basic protein is the antigen for experimental allergic encephalomyelitis (EAE) (Kies, 1965; Kies et al., 1965). EAE may also be induced by peptides derived from the encephalitogenic basic protein (Carnegie and Lumsden, 1966; Lumsden et al., 1966; Nakao et al., 1966; Carnegie et al., 1967; Hashim and Eylar, 1969a,b; Palmer and Dawson, 1969; Westall et al., 1971). EAE represents a form of neuroallergy, and has been considered an experimental model for the human demyelinating disease, multiple sclerosis (review, Mackay et al., 1973). In contrast to the PLP, BP is water-soluble, and the ease with which it can be extracted from myelin probably represents another reason why attention has been focused on this component. The extensive literature on the encephalitogenic basic protein of myelin has been reviewed by a number of authors (Shooter and Einstein, 1971; Einstein, 1972; Eylar, 1972, 1973; Kies, 1973; Eylar et al., 1974; London, 1974).

Isolated myelin BP has an isoelectric point greater than pH 10.5 (Carnegie, 1971) and, from its amino acid composition, a calculated isoinic point between 12 and 13 (Eylar and Thompson, 1969). Thus the isolated basic protein exists as a polycation over virtually the entire pH range. However, reduction of the net charge on the molecule below 14, the figure for the net positive charges at physiological pH (30 at pH 4–5), results in detectable aggregation of the isolated protein (Liebes et al., 1975). These observations suggest that the major interactions of BP within the myelin membrane system will be ionic, although the possibility that specific regions of the protein molecule react hydrophobically will be discussed (see section 6).

The complete amino acid sequence for myelin basic proteins isolated from human (Carnegie, 1971) and bovine (Eylar et al., 1971) sources has been reported, and respective molecular weights of 18,500 and 18,300 daltons have been deduced. The small BP from rat myelin has also been sequenced (Dunkley and Carnegie, 1974), and a molecular weight of 14,100 daltons has been calculated. Incomplete amino acid sequences are available for a number of other species. The sequence data have been reviewed recently by Carnegie and Dunkley (1975). There is considerable conservation of sequence for the basic proteins isolated from different species (Carnegie and Dunkley, 1975) and, allowing for a 40-residue deletion in the rat small basic protein, its sequence shows considerable homology with other myelin basic proteins (Dunkley and Carnegie, 1974). Two basic proteins, P_1 (molecular weight 18,100 daltons) and P_2 (molecular weight 11,100 daltons) have been isolated from rabbit sciatic nerve (Brostoff et al., 1971; Brostoff and Eylar, 1972). The P_1 protein has a blocked N-terminal residue and appears to have substantially the same amino acid sequence as the basic protein from CNS myelin (Brostoff and Eylar, 1972). These workers have also shown that the P_2 protein has a substantially different amino acid composition from the P_1 component of peripheral myelin. A similar pair of basic proteins has been isolated from spinal root myelin of the pig (Uyemnura et al., 1970). The substantial conservation of primary structure in CNS basic proteins and P_1 from the peripheral nervous system suggests an essential role for these basic proteins within the myelin lamellae (see Dunkley and Carnegie, 1974; Carnegie and Dunkley, 1975, for a detailed discussion).

In both human and bovine basic proteins the N-terminal alanine residue is blocked by acetylation (Carnegie, 1969; Hashim and Eylar, 1969). Basic proteins isolated from other species also appear to have blocked N-terminal residues (Dunkley and Carnegie, 1974). A number of posttranscriptional modifications of the residues in the BP have been reported. The presence of mono- and dimethyl arginine at residue 107 (Baldwin and Carnegie, 1971; Carnegie, 1971) and the presence of phosphorylated serine and threonine residues have been reported. Heterogeneity of basic protein species has been demonstrated by electrophoresis and chromatography at alkaline pH (Martenson et al., 1969). This is not due to the extent of arginine methylation (Deibler and Martenson, 1973a,b) but has been attributed to two sites of modification, the phosphorylation of serine and threonine residues and the loss of carboxyl-terminal arginine (Diebler et al., 1975). Deamidation has also been suggested as a source of basic protein heterogeniety (Deibler et al., 1975). Although several possible roles for these modifications have been presented (Carnegie and Dunkley, 1975), their structural and functional significance remains unclear.

5.3.4. Other myelin proteins and enzyme activity

Wolfgram and Kotorii (1968) have commented that the observed amino acid composition of isolated myelin could not be duplicated by any combination of the proteolipid protein and basic protein amino acid compositions. These workers have suggested that at least a third protein with a high acidic amino acid content was required to make up the observed amino acid composition of myelin in central nerve tissue. The so-called "Wolfgram protein" has been identified as the insoluble residue obtained when myelin is extracted with chloroform-methanol (Gonzalez-Sastre, 1970). However, the myelin protein fraction isolated in this way shows considerable heterogeneity (Wiggins et al., 1974). A number of well-resolved high molecular weight components that migrate more slowly than the PLP are found on SDS gel electrophoresis of myelin proteins (Fishman et al., 1975). In recent reviews on myelin biochemistry, Norton (1975, 1976) feels that the case for the Wolfgram protein being a true myelin component is now fairly strong. Doubt, is expressed, however, about the other high molecular weight components detected in myelin gels. Norton (1976) suggests that such components may occur in myelin as a result of contamination by axolemma or oligodendroglial plasma membrane fractions. Although this is possible it would be surprising if compact myelin, in view of its origins, contained only three protein components. We believe that many of the high-molecular-weight protein constituents which run slower than the proteolipid protein band on gel electrophoresis will turn out to be native components of the oligodendroglial cell plasma membrane that are also localized in compact myelin in greatly diluted levels because of the continuity between the two systems. Agrawal et al. (1972) have identified another protein component in myelin gels. This is the DM-20 protein that migrates between the BP and the major PLP and may have properties of a proteolipid species.

Hydrolytic activity for 2′,3′-cyclic nucleotides has been demonstrated in isolated myelin samples (Kurihara and Tsukada, 1967). The proposed localization of this activity in myelin (Olafson et al., 1969) has resulted in 2′,3′-cyclic nucleotide-3′-phosphohydrolase (CNPase) activity being taken as a marker enzyme for myelin. Electrophoretic separation of myelin proteins, followed by a coupled assay for CNPase activity in the gel, has localized the enzyme to one of the high molecular weight bands (Braun and Barchi, 1972). Cholesterol ester hydrolase activity also seems to be established as a myelin-specific enzyme component (Eto and Suzuki, 1972, 1973). Other enzyme activities have also been demonstrated to be associated with purified myelin preparations, including L-alanyl-β-naphthylamidase (Beck et al., 1968; Riekkinen and Clausen, 1969); a non-specific esterase and perhaps lipase activity (Rumsby et al., 1970; Rumsby et al., 1973); a phosphoprotein kinase capable of phosphorylating myelin basic protein (Miyamoto, 1975); and a phosphoprotein phosphatase for myelin basic protein (Miyamoto and Kakiuchi, 1975). The metabolic significance of these enzymes is unclear and for activities other than the cholesterol ester hydrolase and the basic protein kinase and phosphatase must await the discovery of the natural substrates for these experimentally determined activities.

A glycoprotein has been demonstrated to be closely associated with purified myelin (Quarles et al., 1972, 1973a). Modification of the carbohydrate moiety during myelinogenesis has been proposed (Quarles et al., 1973b) and its location at the surface of the myelin-glial membrane has been shown (Poduslo, 1975a; Poduslo et al., 1976). A glycoprotein, P_0, is the major protein component of PNS myelin (Everley et al., 1973; Wood and Dawson, 1973). This component has a similar amino acid composition to the P_2 basic protein (section 5.3.3. and also see Greenfield et al., 1973; Wood and Dawson, 1973), but it remains to be proven that the P_0 glycoprotein is simply a glycosylated derivative of the P_2 protein (Carnegie and Dunkley, 1975).

5.3.5. Developmental changes
The period of myelination in the central nervous system is characterized by changes in the protein composition of the myelin sheath (reviews, Norton, 1975, 1976). The study of these developmental changes is very important as it can provide an indication of which components in myelin are integral constituents of the mature compact lamellae. The major findings are that myelin from immature brain tissue has a very complex protein composition compared with the mature membrane. As development proceeds through myelination and on the pattern alters so that the classic myelin proteins, the BP and PLP, emerge as the major constituents. These studies have been made on myelin from the mouse (Morell et al., 1972), rabbit (Einstein et al., 1970), rat (Adams and Osbourne, 1973; Zgorzalewicz et al., 1974) and man (Savolainen, 1972; Banik et al., 1974; Fishman et al., 1975). Inevitably, there are species differences as well as discrepancies within species due to variation in experimental techniques, modes of myelin isolation, and so on. The general results of this approach emphasize that even in immature myelin basic protein and proteolipid protein are present though at very reduced levels, and that the composition of the myelin material isolated early in development is more in keeping with that of the oligodendroglial cell plasma membrane (Poduslo, 1975b), as has been suggested by Davison and colleagues (Davison et al., 1966; Cuzner and Davison, 1968; Banik and Davison, 1969). The fact that BP and PLP become the major components of the myelin isolated from

mature brain indicates that these proteins are characteristic of the compact mature sheath membrane and must have an important structural role therein. The observation that high molecular weight components present in the immature membrane decrease during development agrees with the above suggestion that oligodendrocyte plasma membrane proteins become diluted out in mature myelin. With respect to this experimental approach, Norton (1975, 1976) reminds us that nervous tissue is heterogeneous in development and probably in myelin composition. The next requirement is a developmental analysis of myelin isolated from different regions of the CNS (cf. Banik and Smith, 1975). Studies on changes in the protein components of peripheral myelin during development have been far fewer and have revealed that the proteins of the system do not apparently change significantly in ratio during development (Zgorzalewicz et al., 1974; Wiggins et al., 1975).

6. Physicochemical studies on compact myelin and its constituents

As summarized in section 1, the application of such techniques as thermal analysis, spectroscopic and fluorescence methods, and monolayer and vesicle studies has greatly improved our understanding of the organization and dynamic nature of lipids and proteins in natural and artificial membrane systems. Some of these physicochemical techniques have been applied to compact myelin and its constituents.

6.1. Studies on compact myelin

Thermal and spectroscopic techniques have provided information about the state of lipid and protein components in intact myelin (Ladbrooke et al., 1968; Jenkinson et al., 1969). Neither hydrated central nerve myelin preparations nor total lipid extracts of myelin in water show any evidence of an endothermic phase transition when heated between $-40°$ and $100°C$. The absence of this feature in both isolated intact myelin and hydrated total myelin lipid fractions is significant because it indicates that the lipid phase in both systems is in a uniform fluid condition and that no lipid components in the membrane are in the rigid crystalline form. This fluidity implies that lipid molecules, and perhaps proteins, can diffuse laterally and rotate in the plane of the membrane unless specific inter- or intralamella forces operate to restrict such movement. The observed mobility of concanavalin A receptors in isolated myelin samples (Matus et al., 1973; Mattheiu et al., 1974) supports the observation of fluidity in the lipid phase. The immobilization of the concanavalin A receptors by glutaraldehyde observed in the same work is in agreement with the known action of the fixative in cross-linking protein and lipid molecules in myelin (Napolitano et al., 1967; Matus et al., 1973; Wood, 1973). The data of Rawlins (1973), showing that tritiated cholesterol injected into sciatic nerves is distributed homogeneously through all lamellae in myelin sheaths, can also be interpreted in terms of the fluid nature of the lipid phase which allows the sterol to equilibrate throughout the sheath. The ability of lipid and protein components in myelin to move in the plane of the membrane would help to explain processes such as myelination and the lengthening and increase in diameter of compact myelin sheaths during growth (Hirano and Dembitzer, 1967). Both of these events necessarily implicate the flow of membrane components within the glial-myelin system.

The maintenance of fluidity in the lipid phase in intact myelin appears to depend on the presence of cholesterol and water in the system (Ladbrooke et al., 1968). These authors also show that in isolated intact myelin 20% of the water is tightly bound to the polar groups of lipids and proteins. This water does not freeze at 0°C. A similar level of structural hydration has been observed for pure phospholipid systems (Chapman et al., 1967; Gary-Bobo et al., 1971). The importance of water in the maintenance of structural fluidity in myelin lamellae is revealed by the finding that lyophilized myelin samples and anhydrous total lipid extracts of myelin give distinct endothermic phase transitions. X-ray diffraction and electron microscope studies show that the removal of water from intact myelin results in the separation of lipid and lipoprotein lamellar phases (Finean, 1961). Dehydration may induce a transition to a rigid crystalline condition in the acyl chains of lipids in the membrane (Ladbrooke et al., 1968; Lecar et al., 1971). Apparently, lyophilization does not remove all the water from myelin. According to Ladbrooke et al. (1968) some water remains associated with protein, which can be removed by more rigorous drying.

From spectroscopic studies on intact myelin and its lipid components, Jenkinson et al. (1969) confirmed that cholesterol influences the mobility of fatty acyl chains in the membrane and thus cholesterol molecules are prevented from having complete isotropic movement. This observation is consistent with the proposed mechanisms for the interaction of cholesterol with lipids to create an intermediate fluid state as described in 2.3.2. Thermotropic phase changes in peripheral nerve were not observed using electron-spin resonance techniques (Schummer et al., 1975), again in keeping with the notion of a fluid lipid phase. However no evidence was presented in this study to indicate that the spin probe reports exclusively on the condition of the lipid phase in the myelin sheath.

Circular dichroism and optical rotatory dispersion studies have been interpreted by Moore and Wetlaufer (1973) as indicating that, despite large particulate light-scattering components, absorption differences between synaptosomal membranes, synaptosomal vesicles, and myelin isolated from rat brain reveal that these membranes contain proteins of differing three-dimensional structure. A further application of ultraviolet absorption techniques suggests that the extinction coefficients of tyrosine and tryptophan residues of myelin proteins are enhanced by structural features within the myelin lamellae (Gent et al., 1973) and that this enhancement is considerably reduced by dissolution of myelin preparations with lysophosphatidylcholine. Infrared spectroscopy of dehydrated myelin samples has provided data to suggest that the protein of the membrane is in an α-helical or random coil form and that there is no detectable β-conformation (Jenkinson et al., 1969).

The intrinsic fluorescence characteristics of isolated myelin preparations in suspension have been examined in depth (Crang et al., 1974; Crang, 1976; Crang and Rumsby, in preparation). The fluorescence spectra suggest that the major emitting tryptophan residues in myelin proteins are in very rigid environments in the membrane. The fluorescence characteristics of myelin differ significantly from published intrinsic protein fluorescence from various other isolated membrane systems, both in the detection of significant tyrosine emission from myelin preparations and a tryptophan emission maximum at 318nm (Wallach et al., 1970; Sonenberg, 1971; Avruch et al., 1972; Crang, 1976). These findings suggest that tyrosine and tryptophan residues of proteins in myelin are in significantly different structural environments from other membrane preparations studied to date. The detection of a large tyrosine fluorescence from isolated myelin samples is very characteristic of the membrane and it indicates that the structural features which are normally responsible for quenching tyrosine fluorescence within proteins are absent around a significant proportion of the tyrosine residues within the myelin lamellae. The tryptophan peak in central myelin samples is blue shifted by some 40 nm from the position of the tryptophan peak in isolated BP (Jones, 1975; Jones and Rumsby, 1975b). This suggests that in myelin the tryptophan residue of the BP is in a more structured environment.

Conclusions. Studies on intact myelin indicate that the lipids of the membrane are in an intermediate fluid condition that appears to depend on the presence of both cholesterol and water. This condition may allow lipid and protein components in myelin some mobility in the plane of the membrane. The protein of myelin has little detectable β-structure but exists in α-helix and random coil form. Ultraviolet absorption and fluorescence studies suggest that tyrosine and tryptophan residues of proteins in myelin exist in very structured environments.

6.2. Studies on myelin lipids

6.2.1. Phosphoglycerides

In compact myelin complex amphipathic lipid molecules are involved in interactions with neighboring lipids and proteins. These intermolecular interactions will be influenced by the nature, charge, and orientation of the polar headgroup of the lipid and by the chain length and degree of fatty acid unsaturation. These structural features of complex lipids, especially the nature of the fatty acids, will also influence what mesomorphic structural state the lipid will tend to adopt at body temperature. Our consideration of the mesomorphic condition of the lipid phase in myelin is restricted to homeothermic species. In poikilotherms the fatty acid composition of myelin lipids has been shown to change in response to environmental temperature changes (quoted in O'Brien, 1970b). The presence of cholesterol in the system and the level of hydration will also have a significant influence on the dynamic state of the lipid phase. Thus an appreciation of the fatty acid composition of individual complex lipid types in myelin is important for an understanding of their structural significance within the membrane. Chain length and the presence of *cis* unsaturated double bonds are two of the main features that influence the temperature at which the gel-liquid crystalline transition occurs for hydrated lipid systems (section 2). The phosphoglycerides of myelin differ notably from the sphingolipids in fatty acid composition (Fig. 8). Sphingomyelin, cerebrosides, and sulfatides contain longer chain length fatty acids on the average than are found in glycerol-based phospholipids such as PC, PE and PS, which are of shorter average chain length (C_{16}–C_{18}). PE and PS contain significant amounts of longer chain, polyunsaturated fatty acids.

Overall molecular length and the cross-sectional area of complex lipids are relevant parameters to the molecular organization of lipid components in myelin. In a membrane this cross-sectional area is influenced considerably by the packing between adjacent components in the lipid phase. Cross-sectional areas for lipids in the fully condensed state, as revealed by monolayer studies, do not necessarily correspond to the state of the lipid in a membrane. Interfacial pressures from 31 to 35 dynes/cm have been estimated for lipids in the outer surface of the erythrocyte membrane (Demel et al., 1975). Similar estimates for myelin are not available. Nevertheless, under such pressures, phospholipids such as those found in myelin would have cross-sectional areas of from 60 to 90Å^2 (Demel et al., 1967) depending on the chain length and degree of unsaturation. The molecular length of the molecule (Figure 9) in the membrane will depend on whether the headgroups are extended or folded over and whether the fatty acids are in the rigid or liquid-crystalline state. X-ray diffraction studies have provided the most accurate results on the dimensions of complex lipids in bilayer form. This area has been reviewed by Shipley (1973). At this stage the width of the lipid phase in compact myelin as revealed by x-ray diffraction (section 3.2.1.) is too narrow to accommodate a bilayer of lipid components, especially myelin sphingolipids, which have their fatty acids in a fully extended all-*trans*

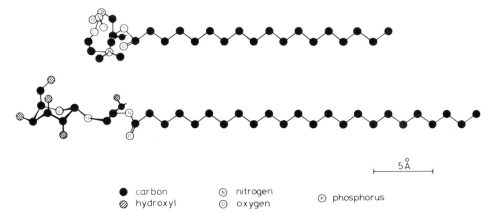

5 Å

● carbon   Ⓝ nitrogen

⊘ hydroxyl   ⊙ oxygen   Ⓟ phosphorus

Fig. 9. Scale diagrams of typical myelin phosphatidylcholine (compact headgroup) and cerebroside (galactose ring extended). Both lipids have their fatty acyl chains drawn in the all-*trans* extended conformation.

configuration. O'Brien (1970b) has invoked the idea of interdigitation of acyl chains across the bilayer to accommodate myelin lipids within the observed bilayer dimensions. However, this model was proposed before the concept of acyl chain fluidity in membranes was developed fully.

A dehydrated, mixed phospholipid fraction from myelin shows a broad endothermic phase transition extending from about 10° to 70°C, with a peak at about 49°C (Ladbrooke et al., 1968). The breadth of the transition, occuring in the absence of water, suggests that some phospholipids are in the liquid-crystalline form below body temperature at 37°C whereas others are still in the crystalline state beyond this point. Primarily this reflects the mixed fatty acid nature of the system, but it can be assumed that those lipids with long chain fatty acids, such as the spingolipids, will remain crystalline at the higher temperatures (Reiss-Husson, 1967). Cholesterol-free total lipid extracts of myelin in water (phospholipid + glycolipid) also show a broad endothermic transition with some components clearly in the liquid-crystalline form below body temperature. As both hydrated, total lipid extracts of myelin and intact myelin dispersed in water give no endothermic phase transition (Ladbrooke et al., 1968), it is evident that the addition of water and cholesterol to the phospholipids and glycolipids of myelin eliminates the phase transition. The observations of Jenkinson et al. (1969) that some degree of order exists in the hydrocarbon region of myelin are compatible with acyl chains being in an intermediate fluid condition where methylene groups up to about C_{10} along a fatty acid chain interact with the rigid sterol molecule (section 2.3.2.). Interactions of cholesterol with sphingomyelin and PC (Darke et al., 1972; Oldfield and Chapman, 1972a) also reduces the molecular motion of the choline portion of the headgroup compared with the mobility of this region in hydrated phospholipid bilayers in the fully liquid-crystalline form (review, Lee et al., 1974).

Plasmalogens, especially ethanolamine plasmalogens, increase in content in nerve tissue over the period of myelination (summaries, Eichberg et al., 1969; and Ansell, 1973) and thus must have some structural significance in compact myelin. Over 80% of the ethanolamine-containing lipids of myelin are the plasmalogenic form (Table 1). The double bond of the alk-1-enyl plasmalogenic group is in the cis-configuration (Ansell, 1973). Unfortunately, this class of lipid has not been widely studied by physicochemical methods compared with diacyl phosphoglycerides. Thus we know little of the structural features or function of these molecules in myelin.

Two reports have examined structural aspects of plasmalogens in relation to membrane organization. Shah and Schulman (1965) used natural choline plasmalogens of mixed hydrocarbon chain composition, while Paltauf et al. (1971) examined the behavior of synthetic choline and ethanolamine 1-alkyl-2-acyl and 1,2-dialkyl phospholipids. It should be noted that the molecules used by Paltauf and coworkers are not true plasmalogens for they lack the alk-1-enyl group although they do contain ether-linked hydrocarbon chains. Replacement of an ester linkage by an ether bond has no substantial effect on the molecular packing of a molecule in expanded or condensed monolayers. At 25 dyne/cm pressure monolayers of natural phosphatidalcholine have an average cross-sectional area per molecule of about 65Å² compared with figures of 55 Å² for dipalmitoyl PC and 75 Å² for PC from egg yolk (Shah and Schulman, 1965). The thermal characteristics of corresponding diester and diether forms were also similar (Paltauf et al., 1971), but here it would be of interest to see the effect of the true plasmalogenic form on the thermal properties of this lipid class. The studies reveal, however, that values for surface potential of plasmalogens and alkyl ether phospholipids are much lower when compared with the corresponding diester form (Shah and Schulman, 1965; Paltauf et al., 1971). The explanation advanced by both groups of workers for this finding is that the dipole on the carbonyl groups of ester links, which is involved in determining the surface potential, is reduced in plasmalogens where there is only one ester-linked fatty acid. The significance of this observation for the organization of plasmalogens in myelin is not clear. Paltauf and colleagues have commented that replacement of the 1-ester group by an ether-linked alkyl chain affects the chemical behavior of the whole phospholipid molecule for it renders the 2-ester bond more resistant to the action of phospholipases. Much more work needs to be carried out on native plasmalogens to define their membrane characteristics in relation to function in myelin.

6.2.2. Sphingolipids

The structural characteristics of sphingomyelin have been studied by Reiss-Husson (1967). At 25°C the fatty acids of a fully hydrated beef brain sphingomyelin are in the rigid crystalline form, but the lipid has an endothermic phase transition to the liquid-crystalline form at 40°C. At this point the bilayer thickness for spingomyelin is 40 Å while the cross-sectional area for the molecule is about 55 Å², as also noted for a nervonic acid ($C_{24:1}$)-containing sphingomyelin (Raper et al., 1966). These latter authors have found that nervonic acid is more characteristic of myelin sphingomyelin, and when this fatty acid is hydrogenated the cross-sectional area of the sphingomyelin is reduced to about 40 Å². This figure is more representative of a stearic acid-containing sphingomyelin. Shipley (1973) noted that the structural behavior of sphingomyelin is similar to PC, which it ressembles in headgroup characteristics, except that the thicker layer spacing and the smaller cross-sectional area are due to the more saturated and longer chain length fatty acids of the sphingomyelin class of molecules.

Cerebrosides and sulfatides account for 20% of the total myelin lipid on a molar basis (Table 1) and these glycolipids show a clearly defined increase in concentration of nerve tissue at the period of myelination (Eichberg et al., 1969). These two classes of lipids are distinctive for myelin. Recently they have also been detected in axolemma fractions (De Vries et al., 1976). Typically, the cerebrosides and sulfatides have a high content of C_{24} chain length fatty acids (Fig. 8). Of these, about 50% have an additional hydroxyl group on the C2 position of the amide-linked fatty acid. To a large extent the fatty acid composition will determine the structural properties of this class of lipids, and it has been shown that a dehydrated glycolipid extract from myelin has its endothermic phase transition between 110° and 140°C (Ladbrooke et al., 1968). As with other lipids, hydration of the polar regions lowers the temperature at which the phase transition occurs so that the gel to liquid-crystalline transition of hydrated cerebrosides is from 55 to 70°C. For hydrated sulfatides the transition occurs at around 75°C (Reiss-Husson, 1967; Clowes et al., 1971; Abrahamsson et al., 1972; summarized in Shipley, 1973). These figures reveal that even in the hydrated state the fatty acids of cerebrosides and sulfatides will be in a rigid all-*trans* arrangement at body temperature (37°C). However, no phase transition is shown by intact myelin or by hydrated total myelin lipids (Ladbrooke et al., 1968). The phase transition for hydrated cerebrosides can be lowered from 55 to 70°C down to around 20°C by the addition of other amphipathic lipids, such as egg PC for example, up to a 1:1 molar ratio (Clowes et al., 1971). Similarly, Oldfield and Chapman (1972a) have shown that cholesterol reduces the gel to fluid transition temperature for cerebrosides (and for sphingomyelin) up to a 1:1 molar ratio. It must be assumed that similar interactions of this type operate in intact myelin so that the fatty acids of the sphingolipids are in an intermediate fluid condition.

Monolayer and x-ray diffraction studies on cerebrosides and sulfatides have defined some of the structural characteristics of these glycolipids (Quinn and Sherman, 1971; Abrahamsson et al., 1972; Oldani et al., 1975). At 25 dyne/cm pressure, pure cerebrosides have a mean cross-sectional area of about 42Å^2 that approximates the value for the lipid in the fully condensed state. This low figure reflects the close packing that is possible for these molecules with their long, saturated fatty acids (Demel et al., 1967). The cross-sectional area of hydroxycerebrosides such as phrenosine is not very different from the cerasine type of cerebroside. Hydroxy fatty-acid containing cerebrosides have an increased surface potential compared with the cerasine form (Quinn and Sherman, 1971). Thus phrenosine has a greater surface charge than cerasine by about 90mV (Oldani et al., 1975). This reveals that small variations in molecular structure, such as the presence of the extra hydroxyl group, although they do not affect the average packing density of the molecules, can cause a change in surface potential. Overall, however, the absence of ester-linked fatty acids means that glycosphingolipids as a class of lipids have lower values for surface potential compared with the phosphoglycerides (Oldani et al., 1975). In the liquid-crystalline state (at 70°C) cerebrosides have a cross-sectional area of about 55Å^2, and the width of the bilayer formed by these lipids is about 50Å (Reiss-Husson, 1967; Abrahamsson et al., 1972). Sulfatide molecules have similar dimensions in the liquid-crystalline state. It is relevant that the bilayer width of these glycolipids in the liquid-crystalline condition, compared with the rigid all-*trans* form, is much more compatible with the dimensions of the lipid phase in compact myelin revealed by x-ray diffraction studies (section 3.2.1. and Fig. 3). Abrahamsson et al., (1972) reported that with natural sodium sulfatide in the gel state water layers of up to 44 Å in thickness between the polar end groups have been observed to be stable for several days.

Sulfatides are characterized by the presence of a sulfate group on C3 of the galactose ring. This group is fully ionized above pH 3.5 (Quinn and Sherman,

1971) and is more strongly acidic than the phosphate or carboxyl groups of phospholipids such as PE and PS (Abramson et al., 1967). The sulfate group is available for interaction with charged groups on other lipid components (Abramson and Katzman, 1968), with proteins (Demel et al., 1973; London et al., 1973; Banik and Davison, 1974; Jones, 1975; Jones and Rumsby, in preparation) and ions, especially calcium (Abramson et al., 1967; Quinn and Sherman, 1971).

Spectroscopic techniques have been used to confirm previous findings that the glycosidic bond in these glycolipids is in the β-configuration (Martin-Lomas and Chapman, 1973). However, a pertinent question that has not yet been solved for the glycosphingolipids is whether the galactose moiety extends out of the plane of the bilayer or whether it is oriented to lie in the plane of the membrane. Fully extended, the galactose group would be about 10 to 12Å in length. Results from optical and capacitance studies have been interpreted to indicate that since each polar region of a mixed cerebroside: PC bilayer is less than 8Å in thickness, galactose headgroups lie parallel to the plane of the membrane. Because the dielectric constant and the conductance are low in the polar regions of the same lipid mixture, it is believed that water is excluded from the polar groups. However, even with the galactose moieties parallel to the membrane, it is still possible for hydroxyl groups to orient at the outermost surface of the membrane. In this position they can hydrogen bond with the aqueous phase without an appreciable penetration of water into the non-polar regions. The data of Oldani et al. (1975) indicates that fatty acid hydroxyl groups in cerebrosides such as phrenosine are available to the surface to contribute to surface potential. Clearly, the sulfate on C3 of the galactose in sulfatides is available for interaction (section 5.4.2.), while labeling studies with galactose oxidase (Poduslo et al., 1976) indicate that the hydroxyl on C6 of the galactose in cerebrosides in myelin and in mixed lipid vesicles (Poduloso, 1975; Linnington and Rumsby, 1977) is accessible to enzymes. The glycosidic bond in these glycolipids in situ can also be reached by glycosidases. Dielectric measurements on myelin (Gent et al., 1970) indicate that much of the water associated with the membrane is more tightly bound than would be expected from interaction between bulk water molecules. The interaction of hydroxyl groups on the sugar moiety of these glycolipids with water molecules, which leads to a structuring of water in the vicinity and beyond, may play an important part in the ordering of the aqueous component in compact myelin. The binding of water to various sugars through hydrogen bonding to hydroxyl groups has been known for some time (quoted in Quinn and Sherman, 1971). However, more work on the physicochemical nature of these glycosphingolipids in membranes must be undertaken to identify their true structural role in the myelin sheath. It may be relevant that incorporation of cerebroside into bilayers of egg PC increases the resistance and stability of the system (Clowes et al., 1971). These authors comment that "this may be of advantage in myelin whose principal function is that of an electrical insulator." In the same study Clowes et al. interpret results to indicate that in a mixed cerebroside:PC bilayer there is little interdigitation of hydrocarbon chains and that this is also true for sphingomyelin mixed with PC.

6.2.3. Cholesterol

Hypotheses about how cholesterol interacts with amphipathic lipids in natural and artifical membrane systems have been summarized in section 2.3.2. Cholesterol is an abundant component of myelin and the thermal data of Ladbrooke et al. (1968) has established that together with water the sterol has a vital role in influencing the structural form of other myelin lipids by putting fatty acyl chains into an intermediate fluid state. Oldfield and Chapman (1972b) note that in this condition a phase transition, detectable only by laser Raman techniques (Lippert and Peticolas, 1971), is still present and that it occurs over a very broad temperature range and is not observed by thermal methods of analysis. It is also a non-cooperative phenomenon. The interaction of many lipid classes found in myelin with cholesterol has been investigated (section 2). Maximum effect occurs at a 1:1 molar ratio of cholesterol with the lipid, and the way in which the sterol affects the fluid nature of a lipid has been summarized in section 2. Thus sphingolipids with their long fatty acid chains of up to C_{24} in length can be brought into an intermediate fluid state. Cerebrosides have no polar headgroup for hydrogen bonding with the hydroxyl of cholesterol as has been proposed for phospholipids (Darke et al., 1972). According to Oldfield and Chapman (1972a) the headgroup may not be of major structural importance, since the principal structural requirement may be an amphipathic molecule with a fatty acid chain length greater than the 12 carbon atoms necessary for optimal stabilisation by van der Waals forces.

6.3. Studies on myelin proteins

The chemistry of the two major proteins of compact myelin—the proteolipid protein and the basic protein—has been described in sections 5.3.2. and 5.3.3. Various physicochemical techniques have been used to obtain information on the nature of these isolated components and their interaction with lipids. The results of such studies are described in this section for they have yielded information that may help us appreciate how such molecules are organized in the myelin sheath. The general extraction and chemical properties of these two major myelin proteins indicate that the basic protein may be located as an extrinsic protein in myelin while the proteolipid protein is more likely to be an intrinsic protein.

6.3.1. The basic protein

Physicochemical studies on isolated BP from myelin give little evidence for significant α-helical or β structure, and it is assumed that the molecule adopts a random coil conformation (Chao and Einstein, 1970; Block et al., 1973; Moscarello et al., 1974; Liebes et al., 1975). However, it has been proposed that the BP adopts a folded conformation in aqueous solution to give a prolate ellipsoid with approximate dimensions of 15×150 Å (Epand et al., 1974; Moscarello et al., 1974). This conformation could incorporate the U-bend suggested by Brostoff and Eylar (1971) and discussed by Eylar (1973) as a possible arrangement for the triproline sequence in the protein (Eylar et al., 1971). The β-bend is compatible with and is necessary to explain the observed axial ratio of about 10:1 of the isolated molecule (Eylar and Thompson, 1969; Chao and Einstein, 1970; Kornguth and Perrin, 1971; Epand et al., 1974; Liebes et al., 1975). The open structure for the isolated BP is consistent with the observed lability of the protein to proteolytic attack (Einstein et al., 1968; Carnegie et al., 1971; Eylar et al., 1971; Brostoff et al., 1974). Spectroscopic studies show that the temperature sensitivity of the absorption spectrum for the single tryptophan residue in purified BP is indistinquishable from that for tryptophan. Intrinsic fluorescence studies on isolated BP reveal that the emitting tyrosine and tryptophan species in the molecule are largely exposed to the solvent (Jones, 1975; Jones and Rumsby, 1975). Resonance energy transfer from tyrosine to tryptophan is inefficient in the protein, suggesting that the residues are all well separated from one another. Urea has little effect on the fluorescence properties of isolated BP, in accord with the proposed open conformation. This similarity of the fluorescence spectrum of BP to that given by free tryptophan has also been noted by Burnett and Eylar (1971).

Most of the work on the structure of basic protein has been performed with the protein in aqueous solution. By contrast, some studies have attempted to examine the structure of the protein in organic solvents. In media such as n-propanol, 2-chlorethanol and trifluoroethanol, for example, the α-helix content of BP is increased considerably (Kornguth and Perrin, 1971; Block et al., 1973; Liebes et al., 1975). The data of Anthony and Moscarello (1971a) and Palmer and Dawson (1969) on the induction of secondary structure in the BP after interaction with negatively charged lipid and detergent molecules emphasizes the fact that the structure of the isolated water-soluble BP may bear little resemblence to the conformation of the protein in myelin. Interaction of the anionic detergent SDS with isolated BP alters the fluorescence spectrum of the protein to indicate that the environment of the residue in the protein (which is normally exposed to the solvent when the protein is in aqueous solution) becomes considerably less polar (Jones and Rumsby, 1975). Initial interaction of SDS with BP is likely to be ionic, but subsequently the hydrocarbon chains of the detergent will tend to form hydrophobic interactions with any non-polar regions on the surface of the protein (Reynolds et al., 1970). In relation to this effect with SDS, it is relevant to note that fluorescence studies on isolated myelin samples (Crang et al., 1974; Jones and Rumsby, 1975; Crang, 1976; Crang and Rumsby, in preparation) that reveal an emission maximum for tryptophan of 318nm as opposed to 348nm for the isolated BP could arise because the tryptophan of BP in myelin exists in a far more structured environment. In myelin this could arise by association and interaction of the BP with lipids (see section 9). Basic proteins are also located in peripheral nerve myelin (section 5.3.3.) but their physicochemical properties have been less well studied than basic proteins from central nerve myelin. It will not be surprising if similarities between P_1 in peripheral and basic protein in central myelin sheaths are not apparent (section 5.3.3. for a recent review, see Carnegie and Dunkley, 1975).

6.3.2. The proteolipid protein

Proteolipid proteins are not located exclusively in myelin. Molecules having similar characteristics are found in white matter and in other membrane systems (Folch-Pi, 1973). The form in which the PLP of central nerve myelin is initially extracted is insoluble in water (Folch-Pi and Lees, 1951; Gonzalez-Sastre, 1970). Spectroscopic studies show that proteolipid proteins dissolved in chloroform:methanol have a high helical content. This is lost when the water-soluble apoprotein is formed, but the change in structure is reversible when these proteins are introduced again into a nonpolar environment (Zand, 1968; Sherman and Folch-Pi, 1970; Colacicco et al., 1972). A myelin protein having similar properties to the PLP has been isolated by Anthony and Moscarello (1971b) and Moscarello et al. (1973). Spectroscopic studies show that it can assume either an α-helical or a β-type structure depending on the conditions in solution. These studies reveal that the PLP has a marked conformational flexibility.

Conclusions. Thermal and spectroscopic studies on the lipid components of myelin indicate that at body temperature the anhydrous sphingolipids are in a crystalline condition. The addition of water and cholesterol up to a 1:1 molar ratio results in a significant reduction in the endothermic transition temperature, so that in intact isolated myelin no thermal transition is observed between 0 and 100°C. Physical studies have identified some of the packing properties of individual myelin lipids. The dimensions of hydrated sphingolipid bilayers above their endothermic transition temperature or in the presence of cholesterol are consistent with the observed bilayer dimensions of myelin. Physical studies on isolated basic protein and proteolipid protein have established details of the conformational changes induced in the basic and proteolipid proteins by interaction with lipids and in different solvent systems. The results indicate that both proteins possess conformational flexibility.

6.4. Interactions between myelin lipids and proteins

Two main experimental approaches have been employed in attempts to evaluate which lipids are associated with the major proteins of the myelin sheath, the proteolipid protein, and the basic protein. In an *"in situ"* approach mild methods are used to disrupt myelin so that complexes which may retain some native lipid-protein interactions can be released and analyzed. The other approach, using model systems, is to purify individual protein and lipid species and to study their interactions in monolayer, bilayer, and vesicle systems. Both approaches have produced important data for a consideration of myelin structure.

6.4.1. The "in situ" approach

The use of reagents that cause dissolution of myelin structure has not found wide application for investigating lipid-protein interaction in myelin, probably because a likelihood exists that membrane components will reorganize as a result of membrane disruption. Nevertheless, there are results from disruptive studies in the literature that may indicate native lipid-protein interactions in myelin. The PLP can be isolated from myelin (and from white matter) by chloroform:methanol mixtures as a complex that comprises 33% lipid and 66% protein (Folch-Pi et al., 1959). Much of this associated lipid can be easily removed by subsequent chromatography, partition, or extraction (Matsumoto et al., 1964; Cavanna and Rapport, 1967; Uda and Nakazawa, 1973). This lipid is thus only loosely associated with the PLP through weak lipid-lipid interactions. When the lipid is removed, the number of free -SH groups and the electrophoretic pattern of the protein does not change (Nicot et al., 1973). Analysis shows that this loosely associated lipid is mainly glycolipid and sphingomyelin (Uda and Nakazawa, 1973). Analysis of the PLP after removal of the loosely bound lipid reveals that some phospholipid with traces of cerebroside remain. This lipid is more tightly associated with the PLP and has been shown to be acidic phospholipid, especially PS with lower levels of phosphoinositides (Pritchard and Folch-Pi, 1963; Tennenbaum and Folch-Pi, 1966). The residual lipid can be largely removed from the proteolipid protein by dialysis of the complex against acidified solvent, which indicates the involvement of ionic interactions between the PLP and this more tightly bound lipid. Cleavage of nonpolar interactions between fatty acyl chains of the residual lipid and the PLP would also be facilitated by this dialysis procedure. When the tightly bound lipid is removed from the PLP, the remaining protein aggregates, as shown by a decrease in the number of free -SH groups (Nicot et al., 1973). Aggregation of the PLP at this stage can be prevented by the presence of detergents which presumably substitute for the more tightly bound lipid.

Far less work using this dissociation approach has been done with other myelin proteins. The polycationic nature of the basic protein (section 5.3.3.) suggests that ionic binding will be the major force for lipid-protein interactions in myelin. Such interactions would not survive acidic extraction procedures. In a milder approach, extracting basic protein from brain tissue with a sodium chloride-ethanol mixture at neutral pH, Saito et al. (1972) have found some carbohydrate and fatty acids (mainly $C_{18:0}$ and $C_{18:1}$) present in the preparation that also contained another acidic protein component. The lipid species present in this extract were not analyzed further.

To date minor protein components of myelin have not been well characterized. These protein components occur in low proportions. The fact that some of these components can be removed with non-ionic detergents such as Triton X-100 (Braun and Barchi, 1972; Lees et al., 1974; Torrie and Rumsby, 1975) suggests that they are mainly intrinsic in nature. Analyses of residual lipid associated with the detergent-extracted complexes are in progress.

Another mild experimental approach to disassembling myelin has been to digest isolated myelin samples with trypsin and acetyl trypsin and to follow lipid loss as the protein of the membrane is attacked (Kies et al., 1965; Raghavan et al.,

1973; Banik and Davison, 1974; Wood et al., 1974). This approach would seem promising at first sight but there have been conflicting results from various research groups with regard to the amount of basic protein digested, the release of lipids, and in interpretation. Banik and Davison (1974) reported that digestion of myelin with trypsin leads to a 15% loss of total protein and a significant loss of PI, PS, and sulfatide. Surprisingly, the myelin material left after digestion with the enzyme was found to be structurally the same as control myelin samples on examination by electron microscopy. On the other hand, Wood et al. (1974), found that the lipids released by trypsin digestion of myelin were only a few percent of the total in the sample and that the composition was no different in the distribution of cholesterol:glycolipid:phospholipid relative to myelin controls. This variation in the results no doubt reflects differences in experimental technique, such as levels of enzymes used, digestion conditions, and the use of myelin preparations of varying purity. Further experiments are required to clarify what occurs in this approach. We have argued that the use of lyophilized resuspended myelin (Raghavan et al., 1973) is not suitable in such work (section 4. 2.). The fact that trypsin can bind to isolated myelin samples would also seem to be a serious potential source of error, and the use of acetylated trypsin is preferred (Banik and Davison, 1974; Wood et al., 1974). In a double enzyme approach, Banik and Davison (1975) have noted that when isolated myelin samples are incubated with phospholipase A (specificity unspecified) and trypsin together, extensive loss of both basic protein and the proteolipid protein and the conversion of phospholipid to the lyso-derivatives occurs. As lipid is removed the PLP apparently becomes exposed to proteolytic attack. This observation is of interest since the isolated PLP has been reported to be resistant to attack by trypsin (Folch-Pi, 1972, 1973).

6.4.2. The model approach

In the model approach purified lipids and proteins, with at least one of the components of the system coming from myelin, are brought together and their interaction studied *in vitro*, using monolayer, vesicle, or black lipid film techniques. The interaction, nature, and purity of the components used in such studies can be carefully controlled and valuable information obtained about the way lipids and proteins may associate in myelin. In these model systems there has been a tendency to utilize synthetic nonmyelin lipids to investigate interactions with myelin proteins. Although this use of accurately defined components potentially gives a precise picture of lipid-protein interaction at a molecular level, the results are not as applicable directly to the complex mixture of lipid components in the intact myelin sheath that are defined in a stoichiometric ratio.

The *in vitro* interaction of purified proteolipid apoprotein with different lipids has been studied in a model system (Braun and Radin, 1969). The apoprotein was found to bind different classes of lipids to form two types of complexes. Anionic lipids such as PS, PI, cerebroside sulfate, and oleic acid bound to the apoprotein to form complexes which became insoluble in the aqueous phase. The minimum amount of acidic lipid needed to precipitate the apoprotein corresponded to about 40% of the basic residues on the protein molecule. A second type of nonprecipitating complex was formed between the proteolipid apoprotein and nonionic lipids such as cholesterol, cerebroside, and PC. The possibility that proteolipid apoprotein can interact with charged and noncharged lipids is suggested by this work. Using monolayer techniques, the binding affinity of the proteolipid apoprotein for acidic lipids such as cerebroside sulfate, PE, and PS has been demonstrated by London et al. (1974). The apoprotein had a lower affinity for cerebroside and sphingomyelin. In the same work it was noted that the apoprotein has a particular affinity for cholesterol under the conditions of the experiment. A subphase pH of 3 was used in the study because the apoprotein shows surface activity above pH 4 and precipitates above pH 7, near its isoelectric point. Interaction of the apoprotein with cholesterol was found to depend on the specific chemical structure of the sterol. Removal of the 3-β-hydroxyl group or the side chain at C17 in cholesterol decreased the interaction markedly (London et al., 1974). Although the use of monolayers of a single lipid species to study lipid interaction with the proteolipid protein are not directly applicable to the in vivo situation in myelin, the results confirm that the proteolipid apoprotein has an affinity for both charged and neutral lipid species, as noted by Braun and Radin (1969). This is likely to be a general characteristic of intrinsic membrane proteins that penetrate the lipid bilayer. Such proteins can be stabilized by headgroup interactions at the surface of the bilayer and by associations in the hydrophobic phase of the membrane. The other major myelin protein, the basic protein, was also studied in these experiments (London et al., 1974) but showed virtually no affinity for less polar lipids such as PE, PC, and cholesterol.

The monolayer technique has also been used by Papahadjopoulos et al. (1975a) to investigate lipid-protein interactions between myelin components. These authors have found that the presence of cholesterol in monolayers of bran PS does not decrease the penetration of proteolipid apoprotein into the lipid phase as much as it did for the other proteins studied. It was also found that proteolipid apoprotein increased the permeability of PS vesicles equally well in the presence or absence of cholesterol, while the presence of the sterol decreased the effect other proteins had on increasing vesicle permeability. The observation that PLP has no cholesterol effect is attributed to the high hydrophobicity of the proteolipid apoprotein molecule. Feinstein and Felsenfeld (1975b) have observed that the proteolipid apoprotein has a higher affinity for the nonpolar fluorescent probes anilino-naphthalene-sulphonic acid (ANS) and toluidinyl-naphthalene-sulfonic acid (TNS) than does the myelin basic protein.

Liposome techniques have been used to provide information about the interaction of the proteolipid apoprotein with lipids. Recently this topic has been studied in some detail (Gould and London, 1972; Vail et al., 1974; Papahadjopoulos et al., 1975a,b). The freeze-fracture technique has been employed by Vail and colleagues to examine the structure of vesicles of PS with and without incorporated apoprotein. Fracture surfaces of the liposomes incubated with the apoprotein showed a particulate appearance while those of liposomes alone were smooth. Particles observed in the fracture surfaces were 115 to 120 Å in diameter. This contrasts with the reported figure of 85 to 90 Å for the diameter of the PLP obtained from negative staining techniques (quoted by Vail et al., 1974). The authors ascribe the increased size seen in freeze-fracture observations to the effects of shadowing and coating during the freeze-fracture process. Since it is now accepted that the fracture plane cleaves through the center of the lipid phase in the membrane, Vail and coworkers conclude that apoprotein can penetrate the bilayer of the PS vesicles, confirming the view that it is an intrinsic protein. It should be noted that the value for the dimensions of the PLP as revealed by negative staining (85 to 90 Å is considerably higher than our calculated figure of 40 Å diameter (see section 9) and would correspond to the dimensions of a molecule with a molecular weight of 275,000 daltons. In our view, negative staining techniques may result in overestimating size due to the probable aggregation of components during sample preparation.

In later work, Papahadjopoulos et al. (1975a) have reported that the incorporation of the apoprotein into PS or PC vesicles increases the permeability of the liposomes to sodium, and that changes in lipid composition, surface charge, and cholesterol content have no appreciable effect on the behavior of the protein. The authors conclude that the interactions between the apoprotein and the lipid vesicles are primarily nonionic, probably hydrophobic. This conclusion has been further substantiated by thermal studies on the interaction of the apoprotein with lipids (Papahadjopoulos et al., 1975a,b). The effect of the apoprotein did not lower the temperature at which the phase transition of the lipids occured. Instead, it decreased the size of the endothermic peak at the phase transition. This result is interpreted (Papahadjopoulos et al., 1975b) as probably being due to the penetration of protein into the bilayer, where interaction with and stabilization of fatty acids occurs to prevent these acyl chains from participating in the thermal phase transition. The proteolipid apoprotein is clearly acting as an intrinsic protein in which "it binds a certain amount of lipid within the bilayer but it leaves the rest of the bilayer largely unperturbed" (Papahadjopoulos et al., 1975a). As revealed in these studies, the intrinsic nature of the PLP is enhanced by the recent reports on the location of particles in the fracture planes of myelin preparations discussed in section 3.2.2.

The myelin basic protein has the general characteristics of an extrinsic protein because of the relative ease with which it can be liberated from myelin by extraction with acid (Kies et al., 1965; Roytta et al., 1974). Thus the protein would be expected to associate with lipid by ionic interactions. Palmer and Dawson (1969) have noted that the BP interacts with acidic phospholipids such as PS and triphosphoinositide (TPI). A binding affinity for TPI was observed that resulted in the complete charge neutralization of both the BP and the lipid. The TPI:BP ratio was constant between pH 7.5 and 10, and it was noted that this ratio is approximately the same as is found in myelin. In a biphasic system, the BP formed complexes with PA, PS, and PI (Palmer and Dawson, 1969) and with sulfatide (Banik and Davison, 1974; Jones, 1975). The two-phase system has been used further by Jones and Rumsby (in preparation) to explore the interaction of the BNPS-skatole cleavage products of the BP with lipids. Ionic interactions of intact BP and of the two peptides formed from the BP with lipids such as sulfatide and PS were all located in the C-terminal fragment (residues 117 – 170), while the other fragment, the N-terminal portion (residues 1–116), did not interact. The work also showed that PS and PI are approximately twice as efficient as sulfatide at causing protein redistribution between phases. The results from this study suggest that the sites for lipid-protein ionic interaction in the BP are located in the C-terminal region of the protein. Interaction of BP with the anionic detergent SDS reveals that the protein is capable of reacting hydrophobically and ionically with amphipathic molecules (Jones, 1975; Jones and Rumsby, 1975a,b; Jones and Rumsby, in preparation). Gould and London (1972) have found that while BP interacts with liposomes of mixed nerve tissue lipids or of PC/PA to increase the rate of leakage of glucose, tryptic peptides of the protein have no such effect.

Basic protein-lipid interactions have been studied extensively using monolayer and vesicle techniques, and the association of the protein with lipid has been found to afford protection of the N-terminal region (residues 20 to 113) of the BP from digestion by trypsin (London and Vossenberg, 1973; London et al., 1973). The degree of protection depended on the lipid component of the monolayer and was in the order CS > myelin acidic lipids > PS = total lipid extract of myelin. It is suggested that the protected N-terminal sequence is involved primarily in hydrophobic interactions with lipid in which particular peptide chain regions of the basic protein penetrate into the hydrocarbon region of the lipid phase (London et al., 1973). This affinity of the BP for acidic lipids has been further confirmed by Demel et al., (1973), where the degree of interaction with monolayer lipids was also found to depend on the fatty acid chain length of the monolayer component species, suggesting that hydrophobic interactions between BP and lipids are also important. These studies show that the BP has markedly less affinity for interaction with neutral lipids such as PC, cholesterol, and cerebrosides. Similar conclusions have been emphasized by Papahadjopoulos et al. (1975a,b), who propose that the BP interaction with membranes involves initial surface binding by ionic forces followed by a partial penetration of the lipid phase. While such conclusions can be drawn from the monolayer studies, similar features have yet to be shown for the BP in myelin. The ease with which the basic protein can be removed by acid extraction reveals that the extent of hydrophobic association in the lipid phase cannot be very significant and that electrostatic interactions essentially determine the location of the BP in the membrane. In the work of Papahadjopoulos et al. (1975b) it is significant to note that interaction of BP and lipid had an effect on the thermal properties of the lipid phase. This effect was interpreted as being due to the protein increasing the mobility of acyl chains in its vicinity after an initial interaction with the membrane by ionic forces. The BP action in making fatty acyl chains more mobile contrasts with the earlier view of Eylar (1973) in which interaction of BP with lipids was seen as inducing rigidity of fatty acyl chains (see also Verkleij et al, 1974). Increased acyl chain mobility could arise because charged lipids interacting with the BP at discrete ionic binding sites along the protein molecule are structured further apart compared with their normal bilayer arrangement. However, the significance of these studies for myelin structure must be weighed against the use of nonmyelin, synthetic PG in these model systems.

Mateu et al. (1973) used x-ray diffraction and electron microscope techniques to study the interaction of basic protein from central and peripheral myelin with acidic phospholipids and sulphatides. Their results show that basic proteins can direct the formation of a lamellar phase having a 150 to 180 Å repeating distance. The structure proposed for the phase consists of BP sandwiched between two types of symmetric lipid bilayers, acidic phospholipid in the liquid crystalline form and sulfatide. The absence of acidic phospholipids or the sulfatide or the use of peptides T and L from the BP give rise to the formation of a more normal lipo-protein phase that lacks the large repeating unit. The authors believe that basic proteins act differently with the two classes of lipid. The affinity of one face of the protein for sulfatide (as has been shown in the monolayer studies discussed above) is indicated. It should be noted that when cholesterol is added to the system, as in myelin, the periodicity of the complex phase is halved and the effect of the BP is removed. Another interesting observation from this work is that the dimensions of the complex phase are 154 Å when BP from central nerve myelin is used, but are increased to 175 Å with BP from peripheral nerve. Such figures approximate those found for myelin from which the BP is purified. However, it should be noted that the major difference in the repeating unit of myelin in the PNS compared with the CNS is in the dimensions of the external apposition region (section 3.2.1.).

Conclusions. Two general experimental approaches have been considered which give information about the interaction of lipid and protein in myelin. An *"in situ"* approach has demonstrated that a proteolipid protein-lipid complex can be isolated that contains strongly bound acidic lipids, notably PS, and weakly bound neutral lipids. The high molecular weight proteins of myelin are considered to be intrinsic components by virtue of their extraction from myelin with nonionic detergents. Reports of the enzymatic disassembly of myelin are inconclusive with respect to any native lipid-protein interactions. A model approach has indicated particular features for the potential interaction of proteins with lipids in well-defined experimental systems. These experiments confirm the ability of the PLP to interact with both acidic and neutral lipids, and demonstrate a considerable penetration of PLP into the hydrocarbon phase so that neighboring hydrocarbon chains are removed from the bulk lipid phase transition in model systems. Freeze-fracture electron microscopy has also revealed that the PLP penetrates into the hydrocarbon fracture plane of PS liposomes. The apparent extrinsic nature of the BP is confirmed. However, an amphipathic character for the primary structure of the BP is suggested by monolayer protection studies and by lipid-interacting properties of the two BNPS-skatole cleavage products. The interaction of the BP with lipid vesicles suggests that this causes fluidization of the hydrocarbon chains of bound lipids. These studies have generally involved homogeneous lipid systems in contrast to the mixed lipid phase of intact myelin. The exact relevance of these model studies to the interaction of BP and PLP in myelin remains to be established.

7. Perturbants of myelin structure

Various investigators have examined the changes that occur in the structure of compact myelin when the membrane system is perturbed by some defined event. The effects of changes in physical conditions such as freezing, thawing, and dehydration, as well as the action of enzymes, ions, detergents, and solvents on myelin have been investigated. These data have yielded information on structural detail within the myelin membrane system.

7.1. Freezing, thawing, and dehydration

Finean and his colleagues have used x-ray diffraction and electron microscopy to examine the effects of freezing, thawing, and dehydration on compact myelin (summarized in Finean, 1961, and examined in detail by Joy and Finean, 1963). Freezing fresh peripheral nerve to below $-9°C$ causes a halving of the repeating unit in the lamellae, and it has been suggested that the two layers that constitute the repeating unit become identical in x-ray scattering power. Fixation of frozen and thawed material with osmium tetroxide results in a distribution of the stain which differs from that found with untreated myelin. In frozen-thawed samples the IPDL region appeared broadened and more intense. Finean (1961) and Joy and Finean (1963) concluded that the structural changes associated with freezing myelin are essentially hypertonicity effects, and that the 60 Å repeating unit that appears in frozen peripheral nerve myelin is related to collapsed regions of membrane layering. It is worth noting that the structural changes brought about by freezing peripheral nerve myelin remain reversible provided that the freezing temperature stays above $-9°C$. If taken below this temperature, an expanded structure is obtained when the myelin is thawed. The important differences between peripheral and central nerve myelin in respect to freezing and thawing behavior have been compared (Finean, 1961). The effects of freezing whole brain tissue on myelin morphology have been examined by Ansari et al. (1975) from an isolation viewpoint (section 4. 1.).

The effects of dehydration on myelin in situ and in isolation, either by air-drying or by lyophilization, have been investigated and the dehydrated structure subjected to analysis. The application of x-ray diffraction and electron microscopy procedures to dehydrated myelin has been examined in depth by Finean (1961). Removal of water from myelin causes a profound structural modification in the organization of the normal lipoprotein structure of the system. The modifications have been interpreted in terms of a collapsed layered structure that forms as water is removed along with separate, diffracting lipid phases of phospholipid (perhaps with cerebroside) and cholesterol which are liberated from the residual lipoprotein structure. Ladbrooke et al. (1968) noted that dehydrated myelin preparations show a clearly defined endothermic phase transition that is not apparent in the fully hydrated (over 30% water) membrane system. These thermal studies reveal that as the concentration of water in myelin is reduced from 30% to 5% an endothermic peak indicative of a phase transition in the system appears and increases in size. Thus removal of water has a considerable effect in disordering the normal lipoprotein organization of the myelin lamellae especially with regard to the release of lipid. Water is therefore a vital feature of the membrane for stability and organization of lipid and protein in the normal lamellar structure.

7.2. Ionic conditions and pH changes

The effect of soaking myelin in hypotonic and hypertonic salt solutions has been examined in some detail because the approach gives essential experimental data for the derivation of electron density profiles of the myelin repeating unit (Finean and Burge, 1963 and section 3.2.1.). Hypotonic salt solutions induce the separation of myelin lamellae at the external apposition (IPDL) as increasingly thick layers of water penetrate the region. Under such conditions, the cytoplasmic apposition region remains tightly apposed but shows a slight contraction in dimensions (Blaurock, 1971).

When compact myelin in intact peripheral nerve is soaked in hypertonic solutions such as Ringer's of 2 to 10 times the physiological concentration, the periodicity of the myelin sheath increases (Finean and Millington, 1957; Joy and Finean, 1963; Blaurock, 1971). This swelling is reversible and the periodicity can, under appropriate conditions, return to normal when the nerve is replaced in isotonic conditions. The increase in myelin periodicity that occurs during hypertonic swelling is less than in hypotonic swelling and has been interpreted in terms of expansion at the cytoplasmic apposition region (see Joy and Finean, 1963; Blaurock, 1971). It should be emphasized that while hypotonic swelling involves penetration by water, hypertonic swelling will involve ionic effects on the membrane. Finean (1961) has discussed this in relation to the presence of a component in peripheral myelin that is critically affected by ionic strengths in the region of 2 to 3. Protein is suggested as the most likely candidate to be influenced by ionic changes, as protein solubility is critically affected by changes in ionic strength. Ionic interactions between lipid and protein would also be affected by increasing ionic strength (e. g., Eng et al., 1968). Finean (1961) has stressed important differences between peripheral nerve myelin and central myelin with regard to hypertonicity effects.

Wolman and colleagues examined the action of hypertonic saline and calcium chloride solutions on homogenized white matter (Wolman and Wiener, 1965; Wolman et al., 1966; Bubis and Wolman, 1968; Wolman, 1971) to make deductions about structural organization in compact myelin. The action of hypertonic sodium chloride in dislodging protein and lipid in the system indicates that ionic associations are involved in lipid-protein complexes in membrane structures in white matter. It is acknowledged, however, that the quantitative date reported must be regarded as approximations as far as myelin is concerned (Wolman, 1971). The use of white matter as a source of myelin (only 50–60% myelin on a dry weight basic—Norton and Autilio, 1966), the probability of proteolytic enzyme activity in homogenates, and the very low gravitational speeds used to separate "aqueous extracts" inevitably introduce uncertainty about the conclusions. The behavior of isolated myelin samples (CNS) in solutions of differing cation valence has been examined by Leitch (1966) and by Leitch et al. (1969). An apparent isoelectric point at about pH 4 was indicated for the isolated myelin particles. Exposure of isolated myelin to mainly hypotonic solutions of increasing valence resulted in no pronounced changes in the ultrastructure of the membrane, although extensive beading of the lamellae in the IPDL region was observed in electron micrographs.

At acid pH, around 2, isolated myelin suspended in water coagulates, beginning to come out of suspension at pH 4 to 4.5 which is around the surface isoelectric point for isolated bovine central nerve myelin samples (Crang, unpublished observations). At alkaline pH beyond 10, the suspension becomes opalescent. The intrinsic fluorescence features of myelin in suspension over a wide pH range have been studied by Crang (1976). Between pH 2 to 10.9, the fluorescence emission from myelin samples shows a gradual increase in the intensity of the 326nm tryptophan-containing component relative to the tyrosine-containing component at 313nm. At acid pH, the 313nm emitting component is enlarged relative to the 326nm component. Myelin preparations appear to be "solubilized" above pH 12.2 by the action of the alkali in disordering the lipoprotein complex and perhaps effecting the hydrolysis of some acyl ester bonds. At pH 12.2, the tyrosine component of the fluorescence spectrum is virtually indiscernible and the tryptophan emission maximum is shifted to the red from 318nm to 326nm. Over the pH range 100,000g, supernatants of myelin suspensions can be recovered following incubation at the required pH for a set time. Little protein is recovered in such supernatants between pH 4 to 10, but levels increase below and above this range. Below pH 4, as would be expected from the known extractability of the basic protein with acid (Kies et al., 1965; Dickinson et al., 1970; Deibler et al., 1972; Sammeck and Brady, 1972), protein is recovered in supernatants and the intrinsic fluorescence of the sample is characteristic of the isolated basic protein of myelin (Jones and Rumsby, 1975). Protein release to the supernatant is most marked above pH 10 as the particulate nature of myelin suspensions is disrupted. The fluorescence emission of the material released at this alkaline pH is significantly different from that of isolated myelin at pH 7.2, indicating destruction of normal myelin. Data of Crang (1976) indicates that some maintenance of lipid-protein interaction is still present in the supernatant fraction above pH 10.

Worthington (1976) has reported briefly on the physical state of intact myelin in situ as a function of pH as revealed by x-ray diffraction. Over the pH range from 5 to 9 the diffraction pattern recorded is typical of normal, fresh nerve myelin. Marked changes in structure, as revealed by modifications in the x-ray pattern, were detected in myelin at pH values below and above this range.

Conclusions. Removal of water from intact myelin results in the formation of a residual lipoprotein structure with additional separate lipid phases that have dissociated from the original single phase complex. Water thus has a vital structural role in maintaining a single phase lamellar structure in normal myelin. The ionic and pH effects indicate that myelin structure is relatively stable over a pH range near neutrality but that the system is labile at acid and alkaline pH when individual components such as the basic protein are extracted or fragmentation of the system occurs. Such results would be expected from a structure in which ionic interactions are important in the maintenance of structural integrity.

7.3. Action of enzymes on myelin

7.3.1. Proteolytic enzymes

The action of proteolytic enzymes, especially trypsin, on myelin in intact nerve tissue has been widely studied (see Mokrasch et al., 1971), and the early work using this approach was followed up by the investigations on the neurokerratin component of nerve tissue (review, Wolman, 1969). Histochemical investigations have involved trypsin digestion of nerve tissue and of the myelin sheath to study the localization of components in the system (review, Adams and Leibowitz, 1969). Trypsin treatment of the myelin sheath in situ results in a loss of protein, patchy disintegration, and the release of lipid. It is well established that the isolated proteolipid protein of myelin is resistant to the action of trypsin but that the basic protein and other high molecular weight proteins are very susceptible to proteolytic attack (Shooter and Einstein, 1971; Eylar, 1972; Folch-Pi, 1972). Recent observations of Banik and Davison (1975) and Marks et al. (1975) indicate that the PLP in situ is subject to proteolytic attack with trypsin and acid proteinase.

More recently, the action of trypsin, acetyl trypsin, and other proteolytic enzymes on isolated central nerve myelin samples has been studied by several groups of investigators to examine the structural organization and lipid-protein interactions in the myelin sheath (McIlwain, 1973; Raghavan et al., 1973; Banik and Davison, 1974, 1975; Roytta et al., 1974; Wood et al., 1974; Marks et al., 1975; Schafer and Franklin, 1975). The results with trypsin, as discussed from a different viewpoint in section 6.4.1., show considerable variation. Observations range from finding a marked digestion of basic protein in isolated calf brain myelin (Raghavan et al., 1973), bovine brain myelin (Roytta et al., 1974), and rat brain myelin (Wood et al., 1974; Marks et al., 1975), through a much lower digestion (Banik and Davison, 1974) to no digestion of the protein at all in isolated myelin preparations treated with chymotrypsin (Schafer and Franklin, 1975). The inconsistency of the results is not entirely unexpected since substantially different experimental procedures were used by the different experimenters. As might be expected from the variation in data, conclusions about the location of various components in myelin also differ. Thus Wood et al. (1974) conclude that the basic proteins of rat brain myelin are located preferentially in the outer layers of the myelin sheath, are not uniformly distributed and do not play any significant part in stabilizing the bulk of the myelin membrane. Banik and Davison (1974) conclude that the basic protein may neither be located exclusively at the IPDL or at the MDL while, from their result showing no digestion of basic protein, Schafer and Franklin (1975) conclude that the basic protein is occluded entirely from the action of chymotrypsin in their isolated myelin samples. Comparison of the effects of trypsin and chymotrypsin may not be entirely valid, since although there is little difference in molecular weight, the enzymes differ in specificity; and it could be argued that sites specific for trypsin (adjacent to basic residues) are more accessible than those for chymotrypsin (adjacent to aromatic residues). In fact, London et al. (1973) have shown in their protection studies (section 6.4.2.) that chymotrypsin has a less effective action on basic protein-lipid complexes at the air-water interface compared with trypsin, which is found to act on the more polar parts of the protein molecule.

Roytta et al. (1974) examined the effect of neutral proteinase- and acid proteinase-rich fractions on isolated myelin. The action of the enzymes differed in that neutral proteinase at pH 7.0 had little effect on the structure of the membrane system, while incubation of myelin with the acid proteinase-rich fraction at pH 3.6 resulted in a complete loss of basic protein after 24 hours with the appearance of breakdown products from the protein. Digestion controls at pH 3.5 and 7.0 also differed since as expected at acid pH, basic protein was completely liberated from the membrane but no breakdown products were detected. At neutral pH there was no significant change in myelin composition. This observation conflicts with the results of Marks et al. (1975) that endogenous neutral proteinase activity in myelin gives a 30 to 46% breakdown of the basic protein and that addition of exogenous neutral proteinase resulted in the further degradation of some proteolipid protein. The picture remains unclear on neutral proteinase activity in isolated myelin since Reikkinen and Rumsby (1972) noted that only low levels of this enzyme activity remain associated with myelin after extensive purification, a view supported by the earlier works of Marks (1971). The action of acid proteinases on isolated myelin basic protein is known to result in the rapid digestion of the protein molecule (Einstein et al., 1968; Sammeck and Brady, 1972).

The action of trypsin on peripheral nerve myelin in situ and in isolation has produced more consistent results. Observations have been related to an understanding of the digestive processes that occur during Wallerian degeneration (Wood and Dawson, 1974a,b; Peterson, 1975, 1976). Isolated peripheral nerve myelin digested with trypsin gives breakdown products that seem to be derived from the major glycoprotein of the membrane (Everley et al., 1973; Wood and Dawson, 1974a,b). The same effects have been noted by Peterson (1975, 1976) in intact peripheral nerve treated with trypsin. The ultrastructural appearance of trypsin-digested peripheral myelin, studied initially by Fernandez-Moran and Finean (1957), has been reexamined recently (Peterson, 1975). Proteolytic digestion alters the appearance of the structure markedly so that splitting at IPDL and MDL regions, formations of granules within lamellae, and the gradual disappearance of IPDL region occurs. Peterson (1975) has interpreted his data to indicate the presence of trypsin-digestible proteins at both the MDL and IPDL sites. In a subsequent analysis of proteins during trypsin treatment of peripheral nerve, Peterson (1976) noted that the glycoprotein P_o and the two basic proteins P_1 and P_2 are all affected by the enzyme.

From the confused state of the results on the action of proteolytic enzymes on myelin it is clear that little structural detail for the membrane can be concluded from this approach until more uniformity of experimental detail yields consistent results.

7.3.2. Lipolytic enzymes

The action of crude phospholipase C on lyophilized reconstituted myelin samples has been described by McIlwain and Rapport (1971), though the preparation also contained sphingomyelinase. The preparation was effective in attaching the bulk of the choline and ethanolamine phospholipid but less of the serine component. Although the implications of the results for structure in myelin could be that the enzymes have access to such lipids, lyophilization of myelin probably causes the formation of separate lipid phases with a residual lipid-protein complex. From the standpoint of phospholipase attack, rehydration of the lyophilized material may leave these separate phases more exposed. The use of purified phospholipases and sphingomyelinase preparations in this approach is essential to ensure absolute specificity and because venoms contain other factors that aid membrane breakdown (Zwaal et al., 1973; Zwaal and Roelofson, 1976). Crude phospholipase C from *Clostridium perfringens* is known to contain a hemolysin component (Sabban et al., 1972). Details about the purification of phospholipases have been summarized by Zwaal and Roelofson (1976). In work on the red cell membrane (Colley et al., 1973) purified phospholipase C had no effect in digesting phospholipid in the intact membrane until sphingomyelinase was added.

More recently, Rapport and coworkers (Cole et al., 1974) have examined the action of purified phospholipase A_2 on lyophilized and resuspended myelin and on lipid extracts from myelin. Incubation with the enzyme for 2 hours at 37°C led to the hydrolysis of 75% of the PC, 74% of the PS, and 53% of the PE. Complete cleavage of all susceptible ester groups could be achieved by increasing the enzyme concentration and hydrolysis time. As with the action of crude phospholipases on myelin, the lysolipids produced by enzymatic hydrolysis remain in the membrane phase and are not released to the medium. This also applies to the action of phospholipases on the red cell membrane (Zwaal et al., 1973). In its action on dried and reconstituted myelin, the pure phospholipase showed a preference for PE compared with the plasmalogenic form, but this specificity was reversed for myelin lipids in aqueous suspension. Rates of hydrolysis of lipids in myelin were slower than in the purified form. This is ascribed to the relative insusceptibility of plasmalogens to phospholipase A_2 attack (section 6.3.1), but could also be due to the nature of the preparation where lipid may be complexed with protein. The exposure of lipid to the action of phospholipase A_2 in lyophilized, resuspended myelin samples can be considered different compared with the situation in native myelin in situ or in freshly isolated form (section 4.2.).

Insufficient data has been presented on the action of lipolytic enzymes on myelin to draw conclusions other than that lipid can be attacked in lyophilized reconstituted myelin samples. The use of purified phospholipases and fresh myelin is obligatory for studies on molecular organization and lipid accessibility in the myelin lamellae.

7.4. Solvents

7.4.1. On compact myelin in situ
The action of a variety of organic solvents on myelin in situ has been studied using physical techniques to gain information about the nature of lipid-protein interactions within the lamellae (Finean, 1961; Rumsby and Finean, 1966a,b,c; Glaisher, 1975). Short chain aliphatic alcohols such as methanol and ethanol have a modifying effect on the myelin lamellae and lipoprotein structure that has been interpreted as a dehydration action with the appearance of a con- tracted, residual lipoprotein lamellar phase and separate lipid phases (section 7.1. and Rumsby and Finean, 1966a). By contrast, alcohols such as butanol and pentanol, which are less miscible with water, cause the rapid and complete breakdown of all organized structure in myelin as revealed by x-ray diffraction and electron microscopy (Rumsby and Finean, 1966b; Glaisher, 1975). The ac- tion of these two alcohols in dissociating lipoprotein complexes has been re- ported for other membrane systems (Maddy, 1964, 1966; Rega et al., 1967; Zwaal and van Deenan, 1968a,b; Maddy and Kelly, 1970; Davidson et al., 1973). Water-immiscible chlorinated hydrocarbon solvents such as chloroform and car- bon tetrachloride are very effective in penetrating myelin and in producing the complete disruption of the system's structure. The general granularity seen in electron micrographs of chloroform-treated myelin might indicate a breakdown of the lamellae into small units (Rumsby, 1963; Rumsby and Finean, 1966a,c). A lamellar structure is preserved in myelin sheaths if nerve tissue is fixed with glutaraldehyde prior to extraction with chloroform:methanol mixture (Napolitano et al., 1967).

7.4.2. On isolated myelin
The gross solubilization effect of chloroform:methanol (1:1 or 2:1, v/v) on myelin has been known for some time, and the solubility of the membrane in this solvent mixture has almost come to be regarded as a criterion of purity (Hulcher, 1963). However, Gonzalez-Sastre (1970) regard the sediment from chloroform:methanol-solubilized myelin as being "Wolfgram protein." Clearly, the myelin basic protein is soluble in chloroform:methanol only by virtue of its interaction with lipids as addition of electrolytes renders the protein insoluble in the solvent (Gonzalez-Sastre, 1970).

The use of solvents to cleave the myelin membrane system into smaller struc- tural units has not been widely investigated because of the difficulty of finding mild solvents to achieve this aim without gross distortion of lipid-protein interactions in the system. Rumsby and Glaisher (1971); Glaisher (1975) have investigated the action of lower aliphatic alcohols, particularly *n*-butanol, on isolated central nerve myelin following the observation that certain aliphatic alcohols completely disrupt the layered structure of myelin. On extraction of aqueous suspensions of purified myelin with *n*-butanol, protein is distributed between the butanol phase and the interfacial precipitate. In contrast to the results of Maddy (1966) and Zwaal and van Deenan (1968a) with the erythrocyte membrane, where most of the protein is made water-soluble, very little protein is rendered water-soluble by this extraction procedure. With myelin, the distribution of proteins between the butanol, aqueous, and interfacial phases was found to depend on ionic strength and pH, and it is suggested that butanol extraction of myelin releases various lipid-protein units from the membrane that then partition according to their composition and association with specific lipids (Rumsby and Glaisher, 1971; Glaisher, 1975). The chemical composition of such units has yet to be determined. The action of *n*-butanol in releasing proteins from membranes has been attributed by Morton (1955) to the fact that the solvent has a balance of hydrophobic and hydrophilic groups and displaces lipid from protein, possibly by competing for binding sites. In myelin the actual mechanism is unclear. The contrasting results of primary alcohols on lyophilized myelin (Adams and Osborne, 1975) can be attributed to the absence of water and thus the native structure in the system.

Conclusions. The action of a variety of organic solvents on intact myelin have been studied by physical techniques. X-ray diffraction studies have demonstrated the rapid breakdown of organized lamellar structure in myelin by chlorinated hydrocarbons and by butanol and pentanol. The dissolution of myelin preparations by butanol suggests that structural breakdown may involve the release of lipid-protein complexes, which may have some relevance to myelin ultrastructure.

7.5. Detergent action on myelin

The action of detergents on myelin has not been widely studied, even though considerable use is made of the fact that the anionic detergent SDS solubilizes myelin proteins for gel electrophoresis (Greenfield et al., 1971; Agrawal et al., 1972; Mehl, 1972; Waehneldt and Mandel, 1972; Chan and Lees, 1974; Lees and Paxman, 1974; Fishman et al., 1975; Morell et al., 1975) and for gel filtration (Reynolds and Green, 1973; Crang and Rumsby, 1975; Wajgt et al., 1975; Crang, 1976). Yet a study of the detailed action of specific detergents on myelin offers an extremely valuable technique for the elucidation of lipid-protein interaction within the membrane. The solubilization of membranes by detergents has been reviewed by Helenius and Simons (1975).

The action of lysophosphatidylcholine (LPC) injected into myelinated peripheral nerves in vivo induces marked structural changes in the myelin sheath that depend on the concentration of the lysolipid (Hall and Gregson,

1971; Gregson and Hall, 1973). The detergent, believed to penetrate via incisural and paranodal extracellular spaces, causes complete disintegration of the myelin sheath in the locality of the injection site. Schwann cells and axons remain undamaged, probably due to their ability to metabolize the LPC. Hall and Gregson (1971) have followed the course of demyelination by electron microscopy. Detergent action on myelin was initially observed as a splitting of the IPDL, a thickening of the MDL, and an expansion of the periodicity. Subsequently, the expanded structure collapses to give a 40 Å layering in which the bands of electron dense material are of identical density. Uniformly dense layers with a 40 Å periodicity have been long recognized as a feature in electron micrographs of air-dried myelin sheaths in peripheral nerve and have been ascribed to lipid phases (Finean, 1961). The limitations of this in vivo system of detergent action make it difficult to elucidate the molecular action of LPC on myelin. The action of LPC on isolated myelin preparations has been studied by Gent and coworkers. In an initial study, Gent et al. (1964) suggested that LPC liberates structural lipoprotein subunits from myelin that appear as single components on electrophoresis and ultracentrifugation. This conclusion has been modified in more recent reports (Gent and Gregson, 1966; Gent et al., 1971a,b,c) which find that interaction of LPC with myelin is a two-stage process resulting in the formation of at least two products of solubilization. These can be separated by density gradient centrifugation (Gent et al., 1971a) and by gel filtration (Gent et al., 1971b). In the initial action of LPC with myelin the detergent is thought to substitute for a protein-lipid component of the membrane system, and examination of the material released in the first stage of solubilization shows that it consists primarily of high-density large-particle-sized components and also includes the basic protein that is predominately recovered in this fraction. The release of basic protein early in the solubilization process may be relevant to the action of the detergent on myelin in situ where the observed structural changes perhaps reflect the disorganization brought about by the removal of basic protein from the membrane. In the second phase of LPC solubilization of myelin, release of low density material including the CNP hydrolase activity occurs as the LPC:membrane ratio is raised. Complement fixing antigen activity, identified as galactocerebroside (Gregson et al., 1971), is associated with this phase. The initial binding of detergent to myelin seems to be accompanied by little release of lipid and protein. As more detergent binds, lipid and protein components are released from the membrane. This is in consistent with the concentration-dependent action of detergents on membranes (Helenius and Simons, 1975). A similar mechanism for the action of SDS on myelin has been proposed by Crang (1976) and Crang and Rumsby (1975).

LPC can be regarded as a natural detergent because it is derived from a native lipid component of the myelin sheath. The single fatty acid on the 1-position of the glycerol backbone will be approximately similar in chain length to that of native myelin PC. McIlwain et al. (1971) have examined the action of lysoPC lipids on lyophilized myelin and have noted that the LP-al C and reduced LPC are significantly more effective in "solubilizing" protein at a detergent concentration of 0.5% than is LPC. These authors have suggested that lipid release (cerebroside and cholesterol) parallels the release of protein. The stoichiometry of the protein release from myelin by LPC reported in the study differs from that of Gent et al. (1971a,b,c). The difference can be ascribed to the use of lyophilized resuspended myelin.

The action of other detergents on myelin has not been widely reported. The behavior of SDS in solubilizing myelin has been examined by Crang and Rumsby (1975) and Crang (1976). The release of detergent-lipid-protein structures followed by detergent-protein and lipid-detergent complexes occurs as the detergent concentration is increased relative to myelin. This agrees with the reported action of SDS in solubilizing other membrane systems (Engleman et al., 1967; Becker et al., 1975). Different myelin phospholipid classes are released as the detergent:myelin ratio is increased. PE is liberated at a low SDS:myelin ratio, PC and sphingomyelin at a slightly higher ratio, and PS and PI at an SDS:myelin dry weight ratio of 13. Release of PE before PC has also been noted for SDS solubilization of the red cell membrane (Kirkpatrick et al., 1974). The subsequent separation of SDS-solubilized myelin fractions by gel exclusion chromatogrphy permits reasonable resolution of basic protein and proteolipid components. The stoichiometry of the solubilization of myelin by SDS, as with LPC, seems to occur in a multistep process. Much higher levels of SDS compared with LPC are required for solubilization of myelin, but it should be noted that SDS effects the monomerization of protein and lipid components which does not seem to be the case for LPC. In relation to the use of SDS in solubilizing myelin for electrophoresis, Crang (1976) has stressed the need for a correct detergent:myelin ratio if reproducibility and accurate quantitation is required. For SDS in water, an SDS:myelin dry weight ratio of at least 7:1 is required for monomerization. In 0.26M sodium chloride, this ratio is decreased to 3.3:1. Inattention to this detail can lead to artifacts in gel electrophoresis and gel exclusion chromatography.

We are presently studying the action of a wide range of ionic and nonionic detergents on freshly isolated myelin as a means of releasing enzyme activity from the membrane for purification and also to investigate the structure of the system. Preliminary results have been reported (Torrie and Rumsby, 1975) and reveal that the levels of protein and lipid removed by nonionic detergents such as Triton X-100 and Lubrol WX are very low when compared with the solubilizing powers of anionic detergents such as SDS and desoxycholate (DOC) or a cationic detergent like cetyltrimethylammonium chloride (CTC). Certain nonionic detergents, such as Tween 20, extract negligible protein from myelin (see Helenius and Simons 1975, for a discussion of nonionic detergent action). The powerful effect of ionic detergents on myelin relative to nonionic detergents is regarded as indicating the significance of ionic interactions in the association of lipid with protein in the membrane system. While ionic detergents produce nearly total extraction of myelin proteins, the specific activity of nonspecific esterase and CNPH in extracts is low, due probably to inactivation of the enzymes (Lees et al., 1974; Torrie and Rumsby, 1975). Nonionic detergents release much more of the activity of these enzymes without serious contamination from the major myelin proteins (Sims and Carnegie, 1975; Torrie and Rumsby, 1975), suggesting that the enzymes are intrinsic membrane components and that strong ionic interactions are involved in holding the basic protein, and perhaps to some extent the proteolipid protein as well, in the myelin membrane system.

Eng et al. (1968) used extraction of myelin with Triton X-100 as the basis for removing different protein fractions from the membrane. These workers used Triton X-100 in 0.5M ammonium acetate to effect protein removal from myelin and found it to be as effective as using an ionic detergent. From our own results (Torrie and Rumsby, 1975) which show that concentrations of Triton X-100 up to 1% in aqueous form extract little protein from myelin, it is clear that the hypertonic ammonium acetate used by Eng et al. has a marked effect in disrupting lipid-protein interactions.

McIlwain et al. (1971) found that anionic, cationic, and nonionic detergents exert different effects on lyophilized myelin that are indicative of how the detergents attack the membrane. Ionic detergents were more potent than nonionic detergents. In this study nonionic detergents such as Triton X-100 and Cutscum had a more effective action in "solubilizing" protein (47% and 20% repectively) compared with the low values reported by others. The use of lyophilized resuspended myelin and a broad definition of solubilization mean that results cannot be compared with those found with freshly isolated myelin.

Conclusions. Detergents differ in their actions on myelin depending on their structure. Ionic detergents are most effective in solubilizing proteins from myelin, and this is taken to indicate the importance of ionic interactions in myelin lamellae. Nonionic detergents in water extract little protein, but certain myelin enzymes are recovered from the membrane and retain activity. Detergents offer a novel way of exploring the molecular arrangement of components in myelin but the mechanics of detergent action on the membrane must be defined precisely. Such information is available for LPC and SDS.

8. Labeling studies on compact and isolated myelin

8.1 Noncovalently reacting probes

Much of our current information on the disposition of both proteins and lipids in natural membranes has come from the application of labeling techniques to different membrane systems (Carraway, 1975; Hubbard and Cohn, 1976). In this approach probe molecules, which possess a distinguishing group so they can be identified after interaction, are reacted with a membrane. Probes interact noncovalently or covalently depending on their structure. Unreacted probe molecules are then removed and the treated membrane is examined to identify probe-binding sites. Results from this approach have been especially valuable in providing information about the asymmetry of lipid and protein components in membranes. These methods are now being applied to myelin.

The interaction of probes that bind noncovalently with myelin has not been widely studied since the information that can be derived is relatively limited. The fluorescent noncovalent probe molecules 1-anilino-8-naphthalene sulphonate (ANS) and toluidinyl-naphthalene sulphonate (TNS) have been reacted with isolated myelin samples and with purified basic and proteolipid proteins (Feinstein and Felsenfeld, 1975a,b). The authors identify the binding sites of such probes as the lipid phase in the isolated membrane system. The action of proteolytic enzymes that remove 6 to 30% of the myelin protein had no effect in reducing the number of dye-binding sites. As purified BP and PLP both show an affinity for the probes (BP higher than PLP), the authors argue that the results suggest that these protein-binding sites in isolated myelin are occluded from interaction with the probe by protein-lipid interactions within the membrane.

A number of the components of the myelin sheath, lipids as well as proteins, are antigenic (Rapport and Graf, 1969; Eylar, 1972; Alving et al., 1974; Niedieck, 1975). Antibodies formed to such components can be used as noncovalently reacting probes to examine the disposition of the antigenic molecules in myelin structure. There are, however, practical problems associated with this approach and these necessitate the use of correct controls to ensure that the antigen-antibody reaction with myelin is specific and is not due merely to the nonspecific binding of gamma globulin (Aarli et al., 1975), and also that components with a very similar structure to the antigen do not react with the antibody (e.g., histones reacting to antibodies to BP). Complement fixation techniques are used to provide an indirect measurement of antigen-antibody reactions. Further, considerable attention must be focused on the type of isolated myelin preparation used to look for antibody binding.

Myelin suspensions can be agglutinated by antisera to galactocerebroside and ganglioside (McMillan et al., 1971; Oxberry and Gregson, 1974; Gregson and Oxberry, 1976), but an antibasic protein serum did not cause agglutination (Oxberry and Gregson, 1974). The antigalactocerebroside antibody activity appears to be the major reactive component of antimyelin sera (Rapport and Graf, 1965; Gregson et al., 1974; Oxberry and Gregson, 1974). A cerebroside antibody is specific for galactocerebroside since it fails to react with glucocerebroside. These results suggest that both galactocerebroside and ganglioside are exposed on the surface of myelin preparations. This observation has also been made for myelin in nerve tissue cultures (Dubois-Daleq et al., 1970). As isolated myelin samples are believed generally to have their external apposition surface exposed (section 4.1.) this places at least some of the cerebroside and ganglioside at this site, whereas the BP is either not at the same site or is occluded for some reason (Oxberry and Gregson, 1974). Myelin samples solubilzed with lysophosphatidylcholine show increased complement fixation with antigalactocerebroside sera (Gregson et al., 1974), presumably as more antigen is exposed with the fragmentation of multilamellar sheets of membrane. McIlwain et al. (1971) believe that in lyophilized myelin redispersed in water less than 15% of the total galactocerebroside determinants available are exposed. Fragmentation of myelin with detergents increases the exposure of the determinant sites. Niedieck (1975) has noted that the sulfatide component of myelin is apparently nonimmunogenic. However, galactosyl diglyceride, a minor component of myelin (section 5.2.4.), is antigenic and thus antibodies to this determinant will be present in antimyelin sera. There have been few studies using the immunologic approach to locate phospholipids in myelin, except for the work of Guarnieri (1975) in which some binding of antisera to PI to myelin has been detected. Some 15% of the available PI was detected by the method, and it was also found that heating the membrane with antisera to 45°C and above increased substantially the exposure of PI sites to the antiserum, presumably as the structure of the membrane was opened up.

Fluorescein-, ferritin- and ^{131}I-labeled antibodies to the basic protein have been prepared in attempts to locate the position of this component in the lamellar structure of compact myelin (Rauch and Raffel, 1964; Kornguth and Anderson, 1965; Tomasi and Kornguth, 1968; McFarland, 1970; Herndon et al., 1973). However, the results have been disappointingly inconclusive, and there is disagreement between the various groups of workers using the approach. This disagreement is perhaps indicative of the problems associated with the use of immunologic techniques and the possibility that histones show a weak antigenic cross-reactivity with the myelin basic protein (Whittingham et al., 1972). More recently, it has been observed that antibody to a hydrophobic myelin protein cross-reacts with the basic protein and the two proteins share a common antigenic determinant (Wood et al., 1975). While most reports on locating BP have dealt with its situation in myelin, Herndon et al. (1973) have, like others, noted that intact myelin shows no uptake of fluorescent antibodies to BP. Labeling was, however, found with disrupted myelin located at the main dense line—the cytoplasmic apposition face.

Guarnieri et al. (1974) and Oxberry and Gregson (1974) have reported that isolated myelin preparations bind antibodies to basic protein. Unlike antigalactocerebroside sera, the anti-BP serum did not cause agglutination of the myelin sample in the work of Oxberry and Gregson, and the authors conclude this is because basic protein hapten sites are either located differently from cerebroside sites, perhaps by being deeper on the surface or are present only at low concentrations, or that agglutination is prevented because the sites are grouped so that cross-linking, and thus agglutination, is prevented. The radioimmunoassay results of Cohen et al. (1975) are also interpreted to indicate that the BP is exposed to anti-BP antibody in isolated myelin samples. At first glance these observations tend to indicate that the BP hapten sites are exposed in isolated myelin preparations. This result is of some concern because of the view expressed in section 4.1. that isolated myelin samples prepared by careful procedures from fresh, nonfrozen brain material preserve the structural features of intact myelin and that separation of the lamellae occurs predominately at the external apposition face. Guarnieri et al. (1974) have claimed the detection of BP haptenic sites indicates that the basic architectural features of myelin change on isolation. We note, however, that the same group (Cohen et al., 1975) use lyophilized myelin samples in their antibody-binding studies and, further, that frozen brain tissue is used as the starting material for isolated preparations. As described in sections 4.2. and 7.1., lyophilization alters the molecular structure of myelin preparations, as does freezing and thawing. An extension of the immunologic approach has been the development of radioimmunoassay techniques for measuring the myelin basic protein in serum. When used with appropriate controls, such methods provide a sensitive technique for assaying the protein (McPherson and Carnegie, 1968; Brostoff et al., 1974; Day and Pitts, 1974; Driscoll et al., 1974; Schmid et al., 1974; Cohen et al., 1975).

The application of histochemical techniques by Adams and colleagues to the problem of locating specific components in myelin can also be considered in this section on noncovalent binding probes. Histochemical staining has shown that myelin proteins can be divided into two categories. One type is rich in tryptophan, resistant to digestion by trypsin, extractable with solvent, and is thought to be proteolipid. The other protein group is basic, sensitive to trypsin, and can be stained with suitable anionic dyes (Adams and Bayliss, 1968). In a more recent report (Adams et al., 1971), a variety of nonspecific anionic stains such as phosphotungstic acid hematoxylin, trypan blue, and amidoblack have been reacted with central and peripheral myelin, and interaction with brain basic proteins has been established. Location of stains in myelin shows that the major part of the myelin BP in peripheral nerve is located at the MDL—the cytoplasmic apposition region. The authors claim that some reaction also occurs at the IPDL. However, due to the relatively unspecific nature of the reaction of such dyes (Dickenson and Aparicio, 1971), this does not provide a positive identification of BP at these sites.

8.2 Covalently binding probes

Various small probe molecules that bind covalently to membranes have been used to study the topography of the myelin sheath. Some of these reagents have been employed nonspecifically, while others are being used much more specifically to identify the location and accessibility of groups in myelin.

Mehl (1972) quotes unpublished work using a carbodiimide reagent $(R-N=C=N-R')$ that can interact with a variety of reactive groups ($-OH$, $-SH$, $-NH_2$) on myelin components. Polymerization of the BP was detected, and Mehl concludes that the BP protein molecules are therefore in relatively close association in the membrane. No experimental data to support this conclusion has appeared, but this report does suggest that the BP in myelin is at least accessible to this small reagent. Carbodiimides do not seem to have achieved much popularity as general covalently-binding probe molecules (Carraway, 1975; Hubbard and Cohn, 1976). Glutaraldehyde also interacts covalently to cross-link proteins and lipids in myelin (Napolitano et al., 1967; Wood, 1973) and thus acts as a probe. Clearly, it can penetrate to both apposition regions in compact myelin either through the lipid phase of the lamellae or along external and cytoplasmic appositions. Wood found cross-linking of myelin protein components together and of protein to lipid. All proteins in myelin—BP, PLP, and high molecular weight components—seem to be cross-linked by glutaraldehyde, though at low glutaraldehyde concentrations the BP and PLP components seem to be affected first.

Wood et al. (1975) have used 4, 4′ diisothiocyano-2, 2′-ditritiostil-benedisulfonic acid (tritiated DIDS) to label exposed amino groups in purified myelin. With normal myelin samples, both BP and PLP acquired label after interaction of the membrane with the probe for several hours. The BP had a lower specific activity than the PLP and the authors interpreted this as indicating BP is less accessible in the membrane for reaction with the probe than the PLP. It is argued that if the BP was exposed it would have a higher specific activity than the PLP. In isolated form, BP has a much higher specific activity than the PLP when treated with DIDS. Presumably this reflects the greater number of $-NH_2$ groups available on the basic protein and its open configuration compared with the PLP molecule, which has only a slight net basic charge (Mokrasch et al., 1971). The basic protein is thought by Wood et al. (1975) to be occluded in the intact myelin membrane. Such results are of considerable interest from the viewpoint of the molecular architecture of the myelin lamellae, since DIDS is believed to be a nonpenetrating probe (Cabantchik and Rothstein, 1974; Hubbard and Cohn, 1976) and the results suggest that both BP, and particularly the PLP, are to some extent accessible to the probe. Again, it should be noted that in this sort of work, as in all labeling experiments (cf. the immunologic work of Cohen et al., 1975), lyophilized and resuspended myelin samples will produce results substantially different from the native, freshly isolated membrane system.

Wood et al. (1975) extended their DIDS labeling studies to myelin isolated from chronic and acute cases of multiple sclerosis (MS). Their results show that overall labeling in myelin from chronic MS tissue is much the same as in normal unaffected material, but in acute MS myelin the basic protein component acquires much more label while the specific activity of the proteolipid protein is decreased. The authors interpret the results as indicating that in acute multiple sclerosis the BP in myelin becomes much more accessible to the reagent as a result of changes in normal membrane structure.

We have studied the covalently binding probes dansyl chloride and trinitrobenzene sulfonic acid (TNBS), both of which interact with $-NH_2$ groups on proteins and lipids (Crang, 1976; Crang and Rumsby, 1977; Rumsby and Grainger, in preparation). When freshly isolated myelin samples are treated with dansyl chloride, both phospholipid and protein are labeled. Some 12% of the myelin PE binds the probe but no distinguishable labeling of PS occurs, suggesting that this lipid class is occluded from the probe by interaction with protein or by headgroup orientation features. With the proteins no labeling of basic protein was detected, in agreement with the view that it may be occluded from interaction with the probe. Label was detected, however, in the PLP and in the high molecular weight components of the lamellae. On a mole-per-mole basis the high-molecular-weight protein components were more highly labeled than PLP. The nonpenetrating TNBS probe also binds to lipid and protein components of myelin.

A different approach to covalently attaching an identifiable label to exposed groups in the myelin sheath has been introduced by Poduslo (1975a), Poduslo and Braun (1975), and Poduslo et al. (1976). In this approach two enzymes, lactoperoxidase and galactose oxidase, have been used to attach radiolabels to accessible myelin groups. The authors perform the labeling reaction on the intact membrane in the dorsal column of the spinal cord. With this system, the arrangement of proteins in the outer surface of the sheath is explored. It should be pointed out that the membrane studied in such work is probably not fully representative of the lamellae in compact myelin and may be similar to the plasma membrane of the oligodendroglial cell, or at least intermediate in composition. Molecules as big as the enzymes used (lactoperoxidase has a molecular weight of 76,000 daltons) would be unlikely to be permeant across the lamellae or to pass through the tight junction arrangement now identified as partially sealing access through the external mesaxon (section 3.2.2.). These results are of great interest in relation to molecular organization in myelin membranes, since they reveal that with lactoperoxidase both PLP and several high molecular weight components are labeled whereas BP is not, though isolated BP can be labeled by lactoperoxidase. The conclusion drawn from this result (Poduslo and Braun, 1975) is that some high-molecular-weight protein components and the PLP (at least partially) are exposed on the outer surface of the membrane. Use of galactose oxidase, followed by reduction with tritiated sodium borohydride, has also revealed the labeling of carbohydrate residues associated with the high-molecular-weight protein fraction (Poduslo, 1975a; Poduslo et al., 1976).

We disagree, however, with the view of Poduslo and Braun (1975) that isolated myelin cannot be used for such labeling work "because the inner and outer membrane surfaces can no longer be distinguished." In our view the weight of evidence supports the opinion that myelin lamellae separate at the weakly-associated extracellular apposition (Finean and Burge, 1963; Lalitha and Worthington, 1975; McIntosh and Robertson, 1976), and this can be seen in electron micrographs of isolated myelin or swollen myelin sheaths. Thus, even though the thickness of myelin sheaths will be reduced during isolation, a common surface will still be presented to the medium—the extracellular surface of the system. Galactose oxidase labeling of isolated myelin samples shows that cerebroside molecules are exposed at the external surface of the lamellae (Linington and Rumsby, 1977). This has been also found by Poduslo (1975a) in the intact dorsal root preparation. It is clear that these labeling techniques are capable of providing exciting data on myelin architecture since the specificity of their interactions is known and can be accurately measured.

Conclusions. The use of probe molecules and modifying reagents to study the localization of the lipid and protein components in the myelin sheath has demonstrated both the problems and potential capabilities of this experimental rationale. Studies involving noncovalent labels clearly demonstrate the inherent problems associated with defining the precise location of these probes within the system. In contrast, covalently labeled components can be separated and their labeling pattern established. This latter approach would appear to be more valuable in the investigation of the surface architecture of myelin, though at this stage it is necessary to make careful assessment of the membrane permeability of the labeling reagent systems and the nature of the myelin preparations used in the studies. Despite these possible limitations, a fairly consistent picture of the surface architecture of myelin is beginning to emerge. Localization of carbohydrate-containing protein and lipid components at the extracellular surface of the myelin sheath has been demonstrated using enzymic techniques and immunologic studies. Location studies using covalently binding probes place the proteolipid protein in a position within the membrane system that is more accessible to a number of reagents than the basic protein.

9. Molecular organization in compact central nerve myelin

In this final section an attempt is made to summarize the important conclusions on molecular organization in myelin and to produce a model for the system. The emphasis in the section, and in the model, is on myelin in central nerve tissue since its chemistry, especially with regard to membrane proteins, is better defined than for peripheral myelin. Nevertheless, we consider that many of the principles will apply to both membrane systems.

9.1. Myelin components

A prerequisite for an appreciation of molecular structure in compact central nerve myelin is that the chemistry and proportions of the individual lipid and protein constituents be clearly defined and also that substantial physical information about the properties of the molecules that make up the membrane be available. This information has now been accummulated for myelin and is summarized in sections 5 and 6. The compositional data for isolated myelin is believed to provide a reasonable indication of the components and their relative proportions within compact cerebral myelin in situ. Developmental studies on myelination (sections 5.2.6. and 5.3.5.) have revealed a number of lipid and protein components that are concentrated in compact myelin and are thus considered to be "myelin-typical" (Norton, 1975, 1976). Although such components may be present in low levels in "early" myelin and/or in the plasma membrane of the oligodendrocyte, they show a significant increase in proportion throughout myelination in relation to other components. They can be regarded as having a fundamental structural role in compact myelin. Within this definition cerebrosides, cerebroside sulfate, ethanolamine plasmalogens, and the proteolipid protein are considered to be myelin-typical. In addition, the basic protein has been shown to be myelin-specific.

With regard to the relative proportions of BP and PLP in compact cerebral myelin, the PLP and BP occur in approximately equimolar proportions. This stoichiometry, which holds for a number of animal species, has been emphasized by Mehl and Wolfgram (1969) and Mehl and Halaris (1970). Although the molecular weight values for the PLP and BP used to derive this ratio have since been shown to be too high by amino-acid analyses (Nussbaum et al., 1974a,b; Carnegie and Dunkley, 1975), their approximate equimolar proportions still hold. Mehl and Halaris (1970) have expressed this observation in terms of lipid-protein subunits of myelin that contain either the BP or the PLP. We consider that the stoichiometry of BP to PLP could have important structural implications in compact myelin, and in a later subsection we will utilize current findings on myelin structure to develop a subunit concept for compact myelin. This subunit consists of a 1:1 molar BP to PLP complex that provides a structural nucleus for binding lipid. The whole comprises a lipo-protein subunit in the membrane.

9.2. The lipid phase of compact myelin

The data from physiochemical studies on myelin (section 6) have established that the lipid phase of compact myelin is in an intermediate liquid-crystalline condition. This state depends on the essential presence of both cholesterol and water in the system (section 2.3.3.). For central nerve myelin from a variety of species, the molar proportions for cholesterol:phospholipid:glycolipid vary between 4:3:2 and 4:4:2 (Table 1, and Norton, 1975, 1976). From these data, Finean (1953) and Vandenheuvel (1963) have suggested that cholesterol and phospholipid form a structural 1:1 molar complex in myelin. From metabolic studies on central nerve myelin, Eng and Smith (1966) have proposed an empirical 1:1 molar relationship for the sum of ethanolamine phosphoglycerides (including plasmalogen) plus sphingolipids to cholesterol. However, because the lipid phase of myelin is in an intermediate fluid state, a principal requirement for cholesterol in the membrane must be to interact with those lipid molecules in the system which, alone in hydrated form, are crystalline at body temperature. These are the sphingolipids, principally the glycosphingolipids. Thus, as has been demonstrated by Oldfield and Chapman (1972a) in model systems, sphingolipid-cholesterol complexes in myelin would appear to be an essential structural feature of the myelin lipid phase. The formation of such complexes would involve some 50% of the total available cholesterol in the membrane system. The rest is presumed to interact with phospholipid. However, insufficient cholesterol remains to complex all the phosphoglycerides in a 1:1 molar ratio. The absence of any detectable phase transition in compact hydrated myelin indicates that acyl chains of uncomplexed phospholipid are in a liquid-crystalline condition (section 2.3.2.). The observation in a model system that interaction with phosphatidyl choline increases the liquid-crystalline state of cerebrosides (Clowes et al., 1971) may be highly relevant to the maintenance of an intermediate fluid condition in the myelin lamellae.

9.3. Localization of basic protein and proteolipid protein in myelin

An exercise that seems not to have been published previously is to relate the molecular weights and characteristics of the two major protein species, the BP and the PLP, to their probable location within the myelin lamellae of the compact sheath. The information for this exercise is readily available in the literature which has been summarized in the preceding sections.

9.3.1. The basic protein

The myelin basic protein has the characteristics of an extrinsic membrane protein (section 5.3.3.). However, five regions of 8 to 10 residues in the BP amino acid sequence lack positively charged groups and are thus capable of penetrating the hydrocarbon region of the bilayer (Eylar, 1970, 1973). The structural significance of these hydrophobic regions has been given an experimental basis by the monolayer penetration studies of London and coworkers, as outlined in section 6.4.2., while the perturbation of the lipid phase by BP in liposome systems has been reported by Papahadjopoulos et al. (1975a,b). In this latter study, however, a significant effect on the fluidity of fatty acyl chains was observed only at protein concentrations far in excess of the levels occurring in myelin.

As an extrinsic membrane protein there are two possible locations for the BP in myelin, either at the extracellular or cytoplasmic apposition surfaces. Evidence from x-ray diffraction and electron microscopy (sections 3.2.1 and 3.2.2) indicates that the cytoplasmic and external apposition regions in compact myelin are maintained by very different structural interactions. The stability of the cytoplasmic apposition region to hypotonic swelling and its separation in hypertonic media suggest that ionic forces are involved in this apposition. This is consistent with a localization of the BP in the cytoplasmic apposition, where extrinsic character and ability to interact ionically with lipids could well provide the necessary intermembrane interactions. Chemical labeling studies using a variety of probe systems (section 8.2.) have provided the most consistent indication for the localization of the BP in myelin even though the consensus from other techniques such as histochemical, immunological, and extraction is equivocal. The inability to detect labeling of BP in intact myelin preparations using nonpermeant labeling systems provides the most conclusive evidence that the BP is localized at the cytoplasmic apposition in the compact sheath (Podulso and Braun, 1975). Also, reaction of freshly isolated central nerve myelin with dansyl chloride does not result in detectable labeling of the BP (Crang, 1976; Crang and Rumsby, 1977). This result further suggests that BP is at the cytoplasmic apposition region where reactive $-NH_2$ groups are occluded from interaction with the dansyl chloride probe. Some authors have suggested that BP may be located partially at both the external and cytoplasmic surfaces. However, we believe the exclusive localization of the BP at one surface is more likely. Support for this view is derived from general observations on other membrane and cellular systems where, for example, the intracellular ionic status of the cell is maintained within strictly defined limits compared with the extracellular fluid and where lipid asymmetry studies (on the red cell) have located negatively charged lipids almost exclusively at the cytoplasmic surface. Also, the biosynthesis of extrinsic proteins for both external and internal surfaces of a plasma membrane system (from which myelin is derived) would appear to require different mechanisms for transport to their membrane location(s).

9.3.2. The proteolipid protein

The proteolipid protein of central nerve myelin has the characteristics of an intrinsic membrane protein (section 5.3.2.) and, considering its lipophilic properties, the lipid phase of the myelin lamellae is the only reasonable site for its location. However, early freeze-fracture studies on intact myelin failed to demonstrate the existence of the classic membrane-associated particles that would indicate the existence of the PLP in the hydrocarbon fracture face. Furthermore, the x-ray diffraction data of Caspar and Kirschner (1971) has been interpreted as indicating that the penetration of the hydrocarbon region of the myelin bilayer by protein could not exceed 10%. However, more recent freeze-fracture studies on myelin and the interaction of the PLP with liposomes have both demonstrated particles at the hydrocarbon fracture face (Fig. 6). These findings are consistent with a localization of the PLP within the lipid phase. In addition, the calculations derived later in this section show that the total amount of PLP present in myelin, assuming the molecule to be spherical, can be completely buried in the lipid phase and yet only involve an overall penetration of the hydrocarbon region of about 9%.

Although hydrophobic interactions between the PLP and lipid predominate, the likelihood of ionic interactions within the membrane are also indicated (sections 6.4.1. and 6.4.2.). Thus partial exposure of the PLP may occur at either the external or the cytoplasmic surfaces. Chemical labeling studies have shown consistently that the PLP is more accessible to a variety of probes than is the basic protein (section 8.2.). In particular, the results of Poduslo and Braun (1975) with intact myelinated nerve preparations have demonstrated that the PLP is partially exposed at the extracellular surface.

9.3.3. Other myelin proteins

The labeling studies discussed in section 8.2 have provided firm evidence on the location of high-molecular-weight protein components in central nerve myelin. The major glycoprotein of myelin has been located at the extracellular surface of the membrane. On a molar basis compared with the PLP the high molecular weight components of myelin incorporate more label (Poduslo and Braun, 1975; Crang, 1976; Poduslo et al., 1976; Crang and Rumsby, 1977). The results reveal that the majority of the high molecular weight protein components are located at the extracellular surface of compact myelin.

9.4. Calculations for the molecular structure of myelin

In order to calculate the occupation of the myelin bilayer by the PLP and BP, the following assumptions are made:

1. Density of isolated myelin = 1.09 g/cm^{-3}(isopycnic with 0.656 M sucrose)
2. Width of hydrated myelin bilayer = 75 Å (Caspar and Kirschner, 1971)
3. Hydration of myelin = 40% (Finean, 1960)
4. Molecular weight of basic protein = 18,400 (Eylar, 1970)

5. Molecular weight proteolipid protein = 23,500 (Nussbaum et al., 1974a,b)
6. Partial specific volume of the basic protein = 0.72 cm$^3 \cdot$g^{-1} (Epand et al., 1974).

9.4.1. Area of myelin bilayer

$$\text{Area of 1 g hydrated myelin bilayer} = \frac{1}{\text{width} \times \text{density}}$$

$$= \frac{1 \times 10^{24}}{75 \times 1.09} \text{ Å}^2$$

$$= 1.223 \times 10^{22} \text{Å}^2$$

9.4.2. Basic protein

$$\text{Weight of one molecule of BP} = \frac{\text{molecular weight}}{\text{Avogadro number}} = \frac{18,400}{6.02 \times 10^{23}} \text{ g}$$

$$= 3.056 \times 10^{-20} \text{g}$$

$$\text{Volume of one molecule BP} = 3.056 \times 10^{-20} \times 0.72 \text{cm}^3$$

$$= 2.201 \times 10^{-20} \text{ cm}^3$$

Assume an axial ratio of 10:1 for the basic protein (Epand et al., 1974) so that length = 20 times the radius.

$$\text{Volume of cylinder} = \pi r^2 h$$
$$\text{for the BP } h = 20r, \text{ so:}$$
$$\text{Volume of BP} = 20\pi r^3$$
$$r^3 = \frac{2.201 \times 10^{-20}}{20\pi} = 3.503 \times 10^{-22}$$
$$r^3 = 7.05 \times 10^{-8} \text{ cm}$$

therefore

$$\text{Diameter of BP} = 14.1 \text{ Å, length BP} = 141 \text{ Å}$$
$$\text{Projected area for BP} = 14.1 \times 141 \text{ Å} = 1988 \text{ Å}^2.$$

9.4.3. Proteolipid protein

$$\text{Weight of one molecule PLP} = \frac{23,500}{6.02 \times 10^{23}} \text{ g}$$

$$= 3.904 \times 10^{-20} \text{g}$$

Volume of one molecule PLP = mass $\times$ partial specific volume. Typically, the range is between 0.70 and 0.75 cm^3 gm^{-1}, so take the mean value of 0.725 for PLP.

Volume of one molecule PLP = $3.904 \times 10^{-20} \times 0.725$ cm^3
$$= 2.830 \times 10^{-20} \text{ cm}^3.$$

Assume the PLP is a spherical molecule, then volume $= \frac{4}{3} \pi r^3$

$$r^3 = \frac{2.830 \times 10^{-20} \times 3}{4\pi}$$

$$= 6.756 \times 10^{-21}$$
$$r = 1.890 \times 10^{-7} \text{ cm}$$

Thus the diameter of the proteolipid protein is 38 Å.
Cross-sectional area of the PLP = $\pi \times 19^2 = 1134$ Å^2.

9.4.4. Occupation of myelin bilayer by protein
1 g wet weight myelin = 600mg dry weight
Protein = 27% dry weight myelin = 162 mg per g
Basic protein = 31% of total protein = 50.2 mg per g
Proteolipid protein = 42% of protein = 68 mg per g

The latter three values are taken from Crang (1976).

9.4.4.1. Basic protein

$$\text{Number of BP molecules per g} = \frac{0.0502 \times 6.02 \times 10^{23}}{18,400}$$

$$= 1.643 \times 10^{18} \text{ molecules}$$

Total area occupied by BP = $1.643 \times 10^{18} \times 1988$ Å^2
$$= 3.266 \times 10^{21} \text{ Å}^2$$

$$\frac{\text{Area occupied by BP}}{\text{Area of bilayer}} = \frac{3.266 \times 10^{21}}{1.223 \times 10^{22}} = 0.267$$

Thus the exclusive location of the basic protein at the cytoplasmic surface, as discussed in section 9.3.1, would result in 27% occupancy of this area.

9.4.4.2. Proteolipid protein

$$\text{Number of PLP molecules per g} = \frac{0.068 \times 6.02 \times 10^{23}}{23,500}$$

$$= 1.742 \times 10^{18} \text{ molecules}$$

$$\text{Total area occupied by PLP} = 1.742 \times 10^{18} \times 1134 \text{ Å}^2$$
$$= 1.9754 \times 10^{21} \text{ Å}^2$$

$$\frac{\text{Area occupied by PLP}}{\text{Area of bilayer}} = \frac{1.9562 \times 10^{21}}{1.223 \times 10^{22}} = 0.16$$

The intrinsic characteristics of the proteolipid locate the molecule within the lipid phase where these results indicate an occupation of a maximum of 16% of the hydrocarbon fracture face. This value represents the maximum projected area for a spherical proteolipid protein molecule. The exposure of PLP at the extracellular surface, indicated by labeling studies (section 9.3.2.), will result in a considerably smaller occupancy of this area by the protein.

9.4.5. Penetration of the myelin bilayer
by proteolipid protein

Assuming the width of the bilayer in myelin is 45 Å (electron density peak-to-peak distance; Caspar and Kirschner, 1971) then to bury one PLP molecule completely within the bilayer perturbs a cylinder of bilayer 38 Å diameter and 45 Å depth.

$$\text{Volume of bilayer perturbed by PLP} = \pi \times 19^2 \times 45 \text{ Å}^3$$
$$= 51035 \text{ Å}^3.$$

$$\frac{\text{Volume of PLP}}{\text{Volume of bilayer perturbed}} = \frac{28300}{51035} = 0.555$$

Projected area of bilayer occupied by PLP = 16%
Total penetration of myelin bilayer by PLP = $0.16 \times 0.555 = 0.089$

Thus, to bury all the proteolipid protein molecules in the myelin bilayer results in a total penetration by protein of 9%. It should be noted that this figure is below the upper limit of 10% protein penetration of the hydrocarbon phase suggested from the x-ray diffraction studies of Caspar and Kirschner (1971).

9.5. Protein-protein association
in compact myelin

With the locations of the two major proteins of compact cerebral myelin proposed in section 9.3., differing structural roles can now be offered for these components in the membrane system. The location of the basic protein at the cytoplasmic apposition surface and its strong ionic interactions with acidic lipid could serve to hold the negatively charged adjacent cytoplasmic surfaces of myelin lamellae in close proximity and thus effect compaction (Davison, 1971). The dimension of the cytoplasmic apposition region, polar to polar centers, as revealed by x-ray diffraction is about 30 Å (section 3.2.1.). The dimensions of the isolated basic protein approximate to 15 Å × 150 Å (Epand et al., 1974) and thus, allowing for interaction between positively and negatively charged groups on protein and lipid, respectively, as well as hydration, the BP can be accommodated within the observed dimensions of the appposition to interact with both polar membrane surfaces. The amphipathic nature of the primary structure of the BP suggested by Eylar (1970, 1973) and its experimental justification by the studies of London and colleagues (section 6.4.2.) has been further established by the work of Jones (1975) and Jones and Rumsby (submitted for publication) on the lipid interacting properties of the two BNPS-skatole cleavage products of the BP. Together these studies suggest that the "amino half" of the BP has the potential to interact hydrophobically with membrane components and the "carboxyl half" can interact ionically with polar membrane components. In BP interactions with the adjacent surfaces of the cytoplasmic apposition region the molecule will occupy 54% of the total cytoplasmic apposition surface area.

A general role for proteolipids in membrane structure can be seen as the provision of a protein-lipid core that can interact hydrophobically in the membrane with other lipids. A major role for the PLP in the lipid-rich myelin lamellae must be considered in these terms.

Studies on the PLP and BP in cerebral myelin indicate that approximately equimolar amounts of the two proteins are present (section 9.1.). To locate the BP in the glial/myelin membrane a structural interaction between these two components can be envisaged. This is considered to involve a specific protein-protein interaction within the bilayer between the BP and the PLP. Specific hydrophobic interactions between nonpolar regions of the BP penetrating the lipid phase and the PLP within the lipid phase may serve to maintain the structural interaction between these two molecules. An association between the BP and PLP may be further stabilized by nonpolar interactions between the PLP and fatty acids of lipids bound ionically to the BP. Two aspects of a model drawn to scale showing how a BP:PLP complex would be localized in myelin are presented in Fig. 10. The diameter of a spherical PLP molecule as calculated in section 9.4.3., is 38 Å. This is large enough to span the hydrocarbon region of the bilayer with its dimensions of about 35 Å. However, partial exposure of the PLP at the external surface has been indicated.

The BP-PLP protein complex proposed above could define an area within the bilayer of 150 Å × 38 Å. The complex would serve as a structural nucleus to bind lipid, as indicated in Fig. 11. Implicit in this model is the location of acidic lipid, particularly cerebroside sulfate and PS, at the cytoplasmic surface. Calculations show there are more than enough positive changes on the BP to interact ionically with all cerebroside sulfate and PS molecules in compact myelin. Labeling studies (section 8.2.) indicate that some cerebroside is indeed located at the

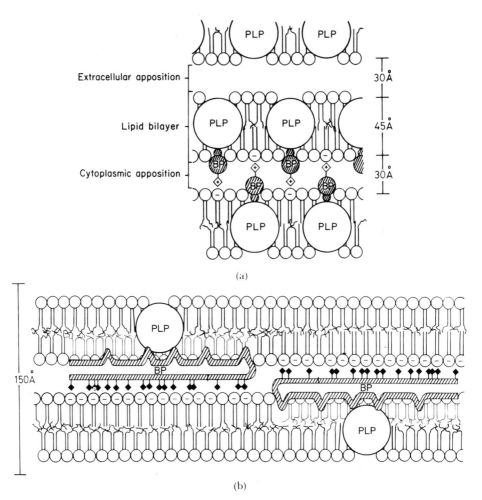

Fig. 10. (a) Cross-section through the proposed proteolipid-basic protein units in compact myelin. Essentially, the proteolipid protein is buried within the bilayer but is slightly exposed at the extracellular surface. The basic protein interacts both hydrophobically and ionically between adjacent cytoplasmic surfaces in compact myelin. (b) Longitudinal section through compact myelin lamellae incorporating the basic protein in the folded conformation, as discussed in the text. This view shows the proposed hydrophobic penetration of the lipid bilayer by the "amino-half" of the basic protein, and the ionic interactions across the cytoplasmic apposition by the basic residues on the "carboxyl-half" of the molecule. Basic amino acid residues are shown as ♦. The basic protein-proteolipid protein interaction within the bilayer is drawn as involving the two central penetrating hydrophobic regions of the basic protein.

external surface in the system. Extrapolation from the erythrocyte membrane where lipid asymmetry has been demonstrated would locate the choline-containing lipids at the external surface as well.

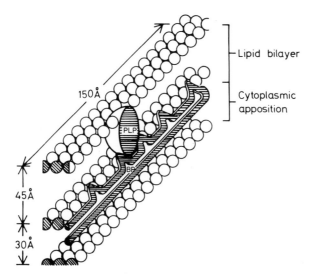

Fig. 11. Cross-sectional perspective view of the proposed proteolipid protein-basic protein unit in compact myelin showing the arrangement of the component lipid molecules. Fatty acyl chains are ommitted for clarity. (See text for further details.)

9.6. Subunits in compact myelin

The proposed PLP-BP protein unit in compact myelin, together with its associated lipid, would constitute a form of structural subunit in the membrane. The area occupied by such subunits within myelin can be calculated as follows:

Dimensions of PLP-BP unit = $150 \times 38 \times 45$ Å (Fig. 11)
Surface area of unit = 150×38 Å = 5700 Å^2
Number of units per g = calculated number of PLP per g
= 1.743×10^{18} units

Area occupied by PLP-BP units = $1.742 \times 10^{18} \times 5700$ Å^2
= 9.930×10^{21} Å^2

$$\frac{\text{Area occupied by units}}{\text{Area of bilayer}} = \frac{9.930 \times 10^{21}}{1.223 \times 10^{22}} = 0.812$$

Thus the proposed PLP-BP units, together with their component lipid molecules, would comprise 80% of the myelin bilayer. To pack these units equally over the bilayer surface would require an equal distance between each PLP-BP unit in the unit cell of which four units comprise 80% of the area (Fig. 12). The distance between each subunit is calculated to be 15 Å. However, the

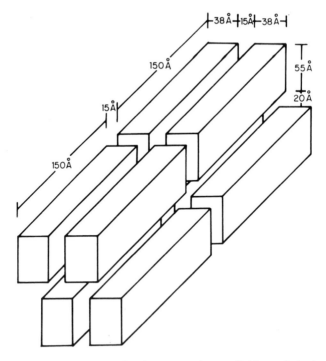

Fig. 12. Calculated packing arrangement for the proposed proteolipid protein-basic protein units in compact myelin so that they occupy 80% of the myelin bilayer surface. (See text for further details.)

exact demarcation of these subunits in myelin lamellae cannot be defined precisely. The intersubunit spaces, some 15 Å in width, will contain lipid components that show little or no affinity for either the basic protein or the proteolipid protein.

The features of myelin structure summarized in this review indicate that the compact membrane has many structural similarities to other naturally-occurring membrane systems and that the physical properties of the component lipids and proteins in terms of their interactions also follow general membrane principles. The existence of essential structural lipid-complexing protein subunits as a feature of compact myelin would induce long-range order and stability in this uniquely lipid-rich membrane structure.

Acknowledgments

We wish to thank Professor A.N. Davison, Dr. J.B. Finean, Dr. N. Gains, Dr. N.A. Gregson, Dr. A.J.S. Jones, and Dr. W.T. Norton for helpful discussions on aspects of the review or for providing copies of their published work, Susan Sparrow for drawing the diagrams, Susan Torrie and Jaki Grainger for battling with the references and Joan Chambers for typing them. The cooperation of the staff of the Morell library, University of York, and especially Margaret Lawty of the interlibrary loans section, is appreciated. Finally, we are grateful to the Medical Research Council, The Wellcome Trust, the Science Research Council and The Multiple Sclerosis Society of Great Britain and Ireland for supporting the research financially over the years.

References

Aarli, J.A., Aparicio, S.R., Lumsden, C.E. and Tonder, O. (1975) Binding of normal human IgG to myelin sheaths, glia and neurones. Immunology 28, 171–185.

Abramson, M.B. (1971) Lipids in water, In: The Chemistry of Biosurfaces. (Hair, M.L., ed.), vol. 1, pp. 45–82. Marcel Dekker, New York.

Abramson, M.B. and Katzman, R. (1968) Ionic interactions of sulphatide with choline lipids. Science 161, 576–577.

Abramson, M.B., Ratzman, R., Curci, R. and Wilson, C.E. (1967) The reactions of sulphatide with metabolic cations in aqueous systems. Biochemistry 6, 295–304.

Abrahamsson, S., Pascher, I., Larsson, K. and Karlsson, K.A. (1972) Molecular arrangements in glycosphingolipids. Chem. Phys. Lipids 8, 152–179.

Adams, C.W.M. and Bayliss, O.B. (1968) Histochemistry of myelin, VII. Analysis of lipid-protein relationships and absence of acid muco-polysaccharides. J. Histochem. Cytochem. 16, 119–127.

Adams, C.W.M. and Leibowitz, S. (1969) The general pathology of demyelinating disease. In: The Structure and Function of Nervous Tissues (Bourne, G.H., ed.), vol. 3, pp. 309–382. Academic Press, New York.

Adams, C.W.M., Bayliss, O.B., Hallpike, J.F. and Turner, D.R. (1971) Histology of myelin. XII. Anionic staining of myelin basic protein for histology electrophoresis and E/M. J. Neurochem. 18, 389–394.

Adams, D.H. and Osborne, J. (1973) A developmental study of the relationship between the protein components of rat CNS myelin. Neurobiology, 3, 91–112.

Adams, D.H. and Osborne, J. (1975) The action of primary alcohols on the structure of isolated rat CNS myelin. Int. J. Biochem. 6. 443–453.

Agrawal, H.C., Burton, R.M., Fishman, M.A., Mitchell, R.F. and Prensky, A.L. (1972) Partial characterisation of a new myelin protein component. J. Neurochem. 19, 2083–2089.

Akers, C.K. and Parsons, D.F. (1970) X-ray diffraction of the myelin membrane. Biophys. J. 10, 116–136.

Akers, C.K. and Parsons, D.F. (1972) Reply to myelin membrane structure as revealed by X-ray diffraction by David Harker. Biophys. J. 12, 1296–1301.

Alving, C.R., Fowble, J.W. and Joseph, K. C. (1974) Comparative properties of four galactosyl lipids as antigens in liposomes. Immunochemistry 11, 475–481.

Ansari, K.A., Hendrickson, H., Sinha, A.A. and Rand, A. (1975) Myelin basic protein in frozen and unfrozen bovine brain; a study of autolytic changes in situ. J. Neurochem. 25, 193–195.

Ansell, G.B. (1973) Phospholipids and the nervous system. In: Form and Function of Phospholipids (Ansell, G.B., Hawthorne, J.N. and Dawson, R.M.C., eds.), pp. 377–422. Elsevier/North-Holland, Amsterdam.

Anthony, J.S. and Moscarello, M.A. (1971a) A conformation change induced in the basic encephalitogen by lipids. Biochem. Biophys. Acta 243, 429–433.

Anthony, J.S. and Moscarello, M.A. (1971b) Conformational transition of a myelin protein. FEBS. Lett. 15, 335–339.

Avruch, J., Carter, J.R. and Martin, D.B. (1972) Insulin-stimulated plasma membranes from rat adipocytes: their physiological and physicochemical properties. Biochim. Biophys. Acta 288, 27–42.

Baldwin, G.S. and Carnegie, P.R. (1971) Isolation and partial characterisation of methylated arginines from the encephalitogenic basic protein of myelin. Biochem. J. 123, 69–74.

342

Bangham, A.D. (1972) Lipid bilayers and biomembranes. Ann. Rev. Biochem. 41, 753–776.

Banik, N.L. and Davison, A.N. (1969) Enzyme activity and composition of myelin and subcellular fractions in the developing rat brain. Biochem. J. 115, 1051–1062.

Banik, N.L. and Davison, A.N. (1973) Isolation of purified basic protein from human brain. J. Neurochem. 21, 489–494.

Banik, N.L. and Davison, A.N. (1974) Lipid and basic protein interaction in myelin. Biochem. J. 143, 39–45.

Banik, N.L. and Davison, A.N. (1975) Cooperative effect of phospholipase A and trypsin on the dissolution of myelin. Fifth ISN Conference, Barcelona, Abstr. No. 335.

Banik, N.L. and Smith, M.E. (1975) Myelin protein in different areas of developing rat CNS. Fifth ISN Conference, Barcelona, Abstract No. 312.

Banik, N.L., Davison, A.N., Ramsey, R.B. and Scott, T. (1974) Protein composition in developing human brain myelin. Dev. Psychobiol. 7, 539–549.

Baumann, N., Bourre, J.M., Jacque, C. and Harpin, M.L. (1973) Lipid composition of quaking mouse myelin: comparison with normal mouse myelin in the adult and during development. J. Neurochem. 20, 753–759.

Beck, C.S., Hasinoff, C.W. and Smith, M.E. (1968) L-alanyl-β-naphthylamidase in rat spinal cord myelin. J. Neurochem. 15, 1297–1301.

Becker, R., Helenius, A. and Simons, K. (1975) Solubilisation of the Semliki Forest virus membrane with sodium dodecyl sulphate. Biochemistry. 14, 1835–1841.

Bischoff, A. and Moor, H. (1967a) Ultrastructural differences between the myelin sheaths of peripheral nerve fibres and CNS white matter. Z. Zellforsch. 81, 303–310.

Bischoff, A. and Moor, H. (1967b) The ultrastructure of the "difference factor" in myelin. Z. Zellforsch. 81, 571–580.

Bischoff, A. and Moor, H. (1969) Myelin ultrastructure revealed by freeze-etching. Med. Biol. Illust. 19, 89–94.

Blank, W.F., Bunge, M.B. and Bunge, R.P. (1974) The sensitivity of the myelin sheath, particularly the Schwann cell-axolemma junction, to lowered calcium levels in cultured sensory ganglia. Brain Res. 67, 503–518.

Blaurock, A.E. (1971) Structure of the nerve myelin membrane: proof of the low resolution profile. J. Mol. Biol. 56, 35–52.

Blaurock, A.E. (1973a) Some comments on 'myelin membrane structure as revealed by x-ray diffraction' by David Harker. Biophys. J. 13, 1261–1262.

Blaurock, A.E. (1973b) X-ray diffraction pattern from a bilayer with protein outside. Biophys. J. 13, 281–289.

Blaurock, A.E. (1976) Myelin X-ray patterns reconciled. J. Mol. Biol. 16, 491–501.

Blaurock, A.E. and Worthington, C.R. (1969) Low angle X-ray diffraction patterns from a variety of myelinated nerves. Biochim. Biophys. Acta. 173, 419–426.

Block, R.E., Brady, A.J. and Joffe, S. (1973) Conformation and aggregation of bovine myelin proteins. Biochem. Biophys. Res. Comm. 54, 1595–1602.

Bowen, D.M., Davison, A.N., and Ramsey, R.B. (1974) The dynamic role of lipids in nervous system. In: Biochemistry of Lipids (Goodwin, T.W., ed.), vol. 4, pp. 141–179. Butterworths, London.

Boxer, D.H., Jenkins, R.E. and Tanner, M.J.A. (1974) The organisation of the major proteins of the human erythrocyte membrane. Biochem. J. 137, 531–534.

Boyd, I.A. and Kalu, K.U. (1973) The relation between axon size and number of lamellae in the myelin sheath for afferent fibres in groups 1, 2 and 3 in the cat. J. Physiol. 232, 31–33.

Brady, R.O. and Quarles, R.H. (1973) The enzymology of myelin. Mol. Cell Biochem. 2, 23–29.

Branton, D. (1967) Fracture faces of frozen myelin. Exp. Cell Res. 45, 703–707.

Branton, D. (1973) The fracture process of freeze-etching. In: Freeze Etching and Techniques and Applications (Benedetti, E.L. and Favard, P. eds.) pp. 107–112. Societe Française de Microscopie Electronique, Paris.

Braun, P.E. and Barchi, R.L. (1972) 2', 3'-cyclic nucleotide -3'-phosphohydrolase in the nervous system. Electrophoretic properties and developmental studies. Brain Res. 40, 437–444.

Braun, P.E. and Radin, N.S. (1969) Interactions of lipids with a membrane structural protein from myelin. Biochem. 8, 4310–4313.

Brazier, M.A.B. (1969) Electrical activity of the nerve fibre and propagation of the nerve impulse. In: The structure and Function of Nervous Tissue (Bourne, G.H., ed.), vol. 2, pp. 409–421. Academic Press, New York.

Bretscher, M.S. (1973) Membrane Structure: Some general principles. Science 181, 622–629.

Bretscher, M.S. (1974) Some aspects of membrane structure. In: Perspectives in Membrane Biology (Estrada, O.S. and Gitler, C., eds.), pp. 3–24, Academic Press, New York.

Brightman, M.W. and Reese, T.S. (1969) Junctions between intimately apposed cell membranes in the vertebrate brain. J. Cell Biol. 40, 648–677.

Brostoff, S., Burnett, P., Lampert, P. and Eylar, E.H. (1971) Isolation and characterization of a protein from sciatic nerve myelin responsible for experimental allergic neuritis. Nature New Biol. 235, 210–212.

Brostoff, S.W. and Eylar, E.H. (1971) Localization of methylated arginine in the A1 protein from myelin. Proc. Nat. Acad. Sci. U.S.A. 68, 765–769.

Brostoff, S.W. and Eylar, E.H. (1972) The proposed amino acid sequence of the P 1 protein of rabbit sciatic nerve myelin. Arch. Biochem. Biophys. 153, 590–598.

Brostoff, S.W., Reuter, W., Hichens, M. and Eylar, E.H. (1974) Specific cleavage of the A1 protein from myelin with Cathepsin D. J. Biol. Chem. 249, 559–567.

Brostoff, S.W., Karkhanis, Y.D., Carlo, D.J., Reuter, W. and Eylar, E.H. (1975) Isolation and partial characterization of the major proteins of rat sciatic nerve myelin. Brain Res. 86, 449–458.

Brunngraber, E.G. (1969) Glycoproteins. In: Handbook of Neurochemistry (Lajtha, A. ed.), vol. 1, pp. 223–244. Plenum Press, New York.

Bubis, J.J. and Wolman, M. (1968) Study of the localisation of the encephalomyelitis-producing protein in the myelin sheath. Acta. Neuropath. 10, 356–358.

Bunge, R.P., (1968) Glial cells and the central myelin sheath. Physiol. Rev. 48, 197–251.

Burnett, P.R. and Eylar, E.H. (1971) Oxidation and cleavage of the single tryptophan residue of the A1 protein from bovine and human myelin. J. Biol. Chem. 246, 3425–3430.

Cabantchik, Z.I. and Rothstein, A. (1974) Effects of proteolytic enzymes on disulfonic stilbene sites of surface proteins. J. Membrane Biol. 15, 227–248.

Capaldi, R.A. (1974) A dynamic model of cell membranes. Scientific American 230, 26–33.

Capaldi, R.A. and Vanderkooi, G. (1972) The low polarity of many membrane proteins. Proc. Nat. Acad. Sci. U.S.A. 69, 930–932.

Carnegie, P.R. (1969) N-Terminal sequence of an encephalitogenic protein from human myelin. Biochem. J. 111, 240–242.

Carnegie, P.R. (1971) Amino acid sequence of the encephalitogenic protein from human myelin. Biochem. J. 123, 57–67.

Carnegie, P.R. and Dunkley, P.R. (1975) Basic proteins of central and peripheral nervous system myelin. In: Advances in Neurochemistry (Agranoff, B.W. and Aprison, M.H., eds.), vol. 1, pp. 95–126. Plenum Press, New York.

Carnegie, P.R. and Lumsden, C.E. (1966) Encephalitogenic peptides from spinal cord. Nature 209, 1354.

Carnegie, P.R., Bencina, B. and Lamoureux, G. (1967) Experimental allergic encephalomyelitis. Isolation of basic proteins and polypeptides from central nervous tissue. Biochem. J. 105, 559–568.

Carraway, K.L. (1975) Covalent labeling of membranes. Biochim. Biophys. Acta 415, 379–410.

Caspar, D.L.D. and Kirschner, D.A. (1971) Myelin membrane structure at 10Å resolution. Nature New Biol. 231, 46–52.

Casu, A., Pala, V., Monacelli R. and Nanni, G. (1971) Phospholipase C from Clostridium perfringens: purification by electrophoresis on acrylamide-agarose gel. Ital. J. Biochem. 20. 166–178.

Cavanna, R. and Rapport, M.M. (1967) An improved preparation of bovine proteolipid. J. Lipid Res. 8, 65–68.

Cerbon, J., (1974) Water in the structure and function of cell membrane. In: Perspectives in Membrane Biology. (Estrado, O.S. and Gitler, C., eds.), pp. 71–84. Academic Press, New York.

344

Chao, L-P, and Einstein, E.R. (1970) Physical properties of the bovine encephalitogenic protein. J. Neurochem. 17, 1121–1132.

Chapman, D. (1968) Biological Membranes, Physical Fact and Function. Academic Press, London.

Chapman, D. (1972) Spectroscopic studies of biological membranes. Ann. N.Y. Acad. Sci. 195, 179–206.

Chapman, D. (1973a) Thermal and spectroscopic studies of membranes and membrane components. In: Biological Horizons in Surface Science (Prince, C.M. and Sears, D.F. eds.), pp. 35–68. Academic Press, London.

Chapman, D. (1973b) Some recent studies of lipids, lipid-cholesterol and membrane systems. In: Biological Membranes (Chapman, D. and Wallach, D.F.H. eds.), pp. 91–144. Academic Press, London.

Chapman, D. (1973c) Physical chemistry of phospholipids. In: Form and Function of Phospholipids (Ansell, G.B., Hawthorne, J.N. and Dawson, R.M.C. eds.), pp. 117–142, Elsevier/North-Holland, Amsterdam.

Chapman, D. (1975) Phase transitions and fluidity characteristics of lipid and cell membranes. Quart. Rev. Biophys. 8, 185–235.

Chapman, D. and Wallach, D.F.H., Eds., (1973) Biological Membranes. Vol. 2. Academic Press, London.

Chapman, D., Williams, R. M. and Ladbrooke, B. D. (1967) Physical studies of phospholipids. VI Thermotropic and lyotropic mesomorphism of some 1,2-diacyl-phosphatidylcholines (lecithins). Chem. Phys. Lipids 1, 445–475.

Chan, D. S. and Lees, M.B. (1974) Gel electrophoresis studies of bovine brain white matter proteolipid and myelin proteins. Biochemistry 13, 2704–2712.

Claude, P. and Goodenough, D. A. (1973) Fracture faces of zonulae occludentes from "tight" and "leaky" epithelia. J. Cell Biol. 58, 390–400.

Clowes, A. W., Cherry, R. J. and Chapman, D. (1971) Physical properties of lecithin-cerebroside films. Biochim. Biophys. Acta. 249, 301–307.

Cogan, U., Shinitzky, M., Weber, G. and Nishida, T. (1973) Microviscosity and order in the hydrocarbon region of phospholipid and phospholipid-cholesterol dispersion determined with fluorescent probes. Biochemistry 12, 521–528.

Cohen, S. R., McKhann, G. M. and Guarnieri, M. (1975) A radioimmunoassay for myelin basic protein and its use for quantitative measurements. J. Neurochem. 25, 371–376.

Colacicco, G., Hendrickson, H. and Joffe, S. (1972) Surface properties of membrane systems. Proc. Nat. Acad. Sci. USA, 69, 1848–1850.

Coleman, R. (1973) Membrane-bound enzymes and membrane ultrastructure. Biochim. Biophys. Acta. 300, 1–30.

Coles, E., McIlwain, D. L. and Rapport, M. M. (1974) The activity of pure phospholipase A_2 from crotalus atrox venom on myelin and on pure phospholipids. Biochim. Biophys. Acta. 337, 68–78.

Colley, L. M., Zwaal, R. F. A., Roelofsen, B. and van Deenen, L. L. M. (1973) Lytic and non-lytic degredation of phospholipids in mammalian erythrocytes by pure phospholipases. Biochim. Biophys. Acta. 307, 74–82.

Cone, R. A. (1972) Rotational diffusion of rhodopsin in the visual receptor membrane. Nature New Biol. 236, 39–43.

Crang, A. J. (1976) D. Phil. Thesis, University of York, England.

Crang, A. J. and Rumsby, M. G. (1975) Solubilization of purified central nerve myelin with the anionic detergent sodium dodecyl sulphate. Fifth I.S.N. Conference, Barcelona, Abstract No. 320.

Crang, A. J. and Rumsby, M. G. (1977) The labeling of lipid and protein components in isolated central nervous system myelin with dansyl chloride. Biochem. Soc. Trans. 5:110–112.

Crang, A. J., Jones, A. J. S. and Rumsby, M. G. (1974) Myelin intrinsic flourescence. Biochem. Soc. Trans. 2, 552–554.

Critchley, D. R. and Vicker, M. G. (1977) Glycolipids as membrane receptors important in growth regulation and cell-cell interactions. In: Dynamic Aspects of Cell Surface Organization. (Poste, G. and Nicolson, G. L., eds.). Cell Surface Reviews, vol. 3. Elsevier/North-Holland, Amsterdam.

Cuzner, M. L. and Davison, A. N. (1968) The lipid composition of rat brain myelin and sub-cellular fractions during development. Biochem. J. 106, 29–34.

Cuzner, M. L., Davison, A. N. and Gregson, N. A. (1965) The chemical composition of vertebrate myelin and microsomes. J. Neurochem. 12, 469–481.

Dalal, K. B. and Einstein, E. R. (1969) Biochemical maturation of the central nervous system. 1. Lipid changes. Brain Res. 16, 441–451.

Darke, A., Finer, E. G., Flook, A. G. and Philips, M. C. (1971) The use of N.M.R. spectra of sonicated phospholipid dispersions in studies of interactions within the bilayer. FEBS. Lett. 18, 326–334.

Darke, A., Finer, E. G., Flook, A. G. and Philips, M. C. (1972) Nuclear magnetic resonance study of lecithin-cholesterol interactions. J. Mol. Biol. 63, 265–279.

Davidson, B., Neary, J. T., Schwartz, S. and Maloof, F. (1973) Partial purification and some properties of thyroid peroxidase solubilised by extraction with n-butanol. Prep. Biochem. 3, 473–493.

Davison, A. N. (1968) Myelination. In: Fortschritte der Padolögie (Linnerwch, F. ed.), vol. 2, pp. 65–87. Springer-Verlag, Berlin.

Davison, A. N. (1970) Lipid metabolism—nerve tissue. In: Comprehensive Biochemistry (Florkin, M. and Stotz, E. H. eds.), vol. 18, pp. 293–329. Elsevier/North-Holland, Amsterdam.

Davison, A. N. and Peters, A. (1970) Myelination. C.C. Thomas, Springfield, Ill.

Davison, A. N., Cuzner, M. L., Banik, N. L. and Oxberry, J. M. (1966) Myelinogenesis in the rat brain. Nature 212, 1373–1374.

Davison, A. N., Mandel, P. and Morgan, I. G. (1972) Functional and structural proteins of the nervous system. Adv. Exp. Med. Biol. 32, 157–273.

Day, E. D. and Pitts, O. H. (1974) Radioimmunoassay of myelin basic protein in sodium sulfate. Immunochemistry 11, 651–659.

Deamer, D. W., Leonard, R., Tardieu, L. R. and Branton, D. (1970) Lamellar and hexagonal lipid phases visualised by freeze-etching. Biochim. Biophys. Acta. 219, 47–60.

Deibler, G. E. and Martenson, R.E. (1973a) Determination of methylated basic amino acids with the amino acid analyser. J. Biol. Chem. 248, 2387–2391.

Deibler, G. E. and Martenson, R. E. (1973b) Chromatographic fractionation of myelin basic protein. J. Biol. Chem. 248, 2392–2396.

Deibler, G. E., Martenson, R. E. and Kies, M. W. (1972) Large scale preparation of myelin basic protein from central nervous tissue of several mammalian species. Prep. Biochem. 2, 139–165.

Deibler, G. E., Martenson, R. E., Kramer, A. J., Kies, M. W. and Miyamoto, E. (1975) The contribution of phosphorylation and loss of COOH-terminal arginine to the microheterogeneity of myelin basic protein. J. Biol. Chem. 250, 7931–7938.

Demel, R. A., van Deenen, L. L. M. and Pethica, B. A. (1967) Monolayer interactions of phospholipids and cholesterol. Biochim. Biophys. Acta. 135, 11–19.

Demel, R. A., London, Y., Geurts van Kessel, W. S. M., Vossenberg, F. G. A. and van Deenen, L. L. M. (1973) The specific interaction of myelin basic protein with lipids at the air-water interface. Biochim. Biophys. Acta. 311, 507–519.

Demel, R. A., Geurts van Kessel, W. S. M., Zwaal, R. F. A., Roelofsen, B. and van Deenen, L. L. M. (1975) Relation between various phospholipase actions on human red cell membranes and the interfacial phospholipid pressure in monolayers. Biochim. Biophys. Acta. 406, 97–107.

Dermietzel, R. (1974) Junctions in the central nervous system of the cat. Cell Tissue Res. 148, 565–576.

Deshmukh, D. S., Flynn, T. J. and Pieringer, R. A. (1974) The biosynthesis and concentration of galactosyl diglyceride in glial and neuronal enriched fractions of actively myelinating rat brain. J. Neurochem. 22, 479–485.

De Vries, G. H. and Norton, W. T. (1974) The fatty acid composition of sphingo lipids from bovine CNS axons and myelin. J. Neurochem. 22, 251–257.

De Vries, G. H., Norton, W. T. and Raine, C. S. (1972) Axons: isolation from mammalian central nervous system. Science 175, 1370–1372.

De Vries, G. H., Saul, R. G., Beals, D. and Hadfield, M. G. (1975) Isolation and characterisation of an axolemma enriched fraction from bovine CNS. Fifth I.S.N. Conference, Barcelona, Abstract No. 323.

De Vries, G. H., Hadfield, M. G. and Cornbrooks, C. (1976) The isolation and lipid composition of myelin-free axons from rat CNS. J. Neurochem. 26, 725–732.

346

Dickinson, J. P. and Aparicio, S. R. (1971) Trypan Blue: Reaction with myelin. Biochem. J. 122, 65–66p.

Dickinson, J. P., Jones, K. M., Aparicio, S. R. and Lumsden, C. E. (1970) Localisation of encephalitogenic basic protein in the intra-period line of lamellar myelin. Nature 227, 1133–1134.

Doggenweiller, C. F. and Frenk, S. (1965) Staining properties of lanthanum on cell membranes. Proc. Nat. Acad. Sci. U.S.A. 53, 425–430.

Dreher, K. D., Schulman, O., Anderson, R. and Roels, O. A. (1967) The stability and structure of mixed lipid monolayers and bilayers. I. Properties of lipid and lipoprotein monolayers on OsO₄ solutions and the role of cholesterol, retinol and tocopherol in stabilizing lecithin monolayers. J. Ultrastruct. Res. 19, 586–599.

Driscoll, B. F., Kies, M. W. and Alvord, E. C. (1974) Successful treatment of experimental allergic encephalomyelitis (EAE) in guinea pigs with homologous myelin basic protein. J. Immunol. 112, 392–397.

Dubois-Dalcq, M., Niedieck, B. and Bayse, N. (1970) Action of anti-cerebroside sera on myelinated nervous tissue cultures. Path. Europ. 5, 331–347.

Dunkley, P. R. and Carnegie, P. R. (1974) Amino acid sequence of the smaller basic protein from rat brain myelin. Biochem. J. 141, 243–255.

Eichberg, J. and Dawson, R. M. C. (1965) Polyphosphoinositides in myelin. Biochem. J. 96, 644–650.

Eichberg, J., Hauser, G. and Karnovsky, M. L. (1969) Lipids of nervous tissue. In: The Structure and Function of Nervous Tissue (Bourne, G. H., ed.), vol. 3, pp. 185–287. Academic Press, New York.

Eichberg, J., Whittaker, V. P. and Dawson, R. M. C. (1964) Distribution of lipids in subcellular particles of guinea pig brains. Biochem. J. 92, 91–100.

Einstein, E. R. (1972) Basic protein of myelin and its role in experimental allergic encephalomyelitis and multiple sclerosis. In: Handbook of Neurochemistry, (Lajtha, A., ed.), vol. 7, pp. 107–129. Plenum Press, New York.

Einstein, E. R., Csejtey, J. and Marks, N. (1968) Degradation of encephalitogen by purified brain acid proteinase. FEBS. Lett. 1, 191–195.

Einstein, E. R., Dalal, K. B. and Csejtey, J. (1970) Biochemical maturation of the central nervous system. II Protein and proteolytic enzyme changes. Brain Res. 18, 35–49.

Elfvin, L. G. (1968) The structure and composition of motor, sensory and autonomic nerves and nerve fibres. In: Structure and Function of Nerve Tissue (Bourne, G. H., ed.). Vol. 1, pp. 325–377. Academic Press, New York.

Eng, L. F. and Noble, E. P. (1968) The maturation of rat brain myelin. Lipids 3, 157–162.

Eng, L. F. and Smith, M. E. (1966) The cholesterol complex in the myelin membrane. Lipids, 1, 269.

Eng, L. F., Chao, F-C., Gerstl, B., Pratt, D. and Tavaststjerna, M. G. (1968) The maturation of human white matter myelin. Fractionation of the myelin membrane proteins. Biochemistry 7, 4455–4465.

Engleman, D. E., Terry, T. M. and Morowitz, H. J. (1967) Characterisation of the plasma membrane of *Mycoplasma laidlawii*. Biochim. Biophys. Acta. 135, 381–390.

Epand, R. M., Moscarello, M. A., Zierenberg, B. and Vail, W. J. (1974) The folded conformation of the encephalitogenic protein of the human brain. Biochemistry 13, 1264–1267.

Eto, Y. and Suzuki, K. (1972) Cholesterol esters in developing rat brain: enzymes of cholesterol ester metabolism. J. Neurochem. 19, 117–121.

Eto, Y. and Suzuki, K. (1973) Cholesterol ester metabolism in rat brain. A cholesterol ester hydrolase specifically localised in the myelin sheath. J. Biol. Chem. 248, 1986–1991.

Evans, M. J. and Finean, J. B. (1965) The lipid composition of myelin from brain and peripheral nerve. J. Neurochem. 12, 729–734.

Everly, J. L., Brady, R. O. and Quarles, R. H. (1973) Evidence that the major protein in rat sciatic nerve myelin is a glycoprotein. J. Neurochem. 21, 329–334.

Eylar, E. H. (1970) Amino acid sequence of the basic protein of the myelin membrane. Proc. Nat. Acad. Sci. U.S.A. 67, 1425–1431.

Eylar, E. H. (1972) The chemical and immunologic properties of the basic A1 protein of myelin. In: Functional and Structural Proteins of the Nervous System (Davison, A. N., Mandel, P. and Morgan, I. G., eds.), pp. 215–240. Plenum Press, New York.

Eylar, E. H. (1973) Myelin-specific proteins. In: Proteins of the Nervous System (Johnson Schneider, D., Angeletti, R. H., Brandshaw, R. A., Grasso, A., Moore, B. W., eds.), pp. 27–66. Raven Press, New York.

Eylar, E. H. and Thompson, N. (1969) Allergic encephalomyelitis: the physicochemical properties of the basic protein encephalitogen from bovine spinal cord. Arch. Biochem. Biophys. 129, 468–479.

Eylar, E. H., Brostoff, S., Hashim, G., Caccam, J. and Burnett, P. (1971) Basic A 1 protein of the myelin membrane: the complete amino acid sequence. J. Biol. Chem. 246, 5770–5784.

Eylar, E. H., Kniskern, P. J. and Jackson, J. J. (1974) Myelin basic proteins. In: Methods in Enzymology. (Fleischer, S. and Packer, L., eds.). vol. 32B, pp. 323–341. Academic Press, New York.

Feinstein, M. B. and Felsenfeld, H. (1975a) Reactions of fluorescent probes with normal and chemically modified myelin. Biochemistry 14, 3041–3048.

Feinstein, M. B. and Felsenfeld, H. (1975b) Reactions of fluorescent probes with normal and chemically modified myelin basic protein and proteolipid. Biochemistry 14, 3049–3056.

Fernandez-Moran, H. and Finean, J. B. (1957) E-M and low angle X-ray diffraction on the nerve myelin sheath. J. Biophys. Biochem. 3, 725–748.

Finean, J. B. (1953) Phospholipid-cholesterol complex in the structure of myelin. Experientia, 9, 17.

Finean, J. B. (1957) Role of water in the structure of peripheral nerve myelin. J. Biophys. Biochem. Cytol. 3, 95–102.

Finean, J.B. (1958) X-ray diffraction studies of the myelin sheath in peripheral and central nerve tissue. Exp. Cell Res. Suppl.5, 18–32.

Finean, J. B. (1960) Electron microscopy and X-ray diffraction studies of the effects of dehydration on the structure of nerve myelin: I peripheral nerve. J. Biophys. Biochem. Cytol. 8, 13–29.

Finean, J. B. (1961) The nature and stability of nerve myelin. Int. Rev. Cytol. 12, 303–336.

Finean, J. B. (1962) The nature and stability of the plasma membrane. Circulation 26, 1151–1162.

Finean, J. B. (1966) The molecular organisation of cell membranes. Prog. Biophys. Mol. Biol. 16, 143–170.

Finean, J. B. (1969) Biophysical contribution to membrane structure. Quart. Rev. Biophys. 2, 1–23.

Finean, J. B. (1972) The development of ideas on membrane structure. Subcell. Biochem. 1, 363–373.

Finean, J. B. (1973) Phospholipids in biological membranes and the study of phospholipid-protein interactions. In: Form and Function of Phospholipids (Ansell, G. B., Hawthorne, J. N. and Dawson, R. M. C., eds), pp. 171–173. Elsevier/North-Holland, Amsterdam.

Finean, J. B. and Burge, R. E. (1963) The determination of the Fourier transform of the myelin layer from a study of swelling phenomena. J. Mol. Biol. 1, 672–682.

Finean, J. B. and Millington, P. F. (1957) Effects of ionic strength of immersion medium on the structure of peripheral nerve myelin. J. Biophys. Biochem. Cytol. 3, 89–94.

Finer, E. G. and Darke, A. (1974) Phospholipid hydration studied by deuteron magnetic resonance spectroscopy. Chem. Phys. Lipids 12, 1–16.

Fishman, M. A., Agrawal, H. C., Alexander, A., Golterman, J., Martenson, R. E. and Mitchell, R. F. (1975) Biochemical maturation of human central nervous system myelin. J. Neurochem. 24, 689–694.

Fleischer, S. and Packer, L., eds. (1974) Methods in enzymology. Vol. 31, part A. Academic Press, New York.

Fluck, D., Henson, A. and Chapman, D. (1969) The structure of dilute lecithin-water systems revealed by freeze-etching and electron microscopy. J. Ultrastruct. Res. 29, 416–429.

Folch-Pi, J. (1972) Proteolipids. In: Functional and Structural Proteins of the Nervous System (Davison, A. N., Mandel, P. and Morgan, I. G., eds.), pp. 171–199. Plenum Press, New York.

Folch-Pi, J. (1973) Proteolipids. In: Proteins of the Nervous System (Johnson Schneider, D., Angelett, R. H., Bradshaw, R. A., Grasso, A., Moore, B. W., eds.), pp. 27–66. Raven Press, New York.

Folch-Pi, J. and Lees, M. (1951) Proteolipids, a new type of tissue lipoproteins. Their isolation from brain. J. Biol. Chem. 191, 807–817.

Folch-Pi, J., Webster, G. R. and Lees, M. J. (1959) The preparation of proteolipids. Fed. Proc. 18, abstr. 898.

348

Fraher, J. P. (1972) A quantitative study of anterior root fibres during early myelination. J. Anat. 112, 99–124.

Fraher, J. P. (1973) A quantitative study of anterior root fibres during early myelination. II Longitudinal variation in sheath thickness and axon circumference. J. Anat. 115, 421–444.

Friede, R. L. (1972) Control of myelin formation by axon calibre. J. Comp. Neurol. 144, 233–252.

Friede, R. L. and Samorajski, T. (1967) Relation between the number of myelin lamellae and axon circumference in fibres of vagus and sciatic nerves of mice. J. Comp. Neurol. 130, 223–232.

Frye, L. D. and Edidin, M. (1970) The rapid mixing of cell surface antigens after formation of mouse-human heterokaryons. J. Cell Sci. 7, 319–335.

Gagnon, J., Finch, P. R., Wood, D. D. and Moscarello, M. A. (1971) Isolation of a highly purified myelin protein. Biochemistry 10, 4756–4763.

Galla, H. J. and Sackman, E. (1975) Chemically induced lipid phase separation in model membranes containing charged lipids: a spin label study. Biochim. Biophys. Acta. 401, 509–529.

Gary-Bobo, C. M., Lange, Y. and Rigaud, J. L. (1971) Water diffusion in lecithin-water and lecithin-cholesterol-water lamellar phases at 22°. Biochim. Biophys. Acta. 233, 243–246.

Gent, W. L. G., and Gregson, N. A. (1966) In homogeneity of lysolecithin-solubilized myelin. Biochem. J. 98, 27p–280p.

Gent, W. L. G., Grant, E. H. and Tucker, S. W. (1970) Evidence from dielectric studies for the presence of bound water in myelin. Biopolymers, 9, 124–126.

Gent, W. L. G., Gregson, N. A., Gammack, D. B. and Raper, J. H. (1964) The lipid-protein unit in myelin. Nature 204, 553–555.

Gent, W. L. G., Gregson, N. A., Lovelidge, C. A. and Winder, A. F. (1971a) Separation of lipid-protein complexes of rat brain myelin by isopycnic centrifugation. Biochem. J. 122. 63p.

Gent, W. L. G., Gregson, N. A., Lovelidge, C. A. and Winder, A. F. (1971b) Resolution of protein components in lysophosphatidyl choline-solubilised rat brain myelin. Biochem. J. 122, 64p.

Gent, W. L. G., Gregson, N. A., Lovelidge, C. A. and Winder, A. F. (1971c) Interaction of lysophosphatidyl choline with central nervous system myelin. Biochem. J. 122, 64p–65p.

Gent, W. L. G., Gregson, N. A., Lovelidge, C. A. and Winder, A. F. (1973) Ultraviolet absorption characteristics of myelin for the central nervous system. J. Neurochem. 21, 697–702.

Gerstl, B., Eng, L. F., Hayman, R. B., Travaststjerna, M. G. and Bond, P. R. (1967) On the composition of human myelin. J. Neurochem. 14, 661–670.

Gigg, R. and Payne, S. (1969) The reaction of glutaraldehyde with tissue lipids. Chem. Phys. Lipids 3, 292–295.

Glaisher, M. P. (1975) D. Phil Thesis, University of York, England.

Glick, M. C. (1974) Isolation and characterisations of surface membrane glycoproteins from mammalian cells. In: Methods in Membrane Biology. (Korn, E. D., ed.), vol. 2, pp. 157–204. Plenum Press, New York.

Gonzalez-Sastre, F. (1970) The protein composition of isolated myelin. J. Neurochem. 17, 1049–1056.

Gonzalez-Sastre, F., Eichberg, J. and Hauser, G. (1971) Metabolic pools of polyphosphoinositides in rat brain. Biochim. Biophys. Acta. 248, 96–104.

Gottlieb, A. M., Inglefield, P. T. and Lange, Y. (1973) Water-lecithin binding in lecithin-water lamellar phases at 20°. Biochim. Biophys. Acta. 307, 444–451.

Gould, R. M. and London, Y. (1972) Specific interaction of central nervous system basic protein with lipids. Biochim. Biophys. Acta. 290, 200–218.

Grant, C. W. M. and McConnell, H. M. (1973) Fusion of phospholipid vesicles with viable Acholiplasma laidlawii. Proc. Nat. Acad. Sci. U.S.A. 70, 1238–1240.

Gray, G. M. (1974) Glycosphingolipids in biological membranes. In: Perspectives in Membrane Biology. (Estrada, O. S. and Gitler, C., eds.), pp. 85–106. Academic Press, New York.

Green, D.E. (1971) Membrane proteins. Science 174, 863–867.

Green, D. E. and Perdue J. F. (1966) Membranes as expressions of repeating units. Proc. Nat. Acad. Sci. U.S.A. 55, 1295–1302.

Greenfield, S., Brostoff, S., Eylar, E. H. and Morell, P. (1973) Protein composition of myelin of the peripheral nervous system. J. Neurochem. 20, 1207–1216.

Greenfield, S., Norton, W. T. and Morell, P. (1971) Quaking mouse: Isolation and characterisation of myelin protein. J. Neurochem. 18, 2119–2128.

Gregson, N. A. and Hall, S. M. (1973) A quantitative analysis of the effects of the intraneural injection of lysophosphatidyl choline. J. Cell Sci. 13, 257–277.

Gregson, N. A. and Oxberry, J. M. (1972) The composition of myelin from the mutant mouse "quaking." J. Neurochem. 19, 1065–1071.

Gregson, N. A. and Oxberry, J. M. (1976) An immunochemical investigation of ganglioside in rat brain myelin. Biochem. Soc. Trans. 4, 310–311.

Gregson, N. A., Kennedy, M. L. and Leibowitz, S. (1971) Immunological reactions with lysolecithin-solubilized myelin. Immunol. 20, 501–572.

Gregson, N. A., Kennedy, M. C. and Leibowitz, S. (1974) The specificity of antigalactocerebroside antibody and its reaction with lysolecithin-solubilized myelin. Immunol. 26, 743–757.

Griffith, O. H., Dehlinger, P. J. and Van, S. P. (1974) The shape of the hydrophobic barrier of phospholipid bilayers. Evidence for water penetration in biological membranes. J. Membrane Biol. 15, 159–192.

Guarnieri, M. (1975) Reaction of anti-phosphatidyl inositol antisera with neural membranes. Lipids 10, 294–298.

Guarnieri, M., Himmelstein, J. and McKhann, G. M. (1974) Isolated myelin quantitatively adsorbs antibody to basic protein. Brain Res. 72, 172–176.

Guidotti, G. (1972) Membrane Proteins. Ann. Rev. Biochem. 41, 731–752.

Hall, S. M. and Gregson, N. A. (1971) The in vivo and ultrastructural effects of injection of lysophosphatidyl choline into myelinated peripheral nerve fibres of the adult mouse. J. Cell Sci. 9, 769–789.

Hall, S. M. and Williams, P. L. (1970) Studies on the "incisures" of Schmidt and Lanterman. J. Cell Sci. 6, 767–791.

Hall, S. M. and Williams, P. L. (1971) The distribution of electron-dense tracers in peripheral nerve. J. Cell Sci. 8, 541–556.

Harker, D. (1972) Myelin membrane structure as revealed by x-ray diffraction. Biophys. J. 12, 1285–1295.

Hashim, G. A. and Eylar, E. H. (1969a) Allergic encephalomyelitis: enzymatic degradation of the encephalitogenic basic protein from bovine spinal cord. Arch. Biochem. Biophys. 129, 635–644.

Hashim, G. A. and Eylar, E. H. (1969b) Allergic encephalomyelitis: isolation and characterization of encephalitogenic peptides from the basic protein of bovine spinal cord. Arch. Biochem. Biophys. 129, 645–654.

Hauser, G. and Eichberg, J. (1973) The subcellular distribution of polyphosphoinositides in myelinated and unmyelinated rat brain. Biochim. Biophys. Acta. 326, 210–223.

Hauser, G., Eichberg, J. and Gonzalez-Sastre, F. (1971) Regional distribution of polyphosphoinositides in rat brain. Biochim. Biophys. Acta. 248, 87–95.

Helenius, A. and Simons, K. (1975) Solubilization of membranes by detergents. Biochim. Biophys. Acta. 415, 29–79.

Henn, F. A., Decker, G. L., Greenawalt, J. W. and Thompson, T. E. (1967) Properties of lipid bilayer membranes separating two aqueous phases: electron microscope studies. J. Mol. Biol. 24, 51–58.

Herndon, R. M., Rauch, H. C. and Einstein, E. R. (1973) Immuno-electron microscopic localization of the encephalitogenic basic protein in myelin. Immunol. Commun. 2, 163–172.

Hildebrand, C. (1971) Ultrastructural and light-microscopic studies of the nodal region in large myelinated fibres of the adult feline spinal cord white matter. Acta Physiol. Scand. Suppl. 364, 43–71.

Hildebrand, C. (1972) Evidence for a correlation between myelin period and number of myelin lamellae in fibres of the feline spinal cord white matter. J. Neurocytol. 1, 223–232.

Hildebrand, C. and Muller, H. (1974) Low angle x-ray diffraction studies of the period of central myelin sheaths during preparation for electron microscopy. Neurobiol. 4, 71–81.

Hirano, A. (1972) The pathology of the central myelinated axon. In: The Structure and Function of Nervous Tissue (Bourne, G. H., ed.), vol. 5, pp. 73–162. Academic Press, New York.

Hirano, A. and Dembitzer, H. M. (1967) A structural analysis of the myelin sheath in the central nervous system. J. Cell Biol. 34, 555–567.

Hirano, A. and Dembitzer, H. M. (1969) The transverse bands as a means of access to the perixonal space of the central myelinated nerve fibre. J. Ultrastruct. Res. 28, 141–149.

Hopwood, D. (1969) Fixatives and fixation: a review. Histochem. J. 1, 323–360.

Horrocks, L. A. (1967) Composition of myelin from peripheral and central nervous systems of the squirrel monkey. J. Lipid Res. 8, 569–576.

Horrocks, L. A. (1968) Composition of mouse brain myelin during development. J. Neurochem. 15, 483–488.

Horwitz, A. F. (1972) Nuclear magnetic resonance studies on phospholipids and membranes. In: Membrane Molecular Biology (Fox, C. F. and Keith, A. D., eds.), pp. 164–191. Sinauer Associates Inc. Stamford, Connecticut.

Hubbard, A. L. and Cohn, Z. A. (1976) Specific labels for cell surfaces. In: Biochemical Analysis of Membranes (Maddy, A. H., ed.), pp. 427–501. John Wiley, New York.

Hughes, R. C. (1973) Glycoproteins as components of cellular membranes. Prog. Biophys. Mol. Biol. 26, 191–268.

Hughes, R. C. (1975) The complex carbohydrates of mammalian cell surfaces and their biological roles. Essays in Biochemistry (Campbell, P. N. and Aldridge, W. N., eds.), vol. 11, pp. 1–36. Academic Press, London.

Hulcher, F. H. (1963) Physical and chemical properties of myelin. Arch. Biochem. Biophys. 100, 237–244.

Inglefield, P. T., Lindblom, K. A. and Gottlieb, A. M. (1976) Water binding and mobility in the phosphatidylcholine/cholesterol/water lamellar phase. Biochim. Biophys. Acta. 419, 196–205.

Ito, T. and Ohnishi, S. (1974) Ca^{2+}-induced lateral phase separations in phosphatidic acid-phosphatidylcholine membranes. Biochim. Biophys. Acta. 352, 29–37.

Jain, M. K. (1975) Role of cholesterol in biomembranes and related lipids. In: Current Topics in Membranes and Transport (Bronner, F. and Kleinzeller, A., eds.), vol. 6, pp. 1–57. Academic Press, New York.

James, R. and Branton, D. (1972) Composition, structure and phase transition in yeast fatty acid auxotroph membranes: Spin labels and freeze-fracture. J. Supramol. Struct. 1, 38–49.

Jenkinson, T. J., Kamat, V. B. and Chapman, D. (1969) Physical studies on myelin II. Biochim. Biophys. Acta. 163, 427–433.

Johnston, P. V. and Roots, B. I. (1964) Brain lipid fatty acids and temperature acclimatation. Comp. Biochem. Physiol. 11, 303–309.

Jones, A. J. S. (1975) D. Phil. Thesis, University of York, England.

Jones, A. J. S. and Rumsby, M. G. (1975a) Intrinsic fluorescence properties of the myelin basic protein. Fifth I.S.N. Conference, Barcelona, Abstr. No. 321.

Jones, A. J. S. and Rumsby, M. G. (1975b) The intrinsic fluorescence characteristics of the myelin basic protein. J. Neurochem. 25, 565–572.

Jost, P. C., Waggoner, A. S. and Griffith, O. H. (1971) Spin labelling and membrane structure. In: Structure and Function of Biological Membranes (Rothfield, L., ed.), pp. 83–144. Academic Press, New York.

Jost, P. C., Griffith, O. H., Capaldi, R. A. and Vanderkooi, G. (1973a) Evidence for boundary lipid in membranes. Proc. Nat. Acad. Sci. U.S.A. 70, 480–484.

Jost, P. C., Capaldi, R. A., Vanderkooi, G. and Griffith, O. H. (1973b) Lipid-protein and lipid-lipid interactions in cytochrome oxidase model membranes. J. Supramol. Struct. 1, 269–280.

Joy, R. T. and Finean, J. B. (1963) A comparison on the effect of freezing and of treatment with hypertonic solutions on the structure of myelin. J. Ultrastruct. Res. 8, 264–282.

Juliano, R. L. (1973) The proteins of the erythrocyte membrane. Biochim. Biophys. Acta. 300, 341–378.

Karlsson, U. (1966) Comparison of the myelin period of peripheral and central origin by electron microscopy. J. Ultrastruct. Res. 15, 451–468.

Kies, M. W. (1965) Chemical studies on an encephalitogenic protein from guinea pig brain. Ann. N.Y. Acad. Sci. 122, 161–170.

Kies, M. W. (1973) Experimental allergic encephalomyelitis. In: Biology of Brain Dysfunction. (Gaul,

G. E., ed.), vol. II, pp. 185–224. Plenum Press, New York.

Kies, M. W., Thompson, E. B. and Alvord, E. C. (1965) The relationship of myelin proteins to experimental allergic encephalomyelitis. Ann. N.Y. Acad. Sci. 122, 148–160.

Kies, M. W., Martenson, R. E. and Deibler, G. E. (1972) Myelin basic proteins. In: Functional and Structural Proteins of the Nervous System. (Davison, A. N., Mandel, P. and Morgan, I. G., eds.), pp. 201–214. Plenum Press, New York.

Kimelberg, H. K. (1977) The influence of membrane fluidity on the activity of membrane-bound enzymes. In: Dynamic Aspects of Cell Surface Organization (Poste, G. and Nicolson, G. L., eds.), Cell Surface Reviews, vol. 3. Elsevier/North-Holland, Amsterdam.

Kirkpatrick, F. H., Gordesky, S. E. and Marinetti, G. V. (1974) Differential solubilization of proteins, phospholipids and cholesterol of erythrocyte membranes by detergents. Biochim. Biophys. Acta. 345, 154–161.

Klemm, H. (1970) Das perineurium als diffusionsbarriere gegenuber peroxydase bei epi-nud endoneuraler applikation, Z. Zellforsch. 108, 431–445.

Korn, E. D. (1968) Structure and function of the plasma membrane. J. Gen. Physiol. 52, 257s–278s.

Korn, E. D. (1969) Cell membranes: structure and synthesis. Ann. Rev. Biochem. 38, 263–288.

Kornberg, R. D. and McConnell, H. M. (1971) Inside-outside transitions of phospholipids in vesicle membranes. Biochemistry 10, 1111–1120.

Kornguth, S. E. and Anderson, W. J. (1965) Localization of a basic protein (immunohistology) in the myelin of various species with the aid of fluorescence and electron microscopy. J. Cell Biol. 26, 157–166.

Kornguth, S. E. and Perrin, J. H. (1971) Circular dichroism and viscometric studies on a basic protein from pig brain. J. Neurochem. 18, 983–988.

de Kruyff, B., Van Dijek, P. W. M., Remel, R. A., Schnijff, A., Brants, F. and van Deenen, L. L. M. (1974) Non random distribution of cholesterol in phosphatidylcholine bilayers. Biochim. Biophys. Acta. 350, 1–7.

Kurihara, T. and Tsukada, Y. (1967) The regional and subcellular distribution of 2′, 3′-cyclic nucleotide 3′-phosphohydrolase in the central nervous system. J. Neurochem. 14, 1167–1174.

Ladbrooke, B. D., Jenkinson, T. J., Kamat, V. B. and Chapman, D. (1968) Physical studies of myelin. Biochim. Biophys. Acta. 164, 101–109.

Laine, R. A., Stellner, K. and Hakomori, S-I. (1974) Isolation and characterisation of membrane glycosphingolipids. In: Methods in Membrane Biology (Korn, E. D., ed.), vol. 2, pp. 205–244, Plenum Press, New York.

Lalitha, A. and Worthington, C. R. (1975) The swelling property of central nervous system membranes using x-ray diffraction. J. Mol. Biol. 96, 625–639.

Lang, R. D. A., Crosby, P. and Robards, A. W. (1976) An inexpensive spray-freezing unit for preparing specimens for freeze-etching. J. Microscop. 108, 101–104.

Lascelles, R. G. and Thomas, P. K. (1966) Changes due to age in internodal length in the sural nerve in man. J. Neurol. Neurosurg. Psychiat. 29, 40–44.

Lebaron, F. N. (1969) The lipid-protein complexes of myelin. In: Structural and Functional Aspects of Lipoproteins in Living Systems (Tria, E. and Scanu, A. M., eds.), pp. 201–226. Academic Press, London.

Lebaron, F. N. (1970) Metabolism of myelin constituents. In: Handbook of Neurochemistry (Lajtha A., Ed.), vol. 3, pp. 561–573, Plenum Press, New York.

Lecar, H., Ehrenstein, F. and Stillman, I. (1971) Detection of molecular motion in lyophilised myelin by NMR. Biophys. J. 11, 140–145.

Lee, A. G., Birdsall, N. J. M. and Metcalf, J. C. (1974) Nuclear magnetic relaxation and the biological membrane. In: Methods in Membrane Biology, (Korn, E. D., ed.), vol. 2, pp. 1–156, Plenum Press, New York.

Leeden, R. W., Yu, R. K. and Eng, L. F. (1973) Gangliosides of human myelin: Sialosylgalactosyl ceramide (G 7) as a major component. J. Neurochem. 21, 829–839.

Lees, M. B. and Paxman, S. A. (1974) Myelin proteins from different regions of the central nervous system. J. Neurochem 23, 825–831.

Lees, M. B., Sandler, S. W. and Eichberg, J. (1974) Effect of detergents on 2′,3′-cyclic nucleotide

3'-phosphohydrolase activity in myelin and erythrocyte ghosts. Neurobiol. 4, 407–413.

Leitch, G. J. (1966) Some ion exchange properties of a myelin extract from bovine optic nerve. Proc. Soc. Exp. Biol. Med. 121, 1253–1256.

Leitch, G. J., Horrocks, L. A. and Samorajski, T. (1969) Effects of cations on isolated bovine optic nerve myelin. J. Neurochem. 16, 1347–1354.

Lenard, J. and Singer, S. J. (1968) Alteration of the conformation of proteins in red blood cell membranes and in solution by fixative used in electron microscopy. J. Cell Biol. 37, 117–121.

Lenaz, G. (1973) Lipid-protein interactions in the structure of biological membranes. In: Membrane Structure and Mechanisms of Biological Energy Transduction, (Avary, J., ed.), pp. 455–526. Plenum Press, London.

Levine, Y. K. (1972) Physical studies of membrane structure. Prog. Biophys. Mol. Biol. 24, 1–74.

Levine, Y. K. and Wilkins, M. H. F. (1971) Structure of oriented lipid bilayers. Nature New Biol. 230, 69–72.

Levine, Y. K., Birdsall, N. J. M., Lee, A. G. and Metcalf, J. C. (1972) ^{13}C Nuclear magnetic resonance relaxation measurements of synthetic lecithin and the effect of spin-labelled lipids. Biochemistry 11, 1416–1421.

Liebes, L. F., Zand, R. and Phillips, W. D. (1975) Solution behaviour, circular dichroism and 220 MHz PMR studies on the bovine myelin basic protein. Biochim. Biophys. Acta. 405, 27–39.

Linington, C. and Rumsby, M. G. (1977) Localization of cerebroside in central nervous system myelin lamellae. An initial approach. Biochem. Soc. Trans. 5:196–198.

Lippert, J. L. and Peticolas, W. L. (1971) Laser Raman investigation of the effect of cholesterol on conformation changes in dipalmitoyl lecithin multilayers. Proc. Nat. Acad. Sci. U.S.A. 68, 1572–1576.

Litman, R. B. and Barrnett, R. J. (1972) The mechanism of the fixation of tissue components by osmium tetroxide via hydrogen bonding. J. Ultrastruct. Res. 38, 63–86.

Livingston, R. B., Pfenninger, K., Moor, H. and Akert, K. (1973) Specialised paranodal and inter-paranodal glial-axonal junctions in the peripheral and central nervous systems: a freeze etching study. Brain Res. 58, 1–24.

London, Y. (1974) Isolation of two basic proteins from peripheral nerve myelin. In: Methods in Enzymology, (Fleischer, S. and Packer, L., eds.), vol. 32B, pp. 341–345. Academic Press, New York.

London, Y. and Vossenberg, F. G. A. (1973) Specific interaction of central nervous system myelin basic protein with lipids. Biochim. Biophys. Acta. 307, 478–490.

London, Y., Demel, R. A., Geurts van Kessel, W. S. M., Vossenberg, F. G. A. and van Deenen, L. L. M. (1973) The protection of A$_1$ myelin basic protein against the action of proteolytic enzymes after interaction of the protein with lipids at the air-water interface. Biochim. Biophys. Acta. 311, 520–530.

London, Y., Demel, R. A., Guerts van Kessel, W. S. M., Zahler, P. and van Deenen, L. L. M. (1974) The interaction of the "Folch-Lees" protein with lipids at the air-water interface. Biochim. Biophys. Acta. 332, 69–84.

Luft, J. H. (1971) Ruthenium red and violet. II. Fine structural localization in animal tissue. Anat. Rec. 171, 369–416.

Lumsden, C. E., Robertson, D. M. and Blight, R. J. (1966) Chemical studies on experimental allergic encephalomyelitis. Peptide as the common denominator in all encephalitogenic 'antigens'. J. Neurochem. 13, 127–162.

Luzzati, V. (1968) X-ray diffraction studies on lipid-water systems. In: Biological Membranes (Chapman, D., ed.), pp. 71–123. Academic Press, London.

Luzzati, V. (1974) X-ray diffraction approach to the structure of biological membranes. In: Perspectives in Membrane Biology (Estrada-O, S. and Gitler, C., eds.), pp. 25–43. Academic Press, New York.

MacBrinn, M. C. and O'Brien, J. S. (1969) Lipid composition of optic nerve myelin. J. Neurochem. 16, 7–12.

Mackay, I. R., Carnegie, P. R. and Coates, A. S. (1973) Immunopathological comparisons between experimental autoimmune encephalomyelitis and multiple sclerosis. Clin. Exp. Immunol. 15, 471–482.

Maddy, A. H. (1964) The solubilisation of the protein of the ox-erythrocyte ghost. Biochim. Biophys. Acta. 88, 448–449.

Maddy, A. H. (1966) The properties of the protein of the plasma membrane of ox erythrocytes. Biochim. Biophys. Acta. 117, 193–200.

Maddy, A. H. and Kelly, P. G. (1970) A mild procedure for the fractionation of ox erythrocyte membrane proteins. FEBS. Lett. 8, 341–344.

Marks, N. (1971) Peptide hydrolases. In: Handbook of Neurochemistry (Lajtha, A., ed.), vol. 3, pp. 133–171, Plenum Press, New York.

Marks, N., Grynbaum, A. and Lajtha, A. (1975) Breakdown of myelin bound proteins *in situ* by endogenous enzymes. Fifth I.S.N. Conference, Abstr. No. 336.

Marsh, D. (1975) Spectroscopic studies of membrane structure. In: Essays in Biochemistry (Campbell, P. N. and Aldridge, W. N., eds.), vol. 11, pp. 139–180. Academic Press, London.

Marsh, D. and Smith, I. C. P. (1972) Interacting spin labels as probes of molecular separation within phospholipid bilayers. Biochem. Biophys. Res. Commun. 49, 916–922.

Marsh, D. and Smith, I.C. P. (1973) An interacting spin label study of the fluidising and condensing effects of cholesterol on lecithin bilayers. Biochim. Biophys. Acta. 298, 133–144.

Martenson, R. E., Deibler, G. E. and Kies, M. W. (1969) Microheterogeneity of guinea pig myelin basic protein. J. Biol. Chem. 244, 4261–4272.

Martenson, R. E., Deibler, G. E. and Kies, M. W. (1970) Rat myelin basic proteins; relationship between size differences and microheterogeneity. J. Neurochem. 17, 1329–*1330*.

Martin-Lomas, M. and Chapman, D. (1973) Structural studies on glycolipids. Chem. Phys. Lipids 10, 152–164.

Mateu, G., Luzzati, L., London, Y., Gould, R. M. and Vossenberg, F. G. A. (1973) X-ray diffraction and electron microscopy study of the interactions of myelin components. J. Mol. Biol. 75, 697–709.

Matsumoto, M., Matsumoto, R. and Folch-Pi, J. (1964) The chromatographic fractionation of brain white matter proteolipids. J. Neurochem. 11, 829–838.

Matthews, M. M. and Duncan, D. (1971) A quantitative study of morphological changes accompanying the initiation and progress of myelin production in the dorsal funiculus of the rat spinal cord. J. Comp. Neurol. 142, 1–22.

Matthieu, J-M., Daniel, A., Quarles, R. H. and Brady, R. O. (1974) Interaction of concanavalin A and other lectins with central nervous system myelin. Brain Res. 81, 348–353.

Matus, A., de Petris, S. and Raff, M. L. (1973) Mobility of concanavalin A receptors in myelin and synaptic membranes. Nature New Biol. 204, 278–280.

McFarland, H. F. (1970) Immunofluorescent study of circulating antibody in experimental allergic encephalomyelitis. Proc. Soc. Exp. Biol. Med. 133, 1195–1200.

McIlwain, D. L. (1973) Non-osmotic swelling in purified bovine myelin. Brain Res. 52, 97–113.

McIlwain, D. L. and Rapport, M. M. (1971) The effects of phospholipase C (*Clostridium perfringens*) on purified myelin. Biochim. Biophys. Acta. 239, 71–80.

McIlwain, D. L., Graf. L. and Rapport, M. M. (1971) Membrane fragments from myelin treated with different detergents. J. Neurochem. 18, 2255–2263.

McIntosh, T. J. and Robertson, J. D. (1976) Observations on the effect of hypotonic solutions on the myelin sheath in the central nervous system. J. Mol. Biol. 100, 213–217.

McIntosh, T. J. and Worthington, C. R. (1973) The choice between the positive and negative structures for nerve myelin. Biophys. J. 13, 492–500.

McIntosh, T. J. and Worthington, C. R. (1974a) Direct determination of the lamellar structure of peripheral nerve myelin at low resolutions. Biophys. J. 14, 363–386.

McIntosh, T. J. and Worthington, C. R. (1974b) The lamellar structure of nerve myelin after rehydration. Arch. Biochem. Biophys. 162, 523–529.

McMillan, P. N., Mickey, D. D., Kaufman, B. and Day, E. D. (1971) The specificity and cross reactivity of antimyelin antibodies as determined by sequential adsorption analysis. J. Immunol., 107, 1611–1617.

McNamee, M. G. and McConnell, H. M. (1973) Transmembrane potentials and phospholipid flip-flop in excitable membrane vesicles. Biochemistry 12, 2951–2958.

McNutt, N. S. (1977) Freeze-fracture techniques and applications to the structural analysis of the mammalian plasma membrane. In: Dynamic Aspects of Cell Surface Organization (Poste, G. and

354

Nicolson, G. L., eds.), Cell Surface Reviews, vol. 3, pp. 75–126. Elsevier/North-Holland, Amsterdam.

McPherson, T. A. and Carnegie, P. R. (1968) Radioimmunoassay with gel filtration for detecting antibody to basic proteins of myelin. J. Lab. Clin. Med. 72, 824–831.

Mehl, E. (1972) Separation and characterization of myelin proteins. Adv. Exp. Med. Biol. 32, 157–170.

Mehl, E. and Halaris, A. (1970) Stoichiometric relation of protein components in central myelin from different species. J. Neurochem. 17, 659–668.

Mehl, E. and Wolfgram, F. (1969) Myelin types with different protein components in the same species. J. Neurochem. 16, 1091–1097.

Miyamoto, E. (1975) Protein kinases in myelin of rat brain: solubilization and characterisation. J. Neurochem. 24, 503–512.

Miyamoto, E. and Kakiuchi, S. (1975) Phosphoprotein phosphatases for myelin basic protein in myelin and cytosol fractions of brain. Biochim. Biophys. Acta. 384, 458–465.

Mokrasch, L. C. (1969) Myelin. In: Handbook of Neurochemistry (Lajtha, A., ed.), vol. 1., pp. 171–193, Plenum Press, New York.

Mokrasch, L. C. (1971a) Historical introduction. In Myelin. (Mokrasch, L. C., Bear, R. S. and Schmitt, F. O., eds.), Neurosciences Research Bulletin, vol. 9, pp. 445–451. Neurosciences Research Program. Brookline, U.S.A.

Mokrasch, L. C. (1971b) Purification and properties of isolated myelin. In: Methods of Neurochemistry (Friede, R. L., ed.), vol. 1. pp. 1–29. Marcel Dekker, New York.

Mokrasch, L. C., Bear, R. S. and Schmitt, F. O., Eds. (1971) Myelin. Neurosciences Research Bulletin, vol. 9., No. (4), pp. 440–598. Neurosciences Research Program, Brookline, Mass.

Moore, W. V. and Wetlaufer, D. B. (1973) Circular Dichroism of nerve membrane fractions: effect of temperature, pH and electrolytes. J. Neurochem. 20, 135–149.

Morell, P., Greenfield, S., Constantino-Ceccarin, E. and Wisniewski, H. (1972) Changes in the protein composition of mouse brain myelin during development. J. Neurochem. 19, 2545–2554.

Morell, P., Lipkind, R. and Greenfield, S. (1973) Protein composition of myelin from brain and spinal cord of several species. Brain Res. 58, 510–514.

Morell, P., Wiggins, R. C. and Gray, M. J. (1975) Polyacrylamide gel electrophoresis of myelin proteins: A caution. Anal. Biochem. 68, 148–154.

Morton, R. K. (1955) Methods of extraction of enzymes from animal tissues. In: Methods of Enzymoloy, (Colwick, S. P. and Kaplan, N. O., eds.), vol. 1. pp. 25–50, Academic Press, New York.

Moscarello, M. A., Gagnon, J., Wood, P. D., Anthony, J. and Epand, R. (1973) Conformational flexibility of a myelin protein. Biochemistry 12, 3402–3406.

Moscarello, M. A., Katona, E., Neumann, A. W. and Epand, R. M. (1974) The ordered structure of the encephalitogenic protein from normal human myelin. Biophys. Chem. 2, 290–295.

Nakao, A., Davis, W. J. and Einstein, E. R. (1966) Basic proteins from the acidic extract of bovine spinal cord. I Isolation and characterisation. Biochim. Biophys. Acta. 130, 163–170.

Napolitano, L. M. and Scallen, T. J. (1969) Observations on the fine structure of peripheral nerve myelin. Anat. Record. 163, 1–6.

Napolitano, L. M., Lebaron, F. and Scaletti, J. (1967) Preservation of myelin lamellar structure in the absence of lipid. A correlated chemical and morphological study. J. Cell Biol. 34, 817–826.

Nicolson, G. L., Poste, G., and Ji, T. H. (1977). The dynamics of cell membrane organization. In: Dynamic Aspects of Cell Surface Organization, vol. 3, Cell Surface Reviews (Poste, G. and Nicolson, G. L., eds.) pp. 1–73, Elsevier/North-Holland, Amsterdam.

Nicot, C., Nguyen Le, T., Lepretre, M. and Alfsen, A. (1973) Study of Folch-Pi apoprotein. I. Isolation of two components, aggregation during delipidation. Biochim. Biophys. Acta. 332, 109–123.

Niedieck, B. (1975) On a glycolipid hapten of myelin. Prog. Allergy 18, 353–422.

Norton, W. T. (1971) Recent developments in the investigation of purified myelin, In: Chemistry and Brain Development (Paoletti, R. and Davison, A. N., eds.), pp. 327–337. Plenum Press, New York.

Norton, W. T. (1972) Myelin. In: Basic Neurochemistry (Albers, W., Siegel, G., Katzman, R. and

Agranoff, B., eds.), pp. 365–386. Little, Brown, Boston, Mass.

Norton, W. T. (1974) Isolation of myelin from nerve tissue. In: Methods of Enzymology (Fleischer, S. and Packer, L., eds)., vol. 31A, pp. 435–444. Academic Press, New York.

Norton, W. T. (1975) Myelin: structure and biochemistry. In: The Nervous System (Tower, D. B., ed.), vol. 1., pp. 467–481, Raven Press, New York.

Norton, W. T. (1977) Isolation and characterisation of myelin. In: 'Myelin' (Morell, P., ed.), Plenum Press, New York.

Norton, W. T. and Autilio, L. A. (1965) The chemical composition of bovine CNS myelin. Ann. N.Y. Acad. Sci. 122, 77–85.

Norton, W. T. and Autilio, L. A. (1966) The lipid composition of purified bovine brain myelin. J. Neurochem. 13, 213–222.

Norton, W. T. and Brotz, M. (1963) New galactolipids of brain: a monoalkylmonoacyl-glyceryl galactoside and cerebroside fatty acid esters. Biochem. Biophys. Res. Commun. 12, 198–203.

Norton, W. T. and Brotz, M. (1967) The glyceryl galactoside derivatives of brain. Fed. Proc. 26, 675.

Norton, W. T. and Poduslo, S. E. (1973a) Myelination in rat brain: method of myelin isolation. J. Neurochem. 21, 749–758.

Norton, W. T. and Poduslo, S. E. (1973b) Myelination in rat brain: changes in myelin composition during brain maturation. J. Neurochem. 21, 759–773.

Nussbaum, J. L., Bieth, R. and Mandel, P. (1963) Phosphatides in myelin sheaths and repartition of sphingomyelin in the brain. Nature 198, 586–587.

Nussbaum, J. L., Rouayrenc, J. F., Jollès, J., Jollès, P. and Mandel, P. (1974a) Amino acid analysis and N-terminal sequence determination of P7 proteolipid apoprotein from human myelin. FEBS. Letts. 45, 295–298.

Nussbaum, J. L., Rouayrenc, J. L., Jollès, J. and Jollès P. (1974b) Isolation and terminal sequence determination of the major rat brain myelin proteolipid P7 apoprotein. Biochem. Biophys. Res. Commun. 57, 1240–1247.

O'Brien, J. S. (1970a) Chemical composition of myelinated nervous tissue. In: Handbook of Clinical Neurology (Vinken, B. and Bruyn, G., eds.), vol. 7, pp. 40–61. Elsevier/North-Holland, Amsterdam.

O'Brien, J. S. (1970b) Lipids and Myelination. In: Developmental Neurobiology (Himwich, W. A., ed.), pp. 262–286. C.C. Thomas, Springfield, Illinois.

O'Brien, J. S. (1971) Lipids. In: Myelin. Neurosciences Research Program Bulletin, vol. 9, no. 4. Neurosciences Research Program. Brookline, Mass.

O'Brien, J. S. and Sampson, E. L. (1965a) Lipid composition of the normal human brain: gray matter, white matter and myelin. J. Lipid Res. 6, 537–544.

O'Brien, J. S. and Sampson, E. L. (1965b) Fatty acid and fatty aldehyde composition of the major lipids in normal human gray matter, white matter and myelin. J. Lipid Res. 6, 545–551.

Ohnishi, S. and Ito, T. (1974) Calcium induced phase separations in phosphatidyl serine-phosphatidylcholine membranes. Biochemistry 13, 881–887.

Olafson, R. W., Drummond, G. I. and Lee, J. F. (1969) Studies on 2', 3'-cyclic nucleotide - 3'-phosphohydrolase from brain. Can. J. Biochem. 47, 961–966.

Oldani, D., Hauser, H., Nichols, B. W. and Phillips, M. C. (1975) Monolayer characteristics of some glycolipids at the air-water interface. Biochim. Biophys. Acta. 382, 1–9.

Oldfield, E. and Chapman, D. (1972a) Molecular dynamics of cerebroside-cholesterol and sphingomyelin-cholesterol interactions for myelin membrane structure. FEBS. Lett. 21, 303–306.

Oldfield, E. and Chapman, D. (1972b) Dynamics of lipids in membranes: heterogeneity and the role of cholesterol. FEBS. Lett. 23, 285–297.

Olive, J. (1973) Cryo-ultramicrotomy and freeze-etching of lipid-water phases, In: Freeze Etching Techniques and Applications, (Benedetti, E. L. and Favard, P., eds.), pp. 187–198. Société Française de Microscopie Electronique, Paris.

Oseroff, O., Robbins, P. W. and Burger, M. M. (1973) The cell surface membrane: biochemical aspects and biophysical probes. Ann. Rev. Biochem. 42, 647–682.

Oxberry, J. M. and Gregson, N. A. (1974) The agglutination of myelin suspensions by specific antisera. Brain Res. 78, 303–313.

Palmer, F. and Dawson, R. M. C. (1969) The isolation and properties of experimental allergic en-

356

cephalitogenic protein. Biochem. J. 111, 629–636.

Paltauf, F., Hauser, H. and Phillips, M. C. (1971) Monolayer characteristics of some 1,2-diacyl, 1-alkyl-2-acyl and 1,2-dialkyl phospholipids at the air-water interface. Biochim. Biophys. Acta. 249, 539–547.

Papahadjopoulos, D. (1973) Phospholipids as model membranes: monolayers, bilayers, and vesicles. In: Form and Function of Phospholipids (Ansell, G. B., Hawthorne, J. N. and Dawson, R. M. C., eds.), pp. 143–169. Elsevier/North-Holland, Amsterdam.

Papahadjopoulos, D. and Kimelberg, H. K. (1973) Phospholipid vesicles (liposomes) as models for biological membranes: their properties and interactions with cholesterol and proteins. In: Progress in Surface Science, (Davison, S. G., ed.), vol. 4, pp. 141–232. Pergamon Press, Oxford.

Papahadjopoulos, D. and Weiss, L. (1969) Amino groups at the surface of phospholipid vesicles. Biochim. Biophys. Acta. 183, 415–426.

Papahadjopoulos, D., Nir, S. and Ohki, S. (1972) Permeability properties of phospholipid membranes: effect of cholesterol and temperature. Biochim. Biophys. Acta. 266, 561–583.

Papahadjopoulos, D., Poste, G., Schaeffer, B. E. and Vail, W. J. (1974) Membrane fusion and molecular segregation in phospholipid vesicles. Biochim. Biophys. Acta. 352, 10–28.

Papahadjopoulos, D., Moscarello, M., Eylar, E. H. and Isac, T. (1975a) Effects of proteins on thermotropic phase transitions of phospholipid membranes. Biochim. Biophys. Acta. 401, 317–335.

Papahadjopoulos, D., Vail, W. J. and Moscarello, M. (1975b) Interaction of a purified hydrophobic protein from myelin with phospholipid membranes: studies on ultrastructure, phase transitions and permeability. J. Membrane Biol. 22, 143–164.

Parsons, D. F. and Subjeck, J. R. (1972) The morphology of the polysaccharide coat of mammalian cells. Biochim. Biophys. Acta. 265, 85–113.

Peters, A. (1968) The morphology of axons of the central nervous system. In: The Structure and Function of Nervous Tissue (Bourne, G. H., ed.), vol. 1, pp. 141–196. Academic Press, New York.

Peters, A. and Proskauer, C. C. (1969) The ratio between myelin segments and oligodendrocytes in the optic nerve of the adult rat. Anat. Record 163, 243 (abstr.)

Peters, A. and Vaughn, J. E. (1970) Morphology and development of the myelin sheath. In: Myelination. (Davison, A. N. and Peters, A., eds.), pp. 3–79. C. C Thomas, Springfield, Ill.

Peters, A., Sanford, L. P. and Webster, H. de F. (1970) The Fine Structure of the Nervous System. Harper and Row, London.

Peters, A., Sanford, L. P. and Webster, H. de F. (1976) The Fine Structure of the Nervous System. Harper and Row, London. In press.

Peterson, R. G. (1975) Electron microscopy of trypsin-digested peripheral nerve myelin. J. Neurocytol. 4, 115–120.

Peterson, R. G. (1976) Myelin protein changes with digestion of whole sciatic nerve in trypsin. Life Sci. 18, 845–850.

Peterson, R. G. and Pease, D. C. (1972) Myelin embedded in polymerized glutaraldehyde-urea. J. Ultrastruct. Res. 41, 115–132.

Phillips, M. C. (1972) The physical state of phospholipids and cholesterol in monolayers, bilayers and membranes. Prog. Surf. Memb. Sci. 5, 139–221.

Phillips, M. C. and Finer, E. G. (1974) The stoichiometry and dynamics of lecithin-cholesterol clusters in bilayer membranes. Biochim. Biophys. Acta. 356, 199–206.

Pieringer, R. A., Deshmukh, D. S. and Flynn, T. J. (1973) The association of galactosyl diglycerides of nerve tissue with myelination. In: Progress in Brain Research (Ford, D., ed.), vol. 40, pp. 397–405. Elsevier/North-Holland, Amsterdam.

Pinto da Silva, P. and Miller, R. M. (1975) Membrane particles on fracture faces of frozen myelin. Proc. Nat. Sci. U.S.A. 72, 404–406.

Poduslo, J. F. (1975a) Distribution of galactose residues in surface membrane glycoproteins and glycolipids of the intact myelin sheath. Fifth I.S.N. Conference, Barcelona, Abstr. No. 319.

Poduslo, S. E. (1975b) The isolation and characterization of a plasma membrane and a myelin fraction derived from oligodendroglia of calf brain. J. Neurochem. 24, 647–654.

Poduslo, J. F. and Braun, P. E. (1975) Topographical arrangement of membrane proteins in the intact myelin sheath. J. Biol. Chem. 250, 1099–1105.

Poduslo, J. F., Quarles, R. H. and Brady, R. O. (1976) External labelling of galactose in surface membrane glycoproteins of the intact myelin sheath. J. Biol. Chem. 251, 153–158.

Poo, M-M. and Cone, R. A. (1974) Lateral diffusion of rhodopsin in the photoreceptor membrane. Nature 247, 438–441.

Pritchard, E. T. and Folch-Pi, J. (1963) Tightly bound proteolipid phospholipid in bovine brain white matter. Biochim. Biophys. Acta. 70, 481–483.

Quarles, R. H., Everly, J. L. and Brady, R. O. (1972) Demonstration of a glycoprotein which is associated with a purified myelin fraction from rat brain. Biochem. Biophys. Res. Commun. 47, 491–497.

Quarles, R. H., Everly, J. L. and Brady, R. O. (1973a) Evidence for the close association of a glycoprotein with myelin in rat brain. J. Neurochem. 21, 1177–1191.

Quarles, R. H., Everly, J. L. and Brady, R. O. (1973b) Myelin-associated glycoprotein: a developmental change. Brain Res. 58, 506–509.

Quinn, P. J. and Sherman, W. R. (1971) Monolayer characteristics and calcium adsorption to cerebroside and cerebroside sulphate oriented at the air-water interface. Biochim. Biophys. Acta. 233, 734–752.

Raghavan, S. S., Rhoads, D. B. and Kanfer, J. N. (1973) The effects of trypsin on purified myelin. Biochim. Biophys. Acta. 328, 205–212.

Ranck, J. L., Mateu, L., Sadler, D. M., Tardieu, A., Gulik-Krzywicki, T. and Luzzati, V. (1974) Order-disorder conformational transitions of the hydrocarbon chains of lipids. J. Mol. Biol. 85, 249–277.

Rand, R. P. and Luzatti, J. V. (1968) X-ray diffraction study in water of lipids extracted from human erythrocytes. Biophys. J. 8, 125–137.

Raper, J. H., Gammack, D. B. and Sloane-Stanley, G. H. (1966) A study of cerebral sphingomyelins in monomolecular films. Biochem. J. 98, 21p.

Rapport, M. M. and Graf, L. (1965) Immunological reactions of myelin in vitro. Ann. N.Y. Acad. Sci. 122, 277–279.

Rapport, M. M. and Graf, L. (1969) Immunochemical reactions of lipids. Prog. Allergy, 13, 273–331.

Rauch., H. C. and Raffel, S. (1964) Immunofluorescent localization of encephalitogenic protein in myelin. J. Immunol. 92, 452–455.

Rawlins, F. A. (1973) A time-sequence autoradiographic study of the in vivo incorporation of (1,2,3-H) cholesterol into peripheral nerve myelin. J. Cell Biol. 58, 42–53.

Reale, E., Luciano, L. and Spitznas, M. (1975) Zonulae occludentes of the myelin lamellae in the nerve fibre layer of the retina and in the optic nerve of the rabbit: a demonstration by the freeze-fracture method. J. Neurocytol. 4, 131–140.

Reese, T. S., Feder, N. and Brightman, M. W. (1971) Electron microscopic study of the blood-brain and blood-cerebrospinal fluid barriers with microperoxidase. J. Neuropathol. Exp. Neurol. 30, 137–138.

Rega, A. F., Weed, R. I., Reed, C. F., Berg, G. G. and Rothstein, A. (1967) Changes in the properties of human erythrocyte membrane proteins after solubilisation by butanol extraction. Biochim. Biophys. Acta. 147, 297–312.

Reiss-Husson, F. (1967) Structure des phase liquida-crystallines de differents phospholipides, monoglycerides, sphingolipides, anhydres ou en presence d'eau. J. Mol. Biol. 25, 363–382.

Revel, J-P. and Hamilton, D. W. (1969) The double nature of the intermediate dense line in peripheral nerve myelin. Anat. Record, 163, 7–16.

Reynolds, J. A. and Green, H. O. (1973) Polypeptide chains from porcine cerebral myelin. J. Biol. Chem. 248, 1207–1210.

Reynolds, J. A., Gallagher, J. P. and Steinhardt, J. (1970) The effect of pH on the binding of N-alkyl sulphates to bovine serum albumin. Biochemistry 9, 1232–1238.

Richards, F. M. and Knowles, J.R. (1968) Glutaraldehyde as a protein cross-linking reagent. J. Mol. Biol. 37, 231–233.

Riekkinen, P. J. and Clausen, J. (1969) Neutral proteinase activity of myelin. Brain Res. 15, 413–430.

Riekkinen, P. J. and Rumsby, M. G. (1972) Aminopeptidase and neutral proteinase activity associated with central nerve myelin preparations during purification. Brain Res. 41, 512–517.

Roozemond, R. C. (1969) The effect of fixation with formaldehyde and glutaraldehyde on the composition of phospholipids extractable from rat hypothalamus. J. Histochem. Cytochem. 17, 482–486.

Rothfield, L. (1971) Biological membranes: an overview at the molecular level, In: Structure and Function of Biological Membranes (Rothfield, L., ed.), pp. 3–9. Academic Press, New York.

Rothstein, A., Cabantchik, Z. I. and Knauf, P. (1976) Mechanism of anion transport in red blood cells: role of membrane proteins. Fed. Proc. 35, 3–10.

Rouser, G., Nelson, G. J., Fleischer, S. and Simon, G. (1968) Lipid composition of animal cell membranes, organelles and organs. In: Biological Membranes: Physical Fact and Function (Chapman, D., ed.), pp. 5–70. Academic Press, New York.

Roytta, M., Frey, H., Riekkinen, P.J., Laaksonen, H. and Rinne, U.K. (1974) Myelin breakdown and basic protein. Exp. Neurol. 45, 174–185.

Rumsby, M. G. (1963) Ph.D. Thesis, University of Birmingham, England.

Rumsby, M. G. (1967) Preparation and characterisation of a glycerogalactolipid fraction from sheep brain. J. Neurochem. 14, 733–741.

Rumsby, M. G. and Finean, J. B. (1966a) The action of organic solvents on the myelin sheath of peripheral nervous tissue. I. Methanol, ethanol, chloroform and chloroform-methanol (2:1, v/v) J. Neurochem. 13, 1501–1507.

Rumsby, M. G. and Finean, J. B. (1966b) The action of organic solvents on the myelin sheath of peripheral nervous tissue. II. Short chain aliphatic alcohols. J. Neurochem. 13, 1509–1511.

Rumsby, M. G. and Finean, J. B., (1966c) The action of organic solvents on the myelin sheath of peripheral nervous tissue. III. Chlorinated hydrocarbons. J. Neurochem. 13, 1513–1515.

Rumsby, M. G. and Glaisher, M. P. (1971) The action of n-butanol on isolated central nerve myelin preparations. Brain Res. 35, 576–579.

Rumsby, M.G. and Rossiter, R.J. (1968) Alkyl ethers from the glycerogalactolipid fraction of nerve tissue. J. Neurochem. 15, 1473–1476.

Rumsby, M. G., Riekkinen, P. J. and Arstila, A. V. (1970) A critical evaluation of myelin purification: Non-specific esterase activity associated with central nerve myelin preparations. Brain Res. 24, 495–516.

Rumsby, M.G., Getliffe, H.M. and Riekkinen, P.J. (1973) On the association of non-specific esterase activity with central nerve myelin preparations. J. Neurochem. 21, 959–967.

Sabban, E., Laster, Y. and Loyter, A. (1972) Resolution of the hemolytic and the hydrolytic activities of phospholipase-C preparation from Clostridium perfringens. Eur. J. Biochem. 28, 373–380.

Saito, M., Nagai, Y. and Tsumita, T. (1972) A novel method for the extraction of encephalitogenic protein from bovine brain. Jap. J. Exp. Med. 42, 473–481.

Salsbury, M. J., Darke, A. and Chapman, D. (1972) Deuteron magnetic resonance studies of water associated with phospholipids. Chem. Phys. Lipids 8, 142–151.

Sammeck, R. and Brady, R. O. (1972) Studies of the catabolism of myelin basic proteins of the rat in situ and in vitro. Brain Res. 92, 441–453.

Samorajski, T. and Friede, R. L. (1968) A quantitative electron microscopic study of myelination in the pyramidal tract of rat. J. Comp. Neurol. 134, 323–337.

Savolainen, H. (1972) Proteins and glycoproteins of human myelin and glial cell membrane with special reference to myelin formation. T.-I.-T. J. Life Sci. 2, 35–38.

Schafer, R. and Franklin, R. M. (1975) Resistance of the basic membrane proteins of myelin and bacteriophage PM2 to proteolytic enzymes. FEBS. Lett. 58, 265–268.

Schmid, G., Thomas, G., Hempel, K. and Gruninger, W. (1974) Radioimmunological determination of myelin basic protein (MBP) and MBP-Antibodies. Europ. Neurol 12, 173–185.

Schnapp, B. and Mugnaini, E. (1975) The myelin sheath: electron microscopic studies with thin sections and freeze fracture. In: Golgi Centennial Symposium Proceedings (Santini, M., ed.), pp. 209–233. Raven Press, New York.

Schoenborn, B. P. (1976) Neutron scattering for the analysis of membranes. Biochim. Biophys. Acta. 457, 41–56.

Schreier-Mucillo, S., Marsh, D., Dugas, H., Schneider, H. and Smith, I. C. P. (1973) A spin probe study of the influence of cholesterol on motion and orientation of phospholipids in oriented multibilayers and vesicles. Chem. Phys. Lipids 10, 11–27.

Schummer, U., Hegner, D., Schnepel, G. H. and Wellhaner, H. H. (1975) Investigations of thermotropic phase changes in peripheral nerve of frog and rat. Biochim. Biophys. Acta. 394, 93–101.

Shah, D. O. and Schulman, J. H. (1965) Binding of metal ions to monolayers of lecithins, plasmalogen, cardiolipin and diacetyl phosphate. J. Lipid Res. 6, 341–349.

Sheltawy, A. and Dawson, R. M. C. (1968) On the phosphatidic acid of myelin. J. Neurochem. 15, 144–146.

Sherman, G. and Folch-Pi, J. (1970) Rotatory dispersion and circular dichroism of brain proteolipid protein. J. Neurochem. 17, 400–409.

Shimshick, E. J. and McConnell, H. M. (1973a) Lateral phase separation in phospholipid membranes. Biochemistry 12, 2351–2360.

Shimshick, E. J. and McConnell, H. M. (1973b) Lateral phase separations in binary mixtures of cholesterol and phospholipids. Biochem. Biophys. Res. Commun. 53, 446–451.

Shipley, G. G. (1973) Recent X-ray diffraction studies of biological membranes and membrane components. In: Biological Membranes (Chapman, D. and Wallach, D. F. H., eds.), pp. 1–89. Academic Press, London.

Shooter, E. M. and Einstein, E. R. (1971) Proteins of the nervous system. Ann. Rev. Biochem. 40, 635–652.

Sima, A. (1974) Relation between the number of myelin lamellae and axon circumference in fibres of ventral and dorsal roots and optic nerve in normal, undernourished and rehabilitated rats; An ultrastructural morphometric study. Acta. Physiol. Scand. Suppl. 410, 1–38.

Sims, N. R. and Carnegie, P. R. (1975) Release of 2′ 3′ cyclic nucleotide 3′ phosphohydrolase from myelin-useful model of membrane protein solubilisation. Proc. Aust. Biochem. Soc. 8, 75.

Singer, S. J. (1971) Molecular organisation of membranes. In: Structure and Function of Biological Membranes (Rothfield, L. I., ed.), pp. 145–222. Academic Press, New York.

Singer, S. J. (1974a) The molecular organisation of membranes. Ann. Rev. Biochem. 43, 805–833.

Singer, S. J. (1974b) Molecular biology of cellular membranes with applications to immunology. Adv. Immunol. 19, 1–66.

Singer, S. J. (1974c) On the fluidity and assymmetry of biological membranes. In: Perspectives in Membrane Biology (Estrado-O, S. and Gitler, C., eds.), pp. 131–148. Academic Press, New York.

Singer, S. J. and Nicolson, G. L. (1972) The fluid mosaic model of the structure of cell membranes. Science 175, 720–731.

Singh, H. and Spritz, N. (1974) Polypeptide components of myelin from rat peripheral nerve. Biochim. Biophys. Acta. 351, 379–386.

Singh, H., Spritz, N. and Geyer, B. (1971) Studies of brain myelin in the 'quaking mouse'. J. Lipid Res. 12, 473–481.

Smith, M. E. (1967) The metabolism of myelin lipids. Adv. Lipid Res. 5, 241–278.

Sonenburg, M. (1971) Interaction of human growth hormone and human erythrocyte membranes studied by intrinsic fluorescence. Proc. Nat. Acad. Sci. U.S.A. 68, 1051–1055.

Soto, E. F., Seminario, L. B. and del Carmen, M. C. (1966) Chemical composition of myelin and other subcellular fractions isolated from bovine white matter. J. Neurochem. 13, 989–998.

Spatz, L. and Strittmatter, P. (1971) A form of cytochrome b5 that contains an additional hydrophobic sequence of 40 amino acid residues. Proc. Nat. Acad. Sci. U.S.A. 68, 1042–1046.

Spohn, M. and Davison, A. N. (1972) Separation of myelin fragments from the central nervous system. In: Research Methods in Neurochemistry (Marks, E.D. and Rodnight, R. eds.), pp. 33–43. Plenum Press, New York.

Staehelin, L. A. (1973) Further observations on the fine structure of freeze-cleaved tight junctions. J. Cell Sci. 13, 763–786.

Steck, T. L. (1974) The organisation of proteins in the red blood cell membrane. J. Cell Biol. 62, 1–19.

Stoffyn, P. and Folch-Pi, J. (1971) On the type of linkage binding fatty acids present in brain white matter proteolipid apoprotein. Biochem. Biophys. Res. Commun. 44, 157–161.

Stolinski, C. and Breathnach, A. S. (1975) Freeze Fracture Replication of Biological Tissue: Techniques, Interpretations and Applications. Academic Press, New York.

Suzuki, K. (1970) Formation and turnover of myelin ganglioside. J. Neurochem. 17, 209–213.

Suzuki, K. (1972) Chemistry and metabolism of brain lipids. In: Basic Neurochemistry (Albers, R.

360

W., Siegel, G. J., Katzman, R. and Agranoff, B. W., eds.), pp. 207–228. Little, Brown, Boston.

Suzuki, K., Poduslo, S. E. and Norton, W. T. (1967) Gangliosides in the myelin fraction of developing rats. Biochim. Biophys. Acta. 144, 375–381.

Suzuki, K., Poduslo, J. F. and Poduslo, S. E. (1968) Further evidence for a specific ganglioside fraction closely associated with myelin. Biochim. Biophys. Acta. 152, 576–586.

Svennerholm, L. (1970a) Gangliosides. In: Handbook of Neurochemistry (Lajtha, A., ed.), vol. 3, pp. 425–452. Plenum Press, New York.

Svennerholm, L. (1970b) Ganglioside Metabolism. In: Comprehensive Biochemistry (Florkin, M. and Stotz, E. H., eds.), vol. 18, pp. 201–227. Elsevier/North-Holland, Amsterdam.

Tani, E., Ikeda, K. and Nishiura, M. (1973) Freeze etching images of central myelinated nerve fibres. J. Neurocytol. 2, 305–314.

Tenenbaum, D. and Folch-Pi, J. (1966) The preparation and characterisation of water-soluble proteolipid protein from bovine brain white matter. Biochim. Biophys. Acta. 115, 141–147.

Thompson, E. B. and Kies, M. W. (1965) Current studies on the lipid and proteins of myelin. Ann. N.Y. Acad. Sci. 122, 129–147.

Tien, H. T. (1971) Bilayer lipid membranes: an experimental model for biological membranes. In: The Chemistry of Biosurfaces (Hair, M. L., ed.), vol. 1, pp. 233–348. Marcel Dekker, New York.

Tien, H. T. and Ting, H. P. (1968) Permeation of water through bilayer lipid membranes. J. Colloid Interface Sci. 27, 702–713.

Tomasi, L. and Kornguth, S. E. (1968) Characterization and immunochemical localization of the basic protein from pig brain. J. Biol. Chem. 243, 2507–2513.

Torrie, S. E. and Rumsby, M. G. (1975) Purification of non specific esterase activity from isolated central nerve myelin preparations. Fifth I.S.N. Conference, Barcelona, Abstr. No. 330.

Traüble, H. (1972) Phase transition in lipids. In: Biomembranes (Kreuzer, F. and Slegers, J. F. G., eds.), vol. 3. pp. 197–227.

Traüble, H. and Eibl, H. (1974) Electrostatic effects on lipid phase transitions: membrane structure and ionic environment. Proc. Nat. Acad. Sci. U.S.A. 71, 214–219.

Uda, Y. and Nakazawa, Y. (1973) Proteolipid of bovine brain white matter. J. Biochem. 74, 545–549.

Urry, D. W. (1972) Protein conformation in biomembranes: Optical rotation and absorption of membrane suspensions. Biochim. Biophys. Acta. 265, 115–168.

Uyemura, K., Tobari, C., and Hirano, S. (1970) Purification and properties of basic proteins in pig spinal cord and peripheral nerve. Biochim. Biophys. Acta. 214, 190–197.

Vail, W. J. Papahadjopoulos D. and Moscarello, M. A. (1974) Interaction of hydrophobic protein with liposomes: Evidence for particles seen in freeze fracture as being proteins. Biochim. Biophys. Acta. 345, 463–467.

Vandenheuvel, F. A. (1963) Biological structure at the molecular level with stereomodel projections. I. The lipids of the myelin sheath of nerve. J. Am. Oil Chem. Soc. 40, 455–472.

Vandenheuvel, F. A. (1965) Structural studies of biological membranes: the structure of myelin. Ann. N.Y. Acad. Sci. 122, 57–76.

Vandenheuvel, F. A. (1971) Structure of membranes and role of lipids therein. Adv. Lipid Res. 9, 161–248.

Vanderkooi, G. (1974) Organisation of proteins in membranes with special reference to the cytochrome oxidase system. Biochim. Biophys. Acta. 344, 307–345.

Veerkamp, J. H. (1972) Lipids in the cell plasma membrane. In: Biomembranes (Manson, L., ed.), vol. 3, pp. 159–180. Plenum Press, New York.

Verkleij, A. D., De Kruyff, B., Ververgaert, P. H. J. Th., Tocanne, J. F. and van Deenen, L. L. M. (1974) The influence of pH, Ca^{2+} and protein on the thermotropic behaviour of the negatively charged phospholipid, phosphatidylglycerol. Biochim. Biophys. Acta. 339, 432–437.

Waehneldt, T. V. and Mandel, P. (1972) Isolation of rat brain myelin monitored by polyacrylamide gel electrophoresis of dodecyl-sulphate-extracted proteins. Brain Res. 40, 419–436.

Wajgt, A., Burton, R. M. and Agrawal, H. C. (1975) Isolation of myelin proteins. Fifth I.S.N. Conference, Barcelona, Abstr. No. 310.

Wallach, D. F. H., Ferber, F., Selin, D., Weidekamm, E. and Fischer, H. (1970) The study of lipid-protein interactions in membranes by fluorescent probes. Biochim. Biophys. Acta. 203, 67–76.

Webster, H. deF. (1971) The geometry of peripheral myelin sheaths during their formation and growth in rat sciatic nerves. J. Cell Biol. 48, 348–367.

Westall, F. C., Robinson, A. B., Caccam, J., Jackson, J. and Eylar, E. H. (1971) Essential chemical requirements for induction of allergic encephalomyelitis. Nature 229, 22–24.

White, D. A. (1973) The phospholipid composition of mammalian tissues. In: Form and Function of Phospholipids (Ansell, G. B., Hawthorne, J. N. and Dawson, R. M. C., eds.), pp. 441–481. Elsevier/North-Holland, Amsterdam.

Whittingham, S., Bencina, B., Carnegie, P. R. and McPherson, T. A. (1972) Properties of antibodies produced in rabbits to human myelin and myelin basic protein. Int. Arch. Allergy Appl. Immunol. 42, 250–263.

Wiggins, R. C., Benjamins, J. A. and Morell, P. (1975) Appearance of myelin proteins in rat sciatic nerve during development. Brain Res. 89, 99–106.

Wiggins, R. C., Joffe, S., Davidson, D. and Del Valle, U. (1974) Characterisation of Wolfgram proteolipid protein of bovine white matter and fractionation of molecular weight heterogeneity. J. Neurochem. 22, 171–175.

Wilkins, M. H. F., Blaurock, A. E. and Engelman, D. M. (1971) Bilayer structure in membranes. Nature New Biol. 230, 72–76.

Williams, P. L. and Hall, S. M. (1969) In vivo observations on mature myelinated nerve fibres of the mouse. J. Anat. 107, 31–38.

Williams, P. L. and Wendell-Smith, C. P. (1971) Some additional parametric variations between peripheral nerve fibre populations. J. Anat. 109, 505–526.

Williams, R. M. and Chapman, D. (1970) Phospholipids, liquid crystals and cell membranes. In: Progress in the Chemistry of Fats and Other Lipids (Holman, R. T., ed.), vol. II, pp. 1–79. Pergamon Press, Oxford.

Woelk, H. and Borri, P. (1973) Lipid and fatty acid composition of myelin purified from normal and MS brains. Eur. Neurol. 10, 250–260.

Wolfgram, F. and Kotorii, K. (1968) The composition of myelin proteins of the central nervous sytem. J. Neurochem. 15, 1281–1290.

Wolman, M. (1969) The nature of neurokeratin. In: Structure and Function of Nervous Tissue (Bourne, G. H., ed.), vol. 2., pp. 241–292. Academic Press, New York.

Wolman, M. (1971) Distribution of various protein fractions in central and peripheral myelin. Exp. Neurol. 30, 309–323.

Wolman, M. and Wiener, H. (1965) Structure of the myelin sheath as a function of concentration of ions. Biochim. Biophys. Acta. 102, 269–279.

Wolman, M., Wiener, H. and Bubis, J. J. (1966) On the nature of the cation-insensitive bonds of myelin. Israel J. Chem. 4, 53–58.

Wood, D. D., Vail, W. J. and Moscarello, M. A. (1975) Re-localisation of the basic protein and N-2 in diseased myelin. Brain Res. 93, 463–471.

Wood, J. G. (1973) The effects of glutaraldehyde and osmium on the protein and lipids of myelin and mitochondria. Biochim. Biophys. Acta. 329, 118–127.

Wood, J. G. and Dawson, R. M. C. (1973) A major myelin glycoprotein of sciatic nerve. J. Neurochem. 21, 717–719.

Wood, J. G. and Dawson, R. M. C. (1974a) Some properties of a major structural glycoprotein of sciatic nerve. J. Neurochem. 22, 627–630.

Wood, J. G. and Dawson, R. M. C. (1974b) Lipid and protein changes in sciatic nerve during Wallerian degeneration. J. Neurochem. 22, 631–635.

Wood, J. G. and Dawson, R. M. C. and Hauser, H. (1974) Effect of proteolytic attack on the structure of CNS myelin membrane. J. Neurochem. 22, 637–643.

Worthington, C. R. (1969) Structural parameters of nerve myelin. Proc. Nat. Acad. Sci. U.S.A. 63, 604–611.

Worthington, C. R. (1971) X-ray analysis of nerve myelin. In: Biophysics and Physiology of Excitable Membranes (Adelman, W. J., ed.), pp. 1–46. Van Nostrand-Reinhold, Princeton.

Worthington, C. R. (1973) X-ray diffraction studies on biological membranes. Curr. Top. Bioenerg. 5, 1–39.

362

Worthington, C. R. (1976) The physical states of nerve myelin as a function of pH using X-ray diffraction. Biophys. J. 16, 49a, Abstr. no. W-PM-G 6.

Worthington, C. R. and Blaurock, A. E. (1969a) A structural analysis of nerve myelin. Biophys. J. 9, 970–990.

Worthington, C. R. and Blaurock, A. E. (1969b) A low angle X-ray diffraction study of the swelling behaviour of peripheral nerve myelin. Biochim. Biophys. Acta. 173, 427–435.

Worthington, C. R. and King, G. I. (1971) Electron density profiles of nerve myelin. Nature, 234, 143–145.

Worthington, C. R. and McIntosh, T. J. (1974) Direct determination of the lamellar structure of peripheral nerve myelin at moderate resolution (7Å). Biophys. J. 14, 703–729.

Worthington, C. R., King, G. I. and McIntosh, T. J. (1973) Direct structure determination of multilayered membrane-type systems which contain fluid layers. Biophys. J. 13, 480–494.

Yee, A. G. and Revel, J-P. (1975) Endothelial cell junctions. J. Cell Biol. 66, 200–204.

Zand, R. (1968) Solution properties and structure of brain proteolipids. Biopolymers, 6, 939–953.

Zgorzalewicz, B., Neuhoff, V. and Waehneldt, T. V. (1974) Rat myelin proteins: compositional changes in various regions of the nervous system during ontogenetic development. Neurobiology 4, 265–276.

Zingsheim, H. P. (1972) Membrane structure and electron microscopy: The significance of physical problems and techniques (freeze-etching). Biochim. Biophys. Acta. 265, 339–366.

Zwaal, R. F. A. and van Deenen, L. L. M. (1968a) The solubilisation of human erythrocyte membranes by n-pentanol. Biochim. Biophys. Acta. 150, 323–325.

Zwaal, R. F. A. and van Deenen, L. L. M. (1968b) Protein patterns of red cell membranes from different mammalian species. Biochim. Biophys. Acta. 163, 44–49.

Zwaal, R. F. A. and Roelofson, B. (1976) Applications of pure phospholipases in membrane studies. In: Biochemical Analysis of Membranes (Maddy, A. H., ed.), pp. 352–377. Chapman & Hall, London.

Zwaal, R. F. A., Roelofsen, B. and Colley, C. M. (1973) Localization of red cell membrane constituents. Biochim. Biophys. Acta. 306, 159–182.

Cell surface changes in phagocytosis $\qquad$ 6

Peter ELSBACH

1. Introduction

The many facets of phagocytosis, i.e., the engulfment of particulate matter by specialized cells, represent events that involve surface membranes. For example, the effectiveness of the most important biological function of the phagocytic cells of higher organisms, namely, the detection, sequestration, killing, and destruction of invading microorganisms is determined by the surface membrane properties of both the phagocyte and the microbe. Membrane modification and interaction are also part of the degranulation phenomenon with its accompanying release of biologically active substances either into the phagocytic vacuole by fusion of granule and vacuole membranes, or, under certain circumstances, into the extracellular environment by fusion with surface membranes without vacuoles having formed. Finally, the efficiency of killing and the extent of structural dissolution of ingested microorganisms are generally determined by the properties of the microbial surface and the ability of the host to alter these properties before ingestion by interaction with serum factors.

Many recent reviews have dealt with membrane alterations in endocytosis. I will attempt to refrain from covering the same territory, referring whenever possible to these earlier reviews for specific segments of the topic. This chapter will be restricted therefore to those surface phenomena accompanying phagocytosis that have not previously been reviewed in detail. The list of references is not meant to be exhaustive and serves to direct the reader to the most recent literature and only to those publications most pertinent to the points this reviewer wishes to emphasize. The following reviews have appeared within the last three years and should be consulted for areas not adequately covered in this chapter: general features of phagocytosis (Stossel, 1974; Elsbach, 1974); macrophage physiology (Steinman and Cohn, 1974); and antimicrobial mechanisms (Klebanoff, 1975).

G. Poste and G.L. Nicolson (eds.) The Synthesis, Assembly and Turnover of Cell Surface Components
© Elsevier/North-Holland Biomedical Press, 1977.

2. Changes involving the membranes of the phagocyte

The successful execution of engulfment and antimicrobial activity requires that the phagocyte be capable of dramatically increasing its mechanical and metabolic work. This ability to respond to specific stimuli with locomotion, formation of phagocytic vacuoles, degranulation, a respiratory burst, and increased membrane synthesis is linked to membrane organization and function. The signals that elicit these various responses are often, but clearly not always, the same for each category. Hence, in many experimental situations one can dissociate one response from another, thereby establishing that they are not part of an obligatory sequence nor all the expression of the same function. On the other hand, certain physiological mechanisms may be common to more than one set of responses. In this section we will examine the role of the surface membrane of the phagocyte as a sensor and modulator of incoming stimuli and as a trigger of various responses.

2.1. Motility

Motility is used here to encompass all aspects of mechanical work by phagocytes, not only those leading to migration of the whole cell, as for example upon chemotactic stimulation, but also those aspects involved with the envelopment and sequestration of particles and the translocation of intracellular organelles.

The electron microscopic discovery in several types of phagocytic cells of a cytoskeleton consisting of microtubules and networks of microfilaments, in contiguity to the surface membrane (Reaven and Axline, 1973), and the observation in other motile cells that the finding of such filaments often indicates the presence of actin and myosin (Pollard and Weihing, 1975), has prompted the search for similar substances in cells capable of engulfment. Indeed, the isolation in partially purified form of actin and myosin from *Acanthamoeba castellanii* (Pollard and Korn, 1973), rabbit lung macrophages (Hartwig and Stossel, 1975), and horse peripheral blood leukocytes (Tatsumi et al., 1973), and of myosin from guinea pig granulocytes (Stossel and Pollard, 1973) lends support to the speculation that a contractile apparatus, resembling that of skeletal muscle, is instrumental in the locomotion and other mechanical work of phagocytic cells. The effects of cytochalasin B, colchicine and vinblastine, agents presumed to have an inhibitory action on phagocytosis and locomotion, at least in part through structural modification of the actin-like filaments and disassembly of microtubules, have also been adduced in support of this concept (Stossel, 1974; Cohn, 1975). Nevertheless, no definitive functional relationship has yet been established between the identification of the chemical elements of a contractile system and the actual events that precede and accompany phagocytosis. The most suggestive circumstantial evidence of the functional importance of the phagocyte's actomyosin apparatus comes from a clinical observation (Boxer et al., 1974). These investigators found that the granulocytes of a child subject to frequent bacterial infection exhibited reduced motility and ingestion in vitro and that monomeric actin isolated from these cells failed to polymerize completely under conditions that permitted total polymerization of actin from normal granulocytes.

If methods can be developed to analyze in the intact cell the interaction between actin, myosin, and an actin-binding macroprotein (Stossel and Hartwig, 1975), that is, the likely means by which the contractile apparatus is regulated, the way will be paved for a renewed study of the multiple signals reaching the surface membrane that induce the phagocyte to react with directed or random motion, or with pseudopod formation, or with engulfment, or with the movement of granules toward the periphery and of phagocytic vesicles toward the center of the cell. In this setting the numerous membrane-modifying factors now known to influence chemotaxis, recognition, and attachment, whether followed or not by engulfment and/or degranulation and metabolic stimulation, may hopefully fall into a more comprehensible scheme than is now the case (Bray, 1976).

Evidence is accumulating that divalent cations, cyclic nucleotides, and agents affecting the levels of these components, together with the complement system, may interact in some fashion to govern all these aspects of the mechanical work associated with phagocytosis.

2.2. Chemotaxis

This complex phenomenon of migration of phagocytes toward a chemical stimulus has been reviewed recently (exhaustively by Grant [1974] and succinctly by Stossel [1974]). I will therefore confine my comments to the possible mechanisms involved in the translation of the chemotactic stimulus into a surface membrane alteration.

Among the many chemotactic factors, those that are generated during host-microbe interaction are obviously of greatest pathophysiological significance. The interconversion of the components of the complement system, triggered by this interaction, yields products with vigorous chemotactic activity (Grant, 1974; Stossel, 1974), in particular the fragments of proteolysis, C3a and C5a, and the complement protein sequence C567 (Müller-Eberhard, 1975). Chemotactic factors interact with the cell surface of phagocytes in unknown fashion. However, a number of interesting clues have been obtained in recent experiments with human granulocytes (Gallin et al., 1975). Incubation of these cells with the established chemotactic factors C5a, kallikrein, plasminogen activator, and dialyzable transfer factor, but not with nonchemotactic proteins, results in a dose-dependent reduction in surface charge that parallels the chemotactic stimulation. Cells incapable of chemotaxis show no or less alteration in surface charge upon incubation with C5a. Whether attachment of chemotactic factors to charged sites on the granulocyte surface accounts for the reduced surface charge is uncertain, since both positively and negatively charged factors produce the effect. Binding of the various agents to granulocytes apparently has not yet been studied. An additional effect of C5a, kallikrein, and transfer factor is on Ca^{2+} fluxes into and out of granulocytes, with resulting net release of this ion (Gallin and Rosenthal, 1974). This effect is again dose-dependent and roughly parallel to the extent of chemotactic activity that is generated. The most pronounced effect of the chemotactic factors on transport of $^{45}Ca^{2+}$ is to cause an increased efflux, but influx is simultaneously inhibited. The authors consider the possibility that the released Ca^{2+} neutralizes negative surface charges. Finally, morphological observations suggest that incubation with C5a is associated with enhanced assembly of microtubules (Goldstein et al., 1973; Gallin and Rosenthal, 1974). Whether a chemotactic factor induces stimulation of random movement (after addition to the incubation medium) or movement in a specific direction (in response to a chemotactic gradient) would then depend on whether the decrease in cytoplasmic Ca^{2+} was sufficiently localized to cause focal polymerization of tubulin, followed in turn by activation of the actomyosin system (Gallin and Rosenthal, 1974).

Why mononuclear phagocytes move slower than polymorphonuclear leukocytes in response to the same chemotactic stimulus is unclear. Polymorphonuclear leukocytes that have been attracted to a given site are thought to promote migration of other polymorphonuclear cells (Borel et al., 1969). In this regard the observation of Turner et al., (1975) is of interest. Oxidized arachidonic acid and other polyenoic lipids, in the absence of serum and chemotactic proteins, are chemotactic in the Boyden chamber. In the inflammatory milieu where the products of the enhanced oxidative metabolism of the polymorphonuclear leukocyte engaged in phagocytosis abound (Root et al., 1975), and where granulocytes disintegrate, oxidation and peroxidation of membrane lipids are likely to occur, since these chemical processes have been shown to take place in the intact cells (Stossel et al., 1974; Smolen and Shohet, 1974). The extracellular generation of such chemotactic molecules might thus provide a mechanism for recruitment of new phagocytes. It would be of interest to know if mononuclear phagocytes and polymorphonuclear leukocytes differ in the speed of their migratory response to oxidized lipid.

2.3. Recognition, attachment, and ingestion

A prominent feature of the surface interaction between the phagocyte and the particle to be phagocytosed after the two have arrived in close proximity to each other is its remarkably discriminatory character. Whether a given particle is "accepted" by the phagocyte, first permitting attachment and then (if certain requirements are met [Rabinovitch, 1967; Jones and Hirsch, 1971; Jones et al., 1972]) also engulfment, is determined by the surface properties of both the particle (microorganism) and the phagocyte. Recognition by granulocytes (and to a lesser extent by mononuclear phagocytes), leading to attachment and phagocytosis of a rather broad spectrum of microorganisms and particles, occurs in vitro in the absence of added serum (Elsbach, 1968; Holland et al., 1972; Elsbach et al., 1973; Beckerdite et al., 1974). However, there is no doubt that recognition of many microorganisms and types of particles, and subsequent events depend prominently upon serum factors that prepare both the particle and the phagocyte for rapid engulfment (opsonization) (Stossel, 1973; Stossel, 1974). Heat-labile components of complement (mainly a fragment (C3b) of C3, activated either by the classical or the properdin pathway [Mantovani et al., 1972; Stossel, 1973; Stossel et al., 1975]) are the most important and nonspecific serum factors leading to attachment of particles to phagocytes. Opsonization of at least some ingestible particles, such as heterologous erythrocytes, is markedly amplified by heat-stable specific IgG antibodies (Mantovani et al., 1972). Recently, much additional evidence has been presented indicating that complement and antibody fulfill cooperative, but quite distinct, functions in the sequential steps of phagocytosis. Portions of C3b and IgG attach to the surfaces of microbes and ingestible particles and to those of phagocytes as well. The physicochemical forces that govern the interaction of these serum proteins with receptors on the various surfaces are incompletely understood. What is known is that not all of C3b needs to be bound to the particle and that only a fragment is opsonically active (Stossel et al., 1975). Further, IgG attaches to the particle (antigen) through the Fab region, and consequently undergoes a conformational change which, in turn, promotes binding of the Fc segment of the antibody to receptors on the surface of the phagocyte (Messner and Jelinek, 1970). Phagocytes (including granulocytes, monocytes, and macrophages from a number of animal species) also possess binding sites for C3b, but not for C3d, that is, C3b inactivated by C3b inactivator (Ehlenberger and Nussenzweig, 1975; Griffin et al., 1975). The receptors on the phagocyte membrane for C3b and for IgG differ in their dependence upon Mg^{2+} for binding of their ligands and in susceptibility to destruction by trypsin (Lay and Nussenzweig, 1968). Several laboratories have now provided further evidence that two classes of functionally distinct receptors exist on the surfaces of polymorphonuclear leukocytes as well as macrophages. Thus, the binding of C3-coated erythrocytes to macrophages results in attachment without much ingestion, whereas bridging between particles and phagocytes when IgG couples with its receptor promotes ingestion as well (Mantovani et al., 1972; Griffin et al., 1975a). It remains to be established, however, how sharp the distinction really is between the apparently different receptors and their functional roles. For example, comparison of activated and nonactivated macrophages has shown that activation stimulates not only ingestion of particles mediated by the Fc receptors, but also initiates ingestion of complement-coated particles (Bianco et al., 1975). It is not yet clear whether activation of macrophages is associated with the genesis of more and/or different receptors or whether activation establishes a previously unexpressed link with a contractile apparatus involved in engulfment. Besides, experiments using as particles erythrocytes coated with immunoglobulins and/or complement, revealing as they are, may not give a complete picture. Thus, uptake of albumin-coated paraffin particles is stimulated by C3 and apparently does not require participation of immunoglobulins (Stossel, 1973).

The dissociation of attachment and ingestion of variously coated erythrocytes presented to cultivated peritoneal mouse macrophages has also been helpful in demonstrating that the stimulus to ingestion is confined only to that portion of the surface membrane to which the particle is attached. Thus, erythrocytes "immunologically" attached to macrophages, but not ingested because of absence of the Fc region of the linking immunoglobulin, are not engulfed when latex particles or opsonized pneumococci are avidly taken up (Griffin and Silverstein, 1974). That uptake of one particle (heat-killed streptococci) does not lead to engulfment of another particle (chylomicra) attached to the phagocyte's surface has also been shown for rabbit peritoneal polymorphonuclear leukocytes (Elsbach, 1965). This discrete responsiveness of the plasma membrane of phagocytes allows for adjustment of the response to the number and size of attached particles (Griffin et al., 1975b). Experiments in support of this concept have been presented by the same authors, who showed that in order for attached erythrocytes to be ingested ligands must be available on the whole surface of the particle for interaction with phagocyte surface receptors. Proteolytic stripping of all particle ligands except those involved in and protected by attachment, as well as blockade of the receptors on the phagocyte surface by antimacrophage IgG prevented ingestion. Griffin et al. (1975b) proposed that "ingestion requires the sequential, circumferential interaction of particle-bound ligand with specific plasma membrane receptors not involved in the initial attachment process."

Uptake of polystyrene and polyvinyltoluene latex particles by *Acanthamoeba* also shows a tight envelopment of the particles by the phagocytic vesicle, thereby virtually excluding all extracellular medium. However, apparently unlike the uptake of erythrocytes by macrophages, phagocytosis by *Acanthamoeba* is triggered by a critical volume of particles attached to the cell surface and, once this volume has been reached, internalization occurs in vesicles of fairly uniform size that contain either one or a few larger or many small particles (Weisman and Korn, 1967); Korn and Weisman, 1967).

2.4. Other membrane modifications during phagocytosis

In addition to the surface-membrane-related events of recognition (interaction with specific ligands) and locomotion (chemotaxis, spreading and pseudopod formation)the actual (or attempt at, i.e., , "frustrated" [Henson1971]) internalization of particles by phagocytes is accompanied by ultrastructural changes indicative of fusing and separating membranes and by biochemical alterations reflecting the synthesis of new membrane.

2.4.1. Fusion of membranes

Endocytosis is associated with three fusion-producing contacts between membranes: (1) the joining of two plasma membrane portions in formation of the initial phagocytic vacuole; (2) the degranulation phenomenon when granules and phagocytic vesicles fuse; and (3) degranulation by fusion of granule membrane with plasma membrane, resulting in release of granule contents into the extracellular environment. Much speculation has been generated to explain fusion (recently reviewed by Korn et al., 1974). Briefly, two concepts underlie the various theories: (1) membrane fusion simply follows the close contact between membranes, either because of surface forces that favor larger rather than smaller membrane-bounded structures, or through charge interactions in which divalent cations (Ca^{2+}) act as a link (Poste and Allison, 1973). These considerations do not readily explain the apparent specificity of the fusion events during phagocytosis, such as the sequential degranulation of two types of neutrophil granules (Bainton, 1973) and the budding off of the phagocytic vesicle from the plasma membrane; (2) membrane fusion is a consequence of a change in membrane lipids, consisting of the accumulation of membrane-labilizing lyso-compounds and permitting merging of membrane components. This concept was first proposed to explain degranulation and phagosome formation during endocytosis by granulocytes (Elsbach and Rizack, 1963; Elsbach et al., 1965), and has since been advocated as a mechanism for membrane fusion in general (Lucy, 1970). The subsequent demonstration in polymorphonuclear leukocytes and alveolar macrophages of phospholipases A that are at least in part granule-associated (Franson et al., 1973; Franson et al., 1974) and also of lysophospholipid reacylation, that is, a diacyl-monoacylphosphatide cycle (review, see Elsbach, 1972) does indeed provide an enzymatic mechanism for the formation of lysophosphatides at fusion sites and their reutilization for diacylphosphatide synthesis with reconstitution of a stable membrane. However, there is actually more evidence against than for involvement of lysophosphatides in fusion (Poste and Allison, 1973; Korn et al., 1974). Thus, high concentrations of exogenous monoacylphosphatides such as lysolecithin are required to induce membrane-lytic and fusion effects. In polymorphonuclear leukocytes little or no net degradation of membrane phospholipids can be demonstrated during phagocytosis (Elsbach et al., 1972a). In alveolar macrophages, turnover of phospholipid is more apparent but not increased during phagocytosis (Franson et al., 1973). Furthermore, the presence of lysophospholipases (Elsbach and Rizack, 1963; Elsbach et al., 1965; Elsbach, 1966, 1967; Victoria and Korn, 1975a,1975b) in phagocytic cells could well prevent accumulation of membrane-lytic concentrations of lysophosphatides. Perhaps most damaging to the proposed role of lysolecithin in promoting membrane fusion are the observations that the dramatic and extensive fusion of several types of mammalian cells by viruses has not provided any evidence of either the formation (Falke et al., 1967; Elsbach et al., 1969) or the participation (Parkes and Fox, 1975) of lyso-compounds. Nonetheless, although failure to find support for a role of deacylation-reacylation cycle in fusion of membranes during phagocytosis has been emphasized here, it must be noted that sensitive techniques for detecting accumulation of lysophosphatides in discrete areas of the cell are not available. Pending the development of such techniques and their application to the problem, the notion that lysophosphatides may play a part in membrane lysis and fusion cannot yet be dismissed entirely.

Aside from the possible role of lysophosphatides as membrane labilizers and as precursors in the resynthesis of major membrane lipids in membrane alterations during phagocytosis, a deacylation-reacylation cycle may also serve to

370

change the fatty acid composition of membrane phospholipids (Smolen and Shohet, 1974). Local changes in the degree of saturation or unsaturation of the lipid bilayer could well be important in influencing the ease with which membranes fuse, whatever the mechanism.

2.4.2. Synthesis of membranes during or after phagocytosis

The stimulus to ingestion of a particle (as discussed earlier in sections 2.1., 2.3., 2.4.1.) may well be the transduction of an appropriate signal, such as the connection of a 7S antibody with a surface membrane receptor, which triggers the assembly of the components of a contractile apparatus into a working unit (Stossel, 1974; Griffin et al., 1975b). The very early, if not immediate, stimulation of biochemical work as engulfment commences would then reflect the energy needs of an activated contractile apparatus that causes the extension of pseudopodia as the initial step in the envelopment of the particle. What other elements in the engulfment process demand biochemical energy? How does the actively engulfing cell furnish the membrane material needed to accommodate ingested particles? Does the phagocyte possess enough redundant surface membrane or is new membrane synthesis required, either to complete formation of the phagocytic vesicles and/or to replace the internalized plasma membrane? Since these questions were recently considered in some detail (Elsbach, 1974) only a synopsis will be presented here of the evidence indicating that new membrane synthesis may accompany or follow engulfment.

Until recently, studies of the questions raised above have been limited by difficulties in reliably measuring the biosynthesis of plasma membrane components other than lipids (De Pierre and Karnovsky, 1973; Chang et al., 1975) Therefore, comparison of (phospho)lipid synthesis by resting and actively engulfing phagocytes has been used as the primary criterion for membrane synthesis (review, Elsbach, 1972).

Coincident with the onset of phagocytosis by human and rabbit polymorphonuclear leukocytes and by rabbit alveolar macrophages, only one pathway of membrane lipid biosynthesis shows substantial stimulation, namely, the conversion of extracellular albumin-bound lysolecithin and lysophosphatidylethanolamine to their corresponding cellular diacyl derivatives (Elsbach, 1968; Elsbach et al., 1969). Polymorphonuclear leukocytes obtain the fatty acid needed for the increased acylation of the lyso-compound from hydrolysis of cellular triglycerides rather than from extracellular free fatty acid (Elsbach and Farrow, 1969; Shohet, 1970), as follows:

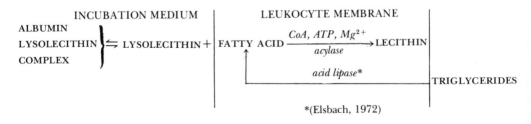

*(Elsbach, 1972)

In the engulfing granulocyte, the acylation of exogenous lysolecithin is stimulated up to threefold over resting values. Since in this type of phagocyte virtually no turnover of phospholipid can be detected during engulfment (Elsbach et al., 1972b), nor any loss of lipid through shedding of membrane material, this enhanced single-step conversion actually appears to represent net addition of phospholipid permitting an increase in lecithin content of as much as 10% in 30 minutes (Elsbach, 1968). That this evidence of increased net phospholipid synthesis during phagocytosis is indicative of formation of new membrane used for the engulfment process, is rendered more likely by the observation that most of the increase in lecithin formed by acylation of medium lysolecithin is recovered in association with the isolated phagosomes (Elsbach et al., 1972b). Furthermore, inhibition of glycolysis, the main source of energy in the polymorphonuclear leukocyte, inhibits both phagocytosis and the increase in acylation (Elsbach, 1972). Alveolar macrophages engaged in phagocytosis show approximately the same stimulation of acylation of albumin-bound lysophospholipids per cell as exhibited by granulocytes. However, because the much larger macrophage contains about eight times more phospholipid per cell than the polymorphonuclear leukocyte, the actual contribution of phospholipid from this source is rather small. Moreover, phospholipid turnover in alveolar macrophages is quite brisk (Franson et al., 1973).

Thus it is uncertain whether the increased incorporation of lysophospholipids during phagocytosis by this phagocytic cell represents net addition of phospholipid or only increased turnover. Actually, observations on another type of macrophage, the cultivated mouse peritoneal macrophage, suggest that appreciable membrane (lipid) synthesis may not take place during engulfment (Werb and Cohn, 1972). Indeed, an initial decrease in plasma membrane can be demonstrated. Several hours after ingestion accumulation of cholesterol and phospholipid occurs. The amount of net increase in lipid is proportional to the quantity of particles ingested and also to the amount of plasma membrane interiorized. The reconstitution of plasma membrane following ingestion then permits the macrophages to phagocytize anew. This study also provided the first indication of plasma membrane protein resynthesis as part of membrane replacement. During phagocytosis plasma membrane 5'-nucleotidase activity fell in proportion to the quantity of phagolysosome formed, that is, the amount of plasma membrane internalized. The phagolysosome-associated enzyme activity is apparently destroyed as time goes on, but plasma membrane enzyme activity begins to reappear some 6 hours after ingestion. At about this time regeneration of a receptor protein for serum lipoproteins that supply the macrophage with cholesterol also starts (Werb and Cohn, 1971).

These important studies not only show that at least the mouse peritoneal macrophage can ingest large numbers of particles (at least one hundred 1.1 μm latex particles per cell) by internalizing preexisting plasma membrane, but also that these cells ultimately replace the membrane used. As pointed out by Werb and Cohn, membrane synthesis by the short-lived polymorphonuclear leukocyte and the long-lived macrophage must have different functions. This may account for the fact that the net synthesis of membrane (lipid) by the polymorphonuclear leukocyte accompanies its single act of phagocytosis, but follows engulfment in the case of the macrophage which prepares itself for renewed phagosome formation.

It must be pointed out that the experimental model studied by Werb and Cohn is a special one in that the particle presented for ingestion was indigestible and hence forced preservation of the phagosome membrane since nondegradable particles remain sequestered indefinitely. Much less demand on new membrane formation might be made under conditions of complete digestion of material taken up. Stimuli might then be generated, leading to recycling of the internalized plasma membrane and its reappearance on the surface, perhaps via the exocytosis and fusion of empty vesicles. Apparently this is the best explanation for the maintenance of surface area of pinocytosing *Acanthamoeba* which, despite enormous and rapid internalization of its whole surface membrane (5 to 50 times per hr), shows no concomitant increase in biochemical turnover of plasma membrane phospholipids (Korn et al., 1974).

Recycling of whole membrane in this fashion would represent a biochemically inexpensive means of restoring surface membrane mass (see review in this volume by Holtzman et al.). Whether such reinsertion of previously internalized plasma membrane occurs during the normal mammalian phagocytic process in a quantitatively important way has not been determined. The demonstration of fusion of granules with the plasma membrane during the exocytosis that accompanies "frustrated phagocytosis" (Henson; 1971; Goldstein et al., 1973) indicates that intracellular membranes, at least under certain experimental conditions, can indeed become part of the phagocyte's surface membrane.

An unexplored redistribution mechanism of components of phagocyte membranes is the transfer of phospholipid molecules by exchange proteins from one intracellular particle to another (reviews, Wirtz, 1974; Kader, 1977). These proteins exhibit specificity for individual phospholipid species. Such exchange provides another means of nonmetabolic phospholipid turnover; the physiological importance of which is still a matter of conjecture. However, the capacity of pure exchange proteins to transfer in vitro substantial amounts of lipid between membranes (vesicles) is consistent with a biological function of considerable magnitude (van den Besselaar et al., 1975). This would be even more likely if experimental support is found for the speculation that phosoholipid exchange proteins can effect net transfer of phospholipid. In view of the demonstration of phospholipid exchange proteins in all tissues examined so far, it would not be surprising if these substances also were present in phagocytic cells.

Bretscher (1976) has speculated on a somewhat similar dynamic system of movement of specific lipids through the cell, permitting "directed" flow of phospholipids in cell membranes. Finally, for membrane synthesis, phagocytes can make use of materials derived from ingested matter. To date, only incorporation of hydrolysis products of bacterial lipids into lipids of polymorphonuclear leukocytes has been documented (Cohn, 1963; Elsbach et al., 1972a; Patriarca et al., 1972). Incorporation of bacterial fatty acids hydrolyzed during phagocytosis into leukocyte phospholipids is quite efficient and proportional to the extent of hydrolysis. Its quantitative significance therefore depends on the number of organisms ingested and the susceptibility of the microbial lipids to hydrolytic attack during phagocytosis (Elsbach et al., 1972a; Patriarca et al., 1972).

2.4.3. Degranulation

Degranulation, the disappearance of the phagocyte's cytoplasmic granules, is the consequence of the merging (fusion) of the granule membranes with other membranes, followed by discharge of the granule contents. Typically degranulation occurs during engulfment, when fusion of granules with forming or enclosed phagocytic vesicles is seen very soon after ingestion begins (Zucker-Franklin and Hirsch, 1964). Degranulation may also take place under certain experimental conditions in the absence of phagocytosis, when granules fuse with the plasma membrane, releasing their contents into the extracellular environment. This occurs during "frustrated phagocytosis" (Henson, 1971), when particles recognized as fit for ingestion are too large to be sequestered. Extracellular release of granule (lysosomal) proteins and enzymes also follows treatment with a range of substances, including bacterial toxins such as staphylococcal leukocidin (Woodin, 1962, 1973) and a streptococcal cell wall substance (Davies et al., 1974), or some cytochalasins which are fungal metabolites (Skosey et al., 1973; Axline and Reaven, 1974; Goldstein et al., 1975). The principle of degranulation appears to be the same whether externally or into the phagosome. What determines the granule's migration and apposition to and fusion with the plasma membrane before it has become a phagocytic vesicle is not known. It has been speculated that the microtubules and microfilaments form a barrier underneath the plasma membrane, thus preventing contact between the plasma membrane and the granules (Skosey et al., 1973; Zurier et al., 1973). Removal of the barrier, either during phagosome formation, or by agents acting on microfilaments and/or microtubules; would permit fusion to occur. Support for this concept has been mustered from observations on the opposite effects of agents that raise levels of cyclic AMP and inhibit enzyme release and those that raise levels of cyclic GMP and stimulate enzyme release (Ignarro, 1974). Since raising cyclic AMP levels tends to inhibit, and elevation of cyclic GMP to stimulate, all forms of motility, including migration (chemotaxis) (Goldstein et al., 1973), engulfment (Cox and Karnovsky, 1973), and degradation (Zurier et al., 1974), it has been proposed that the cyclic nucleotides influence the common effector of all these functions, namely, the contractile apparatus, by a regulatory effect on microtubule assembly (Zurier et al., 1973; Zurier et al., 1974; Stossel, 1974).

Cytochalasin B, as reported for other surface active agents that cause enzyme release (Woodin, 1962, 1964), increases the radiolabeling of phosphatidic acid and inositolphospholipids (Tou and Stjernholm, 1974, 1975). This phenomenon, common to many instances of surface receptor stimulation (Michell, 1975) may represent part of a transmembrane signal transmission system leading to the contractile apparatus. Although these apparently related findings suggest a common underlying mechanism involving the phagocyte's muscular apparatus, a number of issues need resolution before this key role can be accepted. For example, the effects of cytochalasin B on enzyme release by polymorphonuclear leukocytes are not unequivocal. Thus, Skosey et al., (1974) have reported that this agent actually does *not* enhance zymosan-stimulated release of lysosomal enzymes but rather retards and inhibits release. Further, previous interpretations of cytochalasin B effects on enzyme release in the presence of zymosan did not take into consideration that the activity of lysosomal enzymes, discharged into the phagosome or into the extracellular fluid, decreases with time, presumably because of degradation by activated proteolytic enzymes that are also discharged.

Macrophages have been shown to secrete large amounts of several polypeptides, including lysosomal lysozyme, both at rest and when challenged in various ways (Gordon et al., 1974; Unkeless et al., 1974; Cohn, 1975). The secretion mechanism of large quantities of protein is unknown, but its apparent highly selective nature suggests a different process than that of "regurgitation during feeding" (i.e., fusion of granules with plasma membrane) (Henson, 1971). If different modes of enzyme release by phagocytes exist, this need not be limited to the longer-lived and biosynthetically more active macrophages. Hence, in polymorphonuclear leukocytes, the determinants of degranulation may be considerably more diverse than proposed in some of the quoted studies. Further evidence exists that the process of degranulation may be profoundly influenced by factors not directly related to the participation of the contractile apparatus in the event. Phagocytic vesicles of cultured mouse peritoneal macrophages that contain live *Mycobacteria tuberculosis* or *Toxoplasma gondii* do not undergo fusion with either primary or secondary lysosomes. Whereas live *Toxoplasma* parasites prevent degranulation, survive, and divide until rupture of the macrophage ultimately occurs, phagocytic vesicles that hold dead or damaged *Toxoplasma* do fuse with lysosomes (Jones et al., 1972; Jones and Hirsch, 1972). Of particular interest is the observation that phagocytic vacuoles containing live *Toxoplasma*, and apparently unable to coalesce with lysosomes, fused with endoplasmic reticulum and mitochondria. Normally this does not happen, and possibly it reflects the successful recruitment by the parasite of the host's metabolic resources (Jones and Hirsch, 1972). Similar findings have been recorded with virulent *Mycobacteria tuberculosis* (Armstrong and Hart, 1971, 1974). Live bacilli remain intact and capable of multiplication in nonfusing phagocytic vesicles. On the other hand, damaged organisms or viable bacilli pretreated with specific antiserum readily fused with lysosomes. Serum-treated live *Mycobacteria* were not killed, however, despite exposure to the lysosomal contents. These findings suggest that certain microorganisms have in their envelopes, or produce and release, substances that alter the phagosome membrane, thereby interfering with fusion. In the case of *Toxoplasma*, the fact that fusion is possible with other intracellular particles that ordinarily do not fuse with phagosomes suggests that the organisms may cause a very specific membrane modification. That modification of the phagosomal membrane can have an effect on fusion is demonstrated by experiments with concanavalin A (Edelson and Cohn, 1974a,b).

Macrophages treated with concanavalin A form large pinocytic vacuoles that do not fuse with lysosomes. Displacement of the lectin from the membrane receptor by incubation with mannose (but not other sugars) restores the ability to fuse with lysosomes and permits degradative events to take place within the phagolysosome. Once again, the effect on fusion seems to be specific, since formation and sealing of vesicles (fusion of plasma membrane with plasma membrane) continues quite effectively in the presence of concanavalin A, as does engulfment of particles. These and other studies (Ji and Nicolson, 1974) also show that the attachment of the lectin to one side of the membrane (inner face of the phagosome membrane) alters its reactivity on the other side. Another agent that inhibits fusion of phagosomes and lysosomes of mouse peritoneal macrophages is suramin, a drug that accumulates within lysosomes (D'Arcy Hart and Young, 1975). These intriguing observations raise many questions concerning the effects of microorganisms on phagocytic cell membranes and, consequently, on host defense. Clearly, this aspect of phagocyte-microbe interaction needs further exploration.

2.4.4. The surface membrane of phagocytes and stimulation of oxidative metabolism

The increase in oxygen uptake that accompanies phagocytosis, in particular by polymorphonuclear leukocytes, and its implications for the generation of antimicrobial systems have been the subject of intense study for the past two decades. Expertly executed reviews have appeared at short intervals. The interested reader is referred especially to those by Karnovsky (1974), Klebanoff (1975), De Chatelet (1975) and Rossi et al. (1977), which cover this complicated aspect of phagocyte function. Figure 1 presents a simplified scheme of the respiratory burst leading to production of highly reactive molecules by the polymorphonuclear leukocyte. All of these molecules; H_2O_2 (Klebanoff; 1975), superoxide (Babior et al., 1973; Curnutte et al., 1974; Babior et al., 1975; Johnston et al., 1975), the hydroxyl free radical (Johnston et al., 1975), and perhaps singlet oxygen (Krinsky, 1974) may by themselves contribute to the phagocyte's bactericidal activity to a greater or lesser extent, depending on the microorganism exposed. However, in combination with the granule-associated myeloperoxidase of the polymorphonuclear leukocyte and an oxidizable substrate such as a halide, a particularly potent antimicrobial system is formed (Klebanoff, 1974; Sbarra et al., 1977).

Other types of phagocytic cells such as macrophages from the peritoneal cavity and the lung, may also manifest respiratory stimulation during phagocytosis and production of superoxide (Oren et al., 1963; Drath and Karnovsky, 1975). The extent of stimulation varies with the species of animal and the functional history of the cell (activated or nonactivated) (De Chatelet et al., 1975a; Drath and Karnovsky, 1975). Macrophages from some animals, whether activated or not, produce no detectable superoxide. It must be recognized, however, that only superoxide released into the medium (which is probably only a

376

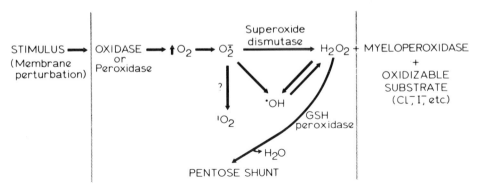

Fig. 1. Oxygen dependent production of antimicrobial system by granulocytes.

small portion of the total produced) is measured in the assay (Barbior et al., 1973).

The respiratory burst is thought to follow a cell surface perturbation. There is considerable support for this belief. Numerous stimuli other than internalization of plasma membrane, degranulation, and phagosome formation can initiate a respiratory burst (Elsbach, 1974; Rossi, et al., 1977). Indeed, most of the agents and conditions that stimulate O_2 consumption and associated release of superoxide (Goldstein et al., 1975) are either known surface active compounds, such as detergents and divalent cation ionophores (Schell-Frederick, 1974; Rossi et al., 1977), or interact with membrane receptors involved in chemotaxis, attachment, or ingestion of particles and binding of lectins (Goldstein et al., 1975; Rossi et al., 1977).

Because the oxidative response on addition of the stimulating agent to a leukocyte suspension is essentially instantaneous, the surface perturbation must trigger an event within the membrane, such as a plasma membrane-associated oxidase, or an exceedingly rapid transfer of the signal to an intracellular catalytic site.

Preliminary evidence for a CN-insensitive "ecto-NADH-oxidase" has been presented by Briggs et al., (1974). Takanaka and O'Brien (1975a,b) also have associated NAD(P)H-oxidase activity with a plasma membrane enzyme. On the other hand, Patriarca et al. (1975), and De Chatelet et al. (1975b) attribute the enzymatic activity that represents the initial step in the respiratory burst solely to the granule-associated NADPH oxidase. This enzyme is also CN-insensitive, exhibits an acid pH optimum, is stimulated by Mn^{2+}, and is postulated to produce a free radical of NADPH which, reacting with O_2, generates O_2^- (Patriarca et al., 1975). NADPH-oxidase activity in granule preparations is increased after phagocytosis. Its specific activity in plasma membrane preparations is low compared to that of granules (Rossi et al., 1977). Hence these investigators invoke a rapid transducing mechanism for a signal propagated from the surface. Recent studies of cytochalasin B-treated polymorphonuclear leukocytes (which are incapable of phagocytosis) exposed to a wide range of soluble and particulate materials show unequivocally that interaction with different surface sites (e.g., receptors specific for Fc, C3b, or C5a, concanavalin A attachment sites, membrane phospholipids) can all increase respiration and superoxide formation (Goetzl and Austin, 1974; Goldstein et al., 1975; Rossi et al., 1977). In one study the results were weighted toward postulating activation of an ectoenzyme (Goldstein et al., 1975; Root et al., 1975), and in the other toward involvement of a granule-associated enzyme (Rossi et al., 1977). At present convincing evidence in support of one or the other possibility is not available.

It may be expected that the reactive molecules produced during the stimulated oxidative metabolism, possibly within or near the phagocyte's membranes, cause peroxidative changes in membrane constituents, not only in the engulfed microorganisms but also in the membranes of the phagocyte itself. Lipid peroxides are formed during ingestion of various particles by mononuclear macrophages, and by neutrophils during ingestion of linoleate-containing particles (Mason et al., 1972; Smolen and Shohet, 1974; Stossel et al., 1974). Lipid peroxides have also been detected in isolated phagosomes. Which of the various radicals are most likely to cause peroxidation appears undetermined. Fong et al. (1973) attribute this role to the hydroxyl free radical and not to the superoxide radical, whereas Tyler (1975) has presented evidence of involvement of superoxide but not of the hydroxyl radical. Lipid peroxidation is much reduced (but not eliminated) in chronic granulomatous disease (Stossel et al., 1974) with its impaired oxidative metabolism (Karnovsky, 1973). Of interest in this context is that Gold et al. (1974) have reported retarded degranulation (measured as enzyme release) in leukocytes of patients with the X-linked variety of this disease. One of the consequences of lipid peroxide formation may be a membrane-labilizing effect that facilitates fusion.

2.4.5. Conclusions

Phagocytes are capable of multiple responses to diverse stimuli that involve mechanical work, morphological transformations, and increased biochemical activity. There is growing evidence that surface perturbations elicit these responses (Elsbach, 1973; Rossi et al., 1977) and that interaction with specific sites determines which response is triggered.

Attempts have been made to find common determinants of at least those responses that require mechanical work—spreading, migration, pseudopod formation, engulfment, and degranulation. Understandably, the recently discovered elements of an intracellular contractile apparatus in phagocytes have been proposed as a likely common effector in these various functions of a motile cell.

The manner in which the signals reaching the surface would be transmitted to these effectors is still a matter of speculation. A currently popular notion is that cyclic nucleotides are regulatory agents of the contractile apparatus. Such a role for substances endowed with so many well-established regulatory functions in

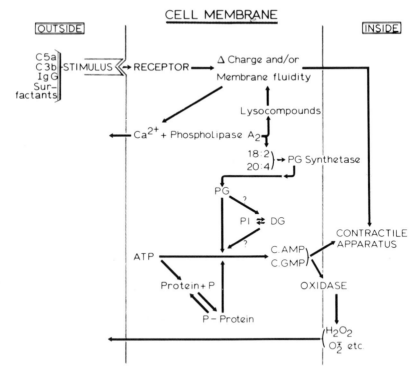

Fig. 2. Key: 18:2 = linoleic acid; 20:4 = arachidonic acid; PG = prostaglandin; PI = phosphatidylinositide; DG = diglyceride.

other physiological events is not difficult to envision. An integrative role of cyclic nucleotides in phagocyte function is further suggested by the parallel effects that altered levels of cyclic nucleotides have not only on motility but also on the respiratory response of stimulated leukocytes.

One possible sequence of events that combines a number of observations is shown in Fig. 2. The effects of various agonists upon the phagocyte's surface membrane might include alterations in charge (Gallin et al., 1975) and perhaps hydrophobic properties (fluidity), as suggested for the effect of complement components on erythrocytes (Hammer et al., 1975). Such effects might be directly transmitted to the contractile apparatus, as has been proposed in the case of local anesthetics (Poste et al., 1975; Nicolson et al., 1976; 1977), thereby inhibiting functions such as spreading and endocytosis (Rabinovitch and De Stefano, 1973; 1974). Alternatively, the displacement, translocation, or release of Ca^{2+} observed in a number of instances of presumably electrostatic ligand-receptor interactions may raise local Ca^{2+} concentrations in proximity to membrane phospholipases A_2. In rabbit polymorphonuclear leukocytes, a portion of the Ca^{2+}dependent phospholipase A_2 may be associated with the plasma membrane (Franson et al., 1974). Activation by Ca^{2+} of this phospholipase A_2 would cause accumulation of lysophospholipids and unsaturated fatty acids that occupy the 2-position of the membrane phospholipids. The lysophospholipids may contribute to changes in membrane lipid fluidity and promote membrane-interaction (fusion) (section 2.4.1). The fatty acid released would include the polyunsaturated fatty acids linoleic, linolenic, and arachidonic acid (Elsbach, 1959), which are precursors of prostaglandin synthesis. Synthesis of PGE1 has been demonstrated in polymorphonuclear leukocytes, and its production is increased tenfold during phagocytosis (McCall and Youlton, 1973). Accumulation of fatty acid precursors might also trigger prostaglandin synthesis. Stimulation by E-type prostaglandins of cyclic nucleotide synthesis has been described in polymorphonuclear leukocytes (Bourne et al., 1971). Regulation of cyclase activity in leukocytes, as in other tissues, might further include synthesis and breakdown of inositolphospholipids (Michel, 1975) and phosphorylation-dephosphorylation of membrane protein (Constantopoulos and Najjar, 1973; Layne et al., 1973). Prostaglandin E_1 activates adenylcyclase of granulocyte membranes and counteracts the inhibition caused by membrane protein (adenylcyclase?) phosphorylation (Constantopoulos and Najjar, 1973). In the scheme shown in Fig. 2, the possibility is considered that prostaglandins might directly affect the interconversion of inositolphospholipids and diglycerides. To our knowledge, such an effect has not yet been demonstrated. Diglycerides may also be a source of fatty acids for prostaglandin synthesis. Finally, activation of the final effectors, the contractile apparatus, and a pyridine-nucleotide oxidase (of still uncertain localization) is the result of the interplay of altered cyclic nucleotide levels (Fig. 2).

In this scheme no function is assigned to Ca^{2+} as a primary regulator for the cytoskeletal apparatus. Generally, such a specific role for Ca^{2+} (Rasmussen, 1970) need not exist (e.g., Borisy et al., 1975). The Ca^{2+} dependence of the actin-activated Mg^{2+} ATPase activity of polymorphonuclear leukocyte myosin is doubtful (Stossel and Pollard, 1973).

It must be recognized, however, that Ca^{2+} exerts so many effects on membranes and their individual components that no serious attempt should be made to attribute to these important ions a single role, even if only a hypothetical one. The purpose of the above comments is to draw attention to yet another possible mechanism by which Ca^{2+} might regulate the transmission of signals from receptor to effector in phagocytic cells.

3. Surface properties of microorganisms as determinants of phagocytosis and postphagocytic events

With the rapid improvement of tools and techniques for the structural and chemical analysis of membranes and their constituents, it has become apparent that the diversity of microbial envelopes is enormous. The study of this diversity has been greatly facilitated by the use of modern microbial genetics which often permits precise identification of the biochemical reason for subtle differences in envelope structure and function (Costerton et al., 1974; Silbert, 1975). Only recently have attempts been made to relate this new information to molecular mechanisms of host defense against infection, even though it has long been known that the degree of virulence of microorganisms for a given animal species is mainly determined by the surface properties of the microbial envelope (Roantree, 1967). Although this statement undoubtedly applies to both gram-positive and gram-negative organisms, the available evidence seems most convincing for gram-negative bacteria. For example, a close correlation has been found between pathogenicity and the presence of specific surface antigens of several species of gram-negative organisms such as *Salmonella* and *E. coli* (Medearis et al., 1968). I will review briefly recent observations indicating further that the surface properties of these gram-negative bacteria exert their effect on the phagocyte-microbe interaction at several levels.

3.1. Properties of the gram-negative microbial envelope

3.1.1. Structure

The progress made during the last decade in exploring the envelopes of gram-negative microorganisms has been extensively reviewed (Bayer, 1974; Osborn et al., 1974; Costerton et al., 1974; Braun, 1975; Cronan and Gelman, 1975).

The envelope of gram-negative bacteria differs from that of other microorganisms because in addition to a cytoplasmic membrane and a peptidoglycan layer, a distinct outer layer is present. This layer has the structural and chemical characteristics of a membrane, quite similar to a typical cytoplasmic membrane, but endowed with several unique components that determine many if not most of the surface properties of gram-negative organisms. These components, the lipopolysaccharides (LPS) (Osborn et al., 1974) and the lipoproteins (Inouye, 1974; Braun, 1975) have a number of structural and functional features that are still incompletely understood. It has been well established, however, that both components play an important role in maintaining the structural integrity of the whole microbial envelope. Thus, the lipoproteins link the peptidoglycans of the cell wall to the outer membrane, presumably through peptide bonding of the protein portion of the lipoprotein with the peptidoglycans and through hydrophobic interaction of the acyl chains with the lipid bilayer of the outer membrane (Costerton et al., 1974; Inouye, 1974; Braun, 1975).

The LPS form an integral part of the outer membrane and determine in large measure the functional integrity of this layer as an important permeability barrier (Leive, 1974). It appears that the anchoring of the lipid portion of the LPS (Lipid A) in the bilayer of the outer membrane is insufficient to preserve the normal LPS-outer membrane relationship, and that divalent ions (both Mg^{2+} and Ca^{2+}) must provide cross-links for additional stability (Costerton et al., 1974; Leive, 1974). The amount and chemical composition of LPS in the envelope appear to be of major importance to the gram-negative microorganism's ability to resist the effects of antibiotics, phages, and other agents (Harold, 1970, Leive, 1974; Costerton et al., 1974). Experimental modification of the LPS with consequent functional changes of the envelopes of *E. coli* and *S. typhimurium* has been brought about by several compounds, notably ethyldiaminetetraacetate (EDTA) (Leive, 1974), but also by numerous others (Costerton et al., 1974). Especially helpful in providing insight into the role of LPS in the barrier function of the outer membrane have been observations on well-defined chemical variations in LPS structure that are genetically determined and that give rise to alterations in outer membrane function (review, Costerton et al., 1974).

3.1.2. Function

3.1.2.1. Biochemical machinery

The biochemical pathways of energy production and macromolecular biosynthesis in all bacteria, including gram-negative bacteria, appear to be exclusively associated with the cytoplasmic membrane. Hence the outer membrane serves no known biosynthetic role. To date, the only enzymatic activity found to be restricted to the outer membrane is a phospholipase A_1 (Scandella and Kornberg, 1971; Albright et al., 1973; McIntyre and Bell, 1975).

Recently much attention has focused on the identification and separation of a range of outer membrane proteins of *E. coli* (Schnaitman, 1970; Haller et al., 1975; Lugtenberg et al., 1975). Whether or not these proteins possess catalytic activity is unknown.

3.1.2.2. Transport

Energy-dependent translocation of solutes by bacteria is solely a function of the cytoplasmic membrane. This complicated topic has recently been reviewed (Simoni and Postma, 1975) and will not be considered here.

3.1.3. Permeability barriers

Both the cytoplasmic and the outer membranes constitute permeability barriers. The outer membrane bars relatively larger molecules, such as antibiotics and other toxic agents, but generally permits the passage of small ions and low molecular weight nutrients. The entry of these needed substances is regulated by the permeases and other transport proteins associated with the cytoplasmic membrane (Simoni and Postma, 1975). The barrier function of the outer membrane is readily impaired when the LPS content or structure is altered (Leive, 1974; Costerton, 1974).

3.2. The role of the microbial envelope in phagocyte-microbe interaction

That surface properties of gram-negative bacteria determine virulence was first suggested by observations on the bactericidal and opsonic effects of serum on various *E. coli* and *Salmonella* mutants (review, Medearis et al., 1968; see also, Nelson and Roantree, 1967; Dlabač, 1968; Krischnapillai, 1971). Similarly, the outcome of the interaction between phagocytes and gram-negative bacteria varies with apparently minor structural and chemical differences in bacterial surface components. Indeed, defined chemical variation of, for example, the LPS of

S.typhimurium and *E.coli* affects bacterial resistance or sensitivity to ingestion, and also to killing and digestion (so-called postphagocytic events)

3.2.1. Ingestion

Medearis et al. (1968) have shown that phagocytosis and the rate of clearance of intraperitoneally administered mutants of *E.coli*, lacking enzymes concerned with the attachment of specific sugars to the LPS chain, is related to the structure of the synthesized LPS. Thus the parent strain of *E.coli* 0111:B4 is virulent to mice and resistant to phagocytosis, both in vivo (macrophages) and in vitro (polymorphonuclear leukocytes), but a mutant lacking uridine diphosphate galactose-4-epimerase, and hence certain sugars in its envelope LPS, is avirulent and highly susceptible to ingestion and killing by phagocytes. Growing of the mutant in media supplemented with galactose, to allow bypass of the enzymatic deficiency and synthesis of complete LPS, renders the mutant strain as virulent and resistant to phagocytosis as the parent strain.

These observations have been extended in a series of recent studies on other gram-negative microorganisms. For example, an inverse relationship appears to exist between susceptibility of mutants of *S. typhimurium* to phagocytosis and killing in vitro by various types of phagocytes (both macrophages and polymorphonuclear leukocytes) on the one hand and the length of the polysaccharide chain of the envelope lipopolysaccharides on the other (Friedberg and Shilo, 1970a,b; Stendahl and Edebo, 1972; Tagesson and Stendahl, 1973). Rottini et al. (1975) and Dri et al. (1975) report that growth at 45°C (at this temperature little synthesis of polysaccharide occurs) renders previously resistant strains of *E.coli* susceptible to phagocytosis by guinea pig macrophages or polymorphonuclear leukocytes. Resistance to phagocytosis could be restored by addition of a partially purified (lipo)polysaccharide isolated from a phagocytosis-resistant *E.coli* strain while a similarly prepared fraction from a phagocytosis-sensitive strain did not restore resistance. Despite considerable effort (Stendahl et al., 1973a,b; Stendahl et al., 1974) it is not yet clear how the length of the lipopolysaccharides influences the susceptibility of a given gram-negative (*S. typhimurium*) mutant to uptake and killing. The prediction that shorter polysaccharide chains might produce more hydrophobic bacterial surfaces and thereby promote phagocytosis, possibly through an easier surface interaction between microorganism and phagocyte (van Oss and Gillman, 1972) has not been realized. In fact, the partition of smooth and rough strains of *S. typhimurium* in a two-phase system ran counter to this prediction (Stendahl et al., 1973a), suggesting that complex alterations occur in the surface properties of gram-negative bacteria when the LPS structure is changed, whether in the core region or in the O-antigen portion of the molecule (Stendahl and Edebo, 1972). Regardless of the nature of these alterations, addition of normal serum only partly reduces the differences in apparent rates of ingestion. Whether the opsonins in hyperimmune serum have the same effects on ingestibility as do bacterial extracts (Dri et al., 1975) has not been reported.

3.2.2. Postphagocytic events

Evidence indicates that those bacterial surface properties which influence the ability of the phagocyte to ingest a microorganism also affect the fate of the organism after ingestion.

To review the factors that determine events after engulfment, it is first necessary to consider a number of assumptions that have ruled the prevailing concepts about the order in which these events take place. Thus it is generally held that killing and digestion do not start until the microorganisms are sequestered within the phagocytic vacuole. This concept derives in large measure from the evidence that the antimicrobial armamentarium resides entirely in the cytoplasmic granules and that attack on the microorganisms must hence await degranulation and discharge of granule contents into the vacuole. Although this view seems amply supported by numerous observations, two considerations suggest that it may be too narrow: (1) very early after the start of phagocytosis by human polymorphonuclear leukocytes, biologically active compounds are released into the extracellular environment (Skosey et al., 1974; Root et al., 1975). Although to my knowledge it has not yet been determined whether the concentrations of antibacterial substances released are sufficient to kill bacteria, this possibility exists and needs further examination; (2) certain granule-associated, catalytically active proteins apparently may occur in association with other membranes. This has been shown for leukocyte alkaline phosphatase (West et al., 1974; Bretz and Baggiolini, 1974) and has been suggested for phospholipase A_2 (Franson et al., 1974). Moreover, in light of certain questions that have been raised concerning the "purity" of various fractions obtained in conventional fractionation studies (De Pierre and Karnovsky, 1973; Chang et al., 1975), it may be premature to accept as unequivocal that the granule preparations are uncontaminated by other membrane fractions. Since the polymorphonuclear leukocyte is relatively free of mitochondria and endoplasmic reticulum, the plasma membrane might also be a site of antimicrobial substances. Hence some microbicidal action against susceptible organisms might also commence upon surface contact.

Another common assumption is that killing precedes digestion, while actually it is likely that degradative attack on bacterial constituents is an essential component of at least some bactericidal mechanisms. It should be emphasized here that except in the case of lysozyme in a few instances of lysozyme-sensitive gram-positive bacteria, the molecular mode of action of none of the phagocyte's antimicrobial systems and agents has been established. It seems probable, however, that most if not all microbicidal events depend on or at least involve envelope alterations (Elsbach, 1973). A review of what is known about the various antimicrobial systems of phagocytes is quite compatible with the notion that effective killing does not occur unless the envelope constituents are modified or destroyed, either by catalytic attack and/or by noncatalytic chemical alteration through interaction with the various highly reactive compounds that are produced by the phagocyte while it ingests (Fig. 1).

Further, there is surprisingly little concrete experimental support for the general idea that the digestive process, so often depicted as an obligatory step in phagocytosis that follows after the organism is killed, is necessarily complete or even very extensive once the rapid-killing phase is over (Elsbach, 1973). On closer scrutiny, it is also not as clear as is frequently assumed that the degradative enzymes of the phagocyte are responsible for destruction of engulfed organisms. This assumption stems, of course, from the lysosome concept and the demonstration of degranulation as an integral feature of phagocytosis. It must be recognized, however, that bacteria possess their own degradative apparatus and that conditions unfavorable to microbial multiplication or survival may result in activation of catabolism (Ballesta and Schaechter, 1971; Wurster et al., 1971; Tomasz, 1974).

Surprisingly few in vitro biochemical studies have been carried out of the degradation of microorganisms during phagocytosis. The time-course of degradation of various radiolabeled microorganisms during killing by polymorphonuclear leukocytes was examined by Cohn (1975) more than a decade ago. In the presence of fresh frozen serum Cohn found rather extensive degradation of $^{32}P_i$ or ^{14}C-glucose-labeled constituents of both gram-positive and gram-negative bacteria at low ratios of bacteria to leukocytes (1–2:1). Since degradation lagged behind killing, these results could be interpreted as typical of "digestion after killing." A more comprehensive examination of the structural and functional alterations found in bacteria exposed to intact granulocytes or purified bactericidal fractions derived from them was recently initiated in the author's laboratory. Using a range of precursors for the purpose of specifically labeling bacterial macromolecules, Elsbach et al. (1973) found large differences in the degradative fates of different bacterial constituents and among different microorganisms during phagocytosis by granulocytes. These experiments were carried out in the absence of serum because of the profound effects serum may have on bacterial viability and envelope structure (Feingold et al., 1968a,b; Beckerdite-Quagliata et al., 1975). Omission of serum did not prevent vigorous killing (at least one log) within 15 minutes, at a ratio of 20 bacteria to one granulocyte. The gram-positive bacteria *Micrococcus lysodeikticus* and *Bacillus megaterium* were rapidly destroyed as is evident from almost complete dissolution of cell wall peptidoglycans of these lysozyme-sensitive organisms, presumably resulting in spheroplast formation and cell rupture. In line with drastic structural disorganization, these bacteria completely lost their ability to carry out macromolecular synthesis within minutes after exposure to the leukocytes. Despite the apparently massive destruction and, as one would expect, close contact of the leukocyte's degradative enzymes with the bacterial fragments, both radiolabeled DNA and protein undergo little breakdown into acid-soluble material. Degradation of RNA is more pronounced but reaches a plateau within 30 minutes. Interestingly, breakdown of phosphatidylglycerol, the main phospholipid of *M. lysodeikticus,* also reaches a plateau after 30 minutes despite fragmentation of the organism. This is unexpected in view of the evidence that exogenous phospholipases A_2 readily hydrolyze the bulk of the glycerol phosphatides of nonsealed erythrocyte ghosts and of protoplasts of *Bacillus subtilis* (Op den Kamp et al., 1972; Zwaal et al., 1975). In this situation the phospholipase evidently "sees" both layers of an asymmetrical lipid bilayer (Bretscher, 1973). Why leukocyte phospholipase A_2 would lack access to most or all cytoplasmic membrane phosphatidylglycerol is not apparent. Disrupted granulocytes, which appear at least as rapidly bactericidal and destructive as intact cells, cause no greater hydrolysis of phospholipids of *M. lysodeikticus,* rendering it unlikely that insufficient phospholipase A_2 had reached the contents of the phagocytic vacuoles of intact leukocytes. These findings suggest shielding of a substantial portion of the microbial lipid from leukocyte phospholipase A_2 action, presumably through interaction with other lipids such as cardiolipin that remain essentially intact (Patriarca et al., 1972) and/or with nonlipid membrane constituents. In any case the mobility of degradable lipid substrates (Devaux and McConnell, 1972) within the membrane of *M. lysodeikticus* must be more limited than suggested for eukaryotic cells by Bretscher (1976), unless they can move in shielding clusters or domains.

Similar experiments with *E. coli* show that these gram-negative bacteria, which are endowed with a far more complex envelope, are even more resistant to degradative attack (Patriarca et al., 1972; Elsbach et al., 1972a; 1973; 1974; Mooney and Elsbach, 1975). This is also supported not only by evidence from morphological observations (Zucker-Franklin et al., 1971) and by data showing very incomplete degradation during prolonged incubation of previously labeled bacterial constituents including lipids, but also by the fact that *E. coli* remain capable of integrated biochemical activity (macromolecular and lipid synthesis; synthesis of β-galactosidase upon presentation with inducer) after loss of ability to multiply (Elsbach, 1973). Such preservation of biochemical function seems to imply that the cytoplasmic membrane must have escaped major destruction. These observations on the interaction of various bacterial species with intact phagocytes do not aid in establishing whether the modification of bacterial constituents that accompanies the killing process is an integral part of the bactericidal event, or merely a concomitant phenomenon, or part of the digestive process that is thought to follow loss of viability.

Obviously, by observing the structural and functional alterations of bacteria ingested by various phagocytes, one examines the resultant of multiple forces acting on the microorganism, often in a synergistic fashion. To gain insight into single vectors involved in phagocyte-microbe interaction, it has been necessary to dissect the antimicrobial systems of the whole cell into individual components. A voluminous literature concerned mainly with the polymorphonuclear leukocyte deals with attempts of this sort. Because an excellent review of the antimicrobial systems of the polymorphonuclear leukocyte has appeared recently (Klebanoff, 1975), the following section will be restricted only to consideration of those aspects that are pertinent to effects on microbial envelopes.

3.2.3. Microbicidal factors and systems of polymorphonuclear leukocytes that can cause microbial surface alterations

The granulocyte's antimicrobial armamentarium can be divided into two categories (Table 1): one depends on the leukocyte's ability to increase its metabolic activity and to produce chemically reactive metabolites on encounter

386

Table 1
Antimicrobial systems of polymorphonuclear leukocytes

1.	*Dependent for formation on metabolic responsiveness of intact leukocytes*	
	(a)	oxygen-dependent (see Fig. 1)
	(b)	oxygen-independent (e.g., drop in intracellular pH).
2.	*Independent of metabolic activity;* evident in broken cell preparations (i.e., preexisting factors)	
	(a)	proteins that are not catalytically active
	(b)	catalytically active proteins:
		lysozyme
		proteases
		phospholipase A
		other degradative enzymes

with an ingestible particle (microorganism); another is independent of active metabolism. Thus the antimicrobial agents in this second category preexist in the mature leukocyte.

3.2.3.1. Antimicrobial systems dependent on the metabolic responsiveness of the leukocyte. Of these systems, by far the most potent and well-studied are those that are generated during the respiratory burst. As shown in Fig. 1, the dramatically increased consumption of O_2 is associated with the increased formation of highly reactive molecules which, probably alone, but particularly in combination with granule-connected myeloperoxidase and an oxidizable substrate, exert lethal effects on a wide range of microorganisms. The molecular mechanism of microbicidal action of these systems is unclear but, in view of the multiple highly reactive intermediates that can be formed, is likely to vary with the prevailing conditions in the phagocyte and with the different substrates provided by different organisms (Klebanoff, 1975; Sbarra et al., 1977). Mechanisms that may be involved in lethal events include oxidation (for example of SH-groups), peroxide formation, deamination, and decarboxylation of aminoacids with aldehyde formation. These chemical reactions are probably mediated by the oxidation of various halides such as chloride (Sbarra et al., 1977; Zgliczynski and Stelmaszynska, 1977) or iodide (Klebanoff, 1975). Substrates in the reactions include bacterial proteins and probably lipids (Shohet et al., 1974; Klebanoff, 1975; Sbarra et al., 1977). Since myeloperoxidase adheres to bacterial surfaces, it seems likely that envelope proteins are important participants in the chemical reactions. Indirect evidence has been presented suggesting that the oxygen-dependent antimicrobial system also produces bacterial (pneumococcal) lipid peroxides (Shohet et al., 1974). These studies, carried out with intact leukocytes that are potently destructive toward pneumococci (Tomasz et al., 1977), permit no conclusion about the possible lethal effects on bacteria of envelope lipid peroxidation. The isolated myeloperoxidase system is bactericidal over a wide pH range (Sbarra et al., 1977). Its antimicrobial effect against *Staphylococcus aureus* and *E. coli* required from 15 to 60 minutes to become apparent (Odeberg,

1976). This contrasts with the much more immediate effects of certain cationic bactericidal proteins (Zeya and Spitznagel, 1966; Odeberg, 1976; Weiss et al., 1975). This lag of the myeloperoxidase system is reduced in the presence of a purified granulocyte elastase that has no bactericidal action by itself (Odeberg, 1976).

The enhanced metabolic activity of the phagocytizing granulocyte also causes a drop in intravacuolar pH. This has been demonstrated convincingly in rabbit granulocytes (review, Klebanoff, 1975). There is little doubt that a low ambient pH within the phagocytic vacuole will contribute to the hostile environment created by the phagocytes. The effects on different microorganisms are clearly multiple and not restricted to those of proton accumulation per se, for example, on surface charge, envelope function (including energy production and transport), and possibly activation of bacterial degradative enzymes. Another important consequence for ingested microorganisms is activation of the leukocyte's acid hydrolases that are released into the vacuole. I am unaware of any published attempts to dissect the diverse effects on microorganisms created by a drop in the pH.

3.2.3.2. Antimicrobial agents of phagocytes active in the absence of metabolic activity. The polymorphonuclear leukocyte has been the main target for the study of these agents. The antimicrobial activity demonstrable in disrupted granulocytes and extracts is predominantly, if not entirely, attributable to a group of so-called cationic proteins (Klebanoff, 1975). Some of these possess enzymatic activity; others have no known catalytic role. In each instance the antimicrobial effects of this group of proteins are accompanied by envelope alterations.

3.2.3.2.1. Noncatalytic cationic proteins. That cationic proteins can exert potently bactericidal effects without apparent catalytic effect is best exemplified by the effects of the basic peptide antibiotics of the polymixin class. These antibiotics interact with the gram-negative bacterial envelope, causing irreversible breakdown of the permeability barriers along with their bactericidal action. The envelope effects are thought to be based on interaction with phosphatidylethanolamine (HsuChen and Feingold, 1973) or with phosphatidic acid and phosphatidylglycerol (Teuber, 1973). Evidence has also been presented indicating penetration of polymixin B, at minimal lethal concentrations, into the bacterial cytoplasm of *S. typhimurium* and *E. coli* B (Schindler and Teuber, 1975).

The group of bactericidal basic proteins isolated from polymorphonuclear leukocytes from various animal species was thought initially to be devoid of catalytic activity (Zeya and Spitznagel, 1966). Their bactericidal effects were found to be associated with suppression of O_2 uptake and leakage of bacterial constituents. Recently, however, several of these proteins in highly purified form have been shown to be enzymatically active (Odeberg and Olsson, 1975; Odeberg et al., 1975). However, as will be discussed below, to date there is no definite evidence that the catalytic and bactericidal properties of these proteins are related. A beginning has also been made with the study of fungicidal cationic proteins of human monocytes (Lehrer, 1975).

3.2.3.2.2. Catalytically active proteins. Lysozyme. This very basic protein (molecular weight approximately 14,500; isoelectric point > 10.0) is capable of degrading peptidoglycans by hydrolyzing the β (1-4) glycosidic linkage between N-acetylglucosamine and N-acetylmuramic acid. The effectiveness of this muramidase action on cell walls of gram-positive organisms depends on at least two factors (Johnson and Campbell, 1972). First, the overall complexity of the peptidoglycan matrix, the extent of cross-linking, and the length of the cross bridges determine if lysozyme can penetrate the peptidoglycan mesh to reach the glycosidic bonds. Second, the degree of O-acetylation of the peptidoglycans is an important determinant of lysozyme sensitivity. In gram-negative bacteria the outer membrane (LPS) generally provides an additional impediment to lysozyme action on cell wall peptidoglycans. It is not surprising therefore that lysozyme resistance is the rule among gram-negative and common among gram-positive bacteria. Lysozyme's ability to kill bacteria is thus limited to rather few, and usually nonpathogenic, organisms. The biological importance of this quantitatively major constituent of the specific (secondary) granules of leukocytes probably lies in its contribution to the combined catabolic action of the complete antimicrobial apparatus of the granulocyte. The biological significance of the continuous selective secretion of large amounts of lysozyme into the extracellular environment by mononuclear phagocytes is unknown (Gordon et al., 1974).

3.2.3.2.3. Antimicrobial cationic proteins with proteolytic activity. A number of the azurophil (primary) granule-associated cationic proteins, specifically those of human polymorphonuclear luekocytes, have been purified to a high degree of homogeneity. Many of these cationic proteins possess esterase and proteolytic activities (Janoff, 1972; 1975; Ohlsson and Olsson, 1973; 1974; Dewald et al., 1975; Feinstein and Janoff, 1975a,b; Odeberg et al., 1975; Rindler-Ludwig and Braunsteiner, 1975). Several of these proteases, for instance a chymotrypsin-like protease, are bactericidal in highly purified form (Odeberg and Olsson, 1975). Others, such as elastase, are not bactericidal alone but may apparently potentiate other isolated bactericidal systems (Odeberg, 1976). Human granulocyte collagenase has no such potentiating effect. The bactericidal action of the cationic protein with chymotrypsin-like properties is expressed toward gram-positive as well as gram-negative bacteria and is thought to be associated with envelope alterations including increased permability. The mode of action of the proteases on the viability of microorganisms is unknown. In fact, the catalytic activity seems unrelated to the bactericidal effect because heat treatment does not affect the latter but eliminates the former. Progress in the purification and characterization of the primary granule-associated cationic proteases of human leukocytes has become very rapid, and new insights into their biological roles should be forthcoming in the near future (Janoff, 1975). In this context it should be noted that the potently bactericidal rodent granulocyte contains little protease activity. Cultivated mouse macrophages, on the other hand, produce and secrete both collagenase and elastase (Werb and Gordon, 1975a,b).

3.2.3.2.4. Cationic phospholipase A$_2$ associated with a bactericidal complex. The

· effect on gram-negative bacteria of another cationic protein species with bactericidal and catalytic activities has been studied by Elsbach and collaborators (Beckerdite et al., 1974; Mooney and Elsbach, 1975; Weiss, et al., 1975, 1976).

As mentioned earlier, disrupted rabbit granulocytes effectively kill *E. coli* and other gram-negative bacteria without causing extensive structural or functional disorganization. However, almost immediately upon exposure of *E. coli* to disrupted granulocytes, an increase in envelope permeability becomes manifest; the bacteria become sensitive to the inhibitory effects on macromolecular synthesis of actinomycin D, an antibiotic (molecular weight 1300) to which the outer membrane of the *E. coli* envelope is normally impermeable. Cytoplasmic β-galactosidase and periplasmic alkaline phosphatase are not released under these conditions, indicating that the effect on permeability is relatively discrete (Beckerdite et al., 1974). The rapid loss of viability and the increased envelope permeability are accompanied by the prompt onset of net hydrolysis of phospholipid and the stimulation of biosynthesis of certain phospholipid species. Incorporation of oleic acid into all phospholipids is increased, suggesting that in particular the turnover of the fatty acid in the 2-ester position (*E. coli* incorporate oleic acid almost exclusively into the 2-position) is stimulated and possibly triggered by the action of a phospholipase A_2 (Mooney and Elsbach, 1975). Leukocyte phospholipase A_2 and permeability-increasing activities are mainly associated with the cytoplasmic granules. These activities show no latency and appear tightly membrane-bound (Franson et al., 1974; Weiss et al., 1975). Solubilization can, however, be achieved readily by acid extraction. Further purification has yielded a fraction that is about equally enriched with respect to bactericidal, permeability increasing, and phospholipase A_2 activities. This fraction is devoid of demonstrable lysozyme, protease, and myeloperoxidase activity (Weiss et al., 1975). All three biological activities of the fraction are equally heat-stable and appear to require binding of the active principle(s) to the envelope for their expression. Resistant microorganisms exhibit minimal or no binding. Both the binding and the three biological activities of this fraction are completely blocked by Mg^{2+} or Ca^{2+} (Weiss et al., 1975). Addition of these ions after the biological effects become apparent produces abrupt cessation of the net hydrolysis of phospholipids, initiates reincorporation of the products of hydrolysis into the bacterial phospholipids and, along a similar time course, reestablishes insensitivity to actinomycin D. This reversal of the envelope effects requires from 15 to 30 minutes at 37°C, depending on dose of and time of exposure to the active principle, and can be produced as long as 2 hours after the effects on permeability and phospholipids became apparent. However, ability to multiply is irreversibly lost within 5 minutes after exposure of *E. coli* populations (2–5 × 10^8 organisms per ml) to μg quantities of the active principle (Weiss et al., 1976). Mutants of *E. coli* devoid of demonstrable phospholipases A (kindly donated by Prof. S. Nojima, Department of Chemistry, National Institute of Health, Tokyo, Japan; cf. Doi and Nojima, 1973) are also sensitive to the bactericidal, permeability, and phospholipid effects of the purified leukocyte fraction (Weiss and Elsbach, 1977). Work in progress is consistent with the conclusion that the three biological activities reside in a complex with a molecular weight of approximately 50,000. Using SDS-polyacrylamide gel electrophoresis, three main components are detected with apparent molecular weights of approximately 50,000, 30,000, and 15,000. Conditions can be created that result in the conversion of at least 70% of the protein in the complex into the lower molecular weight components. The dissociation of the complex is accompanied by a loss of bactericidal and permeability increasing activities, that corresponds to the loss of higher molecular weight protein. Thus the specific biological activity of the remaining complex does not change during dissociation. The low molecular weight components are totally devoid of biological activity toward intact E. coli, but they retain phospholipase A_2 activity. It is remarkable that the potent bactericidal action of the complex is associated with minimal structural and functional disorganization, as judged by the fact that the organisms, despite their loss of ability to multiply, retain their biosynthetic potential. Thus, synthesis of β-galactosidase in response to inducer after the envelope effects have been repaired (upon addition of Mg^{2+}) is the same as that by E. coli populations that were never exposed to the leukocyte fraction. Although the action of the active principle(s) on structure and function of E. coli appears subtle by the experimental criteria that we have applied, it is obvious that numerous structural and metabolic regulatory effects may have gone unrecognized. Any one or more of such effects might account for loss of ability to multiply. Whether any direct relationships actually exist between the bactericidal and envelope effects and the components of the active fraction that cause them cannot be deduced from the available information. One must be aware of the propensity of cationic proteins to form complexes nonspecifically. On the other hand, several observations suggest that the three main components of the complex bear more than an incidental connection, particularly the findings that conversion of complex to the low molecular weight component occurs with preservation of specific biological activity in the remaining complex, that the heat stability of the three biological activities is similar, and that loss of one activity tends to be associated with loss of the others. Furthermore, the effects on phospholipids and permeability seem closely linked. Specifically, increased permeability is associated with hydrolysis of phospholipid, and conversely, restoration of the permeability barrier is associated with removal of membrane-lytic products of phospholipid hydrolysis, either by their reincorporation into diacylphosphatides upon addition of divalent cations or by complex formation with extracellular albumin (Weiss et al., 1976).

What properties of the gram-negative envelope determine susceptibility to the bactericidal and envelope effects of the purified fraction has not yet been established. Both rough and smooth E coli strains have been found to be sensitive. A mutant of S. typhimurium, deficient in uridine diphosphate galactose 4-epimerase and therefore unable to insert galactose into LPS, is highly sensitive to the bactericidal and permeability effects of the fraction. However, this organism is totally resistant after growth in the presence of galactose, which overcomes the defect in LPS synthesis. It is tempting to consider the possibility that specific chemical properties of LPS determine whether or not binding and biological activity of antimicrobial proteins become manifest. Yet it must be recalled that *S. typhimurium* mutants differing in relative resistance to phagocytosis by granulocytes and to the bactericidal effects of various isolated antimicrobial systems of granulocytes, and also to those of serum (Stendahl and Edobo, 1972; Tagesson and Stendahl, 1973). The effects of these very diverse systems (including crude granule preparations, myeloperoxidase-H$_2$O$_2$-halide and D-aminoacid oxidase systems from leukocytes, and antibody plus complement in serum) must thus depend on different kinds of interaction with the microbial envelopes. For example, the properties of the gram negative envelope of *E. coli* that determine susceptibility to the lytic action of serum are not the same as those that determine sensitivity to certain antibacterial cationic proteins of human leukocytes (Odeberg and Olsson, 1975). It appears therefore that no simple determinants can yet be recognized that explain differences in susceptibility of closely related microorganisms to host-defense systems.

It is also unclear how Mg^{2+} and Ca^{2+} protect susceptible microorganisms against so many antimicrobial substances of different chemical composition, including cationic leukocyte proteins (briefly reviewed by Weiss et al., 1975). Except in the case of EDTA after brief exposure (Leive and Kollin, 1967), these divalent cations are generally unable to restore the permeability barrier of *E. coli* once it is impaired, by serum or polymixin for example (HsuChen and Feingold, 1972; Pruul and Reynolds, 1972). The effects on permeability caused by the purified granulocyte fraction are, however, readily reversed by Mg^{2+}, and this reversal is accompanied by such acute interruption of phospholipid hydrolysis (Weiss et al., 1976), that an instantaneous physical effect on the interaction between active principle(s) and bacterial envelope seems most likely. Such an effect is probably not merely an electrostatic one, since Na^+ or K^+ do not prevent or reverse envelope alterations. Rather, Mg^{2+} and Ca^{2+} may modify the envelope by influencing interactions among several molecular species at different sites, for example, LPS (Leive, 1974), peptidoglycans (Rayman and McLeod, 1975), lipids (Verkley et al., 1974), and perhaps other components of the cytoplasmic membrane (Razin, 1969; Scherrer and Gerhardt, 1973; Nakamura and Mizushima, 1975). Apparently, the ions produce a tighter envelope by promoting cross-linking. This could cause conformational changes in attachment sites for antimicrobial substances leading to their detachment, as may have been shown in the case of granulocyte phospholipase A$_2$ (Weiss et al., 1976). Proteins, such as phospholipases, that penetrate into a lipid-water interphase (Verger et al., 1973) may also be "squeezed out" when divalent cations interacting with acidic polar headgroups of phospholipids produce tighter packing of lipid bilayers (Nicolson, 1976). Whether or not the biochemical repair of the envelope follows detachment of antimicrobial substances would then depend on the extent and location of the damage. For example, any effects that are confined mainly to the outer envelope layers would be reversible, but more substantial involvement of the cytoplasmic membrane and hence of the biochemical apparatus would be irreversible. The fact that divalent cations may trigger repair of the *E. coli* envelope effects caused by the purified leukocyte fraction without rescuing the bacteria from the effects on multiplication again illustrates the still undefined relationship between these two effects. Nevertheless, it is necessary to reemphasize the important finding that the leukocyte fraction only acts on *E. coli* when it exists as an apparent complex. The leukocyte phospholipase A $_2$ has no recognizable effect on wild-type *E. coli,* nor do other (pure) phospholipases A $_2$. Actually, many intact natural membranes are quite resistant to hydrolytic attack by phospholipases A. However, in combination with surface-active cationic proteins without known catalytic activity, phospholipase activity toward both natural (Habermann, 1972) and artificial (Mollay and Kreil, 1974) membranes may undergo a manifold enhancement. Interestingly, these cationic proteins occur as major constituents of snake (direct lytic factor) and insect (e.g., melittin) venoms (Habermann, 1972). The basic polypeptide antibiotic, polymixin B, can substitute for the polypeptide native to the venom in the potentiation of the venom phospholipase A_2 (Mollay and Kreil, 1974). Thus there is a precedent for cooperative effects between components of a biologically active mixture. In bee venom it is evident that melittin and the very basic phospholipase A_2(isoelectric point: 10.5; Shipolini et al., 1971) are tightly associated (Mollay and Kreil, 1974). Whether or not such associations between identical subunits may also elicit otherwise unexpressed biological activities remains to be established.

4. Concluding remarks

In only a few years our knowledge of the cell surface changes that occur during phagocytosis has grown vastly. Rapid progress can be discerned especially in the characterization of the phenomena that occur at the surface of the phagocyte (section 2). Our understanding of the molecular events at the microbial envelope that govern the fate of microorganisms exposed to phagocytes has not increased as rapidly. However, important clues have been obtained that should help direct future investigations in this latter area. These clues derive from studies on mutant organisms with presumably single enzymatic defects and from observations on the effects on microbial function of purified antimicrobial systems and substances isolated from phagocytes. To profitably continue comparative studies of the susceptibility of different mutant bacteria, it seems particularly necessary to define the physical and chemical properties of the envelopes of microorganisms incapable of a specific step in the biosynthesis of an envelope component. It is evident that the consequent alterations in the surface properties are not merely those of the defective product.

In light of the complexity of the effects of chemically precisely defined antibiotic agents on microorganisms it must be recognized that the study of pure antimicrobial compounds obtained from phagocytes is not prone to yield unequivocal results. It should be possible, however, to identify the specific structural and functional sequelae of the effects of isolated antimicrobial systems and proteins on well-defined microbial envelopes, thereby providing new insights into phagocyte-microbe interactions and host-defense in general.

Acknowledgments

This study was supported in part by the 1975–1976 Josiah Macy Jr. Foundation Faculty Scholar Award and by Public Health Service grant AM 05472 from the National Institutes of Arthritis, Metabolism and Digestive Diseases.

References

Albright, F. R., White, D.A. and Lennarz, W.J. (1973) Studies on enzymes involved in the catabolism of phospholipids in Escherichia coli. J. Biol. Chem. 248,3968–3977.

Allison, A.C., Davies, P. and dePetris, S. (1971) Role of contractile filaments in macrophage movement and endocytosis. Nature New Biol. 232, 153–155.

Armstrong, J.A. and Hart, P.D. (1971) Response of cultured macrophages to Mycobacterium tuberculosis, with observations on fusion of lysosomes with phagosomes. J. Exp. Med. 134, 713–740.

Armstrong, J.A. and Hart, P.D. (1975) Phagosome-lysosome interactions in cultured macrophages infected with virulent tubercule bacilli. J. Exp. Med. 142,1–16.

Axline, S.G. and Reaven, E.P. (1974) Inhibition of phagocytosis and plasma membrane mobility of the cultivated macrophage by cytochalasin B. J. Cell Biol. 62, 647–659.

Babior, B.M., Kipnes, R.S. and Curnutte, J.T. (1973) Biological defense mechanisms. The production by leukocytes of superoxide, a potential bactericidal agent. J. Clin. Invest. 52, 741–744.

Babior, B.M., Curnette, J.T. and Kipnes, R.S. (1975) Biological defense mechanisms. Evidence for the participation of superoxide in bacterial killing by xanthine oxidase. J. Lab. Clin. Med. 85, 235–244.

Bainton, D.F. (1973) Sequential degranulation of the two types of polymorphonuclear leukocyte granules during phagocytosis of microorganisms. J. Cell Biol. 58, 249–264.

Ballesta, J.P.G. and Schaechter, K. (1971) Effect of shift-down and growth inhibition on phospholipid metabolism of Escherichia coli. J. Bacteriol. 107,251–258.

Bayer, M.E. (1974) Ultrastructure and organization of the bacterial envelope. Ann. N.Y. Acad. Sci. 235,6–28.

Beckerdite, S., Mooney, C., Weiss, J., Franson, R. and Elsbach, P. (1974) Early and discrete changes in permeability of Escherichia coli and certain other gram-negative bacteria during killing by granulocytes. J. Exp. Med. 140, 396–409.

Beckerdite-Quagliata, S., Simberkoff, M. and Elsbach, P. (1975) Effects of human and rabbit serum on viability, permeability and envelope lipids of Serratia marcescens. Infect. Immun. 11,758–766.

Bianco, S., Griffin, F.M. and Silverstein, S.M. (1975) Studies of the macrophage complement receptor. Alterations of receptor function upon macrophage activation. J. Exp. Med. 141, 1278–1290.

Borel, J.F., Keller, H.U. and Sorkin, E. (1969) Studies on chemotaxis. XI. Effect on neutrophils of lysosomal and other subcellular fractions from leukocytes. Int. Arch. Allergy Appl. Immunol. 35, 194–205.

Borisy, G.G., Marcum, J.M., Olmsted, J.B., Murphy, D.B. and Johnson, K.A. (1975) Purification of tubulin and associated high molecular weight proteins from porcine brain and characterization of microtubule assembly in vitro. Ann. N.Y. Acad. Sci. 253, 107–132.

Bourne, H.R., Lehrer, R.I., Cline, M.J. and Melmon, K.L. (1971) Cyclic-3'5'-adenosine monophosphate in the human leukocyte: Synthesis, degradation and effects on neutrophil candidacidal activity. J. Clin. Invest. 50, 920–929.

Boxer, L.A., Hedley-Whyte, E.T. and Stossel, T.P. (1974) Neutrophil actin dysfunction and abnormal neutrophil behavior. New Engl. J. Med. 291, 1093–1099.

Braun, V. (1975) Covalent lipoprotein from the outer membrane of Escherichia coli. Biochim. Biophys. Acta 415, 335–377.

Bray, D. (1976) Spectrin-like proteins. Nature 260, 16.

Bretscher, M.S. (1973) Membrane structure: Some general principles. Science 181, 622–629.

Bretscher, M.S. (1976) Directed lipid flow in cell membranes. Nature 260,21–23.

Bretz, U. and Baggiolini, M. (1974) Biochemical and morphological characterization of azurophil and specific granules of human neutrophilic polymorphonuclear leukocytes. J. Cell Biol. 63, 251–269.

Briggs, R.T., Drath, D.R., Karnovsky, M.J. and Karnovsky, M.L. (1974) Surface localization of NADH-oxidase in polymorphonuclear leukocytes. J. Cell Biol. 63, 36a.

Chang, K., Bennett, V. and Cuatrecasas, P. (1975) Membrane receptors as general markers for plasma membrane isolation procedures. J. Biol. Chem. 250, 488–500.

Clark, R.A. and Klebanoff, S.J. (1975) Neutrophil-mediated tumor cell toxicity: Role of the peroxidase system. J. Exp. Med. 141,1442–1447.

Cohn, Z.A. (1975) Macrophage physiology. Fed. Proc. 34, 1725–1729.

Constantopoulos, A. and Najjar, V.A. (1973) The activation of adenylate cyclase: II. The postulated presence of (A)adenylate cyclase in a phospho(inhibited) form (B) a dephospho(activated) form with a cyclic adenylate stimulated membrane protein kinase. Biochem. Biophys. Res. Commun. 53, 794–799.

Costerton, J.W., Ingram, J.M. and Cheng, K.J. (1974) Structure and function of the cell envelope of gram-negative bacteria. Bacteriol. Rev. 38, 87–110.

Cox, J.P. and Karnovsky, M.L. (1973) The depression of phagocytosis by exogenous cyclic nucleotides, prostaglandins and theophylline. J. Cell Biol. 59, 480–490.

Cronan, J.E., Jr. and Gelman. E.P. (1975) Physical properties of membrane lipids: biological relevance and regulation. Bacteriol. Rev. 39, 232–256.

Curnutte, J.T., Whitten, D.M. and Babior, B.M. (1974) Defective superoxide production by granulocytes from patients with chronic granulomatous disease. New Eng. J. Med. 290, 593–597.

D'Arcy Hart, P. and Young, M.R. (1975) Interference with normal phagosome-lysosome fusion in macrophages, using ingested yeast cells and suramin. Nature 256, 47–49.

Davies, P., Page, R.C. and Allison, A.C. (1974) Changes in cellular enzyme levels and extracellular release of lysosomal acid hydrolases in macrophages exposed to group A streptococcal cell wall substance. J. Exp. Med. 139, 1262–1282.

De Chatelet, L.R. (1975) Oxidative bactericidal mechanisms of polymorphonuclear leukocytes. J. Inf. Dis. 131, 295–203.

De Chatelet, L.R., McPhail, L.C., Mullikin, D. and McCall, C.E. (1975a) An isotopic assay for NADPH oxidase activity and some characteristics of the enzyme from human polymorphonuclear leukocytes. J. Clin. Invest. 55, 714–721.

De Chatelet, L.R., Mullikin, D. and McCall, C.E. (1975b) The generation of superoxide anion by various types of phagocytes. J. Infect. Dis. 131, 443–446.

De Pierre, J.W. and Karnovsky, M.L. (1973) Plasma membranes of mammalian cells. A review of methods of their characterization and isolation. J. Cell Biol. 56, 275–303.

Devaux, P. and McConnell, H.M. (1972) Lateral diffusion in spin-labeled phosphatidylcholine multilayers. J. Am. Chem. Soc. 94, 4475–4481.

Dewald, B., Rindler-Ludwig, R., Bretz, U. and Baggiolini, M. (1975) Subcellular localization and heterogeneity of neutral proteases in neutrophilic polymorphonuclear leukocytes. J. Exp. Med. 141, 709–723.

Dlabač, V. (1968) The sensitivity of smooth and rough mutants of Salmonella typhimurium to bactericidal and bacteriolytic action of serum, lysozyme and to phagocytosis. Folia Microbiol. 13, 439–449.

Doi, O. and Nojima, S. (1973) Detergent-resistant phospholipase A_1 and A_2 in Escherichia coli. J. Biochem. 74, 667–674.

Drath, D.B. and Karnovsky, M.L. (1975) Superoxide production by phagocytic leukocytes. J. Exp. Med. 141, 257–262.

Dri, D., Rottini, G.D., Bellavita, L. and Patriarca, P. (1975) Antiphagocytic activity of the cell wall polysaccharide of E. coli. J. Reticuloend. Soc. Abstr. Suppl. 18, 25a.

Edelson, P.J. and Cohn, Z.A. (1974a) Effects of concanavalin A on mouse peritoneal macrophages. I. Stimulation of endocytic activity and inhibition of phago-lysosome formation. J. Exp. Med. 140, 1364–1386.

Edelson, P.J. and Cohn, Z.A. (1974b) Effects of concanavalin A on mouse peritoneal macrophages.

II. Metabolism of endocytized proteins and reversibility of the effects by mannose. J. Exp. Med. 140, 1387–1403.

Elsbach, P. (1959) Composition and synthesis of lipids in resting and phagocytizing leukocytes. J. Exp. Med. 110, 969–980.

Elsbach, P. (1965) Uptake of fat by phagocytic cells. An examination of the role of phagocytosis. I. Rabbit polymorphonuclear leukocytes. Biochim. Biophys. Acta 98, 402–419.

Elsbach. P. (1966) Phospholipid metabolism by phagocytic cells. I. A comparison of conversion of ^{32}P lysolecithin to lecithin and glycerylphosphorylcholine by homogenates of rabbit polymorphonuclear leukocytes and alveolar macrophages. Biochim. Biophys. Acta 125, 510–524.

Elsbach, P. (1967) Phospholipid metabolism by phagocytic cells. II. Metabolism of lyso phosphatidylethanolamine and lysophosphatidylcholine by homogenates of rabbit polymorphonuclear leukocytes and alveolar macrophages. J. Lipid Res. 8, 359–365.

Elsbach, P. (1968) Increased synthesis of phospholipid during phagocytosis. J. Clin. Invest. 47, 2217–2229.

Elsbach, P. (1972) Lipid metabolism by phagocytes. Semin. Hematol. 9, 227–239.

Elsbach, P., (1973) On the interaction between phagocytes and microorganisms. New Engl. J. Med. 289, 846–852.

Elsbach, P. (1974) Phagocytosis. In: The Inflammatory Process (Zweifach, B.W., Grant, L. and McCluskey, R.T., eds.), pp. 363–408. Academic, New York, N.Y.

Elsbach, P. and Rizack, M.A. (1963) Acid lipase and phospholipase activity in homogenates of rabbit polymorphonuclear leukocytes. Am. J. Physiol. 205, 1154–1158.

Elsbach, P. and Farrow, S. (1969) Cellular triglyceride as a source of fatty acid for lecithin synthesis during phagocytosis. Biochim. Biophys. Acta 176, 438–441.

Elsbach, P., Zucker-Franklin, D. and Sansaricq, C. (1969). Increased lecithin synthesis during phagocytosis by normal leukocytes and by leukocytes of a patient with chronic granulomatous disease. New Engl. J. Med. 280, 1319–1322.

Elsbach, P., Goldman, J. and Patriarca, P. (1972a) VI. Observations on the fate of phospholipids of granulocytes and ingested Escherichia coli during phagocytosis. Biochim. Biophys. Acta 280, 33–44.

Elsbach, P., van den Berg, J.W.O., van den Bosch. H. and van Deenen, L.L.M. (1965) Metabolism of phospholipids by polymorphonuclear leukocytes. Biochim. Biophys. Acta 106,338–347.

Elsbach, P., Pettis, P., Beckerdite, S. and Franson, R. (1973) Effects of phagocytosis by rabbit granulocytes on macromolecular synthesis and degradation in different species of bacteria. J. Bacteriol. 115, 490–497.

Elsbach, P., Beckerdite, S., Pettis, P. and Franson, R. (1974) Persistance of regulation of macromolecular synthesis by Escherichia coli during killing by disrupted rabbit granulocytes. Infect. Immun. 9, 663–668.

Elsbach, P., Patriarca, P., Pettis, P., Stossel, T.P., Mason, R.J. and Vaughan, M. (1972b) The appearance of lecithin-^{32}P, synthesized from lysolecithin-^{32}P, in phagosomes of polymorphonuclear leukocytes. J. Clin. Invest. 51, 1910–1914.

Falke, D., Schiefer, H.G. and Stoffel, W. (1967) Lipoid Analysen bei Riesenzellbildung durch Herpesvirus Hominis. Z. Naturforsch. 22b, 1360–1362.

Feingold, D.S., Goldman, J.N. and Kuritz, H.M. (1968a) Locus of action of serum and the role of lysozyme in the serum bactericidal reaction. J. Bacteriol. 96, 2118–2126.

Feingold, D.S., Goldman, J.N., and Kuritz, H.M. (1968b) Locus of the lethal event in the serum bactericidal reaction. J. Bacteriol. 96, 2127–2131.

Feinstein, G. and Janoff, A. (1975a) A rapid method for purification of human granulocyte cationic neutral proteases: Purification and characterization of human granulocyte chymotrypsin-like enzyme. Biochim. Biophys. Acta 403; 477–492.

Feinstein, G. and Janoff, A. (1975b) A rapid method of purification of human granulocyte cationic neutral proteases: Purification and further characterization of human granulocyte elastase. Biochim. Biophys. Acta 403, 493–505.

Fong, K., McCay, P.B., Poyer, J.L., Keele, B.B. and Misra, H. (1973) Evidence that peroxidation of lysosomal membranes is initiated by hydroxyl free radicals produced during flavin enzyme activity. J. Biol. Chem. 248, 7792–7797.

396

Franson, R., Beckerdite, S., Wang, P., Waite, M. and Elsbach, P. (1973) Some properties of phospholipases of alveolar macrophages. Biochim. Biophys. Acta 296, 365–373.

Franson, R., Patriarca, P. and Elsbach, P. (1974) Phospholipid metabolism by phagocytic cells. Phospholipase A$_2$ associated with rabbit polymorphonuclear leukocyte granules. J. Lipid Res. 15, 380–388.

Friedberg, D. and Shilo, M. (1970a) Interaction of gram-negative bacteria with the lysosomal fraction of polymorphonuclear leukocytes. I. Role of cell wall composition of Salmonella typhimurium. Infect. Immun. 1, 305–310.

Friedberg, D. and Shilo, M. (1970b) Role of cell wall structure of Salmonella in the interaction with phagocytes. Infect. Immun. 2, 279–285.

Gallin, J.T. and Rosenthal. A.S. (1974) The regulatory role of divalent cations in human granulocyte chemotaxis. J. Cell Biol. 62, 594–609.

Gallin, J.T., Durocher, J.R. and Kaplan, A.P. (1975) Interaction of leukocyte chemotactic factors with the cell surface. I. Chemotactic factor-induced changes in human granulocyte surface charge. J. Clin. Invest. 55, 967–974.

Goetzl, E.J. and Austen, K.F. (1974) Stimulation of human neutrophil leukocyte aerobic glucose metabolism by purified chemotactic factors. J. Clin. Invest. 53, 591–599.

Gold, S.B., Hanes, D.M., Stites, D.P. and Fudenberg, H.H. (1974) Abnormal kinetics of degranulation in chronic granulomatous disease. New Engl. J. Med. 291, 332–337.

Goldstein, J., Hoffstein, S., Gallin, J. and Weissmann, G. (1973) Mechanisms of lysosomal enzyme release from human leukocytes: Microtubule assembly and membrane fusion induced by a component of complement. Proc. Nat. Acad. Sci. USA 70, 2916–2920.

Goldstein, J., Roos, D., Kaplan, H.B. and Weissmann, G. (1975) Complement and immunoglobulins stimulate superoxide production by human leukocytes independently of phagocytosis. J. Clin. Invest. 5, 1155–1163.

Gordon, S., Todd, J., and Cohn, Z.A. (1974) In vitro synthesis and secretion of lysozyme by mononuclear phagocytes. J. Exp. Med. 139, 1228–1248.

Griffin, F.M. and Silverstein, S.C. (1974) Segmental response of the macrophage plasma membrane to a phagocytic stimulus. J. Exp. Med. 139, 323–336.

Griffin, F.M., Bianco, C. and Sliverstein, S.C. (1975a) Characterization of the macrophage receptor for complement and demonstration of its functional independence from the receptor for the Fc portion of immunoglobulin G. J. Exp. Med. 141, 1269–1277.

Griffin, F.M., Griffin, J.A., Leider, J.E. and Silverstein, S.C. (1975b) Studies on the mechanism of phagocytosis. I. Requirements for circumferential attachment of particle-bound ligands to specific receptors on the macrophage plasma membrane. J. Exp. Med. 142, 1263–1282.

Habermann, E. (1972) Bee and wasp venoms. Science 177, 314–322.

Haller, I., Hoehn, B. and Henning, U. (1975) Apparent high degree of asymmetry of protein arrangement in the Escherichia coli outer cell envelope membrane. Biochemistry 14, 478–484.

Hammer, C.H., Nicholson, A. and Mayer, M.M. (1975) On the mechanism of cytolysis by complement: Evidence on insertion of C5b and C7 subunits of the C5b,6,7 complex into phospholipid bilayers of erythrocyte membranes. Proc. Nat. Acad. Sci. U.S.A. 72, 5076–5080.

Harold, F.H. (1970) Antimicrobial agents and membrane function. Adv. Microb. Physiol. 4, 45–104.

Hartwig, J.H. and Stossel, T.P. (1975) Isolation and properties of actin, myosin, and a new actinbinding protein in rabbit alveolar macrophages. J. Biol. Chem. 250, 5696–5705.

Henson, P.M. (1971) The immunologic release of constituents from neutrophil leukocytes. II. Mechanisms of release during phagocytosis and adherence to nonphagocytosable surfaces. J. Immunol. 107, 1547–1557.

Holland, P., Holland, N.H. and Cohn, Z.A. (1972) The selective inhibition of macrophage phagocytic receptors by anti-membrane antibodies. J. Exp. Med. 135, 458–475.

Holtzman, E., Schacher, S., Evans, J. and Teichberg, S. Chapter 4, this volume.

HsuChen, C. and Feingold, D.S. (1973). The mechanism of polymyxin B action and selectivity toward biological membranes. Biochemistry 12, 2105–2111.

Ignarro, L.J. (1974) Nonphagocytic release of neutral protease and β-glucuronidase from human neutrophils. Arthr. Rheumat. 17, 25–36.

Inouye, M. (1974) A three-dimensional molecular assembly model of a lipoprotein from the *Escherichia coli* outer membrane. Proc. Nat. Acad. Sci. U.S.A. 71, 2396–2400.

Janoff, A. (1972) Neutrophil proteases in inflammation. Ann. Rev. Med. 23, 177–190.

Janoff, A. (1975) At least three human neutrophil lysosomal proteases are capable of degrading joint connective tissues. Ann. N.Y. Acad. Sci. 256, 402–408.

Ji, T.H. and Nicolson, G.L. (1974) Lectin binding and perturbation of the outer surface of the cell membrane induces a transmembrane organisational alteration at the inner surface. Proc. Nat. Acad. Sci. U.S.A. 71, 2212–2216.

Johnson, K.G., and Campbell, J.N. (1972) Effect of growth conditions on peptidoglycan structure and susceptibility to lytic enzymes in cell walls of Micrococcus sodonensis. Biochemistry 11, 277–286.

Johnston, R.B., Jr., Keele, B.B., Jr., Misra. H.P., Lehmeyer, J.E., Webb, L.S., Baehner, R.L., and Rajagopalan. K.V. (1975) The role of superoxide anion generation in phagocytic bactericidal activity. J. Clin. Invest. 55, 1357–1372.

Jones, T.C. and Hirsch, J.G. (1971) The interaction in vitro of Mycoplasma pulmonis with mouse peritoneal macrophages and L-cells. J. Exp. Med. 133, 231–259.

Jones, T.C. and Hirsch, J.G. (1972) The interaction between Toxoplasma gondii and mammalian cells. II. The absence of lysosomal fusion with phagocytic vacuoles containing living parasites. J. Exp. Med. 136, 1173–1194.

Jones, T.C., Yeh, S. and Hirsch, J.G. (1972a) Studies on attachment and ingestion phases of phagocytosis of Mycoplasma pulmonis by mouse peritoneal macrophages. Proc. Soc. Exp. Biol. Med. 139, 464–470.

Jones, T.C., Yeh, S. and Hirsch, J.G. (1972b) The interaction between Toxoplasma gondii and mammalian cells. I. Mechanism of entry and intracellular fate of the parasite. J. Exp. Med. 136, 1157–1172.

Kader, J.C. (1977) Exchange of phospholipids between membranes. In: Dynamic Aspects of Cell Surface Organization (Poste, G. and Nicolson, G.L., eds.) Cell Surface Reviews, Vol. 3, pp. 127–204. Elsevier/North-Holland, Amsterdam.

Karnovsky, M.L. (1962) Metabolic basis of phagocytic activity. Physiol. Rev. 42, 143–168.

Karnovsky, M. L. (1973) Chronic granulomatous disease—pieces of a cellular and molecular puzzle. Fed. Proc. 32, 1527–1533.

Klebanoff, S.J. (1975) Antimicrobial mechanisms in neutrophilic polymorphonuclear leukocytes. Seminars Hematol. 12, 117–142.

Korn, E.D. and Weisman, R.A. (1967) Phagocytosis of latex beads by Acanthamoeba. II. Electron microscopic study of the initial events. J. Cell Biol. 34, 219–227.

Korn, E.D., Bowers, B., Batzri, S., Simmons. S.R., and Victoria, E.J. (1974) Endocytosis and exocytosis: Role of microfilaments and involvement of phospholipids in membrane fusion. J. Supramolec. Struct. 2, 517–528.

Krinsky, N.I. (1974) Singlet excited O_2 as mediator of antibacterial action of leukocytes. Science 186, 363–365.

Krischnapillai, V. (1971) Uridinediphosphogalactose-4-epimerase deficiency in Salmonella typhimurium and its correction by plasmid borne galactose genes of Escherichia coli K-12: Effects on mouse virulence, phagocytosis and serum sensitivity. Infect. Immun. 4, 177–188.

Lay, W.H. and Nussensweig, V. (1968) Receptors for complement on leukocytes. J. Exp. Med. 138, 991–1007.

Layne, P., Constantopoulos, A., Judge, J.F.X., Rauner, R. and Najjar, V.A. (1973) The occurrence of fluoride stimulated membrane phosphoprotein phosphatase. Biochem. Biophys. Res. Commun. 53, 800–805.

Lehrer, R.I. (1975) The fungicidal mechanisms of human monocytes. I. Evidence for myeloperoxidase-linked and myeloperoxidase independent candidacidal mechanisms. J. Clin. Invest. 55, 338–346.

Leive, L. (1974) The barrier function of the gram-negative envelope. Ann. N.Y. Acad. Sci. 235, 109–129.

Lucy, J.A. (1970) The fusion of biological membranes. Nature 227, 815–817.

Lugtenberg, B., Meijers, J., Peters, R., van der Hoek, P. and van Alphen, L. (1975) Electrophoretic resolution of the "major outer membrane protein" of Escherichia coli K12 into four bands. FEBS Lett. 58, 254–258.

Mantovani, M., Rabinovitch, M. and Nussenzweig, V. (1972) Phagocytosis of immune complexes by macrophages. Different roles of the macrophage receptor sites for complement (C3) and for imunoglobulin (IgG). J. Exp. Med. 135, 780–792.

Mason, R.J., Stossel, T.P. and Vaughan, M. (1972) Lipids of alveolar macrophages, polymorphonuclear leukocytes and their phagocytic vesicles. J. Clin. Invest. 51, 2399–2407.

McCall, E. and Youlten, L.J.F. (1973) Prostaglandin E 1 synthesis by phagocytosing rabbit polymorphonuclear leukocytes: its inhibition by indomethacin and its role in chemotaxis. J. Physiol. 234, 98–100P.

McIntyre, T.M. and Bell, R.M. (1975) Mutants of Escherichia coli defective in membrane phospholipid synthesis. Effect of cessation of net phospholipid synthesis on cytoplasmic and outer membranes. J. Biol. Chem. 250, 9053–9059.

Medearis, D.N., Camitta, B.M. and Heath, E.C. (1968) Cell wall composition and virulence in Escherichia coli. J. Exp. Med. 128, 399–404.

Messner. R.P. and Jelinek, J. (1970) Receptors for human γG globulin on human neutrophils. J. Clin. Invest. 495, 2165–2171.

Michell, R.H. (1975) Inositol phospholipids and cell surface receptor function. Biochim. Biophys. Acta 415, 81–147.

Mollay, C. and Kreil, G. (1974) Enhancement of bee venom phospholipase A_2 activity by melittin, direct lytic factor from cobra venom and polymyxin B. FEBS Lett. 46, 141–144.

Mooney, C. and Elsbach, P. (1975) Altered phospholipid metabolism in Escherichia coli accompanying killing by disrupted granulocytes. Inf. Immun. 11, 1269–1277.

Müller-Eberhard, H.J. (1975) Complement. Ann. Rev. Biochem. 44, 697–724.

Nakamura, K. and Mizushima, S. (1975) In vitro reassembly of the membranous vesicle from Escherichia coli outer membrane components. Role of individual components and magnesium ions in reassembly. Biochim. Biophys. Acta 413, 371–393.

Nelson, B.W. and Roantree, R.J. (1967) Analysis of lipopolysaccharides extracted from penicillin-resistant, serum sensitive Salmonella mutants. J. Gen. Microbiol. 48,179–188.

Nicolson, G.L. (1976) Transmembrane control of the receptors on normal and tumor cells. I. Cytoplasmic influence over cell surface components. Biochim. Biophys. Acta 457, 57–108.

Nicolson, G.L., Smith, J.R. and Poste, G. (1976) Effects of local anaesthetics on cell morphology and membrane-associated cytoskeletal organization in BALB/3T3 cells. J. Cell Biol. 68, 395–402.

Nicolson, G.L., Poste, G. and Ji, T.H. (1977) The dynamics of cell membrane organization. In: Dynamic Aspects of Cell Surface Organization (G. Poste and G. L. Nicolson, eds.). Vol. 3, Cell Surface Reviews, pp 1–73. Elsevier/North-Holland, Amsterdam.

Odeberg, H. (1976) Bactericidal mechanisms in granulocytes. Contribution by granular cationic protein to the antibacterial activity of human polymorphonuclear leukocytes. Acad. thesis, Lund.

Odeberg, H. and Olsson (1975) Antibacterial activity of cationic proteins from human granulocytes. J. Clin. Invest. 56, 1118–1124.

Odeberg, H., Olsson, I. and Venge, P. (1975) Cationic proteins of human granulocytes. IV. Esterase activity. Lab. Invest. 32, 86–90.

Ohlsson, K. and Olsson, I. (1973) The neutral proteases of human granulocytes. Isolation and partial characterization of two granulocyte collagenases. Europ. J. Biochem. 36, 473–481.

Ohlsson, K. and Olsson. I. (1974) The neutral proteases of human granulocytes. Isolation and partial characterization of granulocyte elastases. Europ. J. Biochem. 42, 519–527.

Op den Kamp. J.A.F., Kauerz, M.T. and Van Deenen, L.L.M. (1972) Action of phospholipase A 2 and phospholipase C on Bacillus subtilis protoplasts. J. Bacteriol. 112, 1090–1098.

Oren, R., Farnham. A.E., Saito, K., Milofsky, E. and Karnovsky, M.L. (1963) Metabolic patterns in three types of phagocytic cells. J. Cell Biol. 17, 487–502.

Osborn, M.J., Rick, P.D., Lehmann, V., Rupprecht, E. and Singh, M. (1974) Structure and biogenesis of the cell envelope of gram-negative bacteria. Ann. N.Y. Acad. Sci. 235, 52–65.

Parkes, J.G. and Fox, C.F. (1975) On the role of lysophosphatides in virus-induced cell fusion and lysis. Biochemistry 14, 3627–3725.

Patriarca, P., Beckerdite, S., Pettis, P. and Elsbach, P. (1972) Phospholipid metabolism by phagocytic cells. VII. The degradation and utilization of phospholipids of various microbial species by rabbit granulocytes. Biochim. Biophys. Acta 280, 45–56.

Patriarca, P., Dri, P., Kakinuma, K., Tedesco, F. and Rossi, F. (1975) Studies on the mechanism of metabolic stimulation in polymorphonuclear leukocytes during phagocytosis. I. Evidence for superoxide anion involvement in the oxidation of NADPH 2 Biochim. Biophys. Acta 385, 380–386.

Pesanti, E.L. and Axline, S.G. (1975) Colchicine effects on lysosomal enzyme induction and intracellular degradation in the cultivated macrophage. J. Exp. Med. 141, 1030–1046.

Pollard, T.D. and Korn, E.D. (1971) Filaments of Amoeba proteus. II. Binding of heavy meromyosin by thin filaments in motile cytoplasmic extracts. J. Cell Biol. 48, 216–219.

Pollard, T.D. and Weihing, R.A. (1975) Actin and myosin and cell movement. CRC Crit. Rev. Biochem. 2, 1–65.

Poste, G. and Allison, A.C. (1973) Membrane fusion. Biochim. Biophys. Acta 300, 421–465.

Poste, G., Papahadjopoulos, D. and Nicolson, G.L. (1975) Local anesthetics affect transmembrane cytoskeletal control of mobility and distribution of cell surface receptors. Proc. Nat. Acad. Sci. U.S.A. 72, 4430–4434.

Pruul, H. and Reynolds, B.L. (1972) Interaction of complement and polymixin with gram-negative bacteria. Infect. Immun. 6, 709–717.

Rabinovitch, M. (1967) The dissociation of the attachment and ingestion phases of phagocytosis by macrophages. Exp. Cell Res. 46, 19–28.

Rabinovitch, M. and DeStefano, M.J. (1973) Macrophage spreading in vitro I. Inducers of spreading. Exp. Cell Res. 77, 323–334.

Rabinovitch, M. and DeStefano, M.J. (1974) Macrophage spreading in vitro III. The effect of metabolic inhibitors, anaesthetics and other drugs on spreading induced by subtilisin. Exp. Cell Res. 88, 153–162.

Rasmussen, H. (1970) Cell communication, calcium ion, and cyclic adenosine monophosphate. Science 170, 404–412.

Rayman, M.K. and MacLeod, R.A. (1975) Interaction of Mg^{2+} with peptidoglycan and its relation to the prevention of lysis of a murine Pseudomonad. J. Bacteriol. 122, 650–659.

Razin, S. (1969) Structure and function in mycoplasma. Ann. Rev. Microbiol. 23, 317–356.

Reaven, E.P. and Axline, S.G. (1973) Subplasmalemmal microfilaments and microtubules in resting and phagocytizing cultivated macrophages. J. Cell Biol. 59, 12–27.

Rindler-Ludwig, R. and Braunsteiner, H. (1975) Cationic proteins from human neutrophil granulocytes: Evidence for their chymotrypsin-like properties. Biochim. Biophys. Acta 379, 606–617.

Roantree, R.J. (1967) Salmonella O antigens and virulence. Ann. Rev. Microbiol. 21, 443–466.

Root, R.K., Metcalf, J., Oshino, N. and Chance, B. (1975) H_2O_2 release from human granulocytes during phagocytosis. I. Documentation, quantitation, and some regulating factors. J. Clin. Invest. 55, 945–955.

Rossi, F., Patriarca, P., Romeo, D., and Zabucchi, G. (1977) The mechanism of control of phagocytic metabolism. In: The Reticuloendothelial System in Health and Disease (Reichard, S.H., Escobar, M.R. and Friedman, H. eds.). Plenum Press, New York, 205–223.

Rottini, G., Dri, P., Soranzo, M.R. and Patriarca, P. (1975) Correlation between phagocytic activity and metabolic response of polymorphonuclear leukocytes toward different strains of E. coli. Infect. Immun. 11, 417–423.

Schindler, P.R.G. and Teuber, M. (1975) Action of polymixin B on bacterial membranes: Morphological changes in the cytoplasm and in the outer membrane of Salmonella typhimurium and Escherichia coli B. Antimicr. Agents Chemoth. 8, 95–104.

Sbarra, A.J., Selvaraj, R.J., Paul, B.B., Zgliczynski, J.M., Poskitt, P.K.F., Mitchell, G.W. Jr., and Louis, F. (1977) Chlorination, decarboxylation and bactericidal activity mediated by the $MPO-H_2O_2-Cl$-system. In: The Reticuloendothelial System in Health and Disease (Reichard, S.H., Escobar, M.R. and Friedman, H., eds.) Plenum Press, New York, 191–203.

Scandella, C.J. and Kornberg, A. (1971) A membrane-bound phospholipase A 1 purified from *Escherichia coli*. Biochemistry 10, 4447–4456.

Schell-Frederick, E. (1974) Stimulation of the oxidative metabolism of polymorphonuclear leukocytes by the calcium ionophore A23187. FEBS Letters 48, 37–40.

Scherrer, R. and Gerhardt, P. (1973) Influence of magnesium ions on porosity of the Bacillus megaterium cell wall and membrane. J. Bacteriol. 114, 888–890.

Schnaitman, C.A. (1970) Protein composition of the cell wall and cytoplasmic membrane of Escherichia coli. J. Bacteriol. 104, 890–901.

Shipolini, R.A., Callewaert, G.I., Cottrell, R.C., Doonan, S. and Vernon, C.A. (1971) Phospholipase A from bee venom. Europ. J. Biochem. 20, 459–468.

Shohet, S.B. (1970) Changes in fatty acid metabolism in human leukemic granulocytes during phagocytosis. J. Lab. Clin. Med. 75, 659–667.

Shohet, S.B., Pitt, M., Baehner, R.L. and Poplack, D.G. (1974) Lipid peroxidation in the killing of phagocytized pneumococci. Infect. Immun. 10, 1321–1328.

Silbert, D.F. (1975) Genetic modification of membrane lipid. Ann. Rev. Biochem. 44, 315–339.

Simoni, R.D. and Postma, P.W. (1975) The energetics of bacterial active transport. Ann. Rev. Biochem. 44, 523–554.

Skosey, J.L., Chow, D., Damgaard, E. and Sorensen, L.B. (1973) Effect of cytochalasin B on response of human polymorphonuclear leukocytes to zymosan. J. Cell Biol. 57, 237–240.

Skosey, J.L., Damgaard, E., Chow, D. and Sorensen, L.B. (1974) Modification of zymosan-induced release of lysosomal enzymes from human polymorphonuclear leukocytes by cytochalasin B. J. Cell Biol. 62, 625–634.

Smolen, J.E. and Shohet, S.B. (1974) Remodeling of granulocyte membrane fatty acids during phagocytosis. J. Clin. Invest. 53, 726–734.

Stendahl, O. and Edebo, L. (1972) Phagocytosis of mutants of Salmonella typhimurium by rabbit polymorphonuclear cells. Acta Path. Microbiol. Scand. (B)80, 481–488.

Stendahl, O., Magnusson, K.E., Tagesson, C., Cunningham, R. and Edebo, L. (1973a) Characterization of mutants of Salmonella typhimurium by counter-current distribution in an aqueous two-polymer phase system. Infect. Immun. 7, 573–577.

Stendahl, O., Tagesson, C. and Edebo, M. (1973b) Partition of Salmonella typhimurium in a two-polymer aqueous phase system in relation to liability to phagocytosis. Infect. Immun. 8, 36–41.

Stendahl, O., Tagesson, C. and Edebo, L. (1974) Influence of hyperimmune immunoglobulin G on the physicochemical properties of the surface of Salmonella typhimurium 395MS in relation to interaction with phagocytic cells. Infect. Immun. 10, 316–319.

Steinman, R.M. and Cohn, Z.A. (1974) The metabolism and physiology of the mononuclear phagocytes. In: The Inflammatory Process (Zweifach, B.W., Grant, L. and McCluskey, R.T., eds), pp. 449–510. Academic, New York.

Stossel, T.P. (1973) Quantitative studies of phagocytosis. Kinetic effects of cations and heat-labile opsonin. J. Cell Biol. 58, 346–356.

Stossel, T.P. (1974) Phagocytosis. New Engl. J. Med. 290, 717–723; 774–782; 833–839.

Stossel, T.P. and Pollard, T.D. (1973) Myosin in polymorphonuclear leukocytes. J. Biol. Chem. 248, 8288–8294.

Stossel, T.P., and Hartwig, J.H. (1975) Interaction between actin, myosin and an actin-binding protein from rabbit alveolar macrophages. J. Biol. Chem. 250, 5706–5712.

Stossel, T.P., Mason, R.J. and Smith, A.L. (1974) Lipid peroxidation by human blood phagocytes. J. Clin. Invest. 54, 638–645.

Stossel, T.P., Field, R.J., Gitlin, J.D., Alper, C.A. and Rosen, F.S. (1975) The opsonic fragment of the 3rd component of human complement (C3). J. Exp. Med. 141, 1329–1347.

Tagesson, C. and Stendahl, O. (1973) Influence of the cell surface lipopolysaccharide structure of Salmonella typhimurium on resistance to intracellular bactericidal systems. Acta Path. Microbiol. Scand. (B)81, 473–480.

Takanaka, K. and O'Brien, P. (1975a) Mechanisms of H_2O_2 formation by leukocytes. Evidence for a plasma-membrane location. Arch. Biochem. Biophys. 169, 428–435.

Takanaka, K. and O'Brien, P. (1975b) Mechanisms of H_2O_2 formation by leukocytes. Properties of the NAD(P)H oxidase activity of intact leukocytes. Arch. Biochem. Biophys. 169, 436–442.

Tatsumi, N., Shibata, N., Okamura, Y., Takeuchi, K. and Senda, N. (1973) Actin and myosin A from leukocytes. Biochim. Biophys. Acta 305, 433–444.

Teuber, M. (1973) Action of polymixin B on bacterial membranes. II. Formation of lipophilic complexes with phosphatidic acid and phosphatidylglycerol. Z. Naturforsch. 28, 476–477.

Tomasz, A. (1974) The role of autolysins in cell death. Ann. N.Y. Acad. Sci. 235, 439–447.

Tomasz, A., Beckerdite, S., McDonnell, M. and Elsbach, P. (1977) The activity of the Pneumococcal autolytic system and the fate of the bacterium during ingestion by rabbit polymorphonuclear leukocytes. J. Cell. Physiol. (in press).

Tou, J. and Stjernholm, R.L. (1975) Cytochalasin B: effect on phospholipid metabolism and lysosomal enzyme release by leukocytes. Biochim. Biophys. Acta, 392, 1–11.

Tou, J. and Stjernholm, R.L. (1974) Stimulation of the incorporation of $^{32}P_i$ and myo-[2-^{3}H]inositol into the phosphoinositides in polymorphonuclear leukocytes during phagocytosis. Arch. Biochem. Biophys. 160, 487–494.

Turner, S.R., Campbell, J.A. and Lynn, W.S. (1975) Polymorphonuclear leukocyte chemotaxis towards oxidized lipid components of cell membranes. J. Exp. Med. 141, 1437–1441.

Tyler, D.D. (1975) Role of superoxide radicals in the lipid peroxidation of intracellular membranes. FEBS Lett. 51, 180–183.

Unkeless, J.C., Gordon, S. and Reich, E. (1974) Secretion of plasminogen activator by stimulated macrophages. J. Exp. Med. 139, 834–850.

Van den Besselaar, A.M.H.P., Helmkamp, G.M. and Wirtz, K.W.A. (1975) Kinetic model of the protein-mediated phosphatidylcholine exchange between single bilayer liposomes. Biochemistry 14, 1852–1858.

van Oss, C.J. and Gillman, C.F. (1972) Phagocytosis as a surface phenomenon. I. Contact angles and phagocytosis of non-opsonized bacteria. J. Reticuloendothelial Soc. 12, 283–292.

Verger, R., Mieras, M.C.E. and de Haas, G.H. (1973) Action of phospholipase A at interfaces. J. Biol. Chem. 248, 4023–4034.

Verkley, A.J., DeKruyff, B., Ververgaert, P.H.J.Th., Tocanne, J.F. and van Deenen, L.L.M. (1974) The influence of pH, Ca^2 + and protein on the thermotropic behavior of the negatively charged phospholipid, phosphatidylglycerol. Biochim. Biophys. Acta 339, 432–437.

Victoria, E.J. and Korn, E.D. (1975a) Enzymes of phospholipid metabolism in the plasma membrane of Acanthamoeba castellanii. J. Lipid Res. 16, 54–60.

Victoria, E.J. and Korn, E.D. (1975b) Plasma membrane and soluble lysophospholipases of Acanthamoeba castellanii. Arch. Biochem. Biophys. 171, 255–258.

Weisman, R.A. and Korn, E.D. (1967) Phagocytosis of latex beads by Acanthamoeba I. Biochemical properties. Biochemistry 6, 485–497.

Weiss, J. and Elsbach, P. (1977) The use of phospholipase A-less Escherichia coli mutant to establish the action of granulocyte phospholipase A on bacterial phopholipide during killing by highly purified granulocyte fraction. Biochim. Biophys. Acta 466, 23–33.

Weiss, J., Franson, R., Beckerdite, S., Schmeidler, K. and Elsbach, P. (1975) Partial characterization and purification of a rabbit granulocyte factor that increases permeability of E. coli. J. Clin. Invest. 55, 33–42.

Weiss, J., Schmeidler, K., Beckerdite-Quagliata. S., Franson, R. and Elsbach, P. (1976) Reversible envelope effects during and after killing of E. coli by a highly purified granulocyte preparation. Biochim. Biophys. Acta 436, 154–169.

Werb, Z. and Cohn, Z.A. (1972) Plasma membrane synthesis in the macrophage following phagocytosis of polystyrene latex particles. J. Biol. Chem. 247, 2439–2446.

Werb, Z. and Gordon, S. (1975a) Secretion of a specific collagenase by stimulated macrophages. J. Exp. Med. 142, 346–360.

Werb, Z. and Gordon, S. (1975b) Elastase secretion by stimulated macrophages. Characterization and regulation. J. Exp. Med. 142, 361–376.

West, B.C., Rosenthal, A.S., Gelb, N.A. and Kimball, H.R. (1974) Separation and characterization of human neutrophil granules. Am. J. Pathol. 77, 41–66.

Wirtz, K.W.A. (1974) Transfer of phospholipids between membranes. Biochim. Biophys. Acta 344, 95–117.

Woodin, A.M. (1962) The extrusion of protein from the rabbit polymorphonuclear leukocyte treated with staphylococcal leucocidin. Biochem. J. 82, 9–15.

Woodin, A.M. (1973) The leucocidin-treated leukocyte. In: Lysosomes in Biology and Pathology (Dingle, J.T., ed.), vol. 3, pp. 395–422. Elsevier/North-Holland, Amsterdam.

Wurster, N., Elsbach, P., Rand. J. and Simon, E.J. (1971) Effects of levorphanol on phospholipid metabolism and composition in Escherichia coli. Biochim. Biophys. Acta 248, 282–292.

Zeya, H.I. and Spitznagel, J.K. (1966) Cationic proteins of polymorphonuclear leukocyte lysosomes. II. Composition, properties and mechanism of antibacterial action. J. Bacteriol. 91, 755–762.

Zgliczynski, J.M. and Stelmaszynska. T. (1975) Chlorinating ability of human phagocytosing leukocytes. Eur. J. Biochem. 56, 157–162.

Zucker-Franklin, D. and Hirsch, J.G. (1964) Electron microscopic studies on the degranulation of rabbit peritoneal leukocytes during phagocytosis. J. Exp. Med. 120, 569-575.

Zucker-Franklin. D., Elsbach, P. and Simon, E.J. (1971) The effect of levorphanol, a morphine analog on phagocytosing leukocytes—an electron microscope study. Lab. Invest. 25,415–421.

Zurier, R.B., Hoffstein, S. and Weissmann, G. (1973) Cytochalasin B: effect on lysosomal enzyme release from human leukocytes. Proc. Nat. Acad. Sci. U.S.A. 70, 844–848.

Zurier, R.B., Weissmann, G., Hoffstein, S., Kammerman, S., and Tai, H.H. (1974) Mechanisms of lysosomal enzyme release from human leukocytes. II. Effects of cyclic AMP and cyclic GMP, autonomic agonists, and agents which affect microtubule function. J. Clin. Invest. 53, 297–309.

Zwaal, R.F.A., Roelofsen, B., Comfurius, P. and van Deenen, L.L.M. (1975) Organization of phospholipids in human red cell membranes as detected by the action of various purified phospholipases. Biochim. Biophys. Acta 406, 83–96.

Plasma membrane assembly as related to cell division

7

John G. BLUEMINK and Siegfried W. de LAAT

1. Preface

The term "assembly" used in the title of this chapter might raise the expectation that the discussion will focus mainly on studies dealing with the molecular aspects of membrane formation. As the reader will see, this is not so. On going through the relevant literature it soon becomes clear that those papers dealing with plasma membrane formation in relation to cell division have not contributed much to our understanding of the molecular basis of the membrane assembly process. In this respect, not much progress has been made since the review on this topic by L. Warren in 1969. The biochemical investigations since then have been aimed primarily at unraveling how plasma membrane growth could be correlated in time with certain phases of the cell cycle. The evidence produced by these investigations, which used mainly cultured animal cells, is rather divergent and this probably reflects variation in the design of the experiments and the wide variety of cells used.

Our knowledge of the molecular mechanisms involved in membrane assembly has gained more from investigations using well-defined membrane systems. Such studies will not be discussed in this chapter because they do not deal with the phenomenon of cell division as such (for review, see Singer and Rothfield, 1973). In the context of the present chapter we have aimed at interrelating data obtained from morphological, biochemical, and biophysical experiments on dividing cells as well as cleaving eggs. For the reader who is particularly interested in the molecular aspects of membrane assembly the discussion may remain too much at the cytological level. Nevertheless, answers to certain no less important questions begin to emerge from the literature to be discussed. Such questions include the following: is membrane formation an integral part of the process of cytokinesis? When and where is new membrane inserted? Does it occur all over the cell surface or only in localized areas? If localized, is it followed by rapid intermixing of existing and new membrane areas? Are membrane components inserted in concert or sequentially? Do vesicles play a role?

G. Poste and G.L. Nicolson (eds.)   The Synthesis, Assembly and Turnover of Cell Surface Components
© Elsevier/North-Holland Biomedical Press, 1977.

The kind of information required to answer such questions is basic for understanding the changes occurring in the plasma membrane throughout the cell cycle. Starting from this point it may be possible to sort out the instrumental changes in the regulation of the cell cycle and cell differentiation.

> As cell arises from cell, so membrane arises from membrane: existent membranes have a role not as templates, but as organizational fields within which occurs the coordinated development of further and more complete membrane structures. (Siekevitz, 1972)

2. Introduction

Duplication of an eukaryotic cell or cleavage of an egg takes place as a consequence of "mitosis," or nuclear division. The mechanism by which the cell body is subdivided is called cytokinesis. The process of cytokinesis is causally related to mitosis. In most cells it begins when the process of mitosis has advanced to the metaphase/telophase stage. For an earlier review of the various aspects of cytokinesis in animal cells, see Rappaport (1971).

Cytokinesis involves a drastic change in cell shape which becomes apparent with the formation of a furrow. Firm evidence exists that the time and place of appearance of the furrow are determined by the mitotic event (Rappaport, 1971). During late anaphase, the future plane of furrowing is determined by the position of the mitotic spindle; the plane will appear in a position perpendicular to the middle of the earlier spindle (Figs. 1, 2). Furrow formation begins as a process of constriction. Electron microscopy has provided evidence for the presence of a ring of parallel microfilaments arranged in the proper orientation to provide the structural basis of constriction (Schroeder, 1970, 1972). It is assumed that the microfilaments, in their anchored position to the plasma membrane, are able to slide along each other. This results in an increased extent of overlap between filaments and a decreased span width of the ring. Consequently, a dividing cell changes its shape into that of a dumbbell (Figs. 2, 3). It has been recognized previously that a contractile ring of filaments may initiate furrow formation but cannot complete cytoplasmic cleavage. It is unlikely, if not impossible, that a ring of microfilaments can reduce in diameter till the cell is pinched in two. Apparently another mechanism exists that completes cell separation. Such a mechanism might be membrane (in)growth.

Let us envisage the cell as a sphere with a smooth surface. By changing into a dumbbell its surface/volume ratio increases. After complete cell separation (cell division) the total volume of the daughter cells equals that of the original cell. To package the contents of the original sphere into two separate spheres of equal size, the total surface must increase in area by about 33%. In theory, therefore, there should be about 33% more plasma membrane area after cleavage than before. However, the surface of a dividing cell is not smooth. An extra amount of cell membrane is present in the form of microvilli, ruffles, or protrusions. It remains to be established whether plasma membrane is formed concomitantly with the cytokinetic event or at an earlier time.

After completion of cytokinesis the situation is different for early embryos when compared with tissue cells. The latter are known to grow after cell division till they have attained the size of the mother cell. This implies that soon after cytokinesis, growth-related formation of plasma membrane can be expected. In early embryos, on the other hand, there is no growth during the period of rapid cell division. The volume of the egg is merely subdivided into cells of increasingly smaller dimensions. The plasma membrane formed is necessary only for the increase in intercellular surface area. This implies that in early embryos plasma membrane formation occurs in direct relation to cell division, but not necessarily simultaneously.

In summary, it can be argued that new membrane formation is necessary to complete cell separation both in tissue cells and in early embryos. In tissue cells the process of cell division is followed by cell growth. In eggs new membrane formation is related solely to cell division.

3. Cell surface irregularities: storage sites for plasma membrane?

Light- and electron-microscopy studies of dividing eggs have provided evidence that some time after the onset of contraction there is a strong local expansion of the surface in the region of the furrow. Surface marking experiments in the dividing sea-urchin egg have revealed that cell surface expansion takes place in a localized region around the isthmus of the constricted cell (Dan et al., 1938; Dan and Dan, 1940; Dan, 1954a, 1954b). Experiments using echinoderm eggs have demonstrated that in contrast to the surface inside the region of the furrow, which expands, the egg surface outside the furrow region is stationary (Rappaport and Ratner, 1967). Mechanical constraints applied to the surface next to the furrow were found to impede the stretching and shifting of the surface that normally accompanies division, and to inhibit cell division. The same constraints, when applied to other regions of the egg, do not prevent cell division. This suggests that new surface formation in the region of the furrow might be necessary to complete cell division.

Electron microscopy of dividing eggs of the mollusc *Barnea candida* has shown that during furrow formation microvilli on the surface outside the furrow region remain attached to the vitelline envelope surrounding the egg (Pasteels and de Harven, 1962). This observation suggested that new surface is formed only in the region of the furrow, which represents a band 3.5 μm wide around the egg.

Indirect evidence that plasma membrane formation occurs in the region of the furrow is also available for tissue cells. Electron microscopy (Buck and Krishan, 1964) and micromanipulation experiments with amphibian cells in culture (Rappaport and Rappaport, 1968) have shown that these cells can divide while remaining firmly anchored to neighboring cells or to the substratum by attachment sites all over the surface. On the basis of these observations it has been concluded that the plasma membrane outside the furrow region does not contribute to the

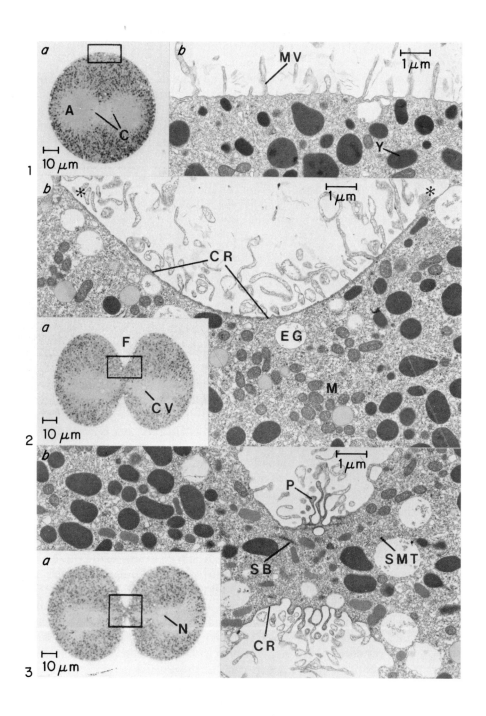

formation of the furrow, and that new surface is probably produced in the region of the furrow.

The evidence for new membrane formation during cytokinesis in amphibian eggs is based on light-microscopy, experimental, and electron-microscopy studies. In several earlier studies, the finding that unpigmented surface formed exclusively in the furrow region was interpreted as representing new membrane formation (Selman and Waddington, 1955; Dan and Kuno-Kojima, 1963; Selman and Perry, 1970; Bluemink, 1971a). Using surface markers, it was demonstrated that in the *Xenopus* egg unpigmented surface grows in area. Stretching of pigmented surface as a means of contributing preexisting membrane to the unpigmented region was not considered responsible (Bluemink and de Laat, 1973).

The above morphological evidence thus suggests that in dividing cells the preexisting plasma membrane outside the furrow region neither shifts nor stretches to allow for the furrow to form, and that concomitant with cytokinesis new surface membrane is produced in the region of the furrow.

Others have suggested (Follett and Goldmann, 1970) that surface extensions of cells, such as microvilli, may function as storage sites of plasma membrane. These structures would supply the membrane needed for furrow formation during cytokinesis. Scanning electron microscopy has revealed cell-cycle-dependent fluctuations in the "roughness" of the cell surface (Porter et al., 1973; Paweletz and Schroeter, 1974; Enlander et al., 1975; Knutton et al., 1975). Knutton et al. (1975) have suggested that during cytokinesis a smoothing out of previously accumulated microvilli takes place. They believe that cytokinesis is essentially a physical process of unfolding. Membrane biogenesis is thought to be constant throughout the cell cycle and not necessarily related to furrow formation. Other authors, however, have not come to a definitive conclusion about the role of

Fig. 1. (a) Meridional section of an *Arbacia* egg at late anaphase, about 30 seconds before furrowing begins. Daughter sets of chromosomes (C) are nearly maximally separated but the egg is still spherical. A, yolkfree zone of mitotic aster. The box frames the field of Fig. 1b. 490 ×. (b) Presumptive furrow region of the same cell shown in Fig. 1a. Neither cleavage furrow nor contractile ring is present. MV, microvillus-like projection; Y, yolk granule. 8,000×.

Fig. 2. (a) Meridional section about 4 minutes after cleavage has begun. Clusters of chromosomal vesicles (CV) characterize the reforming nuclei at this stage. F, cleavage furrow. Box corresponds to Fig. 2b. 490×. (b) Contractile ring (CR) is readily seen in this half-cleaved egg (not the same specimen as Fig. 2a). The CR spans the concave surface (between asterisks). CR prevents cytoplasmic particles, yolk, mitochondria (M), and echinochrome granules (EG) from coming into direct contact with the plasma membrane. 8,000×.

Fig. 3. (a) Meridional section at late cleavage, about 5 minutes after onset. Daughter nuclei (N) are now completely reformed. Box corresponds to Fig. 3b. 490×. (b) Contractile ring (CR) is waning, about to disappear in this egg about 5.5 minutes after it began to cleave. Complex protrusions (P) occur at this stage; the plasma membrane associated with them appears particularly dense (by appearance of osmiophilic membrane precursor material?). Stem-bodies (SB) and a few spindle microtubules (SMT) are remnants of the mitotic apparatus. 8,000×.

(Figures 1, 2, and 3 are courtesy of Dr. Tom Schroeder and reproduced by permission of the Rockefeller University Press, New York.)

408

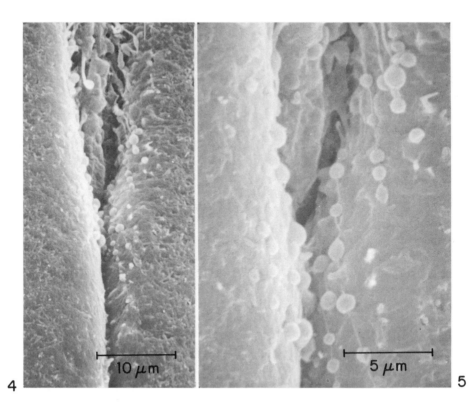

Fig. 4. Scanning electron micrograph of the furrow in the blastodisc of a dividing squid egg (*Loligo pealii*). Note the groups of furrow-related bodies (blebs) aligned on strands in the region of the furrow and the general smooth nature of the edge of the furrow as compared to the surface away from the furrow. Approximately 2,233×.

Fig. 5. Higher magnification of the blebs along the furrow surface. Approximately 4,837×.

(Figures 4 and 5 are courtesy of Dr. John M. Arnold and reproduced by permission of Academic Press Inc., New York.

microvilli (Porter et al., 1973; Paweletz and Schroeter, 1974; Enlander et al., 1975).

Extensive folding of the cell surface in the region of the furrow has been reported for dividing eggs of various species (Selman and Wadington, 1955; Arnold, 1969, 1974; Szollosi, 1970; Fullilove and Jacobson, 1971; Kalt, 1971; Schroeder, 1972; Gipson, 1974), and it has been suggested that cell membrane needed for furrow formation may be derived at least partially from such surface folds.

Folds of a different kind, pleats of cell membrane, have been observed in eggs of *Pomatoceros triqueter* (Gwynn and Jones, 1972). These pleats were present prior to cytokinesis and had disappeared after division, suggesting that unfolding of the plasma membrane may provide the membrane needed to form a furrow. Similar pleats have been seen in the *Drosophila* egg (Fullilove and Jacobson, 1971).

So far these observations suggest that in some cells the area of plasma membrane contained in surface irregularities may provide the extra membrane needed for the cell to duplicate. During which period(s) of the cell cycle the membrane contained in surface irregularities is assembled remains unanswered.

3.1. Surface blebs: sites of plasma membrane formation?

In time-lapse movies both dividing tissue cells and eggs show vigorous surface activity, particularly during the division-related part of the cell cycle. This activity takes the form of a high-frequency fluttering or blebbing of the cell surface. Bleb formation (zeiosis) can be visualized by scanning electron microscopy, and blebs have been observed along the furrow surface in dividing eggs (Bluemink and de Laat, 1973; Arnold, 1974) (Figs. 4,5). In tissue cells they are most prominent during and shortly after mitosis (Porter et al., 1973; Rubin and Everhart, 1973; Paweletz and Schroeter, 1974). Observations on Chinese hamster ovary and lung cells throughout the cell cycle revealed that blebs are predominant at the cell surface during G_1 (Figs. 6,7) and persist into S in a small percentage of cells, but vanish during the latter half of the cell cycle (Porter et al., 1973). Similar observations have been reported for HeLa cells (Robbins and Gonatas, 1964; Paweletz and Schroeter, 1974). In mouse mastocytoma cells, however, bleb formation is not confined to a particular stage of the cell cycle (Knutton et al., 1975).

The transient existence of blebs in many cultured tissue cells is quite familiar to all investigators who use time-lapse cinematography as a tool to analyze cell behavior. Nevertheless, the possible function of these surface protrusions has remained an open question. Transmission electron microscopy of cultured tissue cells has shown blebs in the form of cell surface hernias (Price, 1967). Blebs can be induced experimentally by methods that destabilize the plasma membrane (Korohoda and Stockem, 1975) or by techniques affecting the microfilament network associated with the plasma membrane (Godman et al., 1975; Nicolson et al., 1976).

In one publication the appearance of blebs has been related directly to remodeling of the plasma membrane (Gasko and Danon, 1974). In this study, the blebs seen in rabbit reticulocytes were interpreted as a manifestation of plasma membrane shedding. It was found that the membrane forming the blebs was different from the rest of the membrane in that it was less heavily labeled with gold or ferritin. A similar observation was made by Branton et al. (1974), who freeze-fractured human erythrocyte ghosts and found that protamine-induced blebs were completely free of intramembranous particles and lactose determinants. They concluded that bleb formation in the erythrocyte may result from the contraction of the spectrin filament network, causing mass flow of lipid. The blebs are covered by fluid membrane domains consisting mainly of lipids.

It may be assumed that a structural link exists between intrinsic membrane proteins and the cortical network of microfilaments present in a wide variety of cells (Clarke et al., 1975; Röhlich, 1975; Nicolson, 1976a; Nicolson, et al., 1977).

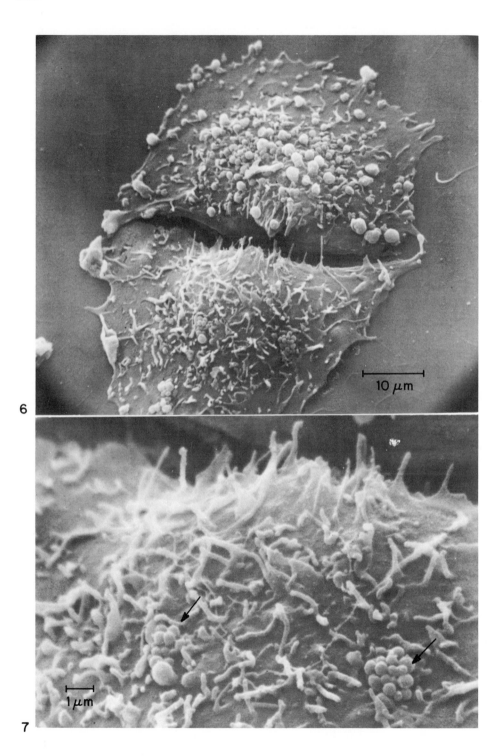

6

7

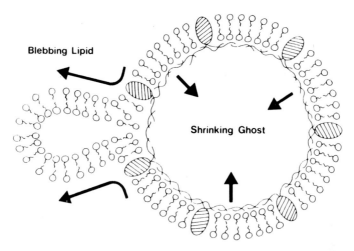

Blebbing Lipid

Shrinking Ghost

Fig. 8. Schematic representation of bleb formation in the erythrocyte membrane. As the spectrin meshwork contracts, the surface area of the ghost falls below minimum for lipid bilayer, and a region of protein-free lipid is squeezed out as a bleb. (Courtesy of Dr. D. Branton and reproduced by permission of Plenum Publishing Co. London.)

Contraction of the microfilament network, as in the case of the contracting spectrin network in the reticulocyte (Branton et al., 1974), might thus generate blebs by squeezing lipid into membrane domains which then detach from the microfilament network (Fig. 8). A bleb would thus represent a plasma membrane region of enriched lipid composition having no microfilaments associated with it. This may explain why blebs are frequently seen at the surface of dividing cells that are under the constraint of microfilament contraction.

To date blebs have not been related to membrane biogenesis. It is conceivable that fluid membrane domains consisting mainly of lipids may give rise to bleb formation. Blebs may then arise when an amount of lipid is inserted in excess of what can be integrated into an overall expansion at any one time and place. In view of their mode of production, such blebs would be different from the ones evoked by damage to the filament network (Godman et al., 1975; Nicolson et al., 1976). However, blebs resulting from excessive incorporation of lipid membrane precursors would be structurally similar to the ones described by Branton et al.

Fig. 6. Scanning electron micrograph of Chinese hamster ovary cells. Two sister cells in mid-G_1 fixed at 3 hours after shake-off. One has developed a large number of blebs; the other shows microvilli predominantly, although there are clusters of small blebs that may represent an early stage in bleb formation. The two cells were connected by a number of slender strands (filopodia) until these were broken in preparation for microscopy. 1,600×.

Fig. 7. The bottom cell of Fig. 6 at higher magnification showing the clusters of small blebs (arrows). 6,600×.

(Figures 6 and 7 are courtesy of Prof. Dr. K. Porter and reproduced by permission of The Rockefeller University Press, New York.)

(1974). In transmission electron microscopy they may exhibit liposome-like characteristics (Fluck et al., 1969; Gebicki and Hicks, 1973). Such blebs have been reported to occur during membrane remodeling (Gasko and Danon, 1974) and during plasma membrane formation (Bluemink and de Laat, 1973).

In summary, before and during the first half of cell division the cell is in a contracted state and the surface is highly irregular. The cell surface can be assumed to be in a more rigid state in this phase of division (Hiramoto, 1970, 1974, 1975; Sawai and Yoneda, 1974). Smoothing out of the surface area of microvilli or other surface projections would be sufficient to compensate for the increase in surface/volume ratio upon cell division. However, there is evidence that the area of the preexisting plasma membrane outside the furrow region remains constant, while the plasma membrane in the region of the furrow increases in area (section 3). Surface irregularities that appear as blebs during late cell division or following cell division may represent fluid membrane regions where lipids are inserted in excess of the amount that can be integrated into an overall expansion. In dividing eggs, the appearance of blebs is confined to the furrow region (Figs. 4,5), while in other cells it is not (Figs. 6,7).

4. Possible rates of membrane formation

During cell division, the cell or the egg must gain in surface area. When stretching of existing plasma membrane is not involved, a corresponding amount of new membrane must be assembled, depending on the shape of the cell and the type of cleavage (holoblastic versus meroblastic cleavage). Two examples can be selected from the literature which illustrate that a high rate of plasma membrane formation may be required. The first example is an insect egg (*Drosophila*). After fertilization the *Drosophila* egg undergoes a sequence of nuclear divisions (one every 10 minutes) without cytoplasmic cleavage (meroblastic blastema). At the twelfth mitotic division (approximately 2.5 hr later) the 4,000 nuclei formed take up their positions in the peripheral cytoplasm. The peripheral cytoplasm is then divided into approximately 4,000 cells simultaneously and the embryo changes from syncytial blastema into a cellular blastoderm. A rough estimate of the area of new membrane required for this formation of approximately 4,000 cells is $2.3 \times 10^6 \ \mu m^2$ (Fullilove and Jacobson, 1971). This amount of new surface is formed in about 30 to 60 minutes at 25°C (Rabinowitz, 1941). From this it can be calculated that approximately $7.5 - 15 \times 10^4 \ \mu m^2$ of membrane area is produced per minute. Another organism that has provided evidence of rapid membrane formation is the egg of the amphibian, *Xenopus laevis*. *Xenopus* eggs are large cells (approximately 1.25 mm in diameter) that undergo complete cell division. In the normally cleaving egg $1.4 \times 10^6 \ \mu m^2$ of intercellular membrane is formed in about 25 minutes at room temperature (Bluemink and de Laat, 1973), which means that intercellular membrane is produced at a rate of approximately $4 - 5.5 \times 10^4 \ \mu m^2$ per minute.

Both examples indicate that in certain cells membrane formation in relation to cell division may occur at a spectacular rate. In tissue culture cells, microvilli have been considered as a possible source of extra plasma membrane (section 3.1), and evidence exists that microvilli diminish in number during cell division (Porter et al., 1973; Knutton et al., 1975). For mouse mastocytoma cells it has been calculated that 33% of the total surface area is in the form of microvilli, an amount equal to 110 μm^2 per cell (Knutton et al., 1975). The proportion of membrane to be elaborated into microvilli is thought to be formed as new membrane during interphase. If this is so, the extra 110 μm^2 used during cell division is formed in the 16-hour period preceding mitosis. The rate at which membrane is formed here is thus only a fraction of that calculated for dividing eggs.

5. Transport and allocation of membrane material

Two alternative hypotheses describe membrane assembly during cytokinesis: (1) the plasma membrane grows by fusion of preassembled cytoplasmic membranous vesicles with the existing plasma membrane; and (2) plasma membrane grows by self-assembly in which individual molecules or macromolecular complexes are inserted directly into existing plasma membrane regions.

5.1 Membrane formation and vesicle fusion

Evidence for the formation of plasma membrane via vesicle fusion derives mainly from electron microscopic studies. An array of vesicles at the site of the advancing furrow has been reported for mammalian cells (Buck and Tisdale, 1962) as well as for dividing eggs (Humphreys, 1964; Zotin, 1964; Agrell, 1966; Schäfer and Bässler, 1967; Thomas, 1968; Schwalm and Bender, 1973). These vesicles were thought to represent preassembled plasma membrane material. Similar to what is seen during cytokinesis in higher plants, the vesicles would be transported to the plane of cell division where they would align and fuse, thus forming the opposed walls of the advancing furrow. However, such a mechanism is highly unlikely. Schroeder (1970) is probably right in stating that aligned vesicles are likely to be artificial, since they only appeared in cells fixed with osmium tetroxide. The idea that vesicles are artifacts and arise secondarily is supported by the observation that similar vesicles arrangements occur in eggs that have been exposed to membrane-active agents (Figs. 9,10) which affect the newly formed membrane in the plane of cleavage (Bluemink, 1971b; Emanuelsson, 1974).

The fusion of prealigned vesicles as a mechanism for furrow formation is also considered unlikely in light of the demonstration that stirring the deeper cytoplasm in the plane of cleavage (Rappaport, 1966; Rappaport and Rappaport, 1968) or replacing it (Hiramoto, 1965) does not obstruct the advancement of the cleavage furrow. As concluded earlier by Rappaport (1971), "the organization, indeed the presence of deeper cytoplasmic structures, is of no demonstrable importance in the process."

Studies on membrane biogenesis in processes other than cell division have also

414

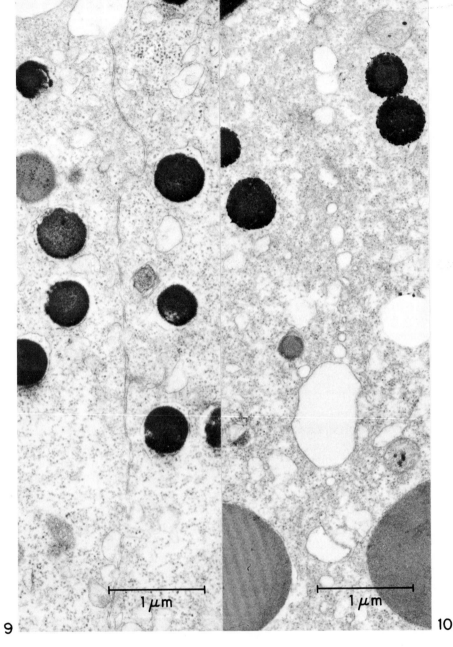

Fig. 9. Ingrowing normal furrow consisting of a double sheet of newly formed cell membrane in the *Xenopus* egg. 26,300×.

Fig. 10. Dividing *Xenopus* egg exposed to 20 μg/ml phospholipase C to remove the cell coat material in the furrow. Instead of the continuous membrane of the ingrown furrow, a track of membrane-bounded vesicles is seen. 26,300×.

(Figures 9 and 10 are reproduced by permission of Springer-Verlag, Heidelberg.)

provided evidence that the formation of membrane can occur without observable preassembly of membrane material in vesicles. This has been reported in the biogenesis and renewal of the cell membrane of photosensitive outer segments of vertebrate visual cells (Young, 1974) and for plasma membrane formation in the unicellular organism *Tetrahymena* (Subbaiah and Thompson, 1974).

This does not mean that vesicles never fuse to contribute material to the forming furrow. The fusion of vesicles with the cell membrane in the furrow has been reported in a variety of cells and eggs (section 5.2). Upon fusion the vesicle membrane is added to the plasma membrane, but vesicles per se are probably not the main source of plasma membrane material. The contents of such vesicles, rather than their membrane, may be important for furrow formation. It is interesting to note that the vesicles in question have frequently been related to the Golgi system which is the site of synthesis of cell-surface carbohydrate-containing material.

5.2 The possible role of Golgi vesicles in membrane biogenesis

Firm evidence exists that vesicles bud off from the Golgi cisternae, migrate through the cytoplasm, and fuse with the plasma membrane to release their contents (Neutra and Leblond, 1966; Reith et al., 1970; Cuminge and Dubois, 1972; Bennet et al., 1974; Flickinger, 1975). The idea that the carbohydrate-containing components of the plasma membrane are synthesized in the Golgi system is widely accepted. The secretion of mucopolysaccharide material in the furrow has been reported for dividing eggs of various species (Motomura, 1958, 1960, 1966, 1967a, 1967b; Zotin, 1964; Matsumoto, 1968a, 1968b). Transmission electron microscopy, in combination with cytochemical techniques, has shown that in the dividing *Xenopus* egg thiamine pyrophosphatase activity, which in turn is linked with carbohydrate metabolism, is associated with the Golgi vesicular system and the furrow surface (Sanders and Singal, 1975). This has been interpreted in support of the view that Golgi vesicles fuse with the plasma membrane in the region of the furrow (Singal and Sanders, 1974a, 1974b).

Transmission electron microscopy of dividing HeLa cells has shown that small membrane-bounded dense bodies occur in association with the Golgi system (Robbins and Gonatas, 1964). During the process of cytokinesis such dense bodies appear in the region of the furrow. Similar bodies have been reported near the cleavage zone in embryonic heart muscle cells (Goode, 1975) and in embryonic heart cells they also occur in the Golgi complex (Manasek, 1969). It may well be that the dense bodies are related in function to similar bodies found near the furrow in dividing eggs of the squid (Arnold, 1969), the axolotl (Bluemink, 1970), and in the chick blastoderm (Gipson, 1974). However, no direct evidence exists for fusion of dense bodies with the membrane in the region of the furrow in tissue cells.

As mentioned earlier, the Golgi vesicles probably transport a surface carbohydrate component which, afer deposition, can be demonstrated with histochemical techniques using lanthanum (Doggenweiler and Frank, 1965) or ruthenium polycationic dyes (Luft, 1971a, 1971b). These dyes form complexes at the carbohydrate-rich external side of the plasma membrane and reveal the so-called glycocalyx, or cell coat, as an amorphic electron-dense layer. By use of these techniques it has been demonstrated that in the *Xenopus* egg such a layer lines the furrow surface exclusively (Bluemink, 1971b; Bluemink and de Laat, 1973; Singal and Sanders, 1974b). Also, when exposed to fluorescein-conjugated concanavalin A (Con A) (O'Dell et al., 1974), the Con A-binding sites were found exclusively in this region. This evidence suggests that carbohydrate-containing material(s) is associated in relatively large amounts with the newly formed cell membrane along the furrow, and not so much with the preexisting plasma membrane outside the furrow.

A possible function for the carbohydrate-containing material deposited along with newly formed cell membrane can be inferred from a study on membrane biogenesis in the giant heliozoan (Vollet and Roth, 1974). In this fresh-water protozoan reextension of axopodia after experimentally induced regression is concurrent with new cell membrane formation. The authors demonstrated that in the absence of the cell coat cell fusion occurred when the new membrane faces came into physical contact. From this experiment it may be inferred that the cell coat is essential to prevent fusion when the risk exists that newly formed membrane faces come into contact. A similar role for the cell coat in fusion of mammalian cells has been proposed by Poste (1972). In the dividing cells and eggs, where the furrow consists of two parallel leaflets of newly-formed membrane, the conditions for fusion are optimal. That fusion may actually occur has been demonstrated in *Xenopus* eggs, where the carbohydrate-rich coat in the furrow was digested away with phospholipase C (Bluemink, 1971b). Eggs treated in this manner do not cleave. Transmission electron microscopy reveals a row of isolated vesicles (Fig. 10) instead of an ingrowing furrow.

Strong evidence for the involvement of the Golgi system in the assembly of cell coat material has come from a study using amoebae (Flickinger, 1975). ^{3}H-mannose was found to be incorporated first into the Golgi system and then transferred by Golgi vesicles to appear finally in association with the plasma membrane. This suggests that a substantial amount of material synthesized in the Golgi system is ultimately incorporated into the carbohydrate-containing components of the cell surface. There was no indication of reutilization of the labeled surface components by the Golgi system. Some authors (Gonatas et al., 1975)have proposed that Golgi vesicles are shuttled to and from the plasma membrane, the vesicles returning to the Golgi system for reutilization (see chapter 4 in this volume by Holtzman et al., p. 165).

The evidence discussed can be summarized as follows. It seems unlikely that furrow formation occurs by the fusion of prealigned vesicles. Nevertheless, light microscopic evidence exists that in many dividing eggs vesicles can fuse with the furrow surface and secrete mucopolysaccharide material. This material is probably synthesized in the Golgi system and transported by vesicles to the region of the furrow where new membrane is formed. Possibly, part of the carbohydrate-containing material released in this way is used to build up a cell coat. One function of the coat may be to prevent the fusion of the two parallel leaflets of newly formed cell membrane which together form the furrow. Not much is known about the role of Golgi-derived vesicles in relation to membrane biogenesis during cell division in tissue cells.

5.3 Membrane formation by self-assembly

The alternative hypothesis is that plasma membrane does not grow by the fusion of vesicles but by self-assembly, that is, individual molecules or macromolecular complexes are inserted directly into the existing plasma membrane from the cytosol. To date, the evidence for cell division-related plasma membrane growth by self-assembly is meager.

Lamellar, osmiophilic dense material has been observed in the cytoplasm (Fig. 11) in dividing amphibian eggs (Bluemink, 1970; Bluemink and de Laat, 1973; Singal, 1975) as well as during blastoderm formation (Figs. 12–14) in the *Drosophila* embryo (Sanders, 1975). It has been suggested that these lamellar-dense bodies are rich in phospholipid and represent pools of membrane precursors. Cytoplasmic dense droplets containing myelin-like material were first observed in relation to new membrane formation in injured amoebae (Szubinska, 1971). Dense droplets found to be in the process of fusing with the plasma membrane were thought to contribute material to the membrane, thereby expanding its area. The dense droplets or lamellar bodies seen in the dividing amphibian egg and in the *Drosophila* embryo during blastoderm formation may have a similar function. In the *Drosophila* embryo they have been found close to or in contact with the furrow region where new membrane is formed. The bodies are very prominent before cleavage (Fig. 12) but become sparse as it progresses (Sanders, 1975). In the *Xenopus* egg the bodies seen in the process of fusing with the plasma membrane may give rise to blebs. It has been suggested (Bluemink and de Laat, 1973) that these blebs are produced by the interaction of the fixation fluid with labile (fluid?) membrane regions. Upon interaction, lipids may erupt from the surface and form liposome-like bodies while mixing with the fixative. The artificially produced blebs can be taken as evidence of the presence of lipid membrane precursors in contact with the plasma membrane (Spornitz, 1973).

Why are similar cytoplasmic inclusions not reported in electron microscope studies of many other dividing cells and eggs? Probably most, if not all, phospholipid is removed from biological material during conventional tissue preparation involving alcohol or acetone dehydration. Dense droplets have been found (Szubinska, 1971; Bluemink and de Laat, 1973) when ruthenium red was used as a phospholipid-complexing agent (Luft, 1971a, 1971b). It may be expected that dense droplets will be found in a variety of tissue cells and eggs upon application of lipid-retaining fixatives (Elbers et al., 1965; Luft, 1971a, 1971b) and preparation procedures (Stratton, 1975). In dividing eggs they may be more easily demonstrated because pooled membrane precursors are present in large amounts in these cells and formation of membrane occurs at an exceptionally high rate,

418

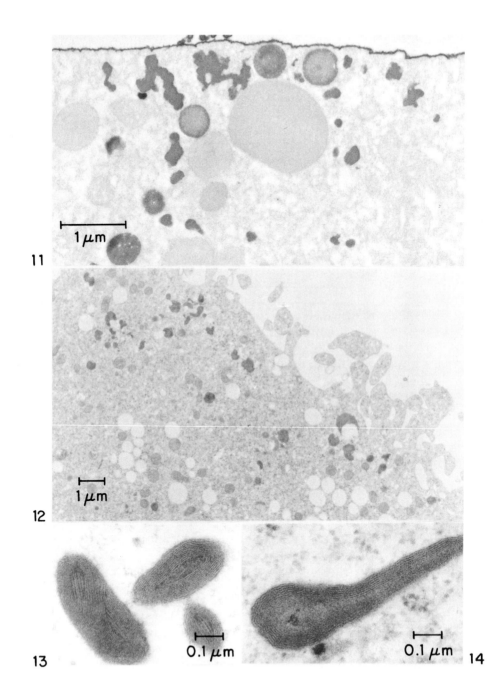

involving large quantities of precursors to be inserted in a short time period (section 4).

To date electron microscopic findings in amphibian eggs and *Drosophila* embryos constitute the only evidence, albeit limited, in support of a role for the self-assembly mechanism in the formation of new membrane during cell division.

6. Freeze-fracture data on membrane biogenesis

Freeze-fracture electron microscopy offers the possibility of studying cell division-related changes in the structure of the plasma membrane in unfixed material. Cell cycle related changes in the density of the intramembranous particles (IMP) have been observed in synchronized mouse L cells (Fig. 15) and Chinese hamster cells (Fig. 16) (Scott et al., 1971). The IMP density was high throughout mitosis and the S- and G_2-phase, but showed a steep decline followed by gradual restoration during the G_1-phase. To explain this fluctuation it has been suggested (one possibility out of three) that during cytokinesis rapid assembly of plasma membrane occurs by intussusception of membrane components other than IMPs. The IMPs found in a variety of biological membranes are thought to represent, at least partially, proteins and/or glycoproteins [or oligomeric protein complexes] complexed with lipids, which float in the lipid bilayer (Marchesi et al., 1972; Pinto da Silva, 1972; Singer and Nicolson, 1972; Bullivant, 1973; Pinto da Silva et al., 1973; Tourtellotte and Zupnik, 1973; Nicolson, 1976a;McNutt, 1977). The other components added may be of a lipid type.

In dividing *Xenopus* eggs the preexisting plasma membrane and the newly formed cell membrane show a remarkable difference in IMP density (Bluemink et al., 1976). In the preexisting plasma membrane region the exposed core on the extracellular side (E-face) showed an IMP density of 1600 to 2200 particles/μm² (Fig. 17), whereas its mirror image, the protoplasmic side, exhibited an IMP density of only 300 particles/μm². In the region of newly formed cell membrane all fracture faces consistently showed a low density of 300 to 500 particles/μm² (Fig. 18). The newly formed cell membrane was also more difficult to fracture. This finding, together with the observation that the trilamellar organization of the newly formed cell membrane always remains indistinct in ul-

Fig. 11. *Xenopus* egg fixed in the presence of ruthenium red showing dense bodies in the cytoplasm (unstained thin section) of the furrow region. 18,000×.

Fig. 12. The cortical region of an uncleaved embryo of *Drosophila melanogaster* showing lamellar dense bodies in the cytoplasma. 7,000×.

Fig. 13. Lamellar dense bodies from the cortical region of the *Drosophila* embryo, fixed with osmium tetroxide followed by glutaraldehyde and acrolein. 75,800×.

Fig. 14. Same as Fig. 13. 78,600×.

(Figures 12–14: Courtesy of Dr. E.J. Sanders and reproduced by permission of Springer-Verlag, Heidelberg.)

420

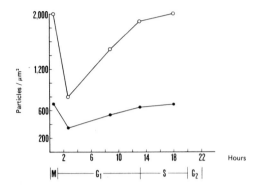

Fig. 15. Density analysis of 70 Å particles associated with fracture faces of the cell membrane of mouse L cells synchronized by chemical technique. Standard error for each point was less than 50 for the outer fracture face and less than 300 for the inner fracture face. O—O, Inner fracture face; ●—●, outer fracture face. (Courtesy of Dr. R. E. Scott and reproduced by permission of Macmillan Ltd., London.)

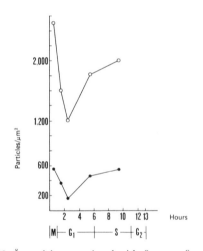

Fig. 16. Density analysis of 70 Å particles associated with fracture faces of the cell membranes of Chinese hamster ovary cells synchronized by the shake-select technique. Standard error for each point was less than 50 for the outer fracture face and less than 200 for the inner fracture face. O—O. Inner fracture face; ●—●, outer fracture face. (Courtesy of Dr. R. E. Scott and reproduced by permission of Macmillan Ltd., London.)

trathin sections (Bluemink and de Laat, 1973), may perhaps be taken as evidence to assume that the degree of disorder in the lipid bilayer is considerable.

Also pertinent to this question are the data on nerve axon elongation (Pfenninger, 1972; Pfenninger and Bunge, 1974), in which new cell membrane is assembled at the anterior end of the ruffled membrane region, the growth cone

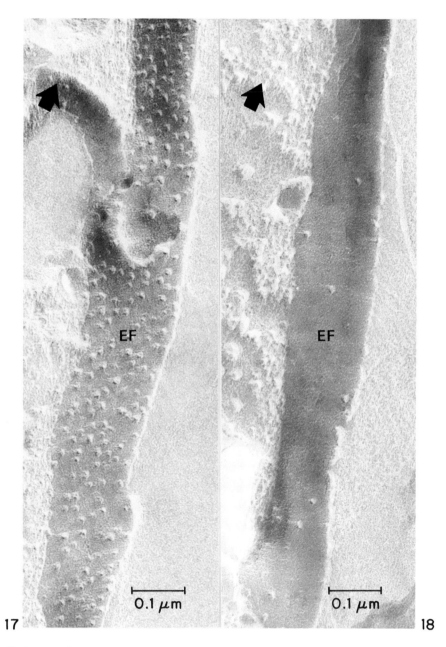

Fig. 17. External fracture face (EF) of preexisting plasma membrane in the dividing *Xenopus* egg. Direction of shadowing indicated by encircled arrowhead. 144,000×.

Fig. 18. External fracture face (EF) of newly formed cell membrane in the dividing egg of *Xenopus laevis*. The low particle density is representative for nascent membrane regions. 144,000×.

(Yamada et al., 1971; James and Tresman, 1972). This membrane differs from the one in the region of the perikaryon by the scarcity of IMPs. A similar paucity of IMPs has been reported for newly formed cell membrane in *Tetrahymena* (Satir et al., 1976). In this organism, dibucaine treatment induces deciliation which is followed by cell membrane repair and the reappearance of cilia. It was found that initially the newly formed cell membrane is devoid of the characteristically arranged IMPs, and IMPs only begin to appear hours later.

The evidence discussed above is consistent in that newly formed plasma membrane appears as a relatively simple lipid bilayer containing only a few IMPs. 3T3 fibroblasts (Furcht and Scott, 1974) show aggregated IMPs during interphase that disaggregate during mitosis. This disaggregation process could be related to an increase in membrane fluidity, as revealed by spin-label experiments (Barnett et al., 1974; see, however, Dodd 1975 and Gaffney 1975). It is conceivable that a transient change in fluidity can be brought about by increasing the proportion of fluid lipids.

The freeze-fracture data could thus be interpreted as indicating that formation of plasma membrane in cell division (Scott et al., 1971; Furcht and Scott, 1974; Bluemink et al., 1976) occurs by rapid insertion of primarily lipid components. This would be consistent with the observations made in ultrathin sections that during cell division-related membrane formation lipid-dense droplets fuse with the plasma membrane (Bluemink and de Laat, 1973; Sanders, 1975). Generally, it seems that membrane components are not added in concert. During the initial phase of rapid membrane growth lipids are inserted in excess of other components (bleb formation), the assembly of other membrane components lagging behind.

Two alternative hypotheses regarding cell division-related plasma membrane assembly have been mentioned. Plasma membrane may grow by self-assembly or by fusion of vesicles consisting of preassembled membrane components. The evidence discussed so far indicates that both possibilities exist, and the alternatives are probably not mutually exclusive.

7. Cell cycle-dependent changes at the cell surface

The term "cell coat" refers to the carbohydrate-rich external part of the plasma membrane present in all mammalian cells (Parsons and Subjeck, 1972). The cell coat consists of a complex of oligosaccharide chains of membrane glycoproteins and glycolipids that have additional glycoproteins and glycolipids adsorbed to it. The amount of glycolipids in the plasma membrane is much smaller than that of glycoproteins. The glycolipids are responsible for much of the antigenic activity of the cell surface, including A, B, and H antigens (Winzler, 1970), whereas the glycoproteins carry the M and N antigens. Due presumably to their hydrophobic properties, the glycolipids are closely associated with the nonpolar part of the plasma membrane and have relatively short sugar chains projecting into the external milieu. The glycoproteins are probably fully extended, forming a branching, "Christmas-tree-"-like structure due to repulsing forces between the negatively charged sialic acid groups.

The protein portion of membrane glycoproteins is synthesized on polyribosomes, while the sugar chains are thought to be built up by membrane-bound glycosyl transferases. The widely accepted view is that the sugars and proteins are assembled in the Golgi system and that complete glycoproteins are transported by Golgi vesicles to the plasma membrane (Neutra and Leblond, 1966; Bennett, 1970; Bennett and Leblond, 1970; Reith et al., 1970; Cuminge and Dubois, 1972; Bennett et al., 1974; and chapters by Morré and Cook in this volume). Genetic control over the final carbohydrate structure is thought to depend on the glycosyl transferases present. Alternatively, Lennarz (1975) has proposed that sugars and proteins may be assembled into membrane glycoproteins by glycosyl transferases present in the plasma membrane, whereas Keenan and Morré (1975) have suggested that the glycosyl transferases in the membrane are responsible only for the building up of the cell coat.

Cell cycle-dependent changes at the cell surface due to plasma membrane biogenesis may be expressed as changes in the molecular arrangement, that is, a redistribution of the coat material or a change in its molecular composition.

7.1 The arrangement of cell coat material

A cell cycle-dependent change in the arrangement of cell coat components has been observed in synchronous cultures of Chinese hamster ovary cells and of Cl_2TSV_5 cells (Garrido, 1975). By means of ultrastructural labeling of surface lectin receptor sites, marked difference in the degree of clustering of receptors for concanavalin A and wheat germ agglutinin were found between interphase and mitotic cells in monolayer cultures. A moderately discontinuous pattern of label occurs during S and G_2, and a strikingly discontinuous pattern during M, while in cells completing division and entering G_1 the cell surface is almost devoid of label for a short time. In early G_1 cells there are many endocytotic vacuoles that contain labeled surface material. In mid-G_1 the pattern again resembles that in S and G_2. Thus, the degree of clustering observed on mitotic cells is far greater than that seen on interphase cells. The clustering is thought not to reflect the inherent pattern but to be due to ligand-induced clustering of the lectin receptor sites produced by the divalent lectin molecules (de Petris and Raff, 1973). The increased clustering observed during mitosis would then reflect a higher mobility of the proteins carrying the receptor sites, since it may be assumed that the probability of receptor cross-linking increases with the freedom for lateral movement of membrane proteins (Garrido, 1975). Three possibilities have been mentioned that could explain this phenomenon (Nicolson, 1974): (1) an increase in the intrinsic fluidity of the lipid bilayer; (2) a structural alteration of the lectin-binding site enabling the complex to diffuse more rapidly; or (3) the participation of peripheral cytoplasmic structures that may interact with specific cell surface components. In light of the freeze-fracture data discussed above (section 5) the first possibility seems to best fit the facts, though, as mentioned, final resolution of this question has yet to be determined and powerful arguments can also be made for alternative (3) (see review by Nicolson, 1976b).

7.2 The composition of the cell coat

In the search for deeper insight into plasma membrane biogenesis in relation to cell division, the analysis of cell cycle-dependent changes in the molecular composition of the cell coat may provide useful information. An elegant and reliable technique is to split the carbohydrate part from the cell surface by mild proteolytic enzyme treatment without disrupting the cells. Biochemical analysis of the released material may then provide information on qualitative and quantitative changes throughout the cell cycle.

Synchronized baby hamster kidney cells grown in monolayer were analyzed for the presence of fucose-containing peptides during mitosis (Glick and Buck, 1973). Material removed with trypsin from the surface of cells arrested in metaphase and analyzed by gel filtration revealed a pattern of fucose-containing glycopeptides not present in control cells. A similar pattern was found by analyzing isolated plasma membranes. The cells in metaphase agglutinated in the presence of concanavalin A, whereas the nonmetaphase control cells did not. Apparently, certain fucose-containing glycopeptides are exposed on the cell surface only during mitosis. In addition it was found that the amount of radioactively labeled fucose incorporated by metaphase-arrested cells was three times higher than that incorporated by control cells.

Synchronized mouse LS cells (cultured in suspension) preincubated with ^{3}H-glucosamine have been used for analyzing isolated glycopeptides by column chromatography (Brown, 1972). It was found that a glycopeptide mixture appeared exclusively during the M phase of the cell cycle. The sudden preferential appearance of glycopeptide has been taken as evidence for the assembly of glycoprotein, that is, a manifestation of cell division-related plasma membrane biogenesis. A similar experimental approach has been used with synchronized mouse 3T3 cells and 3T3 cells transformed by simian virus 40 (SV40) (Onodera and Sheinin, 1970). In these cells, cultured in monolayers, it was found that incorporation of radioactively labeled glucosamine into cell surface glycoprotein was signsificantly enhanced immediately after mitosis, when the cells entered G_1. This result is also consistent with the reported incorporation of labeled glucosamine in Chinese hamster ovary (CHO) cells (Kraemer and Tobey, 1972). The CHO cells, grown in suspension, were synchronized by the isoleucine deficiency method and a thymidine regimen. An increased incorporation of radioactively labeled glucosamine was found to occur in the late G_1 phase, suggesting an increased rate of synthesis and transport to the cell surface in this period.

Sialic acid is a terminal sugar of cell surface glycoproteins and probably the chief contributor to the net negative charge on mammalian cells. Using synchronized human lymphoid cells in suspension culture, it has been found that the amount of sialic acid released by neuraminidase treatment fluctuates during the cell cycle (Rosenberg and Einstein, 1972). In the late G_2 phase, immediately before the onset of mitosis, when the cell surface area remains constant, the amount of sialic acid per cell showed a rapid increase, indicating that at least a portion of the surface sialoproteins is introduced into the plasma membrane just prior to cytokinesis. It was considered unlikely that other membrane components would also assemble in the same period of the cycle. In contrast, results obtained with synchronized CHO cells grown in suspension culture indicate that the amount of sialic acid per unit surface area shows only a small gradual increase throughout the cell cycle (Kraemer, 1967). On the other hand, Mayhew (1966) has reported an increase in the mean electrophoretic mobility of mitotic culture cells as compared with interphase cells due to surface-bound, ionized neuraminic acid carboxyl groups. Kraemer (1967) suggested that this transient elevation of electrophoretic mobility was due to conformational or terminal complex changes of sialoglycolipids or sialoglycoproteins in relation to cell division rather than a change in the quantity of surface sialic acid residues.

A variety of external surface labeling techniques have been used recently to study the appearance and distribution of the receptor-bearing molecules throughout the cell cycle. This approach has been used to demonstrate a cell cycle-dependent change in the cell surface of synchronized HeLa S3 cells (Stein and Berestecky, 1975). Using phenylglyoxal as a surface probe specific for arginine-rich moieties of cell membrane proteins, a sharp decrease in the exposure of positively charged protein was found during the change from M to G_1. This was followed by a 26-fold increase during interphase. According to the authors' interpretation, an arginine-rich protein appears to be exposed during the cell division-related phase of the cell cycle.

Another external labeling technique used to identify membrane glycoproteins and glycolipids exposed at the cell surface involves treatment of cells with galactose oxidase followed by reduction with tritiated sodium borohydride (Gahmberg and Hakomori, 1974). The surface-exposed galactosyl and galactosaminyl residues are labeled, and the exposure rate in different phases of the cell cycle can be determined by measuring the specific radioactivity of the sugars. Gahmberg and Hakomori used synchronized hamster NIL cells, obtained by trypsinization of contact-inhibited confluent cells or by double thymidine block. The presence of a glycoprotein with a molecular weight of 200,000 daltons ("galactoprotein a") was demonstrated and the label in this protein appeared in the G_1 phase, remained throughout the cell cycle, and increased further in the next G_1 phase, suggesting that assembly occurs at G_1. Similar results have been obtained by other investigators (review, Hynes, 1976; Nicolson, 1976b).

It is difficult to interrelate the results of the experiments mentioned in this section (summarized in Table 1) to arrive at a more general conclusion. Differences in the outcome of the experiments may be due to variations in the types of cells used, in culture conditions, in methods of synchronization, and in the molecular species analyzed. It is apparent that of the various cell-cyclic changes

Table 1
Cell cycle dependent changes in plasma membrane protein composition in cells cultured in vitro

Designation of cell type and culture method	Reference	Phase of cell cycle
Mouse LS cells (suspension culture)	Brown, 1972	M
Mouse 3T3 cells and SV40 transformed 3T3 cells (monolayer culture)	Onodera and Sheinin, 1970	G_1
Mouse mastocytoma cell Transformed hamster fibroblasts (suspension culture)	Graham et al., 1973 Pasternak and Graham, 1973	G_1
Chinese hamster ovary cells (suspension culture)	Kraemer, 1967	—
Chinese hamster ovary cells (suspension culture)	Kraemer and Tobey, 1972	G_1 (late)
Hamster NIL cells (monolayer culture)	Gahmberg and Hakamori, 1974	G_1
Baby hamster kidney fibroblasts (monolayer culture)	Glick and Buck, 1973	M
KB cells (suspension culture)	Gerner et al., 1970	M/G_1
HeLa Cells (suspension culture)	Nowakowski et al. 1972	late S
HeLa cells (suspension culture)	Johnsen et al., 1975	M
HeLa S3 cells (monolayer culture)	Stein and Berestecky, 1975	G_1 and S
Human lymphoid cells (suspension culture)	Rosenberg and Einstein, 1972	late G_2
Tetrahymena pyriformis	Baugh and Thompson, 1975	Cell division

---

Main results and conclusions

Synchronization of cells by a selection method. Labeled glucosamine incorporation into glycoprotein released from intact cells by proteolytic enzyme treatment. Appearance of a new glycoprotein during M.

Synchronization of cells by a subculture technique after cells were grown to confluence. Labeled glucosamine incorporation into glycoprotein released from intact cells by proteolytic enzyme treatment showed an increase during G_1.

Synchronized cells obtained by zonal centrifugation after thymidine treatment. Biochemical analysis of isolated plasma membrane ghosts. Gradual increase in plasma membrane proteins and phospholipids, but sudden increase in the carbohydrate (glycoprotein plus glycolipid)) moiety during G_1.

The amount of surface-bound sialic acid (sialoprotein) released by neuraminidase treatment shows only a small gradual increase through the cycle.

Cells synchronized by the isoleucine deficiency method followed by thymidine block. The amount of radioactively labeled glucosamine incorporated into both heparan sulfate and the glycopeptide fraction associated with the cell surface showed increased specific activity in late G_1.

Synchronized cells obtained by trypsinization of contact-inhibited confluent cells or by double thymidine block. Surface labeling by galactose oxidase reduced with tritated sodium borohydride. Labeled glycoprotein appeared at G_1, remained during the rest of the cell cycle and increased further during the next G_1.

Cells arrested in metaphase with vinblastine sulfate or by thymidine block. Trypsin-isolated surface material analyzed by gel filtration revealed a pattern of fucose-containing glycopeptides present exclusively during metaphase. This pattern was also found in isolated cell membrane ghosts. Incorporation of radioactive fucose into mitotic cells was 3 times higher than into control, nonmitotic cells.

Cells synchronized by double thymidine block. The incorporation of radioactive glucosamine (together with labeled choline and leucine) showed a marked increase just after division.

Cells synchronized by double thymidine block. Incorporation of labeled fucose (glycoproteins) into isolated plasma membrane is increased in late S.

Cells synchronized by thymidine-colcemid treatment. Gel electrophoresis of proteins solubilized from ghost material showed that cells in mitosis contain proteins that do not occur in other phases of the cycle.

Cells synchronized by mitotic shake-off plus thymidine block. Surface labeling of arginine-rich membrane proteins by radioactive phenylglyoxal. Labeled membrane protein decreased sharply upon entry into G_1, and thereafter increased 24-fold during G_1 and early S.

Cells synchronized by double thymidine block. The amount of surface-bound sialic acid released by neuraminidase treatment from intact cells showed an increase (sialoproteins) during G_2.

Cells synchronized by heat-shock. Incorporation of labeled precursors. During cytokinesis the fraction containing surface membrane material showed a sudden increase in labeled protein. Peaks in phospholipid labeling generally followed peaks in protein labeling in different cell fractions.

detected at the cell surface only some can be related to plasma membrane biogenesis. However, a number of reports seem to be consistent in indicating that the incorporation of labeled sugars is at a maximum during the M/G$_1$ phase of the cell cycle (Onodera and Sheinin, 1970; Kraemer and Tobey, 1972; Glick and Buck, 1973), and also that changes in plasma membrane composition are pronounced during this period (Brown, 1972; Gahmberg and Hakomori, 1974; Stein and Berestecky, 1975).

7.3 The expression of surface antigens and receptor sites

Indirect information on the assembly of specific classes of glycoproteins and glycolipids into the plasma membrane can also be obtained from studies that analyze cell cycle-dependent changes in the expression of membrane receptor sites.

The major methods used in this approach involve either direct binding assays of fluorescent or radiolabeled molecules, or indirect assays such as immune cytolysis or lectin-mediated cell agglutination. Regardless of the method employed, any firm conclusions about assembly of the antigens or receptor sites should be made with care. The binding of labeled probe molecules will depend not only on the surface density and/or the total number of a particular receptor site per cell but also on the molecular organization of the plasma membrane which will influence the exposure of the site, its distribution, mobility, and so on, and thus in turn, affect the binding capacity of the probe molecule. In indirect assays the possible influence of secondary factors is even more obvious.

The relevant studies can be subdivided into two groups: those dealing with the cyclic expression of immunoglobulins, alloantigens, and blood group determinants on the cell surface (summarized in Table 2), and studies on lectin-binding sites (Table 3). A few studies have also been done on the receptor sites for growth factors (Table 2).

As can be seen from Table 2, various authors provide contradictory data regarding the expression of immunoglobulins during the cell cycle, even for the same cell type. In (human) Burkitt lymphoma cells the binding of ^{125}I-labeled monospecific antibodies against a variety of immunoglobulins showed a peak in early S when the cells were released from a double thymidine block (Takahashi et al., 1969), whereas in a combined autoradiographic and cytophotometric analysis the binding of fluorescein-conjugated IgM antibodies (as well as the HL-A alloantigen) was found to increase gradually throughout interphase (Killander et a., 1974). The major difference between these two studies was the absence of any synchronization procedure in the latter. Observations on Wi-L2 lymphoid cells synchronized by thymidine and/or colcemid blocks revealed maximal binding of fluorescein-conjugated antisera against IgG and IgM in late G$_1$ and S (Buell and Fahey, 1969). If immunoglobulins and HL-A alloantigens are actually expressed in concert (Killander et al., 1974), the absence of differences in HL-A expression between G$_1$ and S cells (Ferrone et al. 1973) would be expected. Ferrone et al. (1973) speculated that malignant mouse and human cells would show a maximal expression of histocompatibility antigens (H-2 and HL-A, respectively) in G_1, whereas in nonmalignant cells, such as Wi-L2, cell cycle-dependent changes in antigen expression would be absent or less marked. This hypothesis was not confirmed, however, by later studies on HL-A expression (Killander et al., 1974).

In mouse mastocytoma cells H-2 antigen expression, measured by an immune cytolysis test, appeared to be maximal in G_2 and M (Pasternak et al., 1971). However, later studies from the same laboratory showed that H-2 antigen accessibility, as determined by direct titration with ^{125}I-labeled antibody, increased during G_1 and remained constant during S and G_2 (Pasternak et al., 1974). In both studies, zonal centrifugation of asynchronous cultures was used to separate the various cell cycle stages. The discrepancy between these results shows that secondary factors such as membrane "fragility," the capacity to repair membrane lesions, and other membrane perturbations may influence the cytolysis test. In Moloney leukemia virus-induced lymphoma cells, synchronized by colcemid treatment, indirect immunofluorescence has shown that the highest proportion of cells positive for H-2 and Moloney virus-determined antigens were in G_1 (Cikes and Friberg, 1971). However, the degree of synchronization obtained makes it difficult to distinguish between G_2, M, and G_1 cells, especially since no data were given on the second cycle after release from the colcemid treatment. The expression of blood group H determinants seems to be restricted to mitotic cells (Kuhns and Bramson, 1968; Thomas, 1971), in contrast to blood group B determinants (Thomas, 1971).

In their detailed analysis of cell cycle-dependent synthesis and exposure of glycolipids, Wolf and Robbins (1974) compared the exposure of the Forssman antigen (ceramide pentahexoside) in trypsinized and nontrypsinized synchronized NIL hamster cells. In mitotic cells all the detectable antigen was exposed. The detectable amount increased during G_1 and S and then remained constant. However, an increasing amount became nonaccessible in nontrypsinized cells as the cells progressed though the cell cycle. These results indicate that antibody-antigen interactions can only provide relatively limited information on the assembly of antigens into the plasma membrane.

Indirect assays have also been used to investigate the presence of a macrophage growth factor (MGF) receptor on L cells (Cifone and Defendi, 1974) and a nerve growth factor (NGF) receptor on neuroblastoma cells (Revoltella et al., 1974). MGF was detectable in trypsin digests of S cells only, whereas NGF receptor sites were exposed during late G_1 and early S.

In conclusion, with few exceptions, the studies described in this section show that the expression and exposure of many antigens at the plasma membrane is cell cycle-dependent. Blood group H determinants, for example, are detectable only during mitosis. However, for immunoglobulins, HL-A and H-2 alloantigens, Forssman antigen, and MGF- and NGF-receptors evidence has been provided that exposure at the cell membrane either starts in, or is restricted to, the G_1 and S phases. Whether this reflects the assembly and incorporation of an-

Table 2
Expression of membrane receptor sites during the cell cycle

Cell type	Reference	Phase of maximal expression	Type of receptor site
Human lymphoid cells	Buell and Fahey, 1969	late G_1, S	IgG, IgM
Human lymphoid cells	Ferrone et al., 1969 Pellegrino et al., 1972	no change	HL-A
Human Burkitt lymphoma cells Human lymphoma cells	Takahashi et al., 1969	early S	Immunoglobulins
Human Burkitt lymphoma cells	Killander et al., 1974	G_2?	IgM, HL-A
Mouse lymphoma cells	Cikes and Friberg, 1971	G_1	H-2
Mouse mastocytoma cells	Pasternak et al., 1971	G_2, M	H-2
Mouse mastocytoma cells	Pasternak et al., 1971	late G_1, S,G_2	H-2
Mouse mastocytoma cells	Thomas, 1971	M no change	blood gr. H blood gr. B
HeLa cells	Kuhns and Bramson, 1968	M	blood gr. H
Hamster NIL cells	Wolf and Robbins, 1974	G_1, S	Forssman antigen
L cells	Cifone and Defendi, 1974	S	MGF
Mouse neuroblastoma cells	Revoltella et al., 1974	late G_1, early S	NGF

Main results and conclusions

Synchronization by single thymidine block + colcemid or colcemid + double thymidine block, respectively. Determination of percentage cells stained with fluorescein-conjugated antisera specific for IgG and IgM. Highest percentage in late G_1 and S, little IgG and IgM detectable immediately before, during and after mitosis.

Synchronization by replating G_0 cells. G_1 and S cells were compared. Neither the sensitivity to HL-A antibody-mediated lysis, nor the extent of activation of the complement system, the degree to which labeled components are bound, or the ability to absorb HL-A alloantibodies showed any significant differences.

Synchronization by double thymidine block. Radioautography of monospecific ^{125}I-labeled antibodies against heavy and light chain classes showed that the immunoglobulin levels reach a peak in early S, gradually decrease during late S and G_2, and reach a minimal level in M.

Analysis of asynchronous cultures by combined cytophotometric and autoradiographic techniques. IgM and HL-A were stained with fluorescein-conjugated antibodies. The amounts of IgM and HL-A were found to increase continuously throughout interphase, so that in absolute terms their content was maximal in G_2. However, the IgM (and HL-A)/total protein ratio remained constant. Probably the surface density of these receptor sites is greater in G_2.

Synchronization by colcemid treatment and subsequent shake-off. Indirect membrane immunofluorescence showed the highest proportion of cells positive for HL-A and Moloney virus-determined antigens to be in G_1. As cells entered S, this proportion dropped and remained low until the cells divided again.

Cell cycle stages separated by zonal centrifugation. H-2 antigen expression as measured by a cytolysis test is maximal in G_2 and M.

Cell cycle stages separated by zonal centrifugation. H-2 antigen accessibility as shown by direct titration with [^{125}I] antibody increased during G_1 and remained constant during S and G_2.

Synchronization by double thymidine block. Cells were stained for blood groups B and H by indirect immunofluorescence. H showed a marked increase in M while B showed no changes. However cells unable to enter a second cycle after release from the thymidine block became B⁻.

Synchronization by single thymidine block. Mixed agglutination tests for blood group H showed a maximal expression in M.

Synchronization by double thymidine block, or double thymidine block plus mitotic collection with colcemid. The total amount of Forssman antigen in cell extracts, assayed by a hemolysis test, increased gradually during the cell cycle. The binding of ^{14}C-anti-Forssman antiserum demonstrated that M cells expose all detectable antigen. In trypsinized cells the detectable amount increased during G_1 and early S, but then remained constant. As cells progressed through the cell cycle a large fraction of the antigen became cryptic.

Synchronization by double thymidine block. The presence of macrophage growth factor (MGF) at the surface of L cells was determined by testing the DNA synthesis-inducing capacity of trypsin digests. MGF is presumably synthesized throughout the cell cycle, inserted into the cell membrane in S, and released into the medium in M.

Synchronization by isoleucine deprivation. The expression of nerve growth factor (NGF) receptor sites at the cell surface was determined by the capacity to form rosettes with NGF-coated erythrocytes. Only late G_1 and early S cells were able to form rosettes.

tigens into the plasma membrane, or involves alterations in the organization of the plasma membrane leading to the "unmasking" of preexisting sites (possibly due to the assembly of membrane components other than antigens), is still unknown.

The results of lectin-binding studies done in several laboratories are summarized in Table 3. Fox et al., (1971) showed that binding of fluorescein-conjugated wheat germ agglutinin (FcWGA) was limited to mitotic 3T3 cells, whereas transformed Py3T3 cells bound FcWGA at all stages of the cell cycle. Other work on lectin-mediated cell agglutination and quantitative measurements of lectin-binding have prompted the hypothesis that lectins reveal surface properties involved in the regulation of growth (Fox et al., 1971; Burger, 1973). It is outside the scope of this review to go into detail about the conflicting results occasionally presented by the various authors, and the reader is referred to recent reviews by Nicolson (1974; 1976b) and by Rapin and Burger (1974). We will confine our comments to a discussion of cell cycle-dependent changes in the expression and accessibility of lectin-binding sites on the cell surface (Table 3).

Noonan and Burger (1973) have reported that at mitosis 3T3 cells bind five times more ^{3}H-concanavalin A (^{3}H-ConA) at 0°C than during any other part of the cell cycle and three times more if the difference in surface area is taken into account (Noonan et al., 1973). Other observations on Fc-ConA binding to hamster embryo cells and chemically transformed hamster cells under non-saturation conditions have revealed a correlation between the percentage of fluorescent cells and their agglutinability (Shoham and Sachs, 1971). The percentage of fluorescent cells decreased in the following order: (1) trypsin-treated normal and transformed interphase cells; (2) transformed interphase and normal mitotic cells; (3) transformed mitotic cells; (4) normal interphase cells. Only the cells of the first two classes were agglutinable by ConA. Further observations by the same investigators on the binding of Fc-WGA to 3T3 and SV-3T3 cells gave similar results. However, no significant differences between normal and transformed cells, or between interphase and mitotic cells could be detected with respect to the number of ^{3}H-ConA molecules bound per unit surface area, both under saturation and nonsaturation conditions, probably indicating that clustering of receptor sites is a necessary condition for the appearance of both visible fluorescence and for agglutination (Shoham and Sachs, 1974a, 1974b). In P815Y mastocytoma cells the binding of ^{125}I-ConA increases in late S and G_2, reaching a maximum in mitosis (Pasternak et al., 1974). In Epstein-Barr virus-transformed human lymphocytes the highest ConA agglutinability is found in M and persists during G_1, decreases at the transition to S, but rises abruptly again when the cells enter metaphase (Smets, 1973). A brief treatment with trypsin increases the agglutinability of G_2 cells. Thus, even in transformed cells, ConA agglutinability is not constant during interphase, but, compared with untransformed cells, the postmitotic alteration of the cell membrane that leads to a decreased agglutinability is postponed. A similar study was made with 3T3 and SV-3T3 cells by Smets and De Ley (1974). In 3T3 cells, high agglutinability was restricted to M cells, whereas in transformed cells it remained high during G_1 and diminished gradually during S and G_2.

Table 3
Expression and accessibility of lectin receptor sites during the cell cycle

Cell type	Lectin binding determined by: immunofluorescence	radiolabeling	Phase of maximum agglutinability	Lectin	Synchronization mode	Reference
3T3	M>I			WGA	replating confluent cells	Fox et al., 1971
Py3T3 cells	M→M					
3T3 cells	M>I		M	WGA/ConA	mitotic shake-off	Shoham and Sachs, 1974a, 1974b
SV-3T3 cells	I<M		I			
3T3 cells		M>I		ConA	replating confluent cells	Noonan and Burger,1973; Noonan et al., 1973
3T3 cells			M>G, S, G2	ConA	mitotic shake-off. FUdR, thymidine, X-ray	Smets and De Ley. 1974
SV-3T3 cells			M, G1>S>G2			
Hamster fibroblasts	M>I	M=I	M	ConA	mitotic shake-off	Shoham and Sachs, 1971
Transformed hamster fibroblasts	I>M	M=I	I	ConA	zonal centrifugation	Pasternak et al., 1974
P815Y mastocytoma cells		M>G2>S, G1				Glick and Buck, 1973
BHK cells			M>I	ConA		
EB-3 Burkitt lymphoma cells			M, G1>S>G2	ConA	medium exhaustion, double thymidine block, dibutyryl cAMP, and theophylline, X-ray, vinblastine	Smets, 1973

KEY: WGA: wheat germ agglutinin; ConA: concanavalin A; M→M: independent of cell cycle stage; M, G_1, S, G_2: cell cycle phases; I: interphase (no distinction mode between G_1, S, and G_2).

In summary, there is general agreement that mitotic cells are more agglutinable by lectins than interphase cells. Shoham and Sachs (1971, 1974a, 1974b) found transformed cells to be an exception to this rule. It is evident that the percentage of visible fluorescent cells in fluorescence-binding assays is determined by many factors, including the number of lectin receptor sites and also probably by their ability to cluster. The number of bound radiolabeled lectin molecules is not necessarily reflected in the agglutinability (Ozanne and Sambrook 1971). Although lectins are valuable tools that can reveal changes in cell surface properties in relation to cell growth control, they have yet to provide much information about the assembly of the membrane receptor sites involved.

7.4 Cyclic release of surface material

Bosmann and Winston (1970) studied the synthesis and secretion of glycoprotein, glycolipid, protein, and lipid in synchronized mouse lymphoma cells grown in suspension. The total incorporation of leucine, choline, fucose, glucosamine, or thymidine was measured by pulse labeling at intervals after release from a thymidine/colcemid regimen. In this study the amount of labeled material released into the medium was measured as a function of the cell cycle. Secreted proteins and glycoproteins were found in increased amounts during late S and G_2, whereas the release of glycolipids and lipids occurred during G_2 and M. The authors considered it unlikely that the released material resulted from cell lysis and suggested that it represented surface material secreted by the cells. The synthesis of glycolipids and lipids found exclusively in G_2 and M has led to the suggestion that this may reflect an increased demand for membrane biogenesis in this period of the cell cycle. That the synthesis appears so late in the cell cycle was considered to be due to the fact that the enzymes required are synthesized in S, when protein synthesis occurs. Thus, in the assembly line process of membrane biogenesis, the lipids and glycolipids may be the last molecules to be incorporated (Bosmann and Winston, 1970).

A similar observation that cell surface material is released in increasing amounts during a limited period of the cell cycle has been made by Kraemer and Tobey (1972). They found that CHO cells grown in suspension release surface-associated heparan sulfate just before mitosis. It is unclear whether this is a mechanism to regulate the expression of cell surface receptor sites or a reflection of membrane renewal related to cell division.

7.5 Studies on isolated plasma membranes

A different biochemical approach to the study of membrane biogenesis is to isolate the plasma membrane and to analyze changes in its composition in relation to the different phases of the cell cycle. This kind of experiment is not easy to perform. The method requires that large "ghosts" be obtained which are minimally contaminated by other cellular components. In addition, the purity of the plasma membrane fraction analyzed should be checked for the presence of appropriate plasma membrane markers and lack of contamination by other fractions. (DePierre and Karnovsky, 1973; Graham, 1975).

Biochemical analysis of ioslated ghosts has been used to study plasma membrane biogenesis in mouse mastocytoma cells and in virally transformed hamster fibroblasts (Graham et al., 1973; Pasternak and Graham, 1973). These authors have postulated three alternative schedules for plasma membrane biogenesis throughout the cell cycle as follows:

1. Unfolding during G_1 till G_2, and assembly ($\times$ 2) during G_2 till G_1; or
2. assembly ($\times$ 1.6) during G_1 till G_2, and assembly ($\times$ 0.4) during G_2 till G_1; or
3. assembly ($\times$ 2) during G_1 till G_2, and unfolding during G_2 till G_1.

The yields of membrane proteins and phospholipids were measured relative to the total cellular protein and found to be relatively constant. This suggests a gradual doubling of plasma membrane during interphase (i.e., growth-related plasma membrane assembly). There was, however, a fluctuation in the carbohydrate ratio, indicating an increase during G_1 followed by a relative decrease. On the basis of these experiments the conclusion has been drawn that membrane biogenesis is a constant process that occurs during interphase, and that cytokinesis is a physical event made possible by membrane unfolding or stretching. Thus, the alternative mentioned under item 3 above was considered to be most likely, although the possibility of some mitosis-related membrane biogenesis (item 2) was not excluded.

The fluctuation observed in the carbohydrate content in this study agrees well with similar changes found in KB cells grown in suspension culture (Gerner et al., 1970). In this study pulse-labeling with glucosamine, choline, and leucine (labeled with different isotopes) at set intervals throughout the cell cycle was used. The plasma membranes were isolated after incubation with the precursors for 1 hour and analyzed for the amount of incorporated labels. An increased incorporation rate of all three isotopes was found just after mitosis, again indicating that plasma membrane biogenesis occurs during G_1, starting in the latter part of M.

These data are also consistent with the increased incorporation of labeled glucosamine reported in various other cell types during G_1 (Onodera and Sheinin, 1970; Brown, 1972; Kraemer and Tobey, 1972). They are at variance, however, with similar studies on HeLa cells grown in suspension (Nowakowski et al., 1972). Isolated plasma membranes from cells synchronized with thymidine were analyzed for the incorporation of ^{3}H-fucose. An increased incorporation was found in the late S phase. This was interpreted as a net increase of incorporation into the plasma membrane. However, although fucose is a main constituent of the blood-group H-antigen-bearing glycolipid, the increased fucose incorporation is not accompanied by an increased expression of the blood group H-antigen (Kuhns and Bramson, 1968; Thomas, 1971).

Purified plasma membranes of HeLa cells grown in suspension culture have also been analyzed for cell cycle-dependent changes in protein composition (Johnsen et al., 1975). Gel electrophoresis of proteins solubilized from ghost material showed that cells in mitosis contained membrane proteins that do not occur in other phases. These changes found prior to and during mitosis were thought to reflect the insertion of new membrane precursors.

In experiments that use cell fractionation techniques to isolate the plasma membrane as a ghost, the possibility must be considered that a considerable proportion of the newly assembled membrane may be lost. Due to this, changes in lipid composition or in the ratio of lipid relative to other membrane components may remain unobserved (section 7.7).

The changes in composition of isolated plasma membrane observed during the M/G_1 period (Gerner et al., 1970; Graham et al., 1973; Pasternak and Graham, 1973; Johnsen et al., 1975) are consistent with the outcome of several other experiments discussed above (section 7.2). However, as can be seen in Table 1, not all membrane components appear to be assembled in this period.

7.6 Cyclic changes in Tetrahymena

Some of the difficulties encountered in cultured tissue cells may be circumvented by using free-living animal cells. Interesting data on plasma membrane formation have been reported recently in studies (Baugh and Thompson, 1975) using *Tetrahymena pyriformis*, an eukaryotic unicellular organism. The culture conditions for this organism are relatively simple. Almost complete synchrony can be achieved by temperature treatment, and essentially all of its lipids and structural proteins are intended for use within the cell and do not appear to be secreted. Synchronously dividing cultures of *Tetrahymena* were pulse-labeled for 10 minutes with precursors of proteins (^{3}H-leucine) and phospholipids (^{32}P) at selected time intervals during the cell cycle. The study was designed to determine the pattern in which synthesized lipids and proteins are distributed to various membrane systems during cell division. To analyze the incorporation of the label into the major phospholipids and proteins, radioactivity was measured in six different cell fractions. Two fractions (cilia and cilia supernatant) contained the cell surface material. The cell surface fractions showed a sudden increase in labeled membrane proteins in relation to the period of furrow formation followed by an increase in labeled phospholipids. A comparison of the lipid labeling pattern with that of the proteins showed that the peaks in phospholipid labeling generally followed the respective peaks of protein labeling. The evidence was considered to exclude the possibility that membrane expansion during cell division is initiated by intussusception of phospholipids alone. However, other evidence exists in *Tetrahymena* (Subbaiah and Thompson, 1974) that lipids and proteins do not necessarily move in concert to sites where new membrane is assembled. These authors favor the concept that membranes in cells are in a dynamic equilibrium and that there is a continuous turnover and exchange of membrane constituents. Membrane biogenesis and expansion are thought to occur by intussusception of molecules rather than of complexes.

In *Tetrahymena* the fractions containing the cell surface material showed a sudden increase in labeled membrane proteins during furrow formation. This observation is consistent with the evidence reported for various other cells (Brown, 1972; Glick and Buck, 1973; Gahmberg and Hakomori, 1974; Johnsen et al., 1975; Stein and Berestecky, 1975) in showing that the protein composition of the plasma membrane changes during, or shortly after, mitosis.

7.7 Lipid metabolism in dividing cells

About 40% of the mass of cell membranes consists of phospho- glyco-, and neutral lipids (cholesterol). Phospholipids are reportedly synthesized in the endoplasmic reticulum of rat liver cells (Wilgram and Kennedy, 1963; Stein and Stein, 1969) and were found to be incorporated first into the rough endoplasmic reticulum of these cells (Dallner et al., 1966). According to one current view, a continuous exchange and turnover of phospholipids within and between intracellular membranes, internal pools, and the plasma membrane is envisaged as occurring (for references, see Siekevitz 1972).

Arguments exist for assuming that nondividing tissue cells may have an intracellular phospholipid pool of about 10 to 15% of that present in the plasma membrane, whereas there is almost no intracellular phospholipid pool in the dividing cell (Warren, 1969). Therefore, rapid incorporation of newly synthesized phospholipids in the plasma membrane can be expected. Incorporation studies using whole cells may provide information on plasma membrane formation throughout the cell cycle. Table 4 presents a summary of the reported biochemical data concerning lipid metabolism in dividing tissue cells. These data are inconsistent, however, with regard to the period(s) of the cell cycle during which maximum incorporation was found. In the majority of studies, lipid metabolism was analyzed in whole cells and the differences are thus difficult to explain. They may be due to variations in the methods used to synchronize cells, differences in cell type, in the molecular species analyzed, variation in the rate of intracellular membrane formation and also in the rate of intracellular transport and/or differences in the cytoplasmic pool sizes of lipid intermediates. Incorporation studies on amoebae have shown that in certain cells a time-lag may exist between the period when phospholipids are synthesized in the endoplasmic reticulum and the time they are incorporated into the plasma membrane (Chlapowski and Band, 1971). To date, the biochemical data do not allow us to conclude with certainty that the biogenesis of plasma membrane lipids is confined to a particular period of the cell cycle.

There are, however, four studies that did not analyze whole cells but studied lipid metabolism of either isolated plasma membranes (Gerner et al., 1970; Graham et al., 1973; Pasternak and Graham, 1973) or various cell fractions (Baugh and Thompson, 1975). In synchronized KB cells pulse-labeled for 1 hour at various times after release from a thymidine block, the rate of incorporation of radioactive choline into the plasma membrane showed a marked increase just after division (Gerner et al., 1970). This indicates a high rate of phos-

Table 4
Lipid metabolism in dividing tissue cells

Designation of cell type	Reference	Phase of cell cycle showing increased activity
Mouse mastocytoma cells	Bergeron et al., 1970	S
Mouse mastocytoma cells	Warmsley and Pasternak, 1970	G_1 and S
Mouse mastocytoma cells Transformed hamster fibroblasts	Pasternak and Graham, 1973 Graham et al., 1973	—
Mouse lymphoma cells	Bosmann and Winston, 1970	G_2 and M
Mouse mastocytoma cells	Lingwood and Thomas, 1975	S and G_2
NIL hamster cells	Wolf and Robbins, 1974	G_1 and S
NIL hamster cells	Hirschberg et al., 1975	G_1 and S
Human KB cells	Gerner et al., 1970	early G_1
Human KB cells	Chatterjee et al., 1973	G_1 and M
Human KB cells	Chatterjee et al., 1975	M
Tetrahymena pyriformis	Baugh and Thompson, 1975	M/G_1

Main results and conclusions

Cells synchronized by double thymidine block. Incorporation of radioactive choline into phospholipids through the cell cycle. Fractionation of whole cells. Total phospholipid phosphorus measured in the lipid fraction doubled during the S phase.

Synchronously growing cells obtained by gradient centrifugation to separate cells according to their position in the cell cycle (note: no separation of M and G_2 cells). The incorporation of radioactive choline into the phospholipid fraction extracted from whole cells was determined, as well as the total phospholipid phosphorus. Phospholipid synthesis is continuous through the cell cycle and shows an increase during G_1 reaching a maximum in the S phase.

Synchronized cells obtained by zonal centrifugation after thymidine treatment. Biochemical analysis of isolated plasma membrane ghosts. Gradual increase in plasma membrane phospholipids and proteins through the cell cycle.

Synchronized cells in suspension under thymidine block followed by colcemid plus deoxycytidine treatment. Incorporation of radioactive fucose and choline into the lipid fraction extracted from whole cells. Glycolipids and choline-containing phospholipids were synthesized during G_2 and M phases only.

Synchronously growing cells obtained by gradient centrifugation to separate cells according to their position in the cell cycle (note: no separation of M and G_2 cells) as well as by double thymidine block. The synthesis of phosphatidylcholine and of glycolipid was analyzed by measuring the incorporation of respectively ^{3}H-choline into acid-soluble, and of ^{23}H-fucose into lipid-extractable material of whole cells. A bimodal incorporation for ^{3}H-choline was found, with maxima at early S and G_2. The incorporation of ^{3}H-fucose was maximal during G_2.

Cells synchronized by double thymidine block followed by mitotic collection using colcemid. The incorporation of radioactive palmitate, glucosamine and galactose was analyzed in the lipid extracted from whole cells. Most glycolipids are synthesized during all phases of the cell cycle, but different glycolipids showed maximum rates in different phases. Ceramide trihexoside and tetrahexoside exclusively during the G_1 and S phases.

Cells synchronized by a double thymidine block followed by mitotic collection. The synthesis of the main phospholipid (PC, PI, PE, PS and SP) was analysed by measuring ^{32}P incorporation into phospholipids extracted from whole cells. The rate of synthesis of any of these phospholipids was relatively constant through the cell cycle; that of glycerophospholipids being slightly higher during the late G_1 and S phases.

Cells in suspension synchronized by double thymidine block. Analysis of isolated cell membranes. The rate of incorporation of radioactive choline, leucine and glucosamine into the plasma membrane was found to be markedly increased just after division (early G_1)

Cells synchronized by double thymidine block. Incorporation of ^{14}C-galactose into the lipid fractions extracted from whole cells showed that glycosphingolipids are synthesized mainly during the M and G_1 phases.

Cells synchronized by double thymidine block. The level of glycosphingolipids in whole cells was measured by determining the glucose content on purified glycosphingolipids, using gas-liquid chromatography. A gradual increase was found after release from thymidine block, followed by a dramatic inrease in almost all glycosphingolipids during the M phase.

Cells synchronized by heat-shock. The incorporation of radioactive phosphorus into lipids, and of radioactive leucine into proteins was measured in 6 different cell fractions. The two fractions containing cilia and cilia supernatant (cell surface material) showed a sudden increase in labeled protein followed by a peak in incorporation of phospholipid during cytokinesis.

pholipid incorporation during G_1. The interpretation of this result requires some caution, however, since it is known that the permeability of the plasma membrane sharply increases in this period (section 9), and an increased rate of incorporation may be due simply to an increased uptake of the precursor rather than a change in the rate of synthesis.

A different reservation may apply to the data obtained from the biochemical analysis of isolated plasma membranes (Graham et al., 1973; Pasternak and Graham, 1973). According to these studies, no sudden change occurs in the lipid/protein ratio of the plasma membrane at any phase of the cell cycle. As stated earlier (section 7.5), experiments using dividing *Xenopus* eggs have shown that newly formed plasma membrane regions are very labile. Thus it should be considered that significant proportions of newly assembled plasma membrane may not be included in the biochemical analysis due to loss during the isolation procedure. This may be particularly true for the lipid portion contained in fluid membrane regions. This reservation does not apply to the study in which cell fractions of dividing cells of *Tetrahymena* were analyzed (Baugh and Thompson, 1975). A sudden increase of labeled phospholipids was observed during cytokinesis in the two fractions containing cell surface material (including the cilia and their supernatant). The sudden increase in phospholipid insertion agrees with the freeze-fracture data on other dividing cells, which suggests that during new membrane formation the insertion of lipid precursors prevails initially (Scott et al., 1971; Bluemink et al., 1976; see also Pfenninger and Bunge, 1974; Satir et al., 1976; and section 6).

In final summary, the results of experiments using radioactive precursors discussed in this and preceding sections are consistent in that maximum incorporation of components into the cell surface occurs during or shortly after mitosis or at both times (Gerner et al. 1970; Onodera and Sheinin, 1970; Kraemer and Tobey, 1972; Glick and Buck, 1973). These findings differ, however, from data in other studies (Bosmann and Winston, 1970; Nowakowski et al., 1972; Rosenberg and Einstein, 1972) which indicate that certain plasma membrane proteins are assembled during S and G_2. However, several other studies that have used techniques for labeling plasma membrane proteins (Gahmberg and Hakomori, 1974; Stein and Berestecky, 1975) or analyzing the composition of surface membrane (Brown, 1972; Baugh and Thompson, 1975; Johnsen et al., 1975) also indicate that the plasma membrane changes during or shortly after mitosis.

In an attempt to interrelate these results the following speculative scheme is presented. It is likely that cells passing from M into G_1 start to form new plasma membrane by rapid insertion of mainly lipid precursors into the existing membrane (section 6). This may raise the fluidity of the existing plasma membrane. Membrane growth by insertion of lipid may remain unobserved when ghost material is analyzed (Graham et al., 1973; Pasternak and Graham, 1973) because of specific loss of the newly introduced material during isolation. Almost concomitantly, surface carbohydrates that stabilize the newly formed cell membrane are incorporated (section 5.2). The high rate of incorporation of labeled precursors found in this period (Gerner et al., 1970; Onodera and Sheinin, 1970; Kraemer and Tobey, 1972; Chatterjee et al., 1973, 1975; Glick and Buck, 1973) may be due partly to facilitated transport across the membrane barrier (section 9). Once the cells are back to their original size; lipid insertion may become less pronounced. In the S/G_2 period when the cell area remains almost constant, the incorporation of more specialized membrane components (section 7.3) such as sialoproteins may become prevalent (Nowakowski et al., 1972; Rosenberg and Einstein, 1972). A sudden change in the extent of exposure of certain membrane glycoproteins and glycolipids upon entry into G_1 may be due to a rearrangement of membrane receptors (Gahmberg and Hakomori, 1974; Stein and Berestecky, 1975) and does not necessarily imply that molecules are withdrawn or inserted.

8. Lipid metabolism in dividing eggs

Dividing eggs have several advantages over other cells for the study of plasma membrane biogenesis. In relation to cell division, generally, a free-living egg forms a closed system relying completely on its own reserves. During early development, assembly of intracellular membranes such as the endoplasmic reticulum or mitochondria is neglible. Cell divisions succeed each other rapidly and demand considerable production of new membrane area.

Sea urchin eggs have been used to study phospholipid synthesis and assembly during cytokinesis. In vivo incorporation of the phospholipid precursor choline into sea urchin eggs and embryos was analyzed in both acid-soluble extracts and in the lipid fraction of total embryo homogenates (Pasternak, 1973). No significant change in amount or distribution of the major phospholipid classes was found during development until the gastrula stage. There was an increased uptake of ^{14}C-choline after fertilization (before first cleavage), and the label was found to be inserted as phosphatidylcholine, at a very slow rate. It was concluded that plasma membrane assembly during early cleavage is probably accompanied by resynthesis of phospholipids derived from degraded yolk phospholipids. These observations are consistent with the earlier findings of Mohri (1964), who detected increased incorporation of labeled phospholipid precursors (^{14}C-acetate and ^{14}C-glycerol) after fertilization in the sea urchin egg.

However, later biochemical studies using sea urchin eggs have led to a different interpretation. Byrd (1975) also investigated the uptake and incorporation of radioactive choline and ethanolamine, but concluded that there was no increase in the rate of phospholipid synthesis at the onset of cleavage. The increased precursor incorporation was thought to be due instead to an increase in plasma membrane permeability after fertilization. Eggs prelabeled with radioactive choline and ethanolamine do not show a significant change in incorporation of these precursors until 9 hours after fertilization. Similar experiments were carried out by Schmell and Lennarz (1973), who were also unable to detect a significant net synthesis of the two major membrane phospholipids, phosphatidylcholine and phosphatidylethanolamine. Their conclusion that the increase in uptake of labeled precursor only reflects facilitated transport across the plasma membrane agrees with that of Byrd. This is plausible, since it is known that both the ion permeability of the plasma membrane (Steinhardt et al., 1972) and the rate of transport of adenine and L-leucine across the plasma membrane (Doree and Guerrier, 1974) increase significantly during fertilization.

In the sea urchin egg the total content of lipid or phospholipid remains constant during early development up to the blastula stage (Mohri, 1964). Probably, membrane phospholipids are present in pools (yolk) already formed during oogenesis, and are translocated from these pools for membrane biogenesis. Whether this involves processes of degradation, phospholipid resynthesis, and membrane preassembly in the cytoplasm is unknown.

Amphibian eggs have also been used to examine post-fertilization events. In eggs of *Bufo arenarum* the total phospholipid content, as well as the content of choline and ethanolamine phosphoglycerides, was found to be unchanged during early development up to the stage of gill circulation (150 hr of development) (Barassi and Bazan, 1974). In vivo labeling of oocytes with ^{32}P made it possible to analyze ^{32}P incorporation into phospholipid during development. The two major phospholipid classes showed an increase in specific activity. The authors concluded that there is no net synthesis of phospholipid during the early phase of development, and that phospholipids are probably transported intracellularly from the yolk to regions where membrane is formed (Wallace, 1963a, 1963b). The increase in ^{32}P-specific activity is thought to be due to an increased turnover of the phosphorylated moiety which, in turn, would be related to de- and rephosphorylation of the phospholipid necessary for intussusception into the membrane. The increased phosphorylation of intracellular choline observed in sea urchin eggs (Pasternak, 1973) is open to a similar interpretation.

9. Membrane assembly as reflected by physical membrane properties

Although our present knowledge is far from complete, it has been stressed in previous sections (7.2, 7.5, 7.6) that the formation of new membrane during the cell cycle involves sequential assembly of different membrane components. Therefore it seems reasonable to assume that cyclic membrane assembly will be reflected in cyclic alterations of physical membrane properties. To date, such causal relationships have only been demonstrated in dividing amphibian eggs. As mentioned in section 6, such eggs are well suited for studies of membrane assembly because new membrane material is inserted exclusively into the cleavage furrow region and, moreover, the new membrane can be distinguished from the preexisting membrane by several morphological criteria.

Rana pipiens eggs dividing in hypertonic solutions form partial cleavage furrows, which then regress so that the surface becomes spherical again. In such eggs, the changes in electrical membrane potential (V_m) and electrical membrane resistance (R_m) are much more pronounced than during normal cleavage

(Woodward, 1968). Furthermore, a substantial increase in the electrical membrane capacitance is observed only when furrowing is inhibited. This leads to the hypothesis that the electrical changes at division are due to the insertion between the blastomeres of new plasma membrane having a selective permeability to K^+-ions and a low resistance compared with the preexisting surface membrane. During normal division, the narrow gap between the blastomeres would restrict current flow through the new membrane.

Takahashi and Ito (1968) have compared the changes in V_m and R_m and electrical coupling in normally dividing eggs and dividing isolated blastomeres in *Triturus pyrrhogaster* eggs. During normal division, R_m decreased drastically with the extension of the furrow, followed by a small increase at the end of the first cleavage. V_m changed concomitantly with R_m, so that a decrease in R_m was accompanied by a negative change in V_m. The coupling ratio (Lowenstein, 1966) was greater than 0.8 during the first two cleavages. In isolated blastomeres no changes in V_m were observed during cleavage. However, R_m dropped to a minimum when the blastomere appeared almost to be separated into two daughter cells. At the end of cleavage, R_m rose again. The coupling ratio attained values of 0.3 to 0.8 at the end of cleavage. It was concluded that during cytokinesis new membrane is inserted into the cleavage furrow, which is more permeable to ions than the preexisting membrane.

The membrane potential in intact *Xenopus laevis* embryos was only sensitive to changes in the external K^+-concentration during the second half of the division cycle, when newly formed membrane is exposed in the furrow (Slack and Warner, 1973). In postmorula stages the K^+-sensitivity of V_m could only be demonstrated when the seal that isolates the intercellular fluid from the bathing medium was broken, so that the newly formed intercellular membranes became exposed. Again, it was suggested that the newly formed membrane after fertilization is highly K^+-permeable.

The formation of new membrane during the first division of *Xenopus laevis* eggs has been studied extensively using both electron microscopic and electrophysiological methods (Bluemink and de Laat, 1973; de Laat et al., 1973, 1974; de Laat and Bluemink, 1974). The changes in V_m and R_m during normal cleavage (Fig. 19) followed a characteristic pattern (Fig. 21), which was qualitatively similar to that found in *Triturus* (Takahashi and Ito, 1968). The onset of the hyperpolarization of V_m and the decrease in R_m coincided with the first ultrastructural signs of new membrane insertion into the cleavage furrow (Bluemink and de Laat, 1973; de Laat et al., 1973; de Laat and Bluemink, 1974). The exposed surface area of new membrane can be increased artificially, either by removal of the vitelline membrane (Fig. 20), which ultimately leads to a separation of the blastomeres, or by exposure to cytochalasin B, resulting in furrow regression and subsequent insertion of new membrane into the original furrow region. Under these conditions the changes in V_m and R_m during the division cycle were far more pronounced (Fig. 22), and the extent of reduction of R_m could be correlated quantitatively with the surface area of the inserted new membrane. The specific resistance of this membrane portion was found to

444

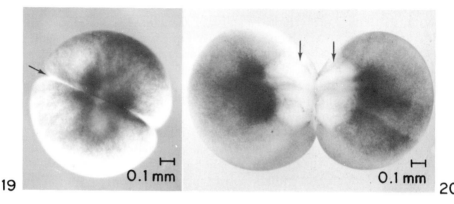

Fig. 19. *Xenopus* egg within the vitelline membrane at the end of the first cleavage. The two blasto-meres are closely apposed and hardly any unpigmented surface (arrow) i.e., new cell membrane, is visible. 32×.

Fig. 20. *Xenopus* egg without vitelline membrane at the end of first cleavage. The two blastometers have moved apart, unpigmented surface (arrows) i.e., new cell membrane, is exposed to the medium. 32×.

(Figures 19 and 20 are reproduced by permission of The Rockefeller University Press.)

be 1 to 2 kOhm · cm² , as against 74 kOhm · cm² for the preexisting surface membrane (de Laat and Bluemink, 1974). Using ion-selective microelectrodes, the ionic permeability properties of the preexisting and the newly formed membrane were analyzed. While the preexisting membrane showed no permselectivity, the relative permeability P_{Na}/P_K of the new membrane was found to be 0.19. It was concluded that the changes in V_m during the normal division cycle were due to the insertion of a fraction of the newly formed, highly K^+-permeable intercellular membrane into the cell surface (de Laat et al., 1974). Similar results were obtained for the dividing *Ambystoma* egg (de Laat et al., 1975). The specific resistances of all intercellular membranes formed at fifth cleavage were found to be constant at about 0.4 kOhm · cm² , when determined in situ. Apparently, the permeability properties of these new membranes remain constant during this period of development (de Laat et al., 1976).

Cell cycle-dependent changes in V_m have also been found in tissue cells. However, a causal relationship with new membrane formation remains to be established. In mouse L cells, V_m was reported to be constant at -10 mV during interphase, but to show a rapid hyperpolarization as the cells entered prophase (Cone, 1969). These results do not agree with those obtained for synchronized CHO cells (Sachs et al., 1974). Here, V_m was low in early G_1 (-14 mV), with hyperpolarization occurring in late G_1 and the beginning of S, and during G_2 it remained constant at -29 mV. This hyperpolarization is probably due to an increased K^+ permeability. Measurements of the specific electroconductivity of Ehrlich ascites tumor cells (Malenkov et al., 1972) revealed a change in membrane conductance during the cell cycle; it was high in G_1, decreased in S, and increased

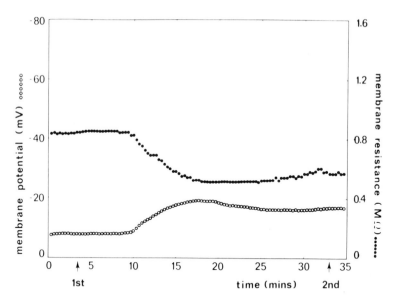

Fig. 21. Changes in membrane potential and membrane resistance during the first cleavage of a *Xenopus* egg within the vitelline membrane. Arrows indicate the onset of the first and second cleavage respectively. (Reproduced by permission of The Rockefeller University Press.)

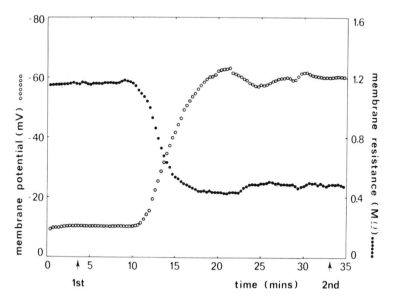

Fig. 22. Changes in membrane potential and membrane resistance during the first cleavage of a *Xenopus* egg devoid of the vitelline membrane. Arrows indicate the onset of the first and second cleavage respectively. (Reproduced by permission of The Rockefeller University Press.)

again in G_2. The authors suggested a relationship between this change and the assembly of new surface components.

In conclusion, the plasma membrane shows drastic changes in ion-permeability properties during the cell cycle, particularly in G_1 which, in amphibian eggs, can be related directly to the assembly of new membrane. It should be noted that transport rates of nutrients have been shown to be cell cycle-dependent (Plagemann and Richey, 1974), but a relationship with the process of membrane assembly has still to be shown. However, the possibility that such permeability changes could influence the incorporation of labeled precursors during the cell cycle, should be kept in mind when interpreting experimental data since changes in incorporation could in fact be due simply to enhanced precursor uptake rather than an increased rate of synthesis.

10. Retrospect

From the literature discussed in this chapter, it becomes evident that wide gaps exist in our knowledge of plasma membrane assembly in cell division. From the available data a few generalizations can be made with regard to: the time of membrane growth; the site of membrane growth; and the mechanism of assembly. It must be emphasized, however, that many of these generalizations are based on incomplete evidence and must therefore contain much speculation.

10.1. Time of membrane growth

The synthesis of membrane lipids, carbohydrates, and proteins must occur prior to membrane assembly. Information on the intracellular sites of synthesis of membrane components is beginning to emerge. The Golgi system and the endoplasmic reticulum are considered to be the sites of synthesis for carbohydrates (Neutra and Leblond, 1966; Bennett, 1970; Bennett and Leblond, 1970; Reith et al., 1970; Cuminge and Dubois, 1972; Bennett et al., 1974; Flickinger, 1975) and lipid components (Wilgram and Kennedy, 1963; Stein and Stein, 1969; Chlapowski and Band, 1971), respectively, while proteins are probably synthesized on free (or bound) polyribosomes (Lodish, 1973; Lowe and Hallinan, 1973; Lodish and Small, 1975). Shortly after synthesis, the membrane components are distributed to the site(s) of membrane assembly (Baugh and Thompson, 1975), and vesicles may form the vehicle of transport for certain precursors (Singal and Sanders, 1974a; Flickinger, 1975; Sanders and Singal, 1975). Probably, the precursors do not move in concert (Subbaiah and Thompson, 1974) and are assembled sequentially.

In eggs, early development is characterized by a period of rapid cell division. No evidence exists that plasma membrane components are synthesized during this period (Barassi and Bazan, 1974; Pechen and Bazan, 1974; Schmell and Lennarz, 1974; Byrd, 1975). Probably they are present in the yolk (Wallace,

1963a, 1963b) from which they are mobilized (Emanuelsson, 1974; Singal and Sanders, 1974b) by means of such processes as degradation, translocation, and reconstitution (Pasternak, 1973; Barassi and Bazan, 1974).

Evidence from immunological (Kuhns and Bramson, 1968; Cikes and Friberg, 1971; Fox et al., 1971; Thomas, 1971; Ferrone et al., 1973; Noonan and Burger, 1973; Smets, 1973; Pasternak et al., 1974; Smets and de Ley, 1974), biochemical (Brown, 1972; Glick and Buck, 1973; Gahmberg and Hakomori, 1974; Baugh and Thompson, 1975; Johnsen et al., 1975; Stein and Berestecky, 1975), biophysical (Takahashi and Ito, 1968; Woodward, 1968; Malenkov et al., 1972; Bluemink and de Laat, 1973; de Laat et al., 1973, 1974; de Laat and Bluemink, 1974; Sachs et al., 1974), and morphological studies (Scott et al., 1971; Furcht and Scott, 1974; Bluemink et al., 1976) indicates that during, or shortly after, mitosis the plasma membrane shows a rapid change in its properties. The changes observed in this period are more dramatic than during any other period of the cell cycle and involve an increased mobility of membrane receptor sites (Webb and Roth, 1974; Garrido, 1975), enhanced incorporation of radioactive precursors (Gerner et al., 1970; Onodera and Sheinin, 1970; Chatterjee et al., 1973, 1975; Glick and Buck, 1973), increased membrane permeability (Takahashi and Ito, 1968; Woodward, 1968; Malenkov et al., 1972; Bluemink and de Laat, 1973; de Laat et al., 1973, 1974; de Laat and Bluemink, 1974; Sachs et al. 1974), and the appearance of membrane areas sparsely seeded with intramembranous particles (Bluemink et al., 1976). These alterations can be taken as evidence for the formation of new membrane areas having properties different from the existing membrane. Probably, the plasma membrane starts to grow in area by the insertion of bulk components, that is, nonspecific membrane precursors such as phospholipids. The insertion of specific membrane components such as special sialoproteins may follow later, after the appropriate synthesizing enzymes have been produced (Bosmann and Winston, 1970: Nowakowski et al., 1972; Rosenberg and Einstein, 1972). In dividing eggs only intracellular membrane is formed, but in dividing tissue cells the plasma membrane grows until the cells resume their original size (Graham et al., 1973; Pasternak and Graham, 1973). New membrane formation in relation to cell division is a postmitotic event, which in dividing Xenopus eggs probably follows approximately 5 to 6 minutes after constriction has been initiated (Fig. 3b). Arguments exist (section 3) for believing that this process is in fact required to complete cytoplasmic cleavage (cytokinesis).

10.2 Site of membrane growth

Furrow formation occurs by contraction of an equatorial ring of microfilaments that pinches the dividing cell into two (Schroeder, 1970, 1972; Selman and Perry, 1970; Sanger, 1975). The constriction process appears to depend on the anchorage of cytoplasmic microfilaments to proteins present in the plasma membrane (Durham, 1974; Clarke et al., 1975; Röhlich, 1975). Thus it seems logical to assume that during the first half of cytokinesis, when surface rigidity is high (Hiramoto, 1970, 1974, 1975), the mobility of proteins in the plasma membrane is restricted. This restriction of mobility could be internal, or it could be due to carbohydrate/carbohydrate or carbohydrate/protein interactions. Lacking such mobility restraints, vectorial forces from microfilaments would redistribute anchorage sites within the plasma membrane to form aggregates (Durham, 1974). Thus if structural stability in the plane of the membrane is required, particularly along the furrow region, it follows that the insertion of membrane precursors during the first half of cytokinesis is not to be expected.

In cultured cells and in small eggs the constriction may proceed till the cell is almost pinched in two (dumbbell stage). This is followed by a release of tension (Fig. 3b), during which the insertion of new membrane may start (section 3). Relaxation by, or concomitant with, putative new membrane formation would cause the cells to flatten against each other (dividing egg) or against the substrate (cultured cells). The extensive surface folds and blebs seen in the furrow (Figs. 4,5) may represent membrane material in excess to what can be inserted into an expanding membrane area (section 3.1).

In eggs, the growth of new surface membrane is presumably restricted to the furrow region and only intercellular membrane is formed. The arguments for this assumption follow. Free-living eggs have a plasma membrane that functions as an effective permeability barrier against external influences. For normal development, the barrier function must be maintained throughout the period of rapid cell division. Random insertion of new membrane area into the existing plasma membrane would perturb the barrier function. The observations on dividing sea urchin (Dan et al., 1938; Dan and Dan, 1940; Dan, 1954a, 1954b; Rappaport and Ratner, 1967; Rappaport and Rappaport, 1968) and amphibian eggs (Bluemink and de Laat, 1973; de Laat et al., 1973; de Laat and Bluemink, 1974) are compatible with the new concept that new membrane is formed (as intercellular membrane) exclusively in the furrow region (section 3).

For cultured tissue cells the situation is different. These cells do not exist naturally as free-living cells. They survive only in a medium that meets their nutrient requirements, which indicates that the barrier function of the plasma membrane in maintaining optimal intracellular conditions is of limited significance. The appearance of blebs and folds all over the surface (Porter et al., 1973; Rubin and Everhart, 1973; Paweletz and Schroeter, 1974; Enlander et al., 1975; Knutton et al., 1975) suggests that the insertion of new membrane is not bound to the region of the furrow (section 3.1). Moreover, cultured tissue cells do not form intercellular membranes, but assemble plasma membrane until they have resumed their original size (Graham et al., 1973). This does not necessarily mean that insertion of new membrane is completely random. Probably, the plasma membrane in cultured tissue cells is topographically heterogeneous in composition (Berlin et al., 1974). Large-scale mosaicism is known to exist in the plasma membrane of hepatocytes, lymphocytes, and thymocytes (Siekevitz, 1972). To that extent the integrity of mosaic areas must have functional value.

10.3. Mechanism(s) of assembly

According to the fluid mosaic membrane model (Singer and Nicolson, 1972), the cell membrane can be conceived of as an oriented, two-dimensional viscous solution of amphipathic proteins (or lipoproteins) and mainly phospholipids in thermodynamic equilibrium. Glycoproteins are structural membrane proteins with their sugar moieties exposed to the outer side. The possibility that complete glycoproteins can be inserted from the cytosol into the plasma membrane is thought to be excluded (Bretscher, 1973). Because of their amphipathic character, it would require a large amount of energy to rotate glycoproteins through the hydrophobic interior of the membrane toward the outer side.

Accepting the idea that glycoproteins are synthesized in the Golgi system (Neutra and Leblond, 1966; Bennett, 1970; Bennett and Leblond, 1970; Cuminge and Dubois, 1972; Bennett et al., 1974), one may imagine that the structural glycoproteins are assembled with their sugar moieties oriented toward the inner side of the vesicular membrane (Hirano et al., 1972). When the Golgi vesicles fuse with the existing plasma membrane, the newly incorporated vesicular membrane will have its inner side out, that is, the glycoproteins will have their sugar moieties exposed to the outer side. Subsequently, mixing of the newly inserted components with those previously present may occur by translational diffusion in the plane of the membrane (Frye and Edidin, 1970). Strong evidence exists that in various types of cells a substantial amount of cell coat material synthesized in the Golgi system is ultimately incorporated into the cell surface (Neutra and Leblond, 1966; Bennett and Leblond, 1969; Bennett, 1970; Cuminge and Dubois, 1970; Reith et al., 1970; Bennett et al., 1974; Flickinger, 1975).

An alternative possibility involves a lipid-mediated assembly process for membrane glycoproteins as proposed by Lennarz (1975). According to this concept glycoproteins are not incorporated as such. The proteins destined for glycoproteins are inserted into the membrane without the hydrophilic oligosaccharide chain, and the complete glycoprotein is then assembled at the plasma membrane itself by a process involving three more steps (Fig. 23): (1) Monosugars are transferred from hydrophilic sugar nucleotides (donors) in the cytoplasm to lipid (polyprenol) phosphate by enzymes associated with the plasma membrane. It is known that some cells possess glycosyltransferase activity associated with the plasma membrane (Roth et al., 1971; Bosmann, 1972, 1974; Roth and White, 1972). (2) More sugar residues are linked to a polyprenol carrier by a phosphodiester-or-pyrophosphoryl bridge, leading to the assembly of a sugar chain (oligosaccharide) linked to hydrophobic lipid, which may exist as such for some time in the interior of the plasma membrane. (3) The preassembled oligosaccharide chain is then transferred to the amino acid residues of an acceptor membrane protein. This may be accompanied by a conformational change in the protein, resulting in the placement of the oligosaccharide residues at the plasma membrane surface.

450

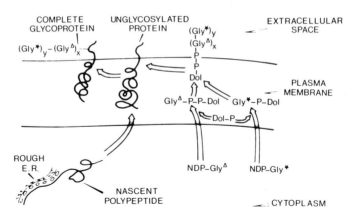

Fig. 23. A model for the glycosylation of plasma membrane glycoproteins via the lipid intermediate pathway. Gly▲ and Gly★ represent two different glycosyl residues (see text for details). (Courtesy of W.J. Lennarz [1975] and reproduced by permission of Science, Washington, D.C.)

No direct evidence exists as to whether such an assembly process is involved in cell-division-related membrane biogenesis, but evidence is available that cell surface glycosyl transferase activity fluctuates during the cell cycle (Bosmann, 1974; Webb and Roth, 1974). 3T3 cells incubated with radioactive UDP-galactose show heavy surface labeling shortly after mitosis. In contrast, interphase 3T3 cells exhibit low labeling (Webb and Roth, 1974). The simplest interpretation is that during mitosis galactosyl transferases and acceptors come close enough to each other for cis-glycosylation to occur. Higher transferase activity might thus reflect increased interaction resulting from increased mobility of ecto-enzymes and acceptors. In mouse lymphoma cells in suspension culture, surface glycosyl transferase activity showed a peak of activity during the S phase (Bosmann, 1974). The results of the two reports do not agree. A comparison is difficult, however, partly because of the difference between the two cell types used.

It is not possible to decide therefore whether such an assembly process is involved in cell-division-related membrane biogenesis. It seems well documented that vesicles function as transport vehicles for surface carbohydrate material (sections 5.1, 5.2). The carbohydrate material is deposited at a time and place to relate it to the formation of new membrane. Available evidence indicates that early membrane biogenesis involves both glycoprotein incorporation and lipid insertion.

It has been suggested that the so-called "peripheral" and "integral" proteins (Singer and Nicolson, 1972) follow different routes for incorporation into the membrane. The erythrocyte membrane in rabbit reticulocytes contains two predominant membrane polypeptides (Lodish, 1973). By lactoperoxidase iodination of whole cells and membranes it was demonstrated that both proteins are localized on the cytoplasmic side of the plasma membrane and that neither protein penetrates to the external surface (Lodish and Small, 1975). It was suggested that the two membrane proteins are released from polyribosomes in the cytoplasm and are translocated to the plasma membrane to be bound on its cytoplasmic side. Direct intussusception of integral membrane proteins could also occur during cell division-related growth of the plasma membrane, but so far no data are available. Integral membrane proteins visible in existing membrane as intramembranous particles (after freeze-fracturing) are not directly seen as such in nascent membrane (Pfenninger and Bunge, 1974; Bluemink et al., 1976; Satir et al., 1976), which suggests that their incorporation follows more slowly.

Molecular insertion of phospholipids, neutral lipids, and lipoproteins may be a basic process in cell division-related membrane biogenesis (Bluemink and de Laat, 1973; Bluemink et al., 1976), at least in certain cells. Increasing evidence exists that in eukaryotic cells phospholipids are transported via the cytosol by carrier proteins (Wirtz and Zilversmit, 1969; Dawson, 1973; Wirtz, 1974). To date, protein-mediated transport has only been demonstrated in the exchange of phospholipids between intracellular membranes. However, shuttling of soluble phospholipids to self-assembling membranes should be considered as a possible method of plasma membrane biogenesis. It has been demonstrated (Wirtz et al., 1975) that upon interaction with an existing membrane transport proteins can release bound phospholipid without binding a membrane phospholipid molecule in return, resulting in net transfer. Possibly, the membrane can govern such a net transfer via the cell membrane surface charge. In experiments in vitro, the transport protein associates more readily with interfaces of increasing negative charge, while dissociation becomes more difficult (Besselaar et al., 1975) that upon interaction with an existing membrane, transport proteins can cytoplasm) to carry phospholipid.

An interesting finding is that changing the concentration of free, soluble bivalent cations (e.g., Mg^{2+}) can overcome the inhibitory effect of the negative surface charge. The possibility has been advanced that neutralization of the negative surface affects the interaction of the protein with the interface, resulting in an increase of the dissociation constant of the protein/liposome complex (Wirtz et al., 1975). This suggestion would provide a mechanism for the onset of postmitotic phospholipid insertion, when we consider the available data on the changes in electrical membrane potential during the cell cycle. In CHO cells (Sachs et al., 1974) and in neuroblastoma cells (de Laat et al., in preparation) the membrane potential showed a marked depolarization in early G_1. Such a depolarization could cause an activation of divalent cations (Jöbsis and O'Connor, 1966; Bissel and Rubin, 1972) that might then lead to a net posphlipid transfer to the plasma membrane. However, the origin of this depolarization remains to be established. At least in amphibian eggs clear evidence exists that the hyperpolarization occurring during the second half of cell division is due to the insertion of new membrane having different permeability properties than the preexisting membrane (Bluemink and de Laat, 1973; de Laat and Bluemink, 1974).

Our knowledge regarding the assembly of new membrane during cell division contains too many gaps to allow us to formulate a unifying molecular theory of the mechanisms involved and their regulation. The morphological alterations resulting from the process of membrane assembly are relatively well documented. However, the biochemical data suffer, among other things, from the difficulties inherent in plasma membrane isolation, including possible specific loss of the newly assembled areas during isolation. Studies on receptor site exposure during the cell cycle, although interesting, have not provided much information on the actual membrane insertion of the receptors. The number of biophysical studies on membrane formation is also limited. Such studies may provide information on the underlying regulatory mechanisms. Finally, the use of different cell types and the unknown effects of metabolically induced cell synchronization often make it difficult to compare the results obtained by different investigators. Nevertheless, we hope this review may serve to suggest critical experiments and to test the various hypotheses for the regulation and molecular mechanisms of membrane assembly during cell division.

Acknowledgment

Grateful acknowledgment is made to Dr. J. Faber for editorial help.

References

Agrell, I.P.S. (1966) Membrane formation and membrane contacts during the development of the sea urchin embryo. Z. Zellforsch. Mikrosk. Anat, 72, 12–21.

Arnold, J.M. (1969). Cleavage furrow formation in telolecithal egg (*Loligo pealii*). I. Filaments in early furrow formation. J. Cell Biol. 41, 894–904.

Arnold, J.M. (1974) Cleavage furrow formation in a telolecithal egg (*Loligo pealii*). III. Cell surface changes during cytokinesis as observed by scanning microscopy. Develop. Biol. 40, 225–232.

Barassi, C.A. and Bazan, N.G. (1974) Metabolic heterogeneity of phosphoglyceride classes and subfractions during cell cleavage and early embryogenesis: Model for cell membrane biogenesis. J. Cell Physiol. 84, 101–114.

Barnett, R.E., Furcht, L.T. and Scott, R.E. (1974) Differences in membrane fluidity and structure in contact-inhibited and transformed cells. Proc. Nat. Acad. Sci. U.S.A. 71, 1992–1994.

Baugh, L.C. and Thompson, G.A. (1975) Studies of membrane formation in *Tetrahymena pyriformis*. IX. Variations in intracellular phospholipid and protein deployment during the cell division cycle. Exp. Cell Res. 94, 111–121.

Bennett, G.(1970) Migration of glycoprotein from Golgi apparatus to cell coat in the columnar cells of the duodenal epithelium. J. Cell Biol. 45, 668–673.

Bennett, G. and Leblond, C.P. (1970) Formation of the cell coat material for the whole surface of columnar cells in the rat intestine, as visualized by radioautography with L-fucose-^{3}H. J. Cell Biol. 46, 409–416.

Bennett, G., Leblond, C.P. and Haddad, A. (1974) Migration of glycoprotein from the Golgi apparatus to the surface of various cell types as shown by radioautography after labeled fucose injection into rats. J. Cell Biol. 60, 258–284.

Bergeron, J.M., Warmsley, A.M. and Pasternak, C.A. (1970) Phospholipid synthesis and degradation during life-cycle of P815Y mast cells synchronized with excess of thymidine B. Biochem. J. 119, 489–492.

Berlin, R.D., Oliver, J.M. Ukena, T.E. and Yin, H.H. (1974) Control of cell surface topography. Nature 247, 45–46.

Besselaar, A.M.H.P. van den, Helmkamp, G.M. and Wirtz, K.W.A. (1975) Kinetic model of the protein-mediated phosphatidylcholine exchange between single bilayer liposomes. Biochemistry 14, 1852–1858.

Bissel, M. J. and Rubin, H. (1972). Growth regulation and Ca^{2+} efflux in chick embryo fibroblasts. J. Cell Biol. 55, 20a.

Bluemink, J.G. (1970) The first cleavage of the amphibian egg. An electron microscope study of the onset of cytokinesis in the egg of *Ambystoma mexicanum*. J. Ultrastruct. Res. 32, 142–166.

Bluemink, J.G. (1971a) Effects of Cytochalasin B on surface contractility and cell junction formation during egg cleavage in *Xenopus laevis*. Cytobiologie 3, 176–187.

Bluemink, J.G. (1971b) Cytokinesis and Cytochalasin-induced furrow regression in the first-cleavage zygote of *Xenopus laevis*. Z. Zellforsch. Mikrosk. Anat. 121, 102–126.

Bluemink, J.G. and de Laat, S.W. (1973) New membrane formation during cytokinesis in normal and Cytochalasin B-treated eggs of *Xenopus laevis*. J. Cell Biol. 59, 89–108.

Bluemink, J.G., Tertoolen, L.G.J., Ververgaert, P.H.J.T. and Verkleij, A.J. (1976) Freeze-fracture electron microscopy of preexisting and nascent cell membrane in cleaving eggs of *Xenopus laevis*. Biochim. Biophys. Acta 443, 143–155

Bosmann, H.B. (1972) Cell surface glycosyl transferases and acceptors in normal and RNA- and DNA-transformed fibroblasts. Biochem. Biophys. Res. Commun. 48, 523–529.

Bosmann, H.B. (1974) Cell plasma membrane external surface glycosyltransferases: activity in the cell mitotic cycle. Biochim. Biophys. Acta 339, 438-441.

Bosmann, H.B. and Winston, R.A. (1970) Synthesis of glycoprotein, glycolipid, protein, and lipid in synchronized L5178Y cells. J. Cell Biol. 45, 23–33.

Branton, D., Elgsaeter, A. and Shotton, D. (1974) Advances in experimental medicine and biology. In: The Cell Surface (Kahan, B.D. and Reisfeld, R.A., eds.) pp. 17–21. Plenum Press. New York.

Bretscher, M.S. (1973) Membrane structure. Some general principles. Science 181, 622–629.

Brown, J.C. (1972) Cell surface glycoproteins 1: Accumulation of glycoprotein on the outer surface of mouse LS cells during mitosis. J. Supramol. Struct. 1, 1–7.

Buck, R.C. and Krishan, A. (1964) Site of membrane growth during cleavage of amphibian epithelial cells. Exp. Cell Res. 38, 426–428.

Buck, R.C. and Tisdale. J.M. (1962) An electron microscopic study of the development of the cleavage furrow in mammalian cells. J. Cell Biol. 13, 117–125.

Buell, D.N. and Fahey, J.L. (1969) Limited periods of gene expression in immunoglobulin synthesizing cells. Science 164, 1524–1525.

Bullivant, S. (1973). Freeze-etching and freeze-fracturing. In: Advanced Techniques in Biological Electron Microscopy (Koehler, J.K., ed.), pp. 67–112. Springer-Verlag. Berlin.

Burger, M.M. (1973) Surface changes in transformed cells detected by lectins. Fed. Proc. 32, 91–101.

Byrd, E.W. (1975) Phospholipid metabolism following fertilization in sea urchin eggs and embryos. Dev. Biol. 46, 309–317.

Chatterjee, S., Sweeley, C.C. and Velicer, L.F. (1973) Biosynthesis of proteins, nucleic acids and glycosphingolipids by synchronized KB cells. Biochem. Biophys. Res. Commun. 54, 585–592.

Chatterjee. S., Sweeley, Ch.C. and Velicer, L.F. (1975) Glycosphingolipids of human KB cells grown in monolayer, suspension, and synchronized cultures. J. Biol. Chem. 250, 61–66.

Chlapowski, F.J. and Band, R.N. (1971) Assembly of lipids into membranes in *Acanthamoeba palestinensis*. J. Cell Biol. 50, 634–651.

Cikes, N. and Friberg, S. (1971) Expression of H-2 and Moloney leukemia virus-determined cell-surface antigens in synchronized cultures of a mouse cell line. Proc. Nat. Acad. Sci. U.S.A. 68, 566–569.

Cifone, M. and Defendi, V. (1974) Cyclic expression of a growth conditioning factor (MGF) on the cell surface. Nature 252, 151–152.

454

Clarke, M., Schatten, G., Mazia, D. and Spudich, J.A. (1975) Visualization of actin fibers associated with the cell membrane in amoeba of *Dictyostelium discoideum*. Proc. Nat. Acad. Sci. U.S.A. 72, 1758–1762.

Cone, C.D. (1969) Electroosmatic interactions accompanying mitosis initiation in Sarcoma cells in vitro. Trans. N.Y. Acad. Sci. Ser. II, 31, 404–427.

Cuminge, D. and Dubois, R. (1972) Etude autoradiographique, au microscope electronique des ébouches gonadiques du poulet après une incorporation de galactose tritié, en culture organotypique: rôle des appareil de Golgi. J. Microscopie, Paris, 14, 299–326.

Dallner, G., Siekevitz, P. and Palade, G.E. (1966) Biogenesis of endoplasmatic reticulum membranes. I. Structural and chemical differentiation in developing rat hepatocyte. J. Cell Biol. 30, 73–96.

Dan, K. (1954a) Further study on the formation of the "new membrane" in the eggs of the sea urchin. *Hemicentrotus (Strongylocentrotus) pulcherrimus*. Embryologia 2, 99–114.

Dan, K. (1954b) The cortical movement in *Arbacia punctulata* eggs through cleavage cycles. Embryologia 2, 115–122.

Dan, K. and Dan, J.C. (1940) Behaviour of the cell surface during cleavage. III. On the formation of new surface in the eggs of *Strongylocentrotus pulcherrinus*. Biol. Bull. (Woods Hole) 78, 486–501.

Dan K. and Kuno-Kojima, M. (1963) A study on the mechanism of cleavage in the amphibian egg. J. Exp. Biol. 40, 7–14.

Dan, K., Dan. J.C. and Yanagita, T. (1938) Behaviour of the cell surface during cleavage. II. Cytologia (Tokyo), 8, 521–531.

Dawson, R.M.C. (1973) Exchange of phospholipids between cell membranes. Sub-Cell. Biochem. 2, 69–89.

DePierre, J.W. and Karnovsky, M.L. (1973) Plasma membranes of mammalian cells. J. Cell Biol. 56, 275–303.

Dodd, N.J.F. (1975) PHA and lymphocyte membrane fluidity. Nature 257, 827–828.

Doggenweiler, C.F. and Frank, S. (1965) Staining properties of lanthanum on cell membranes. Proc. Nat. Acad. Sci. U.S.A. 53, 425–430.

Doree, M. and Guerrier, P. (1974) A kinetic analysis of the changes in membrane permeability induced by fertilization in the egg of the sea urchin *Sphaerechinus granularis*. Develop. Biol. 41, 124–136.

Durham, A.C.H. (1974) A unified theory of the control of actin and myosin in nonmuscle movements. Cell 2, 123–135.

Elbers, P.F., Vervegaert, P.H.J.T. and R. Demel (1965) Tricomplex fixation of phospholipids. J. Cell Biol. 24, 23–30.

Emanuelsson, H. (1974) Localization of serotonin in cleavage embryos of *Ophryotrocha labronica* LaGreca and Bacci. Wilhelm Roux' Arch. Entwicklungsmech. Organismen 175, 253–271.

Enlander, D., Tobey. R.A. and Scott, Th. (1975) Cell cycle-dependent surface changes in Chinese hamster cells grown in suspension culture. Exp. Cell Res. 95, 395–404.

Ferrone, S., Cooper, N.R., Pellegrino, M.A. and Reisfeld, R.A. (1973) Interaction of histocompatibility (HL-A) antibodies and complement with synchronized human lymphoid cells in continuous culture. J. Exp. Med. 137, 55–58.

Flickinger, Ch. J. (1975) The relation between the Golgi apparatus, cell surface, and cytoplasmic vesicles in amoebae studied by electron microscope radioautography. Exp. Cell Res. 96, 189–201.

Fluck, D.J., Henson, A.F. and Chapman, D.J. (1969) The structure of dilute lecithin-water systems revealed by freeze-etching and electron microscopy. J. Ultrastruct. Res. 29, 416–429.

Follett, E.A. and Goldmann, R.D. (1970) The occurrence of microvilli during spreading and growth of BHK21/C13 fibroblasts. Exp. Cell Res. 59, 124–136.

Fox, T.O., Sheppard. J.R., Burger, M.M. (1971) Cyclic membrane changes in animal cells: transformed cells permanently display a surface architecture detected in normal cells only during mitosis. Proc. Nat. Acad. Sci. U.S.A. 68, 244–247.

Frye, L. and Edidin, M. (1970) The rapid intermixing of cell surface antigens after formation of mouse-human heterokaryons. J. Cell Sci. 7, 319–335.

Fullilove, S.L. and Jacobson, A.G. (1971) Nuclear elongation and cytokinesis in *Drosophila montana*. Develop. Biol. 26, 560–577.

Furcht, L.T. and Scott, R.E. (1974) Influence of cell cycle and cell movement on the distribution of intramembranous particles in contact-inhibited and transformed cells. Exp. Cell Res. 88, 311–318.

Gaffney, B.J. (1975) Fatty acid chain flexibility in the membranes of normal and transformed fibroblasts. Proc. Nat. Acad. Sci. U.S.A. 72, 664–668.

Gahmberg, C.G. and Hakomori, S. (1974) Organization of glycolipids and glycoproteins in surface membranes: dependency on cell cycle and on transformation. Biochem. Biophys. Res. Comm. 59, 283–291.

Garrido, J. (1975) Ultrastructural labeling of cell surface lectin receptors during the cell cycle. Exp. Cell Res. 94, 159–175.

Gasko, O. and Danon, D. (1974) Endocytosis and exocytosis in membrane remodelling during reticulocyte maturation. Brit. J. Haematol. 28, 463–470.

Gebicki, J.M. and Hicks, M. (1973) Ufasomes are stable particles surrounded by unsaturated fatty acid membranes. Nature 243, 232–234.

Gerner, E.W., Glick, M.C. and Warren, L. (1970) Membranes of animal cells. V. Biosynthesis of the surface membrane during the cell cycle. J. Cell Physiol. 75, 275–280.

Gipson, I. (1974) Electron microscopy of early cleavage furrows in the chick blastodisc. J. Ultrastruct. Res. 49, 331–347.

Glick, M.C. and Buck, C.A. (1973) Glycoproteins from the surface of metaphase cells. Biochemistry 12, 85–90.

Godman, G.C., Miranda, A.F., Deitsch, A.D. and Tanenbaum, S.W. (1975) Action of Cytochalasin D on cells of established lines. III. Zeiosis and movements at the cell surface. J. Cell Biol. 64, 644–667.

Gonatas, N.K., Steiber, A., Kim, S.U., Graham, D.I. and Avrameas, S. (1975) Internalization of neuronal plasma membrane receptors into the Golgi apparatus. Exp. Cell Res. 94, 426–431.

Goode, D. (1975) Mitosis of embryonic heart muscle cells in vitro; an immunofluoresence and ultrastructural study. Cytobiologie 11, 203–229.

Graham, J.M. (1975) Cellular membrane fractionation. In: New techniques in Biophysics and Cell Biology (Pain, R.H. and Smith, B.J., eds.) vol. 2, pp. 1–42. J. Wiley, London.

Graham, J.M., Sumner, M.C.B., Curtis, D.H. and Pasternak, C.A. (1973) Sequence of events in plasma membrane assembly during the cell cycle. Nature 246, 291–295.

Gwynn, I.ap. and Jones, P.C.T. (1972) Some aspects of cleavage in Pomatoceros triqueter eggs. Z. Zellforsch. Mikrosk. Anat. 123, 486–495.

Hiramoto, Y. (1965) Further studies on cell division without mitotic apparatus in sea urchin eggs. J. Cell Biol. 25, 161–167.

Hiramoto, Y. (1970) Rheological properties of sea urchin eggs. Biorheology 6, 201–234.

Hiramoto, Y. (1974) Mechanical properties of the surface of the sea urchin egg at fertilization and during cleavage. Exp. Cell Res. 89, 320–326.

Hiramoto, Y. (1975) Force exerted by the cleavage furrow of the sea urchin eggs. Develop. Growth Diff. 17, 27–38.

Hirano, H., Parkhouse, B., Nicolson, G.L., Lennox, E.S. and Singer, S.J. (1972) Distribution of saccharide residues on membrane fragments from a myeloma-cell homogenate: its implications for membrane biogenesis. Proc. Nat. Acad. Sci. U.S.A. 69, 2945–2949.

Hirschberg, C.B., Wolf, B.A. and Robbins, P.W. (1975) Synthesis of glycolipids and phospholipids in hamster cells: dependence on cell density and the cell cycle. J. Cell Physiol. 85, 31–40.

Humphreys, W.J. (1964) Electron microscope studies of the fertilised egg and the two-cell stage of Mytilus-edulis. J. Ultrastruct. Res. 10, 244–262.

Hynes, R.O. (1976) Cell surface proteins and malignant transformation. Biochim. Biophys. Acta 458, 73–108.

James, D.W. and Tresman, R.L. (1972) The surface coats of chick dorsal root ganglion cells in vitro. J. Neurocytol. 1, 383–395.

Jöbsis, F.F. and O'Connor, M.J. (1966) Calcium release and reabsorption in the Sartorius muscle of the toad. Biochem. Biophys. Res. Commun. 25, 246–252.

Johnsen, S., Stokke, T. and Prydz, H. (1975) HeLa cell plasma membranes. Changes in membrane protein composition during the cell cycle. Exp. Cell Res. 93, 245–251.

Kalt, M.R. (1971) The relationship between cleavage and blastocoel formation in *Xenopus laevis*. II. Electron microscopic observations. J. Embryol. Exp. Morphol. 26, 51–66.

Keenan, T.W. and Morré, D.J. (1975) Glycosyltransferases: do they exist on the surface membrane of mammalian cells? FEBS Lett. 55, 7–12.

Killander, D., Klein, E. and Levin, A. (1974) Expression of membrane-bound IgM and HL-A antigens on lymphoblastoid cells in different stages of the cell cycle. Eur. J. Immunol. 4, 327–332.

Knutton, S., Sumner, M.C.B. and Pasternak, C.A. (1975) Role of microvilli in surface changes of synchronized P815Y mastocytoma cells. J. Cell Biol. 66, 568–576.

Korohoda, W. and Stockem, W. (1975) Experimentally induced destabilization of the cell membrane and cell surface in *Amoeba proteus*. Cytobiologie 12, 93–110.

Kraemer, P.M. (1967) Configuration change of surface sialic acid during mitosis. J. Cell Biol. 33, 197–200.

Kraemer, P.M. and Tobey, R.A. (1972) Cell-cycle dependent desquamation of heparan sulphate from the cell surface. J. Cell Biol. 55, 713–717.

Kuhns, W.J. and Bramson, S. (1968) Variable behaviour of blood group H on HeLa cell populations synchronized with thymidine. Nature 219, 938–939.

de Laat, S.W. and Bluemink, J.G. (1974) New membrane formation during cytokinesis in normal and Cytochalasin B-treated eggs of *Xenopus laevis*. II. Electrophysiological observations. J. Cell Biol. 60, 529–540.

de Laat, S.W., Luchtel, D. and Bluemink, J.G. (1973) The action of Cytochalasin B during egg cleavage in *Xenopus laevis:* Dependence on cell membrane permeability. Dev. Biol. 31, 163–177.

de Laat, S.W., Buwalda, R.J.A. and Habets, A.M.M.C. (1974) Intracellular ionic distribution, cell membrane permeability and membrane potential of the *Xenopus* egg during first cleavage. Exp. Cell Res. 89, 1–14.

de Laat, S.W., Wouters, W., Marques da Silva Pimenta Guarda, M.M. and da Silva Guarda, M.A. (1975) Intracellular ionic compartmentation, electrical membrane properties, and cell membrane permeability before and during first cleavage in the *Ambystoma* egg. Exp. Cell Res. 91, 15–30.

de Laat, S.W., Bakker, M.I. and Barts, P.W.J.A. (1976) New membrane formation and intercellular communication in the early *Xenopus* embryo. J. Membrane Biol., 27, 109–129.

Lennarz, W.J. (1975) Lipid linked sugars in glycoprotein synthesis. Science 188, 986–991.

Lerner, R.A., Oldstone, M.B.A. and Cooper, N.R. (1971) Cell cycle-dependent immune lysis of Moloney virus-transformed lymphocytes: Presence of viral antigen, accessibility to antibody, and complement activation. Proc. Nat. Acad. Sci. U.S.A. 68, 2584–2588.

Lingwood, C.A. and Thomas, D.B. (1975) Modulation in the rates of incorporation of lipid precursors during the cell cycle. J. Cell Physiol. 86, 635–640.

Lodish, H.F. (1973) Biosynthesis of reticulocyte membrane proteins by membrane-free polyribosomes. Proc. Nat. Acad. Sci. U.S.A. 70, 1526–1530.

Lodish, H.F. and Small, B. (1975). Membrane proteins synthesized by rabbit reticulocytes. J. Cell Biol. 65, 51–64.

Loewenstein, W.R. (1966) Permeability of membrane junctions. Ann. N.Y. Acad. Sci. 137, 441–472.

Lowe, D. and Hallinan, T. (1973). Preferential synthesis of a membrane-associated protein by free polyribosomes. Biochem. J. 136, 825–828.

Luft, J.H. (1971a) Ruthenium red and violet. I. Chemistry, purification, methods of use for electron microscopy and mechanism of action. Anat. Rec. 171, 347–368.

Luft, J.H. (1971b) Ruthenium red and violet. II. Fine structural localization in animal tissues. Anat. Rec. 171, 369–416.

Malenkov, A.G., Voeikov, V.L. and Ovchinnikov, Yu. A. (1972) Electronconductivity changes during the mitotic cycle in Ehrlich ascites tumor cells. Biochim. Biophys. Acta 255, 304–310.

Manasek, F.J. (1969) The appearance of granules in the Golgi complex of embryonic cardiac myocytes. J. Cell Biol. 43, 605–610.

Marchesi, V.T., Tillack, T.W., Jackson, R.L., Segrest, J.P. and Scott, R.E. (1972) Chemical characterization and surface orientation of the major glycoprotein of the human erythrocyte membrane. Proc. Nat. Acad. Sci. U.S.A. 69, 1445–1449.

Matsumoto, M. (1968a) On the mucosubstance in the furrow region of cleaving eggs of *Aquatic oligochaetes*. Zool. Mag. 77, 44–51.

Matsumoto, M. (1968b) Staining properties of mucosubstance in the cleavage furrow of *Tubifex* eggs. Zool. Mag. 77, 81–86.

Mayhew, E. (1966) Cellular electrophoretic mobility and the mitotic cycle. J. Gen. Physiol. 49,

McNutt, N.S. (1977) Freeze-fracture techniques and applications to the structural analysis of the mammalian plasma membrane. In: Dynamic Aspects of Cell Surface Organization (Poste, G. and Nicolson, G.L., eds.), Cell Surface Reviews, vol. 3, pp. 75–125. Elsevier/North-Holland, Amsterdam.

Mohri, H. (1964) Utilization of ^{14}C labelled acetate and glycerol for lipid synthesis during the early development of sea urchin embryos. Biol. Bull. 126, 440–455.

Motomura, I. (1958) Secretion of mucosubstance in the cleaving eggs of the sea urchin. Bull. Mar. Biol. Sta. Asamushi 9, 79.

Motomura, I. (1960) Formation of cleavage plane by the secretion of mucosubstance in the egg of the frog. Sci. Rep. Tohoku Univ. Ser. IV Biol. 26, 53–58.

Motomura, I. (1966) Secretion of a mucosubstance in the cleaving egg of the sea urchin. Acta Embryol. Morphol. Exp. 9, 56–60.

Motomura, I. (1967a) Formation of diastema in the cleaving egg of the sea urchin. Sci. Rep. Tohoku Univ. Ser. IV Biol. 33, 135–142.

Motomura, I. (1967b) Secretion of mucosubstance in the early embryo of an amphibian, *Hynobius lichenatus*. Sci. Rep. Tohoku Univ. Ser. IV Biol. 33, 143–148.

Neutra, M. and Leblond, C.P. (1966) Synthesis of the carbohydrate of mucus in the Golgi complex, as shown by electronmicroscope radioautography of goblet cells from rats injected with glucose-H³. J. Cell Biol. 30, 119–136.

Nicolson, G.L. (1974) The interactions of lectins with animal cell surfaces. Int. Rev. Cytol. 39, 89–190.

Nicolson, G.L. (1976a) Transmembrane control of the receptors on normal and tumor cells. I. Cytoplasmic influence over cell surface components. Biochim. Biophys. Acta 457, 57–108.

Nicholson, G. L. (1976b) Transmembrane control of the receptors on normal and tumor cells. II. Surface changes associated with transformation and malignancy. Biochim. Biophys. Acta 458, 1–72.

Nicolson, G.L., Smith, J.R. and Poste, G. (1976) Effects of local anesthetics on cell morphology and membrane-associated cytoskeletal organization BALB/3T3 cells. J. Cell Biol. 68, 395–402.

Nicolson, G. L., Poste, G. and Ji, T. H. (1977) The dynamics of cell membrane organization. In: Dynamic Aspects of Cell Surface Organization (Poste G., and Nicolson, G. L., eds.), vol. 3. Cell Surface Reviews, pp. 1–73, Elsevier/North-Holland, Amsterdam.

Noonan, K.D. and Burger, M.M. (1973) Binding of ³H-Concanavalin A to normal and transformed cells. J. Biol. Chem. 248, 4286–4292.

Noonan, K.D., Levine, A.J. and Burger, M.M. (1973) Cell cycle-dependent changes in the surface membrane as detected with ³H Concanavalin A. J. Cell Biol. 58, 491–497.

Nowakowski, M., Atkinson, P.H. and Summers, D.F. (1972) Incorporation of fucose into HeLa cell plasma membranes during the cell cycle. Biochim. Biophys. Acta 266, 154–160.

O'Dell, D.S., Tencer, R., Monroy, A. and Brachet, J. (1974) The pattern of concanavalin A-binding sites during the early development of *Xenopus laevis*. Cell Differentiation. 3, 193–198.

Onodera, K. and Sheinin, R. (1970) Macromolecular glucosamine-containing component of the surface of cultivated mouse cells. J. Cell Sci. 7, 337–355.

Ozanne, B. and Sambrook, J. (1971) Binding of radioactively labelled Concanavalin A and wheat germ agglutinin to normal and virus-transformed cells. Nature New Biol. 232, 156–160.

Parsons, D.F. and Subjeck, J.R. (1972) The morphology of the polysaccharide coat of mammalian cells. Biochim. Biophys. Acta 265, 85–113.

Pasteels, J.J. and de Harven, E. (1962) Etude au microscope électronique du cortex de l'oeuf de *Barnea candida* (Mollusque bivalve) et son evolution au moment de la fécondation, de la maturation et de la segmentation. Arch. Biol. 73, 465–490.

458

Pasternak, C.A. (1973) Phospholipid synthesis in cleaving sea urchin eggs: Model for specific membrane assembly. Dev. Biol. 30, 403–410.

Pasternak, C.A. and Graham, J.M. (1973) The assembly of fibroblast plasma membranes. In: Biology of the Fibroblast (Kulonen, E. and Pikkarainen, J. eds.) pp. 261–265. Academic Press, London.

Pasternak, C.A., Warmsley, A.M. and Thomas, D.B. (1971) Structural alterations in the surface membrane during the cell cycle. J. Cell Biol. 50, 562–564.

Pasternak, C.A., Sumner, M.C.B. and Collin, R.C.L.S. (1974) Surface changes during the cell cycle. In: Cell cycle controls (Padilla, G.M., Cameron, I.L. and Zimmerman, A., eds.) pp. 117–124. Academic Press, New York.

Paweletz, N. and Schroeter, D. (1974) Scanning electron microscopic observations on cells grown *in vitro*. II. HeLa cells in mitosis. Cytobiologie 8, 238–246.

Pechen, A.M. and Bazan, N.G. (1974) Membrane ^{32}P-phospholipid labeling in early developing toad embryos. Exp. Cell Res. 88, 432–435.

Pellegrino, M.A., Ferrone, S., Natali, P.G., Pellegrino, A. and Reisfeld, R.A. (1972) Expression of HL-A antigens in synchronized cultures of human lymphocytes. J. Immunol. 108, 573–576.

de Petris, S. and Raff, M.C. (1973) Normal distribution, patching and capping of lymphocyte surface immunoglobulin studied by electron microscopy. Nature New Biol. 241, 257–259.

Pfenninger, K. (1972) Freeze-cleaving of outgrowing nerve fibers in tissue culture. J. Cell Biol. 55, 203a.

Pfenninger, K.H. and Bunge, R.P. (1974) Translational mobility of the membrane intercalated particles of human erythrocyte ghosts, pH-dependent, reversible aggregation. J. Cell Biol. 63, 180–196.

Pinto da Silva, P. (1972) Translational mobility of the membrane intercalated particles of human erythrocyte ghosts, pH-dependent, reversible aggregation. J. Cell Biol. 53, 777–787.

Pinto da Silva, P., Moss, S. and Friedenberg, H.H. (1973) Anionic sites on the membrane intercalated particles of human erythrocyte ghost membranes. Freeze-etch localization. Exp. Cell Res. 81, 127–138.

Plagemann, P.G.W. and Richey, D.P. (1974) Transport of nucleosides, nucleic acid bases, choline and glucose by animal cells in culture. Biochim. Biophys. Acta 344, 263–306.

Porter, K., Prescott, D. and Frye, J. (1973) Changes in surface morphology of Chinese hamster ovary cells during the cell cycle. J. Cell Biol. 57, 815–836.

Poste, G. (1972) Mechanisms of virus-induced cell fusion. Int. Rev. Cytol. 33, 157–253.

Price, Z.H. (1967) The micromorphology of zeiotic blebs in cultured human epithelial (HEp) cells. Exp. Cell Res. 48, 82–92.

Rabinowitz, M. (1941) Studies on the cytology and early embryology of the egg of *Drosophila melanogaster*. J. Morphol. 69, 1–49.

Rapin, A.M.C. and Burger, M.M. (1974) Tumor cell surfaces: general alterations detected by agglutinins. Adv. Cancer Res. 20, 1–91.

Rappaport, R. (1966) Experiments concerning the cleavage furrow in invertebrate eggs. J. Exp. Zool. 161, 1–8.

Rappaport, R. (1971) Cytokinesis in animal cells. Int. Rev. Cytol. 31, 169–213.

Rappaport, R. and Rappaport, B.N. (1968) An analysis of cytokinesis in cultured newt cells. J. Exp. Zool. 168, 187–195.

Rappaport, R. and Ratner, J. H. (1967) Cleavage of sand dollar eggs with altered patterns of new surface formation. J. Exp. Zool. 165, 89–100.

Reith, A., Oftebro, R. and Seljelid, R. (1970) Incorporation of ^{3}H-Glucosamine in HeLa cells as revealed by light and electron microscopic autoradiography. Exp. Cell Res. 59, 167–170.

Revoltella, R., Bertolini, L. and Pediconi, M. (1974) Unmasking of nerve growth factor membrane-specific binding sites in synchronized murine C 1300 neuroblastoma cells. Exp. Cell Res. 85, 89–94.

Robbins, E. and Gonatas, N.K. (1964) The ultrastructure of mammalian cells during the mitotic cycle. J. Cell Biol. 21, 429–464.

Röhlich, P. (1975) Membrane-associated actin filaments in the cortical cytoplasm of the rat mast cell. Exp. Cell Res. 93, 293–298.

459

Rosenberg, S.A. and Einstein, A.B. (1972) Sialic acids on the plasma membrane of cultured human lymphoid cells. J. Cell Biol. 53, 466–473.

Roth, S. and White, D. (1972) Intercellular contact and cell-surface galactosyl transferase activity. Proc. Nat. Acad. Sci. U.S.A. 69, 485–489.

Roth, S., McGuire, E.J. and Roseman, S. (1971) Evidence for cell-surface glycosyl transferases. Their potential role in cellular recognition. J. Cell Biol. 51, 536–547.

Rubin, R.W. and Everhart, L.P. (1973) The effect of cell-to-cell contact on the surface morphology of Chinese hamster ovary cells. J. Cell Biol. 57, 837–844.

Sachs, H.G., Stambrook, P.J. and Ebert, J.D. (1974) Changing in membrane potential during the cell cycle. Exp. Cell Res. 83, 362–366.

Sanders,E. J. (1975) Aspects of furrow membrane formation in the cleaving *Drosophilia* embryo. Cell. Tiss. Res. 156, 463–474.

Sanders, E.J. and Singal, P.K. (1975) Furrow formation in *Xenopus* embryos. Involvement of the Golgi body as revealed by ultrastructural localisation of thiamine pyrophosphatase activity. Exp. Cell Res. 93, 219–224.

Sanger, J.W. (1975) Changing patterns of actin localization during cell division. Proc. Nat. Acad. Sci. U.S.A. 72, 1913–1916.

Satir, B., Sale, W.S. and Satir, P. (1976) Membrane renewal after dibucaine deciliation of *Tetrahymena*. Exp. Cell Res. 97, 83–91.

Sawai, T. and Yoneda, M. (1974) Wave of stiffness propagating along the surface of the newt egg during cleavage. J. Cell Biol. 60, 1–7.

Schäfer, A. and Bässler, R. (1967) Weitere elektronenmikroskopische Beobachtungern am sich teilenden Seeigelei. Verh. Deut. Anat. Ges. 120, 15–23.

Schmell, E. and Lennarz, W.J. (1974) Phospholipid metabolism in the eggs and embryos of the sea urchin *Arbacia punctulata*. Biochemistry 13, 4114–4121.

Schroeder, T.E. (1970) The contractile ring. I. Fine structure of dividing mammalian (HeLa) cells and the effects of cytochalasin-B. Z. Zellforsch. Mikrosk. Anat. 109, 431–449.

Schroeder, T.E. (1972) The contractile ring. II. Determining its brief existence, volumetric changes, and vital role in cleaving *Arbacia* eggs. J. Cell. Biol. 53, 419–434.

Schwalm, F.E. and Bender, H.A. (1973) Early development of the kelp fly, *Coelopa frigida* (Diptera). II. Morphology of cleavage and blastoderm formation. J. Morphol. 141, 235–255.

Scott, R.E., Carter, R.L. and Kidwell, W.R. (1971) Structural changes in membranes of synchronized cells demonstrated by freeze-cleavage. Nature New Biol. 233, 219–220.

Selman, G.G. and Perry, M.M. (1970) Ultrastructural changes in the surface layers of the newt's egg in relation to the mechanism of its cleavage. J. Cell Sci. 6, 207–227.

Selman, G.G. and Waddington, C.H. (1955) The mechanism of cell division in the cleavage of the newt's egg. J. Exp. Biol. 32, 700–733.

Shoham, J. and Sachs, L. (1971) Differences in the binding of fluorescent concanavalin A to the surface membrane of normal and transformed cells. Proc. Nat. Acad. Sci. U.S.A. 69, 2479–2482.

Shoham, J. and Sachs, L. (1974a) Different cyclic changes in the surface membrane of normal and malignant transformed cells. Exp. Cell Res. 85, 8–14.

Shoham, J. and Sachs, L. (1974b) Differences in lectin agglutinability of normal and transformed cells in interphase and mitosis. In: Control of Proliferation in Animal Cells. (Clarkson, B. and Baserga, R., eds.), pp. 297–304. Cold Spring Harbor Laboratory, New York.

Siekevitz, P. (1972) Biological membranes: The dynamics of their organization. Ann. Rev. Physiol. 34, 117–140.

Singal, P.K. (1975) Types and distribution of lamellar bodies in first cleavage *Xenopus* embryos. Cell Tiss. Res. 163, 215–221.

Singal, P.K. and Sanders, E.J. (1974a) An ultrastructural study of the first cleavage of *Xenopus laevis* embryos. J. Ultrastruct. Res. 47, 433–451.

Singal, P.K. and Sanders, E.J. (1974b) Cytomembranes in first cleavage *Xenopus* embryos. Cell Tiss. Res. 154, 189–209.

Singer, S.J. and Nicolson, G.L. (1972) The fluid mosaic model of the structure of cell membranes. Science 175, 720–731.

Singer, S.J. and Rothfield, L.J. (1973) Synthesis and turnover of cell membranes. Neurosci. Res. Bull. 11, 1–86.

Slack, C. and Warner, A. (1973) Intracellular and intercellular potentials in the early *Amphibian* embryo. J. Physiol. 232, 313–330.

Smets, L.A. (1973). Agglutination with Con A dependent on cell cycle. Nature New Biol. 245, 113–115.

Smets, L.A. and De Ley, L. (1974) Cell cycle dependent modulations of the surface membrane of normal and SV40 virus transformed 3T3 cells. J. Cell Physiol. 84, 343–348.

Spornitz, U.M. (1973) Lamellar bodies in oocytes of *Xenopus laevis* and their relation to the mode of fixation. Experientia 29, 589–591.

Stein, O. and Stein, Y. (1969) Lecithin synthesis, intracellular transport, and secretion in rat liver. IX. A radioautographic and biochemical study of choline-deficient rats injected with choline-^{3}H. J. Cell Biol. 40, 461–483.

Stein, M. S. and Berestecky, J. M. (1975) Exposure of an arginine-rich protein at surface of cells in S, G_2 and M phases of the cell cycle. J. Cell Physiol. 85, 243–250.

Steinhardt, R.A., Shen, S. and Mazia, D. (1972) Membrane potential, membrane resistance and an energy requirement for the development of potassium conductance in the fertilization reaction of echinoderm eggs. Exp. Cell Res. 72, 195–203.

Stratton, C.J. (1975) Multilamellar body formation in mammalian lung: An ultrastructural study utilizing three lipid-retention procedures. J. Ultrastruct. Res. 52, 309–320.

Subbaiah, P.V. and Thompson, G.A. (1974) Studies of membrane formation in *Tetrahymena pyriformis*. J. Biol. Chem. 249, 1302–1310.

Szollosi, D. (1970) Cortical cytoplasmic filaments of cleaving eggs: A structural element corresponding to the contractile ring. J. Cell Biol. 44, 192–209.

Szubinska, B. (1971) "New membrane" formation in *Amoeba proteus* upon injury of individual cells. J. Cell Biol. 49, 747–772.

Takahashi, M. and Ito, S. (1968) Electrophysiological studies on membrane formation during cleavage of the amphibian egg. Zool. Magazine 77, 307–316.

Takahashi, M., Yagi, Y., Moore, E. and Pressman, D. (1969) Immunoglobulin production in synchronized cultures of human hematopoietic cell lines. J. Immunol. 103, 834–843.

Thomas, D.B. (1971) Cyclic expression of blood group determinants in murine cells and their relationship to growth control. Nature 233, 317–321.

Thomas, R.J. (1968) Cytokinesis during early development of a teleost embryo *Brachidanio rerio*. J. Ultrastruct. Res. 24, 232–238.

Tourtellotte, M.E. and Zupnik, J.S. (1973) Freeze-fractured *Acholeplasma laidlawii* membranes: Nature of particles observed. Science 179, 84–86.

Vollet, J.J. and Roth, L.E. (1974) Cell fusion by nascent-membrane induction and divalent-cation treatment. Cytobiologie 9, 249–262.

Wallace, R.A. (1963a) Studies in amphibian yolk. III. A resolution of yolk platelet components. Biochim. Biophys. Acta 74, 495–504.

Wallace, R.A. (1963b) Studies on amphibian yolk. IV. An analysis of the main-body component of yolk platelets. Biochim. Biophys. Acta 74, 505–518.

Warmsley, A.M. and Pasternak, C.A. (1970) The use of conventional and zonal centrifugation to study the life cycle of mammalian cells. Phospholipid and macromolecular synthesis in neoplastic mast cells. Biochem. J. 119, 493–499.

Warren, L. (1969) The biological significance of turnover of the surface membrane of animal cells. Curr. Top. Dev. Biol. 4, 197–222.

Webb, G.C. and Roth, S. (1974) Cell contact dependence of surface galactosyltransferase activity as a function of the cell cycle. J. Cell Biol. 63, 796–805.

Winzler, R.J. (1970) Carbohydrates in cell surfaces. Int. Rev. Cytol. 29, 77–125.

Wilgram, G.F. and Kennedy, E.P. (1963) Intracellular distribution of some enzymes catalyzing reactions in the biosynthesis of complex lipids. J. Biol. Chem. 238, 2615–2619.

Wirtz, K.W.A. (1974) Transfer of phospholipids between membranes. Biochim. Biophys. Acta 344, 95–117.

Wirtz, K.W.A. and Zilversmit, D.B. (1969) Participation of soluble liver proteins in the exchange of membrane phospholipids. Biochim. Biophys. Acta 193, 105–116.

Wirtz, K.W.A., Geurts van Kessel, W.S.M., Kamp, H.H. and Demel, R.A. (1976) The protein-mediated transfer of phosphatidylcholine between membranes. Eur. J. Biochem. 61, 515–523.

Wolf, B.A. and Robbins, P.W. (1974). Cell mitotic cycle synthesis of NIL hamster glycolipids including the Forssman antigen. J. Cell Biol. 61, 676–687.

Woodward, D.J. (1968) Electrical signs of new membrane production during cleavage of R. pipiens eggs. J. Gen Physiol. 52, 509–531.

Yamada, K.M., Spooner, B.S. and Wessels, N.K. (1971) Ultrastructure and function of growth cones and axons of cultured nerve cells. J. Cell Biol. 49, 614–635.

Young, R.W. (1974) Biogenesis and renewal of visual cell outer segment membranes. Exp. Eye Res. 18, 215–223.

Zotin, A.J. (1964) The mechanism of cleavage in amphibian and sturgeon eggs. J. Embryol. Exp. Morphol. 12, 247–262.

Biosynthesis and assembly of bacterial cell walls

8

Jean-Marie GHUYSEN

1. Introduction

The distinction between eukaryotic and prokaryotic cells rests upon a limited number of major features (Salton, 1974): the organization of the nucleus, the types of ribosomes, the respiratory and photosynthetic equipment, the absence of nuclear membrane and endoplasmic reticulum in bacteria, their inability (except the mycoplasma) to synthesize sterols, and finally, the occurrence of a whole class of unique heteropolymers in the latter's cell walls and envelope layers.

Electron microscopy has proven to be an invaluable tool for investigating the architecture of the bacterial surface components and for showing the degree of purity of the various preparations of organelles or layers. It has also been an essential technique in understanding the distinctions between the two broad groups of bacteria that are separated by the gram-stain reaction, the gram-positives and the gram-negatives.

Bacteria are surrounded by a delicate plasma membrane that contains or has fixed on it many enzymes and biologically active compounds. Because of the large concentration gradient between the inside and the outside of the plasma membrane in bacteria, water has a strong tendency to flow inward. To preserve the plasma membrane against osmotic disruption, a cell envelope of high tensile strength outside the plasma membrane has evolved. The cell envelopes in gram-positive bacteria differ markedly in structure from those in gram-negative bacteria. Both types, however, perform the same essential function, to keep the cell alive under ordinary hypotonic environmental conditions.

Thin sections of gram-positive bacteria (Fig. 1) usually reveal (1) a well-defined, seemingly homogeneous, thick (15–30 nm) and rigid outer cell wall; (2) an underlying and cell-limiting plasma membrane; and (3) membrane-ous mesosomes that appear to be internal invaginations of the plasma membrane (Ellar et al., 1967). The wall represents about 15 to 30% of the dry weight of the cell. It is the cell supporting structure because of the presence of a rigid, water-

G. Poste and G.L. Nicolson (eds.) The Synthesis, Assembly and Turnover of Cell Surface Components
© *Elsevier/North-Holland Biomedical Press, 1977.*

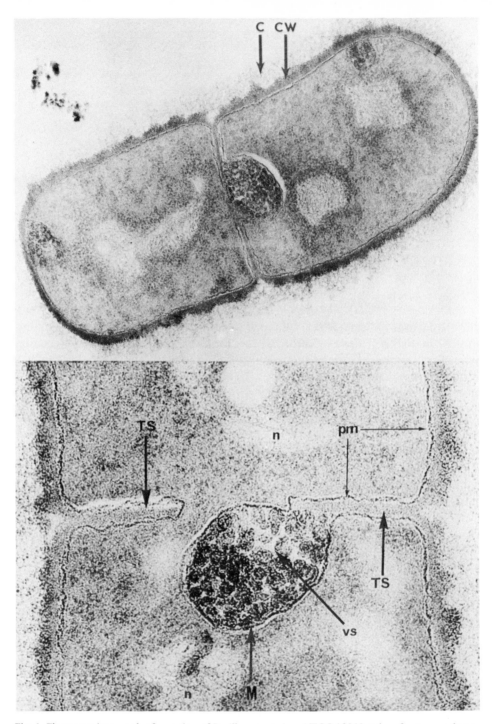

Fig. 1. Electron micrograph of a section of *Bacillus megaterium* ATCC 19213, taken from a synchronously dividing population at 3 hours. CW, cell wall; c, surrounding capsular material; TS, nascent transverse septum; pm, plasma membrane; M, mesosomes; vs, mesosomal vesicles; n, nuclear material. (Ellar et al., 1967.) (Reprinted by courtesy of the American Society for Microbiology.) 32,000 ×: upper part; 64,000: lower part.

insoluble polymer, the peptidoglycan (also called mucopeptide, glycopeptide, or murein).

Thin sections of gram-negative bacteria (Fig. 2) reveal a more complex layering in the cell envelope (Freer and Salton. 1971). The electron-dense peptidoglycan-containing layer is considerably thinner (1–2 nm) than that in the corresponding wall of the gram-positive bacteria, and it occurs sandwiched between the inner plasma membrane and an additional outer membrane that is similar in appearance to the former. Mesosomes are also present but their development is often less conspicuous than in the gram-positive bacteria. In electron micrographs the intermediate, rigid peptidoglycan-containing layer can be seen well separated from the plasma membrane by an electron-transparent zone. Depending on the bacterial species and the preparation techniques used, the peptidoglycan-containing layer may be seen either as a layer that is well separated from the outer membrane by another electron-transparent zone or is firmly connected to its inner segment and is not therefore discernible (unless the cells are submitted to heat or trypsin treatment). The profile of the outer membrane also varies (Freer and Salton, 1971). In some species, it exhibits a typical convoluted, wavy appearance. The structures that include both the peptidoglycan layer and those external to it are generally referred to as the wall of the gram-negative bacteria.

The freeze-fracture technique has amply confirmed the results obtained by thin-section studies. It has also provided valuable additional information by allowing intimate membrane and wall details to be revealed in a state believed to be closer to the "native" one. Instead of showing cross-sectional profiles, this technique shows the topography of outer and inner surfaces of walls and membranes (after etching) or the internal cleavage faces of these structures. Mesosomes are only rarely seen in freeze fractures of unfixed cells, whereas they are seen in high frequency in freeze fractures after fixation with glutaraldehyde (Higgins et al., 1976). In fact, a quantitative agreement was obtained between the increasing numbers of mesosomes seen in central cell fractures and the degree to which these cells had been cross-linked by glutaraldehyde. Moreover, it was observed that with increasing fixation times, the mesosome was progressively displaced from the periphery of the cell, and in the process a "tail" of membrane was created between the septal membrane and the displaced mesosome body. A mechanism has been proposed whereby fixation renders mesosomes observable in freeze fractures. At present, morphology, location, number per cell, and function of mesosomes are still being debated (section 2.2.2.)

This chapter will concentrate on selected aspects of some of the unique structures that are located outside the plasma membrane of the gram-positive and gram-negative bacteria. The goal is to describe the multiplicity of functions they perform in molecular structures and processes of synthesis and assembly. Factors that govern cell surface enlargement, division sites, and changes in shape during the division cycle will not be discussed in detail. These complex factors, which must be integrated with the other cellular processes, are discussed by Daneo-Moore and Shockman in this volume. Capsules that may occur outside

466

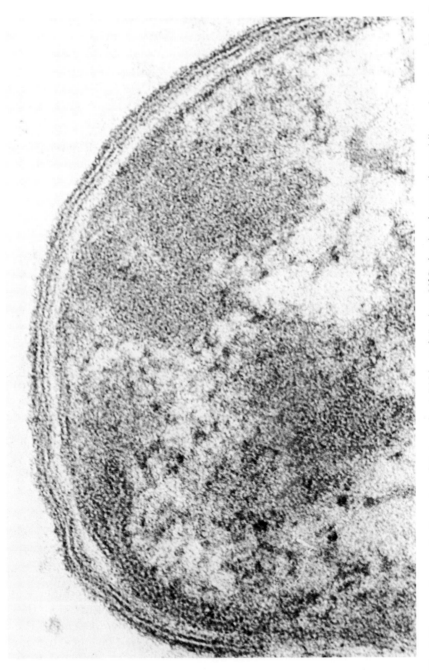

Fig. 2. Electron micrograph of a section of *Proteus vulgaris* P18 heated (5 min at 80°C) showing the complex multilayered structure of the cell envelope. The peptidoglycan is located in the intermediate, electron-dense layer that is sandwiched between the two membrane units (the plasma membrane and the outer membrane, respectively) 240,000 ×

467

the walls in some representatives of both gram-positive and gram-negative bscteria are also beyond the scope of this chapter.

In an attempt to help orient the reader, diagrammatic representations of the location of the various components and layers of the bacterial cell envelopes in both gram-positive and gram-negative bacteria are presented in Figs. 3 and 4. These drawings, however, are purely illustrative and should not be taken literally.

2. Structure and assembly of bacterial walls

2.1. The peptidoglycan in gram-positive and gram-negative bacteria

2.1.1. Isolation
Walls of gram-positive bacteria can be readily obtained reasonably free of cytoplasm and membranous contaminants by mechanical disruption of the cells fol-

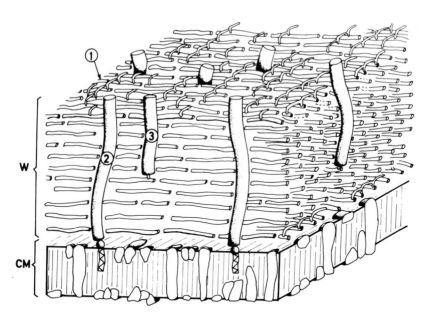

Fig. 3. Schematic representation of the cell envelope of a gram-positive bacterium. CM, cytoplasmic membrane; W, wall. The wall essentially consists of a thick, multilayered peptidoglycan structure ① and of anionic polysaccharides ② and ③. Wall teichoic acids ③ are covalently linked to glycan strands of the peptidoglycan. Lipoteichoic acids ② are anchored in the cytoplasmic membrane. Depending on growth conditions, the wall teichoic acids can be replaced by other anionic polysaccharides (teichuronic acids) that are also linked to the peptidoglycan. Lipoteichoic acids are permanent constituents of the cell envelope, and are not dependent on growth conditions. (Drawing courtesy of Dr. J. Dusart, University of Liège.)

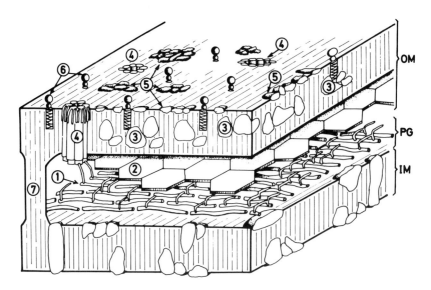

Fig. 4. Schematic representation of the cell envelope of a gram-negative bacterium. IM, inner (cytoplasmic) membrane; PG, mono-layered peptidoglycan; OM = outer membrane; ①, glycan strands of the peptidoglycan monolayer; ②, hexagonally-packed protein that may be closely associated with (but not covalently linked to) the peptidoglycan (Rosenbush, 1974); ④, lipoprotein covalently linked to peptide units of the peptidoglycan. This lipoprotein occurs not only in covalent linkage with the peptidoglycan but also in free form. Bound and free lipoproteins are represented as if they formed cylindrical channels providing the outer membrane with diffusion pores (Inouye, 1974). However, this type of arrangement for the lipoprotein molecules was not supported by more recent experiments (Nakae, 1976). In addition to the lipoprotein and phospholipids (not shown), the outer membrane contains a set of major ③ and minor ⑤ proteins and lipopolysaccharide ⑥. Inner and outer membranes are connected to each other by adhesion sites ⑦ (Bayer, 1968.) (Drawing courtesy of Dr. J. Dusart, University of Liège.)

lowed by differential centrifugation (and sometimes by treatments with proteases and nucleases). Walls isolated from cocci are spherical, whereas walls isolated from bacilli are cylindrical (Salton, 1964) (Fig. 5). Polymers other than peptidoglycan also occur in the walls. Depending on the bacterial species, the wall-associated polymers exhibit extreme variations in chemical nature and in structural complexity. All gram-positive bacteria, however, contain in their walls anionic polymers that are covalently attached to the peptidoglycan. These polymers (teichoic and teichuronic acids; see section 2.3.) are water-soluble. They do not contribute to the rigidity and insolubility of the walls, and can be selectively removed by various means (dilute acid, dilute alkali, hot formamide, dilute periodate in the cold) without affecting the shape and mechanical properties of the residual peptidoglycan.

Crude cell envelopes of gram-negative bacteria can also be obtained by mechanical disruption of the cells and differential centrifugation (Fig. 6a). The protein composition of the isolated cell envelopes is exceedingly complex; about 150 different proteins have been visualized in those of *Escherichia coli* and *Sal-*

Fig. 5. Isolated cell wall of *Bacillus megaterium*. No well-defined fine structure is detectable but wall has a "fibrous" appearance. Latex spheres are 250 nm. (Salton and Williams, 1954.) (Reprinted courtesy of Elsevier/North-Holland, Amsterdam.) 32,000 ×

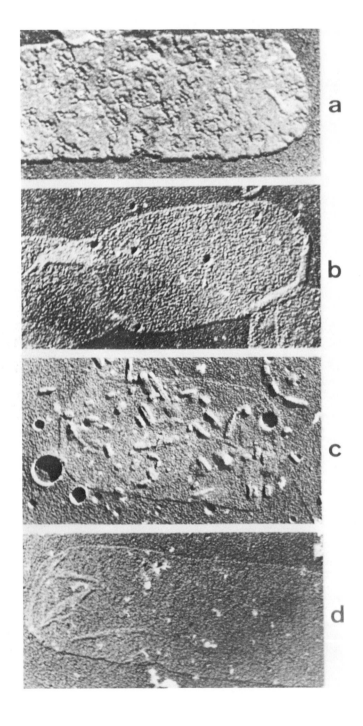

monella typhimurium (Ames and Nikaido, 1976). Extraction of the isolated envelopes, usually with 2% sodium dodecylsulfate (SDS) at 100°C, yields a clear solution from which a small sediment can be recovered by high speed centrifugation (Fig. 6b). Further treatment of the pellet with trypsin (or pepsin) removes lipoprotein molecules (Fig. 6c) (see section 2.5.) which, like the teichoic and teichuronic acids in the walls of gram-positives, are covalently attached to the peptidoglycan of gram-negative bacteria. The residual material is reasonably well purified peptidoglycan material (Martin and Frank, 1962). After shadowing it appears in the electron microscope as a thin, translucent, collapsed cylindrical structure similar in shape to that of the original bacterium (Fig. 6c).

2.1.2. Primary structure

In marked contrast to the eukaryotic cell-supporting exostructures that consist of α-cellulose, hemicellulose, glucan, mannan, or chitin, bacterial peptidoglycans are composed of glycan chains cross-linked by peptide chains (Fig. 7). These peptides contain both D- and L-amino acids. A complete bibliography on this topic may be found in specialized monographs and reviews (Salton, 1964; Weidel and Pelzer, 1964; Martin, 1966; Ghuysen, 1968; Tipper, 1970; Schleifer and Kandler, 1972; Ghuysen and Shockman, 1973; Rogers, 1974).

2.1.2.1. The glycan chains. The glycan moiety of all peptidoglycans examined consists of linear strands of alternate residues of 2-N-acetylamino-2-deoxy-D-glucose (N-acetylglucosamine) and 2-N-acetylamino-3-O-(D-1-carboxyethyl)-2-deoxy-D-glucose (N-acetylmuromic acid). On the basis of extensive chemical investigation carried out in one or two cases (Tipper et al. 1965; Sharon et al., 1966; Jeanloz, 1967) and, more generally, on the basis of the lytic activity of enzymes that have a specificity for 1-4,β bonds or act as endo-β-N-acetylhexosaminidases (Ghuysen, 1968), it is believed that the glycan linkages are uniform in all bacteria and the pyranoside residues of N-acetylglucosamine and N-acetylmuramic acid are linked together by 1-4,β bonds. It is thus a chitin-like structure in which each alternative N-acetylglucosamine residue contains a D-lactyl group ether linked to C-3 (Fig. 8).

Variations occur in the glycan strands, but they do not appear to alter their three-dimensional organization. The C-6 hydroxyl function of N-acetylmuramic acid residues are often substituted by acetyl or phosphodiester groups (Ghuysen, 1968). In at least several species of *Nocardia* and *Mycobacteria*, N-acetylmuramic acid is replaced by N-glycolylmuramic acid (Azuma et al.,

Fig. 6. The cell envelope of *Escherichia coli* B. (a) The isolated wall. (b) The isolated rigid layer, i.e., the peptidoglycan-lipoprotein complex. (c) Same as (b) after treatment with pepsin and before washing. Particles of detached and aggregated lipoprotein still adhere to the sacculus. (d) Peptidoglycan sacculus after treatment of the rigid layer with pepsin and washing. (d) A pure monolayered peptidoglycan or murein layer. (Martin and Frank, 1962.) (Reprinted courtesy of Verlag Zeitschrift für Naturforschung, Tübingen.) 41,600 ×

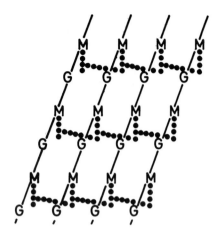

Fig. 7. Schematic representation of a wall peptidoglycan. Glycan chains are composed of N-acetylglucosamine (G) and N-acetylmuramic acid (M). Vertical dots from M represent the amino acid residues of the tetrapeptide subunits. Horizontal dots represent the peptide cross-linking bridges. Five bridging amino acids are shown, corresponding to the peptide bridges of *Staphylococcus aureus* presented in Fig. 11.

1970; Guinand et al., 1970). In spore peptidoglycans, a portion of the muramic acid is replaced by the lactam derivative (Warth and Strominger, 1969). Finally, in a few instances, small amounts of mannomuramic acid were reported to occur along with the glucose derivative (Hoshino et al., 1972). A survey including more than 40 species of gram-positive and gram-negative bacteria, however, failed to reveal the occurrence of anything except glucosamine and glucomuramic acid (Wheat and Ghuysen, 1971).

2.1.2.2. The peptide units (Ghuysen, 1968; Schleifer and Kandler, 1972). In most peptidoglycans that have been examined, every D-lactyl group of the N-acetylmuramic acid is peptide-substituted. In a few species such as *Micrococcus lysodeikticus* (and related *Micrococcaceae*), about 30 to 40% of the total N-acetylmuramic acid is unsubstituted and oligosaccharides as long as octasaccharides with no peptide attached have been identified. All glycans have short tetrapeptide units L-alanyl-D-glutamyl $_\Gamma$L-R$_3$-D-alanine (Fig. 9) (and sometimes tripeptide units L-alanyl-D-glutamyl $_\Gamma$L-R$_3$) linked to their muramyl carboxyl groups. L-alanine at the N-terminus can be replaced by L-serine or glycine. Thus, except for the occasional appearance of glycine at this position, the backbone of all tetrapeptides exhibits an alternating LDLD sequence. The peptide linkages are α except the bond between D-glutamic acid and the L-R$_3$ residue which is in a γ linkage. The α-carboxyl group of D-glutamic acid can be either free, amidated, substituted by a C-terminal glycine or by a glycine amide. Threo-3-hydroxylglutamic acid may occur instead of glutamic acid. Finally, the L-R$_3$ residue may be a neutral amino acid (e.g., L-alanine, L-homoserine), a dicarboxylic amino acid (e.g., L-glutamic acid) or a diamino acid (e.g., L-diaminobutyric acid, L-ornithine, L-lysine, LL-diaminopimelic acid or *meso*-diaminopimelic acid).

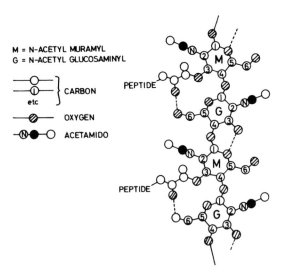

M = N-ACETYL MURAMYL
G = N-ACETYL GLUCOSAMINYL

CARBON
etc

OXYGEN

ACETAMIDO

PEPTIDE

PEPTIDE

Fig. 8. Tetrasaccharide segment of a glycan strand with the chitin-like configuration. The dashed lines represent hydrogen bonds. G, N-acetylglucosamine; M, N-acetylmuramic acid (Tipper, 1970.) (Reprinted courtesy of Iowa State University Press.)

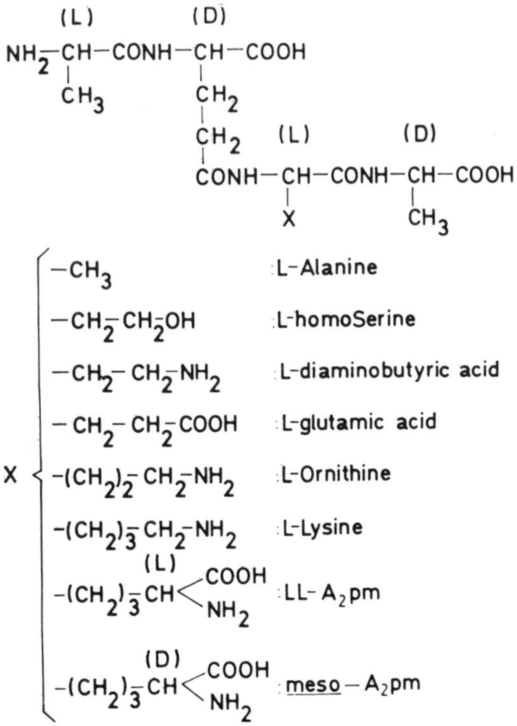

$$
\begin{array}{c}
\text{(L)} \qquad \text{(D)} \\
NH_2-CH-CONH-CH-COOH \\
\quad\ \ |\qquad\qquad | \\
\quad\ \ CH_3\qquad\quad CH_2 \\
\qquad\qquad\qquad\ \ | \\
\qquad\qquad\qquad\ \ CH_2 \quad \text{(L)}\qquad\quad \text{(D)} \\
\qquad\qquad\qquad\ \ | \\
\qquad\qquad\qquad\ \ CONH-CH-CONH-CH-COOH \\
\qquad\qquad\qquad\qquad\qquad | \qquad\qquad\quad | \\
\qquad\qquad\qquad\qquad\qquad X \qquad\qquad\ \ CH_3
\end{array}
$$

X

$-CH_3$	:L-Alanine
$-CH_2-CH_2OH$	:L-homoSerine
$-CH_2-CH_2-NH_2$	:L-diaminobutyric acid
$-CH_2-CH_2COOH$	:L-glutamic acid
$-(CH_2)_2-CH_2-NH_2$	:L-Ornithine
$-(CH_2)_3-CH_2-NH_2$	:L-Lysine
(L) $-(CH_2)_3-CH{<}^{COOH}_{NH_2}$	:LL-A_2pm
(D) $-(CH_2)_3-CH{<}^{COOH}_{NH_2}$	:meso-A_2pm

Fig. 9. General structure of tetrapeptide L-alanyl-γ-D-glutamyl-L-R $_3$-D-alanine subunits. Side chains of amino acids known to occur in the L-R$_3$ position are shown. A$_2$ pm, diaminopimelic acid. (Ghuysen, 1968.)

When *meso*-diaminopimelic acid occurs, both its amino group linked to D-glutamic acid and its carboxyl group linked to D-alanine are located on the same L-center. Finally, the carboxyl group on the D-carbon of *meso*-diaminopimelic acid may be either free or amidated.

2.1.2.3. The interpeptide bridges (Ghuysen, 1968; Schleifer and Kandler, 1972)
Peptide units substituting adjacent glycan chains are covalently linked together by means of "bridges." Based on the composition and location of these interpeptide bridges, peptidoglycans have been classified into four main types. The bridge between two peptide units always extends between the C-terminal D-alanine residue of one tetrapeptide and either the ω-amino group of the L-R$_3$ diamino acid (types I, II and III) or the α-carboxyl group of D-glutamic acid (type IV) of another (tri or tetra) peptide unit. Tripeptides L-alanyl-D-glutamyl-L-R$_3$, which lack the D-alanine residue, are thus necessarily either uncross-linked or located at the C-terminus of a peptide oligomer.

In type I, the bridging consists of a direct N^ω (D-alanyl)-L-R$_3$ peptide bond (Fig. 10). In type II, the bridging is mediated via a single additional amino acid residue (glycine, L-amino acid or D-amino acid) or an intervening short peptide containing up to five amino acid residues. Variations are almost endless (Fig. 11). In type III, the bridging is composed of one or several peptides each having the same amino sequence as the peptide unit attached to muramic acid (Fig. 12). This type of bridging is found in *M. lysodeikticus* where a high proportion of the N-acetylmuramic acid residues in the glycan strands are not peptide substituted as if, at a certain stage of biosynthesis, some peptide units had moved from their muramyl residues into a bridging position. Finally, in type IV, the bridge extends between two carboxyl groups belonging to D-alanine and D-glutamic acid, respectively (Fig. 13). Thus it necessarily involves either a diamino acid residue (which often also has a D configuration) or a diamino acid-containing short peptide.

Fig. 10. Peptidoglycan of chemotype I with *meso*-diaminopimelic acid in the L-R$_3$ position. This structure occurs in the wall of *E. coli*, of probably all other gram-negative bacteria, and of many bacilli. The interpeptide linkage is a D-alanyl-(D)-*meso*-diaminopimelic acid linkage. In *E. coli*, the carboxyl groups are not amidated. The arrow indicates the site of action of the lytic KM endopeptidase from *Streptomyces* strain *albus* G. In some *Bacillaceae*, the α-carboxyl group of D-glutamic acid and/or the carboxyl group of diaminopimelic acid that is not in a peptide bond, are amidated. G, N-acetylglucosamine; M, N-acetylmuramic acid. (Ghuysen, 1968.)

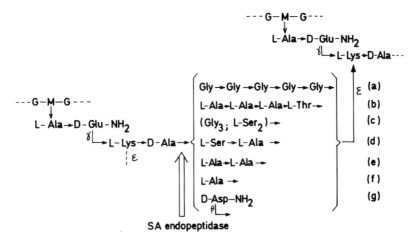

Fig. 11. Peptidoglycans of chemotype II that occur in the walls of (a) *Staphylococcus aureus* Copenhagen; (b) *Micrococcus roseus* R27; (c) *Staphylococcus epidermidis* Texas 26; (d) *Lactobacillus viridescens;* (e) *Streptococcus pyogenes* Group A, type 14; (f) *Arthrobacter crystallopoietes;* (g) *Streptococcus faecalis* (faecium) ATCC 9790; and *Lactobacillus casei* RO94. Arrow indicates the site of action of the lytic SA endopeptidase from *Streptomyces albus* G upon walls (a), (b), (e), and (g). This enzyme has not been tested upon walls (c), (d) and (f). G, N-acetylglucosamine; M, acetylmuramic acid. (Ghuysen, 1968.)

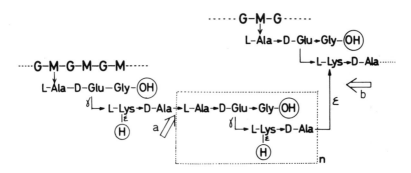

Fig. 12. Peptidoglycan of chemotype III that occurs in the wall of *Micrococcus lysodeikticus*. The site of action of *Myxobacter* ALI enzyme is indicated by arrow a and that of the ML endopeptidase from *Streptomyces albus* G by arrow b. Also shown are the unsubstituted N-acetylmuramic acid residues. G, N-acetylglucosamine, M, N-acetylmuramic acid. (Ghuysen, 1968.)

Recent experiments have shown, in addition to types I to IV, the existence of "atypical" interpeptide bridges extending between the R_3 residue of one peptide to the same R_3 residue of another. To date such atypical bridges have only been found in Mycobacteria. In these organisms cross-linking is mediated through both classical D-alanyl-(D)-*meso*-diaminopimelic acid (Fig. 10) and atypical *meso*-diaminopimelyl-*meso*-diaminopimelic acid linkages occurring in a ratio of about 2:1 (Wietzerbin et al., 1974). The stereochemistry of this atypical bridging is unknown and its physiological significance is still being debated.

476

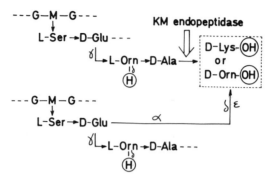

Fig. 13, Peptidoglycan of chemotype IV that occurs in the wall of *Butyribacterium rettgeri*. Arrow indicates the site of action of the KM endopeptidase from *Streptomyces albus* G. G, N-acetylglucosamine; M, N-acetylmuramic acid. (Ghuysen, 1968.)

2.1.3. Evolutionary trends in peptidoglycan structures (Schleifer and Kandler, 1972)

The most frequent type of peptidoglycan is that of type I with *meso*-diaminopimelic acid at the L-R$_3$ position in the peptide unit. Bridging is direct and is mediated via a D-alanyl-(D)-*meso*-diaminopimelic acid bond (Fig. 10). It occurs in the gram-positive bacilli, probably in all gram-negative bacteria, and is also found in highly evoluted prokaryotes such as myxobacteria and blue-green algae. The enormous diversification and complexity of primary structures briefly described above (Figs. 9–13) have only been found in gram-positive bacteria. As Schleifer and Kandler (1972) pointed out, the fact that at present so many different peptidoglycan structures are still conserved by gram-positive bacteria clearly shows that such a diversification was not connected with an appreciable selective advantage or disadvantage. If one accepts the idea that phylogenetic advancement is expressed by simplification and loss of variability (fixation), the complex peptidoglycans of the gram-positive bacteria may represent a primitive stage. Following this view, the directly cross-linked *meso*-diaminopimelic acid-containing peptidoglycan is not only the most successful but would also be the most advanced.

In all gram-positive bacteria shape maintenance and physical protection of the plasma mebrane are almost exclusively, if not entirely, undertaken by the vast amount of multilayered peptidoglycan that is present in their thick wall. Some degree of expansion and contraction has been observed and actually measured based on the impermeability of peptidoglycan to dextran molecules (Marquis, 1968; Ou and Marquis, 1970, 1972). Since these changes, induced by variations of the ionic strength of the medium, are especially sensitive to pH, it has been assumed that they are caused primarily by electrostatic forces within the wall.

In contrast to the gram-positive bacteria (including the bacilli that conserve the multilayered peptidoglycan), all the gram-negative bacteria have developed a peptidoglycan structure that is considered to be a monolayer. This highest level of simplification is such that the supporting function of the peptidoglycan is barely maintained. The outer membrane (linked to the underlying peptidoglycan via a covalent attachment between lipoprotein molecules and the peptidoglycan; see section 2.5.) not only confers the advantage of an additional permeability barrier on the gram-negative bacteria but also reinforces the mechanical strength of the whole cell envelope and further protects the inner plasma membrane against osmotic hazards.

The relative fragility of the *E. coli* K12 peptidoglycan is well demonstrated by the effects of guanidine hydrochloride (Leduc and Van Heijenoort, 1975). The presence of a 6 M concentration of this reagent in the cold causes a suspension of this peptidoglycan to undergo complete clarification. Upon dilution or dialysis of the resulting solution peptidoglycan precipitates; but as shown by electron microscopy, it is recovered as distorted sacculi lacking the initial rod shape. Hence, this fragile peptidoglycan may be converted into a soluble form without breaking covalent linkages. Hydrogen bonding and perhaps electrostatic interactions must contribute significantly to the nondenatured form of this polymer.

The effects of chelating agents on gram-negative cells are also revealing in this respect. When suspended in sucrose solution in the presence of chelating agents such as ethylenediaminetetraacetate (EDTA), exponentially growing cells of *E. coli* (and of other gram-negative bacteria) undergo lysis upon subsequent dilution of sucrose but remain rod-shaped. As shown by Eagon and Carson (1965), Leive (1965a,b), Gray and Wilkinson (1965a,b), Asbell and Eagon (1966), Leive et al. (1968), Roberts et al. (1970), Gilleland et al. (1973), and Leive (1974), the only damage caused by EDTA is removal of divalent cations, which affects the stability of the outer membrane and induces the solubilization of part of its components (lipopolysaccharide together with minor amounts of phospholipids and proteins; see section 2.7). Since the ghosts produced by EDTA and osmotic shock are rod-shaped, it follows that the shape-maintaining properties of the rigid layer are not impaired. The observed lysis, however, suggests that under conditions where the outer membrane has lost its integrity, the very thin peptidoglycan monolayer might not possess enough tensile strength in various areas to resist the high internal osmotic pressure. A thin peptidoglycan monolayer containing areas of low physical strength might also explain the flexibility of some gram-negative bacteria such as the gliding bacteria and *Myxobacteriales* (Verma and Martin, 1967; White et al., 1968). Experimental evidence also supports the view that the peptidoglycan monolayer in gram-negative bacteria is discontinuous in certain areas. These areas would correspond to the adhesion sites that occur between the plasma and the outer membrane (section 2.10.).

Evidence also exists showing that the outer membrane actually contributes to the shape of the wall. Treatment of *E. coli* cells with lysozyme in the absence of EDTA but in the presence of 0.5 M sucrose does not alter their morphology (Birdsell and Cotta-Robles, 1967). However, peptidoglycan is degraded, and dilution in water causes the formation of spherical structures.

Rod-shaped "ghosts" have been prepared from *E. coli* and a number of other bacteria (section 2.8.). Essentially these ghosts consist of a unit membrane derived from the outer membrane. They lack peptidoglycan and have lost virtually all intracellular material. Hence, the outer membrane can maintain its rod shape.

2.1.4. Size

The concept suggested by Weidel and Pelzer (1964) that the peptidoglycan is a bag-shaped macromolecule which completely surrounds the cell is still valid, at least in essence. Studies of primary structure have largely confirmed the netlike arrangement of this polymer. The initial idea, however, might lead to the false impression of extremely long glycan strands interconnected by means of a single, enormous, branched peptide moiety. This is not true, as many terminal groups are present in both the glycan and the peptide moieties.

Average chain lengths for several peptidoglycans have been reported to be of the order of only 10 to 50 disaccharides (Krulwich et al., 1967a, b; Tipper et al., 1967; Kolenbrander and Ensign, 1968; Tipper, 1969). The largest glycan synthesized by a (autolytic-deficient) mutant of *Bacillus licheniformis* contains about 100 to 150 disaccharide units (Ward, 1973; Rogers, 1974). As discussed by Rogers (1974), such a 150-disaccharide unit glycan strand has a length of about 150 nm. Since the length of a bacillus is approximately 1,500 to 3,000 nm, and its circumference is about 2,000 nm, ten or more glycan chains could be stretched in either direction.

The proportion of peptide units that are cross-linked varies from species to species but is invariably rather low (Ghuysen, 1968). The most highly cross-linked peptide is that of *Staphylococcus aureus,* and at least one estimate indicates that the average size does not exceed ten peptide units. The other extreme includes many peptidoglycans of type I (from *E. coli, Proteus vulgaris* and probably most gram-negative bacteria), where approximately equimolar amounts of uncross-linked peptide monomer and cross-linked peptide dimer exist. In spite of the small size of both glycan and peptide moieties, a continuous net of the type shown in Fig. 14 can be built. Such a sheet still may consist of a single macromolecule covering the entire surface of the bacterium. Points of weak mechanical resistance are easily visualized.

2.1.5. Three-dimensional structure

Three-dimensional atomic models of peptidoglycans have been proposed by Tipper (1970), Kelemen and Rogers (1971), Higgins and Shockman (1971), Formanek et al. (1974, 1976) and Oldmixon et al. (1974), respectively. In all of these models the glycan chains are assumed to have the structure of chitin. An α-helical conformation of the peptide units is judged impossible for the obvious reason that the glutamyl bond is γ, and therefore all the carbon atoms of this residue are in the chain. Moreover, in all models proposed, efforts are made to stabilize the peptides by as many hydrogen bonds as possible (like all other known peptide and protein structures) which implies a fairly extended, rather

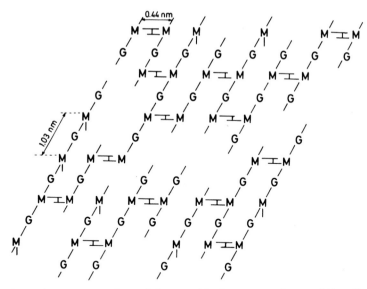

Fig. 14. Schematic representation of the peptidoglycan monolayer of *E. coli*. All of the N-acetylmuramic acid residues (M) are substituted, either by uncross-linked peptide monomers or by cross-linking peptide dimers. Peptide oligomers larger than dimers have not been observed. G, N-acetylglucosamine; M, N-acetylmuramic acid. Note that a continuous net of this type can be built even if the glycans and the peptides are short. The periodicity of the structure is assumed to be about 1.00 and 0.44 nm (see text).

flat conformation. As pointed out by Formanek et al. (1974) for a structure of β-1,4 linked glucose residues, an X-ray reflection corresponding to a periodicity near 1.0 nm can only occur if the sugar residues form a twofold screw axis (Fig. 8). This is actually the case in cellulose and chitin. Moreover, the carbohydrate chains of chitin are packed in a parallel array side by side, with a periodicity of 0.476 nm perpendicular to their long direction. This packing is energetically favored because of the formation of hydrogen bonds between the N-acetyl groups of adjacent carbohydrate chains. Interestingly, X-ray diffraction of dried foils of peptidoglycans from both the gram-negative *Spirillum serpens* and the gram-positive *Lactobacillus plantarum* showed Debye-Scherrer rings indicating periodicity of about 1.00 and 0.44 nm (Fig. 14) (Formanek et al. 1974). In addition, a comparison of the infrared spectra of chitin and peptidoglycan showed that the amide I and amide II bonds have the same frequencies (Formanek et al., 1974, 1976). These experimental data strongly suggest that the structure of the glycan chains in the peptidoglycans may well be similar to that of chitin (and cellulose). In order to fit into a periodic structure of the glycan moiety of 0.44 and about 1 nm, the peptide units must be very flat. Therefore the peptide units were assigned the hypothetical 2.2_7 helix conformation. In this representation, the peptide units form two hydrogen bonds with the sugar residues. Similar models were proposed by Oldmixon et al. (1974) that take into account the ability of walls from gram-positive bacteria to change in volume with ionic conditions (Marquis, 1968).

An interesting conclusion of these studies is that a peptidoglycan monolayer appears to be a highly asymmetric structure. Indeed, on the basis of a chitin-like conformation, the twofold screw axis in the glycan chains of the peptidoglycan necessarily directs all the peptide chains on the same side of the glycan to stack, whereas the O–6 positions of the N-acetylmuramic acid residues are exposed on the other side of the structure sterically unhindered, and therefore are readily available for substitution with acetyl and phosphodiester groups (Fig. 8).

Despite these recent advances, more work remains to be done before a clear representation of the three-dimensional structure of the peptidoglycans can be obtained. In particular, as pointed out by Braun and Wolff (1975), a very dense crystalline, chitin-like arrangement in the *E. coli* rigid monolayer can only exist in certain areas that must be interrupted by less densely packed and less ordered structures to account for the permeability of the peptidoglycan layer, the number of structural repeating units per cell surface area, and the occurrence of the adhesion sites extending between the plasma and the outer membranes. A continuous packing density of the chitin-like arrangement, as proposed by Formanek et al., (1974) would not allow water to pass through and would require that only 30% of the cell surface be covered by peptidoglycan (Braun, 1975).

2.1.6. Peptidoglycan-degrading enzymes (Ghuysen, 1968)

Peptidoglycans can be solubilized by enzymes that hydrolyze bonds either in the glycan strands (endo-N-acetylmuramidases and endo-N-acetylglucosaminidases), in the peptide moiety (endopeptidases hydrolyzing peptide bonds in the interior of the peptide bridges, or those bonds that involve the C-terminal D-alanine residue of the peptide units; see Figs. 10–13), or at the junction between the glycan strands and the peptide units (N-acetylmuramyl-L-alanine amidase). Peptidoglycan solubilization causes cell lysis. Lysis can occur either from within (by autolysins; see below) or externally, when the enzymes are added to cell suspension. Lytic enzymes have been used as the method of choice for the elucidation of the primary structures of all peptidoglycan types. This procedure requires several sequential steps, each involving the use of one specific enzyme so that the original complex heteropolymer can be progressively degraded into small fragments in a controlled manner.

2.1.7. Lysis from within

Bacteria possess enzymes called autolysins, that are capable of hydrolyzing their own peptidoglycan. When such enzymes are permitted to act, cells lose their osmotic protection and autolyze. Autolysins have been found to be localized in the cytoplasm, associated with the membrane, concentrated in the periplasmic region, firmly fixed on the wall, or even excreted in the growth medium. Some are specifically localized in the region of the dividing septum. The first evidence for such localization was suggested by Mitchell and Moyle (1957), who made the intriguing observation that *Staphylococcus* spontaneously becomes osmotically fragile when incubated at 25°C in 1.2 M sucrose, and on further dilution of the sucrose eventually gives rise to hemispheric wall fragments.

The specificities of the autolysins found in various species correspond to those described above: N-acetyl-muramidase, N-acetylglucosaminidase, N-acetylmuramyl-L-alanine amidase, and endopeptidase. The nature and complexity of the autolysin system, however, vary largely among different species. In *S. faecalis* 9790 (Shockman et al., 1967) and *L. acidophilus* strain 63 AM Gasser (Coyette and Ghuysen, 1970b), the only detectable activity is an endo-N-acetylmuramidase. Yet the localization and mode of action of the enzymes seem not to be the same in the two organisms (Higgins et al., 1973). In *D. pneumoniae* (Mosser and Tomasz, 1970; Howard and Gooder, 1975), various bacilli (Brown and Young, 1970), and in *Clostridia* (Tinelli, 1968; Takumi et al., 1971), the major, if not the only activity appears to be an N-acetylmuramyl-L-alanine amidase. In marked contrast, *E. coli* possesses at least six different hydrolases capable of hydrolyzing linkages found in the peptidoglycan. They include N-acetylglucosaminidase, N-acetylhexosaminidase, amidase and cross-bridge splitting peptidases (Weidel and Pelzer, 1964). One of these autolysins was recently discovered in *E. coli* and exhibits a novel and interesting specificity (Taylor et al., 1975; Höltje et al., 1975). In an in vitro system this enzyme has the unique property of degrading the *E. coli* rigid layer peptidoglycan into disaccharide fragments that lack reducing ends and contain muramic acid as an internal (anhydro) structure, apparently 1→6 linked (Fig. 15). Such an in vitro conversion of the glycosidic bond between N-acetylmuramic acid and N-acetylglucosamine into an internal 1→6 anhydro N-acetylmuramyl (hemiacetal) bond is an intramolecular transglycosylation reaction. Therefore this enzyme behaves as a peptidoglycan: peptidoglycan-6-muramyl transferase. A possible role in remodeling the peptidoglycan sacculus during the life cycle of the bacterium can be assigned to this enzyme. Indeed, rupture of a glycan strand could be followed by reattachment of the terminal anhydro N-acetylmuramic acid thus formed to the nonreducing terminal N-acetylglucosamine residue of another glycan fragment, thus conserving the energy of the original sensitive glycosidic bond. As noted by Höltje et al. (1975), however, the in vitro reaction products could be artifacts due to a misdirected intramolecular transfer reaction, whereas the enzyme could serve other fuunctions in vivo. Side reactions similar to that occurring in vitro might also occur to a minor extent in vivo, which would explain the small amounts of nonreducing peptidoglycan fragments that can be detected in lysozyme digests of *E. coli* peptidoglycan (Primosigh et al., 1961).

In view of the ubiquity and multiplicity of autolysins, it could be argued that the many terminal groups found in all peptidoglycans as they are isolated (section 2.1.4.) are artifacts caused by the uncontrolled action of these enzymes during isolation. If special care is not taken to prevent autolysin action, terminal groups can be "artificially" created, and at the limit no insoluble peptidoglycan or wall can be isolated at all. It is striking, however, that the sizes of both the glycan and peptide moieties do not vary to any great extent, among peptidoglycans isolated from cells of which the autolytic systems are markedly different.

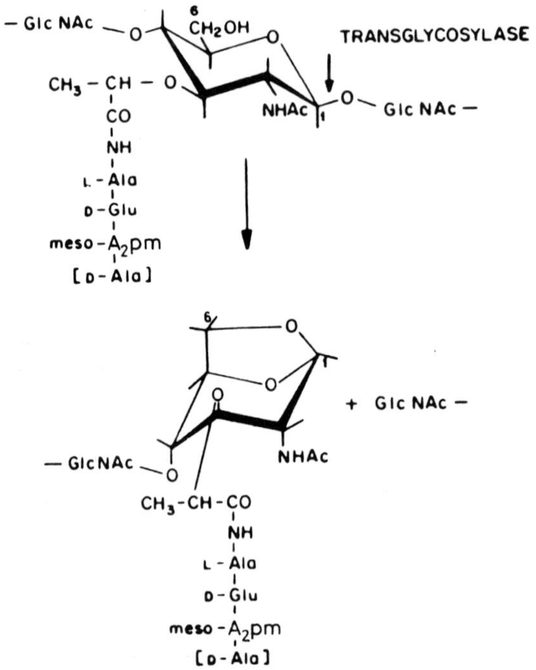

Fig. 15. Conversion of the muramic acid moiety from the $^4C_1$ into the $^1C_4$ conformation upon transglycosylase action by one of the autolysins of *Escherichia coli.* (Höltje et al., 1975.) (Reprinted courtesy of the American Society for Microbiology.)

Thus, the peptide moiety of the peptidoglycan of *L. acidophilus* is essentially a mixture of monomer, dimer, and trimer (Coyette and Ghuysen, 1970a) although the only autolysin actively detected in the cells is an endo-N-acetylmuramidase. Short glycan strands occur in the peptidoglycans of various bacilli (Rogers, 1974), while the main autolysin detected is an N-acetylmuramyl-L-alanine amidase (although hexosaminidase activity has also been detected; Brown and Young, 1970; Fan and Beckam, 1973). *E. coli* has a full complement of autolysins of various types, but its peptidoglycan is not very different in size from those of the same type isolated from bacilli. Finally, except for *M. lysodeikticus,* where a high proportion of the N-acetylmuramic acid is unsubstituted (Leyh-Bouille et al., 1966; Ghuysen, 1968), in all other bacteria the N-acetylmuramic acid is fully substituted (even in those where very active amidases are present in the membranes and walls). From these observations, it seems clear that many of the terminal groups in the peptidoglycan are not artifacts but rather reflect important properties of the peptidoglycan biosynthetic machinery and secondary modifications that occur following wall assembly.

The potentially dangerous autolysins probably play an important role in various cellular events such as remodeling the cell shape throughout the division cycle, cell separation, wall turnover (when it occurs), sporulation, the ability of cells to become competent for transformation, and finally, the excretion of toxins and exoenzymes. Actually, peptidoglycan hydrolases of any specificity could function effectively for all these processes. It has also been proposed that autolysins are actively involved in bacterial growth itself (Higgins and Shockman, 1971). The relationships of peptidoglycan hydrolases to surface growth and cell division are discussed in chapter 9 by Daneo-Moore and Shockman.

2.2. The bacterial membrane systems in gram-positive and gram-negative bacteria

2.2.1. Isolation

Enzymic degradation of the wall peptidoglycan normally causes cell lysis. However, lysis can be prevented for a limited period if the external medium contains a solute to which the cell is impermeable (sucrose) at a concentration that approximately balances the high osmotic pressure of the cell. Under these conditions, bacteria undergo transformation into osmotically fragile bodies. These bodies are either wall-less bacteria (i.e., protoplasts from gram-positives) or bacteria with some defect in their peptidoglycan component (i.e., spheroplasts from gram-negatives).

Plasma membranes of gram-positive bacteria are isolated by using any enzymes that selectively degrade the peptidoglycan and digest the whole wall structure. From the lysate thus obtained, plasma membranes free of insoluble wall residue and of cytoplasmic contaminants can be isolated. Alternatively, the procedure may include protoplast formation followed by osmotic shock.

The isolation of the mesosome intrusions is a more difficult task that has been successfully accomplished with a limited number of gram-positive bacteria (Salton, 1971). Essentially, the procedure is based on controlled and careful extrusion of the mesosome content during protoplasting of the cells, followed by careful differential centrifugation. Fig. 16 shows how the mesosome pocket opens and liberates its membranous content when a gram-positive bacterium is transformed into protoplast in a hypertonic medium (which causes plasmolysis) (Ryter, 1974).

Various methods have been proposed for the physical separation of the outer membrane of the gram-negative bacteria from the plasma membrane and for the selective isolation of both structures (Miura and Mizushima, 1968a,b; Schnaitman, 1970a,b; Osborn et al., 1972a,b). Most methods include the preparation of the total (inner plus outer) membrane fraction through the lysis of spheroplasts obtained by the action of lysozyme and EDTA, although this technique may be damaging to the outer membrane. Extensive loss of its lipopolysaccharide molecules may occur, probably due to the use of EDTA. Recently, Osborn's group showed that loss of lipopolysaccharide from the outer membrane of *Salmonella typhimurium* can be prevented by using an EDTA concentration just sufficient to prevent the aggregation of membranes (Osborn et al., 1972a,b). The two membranes can then be separated from each other by isopyc-

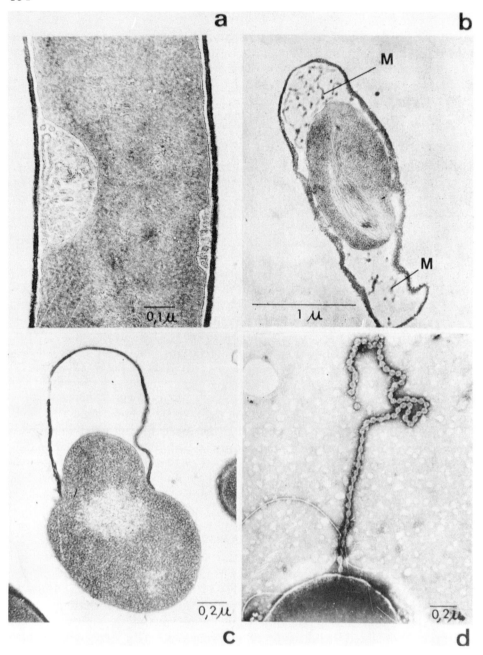

Fig. 16. Protoplast formation and mesosome extrusion (Ryter, 1974). (a) Beginning of plasmolysis. The mesosomic pocket opens. (b) End of plasmolysis. The mesosomic tubules (M) are at the poles of the bacterium between the wall and the plasma membrane. (c) Emerging of the protoplast. (d) Mesosomic tubule still attached to the protoplast. (Reprinted courtesy of Ediscience/McGraw-Hill. Paris.)

nic sucrose density gradient centrifugation of the total membrane fraction. The outer membrane bands at a buoyant density of about 1.22 g/cc, whereas the plasma membrane bands at densities ranging from 1.14 to 1.12 g/cc. The same technique, slightly modified if necessary, can be applied to other gram-negative bacteria. Fig. 17 shows the negatively stained outer and plasma membranes isolated from *E. coli* K12 (Pollock et al., 1974).

Finally, mention should be made of the antibiotic EM49, an octapeptide containing a C:10 or C:11 β-hydroxy fatty acid with no threonine residues, which has the property shared by no other antibiotic, to disrupt the *E. coli* outer-membrane structure and to release from the cells membrane fragments having a density identical to that of the isolated outer membrane (Rosenthal et al., 1976).

2.2.2. *Functions*
The physiological functions fulfilled by the plasma-mesosome membranes involving enzyme activities are numerous and important (Salton, 1971). Membranous organelles such as nuclear membrane, mitochondria, Golgi, and endoplasmic reticulum do not exist in bacteria. The functions associated with these organelles in the eukaryotic cells are packaged in the bacterial multifunctional plasma-mesosome membrane system. At present, a distribution of specific enzymes between plasma membranes and mesosomes has not been established except for the adenosinetriphosphatase (ATPase) connected with oxidative phosphorylation and for the mannosyl-1-phosphorylundecaprenol synthetase involved in mannan synthesis (section 3.2.), which were shown to occur only in the plasma membrane and to be absent from the mesosomes. Concerning all other enzyme activities, including those involved in the biosynthesis of the exocellular cell envelope structures, differences in activity levels of certain enzymes between plasma membrane and mesosomes were reported. The significance of these data, however, has not yet been assessed.

In contrast to the plasma membrane, very few physiological functions involving enzyme activities seem to be performed by the outer membrane of the gram-negative bacteria (Salton, 1971). Enzyme activities associated with terminal electron transport and with the active transport of solutes, which are classic plasma membrane functions, are entirely lacking in the outer membrane. However, phospholipase activities tentatively identified as a mixture of phospholipase A and lysophospholipase, were found associated primarily with the outer membrane of *S. typhimurium* (Osborn et al., 1972a,b) but the role they may play is unclear. UDP-sugar hydrolase, ribonuclease I and endonuclease I activities were found in both plasma and outer membranes (Osborn et al., 1972a,b). Finally, some peptidoglycan hydrolases of unidentified specificity also appear to be confined to the outer membrane (Hakenbeck et al., 1974). Their role remains obscure.

The outer membrane, essentially a phospholipid-lipopolysaccharide-protein structure stabilized by Mg^{2+} cations, contributes to the mechanical stability of the gram-negative bacteria cell envelope and provides them with an additional permeability barrier (Leive, 1974). This barrier function is, at least in part,

486

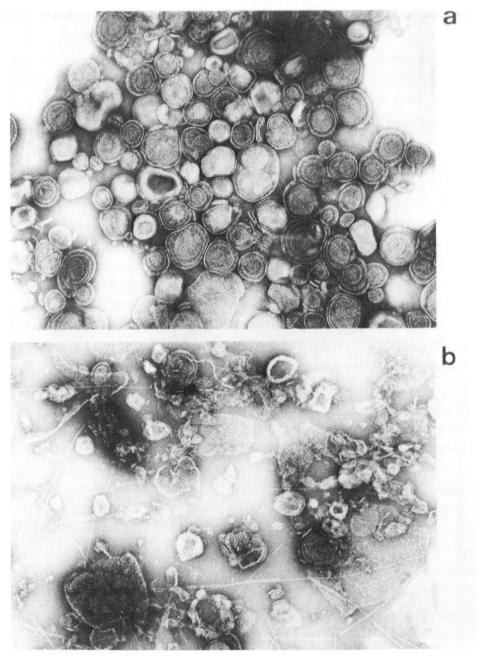

Fig. 17. (Top) Electron micrograph of the outer membrane of *E. coli* K12, mutant strain 44. Note the spherical shape of the membranes containing concentric structures typical of isolated lipopolysaccharide. Bar, 0.5 μm. (Bottom) Electron micrograph of the inner membrane of *E. coli* K 12 mutant strain 44. The plasma membrane fragments are covered with small particles. Pili also occur in the preparation. Bar, 0.5 μm. (Pollock et al., 1974). (Reprinted courtesy of the Federation of European Biochemical Societies.)

related to the lipopolysaccharide component. Treatment with EDTA (Leive, 1965a,b; Voll and Leive, 1970) and mutational alterations of the lipopolysaccharide structure (Tamaki et al., 1971; White et al., 1971) cause a marked increased permeability of the outer membrane to antibiotics, detergents, and other agents. As stressed by Haller et al. (1975), the term membrane should be used as a morphological designation, since it is only meant to describe the unit membrane profile revealed by electron microscopy.

2.2.3. General composition of the outer membrane
of the gram-negative bacteria

In *S. typhimurium, E. coli,* and other gram-negative bacteria (Osborn et al., 1972a,b) the phospholipid composition of the outer membrane is qualitatively similar to that of the plasma membrane: phosphatidylethanolamine and phosphatidylglycerol are the major constituents and small amounts of cardiolipin are also present. Quantitatively, the composition of the two membranes differs markedly. The ratio of phosphatidylglycerol to phosphatidylethanolamine and that of cardiolipin to phosphatidylethanolamine in the outer membrane are reduced to about 50% and 25%, respectively, of those observed in the plasma membrane. Moreover, the outer membrane has an appreciably low ratio of total phospholipid to protein as compared to those of the plasma membrane and other biological membranes. The outer membrane is distinguished by a highly characteristic pattern of protein bands in SDS-polyacrylamide gel elctrophoresis (Schnaitman, 1970a,b; 1971a,b) and by the occurrence in it of several unique components (lipoprotein and lipopolysaccharide; see sections 2.5. and 2.7.). One could assume that the phospholipids occur primarily in a bilayer structure and that the proteins interact with both the polar headgroups and the internal hydrocarbon chains of the lipid layer. This classic view, however, does not explain the many unusual features and properties of the outer membrane. These properties are such that the fluid mosaic model of Singer and Nicolson (1972), in essence a fluid lipid matrix in which the proteins are randomly disposed without forming a long-range ordered system, does not apply to the outer membrane (see section 2.8.).

2.3. Wall and membrane teichoic acids
in gram-positive bacteria

The term teichoic acid was first introduced by Baddiley to designate polymers of either ribitol phosphate or glycerol phosphate in which the repeating units were joined together through phosphodiester linkages (Armstrong et al., 1958). Typical ribitol and glycerol teichoic acids are shown in Fig. 18. This term is intended to include all polymers containing glycerol phosphate or ribitol phosphate residues, associated with the membrane, the wall, or the capsule (Baddiley, 1972). Teichoic acids have been found to occur in gram-positive bacteria and are apparently absent from gram-negative bacteria. They are very effective antigens and have often been identified as group- or type-specific substances.

Wall teichoic acids are either glycerol or ribitol phosphate polymers. They are covalently linked to the peptidoglycan. Membrane-associated teichoic acids are always glycerol phosphate polymers. They are covalently linked to a glycolipid that is part of the plasma membrane, and thus they are also called lipoteichoic acids. Both wall and membrane teichoic acids often occur in the same organism.

2.3.1. Isolation (Coley et al., 1975a)
Wall teichoic acids are usually extracted from purified walls. As they are isolated, however, walls may contain substantial amounts of membrane fragments and lipoteichoic acid. Thus special attention must be paid to the removal of membrane contamination. This can be achieved by extraction with 40 to 80% aqueous phenol. Wall teichoic acid can then be extracted by cold dilute trichloracetic acid, followed by precipitation with organic solvents. Extraction periods of 1 to 2 days or more at 4°C are necessary. In this procedure solubilization is caused by hydrolysis of some of the phosphodiester linkages. Obviously, the compounds thus obtained are of no value for the study of chain length. "Native" teichoic acids can be isolated by selective enzymic degradation of the wall peptidoglycan (section 2.3.2.3.).

Lipoteichoic acids are usually isolated by extraction of whole cells with 40 to 80% aqueous phenol, followed by chromatography on Sepharose 6B. Attention should be paid to the fact that the glycolipid moiety of the lipoteichoic acid is in hydrophobic interaction with the lipids of the plasma membrane. Substantial loss of lipoteichoic acid from the membrane can result from washing with water (Shockman and Slade, 1964). Mg^{2+} ions play a role in maintaining the association. Protoplasts made in the absence of Mg^{2+} ions lack lipoteichoic acid (van Driel et al., 1973). Finally, release of lipoteichoic acids from the membrane through the walls has been observed in several species (Joseph and Shockman, 1975; Markham et al., 1975).

2.3.2. Wall teichoic acids
Wall teichoic acids may exhibit the full range of glycerol phosphate-and ribitol phosphate-containing structures.

2.3.2.1. Structure.
One of the first ribitol teichoic acids that was submitted to thorough structural investigation was found in the walls of a strain of *Staphylococcus aureus* (Baddiley et al., 1962a,b; Ghuysen et al., 1964). Its basic structure is shown in Fig. 18. In all strains examined the phosphodiester linkages are between positions 1 and 5 on ribitol, and N-acetylglucosaminyl residue is present at the D-4 position of each ribitol residue. The glycosidic linkages may be α or β, but both types of linkage are present in the majority of strains. Most of the ribitol residues have a D-alanine ester residue that occupies position D-2. Similar wall ribitol teichoic acids occur in many bacilli and lactobacilli. The sugar substitutes are β-glucopyranosyl in a strain of *B. subtilis* (Armstrong et al., 1961) and are α-glucopyranosyl in *Lactobacillus arabinosis* 17-5 (Archibald et al., 1961).

More complex wall ribitol teichoic acids occur in other species. For example, in *Diplococcus pneumoniae* strain R36A, the wall teichoic acid is not completely characterized but is known to contain choline in addition to glucose, ribitol,

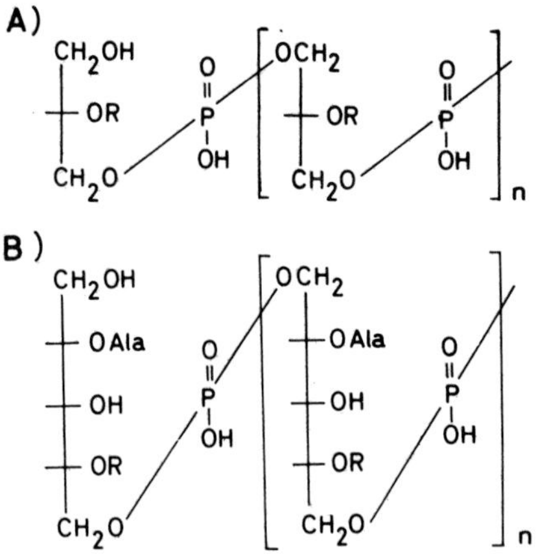

Fig. 18. (a) Typical glycerol teichoic acid (R = H, glycosyl or D-alanyl). (b) Typical ribitol teichoic acid (R = glycosyl; Ala-D-alanyl).

phosphorus, galactosamine, and 2,4,6-trideoxy-2, 4-diaminohexose. This latter teichoic acid has been identified as the serologically somatic antigen known as "C-substance" (Brundish and Baddiley, 1968; Mosser and Tomasz, 1970; Watson and Baddiley, 1974). Wall teichoic acids in which the sugar residues form a part of the main polymer chain are also known. Polymers of glucosylglycerol phosphate (Fig. 19) and galactosylglycerol phosphate occur together as separate molecules in the walls of *Bacillus licheniformis* ATCC 9945 (Burger and Glaser, 1966). Similarly, walls of *Lactobacillus plantarum* N1RD C106 contain a mixture of one polymer of glucosylglycerol phosphate and two polymers of isomeric diglucosylglycerol phosphates (Fig. 20) (Adams et al., 1969). Walls of various bacteria possess teichoic acids which contain sugar-1-phosphate residues. The first example known was *Staphylococcus lactis* 13; glycerolphosphate is attached to the hydroxyl at C-4 on N-acetylglucosamine-1-phosphate, whereas D-alanine is fixed on the hydroxyl at C-6 (Fig. 21) (Archibald et al., 1971). Finally, in *Lactobacillus acidophilus,* the teichoic acid is probably a mixture of (α or β)-1,6-linked polyglucose polymers with monomeric α-glycerol phosphate side chains attached to them on the C_2 or C_4 position (Coyette and Ghuysen, 1970a).

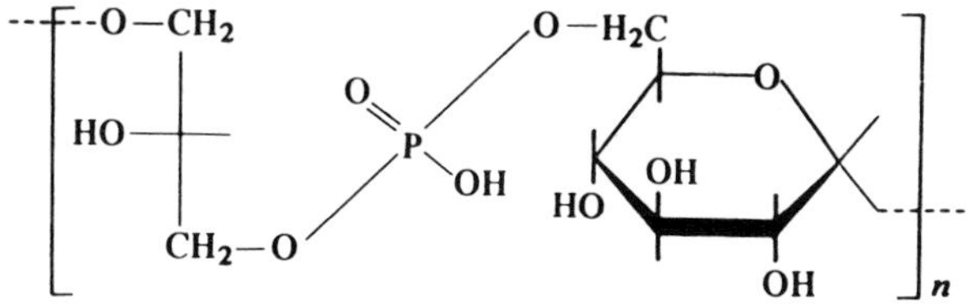

Fig. 19. Glucosylglycerol phosphate teichoic acid from *Bacillus licheniformis* ATCC 9945. (Burger and Glaser, 1966).

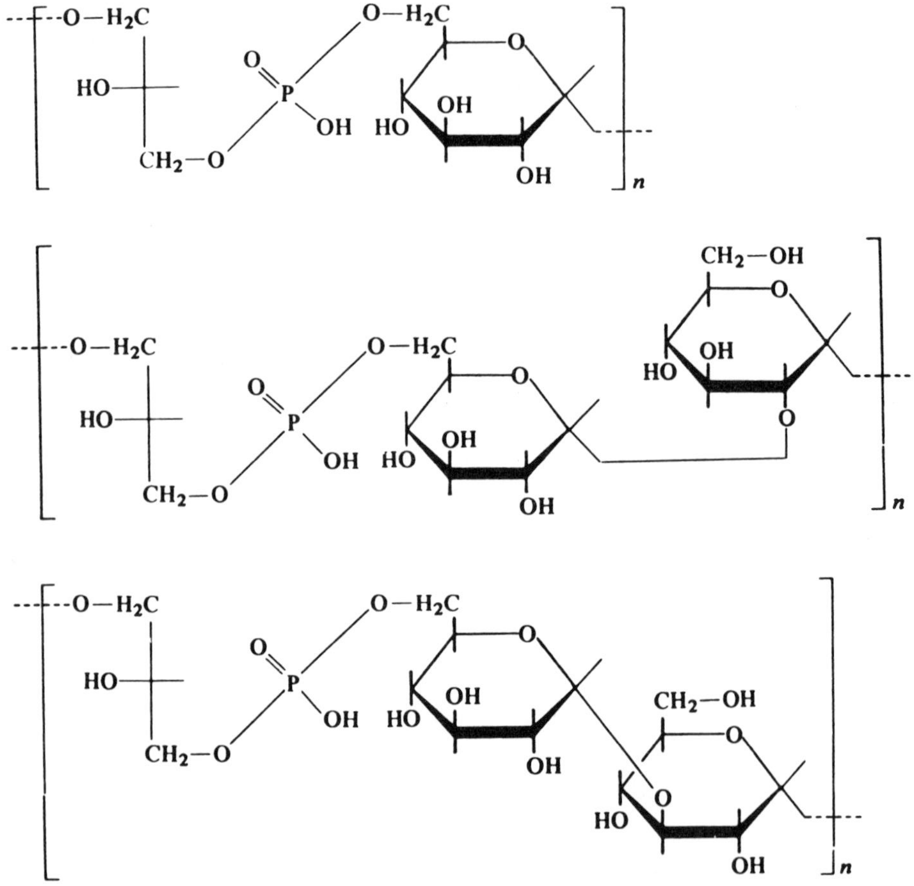

490

Fig. 20. Mixture of teichoic acids in walls of *Lactobacillus plantarum* NIRD C106. (Adams et al., 1969.)

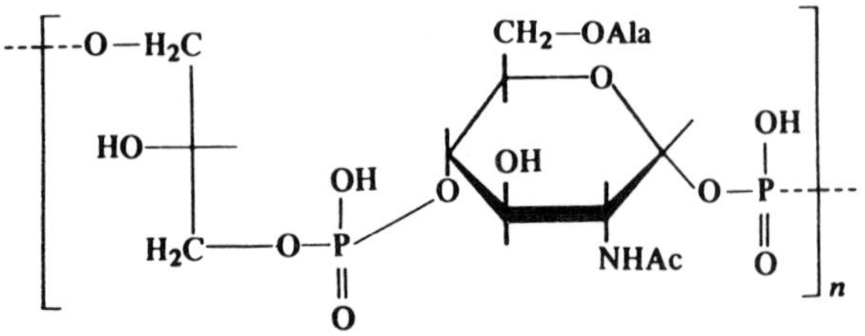

Fig. 21. Teichoic acid containing sugar 1-phosphate linkages from the walls of *Staphylococcus lactis* 13. (Archibald et al., 1971.)

2.3.2.2. Occurrence. The few examples given above illustrate the extreme variability in structure exhibited by wall teichoic acids. Other anionic polymers that lack polyol phosphate, and thus are not strictly teichoic acids, may also occur in the walls, such as polymers containing N-acetylglucosamine 1-phosphate (Archibald and Stafford, 1972) and the so-called teichuronic acids that are acidic polymers containing hexuronic acids. In the walls of *B. subtilis*, the teichuronic acid is a polymer of N-acetylgalactosamine and glucuronic acid (Janczura et al, 1961), and that of *M. lysodeikticus* is a polymer of glucose and 2-acetamido-2-deoxymannuronic acid (Perkins, 1963). An almost endless variety of acidic polysaccharides and teichoic acids constitute the nonpeptidoglycan portion of virtually all the walls of gram-positive bacteria. Growth conditions (especially phosphate and magnesium limitations) deeply influence the composition of the wall in altering the relative proportion of its teichoic acids and anionic polysaccharides (Ellwood and Tempest, 1969). This mobility occurs despite the fact that both of these polymers are covalently linked to the peptidoglycan, and thus implies the presence of an active system for turnover.

2.3.2.3. Linkage with peptidoglycan. Little information is available to date on the exact nature of the bond by which the teichoic acids and acidic polysaccharides are fixed to the walls. Clearly, however, teichoic acids are covalently attached to the glycan moiety of the peptidoglycan. This was proved for the first time by selectively degrading the peptidoglycan of the walls of *S. aureus* Copenhagen with the help of specific enzymes and by isolating the ribitol teichoic acid (Ghuysen et al., 1964). The polymer thus purified had short peptidoglycan fragments still attached to it. By using this technique, an average chain length of about 30 to 40 ribitol phosphate units was estimated.

The fact that muramic acid 6-phosphate occurs in various walls containing teichoic acids (Liu and Gotschlich, 1963; Munoz et al., 1967) strongly suggests that the attachment between wall teichoic acid and peptidoglycan involves a bond between a terminal phosphate group of teichoic acid and C-6 of muramic acid. Good evidence has also been obtained that wall polysaccharides are attached to peptidoglycan through a sugar-1-phosphate linkage between the reducing terminal sugar of the polysaccharide and a N-acetylmuramic acid residue in the peptidoglycan (Knox and Hall, 1965; Knox and Holwood, 1968; Hughes, 1970). So far, however, direct evidence for the nature of the linkage between wall teichoic acid and peptidoglycan has been obtained in only two cases. In *Staphylococcus lactis* 13, the wall glycerol teichoic acid has its terminal glycerol phosphate residue fixed to muramic acid (Button et al., 1966) (Fig. 22). Nearly 40% of the glycan of the peptidoglycan has teichoic acid attached to it, and in that fraction of the wall that contains the two polymers linked to each other each glycan chain (9 disaccharide units) is substituted by only one teichoic acid chain (24 repeating units) (Baddiley, 1972). In the bacteriophage resistant mutant of *S. aureus* H, the wall ribitol teichoic acid (40 repeating units) is also fixed to muramic acid, but in this case the link between the two polymers is mediated via a short oligomer containing 3 or 4 glycerol phosphate residues (Fig. 22) (Heckels

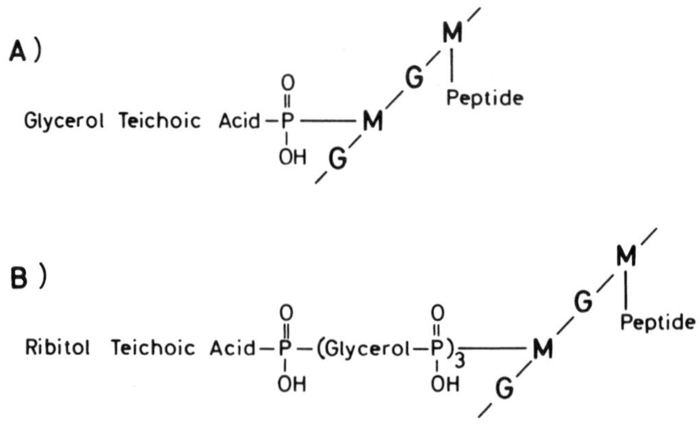

Fig. 22. Wall teichoic acids in *S. lactis* 13 (a) and in *S. aureus* H (b). The attachment to the peptidoglycan is probably on C_6 of N-acetylmuramic acid. G, N-acetylglucosamine; M, N-acetylmuramic acid.

et al., 1975; Coley et al., 1975b, 1976; Hancock and Baddiley, 1976) and perhaps N-acetyl-D-glucosamine (Bracha and Glaser, 1976). It is not known whether this type of link occurs with all ribitol teichoic acids. It is striking, however, that even after trypsin treatment and aqueous phenol extraction small amounts of glycerol phosphate are found in the walls of various bacteria that contain ribitol teichoic acids (Archibald and Stafford, 1972). This observation suggests that an intervening short oligomer of 3 to 4 glycerol phosphate units between peptidoglycan and ribitol teichoic acid may usually occur. Such a structural feature would be consistent with the finding that, at least in some bacteria, the glycerol lipoteichoic acid carrier is involved in the biosynthesis of the wall ribitol teichoic acid (Fiedler and Glaser, 1974; also see section 3.2.).

2.3.2.4. Arrangement. Peptidoglycan, together with the teichoic acids and the acidic polysaccharides covalently attached to it, should be considered as components of the same macromolecule and not as separate polymers (Rogers, 1974). The question arises, however, whether the attachment sites for the nonpeptidoglycan polymers are distributed randomly or nonrandomly within the wall. A nonrandom distribution could give rise to distinct wall areas containing unsubstituted and substituted peptidoglycan, respectively. This problem has not been solved. Clearly at least a portion of the teichoic acid must be localized close enough to the surface for reactions to occur between whole organisms and teichoic acid—specific antibodies (Burger, 1966; Knox and Wicken, 1971, 1973), plant lectins (Birdsell et al., 1975) and bacteriophages (Young, 1967; Coyette and Ghuysen, 1968; Chatterjee, 1969; Doyle et al., 1973; Archibald, 1976). For example, concanavalin A, which interacts specifically and reversibly with the polyglucosyl glycerolphosphate teichoic acid of *B. subtilis* 168 walls, has been used as a probe to study the organization of the surface teichoic acid in this organism (Birdsell et al., 1975). Treatment of whole cells and walls with Con A caused the appearance of a discontinuous, irregular fluffy layer (25–60 nm thick) on the outer profile of the wall. On this basis, the portion of teichoic acid exposed at the cell surface was proposed to be oriented perpendicularly to the long axis of the cell.

2.3.3. Membrane lipoteichoic acids

Membrane lipoteichoic acids are almost always of the classical glycerolphosphate polymer type. They are almost universally present in the gram-positive bacteria, and this presence does not depend on growth conditions as do wall-associated teichoic acids.

2.3.3.1. Structure. Lipoteichoic acids are amphipathic molecules, each having a long, polar glycerolphosphate chain linked to a small hydrophobic lipid portion. The polar portion consists of 1→3 phosphodiester-linked chains of 25 to 35 glycerolphosphate residues variously substituted with glycosyl and D-alanine ester groups (Fig. 18). The lipid moiety is a glycolipid that is probably identical to the free glycolipid of the plasma membrane (Button and Hemmings, 1976). In *S. aureus* H, the lipoteichoic acid (Fig. 23) contains one molecule of gentiobiosyl-glycerol for every 30 glycerol phosphate units. The polyglycerolphosphate chain is linked at its phosphate terminal end to the hydroxyl group at position 6 of the terminal glucose moiety of diacylgentiobiosylglycerol (Duckworth et al., 1975). In *S. faecalis*, this glycolipid has been characterized as a phosphatidylkojibiosyl diglyceride (Toon et al., 1972; Ganfield and Pieringer, 1975). The two portions of the lipoteichoic acids are linked by a phosphodiester bond that involves a sugar hydroxyl group of the glycolipid and the terminal glycerolphosphate residue of the teichoic acid chain. The partial structures of the lipoteichoic acids in other *Streptococci* and in various *Lactobacilli* (Knox and Wicken, 1973; Wicken and Knox, 1975) are also known. Deacylation removes the fatty acyl and D-alanine ester groups and yields glycerolphosphate chains still attached to the glycerol glycoside of the original polymer. The phosphodiester link between the glycerolphosphate chain and the glycolipid is acid-labile. Treatment with cold trichloracetic acid yields teichoic acid devoid of glycolipid.

Not all gram-positive bacteria possess conventional lipoteichoic acids. In *M. lysodeikticus*, *sodonensis*, and *flavus*, the presence of lipoteichoic acids has been

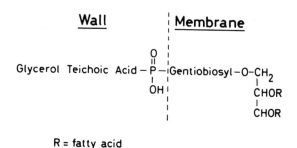

Fig. 23. Lipoteichoic acid in *Staphylococcus aureus* H.

precluded, but there are membrane-associated lipomannans that have the same properties as lipoteichoic acids (Powell et al., 1974, 1975; Schmit et al., 1974; Owen and Salton, 1975a,b,c). Lipomannans have mannose as the only sugar component and also contain glycerol, ester-linked fatty acids, and succinic acid substituents. In *M. Lysodeikticus,* the mannan has a chain of about 60 hexose units linked at its reducing end to a glycerol residue which itself bears 2 fatty acid esters (Powell et al., 1975). This hydrophobic terminal is similar to the linkage of the polyglycerolphosphate chain to lipid in conventional lipoteichoic acids and ensures the anchorage of the mannan to the plasma membrane. Lipomannans occur in both plasma and mesosomal membranes (Owen and Salton, 1975c).

D. pneumoniae is another exception. In this organism, the so-called pneumococcal Forssman antigen can probably be regarded as a lipoteichoic acidlike component. It is localized in the plasma membrane and contains bound lipids and choline (Goebel et al., 1943; Brundish and Baddiley, 1968; Briles and Tomasz, 1973; Höltje and Tomasz, 1975a,b).

2.3.3.2. Lipoteichoic acids as surface components. Despite their inner cellular localization, reaction of whole organisms with antibodies specific for lipoteichoic acids has provided evidence for the exposure of lipoteichoic acids at the cell surface (Wicken and Knox, 1975). Thin sections of *Lactobacillus plantarum* treated with rabbit IgG specific for the glycerolphosphate sequence of the lipoteichoic acid and then with ferritin-labeled goat antibodies to rabbit IgG showed the label extending from the outer surface of the plasma membrane throughout the wall and, in some cases, outside the outer boundary of the cell. Mesosomes were also labeled by the indirect ferritin antibody technique, indicating that both types of membrane carry lipoteichoic acid. However, variations may occur. Thus *L. fermenti,* but not *L. casei,* was agglutinated by antisera to lipoteichoic acid. By use of the same ferritin technique as described above, *L. casei* showed some surface adsorption of the antibody, but this was irregular and less significant than the confluent labeling seen with *L. fermenti* (Knox and Wicken, 1973; van Driel et al., 1973). On the basis of these observations and others, it has been proposed that the long, polar glycerolphosphate chains of lipoteichoic acids probably extend through the network of the wall until, in at least some cases, they come near enough to the outer surface to act as surface antigens. Surprisingly, extracellular lipoteichoic acids have been found in cultures of *S. faecalis* 9790, *Streptococcus mutans* FA-1 and in a variety of other bacterial species (Joseph and Shockman, 1975; Markham et al., 1975). They occur in both the deacylated and micellar (and presumably acylated) forms. Extracellular lipoteichoic acids do not seem to be products of cellular lysis, and the role of excretion of large amounts of such energy-rich polymers remains to be determined. Recent observations, however, suggest they may play a role in regulating the activity of the cell autolysins (section 3.1.6.).

2.3.4. Functions of wall and membrane teichoic acids
Because of the widespread distribution of wall and membrane teichoic acids in gram-positive bacteria, it has long been assumed that these polymers confer

advantages on these bacteria, or that they fulfill one or several important functions.

When *B. subtilis* (and other bacteria) are grown under conditions of phosphate limitation, teichoic acid is no longer present in the wall. The preexisting teichoic acid is removed by wall turnover and is replaced by teichuronic acid (Ellwood and Tempest, 1969). Conversely, when *B. subtilis* is grown under various limitations in the presence of excess phosphorus, the walls always contain teichoic acid but no teichuronic acid. The amount of teichoic acid formed, however, varies depending on the culture limitations. Among the limitations tested, the one that yields the largest amount of wall teichoic acid is magnesium starvation (Ellwood, 1975). The substitutions between teichoic and teichuronic acid are phenotypic responses. They occur rapidly as pointed out by Rogers (1970), "providing either one or another of the groups of negatively charged polymers is present on its surface, the microorganism seems content". Significantly, even under conditions of phosphate limitation when no wall teichoic acid is formed, the membrane lipoteichoic acid is still produced. Thus it seems that cell must possess membrane lipoteichoic acid but that "all that is required in the wall is one or more acidic polymers" (Rogers, 1970).

The most gram-negative bacteria do not possess teichoic acids. Glycerol teichoic acids, however, have been isolated from some strains of *Butyrivibrio fibrisolvens* (Sharpe et al., 1975). Moreover, gram-negative bacteria have in their outer membrane a related class of macromolecules, the lipopolysaccharides, which include in their structures, phosphodiester groups, sugar residues, and basic centers (section 2.7.). These may serve a purpose similar to that of teichoic acids. This idea is further supported by findings that phosphate esters may be constituents of the O-side chain of the lipopolysaccharide in some gram-negative bacteria. Glycerolphosphate was found in one strain of *E. coli* (Jann et al., 1970) and ribitol phosphate occurs in *Proteus mirabilis* (Gmeiner, 1975, 1977; Gmeiner and Martin, 1976) as a side branch of the sugar polymer. Moreover, uronic acids are also frequently found in lipopolysaccharides of *P. mirabilis* as if this organism could satisfy its need for anionic groups in the O-side chain either by phosphodiester groups or by uronic acids. This situation is reminiscent of that in gram-positive bacteria, where a reversible shift from teichoic acid to teichuronic acid, and vice versa, occurs depending on growth conditions.

2.3.4.1. Magnesium uptake.   Teichoic acids and other negatively charged polymers bind divalent cations. Similarly, the succinic acid substituents on lipomannan determine its charged properties and also enable this polymer to bind divalent cations efficiently. Baddiley and his colleagues have suggested that the two regions of anionic polymers (i.e., the wall and the plasma membrane, respectively) would function as an integrated cation-exchange system between the exterior of the cell and the plasma membrane, ensuring to the latter the high concentration of Mg^{2+} required for stability and many enzymic functions (Heptinstall et al., 1970; Hughes et al., 1971; Baddiley, 1972). This view is not only consistent with the cation-binding properties of these polymers but, as stressed by Baddiley (1972), would also explain their structural diversities since a variety of phosphate, hydroxyl, and amino groups could effectively regulate cation binding. Cells could control this activity in part by controlling the amount of ester-bond alanine in the teichoic acids.

The interaction between Mg^{2+} ions and wall teichoic acids has been studied by equilibrium dialysis (Lambert et al., 1975). On the basis of binding data, the apparent association constants and the number of Mg^{2+} binding sites on the teichoic acid molecules were shown to vary depending on the pH, ionic strength, and the presence of other divalent cations. Thus in 10 mM NaCl aqueous solution at pH 5.0, one Mg^{2+} ion is bound for every two phosphate groups in the wall teichoic acid of *Lactobacillus bulchneri* with a $K_{assoc.}$ apparent value of 2.7 $\times 10^3$ M^{-1}. Moreover, the number of Mg^{2+} binding sites was shown to be reduced by the presence of alanine ester substituents. These properties and apparent association constant values presumably reflect an optimum balance between two antagonistic requirements. The first requirement is that wall teichoic acid must possess sufficient affinity and selectivity for Mg^{2+} ions to be able to scavenge them from an excess of competing ions. The second is that Mg^{2+} ions must not be too tightly bound in order to be transferred to the plasma membrane. On the basis of equilibrium dialysis experiments, the Mg^{2+} binding properties of walls of *B. subtilis* W13 containing teichoic acid were compared with those of walls of the same organism containing teichuronic acid (Heckels et al., 1977). Both walls had similar properties and their affinity was greater than that displayed by the isolated polymers in solution.

The relative affinities of various cations for anionic sites in isolated walls were also assessed by a technique involving displacement of one cation by another (Marquis, R.E., personal communication). The affinity series was $H^+ \gg La^{3+} \gg Sr^{2+} > Ca^{2+} > Mg^{2+} \gg K^+ > Na^+ > Li^+$. The total amounts of magnesium that could be displaced with Na^+ or H^+ from the magnesium forms of isolated walls varied from $73\mu mole/g$, dry weight, for walls of the teichoic acid deficient *S. aureus* 52A5 to about 520 $\mu mole/g$ for walls of *B. megaterium* KM. These studies have also shown that interaction of the walls with cations is a complex phenomenon. Thus, for cells grown in usual laboratory media that often contain an excess of monovalent versus divalent cations, there is a mix of small cationic counterions in the wall and monovalent cations may predominate even if the wall has a higher affinity for diavalent cations.

Mutants of bacilli have been used to show that the presence of the proper amounts of negatively charged polymers attached to peptidoglycan in gram-positive bacteria is important (Rogers, 1975, 1976; Rogers et al., 1971, 1974, 1976). Three classes of mutants of *B. subtilis* that have phenotypes with reduced amounts of these polymers in their walls are of particular interest since they do not grow as rods but as deformed cocci. These mutants are the following: (1) the *Rod* A mutant, when grown at 45°C, has a wall containing only about 20% of the amount of teichoic acid present when grown as rods at 30°C (Rogers et al., 1974); (2) the phosphoglucomutase negative mutants, when grown under phosphate limitation, have walls consisting almost entirely of peptidoglycan (Forsberg et al., 1973); and (3) the CDP-glycerol pyrophosphorylase mutants also have markedly reduced amounts of teichoic acids in the walls (Rogers, 1975, 1976). Surprisingly, at first sight, *Rod* B mutants of *B. subtilis* under restrictive conditions have a deformed coccal morphology but virtually normal amounts of wall teichoic acid (Rogers et al., 1971). However, these mutants are grossly deficient in their capability to use Mg^{2+} ions, probably due to a deficient transport system. Thus, it seems likely that the *Rod* type mutants as a group are disturbed in the supply and/or movement of the Mg^{2+} ions, this disturbance being related either to the absence of wall teichoic acids in proper amounts or to a deficient transport system for this cation.

Manipulations of the Mg^{2+} concentration and of the nature and concentration of halogen anions can alter the morphology of at least some of these mutants (Rogers, 1975, 1976). Thus if 10 mM, instead of 1 mM Mg^{2+}, is added to the growth medium the CDP-glycerol pyrophosphorylase mutants change from deformed cocci to rods. Rapid growth of *Rod* B mutant as deformed cocci can be obtained with 10mM Mg^{2+} and 20 mM Cl^-. But if Cl^- is substituted by 15 mM Br^- or 5 − 10 mM I^-, the morphology is that of rods. Moreover, by using a constant concentration of the latter two halogen ions, the morphology can be precisely controlled by the Mg^{2+} concentration. Thus, in some bacteria at least, Mg^{2+} uptake seems to be an important factor related to cell morphology. A further discussion of the effects of changes in the anionic polymer content of walls can be found in the chapter by Daneo-Moore and Shockman in this volume.

2.3.4.2. Other functions. In addition to the suggested role of the negatively charged polymers in wall and membrane as the suppliers of Mg^{2+} ions to bacterial cells, other important functions are fulfilled by both wall teichoic acids and lipoteichoic acids that are related to the autolytic activity of the cells. These functions are discussed in section 3.1.6. Moreover, lipoteichoic acids were shown to serve, at least in some bacteria, as assembly sites for the wall teichoic acids (and perhaps for the assembly of the peptidoglycan). This role is discussed in section 3.2.2.5.).

2.4. Wall proteins in gram-positive bacteria

Walls of most gram-positive bacteria contain some protein. Little is known about their association with the other wall polymers. In some cases, protein and peptidoglycan are covalently bound to one another. Thus, protein A in the walls of *Staphylococcus aureus* is linked to an amino group of the peptide moiety of the peptidoglycan (Sjöquist et al., 1972; Movitz, 1976; Sjödahl, 1977; Lindmark et al., 1977). In other cases, proteins cover the outer surface of the cell from which they can be removed by trypsin. Thus the M protein antigen of group A streptococci is labile to proteolytic enzymes and can be digested without affecting the viability of the cells or the insolubility and electron microscopic appearance of isolated walls (Fox, 1974). Regular surface patterns due to outer protein-containing layers have also been described; a well-characterized one is the so-called T-layer from *Bacillus brevis* (Brinton et al., 1969; Aebi et al., 1973; Howard and Tipper, 1973). T designates the tetragonal symmetry of the surface layer. It is located on the exterior surface of the cell and consists of a single polypeptide chain (molecular weight: 140,000) and is involved in phage binding. It is not linked covalently to the peptidoglycan since it can be released with 2 M guanidine hydrochloride and is cleaved by proteolytic enzymes. The purified protein can reassociate from a subunit state to give rise to planar T-layer sheets. Pronase-treated protein (molecular weight: 125,000) can also reassociate to give rise to hollow cylinders with the diameter of the cells. The ultrastructure of planar and cylindrical reassociates has been studied. *Bacillus polymixa* also possesses a patterned surface layer, containing mainly protein and carbohydrates, which is noncovalently associated with the underlying peptidoglycan structure (Baddiley, 1964; Goundry et al., 1967).

The wall of *Bacillus subtilis* 168, strain Cbl-1, exhibits several unusual features (Leduc et al., 1973). The most striking is the presence of a high molecular weight protein, instead of wall teichoic acid, that accounts for over 50% of the wall material. To our knowledge, this strain appears to be the only one known that contains such a "novel" wall protein. This protein is not covalently linked to the peptidoglycan and can be readily removed by treating the walls with 6M LiCl. It consists of two noncovalently associated chains. Treatment of walls with SDS yields a peptidoglycan that exhibits the classical rod-shaped morphology. Very thin homogeneous ghosts are seen that are reminiscent of the rigid layer of *E. coli*. Like this latter structure, the isolated peptidoglycan layer of the *B. subtilis* Cbl-1 is especially fragile; it can be dissolved with 6M guanidine hydrochloride (Leduc and Van Heijenoort, 1975). The isolated peptidoglycan fraction contains galactosamine and an excess of glucosamine, suggesting that a polysaccharide containing these two hexosamines is bound to the peptidoglycan. The mutant fails to grow on a minimum glucose medium and exhibits a long lag before growth on complex media. The occurrence of lipoteichoic acid in the mutant was not investigated. The role played by the novel wall protein in the physiology of the Cbl-1 strain is unknown. Whether or not this protein may serve some of the same functions as the wall teichoic acids or acidic polysaccharides remains unresolved.

2.5. The wall lipoprotein of the gram-negative enteric bacteria

2.5.1. Localization

The peptidoglycan in *E. coli* (and in other enteric bacteria) has covalently attached lipoprotein, a situation reminiscent of the peptidoglycan-anionic polymer complexes found in the walls of gram-positive bacteria. Lipoprotein, however, occurs not only in covalent linkage with the peptidoglycan but also in one free form. Both covalently bound and free lipoproteins are part of the outer membrane, where they represent one of the major proteins. In *E. coli*, the free form is present in twice the amount of that in the peptidoglycan-linked form (Braun and Sieglin, 1970; Braun and Wolff, 1970; Inouye et al., 1972; Hirashima et al., 1973b). The total number of lipoprotein molecules in cells growing exponentially in a rich medium is about 750,000 per cell, and approximately 250,000 are attached to the peptidoglycan. The number of disaccharide peptide units required to form a monomolecular layer of peptidoglycan in one cell has been estimated to be about 3,500,000. Thus one lipoprotein molecule is statistically linked to every 10th to 12th disaccharide peptide unit (Braun et al., 1974a). Because of the covalent association between lipoprotein and the peptide moiety of the peptidoglycan, one may assume that the side of the glycan stack of the peptidoglycan monolayer that bears the peptide side chains is oriented toward the outer membrane of the cell.

2.5.2. Isolation

The peptidoglycan-lipoprotein complex (or murein-lipoprotein complex) is usually obtained as the residue left after extraction of the isolated cell envelopes with 2 to 4% SDS at 100°C. This complex has been also called the rigid layer, or sacculus. Trypsin selectively cleaves bonds at the junction between the lipoprotein and the peptidoglycan so that the two components can then be separated from each other. As mentioned earlier, the lipoprotein-free peptidoglycan exhibits fragile, rodlike structures similar in shape to that of the intact bacterial cell (Fig. 6).

When isolated *E. coli* cell envelopes are submitted to SDS polyacrylamide gel electrophoresis, only the free form of the lipoprotein migrates into the gel to a position corresponding to a molecular weight of about 7,500 (Inouye et al., 1974). Because of its attachment to peptidoglycan, the bound lipoprotein remains on the top of the gel. If, prior to electrophoresis, the cell envelope is treated with an endo-N-acetylmuramidase (such as lysozyme) that degrades the peptidoglycan into fragments the bound lipoprotein can then penetrate the gel. Since it has an attached peptidoglycan fragment, this substituted lipoprotein migrates to a position corresponding to a higher molecular weight than that of the free protein. If the isolated peptidoglycan-lipoprotein complex is treated with trypsin prior to electrophoresis, then all the lipoprotein migrates as the free form. After double labeling of the total proteins of the cell envelope with L-[^{3}H]arginine and L-[^{14}C]histidine, the lipoprotein (which lacks histidine) is readily characterized as the one that bears only the tritium label. By using these techniques, the quantitative relationship of the two forms of lipoprotein was investigated (Inouye et al., 1974). Recently, a procedure has been described (Hindennach and Henning, 1975) that allows the isolation of the free lipoprotein (and the other major proteins) of the outer membrane of *E. coli*. The method involves differential extraction of cell envelopes with ionic and nonionic detergents with and without Mg^{2+}, and the proteins are finally separated from each other by molecular sieve chromatography in the presence of SDS (yield: 30 mg of pure lipoprotein from 200 g cell paste).

2.5.3. Primary structure

The lipoprotein of *E. coli* contains 58 amino acids (molecular weight: 7,800) and lacks several amino acids including histidine. The amino acid sequence has been determined by Braun and coworkers (Braun and Rehn, 1969; Braun and Sieglin, 1970; Braun and Wolff, 1970; Braun et al., 1970; Braun and Bosh, 1972a,b; Braun 1973; Hantke and Braun, 1973; Braun and Hantke, 1974; Braun et al., 1974a,b; Braun, 1975; Braun and Wolff, 1975). Fig. 24, in which the attachment site to the peptidoglycan is shown, has been drawn in a way that emphasizes the repetitive design of the molecule. Essentially, tripeptides extend at each end of a middle repetitive section and the lipid and the peptidoglycan are attached there.

The lipid is at the N-terminal end of a Cys-Ser-Ser sequence. A fatty acid is bound as an amide to the N-terminal α-amino group of cysteine, and other fatty acids are bound as esters to the hydroxyl groups of S-glyceryl cysteine. The fatty acids bound as amides are mainly palmitic acid (65%), palmitoleic acid (11%) and *cis*-vaccinic acid (11%). The fatty acids bound as esters are mainly palmitic acid (45%), palmitoleic acid (11%), *cis*-vaccinic acid (24%), cyclopropylenehexadecanoic acid (12%) and cyclopropyleneoctadecanoic acid (8%).

The first 15 amino acid residues of the middle section are duplicated almost identically and the next stretch of 7 amino acid residues is almost identically quadruplicated. This amino acid sequence suggests the possible evolution of the molecule from a gene that coded originally for 15 amino acids, which was duplicated and only the C-terminal half was then added four times. The dashes in Fig. 24 represent deletions that may have occurred during evolution.

The attachment site to the peptidoglycan at the C-terminal of the lipoprotein is a Tyr-Arg-Lys sequence and the lipoprotein is fixed by the ϵ-amino group of its C-terminal lysine residue to the carboxyl group at the L-center of *meso*-diaminopimelic acid of the peptidoglycan (i.e., where a D-alanine residue occurs in a conventional tetrapeptide unit) (Fig. 24). The tyrosine, arginine, and lysine residues that accumulate at this attachment site fit especially well into the active center of trypsin. The Arg-Lys bond is preferentially cleaved and the lipoprotein is detached from the peptidoglycan. The released lipoprotein lacks its C-terminal lysine, and this lysine residue is found at the C-terminal end of one of 10th to 12th tetrapeptide L-Ala-D-Glu $\ulcorner$(L)-*meso*-A $_2$pm-(L)-L-Lys units in the peptidoglycan.

These structural studies were carried out on the peptidoglycan-bound lipoprotein. The protein part of the free protein obtained by the procedure of Hindennach and Henning (1975) does not differ substantially from that of the bound form.

2.5.4. Conformation

Circular dichroism of the lipoprotein in aqueous environment reveals an α-helical content of about 80% (Braun, 1975). This high degree of ordered structure explains why the peptide bonds involving arginine or lysine, which are localized in the helical portion of the molecule, are fairly resistant to trypsin action. Another remarkable property of the molecule is that, starting with its

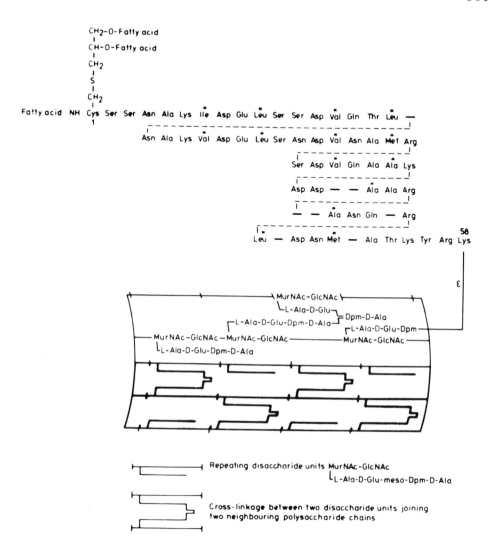

Fig. 24. Structure of the peptidoglycan-lipoprotein complex (rigid layer) of the outer membrane of *E. coli.* The cylindrical section indicates a part of the peptidoglycan with the shape of the rodlike *E. coli* cell. It is not known how the glycan chains span the cell relative to the long axis of the cylinder. In this model they are arbitrarily drawn parallel. The peptidoglycan is composed of roughly 10^6 repeating units, to which approximately 10^5 lipoprotein molecules are covalently bound. The lipoprotein replaces D-alanine on the diaminopimelate residue. Dashes represent hypothetical deletions of amino acids that may have occurred during evolution (see text). Stars indicate the hydrophobic amino acids at every 3.5th position. (Braun, 1975.) (Reprinted courtesy of Elsevier/North-Holland, Amsterdam.)

repetitive segment, every fourth and third amino acid residue, respectively, is hydrophobic. Since a helical turn contains 3.6 residues, all the hydrophobic residues would be localized on one face of the helical rod. Fig. 25 emphasizes the peculiar distribution of the amino acid residues with all the hydrophobic side

502

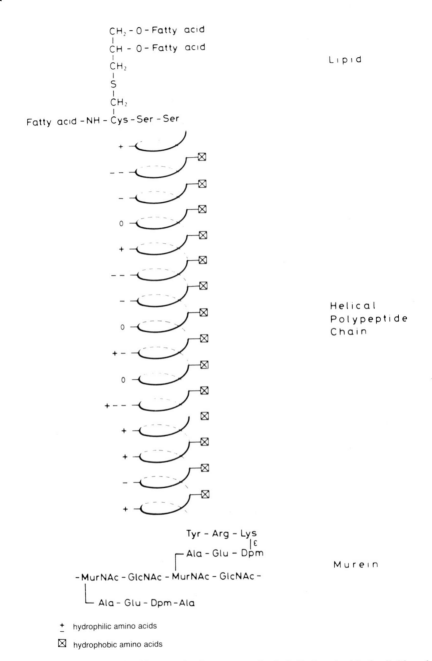

CH₂ – O – Fatty acid
CH – O – Fatty acid
CH₂
S
CH₂
Fatty acid –NH – Cys –Ser –Ser

Lipid

Helical
Polypeptide
Chain

Tyr – Arg – Lys
|ε
Ala – Glu – Dpm
–MurNAc – GlcNAc – MurNAc – GlcNAc –
Ala – Glu – Dpm –Ala

Murein

± hydrophilic amino acids

⊠ hydrophobic amino acids

Fig. 25. Tentative model of the lipoprotein drawn as a single helical rod with the lipid and two peptidoglycan repeating units attached. ±, Hydrophilic amino acids. ×, Hydrophobic amino acids. In aqueous solution, intermolecular interactions along the hydrophobic face must occur, leading to aggregation of molecules which, in fact, is observed. (Braun, 1975.) (Reprinted courtesy of Elsevier/North-Holland, Amsterdam.)

chains on one face of the α-helix and all the hydrophilic side chains on the other.

In a search for an experimental approach to the possible conformations of the lipoprotein, Braun (1975) and Braun et al. (1976a) used the rules that Chou and Fasman (1974) derived from 15 proteins of known conformation, which allow one to predict helix, β, and coiled regions of proteins with known sequence with up to 80% accuracy. The model derived from such a calculation is shown in Fig. 26. The simple helical rod initially proposed on the basis of the amino acid sequence and circular dichroism measurements, is now broken by a β-loop occurring approximately at the middle of the amino acid sequence. Amino acid residues 1 to 4 form a β-loop, 5 to 24 an α-helix (symbolized by a cylinder), 25 to 29 a β-loop, 30 to 47 an α-helix (also symbolized by a cylinder), 48 to 51 a β-loop, 52 to 56 a β-sheet, and residues 57 and 58 a coil. One interesting feature of the model is that both the attachment sites of the lipid and the peptidoglycan protrude from the helical part of the molecule. They may serve as recognition sites for the enzyme system that are responsible for the transfers of the lipid to the polypeptide chain and for the attachment of the lipoprotein to the peptidoglycan (Braun, 1975). The secondary structure of the lipoprotein was also examined by Green and Flanagan (1976) using the method of Lim (1974a,b). The two methods of Chou and Fasman, and of Lim, respectively, agreed moderately well.

2.5.5. Function

One possible function of the bound lipoprotein is to connect the outer membrane with the peptidoglycan layer. Thus, it probably plays an important role in the construction and stabilization of the cell envelope of the *Enterobacteriaceae*. It is noteworthy that *E. coli* mutants lacking different major proteins in the outer membrane or even all of them except the lipoprotein (section 2.8.2.) have been found, suggesting that the lipoprotein may well be an essential structural component of the outer membrane. Lipoprotein is a normal wall constituent of all *Enterobacteriaceae*. Contrary to a previous report that lipoprotein was lacking in at least one enteric bacterium, *Proteus mirabilis*, lipoprotein has been shown to occur in covalent linkage to peptidoglycan in this organism, at least in the stationary phase of growth (Gruss et al., 1975). This lipoprotein, however, differs from that of *E. coli* because the *P. mirabilis* lipoprotein has a lower molecular weight (5,500) and has glycine and phenylalanine as specific components that are absent in lipoproteins of other enteric bacteria. The lipoprotein substitutes every 15th to 20th peptidoglycan unit; treatment with trypsin leaves lysine as the only lipoprotein amino acid attached to the peptidoglycan.

The fact that, at least in *E. coli*, there is twice as much free compared to bound lipoprotein suggests another function for the lipoprotein. By assuming that the lipoprotein molecule is a continuous, single helical rod (Fig. 25) and based on theoretical considerations, Inouye (1974, 1975) proposed an assembly model for the bound and free lipoprotein molecules in the form of cylindrical channels that might provide the outer membrane with passive diffusion pores. In this model, the lipoprotein α-helices in the outer membrane are arranged so that the

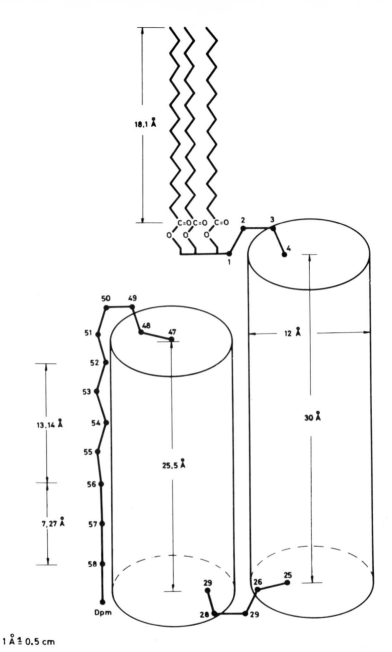

Fig. 26. Three-dimensional structure of the lipoprotein as deduced from the circular dichroism measurements and the amino acid sequence, applying the Chou-Fasman rules. Amino acid residues 1–4 form a β-loop, 5-24 an α-helix with the possible exception of 13-17, which also could be arranged in a β-sheet, 25-29 clearly form a β-loop, 30-47 an α-helix, 48-51 a β-loop, 52-56 a β-sheet and residues 57 and 58 a coil. The arrangement of the amino acid residues in the helical portions is only symbolized by the 2 cylinders. The length of the lipoprotein (48 Å) would span just half the thickness of the outer membrane, suggesting that the lipid portion of the lipoprotein immerses into the inner layer of the lipid bilayer of the outer membrane. (Braun, 1975.) (Reprinted courtesy of Elsevier/North-Holland, Amsterdam.)

hydrophobic residues face outward from the assembly whereas the hydrophilic residues are localized inside. In particular, six α-helical coils may form a super-helix or a coiled-coil with the hydrophobic residues of each of the six α-helices pointed toward the outer surface (structure 4, Fig. 4). This arrangement makes a number of ionic interactions possible between adjacent molecules, a situation that enhances stabilization of the entire assembly. Moreover, according to the model, the whole assembly structure (height: 76 Å) would penetrate through the outer membrane (75 Å thick), providing additional hydrophobic interactions between the outer surface of the assembly and the lipid bilayer of the membrane. Further stabilization of the arrangement could also be achieved by assuming that the three hydrocarbon chains at the amino terminal ends of the lipoproteins are flipped back over the helix and inserted into the bilayer structure (Fig. 4), whereas at the carboxyl terminal end of the assembly, two out of the six lipoprotein molecules would be covalently linked to the peptidoglycan layer, thus preventing the assembly from moving freely in the lipid bilayer (Fig. 4). The channel thus created through the outer membrane would have a diameter of 12.5 Å. On a basis of a twelve (instead of six) lipoprotein assemblies, the diameter would be 35.8 Å. Accordingly, the total number of channels per cell would change from 1.25×10^5 to 0.63×10^5, and the area occupied by these channels would change from 35 to 46% of the total cell surface. Recently, the lipoprotein was obtained in the form of paracrystals by adding acetone to a solution of purified lipoprotein made in Triton X-100 or SDS and in the presence of about 0.01 M $MgSO_4$ (De Martini et al., 1976, Inouye et al., 1976). When examined by the electron microscope after negative staining, these paracrystals showed considerable ultrastructure and the observed patterns were found to be compatible with the tubular superstructure proposed for the lipoprotein assembly in the outer membrane.

If one accepts the idea that one function of the lipoprotein is to provide the outer membrane with passive diffusion pores, then the accessibility of these channels should be controlled, at least in part, by the polysaccharide chains of the lipopolysaccharides (also localized on the outer surface of the outer membrane; see section 2.7.). Inouye suggested that these polysaccharides might cover the entrance of the channels, preventing some molecules from passing through them. Such an arrangement would explain the known effect of EDTA on the permeability barrier of the cell (Leive, 1965a,b; Vol and Leive, 1970). By removing part of the lipopolysaccharide, EDTA would readily expose the entrance of the channels to the outside of the cell. Immunological studies do not contradict this view. Both free and bound forms of lipoprotein are available to lipoprotein-directed antibodies only in those mutants that have incomplete lipopolysaccharides and hence a defective cell surface structure (Braun et al., 1974a). In wild strains, where complete lipopolysaccharides occur, lipoprotein does not function as an antigen and is probably shielded by other surface structures (Braun, 1975; Braun et al., 1976b).

Although Inouye's model is very attractive in many respects, at the present time it should not be taken literally. In particular, if the conformation for the lipoprotein most recently proposed by Braun (Fig. 26) is that occurring in the membrane, then the length of the lipoprotein when broken by a β-loop approximately at the middle of it, is about 48 Å . Thus the lipoprotein would span half of the outer membrane thickness and the lipid moiety of the lipoprotein would be immersed in the inner portion of the outer membrane's lipid bilayer. Exactly how the free and bound forms of lipoprotein are arranged in the outer membrane remain matters for speculation (see also section 2.9).

2.6. Possible occurrence of a patterned inner protein layer in the walls of gram-negative enteric bacteria

In the 1960s, a periodic structure was observed in *E. coli* that apparently was buried deep in the cell envelope because extractions with phenol or detergents were necessary to reveal it (Weidel et al., 1960; Bayer and Anderson, 1965; Boy de la Tour et al., 1965). Subsequently, the occurrence of a regular array was confirmed by using other preparative procedures (de Petris, 1965, 1967; Fischman and Weinbaum, 1967; Bayer and Remsen, 1970; Nanninga, 1970) but the component(s) responsible for the observed pattern could not be identified. Recently, by using a differential heat extraction procedure in SDS, Rosenbush (1974) isolated a protein that might be responsible for the occurrence of this periodic pattern.

2.6.1. Isolation
Essentially, the procedure involves treatment of the cell envelopes with a large excess of SDS at 60°C. Variations in temperature between 30 and 70°C do not affect the procedure, but higher temperatures must be avoided. A turbid solution is obtained and, after sedimentation, the washed pellet consists of classic rod-shaped structures. When compared to the rigid layer obtained after extraction at 100°C in the detergent (i.e., the peptidoglycan-lipoprotein complex studied in the previous section), the rod-shaped structures obtained at lower temperature are less translucent and their surfaces show a regular pattern (Fig. 27a). Further extraction in SDS at 100°C for 5 minutes eliminates this regular array, and the released protein that was apparently responsible for it can be purified from the supernatant fraction and subsequently freed of the detergent.

2.6.2. Properties
The protein-lipoprotein-peptidoglycan complex (obtained after extraction at 60°C with SDS) is composed of about 65% envelope protein, the remaining mass being accounted for almost exclusively by the lipoprotein-peptidoglycan complex. In this bound form the protein is resistant to prolonged incubations with trypsin, in marked contrast to the lipoprotein. It does not bind SDS tightly even after prolonged exposure to a great excess at 60°C, and is insoluble in 5 M guanidine hydrochloride. There are 1.1×10^5 molecules of this polypeptide per cell. Diffraction pattern and electron micrograph measurements (Steven et al.,

1977) revealed that they are arranged in a lattice structure with hexagonal symmetry and a periodicity of 7.5 nm (Fig. 27a). Electron microscopic evidence, the known asymmetry of the peptidoglycan layer itself, and the fact that antibodies specific for the solubilized protein selectively agglutinate uninverted peptidoglycan-envelope protein complexes strongly suggest that the protein occurs solely on the outer face of the peptidoglycan. On the basis of 1.1×10^5 molecules per cell, a 56 nm^2 value for the area of each unit, and an average value of 3 μm^2 for the surface area of the peptidoglycan layer, two subunits per morphological unit has been estimated. The lattice structure appears to be due to the protein. Treatment with SDS at 100°C dissociates the protein from the complex and leaves the latter devoid of the characteristic array. Moreover, the released protein constitutes more than 90% of the components dissociated by boiling in the detergent. Clearly, the protein is not covalently linked to the peptidoglycan (in contrast to the lipoprotein), but the nature of the interactions between the protein and the peptidoglycan-lipoprotein network as well as the neighboring protein subunits is unknown.

The dissociated detergent-free protein has been found homogeneous on the basis of several criteria. It has a molecular weight of 36,500 and exhibits moderate hydrophobicity. Circular dichroism and infrared spectroscopy indicated that a large fraction of the protein exists as β-structure. In contrast to the bound protein, the free protein binds SDS in amounts equivalent to those found with most polypeptides (Helenius and Simons, 1975), and is denatured by guanidinium ions to a random coil conformation. The same protein in an amorphous state has also been obtained by Hindennach and Henning (1975) by using the procedure described earlier (section 2.5.2). The yield is about 120 mg protein from 200 g cell paste. Henning calls this protein I.

2.6.3. Occurrence and possible function

Examination of 10 strains of E. coli revealed strain-specific variations in the binding properties of the protein to the complex (e.g., as a function of the temperature) rather than variations in the amount of protein per cell (Rosenbush, 1974). Proteins corresponding to that described here apparently occur in various gram-negative organisms other than E. coli. Spirillum serpens (Murray, 1963) is remarkable because it possesses both an inner regular structure comparable to that of E. coli and a protein surface lattice. This lattice is probably analogous to those found in various gram-positive bacteria as, for example, the T-layer of Bacillus brevis (section 2.4.). The surface lattices are very different from the inner ones. As previously mentioned, the surface lattices are localized on the exterior surface of the cell, released by 2 M guanidine hydrochloride, cleaved by proteolytic enzymes, and during renaturation they reassemble into highly organized structures.

The peptidoglycan is generally accepted as the main shape-maintaining structure in bacteria. However, it probably does not contain shape-determining information. For this function, one could assume that a protein matrix in the form

508

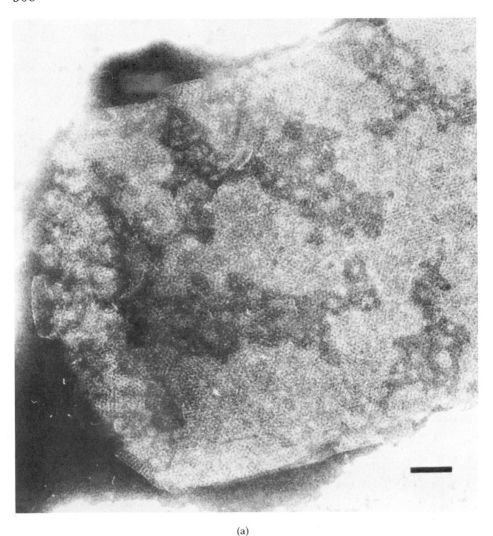

(a)

Fig. 27. Wall peptidoglycan in *Escherichia coli* with its associated envelope protein: (a) Prepared by SDS treatment at 60°C according to the procedure of Rosenbush (1974). Electron microscopy reveals particles covered with a regular pattern. (See text for details.) Bar, 100 nm. Particles stained negatively with potassium phosphotungstate. (Reprinted from Rosenbush (1974) by courtesy of the Journal of Biological Chemistry). (b) Prepared by SDS treatment at room temperature by Dr. M. Bayer (unpublished data). Unfixed cells from a growing culture were quickly frozen and cut in frozen state with the knife of an ultramicrotome. Sections containing large segments of the cells were picked up on an electron microscope grid, treated with 0.5% SDS at room temperature for 5 min, briefly washed and negatively stained. In the process, more than 50% of the dry mass of the wall disappeared. When treated with lysozyme (which degrades the peptidoglycan), the residual structure shown in the figure disintegrates into small spherical units some of which can also be seen in the background. These spherical elements are destroyed by both trypsin and pronase. Note the gap in the tight protein cover. Crystallinity could not be detected in the layer shown in the figure even by employing optical diffraction. (Unpublished figure and data courtesy of Dr. M. Bayer, Institute for Cancer Research, Philadelphia, Pa.)

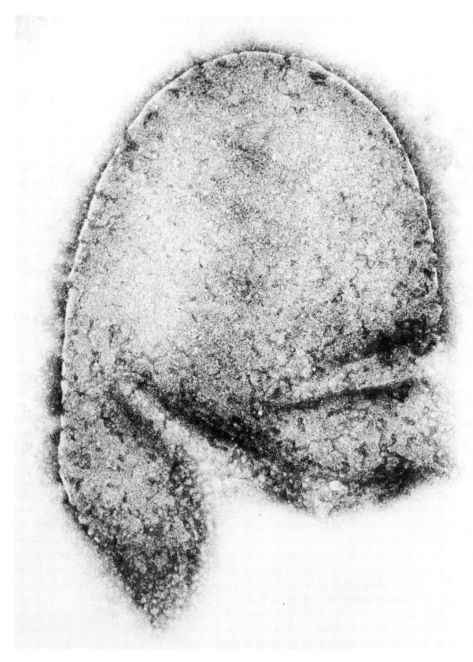

Fig. 27. Part (b).

of a two-dimensional lattice would be the most likely candidate (Henning and Schwartz, 1973). Since surface lattices are located on the outer surface of the cell, and since their absence does not noticeably affect the bacteria, it is not possible to correlate these structures with such a function. The periodic structure that seems to be deeply buried in the envelope of the gram-negative bacteria in close contact with the peptidoglycan structure may have played such a role. However, *E. coli* mutants that lack protein I (and other major proteins except the lipoprotein) in the outer membrane exist and do not show any gross defect (Henning and Haller, 1975). Therefore it is also not possible to assign a shape-determinig function to Rosenbush's protein. At present, the functional significance of this protein in cellular processes is unknown.

Finally, the occurrence of a deep, regular, or crystalline structure in the walls of gram-negative bacteria should not be regarded as being established beyond a doubt. Thus for example, examination of sections of *E. coli* cell wall like that shown in Fig. 27b (in which case treatment with SDS was carried out at room temperature on the electron microscopic grid; see figure legend) failed to reveal any crystallinity in the layer even after employment of optical diffraction (M. Bayer; personal communication).

2.7. The wall lipopolysaccharide of gram-negative enteric bacteria

The lipopolysaccharides, major somatic antigens (O-antigens) of enteric bacteria, are very powerful toxic agents (endotoxins). Their chemical structure, physical structure, synthesis, and genetic control have been studied extensively and numerous reviews have appeared on these topics (Nikaido, 1968, 1970, 1973; Luderitz et al., 1966, 1968, 1971, 1973, 1974; Osborn, 1969, 1971; Osborn et al., 1971, 1974; Rothfield and Romeo, 1971; Mäkelä and Stocker, 1969).

2.7.1. Isolation

Although the lipopolysaccharides are localized exclusively in the outer membrane, the most common procedure used for their isolation is phenol-water extraction of the intact cells (45% aqueous phenol at 65–68°C). Cooling the extracts causes partitioning of the extracted substances between phenol layer and aqueous layer. Usually the lipopolysaccharide occurs in the aqueous layer in the form of large aggregates. It can be recovered and purified by ultracentrifugation.

Lipopolysaccharides produced by certain mutants have lost part or most of their hydrophilic polysaccharide moiety. They are hydrophobic. In these cases an extraction method with phenol-petroleum ether-chloroform has been developed that generally gives rise to homogeneous preparations (Galanos et al., 1969).

The lipopolysaccharides contain many ionic groups and thus they usually carry a mixture of various cations (including alkali, alkaline earths, and heavy metal ions) and amines such as spermine, spermidine, and putrescine. Further purification can be achieved by electrodialysis. After such treatment, the lipopolysaccharides are acidic and water-insoluble. They can then be solubilized in a controlled manner by using any base.

2.7.2. Structure

Lipopolysaccharides are complexes of polysaccharide chains covalently linked to a unique glucosamine-containing lipid known as lipid A. An important feature of these molecules is the occurrence of three different regions (Fig. 28). Typically, the polysaccharide moiety is composed of two distinct parts, the so-called superficial O-antigen chains (region I) and the "core" (region II), which is characterized by two unique sugars in it, L-glycero-D-mannoheptose and 2-keto-3-deoxyoctonate or KDO (Fig. 29). Region II is, in turn, linked to lipid A or region III (Fig. 28). Detailed studies have been performed mainly on lipopolysaccharides of *Salmonella, E. coli* and *Shigella flexneri*. The following description is related to the lipopolysaccharide of *S. typhimurium* which is especially well known (Fig. 28) (Luderitz et al., 1973, 1974; Hämmerling et al., 1973).

Structural studies were made possible because of the isolation of various types of mutants that were defective at various stages of the lipopolysaccharide synthesis (Nikaido, 1968; Mäkelä and Stocker, 1969; Stocker and Mäkelä, 1971). However, the innermost part of the molecule, the KDO-lipid A portion, seems to be indispensable for the survival of the bacteria. So far, mutants defective in the

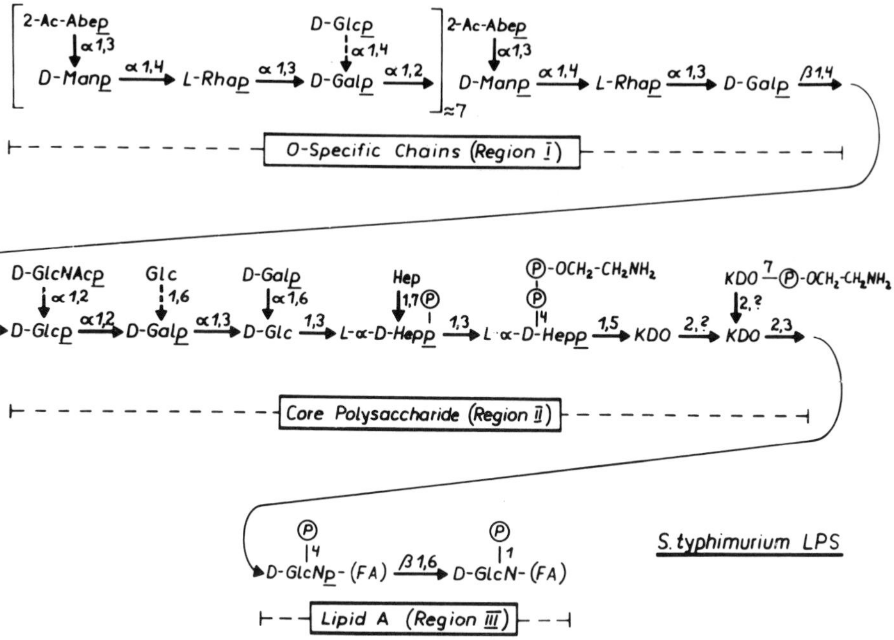

Fig. 28. Chemical structure of the lipopolysaccharide of *Salmonella typhimurium* S form. (Lüderitz et al., 1974). (Reprinted courtesy of Avicenum, Czechoslovak Medical Press, Prague.)

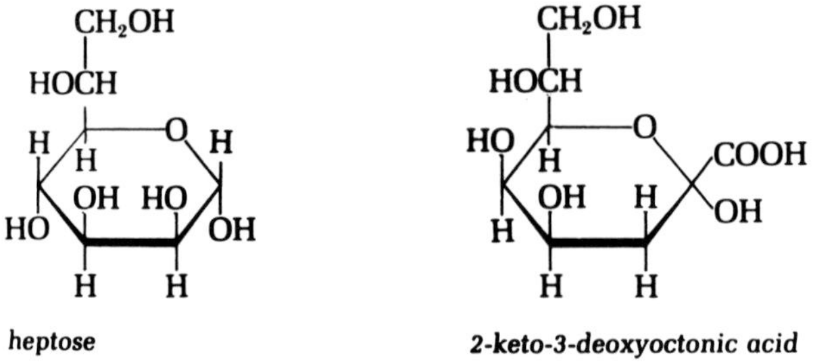

heptose   2-keto-3-deoxyoctonic acid

Fig. 29. Structures of the C_7 and C_8 constituents of the lipopolysaccharide core region.

synthesis of this portion (except conditional mutants) have never been isolated (Osborn et al., 1974). The O-side chain and the greatest part of the polysaccharide core are not indispensable. Many different mutants lacking these components to various extents are known. Wild-type organisms usually produce complete lipopolysaccharides with O-side chains. They form smooth colonies on solid media (S strains) and they react with anti-O-antibodies. Mutants without O-side chains grow as rough-surface colonies on solid media (R strains). They do not react with anti-O-antibodies. The smooth appearance of the S colonies is due presumably to the hydrophilic and water-retaining properties of the O-side chains which cover the surface of the cells. In addition to the S and R types, SR mutants are known that can be regarded as intermediates between the S and the R mutants. These SR mutants are defective in the polymerization of O-repeating units, so their lipopolysaccharides have very short unpolymerized O-side chain units (Naide et al., 1965; Mäkelä, 1966; Nikaido et al., 1966; Yuasa et al., 1970).

Lipid A (region III) at one end of the molecule is an unusual phospholipid because it contains no glycerol (Fig. 30) (Luderitz et al., 1974). Its backbone is a disaccharide of D-glucosamine (GlcN-β-1,6-GlcN) with the specific 3-D-hydroxymyristic acid in an amide linkage to the amino groups. The glucosamine disaccharide is substituted at position 3' with the KDO trisaccharide of the core (region II) and at position 4' and 1 with phosphate residues. These latter residues may be used to link 3 to 4 individual lipid A molecules through pyrophosphate bridges. The remaining hydroxyl groups of glucosamine (positions 6', 3 and 4) are esterified by long-chain saturated fatty acids (lauric, palmitic, and 3-D-myristoxymyristic acid). Mild acid hydrolysis—for example, autolysis at 100°C of the acidic lipolysaccharide in water (pH 3–4)—cleaves the ketosidic linkage between KDO and lipid A and releases the polysaccharide-free lipid A as a water-insoluble compound. Neutralization produces a viscous and opalescent solution (Lüderitz et al., 1974). Lipid A is immunogenic, when administered in a suitable form, that is, incorporated into liposomes (Galanos et al., 1971) or exposed on bacterial cells (Lüderitz et al., 1973). Lipid A is also the endotoxic center of the lipopolysaccharides. This has been demonstrated from isolated lipid A solubilized by complexing it with proteins and other solubilizing carriers (Lüderitz et al., 1973).

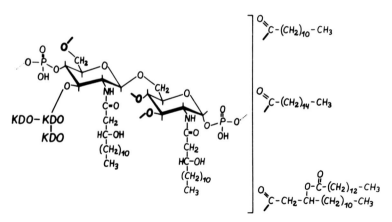

Fig. 30. Chemical structure of a *Salmonella* lipid A unit with an attached KDO trisaccharide. The three acyl residues shown are linked in an unknown distribution to the hydroxyl groups of the glucosamine residues. (Lüderitz et al., 1974). (Reprinted courtesy of Avicenum, Czechoslovak Medical Press, Prague.)

The O-specific polysaccharide (region I) at the other end of the molecule contains (in *S. typhimurium*) five sugar constituents (abequose, mannose, rhamnose, galactose, and glucose) that form the repeating, branched pentasaccharide units of one chain. The great variety of lipopolysaccharides that occur in the various strains, species, and bacterial groups is determined by the nature and linkages of the sugar constituents of region I. Often the chains are composed of repeating oligosaccharide units, as in *S. typhimurium*. These O-specific chains determine the serological specificity of the lipopolysaccharides and the species that contain them.

The core (region II) extends between the O-specific chains and lipid A. Heptose and KDO accumulate in its inner portion. Both residues occur as branched trisaccharides that are substituted by phosphate, phosphorylethanolamine, and pyrophosphorylethanolamine residues, causing an accumulation of charged groups in that part of the molecule. The outer portion of the core is more conventional and consists of a branched polysaccharide of glucose, galactose, and N-acetylglucosamine. Region II carries antigenic determinants that are hardly detectable in complete lipopolysaccharides but become immunologically active in mutants with defective lipopolysaccharides. The structure of region II is invariant within the various *Salmonella* serotypes. The core structures in other *Enterobacteriaceae*, however, may be different. Various distinct groups of *E. coli* with different core structures have been isolated by serological methods. Finally, in gram-negative bacteria other than the *Enterobacteriaceae*, the core structures may lack heptose or KDO, or even both (Jann et al., 1973). Core-defective R mutants in *Salmonella* are known (Fig. 31). The terminal sugar units are the main determinants responsible for the serological specificity of the corresponding lipopolysaccharides. All the R lipopolysaccharides, even those with very profound defects such as the Re mutants, which lack heptose and contain only KDO and lipid A, are potent endotoxins—a finding consistent with the idea that lipid A is the endotoxic center of all lipopolysaccharides (Lüderitz et al., 1973).

2.7.3. Localization and arrangement

The lipopolysaccharides are asymmetrically distributed in the outer membrane with preferential localization on its outer face (Mühlradt and Golecki, 1975). Their surface density is relatively constant ($0.7-1.0 \times 10^5$ molecules per μm^2) and only about 25% of the cell surface is occupied by the lipid portion of the lipopolysaccharides (Mühlradt et al., 1974).

The lipopolysaccharides exhibit surface mobility. Lateral diffusion is low with a constant of about 3×10^{-13} cm²/s (Mühlradt et al., 1974). This value is orders of magnitude lower than the diffusion constant values of phospholipids in most biological membranes ($D \approx 5 \times 10^{-9}$ cm²/s) (Lee et al., 1973) and of animal cell-surface antigens ($D \approx 5 \times 10^{-11}$ cm²/s) (Frye and Edidin, 1970). Trans-membrane movement, flip-flop transition, can also occur but only in an area where the peptidoglycan layer is defective and the temperature is sufficiently high (25–37°C) to allow it (Mühlradt and Golecki, 1975). For example, by using ferritin-conjugated antibodies directed against the polysaccharide moiety of the lipopolysaccharide, the walls of *S. typhimurium* from which the peptidoglycan had been removed by lysozyme digestion at 0°C were shown to carry the label only on the outer face of the membrane. When removal of the peptidoglycan was performed at physiological temperatures (25–37°C), lipopolysaccharide was localized on both membrane faces.

Divalent cations are essential for integration of the lipopolysaccharide molecules into the outer membrane. The outer membrane is able to withstand treatment with nonionic detergent (Triton X-100) if enough Mg^{2+} ions are present (De Pamphilis and Adler, 1971; Schnaitman, 1971b), but treatment with EDTA in the absence of detergent causes the selective release of large amounts of lipopolysaccharide (Leive, 1965a; Leive et al., 1968). About 50% of the total lipopolysaccharide molecules are freed and the released material has an overall composition of 85 to 90% lipopolysaccharide, 5 to 10% protein and 5% phospholipid. Clearly, such a release is not a peeling apart of the outer membrane. Newly synthesized lipopolysaccharide molecules are essentially nonreleasable. Subsequently they become evenly distributed between releasable and nonreleasable fractions. Actually, the two fractions are in constant equilibrium (Levy and Leive, 1968). These phenomena, however, are not yet understood.

The lipopolysaccharide molecules play an important role in the permeability barrier of the outer membrane (section 2.9.). They exert a masking effect on various proteins localized on the outer cell surfaces (section 2.8.3.); yet agents such as colicin, phages, and antibodies directed against components other than the lipopolysaccharide can bind to and affect the bacterial cell surface. These observations suggest that the lipopolysaccharide molecules are not evenly distributed on the exterior of the cell. In this context, one may mention that according

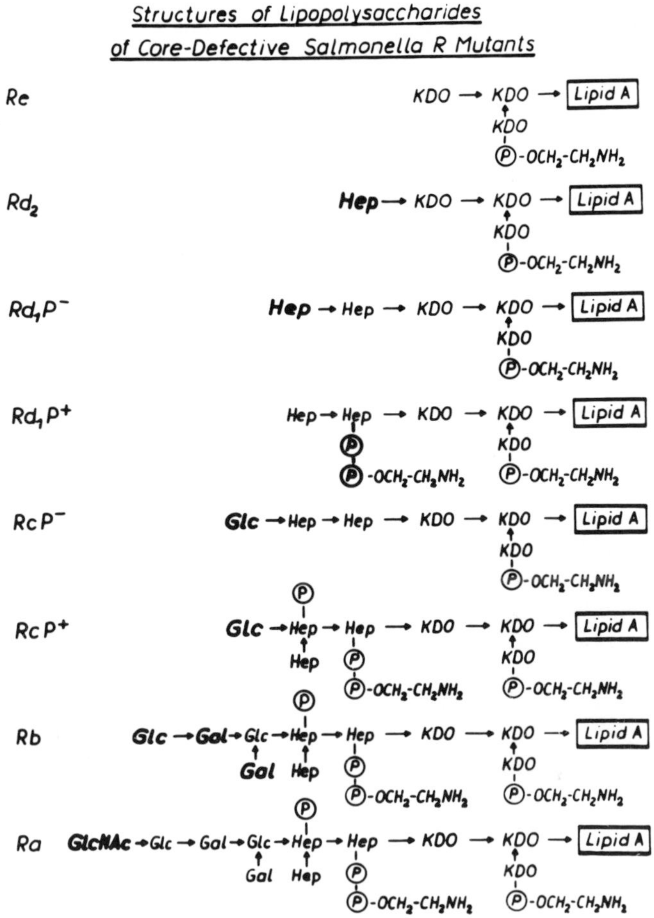

Structures of Lipopolysaccharides
of Core-Defective Salmonella R Mutants

Fig. 31. Chemical structure of the defective lipopolysaccharides from R mutants of chemotype Ra through Re. (Lüderitz et al., 1974). (Reprinted courtesy of Avicenum, Czechoslovak Medical Press, Prague.)

to recent reports, lipopolysaccharide in its native state might occur in close association with protein. In at least one case (*E. coli* 0111-B $_4$) where mild detergent extraction was used, the lipopolysaccharide was quantitatively recovered as an apparent covalently linked polysaccharide-protein complex (Wu and Heath, 1973). It has also been hypothesized (Verkleij et al., 1977) that the many particles seen on the outer fracture face of the outer membrane by freeze fracturing electron microscopy of *E. coli* K12, would consist of lipopolysaccharide aggregates stabilized by divalent cations and complexed with protein and/or phospholipid.

The lipopolysaccharide molecules also exert a masking effect on the phospholipid head groups present in the outer membrane. Whole cells of *S.*

typhimurium were treated with phospholipase C (which should hydrolyze the phosphatidylethanolamine head groups) or with CNBr-activated dextran (which should undergo coupling with these groups). Results did not indicate the presence of any accessible head groups on the outer surface of strains producing lipopolysaccharides of S or R_c type. In contrast, with strains that produce less complete R_d or R_e lipopolysaccharides and reduced amounts of proteins, both methods showed the presence of exposed phosphatidylethanolamine head groups (Kamio and Nikaido, 1976). Resistance of phospholipids in whole cells of wild type of *E. coli* K12 to exogenous phospholipases was also attributed by van Alphen et al. (1977b) to shielding by some other outer membrane components. Proteins b and d, the heptose-bound glucose of lipopolysaccharide and divalent cations would be involved in the phenomenon.

2.7.4. Functions

Available data is limited: (1) Clearly, lipopolysaccharide is an important constituent of the outer permeability barrier of the cell. (2) The presence of at least the glycolipid found in the Re-mutants (Fig. 31) seems to be essential for the survival of the cell, but the function of this apparently indispensable component is not clear. It has been hypothesized that at least this portion of the lipopolysaccharide is needed for the proper assembly of the outer membrane. (3) Wild strains, as they are isolated, are always smooth and contain lipopolysaccharides with O-side chains. When kept as pure cultures, R mutants appear and sometimes overgrow the parent organism. The well-known *E. coli* K12, B, and C strains, for example, are rough strains. O-side chains enable pathogenic organisms to escape phagocytosis in host animals that do not possess the proper antibody. Complete lipopolysaccharides may play a similar role in nature, such as preventing phagocytosis by protozoa. (4) Finally, the lipopolysaccharides exhibit overall structural features which resemble those of the teichoic acids to some extent (see section 2.3.4.). These two types of polymers, though especially devised for gram-positives and gram-negatives, respectively, might, at least in part, be functionally related.

2.8. Protein arrangement in the outer membrane
of Escherichia coli *and other gram-negative bacteria*

2.8.1. The membrane ghost

When treated with lysozyme (which destroys the peptidoglycan), *E. coli* cells plasmolyzed in sucrose become osmotically fragile. Nevertheless they retain their rod shape, and it is only upon sucrose dilution that the cells round up into spheres (Birdsell and Cota-Robles, 1967). Sphere formation is due to osmotic pressure and/or surface tension phenomena. These phenomena can be avoided by disrupting the membrane or by making it leaky. Under these conditions, the ghosts obtained remain rod-shaped upon sucrose dilution (Henning et al., 1973a). Rod-shaped ghosts can be obtained by treating sucrose-plasmolyzed *E.*

coli cells with 1% Triton X-100 or Brij 18 to disrupt the plasma membrane. Subsequent steps usually include treatments with urea, trypsin, $MgSO_4$, and lysozyme (Henning et al., 1973a).

Rod-shaped ghosts (Fig. 32) consist almost exclusively of an outer membrane unit (and sometimes of inner membranous material; this material, however, does not contribute to shape maintenance). Peptidoglycan is lacking (Henning et al., 1973a; Haller and Henning, 1974; Schweizer et al., 1975). The phospholipid (20–30%) of the ghosts is almost entirely phosphatidylethanolamine. Very little phosphatidylglycerol and only traces of cardiolipin are present.

The rod shape of the ghosts is resistant to pronase, trypsin, chymotrypsin, EDTA, Triton X-100 and Brij 58; however, SDS (1%) causes complete solubilization. Ghosts delipidated by chloroform methanol extraction retain their rod shape (Henning et al., 1973b). They are distorted and exhibit few local breaks but are otherwise indistinguishable from normal ghosts. They contain about 70% protein and a substantial but imprecisely known amount of lipopolysaccharide (25–30%). The proteins have been classified into major and minor proteins (for a recent review, see Henning, 1975).

2.8.2. The major ghost proteins

Trypsinized ghosts from *E. coli* contain four major proteins which, on the basis of SDS polyacrylamide gel electrophoresis, exhibit the following molecular weights (Henning et al., 1973a,b; Garten and Henning, 1974): protein I (mol. wt. $\approx$ 38,000), protein II (mol. wt. $\approx$ 28,000), protein III (mol. wt. $\approx$ 17,000) and protein IV (mol wt. $\approx$ 7,000–10,000). Protein I is Rosenbush's protein (section 2.6), but protein I consists of two main components (Ia and Ib), both of which contain six separable isoelectric species (Schmitges and Henning, 1976). These components are almost identical in their primary structure, and represent essentially the same polypeptide. Protein II, probably an homogeneous polypeptide, is derived from a protein II (mol wt. about 33,000) by the action of trypsin, which is usually used during preparation of the ghosts. Evidence for the structural gene of protein II* has been obtained (Henning et al., 1976; Datta et al., 1976). Protein III occurs in small amounts that vary from preparation to preparation. Protein IV is Braun's lipoprotein (section 2.5.). The molar ratio of polypeptides I, II*, and IV in nontrypsinized ghosts about 1:1:8–10 (Henning et al., 1973b). Henning's nomenclature is used for the major proteins. Protein I is probably identical with protein I of Schnaitman (1974), with protein A_1 of Bragg and Hou (1972), with protein B of Koplow and Goldfine (1974), and with the Salmonella 35K protein of Ames (1974). Protein II is probably identical with protein B* of Reithmeier and Bragg (1974). Protein II* is probably identical with protein d of Lugtenberg et al. (1975), with protein 3a of Schnaitman (1974) and with protein B of Reithmeier and Bragg (1974).

As previously mentioned (section 2.5 and 2.6), preparative isolation of all major membrane proteins has been accomplished (Hindennach and Henning, 1975). The yields (per 200 g cell paste) were $\approx$ 120 mg protein I, $\approx$ 110 mg protein II*, $\approx$

518

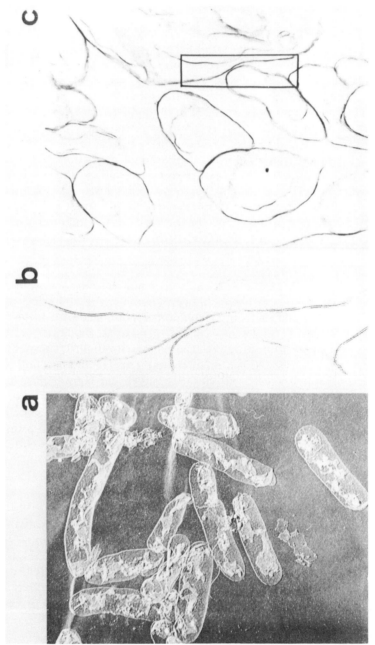

Fig. 32. Shadowed and sectioned ghosts of *Escherichia coli* K12. Insert b is a 2.5-fold further magnification of c. Each bar represents 1 μm. (Henning et al., 1973a.) (Reprinted courtesy of the Federation of European Biochemical Societies.)

50 mg protein III, and 30 mg protein IV. A comparison of all major proteins can be found in recent papers by Garten et al. (1975a,b).

Cross-linking of nontrypsinized ghosts with bifunctional diimidoester cross-linking reagents, with spanning distances from about 0.3 nm to 1.8 nm, produces a macromolecule that shows the size and shape of the cell and resists boiling SDS (Haller and Henning, 1974). Before treatment with SDS, these sacs contain 60 to 70% protein. After treatment, they consist of virtually pure protein. Ammonolysis of the cross-linked sacs shows that the four major proteins (I to IV) found in normally prepared ghosts have been cross-linked. Since reagents spanning only about 0.3 nm can stabilize the ghosts into covalently closed structures, it is clear that these proteins are very densely packed. Ghost membranes do not fit the fluid mosaic membrane structure model of Singer and Nicolson (1972) but rather constitute an extreme case of the types of models discussed by Capaldi and Green (1972). The fact that protein III always occurs in small and variable amounts and that the molar ratio of polypeptides I, II*, and IV is 1:1:8–10 suggests that cross-linking requires repeating I-II* sequences. Protein II* involvement in cross-linking poses an interesting problem because trypsinized ghosts cannot be cross-linked with the diimidoesters (but they can be cross-linked with glutaraldehyde) (Haller and Henning, 1974). This may be due to the removal of about 30% of the amino acid residues of protein II* by trypsin action, causing its transformation into protein II.

Because of the noncovalent, long-range ordered structure exhibited by the ghost membrane, the outer membrane might play a role in shape determination and the major membrane proteins might belong to a self-assembly system participating in or responsible for cellular shape. However, this is not the case (Henning and Haller, 1975). Rod-shaped mutants that lack one or another major protein, or all of them but lipoprotein IV, did not show any apparent gross fundamental defects when isolated. In the mutants where all major proteins (except protein IV) were lacking, no detectable increase occurred in any other single protein species and the integrity of the outer membrane seemed to be perfectly maintained. It is not known whether the space thus made available has been "filled up" with phospholipid or other components. Whatever the situation, it seems obvious that an assembly of the major proteins is not required for the generation or for the shape maintenance of the outer membrane.

Ghosts morphologically identical to those of *E. coli* were prepared from *Serratia marcescens, Proteus mirabilis, Salmonella typhimurium, Spirillum serpens* and *Pseudomonas aeruginosa* (Henning et al., 1973a,b). Sometimes the procedure had to be modified slightly; that used for *S. serpens* (Schweizer et al., 1975) includes treatment with 2 M guanidine hydrochloride to remove the protein cell surface lattice. Except for *Pseudomonas aeruginosa,* the band patterns in SDS gels are identical (*Salmonella*) or at least similar to that of *E. coli.* The ghosts membranes of *S. serpens* contain three major proteins (I:mol. wt., 40,000, II:mol. wt., 26,000, and III:mol. wt., 21,000). In contrast to *E. coli,* all the proteases used cause desintegration of the *S. serpens* ghosts (Schweizer et al., 1975).

2.8.3. The minor ghost proteins
(Haller et al., 1975)

One step in the preparation of the E. coli ghosts is trypsin treatment, which removes a heterogeneous set of proteins and transforms protein II* into protein II. It has been found that most of these proteins in nontrypsinized ghosts can be covalently linked to each other by oxidation with $CuSO_4$-O-phenanthroline or ferricyanide-ferrocene. The resulting rod-shaped "oxidation containers" apparently held together by disulfide bridges are virtually pure protein. They retain the size and shape of the cell even when treated with hot SDS. When reduced, they are soluble in SDS and contain a set of about 30 different polypeptide chains. Remarkably, the four major proteins (I, II, III, and IV) are not among the "oxidation proteins."

The oxidation proteins are mainly, if not exclusively, localized on the outer surface of the ghosts, and the same localization and arrangement was found in cells. The trypsin-sensitive portion of protein II* is also exposed at the outer surface of the membrane. Protein II* was found together with lipopolysaccharide to stoichiometrically inhibit the F-pilus mediated conjugation. All data available indicate that protein II* acts as a receptor in conjugation (Schweizer and Henning, 1977; van Alphen et al., 1977a). Asymmetry of membrane architecture, especially with regard to localization of proteins and phospholipids, exists in various systems (Blaurock and Stoeckenius, 1971; Fukui et al., 1971; Bretscher, 1972; Zwaal et al., 1973). The outer membrane of E. coli (and other gram-negative bacteria) may be one of the most extreme cases of asymmetric protein distribution. At present, however, a model cannot be drawn for the outer membrane since the arrangements of phospholipid and lipopolysaccharide are unknown. Interestingly, it has been shown with Salmonella typhimurium that native cells of deep rough mutants, but not the wild-type cells, can be oxidized with ferrocene, suggesting that in the wild type the lipopolysaccharides protect the external face of the outer membrane from the required lipophilic oxidant (Haller et al., 1975).

2.8.4. Outer membrane proteins involved in
DNA replication and cell morphogenesis

Recent findings have shed new light on the involvement of some of the outer membrane proteins in DNA synthesis, cell elongation, and septation. Thus, protein G (mol. wt., 15,000) appears to be a structural protein that is specific for cell elongation (James, 1975); protein D (mol. wt. 80,000) appears to be synthesized just before initiation of a round of DNA synthesis (Gudas et al, 1976); nalidixic acid, which inhibits DNA synthesis, causes the inhibition of the envelope protein D and stimulates the synthesis of a protein X (Gudas and Pardee, 1976; Gudas et al., 1976); another protein (mol. wt., 76,000) is only synthesized during a brief period near the time of bacterial division (Churchward and Holland, 1976). R plasmid RM98 apparently mediates an increase of this 76,000 dalton protein in the outer membrane (and the specific loss of a 36,000 dalton protein (Iyer et al., 1976). The relationships of outer membrane proteins and cell cycle are discussed in this volume by Daneo-Moore and Shockman.

2.9. The outer membrane as a permeability barrier

The outer membrane behaves as a fine molecular sieving barrier on the surface of the gram-negative bacteria with an exclusion limit (for uncharged oligosaccharides) of about 650 to 900 molecular weight (Nikaido, 1973, 1976; Nakae and Nikaido, 1976; Dacad and Nikaido, 1976). This limit is much smaller than that of the multilayered peptidoglycan-containing walls of gram-positive bacteria. An upper limit of 100,000 has been estimated in *B. megaterium* (Scherrer and Gerhardt, 1971). According to more recent estimates, the apparent exclusion threshold of the walls of *B. subtilis* and *B. licheniformis* would correspond to molecules with a diffusion radius of 2.5 nm (i.e., molecular weight ca. 70,000). Isolated, resealed outer membrane vesicles can be penetrated by saccharides with molecular weights higher than 1,000 (Nakae and Nikaido, 1976). This penetration seems to occur through "cracks" that are different from the physiological "pores" and are probably due to incorrect resealing. Such resealing may be a reflection of the high degree of protein-to-protein interaction that characterizes the outer membrane (section 2.8.). Since phospholipid bilayers are essentially impermeable to sugars larger than pentoses, and since mixed bilayers of both phospholipid and lipopolysaccharide have similar permeability properties, proteins are the best candidates responsible for outer membrane permeability. Reconstruction of sucrose-permeable vesicles from lipopolysaccharide, phospholipids, and proteins of the outer membrane may lead to the final identification of the permeability-conferring protein (Nakae and Nikaido, 1976). A protein complex participating in selective membrane permeability has been isolated. The active fractions contained three major protein species and the Braun lipoprotein was not one of them (Nakae, 1976). Peptidoglycan-associated outer membrane proteins were also isolated and compared. They would form, or be part of hydrophilic channels through the outer membrane (Lugtenberg et al., 1977). One of them, protein c, acts as a specific phage receptor in *E. coli* (Verhoef et al., 1977). A gene *meo* A, resonsible for the lack of this protein in some *coli* strains was mapped at 48 min on the linkage map (Verhoef et al., 1977).

Whereas hydrophobic substances mainly penetrate by dissolving into the hydrocarbon interior of the outer membrane (Nikaido, 1976), hydrophilic molecules that are too large to cross the pores must also pass through the outer membrane to support cell growth. Thus, other permeation mechanisms must exist. Recent studies on transport of Fe^{3+} ions have revealed complex functional interplay between specific receptor sites on the outer membrane, the active transport systems of the plasma membrane (Kaback, 1973), and specific binding proteins in the periplasmic space (i.e., the region between the outer and plasma membranes) (Rosen and Heppel, 1973).

E. coli K12 has a low affinity system for iron uptake that satisfies the cell requirement under conditions where a high concentration of Fe^{3+} ions is present in the medium (Frost and Rosenberg, 1973). It has also three high affinity systems through which Fe^{3+} ions are taken up either as complexes with citrate (Frost and Rosenberg, 1973; Hancock et al., 1976), enterochelin (Langman et al., 1972), or ferrichrome (Pollack et al., 1970; Luckey et al., 1972). The protein specified by the *ton* A gene (at 3 min on the linkage map) is involved in the ferrichrome-dependent iron uptake. This protein has been isolated by Braun and coworkers (Braun and Wolff, 1973; Braun et al., 1973; Braun and Hantke, 1974; Braun et al., 1974b; Hantke and Braun, 1975a,b). It consists of a single polypeptide (mol. wt., 85,000 daltons), and is localized in the outer membrane, where it also serves as receptor for phages T_5, T_1 and $\phi80$ and for colicin M; whereas T_5 binds irreversibly to the protein and releases its DNA, productive infection by T_1 and $\phi80$ and effective peptidoglycan degradation (and spheroplast formation) by colicin M require in addition, another function, the *ton* B function. This function has not yet been identified biochemically but it may be a "binding protein" for the three types of iron complexes, serving as a shuttle between the outer membrane and the plasma membrane (Boos, 1974). Finally a *feu* function (specified by a gene at 60 min), also localized in the outer membrane, is involved in the enterochehin-dependent iron uptake (Hantke and Braun, 1975a). It might be regarded as an analog of the *ton* A protein for the ferrichrome transport. Thus it appears that the transport of the large iron complexes (molecular weight higher than 700) through the outer membrane depends on the *feu* and the *ton* A functions (specific receptor sites) together with the *ton* B function (periplasmic binding protein).

Other receptor-dependent high affinity transport systems also exist in the outer membrane of *E. coli* such as the receptor protein for vitamin B_{12}, which is also utilized by the E colicins and by phage BF_{23} (Di Masi et al., 1973) and the maltose receptor system which is also utilized by phage λ (Schwartz, 1975, 1976; Schwartz and Le Minor, 1975; Schwartz et al., 1976; Szmelcman and Hofnung, 1975; Szmelcman et al., 1976; Ryter et al., 1975). Both vitamin B_{12} and maltose systems depend on additional periplasmic "binding proteins." These findings suggest that many, if not all, of the receptor proteins evolved as high-affinity binding components of the outer membrane. Initially devised to take up substrates present in the cell environment, these proteins were alternatively utilized by toxic agents such as phages and colicins in order to enter the bacteria.

Because of the presence of the outer membrane, many enzymes that are typically extracellular in gram-positive bacteria are "periplasmic" in the gram-negative bacteria (Heppel, 1967). The retention of these enzymes also poses an interesting problem because it may involve a complex mechanism. Indeed, mutants are known that are leaky (i.e., able to degrade substrates in the medium by an enzyme that is normally periplasmic) for only one or two periplasmic enzymes but not for all of them (Lopes et al., 1972). The molecular basis of this selective mechanism is not understood. At least one of these "leaky" mutants is known to possess an altered outer membrane protein (Lopes et al., 1972).

*2.10. Adhesion zones between outer
and plasma membranes*

Sections of plasmolyzed *E. coli*, *Salmonella*, and other gram-negative bacteria reveal that not all the plasma membrane is retracted from the more rigid outer membrane, but that both the outer and plasma membranes remain attached to each other at many distinct areas (Cota-Robles, 1963; Bayer, 1968, 1974, 1975). There are about 200 to 400 such adhesion zones per growing cell, each of them measuring approximately 25 to 50 nm in cross-sections (Fig. 33). Following deplasmolysis both membranes return to their normal positions more or less in contact with each other. Adhesion zones are not seen in stationary cells. They may play a part in cell growth and the question then arises of what function(s) could be ascribed to them.

2.10.1. Export sites of lipopolysaccharides

Bayer (1974), Mühlradt et al. (1973), and Kulpa and Leive (1976) have shown independently that these adhesion sites are physical channels through which the lipopolysaccharides are translocated across the peptidoglycan layer to the outer membrane once they have been synthesized at the plasma membrane (see section 3.4.4.).

2.10.2. Phage adsorption

Phage adsorption is not a "function." However, most cell surface structures of gram-negative bacteria are utilized by phages in order to inject their nucleic acid inside the bacterium. This is true for the following: (1) the F-pili that provide the adsorption sites for two classes of pili-specific phages, the filamentous DNA phages that adsorb to the tip and the icosahedral RNA phages that adsorb along the length of the pili (Curtis et al., 1969), (2) the flagellae on which phages such as PBS1 (Raimondo et al., 1968) and Xl (Meynell and Aufreiter, 1969) are initially attached, and (3) the outer membrane, which shows a wide spectrum of phage receptors assimilated with lipopolysaccharides and proteins (Bayer, 1974). Regardless of the strain used, localization of the phage receptors by the plasmolysis technique reveals that at low multiplicities of infection (maximum 100) phages are positioned preferentially over the adhesion zones. These phages are frequently seen with empty heads suggesting that release of the viral nucleic acid also occurs at these zones. Finally many, if not all, of the adhesion sites are able to adsorb all phages tested. Since a limited number of adhesion sites are available and many different phages can be adsorbed on these sites, each site appears to have more than one type of receptor.

2.11. Conclusions

Recently the gap that existed between the different conceptual approaches to bacterial wall structure, those of the electron microscopists, biochemists and physiologists, began to narrow and a unified molecular structural view emerged.

Walls exist in both gram-positive and gram-negative bacteria to preserve the plasma membrane against osmotic disruption. Gram-positive bacteria probably represent a primitive stage of evolution. They have solved the problem by synthesizing a vast amount of solid, multilayered peptidoglycan which, depending on

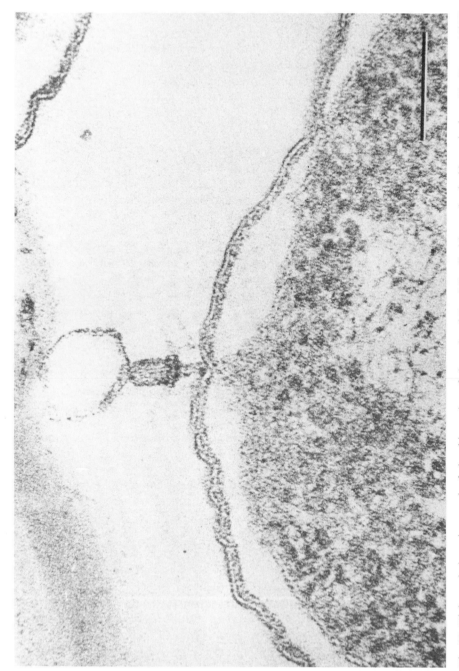

Fig. 33. High-resolution micrograph of ultrathin section of plasmolyzed *E. coli* B cell with an adsorbed bacteriophage T₂. (Bayer, 1975.) (Reprinted courtesy of Plenum Press, New York.) 288,000 ×

the species, exhibits many different structures. The thick walls of the gram-positive bacteria are highly permeable and do not pose any serious problems to the penetration of substances necessary for cell growth. Highly specialized acidic polymers such as teichoic acids, teichuronic acids, and anionic polysaccharides are embedded in the solid peptidoglycan matrix. They are covalently attached to the peptidoglycan or anchored in the underlying plasma membrane. At least a portion of them is localized very close to or over the cell surface. These non-peptidoglycan components seem to be involved in magnesium uptake, possibly functioning as a highly selective cation-exchange system between the exterior of the cell and the plasma membrane. Other important metabolic roles have also been assigned to these polymers, especially in regulating the autolytic activities of the cell.

Gram-negative bacteria have reached a much more advanced stage. They adopted the same simple structure for their peptidoglycan which, in addition, evolved as a monolayer. This simplification was due to the development of an outer membrane which, in addition, gave these bacteria the appreciable advantage of having an additional fine, molecular sieving barrier on their surface.

Both the peptidoglycan and the outer membrane contribute to the mechanical strength of the cell envelope. The plasma membrane, the peptidoglycan layer, and the outer membrane are closely associated. The peptidoglycan and the outer membrane are covalently attached to each other through part of the lipoprotein molecules. The plasma membrane and the outer membrane are also connected to each other through adhesion zones that extend through the peptidoglycan layer and are sites of export for at least some of the specific constituents (lipopolysaccharides) of the outer membrane. The plasma membrane and the outer membrane are also connected to each other through adhesion zones that extend through the peptidoglycan layer and are sites of export for at least some of the specific constituents (lipopolysaccharides) of the outer membrane. Many molecules required for cell growth cannot cross the pores of the outer membrane; consequently, specialized permeation mechanisms evolved in which "periplasmic" binding proteins serve as shuttles between specific receptor sites in the outer membrane and active transport systems in the plasma membrane. The outer membrane is a unique structure. Not only does it contain components that are not found elsewhere in nature but it also exhibits a noncovalent long-range ordered structure which does not fit the classical fluid mosaic model of membrane structure and may represent the most extreme case of asymmetric protein distribution ever encountered.

Finally, teichoic acids and anionic polysaccharides in gram-positive bacteria, lipopolysaccharides and proteins in the gram-negative bacteria—in fact virtually all the specialized components of the bacterial envelope that are distributed on the exterior of the cells and were devised to fulfill important physiological functions—are also utilized by toxic agents such as phages and/or colicins to enter the cell.

3. Biosynthesis and mode of assembly of bacterial walls

Bacterial walls are exocellular structures, and wall biosynthesis is therefore a three-stage process, each stage occurring at a different cell site: in the cytoplasm, on the membrane, and within the wall. Biosynthesis will be described for the four following main wall components: peptidoglycan, teichoic acid, lipoprotein, and lipopolysaccharide. The wall polymers are assembled from cytoplasmic precursors on specific "centers" that are functionally specialized portions of the plasma membrane. Each of these assembly centers contains (1) an "acceptor," or "carrier" to which the various units of a given wall polymer are transferred from the proper cytoplasmic precursors; (2) a series of transferases that function in a specific sequence so that the transfer reactions are catalyzed in the right order; and (3) phospholipids, of defined structure, which may act as ligands or affectors for some enzymes of the assembly centers. Normal wall or cell envelope synthesis requires an intimate functional integration of the various assembly centers involved.

3.1. Biosynthesis of peptidoglycan

3.1.1. The strategy
Contributions from the research groups of Strominger, Park, and Neuhaus, respectively, were essential in unraveling the processes involved in peptidoglycan synthesis (Strominger, 1970). In essence, the peptidoglycan precursors made on a uridylic acid cytoplasmic carrier (stage 1) are transferred from uridylic acid to the membrane assembly centers (stage 2), and from these to the expanding wall peptidoglycan where attachment occurs (stage 3). At some point during stage 3 peptidoglycan becomes insoluble.

Reactions of stage 1 lead to the synthesis of two nucleotide precursors: UDP-N-acetylglucosamine and UDP-N-acetylmuramyl-L-alanine-D-glutamyl-L-R$_3$-D-alanyl-D-alanine (Fig. 34). The peptide moiety of the latter is not a tetrapeptide as usually found in the completed wall peptidoglycans, but a pentapeptide that ends in a D-Ala-D-Ala sequence. Reactions of stage 1 are catalyzed by soluble, cytoplasmic enzymes.

Reactions of stage 2 ensure the assembly of the activated precursors and their transfer through the plasma membrane. The membrane carrier of the assembly centers involved in peptidoglycan synthesis is a C$_{55}$-polyisoprenoid alcohol phosphate (undecaprenyl phosphate). At the end of the process the uridylic acid carriers are regenerated into the cytoplasm pool. Reactions of both stage 1 and stage 2 are linked. They function as two integrated cycles: the nucleotide cycle and the lipid cycle, respectively.

Reactions of stage 3 occur on the outer face of the plasma membrane. They are also catalyzed by membrane-bound enzymes, and result in the insolubilization of the newly synthesized soluble peptidoglycan by incorporation into the preexisting wall peptidoglycan. One of the reactions involved is a transpeptida-

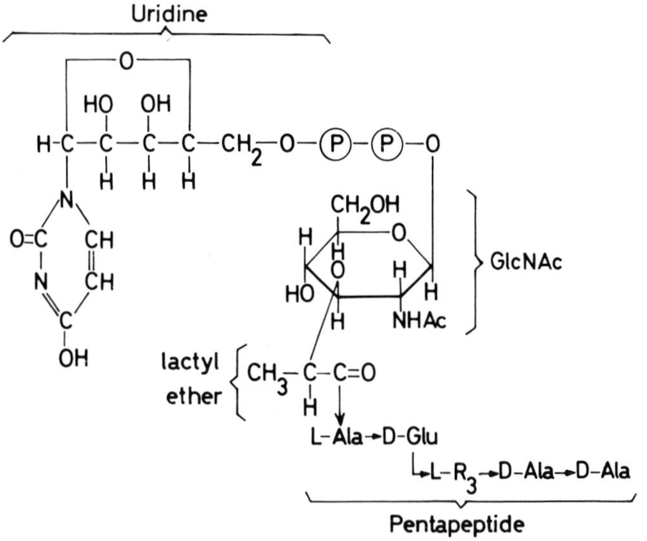

Fig. 34. The completed nucleotide precursor UDP-N-acetylmuramylpentapeptide.

tion. It causes peptide cross-linking between "new" polymers as well as between the "new" and the "old" polymers with concomitant release of one of the C-terminal D-alanine residues originally present in the nucleotide precursor. The released D-alanine can be reutilized by the cell either for peptidoglycan synthesis or for other processes by means of a specific transport system. In *E. coli* the transport for D-alanine, L-alanine, and glycine has been resolved. It is mediated by at least two enzymes. The system for D-alanine and glycine are related and differ from that of L-alanine (Neuhaus et al., 1972).

Figure 35 shows the main reactions involved in the synthesis of a type I peptidoglycan where additional amino acid residues do not occur in the interpeptide bridging.

In all peptidoglycans other than type I, the interpeptide bridges are not direct but consist of one or several additional amino acid residues (section 2.1.2.3.). The incorporation of those bridging residues onto the pentapeptide units L-Ala-D-Glu-L-R $_3$-D-Ala-D-Ala takes place during stage 1 or, usually during stage 2. Mechanisms also exist that are responsible for the variations encountered in both the glycan and the peptide moieties of the peptidoglycans of various bacterial species. These variations are of minor importance in terms of the general structure of the polymer (section 2.1.2.). In most cases the mechanisms involved are poorly understood or unknown. It is known, however, that in *S. aureus,* conversion of D-glutamic acid into D-isoglutamine (Fig. 11) is accomplished by amidation in the presence of ATP and ammonium ions at the level of the lipid cycle (Siewert and Strominger, 1968). Similarly, in *M. lysodeikticus,* substitution of the same α-carboxyl group of D-glutamic acid by a glycine residue (Fig. 12) is carried out in the presence of ATP at the lipid cycle level (Katz et al., 1967).

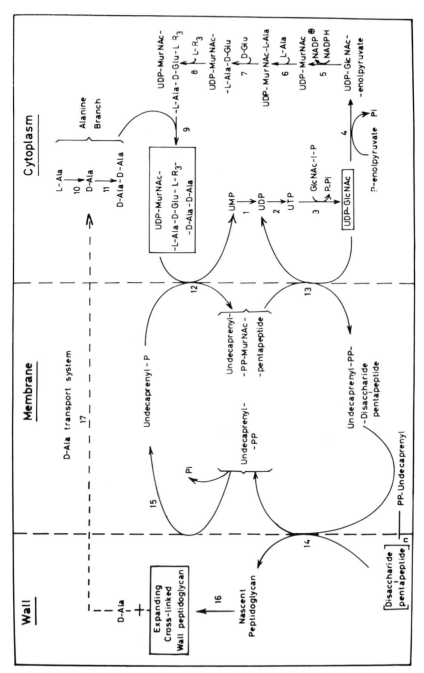

Fig. 35. Schematic representation of the biosynthesis of a peptidoglycan of type I. The three stages—cytoplasmic, membrane-bound, and wall-bound—are separated by the dashed vertical lines. GlcNAc = N-acetylglucosamine; MurNAc = N-acetylmuramic acid; L-R₃ = for example, *meso*-diaminopimelic acid. For peptidoglycans of types II, III and IV, interpeptide chains are incorporated at the cytoplasmic level or, more often, at the lipid cycle (see text). Structure of the undecaprenyl-PP is given in Fig. 36.

3.1.2. The membrane-bound lipid carrier
The lipid carrier of the peptidoglycan assembly centers in *S. aureus* has been isolated to a very high degree of purity, allowing its structure to be ascertained by mass spectrometry (Higashi et al., 1967, 1970b). It is a C_{55}-isoprenoid alcohol containing 11 isoprene units, with the chain ending in an alcoholic function (Fig. 36). In order to be functional, the carrier must be phosphorylated. In *S. aureus,* phosphorylation is achieved by a membrane-bound isoprenyl alcohol phosphokinase in the presence of ATP (Higashi et al., 1970a; Higashi and Strominger, 1970). The active enzyme is composed of a protein fraction and a phospholipid (phosphatidylglycerol). The enzyme-lipid complex is soluble and stable in several organic solvents. The protein fraction is inactive and insoluble both in water and in organic solvents. Its activity and solubility in organic solvents can be restored by addition of the phospholipid fraction. In other organisms, phosphorylation of the lipid carrier is achieved by a different route. In *Salmonella newington,* a particulate enzyme catalyzes the synthesis of C_{55}-lipid pyrophosphate from farnesyl pyrophosphate and isopentenyl pyrophosphate which, in turn, is derived from mevalonate (Christenson et al., 1969; Rothfield and Romeo, 1971). The C_{55}-lipid pyrophosphate is then dephosphorylated into inorganic phosphate and active C_{55}-isoprenoid alcohol phosphate carrier. Bacitracin prevents the action of the membrane-bound pyrophosphatase.

Lipid intermediates are key compounds in the biosynthesis of complex polysaccharides in prokaryotes and in eukaryotes. They have been well characterized in the biosynthesis of the peptidoglycans and the O-antigen lipopolysaccharides (section 3.4.). Their participation has also been demonstrated in the synthesis of a capsular polysaccharide (Troy et al., 1971), the membrane-bound lipomannan of *M. lysodeikticus* (section 3.2.3.), and mannolipids in *Mycobacteria* (Takayama and Goldman, 1970). These lipid intermediates generally contain a pyrophosphate or sometimes a monophosphate linkage between carbohydrate and lipid moieties, and in bacteria the latter is the polyisoprenoid alcohol with 55 carbon atoms. Similar structures are also involved in the synthesis of the yeast mannan (Tanner et al., 1971; Sentandreu and Lampen, 1971) and of mammalian glycolipids and glycoproteins (Oliver et al., 1975; also see the chapter by Cook in this volume). Mammalian glycolipids consist of a moiety with the properties of dolichol (or a similar compound), linked through a phosphate group to a carbohydrate residue. Dolichol is a mixture of very long-chain polyprenols

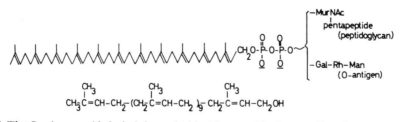

Fig. 36. The C_{55}-isoprenoid alcohol charged with either peptidoglycan or O-antigen precursors (see section 3.4.3.) *via* a pyrophosphate bridge.

(C_{80}–C_{105}) that are widely distributed in animal cells (Butterworth and Hemming, 1968).

3.1.3. Stage 1. The nucleotide cycle (Figure 35)

3.1.3.1. Synthesis of the precursor UDP-N-acetylglucosamine.
UTP, produced by phosphorylation of UMP and UDP by ATP (reactions 1 and 2), reacts with α-N-acetylglucosamine-1-phosphate to yield inorganic pyrophosphate and the nucleotide precursor UDP-N-acetylglucosamine (reaction 3). The reaction is catalyzed by UDP-N-acetylglucosamine pyrophosphorylase and is analogous to reactions leading to UDP-glucose and other compounds of this type.

3.1.3.2. Synthesis of the precursor UDP-N-acetylmuramylpentapeptide.
The next two reactions of the cycle consist of the transfer of a three-carbon unit from 2-phosphoenolpyruvate with formation of the pyruvate enol ether of UDP-N-acetyl-D-glucosamine (reaction 4), followed by reduction into UDP-N-acetylmuramic acid (reaction 5). The reactions are catalyzed by a transferase and a UDP-N-acetylglucosaminepyruvate reductase, respectively (Gunetileke and Anwar, 1968). The transferase is the specific target of the antibiotic phosphonomycin. The mode of action of phosphonomycin at the molecular level is known (Kahan et al., 1974).

Completion of the UDP-N-acetylmuramylpentapeptide is carried out by stepwise addition of L-alanine (reaction 6), D-glutamic acid (reaction 7), the L-R_3 residue (reaction 8) and, finally, a preformed D-alanyl-D-alanine dipeptide (reaction 9). Each step is catalyzed by a separate enzyme requiring ATP and either Mg^{2+} or Mn^{2+}. The synthesis is not directed by a nucleic acid template, and the constancy of peptidoglycan composition depends on the specificity of the relevant enzymes for their substrates (Strominger, 1970).

The synthesis of D-Ala-D-Ala (reactions 10 and 11) and its addition to the nucleotide tripeptide UDP-N-acetylmuramyl-L-Ala-D-Glu⌐L-R_3 (reaction 9) are carried out by three enzymes: (1) alanine racemase; (2) D-Ala:D-Ala ligase (ADP) or D-Ala-D-Ala synthetase; and (3) UDP-N-acetylmuramyl-L-Ala-D-Glu⌐L-R_3; D-Ala D-Ala ligase (ADP) or D-Ala-D-Ala adding enzyme (Neuhause et al., 1972). Several D-alanine antagonists are known, and among them D-cycloserine is of great importance. It behaves as a competitive inhibitor of both the racemase (with either D-Ala or L-Ala as substrate) and the synthetase, with K_i values considerably smaller than the K_m values for the substrates, D-cycloserine does not inhibit the D-Ala-DAla-adding enzyme.

In *E. coli* and *B. subtilis,* the alanine racemase is inhibited by both D- and L-cycloserine (Neuhaus et al., 1972; Johnston et al., 1966), whereas in *S. aureus* (Roze and Strominger, 1966) and *S. faecalis* (Lynch and Neuhaus, 1966; Wood and Gunsalus, 1951), this enzyme is inhibited by D- but not L-cycloserine. It has been suggested that some racemases have a single site for both D- and L-alanine and that others have two distinct binding sites. Racemases may also occur in two forms, one of which would bind L-alanine and the other D-alanine (Johnston and Diven, 1969).

The synthetase

$$(2 \text{ D-Ala} + \text{ATP} \xrightarrow[\text{K}^+]{\text{Mg}^{2+}} \text{D-Ala-D-Ala} + \text{ADP} + \text{Pi})$$

is inhibited by D-cycloserine, D-Ala-D-Ala and analogs of D-Ala-D-Ala (Neuhaus et al., 1972). Inhibition is specific for dipeptides. Additions to the N-terminal residue decrease the inhibitory activity (e.g., D-norvalyl-D-alanine is not an inhibitor), whereas additions to the C-terminal residue sometimes enhance the effectiveness of the peptides as inhibitors (e.g., D-alanyl-D-norvaline is a better inhibitor than D-Ala-D-Ala). Multiple binding sites on the enzyme exist, and it is believed that the function of these sites is to control the rate of dipeptide formation.

The specificity pattern of the D-Ala-D-Ala-adding enzyme for the addition of D-Ala-D-Ala analogs has also been studied. Remarkably, the substrate profile of this enzyme complements that of the D-Ala-D-Ala synthetase (Neuhaus and Struve, 1965). For example, the synthetase can incorporate D-norvaline (and other D-amino acids) in the C-terminal end of a dipeptide but not in the N-terminal end. The adding enzyme, in turn, binds and is inhibited by D-norvalyl-D-alanine but not by D-alanyl-D-norvaline. This combination of specificities also contributes to the relative accuracy of synthesis of the complete nucleotide precursor. Further, it accounts for the growth-inhibitory action exerted by some D-amino acids. Modification of peptidoglycan structure as a result of growth in unbalanced media has been investigated by Schleifer (Schleifer, 1975; Schleifer et al., 1976).

3.1.3.3. Incorporation of interpeptide chains (not shown in Figure 35). In *Lactobacillus viridescens*, the interpeptide chain consists of one L-alanine residue. L-alanine is incorporated to the ε-amino group of L-lysine of the nucleotide UDP-N-acetylmuramylpentapeptide by transfer from L-alanyl-tRNA (Plapp and Strominger, 1970). The reaction is catalyzed by a soluble enzyme of poor specificity that is able to transfer not only L-alanine but also, although less efficiently, L-serine, L-cysteine, and probably glycine from the corresponding tRNAs. *L. viridescens* is unusual. In all other bacteria examined (section 3.1.4.3.), incorporation of the interpeptide bridges occurs at the lipid intermediate level (Strominger, 1970).

3.1.3.4. Regulation. Osmotically fragile temperature-sensitive mutants of *E. coli* K12 impaired at the level of UDP-N-acetylglucosamine enolpyruvate reductase (proposed symbol : MurB), L-alanine-adding enzyme (MurC), *meso*-diaminopimelic acid-adding enzyme (MurE) or D-Ala-D-Ala adding enzyme (MurF), respectively, were isolated (Matsuzawa et al., 1969; Lutgenberg, 1971; Lutgenberg and de Haan, 1971; Wijsman, 1972). Mutants impaired at the level of UDP-N-acetylglucosamine-2-phosphoenolpyruvate transferase (MurA) and of the D-Glu-adding enzyme (MurD) have not yet been obtained. The MurC, E and F genes are localized extremely close to each other (at 1–1.5 min of the chromosome map). They might form or be part of an operon. The MurB gene is located at 77 minutes. Mutants with impaired L-Ala racemase (alr, at 3 min) and D-Ala-D-Ala ligase (ddl, at 17 min) activities have also been obtained.

D-cycloserine causes an enormous accumulation of UDP-N-acetylmuramyl-tripeptide in *E. coli* K12, but penicillin, which inhibits one stage 3 reaction of biosynthesis (section 3.1.5.), fails to cause the accumulation of UDP-N-acetylmuramylpentapeptide. Similarly, mutants altered in one of the membrane-bound enzymes that could accumulate this nucleotide apparently do not exist. These observations suggest that in *E. coli* K12 UDP-N-acetylmuramylpentapeptide may regulate its own biosynthesis by feedback inhibition (Lutgenberg, 1971). This phenomenon is certainly not frequent. *S. aureus* lacks this type of regulation, since penicillin causes a large accumulation of nucleotides. This accumulation was one of the key observations leading to our current knowledge of penicillin's mode of action (Wise and Park, 1965; Tipper and Strominger, 1965).

The amount of D-Ala-D-Ala available for condensation with UDP-N-acetylmuramyl-L-Ala-D-Glu⌐L-R$_3$ is submitted to complex regulation. In addition to the occurrence of several product-binding sites on the synthetase the phenomenon also includes (1) regulation of the size of the intracellular pools of D- and L-alanine (Wargel et al., 1971); (2) competition for available D-alanine by other reactions that utilize it (e.g., teichoic acid synthesis); (3) D-Ala is a specific inducer of L-alanine dehydrogenase (in *B. subtilis*) and hence limits the amount of L-alanine available to the racemase (Berberich et al., 1968). In other bacteria (*Pseudomonas aeruginosa*), D-Ala induces a D-amino acid dehydrogenase which limits the level of D-Ala (Marshall and Sokatch, 1968); (4) in *E. coli*, the synthesis of the racemase is repressed by high concentrations of alanine (Neuhaus et al., 1972); and (5) the kinetic properties of the racemases in both *E. coli* and *S. faecalis* are such that the intracellular L-Ala pool must be larger than the D-Ala pool for the reaction to proceed in the L→D direction with the required velocity (Neuhaus et al., 1972).

*3.1.4. Stage 2. The lipid cycle in the assembly centers
of the plasma membrane (Figure 35)*
The first two reactions consist of translocating the peptidoglycan precursors from the hydrophilic environment of the cytoplasm to the hydrophobic environment of the membrane. This interchange of carriers leads to the synthesis of disaccharide-peptide β-1,4-N-acetylglucosaminyl-N-acetylmuramylpeptide units, a mechanism that ultimately is responsible for the alternating sequence of N-acetylglucosamine and N-acetylmuramic acid in the glycan strands of the completed wall peptidoglycan (Neuhaus, 1971).

3.1.4.1. Transphosphorylation (Neuhaus, 1971). First, phospho-N-acetylmuramylpentapeptide is translocated from UDP-N-acetylmuramyl-pentapeptide to the membrane undecaprenyl phosphate carrier (reaction 12). Undecaprenyl-PP-N-acetylmuramylpentapeptide is formed and UMP is generated. The translocase has been solubilized and partially purified (Heydanek and Neuhaus, 1969). The transphosphorylation reaction proceeds without loss of energy: UDP-MurNAc (pentapeptide) + undecaprenyl-P $\rightleftharpoons$ undecaprenyl-PP-MurNAc (pentapeptide) + UMP. The equilibrium may be reached by either route and the K_{eq} value is about 0.25. Forward and reverse reactions require Mg^{2+}. Potassium ions stimulate the forward reaction and UMP inhibits it. An enzyme-P-N-acetylmuramyl (pentapeptide) complex is transitorily formed (with release of UMP; exchange reaction), and then P-N-acetylmuramyl (pentapeptide) is transferred from the enzyme to the undecaprenyl phosphate carrier (transfer reaction). The translocase has a high specificity for the uracyl moiety. Fluorouracil causes great accumulation of FUDP-N-acetylmuramyl-pentapeptide. A C-terminal D-Ala-D-Ala sequence is also an essential feature of the substrate. The specificity profiles of both exchange and transfer reactions complement those of the earlier reactions in the synthesis.

3.1.4.2. Transglycosylation (Chatterjee and Park, 1964; Meadow et al. 1964; Struve and Neuhaus, 1965).   N-acetylglucosamine is, in its turn, translocated from UDP-N-acetylglucosamine to undecaprenyl-PP-N-acetylmuramylpentapeptide (reaction 13). Undecaprenyl-PP-disaccharidepentapeptide is formed and UDP is generated. This transglycosylation differs from the preceding reaction in that only N-acetylglucosamine and not the terminal phosphate of UDP is transferred to the lipid intermediate. Tunicomycin prevents formation of disaccharide lipid intermediate by blocking the transfer of N-acetylglucosamine onto undecaprenyl muramylpentapeptide pyrophosphate (Bettinger and Young, 1975).

3.1.4.3. Incorporation of interpeptide chains (not shown in Figure 35).   With most type II peptidoglycans (section 2.1.2.3.), incorporation of interpeptide chains occurs within the peptidoglycan assembly centers by extension of the length of the side chain of the L-R$_3$ residue of the lipid intermediate through substitution of the ω-amino group by one or several amino acid residues. These transfers are catalyzed by particulate enzymes and differ completely from mRNA-coded protein synthesis on ribosomes.

tRNAs participate in the incorporation of glycine and L-amino acid residues. Thus for example, in *S. aureus,* five glycine residues from glycyl-tRNA are sequentially added to the ε-amino group of L-lysine on the undecaprenyl-PP-disaccharide-peptide intermediate (Matsuhashi et al., 1967; Kamiryo and Matsuhashi, 1969; Thorndike and Park, 1969). Four species of tRNA-Gly exist in *S. aureus* and support glycine incorporation in the peptidoglycan. Three of them participate in template-directed polypeptide synthesis. The fourth one is apparently peptidoglycan specific; it may be a unique gene product. A similar mechanism occurs in *Arthrobacter crystallopoietes* for the incorporation of one L-alanine residue (from L-alanyl-tRNA) (Roberts et al., 1968) and in *Staphylococcus epidermidis* for the incorporation of 3 glycine and 2 L-serine residues (from glycyl-tRNA and L-seryl-tRNA) (Petit et al., 1968).

Contrary to the aforementioned examples, the interpeptide bridges in *S. faecalis* and *L. casei* consist of a single iso-asparaginyl residue that has the D-configuration. In these cases tRNAs do not participate in the incorporation

(Staudenbauer and Strominger, 1972; Staudenbauer et al., 1972). D-aspartic acid is activated as β-D-aspartyl-phosphate by a membrane-bound enzyme and is then transferred to the ε-amino group of L-lysine of the lipid intermediate. Finally, amidation of the α-carboxyl group is achieved in the presence of NH_3 and ATP.

3.1.4.4. Transfer of disaccharide peptide (Figure 35). The transfer of the disaccharide peptide units from the undecaprenyl-PP-disaccharide-peptide to an appropriate acceptor generates undecaprenyl pyrophosphate (reaction 14) which, in turn, is dephosphorylated by a membrane-bound pyrophosphatase (reaction 15). This latter reaction yields inorganic phosphate and the initial C_{55}-isoprenoid alcohol phosphate carrier which can begin a new cycle. As previously mentioned, bacitracin is an inhibitor of the pyrophosphatase (Siewert and Strominger, 1967; Siewert, 1969; Storm, 1974). Transfer of the disaccharide peptide from the lipid carrier and wall peptidoglycan expansion involves two reactions: one insures elongation of the glycan chains by transglycosylation and the other insures peptide cross-linking by transpeptidation. These reactions are part of stage 3 of biosynthesis and are discussed in the following section.

3.1.5. Stage 3. Wall peptidoglycan expansion
The first attempts to study the mechanism of wall peptidoglycan expansion were made by Mirelman, Sharon, and coworkers (Mirelman and Sharon, 1972; Mirelman et al., 1972). *S. aureus* and *M. lysodeikticus* were used as models for these studies (section 3.1.5.3.). Since additional amino acid residues occur in the interpeptide bridges in these bacteria (Figs. 11 and 12), the processes involve a complex series of steps. Since bacilli that have a type I peptigolycan with a direct cross-linkage between the peptide units (Fig. 10) are simpler models, studies with these bacteria will be described first.

3.1.5.1. Synthesis of nascent peptidoglycan in bacilli. The elegant studies of Ward and Perkins (1973, 1974) on the biosynthesis of peptidoglycan by cell-free membrane preparations from a poorly lytic mutant of *Bacillus licheniformis* have shown that chains consisting of multiple disaccharide peptide units grow by addition of the new disaccharide peptide units at the reducing terminal of the lengthening chain. In the process the reducing terminal end (i.e., N-acetylmuramic acid) of the growing chain is transferred from its link with the membrane to the nonreducing terminal residue (i.e., N-acetylglucosamine) of the new disaccharide peptide unit which is itself linked to the membrane. In their original papers Ward and Perkins interpreted their observations as polymerization of nascent glycan chains on the undecaprenyl phosphate carrier (Fig. 37), but the nature of the acceptor was not characterized. Regardless of the exact mechanism, the nascent peptidoglycan thus formed emerges on the exterior of the plasma membrane, which has actually been seen using isolated protoplasts of *B. megaterium* (Fritz-James, 1974) and of *B. licheniformis* (Elliott et al., 1975a,b).

The synthesis of an uncross-linked, linear peptidoglycan by purified plasma

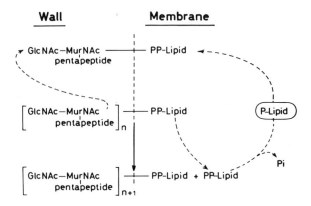

Fig. 37. Synthesis of the nascent peptidoglycan on the undecaprenyl phosphate carrier. Elongation of the glycan chain by transglycosylation on the lipid intermediate. The reaction shows the addition of a new disaccharide peptide unit.

membranes depends strictly on the presence of exogenously added UDP-N-acetylglucosamine and UDP-N-acetylmuramylpentapeptide precursors. It is entirely prevented by the presence of lysozyme (which degrades the polymer as it is being formed) but is unaffected by penicillin.

3.1.5.2. Insolubilization of nascent peptidoglycan by attachment to preexisting wall peptidoglycan and by autopolymerization in bacilli. The attachment of the nascent peptidoglycan to the preexisting wall peptidoglycan in the same poorly lytic mutant of *B. licheniformis* was studied using a cell-free, wall plus membrane preparation (Ward, 1974). Attachment is achieved by transpeptidation (Fig. 38). In this reaction, the penultimate C-terminal D-alanine of a pentapeptide unit acting as donor is transferred to the proper N-terminal group of another peptide unit acting as acceptor. Interpeptide bonds are formed in the absence of exogenous energy and equivalent amounts of D-alanine residues are liberated from the donor peptides (Tipper and Strominger, 1968). The reaction is catalyzed by a membrane-bound transpeptidase and is inhibited by the β-lactam antibiotics (penicillins and cephalosporins) (Wise and Park, 1965; Tipper and Strominger, 1965).

The direction of the transpeptidation reaction with the wall plus membrane preparation from *B. licheniformis* was established by using a nucleotide precursor UDP-N-acetylmuramylpentapeptide in which the free amino group of *meso*-diaminopimelic acid (i.e., the potential acceptor group for transpeptidation) had been blocked by an [14C]acetyl group (Ward and Perkins, 1974). Incorporation of the resulting acetylated nascent peptidoglycan in the wall material occurred, demonstrating that the nascent peptidoglycan must act as donor (through its C-terminal D-Ala-D-Ala sequences) and that the preexisting peptidoglycan must act as acceptor (through the amino group located on the D-center of *meso*-diaminopimelic acid). In vivo studies confirmed the overall process, that attachment of the nascent peptidoglycan is mainly performed by transpeptidation whereas addition of newly synthesized material to the glycan strands of the preexisting wall peptidoglycan by transglycosylation could be at best a minor pathway of peptidoglycan synthesis in *B. licheniformis.*

The system involved in peptide crosslinking contains more than one single enzyme and consists of at least two antagonistic activities. In addition to the transpeptidase activity which catalyses peptide bound formation, bacteria also contain a DD-carboxypeptidase activity which simply hydrolyses the C-terminal D-Ala-D-Ala sequences of pentapeptide units without performing any transfer reaction. As the transpeptidase is sensitive to penicillin, so too is the DD-carboxypeptidase activity. The DD-carboxypeptidase activity is probably there to limit the number of pentapeptide units made available to the transpeptidase activity and therefore, to control the extent of peptide crosslinking. The specific inhibition of the DD-carboxypeptidase activity in *B. subtilis* and *B. stearothermophilus* does not cause detectable damage to the cell (Blumberg and Strominger, 1971; Yocum et al., 1974), suggesting that the transpeptidase is the enzyme that is physiologically important.

Cells incubated in the presence of β-lactam antibiotics released uncross-linked peptidoglycan material in the medium (Tynecka and Ward, 1975). A similar observation was made with *Brevibacterium divaricatum* (Keglevic et al., 1975). Again, these observations showed that glycan elongation on the plasma membrane by transglycosylation was insensitive to these antibiotics, whereas attachment of the nascent peptidoglycan to the preexisting wall peptidoglycan was inhibited under these conditions.

Quantitation of the nascent peptidoglycan attachment by the wall plus membrane preparation of *B. licheniformis* showed that the acetylated nascent peptidoglycan was utilized less efficiently than the natural compound, only about 23% as much being incorporated under identical conditions. This reduced degree of cross-linking could suggest that with the natural precursor not only attachment to the "old" wall peptidoglycan occurred but many crosslinks were also formed between the new chains themselves (Fig. 38), a process that was precluded by the use of the acetylated precursor. This assumption is probably correct, as shown by the remarkable study by Rogers and coworkers (Elliott et al., 1975a,b) on the reversion of protoplasts of *B. licheniformis* (MH-1 and 6346His⁻) to normally rod-shaped bacilli. In liquid suspension growing protoplasts excrete the soluble, nascent peptidoglycan they produce. Obviously, at some stage of the extension, detachment from the carrier must occur. The mechanism of this reaction is not known. When incubated on the surface of a medium containing 2.5% agar, however, protoplasts of *B. licheniformis* can reverse successfully to normal bacilli. Hence insolubilization of the nascent peptidoglycan by peptide cross-linking (and incorporation of the other wall polymers synthesized simultaneously) can be achieved by cells that completely lack preexisting walls at the start of the process. The operation, however, is difficult and immobilization of the protoplasts by attachment to the agar in a medium of low fluidity is necessary.

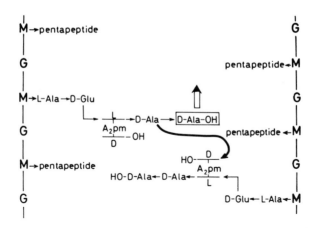

Fig. 38. Formation of a cross-linkage by transpeptidation between two nascent strands of peptidoglycan (adapted from Perkins et al., 1975). In the process, one pentapeptide serves as peptide donor (through the C-terminal D-Ala-D-Ala sequence) and another pentapeptide serves as peptide acceptor (through the N-terminal group of the L-R$_3$ residue; here a *meso*-diaminopimelic acid residue). Attachment of the nascent peptidoglycan to the preexisting wall peptidoglycan is also catalyzed by transpeptidation. (See text for details.)

In the experiments reported here (Elliott et al., 1975a,b) protoplasts were incubated on the surface of a medium containing 2.5% agar, N-acetyl [^{14}C]glucosamine to measure peptidoglycan synthesis, [2-^{3}H]glycerol to measure teichoic acid synthesis, and [^{3}H]tryptophan to measure protein synthesis. Early stages of peptidoglycan synthesis were visualized by use of ferritin-conjugated antibody. Cells still attached to the agar were also studied by freeze-etching, and cells scraped off the agar were used for biochemical work. A study of the sequence of events occurring during reversion of protoplasts to bacilli showed that at early stages the glycan chains were very short and the peptides poorly cross-linked. Reaction with ferritin-labeled peptidoglycan antibody showed patches of long, thin, flexible fibrils emerging from large, misshapen membranous bodies. Subsequently, the surface became covered increasingly with fibrils emerging from an increasing number of points, and after several hours a layer of loosely organized material was seen surrounding the cells using osmium fixatives or freeze-etching (Fig. 39). At this stage, however, the cells still rounded up when removed from the agar and suspended in isoosmolal solutions. As reversion proceeded, cross-linking reached its maximal value before the glycan chains reached their normal, final average length. After complete reversion, the wall material was rod-shaped and identical to that of the original bacilli. It contained peptidoglycan, teichoic acid, and teichuronic acid in about the same proportions. During the process, soluble peptidoglycan was found in the reversion medium in amounts that decreased as the reversion proceeded. Soluble products, however, were not formed by reverting protoplasts of an autolysin-deficient mutant.

538

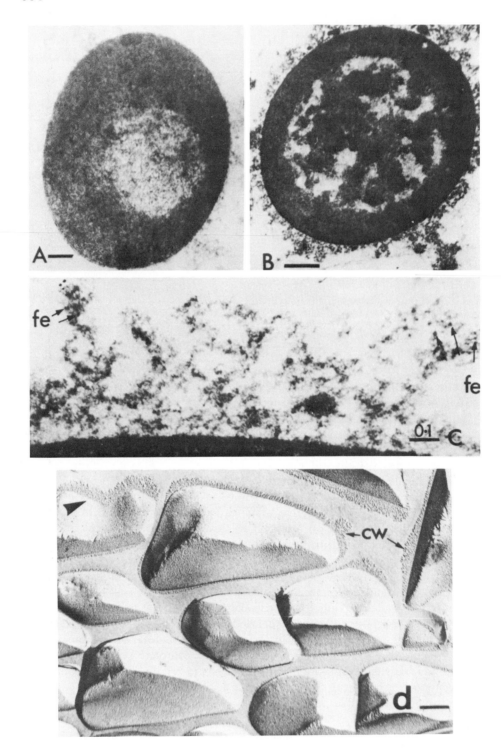

Reversion of fused protoplasts to the bacillary form has opened a new tool in bacterial genetics. While transformation, conjugation, transduction, sexduction and transfection are unidirectional processes transferring DNA, and only DNA, from a donor into a recipient bacterium, fusion occurring in mixed populations of protoplasts derived from two parental strains which are both nutritionally-complementing and polyauxotrophic, followed by wall regeneration can lead to the formation of prototrophic bacteria. Recently such a process was applied successfully to *B. subtilis* by Schaeffer et al. (1976) and to *B. megaterium* by Fodor and Alföldi (1976).

B. megaterium expands its wall peptidoglycan in a way similar, if not identical, to that described for *B. licheniformis* (Reynolds, 1971; Schrader et al., 1974; Schrader and Fan, 1974; Taku et al., 1975; Fuchs-Cleveland and Gilvary, 1976). Isolated membranes catalyzed glycan chain elongation and also partial attachment of these nascent chains to each other by transpeptidation. Partially cross-linked peptidoglycan not attached to preexisting wall peptidoglycan and exhibiting an average molecular weight greater than 6×10^6 was shown to be formed (Schrader et al., 1974). Since peptidoglycans of smaller molecular weights were obtained in the presence of penicillin, transpeptidation was thus partially responsible for the large size of the product.

Toluenized cells of *B. megaterium,* where wall and membrane remain in close contact, catalyzed both peptidoglycan synthesis and its attachment to preexisting wall peptidoglycan by cross-linking (Schrader and Fan, 1974). Toluene treatment seems to preserve remarkably the cross-linking machinery of the plasma membrane. Extraction with LiCl of the toluene-treated cells resulted in a greatly diminished capability of peptidoglycan synthesis. Recovery of the original activity could be obtained by readding the LiCl solubilized proteins to the deficient cells (Taku et al., 1975). One of these protein factors, a glycoprotein with a molecular weight of about 42,000 to 52,000 was purified 124-fold (Taku and Fan, 1976a,b). At present, the nature and exact physiological role of these protein factors in peptidoglycan synthesis are still unknown.

In vivo studies of the peptide cross-linking reaction in *B. megaterium* and its sensitivity to penicillin were performed by Fordham and Gilvarg (1974). One interesting result showed that in the absence of penicillin the reaction continued for many minutes after incorporation of precursor subunits into peptidoglycan.

3.1.5.3. Wall peptidoglycan expansion gram-positive cocci. Synthesis of a nascent

Fig. 39. Reversion of protoplasts of *Bacillus licheniformis.* Sections of reverting protoplasts after 6 h of incubation: (a) Treated first with unconjugated antibody against peptidoglycan followed by reaction with ferritin-conjugated antibody, showing absence of ferritin particles. (b) Low magnification view of section after direct reaction with ferritin-labeled antibody. (c) Appearance examined at higher magnification (fe, ferritin particles). (a,b and c) Cells removed from the agar and suspended in liquid medium. Bars represent 0.5 μm unless otherwise indicated. (d) Freeze-etched cells examined in situ, showing the shape and the fringe of variable length around the cells (CW). Arrow shows the direction of shadowing of replicas. The bar represents 0.5 μm. (Elliott et al., 1975.) (Reprinted courtesy of the American Society for Microbiology.)

peptidoglycan and its attachment to the "old" wall peptidoglycan by transpepti- dation are also essential features of peptidoglycan expansion in gram-positive cocci. Crude wall preparations of *S. aureus* containing strongly associated mem- brane fragments were utilized both as enzyme sources and as acceptors for peptidoglycan synthesis (Mirelman and Sharon, 1972). These preparations also contain tRNA and glycyl-tRNA synthetase, so that they incorporate glycine into the preexisting wall peptidoglycan. Both cross-linking of the newly synthesized peptidoglycan by transpeptidation and glycine incorporation in the wall fraction were strongly inhibited by penicillin.

Autoplasts of *Streptococcus faecalis* 9790 produced by the action of the native autolytic N-acetylmuramidase were grown in the presence of heat-inactivated cell walls (to bind autolytic enzyme and thus reduce the level of peptidoglycan hydrolysis) and of tetracycline (to inhibit further autolysin synthesis) (Rosenthal and Shockman, 1975). Autoplasts secreted soluble, infrequently peptide cross- linked glycan chains. Transfer of this material to the exogenously added walls did not occur, suggesting that a close connection between wall and membrane is necessary for the wall to act as acceptor in transpeptidation. The transpeptida- tion reaction was studied with whole cells continuously labeled during exponen- tial growth. The pattern of distribution of monomers, dimers, and trimers sug- gested that cross-linking between peptides was not a random condensation pro- cess but proceeded by a monomer addition mechanism (Dezélée and Shockman, 1975; Oldmixon et al., 1976).

In some cocci, transpeptidation is not the only mechanism of precursor incor- poration. Thus in *M. lysodeikticus*, some incorporation of precursors in the wall peptidoglycan still occurs in the presence of penicillin at concentrations that completely block transpeptidation. The amounts of precursor attached under these conditions is about 30% of that incorporated in the absence of penicillin. Presumably, this penicillin-insensitive incorporation results from elongation of the preexisting wall glycan strands by transglycosylation (Fig. 40). With *M. lysodeikticus*, additional complications also arise because of its complex interpep- tide bridges (Fig. 12). The linear uncross-linked glycan strands secreted by this bacterium (Mirelman et al., 1972; 1974a,b,c; 1975; Mirelman and Bracha, 1974) in the presence of penicillin contain about 150 disaccharide units. About half of these strands are substituted by the hexapeptide L-Ala-D-Glu(Gly) ⌐ L- Lys-D-Ala-D-Ala, whereas the others have their muramic acid residues with free carboxyl group. Free hexapeptide is also secreted in the medium and occurs in amounts equivalent to that of the unsubstituted muramic acid residues. The secretion of this free peptide must be the result of a penicillin-insensitive amidase acting on the nascent peptidoglycan. The latter enzyme is presumably involved in the translocation of some peptide units from their muramyl residues into a bridging position. Incorporation of the liberated peptide units in that position requires formation of D-Ala-L-Ala linkages. Evidence has been obtained that this link is probably also made by a transpeptidation reaction in which the C-terminal D-alanine of one hexapeptide that is attached to the glycan is trans- ferred to the N-terminal L-alanine of a free hexapeptide. Amidase action and

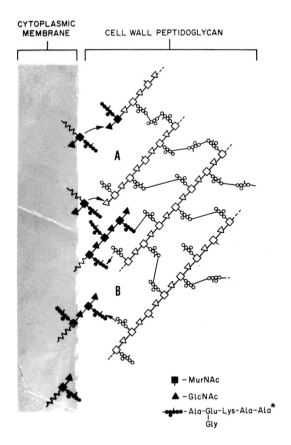

CYTOPLASMIC
MEMBRANE   CELL WALL PEPTIDOGLYCAN

A

B

■ – MurNAc

▲ – GlcNAc

–•••– Ala-Glu-Lys-Ala-Ala*
                    |
                   Gly

Fig. 40. Model of growth of *M. lysodeikticus* wall peptidoglycan. Newly synthesized strands are attached to preexisting wall peptidoglycan by two mechanisms. The main one is transpeptidation to an amino group on a preexisting peptide side chain, with concomitant release of terminal D-alanine from the newly synthesized peptidoglycan chain (see B on Fig.). The second mechanism is the attachment by transglycosylation of an oligosaccharide-peptide intermediate to a nonreducing end of a preexisting glycan chain (see A on Fig.). Symbols representing preexisting glycan strands are: GlcNAC-MurNAc, △—☐; and the peptide

Ala-Glu-Lys-Ala-Ala, oo○oo.
        | ○
       Gly

Newly synthesized peptidoglycan is in black. The strands incorporated either singly or in polymerized form are depicted as lipid bound, although it is possible that the lipid moiety is removed before incorporation. (Mirelman et al., 1974c.) (Reprinted courtesy of the New York Academy of Sciences.)

D-Ala-L-Ala peptide bond formation must occur in strict coordination. By repetition of the process, interpeptide bridges of increasing size can be achieved. The penicillin susceptibility of the formation of D-Ala-L-Ala crosslinkages is about 50-fold less than that of the N^ϵ-(D-Ala)-L-Lys linkages. Fig. 41 shows the proposed sequence for the release of hexapeptide units (through amidase action at sites 2, 4, 6, 8), the formation of a cross-linking peptide chain composed of four

542

pentapeptide units (through D-Ala-L-Ala formation at sites 9, 7, 5, 3), and the closure of the cross-linking bridge (through N$^\epsilon$-(D-Ala)-L-Lys formation at site 1) (Ghuysen, 1968). Two transpeptidases exhibiting different penicillin susceptibility may exist. Alternatively, only one transpeptidase might be responsible for both peptide cross-linking reactions and the variations in penicillin susceptibility might be ascribed to differences in location of the reactions in the cell.

The direction of transpeptidation in *Gaffkya homari* (Hammes, 1976; Hammes and Kandler, 1976) was shown to proceed in such a way that the preexisting wall peptidoglycan functions as carbonyl donor and the newly synthesized peptidoglycan strands function as amino acceptors. This situation is entirely different from that described above for *B. licheniformis* (see section 3.1.5.2.). The mechanism is especially complex. Indeed, these peptide units of the nascent peptidoglycan function as carbonyl donor and the newly synthesized peptidoglycan strands function as amino acceptors. This situation is entirely different units must remain present in this nascent peptidoglycan in such a way that through the active transpeptidation of the tetrapeptide units, these pentapeptide units are passively incorporated into the wall peptidoglycan where they can serve as donor units for further expansion of the polymer. In vivo, the newly synthesized peptidoglycan strands which undergo attachment probably contain equimolar amounts of pentapeptide and tetrapeptide units. Such a ratio would allow a maximum degree of crosslinking of 50% and oligomers larger than dimers could not be formed. In fact, such structural features are those found in the wall peptidoglycan of *G. homari*. Whereas the specific inhibition of the DD-carboxypeptidase activity in various bacilli by penicillin does not cause detectable damage to the cell (see section 3.1.5.2), the specific inhibition of the DD-

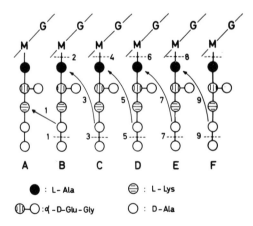

Fig. 41. Proposed biosynthetic sequence of the modification of the peptidoglycan in *Micrococcus lysodeikticus*. Reactions 2, 4, 6 and 8 are the result of hydrolysis by an N-acetylmuramyl-L-alanine amidase. Reactions 1, 3, 5, 7 and 9 are transpeptidations. Reaction 1 leads to formation of N$^\epsilon$-(D-Ala)-L-Lys linkage. Reactions 3, 5, 7 and 9 lead to formation of L-Ala-D-Ala linkages. (Ghuysen et al., 1968.) (Reprinted courtesy of the American Chemical Society.)

carboxypeptidase activity in *G. homari* prevents peptidoglycan incorporation (Hammes, 1976). Hence, an unified view on the nature of the enzyme on which penicillin binding causes cessation of cell growth (and cell lysis) cannot be proposed. The concept of a penicillin "killing site" may well escape a precise definition.

3.1.5.4. Effect of the microenvironment of the assembly centers of peptidoglycan synthesis. Virtually nothing is known about the functioning of the enzymes in the peptidoglycan assembly centers. Mention should be made, however, of recent experiments by Neuhaus and coworkers who introduced a new and promising methodology for studying the effects of the microenvironment on, and the dynamics of, the processes that lead to the synthesis of the nascent peptidoglycan. The synthesizing system of *Gaffkya homari* (Hammes and Neuhaus, 1974a,b) was utilized to synthesize a spin-labeled nascent peptidoglycan from UDP-N-acetylglucosamine and UDP-N-acetylmuramyl-(TEMPO-ϵN)-pentapeptide (Johnston et al., 1975; Johnston and Neuhaus, 1975). Spin labels are sensitive probes to study the microenvironments of intermediates in membrane-catalyzed reactions, since the electron-spin resonance spectrum of the spin label is a function of the motion that the probe experiences and of the polarity of the medium surrounding the probe. In the above system 188 pmoles of lysozyme-sensitive polymer were synthesized in 15 minutes. The spectrum of the spin-labeled lipid intermediate indicated that it was weakly immobilized relative to the spin-labeled nucleotide. The spectrum of the spin-labeled nascent peptidoglycan had a line-broadening characteristic of nitroxide-nitroxide interactions due to the high concentration of spin labels. Incubation with lysozyme resulted in marked sharpening of the spectrum. Both spin-labeled lipid intermediate and spin-labeled peptidoglycan were found to be accessible to vancomycin and ristocetin. The mechanism of action of these two antibiotics is related to their ability to complex with the C-terminal D-Ala-D-Ala moieties that are present at various phases of peptidoglycan synthesis (Perkins and Nieto, 1974).

3.1.5.5. Peptidoglycan expanson in E. coli. In their studies with *E. coli* W7 and W945T3282, Braun and coworkers (Braun and Bosch, 1973; Braun et al., 1974a; Braun and Wolff, 1975) observed that the only detectable radioactive peptidoglycan in the plasma membrane that could be chased from the plasma membrane into the rigid layer was the disaccharide peptide unit linked to the lipid carrier. The technique failed to reveal any formation of glycan chains before transfer from the lipid. Thus, peptidoglycan extension in *E. coli* might proceed by direct insertion of newly synthesized disaccharide-peptide monomer units. The results obtained also indicated that the rigid layer was not enlarged at growing points distributed at random throughout the rigid layer. Instead, during a 1-minute pulse, the diaminopimelate label was shown not to be incorporated in those regions where lipoprotein was attached to peptidoglycan; areas comprising 30% of the peptidoglycan remained largely preserved from growth. Prevention of peptidoglycan unit insertion at lipoprotein attachment sites might be due to steric reasons since, for example, lysozyme is also sterically hindered by lipoprotein and leaves about three peptidoglycan units attached to each lipoprotein molecule. In this process, attachment of lipoprotein to the newly synthesized peptidoglycan portions would occur subsequently.

Assuming that the above model is correct, acceptor sites in the preexisting glycan chains (i.e., nonreducing N-acetylglucosamine) could be created by any type of endo-N-acetylmuramidase autolysin. Following insertion or concomitant with it, closure of bridges between peptide units belonging to adjacent strands would be achieved by transpeptidation. Virtually nothing is known about the transglycosylation reaction; however, endo-N-acetylmuramidases exhibiting potential biosynthetic activities are known. Hen egg white lysozyme, for example, is able to catalyze transglycosylation reactions (Chipman et al., 1968; Pollock and Sharon, 1970). The same endo-N-acetylmuramidase, occurring perhaps in two forms, might catalyze both creation of acceptor sites in the glycan strands by hydrolysis and insertion of new units by transglycosylation. The recently discovered peptidoglycan:peptidoglycan-6-muramyl transferase (section 2.1.7.) might also fulfill such a function.

A different view is expressed by Mirelman, Yashow-Gan, and Schwarz (1976) who showed experimentally that the mechanism of wall peptidoglycan expansion in E. coli (a thermosensitive division mutant PAT84) may be very similar to that in bacilli. E. coli PAT84, made permeable to nucleotide precursors by ether treatment, catalyzes peptidoglycan synthesis and its attachment to the preexisting wall peptidoglycan. Synthesis during a bacterial division cycle was investigated by initiating synchronous cell division with a shift down from restrictive (42°C) to a permissive (30°C) growth temperature. An abrupt increase in cell number occurred 20 to 30 minutes after the temperature shift down indicating that in these cells a synchronous triggering of septum and polar cap formation had occurred. At various times after the temperature shift down cells were collected, treated with ether, and utilized as representatives of the various stages during cell division. Nonseptated filaments formed at the restrictive temperature were similarly investigated. In these experiments newly synthesized, mostly peptide uncross-linked macromolecular peptidoglycan was estimated to be insoluble in trichloracetic acid but soluble in hot sodiumdodecylsulfate. Only the peptidoglycan that became covalently linked to preexisting wall peptidoglycan (i.e., the sacculus in Fig. 42) was insoluble in both reagents (Fig. 42).

Of interest was the finding that the amount of pentapeptide units in the SDS-soluble peptidoglycan intermediate was under the control of a DD-carboxypeptidase activity. Release of D-alanine residues (by both transpeptidase and DD-carboxypeptidase), covalent attachment of new peptidoglycan and formation of peptide cross-linkages were all completely inhibited by high concentrations of ampicillin (50 μg/ml). Low dose levels of ampicillin (0.5 μg/ml), however, specifically inhibited the DD-carboxypeptidase activity of cells grown at permissive temperature without affecting transpeptidation. Peptidoglycan incorporation in the preexisting wall was increased and septum formation was inhibited. Similarly, the DD-carboxypeptidase activity of nonseptated filaments

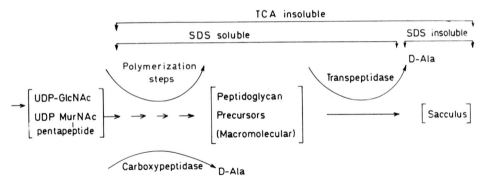

Fig. 42. Schematic model of principal reactions which participate in the synthesis and attachment of newly synthesized peptidoglycan strands to preexisting ones in the *E. coli* cell wall. The SDS-insoluble material accounts only for newly synthesized peptidoglycan bound covalently to the preexisting cell wall. The macromolecular peptidoglycan intermediates remain in the cell upon treatment with TCA, and are predominantly uncross-linked at its peptide side-chains (Mirelman et al., 1976.) (Reprinted courtesy of the American Chemical Society.)

grown at restrictive temperature was lower than that of the transpeptidase. Very little SDS-soluble peptidoglycan could be detected in these filaments and the extent of cross-linkage was increased when compared to that of cells grown at permissive temperature. Thus an increased ratio of transpeptidase to DD-carboxypeptidase, induced either by low ampicillin concentration at permissive temperature or by a shift up to nonpermissive temperature, appeared to result in an increased amount of substrate donor available for transpeptidation. Consequently, an increased rate of peptidoglycan incorporation in the rigid layer and a greater extent of peptide cross-linking were obtained. Also, under conditions where the extent of transpeptidation was not controlled by DD-carboxypeptidase activity, triggering of formation of new septa was prevented. Regardless of the exact situation, it is clear that a balance of the relative activities of transpeptidase and DD-carboxypeptidase is important for cell division and morphology. On the basis of these observations, and more recently, of the effects caused by cephalexin and nalidixic acid on peptidoglycan synthesis at the permissive temperature (Mirelman et al., 1977), the attractive working hypothesis was made that in *E. coli*, the nascent peptidoglycan would act as carboxyl donor for transpeptidation reactions with two types of acceptors. Reaction with the preexisting wall peptidoglycan would lead to wall elongation, whereas reaction with a modified nascent peptidoglycan previously deprived of donor sites by DD-carboxypeptidase action and accumulated at the equator of the cell (i.e., where it is synthesized) would lead to septum formation. The specific inhibition of the DD-carboxypeptidase responsible for the destruction of the donor sites of the nascent peptidoglycan (and for its accumulation) and/or the specific inhibition of the transpeptidase which uses this modified nascent polymer as amino acceptor would therefore prevent cell septation and would cause cell filamentation.

Recently, the work of Spratt and Pardee (1975) and Spratt (1975, 1977) has shed new light on the subtle processes that control cell wall growth and septation in *E. coli.* These authors have shown that a set of membrane-bound proteins in this organism are apparently involved specifically in cell division, cell elongation, and cell shape. [^{14}C]benzylpenicillin and [^{14}C]penicillin FL-1060 were bound to the plasma membrane selectively solubilized with Sarkosyl NL-913. Penicillin FL-1060, the amidino-penicillanic acid (Lund and Tybring, 1972), differs from other β-lactam antibiotics in that penicillin FL-1060 at low concentration pro- duces ovoid shaped gram-negative cells without inhibiting cell division or caus- ing lysis. Penicillin FL-1060 is known to interfere with an event that starts at the beginning of the C period of the cell division cycle in *E. coli* and is responsible for normal cell elongation (James et al., 1975). Six main penicillin-binding proteins (numbered 1 to 6 on the basis of decreasing molecular weights) and a series of other minor binding proteins were separated on sodium dodecylsulfate polyac- rylamide slab gels and detected by fluorography. Evidence that protein n°2 (molecular weight 66,000) was involved in rod-shape maintenance was strong. Indeed [^{14}C]penicillin FL-1060, which specifically causes ovoid cells, bound exc- lusively to protein n°2. Moreover, a mutant (B-6) was isolated that grew as round cells and failed to bind β-lactam antibiotics to protein n°2. About 10 copies of protein n°2 exist per cell. Weaker evidence indicated the probable involvement of protein n°3 (mol. wt., 60,000) in cell division and of protein n°1 (mol. wt. 91,000) in cell elongation. Bulge formation, typical of the early stages of penicil- lin intoxication, might result from inhibition of proteins 2 and 3 since the β-lactam antibiotics that caused bulges in the strain used also showed higher affinities for protein 2 and 3 than for protein 1. Finally, since the peptidoglycan produced under conditions where either protein n°2 was inhibited (by ovoid cells) or protein n°3 was inhibited (by filaments) was apparently not defective, and since inhibition of protein n°1 stopped cell elongation, protein n°1 was proposed as the "main" transpeptidase responsible for the attachment to the wall of the newly synthesized peptidoglycan. The significance of this work, however, remains incomplete as long as the possible correlation between those multiple penicillin-binding proteins and the known multiple penicillin-sensitive enzymes remains unknown.

Finally, Goodell and Schwarz (1975) have shown that in addition to the pro- tein 2 of Spratt and Pardee the peptidoglycan itself may play a role in cell morphogenesis. Spherical *E. coli* cells retaining an apparently mechanically in- tact but spherical peptidoglycan were produced by mutation and by treatment with penicillin FL-1060. Under proper conditions these spherical cells were seen to reshape themselves into rods. During this process the areas of the cell en- velope that had been the ends of the original rods became the ends of the newly formed rods. In contrast, osmotically sensitive spheroplasts lacking the rigid layer were unable to revert to rods although they were able to synthesize a new but spherical rigid layer. Loss of the original rigid layer in *E. coli* might thus be paralleled by loss of the ability to revert to a rod, that is, to retain cell polarity.

3.1.5.6. The peptide cross-linking enzyme system   Transpeptidation is a key reaction in peptidoglycan expansion. The complexity of the enzyme system involved in the peptide cross-linking reaction varies greatly depending on the bacteria (Ghuysen, 1977a,b). The most simple systems appear to occur in *Streptomyces*, gram-positive eubacteria that form a characteristic mycelium. The enzyme system seems to consist, of one membrane-bound enzyme in *Streptomyces* strain *rimosus* (a transpeptidase that exerts a low DD-carboxypeptidase activity) and two membrane-bound enzymes in *Streptomyces* strains K15 and R61 (a transpeptidase that exerts a low DD-carboxypeptidase activity, and a DD-carboxypeptidase that shows a low transpeptidase activity). Strains R61 and K15 also possess two other DD-carboxypeptidases performing a low transpeptidase activity; one of them can be released from the cells during protoplast formation by lysozyme treatment and the other is excreted in the culture medium during growth. Membrane-bound, lysozyme-releasable and exocellular DD-carboxypeptidases are immunologically related to each other (Ghuysen, 1977b). The membrane-bound peptide cross-linking enzyme system in *E. coli* is exceedingly complex (Pollock et al., 1974; Nguyen-Distèche et al., 1974a,b; Mirelman et al., 1976; Ghuysen, 1977a). Fractionation of the extracted complex suggested the occurrence of (1) a transpeptidase whose main function is to catalyze transpeptidation, that is, dimerization between two peptide units. An in vitro specific test for the transpeptidase has been devised. It consists of a mixture containing a low concentration of radioactive pentapeptide L-Ala-D-Glu-(L)-*meso*-diamino-pimelyl-(L)-D-[^{14}C]Ala-D-[^{14}C]Ala (acting mainly as donor) and a high concentration of amidated tetrapeptide L-Ala-D-Glu(amide)-(L)-*meso*-diaminopimethyl-(L)-D-Ala (acting exclusively as acceptor). D-[^{14}C]Ala is liberated and the monoamidated octapeptide dimer is formed (Fig. 43); (2) a DD-carboxypeptidase whose main function is to hydrolyze the C-terminal D-Ala-D-Ala peptide bond of pentapeptide units (whether they occur free, as disaccharide peptide, in the form of the nucleotide precursor, or at the C-terminal end of a peptidoglycan polymer) without concomitant transpeptidation; and (3) an endopeptidase whose main function is to hydrolyze the dimers formed by transpeptidation into monomers. Since the interpeptide bond D-Ala-(D)-*meso*-diaminopimelic acid made by transpeptidation extends between two D-centers and is in α-position to a free carboxyl group (Fig. 43), the endopeptidase that hydrolyzes it has the specificity of a DD-carboxypeptidase. Although they evolved to perform one of these specific activities with high efficiency, each of these enzymes seems also to be able to perform the other activities with low efficiency. The situation becomes more complicated since this multienzyme (transpeptidase-DD-carboxypeptidase-endopeptidase) complex probably occurs in two distinct forms, one of them sensitive to very low dose levels and the other to higher dose levels of ampicillin.

As discussed earlier (section 3.1.5.3.), the peptide cross-linking enzyme system in *M. lysodeikticus* is also exceedingly complex since its functioning probably implies the involvement of two transpeptidases and a N-acetyl-muramyl-L-alanine amidase. In addition, *M. lysodeikticus* contains a periplasmic DD-carboxypeptidase activity of undetermined physiological function (Linder and Salton, 1975).

Bacteria may contain LD-carboxypeptidase-transpeptidase enzyme systems.

548

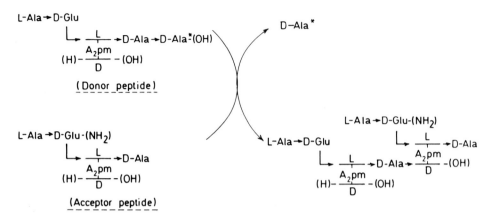

Fig. 43. Mono-amidated octapeptide dimer formed *in vitro* by the *E. coli* peptide cross-linking enzyme system from pentapeptide L-Ala-D-Glu $\ulcorner$(L)-*meso*-diaminopimelyl-(L)-D-Ala-D-Ala and amidated tetrapeptide L-Ala-D-Glu (amide) $\ulcorner$(L)-*meso*-diaminopimelyl-(L)-D-Ala.

These systems liberate the C-terminal D-alanine residues from tetrapeptides L-Ala-D-Glu$\ulcorner$L-R$_3$-D-Ala (that are formed by action of a DD-carboxypeptidase) either without concomitant transpeptidation (LD-carboxypeptidase) or with concomitant transfer of the tripeptide L-Ala-D-Glu$\ulcorner$L-R$_3$ to a proper amino group acceptor ("atypical" transpeptidase). *E. coli* possesses an LD-carboxypeptidase which is periplasmic (Strominger, 1970). Its activity is not constant throughout the division cycle (in cells of D11/lac $^+$pro$^+$), detectable activity being highest at the time of division by a factor of three (Beck and Park, 1976). The corresponding atypical transpeptidase has not been detected but its involvement in the covalent attachment of the lipoprotein to the peptidoglycan has been postulated (section 3.3.). *Streptococcus faecalis* possesses a membrane-bound "atypical" transpeptidase that is inhibited by various β-lactam antibiotics (Coyette et al., 1974). Its physiological role is unknown. Such "atypical" transpeptidases might catalyze the synthesis of "atypical" L-R$_3$L-R$_3$ interpeptide bridges described earlier (section 2.1.2.3.).

Multiple penicillin binding sites exist in both gram-negative and gram-positive bacteria. These binding sites occur in the plasma membrane and, to all appearances, they are distinct proteins. The fluorography technique has revealed the presence of at least 6 penicillin binding sites in *E. coli* (see section 3.1.5.5.), 5 in *S. typhimurium* (Shepherd et al., 1977), in *B. subtilis* and *B. megaterium,* 4 in *B. stearothermophilus* and 3 in *B. cereus* (Blumberg and Strominger, 1974). In most cases, the physiological role and the possible enzymatic function of these binding sites remain to be discovered. However, proteins 5 and 6 in *E. coli* were shown to be DD-carboxypeptidases-transpeptidases (Spratt and Strominger, 1976). Similarly, proteins 1, 4 and 5 in *Salmonella typhimurium* were shown to be DD-carboxypeptidases-transpeptidases, protein 4 having the greatest capacity for carrying out transpeptidation *in vitro* (Shepherd et al., 1977). It has also been suggested that in *B. subtilis,* the penicillin binding component n°2 would be the most likely target for killing by penicillins (Buchanan and Strominger, 1976).

Extraction and characterization of a membrane-bound transpeptidase pose technical problems because experiments of this type cannot be attempted unless the transpeptidase activity can be directly and specifically estimated with the help of a proper system of donor and acceptor peptides. The nucleotide precursors UDP-N-acetylglucosamine and UDP-N-acetylmuramylpentapeptide are not substrates of the transpeptidase. Nascent peptidoglycans are poorly characterized, both chemically and physically, and preclude precise kinetic studies. Defined substrate systems that allowed these enzymes to be estimated independently of the preceding biosynthetic reactions have been developed and their use has allowed several transpeptidases to be isolated and studied (Ghuysen, 1977a,b). Several strains of *Streptomyces* were shown to excrete enzymes during growth that appeared to be soluble forms of the membrane-bound transpeptidases. Some of these extracellular enzymes were purified to protein homogeneity. Each extracellular enzyme was shown to be capable of performing both DD-carboxypeptidase and transpeptidase activities, and the ratio between these activities could be modified by changing the environmental conditions (e.g., pH, polarity of the incubation medium, ionic strength, acceptor and donor concentrations) (Frère et al., 1973; Ghuysen et al., 1973). Membrane-bound DD-carboxypeptidases-transpeptidases were also solubilized and partially purified. Both membrane-bound and exocellular enzymes were used as models to study the transpeptidation reaction and the interaction with penicillin (Ghuysen, 1977a,b). β-Lactam antibiotics are actually substrates of these enzymes. Enzyme-antibiotic complexes are formed that subsequently undergo spontaneous breakdown, and the half-lives of these complexes vary from minutes to days depending on the enzyme and the antibiotic, so that the velocity with which the antibiotic is degraded during the interaction may vary greatly. By using the exocellular DD-carboxypeptidase-transpeptidase of *Streptomyces* strain R61, benzylpenicillin was shown to be fragmented into phenylacetylglycine (Frère et al., 1975) and N-formyl-D-penicillamine (Frère et al., 1976).

3.1.6. Regulation of the autolysins

Wall growth and the changes that occur in the shape of the peptidoglycan network during the life cycle are complex phenomena. They are discussed by Daneo-Moore and Shockman elsewhere in this volume. The cell autolysins are actively involved in these processes. Obviously, the peptidoglycan hydrolases must be prevented from uncontrolled action; a "barrier" exists between these enzymes and their substrate.

3.1.6.1. The "barrier" between autolysins and peptidoglycan in E. coli.

E. coli possesses a set of different types of membrane-bound peptidoglycan hydrolases despite which peptidoglycan breakdown will not occur if intact cells from either the exponential or stationary phases are maintained in buffer at 37°C (Hartmann et al., 1974). The characteristics of the barrier are unknown but its maintenance clearly depends on the integrity of the cell envelope. Breaking the barrier, and triggering hydrolase activity, can be achieved in many different ways, as follows:

(1) mechanical opening of the cells; (2) changing osmotic pressure (e.g., treatment of the cells with 1 M NaCl at 0°C or with 20% sucrose); (3) removal of divalent cations by EDTA; (4) interaction with penicillin (for further details on the effects of penicillin on cell surface growth and division, see Daneo-Moore and Shockman, chapter 9 in this volume) and (5) treatment of the cells with 5% trichloracetic acid. The latter treatment causes optimal activation of the peptidoglycan hydrolase complex, and this shows that the membrane-bound hydrolases in their native environment are well protected against otherwise efficient denaturing agents. Despite the presence of this barrier, localized hydrolase action appears to be triggered at a given stage of the cell life cycle. Presumably "enzyme action would be permitted for a defined time period in a specific cell area by controlled local breakdown and reestablishment of the barrier" (Hartmann et al., 1974). Experimentally, peptidoglycan hydrolase activities in synchronized cultures of *E. coli* B/r were shown to increase discontinuously during the cell cycle with a maximal activity occurring at the same time that the rate of wall synthesis was maximal (Hakenbeck and Messer, 1977). The transpeptidase-DD-carboxypeptidase-endopeptidase system in *E. coli* also seems to be controlled by some "barrier." Thus, the lower DD-carboxypeptidase activity exhibited by the ether-treated filaments of *E. coli* (obtained at the nonpermissive temperature; section 3.1.5.5.) is not due to temperature sensitivity of the enzyme. Mechanical disruption of the filaments by sonication causes a great increase in DD-carboxypeptidase activity, suggesting that this enzyme is partially inactive in its natural environment in the intact filament (Mirelman et al., 1975).

3.1.6.2. The "barrier" between autolysins and peptidoglycan in gram-positive bacteria. The barrier between autolysins and peptidoglycan in gram-positive bacteria can be partially described in biochemical terms. In *S. faecalis*, the occurrence of two interconvertible forms of the autolytic enzyme, one active and the other latent, provides a possible mechanism for the mode of control of the wall autolysin (Higgins and Shockman, 1971; Joseph and Shockman, 1976). Various defined wall components are also involved in this control; they include the lipoteichoic acids, the wall teichoic acids, the wall teichuronic acids and specialized autolysin modifiers.

Membrane lipoteichoic acids isolated from several bacterial species (with common backbone of polyglycerol phosphate, but differing in sugar substituents and in glycolipids) and the pneumococcal Forssman antigen (section 2.3.3.) have been examined for effects on various wall autolysins (endo-N-acetyl-muramidases of *S. faecalis* and *L. acidophilus*; N-acetylmuramyl-L-alanine amidases of *B. subtilis* and *D. pneumoniae*) (Cleveland et al., 1975). The conventional lipoteichoic acids inhibit both N-acetylmuramidases and the *B. subtilis* amidase at relatively low concentrations, but fail to inhibit the pneumococcal amidase. Conversely, the pneumococcal Forssman antigen inhibits the pneumococcal amidase but has no effect on the three other autolysins. Deacylation of the lipoteichoic acids results in loss of lytic inhibitory activity on the sensitive autolysins, and similarly the pneumococcal choline-containing wall teichoic acid is 700-fold less active with the pneumococcal amidase than the Forssman antigen. Further work is required to unravel the specificities of the observed effects, but the occurrence of the lipoteichoic acids in both acylated and deacylated forms implicates a role for these polymers in the in vivo regulation of autolytic activity. It should also be mentioned that autolysis of intact cells of *S. faecalis* was found to be inhibited to a greater extent by phospholipids than by lipoteichoic acid (Cleveland et al., 1976). This observation suggests a possible difference in the accessibility of native autolysin to these various substances.

In vivo, the specificity of at least some autolysins can be modulated by specific modifiers. Thus, both the N-acetylmuramyl-L-alanine amidase autolysin and a specific modifier protein were purified from *B. subtilis* ATCC 6051 (Herbold and Glaser, 1975a,b). The modifier protein combines stoichiometrically with the enzyme and stimulates its activity threefold. The major effect of the modifier, however, is not to alter the level of enzyme activity but rather to change the pattern of cleavage from a random one, when only the enzyme hydrolyzes the wall, to a more sequential, specific pattern in the presence of the modifier protein.

Wall teichoic acids have been assigned a role as "allosteric" (or "allotopic") ligands for wall autolytic enzymes (Tomasz and Westphal, 1971; Herbold and Glaser, 1975a,b). These ligands appear to be responsible for the tight binding of at least some autolysins to their wall substrate. One revealing example of this type of interaction is found in *D. pneumoniae*. Walls from pneumococci grown on choline-containing medium have a choline-containing teichoic acid and are substrates of the N-acetylmuramyl-L-alanine amidase autolysin. Removal of the wall teichoic acid by treatment with periodate or with formamide renders the walls resistant to the amidase (Mosser and Tomasz, 1970). Moreover, pneumococci grown on the choline analog, ethanolamine, incorporate ethanolamine in the teichoic acid where it replaces choline, resulting in synthesized walls that are resistant to the homologous autolysin (Höltje and Tomasz, 1975b). In vitro methylation of these autolysin-resistant, ethanolamine-containing walls under conditions where virtually all the ethanolamine residues are converted to choline results in recovery of amidase sensitivity. Clearly, the autolysin-catalyzed hydrolysis of the amide bonds on the peptidoglycan requires interaction between the enzyme and the choline residues of the wall teichoic acid. The wall choline-containing teichoic acid activates the amidase, whereas the membrane choline-containing lipoteichoic acid inhibits it. Thus, regulation of the enzyme activity appears to depend on the relative amount, activity, or localization of these two ligands.

Another example of the ligand properties of wall teichoic acids toward autolysins is *B. subtilis* 6051. In this bacterium the teichoic acid is a polyglycerolphosphate polymer containing α-linked glucosyl residues, and teichoic acid is also required for the action of the modifier protein on the amidase autolysin. Herbold and Glaser (1975a,b) have suggested that the apparent absence of wall hydrolytic enzymes in teichoic acid-negative mutants may not be due to the lack of enzyme synthesis by these cells, but may instead reflect the fact that in the

552

absence of teichoic acid the enzymes are not bound tightly to the wall and are probably lost from the cell *in vivo* and during cell breakage and wall isolation.

The studies mentioned above led to an interesting observation concerning the mode of action of penicillin (and other antibiotics inhibiting peptidoglycan synthesis) at the cellular level. Working with a mutant pneumococcus defective in autolysin, Tomasz and Waks (1975) noted: (1) growth of the mutant was inhibited by penicillin at the same concentration as the one that induced lysis of the wild type; (2) exogenous, wild-type autolysin alone had no effect on the growth of the mutant but caused its lysis if penicillin (at the minimum growth inhibiting concentration) was added simultaneously to the medium; (3) penicillin and other inhibitors of peptidoglycan synthesis caused the escape into the medium of a trichloracetic acid-precipitable, choline-containing polymer; and (4) growth inhibition of the mutant, its sensitization to exogenous penicillin, and release of the choline-containing polymer all showed the same dose response as that of the penicillin-induced lysis of the wild type. One probable interpretation of these results is that inhibition of peptidoglycan synthesis destabilizes the complex between the autolysin and its inhibitor (lipoteichoic acid) and triggers autolysin activity.

Finally in *B. licheniformis,* it seems that the wall teichuronic acid (and not the wall teichoic acid) controls lytic activity. Cells grown under certain conditions, contain both polymers in their walls. Removal of teichoic acid from isolated walls had no effect on autolysin sensitivity, whereas removal of teichuronic acid made the walls resistant to autolysin. Moreover a novobiocin-resistant mutant lacking teichuronic acid was isolated (Robson and Baddiley, 1977). It was defective in its cell morphology, the isolated walls had an increased resistance to autolysin and did not possess such autolysin. It was concluded that teichuronic acid was necessary for both the binding of the autolysin to the wall and for the bound enzyme to hydrolyse the peptidoglycan. It was also suggested that small variations in the localization of teichuronic acid in the wall might result in different rates of autolysis in specific regions and that cell separation might be achieved by discontinuities in the distribution of this polymer across the septal wall (Robson and Baddiley, 1977).

*3.2. Biosynthesis of teichoic acids
and other anionic polysaccharides*

3.2.1. Synthesis of lipoteichoic acids
Little is known about the synthesis of lipoteichoic acids. Attempts to demonstrate lipoteichoic acid formation from CDP-glycerol in a variety of microorganisms failed. Studies with *S. aureus* H (Glaser and Lindsay, 1974) suggested that phosphatidylglycerol was a precursor of lipoteichoic acid, not only of the hydrophobic end of the molecule but also of at least a portion of the polyglycerol phosphate chain. Data were also consistent with the assumption that only the glycerol-phosphate moiety of phosphatidylglycerol was used in the synthesis of lipoteichoic acid. Work on the biosynthesis of the phosphatidylkojibiosyl dig-

lyceride, the phosphoglycolipid moiety of the lipoteichoic acid in S. *faecalis*, is in progress (Pieringer and Ganfield, 1975).

Attempts to demonstrate the incorporation of D-alanine into isolated lipid- and D-alanine-free teichoic acids were unsuccessful. Thus, it was thought that D-alanine incorporation might occur only when the polymer was associated with the membrane, perhaps in a particular conformation or environment. Such a possibility was supported by other observations showing that isolated teichoic acid has lost properties that it performs when it is integrated in the cell envelope. For example, bacteriophages adsorb to teichoic acid bound with the wall but not to isolated teichoic acid as if adsorption requires a given conformation of the receptor or a special orientation of the sugar residues that are imparted by the binding of teichoic acid to the cell envelope (Young, 1967; Coyette and Ghuysen, 1968). Incorporation of D-alanine into membranes was therefore investigated (Reusch and Neuhaus, 1971). *Lactobacillus casei* was chosen for these studies because its membrane contains a simple lipoteichoic acid consisting of a linear polyglycerol phosphate with D-alanine esterified at position 2 of the glycerol moiety (Kelemen and Baddiley, 1961). In the presence of Mg^{2+} ions and ATP, D-alanine incorporation appeared to proceed through a two-step reaction involving two cytoplasmic, soluble enzymes, a D-alanine activating enzyme (E_1) and a D-alanine:membrane acceptor ligase (E_2) (Linzer and Neuhaus, 1973; Neuhaus et al., 1974):

$$\text{D-Ala} + E_1 + \text{ATP} \rightleftharpoons E_1 \cdot \text{AMP-D-Ala} + \text{PP}$$

$$E_1 \cdot \text{AMP-D-Ala} + \text{MEMBRANE ACCEPTOR} \xrightarrow{E_2} E_1 + \text{AMP} + \text{D-Ala-MEMBRANE ACCEPTOR}$$

The acceptor has not been characterized. It may be the lipoteichoic acid itself. The system is exceedingly complex as has been revealed particularly by a study of the membrane glycerol teichoic acid of a stabilized L-form of *Streptococcus pyogenes* (Chevion et al., 1974). The glucose-containing teichoic acid from this L-form lacks D-alanyl esters and is shorter in length than that of the parental coccus (13 *versus* 25 units) (Slabyi and Panos, 1973). Like the parental coccus, the stabilized L-form contains both D-alanine activating and ligase enzymes and the L-form enzymes successfully catalyze D-alanine incorporation into the membranes of the parental coccus. No incorporation was observed, however, when the L-form membranes were used as substrates, demonstrating that the L-form membranes lack functioning D-alanine acceptor activity. The exact reason for this situation is unclear. It may be due to an alteration of the lipoteichoic acid acceptor (which appears to be shorter than that of the parental strain), or to a change in its conformation, the absence of one enzyme or several enzymes associated with the membrane or, finally, to changes in the complex lipid and fatty acids of the L-form membrane.

3.2.2. Synthesis of wall teichoic acids
Wall teichoic acids are synthesized through a three-stage process which, in essence, is similar to that described for peptidoglycan synthesis.

3.2.2.1. The nucleotide precursors. CDP-glycerol is the precursor of the wall glycerol teichoic acids and CDP-ribitol is the precursor of the wall ribitol teichoic acids (Fig. 44). CDP-glycerol is synthesized from D-glycerol-1-phosphate (which arises from glycolysis) and CTP. The reaction is catalyzed by a CDP-glycerol pyrophosphorylase: D-glycerol-1-P + CTP $\rightleftharpoons$ CDP-glycerol + PP. Similarly, CDP-ribitol is synthesized from D-ribitol-5-phosphate and CTP by a CDP-ribitol pyrophosphorylase. D-ribitol-5-phosphate is formed by enzymic reduction of D-ribulose-5-phosphate by NADH. CDP-glycerol and CDP-ribitol are known to occur in gram-positive bacteria and were actually characterized by Baddiley and his colleagues before the polymers for which they serve as biosynthetic precursors were discovered (review, Baddiley, 1972).

An interpendence of the peptidoglycan and teichoic acid synthesizing systems exists at the level of the relevant nucleotides. Such an early reciprocal control is suggested by the fact that a precursor of one wall polymer can interact with an enzyme involved in the synthesis of the second wall polymer. The UDP-N-acetylglucosamine pyrophosphorylase and the phosphoenolpyruvate UDP-N-acetylglucosamine enolpyruvyl transferase involved in peptidoglycan synthesis and the CDP-glycerol pyrophosphorylase involved in teichoic acid synthesis appear to be possible control points in the synthesis of the wall of *B. licheniformis* (Anderson et al., 1972, 1973) (Fig. 45). UDP-N-acetylmuramylpentapeptide inhibits all three enzymes, influencing both peptidoglycan and teichoic acid syntheses. CDP-glycerol can affect peptidoglycan synthesis by inhibiting UDP-N-acetylglucosamine pyrophosphorylase. The only stimulating effects observed are

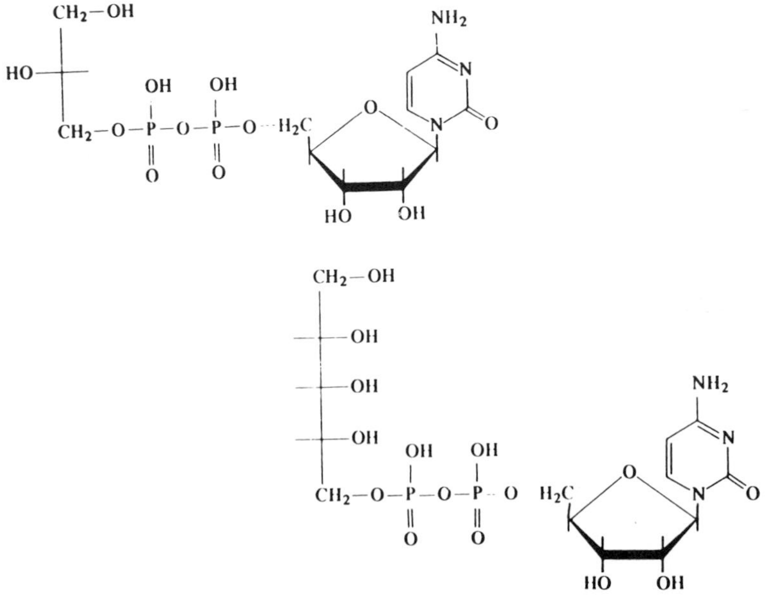

Fig. 44. Cytidine diphosphate glycerol and cytidine diphosphate ribitol.

on CDP-glycerol pyrophosphorylase by low concentrations of UDP-N-acetylglucosamine and UDP-N-acetylmuramylpentapeptide. Finally, the pyrophosphorylases investigated are inhibited by their reaction products. The relative importance of these possible regulatory processes, remains to be determined.

3.2.2.2. Polymerization and transglycosylation reaction. The synthesis of those teichoic acids where the backbone is a simple polymer of polyglycerol- or polyribitol phosphate units (Fig. 18) proceeds through sequential transfers of polyol phosphate residues from CDP-glycerol or CDP-ribitol to a membrane lipid carrier. These reactions are transphosphorylations where CMP residues are released. Incorporation of the side-chain sugars onto the polyolphosphate backbone is performed by transfer of glycosyl residues from UDP-glucose or UDP-N-acetylglucosamine to the free hydroxyl groups of the polyol. These reactions are transglycosylations; UDP residues are released. Transphosphorylation and transglycosylation reactions can occur concomitantly, and bivalent cations are required for optimal synthesis. The first successful syntheses of this type were carried out with particulate preparations from various bacilli and staphylococci (Glaser, 1964). Preformed poly(glycerol or ribitol) phosphate polymers can be used as substrates by the transglycosylases. For example, glucosyl transfer from UDP-glucose to polyribitol phosphate backbone was achieved with a particulate enzyme preparation from *B. subtilis* W23. All the linkages formed had the β-configuration (Chin et al., 1966). Similarly, insertion of both α- and β-N-acetylglucosamine residues on a polyribitol phosphate backbone from

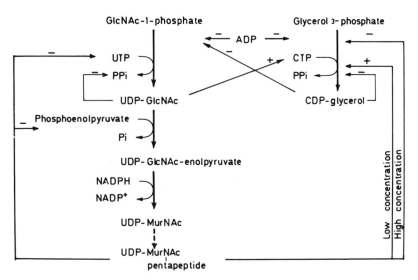

Fig. 45. Interrelations in the synthesis of UDP-GlcNAc, UDP-MurNAc-pentapeptide and CDP-glycerol. Stimulatory effects are indicated as positive, inhibitory effects as negative. A$_2$pm, diaminopimelate. (Anderson et al., 1973.) (Reprinted courtesy of the Biochemical Society.)

UDP-N-acetylglucosamine was achieved by enzyme preparations from *S. aureus* (Nathenson and Strominger, 1966).

The synthesis of those teichoic acids where the sugar residues are part of the polymeric chain (Figs. 19 to 21) is more complex. For synthesis of the teichoic acid of *S. lactis* 13 (Fig. 21), both CDP-glycerol and UDP-N-acetylglucosamine must be incubated simultaneously with the cell-free enzyme preparation. UDP-N-acetylglucosamine donates an N-acetylglucosamine-1-phosphate residue to a membrane lipid carrier and the intermediate thus formed accepts a glycerol phosphate residue from CDP-glycerol. Both reactions are transphosphorylations and UMP and CMP are released. The product formed possesses the complete repeating unit of the polymer. In the process, the glycerol phosphate moiety is provided by CDP-glycerol and the N-acetylglucosamine-1-phosphate moiety is provided by UDP-N-acetylglucosamine (Fig. 46). For synthesis of the teichoic acid of *B. licheniformis* ATCC-9945 (Fig. 19), glucose incorporation proceeds by transglycosylation from UDP-glucose to a membrane lipid carrier with release of UDP, and glycerol phosphate incorporation proceeds by transphosphorylation from CDP-glycerol with release of CMP. As a result, the intermediate contains glucose and glycerol phosphate in the form of the repeating unit of the polymer (review, Baddiley, 1972.)

3.2.2.3. The membrane carrier. The polyglycerol phosphate and polyribitol phosphate polymers being synthesized are linked to an endogenous acceptor. The polyglycerol phosphate polymerase (from *B. subtilis*) and the polyribitol phosphate polymerase (from *S. aureus*) involved in these polymerizations were extracted from the relevant bacterial membranes and purified to the stage where they depended almost entirely on the addition of a heat-stable component for activity (Mauck and Glaser, 1972, 1973; Fiedler and Glaser, 1974; Fiedler et al., 1974). This component was also isolated. It was not an activator of the reaction, but behaved as an acceptor for the growing polymer. With the *S. aureus* system

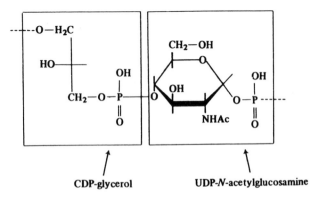

Fig. 46. Transfer of residues from nucleotide precursors to teichoic acid from *Staphylococcus lactis* 13. (Baddiley, 1972.)

the length of the complete polyribitol phosphate formed on this acceptor was about 35 units, in fair agreement with a chain length of 40 to 50 units for the teichoic acid in the completed wall (Ghuysen et al., 1964). In addition to the acceptor the *S. aureus* polyribitol phosphate polymerase also required a phospholipid for full activity. Cardiolipin was found to be a very effective activator, but a variety of other phospholipids (e.g., phosphatidylglycerol, phosphatidylethanolamine and phosphatidylcholine) could substitute for cardiolipin with diminished efficiency (Fiedler et al., 1974).

Transfer of glycerol phosphate and ribitol phosphate units to the acceptor is a direct process that does not involve an additional intermediate carrier. Both *S. aureus* and *B. subtilis* polymerases did not appear to require any addition of undecaprenol phosphate (i.e., the lipid carrier involved in peptidoglycan synthesis) and formation of undecaprenol pyrophosphate glycerol or ribitol could not be detected in these systems. The carrier was subsequently identified as a (polyglycerol) lipoteichoic acid. Hence, the nascent teichoic acid that ultimately undergoes insertion in the wall peptidoglycan grows on a preformed lipoteichoic acid; lipoteichoic acid serves as the assembly center for the wall teichoic acid. The growing site for the future wall teichoic acid on the lipoteichoic acid is probably a glycerol residue located at the hydrophobic end of the carrier where all the D-glucose and fatty acids accumulate (Fiedler and Glaser, 1974). Thus the hydrophobic end of the lipoteichoic acid carrier would serve not only to anchor the carrier into the membrane but would also be the active site of the molecule. Such a model could explain why a polyglycerol phosphate polymer that lacks the hydrophobic end of the lipoteichoic acid cannot substitute for the carrier in the teichoic acid polymerase reaction.

Lipoteichoic acid carriers do not exhibit a high specificity with respect to the type of wall teichoic acid for which they serve as carriers. Lipoteichoic acid carriers that are active with the polyribitol phosphate polymerase of *S. aureus* have been extracted from various gram-positive bacteria. Similarly, lipoteichoic acid isolated from *S. aureus* also acts as a carrier for the polyglycerol phosphate polymerase of *B. subtilis*.

Whether the lipoteichoic acid that acts as carrier for wall teichoic acid synthesis is distinct or not from the bulk of lipoteichoic acid which occurs in the plasma membrane of the gram-positive bacteria is not definitely established. Thus, the purified lipoteichoic acid carrier of *S. aureus* H was considered by Fiedler and Glaser (1974) to be an acylated polyglycerol phosphate with a chain length of 12 to 14 units and one glucose molecule per chain. Duckworth et al. (1975), on the other hand, assigned to the lipoteichoic acid isolated from the same organism a chain of 28 to 30 units linked to the terminal glucose moiety of diacetylgentiobiosylglycerol. The proportions of glycerol, phosphorus, glucose and fatty acids reported by Fiedler and Glaser, and by Duckworth and associates were virtually identical. The difference between the reported properties of the two preparations might be resolved on the basis that the lipoteichoic acid carrier actually contains two molecules of glucose per chain, and the published evidence does not exclude this possibility.

558

Finally, the high levels of D-alanine ester residues in wall teichoic acid also pose the question of D-alanine incorporation in this polymer. Presumably, such incorporation might occur when the growing wall teichoic acid is linked to its lipoteichoic acid carrier. The exact mechanism involved is unknown.

3.2.2.4. The direction of chain extension. The synthesis of the wall polyglycerol phosphate chains in *B. subtilis* was studied by pulse-labeling techniques (Kennedy and Shaw, 1968). The individual units of glycerol phosphate were shown to be transported from the precursor CDP-glycerol so that the newly added glycerol on the carrier was susceptible to oxidation with periodate, yielding formaldehyde. The most recently introduced glycerol phosphate unit was at the glycol end of the growing chain (Fig. 47). Thus, chain extension in teichoic acid synthesis, at least in *B. subtilis,* is similar to glycan chain extension in glycogen synthesis where the glucose units are directly transferred from nucleotide diphosphate sugar to the nonreducing end of the chain. These processes are entirely different from those of peptidoglycan synthesis (section 3.1.5.) and O-antigen synthesis (section 3.4.) that occur through undecaprenol phosphate intermediates where chain growth proceeds by extension from the reducing end of the oligosaccharide. The latter processes are comparable to the synthesis of peptide chains in proteins that occurs at the carboxyl end by transfer of the growing oligopeptide from a peptidyl tRNA to an aminoacyl-RNA.

3.2.2.5. Attachment of wall teichoic acid to peptidoglycan. When the growing (wall) teichoic acid linked to its lipoteichoic acid carrier has reached the correct length, the enzyme reaction stops and the polymer is translocated to the proper N-acetylmuramic acid residue of the peptidoglycan chain (section 2.3.2.3.). In the process, a chain of 3 to 4 glycerolphosphate residues is inserted as the link between the wall teichoic acid and the 6-position of an N-acetylmuramic acid residue of the peptidoglycan (Heckels et al., 1975; Coley et al., 1976).

With cell-free membrane preparations, teichoic acid synthesis leads to products that are not attached to peptidoglycan. Thus the question arose whether the teichoic acid in vivo is inserted to a preexisting wall peptidoglycan or to a concomitantly synthesized peptidoglycan. On the basis of elegant double-label experiments, Mauck and Glaser (1972) showed that in *B. subtilis* cells newly synthesized teichoic acid (and teichuronic acid) were linked only to peptidoglycan which had been synthesized simultaneously.

More recent experiments by Wyke and Ward (1975) performed with a cell-free membrane plus wall preparation of *B. licheniformis* largely confirmed these

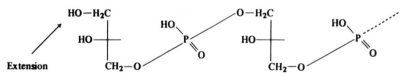

Fig. 47. Growth of a poly(glycerol phosphate) (Baddiley, 1972.)

results. Such membrane plus wall preparations synthesized teichoic acid from CDP-ribitol as well as peptidoglycan from UDP-N-acetylglucosamine and UDP-N-acetyl-muramylpentapeptide. The following observations were made: (1) Addition of penicillin to the incubation mixtures at concentrations that completely inhibited attachment of the nascent peptidoglycan to the preexisting wall peptidoglycan by transpeptidation had no effect on the synthesis of the nascent teichoic acid but caused an 81% decrease in the amount of teichoic acid linked covalently to the wall. (2) The same result was obtained when teichoic acid synthesis was carried out in the absence of added peptidoglycan precursors in the reaction mixtures. Since, as shown by Tynecka and Ward (1975), the nascent peptidoglycan synthesized by *B. licheniformis* under conditions where transpeptidation is inhibited lacks wall-associated teichoic acid, it follows that, at least with this cell-free system, the incorporation of the majority of the newly synthesized teichoic acid to the wall peptidoglycan requires the concomitant synthesis of both teichoic acid and cross-linked peptidoglycan. These observations suggest that the nascent teichoic acid made on the lipoteichoic acid assembly center could not undergo attachment to the nascent peptidoglycan made on the undecaprenol phosphate assembly center unless the latter polymer would exhibit the proper orientation or correct alignment that would be imparted by its concomitant fixation on the "old" wall peptidoglycan by transpeptidation (Fig. 48). Obviously, perfect coordination among all these events would be essential.

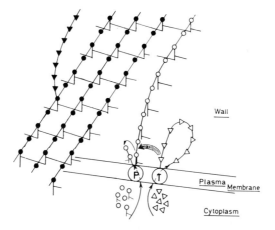

Fig. 48. Model of wall assembling in Gram-positive bacteria. In black: preexisting wall crosslinked peptidoglycan-teichoic acid complex. ●—🐓 = peptidoglycan; ▲—▲ = teichoic acid. In white: nascent peptidoglycan (○—○) and nascent teichoic acid (△—△). ○ and 🐓: peptidoglycan precursors. △: teichoic acid precursors.Ⓟ: undecaprenol phosphate-containing assembly center for peptidoglycan synthesis. Elongation proceeds at the reducing terminal of the growing chain.Ⓣ: lipoteichoic acid-containing assembly center for wall teichoic acid synthesis. Elongation proceeds at the nonreducing terminal of the growing chain. The model shows the attachment of the nascent teichoic acid on a N-acetylmuramic acid residue (arrow) of the simultaneously synthesized peptidoglycan, which itself undergoes attachment to the preexisting wall peptidoglycan by peptide cross-linking. Attachment of nascent ribitol teichoic acid to nascent peptidoglycan involves an intervening 3-glycerol phosphate residues chain unit. A possible mechanism for the insertion of this link is shown in Fig. 49.

The integration and reciprocal control of peptidoglycan and wall teichoic acid syntheses at the level of the corresponding assembly centers is a major problem that remains to be solved. Recent experiments by Bracha and Glaser (1976) with *S. aureus* H, and by Hancock and Baddiley (1976) with the same and other organisms, have shed new light on the processes involved. Fig. 49 shows a possible mechanism (Hancock and Baddiley, 1976) for synthesis of the 3-glycerol phosphate residue-containing chain that links wall teichoic acid to peptidoglycan and for incorporation of teichoic acid into wall material. Peptidoglycan synthesis on the undecaprenyl phosphate carrier and wall teichoic acid synthesis (here, a polyribitol phosphate polymer) on the lipoteichoic acid carrier are represented in the upper part and the lower part, respectively, of Fig. 49. It is suggested that the acceptor for the 3-glycerol phosphate residue-containing linkage unit is the undecaprenyl-PP-disaccharide peptide intermediate. Subsequently, the completed wall teichoic acid is transferred from the lipoteichoic acid carrier and the complete unit thus formed is incorporated into a growing peptidoglycan chain and then into the wall. At present, a choice cannot be made between addition of the 3-glycerol phosphate residues linkage unit to peptidoglycan at the lipid intermediate stage or to the growing glycan chain. The mechanism proposed by Bracha and Glaser (1976) differs from that represented in Fig. 49: the oligomer which links the polyribitol phosphate to the wall peptidoglycan would contain not only glycerol phosphate but also N-acetyl-D-glucosamine.

Finally, the dynamics of the wall teichoic acid synthesis in *B. subtilis* has been studied by Archibald (1976). On the basis of the development of bacteriophage-binding properties as a result of the pulsed incorporation of teichoic acid, the newly synthesized receptors appeared to be incorporated at the inner surface of the wall and became exposed at the outer surface during subsequent growth.

3.2.3. Synthesis of acidic polysaccharides
other than teichoic acids

The synthesis of most of these wall polyanions has not been studied extensively, but from what is known the process appears to resemble that described for wall teichoic acids. During logarithmic growth in phosphate-rich medium *B. subtilis* synthesizes wall teichoic acid. Under conditions of phosphate limitation, teichoic acid synthesis stops and teichuronic acid synthesis is activated. Wall insertion of teichuronic acid induced by phosphate deprivation was found to proceed exactly as wall insertion of teichoic acid in a rich medium; teichuronic acid was linked only to peptidoglycan strands synthesized at the same time as teichuronic acid was produced (Mauck and Glaser, 1972). The assembly center for teichuronic acid synthesis might be the lipoteichoic acid carrier involved in wall teichoic acid synthesis, although this has not yet been proven.

Another interesting example is that of the D-glucose and N-acetyl-D-mannosaminuronic acid-containing polymer that occurs in the walls of *M. lysodeikticus* (section 2.3.2.1.). A particulate enzyme fraction isolated from this organism was shown to synthesize this polymer from the nucleotides UDP-

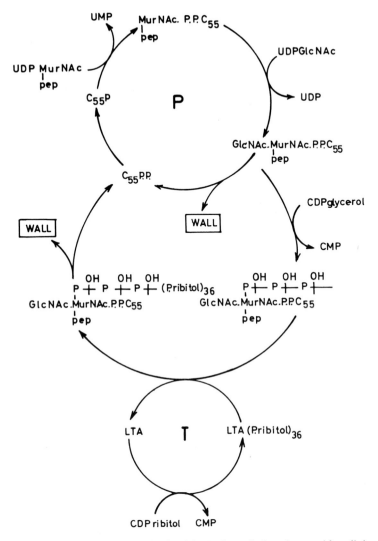

Fig. 49. Possible mechanism for the synthesis of the 3-glycerol phosphate residues linkage unit that links wall ribitol teichoic acid to peptidoglycan and for the incorporation of teichoic acid into wall material (Hancock and Baddiley, 1976.)

Ⓟ : undecaprenol phosphate-containing assembly center for peptidoglycan synthesis.

Ⓣ : lipoteichoic acid-containing assembly center for wall teichoic acid synthesis.

(Reprinted courtesy of the American Society for Microbiology)

glucose, UDP-N-acetylhexosaminuronic acid, and UDP-N-acetylglucosamine (Page and Anderson, 1972; Anderson et al., 1972). Glucose and N-acetylhexosaminuronic acid were incorporated in approximately equimolar amounts, and N-acetylglucosamine was incorporated only to the extent of one residue or less for each 15 glucose residues. Wall polysaccharide contains the two former sugars in about equal amounts. Interestingly, the polysaccharide-synthesizing system required the simultaneous presence of a high molecular weight, heat-stable factor. Although the data suggested that this factor might play an acceptor role, its nature is still unknown. At present, it is also not known whether the membrane-associated succinylated lipomannan might play a role in the synthesis of this wall polysaccharide analogous to that of the lipoteichoic acid carrier in wall teichoic acid synthesis.

Synthesis of the membrane-bound succinylated mannan in *M. lysodeikticus* was elucidated by the important studies of Lennarz and coworkers who established that the carrier lipid mannosyl-1-phosphorylundecaprenol is an obligatory intermediate (Scher et al., 1968; Lahav et al., 1969; Scher and Lennarz, 1969; Lennarz and Scher, 1972). Synthesis involves transfer of mannosyl groups from GDP-mannose to the carrier lipid and is followed by subsequent transfer of hexose from mannosyl-1-phosphorylundecaprenol to mannan acceptor, the majority of the mannose units thus incorporated being located on the nonreducing termini of mannan (Scher and Lennarz, 1969). Thus, both peptidoglycan and mannan in this organism are synthesized on the same undecaprenol phosphate-containing assembly centers. Mannan has been localized in the plasma membrane and in the mesosomal membranes of *M. lysodeikticus* (Owen and Freer, 1972). It was therefore tempting to explore the possibility that its synthesis might be localized in the mesosomes, providing a unique enzymatic function for these structures. Owen and Salton (1975b,c) found that unlike the plasma membrane isolated mesosome vesicles could not catalyze the transfer of mannose from GDP-mannose into mannan because of the inability of the vesicles to synthesize the intermediate mannosyl-1-phosphorylundecaprenol. This situation was not attributed to the absence of undecaprenol phosphate in the mesosome vesicles but rather to the absence of the enzyme mannosyl-1-phosphorylundecaprenol synthetase in these organelles. Since the mannan located on both mesosomal and plasma membranes can accept mannosyl units, it was suggested that the juxtaposition of mesosomal vesicles and mesosomal sacculus (i.e., the region of the plasma membrane that surrounds the mesosomal vesicles) in vivo might allow the mannan located on the mesosomal vesicles to accept mannosyl units either from the carrier lipid of the sacculus membrane or by direct carrier lipid exchange between the two membrane systems (Owen and Salton, 1975c). Thus it is significant that newly synthesized bactoprenol has been shown to be localized mainly in the septal region of *Lactobacillus* cells (Thorne et al., 1974) that have just undergone division, and that in many bacteria mesosomes are closely associated with the growing cross wall.

3.3. Biosynthesis of the lipoprotein of the outer membrane of E. coli

Biosynthesis of the outer membrane lipoprotein is a sequential process. The free form is synthesized first and the newly synthesized free form is effectively incorporated into the outer membrane without accumulating in the inner membrane (Inouye et al., 1972, 1974; Lee and Inouye, 1974). Part of the free form is then converted as the peptidoglycan-bound form. A dynamic equilibrium may exist between the two forms (Inouye et al., 1972).

3.3.1. Synthesis of the free form

In vivo biosynthesis of the free lipoprotein presents several striking features: (1) the extraordinarily long half-life of the relevant mRNA (Hirashima and Inouye, 1973; Hirashima et al., 1973a); (2) the extremely strong resistance of the synthesizing enzyme system to puromycin (Hirashima et al., 1973a); and (3) the exclusive synthesis of lipoprotein in cells deprived of histidine (Hirashima et al., 1973a). A cell-free system was devised so that the lipoprotein synthesis was directed by the purified relevant mRNA. (Hirashima et al., 1974). The translation product was characterized as the protein moiety of the lipoprotein by immunoprecipitation and peptide mapping. Using slab-gel electrophoresis the mRNA purified from exponentially growing cells gave rise to two closely adjacent bands migrating similar to 7s RNA. Band 1 consisted of about 250 and band 2 of about 230 nucleotides. Whether or not both of these RNAs have the lipoprotein-specific mRNA activity is not known. Nevertheless, since lipoprotein contains 58 amino acid residues, at least 180 nucleotides are required for its synthesis (i.e., 3 × 58; + 1 for initiation codon; + 1 for termination codon) and thus the mRNA should contain 50 to 70 untranslated nucleotides. These nucleotides might be used for the ribosome recognition site at the 5' end of the mRNA and for the termination signal at the 3' end. Whether the amino-terminal end of the protein is modified in the cell-free system is also unknown. When radioactive palmitic acid was added to the incubation mixture, its incorporation into product could not be detected. Since in a cell-free protein synthesizing system the initiator amino acid, N-formyl-methionine, is known to remain at the N-terminal of the product, perhaps N-formyl-methionine should first be removed from the synthesized protein portion of lipoprotein to allow palmitic acid to undergo attachment. Recent results have suggested that the synthesis of the apoprotein portion of the lipoprotein proceeds independently of the attachment of diglyceride to the SH group of the N-terminal cysteine (Jung-Ching Lin and Wu, 1976) and that the protein is transferred to the outer membrane and linked to the peptidoglycan (see section 3.3.2) at least in the absence of glyceride-linked portion of the lipoprotein lipids (Suzuki et al., 1976). With the cell-free system, mRNA and ribosomes were required for lipoprotein synthesis. Since the mRNA was relatively small, the size of the polyribosomes involved in the synthesis was also expected to be small. This prediction proved to be right. Fractionation of the polyribosomes showed that lipoprotein synthesis was carried out on relatively small polyribosomes, possibly tri- or tetra-ribosomes (Hirashima and Inouye, 1975).

Many intriguing questions remain: (1) Are the polyribosomes membrane-

Braun, V., Bosch, V., Hantke, K. and Schaller, K. (1974a) Structure and biosynthesis of functionally defined areas of the *Escherichia coli* outer membrane. Ann. N.Y. Acad. Sci. 235, 66–82.

Braun, V., Schaller, K. and Wabl, M.R. (1974b) Isolation, characterization and action of colicin M. Antimicrob. Ag. Chemother. 5, 520–533.

Braun, V., Rotering, H., Ohms, J.P. and Hagenmaier, H. (1976a) Conformational studies on murein-lipoprotein from the outer membrane of *Escherichia coli*. Eur. J. Biochem. 70, 601–610.

Braun, V., Bosch, V., Klumpp, E.R., Neff, I., Mayer, H. and Schlecht, S. (1976b) Antigenic determinants of murein lipoprotein and its exposure at the surface of *Enterobacteriaceae*. Eur. J. Biochem. 62, 555–566.

Bray, D. and Robbins, P.W. (1967) The direction of chain growth in *Salmonella anatum* O-antigen biosynthesis. Biochem. Biophys. Res. Commun. 28, 334–339.

Bretscher, M.S. (1972) Asymmetrical lipid bilayer structure for biological membranes. Nature New Biol. 236, 11–12.

Briles, E.B. and Tomasz, A. (1973) Pneumococcal Forssman Antigen. A choline-containing lipoteichoic acid. J. Biol. Chem. 248, 6394–6397.

Brinton, C.C., McNary, J.C. and Carnahan, J. (1969) Purification and *in vitro* assembly of a curved network of identical protein sub-units from the outer surface of a *Bacillus*. Bact. Proc., p. 48.

Brown, W.C. and Young, F.E. (1970) Dynamic interaction between cell wall polymers, extracellular protease and autolytic enzymes. Biochem. Biophys. Res. Commun. 38, 564–568.

Brundish, D.E. and Baddiley, J. (1968) *Pneumococcal* C-substance, a ribitol teichoic acid containing choline phosphate. Biochem. J. 110, 573–582.

Buchanan, Ch.E. and Strominger, J.L. (1976) Altered penicillin-binding components in penicillin-resistant mutants of *Bacillus subtilis*. Proc. Nat. Acad. Sci. USA 73, 1816–1820.

Burdett, I.D.J. and Murray, R.G.E. (1975) Electron microscope study of septum formation in *Escherichia coli* strains B and B/r during synchronous growth. J. Bacteriol. 119, 1039–1056.

Burger, M.M. (1966) Teichoic acids: antigenic determinants, chain separation, and their location in the cell wall. Proc. Nat. Acad. Sci. U.S.A. 56, 910–917.

Burger, M.M. and Glaser, L. (1966) The synthesis of teichoic acids. V. Polyglucosylglycerol phosphate and polygalactosylglycerol phosphate. J. Biol. Chem. 241, 494–506.

Butterworth, P.H.W. and Hemming, F.W. (1968) Intracellular distribution of the free and esterified forms of dolichol in pig liver. Arch. Biochem. Biophys. 128, 503–508.

Button, D. and Hemmings, N.L. (1976) Lipoteichoic acid from *Bacillus licheniformis* 6346 MH–1. Comparative studies on the lipid protein of the lipoteichoic acid and the membrane glycolipid. Biochemistry 15, 989–995.

Button, D., Archibald, A.R. and Baddiley, J. (1966) The linkage between teichoic acid and glycosaminopeptide in the walls of a strain of *Staphylococcus lactis*. Biochem. J. 99, 11c–14c.

Capaldi, R.A. and Green, D.E. (1972) Membrane proteins and membrane structure. FEBS Lett. 25, 205–209.

Chatterjee, A.N. (1969) Use of bacteriophage-resistant mutants to study the nature of the bacteriophage receptor site of *Staphylococcus aureus*. J. Bacteriol. 98, 519–527.

Chatterjee, A.N. and Park, J.T. (1964) Biosynthesis of cell wall mucopeptide by a particulate fraction from *Staphylococcus aureus*. Proc. Natl. Acad. Sci. U.S.A. 51, 9–16.

Chevion, M., Panos, Ch., Linzer, R. and Neuhaus, F.C. (1974) Incorporation of D-alanine into the membrane of *Streptococcus pyogenes* and its stabilized L-form. J. Bacteriol. 120, 1026–1032.

Chin, T., Burger, M.M. and Glaser, L. (1966) Synthesis of teichoic acids VI. The formation of multiple wall polymers in *Bacillus subtilis* W23. Archiv. Biochem. Biophys. 116, 358–367.

Chipman, D.M., Pollock, J.J. and Sharon, N. (1968) Lysozyme-catalyzed hydrolysis and transglycosylation reactions of bacterial cell wall oligosaccharides. J. Biol. Chem. 243, 487–496.

Chou, P.Y. and Fasman, G.D. (1974) Prediction of protein conformation. Biochemistry 13, 222–245.

Christenson, J.G., Gross, S.K. and Robbins, P.W. (1969) Enzymatic synthesis of the antigen carrier lipid. J. Biol. Chem. 244, 5436–5439.

bound? It is well established that in eukaryotic cells, secreted proteins are synthesized on membrane-bound polysomes. Randall and Hardy (1977) showed that the major products of protein synthesis in vitro by membrane-bound polysomes from *E. coli* included proteins of the outer membrane and the maltose-binding protein, a typical secreted periplasmic protein. On the contrary, the major product synthesized in vitro by free polysomes of *E. coli* was characterized as the ribosomal elongation factor Tu, a cytoplasmic protein for which a peripheral location at the inner surface of the plasma membrane has been postulated (Jacobson and Rosenbusch, 1976a,b; Jacobson et al., 1976). Randall and Hardy (1977) also found that the activity of membrane-bound polysomes in vitro was more resistant to puromycin than was the activity of free polysomes and in addition, the mRNA associated with membrane-bound polysomes was more stable than the bulk of cellular mRNA. (2) Which structural feature of the lipoprotein is responsible for the extreme distribution coefficient that the lipoprotein exhibits in favor of the outer membrane? (3) What is the molecular basis for the extreme resistance of the lipoprotein synthesizing system toward puromycin? In the presence of 800 μg/ml of puromycin, lipoprotein synthesis was inhibited by only 30% (Hirashima et al., 1973a). One could hypothesize that because of some sort of compartmentalization membrane-bound ribosomes would behave differently or would be less accessible to puromycin. (4) What is the molecular basis for the very long half-life of the lipoprotein mRNA? The half-life for this mRNA is 11.5 minutes, a value that can be compared with 2 minutes for the cytoplasmic proteins mRNAs and 5.5 minutes for the envelope proteins mRNAs. Obviously, this small lipoprotein mRNA is worth sequencing.

3.3.2. Attachment of the free form to the peptidoglycan

The following mechanism is hypothetical. Attachment of lipoprotein to the underlying peptidoglycan might proceed by a transpeptidation reaction in which the free lipoprotein molecules would act as the acceptors and tetrapeptide units L-Ala-D-Glu $_r$*meso*-A $_2$pm-D-Ala of the peptidoglycan would act as the donors. The reaction would be catalyzed by an LD-carboxypeptidase-transpeptidase system as shown in Fig. 50. Tripeptide units L-Ala-D-Glu$_r$*meso*-A $_2$pm would be the products of the hydrolysis pathway of the reaction (such tripeptide units are actually present in the completed wall peptidoglycan), and the ratio between the tripeptide units and peptidoglycan-bound lipoproteins might depend on the availability of water to the enzyme complex. The tetrapeptide units assumed to be involved as donors in the reaction are themselves products of the hydrolysis pathway of the membrane-bound DD-carboxypeptidase-transpeptidase system (section 3.1.5.6.).

The attachment process in *E. coli* is unaffected by benzylpenicillin and penicillin FL-1060. Penicillin FL-1060 has stimulated interest. Under the influence of this antibiotic peptidoglycan is at most 50% inhibited, as if the inhibition of an enzymatic step of peptidoglycan synthesis could be partially bypassed by a second enzyme, less efficient but resistant to penicillin FL-1060. The rate of lipo-

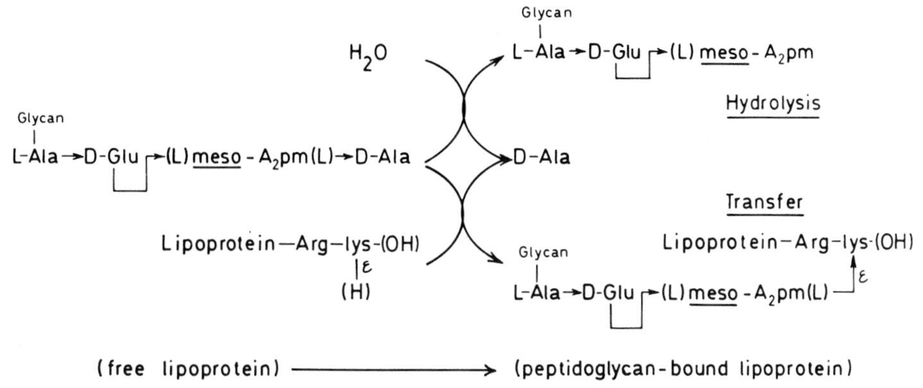

H_2O

Glycan
|
L–Ala→D-Glu $\rightarrow$(L) <u>meso</u> - A_2pm

Hydrolysis

Glycan
|
L–Ala→D-Glu $\rightarrow$(L) <u>meso</u> - A_2pm(L)→D-Ala

D-Ala

Transfer

Lipoprotein—Arg—lys-(OH)
|ε
(H)

Glycan
|
L–Ala→D-Glu $\rightarrow$(L) <u>meso</u> - A_2pm(L)

Lipoprotein–Arg–lys·(OH)
|ε

(free lipoprotein) ⟶ (peptidoglycan-bound lipoprotein)

Fig. 50. Possible mechanism for the enzymatic attachment of lipoprotein to peptidoglycan.

protein attachment is far less affected, and therefore increasing amounts of lipoprotein become attached with time. When the stationary growth phase has been reached, the lipoprotein content of the peptidoglycan has doubled (Braun and Wolff, 1975).

Recent studies by James and Gudas (1976) have shown that the incorporation of free lipoprotein into the outer membrane of E. coli is a cell cycle-specific process with a maximal rate of incorporation occurring at the time of septation. Since no corresponding increase in the rate of incorporation of bound lipoprotein during the cell cycle was observed, an appreciable decrease in the ratio of bound to free lipoprotein must occur shortly before septation. This observed phenomenon may be relevant to the separation of outer membrane and peptidoglycan layer observed at an early stage in septation. Burdett and Murray (1975) have shown that the ingrowing septum consists of a fold of plasma membrane plus peptidoglycan, a double structure composed of two opposed lamellae separated by an electron-transparent gap. Clearly, the outer membrane is absent and only enters the septum during cell separation.

The lipoprotein attachment process presents other intriguing peculiarities. E. coli cells were pulse-labeled for 5 minutes with [³H]arginine and then chased with nonradioactive arginine for several cell-doubling times. After one cell-doubling time, 40% of the newly synthesized lipoprotein was attached to peptidoglycan, and this percentage remained constant following 2 and 3 cell-doubling times (Braun and Wolff, 1975). These observations are not understood. They could suggest that there is a reversible conversion of the free and bound forms of the lipoprotein and that the newly synthesized lipoprotein mixes with a large pool of the free form. Alternatively, 40% of the newly synthesized free form may become irreversibly attached to the peptidoglycan in one doubling time, whereas the rest would undergo compartmentalization in some areas of the cell envelope where it would no longer become attached to peptidoglycan

(Braun, 1975). As mentioned earlier the peptidoglycan units newly inserted into the preexisting wall peptidoglycan appear very slowly at lipoprotein attachment sites (section 3.1.5.6.). Peptidoglycan extension is confined to areas without attached lipoprotein. The enzyme system responsible for peptidoglycan extension may be hindered sterically at these attachment sites. Regardless of the precise situation, it seems clear that the redistribution of the bound lipoproteins over the peptidoglycan sacculus must result from the incorporation of lipoprotein molecules into the newly synthesized peptidoglycan region (Braun and Wolff, 1975).

3.4. Lipopolysaccharide biosynthesis

Biosynthesis of lipopolysaccharide is peculiar because the core oligosaccharide is made by stepwise extension of lipid A whereas the O-side chains are independently synthesized. Joining of the two preformed macromolecules then occurs and the resulting complete structure that has not yet left the plasma membrane is translocated to the outer membrane. The enzymes that perform the O-side chain syntheses are localized exclusively in the plasma membrane (Osborn et al., 1972a,b). Pulse-chase experiments with radioactive mannose (a specific precursor of the O-antigen) also showed that about 70% of the mannose incorporated into the O-antigen chains during a 1-minute pulse with [^{14}C]mannose was in the plasma membrane and that this pulse-labeled lipopolysaccharide was completely chased in the outer membrane during a subsequent 2-minute chase with non-radioactive mannose (Osborn et al., 1974).

The core-synthesizing enzymes are not firmly bound to the plasma membrane. They are released in soluble form during membrane preparation and thus cannot be used as markers for the synthesis site. Pulse-chase experiments, however, demonstrated that the plasma membrane is also the place where core assembly proceeds (Osborn et al., 1974).

Mutants have been extremely useful for unraveling the structure of lipopolysaccharides (section 2.7.2.). They are also extremely useful in devising cell-free systems for the study of biosynthesis. While it would be impossible to show in vitro transfers of lipopolysaccharide constituents in wild-type cells because of lack of empty sites for incorporation, the cell envelope fractions of mutants containing all its lipopolysaccharide in an incomplete form are ideal acceptor substrates for incorporation experiments. Only the principles of the biosynthesis pathways are presented here. Details and a complete reference list can be found in an excellent review by Nikaido (1973).

3.4.1. The sugar nucleotide precursors
The sugar nucleotide precursors are synthesized by soluble, cytoplasmic enzymes. The pathways for the synthesis of the various sugar nucleotide precur-

sors in *S. typhimurium* are known (Nikaido, 1973). In particular, the nucleotide precursor for KDO incorporation is CMP-KDO (Heath et al., 1966). KDO arises from arabinose through the action of a KDO-8-P synthetase according to the reaction D-arabinose-PP + phosphoenolpyruvate $\rightarrow$ KDO-8-P + P (Rick and Osborn, 1972). The nucleotide diphosphate derivative of L-glycero-D-mannoheptose has not yet been isolated, but because a transketolase is needed for heptose synthesis it is thought that sedoheptulose-7-phosphate is an obligatory precursor (Eidels and Osborn, 1971). Radioactive sedo-heptulose-7-phosphate was shown to label specifically the L-glycero-D-mannoheptose residues of the lipopolysaccharide of a transketolase-negative mutant of *S. typhimurium* (Eidels and Osborn, 1974). Synthesis of the nucleotide of L-glycero-D-mannoheptose would proceed through the following pathway:

$$\text{SODOHEPTULOSE}-7-\text{P} \xrightleftharpoons{\text{isomerase}} \text{D}-\text{GLYCERO-MANNOHEPTOSE}-7-\text{P} \xrightleftharpoons{\text{mutase}}$$

$$\text{D-GLYCERO-D-MANNOHEPTOSE}-1-\text{P} \xrightarrow{\text{NDP-heptose synthetase + NTP}}$$

$$\text{NDP-D-GLYCERO-D-MANNOHEPTOSE} + \text{PP} \xrightleftharpoons{\text{epimerase}}$$
$$\text{NDP-L-GLYCEROL-D-MANNOHEPTOSE}$$

(NTP = NUCLEOTIDE TRIPHOSPHATE; NDP = NUCLEOTIDE DIPHOSPHATE).

3.4.2. Synthesis of the core-lipid A moiety

Very little is known about the synthesis mechanism of lipid A (for the structure of lipid A, see Fig. 30). Presumably, specific transferases move each fatty acid to the correct location on the glucosamine disaccharide skeleton. Since 3-hydroxytetradecanoic acid is a specific component of lipid A, its transfer to lipid A preparations were attempted but were unsuccessful. 3-Hydroxytetradecanoic acid, could, however, be transferred from its acyl carrier protein derivative to lysophosphatidylethanolamine, but no evidence was obtained that the resulting product would transfer the fatty acid to lipid A (Taylor and Heath, 1969).

The synthesis of the heptose and KDO containing parts, the inner part of the core lipopolysaccharide (Fig. 28), is also obscure. The first step is probably the transfer on lipid A of a KDO residue from CMP-KDO (Heath et al., 1966). A glucosamine oligosaccharide in which the amino groups were substituted by 3-hydroxytetradecanoic acid residues was shown to be a very good acceptor. This observation might suggest that KDO is transferred before the glucosamine residues are O-acylated. The mechanisms of addition of the two remaining KDO residues and of the heptose moieties are unknown. Phosphorylation is catalyzed at the expense of ATP by an enzyme that can be released by washing EDTA-lysozyme treated cells (Mühlradt et al., 1968; Mühlradt, 1971). Based on the acceptor efficiency of various defective lipopolysaccharides the transfer of the first glucose residue of the outer core might precede phosphorylation of the inner core.

The synthesis of the lipopolysaccharide outer core (Fig. 28) has elicited much interest (Romeo et al., 1970; Hinckley et al., 1972; Müller et al., 1972). It proceeds through a series of conventional glycosyl transferase reactions, each residue being transferred from its proper nucleotide precursor directly to the nonreducing terminal of the growing polymer. Concomitantly, the corresponding nucleotide diphosphate is released. The involvement of undecaprenylphosphate or of other intermediate carriers between the sugar nucleotides and the outer core of the lipopolysaccharide has been excluded. Purified defective lipopolysaccharides (from R_C mutants; see Fig. 31) were completely inactive as acceptors, but activity could be restored by readding the phospholipids extracted from the cell envelope. Actually, the (incomplete) lipopolysaccharide-phospholipid complex but not the (incomplete) lipopolysaccharide alone is the acceptor for the glycosyl transfer reactions (in the presence of Mg^{2+}). Ternary lipopolysaccharide-phospholipid-transferase complexes were isolated by isopyknic sucrose gradient centrifugation and were shown to interact with the relevant nucleotide precursors with transfer of the glycosyl residues. Several transferases were isolated and purified to homogeneity (mol. wt., approximately 20,000). The galactosyl transferase contains glucosamine, perhaps neutral sugars, and a lipid of unknown structure that is probably responsible for the tendency of the transferase to aggregate in an aqueous environment. The efficiency of various phospholipids in transferase reactions has been studied. Phosphatidylethanolamine, phosphatidylglycerol and phosphatidic acid, are active, and the fatty acid composition also exerts a strong influence.

A model for the assembly of the outer core has been proposed by Rothfield and Romeo (1971). Lipid A is first synthesized and anchored in the plasma membrane by insertion of its fatty acid chains into the membrane phospholipids. Lipid A with the added inner core has surface mobility. When it encounters a transferase a specific interaction occurs, resulting in the formation of a ternary complex. If the proper sugar nucleotide is unavailable, the complex is blocked (complexes of this type are released from the cells by EDTA treatment). If the proper sugar nucleotide is available, the transfer reaction occurs, the complex dissociates, and the released lipopolysaccharide is then ready to bind with the transferase thus catalyzing the next step in the sequence (Fig. 51). In this model, the incomplete lipopolysaccharide is assumed to move from one transferase to the next. Consequently, the enzymes must be aligned in correct order. It may be that the transferases interact specifically with each other or that the multienzyme system contains a structural protein to which the transferases are attached in the required sequence.

3.4.3. Synthesis of the O-side chains

The O-side chains are assembled independently of the other portions of the

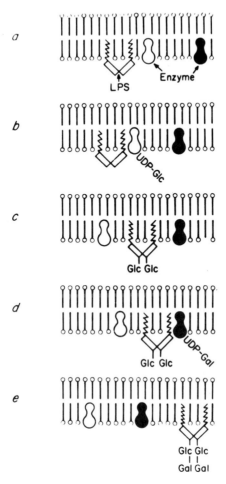

Fig. 51. Mechanism of core oligosaccharide extension on the surface of the membrane. The sequence of events beginning at (a) and ending at (e) are shown, which results in the successive transfer of glucose and galactose residues. It is based on the proposal of Rothfield and Romeo (1971). In essence, incomplete lipopolysaccharide molecules are supposed to move along the phospholipid bilayer membrane, and to accept sugar residues when they encounter transferase enzymes which are half buried in the bilayer. Possibly, transferase enzymes are arranged in the correct order through the formation of a protein aggregate within the membrane; this possibility is not shown here. (Nikaido, 1973.) (Reprinted courtesy of Marcel Dekker, New York.)

lipopolysaccharide. Their synthesis involves a membrane-bound multienzyme system and the same undecaprenyl-phosphate lipid carrier (also called glycosyl carrier lipid or P-GCL) as that involved in peptidoglycan synthesis (Bray and Robbins, 1967; Robbins and Wright, 1971; Lennarz and Scher, 1972; Osborn, 1971). The reaction sequence is analogous to that leading to the formation of the nascent peptidoglycan (section 3.1.4.). The first reaction is a transphosphorylation. Galactose 1-phosphate is transferred from UDP-galactose to the undecaprenyl-phosphate carrier with formation of galactose-PP-undecaprenyl and release of UMP. The next sugar residues are then added sequentially by a series of transglycosylation reactions (with, in each case, the release of the corresponding nucleotide diphsophate) until the O-side chain repeating unit is completed. Through a series of lipid cycles, where each newly synthesized repeating unit linked to the lipid accepts the growing chain made during the preceding runs, a long-chain polymeric intermediate is synthesized while remaining linked at its reducing terminal to the lipid carrier (see Fig. 52, and compare with Fig. 37 for synthesis of the nascent peptidoglycan). After each transfer, the PP-lipid carrier is liberated and the active monoester coenzyme is regenerated by a pyrophosphatase. Like peptidoglycan synthesis, O-side chain synthesis is sensitive to bacitracin.

3.4.4. Assembly of the complete lipopolysaccharide and its transfer to the outer membrane

Transfer of the polymerized O-side chains to the lipopolysaccharide core is catalyzed by an O-antigen: lipopolysaccharide ligase (Cynkin and Osborn, 1968). The mechanism is far from being understood and probably involves a series of reactions. Sonicated cell envelope fraction from *S. typhimurium* can transfer in vitro incomplete repeating units (Nikaido, 1973). This situation never occurs in intact cells, suggesting that control mechanisms exist that preclude premature transfers.

After the joining of the two preformed macromolecules, the resulting complete lipopolysaccharide molecules are translocated into the outer membrane through the adhesion sites described in section 2.10. Once they have emerged from these sites, the lipopolysaccharides are distributed over the entire surface of the cell with preferential localization on the outside face of the outer membrane (Kulpa and Leive, 1972, 1976; Bayer, 1974, 1975; Mühlradt et al., 1973, 1974; Mühlradt, 1976). In his study, Bayer made use of a phenomenon known as "lysogenic conversion" where the O-antigen composition of the cell can be rapidly changed as a result of infection with a given bacteriophage. As a consequence of infection with phage $\epsilon15$, the original O-antigen 10 of *Salmonella anatum* is rapidly diluted by increasing amounts of O-antigen 15. This conversion provides a natural marker for newly synthesized O-antigens either by using immunologic techniques or because the new antigen serves as a receptor for phage $\epsilon34$ but not for phage $\epsilon15$ (i.e., the converting phage deprives itself of new adsorption sites while it induces the synthesis of receptor for $\epsilon34$). Mühlradt and his colleagues used a *Salmonella typhimurium* strain lacking UDP-galactose-4-epimerase. This mutant bears incomplete lipopolysaccharide, and produces the wild-type lipopolysaccharide only after addition of galactose to the medium. After this addition, the preexisting incomplete lipopolysaccharide molecules remain unfinished and only the newly synthesized molecules carry the

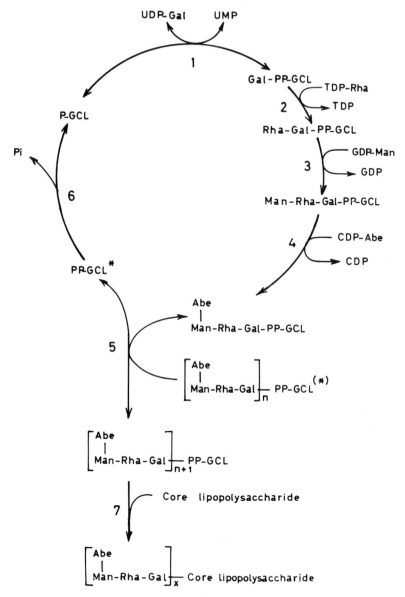

Fig. 52. Pathway of biosynthesis of O-antigen in *S. typhimurium* (reactions 1–6) and transfer of the completed O-side chain to the preformed core lipopolysaccharide (reaction 7). Note that the growing chain made during the preceding runs
Abe
[Man-Rha-Gall]$_n$ is transferred from the lipid carrier to the nonreducing end of the newly synthesized repeating unit linked to the lipid.

572

wild-type specificity, providing a means for their specific localization by using ferritin-conjugated antibodies. Finally, Kulpa and Leive used a mutant of *E. coli* that also lacks UDP-galactose-4-epimerase, and thus produces a lipopolysaccharide with less carbohydrate than the parent unless galactose is added exogenously. Because outer membrane fragments containing carbohydrate-rich lipopolysaccharide are denser than fragments containing carbohydrate-poor lipopolysaccharide, the "insertion" points through which newly synthesized lipopolysaccharide enters the outer membrane can be isolated by isopyknic centrifugation. This technique should make it possible to determine how these areas differ in composition and function from the remainder of the outer membrane.

The three methods used, each with very different limitations, have supported the same conclusions, at least on a qualitative level. In *Salmonella typhimurium*, newly synthesized lipopolysaccharides were seen to appear in about 200 patches on the exterior of each cell 30 seconds after addition of galactose. These patches always occurred over the adhesion sites, and after 2 to 3 minutes of growth the labeled molecules were seen evenly spread over the entire cell surface. In both *S. anatum* and *E. coli* the process was different in that export occurred at a limited number of sites (10–50 per cell) and complete spreading over the cell surface took 10 to 20 minutes.

It should be emphasized that the translocation and integration processes undergone by the lipopolysaccharide and the lipoprotein may also be used for all the other constituent elements of the outer membrane. The outer membrane is assembled from components that are synthesized elsewhere. For example, the outer membrane phospholipids are also synthesized in the plasma membrane and are secondarily translocated to the outer membrane (White et al., 1971). Translocation of pulse-labeled phosphatidylethanolamine from the plasma to the outer membrane is essentially complete within about 7 minutes, and the extent of translocation corresponds to the theoretical value from the known distribution of the phospholipids between the two membranes (section 2.2.3.). At present, the molecular bases for these unidirectional translocations into the outer membrane are not understood. Obviously they represent one of the major areas for future research.

4. Conclusions

An attempt has been made in this chapter to assess the present state of progress in the field of assembly and synthesis of bacterial walls. Thus, it might be appropriate to try to predict how research in this area might develop within the next decade or so. There are number of questions to be answered.

Undoubtedly, the exact functioning, reciprocal control, and coordination of the multienzyme assembly centers for the various wall polymers in the microenvironment of the plasma membrane, the exact functioning, regulation, and triggering of the autolytic system in a specific cell area for a defined time period during the cell life cycle, and the integration of all these activities in terms of cell shape expression and determination are essential topics that will continue to stimulate increasing interest.

The operating units in the various wall assembly centers of the plasma membrane are obviously enzymes. Some of these enzymes are known to be the targets of important antibiotics. Thus, β-lactam antibiotics prevent the functioning of the DD-carboxypeptidase-transpeptidase enzyme systems. Other enzymes may be potential targets for new antimicrobial agents. At present, however, none of these membrane-bound enzymes has been fully characterized or even purified to strict homogeneity. Studies of these fascinating enzymes with respect to amino acid sequence, conformation (both in their natural environment and outside), distribution of hydrophobic and hydrophilic portions, mechanisms of the reactions catalyzed, structure and conformation of active centers, requirement for phospholipid, regulation, and so on leave unanswered questions. In light of our present knowledge, future advances must lead to substantial and gratifying developments and, perhaps, to entirely unexpected discoveries.

All the wall components of the gram-negative bacteria, except the peptidoglycan, are part of a membrane unit which is the outermost structure of the cell. The unidirectional translocation of these components from the plasma membrane where they are made to the exterior of the cell where they are eventually assembled and form a relatively fluid and extremely asymmetric structure, is obviously the expression of a very high distribution coefficient that these components exhibit in favor of the outer membrane. The molecular basis of this astonishing behavior is entirely unknown. Is this a purely physical phenomenon? Do carriers serve as shuttles between the two membranes involved in the process? What is the exact nature of the mixed or fused regions that form the adhesion sites between the plasma and the outer membranes. Do these sites, which appear to be export sites for the lipopolysaccharide, fulfill the same function for the other components? Satisfying answers to any of these questions, and others, cannot be ventured at present.

References

Adams, J. B., Archibald, A.R., Baddiley, J., Coapes, H.E. and Davison, A.L. (1969) Teichoic acids possessing phosphate-sugar linkages in strains of *Lactobacillus plantarum*. Biochem. J. 113, 191–193.

Aebi, U., Smith, P.R., Dubochet, J., Henry, C. and Kellenberger, E. (1973) A study of the structure of the T-layer of *Bacillus brevis*. J. Supramol. Struct. 1, 498–522.

574

Ames, G.F.L. (1974) Resolution of bacterial proteins by polyacrylamide gel electrophoresis on slabs. Membrane, soluble and periplasmic fractions. J. Biol. Chem. 249, 634–644.

Ames, G.F.L. and Nikaido, K. (1976) Two-dimensional gel electrophoresis of membrane proteins. Biochemistry 15, 616–623.

Anderson, J.S., Page, R.L. and Salo, W.L. (1972) Biosynthesis of the polysaccharide of *Micrococcus lysodeikticus* cell walls. II. Identification of uridine diphospho-N-acetyl-D-mannosaminuronic acid as a required substrate. J. Biol. Chem. 247, 2480–2485.

Anderson, R.G., Hussey, H. and Baddiley, J. (1973a) The mechanism of wall synthesis in bacteria. The organization of enzymes and isoprenoid phosphates in the membrane. Biochem. J. 127, 11–25.

Anderson, R.G., Douglas, L.J., Hussey, H. and Baddiley, J. (1973b) The control of synthesis of bacterial cell walls. Interaction in the synthesis of nucleotide precursors. Biochem. J. 136, 871–876.

Archibald, A.R. (1976) Cell wall assembly in *Bacillus subtilis:* Development of bacteriophage-binding properties as a result of the pulsed incorporation of teichoic acid. J. Bacteriol. 127, 956–960.

Archibald, A.R. and Staford, G.H. (1972) A polymer of N-acetyl-glucosamine 1-phosphate in the wall of *Staphylococcus lactis* 2102. Biochem. J. 130, 681–690.

Archibald, A.R., Baddiley, J. and Buchanan, J.G. (1961) The ribitol teichoic acid from *Lactobacillus arabinosus* walls: isolation and structure of ribitol glucosides. Biochem. J. 81, 124–134.

Archibald, A.R., Baddiley, J., Heckels, J.E. and Heptinstall, S. (1971) Further studies on the glycerol teichoic acid of walls of *Staphylococcus lactis* 13. Location of the phosphodiester groups and their susceptibility to hydrolysis with alkali. Biochem. J. 125, 353–359.

Armstrong, J.J., Baddiley, J., Buchanan, J.G., Carss, B. and Greenberg, G.R. (1958) Isolation and structure of ribitol phosphate derivatives (teichoic acids) from bacterial cell walls. J. Chem. Soc., 4344–4345.

Armstrong, J.J., Baddiley, J. and Buchanan, J.G. (1961) Further studies on the teichoic acid from *Bacillus subtilis* walls. Biochem. J. 80, 254–261.

Asbell, M.A. and Eagon, R.G. (1966) Role of multivalent cations in the organization, structure and assembly of the cell wall of *Pseudomonas aeruginosa.* J. Bacteriol. 92, 380–387.

Azuma, I., Thomas, D.W., Adam, A., Ghuysen, J.M., Bonaly, R., Petit, J.F. and Lederer, E. (1970) Occurrence of N-glycolylmuramic acid in bacterial cell walls. A preliminary survey. Biochim. Biophys. Acta 208, 444–451.

Baddiley, J. (1964) Teichoic acids and bacterial cell wall. Endeavour 23, 33–37.

Baddiley, J. (1972) Teichoic acids in cell walls and membranes of bacteria. Essays in Biochem. 8, 35–77.

Baddiley, J., Buchanan, J.G., Martin, R.O. and RajBhandary, U.L. (1962a) Teichoic acids from the walls of *Staphylococcus aureus* H. 2. Location of phosphate and alanine residues. Biochem. J. 85, 49–56.

Baddiley, J., Buchanan, J.G., RajBhandary, U.L. and Sanderson, A.R. (1962b) Teichoic acid from the walls of *Staphylococcus aureus* H. 1. Structure of the N-acetylglucosaminylribitol residues. Biochem. J. 82, 439–448.

Bayer, M.E. (1968) Areas of adhesion between wall and membrane of *Escherichia coli.* J. Gen. Microbiol. 53, 395–404.

Bayer, M.E. (1974) Ultrastructure and organization of the bacterial envelope. Ann. N.Y. Acad. Sci. 215, 6–28.

Bayer, M.E. (1975) Role of adhesion zones in bacterial cell surface, function and biogenesis. In: Membrane Biogenesis (Tzagoloff, A., ed.), pp. 393–427. Plenum Press, New York.

Bayer, M.E. and Anderson, T.F. (1965) The surface structure of *Escherichia coli.* Proc. Nat. Acad. Sci. U.S.A. 54, 1592–1599.

Bayer, M.E. and Remsen, C.C. (1970) Structure of *Escherichia coli* after freeze-etching. J. Bacteriol. 101, 304–313.

Beck, B.D. and Park, J.T. (1976) Activity of three murein hydrolases during the cell division cycle of *Escherichia coli* K 12 as measured in toluenized cells. J. Bacteriol. 126, 1250–1260.

Berberich, R., Kaback, M. and Freese, E. (1968) D-Amino acids as inducers of L-alanine dehydrogenase in *Bacillus subtilis*. J. Biol. Chem. 243, 1006–1011.

Bettinger, G.E. and Young, F.E. (1975) Tunicamycin, an inhibitor of *Bacillus* peptidoglycan synthesis: a new site of inhibition. Biochem. Biophys. Res. Commun. 67, 16–21.

Birdsell, D.C. and Cota-Robles, E.H. (1967) Production and ultrastructure of lysozyme and ethylenediaminete-traacetate-lysosome spheroplasts of *Escherichia coli*. J. Bacteriol. 93, 427–437.

Birdsell, D.C., Doyle, R.J. and Morgenstern, M. (1975) Organization of teichoic acid in the cell wall of *Bacillus subtilis*. J. Bacteriol. 121, 726–734.

Blaurock, A.E. and Stoeckenius, W. (1971) Structure of the purple membrane. Nature New Biol. 233, 152–155.

Blumberg, P.M. and Strominger, J.L. (1971) Inactivation of D-alanine carboxypeptidase by penicillins and cephalosporins is not lethal in *Bacillus subtilis*. Proc. Nat. Acad. Sci. USA 68, 2814–2817.

Blumberg, P.M. and Strominger, J.L. (1974) Interaction of penicillin with the bacterial cell: Penicillin binding proteins and penicillin-sensitive enzymes. Bacteriol. Rev. 38, 291–335.

Boos, W. (1974) Bacterial transport. Ann. Rev. Biochem. 43, 123–146.

Boy de la Tour, E., Bolle, A. and Kellenberger, E. (1965) Nouvelles observations concernant l'action du laurylsulfate de sodium sur la paroi et la membrane d'*Escherichia coli*. Pathol. Microbiol. 28, 229–234.

Bracha, R. and Glaser, L. (1976) *In vitro* system for the synthesis of teichoic acid linked to peptidoglycan. J. Bacteriol. 125, 872–879.

Bragg, P.D. and Hou, C. (1972) Organization of proteins in the native and reformed outer membrane of *Escherichia coli*. Biochim. Biophys. Acta 274, 478–488.

Braun, V. (1973) Molecular organization of the rigid layer and the cell wall of *Escherichia coli*. J. Infect. Diseases 128 (suppl.), 9–16.

Braun, V. (1975) Covalent lipoprotein from the outer membrane of *Escherichia coli*. Biochim. Biophys. Acta 415, 335–377.

Braun, V. and Bosch, V. (1972a) Repetitive sequences in the murein-lipoprotein of the cell wall of *Escherichia coli*. Proc. Nat. Acad. Sci. U.S.A. 69, 970–974.

Braun, V. and Bosch, V. (1972b) Sequence of the murein-lipoprotein and the attachment site of the lipid. Eur. J. Biochem. 28, 51–69.

Braun, V. and Bosch, V. (1973) *In vivo* biosynthesis of murein-lipoprotein of the outer membrane of *Escherichia coli*. FEBS Lett. 34, 302–306.

Braun, V. and Hantke, K. (1974) Biochemistry of bacterial cell envelopes. Ann. Rev. Biochem. 43, 89–121.

Braun, V. and Rehn, K. (1969) Chemical characterization, spatial distribution and function of a lipoprotein (murein-lipoprotein) of the *Escherichia coli* cell wall. The specific effect of trypsin on the membrane structure. Eur. J. Biochem. 10, 426–438.

Braun, V. and Sieglin, U. (1970) The covalent murein-lipoprotein structure of the *Escherichia coli* cell wall. The attachment site of the lipoprotein on the murein. Eur. J. Biochem. 13, 336–346.

Braun, V. and Wolff, H. (1970) The murein-lipoprotein linkage in the cell wall of *Escherichia coli*. Eur. J. Biochem. 14, 387–391.

Braun, V. and Wolff, H. (1973) Characterization of the receptor protein for phage T5 and colicin M in the outer membrane of *Escherichia coli* B. FEBS Lett. 34, 77–80.

Braun, V. and Wolff, H. (1975) Attachment of lipoprotein to murein (peptidoglycan) of *Escherichia coli* in the presence and absence of penicillin FL 1060. J. Bacteriol. 123, 888–897.

Braun, V., Rehn, K. and Wolff, H. (1970) Supramolecular structure of the rigid layer of the cell wall of *Salmonella, Serratia, Proteus,* and *Pseudomonas fluorescens*. Number of lipoprotein molecules in a membrane layer. Biochemistry 9, 5041–5049.

Braun, V., Schaller, K. and Wolff, H. (1973) A common receptor protein for phage T5 and colicin M in the outer membrane of *Escherichia coli* B. Biochim. Biophys. Acta 323, 87–97.

577

Churchward, G. and Holland, J.B. (1976) Envelope synthesis during the cell cycle in *E. coli* B/r. J. Mol. Biol., 105, 245–261.

Cleveland, R.F., Daneo-Moore, L., Wicken, A.J. and Shockman, G.D. (1976) Effect of lipoteichoic acid and lipids on lysis of intact cells of *Streptococcus faecalis*. J. Bacteriol. 127, 1582–1584.

Cleveland, R.F., Höltje, J.V., Wicken, A.J., Tomasz, A., Daneo-Moore, L. and Shockman, G.D. (1975) Inhibition of bacterial wall lysins by lipoteichoic acids and related compounds. Biochem. Biophys. Res. Commun. 67, 1128–1135.

Coley, J., Duckworth, M. and Baddiley, J. (1975a) Extraction and purification of lipoteichoic acids from gram-positive bacteria. Carbohydrate Res. 40, 41–52.

Coley, J., Heckels, J.E., Archibald, A.R. and Baddiley, J. (1975b) The linkage between wall teichoic acid and peptidoglycan in *Staphylococcus aureus* H. Paper presented at 74th Meet. Soc. Gen. Microbiol., Newcastle-upon-Tyne, England.

Coley, J., Archibald, A.R. and Baddiley, J. (1976) A linkage unit joining peptidoglycan to teichoic acid in *Staphylococcus aureus* H. FEBS Lett. 61, 240–243.

Cota-Robles, E. (1963) Electron microscopy of plasmolysis in *Escherichia coli*. J. Bacteriol. 85, 499–503.

Coyette, J. and Ghuysen, J.M. (1968) Structure of the cell wall of *Staphyloccus aureus*. strain Copenhagen. IX. Teichoic acid and phage adsorption. Biochemistry 7, 2385–2389.

Coyette, J. and Ghuysen, J.M. (1970a) Structure of the walls of *Lactobacillus acidophilus* strain 63 AM Gasser. Biochemistry 9, 2935–2944.

Coyette, J. and Ghuysen, J.M. (1970b) Wall autolysin of *Lactobacillus acidophilus* strain 63 AM Gasser. Biochemistry 9, 2952–2955.

Coyette, J., Perkins, H.R., Polacheck, I., Shockman, G.D. and Ghuysen, J.M. (1974) Membrane-bound DD-carboxypeptidase and LD-transpeptidase of *Streptococcus faecalis* ATCC 9790. Eur. J. Biochem. 44, 459–468.

Curtis, R., Caro, L.G., Allison, D.P. and Stallions, D.R. (1969) Early stages of conjugation in *Escherichia coli*. J. Bacteriol. 100, 1091–1104.

Cynkin, M.A. and Osborn, M.J. (1968) Enzymatic transfer of O-antigen to lipopolysaccharide. Proc. Fed. Amer. Soc. Exp. Biol. 27, 293.

Datta, D.B., Kramer, C. and Henning, U. (1976) Diploidy for a structural gene specifying a major protein of the outer cell envelope membrane from *Escherichia coli* K12. J. Bacteriol. 128, 834–841.

Decad, G.M. and Nikaido, H. (1976) Outer membrane of Gram-negative bacteria. XII. Molecular-sieving function of cell wall. J. Bacteriol. 128, 325–336.

De Martini, M., Inouye, S. and Inouye, M. (1976) Ultrastructure of paracrystals of a lipoprotein from the outer membrane of *Escherichia coli*. J. Bacteriol. 127, 564–571.

De Martini, M., Inouye, S. and Inouye, M. (1976) Ultrastructure of paracrystals of a structural lipoprotein of the outer membrane of *Escherichia coli*. J. Mol. Biol., in press.

De Pamphilis, M.L. and Adler, J. (1971) Attachment of flagellar basal bodies to the cell envelope: specific attachment to the outer, lipopolysaccharide membrane and the cytoplasmic membrane. J. Bacteriol. 105, 396–407.

de Petris, S. (1965) Ultrastructure of the cell wall of *Escherichia coli*. J. Ultrastruct. Res. 12, 247–262.

de Petris, S. (1967) Ultrastructure of the cell wall of *Escherichia coli* and chemical nature of its constituent layers. J. Ultrastruct. Res. 19, 45–83.

Dezélée, Ph. and Shockman, G.D. (1975) Studies of the formation of peptide cross-links in the cell wall peptidoglycan of *Streptococcus faecalis*. J. Biol. Chem. 250, 6806–6816.

Di Masi, D.R., White, J.C., Schnaitman, C. and Bradbeer, C. (1973) Transport of vitamin B12 in *Escherichia coli*: common receptor sites for vitamin B12 and the E colicins on the outer membrane of the cell envelope. J. Bacteriol. 115, 506–513.

Doyle, R.J., Birdsell, D.C. and Young, F.E. (1973) Isolation of the teichoic acid of *Bacillus subtilis* 168 by affinity chromatography. Prep. Biochem. 3, 13–18.

578

Duckworth, M., Archibald, A.R. and Baddiley, J. (1975) Lipoteichoic acid and lipoteichoic acid carrier in *Staphylococcus aureus H*. FEBS Lett. 53, 176–179.

Eagon, R.G. and Carson, K.J. (1965) Lysis of cell walls and intact cells of *Pseudomonas aeruginosa* by ethylenediaminetetraacetic acid and by lysozyme. Can. J. Microbiol. 11, 193–201.

Eidels, L. and Osborn, M.J. (1971) Lipopolysaccharide and aldoheptose biosynthesis in transketolase mutants of *Salmonella typhimurium*. Proc. Nat. Acad. Sci. U.S.A. 68, 1673–1677.

Eidels, L. and Osborn, M.J. (1974) Phosphoheptose isomerase, first enzyme in the biosynthesis of aldoheptose in *Salmonella typhimurium*. J. Biol. Chem. 249, 5642–5648.

Ellar, D.J., Lundgren, D.G. and Slepecky, R.A. (1967) Fine structure of *Bacillus megaterium* during synchronous growth. J. Bacteriol. 94, 1189–1205.

Elliott, T.S.J., Ward, J.B. and Rogers, H.J. (1975a) The formation of cell wall polymers by reverting protoplasts of *Bacillus licheniformis* 6346 His⁻. J. Bacteriol. 124, 623–632.

Elliott, T.S.J., Ward, J.B., Wyrick, P.B. and Rogers, H.J. (1975b) An ultrastructural study of the reversion of protoplasts of *Bacillus licheniformis* 6346 His- to bacilli. J. Bacteriol. 124, 905–917.

Ellwood, D.C. (1975) The effect of growth conditions on the chemical composition of microbial walls and extracellular polysaccharides. Paper presented at 74th Meet. Soc. Gen. Microbiol., Newcastle-upon-Tyne. England.

Ellwood, D.C. and Tempest, D.W. (1969) Control of teichoic acid and teichuronic acid biosynthesis in chemostat cultures of *Bacillus subtilis* var. *niger*. Biochem. J. 111, 1–5.

Fan, D.P. and Beckman, B.E. (1973) *Micrococcus lysodeikticus* bacterial walls as a substrate specific for the autolytic glycosidase of *Bacillus subtilis*. J. Bacteriol. 114, 804–813.

Fiedler, F. and Glaser, L. (1974) The synthesis of polyribitol phosphate. I. Purification of polyribitol phosphate polymerase and lipoteichoic acid carrier. J. Biol. Chem. 249, 2684–2689.

Fiedler, F., Mauck, J. and Glaser, L. (1974) Problems in cell wall assembly. Ann. N.Y. Acad. Sci. 235, 198–209.

Fischman, D.A. and Weinbaum, G. (1967) Hexagonal pattern in cell walls of *Escherichia coli B*. Science 155, 472–474.

Fitz-James, Ph. (1974) In the discussion following the paper of Mirelman, D., Bracha, R. and Sharon, N. Ann. N.Y. Acad. Sci. U.S.A. 235, 345–346.

Fodor, K. and Alföldi, L. (1976) Fusion of protoplasts of *Bacillus megaterium*. Proc. Nat. Acad. Sci. USA 73, 2147–2150.

Fordham, W.D. and Gilvarg, Ch. (1974) Kinetics of crosslinking of peptidoglycan in *Bacillus megaterium*. J. Biol. Chem. 249, 2478–2482.

Formanek, H., Formanek, S. and Wawra, H. (1974) A three-dimensional atomic model of the murein layer of bacteria. Eur. J. Biochem. 46, 279–294.

Formanek, H., Schleifer, K.H., Seidl, H.P., Lindemann, R. and Zundel, G. (1976) Three-dimensional structure of peptidoglycan of bacterial cell walls: Infra red investigations. FEBS Letters 70, 150–154.

Forsberg, C.W., Wyrick, P.B., Ward, J.B. and Rogers, H.J. (1973) Effect of phosphate limitation on the morphology and wall composition of *Bacillus licheniformis* and its phosphoglucomutase-deficient mutants. J. Bacteriol. 113, 969–984.

Fox, E.N. (1974) M. proteins of group A *Streptococci*. Bact. Rev. 38, 57–86.

Freer, J.H. and Salton, M.R.J. (1971) The anatomy and chemistry of Gram-negative cell envelopes. In: Microbial Toxins (Weinbaum, G., Kadis, S. and Ajl, S.J., eds.), vol. 4, pp. 67–126. Academic Press, New York.

Frère, J.M., Ghuysen, J.M., Perkins, H.R. and Nieto, M. (1973) Kinetics of concomitant transfer and hydrolysis reactions catalyzed by the exocellular DD-carboxypeptidase-transpeptidase of *Streptomyces* R61. Biochem. J. 135, 483–492.

Frère, J.M., Ghuysen, J.M., Degelaen, J., Loffet, A. and Perkins, H.R. (1975) Fragmentation of benzylpenicillin as a result of its interaction with the exocellular DD-carboxypeptidase-transpeptidase of *Streptomyces* R61 and R39. Nature 258, 168–170.

Frère, J.M., Ghuysen, J.M., Vanderhaeghe, H., Adriaens, P., Degelaen, J. and De Graeve, J. (1976) The fate of the thiazolidine ring during fragmentation of penicillin by the exocellular DD-carboxypeptidase-transpeptidase of *Streptomyces* R61. Nature 260, 451–454.

Frost, G.E. and Rosenberg, H. (1973) The inducible citrate-dependent iron transport system in *Escherichia coli* K12. Biochim. Biophys. Acta 330, 90–101.

Frye, L.D. and Edidin, M. (1970) The rapid intermixing of cell surface antigens after formation of mouse-human heterokaryons. J. Cell Sci. 7, 319–335.

Fuchs-Cleveland, E. and Gilvarg, Ch. (1976) Oligomeric intermediate in peptidoglycan biosynthesis in *Bacillus megaterium*. Proc. Nat. Acad. Sci. USA 73, 4200–4204.

Fukui, Y., Nachbar, M.S. and Salton, M.R.J. (1971) Immunological properties of *Micrococcus lysodeikticus* membranes. J. Bacteriol. 105, 86–92.

Galanos, C., Lüderitz, O. and Westphal, O. (1969) A new method for the extraction of lipopolysaccharides. Eur. J. Biochem. 8, 245–249.

Galanos, C., Lüderitz, O. and Westphal, O. (1971) Preparation and properties of antisera against the lipid-A component of bacterial lipopolysaccharides. Eur. J. Biochem. 24, 116–122.

Ganfield, M.C.W. and Pierringer, R.A. (1975) Phosphatidylkojibiosyl diglyceride. The covalently linked lipid constituent of the membrane lipoteichoic acid from *Streptococcus faecalis (faecium)* ATCC 9790. J. Biol. Chem. 250, 702–709.

Garten, W. and Henning, U. (1974) Cell envelope and shape of *Escherichia coli* K12. Isolation and preliminary characterization of the major ghost-membrane proteins. Eur. J. Biochem. 47, 343–352.

Garten, W., Hindennach, J. and Henning, U. (1975a) The major proteins of *Escherichia coli* outer cell envelope membrane. Characterization of proteins II* and III, comparison of all proteins. Eur. J. Biochem. 59, 215–221.

Garten, W., Hindennach, J. and Henning, U. (1975b) Cyanogen bromide fragments of protein I, composition and order. Eur. J. Biochem. 60, 303–307.

Ghuysen, J.M. (1968) Use of bacteriolytic enzymes in determination of wall structure and their role in cell metabolism. Bact. Rev. 32, 425–464.

Ghuysen, J.M. (1977a) The bacterial DD-carboxypeptidase-transpeptidase enzyme system. A new insight into the mode of action of penicillin. E.R. Squibb Lectures on Chemistry of Microbial Products. (W.E. Brown, series ed.) University Tokyo Press, Tokyo, 162 p.

Ghuysen, J.M. (1977b) The concept of penicillin target: From 1965 until to-day. The 13th Marjory Stephenson Lecture. J. Gen. Microbiol. 101, part I, in press.

Ghuysen, J.M. and Shockman, G.D. (1973) Biosynthesis of peptidoglycan. In: Bacterial Membranes and Walls (Leive, L., ed.), Vol. 1, pp. 37–130, Marcel Dekker, New York.

Ghuysen, J.M., Tipper, D.J. and Strominger, J.L. (1964) Structure of the cell wall of *Staphylococcus aureus*, strain Copenhagen. IV. The teichoic acid-glycopeptide complex. Biochemistry 4, 474–485.

Ghuysen, J.M., Bricas, E., Lache, M. and Lehy-Bouille, M. (1968) Structure of the cell walls of *Micrococcus lysodeikticus*. III. Isolation of a new peptide dimer, N^α-[L-alanyl-γ-(α-D glutamylglycine)]-L-lysyl-D-alanyl-N^α-[L-analyl-(α-D-glutamyl-glycine)]-L-lysyl-D-alanine. Biochemistry 7, 1450–1460.

Ghuysen, J.M., Lehy-Bouille, M., Campbell, J.N., Moreno, R., Frère, J.M., Duez, C., Nieto, M. and Perkins, H.R. (1973) The structure of the wall peptidoglycan of *Streptomyces* R39 and the specificity profile of its exocellular DD-carboxypeptidase-transpeptidase for peptide acceptors. Biochemistry 12, 1243–1251.

Gilleland, H.E., Stinnett, J.D., Roth, I.L. and Eagon, R.G. (1973) Freeze-etch study of *Pseudomonas aeruginosa:* localization within the cell wall of an ethylenediaminetetraacetate-extractable component. J. Bacteriol. 113, 417–432.

Glaser, L. (1964) The synthesis of teichoic acids II. Polyribitol phosphate. J. Biol. Chem. 239, 3178–3186.

Glaser, L. and Lindsay, B. (1974) The synthesis of lipoteichoic acid carrier. Biochem. Biophys. Res. Commun. 59, 1131–1136.

Gmeiner, J. (1975) Identification of ribitol phosphate as a constituent of the lipopolysaccharide from *Proteus mirabilis* strain D52. Eur. J. Biochem., 58, 627–629.

Gmeiner, J. (1977) The ribitol-phosphate-containing lipopolysaccharide from *Proteus mirabilis,* strain

D52. Investigations on the structure of O-specific chains. Eur. J. Biochem. 74, 171–180.

Gmeiner, J. and Martin, H.H. (1976) Phospholipid and lipopolysaccharide in *Proteus mirabilis* and its stable protoplast L-form. Difference in content and fatty acid composition. Eur. J. Biochem. 67, 487–494.

Goebel, W.F., Shedlovsky, T., Lavin, G.I. and Adams, M.H. (1943) The heterophile antigen of *Pneumococcus*. J. Biol. Chem. 148, 1–15.

Goodell, E.W. and Schwarz, U. (1975) Sphere-rod morphogenesis of *Escherichia coli*. J. Gen. Microbiol. 86, 201–209.

Goundry, J., Davison, A.L., Archibald, A.R. and Baddiley, J. (1967) The structure of the cell wall of *Bacillus polymyxa* (NCIB 4747). Biochem. J. 104, 1c–2c.

Gray, G.W. and Wilkinson, S.G. (1965a) The action of ethylenediaminetetraacetic acid on *Pseudomonas aeruginosa*. J. Appl. Bacteriol. 28, 153–164.

Gray, G.W. and Wilkinson, S.G. (1965b) The effect of ethylenediaminetetraacetic acid on the cell walls of some Gram-negative bacteria. J. Gen. Microbiol. 39, 385–399.

Green, N.M. and Flanagan, M.T. (1976) The prediction of the conformation of membrane proteins from the sequence of amino acids. Biochem. J. 153, 729–732.

Gruss, P., Gmeiner, J. and Martin, H.H. (1975) Amino-acid composition of the covalent rigid-layer lipoprotein in cell walls of *Proteus mirabilis*. Eur. J. Biochem. 57, 411–414.

Gudas, L.J. and Pardee, A.B. (1976) DNA synthesis inhibition and the induction of protein X in *Escherichia coli*. J. Mol. Biol. 101, 459–477.

Gudas, L.J., James R. and Pardee, A.B. (1976) Evidence for the involvement of an outer membrane protein in DNA initiation. J. Biol. Chem., 251, 3470–3479.

Guinand, M., Vacheron, M.J. and Michel, G. (1970) Structure des parois cellulaires des *Nocardia*. I. Isolement et composition des parois de *Nocardia kirovani*. FEBS Lett. 6, 37–39.

Gunetileke, K.C. and Anwar, R.A. (1968) Biosynthesis of uridine diphsopho-N-acetylmuramic acid. II. Purification and properties of pyruvate-uridine diphospho-N-acetylglucosamine transferase and characterization of uridine diphospho-N-acetylenolpyruvylglucosamine. J. Biol. Chem. 243, 5770–5778.

Hakenbeck, R. and Messer, W. (1977) Activities of murein hydrolases in synchronized cultures of *Escherichia coli*. J. Bacteriol. 129, 1239–1244.

Hakenbeck, R., Goodell, E.W. and Schwarz, U. (1974) Compartmentalization of murein hydrolases in the envelope of *Escherichia coli*. FEBS Lett. 40, 261–264.

Haller, I. and Henning, U. (1974) Cell envelope and shape of *Escherichia coli* K12. Crosslinking with dimethyl imidoesters of the whole cell wall. Proc. Nat. Acad. Sci. U.S.A. 71, 2018–2021.

Haller, I., Hoehn, B. and Henning, U. (1975) Apparent high degree of asymmetry of protein arrangement in the *Escherichia coli* outer cell envelope membrane. Biochemistry 4, 478–484.

Hämmerling, G., Lehmann, V. and Lüderitz, O. (1973) Structural studies on the heptose region of *Salmonella* lipopolysaccharides. Eur. J. Biochem. 38, 453–458.

Hammes, W.P. (1976) Biosynthesis of peptidoglycan in *Gaffkya homari*. The mode of action of penicillin G and micillinam. Eur. J. Biochem. 70, 107–113.

Hammes, W.P. and Kandler, O. (1976) Biosynthesis of peptidoglycan in *Gaffkya homari*. The incorporation of peptidoglycan into the cell wall and the direction of transpeptidation. Eur. J. Biochem. 70, 97–106.

Hammes, W.P. and Neuhaus, F.C. (1974a) Biosynthesis of peptidoglycan in *Gaffkya homari*: role of the peptide subunit of uridine disphosphate-N-acetylmuramyl-pentapeptide. J. Bacteriol. 120, 210–218.

Hammes, W.P. and Neuhaus, F.C. (1974b) On the mechanism of action of vancomycin: inhibition of peptidoglycan synthesis in *Gaffkya homari*. Antimicrob. Ag. Chemother. 6, 722–728.

Hancock, I. and Baddiley, J. (1976) *In vitro* synthesis of the unit that links teichoic acid to peptidoglycan. J. Bacteriol., 125, 880–886.

Hancock, R.E.W., Hantke, K. and Braun, V. (1976). Iron transport in *Escherichia coli* K12: Involvement of the colicin B receptor and of a citrate-inducible protein. J. Bacteriol. 127, 1370–1375.

Hantke, K. and Braun, V. (1973) Covalent binding of lipid to protein. Diglyceride and amide-linked fatty acid at the N-terminal end of the murein-lipoprotein of the *Escherichia coli* outer membrane. Eur. J. Biochem. 34, 284–296.

Hantke, K. and Braun, V. (1975a) Membrane receptor dependent iron transport in *Escherichia coli.* FEBS Lett. 49, 301–305.

Hantke, K. and Braun, V. (1975b) A function common to iron-enterochelin transport and action of colicins B, I, V. in *Escherichia coli.* FEBS Lett. 59, 277–281.

Hartmann, R., Bock-Hennig, S.B. and Schwarz, U. (1974) Murein hydrolases in the envelope of *Escherichia coli.* Properties *in situ* and solubilization from the envelope. Eur. J. Biochem. 41, 203–208.

Heath, E.C., Mayer, R.M., Edstrom, R.D. and Beaudreau, C.A. (1966) Structure and biosynthesis of the cell wall lipopolysaccharide of *Escherichia coli.* Ann. N.Y. Acad. Sci. U.S.A. 133, 315–333.

Heckels, J.E., Archibald, A.R. and Baddiley, J. (1975) Studies on the linkage between teichoic acid and peptidoglycan in a bacteriophage-resistant mutant of *Staphylococcus aureus H.* Biochem. J. 149, 637–647.

Heckels, J.E., Lambert, P.A. and Baddiley, J. (1977). Binding of magnesium ions to cell walls of *Bacillus subtilis* W23 containing teichoic acid or teichuronic acid. Biochem. J. 162, 359–365.

Helenius, A. and Simons, K. (1975) Solubilization of membranes by detergents. Biochim. Biophys. Acta 415, 29–79.

Henning, U. (1975) Determination of cell shape in bacteria. Ann. Rev. Microbiol. 29, 45–60.

Henning, U. and Haller, I. (1975) Mutants of *Escherichia coli* K12 lacking all "major" proteins of the outer cell envelope membrane. FEBS Lett. 55, 161–164.

Henning, U. and Schwarz, U. (1973) Determinants of cell shape. In: Bacterial Membranes and Walls (Leive, L., ed.), pp. 413–438. Marcel Dekker, New York.

Henning, U., Höhn, B. and Sonntag, I. (1973a) Cell envelope and shape of *Escherichia coli* K12. The ghost membrane. Eur. J. Biochem. 39, 27–36.

Henning, U., Rehn, K. and Hoehn, B. (1973b) Cell envelope and shape of *Escherichia coli* K12. Proc. Nat. Acad. Sci. U.S.A. 70, 2033–2036.

Henning, U., Hindennack, I. and Haller, I. (1976) The major proteins of the *Escherichia coli* outer cell evelope membrane: Evidence for the structural gene of protein II*. FEBS Letters 61, 46–48.

Heppel, L.A. (1967) Selective release of enzymes from bacteria. Treatments affecting the bacterial wall remove certain enzymes and transport factors from living cells. Science 156, 1451–1455.

Heptinstall, S., Archibald, A.R. and Baddiley, J. (1970) Teichoic acids and membrane function in bacteria. Nature 225, 519–521.

Herbold, D.R. and Glaser, L. (1975a) *Bacillus subtilis* N-acetylmuramic acid L-alanine amidase. J. Biol. Chem. 250, 1676–1682.

Herbold, D.R. and Glaser, L. (1975b) Interaction of N-acetylmuramic acid L-alanine amidase with cell wall polymers. J. Biol. Chem. 250, 7231–7238.

Heydanek, M.G. and Neuhaus, F.C. (1969) The initial stage in peptidoglycan synthesis. IV. Solubilization of phospho-N-acetylmuramyl-pentapeptide translocase. Biochemistry 4, 1474–1481.

Higashi, Y. and Strominger, J.L. (1970) Biosynthesis of phosphatidylglycerol and candidipin as cofactors for isoprenoid alcohol phosphokinase. J. Biol. Chem. 245, 3691–3696.

Higashi, Y., Strominger, J.L. and Sweeley, Ch.C. (1967) Structure of a lipid intermediate in cell wall peptidoglycan synthesis: a derivative of a C 55 isoprenoid alcohol. Proc. Nat. Acad. Sci. U.S.A. 57, 1878–1884.

Higashi, Y., Siewert, G. and Strominger, J.L. (1970a) Biosynthesis of the peptidoglycan of bacterial cell walls. XIX. Isoprenoid alcohol phosphokinase. J. Biol. Chem. 245, 3683–3690.

Higashi, Y., Strominger, J.L. and Sweeley, Ch.C. (1970b) Biosynthesis of the peptidoglycan of bacterial cell walls. XXI. Isolation of free C 55-isoprenoid alcohol and of lipid intermediates in peptigolycan synthesis from *Staphylococcus aureus.* J. Biol. Chem. 245, 3697–3702.

Higgins, M.L. and Shockman, G.D. (1971) Procaryotic cell division with respect to wall and membranes. CRC Critical Rev. Microbiol. 1, 29–72.

Higgins, M.L., Coyette, J. and Shockman, G.D. (1973) Sites of cellular autolysis in *Lactobacillus acidophilus.* J. Bacteriol. 116, 1375–1382.

582

Higgins, M.L., Tsien, H.C. and Daneo-Moore, L. (1976) Organization of mesosomes in fixed and unfixed cells. J. Bacteriol. 127, 1519–1523.

Hinckley, A., Müller, E. and Rothfield, L. (1972) Reassembly of a membrane-bound multienzyme system. I. Formation of a particule containing phosphatidylethanolamine, lipopolysaccharide, and two glycosyltransferase enzymes. J. Biol. Chem. 247, 2623–2628.

Hindennach, I. and Henning, U. (1975) The major proteins of the *Escherichia coli* outer cell envelope membrane. Preparative isolation of all major membrane proteins. Eur. J. Biochem. 59, 207–213.

Hirashima, A. and Inouye, M. (1973) Specific biosynthesis of an envelope protein of *Escherichia coli*. Nature 242, 405–407.

Hirashima, A. and Inouye, M. (1975) Biosynthesis of a specific lipoprotein of the *Escherichia coli* outer membrane on polyribosomes. Eur. J. Biochem. 60, 395–398.

Hirashima, A., Childs, G. and Inouye, M. (1973a) Differential inhibitory effects of antibiotics on the biosynthesis of envelope proteins of *Escherichia coli*. J. Mol. Biol. 79, 373–389.

Hirashima, A., Wu, H.C., Venkateswaran, P.S. and Inouye, M. (1973b) Two forms of a structural lipoprotein in the envelope of *Escherichia coli*. Further characterization of the free form. J. Biol. Chem. 248, 5654–5659.

Hirashima, A., Wang, S. and Inouye, M. (1974) Cell-free synthesis of a specific lipoprotein of the *Escherichia coli* outer membrane directed by purified messenger RNA. Proc. Nat. Acad. Sci. U.S.A. 71, 4149–4153.

Höltje, J.V. and Tomasz, A. (1975a) Lipoteichoic acid: a specific inhibitor of autolysin activity in *Pneumococcus*. Proc. Nat. Acad. Sci. U.S.A. 72, 1690–1694.

Höltje, J.V. and Tomasz, A. (1975b) Specific recognition of choline residues in the cell wall teichoic acid by the N-acetylmuramyl-L-alanine amidase of *Pneumococcus*. J. Biol. Chem., 250, 6072–6076.

Höltje, J.V., Mirelman, D., Sharon, N. and Schwarz, U. (1975) A novel type of murein transglycosylase in *Escherichia coli*. J. Bacteriol. 124, 1067–1076.

Hoshino, O., Zehavi, U., Sinay, P. and Jeanloz, R.W. (1972) The isolation and structure identification of a disaccharide containing manno-muramic acid from *Micrococcus lysodeikticus* cell wall. J. Biol. Chem. 247, 381–390.

Howard, L.V. and Gooder, H. (1974) Specificity of the autolysins of *Streptococcus (Diplococcus) pneumoniae*. J. Bacteriol. 117, 796–804.

Howard, L. and Tipper, D.J. (1973) A polypeptide bacteriophage receptor: modified cell wall protein subunits in bacteriophage-resistant mutants of *Bacillus sphaericus* strain P-1. J. Bacteriol. 113, 1491–1504.

Hughes, R.C. (1970) The cell wall of *Bacillus licheniformis* NCTC 8346. Linkage between the teichuronic acid and mucopeptide components. Biochem. J. 117, 431–439.

Hughes, A.H., Stow, M., Hancock, I.C. and Baddiley, J. (1971) Function of teichoic acids and effect of novobiocin on control of Mg^{2+} at the bacterial membrane. Nature New Biol. 229, 53–55.

Inouye, M. (1974) A three-dimensional molecular assembly model of a lipoprotein from the *Escherichia coli* outer membrane. Proc. Natl. Acad. Sci. U.S.A. 71, 2396–2400.

Inouye, M. (1975) Biosynthesis and assembly of the outer membrane proteins of *Escherichia coli*. In Membrane Biogenesis (Tzagoloff, A., ed.), pp. 351–391, Plenum, New York.

Inouye, M., Shaw, J. and Shen, C. (1972) The assembly of a structural lipoprotein in the envelope of *Escherichia coli*. J. Biol. Chem. 247, 8154–8159.

Inouye, M., Hirashima, A. and Lee, N. (1974) Biosynthesis and assembly of a structural lipoprotein in the envelope of *Escherichia coli*. Ann. N.Y. Acad. Sci. 235, 83–90.

Inouye, S., Takeishi, K., Lee, N., De Martini, M., Hirashima, A. and Inouye, M. (1976) Lipoprotein from the outer membrane of *Escherichia coli*: Purification, paracrystallization and some properties of its free form. J. Bacteriol. 127, 555–563.

Iyer, R., Darby, V. and Holland, I.B. (1977) Specific loss of the major outer membrane protein in *Escherichia coli* B/r mediated by the R plasmid RM98. J. Biol. Chem., in press.

Jacobson, G.R. and Rosenbusch, J.P. (1976a) Abundance and membrane association of elongation factor Tu in *Escherichia coli*. Nature 261, 23–26.

Jacobson, G.R. and Rosenbusch, J.P. (1976b) A functionally active tryptic fragment of *Escherichia coli* elongation factor Tu. Biochemistry 15, 5105–5110.

Jacobson, G.R., Takacs, B.J. and Rosenbusch, J.P. (1976) Properties of a major protein released from *Escherichia coli* by osmotic shock. Biochemistry 11, 2297–2302.

James, R. (1975) Identification of an outer membrane protein of *Escherichia coli* with a role in the coordination of deoxyribonucleic acid replication and cell elongation. J. Bacteriol. 124, 918–929.

James, R. and Gudas, L.j. (1976) Cell cycle-specific incorporation of lipoprotein in the outer membrane of *Escherichia coli*. J. Bacteriol. 125, 374–375.

James, R., Haga, J.Y. and Pardee, A.R. (1975) Inhibition of an early event in the cell division cycle of *Escherichia coli* by FL-1060, an amidinopenicillanic acid. J. Bacteriol. 122, 1283–1292.

Janczura, E., Perkins, H.R. and Rogers, H.J. (1961) Teichuronic acid: a mucopolysaccharide present in wall preparations from vegetative cells of *Bacillus subtilis*. Biochem. J. 80, 82–93.

Jann, B., Jann, K. and Beyaert, G.O. (1973) 2-Amino-2,6-dideoxy-D-glucose (D-quinovosamine): a constituent of the lipopolysaccharides of *Vibrio cholerae*. Eur. J. Biochem. 37, 531–534.

Jann, B., Jann, K., Schmidt, G., Ørskov, I. and Ørskov, F. (1970) Immunochemical studies of polysaccharide surface antigens of *Escherichia coli* 0100:K (B):H2. Eur. J. Biochem. 15, 29–39.

Jeanloz, R.W. (1967) The chemical structure of the cell wall of Gram-positive bacteria. Pure Appl. Chem. 14, 57–75.

Johnston, M.M. and Diven, W.F. (1969) Studies on amino acid racemases. I. Partial purification and properties of the alanine racemase from *Lactobacillus fermenti*. J. Biol. Chem. 244, 5414–5420.

Johnston, L.S. and Neuhaus, F.C. (1975) Initial membrane reaction in the biosynthesis of peptidoglycan. Spin-labelled intermediates as receptors for vancomycin and ristocetin. Biochemistry 4, 2754–2760.

Johnston, R.B., Scholz, J.J., Diven, W.F. and Shepard, S. (1966) Some further studies on the purification and mechanism of the action of alanine racemase from *Bacillus subtilis*. In: Pyridoxal catalysis: enzymes and model systems (Snell, E.E., Braunstein, A.E., Severin, E.S. and Torshinsky, Y.M., eds.), pp. 537–545. Wiley-Interscience, New York.

Johnston, L.S., Hammes, W.P., Lazar, H.A. and Neuhaus, F.C. (1975) Spin-labelled intermediates as targets of antibiotic action in peptidoglycan synthesis. In: Topics in Infectious Diseases: Drug Receptor Interactions in Antimicrobial Chemotherapy (Drews, J. and Hahn, F.E., eds), vol. 1, pp. 269–284, Springer-Verlag, Berlin.

Joseph, R. and Shockman, G.D. (1975) Cellular localization of lipoteichoic acid in *Streptococcus faecalis*. J. Bacteriol. 122, 1375–1386.

Joseph, R. and Shockman, G.D. (1976) Autolytic formation of protoplasts (autoplasts) of *Streptococcus faecalis*: Location of active and latent autolysin. J. Bacteriol. 127, 1482–1483.

Jung-Ching Lin, J. and Wu, H.C.P. (1976) Biosynthesis and assembly of envelope lipoprotein in a glycerol-requiring mutant of *Salmonella typhimurium*. J. Bacteriol. 125, 892–904.

Kaback, H.R. (1973) Bacterial transport mechanisms. In: Bacterial Membranes and Walls (Leive, L., ed.), pp. 241–292, Marcel Dekker, New York.

Kahan, F.M., Kahan, J.S., Cassidy, P.J. and Kropp, H. (1974) The mechanism of action of fosfomycin (phosphonomycin). Ann. N.Y. Acad. Sci. 235, 364–386.

Kamio, Y. and Nikaido, H. (1976) Outer membrane of *Salmonella typhimurium*: Accessibility of phospholipid head groups to phospholipase C and cyanogen bromide activated dextran in the external medium. Biochemistry 15, 2561–2569.

Kamiryo, T. and Matsuhashi, M. (1969) Sequential addition of glycine from glycyl-tRNA to the lipid-linked precursors of cell wall peptidoglycan in *Staphylococcus aureus*. Biochem. Biophys. Res. Commun. 36, 215–222.

Katz, W., Matsuhashi, M., Dietrich, C.P. and Strominger, J.L. (1967) Biosynthesis of the peptidoglycans of bacterial cell walls. IV. Incorporation of glycine in *Micrococcus lysodeikticus*. J. Biol. Chem. 242, 3207–3217.

Keglević, D., Ladešić, B., Hadžija, J., Tomašić, Z., Valinger, M., Pokotny, M. and Maumski, R. (1974) Isolation and study of the composition of a peptidoglycan complex excreted by the biotin-requiring mutant of *Brevibacterium divaricatum* NRRL-2311 in the presence of penicillin. Eur. J. Biochem. 42, 389–400.

584

Kelemen, M.V. and Baddiley, J. (1961) Structure of the intracellular glycerol teichoic acid from *Lactobacillus casei* ATCC 7469. Biochem. J. 80, 246–254.

Kelemen, M.V. and Rogers, H.J. (1971) Three-dimensional molecular models of bacterial cell wall mucopeptides (peptidoglycans). Proc. Nat. Acad. Sci. U.S.A. 68, 992–996.

Kennedy, L.D. and Shaw, D.R.D. (1968) Direction of polyglycerol phosphate chain growth in *Bacillus subtilis*. Biochem. Biophys. Res. Commun. 32, 861–865.

Knox, K.W. and Hall, E.A. (1965) The linkage between the polysaccharide and mucopeptide components of the cell wall of *Lactobacillus casei*. Biochem. J. 96, 302–309.

Knox, K.W. and Holmwood, K.J. (1968) Structure of the cell wall of lactobacilli. Role of muramic acid phosphate in *Lactobacillus fermenti*. Biochem. J. 108, 363–368.

Knox, K.W. and Wicken, A.J. (1971) Serological properties of the wall and membrane teichoic acids from *Lactobacillus helveticus* NCIB 8025. J. Gen. Microbiol. 63, 237–248.

Knox, K.W. and Wicken, A.J. (1973) Immunological properties of teichoic acids. Bact. Rev. 37, 215–257.

Kolenbrander, P.E. and Ensign, J.C. (1968) Isolation and chemical structure of the peptidoglycan of *Spirillum serpens* cell walls. J. Bacteriol. 95, 201–210.

Koplow, J. and Goldfine, H. (1974) Alternations in the outer membrane of the cell envelope of heptose-deficient mutants of *Escherichia coli*. J. Bacteriol. 117, 527–543.

Krulwich, T.A., Ensign, J.C., Tipper, D.J. and Strominger, J.L. (1967a) Sphere-rod morphogenesis in *Arthrobacter crystallopoietes*. I. Cell wall composition and polysaccharides of the peptidoglycan. J. Bacteriol. 94, 734–740.

Krulwich, T.A., Ensign, J.C., Tipper, D.J. and Strominger, J.L. (1976b) Sphere-rod morphogenesis in *Arthrobacter crystallopoietes*. II. Peptides of the cell wall peptidoglycan. J. Bacteriol. 94, 741–750.

Kulpa, Ch.F.Jr. and Leive, L. (1972) Mode of insertion of lipopolysaccharide into the outer membrane of *Escherichia coli*. In Membrane Research (Fox, C.F., ed.), pp. 155–160, Academic Press Inc., New York.

Kulpa, Ch.F.Jr. and Leive, L. (1976) Mode of insertion of lipopolysaccharide into the outer membrane of *Escherichia coli*. J. Bacteriol. 126, 467–477.

Lambert, P.A., Hancock, I.C. and Baddiley, J. (1975) The interaction between magnesium ions and teichoic acid. Paper presented at 74th Meet. Soc. Gen. Microbiol., Newcastle-upon-Tyne, England. (abstr.)

Langman, L., Young, I.G., Frost, G.E., Rosenberg, H. and Gibson, F. (1972) Enterochelin system of iron transport in *Escherichia coli*: mutations affecting ferric-enterochelin esterase. J. Bacteriol. 112, 1142–1149.

Lahav, M., Chiu, T.H. and Lennarz, W.J. (1969) Studies on the biosynthesis of mannan in *Micrococcus lysodeikticus*. II. The enzymatic synthesis of mannosyl-1-phosphoryl-undecaprenol. J. Biol. Chem. 244, 5890–5898.

Leduc, M., Van Heijenoort, J., Kaminski, M. and Szulmajster, J. (1973) An unusual type of bacterial cell wall in a mutant of *Bacillus subtilis* 168. Eur. J. Biochem. 37, 389–400.

Leduc, M. and Van Heijenoort, J. (1975) Conversion of bacterial cell wall peptidoglycan into a soluble form by action of guanidine hydrochloride. 10th FEBS Meeting, Paris, Abstr. No. 1055.

Lee, N. and Inouye, M. (1974) Outer membrane proteins of *Escherichia coli*: biosynthesis and assembly. FEBS Lett. 39, 167–170.

Lee, A.G., Birdsall, N.J.M. and Metcalfe, J.C. (1973) Measurement of fast lateral diffusion of lipids in vesicles and in biological membranes by ^{1}H nuclear magnetic resonance. Biochemistry 12, 1650–1659.

Leive, L. (1965a) Release of lipopolysaccharide by EDTA treatment of *Escherichia coli*. Biochem. Biophys. Res. Commun. 21, 290–296.

Leive, L. (1965b) A nonspecific increase in permeability in *Escherichia coli* produced by EDTA. Proc. Nat. Acad. Sci. U.S.A. 53, 745–750.

Leive, L. (1974) The barrier function of the Gram-negative envelope. Ann. N.Y. Acad. Sci. U.S.A. 235, 109–130.

Leive, L., Shovlin, V.K. and Mergenhagen, S.E. (1968) Physical, chemical, and immunological properties of lipopolysaccharide released from *Escherichia coli* by ethylenediaminetetraacetate. J. Biol. Chem. 243, 6384–6391.

Lennarz, W.J. and Scher, M.G. (1972) Metabolism and function of polyisoprenol sugar intermediates in membrane-associated reactions. Biochim. Biophys. Acta 265, 417–442.

Levy, S.B. and Leive, L. (1968) An equilibrium between two fractions of lipopolysaccharide in *Escherichia coli*. Proc. Nat. Acad. Sci. U.S.A. 61, 1435–1439.

Leyh-Bouille, M., Ghuysen, J.M., Tipper, D.J. and Strominger, J.L. (1966) Structure of the cell wall of *Micrococcus lysodeikticus* I. Study of the structure of the glycan. Biochemistry 5, 3079–3090.

Lim, V.I. (1974a) Structural principles of the globular organization of protein chains. Stereochemical theory of globular protein secondary structure. J. Mol. Biol. 88, 857–872.

Lim. V.I. (1974b) Algorithms for prediction of α-helical and β-structural regions in globular proteins. J. Mol. Biol. 88, 873–894.

Linder, R. and Salton, M.R.J. (1975) D-alanine carboxypeptidase activity of *Micrococcus lysodeikticus* released into the protoplasting medium. Eur. J. Biochem. 55, 291–297.

Lindmark, R., Movitz, J. and Sjöquist, J. (1977) Extracellular protein A from a methicillin-resistant strain of *Staphylococcus aureus*. Eur. J. Biochem. 74, 623–628.

Linzer, R. and Neuhaus, F.C. (1973) Biosynthesis of membrane teichoic acid. A role for the D-alanine activating enzyme. J. Biol. Chem. 248, 3196–3201.

Liu, T.Y. and Gotschlich, E.G. (1963) The chemical composition of *pneumococcal* C polysaccharide. J. Biol. Chem. 238, 1928–1934.

Lopes, J., Gottfried, S. and Rothfield, L. (1972) Leakage of periplasmic enzymes by mutants of *Escherichia coli* and *Salmonella typhimurium:* isolation of "periplasmic leaky" mutants. J. Bacteriol. 109, 520–525.

Luckey, M., Pollack, J.R., Wayne, R., Ames, B.N. and Neidlands, J.B. (1972) Iron uptake in *Salmonella typhimurium:* utilization of exogenous siderochromes as iron carriers. J. Bacteriol. 111, 731–738.

Lüderitz, O., Staub, A.M. and Westphal, O. (1966) Immunochemistry of O and R antigens of *Salmonella* and related *Enterobacteriaceae*. Bact. Rev. 30, 192–255.

Lüderitz, O., Jann, K. and Wheat, R. (1968) Somatic and capsular antigens of Gram-negative bacteria. In: Comprehensive Biochemistry (Florkin, M. and Stotz, E.H., eds.), Vol. 26A, pp. 105–228. Elsevier, Amsterdam.

Lüderitz, O., Westphal, O., Staub, A.M. and Nikaido, H. (1971) Isolation and chemical and immunological characterization of bacterial lipopolysaccharides. In: Microbial Toxins (Weinbaum, G., Kadis, S. and Ajl, S.J., eds.), Vol. 4, pp. 145–233. Academic Press, New York.

Lüderitz, O., Galanos, Ch., Lehmann, V., Nurminen, M., Rietschel, E.T., Rosenfelder, G., Simon, M. and Westphal, O. (1973) Lipid A: chemical structure and biological activity. J. Inf. Dis. 128, S17-S29.

Lüderitz, O., Galanos, Ch., Lehmann, V. and Rietschel, E.T. (1974) Recent findings on the chemical structure and biological activity of bacterial lipopolysaccharides. J. Hyg. Epidem. Microbiol. Immun. 4, 381–390.

Lugtenberg, B., Meijers, J., Peters, R., van der Hoek, P. and van Alphen, L. (1975) Electrophoretic resolution of the "major outer membrane protein" of *Escherichia coli* K12 into four bands. FEBS Lett. 58, 254–259.

Lugtenberg, B., Bronstein, H., Van Selm, N. and Peters, R. (1977) Peptidoglycan-associated outer membrane proteins in Gram-negative bacteria. Biochim. Biophys. Acta 465, 571–578.

Lugtenberg, E.J.J. (1971) "*Escherichia coli* mutants impaired in the synthesis of murein". Ph.D. thesis, Rijksuniversiteit te Utrecht, Holland.

Lugtenberg, E.J.J. and de Haan, P.G. (1971) A simple method for following the fate of alanine-containing components in murein synthesis in *Escherichia coli*. Antonie van Leeuwenhock 37, 537–552.

Lund, F. and Tybring, L. (1972) 6β-Amidinopenicillanic acids—a new group of antibiotics. Nature New Biol. 236, 135–137.

Lynch, J. and Neuhaus, F.C. (1966) On the mechanism of action of the antibiotic O-

586

carbamyl-D-serine in *Streptococcus faecalis*. J. Bacteriol. 91, 449–460.

Mäkelä, P.H. (1966) Genetic determination of the O-antigens of *Salmonella* groups B(4,5,12) and C₁(6,7). J. Bacteriol. 91, 1115–1125.

Mäkelä, P.H. and Stocker, B.A.D. (1969) Genetics of polysaccharide biosynthesis. Ann. Rev. Genet. 3, 291–322.

Markham, J.L., Knox, K.W., Wicken, A.J. and Hewett, M.J. (1975) Formation of extracellular lipoteichoic acid by oral streptococci and lactobacilli. Infect. Immun. 12, 378–386.

Marquis, R.E. (1968) Salt-induced contraction of bacterial cell walls. J. Bacteriol. 95, 775–781.

Marshall, V.P. and Sokatn, J.R. (1968) Oxidation of D-amino acids by a particulate enzyme from *Pseudomonas aeruginosa*. J. Bacteriol. 95, 1419–1424.

Martin, H.H. (1966) Biochemistry of bacterial cell walls. Ann. Rev. Biochem. 35 (p. 2), 457–483.

Martin, H.H. and Frank, H. (1962) Quantitative Bausteinanalyse der Stützmembran in der Zellwand von *Escherichia coli* B. Z. Naturforsch. 17b, 190–196.

Matsuhashi, M., Dietrich, C.P. and Strominger, J.L. (1967) Biosynthesis of the peptidoglycan of bacterial cell walls. The role of soluble ribonucleic acid and of lipid intermediate in glycine incorporation in *Staphylococcus aureus*. J. Biol. Chem. 242, 3191–3206.

Matsuzawa, H., Matsuhashi, M., Oka, A. and Sugino, Y. (1969) Genetic and biochemical studies on cell wall peptidoglycan synthesis in *Escherichia coli*. Biochem. Biophys. Res. Commun. 36, 682–689.

Mauck, J. and Glaser, L. (1972) An acceptor-dependent polyglycerolphosphate polymerase. Proc. Nat. Acad. Sci. U.S.A. 69, 2386–2390.

Mauck, J. and Glaser, L. (1973) On the mode of *in vivo* assembly of the cell wall of *Bacillus subtilis*. J. Biol. Chem. 247, 1180–1187.

Meadow, P.M., Anderson, J.S. and Strominger, J.L. (1964) Enzymatic polymerization of UDP-acetylmuramyl-L-Ala-D-Glu-L-Lys-D-Ala and UDP-acetylglucosamine by a particulate enzyme from *Staphylococcus aureus* and its inhibition by antibiotics. Biochem. Biophys. Res. Commun. 14, 382–387.

Meynell, G.G. and Aufreiter, E. (1969) Functional independence of F and I sex pili. Nature New Biol. 223, 1069.

Mirelman, D. and Bracha, R. (1974) Effect of penicillin on the *in vivo* formation of the D-alanyl-L-alanine peptide cross-linkage in cell walls of *Micrococcus luteus*. Antimicrob. Ag. Chemother. 5, 663–666.

Mirelman, D. and Sharon, N. (1972) Biosynthesis of peptidoglycan by a cell wall preparation of *Staphylococcus aureus* and its inhibition by penicillin. Biochem. Biophys. Res. Commun. 46, 1909–1917.

Mirelman, D., Bracha, R. and Sharon, N. (1972) Role of the penicillin-sensitive transpeptidation reaction in attachment of newly synthesized peptidoglycan to cell walls of *Micrococcus luteus*. Proc. Nat. Acad. Sci. U.S.A. 69, 3355–3359.

Mirelman, D., Bracha, R. and Sharon, N. (1974a) Inhibition by penicillin of the incorporation and crosslinking of L-lysine in intact cells of *Micrococcus luteus*. FEBS Lett. 39, 105–110.

Mirelman, D., Bracha, R. and Sharon, N. (1974b) Penicillin-induced secretion of a soluble, uncrosslinked peptidoglycan by *Micrococcus luteus* cells. Biochemistry 13, 5045–5053.

Mirelman, D., Bracha, R. and Sharon, N. (1974c) Studies on the elongation of bacterial cell wall peptidoglycan and its inhibition by penicillin. Ann. N.Y. Acad. Sci. U.S.A. 235, 326–347.

Mirelman, D., Kleppe, G. and Jensen, H.B. (1975) Studies on the specificity of action of bacteriophage T4 lysozyme. Eur. J. Biochem. 55, 369–373.

Mirelman, D., Yashouv-Gan, Y. and Schwarz, U. (1976) Growth pattern of peptidoglycan: biosynthesis in a thermosensitive division mutant of *Escherichia coli*. Biochemistry, 15, 1781–1790.

Mirelman, D., Yashouv-Gan, Y. and Schwarz, U. (1977) Regulation of murein biosynthesis and septum formation in filamentous cells of *Escherichia coli* PAT 84. J. Bacteriol. 129, 1593–1600.

Mitchell, P. and Moyle, J. (1957) Autolytic release and osmotic properties of "protoplasts" from *Staphylococcus aureus*. J. Gen. Microbiol. 16, 184–194.

Miura, T. and Mizushima, S. (1968a) Separation by density gradient centrifugation of two types of membranes from spheroplast membrane of *Escherichia coli*. Biochim. Biophys. Acta 150, 159–161.

Miura, T. and Mizushima, S. (1968b) Separation and properties of outer and cytoplasmic membranes in *Escherichia coli*. Biochim. Biophys. Acta 193, 268–276.

Mosser, J.L. and Tomasz, A. (1970) Choline-containing teichoic acid as a structural component of *pneumococcal* cell wall and its role in sensitivity to lysis by an autolytic enzyme. J. Biol. Chem. 245, 287–298.

Movitz, J. (1976) Formation of extracellular protein A by *Staphylococcus aureus*. Eur. J. Biochem. 68, 291–299.

Mühlradt, P.F. (1971) Biosynthesis of *Salmonella* lipopolysaccharide. Studies on the transfer of glucose, galactose, and phosphate to the core in a cell free system. Eur. J. Biochem. 18, 20–27.

Mühlradt, P.F. (1976) Topography of outer membrane assembly in *Salmonella*. J. Supramol. Structure 5, 103–108.

Mühlradt, P.F. and Golecki, J.R. (1975) Asymmetrical distribution and artifactual reorientation of lipopolysaccharide in the outer membrane bilayer of *Salmonella typhimurium*. Eur. J. Biochem. 51, 343–352.

Mühlradt, P.F., Risse, H.J., Lüderitz, O. and Westphal, O. (1968) Biochemical studies on lipopolysaccharides of *Salmonella* R mutants: 5. Evidence for a phosphorylating enzyme in lipopolysaccharide biosynthesis. Eur. J. Biochem. 4, 139–145.

Mühlradt, P.F., Menzel, J., Golecki, J.R. and Speth, V. (1973) Outer membrane of *Salmonella*. Sites of export of newly synthesized lipopolysaccharide on the bacterial surface. Eur. J. Biochem. 35, 471–481.

Mühlradt, P.F., Menzel, J., Golecki, J.R. and Speth, V. (1974) Lateral mobility and surface density of lipopolysaccharide in the outer membrane of *Salmonella typhimurium*. Eur. J. Biochem. 43, 533–539.

Müller, E., Hinckley, A. and Rothfield, L. (1972) Studies of phospholipid-requiring bacterial enzymes. III. Purification and properties of uridine diphosphate glucose: lipopolysaccharide glucosyltransferase. J. Biol. Chem. 247, 2614–2622.

Muñoz, E., Ghuysen, J.M. and Heymann, H. (1967) Cell walls of *Streptococcus pyogenes* group A, type 14. C polysaccharide-peptidoglycan and G polysaccharide-peptidoglycan complexes. Biochemistry 6, 3659–3670.

Murray, R.G.E. (1963) On the cell wall structure of *Spirillum serpens*. Can. J. Microbiol. 9, 381–392.

Naide, Y., Nikaido, H., Mäkelä, P.H., Wilkinson, R.G. and Stocker, B.A.D. (1965) Semirough strains of *Salmonella*. Proc. Nat. Acad. Sci. U.S.A. 53, 147–153.

Nakae, T. (1976) Outer membrane of *Salmonella*. Isolation of protein complex that produces transmembrane channels. J. Biol. Chem. 251, 2176–2178.

Nakae, T. and Nikaido, H. (1976) Outer membrane as a diffusion barrier in *Salmonella typhimurium*. Penetration of oligo- and polysaccharides into isolated outer membrane vesicles and cells with degraded peptidoglycan layer. J. Biol. Chem. 250, 7359–7365.

Nanninga, N. (1970) Ultrastructure of the cell envelope of *Escherichia coli* B after freeze-etching. J. Bacteriol. 101, 297–303.

Nathenson, S.G. and Strominger, J.L. (1966) Enzymatic synthesis and immunochemistry of N-acetylglucosaminylribitol linkages in the teichoic acids of *Staphylococcus aureus* strains. J. Biol. Chem. 237, PC3839–PC3841.

Neuhaus, F.C. (1971) Initial translocation reaction in the biosynthesis of peptidoglycan by bacterial membranes. Accounts Chem. Res. 4, 297–303.

Neuhaus, F.C. and Struve, W.G. (1965) Enzymatic synthesis of analogs of the cell-wall precursor. I. Kinetics and specificity of uridine diphospho-N-acetylmuramyl-L-alanyl-D-glutamyl-L-lysine: D-alanyl-D-alanine ligase (adenosine diphosphate) from *Streptococcus faecalis* R. Biochemistry 4, 120–131.

Neuhaus, F.C., Carpenter, C.V., Lambert, M.P. and Wargel, R.J. (1972) D-cycloserine as a tool in studying the enzymes in the alanine branch of peptidoglycan synthesis. In: Proceedings of Symposium on Molecular Mechanisms of Antibiotic Action on Protein Biosynthesis and Membranes (Muñoz, E., Ferrandiz, F. and Vazquez, D., eds), pp. 339–362. Elsevier/North-Holland, Amsterdam.

Neuhaus, F.C., Linzer, R. and Reusch, V.M., Jr. (1974) Biosynthesis of membrane teichoic acid: role of the D-alanine-activating enzyme and D-alanine: membrane acceptor ligase. Ann. N.Y. Acad. Sci. 235, 502–518.

Nguyen-Distèche, M., Ghuysen, J.M., Pollock, J.M., Puig, J., Reynolds, P.E., Perkins, H.R., Coyette, J. and Salton, M.R.J. (1974a) Enzymes involved in wall peptide crosslinking in Escherichia coli K12, strain 44. Eur. J. Biochem. 41, 447–455.

Nguyen-Distèche, M., Pollock, J.J., Ghuysen, J.M., Puig, J., Reynolds, P.E., Perkins, H.R., Coyette, J. and Salton, M.R.J. (1974b) Sensitivity to ampicillin and cephalothin of enzymes involved in wall peptide crosslinking in Escherichia coli K12, strain 44. Eur. J. Biochem. 41, 457–463.

Nikaido, H. (1968) Biosynthesis of cell wall lipopolysaccharide in Gram-negative enteric bacteria. Adv. Enzymol. 31, 77–124.

Nikaido, H. (1970) Lipopolysaccharide in the taxonomy of enterobacteriaceae. Int. J. Syst. Bacteriol. 20, 383–406.

Nikaido, H. (1973) Biosynthesis and assembly of lipopolysaccharide and the outer membrane layer of Gram-negative cell wall. In: Bacterial Membranes and Walls (Leive, L., ed.), pp. 131–209, Marcel Dekker, Inc., New York.

Nikaido, H. (1976) Outer membrane of Salmonella typhimurium. Transmembrane diffusion of some hydrophobic substances. Biochim. Biophys. Acta 433, 118–132.

Nikaido, H., Naide, Y. and Mäkelä, P.H. (1966) Biosynthesis of O-antigenic polysaccharides in Salmonella. Ann. N.Y. Acad. Sci. U.S.A. 133, 299–314.

Oldmixon, E.H., Glauser, S. and Higgins, M.L. (1974) Two proposed general configurations for bacterial cell wall peptidoglycans shown by space-filling molecular models. Biopolymers 13, 2037–2060

Oldmixon, E.H., Dézelée, Ph., Ziskin, M.C. and Shockman, G.D. (1976) A mechanism of forming peptide crosslinks in the cell wall peptidoglycan of Streptococcus faecalis. Eur. J. Biochem. 68, 271–280.

Oliver, C.J.A., Harrison, J. and Hemming, F.W. (1975) The mannosylation of dolichol-diphosphate oligosaccharides in relation to the formation of oligosaccharides and glycoproteins in pig-liver endoplasmic reticulum. Eur. J. Biochem. 58, 223–229.

Osborn, M.J. (1969) Structure and biosynthesis of the bacterial cell wall. Ann. Rev. Biochem. 38, 501–538.

Osborn, M.J. (1971) The role of membranes in the synthesis of macromolecules. In: Structure and Function of Biological Membranes (Rothfield, L.I., ed.), pp. 343–400. Academic Press, New York.

Osborn, M.J. and Rothfield, L.I. (1971) Biosynthesis of the core region of lipopolysaccharide. In: Microbial Toxins (Weinbaum, G., Kadis, S. and Ajl, S.J., eds), pp. 331–350. Academic Press, New York.

Osborn, M.J., Gander, J.E., Parisi, E. and Carson, J. (1972a) Mechanism of assembly of the outer membrane of Salmonella typhimurium. Isolation and characterization of cytoplasmic and outer membrane. J. Biol. Chem. 247, 3962–3972.

Osborn, M.J., Gander, J.E. and Parisi, E. (1972b) Mechanism of assembly of the outer membrane of Salmonella typhimurium. Site of synthesis of lipopolysaccharide. J. Biol. Chem. 247, 3973–3986.

Osborn, M.J., Rick, P.D., Lehmann, V., Rupprecht, E. and Singh, M. (1974) Structure and biogenesis of the cell envelope of Gram-negative bacteria. Ann. N.Y. Acad. Sci. 235, 52–65.

Ou, L.T. and Marquis, R.E. (1970) Electromechanical interactions in cell walls of Gram-positive cocci. J. Bacteriol. 101, 92–101.

Ou, L.T. and Marquis, R.E. (1972) Coccal cell wall compactness and the swelling action of denaturants. Can. J. Bacteriol. 18, 623–629.

Owen, P. and Freer, J.H. (1972) Isolation and properties of mesosomal membrane fractions from Micrococcus lysodeikticus. Biochem. J. 129, 907–917.

Owen, P. and Salton, M.R.J. (1975a) A succinylated mannan in the membrane system of Micrococcus lysodeikticus. Biochem. Biophys. Res. Commun. 63, 875–880.

Owen, P. and Salton, M.R.J. (1975b) Isolation and characterization of a mannan from mesosomal membrane vesicles of Micrococcus lysodeikticus. Biochim. Biophys. Acta 405, 214–234.

Owen, P. and Salton, M.R.J. (1975c) Distribution of enzymes involved in mannan synthesis in plasma membranes and mesosomal vesicles of *Micrococcus lysodeikticus*. Biochim. Biophys. Acta 405, 235–247.

Page, R.L. and Anderson, J.S. (1972) Biosynthesis of the polysaccharide of *Micrococcus lysodeikticus* cell walls. I. Characterization of an *in vitro* system for polysaccharide biosynthesis. J. Biol. Chem. 247, 2471–2479.

Perkins, H.R. (1963) A polymer containing glucose and aminohexuronic acid isolated from the cell walls of *Micrococcus lysodeikticus* Biochem. J. 86, 475–483.

Perkins, H.R. and Nieto, M. (1974) The chemical basis for the action of the vancomycin group of antibiotics. Ann. N.Y. Acad. Sci. U.S.A. 235, 348–363.

Perkins, H.R., Ghuysen, J.M., Frère, J.M. and Nieto, M. (1975) Antibiotic action on enzymes involved in peptidoglycan synthesis. In: Protein-Ligand Interactions, pp. 372–384. W. de Gruyter and Co., Berlin.

Petit, J.F., Strominger, J.L. and Söll, D. (1968) Biosynthesis of the peptidoglycan of bacterial cell walls. VII. Incorporation of serine and glycine into interpeptide bridges in *Staphylococcus epidermidis*. J. Biol. Chem. 243, 757–768.

Pieringer, R.A. and Ganfield, M.C.W. (1975) Phosphatidylkojibiosyl diglyceride: metabolism and function as an anchor in bacterial cell membranes. Lipids 10, 421–426.

Plapp, R. and Strominger, J.L. (1970) Biosynthesis of the peptidoglycan of bacterial cell walls. XVII. Purification and properties of L-alanyl transfer ribonucleic acid-uridine diphosphate-N-acetylmuramyl-pentapeptide transferase from *Lactobacillus viridescens*. J. Biol. Chem. 245, 3675–3682.

Pollack, J.R., Ames, B.N. and Neilands, J.B. (1970) Iron transport in *Salmonella typhimurium:* mutants blocked in the biosynthesis of enterobactin. J. Bacteriol. 104, 635–639.

Pollock, J.J. and Sharon, N. (1970) Studies on the acceptor specificity of the lysozyme-catalyzed transglycosylation reaction. Biochemistry 9, 3913–3925.

Pollock, J.J., Nguyen-Distèche, M., Ghuysen, J.M., Coyette, J., Linder, R., Salton, M.R.J., Kim, K.S., Perkins, H.R. and Reynolds, P.E. (1974) Fractionation of the DD-carboxypeptidase-transpeptidase activities solubilized from membranes of *Escherichia coli* K12, strain 44. Eur. J. Biochem. 41, 439–446.

Powell, D.A., Duckworth, M. and Baddiley, J. (1974) An acylated mannan in the membrane of *Micrococcus lysodeikticus*. FEBS Lett. 41, 259–263.

Powell, D.A., Duckworth, M. and Baddiley, J. (1975) The presence in Micrococci of a membrane-associated lipomannan. Paper presented at 74th Meet. Soc. Gen. Microbiol., Newcastle-upon-Tyne, England.

Primosigh, J., Pelzer, H., Maas, D. and Weidel, W. (1961) Chemical characterization of mucopeptides released from the *E. coli* B cell wall by enzymic action. Biochim. Biophys. Acta 46, 68–80.

Raimondo, L.M., Lundh, N.P. and Martinez, R.J. (1968) Primary adsorption site of phage PBSI: the flagellum of *Bacillus subtilis*. J. Virol. 2, 256–264.

Randall, L.L. and Hardy, S.J.S. (1977) Synthesis of exported proteins by membrane-bound polysomes from *Escherichia coli*. Eur. J. Biochem. 75, 43–53.

Reithmeir, R.A.F. and Bragg, P.D. (1974) Purification and characterization of a heat modifiable protein from the outer membrane of *Escherichia coli*. FEBS Lett. 41, 195–198.

Reusch, V.M., Jr. and Neuhaus, F.C. (1971) D-alanine-membrane acceptor ligase from *Lactobacillus casei*. J. Biol. Chem. 246, 6136–6143.

Reynolds, P.E. (1971) Peptidoglycan synthesis in *Bacilli*. II. Characteristics of protoplast membrane preparations. Biochim. Biophys. Acta 237, 255–272.

Rick, P.D. and Osborn, M.J. (1972) Isolation of a mutant of *Salmonella typhimurium* dependent on D-arabinose-5-PO$_4$ for growth and synthesis of 3-deoxy-D-mannoctulosonate (ketodeoxyoctonate). Proc. Nat. Acad. Sci. U.S.A. 69, 3756–3760.

Robbins, P.W. and Wright, A. (1971) Biosynthesis of O-antigens. In: Bacterial Endotoxins (Weinbaum, G., Kadis, S. and Ajl, S.J., eds), vol. 4, pp. 351–368. Academic Press, New York.

Roberts, N.A., Gray, G.M. and Wilkinson, S.G. (1970) The bactericidal action of ethylenediaminetetraacetic acid on *Pseudomonas aeruginosa*. Microbios 7–8, 189–208.

Roberts, W.S.L., Petit, J.F. and Strominger, J.L. (1968) Biosynthesis of the peptidoglycan of bacterial cell walls. VIII. Specificity in the utilization of L-alanyl transfer ribonucleic acid for interpeptide bridge synthesis in *Arthrobacter crystallopoietes*. J. Biol. Chem. 243, 768–773.

Robson, R.L. and Baddiley, J. (1977) Role of teichuronic acid in *Bacillus licheniformis:* Defective autolysis due to deficiency of teichuronic acid in a novobiocin-resistant mutant. J. Bacteriol. 129, 1051–1058.

Rogers, H.J. (1970) Bacterial growth and the cell envelope. Bact. Rev. 34, 194–214.

Rogers, H.J. (1974) Peptidoglycans (mucopeptides): structure, function, and variations. Ann. N.Y. Acad. Sci. U.S.A. 235, 29–51.

Rogers, H.J. (1975) Biosynthesis of mucopeptide and teichoic acid in cell division mutants of *Bacillus subtilis* and *Bacillus licheniformis*. Paper presented at 74th Meeting Soc. Gen. Microbiol., Newcastle-upon-Tyne, England.

Rogers, H.J. (1977) Envelope growth and synthesis in rod mutants and protoplasts of *Bacilli*. Microbiology, in press.

Rogers, H.J., McConnell, M. and Hughes, R.C. (1971) The chemistry of the cell walls of *rod* mutants of *Bacillus subtilis*. J. Gen. Microbiol. 66, 297–308.

Rogers, H.J., Thurman, P.F., Taylor, C. and Reeve, J.N., (1974) Mucopeptide synthesis by *rod* mutants of *Bacillus subtilis*. J. Gen. Microbiol. 85, 335–350.

Rogers, H.J., Thurman, P.F. and Buxton, R.S. (1976) Magnesium and anion requirements of *rod B* mutants of *Bacillus subtilis*. J. Bacteriol. 125, 556–564.

Romeo, D., Hinckley, A. and Rothfield, L. (1970) Reconstitution of a functional membrane enzyme system in a monomolecular film. II. Formation of a functional ternary film of lipopolysaccharide, phospholipid and transferase enzyme. J. Mol. Biol. 53, 491–501.

Rosen, P. and Heppel, L.A. (1973) Present status of binding proteins that are released from Gram-negative bacteria by osmotic shock. In: Bacterial Membranes and Walls (Leive, L., ed.), pp. 209–239. Marcel Dekker, New York.

Rosenbusch, J.P. (1974) Characterization of the major envelope protein from *Escherichia coli*. J. Biol. Chem. 249, 8019–8029.

Rosenthal, R.S. and Shockman, G.D. (1975) The synthesis of peptidoglycan in the form of soluble, glycan chains by growing protoplasts (autoplasts) of *Streptococcus faecalis*. J. Bacteriol. 124, 419–423.

Rosenthal, K.S., Swanson, P.E. and Storm, D.R. (1976) Disruption of *Escherichia coli* outer membranes by EM49. A new membrane active peptide antibiotic. Biochemistry 15, 5783–5792.

Rothfield, L. and Romeo, D. (1971) Role of lipids in the biosynthesis of the bacterial cell envelope. Bact. Rev. 35, 14–38.

Roze, U. and Strominger, J.L. (1966) Alanine racemase from *Staphylococcus aureus:* conformation of its substrates and its inhibitor, D-cycloserine. Mol. Pharmacol. 2, 92–94.

Ryter, A. (1974) Anatomie de la cellule procaryote. In: Microbiologie générale, Institut Pasteur, Bactéries-Bactériophages, pp. 45–73, Ediscience/McGraw-Hill, Paris.

Ryter, A., Shuman, H. and Schwartz, M. (1975). Integration of the receptor for bacteriophage lambda in the outer membrane of *Escherichia coli:* Coupling with cell division. J. Bacteriol. 122, 295–301.

Salton, M.R.J. (1964) The Bacterial Cell Wall. Elsevier/North-Holland, New York.

Salton, M.R.J. (1971) Bacterial membranes. CRC Crit. Rev. Microbiol. 1, 161–197.

Salton, M.R.J. (1974) Molecular bacteriology. In: Molecular Microbiology (Kwapinski, J.B.G., ed.), pp. 387–421. John Wiley, New York.

Salton, M.R.J. and Williams, R.C. (1954) Electron microscopy of the cell walls of *Bacillus megaterium* and *Rhodospirillum rubrum*. Biochim. Biophys. Acta 14, 455–458.

Schaeffer, P., Cami, B. and Hotchkiss, R.D. (1976) Fusion of bacterial protoplasts. Proc. Nat. Acad. Sci. USA. 73, 2151–2155.

Scher, M. and Lennarz, W.J. (1969) Studies on the biosynthesis of mannan in *Micrococcus lysodeikticus*. I. Characterization of mannan-^{14}C formed enzymatically from mannoysl-1-phosphoryl-undecaprenol. J. Biol. Chem. 244, 2777–2789.

Scher, M., Lennarz, W.J. and Sweeley, C.C. (1968) The biosynthesis of mannosyl-1-phosphoryl polyisoprenol in *Micrococcus lysodeikticus* and its role in mannan synthesis. Proc. Nat. Acad. Sci. U.S.A. 59, 1313–1320.

Scherrer, P. and Gerhardt, P. (1971) Molecular sieving by the *Bacillus megaterium* cell wall and protoplast. J. Bacteriol. 107, 718–735.

Schleifer, K.H. (1975) Chemical structure of the peptidoglycan, its modifiability and relation to the biological activity. Z. Immun.-Forsch. 149, S104–117.

Schleifer, K.H. and Kandler, O. (1972) Peptidoglycan types of bacterial cell walls and their taxonomic implications. Bact. Rev. 36, 407–477.

Schleifer, K.H., Hammes, W.P. and Kandler, O. (1976) Effect of endogenous and exogenous factors on the primary structures of bacterial peptidoglycan. Adv. Microbial Physiol. 13, 245–292.

Schmit, A.S., Pless, D.D. and Lennarz, W.J. (1974) Some aspects of the chemistry and biochemistry of membranes of Gram-positive bacteria. Ann. N.Y. Acad. Sci. 235, 91–104.

Schmitges, Cl.J. and Henning, U. (1976). The major proteins of the *Escherichia coli* outer cell-envelope membrane. Heterogeneity of protein I. Eur. J. Biochem. 63, 47–52.

Schnaitman, C.A. (1970a) Examination of the protein composition of the cell envelope of *Escherichia coli* by polyacrylamide gel electrophoresis. J. Bacteriol. 104, 882–889.

Schnaitman, C.A. (1970b) Protein composition of the cell wall and cytoplasmic membrane of *Escherichia coli*. J. Bacteriol. 104, 890–901.

Schnaitman, C.A. (1971a) Solubilization of the cytoplasmic membrane of *Escherichia coli* by Triton X-100. J. Bacteriol. 108, 545–552.

Schnaitman, C.A. (1971b) Effect of ethylenediaminetetraacetic acid, Triton X-100, and lysozyme on the morphology and chemical composition of isolated cell walls of *Escherichia coli*. J. Bacteriol. 108, 553–563.

Schnaitman, C.A. (1974) Outer membrane proteins of *Escherichia coli*. III. Evidence that the major protein of *Escherichia coli* 0111 outer membrane consists of four distinct polypeptide species. J. Bacteriol. 118, 442–453.

Schrader, W.P. and Fan, D.P. (1974) Synthesis of crosslinked peptidoglycan attached to previously formed cell wall by toluene-treated cells of *B. megaterium*. J. Biol. Chem. 249, 4815–4818.

Schrader, W.P., Beckman, B.E., Beckman, M.M., Anderson, J.S. and Fan, D.P. (1974) Biosynthesis of peptidoglycan in the one million molecular weight range by membrane preparations from *B. megaterium*. J. Biol. Chem. 249, 4807–4814.

Schwartz, M. (1975) Reversible interaction between coliphage lambda and its receptor protein. J. Mol. Biol. 99, 185–201.

Schwartz, M. (1976) The adsorption of coliphage lambda to its host: Effects of variations in the surface density of receptor and in phage-receptor affinity. J. Mol. Biol. 103, 521–526.

Schwartz, M. and Le Minor, L. (1975) Occurrence of the bacteriophage lambda receptor in some *Enterobacteriaceae*. J. Virology 15, 679–685.

Schwartz, M., Kellermann, O., Szmelcman, S. and Hazelbauer, G.L. (1976) Further studies on the binding of maltose to the maltose-binding protein of *Escherichia coli*. Eur. J. Biochem. 71, 167–170.

Schweizer, M. and Henning, U. (1977) Action of a major outer cell envelope membrane protein in conjugation of *Escherichia coli* K12. J. Bacteriol. 129, 1651–1652.

Schweizer, M., Sonntag, I. and Henning, U. (1975) Outer cell envelope membrane and shape of *Spirillum serpens*. J. Mol. Biol. 93, 11–21.

Sentandreu, R. and Lampen, J.O. (1971) Participation of a lipid intermediate in the biosynthesis of *Saccharomyces cerevisiae* LK2G12 mannan. FEBS Lett. 14, 109–113.

Sharon, N., Osawa, T., Flowers, H.M. and Jeanloz, R.W. (1966) Isolation and study of the chemical structure of a disaccharide from *Micrococcus lysodeikticus* cell walls. J. Biol. Chem. 241, 223–230.

Sharpe, M.E., Brock, J.H., Wicken, A.J. and Knox, K.W. (1975) Isolation of glycerol teichoic acids from a Gram-negative bacterium. Proc. 75th Ann. Meet. Amer. Soc. Microbiol., New York, Abst. No. K118.

Shepherd, S.T., Chase, H.A. and Reynolds, P.E. (1967) The separation and properties of two penicillin-binding proteins from *Salmonella typhimurium*. Eur. J. Biochem., in press.

Shockman, G.D. and Slade, H.D. (1964) The cellular location of the *streptococcal* group D antigen. J. Gen. Microbiol. 32, 297–305.

Shockman, G.D., Thompson, J.S. and Conover, M.J. (1967) The autolytic enzyme system of *Streptococcus faecalis*. II. Partial characterization of the autolysin and its substrate. Biochemistry 6,

1054–1065.

Siewert, G. (1969) Inhibition of bacterial peptidoglycan synthesis by bacitracin. In: Inhibitors as Tools in Cell Research (Bücher, Th. and Sies H., eds.), pp 210–216. Colloquium des Gesellschaft für Biologische Chemie, Springer-Verlag, Berlin.

Siewert, G. and Strominger, J.L. (1967) Bacitracin: an inhibitor of the dephosphorylation of lipid pyrophosphate, an intermediate in biosynthesis of the peptidoglycan of bacterial cell walls. Proc. Nat. Acad. Sci. U.S.A. 57, 767–773.

Siewert, G. and Strominger, J.L. (1968) Biosynthesis of the peptidoglycan of bacterial cell walls. XI. Formation of the isoglutamine amide group in the cell walls of *Staphylococcus aureus*. J. Biol. Chem. 243, 783–791.

Singer, S.J. and Nicolson, G.L. (1972) The fluid mosaic model of the structure of cell membranes. Cell membranes are viewed as two-dimensional solutions of oriented globular proteins and lipids. Science 175, 720–731.

Sjödahl, J. (1977) Repetitive sequences in protein A from *Staphylococcus aureus*. Eur. J. Biochem. 73, 343–351.

Sjöquist, J., Movitz, J., Johansson, I.B. and Hjelm, H. (1972) Localization of protein A in the bacteria. Eur. J. Biochem. 30, 190–194.

Slabyj, B.M. and Panos, C. (1973) Teichoic acid of a stabilized L-form of *Streptococcus pyogenes*. J. Bacteriol. 114, 934–942.

Spratt, B.G. (1975) Distinct penicillin binding proteins involved in the division, elongation and cell shape in *Escherichia coli* K 12. Proc. Nat. Acad. Sci. U.S.A. 72, 3117–3127.

Spratt, B.G. (1977) Properties of the penicillin-binding proteins of *Escherichia coli* K12. Eur. J. biochem. 72, 341–352.

Spratt, B.G. and Pardee, A.B. (1975) Penicillin-binding proteins and cell shape in *Escherichia coli*. Nature 254, 516–517.

Spratt. B.G. and Strominger, J.L. (1976) Identification of the major penicillin-binding proteins of *Escherichia coli* as D-alanine carboxypeptidase IA. J. Bacteriol. 127, 660–663.

Staudenbauer, W. and Strominger, J.L. (1972) Activation of D-aspartic acid for incorporation into peptidoglycan. J. Biol. Chem. 247, 5095–5102.

Staudenbauer, W., Willoughby, E. and Strominger, J.L. (1972) Further studies of the D-aspartic-activating enzyme of *Streptococcus faecalis* and its attachment to membrane. J. Biol. Chem. 247, 5289–5296.

Steven, A.C., ten Heggeler, B., Müller, R., Kistler, J. and Rosenbusch, J.P. (1977) The ultrastructure of a periodic protein layer in the outer membrane of *Escherichia coli*. J. Cell Biol. 72, 292–301.

Stocker, B.A.D. and Mäkelä, P.H. (1971) Genetic aspects of biosynthesis and structure of *Salmonella* lipopolysaccharide. In: Microbial Toxins (Weinbaum, G., Kadis, S. and Ajl, S.J., eds), vol. 4, pp. 369–438. Academic Press, New York.

Storm, D.R. (1974) Mechanism of bacitracin action: a specific lipid-peptide interaction. Ann. N.Y. Acad. Sci. U.S.A. 235, 387–398.

Strominger, J.L. (1970) Penicillin-sensitive enzymatic reactions in bacterial cell wall synthesis. The Harvey Lectures 64, 171–213. Academic Press, New York.

Struve, W.G. and Neuhaus, F.C. (1965) Evidence for an initial acceptor of UDP-N-acetylmuramyl-pentapeptide in the synthesis of bacterial mucopeptide. Biochem. Biophys. Res. Commun. 18, 6–12.

Suzuki, H., Nishimura, Y., Iketani, H., Campisi, J., Hirashima, A., Inouye, M. and Hirota, Y. (1976) Novel mutation that causes a structural change in a lipoprotein in the outer membrane of *Escherichia coli*. J. Bacteriol. 127, 1494–1501.

Szmelcman, S. and Hofnung, M. (1975) Maltose transport in *Escherichia coli* K12: Involvement of the bacteriophage lambda receptor. J. Bacteriol. 124, 112–118.

Szmelcman, S., Schwartz, M., Silhavy, T.J. and Boos, W. (1976). Maltose transport in *Escherichia coli* K12. A comparison of transport kinetics in wild-type and λ-resistant mutants with the dissociation constants of the maltose-binding protein as measured by fluorescence quenching. Eur. J. Biochem. 65, 13–19.

593

Takayama, K. and Goldman, D.S. (1970) Enzymatic synthesis of mannosyl-1-phosphoryl-decaprenol by a cell-free system of *Mycobacterium tuberculosis*. J. Biol. Chem. 245, 6251–6257.

Taku, A. and Fan, D.P. (1976a) Purification and properties of a protein factor stimulating peptido-glycan synthesis in toluene- and LiCl-treated *Bacillus megaterium* cells. J. Biol. Chem., 126, 48–55.

Taku, A. and Fan, D.P. (1976b). Identification of an isolated protein essential for peptidoglycan synthesis as the N-acetylglucosaminyl transferase. J. Biol. Chem. 251, 6154–6156.

Taku, A., Gardner, H.L. and Fan, D.P. (1975) Reconstitution of cell wall synthesis in toluene- and LiCl-treated *Bacillus megaterium* cells by addition of a soluble protein extract. J. Biol. Chem. 250, 3375–3380.

Takumi, K., Kawata, T. and Hisatsune, K. (1971) Autolytic enzyme system of *Clostridium botulinum*. II. Mode of action of autolytic enzymes in *Clostridium botulinum* type A. Japan J. Microbiol. 15, 131–141.

Tamaki, S., Sato, T. and Matsuhashi, M. (1971) Role of lipopolysaccharides in antibiotic resistance and bacteriophage adsprption of *Escherichia coli* K12. J. Bacteriol. 105, 968–975.

Tanner, W., Jung, P. and Behrens, N.H. (1971) Dolichol monophosphates: mannosyl acceptors in a particulate *in vitro* system of *S. cerevisiae*. FEBS Lett. 16, 245–248.

Taylor, A., Das, C.B. and Van Heijenoort, J. (1975) Bacterial cell wall peptidoglycan fragments produced by phage λ or Vi II endolysin and containing 1,6-anhydro-N-acetylmuramic acid. Eur. J. Biochem. 53, 47–54.

Taylor, S.S. and Health, E.C. (1969) The incorporation of β-hydroxy fatty acids into a phospholipid of *Escherichia coli* B. J. Biol. Chem. 244, 6605–6616.

Thorndike, J. and Park, J.T. (1969) A method for demonstrating the stepwise addition of glycine-tRNA into the murein precursor of *Staphylococcus aureus*. Biochem. Biophys. Res. Commun. 35, 642–647.

Thorne, K.J.I., Swales, L.S. and Barker, D.C. (1974) An investigation by autoradiography and electron microscopy of the localization of prenols in *Lactobacillus casei*. J. Gen. Microbiol. 80, 467–473.

Tinelli, R. (1968) Mise en évidence d'enzymes autolytiques dans les parois de différentes Sporulales. C.R. Acad. Sci. 266, 792–974.

Tipper, D.J. (1969) Structures of cell-wall peptidoglycans of *Staphylococcus epidermidis* Texas 26 and *Staphylococcus aureus* Copenhagen. II. Structure of neutral and basic peptides from hydrolysis with myxobacter AL-1 peptidase. Biochemistry 8, 2192–2212.

Tipper, D.J. (1970) Structure and function of peptidoglycans. Int. J. Syst. Bacteriol. 26, 361–377.

Tipper, D.J. and Strominger, J.L. (1968) Biosynthesis of the peptidoglycan of bacterial cell walls. XII. Inhibition of cross-linkage by penicillins and cephalosporins: studies in *Staphylococcus aureus in vivo*. J. Biol. Chem. 243, 3169–3179.

Tipper, D.J., Ghuysen, J.M. and Strominger, J.L. (1965) Structure of the cell wall of *Straphylococcus aureus*, strain Copenhagen. III. Further studies of the disaccharides. Biochemistry 4, 468–473.

Tipper, D.J., Strominger, J.L. and Ensign, J.C. (1967) Structure of the cell wall of *Staphylococcus aureus*, strain Copenhagen. VII. Mode of action of the bacteriolytic peptidase from *Myxobacter* and the isolation of intact cell wall polysaccharides. Biochemistry 6, 906–920.

Tomasz, A. and Waks, S. (1975) Mechanism of action of penicillin. Triggering of the pneumococcal autolytic enzyme by inhibitors of cell wall synthesis. Proc. Nat. Acad. Sci. U.S.A. 72, 4162–4166.

Tomasz, A. and Westphall, M. (1971) Abnormal autolytic enzyme in a Pneumococcus with altered teichoic acid composition. Proc. Nat. Acad. Sci. U.S.A. 68, 2627–2630.

Toon, P., Brown, P.E. and Baddiley, J. (1972) The lipid-teichoic acid complex in the cytoplasmic membrane of *Streptococcus faecalis* N.C.I.B. 8191. Biochem. J. 127, 399–409.

Troy, F.A., Frerman, F.E. and Heath, E.C. (1971) The biosynthesis of capsular polysaccharide in *Aerobacter aerogenes*. J. Biol. Chem. 246, 118–133.

Tynecka, Z. and Ward, J.B. (1975) Peptidoglycan synthesis in *Bacillus licheniformis*. The inhibition of crosslinking by benzylpenicillin and cephaloridine *in vivo* accompanied by the formation of soluble peptidoglycan. Biochem. J. 146, 253–267.

van Alphen, L., Havekes, L. and Lugtenberg, B. (1977a) Major outer membrane protein d of

Escherichia coli K12. Purification and *in vitro* activity of bacteriophage k3 and f-pilus mediated conjugation. FEBS Lett. 75, 285–290.

van Alphen, L., Lugtenberg, B., van Boxtel, P. and Verhoef, K. (1977b) Architecture of the outer membrane of *Escherichia coli* K12. I. Action of phospholipases A2 and C on wild type strains and outer membrane mutants. Biochim. Biophys. Acta 466, 257–268.

van Driel, D., Wicken, A.J., Dickson, M.R. and Knox, K.W. (1973) Cellular location of the lipoteichoic acids of *Lactobacillus fermenti* NCTC6991 and *Lactobacillus casei* NCTC6375. J. Ultrastruct. Res. 43, 483–497.

Verhoef, C., de Graaff, P.J. and Lugtenberg, E.J.J. (1977) Mapping of a gene for a major outer membrane protein of *Escherichia coli* K 12 with the aid of a newly isolated bacteriophage. Molec. Gen. Genet. 150, 103–105.

Verkleij, A., van Alphen, L., Bijvelt, J. and Lugtenberg, B. (1977) Architecture of the outer membrane of *Escherichia coli* K12. II. Freeze fracture morphology of wild type, and mutant strains. Biochim. Biophys. Acta 466, 269–282.

Verma, J.P. and Martin, H.H. (1967) Über die Oberflackenstruktur von Myxobakterien. I. Chemie und Morphologie der Zellwande von *Cytophaga hutchinsonii* und *Sporacytophaga myxococcoides*. Arch. Mikrobiol. 59, 355–380.

Voll, M.J. and Leive, L. (1970) Release of lipopolysaccharide in *Escherichia coli* resistant to the permeability increase induced by ethylenediaminetetraacetate. J. Biol. Chem. 245, 1108–1114.

Ward, J.B. (1973) The chain length of the glycans in bacterial cell walls. Biochem. J. 133, 395–398.

Ward, J.B. (1974) The synthesis of peptidoglycan in an autolysin-deficient mutant of *Bacillus licheniformis* NCTC 6346 and the effect of β-lactam antibiotics, bacitracin and vancomycin. Biochem. J. 141, 227–241.

Ward, J.B. and Perkins, H.R. (1973) The direction of glycan synthesis in a bacterial peptidoglycan. Biochem. J. 135, 721–728.

Ward, J.B. and Perkins, H.R. (1974) Peptidoglycan biosynthesis by preparations from *Bacillus licheniformis:* crosslinking of newly synthesized chains to preformed cell wall. Biochem. J. 139, 781–784.

Wargel, R.J., Shadur, C.A. and Neuhaus, F.C. (1971) Mechanism of D-cycloserine action: transport mutants for D-alanine, D-cycloserine, and glycine. J. Bacteriol. 105, 1028–1035.

Warth, A.D. and Strominger, J.L. (1969) Structure of the peptidoglycan of bacterial spores: occurrence of the lactam of muramic acid. Proc. Nat. Acad. Sci. U.S.A. 64, 528–535.

Watson, M.J. and Baddiley, J. (1974) The action of nitrous acid on C-teichoic acid (C-substance) from the walls of *Diplococcus pneumoniae*. Biochem. J. 137, 399–404.

Weidel, W. and Pelzer, H. (1964) Bagshaped macromolecules—a new outlook in bacterial cell walls. Adv. Enzymol. 26, 193–232.

Weidel, H., Frank, H. and Martin, H.H. (1960) The rigid layer of the cell wall of *Escherichia coli* strain B.J. Gen. Microbiol. 22, 158–166.

Wheat, R.W. and Ghuysen, J.M. (1971) Occurrence of glucomuramic acid in Gram-positive bacteria. J. Bacteriol. 105, 1219–1221.

White, D., Dworkin, M. and Tipper, D.J. (1968) Peptidoglycan of *Myxococcus xanthus:* structure and relation to morphogenesis. J. Bacteriol. 95, 2186–2197.

White, D.A., Albright, F.R., Lennarz, W.J. and Schnaitman, C.A. (1971) Distribution of phospholipid-synthesizing enzymes in the wall and membrane subfractions of the envelope of *Escherichia coli*. Biochim. Biophys. Acta 249, 636–642.

Wicken, A.J. and Knox, K.W. (1975) Lipoteichoic acids: a new class of bacterial antigen. Membrane lipoteichoic acids can function as surface antigens of Gram-positive bacteria. Science 187, 1161–1167.

Wietzerbin, J., Das, B.C., Petit, J.F., Lederer, E., Leyh-Bouille, M. and Ghuysen, J.M. (1974) Occurrence of D-alanyl-(D)-*meso*-diaminopimelic acid and *meso*-diaminopimelyl- *meso*-diaminopimelic acid interpetide linkages in the peptidoglycan of Mycobacteria. Biochemistry 13, 3471–3476.

Wijsman, H.J.W. (1972) A genetic map of several mutations affecting the mucopeptide layer of *Escherichia coli*. Genet. Res. 20, 65–74.

Wise, E.M. and Park, J.T. (1965) Penicillin: its basic site of action as an inhibitor of a peptide crosslinking reaction in cell wall mucopeptide synthesis. Proc. Nat. Acad. Sci. U.S.A. 54, 75–81.

Wood, W.A. and Gunsalus, I.C. (1951) D-alanine formation: a racemase in *Streptococcus faecalis*. J. Biol. Chem. 190, 403–416.

Wu, M.C. and Heath, E.C. (1973) Isolation and characterization of a lipopolysaccharide protein from *Escherichia coli*. Proc. Nat. Acad. Sci. U.S.A. 70, 2572–2576.

Wyke, A.W. and Ward, J.B. (1975) The synthesis of covalently-linked teichoic acid and peptidoglycan by cell-free preparations of *Bacillus licheniformis*. Biochem. Biophys. Res. Commun. 65, 877–885.

Yocum, R., Blumberg, P.M. and Strominger, J.L. (1974). Purification and characterization of the thermophile D-alanine carboxypeptidase from membranes of *Bacillus stearothermophilus*. J. Biol. Chem. 249, 4863–4871.

Young, F.E. (1967) Requirement of glycosylated teichoic acid for adsorption of phage in *Bacillus subtilis* 168. Proc. Nat. Acad. Sci. U.S.A. 58, 2377–2384.

Yuasa, R., Nakane, K. and Nikaido, H. (1970) Structure of cell wall lipopolysaccharide from *Salmonella typhimurium*. Structure of lipopolysaccharide from a semirough mutant. Eur. J. Biochem. 15, 63–71.

Zwaal, R.F.A., Roelafsen, B. and Colley, C.M. (1973) Localization of red cell membrane constituents. Biochim. Biophys. Acta 300, 159–182.

The bacterial cell surface in growth and division*

9

L. DANEO-MOORE and G. D. SHOCKMAN

1. The role of surface components in bacterial growth and division

Assembly of macromolecular components into a distinct and characteristic cell surface is one of the most complex processes to be investigated thus far. Assembly of a three-dimensional supramolecular surface structure requires the synthesis of a series of complex macromolecules and also precise, predictable, and well-regulated mechanisms to (1) link one macromolecule to another; (2) modify assembled structures in precise and predictable ways; and (3) provide energy for linkage and modification at or on the outside of the cellular permeability barrier. During growth and division, the three-dimensional supramolecular surface structure is duplicated along with all other cellular components, but the surface does not play a passive role during this time. First, in order to integrate and regulate cell surface growth with the synthesis of other cellular components, and in particular with informational macromolecules, the surface must receive data and respond to this information. Second, as a cell-limiting structure, the bacterial surface itself generates information. For example, in its role as a surface, it receives and transmits data from the environment to the rest of the cell. Other, more specific roles in the segregation of the bacterial genome and in cell division have also been suggested.

The principal aim of this chapter is to analyze the evidence available on the specific mechanisms by which bacterial surfaces interact with other cellular elements during cell growth and cell division.

Bacterial cell surfaces are complex in terms of structural components, as detailed in chapter 8 of this volume by Ghuysen. It is commonly thought that the order of complexity of bacterial surfaces is somewhat lower than that of eukaryotic cell surfaces, including mammalian cells. An apparent lower overall order of complexity at both a genetic and a structural level and the availability of

---

*Literature review on this chapter was completed in March 1976.

G. Poste and G.L. Nicolson (eds.)   The Synthesis, Assembly and Turnover of Cell Surface Components
© Elsevier/North-Holland Biomedical Press, 1977.

a multitude of genetic, physiological, and biochemical techniques have been frequently stated as advantages for the use of bacteria in studies of many facets of molecular biology and biochemistry. Frequently, the aspect of relative simplicity in rationalizing the use of bacteria as models for studies directly transferable to higher cells is overemphasized. Differences exist between bacterial and mammalian cell surfaces, and some of these differences may turn out to be exceptions to the rule. Such exceptions, however, may also provide useful experimental tools.

A few words about bacteria and their surfaces are needed to introduce the problems that bacterial cells encounter during the processes involved in increasing their surface area to enclose an increasing volume of protoplasm, and of dividing. The concentration of a large number of molecules in a small volume results in internal osmotic pressures as high as 20 atmospheres. Thus, in contrast to other cells, bacteria must maintain the structural integrity of their strong and protective wall as the cell increases in volume.

Bacteria are completely surrounded by a structurally rigid layer of a cross-linked polymer, the peptidoglycan, a situation that has been likened to the presence of a continuous, exoskeleton- or corsetlike macromolecule. The presence of this macromolecule does not prevent the cells from possessing a variety of other surface macromolecules that interact with the environment at a number of levels, resulting in cellular responses to attractants and repellents, adherence of bacteria (e.g., to various specific eukaryotic cell surfaces), and the antigenic properties of many bacteria. Also, the relationship of other structures to this surface marker can be used to great advantage in studies of structure-function relationships.

In addition to maintaining osmotic integrity, the bacterial cell wall maintains the characteristic size and shape for that particular bacterial species and set of growth conditions while it is itself being enlarged. Normally, a substantial portion of the increase in cell volume is accomplished by extension of the surface area in one, rather than two or three dimensions at any one time, at least in those bacterial species in which this aspect has been relatively well studied. For example, when growing under a defined set of conditions, cylindrically shaped organisms such as *Escherichia coli* or *Bacillus subtilis* usually maintain a constant diameter while increasing in length. Thus, the shape of the cell changes during the cell division cycle. However, to produce two daughter cells that are essentially equivalent in size and shape at birth to the mother cell requires surface extension in a second dimension for the formation of a completed cross wall. This relationship holds for bacterial species that continuously divide in a different plane from the previous division (e.g., Staphylococci) as well as for those which normally divide in only one plane (e.g., Streptococci, Bacilli, etc.).

All of the data obtained so far are consistent in that they clearly demonstrate that preexisting cell wall peptidoglycan can maintain the shape of a bacterial cell. For example, isolated and purified peptidoglycans retain the rod, coccal, or other shape of the bacterial species from which they were isolated (e.g., Fig. 7 in chapter 8, Ghuysen, this volume). However, as a cell grows in volume, the peptidoglycan assembly process must result in a product that varies in size and shape in a rather precisely defined and systematic fashion. Thus, as clearly indicated by Henning (1975), the expression of, or *determination* of, shape within genetically defined and environmentally influenced limits can be distinctly differentiated from shape *maintenance*. The processes by which growing cells assemble a peptidoglycan of continuously varying shape seem more complex than the fairly simple self-assembly systems used in the formation of virus particles. As discussed further on, for example, peptidoglycan assembly is integrated with the synthesis of other cellular components so that, in the case of *E. coli* and related rod-shaped species, the diameter and the length of the peptidoglycan cylinder can change with growth rate.

The roles of the cell wall in bacterial growth and division have been demonstrated in many ways including studies of mutants with one or more defects that result in morphological or other alterations in these processes (review, Slater and Schaechter, 1974). Perhaps the most striking examples of the function of wall formation in growth and division are the consequences of removal of an intact and functional wall. Bacterial walls can be removed or their structural integrity damaged by enzymatic hydrolysis of the wall peptidoglycan (the polymer responsible for all the properties mentioned above) or by inhibiting the synthesis or assembly of the peptidoglycan in growing cultures (Ghuysen, this volume). Such treatments result in osmotically fragile forms that, in addition, no longer possess the characteristic shape of the bacterial species from which they were derived and no longer increase in size and divide in a well-organized, regulated manner. For example, lysozyme treatment of *Bacillus subtilis* in an osmotically protective liquid environment results in osmotically fragile, spherical bodies (protoplasts) which, on transfer to an appropriate solid medium, can form colonies (L-forms) consisting of cellular units of a vast variety of sizes and shapes (Landman, et al., 1968). Morphological and other studies of L-forms suggest that the precise processes that result in compartmentalization of genomes into two daughter cells no longer occur. Instead, a disorganized process takes place, frequently resulting in the production of units that lack DNA or other essential cellular components. When L-forms are placed under conditions resulting in the assembly of an intact wall (Landman et al., 1968; Elliott et al., 1975), it is only after a morphologically recognizable wall and rod shape has been attained that some regularity of the cell division process can be observed.

The most important component of the three-dimensional information containing system of the bacterial cell surface related to chromosome segregation and cell division is the cell wall peptidoglycan. In the absence of a mitotic apparatus, the precise regular and predictable sequence of events from chromosome initiation, cell enlargement, and the partitioning of the two new completed genomes into two equal daughter cells appears to require the presence of the principal structural wall polymer, the peptidoglycan. In one way or another the other surface polymers and structures relate closely to the peptidoglycan and sometimes play obviously important roles in the transfer of information. Here, we will emphasize the role of the wall peptidoglycan in surface growth and division and consider the other surface components only as they relate directly to the questions at hand.

2. Biosynthesis and assembly of cell walls

The enzymatic reactions resulting in the final assembly of the cell wall occur at the interface between the outer surface of the membrane and the wall itself. Our current state of knowledge of the series of biochemical reactions involved is discussed in detail in the previous chapter by Ghuysen. Substantially less information is available concerning these processes than of the earlier steps in the biosynthesis of each of the wall polymers. For the assembly of cell walls in growing and dividing cells, the following overall, generalized picture can be deduced.

Disaccharide peptide units are polymerized into chains of multiple units via a "headward elongation" process (Lipmann, 1968) at the reducing end (Ward and Perkins, 1973), while connected to the membrane by a linkage that is labile to dilute acid at 60°C (Ward and Perkins, 1973). Linkage through a pyrophosphate bond would be acid-labile.

The next step in the assembly of insoluble cell wall peptidoglycan appears to be the insertion of nascent, linear, and uncross-linked peptidoglycan chains into a preexisting wall acceptor. The consensus of data, obtained with several different systems and bacterial species, suggests that new chains are inserted into preexisting wall via transpeptidation reactions. In several systems, transpeptidation appears to be the sole mechanism, while some penicillin-resistant incorporation of precursors into the wall of *Micrococcus lysodeikticus (M. luteus)* seems to occur via transglycosylation. Fiedler and Glaser (1973) proposed that new peptidoglycan chains are transpeptidated into preexisting wall before they are released from the undecaprenol pyrophosphate carrier in the membranes (Fig. 37 in Ghuysen, this volume). Such a mechanism seems to be attractive since it precludes the possible loss of peptidoglycan to the growth medium. The release of soluble uncross-linked or infrequently cross-linked glycan chains to the growth medium when transpeptidation is inhibited (Keglevic et al., 1974; Mirelman et al., 1974b; Tynecka and Ward, 1975) or when protoplasts are grown in the absence of a wall acceptor (Rosenthal and Shockman, 1975 a,b) are consistent with the hypothesis of Fiedler and Glaser.

Synthesis of teichoic acid chains occurs concurrently with the assembly of the peptidoglycan. Chains of teichoic acids appear to be assembled by the transfer of glycerol- or ribitol phosphate units from the corresponding CDP derivatives directly to a lipoteichoic acid carrier. While certain lipids, such as cardiolipin, were found to activate the polymerization, the process seems to be a direct transfer from the CDP derivatives and does not require an intermediate carrier such as undecaprenol phosphate. Chain extension, at least for the polyglycerol phosphate polymer of *B. subtilis* (Kennedy and Shaw, 1968) and the polymer of N-acetylglucosamine-1-phosphate of *S. lactis* N.C.T.C. 2102 (Brooks and Baddiley (1969), appears to take place by a "tailward elongation" (Lipman, 1968) at the glycol end of growing chains. Attachment of newly completed teichoic acid chains to newly made peptidoglycan apparently occurs at about the same time as the glycan chains are incorporated into the cell wall (Fig. 37, Ghuysen, this volume). The recent data of Wyke and Ward (1975), Bracha and Glaser (1976), and Hancock and Baddiley (1976) obtained with cell-free wall plus membrane preparations are consistent with the addition of teichoic acid chains to the peptidoglycan concomitant with, or just after, peptide cross-linking.

While the biochemistry of the assembly of these two wall polymers seems to be fairly well defined, substantially less is known about the factors that regulate their integrated assembly into a recognizable cell wall in vivo. First, the series of enzymatic reactions for synthesis of each polymer, for the attachment of teichoic acid chain to the peptidoglycan, and for transpeptidation into an insoluble product are required. More significantly, these reactions must occur in a manner that yields a wall having the characteristic morphological and physical properties produced by a particular bacterial species growing in a particular manner. Obviously then, the product(s) of any disrupted cell preparation, even those that may closely resemble bacterial cells, will differ from the variety of products made by the same growing bacterial species. Thus, despite difficulties of interpretation, studies of cell wall assembly at the cellular level are essential.

Several questions may be asked about wall chemistry and synthesis in growing cells; only a few have been satisfactorily answered so far. We shall now consider some of these questions.

3. The contribution of surface constituents to cellular mass or volume

Accurate determinations of the contribution of cell wall to cellular mass and/or volume have been made in only a few instances. The paucity of data on this aspect may be attributed to several factors. Investigators of cell wall composition and structure have been more interested in obtaining highly purified samples of wall rather than quantitative recoveries. Less direct estimates, such as those that can be calculated by comparing total cellular content with wall content of specific wall components (e.g., diaminopimelic acid [DAP], muramic acid, sugars found only in the wall), have been made only infrequently. Quantitative estimates of cell volume based on ultrastructural observations are equally rare.

Qualitative inspections of electron micrographs, such as the thin sections shown in Figs. 1, 2 and 16 of chapter 8 by Ghuysen, and approximation of wall thickness relative to cell size indicates that the cell wall can account for a substantial fraction of cell volume or mass, especially in Gram-positive bacteria. For example, wall thicknesses have been estimated to average about 28 and 37 nm in cells from exponential phase and valine- or threonine-deprived cultures of *S. faecalis*, respectively (Higgins and Shockman, 1970b). Estimates of cell volume of 0.55 to 0.7 μm^3 based on particle counter and ultrastructural measurements

(Daneo-Moore and Higgins, unpublished data) yield approximations of about 15 to 20% of the cellular volume as cell wall (calculated as a sphere).

The estimates of cell volume occupied by wall in *S. faecalis* can be compared with weight estimates based on chemical analyses of specific cell wall components (e.g., rhamnose and hexosamine) in cells and cell walls, confirmed by quantitative recoveries of walls (Shockman et al., 1958; Shockman, 1959a; Toennies et al., 1963). Cell walls accounted for about 25% of the dry weight of exponential phase cells (doubling time, or t_d = 33 min) compared with 38 and 44% for stationary phase (20–40 hr) valine- or threonine-deprived cells, respectively.

On the basis of permeability to solutes of various molecular sizes, Scherrer and Gerhardt (1964, 1971) estimated that the fraction of total cell volume outside the permeability barrier in *B. megaterium* is about 20% of the total cell volume, in good agreement with approximations of 20 to 25% of wall by weight for this organism (Salton, 1964).

In other Gram-positive bacterial species, estimates based primarily on quantitative recovery ranged from 20 to 50% of cellular dry weight (Salton, 1964). Greater difficulties in quantitatively isolating relatively pure walls and in precisely determining the limits of the wall of Gram-negative species resulted in little reliable data. Salton (1964) estimated that the wall of Gram-negative species probably accounts for less than 20% of their dry weight.

4. The relationship of chemical composition and structure of the wall to cell shape, changes in shape, and mode of division

Various comparisons have been made among: (1) different bacterial species; (2) coccal and rod or filamentous forms of dimorphic species; (3) the same species grown under different nutritional or physiological conditions and in the presence of inhibitors; and (4) morphological and other mutants and the wild type of a particular species. The data obtained so far have served to eliminate the more obvious and easily detected correlations. Since the peptidoglycan is primarily responsible for cell shape, we will first consider changes in peptidoglycan structure.

4.1. Overall chemical composition and structure of the peptidoglycan

As summarized below, numerous comparisons of peptidoglycan composition and structure have been made. However, the chemical composition and chemical structure of the bulk of the cell wall peptidoglycan does not appear to reflect the variety of shapes or changes in shape of cells in which they are found.

4.1.1. Comparisons between species
As documented in the preceding chapter (Ghuysen), the overall chemical composition of walls and peptidoglycans from a wide variety of species has been

determined. For a wide variety of species, the chemical structure of at least the major portion of peptidoglycans has also been determined and classified into a series of chemotypes. All Gram-negative species examined so far, and several Gram-positive species, have peptidoglycans of chemotype I containing *meso*-diaminopimelic acid (DAP) (Fig. 10 in chapter 8, Ghuysen). This assortment includes both rod- and coccal-shaped species such as *E. coli* and *Neisseria,* respectively. In addition, the walls of some gram-positive rods such as a variety of species of *Bacilli* and *Lactobacillus plantarum* and the filamentous branching forms of *Streptomyces (Actinomadura)* strain R39 have peptidoglycans of chemotype I containing DAP. A similar lack of correlation can be found among other gram-positive bacteria. For example, walls of both *Streptococcus faecalis* ATCC 9790 *(S. faecium)* and *Lactobacillus acidophilus* strain 63AM Gasser have the same overall peptidoglycan structure (chemotype II containing lysine and D-isoasparagine in its cross bridge). Thus, despite the variations in peptidoglycan structure encountered, the implications of their constancy and variation in bacterial taxonomy (Schleifer and Kandler, 1972; Ghuysen, this volume) and the interesting hypothesis of Previc (1970) regarding a role for the second carboxyl group of DAP in shape maintenance, it seems unlikely that the chemical structure of the bulk of the cell wall peptidoglycan plays a role in shape maintenance. It should be stated here that all analyses of peptidoglycans made so far considered only major constituents. While minor, but perhaps important, variations in amino acid sequence in the peptidoglycan of some species are known to exist, they have not been carefully studied.

4.1.2. Comparisons of spherical and rod-shaped forms of dimorphic species

Only a few comparisons have been made. Differences in the composition of the peptidoglycan present in vegetative cells and microcysts of *Myxococcus xanthus* were observed (White et al., 1968). Although the amino acid and amino sugar ratios were about the same in both forms, evidence suggested that the peptidoglycan of the microcysts was more cross-linked than that of the vegetative cells. Ensign and collaborators (Krulwich et al., 1967) compared the sphere and rod forms of *Arthrobacter crystallopoites*. They found the overall peptidoglycan composition and the chemical structure of the isolated disaccharide-peptide structural unit to be the same in both forms. However, they observed that the polydisperse glycan chain lengths were longer (114–135 hexosamines per chain) in the rod form than in the spherical form (14–63 hexosamines per chain) (Krulwich et al., 1967). A decrease in autolytic N-acetylmuramidase activity was associated with the conversion of spheres to rods (Krulwich and Ensign, 1968).

In a strain of *A. crystallopoites* isolated from a culture of *A. crystallopoites* ATCC 15481, Previc and Lowell (1975) found that the peptidoglycan of the coccal form contained lysine while that of the rod form contained DAP. The shift from coccal to rod form was accompanied by a gradual loss of lysine and gain in DAP. A shift back to spheres was accompanied by a return to the amino acid composition of the original spheres. The parent ATCC 15481 strain contains both lysine and DAP in its peptidoglycan, and the ratio of these two amino acids did not change much with varying growth conditions. Also, other derived strains showed individual variations in peptidoglycan composition. Thus, although this change in peptidoglycan composition is similar to the change from lysine in vegetative cell peptidoglycan to DAP in the spore peptidoglycan of *Bacillus sphaericus* (Linnett and Tipper, 1974), its significance is not yet apparent.

4.1.3. Comparisons of the same species under different growth conditions

Little information is available about peptidoglycan composition and structure with changes in growth rate of exponential phase cultures. In general, significant changes in the molar ratios of amino acids and amino sugars have not been observed in walls from cells grown in different growth media (e.g., Young, 1965; Bleiweis et al., 1976) or to different growth stages. For example, the content and ratios of amino acid in walls of exponential phase and the thickened walls of stationary-phase, threonine-deprived cultures of *S. faecalis* were nearly the same (Toennies et al., 1959), suggesting not only little or no change in peptidoglycan composition but also little alteration in relative amounts of peptidoglycan in the walls.

A similar lack of change in composition or amount of peptidoglycan in thickened walls of *B. subtilis* was observed by Hughes et al. (1970). Also consistent with these observations are those of Katz and Martin (1970), which indicate that some L-forms of *Proteus mirabilis* contain a peptidoglycan of the same composition and the same degree of peptide cross-linking (see below) as its normal rod-shaped parent. Similarly, the overall amino-acid composition, and degree of peptide cross-linking of the peptidoglycan, of *E. coli* treated with concentrations of penicillin that resulted in morphological changes were indistinguishable from those in untreated cultures (Schwarz et al., 1969; Schwarz and Leutgeb, 1971).

Ellwood and Tempest (1972) reported relatively small quantitative differences in molar ratios of amino acids and amino sugars in walls of *B. subtilis* W23 grown in the chemostat at slow growth rates (dilution rate of 0.1–0.3 hr^{-1}) when the pH, temperature, or growth-limiting nutrient was varied.

Johnson and Campbell (1972) showed a substantial difference in cell wall composition and peptidoglycan structure between cultures of *Micrococcus sodonensis* (ATCC 11880) grown to the stationary phase in a rich trypticase soy broth (TCS-walls) and a poor chemically defined medium (CD-walls). These detailed studies showed that the TCS-walls (1) had 40% of their muramic acid residues substituted by peptides compared with 13% for the CD-walls; (2) had fewer amino acids in the peptide cross bridges than the CD-walls; and (3) were more resistant to dissolution by lysozyme. In addition, the studies revealed a heterogeneity of peptide structure in walls from both cultures that was less pronounced in the TCS-walls. On the basis of these studies the authors proposed that some areas of the peptidoglycan in intact walls are more highly ordered than others. Lysozyme dissolution of up to 75% of CD-walls left a residue in which recognizable coccal walls were less dense when viewed in negatively stained preparations in the electron microscope. These observations suggested that the lysozyme sensitive and resistant portions of the peptidoglycan do not exist as separate "islands" in the wall, and that the two may "exist as interwoven peptidoglycan nets, one of which is resistant to lysozyme because of extensive peptide cross bridging and increased levels of O-acetylation."

Changes in an amino acid present in peptidoglycan have been induced by certain alterations in growth conditions. For example, hydroxylysine can replace L-lysine in the peptidoglycan of S. faecalis when the cultures are starved for lysine in the presence of hydroxylysine (Shockman et al., 1965). Growth of some species in the presence of high concentrations of certain amino acids can result in the substitution of the fed amino acid for one normally present in the peptidoglycan. Examples of this are growth of M. lysodeikticus in the presence of D-serine (Whitney and Grula, 1964) and Staphylococcus aureus in the presence of DL-serine (Schleifer, 1969). In both cases, substitution of amino acids in the peptide crossbridge (the more variable part of the peptidoglycan; Schleifer and Kandler, 1972) were observed.

4.1.4. Other variations in peptidoglycan composition

As discussed by Ghuysen in chapter 8, the presence of several minor components in peptidoglycans have been noted. To date, none of these minor components has been correlated with cell shape or division. They are, however, worth mentioning since one or more could prove to be important in the future. Such variations include (1) the presence of small amounts of mannomuramic acid in place of the normally found glucomuramic acid in walls of M. lysodeikticus; (2) the presence of acetyl or phosphodiester groups on the C-6 hydroxyl position of muramic acid; (3) the replacement of the N-acetyl group of muramic acid by an N-glycolyl group; (4) the absence of an N-acetyl substituent on muramic acid or glucosamine; (5) the presence of muramic acid lactam in spore peptidoglycan; (6) the substitution of one amino acid for another in the peptide side chain or peptide cross-bridge induced by alterations in growth conditions, as mentioned earlier; (7) the presence of amide or other substituents or otherwise free carboxyl groups; and (8) the presence of quantitatively small but perhaps important variations in peptidoglycan sequence. A few examples may illustrate the possible, but as yet undetermined, consequences of such variations, particularly if they are concentrated in specific topological surface locations.

It is known that the peptide cross-bridge in S. faecalis is mediated via a D-isoasparagine moiety. However, only about 70% of the amino groups of lysine are substituted by this amino acid (Ghuysen et al., 1967). Furthermore, the presence of an LD-transpeptidase activity in membranes of this species (Coyette et al., 1974) provides an enzymatic mechanism by which a bond can be formed between the α carboxyl group of the L-lysine residue and the amino group of a D-isoasparagine on a different glycan chain. At present, the existence of such a linkage in cell walls of this species is unknown. Similarly, the occurrence of the transpeptidation product of the reaction of DD-DAP with the carboxyl group of D-alanine catalyzed by membrane preparations of *B. megaterium* (Wickus and Strominger, 1972) has not been detected in the walls of this species.

Another interesting series of observations that could be related to cell shape and mode of division were made by Johnson and McDonald (1974). They compared the peptidoglycan structure in cell walls of a "filamentous" strain of *Streptococcus cremoris* HP with that of the parent strain. Differences in the chemical structure of the peptidoglycan in the two strains were not observed. Both strains contained peptidoglycans with many unsubstituted muramic acid residues and relatively long peptide cross-bridges. However, about 50% and 15% of the glucosamine and muramic acid residues, respectively, were not N-acetylated. Similar to several other bacterial species (Perkins, 1965; Shockman et al., 1967b; Hayashi et al., 1973), the presence of non-N-acetylated glucosamine residues was correlated with resistance to hen egg-white lysozyme. The presence of nonacetylated hexosamines in the wall infers the existence and action of de-N-acetylase(s) (Araki et al., 1971). These investigators proposed that de-N-acetylation could modify the susceptibility of the peptidoglycan substrate to the action of autolytic peptidoglycan hydrolases at specific topological areas of the wall. Such an interpretation is similar to the proposed role for choline in the wall teichoic acid of the pneumococcus (Tomasz, 1968), and wall teichoic acid in the action of the amidase-modifier system in *B. subtilis* (Herbold and Glaser, 1975a,b), as discussed in section 5.3.

Although studies of the overall composition of peptidoglycans have yielded little insight into the mechanism of shape change and mode of division, minor variations in peptidoglycan structure, such as the compositional changes mentioned here and other known and possible modifications to the peptidoglycan and/or wall polymers mentioned in the following sections, may be important.

4.2. Overall size of peptidoglycan molecules

4.2.1. Length of glycan chains
Due to the presence of peptidoglycan hydrolase activities (section 5), nearly all estimates of either glycan chain lengths or extent of peptide cross-linking in isolated walls must be considered to be minimal values and, in terms of their relationship to wall structure, to be questionable. From the results of Ward (1973) it seems that average glycan chain lengths in wall may be substantially longer than was thought previously. By measuring the ratio of total amino sugar alcohol to total hexosamine, he estimated chain length to be about 80 disaccharide units in walls of a lytic defective mutant of *B. licheniformis,* isolated with all known precautions to prevent enzymatic degradation. From the ratio of muramitol to muramic acid Ward was also able to determine what he considers to be the length of glycan chains biosynthesized, before the action of endogenous β-N-acetylglucosaminidase, to be about 140 disaccharides long. Thus, it seems that long glycan chains are hydrolyzed to shorter lengths when, or after, they are inserted into the wall. Such observations are consistent with the idea of

modifications to peptidoglycans after incorporation into the wall as discussed later in section 7. Similarly, the polydispersed nature of glycan chains could be due either to hydrolysis during wall isolation or to true variations in chain lengths in walls, depending perhaps on topological location or age. To date, experimental difficulties have prevented a clear assessment of the relationship between glycan chain length and morphology, except perhaps in the case of *A. crystallapoites* mentioned earlier.

4.2.2. Peptide cross-links

Estimates of the extent of peptide cross-linking vary with species and, with the possible exception of *S. aureus,* tend to be low. Analyses of the distribution of peptide cross-links in walls of *S. faecalis* grown under a variety of conditions (Dezéllée and Shockman, 1975) showed that in the portion of the peptidoglycan (about 80% of the total) that is not linked to other wall polymer after extensive muramidase hydrolysis of glycan chains (1) less than 35% disaccharide peptides are in units larger than peptide cross-linked trimers even after "ageing" for one full cell generation; (2) a substantial increase in number of peptide-cross-linked units occurred after incorporation of precursors into the insoluble wall (section 7); and (3) the observed distribution pattern of monomers, dimers, trimers, and so on, was not consistent with a random distribution pattern but closely resembled the distribution to be expected from a monomer addition mechanism (Oldmixon et al., 1976). These analyses brought out the concept that peptidoglycans may also "grow" in the peptide chain direction, so that one can picture an increase of mean peptide chain length with, for example, time. Such a concept, along with that of assembly according to the monomer addition model, suggested that mean peptide chain lengths are short, even in walls considered to be extensively cross-linked. For example, 50 and 75% cross-linking would be equivalent to a mean peptide chain length of 2 and 3, respectively. Fordham and Gilvarg (1974) also observed a continuation of peptide cross-linking for many minutes after incorporation of precursor subunits into the peptidoglycan of *B. megaterium.*

Although mean peptide chain lengths tend to be short, an occasional cross-bridge between two glycan strands would be sufficient to construct a very large macromolecule. For example, for a peptidoglycan with a mean glycan chain length of 20 disaccharide units, a minimum of only 10% of the terminal amino (or carboxyl) groups need to be engaged in cross-links to link together 3 or more glycan chains. Obviously, longer glycan chains would require still fewer cross-links. Additional peptide bridges presumably would simply increase the strength and rigidity of the polymer.

Attempts have been made to correlate the extent of peptide cross-linking and cell shape and division. Gross correlations, in terms of the shape or mode of division of various species, for example, have not indicated any striking differences. For example, cocci that divide in more than one plane can have a relatively high *(S. aureus)* or low *(M. lysodeikticus)* degree of cross-linking.

An interesting situation was encountered with Rod⁻ mutants of *B. subtilis* (Rogers et al., 1971). When grown on media of low salt content, these mutants appear as rounded forms that show gross distortions and disorganization of their walls. One class of these mutants (B) appear as normal rods when grown on media containing 0.8 to 1.0 M NaCl or KCl. Initial observations suggested that the peptidoglycan of one of these mutants (rod-4) was less cross-linked than that of the parent strain. However, when the parent organism was grown (as rods) in the same minimal medium, the same very low degree of cross-linking was observed. Thus, in the parent strain, a very low degree of peptide cross-linking in cells grown in normal medium (t_d = 70 min) compared with a higher extent of cross-linking in cells grown in a casein hydrolysate medium (t_d = 36 min). Correlations of changes in cell shape by means of inhibiting cross-linking with various β-lactam antibiotics has not yet shown the correlations expected. The filamentous cells and bulges in cells and sacculi of *E. coli* could not be correlated with the degree of peptide cross-linking (Schwarz et al., 1969).

4.3 Polymers covalently attached to peptidoglycans

4.3.1. Gram-positive bacteria

Removal of the various negatively charged or neutral polysaccharides from isolated cell walls of gram-positive bacteria by gentle means such as dilute acid (Knox and Hall, 1965; Pavlik and Rogers, 1973), dilute alkali (Hughes and Tanner, 1968), dilute periodate (Garrett, 1965), or N,N-dimethyl hydrazine treatment (Anderson et al., 1969) usually leaves a peptidoglycan residue that retains the wall shape. Thus, the various polysaccharides linked covalently to the peptidoglycan of gram-positive species do not appear to contribute significantly to the shape of an already existing wall. However, the question of the influence of the synthesis and attachment of such polymers to the peptidoglycan during growth and division on shape determination is not answered by these types of experiments.

With at least several *Bacillus* species the presence of one or another negatively charged polymer in the wall seems sufficient to insure maintenance of shape and a regular pattern of cell division. As discussed below (section 4.4.1), a change from one nonpeptidoglycan polymer to another, such as the change from a teichoic acid in Mg^{2+} limited cultures to a teichuronic acid polymer in phosphorus-limited cultures, does not seem to greatly affect cell size and shape. In growing cells, however, while the presence of either a phosphate or uronic acid containing polymer in the wall is not essential to maintenance of cellular integrity it does seem to be important for the maintenance of the rod shape and a normal pattern of cell division. For example, some conditional Rod⁻ mutants of *Bacillus subtilis* when grown as irregularly shaped spheres, have been shown to possess a much smaller proportion of wall teichoic acid than the wild type. However, other Rod⁻ mutants (called *rod*B) have essentially the same composition and ratio of peptidoglycan to teichoic acid in their walls as those of the wild type, whether growing as deformed cocci or rods (summarized in Rogers, 1977). The similarities in wall composition of *rod*B and wild type included length of glycan chains and extent of peptide crosslinking (Rogers et al., 1971). Recently, the *rod*B mutants have been shown to have a requirement for high concentrations of Mg^{2+}, and their morphology can be controlled by Mg^{2+} concentration under certain conditions (Rogers et al., 1976). Each of these mutant phenotypes does not seem to be a result of multiple mutations although, as with many bacterial morphological mutants, they exhibit pleiotropic characteristics.

Thus, in *B. subtilis* deformed morphology and, since the coccal forms are actually extremely irregularly-shaped subdivided groups of organisms, mode of division can occur either in the presence or near absence of negatively charged polymers in the wall. However, studies of growth of two phosphoglucomutase deficient mutants of *B. licheniformis*, which lack teichuronic acid and glucose in their walls and are also poorly lytic under conditions of Mg^{2+} and PO_4^{3-} limitations (Forsberg et al., 1973), provide an additional insight. When grown in batch culture in minimal medium, or in Mg^{2+} limited chemostat culture at dilution rates of 0.2 hour^{-1}, the Lyt$^-$5 strain grew as rods or chains of rods. Under such conditions cells could satisfy their requirement for an anionic polymer by incorporating teichoic acid into their walls. When grown under PO_4^{3-} limitation, cells of Lyt$^-$5 grew as ovoid cells with multiple septa in more than one plane, and had multilayered walls of irregular thickness that tended to peel away from the cells. The walls of PO_4^{3-} limited cells contained about 18% of the phosphorus and about 13% of the glycerol, but had substantially higher levels of glucosamine, muramic acid, DAP, glutamate, and alanine than the walls of cells from Mg^{2+} limited cultures. In both PO_4^{3-} and Mg^{2+} limited cells of this mutant, glucuronic acid, glucose and galactose were absent and galactosamine was very low. However, bypassing the phosphoglucomutase requirement by supplying galactose in the growth medium, resulted in normal, rod-shaped cells. Thus, it appears that when *B. licheniformis* is unable to synthesize either wall teichoic acid (because of PO_4^{3-} limitation) or teichuronic acid (because of the absence of phosphoglucomutase), it synthesizes a wall containing more peptidoglycan that is deformed in shape and mode of division.

The effects of the absence of a negatively charged polymer in walls is also illustrated by the series of mutants of *S. aureus* that have a defect in their wall teichoic acid isolated by Park and collaborators (summarized in Park et al., 1974). Their type III mutants totally lack wall teichoic acid and exhibit a series of properties which differ from the parent. The cells of strain 52A5: (1) are larger in size and mass; (2) often separate incompletely so that cells (and cell walls) remain attached to each other in large clumps; (3) are resistant to bacteriophage; (4) grow more slowly; (5) do not accumulate nucleotide precursors of peptidoglycan in the presence of penicillin; and (6) are particularly susceptible to salt-stimulated cellular autolysis. This same phenotype is exhibited by mutants of this class which lack CDP-ribitol, phosphoribitol transferase or both. In all three cases these investigators believe that a single mutational event, probably in a structural gene, seems responsible for both the absence of ribitol teichoic acid in the walls and the pleiotropic phenotype. Thus, although wall teichoic acids do not appear to be essential to S. *aureus,* its presence may confer a selective advantage particularly because of the faster growth rate and facilitated separation of the wild type.

A similar association of the absence of a nonpeptidoglycan wall polymer and the mode of cell separation was recently described by Yamada et al. (1975) for two mutants of *M. lysodeikticus* that lack the teichuronic acid normally found in their wall. When grown on agar, these two mutants formed regular packets of 4 to 64 cells in contrast to the irregular clusters formed by the parent strain (Fig. 1). Analyses revealed about a 30% increase in amino acids and amino sugars, with maintenance of the same molar ratios to each other as in the parent, accompanied by a decrease in hexose content. Cells and walls of the mutants remained sensitive to hen egg-white lysozyme which, along with the constancy of molar ratios of amino acid and amino sugar, suggested the absence of a change in the peptidoglycan. The very low hexose level and the lack of reaction with antiserum to *M. lysodeikticus* (presumably directed against the teichuronic acid) indicated the absence of this wall polysaccharide. Since the levels of peptidoglycan constituents present in the wall do not account for the entire wall weight, another undetected polymer may be present in the wall. Both mutants required an increased concentration of Mg^{2+} for growth comparable to the wild type. The availability of a transformation system in this species should help to characterize genetically the mutation(s) involved. These results are consistent with a role for anionic wall polymers in cell separation (and for binding cations) as well as in possible orientation of future division planes.

Another example of the multiple effects of a change (in this case very small) in one of the wall teichoic acid constituents has been observed in *Streptococcus (Diplococcus) pneumoniae.* Choline, a component of the complex and as yet incompletely characterized wall teichoic acid (C polysaccharide; Brundish and Baddiley, 1968; Mosser and Tomasz, 1970; Watson and Baddiley, 1975), and of the Forssman (or F) antigen (Brundish and Baddiley, 1968; Briles and Tomasz, 1973), is required for growth of the pneumococcus. In the absence of choline, ethanolamine can satisfy the dietary requirement (Tomasz, 1968) and is incorporated into a polymer that has several properties in common with those of the C polysaccharide. When grown in a medium containing ethanolamine, the cells: (1) lose the capacity to undergo transformation; (2) grow in very long chains; (3) resist dissolution by deoxycholate; (4) fail to lyse when grown in the presence of penicillin, cycloserine, or phosphonomycin, although growth is inhibited by these antibiotics; (5) fail to autolyze after the end of the exponential growth phase; and (6) possess cell walls that are not hydrolyzed by the pneumococcal autolytic N-acetylmuramyl-L-alanine amidase. All these changes have been interpreted as being due to the replacement of choline by ethanolamine (which differ by only three methyl groups) in the wall teichoic acid. In fact, methylating ethanolamine-containing walls was shown to restore both susceptibility of walls to the pneumococcal amidase and the ability of walls to convert an inactive form of the enzyme to the active form (Höltje and Tomasz, 1975b). However, choline is also thought to be a component of the F antigen, and the recently obtained evidence that the F antigen is an effective inhibitor of the action of the amidase (Höltje and Tomasz, 1975a) provides alternate interpretations. It seems possible that one or more of the effects observed could be due to the presence of ethanolamine in F antigen preparations.

At this time, it is difficult to reach firm conclusions regarding the role of negatively charged polymers in cell shape and division. Probably, the presence of a negatively charged polymer is highly desirable for the maintenance of orderly division and shape. For bacilli and staphylococci, the presence of either phosphate or carboxyl groups appears to be satisfactory. For *M. lysodeikticus,* loss of the teichuronic acid results in a change in division symmetry, which could be interpreted as either a gain or a loss in degree of complexity. Loss of genetic information could cause the loss of a function that has either a positive (e.g., a topological marker) or a negative (e.g., a hydrolytic activity) effect on degree of order. Also, somewhat more subtle effects may be exerted by small changes in a nonpeptidoglycan wall polymer, such as that shown by the choline-containing polymer of pneumococci.

4.3.2. Gram-negative bacteria

The only polymer known to be covalently attached to the thin layer of peptidoglycan in gram-negative bacteria is a lipoprotein (chapter 8 by Ghuysen, this volume for a description of the relationship of the peptidoglycan and lipoprotein). Removal of covalently linked lipoprotein from the isolated peptidoglycan-lipoprotein complex (rigid layer) of *E. coli* by trypsin treatment still leaves a rodlike structure similar in shape to that of the bacterial cell (Fig. 6 in Ghuysen, chapter 8). Thus, similar to the situation with negatively charged polymers in gram-positive species, lipoprotein does not appear to be required for the maintenance of the shape of an already assembled wall. However, the sequence and topology of attachment of lipoprotein molecules to the peptidoglycan may be important in the synthetic sequence for shape determination.

In the biosynthetic and assembly sequence of the lipoprotein-peptidoglycan complex (for details, see Braun, 1975 and Ghuysen, this volume), the lipoprotein is first incorporated into the outer membrane and a portion of the free form is then covalently linked to the peptidoglycan layer, presumably by an LD-transpeptidase activity which is relatively resistant to various penicillins including amidinopenicillanic acid (FL-1060). An LD-carboxypeptidase whose activity varies during the cell cycle was recently found to be located in the periplasm (Beck and Park, 1976). If this enzyme could also carry out transpeptidation reactions it would be capable of linking the lipoprotein to the peptidoglycan. For some unknown reason, lipoprotein synthesis occurs in the presence of puromycin concentrations which inhibit the synthesis of cytoplasmic and several other membrane proteins. While several questions concerning synthesis, cellular localization, and attachment of the lipoprotein are not yet answered, the following points seem relevant (Braun, 1975; Braun and Wolff, 1975). Lipoprotein has been found to be present in *E. coli* and a variety of other related gram-negative, rod-shaped species. The presence of lipoprotein or its functional equivalent in

612

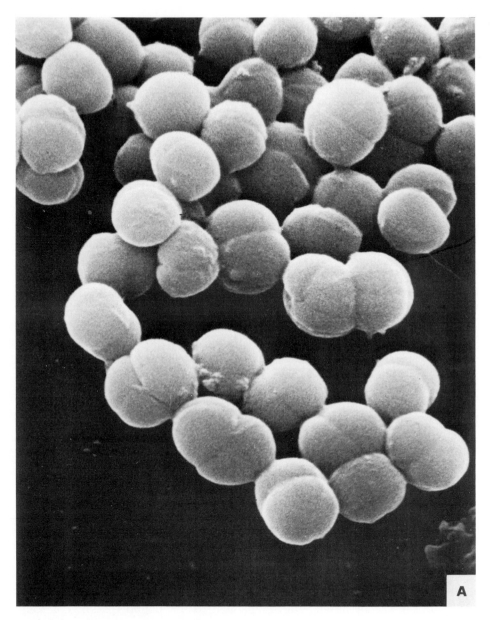

Fig. 1. Scanning electron micrographs of *M. lysodeikticus* and the packet-forming mutants. (a) Parent strain AH-3 collected from BHI agar plates. (b) Packet-forming mutant strain AH-47 collected from BHI agar plates. (Reproduced with permission from Yamada et al. [1975]).

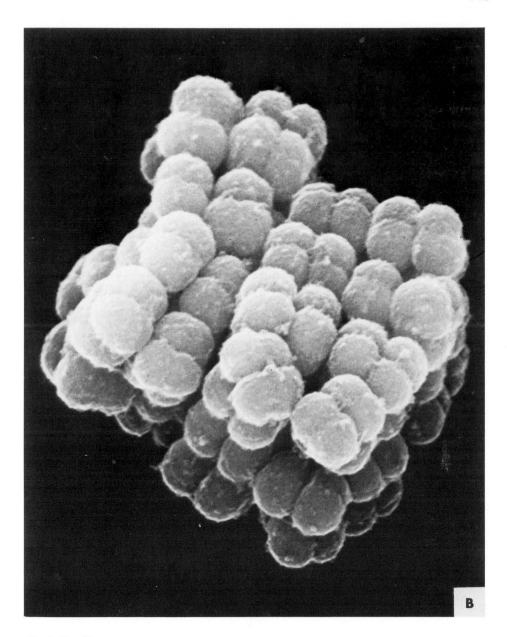

Fig. 1. Part (b).

other gram-negative bacteria remains to be investigated. Mutants of *E. coli* that lack all the quantitatively major proteins except the lipoprotein have been found (Henning and Haller, 1975). Lipoprotein molecules do not appear to be attached to peptidoglycan in clusters, and quantitative assessments indicate that the total lipoprotein could be concentrated on half of the total peptidoglycan at most. Pulse and chase experiments showed that the incorporation of new peptidoglycan preferentially occurs in regions that lack attached lipoprotein and account for 70% of the peptidoglycan. The slow appearance of new peptidoglycan at lipoprotein attachment sites suggests that lipoprotein may be linked to the peptidoglycan layer after peptidoglycan assembly. However, only 40% of newly synthesized lipoprotein becomes attached to the peptidoglycan. Inhibition of peptidoglycan synthesis by about 50% and the appearance of ovoid cells as a consequence of FL 1060 treatment results in a doubling in the amount of the bound lipoprotein observed when the cells reached the stationary phase. This increased ratio did not seem to be due to peptidoglycan turnover.

During nascent cross-wall formation in several strains of *E. coli*, Burdett and Murray (1974a,b) observed the presence of a gap separating the two transverse lamellae of peptidoglycan (Fig. 2). The outer membrane did not appear to participate in septation until cells began to separate. Although the kind of material present in the "gap" is not known, Burdett and Murray postulated that a portion of the covalently attached lipoprotein might be located there. Recently, James and Gudas (1976) observed that the maximal rate of incorporation of free lipoprotein into the outer membrane occurs at about the time of septation. A similar increase in the rate of incorporation of peptidoglycan-bound lipoprotein was not seen. The overall effect would be a decrease in the ratio of bound to free lipoprotein. Thus, it is tempting to postulate a role for the lipoprotein in wall enlargement and division. Such hypotheses are presented in section 9.5.

From the discussions presented above it seems clear that clues to cell shape and cell division do not lie in the overall composition and chemical structure of peptidoglycan and other cell wall polymers. However, it remains possible that topologically localized differences in wall structure could exist. In addition to the possibilities mentioned in section 4.1.4, these differences could take the form of (1) length of glycan chains; (2) presence of unusual glycan or peptide linkages; (3) sites of attachment of teichoic acids, lipoprotein, or other wall polymers; or (4) exact structure of covalently linked polymers. These factors remain to be unraveled in the various systems under investigation.

4.4. Turnover of surface components

4.4.1. Cell wall turnover
In studies of the synthesis of any macromolecule loss of product via turnover is an important factor. Turnover is an even more essential consideration for the interpretation of studies of cell surface components, particularly of one that is essential for maintenance of the osmotic integrity and shape of the cell, and whose shape must change during the cell cycle. For example, turnover of wall or

615

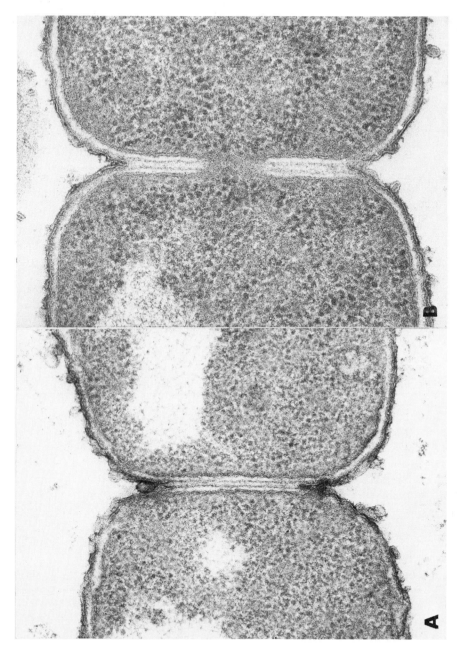

Fig. 2. Septation in *E. coli* CRT 97 showing partial (a) and complete (b) septa; RK fixation was used. (Reproduced with permission from Burdett and Murray [1974a]).

membrane components should be (but is not always) considered as an important factor in all studies of topological location and number of cell surface enlargement sites. All of the wall turnover studies performed so far have measured loss of (usually labeled) material from cells or cell walls. Thus, they have not measured the dynamics of the wall in terms of the movement of apparently conserved molecule within the wall, or from one cell fraction to another.

To our knowledge, experimental evidence for turnover of cell wall polymers was first presented by Chaloupka et al. (1962 a,b). Indications of the dynamic state and turnover of walls and, in this case a change in wall composition, came from chemostat experiments that utilized magnesium or phosphorus concentration to limit growth rate (Tempest et al., 1968; Ellwood and Tempest, 1969; Ellwood, 1971). Growth of *Bacillus subtilis* var. *niger* in phosphorus-limited medium resulted in cells with walls that contained very low amounts of phosphorus and glucose but substantial levels of glucuronic acid and galactosamine. Cells grown at the same dilution rate (0.2 hr^{-1}) in magnesium-limited cultures had walls that contained phosphorus and glucose but not glucuronic acid and galactosamine. Apparently, the walls of magnesium-limited cultures contained teichoic acid, and walls from phosphorus-limited cultures contained teichuronic acid. This difference in wall composition could have resulted from a gradual selection of genotypes by the nutritional limitation, or from a phenotypic shift in synthetic ability, accompanied by the dilution out of the other wall polymer. By following the kinetics of change in polymers after a shift from magnesium to phosphorus limitation, and vice versa, not only was the possibility of selecting a particular genotype eliminated but a shift in nutrient limitation in either direction resulted in the loss of the preexisting polymer at rates substantially faster than the rate predicted from dilution. This was interpreted as a clear indication that during the changes in nutritional limitation the existing polymer was lost by turnover. Evidence was obtained for the presence of both teichoic acid and peptidoglycan, or their degradation products, in the supernatant culture medium. Thus both wall polymers, which are covalently linked to each other in the wall, appeared to turn over.

The observations described above were also taken as an indication that growth conditions affected wall composition and that wall composition was not an invariant property of the bacterial species. The latter conclusion has ramifications not yet fully appreciated by many microbiologists or by many cell biologists. While it seems that bacteria like to have a negatively charged molecule in their surface, molecules containing either phosphate or carboxyl groups seem, at least in some cases, to be satisfactory. Furthermore, a major change in composition will certainly result in a change in the reactive groups exposed on the cell surface that are frequently identified by their ability to combine with specific antibodies or lectins and are important in the attachment of bacterial viruses. The implications of such changes in the use of probes for investigations of cell surfaces and for serologically based identification of microbial species seem obvious.

Mauck and Glaser (1970) and Mauck et al. (1971) showed that during exponential growth of *B. subtilis* W-23 both peptidoglycan and wall teichoic acid turned over at the same rate, which was equivalent to a loss of nearly 50% of wall material per generation. The products of wall turnover that accumulated in the growth medium were expected to result from the action of an N-acetylmuramyl-L-alanine amidase. In the absence of detectable amidase in the growth medium, it was postulated that enzyme bound to the cell surface catalyzed removal of wall material (Fiedler and Glaser, 1973). Furthermore, analysis of turnover products accumulated in the medium indicated that the enzyme hydrolyzed nearly all amide bonds on the released glycan chains. Such action is consistent with that expected from an amidase-modifier complex, carefully studied in another strain of B. subtilis (Herbold and Glaser, 1975,a,b). A lag of about ½ generation before turnover of new wall (pulse labeled for about ⅓ of a generation) but not before turnover of walls labeled for about two generations was observed. However, samples were not taken before about ½ generation after the chase so that a shorter lag before turnover of walls labeled for about 2 generations would not have been detected. The rate of turnover appeared to be proportional to growth rate over a range-generation time of 0.5 to 2 hours. The rate and extent of turnover was decreased in the presence of 10 μg/ml of actinomycin D. Mauck and Glaser also observed that the peptidoglycan of B. megaterium KM turned over. Turnover was at a rate of 30 to 35% per generation in both a rich medium and a chemically defined medium, but 3 different experiments in the medium used by Pitel and Gilvarg (1970) yielded values of 35, 15, and 0% turnover per generation. It should be noted here that Pitel and Gilvarg (1970) did not observe a detectable rate of turnover in a Lys⁻ DAP⁻ mutant of B. megaterium. The rate of peptidoglycan turnover in B. megaterium KM decreased as the cells approached stationary phase and after treatment with chloramphenicol and penicillin (100 μg/ml of each with the penicillin added 20 min after chloramphenicol). These results and those obtained with actinomycin D treatment of B. subtilis W-23 were interpreted as indicating that turnover occurs only during cell wall "growth." Since cell wall synthesis (wall thickening) occurs after inhibition of protein or RNA synthesis, cell wall "growth" should probably be interpreted as cell wall enlargement (see section 4.5.).

Immunity of newly made wall to turnover is an interesting problem. The experiments described above were performed with cultures in which cells in all stages of a division cycle are present at any one instant. Thus it would be difficult to relate the lag before turnover seen by Mauck et al. and by others (Boothby et al., 1973; Pooley, 1976a,b) to specific events in the cell division cycle. This problem will be considered again further on.

Cell walls of five different strains of S. aureus were shown to turnover all at rates of about 15% per generation (Wong et al., 1974). In S. aureus H, the rate of turnover of both peptidoglycan and teichoic acid were indistinguishable and newly made wall turned over at about the same rate as old wall. Turnover continued at an apparent first order rate until about 94% of the wall peptidoglycan was lost. Inhibition of protein synthesis by chloramphenicol (100 μg/ml) or amino acid starvation completely prevented turnover. Turnover at the same rate in S. aureus H, and in mutants deficient in wall teichoic acid (tar-1) or those which lack glucosamine in their wall teichoic acid (52A4), suggests that the presence of teichoic acid in the wall is not essential for the process. A mutant, RUS 3, which autolyzes at 10% of the parent strain rate, was isolated (Chatterjee et al., 1976). The mutant grows as large clusters and does not undergo peptidoglycan turnover. These results are consistent with the assumption that the only autolytic activity detected in *S. aureus* H, an amidase (Singer et al., 1972), plays an active role in turnover. Thus, several aspects of wall turnover in *S. aureus* were similar to those observed in *B. subtilis*, but differed in that turnover did not slow down when cells passed the exponential growth phase and newly made wall was not immune to turnover. During autolysis of isolated walls of *S. aureus* old wall was preferentially lost (Gilpin et al., 1974). Loss of peptidoglycan at a first order rate, until over 90% of the wall was lost in both *B. subtilis* W-23 (Mauck et al., 1971) and *S. aureus* (Wong et al., 1974), suggests that in both organisms random degradation of peptidoglycan occurs over nearly the entire wall surface.

Peptidoglycan turnover in *Lactobacillus acidophilus* strain 63AM Gasser (Boothby et al., 1973; Daneo-Moore et al., 1975) showed several properties that differ from the systems described previously. During exponential growth at doubling times of about 1 to 2.5 hours, peptidoglycan turned over at a first order rate equivalent to a loss of 25 to 35% per generation in the absence of detectable turnover of cellular proteins. Turnover slowed and stopped when cells approached the stationary phase and was completely prevented by valine deprivation or chloramphenicol treatment, even when chloramphenicol was added 1 hour after the start of the chase. However, thickened wall, made during valine starvation, was as susceptible to turnover as wall made during exponential growth when cells were regrown in the presence of valine. In fact, both the lag before turnover and rates of turnover of the portions of wall made during exponential growth and valine starvation were very similar. Since only an N-acetylmuramidase activity seems to be present in this species (Coyette and Ghuysen, 1970), turnover was attributed to the action of this enzyme. However, a substantial rate of peptidoglycan turnover was observed in a mutant that autolyzed at less than 20% of the rate of the wild type. Ultrastructural observations showed a correlation between the sloughing off of pieces of wall, primarily from the external surface of the cylindrical portion of the wall (Fig. 3d), and wall turnover. The polar caps tended to remain smooth and intact; this was particularly apparent in thick-walled cells recovering from valine starvation (Fig. 3c).

A surprising and extremely difficult to explain feature of peptidoglycan turnover in this species is that a lag of 0.8 to 2 generations before turnover was observed in cells that contained peptidoglycan labeled for 6 or more generations, while peptidoglycan pulse labeled for 0.2 generations or less failed to turn over for periods exceeding 2 generations. Neither the lag before turnover of extensively labeled peptidoglycan nor the absence of turnover of short pulses could be interpreted to be related to events in a cell cycle in these randomly dividing cultures, or for that matter to a "culture age" related event. It was presumed, for example, that pulse-labeled wall would eventually "age" as the culture grew, especially into the second (or third) cell generation. Two explanations could account for these observations. The first is that the possible presence of turnover during the early stages of a chase could have been masked and balanced out by the continued incorporation of intermediates from an intracellular pool into peptidoglycan. This event, for which some evidence exists (Dezéllée and Shockman, unpublished observations; see discussion in section 7), would explain the lag before turnover of extensively labeled peptidoglycan. Alternately, the results obtained could be explained by the hypothesis that precursors are incorporated into two products: one that is rapidly labeled and does not turn over (or turns over slowly), and a second that is labeled more slowly and turns over more rapidly.

A second surprising aspect of turnover of the *L. acidophilus* peptidoglycan was based on the observation that after 4 or more generations of exponential growth turnover slowed and stopped, with a minimum of 10 and a maximum of 20% of the peptidoglycan still present, and apparently immune to turnover. A similar fraction of peptidoglycan immune to turnover was observed in a strain of *B. megaterium*, which turns over its peptidoglycan at a rate of about 15% per generation (Chaloupka and Kreckova, 1971). Quantitation of the length of pulse required to observe turnover showed that it was only observed after a pulse exceeding 12% of a generation (6–7 min) and that with longer pulses the rate of turnover increased with pulse time. Thus the fraction of wall immune to turnover synthesized during a pulse of 12% of a generation was in good agreement with the fraction immune to turnover as determined from prolonged periods of chase. The correlation of these two sets of data is consistent with the hypothetical presence of two wall peptidoglycan products, as mentioned earlier.

Based on the ultrastructural observations of loss of less wall from the polar regions of *L. acidophilus* (Boothby et al., 1973) and of two species of *Bacilli* (Frehel et al., 1971), and the observations of Fan et al. (1972) that the poles of *B. subtilis* are less susceptible to dissolution by the autolytic amidase of *B. subtilis*, it is tempting to assume that polar or septal wall is resistant to turnover and is perhaps assembled by a kinetically different process. It should be noted that 10 to 20% of the wall would probably be sufficient to cover the poles of rod-shaped bacterial species. There are alternate explanations (see below) and such an assumption would require more direct experimental evidence.

Pooley (1976 a,b) examined wall turnover of *B. subtilis* 168 and of a mutant of this strain Ni 15, which showed a substantially reduced rate of wall turnover but grew at the same rate and to the same extent as the parent strain. He used ^{14}C-N-acetylglucosamine (GlcNAc) to label both the peptidoglycan and teichoic acid polymers in the wall. Specific incorporation of GlcNAc into walls and not into other cellular fractions permitted direct quantitation of both the appearance of nondialyzable wall turnover products in the growth medium (released from cells) and radioactivity remaining in cells (in the wall), or after cold TCA and proteinase treatment, in the peptidoglycan fraction. Upon chasing an extensively labeled culture, loss of ^{14}C-GlcNAc from cells began slowly but, after about 2 generations of chase, appeared to follow first order kinetics. The amounts of nondialyzable labeled products found in the medium accounted for the ob-

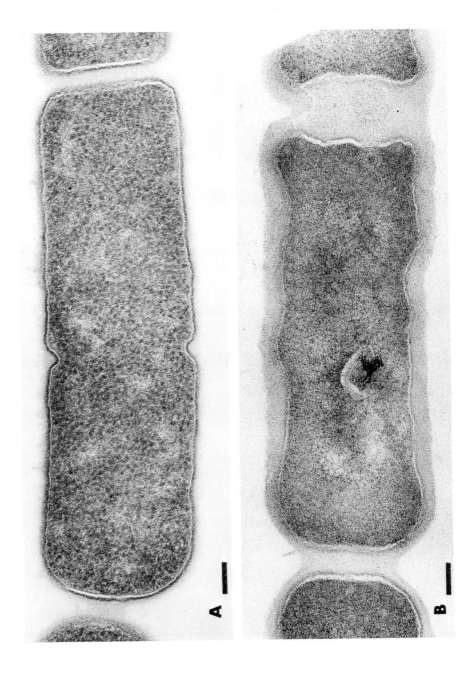

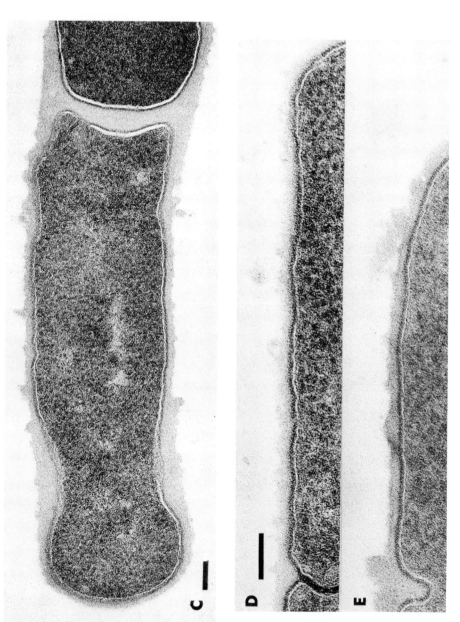

Fig. 3. Parts (a)-(e): See legend on p. 622.

served losses from the cells. Quantitation of ^{14}C-GlcNAc labeled nondialyzable products in the medium in continuously labeled cultures showed that amounts of these products increased exponentially and represented a constant 11 to 12% of the label in (the wall fraction of) the cells. Similarly, the amounts of released products in chased cultures increased exponentially for over 1 generation. These results suggested that as cells grow they continuously lose a constant fraction (about 8%) of their wall substance. Thus, quantitation of turnover by measuring the amount of label remaining in cells after a chase does not yield a true reflection of actual turnover rates. Chase of a culture pulsed for 0.15 generations (3 min) showed a lag of about 1½ generations before a linear decrease in remaining label was seen. The long lag (1½ generations) before loss of label from the pulsed culture suggests that the loss of a constant fraction of wall via turnover is from a portion of the wall that had matured for 1½ generations. A change in dissolution rate of wall substrate by the amidase does not explain this aging process.

Pooley postulated that a layer of wall assembled on the inside of the wall at or near the membrane must pass out through the thickness of the wall to reach the outer surface, where it is then accessible to amidase hydrolysis and loss from the cell. His data suggest that this migratory process takes about 1½ generations. Evidence in support of this hypothesis came from an experiment in which a culture of the Ni 15 strain was pulsed for 0.2 generations (4 min) with ^{14}C-GlcNAc and then chased. In this case SDS-inactivated walls were isolated from cells 0, ½, and 2 generations after the chase, and the kinetics of label release and dissolution of the bulk of the walls by added autolysin from the parent strain were compared (Fig. 4). Amidase action on walls isolated from cells at the start of the chase resulted in an immediate loss of ^{14}C, which closely

---

Fig. 3. (See pp. 620–621.) Electron micrographs of thin sections of *Lactobacillus acidophilus*.

(a) Section of a cell taken from a sample of exponentially growing culture. The wall is thin, densely staining only at the very inner portion, and the remainder appears loosely organized with a very irregular outer profile. (b) Section of a cell taken from a culture deprived of valine for 5 hours. The walls on such cells are thickened, have a fairly smooth and regular outer profile and an irregular, wavelike inner profile, suggesting a lack of uniformity of the wall thickening process on the cell surface. (c) Section of a cell taken from a culture deprived of valine for 5 hours and then allowed to resume growth in a complete medium until the cell mass had increased by 50%. The wall on this cell remains thickened only at a restricted area of the cylindrical portion of the wall and at the pole on the right. Note that, like the wall on the starved cell, the diameter of the cell is about the same over its entire length, and the membrane underlying the thickened section of wall appears to be pushed into the cytoplasm. The smoother outer contour of the thickened wall appears to be lost, except at the cell poles, and pieces of wall appear to be sloughing off from the wall surface. (d) Higher magnification of a section from an exponential phase culture. The thin, irregular outer contour and apparent sloughing off of wall pieces can be seen. (e) Higher magnification of a section from a culture recovering from 5 hours of valine deprivation for 0.5 mass doublings. Most of the cylindrical portion of the wall has been thinned out. Wall still thickened can be seen at extreme left over the nascent cross wall, and slightly thickened and smoother contoured wall on the pole at the right. Bars in (a) through (d) equal 100 nm. The bar in (d) also applies to (e). (Reproduced with permission from Boothby et al. [1973]).

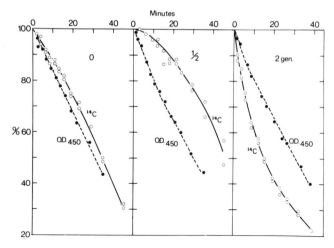

Fig. 4. Variation in susceptibility of cell wall material of different ages to autolytic attack in isolated SLS-wall preparations. *B. subtilis* Ni 15 was grown in 12 ml of medium containing 50 μM NAG to an O.D. of 0.19, before being filtered. The cells were resuspended in 12 ml of medium containing 10 μM ^{14}C-NAG (spec. act. 0.17 μC/μM). After 0.2 generations (4 min) 4 ml of the culture was filtered, the cells mixed with carrier cells, and SLS-walls prepared. The remaining 8 ml was filtered (at 4½ min), the cells were resuspended in 25 ml of unlabeled medium, and growth was allowed to continue. Samples (10 ml) were filtered at ½ generation (10½ min) and 2 generations (41 min) after resuspension in nonradioactive medium. An equal amount of carrier cells (ca. 0.15g dry wt.) was added to all samples and SLS walls prepared. 6 mg of walls from all samples, (0, ½ and 2 generations) were lysed by the addition of 0.16 ml of a preparation of soluble autolysin (20 mg autolyzed wall/ml). The insoluble radioactivity (○) and O.D. $_{450}$ (●) as a percent of zero time values, are plotted against time. 100% radioactivity was 1.25, 2.1, and 2.2 × 10^4 dpm/mg of wall for the samples isolated at zero time, at ½, and at 2 generations, respectively. (Reproduced with permission from Pooley [1976b]).

paralleled the decrease in wall turbidity. After ½ generation of chase, a loss of ^{14}C during the early stages of wall dissolution was very much slower than the loss of wall turbidity. This slow phase was followed by an increasing rate of ^{14}C loss. Comparisons of these two parts of the experiment suggest that after ½ generation of chase the pulse-labeled portion of the wall was not as immediately available for hydrolysis by the amidase as it was just after its incorporation into the wall. The labeled portion of wall was sequestered even in isolated walls. After 2 generations of chase the situation was reversed. In this case, release of ^{14}C was substantially faster than loss of wall turbidity, suggesting that the pulse-labeled wall substrate was even more available for immediate hydrolysis by the amidase than it was just after it was incorporated into the wall at the start of the chase. A topological location of newly incorporated, pulse-labeled wall as a layer on the inner wall surface, followed by the synthesis of additional layers of unlabeled wall to "cover up" the labeled wall during the chase, accompanied by loss of outer wall surface by turnover, would eventually result in the pulse-labeled and chased wall appearing on the outer wall surface and being subject to immediate hydrolysis by the amidase. In intact and growing cells, a layer of wall on the inner wall surface would not be available to bind and to be hydrolyzed by an externally present (or added) enzyme. However, the inner wall surface would be available in isolated wall preparations. Thus, newly assembled wall appears to migrate from the inner to the outer wall surface.

Based on the observation that hydrolysis of wall 2 generations old was more rapid and immediate than that of newly incorporated wall, and on other observations, Pooley proposed that older wall occupies a greater fraction of the total external wall surface than it did when it was newly assembled on the inner wall surface. The slow and gradual "spreading out" of a layer of wall as it is "pushed" out toward the cell surface by more recently assembled wall would suggest a considerable, but apparently slow, mobility within the wall. This "spreading" hypothesis does not take into account the possible effects of modification to peptidoglycan after it is incorporated into the wall, as discussed in section 7 (some of which could be related to, and be an integral part of, the migration and spreading processes), nor the effects on apparent rate of hydrolysis by autolytic peptidoglycan hydrolases of factors such as the modifier protein of Herbold and Glaser (1975a,b), the so-called "allosteric" effects of various wall polymers, or changes in susceptibility of a portion of the peptidoglycan to enzymatic hydrolysis and dissolution depending on the exact chemical structure of that portion of the peptidoglycan. Rates of dissolution or loss of insoluble wall are not always equivalent to rates of enzymatic hydrolysis of susceptible bonds. For example, in B. subtilis, post-incorporation modifications over 2 generations could include hydrolysis of amide (or other) bonds so that hydrolysis of the same number of amide bonds on the external wall surface could result in the more rapid solubilization of labeled wall fragments. It should be noted here that the products of amidase action (N-acetylmuramic acid residues with unsubstituted carboxyl groups and N-terminal L-alanine) have not been found in readily detected amounts in isolated walls of B. subtilis (Warth and Strominger, 1971). However, it is difficult to eliminate the possibility of loss of such unsubstituted groups via a small amount of autolytic wall hydrolysis during wall isolation by the methods normally used. Also, these methods may fail to preserve loosely attached wall fragments. Additional experimental data are required.

Experiments of Archibald and Coapes (1976) generally agree with Pooley's idea that the cell wall of B. subtilis is first assembled as a layer on the inside of the wall, which is then pushed out toward the outer surface by subsequently assembled layers. When cultures of B. subtilis W23 were shifted from phosphate to potassium limitation, the ability to adsorb phage SP50 increased exponentially with wall teichoic acid content. However, after a shift in the reverse direction (K^+ to PO_4^{3-} limitation), maximum phage-binding capacity was retained until the wall teichoic acid content fell to a low level. Since phage bind to the outside of the wall, these results were interpreted as consistent with the assembly of unexposed teichoic acid containing wall that was later exposed on the surface by the combination of turnover and synthesis of additional wall layers. In addition, Archibald and Coapes observed that 2 to 2½ hours after the shift from phosphate to potassium limitation, phage particles adsorbed to the cylindrical portion of the wall and not to the poles. At earlier times (1 to 1½ hr) some cells were seen with phage mainly at regions of the cylinder where cross-wall formation might be expected to occur. Four to five hours after the reverse transition, phage bound to the entire surface were seen, while samples at later times showed phage mainly at one or both poles. These results were interpreted as indicating that incorporation of new material proceeded less rapidly at the poles. However, a contributing factor, could be that wall turnover is slower at the poles. Both interpretations are in agreement with other types of information (Frehel et al., 1971; Fan et al., 1972; Fan and Beckman, 1973) that are consistent with a difference between cylindrical and polar wall in *B. subtilis*. However, density shift experiments (Fan et al., 1974) suggest that polar wall of *B. subtilis* turns over extensively.

Wall turnover has been observed in a variety of Gram-positive and Gram-negative bacterial species of both coccal and rod shapes. In addition to the examples cited above, wall turnover has been observed in other species and strains of *Bacilli* (Chaloupka et al., 1962 a,b; 1964; Chaloupka, 1967; Chaloupka and Kreckova, 1971), in *Neisseria gonorrhoeae* (Hebeler and Young, 1976), and in a DAP-dependent mutant of *E. coli* (Chaloupka and Strnadova, 1972). However, there are also many species and strains in which wall turnover does not seem to occur. This assortment includes the Lys⁻, DAP⁻ mutant of *B. megaterium* and the Ni 15 mutant of *B. subtilis* 168 mentioned above, a mutant of *B. subtilis* W23 that contains a normal complement of lytic enzyme but in which wall turnover equivalent to a loss of only 5% of the wall per generation was observed (Mauck and Glaser, 1971, unpublished results quoted in Fiedler and Glaser, 1973), *E. coli* (Van Tuibergen and Setlow, 1961; Rothfield and Pearlman-Kothencz, 1969), *S. faecalis* (Boothby et al., 1973), and *Streptococcus mutans* (Mychajlonka and Shockman, 1976). Evidence suggesting that the wall of *D. pneumonia* is conserved and does not turn over, at least at any substantial rate, has also been obtained (Briles and Tomasz, 1970; Tomasz et al., 1975). With *S. faecalis,* turnover of pulses or of extensively labeled peptidoglycan were below detectable levels during either exponential growth or recovery from amino acid starvation. Furthermore, turnover could not be induced either by growth in a low concentration of penicillin or during growth after wall damage inflicted by a brief exposure of cells to a low concentration of lysozyme. Thus, it seems clear that wall turnover is not an essential feature of peptidoglycan synthesis in vivo, or for cell surface enlargement. In those species in which it occurs, the role of wall turnover in the economy of the cell and the possible selective advantage that turnover may confer remain to be revealed.

At present, the diversity and paucity of informaton on the various wall synthetic and turnover systems make it extremely difficult to reach even tentative conclusions concerning the general properties of turnover; where, when, and in what species it occurs, and especially its relationship to cell shape, shape change, surface growth and division. Undoubtedly, as first pointed out by Chaloupka et al. (1962a), the existence of wall turnover in some species indicates that assembled wall is not inert but, as discussed above, and in sections 4.4. and 7, undergoes metabolic changes. Clearly, it seems that amidase-catalyzed turnover in various species of *Bacilli,* and N-acetylmuramidase action in *L. acidophilus,* removes material from the outer wall surface. This occurs despite the apparent absence of detectable autolytic enzyme activity in the supernatant growth medium (Mauck et al., 1971; Brown et al., 1976). This topology requires action of a hydrolytic enzyme at some distance from the outside of the cellular permeability barrier in a highly regulated manner related to the rates of growth and cellular peptidoglycan synthesis. The regulatory mechanisms involved are obviously efficient in not permitting loss of osmotic protection and consequent cell lysis.

Of interest are recent observations (Pooley, personal communication) that growth of the Ni 15 mutant of *B. subtilis* (which exhibits a very low turnover rate) in the presence of a crude preparation of the amidase results in about a 50% increase in overall rate of peptidoglycan synthesis. An apparent relationship between overall rate of wall assembly and wall loss via turnover suggests that growing cells of *B. subtilis* are capable of regulating the thickness of their walls by some unknown mechanism.

The ability of cells to maintain a desired wall thickness, especially in relation to their rate of wall turnover, suggests that wall turnover may be more closely related to wall thickening than it is to wall surface enlargement. As discussed in section 4.5, in Streptococci, wall thickening over much of the cell surface occurs concomitantly with surface enlargement, which is largely localized to the septal region. Wall turnover has not been detected in *S. faecalis* (Boothby et al., 1973). Perhaps success in observing localized surface enlargement in streptococci can be, at least in part, attributed to a relatively large fraction of overall wall assembly dedicated to enlargement in the absence of turnover. A substantial decrease in the fraction of wall assembly directed toward enlargement due to wall loss via turnover, for example, would present increased difficulties in the detection of localized processes. Thus, it seems possible that the occurrence of a rapid rate of wall turnover could obscure the detection of a small fraction of wall assembly directed toward localized surface enlargement, not only because of lack of conservation of assembled wall but also because of its consumption of biosynthetic activity.

The occurrence of wall turnover and the consequent changes in topological localization within walls must be considered along with secondary modifications of the chemical structure of the wall in the interpretation of many types of investigations of wall assembly, wall topology, and morphogenetic changes in wall shape.

4.4.2. Turnover of other surface components
Polymers covalently linked to the peptidoglycan, such as the teichoic acids, teichuronic acids, and neutral polysaccharides of Gram-positive bacteria, and peptidoglycan-bound lipoprotein of Gram-negative species would be expected to turn over with the peptidoglycan. This has been shown to be the case in *B. subtilis* (Ellwood and Tempest, 1969; Mauck et al., 1971). Lipopolysaccharide appears to be lost from the outer membrane during exponential growth of *E. coli*

(Rothfield and Pearlman-Kothencz, 1969), apparently as a protein-lipopolysaccharide complex, in the absence of loss of the underlying peptidoglycan. Although some information is available about the turnover of individual lipids, little information seems to exist on the turnover of total membrane lipids or proteins. In general, the rates of membrane lipid and protein turnover are low (summarized by Mindich, 1973). However, changes in amounts of specific membrane lipids and proteins are thought to be important during the cell cycle. Recently evidence has been obtained that the membrane-associated lipoteichoic acids of several bacterial species are excreted into the growth medium (Joseph and Shockman, 1975; Markham et al., 1975). In some species, such as *S. mutans* and several *Lactobacillus* species, large amounts of fully acylated lipoteichoic acid are excreted even during exponential growth. Loss of acylated lipoteichoic acid indicates that the glycolipid end of the lipoteichoic acid (Wicken and Knox, 1975) complete with its fatty acids, presumably anchored in the membrane, is lost from the cell. In other species, such as *S. faecalis*, essentially all the material isolated from the growth medium is as the deacylated form. During exponential growth of this species about 15% of the total of polyglycerol phosphate polymer was found in the growth medium. In stationary phase cultures of either *S. faecalis* or *S. mutans*, a much greater fraction of the total polyglycerol phosphate polymers were found extracellularly. The significance of these observations remains to be determined. The roles for lipoteichoic acid as a carrier in wall teichoic acid synthesis (Fiedler and Glaser, 1974 a,b,c) and as an inhibitor of autolytic enzyme activity (Cleveland et al., 1975) suggest that excretion of these macromolecules may play an important regulatory role in surface growth and division. However, results of preliminary experiments concerned with the kinetics of synthesis and excretion of lipoteichoic acids are not consistent with the excretion of a once used carrier in the assembly of wall polymers.

*4.5. Cellular localization of assembly
of peptidoglycan and other cell wall polymers*

The transfer of peptidoglycan precursors synthesized on the membrane to the wall acceptor must involve a very intimate association of these two fairly fixed structures. Although it is possible that there may be specific membrane units more intimately associated with the wall, in which the enzymes, substrates, carriers, and acceptors are concentrated (Mirelman et al., 1971), it seems probable that such sites are dispersed over the entire membrane surface. Similarly, since new teichoic or teichuronic acid chains are added only to concurrently assembled chains of peptidoglycan in *B. subtilis* (Mauck and Glaser, 1972), at about the same time the peptidoglycan chains are inserted into the wall, the synthesis of nonpeptidoglycan wall polymers, covalently linked to peptidoglycan must be closely integrated with peptidoglycan assembly. The assembly of wall polymers not covalently linked to peptidoglycan, or found in both a covalently linked and a free form, such as the lipopolysaccharide and lipoprotein of Gram-negative species, may occur in a more independent manner than peptidoglycan assembly.

Good evidence that the capacity to assemble a full assortment of wall polymers exists in a large number of sites distributed over the entire surface of the protoplast membrane of gram-positive bacteria is the ability of many Gram-positive species of different shapes or modes of division to thicken their walls. Wall thickening has been most prominently observed in cells after inhibition of protein synthesis and occurs over the entire cell surface (e.g., Fig. 3), including the poles of rod-shaped species (e.g., Hughes et al., 1970; Chung, 1971; Frehel et al., 1971; Boothby et al., 1973). Expression of the wall-thickening capacity is not limited to situations where wall synthesis is uncoupled from the synthesis of other cellular macromolecules. Wall thickening has been shown to occur in rapidly growing cultures of *S. faecalis* (Higgins and Shockman, 1970a) and *B. megaterium* (Ellar, et al., 1967). Other evidence consistent with the assembly of wall polymers at many sites on the wall surface was obtained in experiments on the reversion of protoplasts to bacterial forms. With both *B. subtilis* (Landman et al., 1968) and *B. licheniformis* (Elliott et al., 1975), reversion of protoplasts occurs by the assembly of an ultrastructurally recognizable wall all around pleomorphic cells well before typical rod-shaped bacilli were observed.

The potential capacity to assemble wall polymers at a large number of sites on all or most of the wall surface favors the view that cell surface enlargement occurs by diffuse intercalation. However, such a process would be inconsistent with the integration of cell surface enlargement with chromosome replication, segretion, and cell division (see section 9.7 for a more complete discussion). On the basis of occurrence of wall thickening along with topologically localized wall assembly directed toward surface enlargement in streptococci (section 10.2), the hypothesis can be made that perhaps both types of processes occur in all, or at least in some, Gram-positive bacteria. The relative fractions of total wall synthetic activity directed toward each type of assembly would then contribute to the experimentally obtained data. For example, in a species or strain that exhibits a great deal of wall turnover, one would expect that a greater fraction of wall synthetic activity would be devoted to wall thickening than to wall surface enlargment. The recent results of Pooley (section 4.4.1) would be consistent with such a hypothesis. This interpretation would explain the conflicting data of Hughes and Stokes (1971), who observed localized wall insertion in a lytic defective strain of *B. licheniformis* (which probably has a low rate of wall turnover) by an immunofluorescence technique, and the data of several other investigators (e.g., Chung and Hawirko, 1964; Mauck et al., 1972; de Chastellier et al., 1975 a,b; Archibald and Coapes, 1976) who have used several different techniques with strains of *Bacilli* that probably exhibit turnover, and who conclude that wall growth is not localized.

5. Autolytic peptidoglycan hydrolases

A variety of bacterial species have been shown to possess one or more enzyme activities that are capable of hydrolyzing specific bonds in their own peptidoglycan. Hydrolysis of bonds that are not directly involved in the formation of the two- or three-dimensional peptidoglycan network would simply result in (localized) modification of the peptidoglycan structure. For example, hydrolysis of the bond between terminal and penultimate D-alanine residues on peptide side chains that are not cross-linked by a DD-carboxypeptidase (carboxypeptidase I) or between D-alanine and L-lysine or DAP (on the L center of DAP) by an LD-carboxypeptidase (carboxypeptidase II) would not affect the structural integrity of the peptidoglycan. The sole effect of the action of these enzymes on a susceptible bond would be to remove a potential substrate for the action of other enzymes (e.g., the respective transpeptidases). Thus, such activities could be important in the topology of post-incorporation modifications of peptidoglycans (section 7). In contrast, hydrolysis of a sufficient number of bonds that are directly involved in the maintenance of the two- or three- dimensional integrity of the wall in an intact cell will result in loss of osmotic protection and, lacking a sufficient concentration of a nonpermeable solute, in cellular autolysis. For this reason, such autolytic activities (autolysins) are potentially lethal and require close regulation. In a number of bacterial species the capacity to autolyze is maximal, or near maximal, during exponential growth (e.g., Mitchell and Moyle, 1957; Shockman, 1965; Young, 1966; Coyette and Ghuysen, 1970). This correlation was the basis of the suggestions (Shockman et al., 1958; Weidel and Pelzer, 1964; Shockman, 1965) that such activities may play some role in cell wall growth.

The capacity to autolyze is not usually visibly expressed by cells when they are growing. Expression may be observed under conditions involving some sort of unbalanced growth, frequently inhibition of continued peptidoglycan synthesis. For example, the addition of specific inhibitors of peptidoglycan synthesis, such as one of the penicillins, vancomycin, or D-cycloserine, or deprivation of a nutritionally required peptidoglycan precursor such as lysine, DAP, alanine, glutamate, glucose, or an amino sugar may result in autolysis of a culture when conditions of pH, salt, concentration, and so on are suitable (Shockman, 1965). Other conditions, such as treatment with detergents (e.g., deoxycholate) or oxygen deprivation for a highly aerobic organism, may "activate" cellular autolysis in some species. Although inhibition of continued peptidoglycan synthesis in growing cultures frequently leads to expression of autolytic capacity, the connection is not a direct one. Clearly, it is not simply a matter of the cell growing out of its protective cell wall. While specific inhibition of protein synthesis results in very rapid development of resistance to autolysis (Shockman et al., 1961 a,b; Shockman, 1965; Pooley and Shockman, 1970; Sayare et al., 1972), concomitant inhibition of both protein and peptidoglycan synthesis in S. faecalis, as by deprivation of an amino acid (e.g., lysine) required for both, can result in the onset of cellular autolysis (Shockman et al., 1958, 1961 a,b). Rogers (1967), in studies of the killing of staphylococci, and more recently Rogers and Forsberg (1971) with lytic defective mutants of B. licheniformis and Tomasz et al. (1970) with lytic defective mutants and nutritionally induced lytic defects in the autolytic system of pneumococci, also concluded that inhibition of peptidoglycan synthesis does not necessarily result in cellular lysis and/or cell death.

5.1. Potential roles of autolysins in cell wall assembly, surface enlargement, morphogenesis, and cell division

Several roles for autolytic peptidoglycan hydrolases in cell wall assembly, surface enlargement, and cell division have been proposed (summarized in Higgins and Shockman, 1971; Ghuysen and Shockman, 1973). These potential roles can be grouped as follows: (1) cell wall turnover, (section 4.4.1); (2) cell separation, the final, but not necessarily required, separation of the two daughter cells into two completely distinct units; (3) hydrolytic action to provide new acceptor sites for the addition of peptidoglycan precursors; (4) the potential biosynthetic capacity of a hydrolytic activity; (5) a role in cell division, that is, in compartmentalization into new cell units separated by both wall and membrane; and (6) the hydrolysis of bonds in the peptidoglycan in selected topological areas of the wall so that changes in cell shape may occur ("remodeling function," see also section 7). There is some indirect, experimental evidence for participation of one or more autolytic peptidoglycan hydrolases in each of the processes mentioned above.

Autolytic peptidoglycan hydrolases of different specificities have been detected in various bacterial species (Ghuysen and Shockman, 1973; Ghuysen, 1976). A variety of techniques, including analyses of the products of wall hydrolysis by the autolytic peptidoglycan hydrolase and by other added enzymes, show the presence of only one activity, a glycan-chain splitting, N-acetylmuramyl glycan hydrolase (muramidase) in isolated walls or cells of *S. faecalis* (Ghuysen et al., 1967; Shockman et al., 1967b; Dezéllée and Shockman, 1975; Rosenthal and Shockman, 1975; Rosenthal et al., 1975a,b). Since many of the experiments cited used highly sensitive, double radioactive labeling techniques, products of the activity of any enzyme that would separate the amino sugars in the glycan strands from the amino acids in the peptides (e.g., amidases or peptidases) would have been detected. In contrast to the single activity, and apparently to the presence of a muramidase (an enzymatic specificity that has only been found in a few other bacterial species), many bacteria possess peptidoglycan hydrolases of more than one specificity. The most frequently described and investigated activity is the N-acetylmuramyl-L-alanine amidase (amidase). This activity has been found in species in which cell surface growth and cell division has been most extensively investigated, such as *E. coli, B. subtilis* and other *Bacilli, S. pneumoniae,* and *S. aureus,* and has often been accompanied by other peptidoglycan hydrolase activities. The presence of peptidoglycan hydrolases of multiple specificities in the same species has made it difficult to unravel the potential role of each.

Pleiotropic effects and other evidence could suggest that autolysins may have multiple roles. Furthermore, the absolute necessity for the presence of an autolytic peptidoglycan hydrolase remains to be proven. For example, several bacterial species including some streptococci do not seem to be killed by, or be induced to autolyze by, antibiotics that inhibit cell wall synthesis. Similarly, the lytic-defective mutant of the pneumococcus isolated by Lacks (1970) is resistant to both penicillin and deoxycholate-induced lysis but appears to be normal in other respects (Tomasz et al., 1970). In contrast, a lytic-defective mutant of *S. aureus* appears to show identical kinetics of killing by penicillin G (Chatterjee et al., 1976). However, inability to detect the presence of a potentially lethal activity in cells may be only a reflection of the kind and extent of its regulation.

5.1.1 Role in cell wall turnover

Amidase activities in *Bacilli* (Mauck et al., 1971) and *S. aureus* (Chatterjee et al., 1976) and a muramidase activity in *L. acidophilus* (Boothby et al., 1973) appear to be responsible for wall turnover. However, turnover does not appear to be essential to wall surface growth (section 4.4.1).

5.1.2. Role in cell separation

This process need not be well regulated since immediate or precise timing of cell separation is not essential to growth and division. Despite the relatively trivial nature of cell separation, evidence has been obtained for a role of presumably amidase activities in cell separation in various *Bacillus* species (Fan, 1970; Forsberg and Rogers, 1971), *S. aureus* (Chatterjee et al., 1969, 1976), *S. pneumoniae* (Tomasz, 1968) as well as in other bacterial species. Lack of a requirement for autolysin activity in cell separation was obtained in studies of the DOC⁻ mutant of pneumococcus (Lacks, 1970; Tomasz and Waks, 1975). This mutant contains a very reduced level of amidase activity but appears to grow at the same rate and in the same manner as the wild type. It does not autolyze in the presence of deoxycholate after it reaches the end of the exponential growth phase, or when growing cultures are exposed to an antibiotic which inhibits peptidoglycan synthesis. All of these characteristics are exhibited by the wild type. Also, the DOC⁻ mutant apparently does not grow in long chains of cells. Growth in very long chains of cells, in addition to the other characteristics listed above (plus lack of competence for transformation), is exhibited by the parent pneumococcal strain when grown in the presence of ethanolamine in place of choline (Tomasz, 1968). Walls from ethanolamine grown cells are resistant to hydrolysis by crude amidase preparations (Mosser and Tomasz, 1970). However, susceptibility to amidase is restored by methylation of the walls, and conversion of the ethanolamine residues to choline (Höltje and Tomasz, 1975b). Thus, although amidase action appears to require an interaction with choline residues in walls, action of the amidase does not appear to be required for reasonably good cell separation by the DOC⁻ mutant. Of course, it is difficult to demonstrate the total absence of any enzymatic activity, including that of an autolysin, so that it could be supposed that the DOC⁻ mutant still possessed residual, but undetected, levels of an autolysin, either the amidase or an enzyme of another specificity, sufficient to restore cell separation. Since the DOC⁻ characteristic (apparent amidase defect) in the strains used was introduced into the wild type by transformation, amidase activity may not be responsible for cell separation unless the transfer was of a structurally defective enzyme that possesses some residual activity. On the other hand, the activity of an autolysin of a different, as yet undetected, specificity would also have to possess the property of lack of activity on ethanolamine-containing walls.

Considerable light has been shed on the role of autolysins in cell separation of *B. subtilis* 168 by the recent work of Fein and Rogers (1976). Mutants that produce 5 to 10% of both N-acetyl-L-alanine amidase and N-acetylglucosaminidase activities were isolated. The mutants grew at the same rate as the wild type and had walls identical in chemical composition and in susceptibility to the action of both autolytic activities of the wild type. However, several properties of these mutants differed from those of the wild type, including growth in long chains of unseparated bacilli that were nonmotile and lacked flagella. The mutations were transferred via transformation to a well-recognized strain. Comparisons of the transformants with a constructed, isogenic wild-type strain, the ready reversion of the mutants to the wild-type phenotype, and other data, were consistent with attributing the observed phenotype to a single mutation. However, these investigators carefully state that multiple mutations have not been rigorously ruled out. Since the mutation resulted in deficiencies of two enzyme activities, they proposed that these mutations are either regulatory or result in a change in export of the enzymes. Since growth of the mutant in the presence of the wild type resulted in shorter chains, they believe it seems unlikely that the phenotype of the mutant is due to the hyperproduction of inhibitor(s) of the autolysins.

In terms of a cell division cycle, a role for autolysin in the final stages of cell division or cell separation would require activity toward, and at, the very end of the cell cycle. Since most cells begin separation at the periphery of the nearly completed cross wall, hydrolytic action should commence centripetally from the outside in. The observation of Fan (1970) that chains of septated cells of *B. subtilis* can be separated into individual cells by the addition of crude autolysin of *B. subtilis* or hen egg-white lysozyme, and, of Higgins et al. (1973), that autolysis of *L. acidophilus* in buffer is preceded by an increase in cell number and the earliest change observed by electron microscopy was loss of wall from the periphery of cross walls at all stages of development, are consistent with such a role. Both observations suggest that cross walls are substantially more susceptible to hydrolysis by their respective autolysins, and in the case of *B. subtilis* to hen egg-white lysozyme, than peripheral wall. Attempts at correlating cellular autolytic activity and the chromosome replication and cell division cycles are discussed in section 9.5.

5.1.3. Role in the provision of new acceptor sites and potential biosynthetic capacity

Enzymes of only two specificities, N-acetylmuramidases, and some bridge-splitting peptidases would provide new acceptor sites consistent with current knowledge of the insertion mechanisms of new glycan chains into peptidoglycan by transpeptidation or transglycosylation (see chapter 8, this volume, for details). Although not yet experimentally established, it is possible that the hydrolytic activities observed are the first half of respective transfer reactions. Hen egg-white lysozyme has been shown to carry out transglycosylations (Sharon and

Seifter, 1964; Chipman et al., 1968; Pollock and Sharon, 1970) and some enzymes can carry out both carboxypeptidation and transpeptidation reactions (see chapter 8 for details). In this regard, because of the chemical structure of the peptidoglycan substrate, both the KM endopeptidase of *Streptomyces albus G* and the *E. coli* endopeptidase can be regarded as carboxypeptidases (Ghuysen and Shockman, 1973). The potential roles of transfer reactions in possible postincorporation modifications of peptidoglycan structure are discussed in section 7. Clearly, the frequently found amidase and N-acetylglucosaminidase activities cannot provide suitable acceptor ends.

A sufficient number of acceptor sites, in the form of nonreducing N-acetylglucosamine and C (or N) terminal amino acids seem to be present in walls. The occurrence of peptidoglycan and wall assembly in the absence of an association with autolytic activity during cell wall thickening (Pooley and Shockman, 1970) is inconsistent with an overall lack of acceptor sites. Thus, a role of providing new acceptor sites would probably be more important in terms of surface topology, and the number, location, and suitability of such sites to result in surface enlargement, division, and shape determination.

5.1.4. Role in cell division
The technical problems of distinguishing between cell separation and division have been mentioned earlier. In addition, there are problems in correlating the timing of various events in the cell division cycle (see section 9.1). In rod-shaped species, the time at which cross walls are formed and separated, relative to cylindrical elongation, to become the poles of two daughter cells, will contribute to the determination of cell shape, and the sequence of changes in surface area to volume ratios. In addition, the geometry of cross-wall formation, that is, the extent of cross-wall separation relative to centripetal penetration of the cross wall, will affect the degree of curvature of nascent poles and thus contribute to the morphopoetic sequence observed. This aspect then overlaps with shape change and remodeling.

5.1.5. Role in remodeling functions
Two aspects of remodeling can be considered. Both relate to the postincorporation modifications to peptidoglycan structure considered in section 7 below. Since the peptidoglycan is the shape-maintaining polymer, localized changes in its structure and organization would be expected to result in changes in cell shape. The occurrence of such changes in topologically oriented portions of the wall at precise times in the cell cycle could contribute to the observed sequence of shape change, cell division, and cell separation. One type of role could simply be the hydrolysis of specific bonds at specific topological locations. Such changes could result in a modification of wall morphology. The other role could involve potential biosynthetic (transferase) capacities.

The classification of potential roles discussed above is obviously artificial. There are overlaps. In addition, any single enzyme activity could easily play more than one role. The pleiotropy of some mutants that appear to be defective in more than one function is consistent with multiple roles as are the multiple effects of nutritionally replacing choline with ethanolamine in the pneumococcus. Perhaps the most fruitful and necessary approach would be investigations of mutants known to have a defect in the structural gene(s) specifying an autolytic enzyme protein. To date, unequivocal evidence for the existence of such mutants has not been obtained.

5.2. Cellular localization of autolytic enzyme activities

Obviously, the location of the insoluble and relatively large wall substrate on the outside of the permeability barrier results in requirements for the enzyme to be exocellular at some stage and to have a certain degree of mobility to seek out susceptible bonds. Thus enzyme protein, synthesized on cytoplasmic ribosomes, must cross the membrane at some stage. In a few instances, such as the amidase of *S. aureus* (Wadstrom and Vesterberg, 1971), peptidoglycan hydrolases have been found in culture fluids. In most cases, including the amidase of *B. subtilis*, which is responsible for wall loss via turnover from the outside of the cell (section 4.4.1), little or no extracellular activity has been detected (Brown et al., 1976). Autolytic enzyme molecules lost from the cell may have a short half-life. Otherwise enzyme activity lost from one or a few cells in a culture via either excretion or cell lysis could result in lysis of all the cells in the cultures, especially if the enzyme can attack walls on growing cells from the outside, as appears to be the case for *B. subtilis*. Even with *S. faecalis*, where the autolysin appears to be unable to bind to the external wall surface (Joseph and Shockman, 1976), detection of very low levels of extracellular autolysin required the presence of radiolabeled walls in the medium (Cornett, personal communication). Such small amounts of extracellular activity could be important in cell separation.

Several autolysins, including those of *S. faecalis* (Shockman et al., 1967b); *L. acidophilus* (Coyette and Ghuysen, 1970), *S. aureus* (Singer et al., 1972), and *B. subtilis* (Young, 1966; Brown et al., 1970; Fan and Beckman, 1972), have been found to bind tightly to cell walls so that walls are isolated as a wall-enzyme complex after cell disruption. Spheroplasts of *B. subtilis* and *B. megaterium* appear to release their lytic enzyme into the medium (Glaser, 1973). Similarly, when protoplasts of *S. faecalis* prepared by the action of their own autolysin were then grown in an osmotically stabilized medium, nearly all of the autolysin activity was found to be excreted into the growth medium as the proteinase-activatable, latent form (Joseph and Shockman, unpublished observations). L-forms of *B. licheniformis* synthesized an amidase located in the membrane fraction (Forsberg and Ward, 1972). Addition of walls resulted in removal of this activity from the membrane. In all these cases the autolysins were localized on the outside of the membrane, either bound to membrane or in the wall fraction. In some cases the presence of autolysin activity bound to walls after cell disruption may not reflect its actual location in intact cells. For example, the muramidase activity of *L. acidophilus* was thought to be tightly bound to the wall. However, disruption of cells in 0.1M sodium phosphate, pH 7.8, resulted in a four- to fivefold increase in autolysin activity bound to walls. Furthermore, the addition of inactivated walls to the 12,000 × g soluble fraction after cell disruption in buffer resulted in the detection of additional autolysin activity that accounted for about one fourth of the total activity detected (Coyette and Shockman, 1973). In this organism it seems likely that nonwall-bound activity is present that can be detected only after binding to walls. Additional studies showed the presence of fewer binding sites for the enzyme on walls than the amount of activity present in cells.

A specific topological localization of autolytic enzyme activity has been demonstrated in only one case. In exponential phase cells of *S. faecalis*, several types of evidence strongly indicated that the muramidase activity was concentrated in recently synthesized portions of the wall (Shockman et al., 1967a) which, in turn, were shown to be at the septal region of the cell (Higgins et al., 1970). The extraordinarily high affinity of the enzyme for binding to walls (Pooley et al., 1970) was crucial to such experiments. Removal of autolysin activity from walls requires exposure to salt concentrations in excess of 4 M or to 0.01 M NaOH (Pooley et al., 1970). Cultures were either exposed to pulses of ^{14}C-lysine or grown in the presence of this label and then chased for increasing periods of time. Walls were isolated and the release of radioactivity and dissolution of the walls was followed. In all cases, the data obtained indicated that the active form of the wall-bound autolysin dissolved recently made wall before it attacked older, labeled or unlabeled wall (Shockman et al., 1967a; Pooley and Shockman, 1969). Controls for these experiments included the addition of isolated autolysin to inactivated pre- or postlabeled walls, and studies of release of the ^{14}C label from mixed wall suspensions. Lack of selective release of label in these series of controls showed that the location of the enzyme rather than increased susceptibility of a portion of the substrate was the important factor. Additional controls were studies of the release of the ^{14}C label from pre- and postlabeled walls in the presence of trypsin to activate the latent form of the enzyme. Selective loss of pre- and postlabels was not observed, suggesting that latent enzyme activity was randomly localized in the wall. However, recently obtained data discussed further on indicates that latent autolysin activity may also be localized.

Our interpretation of the above observations has been criticized on the basis that "... rearrangement during cell breakage cannot be ruled out entirely" (Glaser, 1973). However, in the experiments briefly described above, both localized susceptibility of the wall substrate to hydrolysis and a possible rearrangement of autolytic enzyme localization were ruled out experimentally. In the absence of a difference in substrate susceptibility, rearrangement would be expected to favor the observation of random, rather than selective wall loss and probably accounts for the random wall loss observed for the action of latent autolysin activity after trypsin activation. Recent experiments (Joseph and Shockman, 1976) that examined the localization of both active and latent autolysin in intact cells actually demonstrated that latent activity is also localized. Release of a D-^{14}C-alanine from pre- and postlabeled intact cells incubated in an osmotically protective buffer was consistent with localization of the active form of the muramidase in newly assembled wall. Furthermore, evidence was also obtained for the association of the latent form with recently made wall, although the correlation between latent activity and new wall was not as close as that observed either in the recent or earlier experiments for the active form. Along with the radiolabeling data ultrastructural observations of autolysin action in cells incubated in an osmotically protective environment showed that the septally located wall was lost before peripheral wall, in both the absence and presence of trypsin. Thus, in cells, the latent as well as the active form appears to be associated with nascent septa. In our earlier experiments, failure to detect an association of latent activity with new wall was probably due to "rearrangement," that is, release and random rebinding of latent activity during cell breakage and the extensive washing procedure used in those experiments. The affinity of the latent form for walls appears to be somewhat less than that of the active form. Additionally, autolytic loss of a small amount of septally associated wall containing a relatively high concentration of autolysin could easily have resulted in a localized saturation of binding sites and consequent release of some enzyme.

Localization of the action of amidase activity in S. pneumoniae has been demonstrated. In this case, however, localized susceptibility of the wall substrate appeared to be responsible (Mosser and Tomasz, 1970).

5.3. Regulation of peptidoglycan hydrolase activities

Different bacterial species seem to have developed a variety of mechanisms for regulating their peptidoglycan hydrolases. Thus it seems best to discuss separately some of the systems that have been relatively well studied.

5.3.1. The N-acetylmuramidase system of S. faecalis

The muramidase of S. faecalis appears to be synthesized in a latent form that can be activated by treatment of isolated wall-enzyme complex with any of a series of proteinases (Shockman et al., 1967b). Latent autolysin can also be activated by trypsin in intact cells incubated in phosphate or acetate buffers (Joseph and Shockman, 1974). Several types of experimental evidence are consistent with the passage of latent autolysin through the permeability barrier and its binding probably to the inner surface of the wall prior to activation (Pooley and Shockman 1969; 1970; Shockman and Cheney 1969; Joseph and Shockman, 1974; 1976). The requirement for transport and activation on the wall, plus the high affinity of the enzyme for walls and the inability of enzyme to bind to the external surface of cells, would help to prevent enzyme action on neighboring cells. The nature of the S. faecalis wall substrate, in terms of physiological age, degree of N- or O-acetylation, and presence or absence of accessory wall polymers, was found to affect the ease of enzymatic hydrolysis only to a small extent (Shockman et al., 1967b).

A role in cell surface growth and/or cell division would require that the activity of the enzyme be increased and decreased rapidly enough to exert its selective action during a relatively small fraction of a cell division cycle. While proteinase activation could serve to increase activity rapidly, separate mechanisms would be required to decrease activity and to regulate the synthesis, activity, or cellular localization of the proteinase. The capacity of cells to autolyze decreased with a half-life of 4 to 6 minutes after inhibition of protein synthesis in rapidly growing exponential phase cultures with tetracycline or chloramphenicol (Sayare et al., 1972), while the cellular content of both the latent and active forms of the autolysin remained virtually unchanged for many hours (Pooley and Shockman, 1970). This speed of inhibition could not be attributed to a mechanism that involved proteinase activation, or synthesis, or to loss of enzyme protein. Consistent with this view was the speed of recovery of the capacity of cells to autolyze after inhibition of protein synthesis (Sayare et al., 1972). During the early stages of reversal, autolytic capacity increased with doubling time of 20 minutes while cellular mass and protein synthesis had doubling times of about 80 minutes. These and other observations led to the hypothesis that the activity of the autolysin was influenced by a regulatory ligand.

Recently, evidence was obtained that fully acylated lipoteichoic acids, isolated and purified from several bacterial species, inhibited the hydrolysis of isolated cell walls by the autolysin (Cleveland et al., 1975). Deacylated lipoteichoic acid did not inhibit, suggesting that the fatty acids of the glycolipid moiety play a role. The succinylated lipomannan isolated from membranes of *M. lysodeikticus* (Owen and Salton, 1975; Pless et al., 1975) and Forssman antigen preparations of the pneumococcus (Briles and Tomasz, 1973) also failed to inhibit. Lipoteichoic acids also inhibited the action of the muramidase of *L. acidophilus* on walls of the same species and the action of the highly purified amidase of *B. subtilis* (Herbold and Glaser, 1975a) on walls of that species. Evidence was also obtained that certain lipids, notably diphosphatidyl glycerol, also inhibited the *S. faecalis* muramidase. Again, deacylation completely abolished inhibitory activity (Cleveland et al., 1976).

Lipoteichoic acids and diphosphatidyl glycerol are compounds associated with the membrane. The intimate relationship of autolysin, wall substrate, and membrane, together with the rapid mobility of lipids in membranes could provide an appropriately organized arrangement for the rapid, precise, and topological control of autolysin action.

Evidence consistent with a role for N-acetylmuramidase activity in the cell cycle of *S. faecalis* was obtained from studies of both synchronized cell populations and balanced exponential phase populations fractionated on exponential sucrose gradients (Hinks et al., 1974, 1976; Daneo-Moore, unpublished observations). With rapidly growing cultures ($t_d = 32$ min), smaller and younger cells had a somewhat greater capacity to autolyze than larger and older cells. Studies of more slowly growing cultures ($t_d = 75$–80 min) consistently showed a dramatic decrease in autolytic capacity beginning about 15 to 25 minutes before cell division (after the end of C and at about the beginning of D; section 9.2). The results of these and other experiments have been interpreted as being consistent with the participation of autolysin activity in cell surface enlargement and division in a closely regulated fashion (Shockman et al., 1974).

5.3.2. The amidase system of D. pneumoniae

The pneumococcal amidase appears to be regulated via the presence of choline residues on the wall teichoic acid (Tomasz et al., 1975), which appear to be involved in the conversion of the inactive "E form" of the enzyme to the active "C form" and in the provision of a wall substrate that can be hydrolyzed by the enzyme. Removal of choline residues by a teichoic acid phosphorylcholine esterase activity (Höltje and Tomasz, 1974) found in pneumococci could also play a role in regulating sites of autolysin activity on the wall. In addition, choline-containing Forssman antigen preparations inhibit the action of the amidase on pneumococcal walls (Höltje and Tomasz, 1975a).

5.3.3. The amidase system of B. subtilis

The activity of the amidase system of B. subtilis ATCC 6051 is modulated by a modifier protein (Herbold and Glaser, 1975a,b). This modifier protein stoichiometrically combines with the enzyme and changes the pattern of wall hydrolysis from one that is random to a sequential hydrolysis of amide bonds on individual glycan chains. In addition, wall teichoic acid seems to be required for binding the enzyme to the wall and the functional interaction of enzyme and modifier protein. A modifier protein appears to be lacking in a different strain of B. subtilis, W23, raising the possibility that this function may not be required in the amidase system of all bacilli. Also, action of the amidase of B. subtilis 6051 was inhibited by membrane-associated lipoteichoic acid (Cleveland et al., 1975), and tight binding of the enzyme to walls occurred only to teichoic acid-containing walls (Herbold and Glaser, 1975b).

5.3.4. The peptidoglycan hydrolases of E. coli

E. coli is known to contain a variety of enzyme activities capable of hydrolyzing bonds in its peptidoglycan. These activities include the following: LD-carboxypeptidase, DD-carboxypeptidase-endopeptidase, N-acetylmuramyl-L-alanine amidase, N-acetylglucosaminidase and a novel type of N-acetylmuramidase or transglycosylase that hydrolyzes the β, 1→4 linkage between N-acetylmuramic acid and N-acetylglucosamine and converts it into an intramolecular 1→6 anhydro-N-acetylmuramyl bond (Taylor et al., 1975; Höltje et al., 1975). Some of these hydrolytic activities, such as DD-carboxypeptidase, endopeptidase, and transpeptidase appear to be associated with the inner, cytoplasmic membrane (Hackenbeck et al., 1974; Pollock et al., 1974), while others seem to be associated with the outer membrane (an activity with an unspecified specificity; Hackenbeck et al., 1974); the periplasm (LD-carboxypeptidase; Beck and Park, 1976), either periplasmic or associated with the outer membrane (amidase; Van Heijenoort, et al., 1975) or bound loosely to the envelope (transglycosylase; Höltje et al., 1975). While one or more of these activities may play an important and well-regulated role in surface growth and division of intact cells, their overt expression cannot be detected unless the function of the surface components is perturbed. For example, unlike S. faecalis and other species, cells from exponential phase cultures of E. coli will not autolyze in buffers unless the cells have been grown in the presence of an inhibitor of peptidoglycan assembly. Hartmann et al. (1974) found that the action of peptidoglycan hydrolases could be observed only after pretreatments with any of a series of reagents such as 5% TCA, sodium chloride (0.25–2.5 M), 20% sucrose, or chelating agents. Similarly, in their studies of various activities during the cell division cycle of *E. coli*, Beck and Park (1976) and Mirelman et al (1976) treated cell suspensions with toluene and ether, respectively. On the basis of their observations, Hartmann et al. (1974) propose that a "barrier" prevents access of the hydrolases to the peptidoglycan substrate. The nature of this barrier is unknown, but would be effective in preventing the action of both the peptidoglycan hydrolyzing activity that appears to be associated with outer membrane and the endopeptidase and DD-carboxypeptidase activities associated with the inner membrane (Hackenbeck et al., 1974). A similar "barrier" seems to exist for the pneumococcal amidase which requires activation by, for example, deoxycholate, for expression (Tomasz, 1968).

Currently, various bacterial autolytic systems are being investigated. The mechanisms of regulation of each appear to differ, and the features common to all or most of these systems have not yet been established. These common characteristics will be most important in determining the underlying principles governing the roles of such activities in the morphogenetic changes that occur during bacterial growth and division.

6. Inhibition of cell wall assembly

6.1. Morphological studies

Complete inhibition of cell wall synthesis in growing bacterial cultures usually results in osmotically fragile forms which, in the absence of an osmotic stabilizer, will lyse. However, cellular lysis is not a direct consequence of inhibition of cell wall synthesis, but it is a result of the uncontrolled action of autolytic enzyme systems (section 5). Information on cell surface growth can be obtained by observations made: (1) between drug addition and lysis, (2) on cultures treated with sublethal levels of the drug, or (3) on species, mutants, or under growth conditions that either do not result in osmotic fragility or are osmotically protective. Although a number of antibiotics such as phosphonomycin, cycloserine, and vancomycin are known to inhibit peptidoglycan synthesis, the following discussion will center on the well-studied effects of the β-lactam antibiotics.

Lederberg (1956) observed that an early stage in the process of conversion of growing cultures of *E. coli* into spherical bodies by high concentrations of penicillin in sucrose-containing media was the appearance of subterminal or central swelling. The swelling enlarged progressively until completely spherical bodies were seen. Bayer (1967) showed that treatment of rapidly growing cultures of *E. coli* for 15 to 30 minutes with high concentrations of penicillin resulted in the bursting of the surface at a large number of random sites. In some cells the poles were less affected or free from lesions. Schwarz et al. (1969) observed two different types of effects depending on penicillin G concentration in cultures of *E. coli* grown out from overnight cultures. At lower concentrations (30–50 units per ml) cell division was inhibited, but the cells continued to elongate and bulges developed on the wall. The first bulge always developed in the region where cells normally would have divided, and additional bulges appeared about one cell length distant from the first. On continued incubation, cells were observed to burst at the bulges. At higher penicillin concentrations (200 units or more per ml) cell elongation was inhibited, and on further incubation short cells or walls were split in the middle. The morphological effects were paralleled by a decreasing rate of incorporation of DAP into insoluble peptidoglycan. With increasing penicillin concentration, the rate of peptidoglycan synthesis decreased sharply up to 30 units of penicillin per ml, and less steeply with higher concentrations. The biphasic nature of this decrease was thought to be consistent with the presence of two penicillin-sensitive targets, one presumed to be engaged in elongation and the other in cross-wall formation. Recently, three DD-carboxypeptidase activities, each of which can also carry out transpeptidase reactions with certain substrates, and which differ in their sensitivity to β-lactam antibiotics, endopeptidase activity, and ability to bind [^{14}C]penicillin G, have been described (Tamura et al., 1976). The number and location of penicillin-induced bulges in cells of *E. coli* of various sizes was used in the development of a unit cell concept (Donachie and Begg, 1970).

Burdett and Murray (1974a) examined the effects of ampicillin (2 to 5 μg/ml) and cephalothin (0.1 to 5 μg/ml) on the morphology of *E. coli* B/r and a mutant, PAT 84, known to grow as nonseptated filaments at 42° C and to septate and separate into individual cells rapidly 15 to 25 minutes after a shift from 42°C to 30°C (Hirota et al., 1968). These ultrastructural studies employed a fixation procedure that retained septa in *E. coli* B/r. As shown previously for cultures treated with penicillin G (Schwarz et al., 1969), prominent bulges at or near the middle of cells, or more rarely toward one pole, were seen (Figs. 5 b and c), after treatment of *E. coli* B/r for 1 hour with ampicillin. Lysis of cells was seen to proceed via rupture of the wall at the bulges. Very few bulges were seen after treatment with low concentrations (0.1–1.0 μg/ml) of cephalothin and septal areas appeared as furrows (Fig. 5e). Treatment with higher cephalothin concentrations (5 μg/ml or more) showed central enlargement and bulge formation prior to lysis. Addition of chloramphenicol concurrently with either β-lactam antibiotic suppressed bulge formation (Fig. 5d). At 42°C very few bulges were seen in PAT 84. A shift to conditions that permit septation resulted in bulges at one or two sites along the filaments, or occasionally at poles, and cell lysis after a delay of about 20 minutes. In the vicinity of bulges (Fig. 5g) and at the presumed septal area in completely lysed cells the peptidoglycan layer was not seen. Studies with synchronized cultures of *E. coli* B/r (Burdett and Murray, 1974b) showed that ampicillin (2.5 μg/ml) inhibited division when added at any stage of the cell cycle. With cultures beginning their synchronized increase in cell numbers at about 40 minutes and completing it at about 55 minutes, bulges at or near the center were seen in the greatest number of cells at 20 to 25 minutes and 60 to 65 minutes, or when cells would be expected to form cross walls before dividing (e.g., at the beginning of D; see section 9.2). In contrast, at 35 to 45 minutes cylindrical cells without bulges or visible septa were seen. On the basis of these observations, Burdett and Murray proposed a possible sequence of structural events during the D period. Although the penicillin-induced bulges were many orders of magnitude wider than a septum, the authors proposed that the bulges may indicate an extensive area in which the wall structure is highly labile, perhaps because of hydrolytic activity.

In gram-positive bacteria a variety of other effects of penicillin treatment have been observed. Fitz-James and Hancock (1965) showed that penicillin G (100–1000 units/ml) and methicillin (0.01 to 1.0 mg/ml) treatment of *B. megaterium* resulted 20 minutes later in an increase in cell length and disorganization at nascent septa. After 60 minutes bulbous rods and evidence of cellular lysis were seen. Ten minutes after penicillin treatment in osmotically stabilizing growth media, the development of distorted cross wall containing randomly organized fibrous material was observed. Highton and Hobbs (1971, 1972) studied the effects of penicillin G on growing cultures of *B. licheniformis* and *B. cereus*. In both species penicillin treatment resulted in distorted and thickened cross walls. In addition, in *B. licheniformis,* wall accumulation on the surface resulted in localized areas of thickening and cells which grew in twisted coils and chains.

Treatment of *S. aureus* with growth inhibitory concentrations of penicillin G (Murray et al., 1959; Suganuma, 1962) resulted in the formation of larger cells with thinner and less regular peripheral walls and septa that were thickened, blunted, and misshapen. In contrast, treatment of *S. aureus* cultures with penicillin at one third of the minimal inhibitory concentration (Lorian, 1975) resulted in abnormally large cells containing many completed and partly completed cross walls. These cross walls were at various angles to each other and were often thickened and distorted. Thus, these very large cells can be considered to be clusters of divided but unseparated cells. Similar morphological changes were seen in staphylococci exposed to subinhibitory concentrations of oxacillin or cephaloridine and in *N. gonorrhoeae* treated with penicillin G (Lorian 1975; Lorian and Atkinson, 1975; 1976).

Treatment of *S. lactis* (Dring and Hurst, 1969) with growth inhibitory concentrations of penicillin G resulted in the production of rod-shaped forms. Similar effects were seen on exposure of *S. bovis* to one fourth the minimal inhibitory concentration of penicillin (Lorian and Atkinson, 1976). Cross walls were lacking, but slight constrictions and localized wall thickening were seen along the lengths of filaments.

Several attempts have been made to interpret the ultrastructural effects of penicillin with the mode and topology of wall assembly. Such interpretations have ranged from indications of localized surface growth in *E. coli* (Schwarz et al., 1969; Donachie and Begg, 1970) to indications of diffuse intercalation at many sites in Bacilli (Highton and Hobbs, 1971; 1972). Equating observed morphological or lytic effects of penicillin with wall enlargement sites has certain

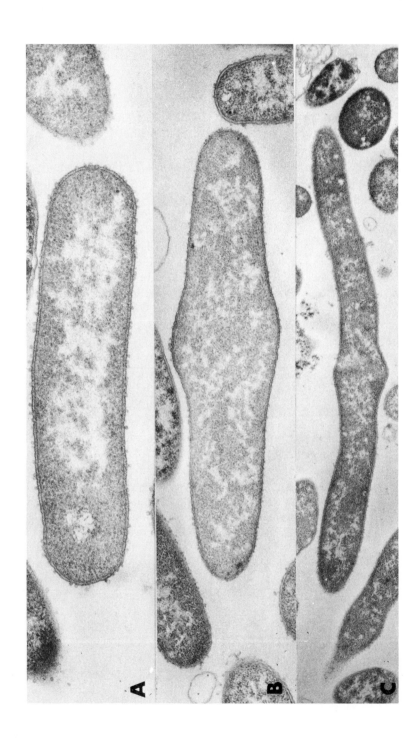

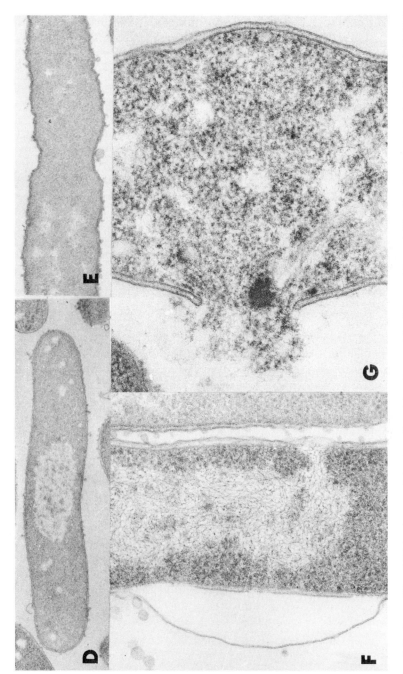

Fig. 5. Sections of *E. coli* B/r in glucose minimal medium (a) and treated for 1 hour with ampicillin (5 μg/ml) showing bulges (b, c, g); CAM (2 μg/ml, d); cephalothin (1 μg/ml, e; note furrow at division site); ampicillin (5 μg/ml) plus CAM (2 μg/ml, f). Acrolein-glutaraldehyde fixation was used. (Reproduced with permission from Burdett and Murray [1974a]).

drawbacks: (1) It is now clear that in at least some cases penicillin does not inhibit peptidoglycan synthesis. In several cases, apparent decreases in peptidoglycan synthesis appear to be merely due to the synthesis of a less cross-linked and therefore more soluble peptidoglycan product in the presence of the drug (Keglevic et al., 1974; Mirelman et al., 1974a; Rosenthal and Shockman, 1975a,b; Tynecka and Ward, 1975). The presence of soluble peptidoglycan products has not always been detected by the methods used. Also, retention of soluble, not fully incorporated, peptidoglycan by cells could be highly variable. (2) The normal process of wall turnover probably continues in the presence of penicillin and strongly influences the morphological effects observed. (3) A significant period of biosynthetic and/or hydrolytic enzyme activity is required before a morphological or lytic effect induced by penicillin can be visualized, even by electron microscopy. (4) Even in species such as *S. faecalis,* where wall surface enlargement is largely localized, wall thickening occurs over the entire surface. Peptidoglycan assembly resulting in wall thickening is also affected by penicillin and other antibiotics that inhibit cell wall synthesis (Toennies and Shockman, 1958; Shockman, 1959b). (5) It has been clearly established that β-lactam antibiotics will inhibit the activity of several enzymes. The relative degree of inhibition of each enzyme will depend on both the β-lactam antibiotic used and its concentration. For example, with membrane preparations of *S. faecalis,* the DD-carboxypeptidase is more susceptible than the atypical LD-transpeptidase to penicillin G, penicillin V, ampicillin, and carbenicillin but is more resistant than the LD-transpeptidase to oxacillin, cloxacillin, cephalothin, cephalexin, cephaloglycin, and cephalosporin C (Coyette et al., 1974). In addition, the relationship of the susceptibilities of these enzymes to the minimal growth inhibitory concentration of each drug is complex. (6) In several bacterial species the presence of multiple components that bind penicillin in a readily detected manner has been observed (Blumberg and Strominger, 1972, 1974; Spratt and Pardee, 1975).

Recent findings require considerable reinterpretation of earlier observations of the effects of β-lactam antibiotics on cell wall growth and division. In this respect, two types of recent experiments are significant: recent measurements of the activities of various enzymes involved in peptidoglycan assembly and degradation during the cell division cycle (Beck and Park, 1976; Hinks et al., 1976; Mirelman et al., 1976), and observations indicating that 6 distinguishable penicillin-binding proteins exist in membrane preparations of *E. coli,* each of which reacts differently to various penicillins, which in turn have varying morphological effects (James, 1975; James et al., 1975; Spratt 1975; Spratt and Pardee, 1975).

6.2. Activities of enzymes of peptidoglycan synthesis, modification, and degradation

Mirelman et al., (1976) examined rates of peptidoglycan synthesis and carboxypeptidase and transpeptidase activities in ether-permeabilized cells of *E. coli* PAT 84 synchronized for division by a shift to the permissive growth temperature. Filaments grown at 42°C contained peptidoglycan that was markedly more cross-linked than that present in the cells grown at the permissive temperature. Evidence that the peptidoglycan in filamentously growing *E. coli* is more cross-linked than that in septated cells was also obtained by Kamiryo and Strominger (1974). Mirelman et al. also observed that a fraction of the DAP-labeled peptidoglycan found in TCA-insoluble cell precipitates was in the form of a high molecular weight product released in soluble form by a hot SDS treatment. The soluble material was mostly uncross-linked peptidoglycan. Following the rate of incorporation of pulses of radioactive DAP into both the SDS- and TCA-insoluble fractions during a shift from 30 to 42°C and back to 30°C showed that the rate of peptidoglycan synthesis was slower at 42° than at 30°. After the shift from 42°C back to 30°C a further drop in incorporation was observed followed by a rise between 15 and 30 minutes before the rapid rise in cell numbers. The relatively small fraction of SDS-soluble peptidoglycan detected in filaments growing at 42°C increased at about 20 minutes after the shift to 30°C. Ether-treated cells incorporated labeled nucleotide-linked precursors into both SDS- and TCA-insoluble peptidoglycan, and the amounts incorporated into the latter were always higher than those incorporated into the former. During the temperature shifts, the patterns of incorporation into the two fractions were similar to that exhibited by the intact cells.

Low concentrations of ampicillin (0.1 and 0.5 µg/ml) reproducibly increased the incorporation of precursors into both TCA- and (TCA and) SDS-insoluble material by a small amount. Low concentrations of 6-aminopenicillanic acid (1.0 µg/ml) slightly increased incorporation of precursors into SDS-insoluble material by ether-treated cells. In contrast, amidinopenicillanic acid (FL-1060) had no effect on incorporation even at high concentrations (50 µg/ml) and cephalothin, at concentrations higher than 10 µg/ml, inhibited incorporation. Similarly, concentrations of ampicillin over 1 µg/ml increasingly inhibited incorporation into SDS-insoluble material. At a concentration of 50 µg/ml, ampicillin nearly completely inhibited incorporation into SDS-insoluble material but inhibited incorporation into TCA-insoluble material by only 50%. Approximations were made of the contribution of ampicillin-sensitive carboxypeptidase activity and transpeptidase activity, insensitive to 0.5 µg/ml of ampicillin, to the release of the terminal D-alanine from UDP-muramylpentapeptide. A relative decrease in ratio of transpeptidase to carboxypeptidase activity in the dividing cells as compared to the filaments of 42°C was interpreted as a possible mechanism for providing more pentapeptide available as substrate for incorporation into peptidoglycan in the filaments which, in turn, were found to contain more peptide cross-links. This interpretation was confirmed by the stimulation of incorporation of precursors into TCA- and SDS-insoluble material by di- and tripeptides of D-alanine, which are known to be substrates of the carboxypeptidase. The lower activity of the carboxypeptidase in ether-treated filaments was not due to a lack of enzyme since sonic disruption caused a large rise in carboxypeptidase activity. Thus, the enzyme was present in substantial amounts but was not fully active in the ether-treated cells. Apparently, carboxypeptidase activity may regulate both the extent of peptide cross-linking and the actual amount of precursors incorporated into peptidoglycan.

A complementary set of data was obtained by Beck and Park (1976) who examined the activities of carboxypeptidase I (DD-carboxypeptidase), carboxypeptidase II (LD-carboxypeptidase) and N-acetylmuramyl-L-alanine amidase in toluene-treated cells of *E. coli* K12, fractionated according to size by gradient fractionation and in a temperature-sensitive division mutant (BUG 6). In toluene-treated *E. coli* K12, carboxypeptidase I and amidase activities were about the same in cells of all sizes. However, carboxypeptidase II activity varied and was about threefold higher in recently divided, or about to divide, cells than it was in cells of intermediate size. Similarly, about 15 minutes after a shift of BUG-6 from 42°C (filaments) to 32°C, the activity of carboxypeptidase II in toluene-treated cells increased about tenfold in conjunction with an increase in cell number. Apparently the increase in carboxypeptidase II activity in toluene-treated cells was not a reflection of increased amounts of enzyme protein since enzyme activity found in completely disrupted cells was about the same in both dividing and nondividing cells. Currently, the function of carboxypeptidase II, which seems to be a periplasmic enzyme, is not known. However, an enzyme of such specificity and cellular location, acting as a transpeptidase, could link lipoprotein to peptidoglycan. Alternately, removal of penultimate D-alanine residues would provide free α-carboxyl groups of lysine to which lipoprotein molecules could be transferred by the action of another enzyme activity.

In addition to the enzyme activities mentioned above, evidence obtained with various strains of *E. coli* suggest that toward the end of the cell cycle (D period; section 9.2.) (1) the rate of incorporation of pulse-labeled precursors into the peptidoglycan increases (Hoffmann et al., 1972). In contrast, however, Cooper (1976) observed an increase in rate of peptidoglycan synthesis during the first 75% of the division cycle and a constant (or possibly decreased) rate in the last 25% of the cycle; (2) the activity of membrane-bound peptidoglycan hydrolase activity increases (Hackenbeck and Messer, 1974); and (3) the sensitivity to killing by a high concentration of penicillin (Hoffmann et al., 1972; Cooper, 1976) or to the production of bulges induced by low concentrations of penicillin or ampicillin increases (Hoffmann et al., 1972; Burdett and Murray, 1974b). In slowly growing (t_d of about 75 min) synchronized cultures of a gram-positive coccus (*S. faecalis*), Hinks et al. (1976) observed a constant and linear rate of peptidoglycan synthesis in middle and latter portions of the cell cycle with little or no synthesis toward the end of the cycle. They also observed decreased cellular autolytic capacity toward the end of the cell cycle of *S. faecalis*.

The inhibition of specific enzymes involved in peptidoglycan assembly by the various β-lactam antibiotics has not yet been correlated with the series of membrane proteins which bind penicillin (Blumberg and Strominger, 1972, 1974; Spratt, 1975; Spratt and Pardee, 1975). However, the ability of these proteins to bind various members of this group of antibiotics, correlated with the morphological and membrane protein changes they appear to induce, sheds some light on the regulation of cell surface assembly and division.

6.3. Membrane proteins that bind
or are affected by penicillin

In *E. coli,* Spratt and Pardee (1975) showed that penicillin G was firmly bound by six inner (cytoplasmic) membrane proteins of different molecular weights. Amidinopenicillanic acid (FL-1060) was bound nearly exclusively by only one protein, "protein 2," of an apparent molecular weight of 66,000. Binding to protein 2 accounts for only 0.5% of the total penicillin G bound. FL-1060 differs from many other β-lactam antibiotics as follows: (1) It does not inhibit cell division to cause filament formation but instead causes the conversion of rods into oval-shaped cells which do not become osmotically fragile for a relatively long period (Lund and Tybring, 1972; Greenwood and O'Grady, 1973a,b; Park and Burman, 1973; Goodell and Schwarz, 1974); and (2) it fails to inhibit DD-carboxypeptidase, endopeptidase, and transpeptidation activities (Park and Burman, 1973; Matsuhashi et al., 1974). FL-1060 reduces the incorporation of DAP into a cellular macromolecular component (peptidoglycan). However, the maximum extent of inhibition, even at very high concentrations, was 35 to 75%, in contrast to about 95% inhibition by ampicillin (Park and Burman, 1973). Except for 6-aminopenicillanic acid (and FL-1060), β-lactam antibiotics that affected cell division failed to compete for binding to protein 2 or bound with low affinity. Within a narrow concentration range, 6-aminopenicillanic acid can also produce ovoid cells and showed a relative affinity for protein 2. Isolation of an ovoid-shaped mutant that failed to bind β-lactam antibiotics to protein 2 was also taken as evidence that protein 2 is involved in determining cell shape.

Cephalexin and penicillin G, which inhibit cell division, appear to bind preferentially to protein 3. A temperature-sensitive cell division mutant in which protein 3 was thermosensitive was isolated, consistent with the suggestion that protein 3 is involved with cell division. Cephaloridine, which at its lowest effective concentration causes cellular lysis, showed preferential binding to protein 1, suggesting that this protein is involved with cell elongation. By the criteria used in these experiments, proteins 5 and 6, which appear to be responsible for binding a very substantial fraction of the total binding of penicillin G bound by membranes, appear to play little or no role in cell shape and division.

The binding of FL-1060 to protein 2 was correlated with its effects on synchronously dividing cultures of *E. coli* B/r (James et al., 1975) and with the synthesis of an outer membrane protein (James, 1975). FL-1060 was found to have no direct effect on DNA synthesis but to block an early event leading to cell division at a time corresponding to the beginning of C (section 9.2). This was interpreted as the inhibition of the process responsible for cell elongation that parallels chromosome replication. The mechanism of this inhibition appears to occur via inhibiting the synthesis of an outer envelope protein, protein G, with a molecular weight of 15,000 daltons. Protein G accumulates when cells elongate after inhibition of DNA synthesis by nalidixic acid, or, at the nonpermissive temperature in a conditional *dna*B mutant, but not when cells elongate after treatment with penicillin G or cephalexin. Ovoid cells, produced by FL-1060 treatment, did not accumulate protein G after nalidixic acid treatment. Thus, protein G was proposed to be a structural protein of cell elongation. Protein G has certain characteristics in common with the free form of lipoprotein of the outer membrane, such as inhibition of its synthesis by chloramphenicol but not by puromycin, and the absence of histidine, but differs in molecular weight and the presence of proline, which is absent in the lipoprotein.

In contrast to its inhibitory effect on the synthesis of protein G, FL-1060 treatment increased the accumulation of another outer membrane protein with a molecular weight of 80,000, protein D (Gudas et al., 1976). A rapid increase in protein D to levels ten times higher than in controls was observed to begin about 30 minutes after FL-1060 addition. Cephalexin, which has a different morphological effect than FL-1060 (see above), also stimulated protein D accumulation but did so immediately. After 90 minutes, levels were threefold higher than the control. In synchronized cultures, FL-1060 or cephalexin addition at the beginning of synchronization did not affect the time at which protein D increased, but synthesis did not terminate as it did in untreated cultures. The synthesis of protein D also reacted differently than the synthesis of protein G to inhibition of DNA synthesis with nalidixic acid. Nalidixic acid addition caused a decrease in protein D synthesis, even after pretreatment with FL-1060, which had stimulated the rate of protein D synthesis. Thus, proteins D and G each react to the two types of inhibitors in an opposite manner. Nalidixic acid causes an increase in protein G and a decrease in protein D, while FL-1060 causes a decrease in protein G and an increase in protein D.

The biochemical, physiological, and morphological changes induced by inhibitors of cell wall assembly discussed above have recently been proven useful in attempts to unravel the sequence of events occurring during the cell division cycle. A speculative attempt to integrate these and other observations is presented in section 9.5.

7. Postincorporation modifications of walls

Conceptually, "remodeling" (Rogers, 1965; 1970) of walls resulting from the hydrolysis of specific bonds at selected topological locations on the cell surface (by autolysins) has been a well accepted concept. Such a process would allow realignment and rearrangement of preexisting peptidoglycan and also provide new acceptor ends (Rogers, 1970; Higgins and Shockman, 1971). A much more slowly developing, and more difficult to accept, idea is that the synthesis of new covalent linkages may also occur in preexisting, insoluble wall outside of the cellular permeability barrier. Recent data indicate that after incorporation of precursors into an insoluble wall the number of peptide cross-links increase.

Fordham and Gilvarg (1974) used the reaction of nitrous acid with unprotected amino groups in intact walls to show that in exponentially growing cultures of Lys⁻, and DAP⁻ mutant of B. megaterium, cross-linking of DAP did not occur concomitant with its incorporation into the SDS-insoluble residue of cells.

After a 5-minute pulse, 30% of the DAP was cross-linked and this fraction increased to 48% during a 20-minute chase. Similarly, a 1-minute pulse, which gave 15% of the DAP in cross-links, was followed by an increase to 33% of the DAP in cross-links during 4 minutes of chase. Although like many other published experiments employing pulse-chase labeling methods (see below) these experiments suffered from the complication of continued incorporation of substantial amounts of DAP into the peptidoglycan during the chase, it seems clear that disaccharide-peptide units incorporated into insoluble peptidoglycan initially contained free NH_2 groups which, on subsequent growth became blocked. Furthermore, penicillin G (1 μg/ml) permitted rapid incorporation of DAP into the peptidoglycan for at least 30 minutes but greatly inhibited cross-linking. In contrast to earlier observations with S. aureus (Tipper and Strominger, 1965; Wise and Park, 1965), removal of penicillin was followed by cross-linking of previously incorporated but uncross-linked subunits, in the absence of detectable growth.

The formation of peptide cross-links in cultures of S. faecalis, grown under a variety of conditions, also appears to continue for an unexpectedly long time (Dezéllée and Shockman, 1975). In these experiments the distribution of doubly labeled disaccharide-peptide monomers and peptide cross-linked dimers, trimers, and higher oligomers in "free" peptidoglycan fragments were determined. A 1-minute pulse followed by a chase of 32 minutes (about 1 generation) showed that the fraction of disaccharide peptide monomer decreased from about 22% to 8% of the total, while the trimer and higher oligomer fractions increased from 24% and 17% to 31% and 32%, respectively (Fig. 6). Calculation of these data in terms of percent cross-linking showed that about 46% crosslinking after the 1-minute pulse increased rather slowly to 61% after 32 minutes of chase. These experiments were interpreted as showing two distinct phases of peptide cross-linking: a rapid phase, approximately concomitant with incorporation into insoluble wall, and a second, slower phase. The second phase could well be important in "remodeling."

Similar to the results reported for B. megaterium (Fordham and Gilvarg, 1974), continued incorporation of labeled precursors into peptidoglycan during a chase has been noted in S. faecalis (Boothby et al., 1973; Dezéllée and Shockman, unpublished observation). For example, a chase following the incorporation of [³H]lysine showed an immediate cessation of the incorporation of lysine into cellular proteins along with a continued increase in radioactivity in the peptidoglycan fraction (Dezéllée and Shockman, unpublished observations). Similarly, Mirelman et al. (1974a) observed that a 10-minute pulse of [¹⁴C]lysine resulted in no increase in radioactivity incorporated into the cells but a very substantial increase in radioactivity in the isolated and SDS-treated wall fraction of M luteus (lysodeikticus). These observations, with three different bacterial species, suggest the presence of a large cellular pool of wall precursors. Since cells from exponentially growing cultures of Gram-positive bacteria usually have only a relatively small pool of nucleotide precursors (Shockman et al., 1965; Lynch and Neuhaus, 1966), it seems possible that a significant portion of this pool may be in

650

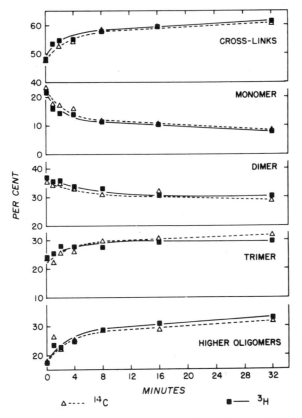

Fig. 6. Frequencies of disaccharide-peptide monomer, and peptide cross-linked dimer, trimer, and higher oligomers, and degree of cross-linking of the free peptidoglycan fragments from walls of cultures pulsed for 1 minute and chased for intervals ranging from 1 to 32 minutes. (Reproduced with permission from Dézelée and Shockman [1975]).

the form of oligomeric (glycan chain?) precursors in or on the membrane. In contrast, Braun and Wolff (1975) found no evidence for oligomeric precursors of peptidoglycan in *E. coli* W7. Instead, a large pool of diaminopimilate appeared to be responsible for continued incorporation during a 1 generation chase.

A hypothetical model for the "remodeling" and reorientation of wall peptidoglycan can be constructed based in part on the observations discussed above, the hypothesis first expressed by Ghuysen (1968) to explain the observed peptidoglycan structure of *M. lysodeikticus* recently presented in a modified form by Fiedler and Glaser (1973), and recent information on the sequence of wall assembly such as that of Mirelman et al. (1972, 1974b) and Ward and Perkins (1974). The model proposed here would involve many reactions known to occur in bacteria. However, not all of the reactions are known to take place in any single bacterial species. Hydrolases, such as glycan-chain-splitting muramidases and/or transglycosylases, glucosamindases, and endopeptidases, which in some instances could be carboxypeptidases, would be involved. Also, transfer enzymes such as transglycosylases, transpeptidases, and possibly a postulated trans-amidase (Wong et al., 1974) would play important roles.

Glycan chains synthesized on the membrane carrier would be inserted into the preexisting wall acceptor via transpeptidation (e.g., the early, fast phase). Initial insertion would be followed by additional transpeptidations to disaccharide peptide units on the same or different glycan chains. Both the initial insertion and the earlier phase of the subsequent transpeptidation reactions would not have to be highly oriented. Initial incorporation, and perhaps release of the long glycan chains from the membrane carrier, would be followed by a series of topologically oriented (and information-containing) reactions, such as the following: (1) hydrolysis of glycan chains at pre-selected intervals. This series of hydrolyses could account for the observed overall polydisperse nature of glycan chains and could provide for definitive chain lengths at specific topological sites at specified times in the cell cycle; (2) transglycosylation reactions. Such an activity could complement glycan chain hydrolytic activity by increasing the length of some glycan chains. Conservation of energy in a 1-6 anhydro-N-acetylmuramyl bond (Höltje et al., 1975) could provide a mechanism by which glycan strands could be linked without prior activation, such as by the coupled hydrolysis of another glycosidic bond; (3) additional transpeptidation reactions, between previously free amino and carboxy groups on glycan chains not previously linked together by peptide bonds. Decreased length of glycan chains, combined with the linkage of additional chains could serve to change the shape of a portion (an "island"?) of peptidoglycan from, for example, long in the glycan direction and thin in the peptide direction to shorter in glycan length and wider in that of the peptide; and (4) endo- (or carboxy-) peptidases could then hydrolyze some of the peptide bonds that had formed, permitting "movement" of peptidoglycan. Transpeptidations could also result in "movement" by exchanging a bond between two different disaccharide peptide units, each of which could be on the same glycan strand (Fiedler and Glaser, 1973) to permit "sliding" of adjacent glycan strands (Fig. 7). Alternately, one reactant could be on an entirely different but adjacent glycan strand, permitting a change in shape. A similar function has been proposed for the amidase of S. aureus (Wong et al. 1974; Chatterjee and Doyle, 1977), although evidence for a transamidase function of such enzymes is lacking.

This model provides a hypothetical picture for a series of remodeling reactions, all of which can be catalyzed by known enzymatic activities outside the cellular permeability barrier in the absence of an energy supply and in a relatively insoluble macromolecular structure. Such possible reactions are visualized as occurring in only very small areas of the wall, but could easily be involved with changes in topological location of peptidoglycan moieties with time. However, it seems unlikely that such processes could result in rapid dispersion of newly incorporated materials over nearly the entire surface of the cell such as that suggested by Ryter et al. (1973).

A separate class of postincorporation modifications to assembled peptidogly-

652

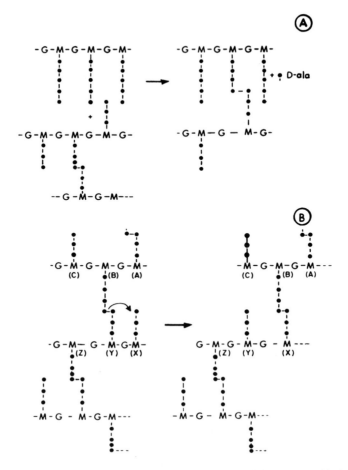

Fig. 7. Growth of cell wall. (a) Insertion of new peptidoglycan strands by transpeptidation and loss of terminal D-alanine. (b) A hypothetical transpeptidation to move peptidoglycan strands relative to each other. Transpeptidation reaction could also occur between strands rather than along a strand. − ○ = amino acids linked to muramic acid in the following order: L-Ala-D-isoGlu-DAP-D-Ala-D-Ala. Horizontal bonds are between D-Ala of one chain and DAP of another chain. (Reproduced with permission from Fiedler and Glaser [1973]).

can results from reactions that add or remove substituents such as acetyl or amide groups (section 4.1.4). In contrast to the lysozyme resistance of walls of *S. aureus*, which contain O-acetyl residues on muramic acid (Ghuysen, 1968), peptidoglycan synthesized in vitro by a crude cell wall preparation from *S. aureus* H was totally degraded by lysozyme or *B. subtilis* amidase, presumably because it lacks O-acetyl groups (Bracha and Glaser, 1976). This observation may be consistent with O-acetylation after wall assembly. Modification, such as O-acetylation, de-N-acetylation, amidation, or deamidation are known to affect not only susceptibility of peptidoglycans to hydrolases but may also change the ability of individual subunits to act as substrates for other enzymatic activities such as transpeptidases (chapter 8 by Ghuysen, this volume).

A third class of postincorporation modification reactions could result from changes in the nonpeptidoglycan polymers of the surface. As discussed in section 4.3, polymers covalently linked to the peptidoglycan may affect cell shape and cell division. A good example of a possible postincorporation modification of a nonpeptidoglycan polymer is in the pneumococcal system. Loss of choline residues from the wall teichoic acid (and possibly from the Forssman antigen) via the action of the phosphorylcholine esterase (Höltje and Tomasz, 1974) could affect the ability of the pneumococcal amidase to be activated or to bind to and hydrolyze bonds at specific topological locations in the wall. Similarly, in *E. coli*, the possible addition of free lipoprotein molecules in the outer membrane to specific areas of the peptidoglycan layer, or at specific times in the cell cycle (James and Gudas, 1976), could affect one or more properties of the peptidoglycan layer. Postincorporation modifications to surface polymers not covalently linked to the peptidoglycan could also affect morphogenesis and division. For example, modifications to one or more outer or inner membrane proteins of *E. coli* which appear to change in amounts during the cell cycle (sections 6 and 10.5), could be important. An additional example would be deacylation of lipoteichoic acid in a gram-positive species. Deacylated LTA has been observed to be present in *S. faecalis* (Joseph and Shockman, 1975). The presence of fatty acids on lipoteichoic acid seem to be required for their action as an inhibitor of various autolytic enzyme systems (section 5.3) and for their postulated function as a carrier in the synthesis of wall teichoic acids (discussed by Ghuysen in chapter 8).

8. Regulation of macromolecular synthesis during growth and division

Bacteria have the ability to regularly and precisely apportion newly replicated chromosomes equally into daughter cells of well-defined and characteristic size and shape in the absence of an elaborate mitotic apparatus such as that found in eukaryotic cells. As discussed above, an intact, exoskeleton-like cell wall appears to be essential to the precise timing and organization of the processes involved. Thus, the view has evolved that the mode of assembly of the rigid wall polymer is not only involved in the replication of cells appropriate in size and shape, but also participates in the processes which result in chromosome segregation. In the following sections several models of bacterial cell growth and division will be examined primarily from the viewpoint of the role of the cell surface in the mechanisms of the processes that occur. Attempts to unify existing data, obtained with a variety of bacterial species of various shapes and modes of division, result in the development of highly speculative models. An essential feature of the proposed models is the presence of a primitive type of topologically or

654

geometrically oriented memory unit in the cell surface that interacts directly or indirectly with the genome.

8.1. Regulation of synthesis of informational macromolecules

The biosynthesis of a linear macromolecule occurs by a process in which successive subunits are sequentially added at a single site to form a growing chain. The addition of each subunit to the growing macromolecule is thought to take a constant period of time, usually called step time. Therefore:

$$\frac{dx}{dt} = k \tag{1}$$

where dx/dt is the rate of synthesis in macromolecule x and k is the rate constant for the reaction, which is clearly linear in this case.

Bacteria divide by fission into two equal daughter cells, and an exponential increase in both cell numbers and cellular substance can be maintained indefinitely. There are two ways that the macromolecular components of a bacterial culture can increase at an exponential rate. The number of growth sites for each species of linear macromolecule must either increase continuously (exponentially) throughout the cell division cycle, or, at some point during the cell cycle, the growth sites must double in number. One of these two processes must occur for each macromolecular species in cells, since during balanced exponential growth each extensive (e.g., additive) property of the cell culture increases exponentially at the same rate (Campbell, 1957). Thus, the overall rate of synthesis of each type of macromolecule in a culture depends on the number of growth sites for that type of macromolecule.

Studies by Schaechter, Kjeldgaard, Kurland, and Maaløe (discussed in Maaløe and Kjeldgaard, 1966) of cell composition and the synthesis of DNA, RNA, and protein helped to establish certain principles for the regulation of synthesis of the various classes of informational macromolecules. They showed that, at a constant temperature, cell size and macromolecular composition of cultures of *Salmonella typhimurium* varied with growth rate. The amount of RNA per mg dry weight increased substantially and almost proportionately with growth rate. These investigators also demonstrated that the rate of protein synthesis was approximately commensurate with stable (ribosomal) RNA content at all growth rates. In later studies, Ecker and Schaechter (1963) showed that the number of ribosomes appeared to be proportional to growth rate. Their results are consistent with a constant efficiency of RNA (and ribosomes) in protein synthesis. Thus, the overall rate of cellular protein synthesis depends on the number of ribosomes present in a culture. At rapid growth rates more ribosomes are present, and therefore more protein can be made per unit time.

These relationships have been analyzed quantitatively by Koch (1970; Koch and Deppe, 1971) who noted that in balanced, exponential phase bacterial cultures:

$$\frac{dp}{dt} = \alpha \cdot p = r \cdot k_p \tag{2}$$

where α is the exponential growth rate constant, and k_p is the rate constant for the synthesis of proteins (p) dependent on stable RNA (r). If this equation is rearranged, one obtains:

$$r/p = \alpha \cdot 1/k_p \tag{3}$$

Since RNA (r), and protein (p) contents can be determined at various growth rates α, one can ask whether the ratio of r/p does or does not increase linearly with α. If the ratio of r/p were to increase linearly with α then $1/k_p$, which is the slope in equation 3, would be constant at all growth rates. Consequently k_p, the rate constant for the synthesis of protein per unit of stable RNA would also be constant. Since k_p can also be defined as the efficiency of stable RNA in protein synthesis, a linear relationship between r/p and α would indicate a constant stable RNA and, by inference, a constant efficiency of ribosomal RNA in protein synthesis.

In practice, when r/p values are analyzed as a function of α, proportionality is observed (reviewed by Koch, 1970). However, if the straight line obtained is extrapolated to 0 growth rate, a substantial y intercept is obtained. The existence of such an intercept has been interpreted in two ways. Dennis and Bremer (1974) have shown that at α values between 0.94 and 1.55 hours^{-1} the relationship between ribosomal protein/total protein increases linearly with α, with an apparent 0 intercept. Koch and Deppe (1971) have shown that at very slow growth rates ribosomal efficiency is decreased. From equation 3, it is obvious that since ribosomal efficiency ($1/k_p$) is the slope; an increased slope indicates a decreased efficiency. Data obtained from *S. faecalis* cultures (Lancy, 1976) tend to suggest that both views may hold, since a much reduced but still substantial intercept is obtained for r/p after subtracting the estimated amounts of non-ribosomal RNA from stable RNA.

Schaechter et al. (1958) also observed that in bacterial cultures the ratio of protein per nucleus or "genome equivalent" was approximately constant at all growth rates. Analogous with the constant ribosome efficiency model, it was proposed that replication of the bacterial chromosome occurs at a constant rate per chromosome growth site. An increase or decrease in growth rate would increase and decrease the number of chromosome growth sites respectively, and therefore the frequency of chromosome replication initiations (Maaløe and Kjeldgaard, 1966). A relatively constant step time of DNA polymerization in chromosome replication was observed in a variety of bacteria, and the effects of alterations in growth rate on frequency of chromosome replication initiation forks have been equally well documented (Helmstetter and Cooper, 1968).

Thus, the pioneer work from Maaløe's laboratory produced two important "theorems" for regulation of macromolecular synthesis in bacterial cultures: (1) the constant efficiency of ribosomes in protein synthesis, and the dependence of synthesis on the number of ribosomes present at any growth rate; and (2) the view that if replication of a macromolecule occurs at a constant rate, then alterations in growth rate must affect the frequency of initiation of that macromolecular species. Actually, in both situations growth rate affects the frequency of initiation of polymerization reactions, since an increase in the number of ribosomes operating at a constant efficiency also increases the frequency of initiations of the amino acid polymerization reaction into proteins.

The principle that the overall rate of biological polymerization reactions is governed by the frequency of initiations has been important in many areas of biology, including our current ideas of chromosome replication regulation and its relationship to cell division (sections 9.2 to 9.5). Based on generalizations about the synthesis of informational macromolecules, attempts have been made to apply the same ideas to the synthesis and assembly of three-dimensional surface components.

8.2. Network analysis

8.2.1. General principles

The use of measurements on undisturbed exponential phase cultures in balanced growth for the deduction on regulatory processes has been mentioned previously. Although the information is limited by the degree of sophistication used in the quantitative isolation and identification of a particular class of macromolecule, this approach may be used in studies of the relationship of surface growth to other cellular processes. The interrelationship between various cellular biosynthetic processes, such as the dependence of protein synthesis on ribosomes (equations 2 and 3, above), can be analyzed quantitatively using the network theorem of Dean and Hinshelwood (1966; Dean, 1973).

For network analysis, all macromolecular processes are considered to belong to or depend on a closed network of biosynthetic reactions consisting of a finite number of elements (Dean and Hinshelwood, 1966; Dean, 1973). Within the closed network a rate-limiting reaction must exist. In a culture in balanced growth, this reaction(s) responds to the environment and in turn determines the overall growth rate. Therefore, at any growth rate the exponential rate constant for each biosynthetic reaction must depend ultimately on the specific rate-determining network reaction(s). Dean and Hinshelwood proposed that in cultures in balanced growth, for a network of reactions containing n elements, the rate-determining reaction must be proportional to the overall growth rate of the culture raised to the $(n-1)$ power. The conceptual framework for this approach was expanded by Koch (1970; 1971).

A typical possible closed network is shown in Fig. 8a. For a network containing three elements, DNA (d), RNA (r), and protein (p), by analogy to equations 2 and 3, three proportionality constants can be recognized:

$$k_p = \frac{\alpha}{r/p} \tag{4}$$

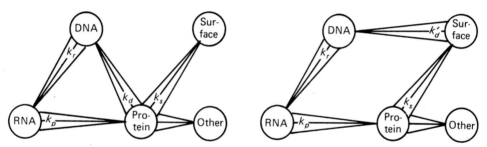

Fig. 8. Closed networks of biosynthetic reactions. (a) The network contains three elements: DNA, RNA, and protein. All other biosynthetic reactions, including synthesis of surface material, depend on one element of the network (e.g., protein). The rate constants for the synthesis of each network element depending on another are given as k_r, k_p, and k_d. The rate constant for the synthesis of surface dependent on protein is given as k_s. (b) The network contains four elements, DNA, RNA, protein, and surface. The surface element has been inserted into the network between protein and DNA. This insertion results in two new rate constants, k_s and k'_d.

$$k_d = \frac{\alpha}{p/d} \tag{5}$$

$$k_r = \frac{\alpha}{d/r} \tag{6}$$

In many bacterial cultures where growth rate is limited by nutritional factors, two constants are observed at all growth rates. They are k_p and the ratio of p/d (Lancy, 1976). Combining equations 4, 5, and 6 yields:

$$\alpha^3 = r/p \cdot k_p \cdot p/d \cdot k_d \cdot d/r \cdot k_r \tag{7}$$

Simplifying equation 7 and dividing by equation 5 gives:

$$\alpha^2 = k_r \cdot k_p \cdot d/p \tag{8}$$

Thus in an $n = 3$ element network (Fig. 8a), when a constant k_p and a constant ratio of d/p are obtained at all growth rates, k_r should be proportional to α^2 (α^{n-1}). In turn, k_r should represent the reaction responsive to the environment and the one which determines the overall growth rate of the culture. Many observations indicate that the rate constant for stable RNA synthesis per unit DNA is affected by the nutritional environment (Maaløe and Kjeldgaard, 1966).

Data obtained from a variety of organisms appear to fit a network of three elements, but do not per se prove that the three-element network is correct. Cellular constituents exist other than DNA, RNA, or protein. In particular, surface constituents contribute substantially to overall cell composition (section 3).

A closed network containing three elements may generate surface biosynthetic reactions from any one of them (Fig. 8a). Alternatively, one or more surface biosynthetic reactions may be an integral part of the network which would then consist of at least four (rather than three) elements (Fig. 8b). The two possibilities can be distinguished experimentally by examining the effects of growth rate limitations on the ratios of any two adjacent elements in the network, and therefore on the rate constants for the reactions. For example, in the two hypothetical networks shown in Fig. 8, one would compare the adequacy of a single rate constant k_d (from the ratio of d/p) to two rate constants, k_s and k'_d (from the ratios of s/p and d/s, where s is a hypothetical surface constituent. Such an analysis was done with $S.$ $faecalis$ under conditions when growth rate was set by: (1) limitation of an essential amino acid, or (2) by the addition of osmotically active solutes (Lancy, 1976; Lancy, et al., 1977).

8.2.2. Experimental evidence for a constant efficiency of protein in surface biosynthesis

Analyses of the cell composition of $S.$ $faecalis$ grown at a series of growth rates were used to obtain information concerning possible roles for various surface components in the synthesis of informational macromolecules. In addition to DNA, RNA, and protein content, content of peptidoglycan, rhamnose (a constituent of the polysaccharide linked covalently to the cell wall peptidoglycan), and of phosphorus and glycerol in the chloroform-methanol soluble lipid fraction were determined (Lancy, 1976). The results obtained were inserted into a four-element network (Fig. 8b) in which the synthesis of DNA is considered to depend on a surface component, and synthesis of the surface component depends on protein. The equations for the two postulated relationships are:

$$k_s = \frac{\alpha}{p/s} \tag{9}$$

and

$$k'_d = \frac{\alpha}{s/d} \tag{10}$$

where s represents a surface component, k_s is the rate constant for the synthesis of the surface component per unit of cellular protein and k'_d is the rate constant for the synthesis of DNA per unit of surface constituent.

When either peptidoglycan, rhamnose, or lipid glycerol was the surface component measured in cultures in which growth rate was controlled by a nutritional restriction, the ratios of p/s increased linearly with increasing growth rates (Fig. 9a,b,c). For lipid the intercept was not significantly different from 0. This linear relationship suggests that the rate of synthesis of lipid is proportional to protein content and therefore occurs at a constant efficiency at all growth rates measured.

In $S.$ $faecalis$, protein content per unit of dry weight is constant at all growth rates (Lancy, 1976). Thus, it could be argued that the relationship of protein to the rate of synthesis of surface components is actually merely a reflection of cellular mass. However, two factors make this interpretation unlikely. First is the case of rhamnose content, where extrapolation to zero growth rate would intercept the y-axis (Fig. 9b). According to Koch and Deppe (1971), this would suggest that efficiency of synthesis for this constituent is constant at relatively fast growth rates but decreases precipitously at very slow growth rates. Since this is not observed for lipid glycerol, which intercepts the y-axis at zero, it is difficult to conceive of a nonspecific parameter such as dry weight as the controlling factor. Second, a relationship between one surface element and DNA synthesis has been demonstrated (section 8.2.3.).

Assuming that protein content per unit dry weight is also constant in other bacteria, we have looked for possible linear relationships between dry weight (or protein) per unit of surface constituent in published data.

In chemostat cultures, Ellwood (1970) examined the amounts of wall that could be recovered from *B. subtilis* var. *niger* grown in magnesium- and phosphorus-limited cultures at different growth rates. With both types of nutritional limitations, the amount of wall per unit dry weight increased as the growth rate decreased (Table 1). A further reduction of growth (dilution) rate to 0.05 hour^{-1} for the phosphorus-limited culture yielded cells with about 37% of their

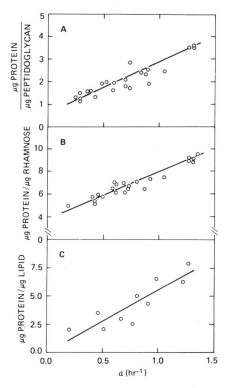

Fig. 9. Effects of growth rate on the ratio of protein to surface elements. Ratio of μg protein to: (a) μg peptidoglycan; (b) μg rhamnose; and (c) lipid glycerol; each plotted as a function of growth rate in glutamate-restricted cultures (Lancy, 1976). Statistical analyses were carried out on a PDP8/M minicomputer (Digital Equipment Corp., Maynard, Mass.).

dry weight as walls. Although the composition of walls in the Mg^{2+} and phosphorus-limited cultures was different at each growth rate, the relative amount of wall substance was very similar. The reciprocal relationship (i.e., the ratio of cellular dry weight to wall weight) increased with growth rate in an approximately proportional manner. An observation that does not seem consistent with the correlation of growth rate and cell wall content was that at a constant dilution rate of 0.3 $hour^{-1}$ the percent content of wall of *B. subtilis* var. *niger* varied from about 13% for glucose-limited cultures to about 21% for sulfur- or nitrogen-limited cultures.

Sud and Schaechter (1964) observed in *B. megaterium* that cellular content (μg/mg dry weight of cells) of two wall components, hexosamine and DAP, and a membrane component, lipid phosphorus, increased as growth rate decreased. Their data also indicated that the relative proportions of the three surface components are fairly constant at all the growth rates examined, suggesting little or no change in wall composition or in wall-to-membrane ratio. Again, the reciprocal relationship (e.g., the ratio of cellular dry weight to the amount of each surface constituent) increased linearly with growth rate.

The amount of cell wall (and in this case wall content) of *Aerobacter aerogenes* was found to decrease with increasing growth rate (Table 1; Tempest and Ellwood, 1969) for both Mg^{2+} and carbon-(glucose) limited cultures. In *Aerobacter aerogenes*, the content of a major wall component, the lipopolysaccharide changed with both growth rate and nutritional limitation. Ratios of KDO to heptose were substantially different. For other gram-negative organisms, data are not readily available.

Table 1.
Influence of Growth Rate and Growth-Limiting Component of the Medium on the Average Wall Content of *Bacillus subtilis* var. *niger* and *Aerobacter aerogenes* Organisms grown in a Chemostat Culture[a]

Dilution rate (hr^{-1})	Bacterial wall content (%, w/w)				
	Bacillus subtilis var. *niger*[b]		*Aerobacter aerogenes*[c]		
	Mg^{2+}-limited	PO_4^3-limited	Mg^{2+}-limited	C-limited	
0.05	—	36.8 37.7	—	—	
0.1'	26.7 27.1	22.8 23.1	19.1 20.5	20.0 21.0	
0.2	23.2 23.9	20.8 21.1	—	—	
0.3	18.2 18.7	17.0 17.5	15.0	14.6	
0.7	—	—	13.0	11.0	

[a] The organisms were grown in 0.5 1. chemostats (Herbert et al., 1965) in simple salts media. The temperature was maintained constant at 35° and the pH value at 7.0 (for the bacilli) or 6.5 (for *A. aerogenes*). Bulk samples were collected from the overflow line into an ice-cooled receiver; organisms were separated from the extracellular constitutents by centrifugation and washed with deionized water. Bacterial pastes were disrupted in a Braun MSK homogenizer, and the walls quantitatively separated from the cytoplasmic constituents by differential centrifugation (Ellwood and Tempest, 1972).
[b] Data of Ellwood (1970).
[c] Data of Tempest and Ellwood (1969).

8.2.3. Role of cell surface in DNA synthesis

While the dependence of surface synthesis on protein content could be outside the network (Fig. 8a), evidence consistent with a role of the surface in chromosome replication is presented in section 9.7. Additionally, with *S. faecalis*, evidence was obtained that lipid phosphorus, but not rhamnose or peptidoglycan, fits into the network shown in Fig. 8b, and provides a determinant for the synthesis of DNA and the other elements of the network (Lancy, 1976). Data from both nutritional and osmotically restricted cultures fit the four-element network shown in Fig. 8b better than a three-element network.

Study of possible regulatory reactions in undisturbed cultures is limited by the degree of resolution of the method used. For example, the constant ribosome efficiency theorem fits best when measurements are made on ribosomal rather than total RNA (Ecker and Schaechter, 1968). Nevertheless, this approach can be useful in the identification and, in some cases, the elimination of potential regulatory elements. It is clear from studies of *S. faecalis* (Lancy, 1976) that k'_d, the rate constant for the synthesis of DNA/unit membrane lipid, is important in the regulation of growth rate. Also, overall wall or polysaccharide composition did not appear to belong in the network, suggesting that the amounts of these elements may not be an important regulatory factor in chromosome replication. Finally, when growth rate is set by limiting the concentration of an essential nutrient, the synthesis of some surface constituents occurs at a constant efficiency per unit protein or per unit mass. Conceivably, this relationship could result in the observation of a relationship between chromosome replication and cell mass (section 9.3).

9. The cell division cycle in bacteria

9.1. Methods used to study the bacterial cell division cycle

The small size of bacteria makes it extremely difficult to quantitate growth and division in individual cells. A few measurements of length, volume, and mass changes of individual cells have been made. However, most information on the behavior of individual cells has been derived from studies of cell populations. Nearly all methods have been criticized because of potential artifacts, usually related to changes induced by pretreatments, or to potential differences in interpretation. Nearly all of the methods employed have been discussed by Helmstetter (1969) and/or Mitchison (1971). A brief description of some of the approaches employed is given here to provide a basis for the interpretation of subsequently presented information.

9.1.1. Synchronization methods

The events occurring during the bacterial cell cycle are probably most easily visualized in synchronously dividing cultures. Three methods of synchronization have been used: (1) size selection, (2) age selection, and (3) induced synchrony.

Size selection, obtained by filtration (Maruyama and Yanagita, 1956; Abbo and Pardee, 1960; Lark and Lark, 1960; Rudner et al., 1965) or by gradient centrifugation (Mitchison and Vincent, 1965; Donachie and Masters 1966; Kubitschek, 1968b) appears to be the simplest method for achieving synchronous cell divisions. However, since selective filtration methods do not appear consistently to yield well-synchronized cultures, gradient centrifugation methods are most commonly used. Sedimentation velocity, rather than sedimentation equilibrium, is utilized. Gradient centrifugation in a zonal rotor has been used successfully in the size fractionation of yeast (Tauro and Halvorson, 1966; Sebastian et al., 1971) and bacteria (Beck and Park, 1976). The smallest cells on the gradient are selected for synchronous growth for three reasons.

First, an unfractionated exponential phase population contains a predominance of recently divided (and therefore smaller) cells, because division by fission generates two daughter cells for each dividing mother cell. Second, cells at the top of a gradient have been exposed to the lowest concentration of gradient solutes and are presumed to be least disturbed. Third, usually a greater homogeneity of cell sizes exists at the top of a gradient. At least some of the size heterogeneity in gradients is due to the statistics of the cell division process, since variations are found in age, as well as size distribution of cells at division (Powell, 1956; Harvey et al., 1967; Koch and Blumberg, 1976).

Synchronization by sucrose gradient fractionation is not a rapid procedure. Under the best circumstances manipulations usually require a minimum of 30 minutes. Usually, during synchronous regrowth the first cell division shows some atypical features. It appears that in addition to effects of manipulation time, exposure to osmotically active solutes and removal of these solutes has some direct effects on the cell division process, especially in the case of Gram-negative bacteria (Wu and Pardee, 1973). For yeasts and for the Gram-positive enterococcus, S. faecalis, the effects of gradient centrifugation are less severe.

Age selection methods were pioneered by Helmstetter who developed a technique usually referred to as the "baby machine" (Helmstetter and Cummings, 1963, 1964). The baby machine is used as an age classification method (section 9.1.2), but can also be used to select cells of relatively uniform age that can then be grown out as synchronous cultures. Cells from an exponentially growing bacterial culture are trapped on a membrane filter, the filter is turned upside down, loosely attached cells are washed off, and the filter is then eluted with prewarmed, conditioned medium. Cells originally caught on the filter divide, and one of the newly divided daughter cells is eluted from the filter and collected. Thus, only "baby cells" come off the filter. There are two principal disadvantages to this age selection method: (1) its reliance on cells that have divided, possibly atypically, on the filter; and (2) when the method is used for preparative purposes, the time required to collect a desired aliquot of baby cells increases with increasing doubling time of the culture. Thus, cells from slowly growing cultures must be collected over a longer period of time than cells from rapidly dividing cultures, unless small samples are collected and used immediately.

Induced (or forced) division synchrony methods are generally of two types. In the first, cells are exposed to successive cycles of growth at 37°C and 25°C or at

43°C and 33°C, resulting in some degree of synchrony (Lark and Maaløe, 1954). Controversy exists about the physiological basis of heat shock synchrony. It is generally thought that heat shock delays divisions at a terminal stage, possibly by destruction of a required component or alteration of a necessary structural configuration (Smith and Pardee, 1970; Lomnitzer and Ron, 1972; Wu and Pardee, 1973). In the second group of methods, cells from a randomly dividing exponential culture are all brought to a specific physiological stage of the cell division cycle, and are then released from the induced state. The stage may be the stationary phase of growth (Masters et al., 1964; Masters and Pardee, 1965; Cutler and Evans, 1966), a block in DNA synthesis (Barner and Cohen, 1956; Burns, 1959), amino acid starvation or any method that inhibits protein synthesis (Matney and Suit, 1966; Stonehill and Hutchison, 1966; Inouye and Pardee, 1970), or a double block, first of protein synthesis, then of DNA synthesis (Jones and Donachie, 1973). Induction of sporulation followed by regrowth has also been used to obtain synchrony of DNA replication in bacilli (Yoshikawa and Sueoka, 1963). Induced synchrony methods rarely affect only the process they are supposed to inhibit, and should therefore be restricted to studies of that process.

Problems in evaluating and comparing currently available data include: (1) lack of consistency in the manner by which data are expressed. Data are not always shown on a per cell basis; (2) variations in interpretations of the beginning and end of a cell cycle based on similar data; and (3) resolution of cell number data and precision of analytic determinations within a cell cycle, and therefore comparisons of the two types of data.

9.1.2. Size and age classification methods
Some size or age classification methods are technologically very similar to selection methods. Frequently, they are considered to be a necessary complement of selection methods (Cooper and Weinberger, 1977). For example, the baby machine described above is also used in age classifications. Exponentially growing cultures are pulsed with a radioactive precursor, and the cells are then caught on a filter and one of each pair of pulsed cells is eluted. The content of label per cell in the eluate from small successive samples is used to deduce the stage of the cycle during which cells were engaged in the biosynthetic process under study. The method is based on the fact that in any cell cycle the first cells emerging from the baby machine are those closest to division at the time of filtration. A possible source of difficulty in studies using this approach is that only one of two nascent cells may be eluted from the filter.

Other classification methods such as cell size classification on gradients are used. If the distribution of cell sizes can be related to specific cell age classes, then it is possible to relate gradient fractions to the cell division cycle. For example, evidence exists for a continuous exponential increase in stable RNA or total protein content per cell during the bacterial cell division cycle (Dennis, 1972), so that one of these measurements can be used to classify cells obtained from a gradient according to age, and to relate other parameters to cell age. Based on theoretical considerations Koch and Schaechter (1962) proposed a useful relationship between cell size and cell age. However, as pointed out by Koch (1966), the use of particle volume or cell length as an index of cell age is controversial, since some disagreement exists about the kinetics of volume increase during the cell division cycle (Harvey et al., 1967; Kubitschek, 1968a). An additional problem is analysis of the gradient portions containing small numbers of cells.

A method that may also be considered to be an age classification is the analysis of frequency distributions in randomly dividing cell populations growing at different rates. This approach is based on the age distribution function, which is theoretically invariant, and can be defined as:

$$1 + y = 2^{(1-x)} = 2^{t_x/t_d} \qquad (11)$$

where y is the frequency of cells at age x (varying from 0 to 1). Alternatively, if a cell cycle event (x) is distinguishable and can be shown to affect a fraction of the cell population, then the timing of the event (t_x) in the cell division cycle can be established from the frequency of cells in the population (y) which exhibit the event in question (Powell, 1956; Cook and James, 1964; Painter and Marr, 1968). For example, from the frequency of cells exhibiting a visible septum, the septum separation time $(T$ or $t_s)$ can be determined in exponential phase population (Woldringh, 1976).

The age distribution equation can also be used to determine the timing of a cell cycle event that depends on the synthesis of a specific class of cellular macromolecules. For example, from residual divisions obtained after inhibition of DNA synthesis with a specific antibiotic, one can determine the time between completion of DNA synthesis and cell division (Clark, 1968).

9.1.3. Other methods

In addition to their use in timing cell cycle events, antibiotics can be used to block cells in specific stages of the cell cycle. For example, inhibition of DNA synthesis will prevent completion of chromosome replication and result in the accumulation of cells in a stage in which the chromosome has not completed its replication. This approach can be dangerous, since inhibition of chromosome replication may permit progress of other processes important for cell division. Sequential blocks employing a series of different antibiotics have also been used (Jones and Donachie, 1973).

A major problem in interpreting the effects of antibiotic inhibitors is based on the interdependence of cellular synthesis of macromolecules (section 8.2) and on the relative specificity of each inhibitor. These factors result in both time- and concentration-dependent phenomena. For example, inhibition of RNA synthesis will often be followed quickly by inhibition of protein synthesis. Less rapid and perhaps less direct effects on DNA and peptidoglycan synthesis can also be observed. At relatively high concentrations of even the more specific inhibitors, inhibition of the secondary process may occur very soon after inhibition of the primary target, substantially decreasing overall specificity and confusing interpretation. Usually antibiotics are considered to inhibit the synthesis of an entire class of macromolecules. There are reports, however, that indicate the synthesis of specific types of molecules within that class are less sensitive to inhibition than is the class as a whole. For example, the synthesis of certain membrane proteins have been reported to be less susceptible to chloramphenicol or puromycin inhibition (Vambutas and Salton, 1970; Hirashima, Childs and Inouye, 1973). Also, it is clear that several enzyme activities which are inhibited by the various β-lactam antibiotics vary in their susceptibility to individual members of this group (e.g., section 6.1).

Studies of macromolecular synthesis in undisturbed, exponential phase cultures can be supplemented by an examination of various processes during growth rate shifts. For example, as suggested by the studies of Maaløe and collaborators, an upward shift in growth rate results first in an increase in the rate of RNA, then of protein, and finally of DNA synthesis (Kjeldgaard et al., 1958). Similarly, a downward shift in growth rate yields a period of no net RNA synthesis. These results are expected in order for the culture to achieve its required new balanced growth composition.

Cell cycle events can be studied by analysis of cell composition, that is, by determining the content of macromolecules per cell at various growth rates or during growth rate shifts. The relationships derived are frequently empirical and are best described in the context of two models of cell growth and division, those of Cooper and Helmstetter (1968) and Donachie (1968), discussed in sections 9.2. and 9.3.

Mutants, particularly conditional mutants, have also been used to study events in the cell cycle. Obviously it is important to know the exact genetic and physiological characterization of the defective function. This has not been a major problem in the use of mutants defective in initiation or synthesis of DNA. However, the pleiotropy frequently exhibited by mutants (and nutritional phenocopies) defective in one of the more terminal stages of the cell cycle or in cell surface assembly serve as indicators of the complexities encountered in interpreting results obtained. Slater and Schaechter (1974) provide a discussion of these problems.

Radioautography has been used to study segregation of cellular component structures such as DNA (Von Tubergen and Setlow, 1961) into progeny. When labeled cells are grown under conditions leading to chain formation, sequential relationships of progeny cells can be followed (Eberle and Lark, 1966; Lin et al., 1971). Alternately, cells are pulsed with a radioactive precursor and chased with an excess of unlabeled precursor for various time periods. The segregation of the label into various cells can be examined by light or electron radioautography. Serious problems encountered in this type of study include lack of label specificity, various defects in the chase, and turnover of cell constituents (Van Tubergen and Setlow, 1961; Shockman et al., 1974).

9.2. The Cooper-Helmstetter model of chromosome replication and the division cycle

This model proposes that replication of the bacterial genome during the division cycle of E. coli B/r growing with doubling times between 20 and 60 minutes can be described by two constants, C and D. C is the time required for a replication point (fork) to transverse the genome, and D the time between the end of a round of chromosome replication and division (Cooper and Helmstetter, 1968). The model is shown schematically in Fig. 10 for cells with $C = 40$ minutes, $D = 20$ minutes, and doubling times between 60 and 20 minutes (at 37°C). The essential aspect of this model may be summarized as follows: Within this range of growth rates, C is a fixed 40 minutes, and D is a fixed 20 minutes. Thus, the frequency of division at any growth rate depends on the timing of C initiations. At doubling times (t_d) faster than 60 minutes, a second cycle of chromosome replication is initiated with a frequency equal to t_d. At doubling times equal to or faster than 40 minutes, chromosome replication occurs continuously during each cell division cycle, and multiple chromosomal forks are present. The presence of multiple chromosome replicating forks was demonstrated at fast growth

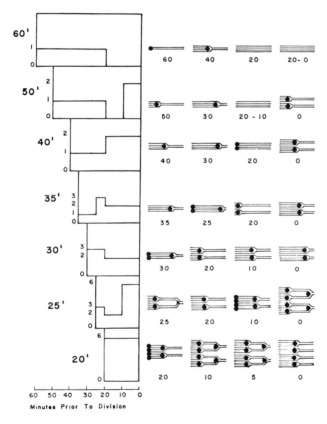

Fig. 10. Schematic illustration of the Cooper-Helmstetter model. The rate of DNA synthesis during the division of *E. coli* B/r growing with various doubling times is shown at left, assuming $C = 40$ minutes and $D = 20$ minutes. The postulated chromosome configurations in these cells are shown at right. The black dot indicates a replication point, and the numbers indicate the time in minutes prior to cell division at which the chromosome configuration is present in the cells. (Reproduced with permission from Cooper and Helmstetter [1968]).

rates (Sueoka and Yoshikawa, 1965) and the rate of DNA synthesis doubled at the expected time in the division cycle in synchronous cultures (Helmstetter, 1967). A less direct test of the model can be made by determining the time interval required to observe the beginning of the final faster rate of cell division when cultures are shifted from a slow to a fast growth rate. In bacterial cultures in which cell division depends on the frequency of C initiation, the final division rate was seen $C + D$ minutes after a shift to a faster growth rate.

In rapidly dividing cultures of $E.\ coli$ $(t_d = 20$ to 60 min) the rate of chromosome replication was found to be constant by autoradiographic study (Cairns, 1963), by determination of thymidine incorporated during short pulses in synchronous culture (Clark and Maaløe, 1967), and by measuring the amount of radioactive thymidine pulse incorporated into exponential phase cells of different ages eluted from the baby machine (Helmstetter, 1967). In the latter two studies, abrupt increases in the rate of incorporation of thymidine into DNA were observed at the times expected for initiation of a new round of chromosome replication. Compared with the replication time of about 42 minutes for the $E.\ coli$ chromosome (Cooper and Helmstetter, 1968), chromosome replication times of 32 to 48, 50, and 52 minutes were obtained for cultures of $S.$ $typhimurium$ (Spratt and Rowbury, 1971; Cooper and Ruettinger, 1973), $B.\ subtilis$ (Ephrati-Elizur and Borenstein, 1971), and $S.\ faecalis$ (Higgins, et al., 1974), respectively.

An increase in C at slow growth rates of some strains of $E.\ coli$ has been reported (Lark, 1966; Helmstetter, 1969; Gudas and Pardee, 1974). The magnitude of these increases seems to vary with strain (Helmstetter and Pierucci, 1976) and/or with methods of measuring C (Kubitscheck and Freedman, 1971; Kubitscheck, 1974; Chandler et al., 1975; Helmstetter and Pierucci, 1976). C can be experimentally manipulated by altering the thymine concentration (and thus step time) in the growth medium for thymine requiring strains of $E.\ coli$ (Pritchard and Zaritsky, 1970). In several bacterial strains at slow growth rates, a gap (G) is observed between cell division and the initiation of chromosome replication. Thus, closely analogous with the eukaryotic cell division cycle, the chromosome replication phase in bacteria can be preceded by a $G1$ phase.

Completion of chromosome replication is usually followed by a fixed time interval before division $(D;$ Cooper and Helmstetter, 1968). In various bacteria, D ranges between 14 and 25 minutes (Clark, 1968; Cooper and Helmstetter, 1968; Spratt and Rowbury, 1971; Sargent, 1973; Higgins et al., 1974). In the Cooper-Helmstetter model, D is defined as the time between completion of chromosome replication and cell division. In a strain of $E.\ coli$ B/r, where C increases at slow growth rates, D was found to increase proportionally, so that $D = \frac{1}{2}C$ (Helmstetter, 1969). Later work using several strains of $E.\ coli$ has not confirmed this empirical relationship (Helmstetter and Pierucci, 1976). Measurements of C and D in various strains of $E.\ coli$ B/r indicate that at a growth rate of 0.5 hours^{-1}, C and D can vary between 80 to 60 minutes and 40 to 20 minutes respectively (Helmstetter and Pierucci, 1976).

Analogous to the eukaryotic cell division cycle, the D phase of the cell cycle in bacteria can be subdivided into at least two phases: a $D1$ phase that requires macromolecular synthesis and a $D2$ phase in which cells are committed to, but have not actually effected, complete daughter cell separation (Perret, 1958; Clark, 1968; Higgins et al., 1974). In $E.$ $coli$, a terminal stage of the cell division cycle has been referred to as T time (Clark, 1968; Gudas and Pardee, 1974; Woldringh, 1976). T represents the time between formation of a visible cell septum and actual cell separation. An increase in T has been observed by Woldringh (1976) at growth rates below 60 minutes.

Discrepancies between reported D values can be expected, since physiological separation or even completion of a visible cell septum are not identical to the formation of two distinct, countable daughter cells (Clark, 1968; Onken and Messer, 1973; Sargent, 1973). As discussed in section 9.1, measurements based on cell counts are subject to several reservations.

A more serious complication arises from the use of terms such as C and D to define the cell division cycle. Operationally, C was defined originally as the time required for a chromosome growth point to progress from origin to terminus. D was defined as the time between completion of that process and final cell division. Recent experimental evidence (discussed in sections 9.4 and 9.5) indicated that surface events leading to division are initiated well before completion of chromosome replication time. These observations indicate that the relationship between C and cell division is not a simple sequential process. In some sections of this review, C and D are used as simple additive terms because some equations were derived in that context (sections 9.3, 9.4, and 9.5).

Qualifications of specific measurements of C and D in bacteria do not detract from the merits of the model for bacterial cell division proposed originally by Cooper and Helmstetter (1968). However, two important questions are raised by their model: (1) the mechanism regulating onset of chromosome replications, and (2) the nature of the relationship between chromosome replication time and events resulting in cell surface growth and division.

9.3. Models for the initiation of chromosome replication

A model for the relationship between onset of C and culture doubling time was proposed by Pritchard (1968) and by Donachie (1968). In this model, chromosome initiations are considered to occur at a constant cell mass or multiples of that mass. The mean mass per cell unit of $S.$ *typhimurium* cultures increases as an exponential function of the average growth rate constant μ, where $\mu = 1/t_d$ (Schaechter, Maaløe and Kjelgaard, 1958). Since at any growth rate the mass per newborn cell is proportional to the mean mass per cell of the total population, the relationship of the size of newborn cells to the growth rate (μ) is also assumed to be exponential. The relative mass of newborn cells at various growth rates is shown in Fig. 11. The cell age at chromosome initiation at each growth rate can be calculated from the model in Fig. 10, and therefore the mass/cell at chromosome initiation can also be calculated. Donachie (1968) assumed that cell mass increases exponentially during a division cycle, and found that the cell mass at chromosome initiation was discontinuous with growth rate, being one at $t_d = 60$ minutes or more, two for cells doubling between less than 60 and 30 minutes, and four for cells doubling faster than 30 minutes (Fig. 11b). These discontinuous values were equal to the number of chromosome initiations/cell/cycle (Fig. 11b). This relationship is used as the basis for some current models of mechanisms for initiation of chromosome replication.

A quantitative expression of this relationship was derived by Pritchard et al. (1969) For mean cell mass $(\bar{M})$ at any growth rate:

$$\bar{M} = K_1 \, 2^{(C + D)\mu} \quad \text{or} \quad \ln \bar{M} = (C + D)\alpha + \ln K_1 \tag{12}$$

where K_1 is a constant and $\alpha = \mu \ln 2$. Assuming that cell density is unchanged at various growth rates, for mean cell volume $(\bar{V})$:

$$\bar{V} = K_2 \, 2^{(C + D)\mu} \quad \text{or} \quad \ln \bar{V} = (C + D)\,\alpha + \ln K_2 \tag{13}$$

where K_2 is a constant.

Two general classes of models have been proposed to explain these observations. One model is based on negative and the other on positive control mechanisms. In the first, an inhibitor of chromosome initiation is considered to be diluted out by mass to permit onset of chromosome replication (Pritchard, 1968; Pritchard et al., 1969). In the second, an activator of chromosome initia-

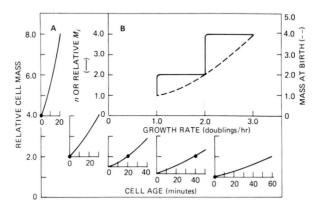

Fig. 11. Donachie's (1968) model for the relationship of cellular mass and time of initiation of chromosome replication. (a) Individual cells from cultures at each of five representative doubling times of 20, 30, 40, 50, and 60 minutes are shown increasing exponentially in mass during their respective cell cycles. The relative mass at birth (time 0) is taken from the plot in (b), where the relative mass at birth is assumed to increase with growth rate in a manner proportional to $\bar{M}$, the average mass/cell of the entire population (see Fig. 13a). In (a) the solid circles indicate the cell age in minutes at chromosome initiation as derived from the Cooper-Helmstetter model (see Fig. 10). (b) Both the relative mass at chromosome initiation (M_i) and the number of chromosome origins per cell per cycle (n) coincide. Therefore, the ratio of $M_i)$ to chromosome origins per cell is constant at all growth rates.

tion increasing with cell mass achieves a critical concentration to initiate chromosome replication (Maaløe and Kjelgaard, 1966). These models are discussed by
Pritchard (1974) and by Fantes et al., (1975). In spite of extensive work, at
present the two models have not been distinguished experimentally (e.g., Engberg et al. 1975). A mechanistic version of a positive control model was proposed
by Jacob et al. (1963), in which growth of cell membrane acts as a segregation
system during chromosome replication and new chromosome initiations occur
when membrane content doubles.

In Donachie's model, an exponential increase in cell volume is assumed during the cell division cycle. In *E. coli,* however, the evidence for an exponential
increase in cell volume during a cell division cycle has been questioned (Harvey
et al., 1967; Kubitscheck, 1968a). Also, in a study of *E. coli* B/r growing at three
different rates (t_d between 27 and 80 min) Ward and Glaser (1971) observed a
linear increase in cell volume with time during synchronous division, with a
doubling in rate occurring at about the time of chromosome replication initiation. Interruption of chromosome replication failed to affect the timing of volume increase and, after a transient interruption of DNA synthesis, initiation of a
new round of chromosome replication occurred at cell volumes smaller than
anticipated. Also, cell mass or volume did not appear to increase exponentially in
single-cell studies of a Gram-positive streptococcus (Mitchison, 1961).

Using equation 12, the expected increase in $\bar{M}$ can be calculated from α and
$C+D$. From the known $(C+D)$ value of 75 minutes (Shockman et al., 1974), the
expected increase in $\bar{M}$ of *S. faecalis* can be estimated for doubling times faster
than 75 minutes. (For convenience, the data are plotted as a function of α the
exponential growth rate constant [$\alpha = 0.69\mu$; μ is frequently referred to as the
growth rate, and represents $1/t_d$ in hours]). Fig. 12a shows that the data do not
fit the expected relationship, and that $\bar{M}$ increases with growth rate far less than
would be anticipated from the model. Furthermore, if the assumption is made
that cultures of *S. faecalis* increase in $\bar{M}$ with growth rate according to equation
12, the best fit obtained gives a calculated slope of about 26 minutes. This value is
close to D and is much smaller than the $(C + D)$ for *S. faecalis* (Higgins et al.,
1974). Thus, the model of Donachie (1968) does not fit the observed data for *S.
faecalis.*

Mean cell length measurements obtained by Sargent (1975a) for *B. subtilis* can
be converted readily to $\bar{V}$, since in this strain cell diameter does not change with
growth rate. A value of 0.55 μm for cell diameter of *B. subtilis* was kindly provided by M. L. Higgins (personal communication). The calculated volume data
for *B. subtilis* are plotted against growth rate in Fig. 12b. In *B. subtilis, C* has been
estimated to be 55 minutes (Ephrati-Elizur and Borenstein, 1971) and cell separation time to be about 20 minutes (Sargent, 1975a). The chromosome replication time has also been estimated to be about 80 minutes (Sargent, 1975c) and
time between nuclear segregation and cell division to be about 45 minutes
(Sargent, 1974). Since $C + D$ minutes is uncertain for this strain of *B. subtilis,* two
values were used: one assumes that $C + D = 75$ minutes, the second, that $C + D$
$= 135$ minutes. The data fail to fit an exponential increase of $\bar{V}$ with α. The best

fit exponential line yields a slope of about 45 minutes. Data for another gram-positive rod, *B. megaterium* (Fig. 12c,d), also failed to exhibit a good fit to an exponential function of either $\bar{M}$ (Herbert, 1961) or $\bar{V}$ (Sud and Schaechter, 1964). The calculated $C + D$ times from the best slopes are 59 and 20 minutes, respectively. For the $\bar{M}$ data a slope of 151 minutes can be obtained, but only at growth rates above 1.0 hour^{-1}. Thus it appears that in instances where $\bar{M}$ or $\bar{V}$ have been determined in Gram-positive organisms the predicted exponential relationship with α does not hold.

A reasonably good fit to an exponential equation can be found for $\bar{M}$ data for *Aerobacter aerogenes* (Herbert, 1961) (Fig. 13b), and *S. typhimurium* (Schaechter et al., 1958) (Fig. 13a), for $\bar{V}$ data for *E. coli* (Kubitcheck, 1974) (Fig. 13c) and for V of $\bar{M}$ data for the blue-green algae *Anacystis nidulans* (Fig. 13d and 13e; Mann

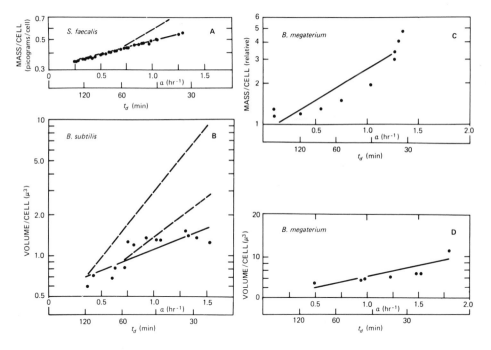

Fig. 12. Average mass or volume per cell at various growth rates for three gram-positive bacteria. The logarithm of the various parameters are plotted against α, the exponential growth rate constant in hours^{-1}. The slope of the line above an α value $= C + D$ minutes should be equal to $C + D$ minutes in Donachie's model (1968; Fig. 11). In each case, the best exponential regression line through the points is given (solid line). The dashed lines are the expected slope from Donachie's model. (a) *S. faecalis*, $\bar{M}$ in picograms/cell obtained from absorbance at 675 nm converted to dry weight per cell particle counts (corrected for chains). (b) *B. subtilis*, $\bar{V}$ in μm^3. Data obtained by Sargent (1975) on cell length were converted to $\bar{V}$, assuming a diameter of 0.55 μm and no change in cell diameter with growth rate. Two dashed lines are given, one based on a $C + D$ value of 75, and one of 135 minutes. (c) *B. megaterium*, $\bar{M}$ as obtained from Sud and Schaechter (1958). Independent estimates of $C + D$ are not available for *B. megaterium*. (d) *B. megaterium*, $\bar{V}$ as obtained from Herbert (1961).

672

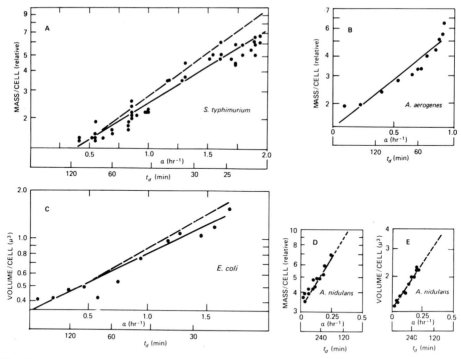

Fig. 13. Average mass or volume per cell at various growth rates for four gram-negative bacteria. For symbols see Fig. 12. (a) *S. typhimurium* (data from Schaechter et al., 1958). (b) *A. aerogenes* (data from Herbert, 1961). (c) *E. coli* B/r (data from Kubitschek, 1974). (d) and (e) $\bar{M}$ and $\bar{V}$ data for *A. nidulans* (Mann and Carr, 1974).

and Carr, 1974). The $C + D$ times calculated from the slope of these plots are 83, 62, 51, and 178 minutes for *A. aerogenes*, *S. typhimurium*, *E. coli*, and *A. nidulans*, respectively. Whereas the $C + D$ times calculated from these data for *A. aerogenes* are somewhat longer than the 60 to 65 minutes, $C + D$ times for *E. coli* as determined by other methods, the calculated $C + D$ values for *E. coli*, *S. typhimurium*, and *A. nidulans* values appear to agree with published data (Spratt and Rowbury, 1971; Kubitscheck, 1974; Mann and Carr, 1974). An interesting point, also made by Kubitscheck (1974), is that in organisms in which an exponential relationship holds, it applies at growth rates slower than $C + D$ min. Therefore, either $C + D$ does not stretch at doubling times in excess of $C + D$ minutes, or the slope of the relationship between $\ln \bar{M}$ or $\bar{V}$ and growth rate has a fortuitous analogy to $C + D$ minutes in some Gram-negative microorganisms. The point that the relationship may be fortuitous has been made by Donachie et al. (1973). Sedgwick and Paulton (1974) have come to a similar conclusion based on changes in $\bar{M}$ at various growth rates in *B. subtilis* and *M. flavus*.

Less direct information indicates that timing of events in the cell division cycle, rather than cell mass, may be the critical event in the initiation of chromosome replication (Helmstetter, 1974a,b). The last observation, made using a Gram-negative rod, raises the possibility that stage of surface growth is a critical element in initiation of chromosome replication in bacteria. This model evolves from the original view of Jacob et al. (1963). A correlation has been observed in *E. coli* between initiation of new rounds of chromosome replication and formation of a cross wall visible in the light microscope (Gudas and Pardee, 1974). Gudas et al. (1976) recently obtained evidence for the synthesis of a protein (protein D) of the outer membrane of *E. coli* B/r, which is synthesized between the end of one round of chromosome replication and the start of the next. Protein D has a high affinity for DNA, a molecular weight of 80,000, and its formation depends on both DNA and peptidoglycan metabolism. These investigators propose that protein D combines with an "initiation (I) structure" and that initiation of chromosome replication would occur as a consequence of the DNA origin binding to protein D. The location of protein D in the *outer* membrane poses some problem but could provide for an additional function of the bridges seen between inner and outer membrane (Bayer, 1967; 1975) in *E. coli*. Evidence for the presence of a similar protein was also reported by Churchward and Holland (1976a,b).

The role of surface growth in initiation of chromosome replication will be discussed further in section 9.7. From the evidence presented here, it seems probable that the exponential increase in cell mass with growth rate observed by Schaechter et al. (1958) is restricted to only a few bacterial species and is the end result of another, more fundamental mechanism common to all bacteria. Consequently, at their simplest level, models for the initiation of chromosome replication that are based on changes in the relative cell concentration of positive or negative control elements do not seem to hold for all prokaryotes.

9.4. Relationship of chromosome replication to cell division

In the Cooper-Helmstetter model, initiation of C could determine the time of cell division, since division would occur after $C + D$ minutes (Fig. 10). However, cell division appears to be determined by factors more complex than initiation of chromosome replication. First, at a doubling time of about 60 minutes, inhibition of DNA synthesis in one strain of *B. subtilis* does not prevent cell division for 138 minutes (Donachie et al., 1971). Cells arising from these divisions are normal in size, but only 50% of the daughter cells contain DNA. In another strain, divisions continued for 60 minutes after interruption of chromosome replication, but the dividing cells contained nuclei (Sargent, 1975a). Anucleate cells were obtained in a temperature-sensitive mutant of *B. subtilis* (Sargent, 1975b). Second, in *E. coli*, interruption of chromosome replication for 40 minutes did not result in delayed division (Jones and Donachie, 1973). Division occurred as early as 5 minutes after resumption of chromosome replication. In a number of chromosome initiation mutants, cell division does not seem to be coupled to chromosome completion (Mendelson and Gross, 1967; reviewed in Slater and Schaechter, 1974;

Pardee and Rozengurt, 1975). Third, a shift to a medium containing a full complement of thymine of cultures of *E. coli* grown in low concentrations of thymine that decreased the velocity of chromosome replication resulted in the immediate onset of a new division (Meacock and Pritchard, 1975). A reverse shift (from high to low thymine) was accompanied by a 20-minute lag before the new division rate was observed. Thus, a downward shift in chromosome replication velocity did not immediately affect division. Overall, these observations indicate that the onset of D in *E. coli* is not necessarily coupled to completion of the chromosome, but that division normally requires completion of the chromosome replication time.

Evidence that inhibition of chromosome replication can prevent division in several bacterial species was also obtained from experiments that utilized bacteria in randomly dividing exponential phase, or in synchronous cultures. These observations are based on studies using inhibitors of DNA synthesis, thymine starvation, or mutants with a temperature-sensitive DNA replication system (reviewed by Slater and Schaechter, 1973). Donachie (1973) proposed that all these observations can be integrated if the assumption is made that there are two "clocks," both of which cover a series of events leading to division, and are initiated at or near the time of chromosome initiation (Fig. 14a). These parallel clocks cover chromosome replication and a second series of events requiring protein synthesis. From studies using inhibitors of the synthesis of specific classes of macromolecules in sequence to each other, Jones and Donachie (1973) suggested that the chromosome replication "clock" and another protein synthesis-requiring "clock" can run independently of each other for all but 5 minutes after completion of C. Synthesis of a critical division protein is thought to occur at that time. In addition, Marunouchi and Messer (1973) suggested that in *E. coli* B/r protein synthesis is required to complete chromosome replication. Differences were observed, however, in the requirement for protein synthesis of various *E. coli* strains (Ron et al., 1975). It has also been suggested that amino acid starvation or addition of chloramphenicol or rifampin to inhibit protein synthesis reduce the chromosome replication velocity in some strains of *E. coli* (Pato, 1975).

A modification of Donachie's model states that the crucial C-dependent division event need not be triggered by completion of chromosome replication in all bacteria (Fig. 14b), and this modification could account for the observation that cell division can occur in the absence of chromosome replication in *B. subtilis* (Donachie et al., 1971; Sargent, 1975b).

A second modification was independently suggested by Shockman et al. (1974) for *S. faecalis* (Fig. 14c). In this modification, at about the time of initiation of chromosome replication a separate series of timed events that lead to the synthesis of a complete new unit of cell surface are also initiated. This series of surface biosynthetic events takes a constant period of time (W). At generation times of about 45 minutes or slower a single W cycle occurs per cell. At faster growth rates, W cycles are initiated at intervals which depend on t_d so that overlapping W cycles occur. These W cycles could give rise to the overlapping

675

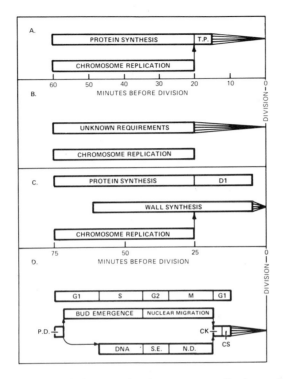

Fig. 14. Models of the relationship between the chromosome replication cycle and other cellular reactions leading to division. (a) Donachie's model. A start signal initiates chromosome replication and a second series of reactions required for division. A trigger at completion of C is required to bring the two pathways together. T.P. indicates a hypothetical termination protein. (b) Modified Donachie model for *B. subtilis*. Essentially the same as the model in (a), but a trigger is not required for division. (c) Shockman et al., (1974) model for *S. faecalis*. A start signal initiates C and, somewhat later, a second reaction called W. Both are completed after a fixed period of time. Protein synthesis is also required. (d) The model of Hartwell et al. (1974) for cell division in yeast, G1; S, G2, and M are phases of the eukaryotic cell division cycle. P.D. = plaque duplication, S.E. = spindle elongation, N.D. = nuclear division, CK = cytokinesis, and C.S. = cell separation.

sequence of secondary surface growth zones observed in *S. faecalis* growing at t_d ≈ 32 minutes. In other details this model resembles Donachie's model, since both completion of the chromosome and protein synthesis are required for cell division in *S. faecalis* (Higgins et al., 1974). An important feature of this scheme is that surface growth zones are spatially fixed and can be visualized as capable of synthesizing only a fixed amount of surface per growth zone. Thus, this model provides a "clock" mechanism that is partially independent of chromosome replication. Recent data on the synthesis of membrane proteins during the cell cycle (sections 6.3 and 9.5) may provide a basis for the further development of the models presented in Figs. 14a and 14c and their eventual integration into a single model.

The model proposed by Donachie for regulation of cell division events in bacteria is similar to one postulated by Hartwell (1974) for yeast. Hartwell suggested that DNA replication and surface division events occur as a result of a master "clock" that operates at the initiation level of each series. After a "start" signal the two sequences proceed independently of each other. Prior to cell division one crucial event requires completion of DNA synthesis (Fig. 14d). It is attractive to think that a requirement for such a terminal event in both yeast and some bacteria may have an evolutionary basis. Such an event could act as a checkpoint to insure that all necessary structures for the replication of daughter cells are passed on from mother to daughter.

In addition to the models shown in Fig. 14, Sargent (1975b) proposed an alternative clock which runs for one generation plus D minutes for $B.$ $subtilis.$

9.5. Regulation of cell surface enlargement in E. coli during the cell cycle

In recent years substantial advances have been made in our knowledge of the biochemistry of changes in the bacterial cell surface. Certain enzyme activities and proteins related to changes in the cell surface have also been related to the cell division cycle. The bulk of this work was discussed in detail in sections 5.3.4 and 6. Here an attempt is made to synthesize a number of the observations made on $E.$ $coli$ into an admittedly selective and speculative sequence with respect to the cell cycle.

Exposure of exponentially growing cultures of $E.$ $coli$ to FL-1060 resulted in continued mass and surface area increases, without an immediate increase in osmotic fragility (section 6.3). The cells gradually change in shape, so that at 50 minutes they have a rounded appearance, and in cultures with a t_d of about 60 minutes division ceases at about 60 minutes, indicating that FL-1060 affects an early event in the cell cycle (e.g., $C + D$ minutes before division; James et al., 1975). The suggestion was made that FL-1060 prevents the occurrence of an early morphogenetic event in the cell division cycle that results in growth of $E.$ $coli$ as a cylinder.

FL-1060 treatment does not appear to affect DNA synthesis (James, 1975). Inhibition of DNA synthesis, on the other hand, results in growth in the form of long cylindrical filaments. When mutants with a temperature-sensitive defect in DNA synthesis are returned to the permissive temperature, the filaments fragment into normal sized cells (for details, review Slater and Schaechter, 1975). FL-1060 treated cells can form filaments on subsequent treatment with nalidixic acid (an inhibitor of DNA synthesis) when the FL-1060 treatment has been administered for 0 or 30 minutes, but cannot be induced to form filaments after 60 minutes or more. These observations suggest that initiation of a cylindrical growth zone can yield a limited amount of actual or potential "filament growth," a point that is expanded on in section 10.4. Eventually, however, FL-1060 treated cells blocked at initiation of cylindrical growth can no longer form filaments. Therefore, the data are also consistent with the view that initiation of cylindrical growth and DNA synthesis do not depend on each other but are instead regulated by a third mechanism.

The early effects of FL-1060 appear to occur via its selective binding to binding protein 2, one of six inner membrane proteins capable of binding penicillin G (Spratt, 1975; Spratt and Pardee, 1975). Estimates of about 10 molecules of protein 2 per cell are consistent with a catalytic rather than a structural role. At present, possible reactions catalyzed by protein 2 are not known. It is unlikely that protein 2 catalyzes either a DD-carboxypeptidase, endopeptidase, or transpeptidase reaction since these three activities are not inhibited by FL-1060 (section 5.3.).

Exposure of cells to FL-1060 reduces the appearance in the outer membrane of protein G (James, 1975). James suggests that protein G may be a structural protein for cell elongation and is inserted into the outer membrane from initiation of cylindrical growth to termination of chromosome replication. Synthesis of protein G could represent all or part of the requirement for protein synthesis for cell division during chromosome replication proposed by Donachie (Fig. 14a).

Cell surface expansion per se does not require the action of binding protein 2, since cells treated with FL-1060 continue to synthesize peptidoglycan (at a slightly reduced rate). In FL-1060 treated cells, lipoprotein continues to be inserted into the outer membrane, and perhaps because of the decreased rate of peptidoglycan synthesis more lipoprotein linked to peptidoglycan was found in FL-1060 treated cells (Braun and Wolff, 1975). In contrast, cephaloridine causes immediate cell lysis when added to cultures at its lowest effective concentration (Spratt, 1975). Cephaloridine has a relatively high binding affinity for another inner membrane protein, protein 1. Spratt suggests that protein 1, possibly acting as a transpeptidase, introduces new precursors into growing peptidoglycan chains and performs the expected functions required for cell surface expansion.

Completion of chromosome replication has been found to be accompanied by several changes in surface composition including: (1) the loss of protein G from the outer membrane; (2) an increased rate of lipoprotein incorporation in the outer membrane; (3) the appearance of protein D in the outer membrane; and (4) an increased activity of several peptidoglycan hydrolases. Inhibiton of DNA synthesis is accompanied by a continued accumulation of protein G, suggesting that on completion of chromosome replication protein G is no longer inserted into the outer membrane. On the other hand, the rate of incorporation of free lipoprotein into the outer membrane increases at the time of septation (James and Gudas, 1976) and, since a corresponding increase in incorporation rate of peptidoglycan-bound lipoprotein was not observed, the ratio of free to bound lipoprotein increases appreciably at septation. On the basis of these observations, and those of Burdett and Murray (1974a, b), it has been proposed that free lipoprotein could be involved in the separation of the outer membrane from the peptidoglycan layer at an early stage of septation. Currently, precise information on the topological distribution of free and bound lipoprotein in *E. coli* is unavailable.

At present we are unable to postulate a role for outer membrane protein D in the cell cycle (see section 6.3.). According to Gudas et al. (1976), protein D appears to increase tenfold between rounds of chromosome replication and is very stable when inserted into the membrane. A protein of similar molecular weight described by Churchward and Holland (1976a, b) was synthesized only during a brief period near the time of division, but in their studies about 70% of newly synthesized protein was lost from the envelope during the succeeding generation. Gudas et al. (1976) suggest that protein D acts as an attachment site between DNA and the outer membrane at a specific time in the cell cycle.

Currently, there are two possible explanations for the observed increase in rate of protein D appearance in the outer membrane of FL-1060 treated cells (Gudas et al., 1976). Proteins D and G may be mutually replaceable so that if one is no longer inserted, the other is. A more likely explanation, however, is that protein D is a transient surface element that is inserted into the outer membrane at completion of chromosome replication and may be required for cell division and initiation of the next chromosome replication cycle (Fig. 14a). A block in initiation of cylindrical surface elongation could prevent the normal cessation of protein D insertion into the outer membrane after initiation of a new round of chromosome replication.

Unlike FL-1060, benzylpenicillin or cephalexin appear to act at a terminal stage in the cell cycle. When cultures are grown in concentrations of cephalexin or benzylpenicillin low enough to just inhibit cell division, the binding of [^{14}C]benzylpenicillin to another of the previously mentioned six inner membrane proteins, protein 3, is reduced. This protein was shown to be the demonstrable, thermolabile, penicillin-binding protein in a temperature-sensitive division mutant isolated by Spratt that forms filaments at the restrictive temperature. In another temperature-sensitive division mutant, PAT-84, low doses of penicillin partially inactivated DD-carboxypeptidase activity at the restrictive temperature (Mirelman et al., 1976). It has been suggested that the action of a penicillin- (and ampicillin) sensitive DD-carboxypeptidase (or endopeptidase) is required for a terminal event in the cell cycle of *E. coli* (Mirelman et al., 1976). Incorporation of newly synthesized peptidoglycan into preexisting wall was stimulated by the low dose of penicillin in the temperature-sensitive division mutant, perhaps indicating that a block of a terminal cell cycle function (possibly protein 3) provides excess precursors for an early cell surface growth function (possibly mediated by protein 1).

Matsuhashi and collaborators (1977, and personal communication) have recently isolated and investigated mutants of *E. coli* K–12 that are defective in D-alanine carboxypeptidase, transpeptidase and endopeptidase activities. Two strains that carry a mutation at 68 min on the *E. coli* chromosome (*dac* B) had decreases in all 3 activities and were considered to be defective in the D-alanine carboxypeptidase activity IB of Tamura et al. (1974). These two strains also lacked penicillin binding protein 4 but grew normally. Similarly, *dac*A mutants which appear to be defective in D-alanine carboxypeptidase IA activity (Tamura et al., 1974) also grew normally as did a double (*dac*A, *dac*B) mutant. These results were interpreted as suggesting that these two enzymes are not essential activities unless: (1) the residual levels of activities remaining are sufficient, (2) the mutant enzymes are inactive in vitro but are active in vivo, or (3) that alternative reactions that can bypass these activities are present.

During the normal cell division cycle of *E. coli*, three distinct cycle phases can therefore be recognized in surface growth: (1) initiation of cylindrical growth accompanied by insertion of protein G and other proteins into the outer membrane; (2) cessation of cylindrical growth at the end of *C*, accompanied by loss of protein G from the outer membrane, and possibly by insertion of protein D into the outer membrane, and an increase in the ratio of free to bound lipoprotein; and (3) cross-wall formation that may require penicillin-binding protein 3. Surface expansion can occur via a reaction associated with binding protein 1, and does not require the initiation event indicated under phase (1) above.

9.6. Mathematical analyses of the regulation of cylindrical surface extension and cell division in rod-shaped bacteria

Direct observations of rod-shaped organisms were used to determine the rates of cell elongation and the timing of cell division (Schaechter et al., 1962; Powell and Errington, 1963; Errington et al., 1965) with the conclusion that, on the average, cells increased in length exponentially. Individual cells showed large variations in their pattern of elongation, but it appeared that on the average large cells grew more rapidly than smaller ones. A doubling of the linear rate of cylindrical elongation at a specific time in the cell cycle has also been suggested.

To investigate the regulation of cell size during the cell division cycle, bacteria can be examined at various growth rates or during growth rate shifts. At any growth rate, rod-shaped bacteria extend only in length (Harvey et al., 1967). In addition, fast growing cells of at least Gram-negative enteric bacterial species have a larger diameter than slow growing cells (Schaechter et al., 1958).

Attempts have been made to analyze mathematically measurements of the average lengths of cells growing at various rates to gain insight into events occurring during the cell cycle. Usually attempts have been made to fit data to an equation of the general form:

$$\bar{L} = k \cdot 2^{x/t_d} \tag{14}$$

where $\bar{L}$ is average cell length, k is a constant that may include a term for rate of elongation, the doubling time, or both, and x is a cell age term related to a cell cycle event (Zaritsky and Pritchard, 1973) or to a specific time in the cell cycle when the rate of cell elongation doubles (Zaritsky and Pritchard, 1973; Sargent, 1975a; Grover et al., 1977). Equation 14 can also be expressed as:

$$\ln \bar{L} = \ln K + x \cdot \alpha \tag{15}$$

If $\bar{L}$ increases exponentially with increasing α, then various models of cell elongation can be evaluated using the numerical values for k and x derived from experimental data (Grover et al., 1977). An obvious constraint is the constant, k, since models of cell elongation can be constructed in which k would not be constant (Sargent, 1975a; Grover et al., 1977).

For *E. coli*, the data are thought to be consistent with a model in which the rate of cell elongation per growth site is linear, is directly proportional to the growth rate, and doubles 17 minutes before division (Grover et al., 1977). These investigators, using lengths of a newly divided cell, proposed that for a newborn cell, L_0:

$$L_0 = k(1 + x/t_d) \qquad (16)$$

where $k = 1.47$ μm and $x = 15.3$ minutes. The data used to derive this relationship showed an increase in L_0, with growth rate from 1.6 μm (at $t_d = 160$ minutes) to 2.4 μm (at $t_d = 24$ minutes), and would equally well fit a linear or an exponential relationship. Thus, although the conclusions of Grover et al. (1976) seem interesting, they may be premature. The idea that cell diameter is regulated by the number of chromosome replication positions in a cycle, whereas cell length is regulated by chromosome termini, has also been expressed (Zaritsky and Pritchard, 1973; Zaritsky, 1975). The average cell diameter of *E. coli* approximately doubled between doubling times of 160 and 24 minutes (Grover et al., 1976).

The average length of *B. subtilis* increases with increasing growth rate in a clearly biphasic manner (Sargent, 1975a; Fig. 15). The data do not fit equation 14 (or 15), indicating that in *B. subtilis* the rate of elongation is regulated in a complex manner. The break at doubling times between 50 and 60 minutes (Fig. 15) suggests that the average cell length may become constant at rapid growth rates, when chromosome replication times overlap (e.g., when the doubling time approaches C).

Data concerning assembly of division poles can be derived from the frequency of visible "septal" sites in cell populations at various growth rates. Two studies are available using different strains of *B. subtilis* (Paulton, 1970; Sargent, 1975a). From the frequency of visible septal sites ($\bar{s}$), the average number of successive division sites per cell (or potential septal growth zones; $\bar{p}$) can be derived, assuming symmetrical distributions, since:

$$\bar{s} = 2^{\bar{p}} - 1 \qquad (17)$$

If the assumption is made that at each generation the number of potential septal growth zones is doubled, then from equation 17:

$$\bar{p} = \ln (\bar{s} + 1)/\ln 2 \qquad (18)$$

So that $\bar{p}$ can be obtained from the septum frequency data. An independent derivation of $\bar{s}$ can be obtained by use of the age distribution equation (equation 14) on the assumtion that p is an integer (Lapidus, 1971). Thus:

$$t_s/t_d = \ln (\bar{s} + 1)/\ln 2 \qquad (19)$$

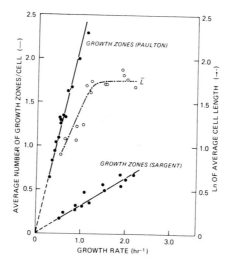

Fig. 15. Average number of potential septal growth zones per cell and average cell length of *B. subtilis* at various growth rates. The average number of potential septal growth zones per cell (O———O) was calculated from the frequency of septa per cell according to equation 20. The upper line was obtained by Paulton (1969) from cells treated with tannic acid; the slope of this curve gives t_s, the time from visible appearance of the potential septal growth zone to cell division, and is 138 minutes. The lower curve was obtained from data of Sargent (1975a) from heat-fixed preparations stained with 0.01% crystal violet. In this case t_s = 19 minutes. The average length $(\overline{L})$ was also obtained by Sargent (1975a). The log of $\overline{L}$ is plotted and shows a definite lack of linearity with increasing growth rate.

where t_s is the time required to complete a septal site from its detectable inception to division.

Combining equations 18 and 19 yields:

$$\bar{p} = t_s/t_d \qquad (20)$$

A plot of $\bar{p}$ versus growth rate ($\mu = 1/t_d$) provides a straight line with slope = t_s.

From the data of Paulton (1970) and Sargent (1975a), $\bar{p}$ was derived and plotted against μ (Fig. 15). A straight line was obtained at values for t_s of 138 and 19.6 minutes, respectively. At this time it is not clear whether the t_s values observed are due to differences between the two strains of *B. subtilis*, growth media or to the methods used to determine $\bar{s}$. In either case it seems clear that formation of a division pole in *B. subtilis* requires a constant time irrespective of growth rate.

A more important inference about surface growth in *B. subtilis* can be made from the observation that while $\bar{p}$ increases linearly with growth rate (Fig. 15), cell length, volume, or surface area do not increase above growth rates of about 1.5 hours^{-1} (Fig. 15). The apparent independence of potential division sites and parameters related to cell volume or cell length suggest that, in *B. subtilis*, formation of division poles is regulated independently of formation of volume or of cylindrical surface. The result of a shift upward during thymine starvation of *B. subtilis* is consistent with this interpretation (Sargent, 1975).

In *E. coli* K12, the cell separation time, t_s, decreases slightly with increased growth rate (Perret, 1958), according to the empirical equation:

$$t_s = 34 + 0.088\, t_d \text{ (min)} \tag{21}$$

This observation agrees qualitatively with the observation of Woldringh (1976) with *E. coli* B/r, although some strains vary in the magnitude of this effect (see also Gudas and Pardee, 1974). Since t_s is not constant at all growth rates, it follows that the number of potential division sites in Gram-negative rods does not increase as a linear function of the growth rate. It is possible that, in Gram-positive rods, extension of cylindrical surface and formation of division poles are regulated independently of each other, whereas in Gram-negative rods the two processes are more closely linked. A more trivial explanation may be that in some organisms under certain growth conditions cell separation time is stretched at slow growth rates.

The discussion above briefly summarizes the essential features of an important current model for regulatory mechanisms in the initiation of C and D. The relationship between cell mass at the initiation of chromosome replication and the number of chromosome origins at various growth rates is interesting but does not seem to be general. This lack of generality suggests that an element or elements other than mass (but related to mass in some bacteria) may regulate chromosome initiation at various growth rates.

The currently accepted operational definition of D as the time between completion of C and cell division fails to include several surface events required for cell division. At least in some bacteria these events appear to be initiated early in the cell division cycle, and completion of chromosome replication is not a necessary requirement for cell division in all bacterial species. In some species events connected with division can occur in the absence of chromosome replication, but division requires completion of both C and D. Finally, a relationship proposed by Zaritsky and Pritchard (1973) between average cell length at division and average number of chromosome *termini* and growth rates requires further analysis.

9.7. The role of surface growth in chromosome replication and segregation

Bacteria lack both a nuclear membrane and a mitotic apparatus. The model of Jacob et al. (1963) was proposed to provide a method for the orderly segregation of the single bacterial chromosome to the two daughter cells. This model specifies that replication of the bacterial chromosome is initiated at a specific point (the origin), and that duplication of the two DNA strands occurs on a structure that is integrated with the bacterial membrane. The membrane was suggested to have two roles in this process: (1) to provide the enzymes and regulatory elements needed for initiation, replication and termination of DNA synthesis; and (2) to segregate the newly replicated chromosomes. Membrane growth between the attached nucleoids would in itself result in segregation (Fig. 16a), so that two resultant chromosomes would be ready for new initiation after cell division (Fig. 16a). The model of Jacob et al. requires that the chromosomes remain attached to the membrane at all stages of the normal cell division cycle. A second model for chromosome segregation (Clark, 1968) proposed that new surface growth occurring to one side (rather than both sides) of the nucleoid serves to segregate attached daughter nucleoids (Fig. 16b).

A third model can be suggested that does not require the attachment of the chromosome to the membrane throughout the cell cycle. In this model the bacterial chromosome is attached to a surface structure during replication. At completion of replication, however, the chromosome dissociates from its original surface site. The two newly replicated chromosomes then attach themselves to two new surface sites at which a new cycle of replication can be initiated. In common with the model of Jacob et al., surface growth is required to generate new surface attachment sites that are important for initiation of chromosome replication (Fig. 16c). In a less extreme form, this model was also proposed by Sargent (1974).

Numerous studies, particularly with bacilli, provide ample evidence for attachment of DNA to the cytoplasmic membrane at its origin and terminus (Sueoka and Quinn, 1968; Snyder and Young, 1969; Fielding and Fox, 1970; Yamaguchi and Yoshikawa, 1975) and its growing forks (Ganesan and Lederberg, 1965; Smith and Hanawalt, 1967; Ivarie and Péne, 1970; Yamaguchi, et al., 1971). Also, the series of enzymes required for chromosome replication have been shown to be closely associated with the membrane. The subject of DNA attachment to membranes has been reviewed extensively (Leibowitz and Schaechter, 1975). Furthermore, genetic evidence (Slater and Schaechter, 1974) has been obtained that correlates temperature-sensitive defects in chromosome initiation to structural alterations in bacterial membrane proteins. At least one antibiotic affecting DNA synthesis, nalidixic acid, is thought to affect membranes.

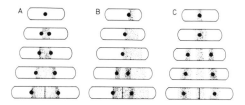

Fig. 16. Models of nuclear segregation in bacteria. (a) The model of Jacob et al. (1963). All surface growth occurs between the nuclei. (b) The model of Clark (1968) in which nuclei are rigidly attached to a particular site and growth occurs only to one side of the nucleus. (c) An alternative model in which nuclei move to a site 25% of the cell length from the poles. Surface growth occurs to both sides of the nucleus. Most intense stippling indicates newest cell envelope. (Reproduced with permission from Sargent [1974].)

684

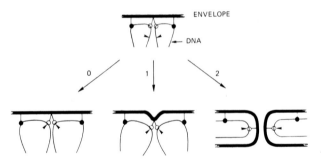

Fig. 17. Schematic illustration of the progression toward initiation of DNA synthesis in three basic classes of *E. coli* cells. The drawing at the top of the figure shows a portion of an *E. coli* cell containing two chromosomal initiation segments. The chromosomes are attached to the cell envelope by the filled circles, and the envelope-attached replication apparatus is indicated by the open circle. The site of initiation of chromosome replication (the origin) is indicated by arrows. The three diagrams in the lower portion of the figure indicate the eventual appearance of class 0, 1, and 2 cells at the time of initiation of chromosome replication. Class 0, 1, and 2 cells initiate chromosome replication respectively before onset of *D*, during *D*, and after *D*. (Reproduced with permission from Helmstetter [1974b]).

Initiation of chromosome replication in bacteria requires protein synthesis (Maaløe and Hanawalt, 1961). Helmstetter's (1974a) data suggested that inhibition of protein synthesis by high concentrations of chloramphenicol (200 μg/ml) failed to prevent chromosome initiation in *E. coli* B/r in cultures growing at such rates that chromosome initiation occurred during *D* (i.e., 40–50 min, Fig. 17). At all growth rates in which chromosome initiation occurred at times in the cycle other than *D*, chloramphenicol successfully prevented initiation. Helmstetter (1974b) used these findings as the basis for a model in which initiation of chromosome replication depends on elongation of the envelope between the origins of sister chromosomes bound to the envelope. Initiation of chromosome replication during the *D* period would be insensitive to chloramphenicol or rifampicin inhibition because envelope elongation between sister chromosomes would continue in the presence of these inhibitors after cells have entered the *D* period (when protein synthesis is not required), but not before entry in the *D* period (when protein synthesis is required). In this model, the timing of chromosome replication initiation is controlled by the rate of envelope elongation between the chromosome origin and the chromosome replication apparatus. In cells, in which chromosome initiation is insensitive to chloramphenicol or rifampicin, septum formation per se is not thought to provide the required surface growth because concentrations of penicillin permitting surface elongation but not septation did not affect chromosome initiation. A requirement of this model is that surface expansion must occur with equal efficiency during both the *C* and *D* periods.

The attachment of the chromosome to a membrane or other surface site throughout the cell cycle requires conservation of a portion of the surface during both chromosome replication and segregation. In turn, permanent attachment of a chromosome to a surface component that occupies a fixed position in the cell requires that a given strand of DNA always segregates toward one pole at division. Nonrandom segregation of DNA strands was suggested from studies of *B. subtilis* (Eberle and Lark, 1966), *L. acidophilus* (Chai and Lark, 1967), and *E. coli* B/r (Pierucci and Zuchowski, 1973). Pierucci and Zuchowski also provide a critical review of the relevant literature. Nonrandom DNA segregation observed in *E. coli* was thought to be consistent with the hypothesis that only one strand (called the positive strand) becomes attached to a fixed cell site, and that this attachment becomes permanent the first time the strand is used as a template.

Cooper and Weinberger (1977) reviewed and expanded the data for the *E. coli* system and provide a new, probabilistic interpretation. They suggest that chromosome segregation depends on a finite probability, characteristic of the medium (and therefore of the growth rate), that a strand will segregate in the same direction as it did previously. The degree of nonrandom segregation was found to depend on the medium in which the cells were grown during chromosome segregation after the period of DNA labeling. At present, compelling evidence is lacking for acceptance of a model in which DNA remains attached to a surface growth zone throughout its replication cycle.

If, at completion of replication, the chromosome migrates or "jumps" to a new cell location, orderly segregation of chromosome attachment sites would still be required. In this model, however, the surface site that segregates need not be a membrane component. It could be any surface element capable of generating new spatially and/or temporally oriented, chromosome attachment (and initiation) sites. Preliminary evidence for the attachment of the origin and terminus of the *B. subtilis* chromosome to detergent-treated walls has been reported (Streips et al., 1976).

Several lines of evidence are consistent with the chromosome jump model. Evidence from Worcel's laboratory (Worcel and Burgi, 1974) indicates that during amino acid starvation the *E. coli* chromosome progressively dissociates from the membrane. During amino acid starvation, ongoing chromosome replication cycles are completed but initiation of a new cycle is prevented (section 9.3.). During regrowth after amino acid starvation, reattachment of the chromosome to a membrane structure preceded reinitiation of replication (Worcel and Burgi, 1974). These results are in good agreement with ultrastructural evidence (Daneo-Moore and Higgins, 1972) which indicated that, during amino acid starvation, the nucleoid of *S. faecalis* becomes centrally visible, presumably away from surface attachment sites. Similar results were obtained in *E. coli* after antibiotic treatment by Zusman et al., (1972). Further indirect evidence consistent with this model was obtained from studies of the number and location of the membranous invaginations called mesosomes in *S. faecalis* (Higgins and Daneo-Moore, 1972; Higgins et al., 1974). Mesosomes were always found in association with the nucleoid and with current or future division sites but never in between (Fig. 18). In small cells with a single nascent cross wall, mesosomes attached to the central septum were seen. However, in larger cells, well before completion of the nascent cross wall, the single central mesosomes were no longer seen, and two mesosomes attached to the two new subequatorial secondary cross-wall sites were observed. Thus, in this species, mesosomes also appear to duplicate and "jump" from one surface location to another. Recent evidence suggests that the assembly of mesosomes, as visualized at specific sites in fixed cells of *S. faecalis,* occurs from a pool of precursors that are not visible in the absence of fixation (Higgins et al., 1976).

The data cited above do not provide information concerning events that occur during a normal cell division cycle. Also, Ryder and Smith (1974), Scheefers-Borchel and Vielmetter (1975), and Meyer et al. (1975) have questioned the findings of Worcel and Burgi (1974) on technical grounds. However, Jones and Donachie (1974) have shown that release of near-completed chromosomes from membranes requires a short period of protein synthesis. In addition, Sargent's (1974) observations of the position of the bacterial nucleoid in cells from growing cultures of *B. subtilis* of various lengths were also consistent with the "chromosome jump" model. He observed that nucleoids clustered at two locations: in mononucleate cells, nucleoids were most frequently found very near the center of the cells; in binucleate cells, most nucleoids were seen at intermediate positions. Interpretation of this data is most consistent with the view that nucleoids of *B. subtilis* jump from a central position to one near the center of the future cell. Sargent interpreted the data conservatively, and suggested that chromosome origins are segregated with new surface. The replication apparatus remains in a central location until replication is completed. It is the change in this complex that would result in the observed jump.

Clearly, the mechanism of nucleoid segregation in bacteria requires further study. However, in either the model of Jacob et al. or in the nucleoid jump model segregation of a surface element is required. The principal difference between the two is that in the model of Jacob et al. the membrane itself is the organelle responsible for segregation, whereas in the nucleoid jump model this function could be performed by either the membrane or a more rigid wall element. Although some evidence exists for the conservation and segregation of at least a small fraction of a possibly specialized region of the membrane during several generations of growth (Green and Schaechter, 1972), this point remains controversial. The high degree of mobility of lipids and proteins within membranes substantially reduces the probability that membrane growth per se would provide a mechanism for the timed and spatially oriented segregation of the chromosome. Perhaps the greatest problem that the model of Jacob et al. would encounter would be in the segregation of chromosomes in cells that divide in more than one plane (section 10.3).

The nuclear jump model permits the participation of other envelope components in the designation of chromosome attachment sites. In this respect evidence of the association of an outer membrane protein with DNA in *E. coli* (Gudas et al., 1976) may be of particular significance. Insertion of this protein into the outer membrane appears to occur at a specific time in the cell cycle (section 9.5). By its very nature the cell wall, and especially the peptidoglycan, has features that make it admirably suited to be a cellular element capable of segregation. In Gram-positive species the wall and underlying membrane are in intimate association. Membrane lipids, proteins, and lipoteichoic acids are sometimes difficult to dissociate from isolated cell wall preparations (Wicken and Knox, 1975; Mirelman et al., 1976). Such intermingling could provide an indirect vehicle for a wall-DNA relationship. Similarly, in Gram-negative organisms the adhesion sites seen between inner membrane, peptidoglycan layer, and outer membrane (Bayer, 1975) could provide a means of contact between DNA and more distal envelope components.

Due to turnover (section 4.4.), the absence of a truly localized assembly process (section 4.5.), and postincorporation modifications and remodeling (section 7), it seems unlikely that the assembly of peptidoglycan per se is important in chromosome segregation. The results of network analysis (section 8.2.) also indicate that peptidoglycan content is not a determining factor in DNA synthesis. However, it seems possible that the segregation of a specific cell surface marker, either as part of the peptidoglycan or covalently linked to it, could serve as a focus for the condensation of membrane components essential for chromosome segregation. Such a marker would have to accompany, and be specifically related to, cell surface enlargement. Each of the surface growth models presented in the next section provide a type of surface primordium that is duplicated and segregated to daughter cells. Their common feature is a ring of surface at which surfaces made during two different cell generations meet. In each case the frequency of primordia duplication depends on growth rate. It is proposed that these primordia are closely linked to chromosome attachment and segregation sites.

10. Surface growth models

10.1. Topology of surface growth

The number and location of cell surface enlargement sites in a number of Gram-positive and Gram-negative bacterial species have been investigated using a variety of methods. Each of the techniques employed appears to suffer one or more drawbacks that either have not, or cannot, be adequately controlled (summarized in Higgins and Shockman, 1971; Shockman et al., 1974; Braun, 1975). Nearly every investigation of the topology of surface growth of rod-shaped species has been interpreted in terms of differentiating between random (or dispersed) growth at a large number of sites and nonrandom (conservative or semiconservative) growth at a highly restricted (one to three) number of sites. Questions that have not been asked include possibilities of: (1) growth at a moderate but definitive number of sites with the exact number related to growth rate for their initiation; or (2) a difference between the initiation, assembly and/or timing of cylindrical and polar wall. Thus, for both a variety of Gram-positive and Gram-negative rod-shaped bacterial species there is a significant lack of agreement among the many attempts to localize wall and membrane enlargement sites.

In contrast, for cocci that divide in one plane (streptococci) there is a notable degree of agreement that cell surface enlargement occurs at one to three sites per nascent diplococcus. To our knowledge, all the results, from the earliest demonstration via the immunofluorescence technique (Cole and Hahn, 1962) to the present, consistently agree that new surface is added at the coccal equator and that the surface of each resultant daughter cell consists of one pole assembled during the current cell generation and the other during a past generation. This model may be termed semiconservative.

In searching for a unifying principle one can consider that, in terms of surface growth, streptococci are more primitive than bacteria of other shapes and modes of division (Higgins and Shockman, 1971). They possess only the ability to assemble the bare essentials for cell division, including a mechanism for placing at least one copy of one chromosome in each daughter cell. Thus, in the absence of a secondary assembly mechanism that could produce additional cell surface between cross walls, such as by the assembly of a cylindrical section of wall, cocci can only increase their surface area by constructing and separating cross walls. Increasing surface area, as cell volume and mass increases, via the assembly of a cylindrical section of wall is more efficient than the assembly and separation of cross walls (Previc, 1970). In the rod then, one could postulate a separation of functions, with one function for cylindrical elongation dedicated to increasing surface area and the second designed for septation to effect cell division.

We will first consider the most primitive streptococcal system and then attempt to apply this information to more complex systems such as cocci that divide in more than one plane and rod-shaped species.

10.2. Surface growth model for cocci that divide in one plane

Ultrastructural observations of *S. faecalis* demonstrated the presence of one to three raised rings of wall material per nascent diplococcus (Shockman and Martin, 1968), which encircled cells at their widest diameter and, in cells in all stages of the divisions cycle, remained at an approximately constant distance from each pole. These raised bands serve as wall surface markers which, in conjunction with a second surface marker, nascent septa, provide a mechanism for studying wall surface growth by electron microscopy (Higgins and Shockman, 1970a,b; 1971; Higgins et al., 1970; 1971; Shockman et al., 1974). The diagrammatic scheme shown in Fig. 18 was developed from these observations. Clearly, the wall bands appear to mark the area of the surface at which wall surfaces assembled during two different cell generations meet. This point was probably best shown by observations of cells that were permitted to regrow after having thickened their walls during a period of amino acid starvation (Higgins et al., 1971). During regrowth, thickened wall was conserved and remained covering the polar caps while new, thinner wall was inserted in between. The conservation of

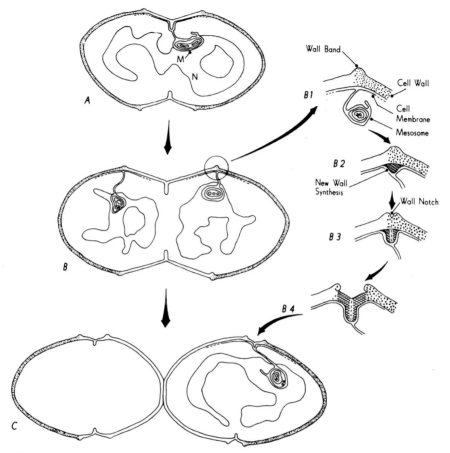

Fig. 18. Diagrammatic representation of the cell division cycle for *Streptococcus faecalis*. The model proposes that linear wall elongation is a unitary process resulting from wall synthetic activity at the leading edges of the nascent cross wall. The diplococcus in A is in the process of growing new wall at its cross wall and segregating its nuclear material to the two nascent daughter cocci. In rapidly growing exponential phase cultures before completion of the central cross wall, new sites of wall elongation are established at the equators of each of the daughter cells at the junction of old, polar wall (stippled) and new equatorial wall beneath a band of wall material that encircles the equator (B). Beneath each band a mesosome is formed, while the nucleoids separate and the mesosome at the central site is lost. The mesosome appears to be attached to the plasma membrane by a thin membranous stalk (B1). Invagination of the septal membrane appears to be accompanied by centripetal cross-wall penetration (B2). A notch is then formed at the base of the nascent cross wall, creating two new wall bands (B3). Wall elongation at the base of the cross wall pushes the newly made wall outward. At the base of the cross wall, the new wall peels apart into peripheral wall, pushing the wall bands apart (B4). When sufficient new wall is made so that the wall bands are pushed to a subequatorial position (e.g., from C to A to B), a new cross-wall cycle is initiated. Meanwhile the initial cross wall penetrates centripetally into the cell, dividing it into two daughter cocci. At all times the body of the mesosomes appears to be associated with the nucleoid. Doubling the number of mesosomes seems to precede completion of the cross wall by a significant interval. Nucleoid shapes and mesosome position are based on reconstruction projections of serially sectioned cells. (Reproduced with permission from Higgins and Shockman [1971]).

the thickened wall during regrowth could be taken as further evidence of the absence of wall turnover in this species.

This model has been extensively discussed elsewhere (Higgins and Shockman, 1971; Shockman et al., 1974). However, several important features of the model should be mentioned. New, nascent septal surface enlargement sites are initiated below equatorial wall bands. An external notch in the wall band is followed by the appearance of two wall bands. Wall assembled in the area of the nascent septum results in the gradual separation of the two bands. Thus, it seems clear that new wall enlargement takes place between the bands. The next question is: Where in this area of the surface, which starts out small in young cells and is substantial in cells about to divide, does wall enlargement occur?

Several observations favor the hypothesis that all or most of wall assembly devoted to cell surface enlargement occurs in the vicinity of the nascent septum. In small cells, nascent cross walls are less than two new wall layers in thickness. As cells enlarge, the nascent cross walls thicken. In addition, a gradient of wall thickness that extends from nascent cross walls to the wall bands was seen. Also, polar wall, assembled in a previous generation is significantly thicker than the thickest wall assembled in the current generation in the vicinity of the wall bands. Thus, in rapidly growing cells, although sites over the entire coccal surface assemble wall to increase wall thickness, thickening appears to occur most rapidly in the vicinity of nascent cross walls. It would be difficult to account for the observed gradient of wall thickness by mechanisms that would require wall enlargement at other specific sites (e.g., at the wall bands).

In addition to wall thickening, observations of the autolytic enzyme system action are consistent with a septal site of wall enlargement. The earliest cell damage observable by electron microscopy of exponential phase cells either undergoing cellular autolysis in buffer (Higgins et al., 1970) or protoplast formation (Joseph and Shockman, 1976) was loss of wall from the sides and tips of nascent cross walls. Wall dissolution continued until nascent cross walls were lost. After this stage, wall loss continued along nearby peripheral wall toward the poles. Even at reasonably advanced stages of peripheral wall loss near nascent cross walls, little damage was seen to polar or more distant peripheral wall. Also, isotope labeling experiments clearly demonstrated that autolytic enzyme activity is closely associated with newly assembled wall in exponential-phase cells (section 5.2). Together, these two types of observations strongly suggest that new wall is assembled primarily at nascent septa.

On the basis of earlier observations, it was postulated that wall enlargement occurs by the assembly of wall at the leading edges and tips of nascent cross walls. In its strictest and most limited sense this hypothesis has been modified. Currently, as a result of careful measurements of many central cell sections, and the mathematical rotation of these data to produce three-dimensional reconstructions of cells (Higgins, 1976; Higgins and Shockman, 1976), we believe that addition of wall precursors to most or all of nascent cross walls occurs. Also, precursor addition accompanying the conversion of nascent cross wall into peripheral wall would explain an observed increase in wall area and would be consistent with a minimum rearrangement of assembled wall. Wall assembly exclusively at the leading edges and tips of cross walls would require a substantial increase in hoop diameter as such wall segments are "pushed" to the periphery of the cell. Thus, even in streptococci, wall surface enlargement (as well as wall thickening) occurs by means of intercalation of new wall precursors into previously assembled wall, even though surface enlargement occurs in a rather localized area.

The overlapping sequence of wall surface enlargement seen in S. faecalis deserves further comment. All ultrastructural observations made so far were of cells from rapidly growing exponential phase cultures (t_d = 32 min). In small, and presumably young, cells one nascent cross wall and two wall bands were seen (e.g., Fig. 18a). In large, and presumably older, cells three nascent cross walls and four wall bands were seen. In all cases the two subequatorially located nascent cross-wall sites were at a less advanced stage of development than the centrally located growing (or completed) cross wall. The initiation and progression of secondary wall surface enlargement sites before completion of the primary site led to the hypothesis that wall surface enlargement of streptococci is governed by the same rules that govern the synthesis of linear macromolecules (section 8.1.). The proposed scheme (Shockman et al., 1974) postulates that each (septal) surface enlargement site(s) is capable of assembling a fixed area of surface in a fixed time interval (W). Thus, the number of e sites would be proportional to growth rate. At growth rates faster than W, one to three overlapping sites are observed (Fig. 19, bottom), and at growth rates slower than W one would expect to see one e site per cell (e.g., Fig. 19, top). A shift upward from a slow to fast growth rate would be expected to result in the rapid initiation of new secondary e sites (Fig. 19, middle). Although enumeration of numbers of e sites in slowly growing cells has not yet been accomplished, other data consistent with this postulate are available. For example, the rather small increase in mass per cell at doubling times between 120 and 30 minutes (Fig. 12a) are consistent with the above postulate.

To provide a basis for the speculative models presented in the following sections several points should be emphasized. In species of other shapes or modes of division, the mechanism by which poles are assembled in cocci that divide in one plane could be modified. The presence of raised bands at the junction of wall surface made during two different cell cycles in streptococci suggests that this portion of the cell surface differs somehow from the rest of the surface. The constancy of topological band location, the duplication of bands at the beginning of each cell cycle, and the relative movement of bands and nascent cross walls suggest that this different area of marked surface is related to a primitive system of spatial memory. Although such spatial markers have not yet been seen in bacteria of other shapes or modes of division, the fundamental principles of the streptococcal model have been used to develop the speculative models outlined in the next two sections.

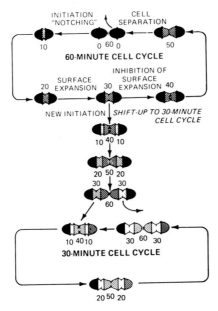

Fig. 19. Two imaginary cell cycles for *S. faecalis* growing with doubling times of 60 (*top*) and 30 (*bottom*) minutes. Midportion of the illustration shows the results of an imaginary growth rate shift. The numbers indicate approximate ages of new surfaces in minutes. The age of the various cell surfaces progresses from ■ to ▨, ▨, and □.

*10.3. Surface growth model for cocci
that divide in more than one plane*

Cocci that divide in more than one plane fall into two groups. In some genera, such as *Sarcina* and *Gaffkya*, regular arrays of packets of four or cubical arrays of eight cells are seen. In others, such as *Staphylococci* and *Micrococci*, irregular clusters of cells are seen. In either group, successive divisions occur in different planes. Of particular interest is the mutant of *M. lysodeikticus* (Fig. 1) which, instead of growing in irregular clusters, grows as regular three-dimensional packets of 64 or more cells (Yamada et al., 1975). The existence of this mutant suggests that the mechanism which results in a random arrangement is lost in this mutant. Since the parent strain appears to divide at right angles to the previous division, the random arrangement could be due to postdivisional processes. Similarly, other ultrastructural observations of staphylococci (e.g., Sugunuma, 1962; Giesbrecht, 1972; Cole et al., 1974) suggest that successive planes of division are at about right angles to the previous plane.

Cell division of staphylococci appears to begin with the formation of a nascent cross wall. In contrast to *S. faecalis*, however, the cross wall seems to completely bisect the coccus in the absence of a peeling apart of the cross wall at its base

(Giesbrecht, 1972; Cole et al., 1974). Only later do cross walls separate to form new poles. However, analogous to streptococci, and excepting wall lost via turnover, the surface of a new coccus of a genus such as *Sarcina* that divides in three planes will consist of one half made in the current generation and one-half made in previous generations. From the geometry of dividing micrococci (e.g., Fig. 1) it is possible to surmise that in these organisms formation of a flat cross-wall annulus occurs first, and then the completed cross wall is expanded into a completed pole. As is suggested in the streptococcus, expansion could occur by autolytic attack of the newly formed cross wall, followed by insertion of new wall at newly created polymerization sites. This mechanism would differ from the mechanism proposed for the streptococcus in the order of sequence of events. In the streptococcus, most of the new hemisphere is generated and then the cross wall is completed. In the other cocci a cross wall would be formed, and then the flattened cross wall would expand to form the hemispherical pole.

In contrast to streptococci, in cocci that divide in more than one plane (Fig. 20), the plane of division during the previous cell cycle (n-1) is at right angles to that of the ongoing division (n). Therefore, the old hemisphere consists of wall surface made during more than one generation. Similarly, going back two generations, and again assuming division at right angles to the two previous planes, the old hemisphere would consist of ½ surface made in the n-1 generation, and ¼ surface made in each of the n-2 and n-3 generations. Thus, the cell surface of a newly divided *Sarcina* consists of: (1) a hemisphere made in the current (n) generation; (2) one-fourth of a cell surface made in the n-1 generation; and (3) two one-eighth cell surface segments, made in the n-2 and in the n-3 (or older) generations, respectively.

In a system such as that described above, wall surfaces made during each generation are at right angles to each other. The generation of two equal and symmetrical daughter cells in regular packets of eight or more cells requires the plane of division to occur at the junction of surface made in the n-2 and n-3 or older generations. Somehow the cell can "remember" that it had divided in that plane two generations earlier. This system would use equatorial memory "bands" because they provide contact with surface made during three rather than two generations.

A similar model could be proposed for *Staphylococci* and *Micrococci* since divisions in these organisms occur at right angles to each other and symmetrical daughter cells are formed. Any other mode of establishing a division plane would produce asymmetrical cells with great frequency, a phenomenon that has not been seen very often. Probably, *Staphylococci* and *Micrococci* use a three-dimensional method for orientation of plane of division, with variations during the postdivisional cell separation events.

Division of cells in more than one plane presents a problem for some of the mechanisms proposed for the segregation of bacterial chromosomes to the two new daughter cells (section 9.7). Models such as those proposed by Jacob et al. (1963) that require continued attachment of the chromosome to the surface and segregation, via growth of the surface between the chromosomes (Fig. 16), could

694

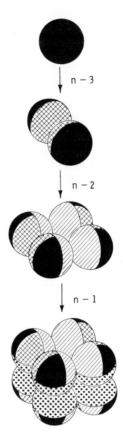

Fig. 20. Model for surface growth of cocci that divide in three planes, showing formation of a packet of 8 cells produced from 3 successive divisions of a single cell. Although, according to this model, any cell would have sections of surface assembled during 3 previous generations, for purposes of illustration the original cell is shown as ■. The first (n-3) division would produce two cells whose surface of which would be ½ old (■) and ½ "new" (⊠). The second (n-2) division, at 90° to the previous division, would produce four cells having ½ new surface (◨) assembled in the n-2 generation and ¼ from each of the two previous generations (■ and ⊠). The third (n-1) division, at 90° to both of the previous two divisions, would produce eight cells having ½ new surface (⊡), and the old surface would be ¼ from the n-2 division (◨) and ⅛ from each of the n-3 division (▨) and the initial cell (■). The next (n) division would occur in the same plane as the n-3 division between the surfaces indicated as ■ and ⊠.

A chromosome attached to the cell surface and segregating as new surface is inserted between its origins (Jacob et al., 1963) could do so during both the n-3 and n-2 generations, if it were located originally in the upper or lower portion of the cell shown at the top of the figure. However, the chromosome would find itself in a position unsuitable for segregation during the n-1 division. A more attractive model would be for the chromosome to switch planes at each division. For example, in the eight-cell packet the chromosome would "jump" from the junction between ■, ⊠, and ◨ to the junction between ◨, ⊡, and □.

not function in bacteria that divide in three planes (Fig. 20). The speculative nuclear jump model presented earlier could provide duplicated and segregated surface sites to which both of the newly replicated chromosomes could attach.

10.4. Surface growth models for rods

In its geometrical configuration, a rod-shaped bacterium may be considered as a cylindrical portion of surface inserted between two poles. In all models for the surface growth of rods, each daughter cell must contain one conserved pole from an earlier cell division and one pole formed in the current cell division cycle. The location of the future septum and the subsequent formation of a pole need not be linked totally to cylindrical enlargement in either topology or in response to regulatory signals.

Within the constraints imposed by the models for growth of cocci dividing in one and more than one plane (Figs. 19,20), two general models for growth of rod-shaped organisms can be constructed. The models maintain the requirement that daughter cells are formed with one pole conserved from a previous generation. They also accomplish a shift upward in growth rate without altering the postulated "rules" for zonal growth and for the location of septal sites imposed upon the cells during the preshift steady state. Longer daughter cells are obtained after the upward shift in growth rate; changes in diameter are not considered. The potential of both models may be visualized by following each through a shift upward in growth rate (Figs. 21, 22).

At slow growth rates, the rod shown in Fig. 21 or 22 contains the equivalent of a "band" or "ring" with the potential to generate a given amount of both cylindrical and cross-wall surface. In both models, cross walls are formed after the elongation potential has been expressed. The model represented in Fig. 21 superficially resembles the streptococcal model (Fig. 19) in that the future division site is located at the initiation site of a new growth zone. Initiation of surface growth occurs after cleavage and duplication of the central wall band. New surface is inserted, primarily from a central zone, until the full growth potential of the zone has been expressed. At that time, a cross wall grows into the cytoplasm, and the two daughter cells become physiologically separate entities. Cleavage between the septal wall generates two new poles, one for each daughter cell. Mechanistically, initiation of an elongation site has two consequences: (1) generation of two future elongation sites; and (2) completion of the growth site by formation of a cross wall and, eventually, of a new pole.

The model shown in Fig. 22 is expanded from the unit cell model of Donachie and Begg (1970). Here, the future septal site is established after the initiation of the growth zone and may or may not be spatially associated with it. At slow growth rates the newly divided rod contains a band located near its most recently formed division pole. This band contains the potential for both elongation and cross-wall formation, but in this model the two potentials can be expressed separately. Elongation is always initiated to one side of a band.

At slow growth rates, the band is pushed toward the center of the cell by

696

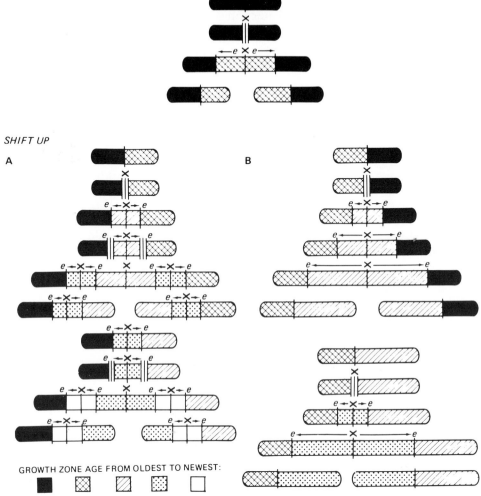

SHIFT UP

A

B

GROWTH ZONE AGE FROM OLDEST TO NEWEST:

Fig. 21. Model for symmetrical surface growth of rod-shaped bacteria. In this model new surface is inserted after splitting and duplicating a central wall "band". At a slow growth rate (*top*), new cylindrical surface (e) is inserted between the newly duplicated bands from a potential cross wall, site *x*. After insertion of a unit cell area of cylindrical surface, the cross wall is formed, closed, and the cell divides in two. The wall bands generated at initiation of new surface enlargement move to the center of each future daughter cell. On shift up two possibilites are considered. In (a) two new growth sites (e) are formed before completion of the ongoing site by splitting the two wall bands. At division, after the shift, the daughter cells are longer than at birth. At the new growth rate, secondary growth sites will continue to be initiated prior to completion of the primary site. In (b) the growth rate shift results in the insertion of more than a unit cell area of surface per site. The cells dividing after the shift are equal in size and longer than at birth. However, subsequent divisions will produce two cell sizes, one intermediate and one large, in a ratio of $1 : 2^{(n-1)}$, where n represents the number of postshift generations. The order of decreasing surface age is: ■, ⊠, ▨, ⊡ and □.

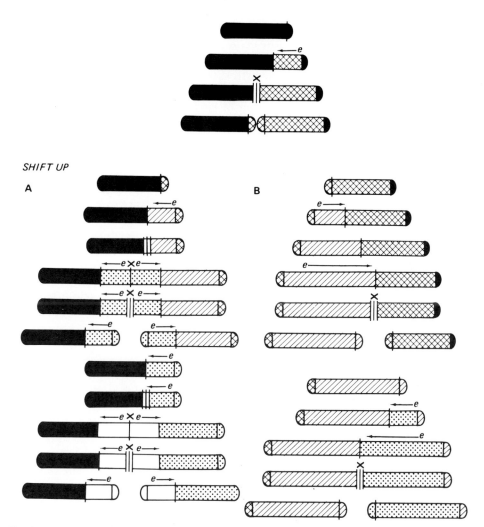

Fig. 22. Model for asymmetrical surface growth of rod-shaped bacteria. In this model new surface is inserted on one side of a band. At slow growth rates (*top*) the cell initiates a new growth zone (*e*) adjacent to a band near the cell pole. The band is pushed to the cell center by formation of a new cylindrical surface. After insertion of a new unit cell area of cylindrical surface, the band is in the center of the cell. Formation of a cross wall (*x*) is initiated by splitting and duplication of the band and the formation of two new poles. Following a shift up in growth rate two possibilities are considered. In (a) two new growth zones are initiated on both sides of the band before initiation of cross walls. Sites are those where elongation would occur in the next division cycle at the slow growth rate. After completion of assembly of the initial unit cell cylinder, the central band splits and generates two new poles. The daughter cells are longer than at birth. In subsequent cycles the cells will initiate secondary growth zones before completion of the primary unit cylinder. In (b) following the shift in growth rate, a longer cylindrical section is inserted. At divison, daughter cells will be unequal in size. In this and future generations small cells will be generated in a ratio of $1 : 2^{(n)}$ to larger cells, where n represents the number of generations after the shift. The order of decreasing surface age is ■, ▨, ▨, ▨ and □.

formation of new cylindrical wall (Fig. 22, top). When a new cylinder is completed, the band (which is now in the center of the cell) is split into two, generating two future growth zones, a new cross wall and, eventually, two new poles.

Upon a shift upward to a faster growth rate (at a constant temperature), three possibilities have been considered: (1) the frequency of initiation of new growth zones will be increased, but the amount of surface generated per growth zone will remain unchanged; (2) the frequency of initiation of new growth zones will be unchanged, but the amount of surface generated per growth zone will be increased; and (3) both the frequency of initiation of growth zones and the amount of surface per growth zone will be increased. The first and second possibilities are illustrated in Figs. 21 and 22, as a and b, respectively. In the models shown in Fig. 21a and 22a, as long as new growth zones are initiated at a frequency determined by a specific ratio of length of new surface to number of unit-sized cells contained at that stage of growth, daughter cells of equal size are obtained during the transition to the new growth rate as well as during the two steady states.

In the models shown in Figs. 21b and 22b daughter cells of equal size will be observed during balanced growth at a fixed rate, but these models do not generate daughter cells of equal size during the shift up. Likewise, the combination of the two possibilities, item 3 above, will fail to generate daughter cells of equal size (as long as new bands are generated on both sides of each old band).

The model shown in Fig. 22a tends to become symmetrical at fast growth rates, as was suggested by Donachie and Begg (1970) and Higgins and Shockman (1971). From a mechanistic viewpoint, in this model, a shift upward in growth rate requires that initiation of new elongation zones occur to both sides of a band. As long as the band is not split into two bands, both sides of the band are topologically identical to one side of the asymmetrically located band at the top of Fig. 22. This model permits the generation of future bands as elongation is initiated, and preserves the original band concerned with the formation of a cross wall and pole. The models shown in Figs. 21 and 22 do not exclude independent regulation of elongation and cross wall formation. The model shown in Fig. 22a has the further feature that the cross wall is no longer situated adjacent to the portion of cylindrical surface from which it was derived.

Other models can be constructed if the constraints of the streptococcal model are changed. The effect of growth rate shifts on the length of daughter cells is a test of the models. Recent evidence from studies of growth rate transitions in *S. typhimurium* indicate that, assuming discrete growth zones, the amount of longitudinal surface synthesized per growth zone is independent of the growth rate (Case, 1976; Case and Marr, 1976).

An interesting model for the elongation of rod-shaped bacteria has been proposed by Mendelson (1976). The model arises from the observation that in a multiple mutant of *B. subtilis* a constraint was generated by the spore coat during outgrowth from spores. This constraint resulted in clones that grew as closed circles in a double helix configuration. The suggestion was made that rods normally elongate helically, causing the poles to rotate away from one another during growth. In this model, new cell surface would be inserted along a helical path.

11. Summary, conclusion, and epilogue

In this chapter we have attempted to summarize currently available information on the assembly of bacterial cell surfaces during growth and division, and to integrate this data with available information on the regulation of the bacterial cell division cycle. Clearly each of the complex processes of cell surface assembly, chromosome replication and its regulation, cell division, and the morphogenetic sequence leading to cell division are all interdependent. Despite the complexities of each process and these interdependencies substantial progress has been made, and some of the features common to all the different systems under investigation are becoming apparent. Thus, such investigations of growth and division of bacterial cells are proving to be model systems for the investigation of morphogenesis at the cellular and subcellular level that are capable of slowly yielding significant data.

Obviously the eventual aim is to describe each event, and the mechanisms regulating it, on a biochemical and molecular level. Currently, we remain far from that goal. However, by artificially subdividing the overall problem into several parts, each of which is experimentally approachable, considerable insight has been gained, and some of the simpler and perhaps naive concepts have been eliminated. For example, while it is clear that the chromosome replication cycle is closely related to the cell division cycle, its relationship is not a simple sequential one.

A major problem in the search for "common principles" has been the integration and correlation of the various ultrastructural, physiological, biochemical, molecular, and genetic approaches. In addition to utilizing different techniques (which makes it difficult to obtain a directly comparable, complete set of data on a single species or strain) each approach utilizes a different philosophy of thought. Despite such difficulties, we have chosen to present speculations that we think are based on such "common principles." A substantial portion of these speculations are based on the requirement that for precise, rapid, and orderly growth and division, bacterial cells must possess a relatively rigid, three-dimensional, protective and perhaps information-containing polymer in their wall (the peptidoglycan). Such speculations include the idea that surface structural elements common to bacteria of different shapes and planes of division govern cross-wall formation and the generation of future poles. Specialized processes, or variations of the common mechanism, would then be required to form cylindrically (or other) shaped sections of surface. Also included is the speculation that the relatively precise segregation of a surface component is essential to provide sites to which newly duplicated chromosomes can attach and segregate to daughter cells (nuclear "jump" model).

The ideas and speculations presented may provide a basis for future investigations not only of bacterial cells but also of eukaryotic cells, which presumably are involved in similar but more complex processes.

Acknowledgments

The work done in the authors' laboratories was supported by research grants AI 05044 and DE 03487 from the National Institutes of Health, and BMS 70-00564 from the National Science Foundation.

We wish to express our gratefulness to the following individuals: A. R. Archibald, J. Baddiley, M. Bayer, B. D. Beck, W. C. Brown, I. D. J. Burdette, D. Button, J. Chaloupka, D. Carson, M. Case, R. Cleveland, S. Cooper, J. Cornett, R. J. Doyle, D. C. Ellwood, L. Glaser, C. E. Helmstetter, M. L. Higgins, E. T. Hinks, R. Hinks, I. B. Holland, S. Holt, F. Jackson, R. James, R. Kessler, A. Koch, H. E. Kubitschek, I. R. Lapidus, P. Lancy Jr., R. Marquis, M. Matsuhashi, N. H. Mendelson, D. Mirelman, M. Mychajlonka, R. G. E. Murray, A. B. Pardee, J. T. Park, L. Pelta, O. Pierucci, H. M. Pooley, F. Robb, H. J. Rogers, E. Z. Ron, R. J. Rowbury, M. G. Sargent, M. Schaechter, D. J. Tipper, J. B. Ward, K. Weightman, A. J. Wicken, C. L. Woldringh, F. E. Young and A. Zaritsky. All have been most helpful in providing data, graphs, electron micrographs, or actual manuscripts to us before their publication, as well as supplying acute, even needlelike, comments and discussions on our manuscript. For all their helpfulness, the authors hastily absolve them from any omissions, excursions, or outright blunders the perspicacious reader may discern. Unhappily, the authors (and the editors) must bear the responsibility for such errors.

References

Abbo, F. E. and Pardee, A. B. (1960) Synthesis of macromolecules in synchronously dividing bacteria. Biochim. Biophys. Acta 39, 478–485.

Adler, H. I., Fisher, W. D. and Hardigree, A.A. (1969) Cell division in *Escherichia coli*. Trans. N. Y. Acad. Sci. 31, 1059–1070.

Anderson, J. C., Archibald, A. R., Baddiley, J., Curtis, M. J. and Davey, N. B. (1969) The action of dilute aqueous NN-Dimethylhydrazine on bacterial cell walls. Biochem. J. 113, 183–189.

Araki, Y., Fukuoka, S., Oba, S. and Ito, E. (1971) Enzymatic deacetylation of N-acetylglucosamine residues in peptidoglycan from *Bacillus cereus* cell walls. Biochem. Biophys. Res. Commun. 45, 751–758.

Archibald, A. R. and Coapes, H. E. (1976) Bacteriophage SP50 as a marker for cell wall growth in *Bacillus subtilis*. J. Bacteriol. 125, 1195–1206.

Barner, H. D. and Cohen, S. S. (1956) Synchronization of division of a thymineless mutant of *Escherichia coli*. J. Bacteriol. 72, 115–123.

Bayer, M. E. (1967) The cell wall of *Escherichia coli:* Early effects of penicillin treatment and deprivation of diaminopimelic acid. J. Gen. Microbiol. 46, 237–246.

Bayer, M. E. (1975) Role of adhesion zones in bacterial cell-surface function and biogenesis. In: Membrane Biogenesis (Tzagoloff, A., ed.), pp. 393–427, Plenum Press, New York.

Beck, B. D. and Park, J. T. (1976) Activity of three murein hydrolases during the cell division cycle of *Escherichia coli* K-12 as measured in toluenized cells. J. Bacteriol., 126, 1250–1260.

Bleiweis, A. S., Taylor, M. C., Deepak, J., Wetherell, J. R. and Brown, T. A. (1976) The comparative chemical compositions of cell walls of *Streptococcus mutans*. J. Dent. Res. 55, A103–A108.

Blumberg, P. M. and Strominger, J. L. (1972) Five penicillin-binding components occur in *Bacillus subtilis* membranes. J. Biol. Chem. 247, 8107–8113.

Blumberg, P. M. and Strominger, J. L. (1974) Interaction of penicillin with the bacterial cell: Penicillin-binding proteins and penicillin-sensitive enzymes. Bacteriol. Rev. 38, 291–335.

Boothby, D., Daneo-Moore, L., Higgins, M. L., Coyette, J. and Shockman, G. D. (1973) Turnover of bacterial cell wall peptidoglycans. J. Biol. Chem. 248, 2161–2169.

Bracha, R. and Glaser, L. (1976) An *in vitro* system for the synthesis of teichoic acid linked to peptidoglycan. J. Bacteriol. 125, 872–879.

Braun, V. (1975) Covalent lipoprotein from the outer membrane of *Escherichia coli*. Biochim. Biophys. Acta 415, 335–377.

Braun, V. and Wolff, H. (1975) Attachment of lipoprotein to murein (peptidoglycan) of *Escherichia coli* in the presence and absence of penicillin FL 1060. J. Bacteriol. 123, 888–897.

Briles, E. B. and Tomasz, A. (1970) Radioautographic evidence for equatorial wall growth in a gram-positive bacterium. J. Cell Biol. 47, 786–790.

Briles, E. B. and Tomasz, A. (1973) Pneumococcal Forssman antigen. J. Biol. Chem. 248, 6394–6397.

Brooks, D. and Baddiley, J. (1969) The mechanism and direction of chain extension of a poly-(N-acetylglucosamine 1-phosphate) from walls of *Staphylococcus lactis* NCTC 2102. Biochem. J. 113, 635–642.

Brown, W. C., Fraser, D. K. and Young, F. E. (1970) Problems in purification of a *Bacillus subtilis* autolytic enzyme caused by association with teichoic acid. Biochim. Biophys. Acta 198, 308–315.

Brown, W. C., Wilson, C. R., Lukehart, S., Young, F. E. and Shiflett, M. A. (1976) Analysis of autolysins in temperature-sensitive morphological mutants of *Bacillus subtilis*. J. Bacteriol. 125, 166–173.

Brundish, D. E. and Baddiley, J. (1968) Pneumococcal C-substance, a ribitol teichoic acid containing choline phosphate. Biochem. J. 110, 573–582.

Burdett, I. D. J. and Murray, R. G. E. (1974a) Septum formation in *Escherichia coli:* Characterization of septal structure and the effects of antibiotics on cell division. J. Bacteriol. 119, 303–324.

Burdett, I. D. J. and Murray, R. G. E. (1974b) Electron microscope study of septum formation in *Escherichia coli* strains B and B/r during synchronous growth. J. Bacteriol. 119, 1039–1056.

Burns, V. W. (1959) Synchronized cell division and DNA synthesis in a *Lactobacillus acidophilus* mutant. Science 129, 566–567.

Cairns, J. (1963) The bacterial chromosome and its manner of replication as seen by autoradiography. J. Mol. Biol. 6, 208–213.

Campbell, A. (1957) Synchronization of cell division. Bacteriol. Rev. 21, 263–272.

Case, M. J. (1976) Nutritional regulation of cell size in *Salmonella typhimurium*. Ph. D. thesis. University of California, Davis.

Case, M. J. and Marr, A. G. (1976) Regulation of cell size and division in *Salmonella typhimurium*. Abstr. Ann. Meet. Am. Soc. Microbiol., p. 127.

Chai, N. C. and Lark, K. G. (1970) Effect of Actinomycin D on the transfer of ribonucleic acid from nucleus to cytoplasm in *Lactobacillus acidophilus*. J. Bacteriol. 101, 1005–1013.

Chaloupka, J. (1967) Synthesis and degradation of surface structures by growing and nongrowing *Bacillus megaterium*. Folia Microbiol. 12, 264–273.

Chaloupka, J. and Kreckova, P. (1971) Turnover of mucopeptide during the life cycle of *Bacillus megaterium*. Folia Microbiol. 16, 372–382.

Chaloupka, J. and Strnadova (1972) Turnover of murein in a diaminopimelic acid dependent mutant of *Escherichia coli*. Folia Microbiol. 17, 446–455.

Chaloupka, J., Kreckova, P. and Rihova. L. (1962a) The mucopeptide turnover in the cell walls of growing cultures of *Bacillus megaterium* KM. Experientia 18, 362–364.

Chaloupka, J., Kreckova, P. and Rihova, L. (1962b) Changes in the character of the cell wall in growth of *Bacillus megaterium* cultures. Folia Microbiol. 7, 269–274.

Chaloupka, J., Rihova, L. and Kreckova, P. (1964) Degradation and turnover of bacterial cell wall mucopeptides in growing bacteria. Folia Microbiol. 9, 9–15.

Chandler, M., Bird, R. E. and Caro, L. (1975) The replication time of the *Escherichia coli* K12 chromosome as a function of cell doubling time. J. Mol. Biol. 94, 127–132.

Chatterjee, A. N. and Doyle, R. J. (1977) A proposed generalized functional role for bacterial N-acetylmuramyl-L-alanine amidases. J. Theoret. Biol., in press.

Chatterjee, A. N., Mirelman, D., Singer, H. J. and Park, J. T. (1969) Properties of a novel pleiotropic bacteriophage-resistant mutant of *Staphylococcus aureus* H. J. Bacteriol. 100, 846–853.

Chatterjee, A. N., Wong, W., Young, F. E. and Gilpin, R. W. (1976) Isolation and characterization of a mutant of *Staphylococcus aureus* deficient in autolytic activity. J. Bacteriol. 125, 961–967.

Chipman, D. M., Pollock, J. J. and Sharon, N. (1968) Lysozyme-catalyzed hydrolysis and transglycosylation reactions of bacterial cell wall oligosaccharides. J. Biol. Chem. 243, 487–496.

Chung, K. L. (1971) Thickened cell walls of *Bacillus cereus* grown in the presence of chloramphenicol: their fate during cell growth. Can. J. Microbiol. 17, 1561–1565.

Chung, K. L. and Hawirko, R. Z. (1964) Cell wall replication: 1. Cell wall growth of *Bacillus cereus* and *Bacillus megaterium*. Can. J. Microbiol. 10, 43–48.

Churchward, G. G. and Holland, I. B. (1976a) Induced synthesis of an envelope protein by thymine starvation of *E. coli* B/r. FEBS Lett. 62, 347–350.

Churchward, G. and Holland, I. B. (1976b) Envelope synthesis during the cell cycle in *E. coli* B/r. J. Mol. Biol. 105, 245–262.

Clark, D. J. (1968) The regulation of DNA replication and cell division in *E. coli* B/r. Cold Spring Harbor Symp. Quant. Biol. 33, 823–838.

Clark, D. J. and Maaløe, O. (1967) DNA replication and the division cycle in *Escherichia coli*. J. Mol. Biol. 23, 99–112.

Cleveland, R. F., Holtje, J.-V., Wicken, A. J., Tomasz, A., Daneo-Moore, L. and Shockman, G. D. (1975) Inhibition of bacterial wall lysins by lipoteichoic acids and related compounds. Biochem. Biophys. Res. Commun. 67, 1128–1135.

Cleveland, R. F., Wicken, A. J., Daneo-Moore, L. and Shockman, G. D. (1976) Inhibition of wall autolysis in *Streptococcus faecalis* by lipoteichoic acid and lipids. J. Bacteriol. 127, 1582–1584.

Cole, R. M., Chatterjee, A. N., Gilpin, R. W. and Young, F. E. (1974) Ultrastructure of teichoic acid-deficient and other mutants of *Staphylococci*. Ann. N. Y. Acad. Sci. 236, 22–53.

Cole, R. M. and Hahn, J. J. (1962) Cell wall replication in *Streptococcus pyogenes*. Science 135, 722–724.

Cook, J. R. and James, T. W. (1964) Age distribution of cells in logarithmically growing cell populations. In: Synchrony in Cell Division and Growth (Zeuthen, E., ed.), pp. 485–495, Interscience, New York.

Cooper, S. and Helmstetter, C. E. (1968) Chromosome replication and the division cycle of *Escherichia coli* B/r. J. Mol. Biol. 31, 519–540.

Cooper, S. and Weinberger, M. (1977) Medium dependent variation of DNA segregation in *Escherichia coli*. J. Bacteriol. 130, 118–127.

Cooper, S. and Ruettinger, T. (1973) Replication of deoxyribonucleic acid during the division cycle of *Salmonella typhimurium*. J. Bacteriol. 11, 966–973.

Coyette, J. and Ghuysen, J.-M. (1970) Wall autolysin of *Lactobacillus acidophilus* strain 63 AM Gasser. Biochemistry 9, 2952–2955.

Coyette, J., Perkins, H. R., Polacheck, I., Shockman, G. D. and Ghuysen, J.-M. (1974) Membrane-bound DD-carboxypeptidase and LD-transpeptidase of *Streptococcus faecalis* ATCC 9790. Eur. J. Biochem. 44, 459–468.

Coyette, J. and Shockman, G. (1973) Some properties of the autolytic N-acetylmuramidase of *Lactobacillus acidophilus*. J. Bacteriol. 114, 34–41.

Cutler, R. G. and Evans, J. E. (1966) Synchronization of bacteria by a stationary-phase method. J. Bacteriol. 91, 469–476.

Daneo-Moore, L., Coyette, J., Sayare, M., Boothby, D. and Shockman, G. D. (1975) Turnover of the cell wall peptidoglycan of *Lactobacillus acidophilus*. J. Biol. Chem. 250, 1348–1353.

Daneo-Moore, L. and Higgins, M. L. (1972) Morphokinetic reaction of *Streptococcus faecalis* (ATCC 9790) cells to the specific inhibition of macromolecular synthesis: Nucleoid condensation of the inhibition of protein synthesis. J. Bacteriol. 109, 1210–1220.

Dean, A. C. R. (1973) Control mechanisms of bacterial cells. Pure Appl. Chem. 36, 317–324.

Dean, A. C. R. and Hinshelwood, C. (1966) Growth, Function and Regulation in Bacterial Cells. Clarendon Press, Oxford.

De Chastellier, C., Frehel, C. and Ryter, A. (1975a) Cell wall growth of *Bacillus megaterium*: Cytoplasmic radioactivity after pulse-labeling with tritiated diaminopimelic acid. J. Bacteriol. 123, 1197–1207.

De Chastellier, C., Hellio, R. and Ryter, A. (1975b) Study of cell wall growth of *Bacillus megaterium* by high-resolution autoradiography. J. Bacteriol. 123, 1184–1196.

Dennis, P. P. (1972) Regulation of ribosomal and transfer ribonucleic acid synthesis in *Escherichia coli* B/r. J. Biol. Chem. 247, 2842–2845.

Dennis, P. O. and Bremer, H. (1974) Differential rate of ribosomal protein synthesis in *E. Coli* B/r. J. Mol. Biol. 84, 407–422.

Dezélée, P. and Shockman, G. (1975) Studies of the formation of peptide cross-links in the cell wall peptidoglycan of *Streptococcus faecalis*. J. Biol. Chem. 250, 6806–6816.

Donachie, W. D. (1968) Relationship between cell size and time of initiation of DNA replication. Nature 219, 1077–1079.

Donachie, W. D. (1973) Regulation of cell division in bacteria. Brit. Med. Bull. 29, 203–207.

Donachie, W. D. and Begg, R. J. (1970) Growth of the bacterial cell. Nature 227, 1220–1224.

Donachie, W. D. and Masters, M. (1966) Evidence for polarity of chromosome replication of F⁻ strains of *Escherichia coli*. Genet. Res., Camb. 8, 119–124.

Donachie, W. D., Jones, N. C. and Teather, R. (1973) The bacterial cell cycle. Symp. Soc. Gen. Microbiol. 23, 9–44.

Donachie, W. D., Martin, D. T. M. and Begg, R. J. (1971) Independence of cell division and DNA replication in *Bacillus subtilis*. Nature New Biol. 231, 274–276.

Dring, G. J. and Hurst, A. (1969) Observations on the action of benzylpenicillin on a strain of *Streptococcus lactis*. J. Gen. Microbiol. 55, 185–193.

Eberle, H. and Lark, K. G. (1966) Chromosome segregation in *Bacillus subtilis*. J. Mol. Biol. 22, 183–186.

Ecker, R. E. and Schaechter, M. (1963) Ribosome content and the rate of growth of *Salmonella typhimurium*. Biochim. Biophys. Acta 76, 275–279.

Ellar, D. J., Lundgren, D. G. and Slepecky, R. A. (1967) Fine structure of *Bacillus megaterium* during synchronous growth. J. Bacteriol. 94, 1189–1205.

Elliott, T. S. J., Ward, J. B., Wyrick, P. B. and Rogers, H. J. (1975) Ultrastructural study of the reversion of protoplasts of *Bacillus licheniformis* to *Bacilli*. J. Bacteriol. 124, 905–917.

Ellwood, D. C. (1970) The wall content and composition of *Bacillus subtilis* var. *niger* grown in a chemostat. Biochem. J. 118, 367–373.

Ellwood, D. C. (1971) The anionic polymers in the cell wall of *Bacillus subtilis* var. *niger* grown in phosphorus-limiting environments supplemented with increasing concentrations of sodium chloride. Biochem. J. 121, 349–351.

Ellwood, D. C. and Tempest, D. W. (1969) Control of teichoic acid and teichuronic acid biosyntheses in chemostat culture of *Bacillus subtilis* var. *niger*. Biochem. J. 111, 1–5.

Ellwood, D. C. and Tempest, D. W. (1972) Effects of environment on bacterial wall content and composition. Adv. Microbial. Physiol. 7, 83–117.

Engberg, B., Hjalmarsson, K. and Norstrom, K. (1975) Inhibition of cell division in *Escherichia coli* K-12 by the R-factor R1 and copy mutants of R1. J. Bacteriol. 124, 663–640.

Ephrati-Elizur, E. and Borenstein, S. (1971) Velocity of chromosome replication in thymine-requiring and independent strains of *Bacillus subtilis*. J. Bacteriol. 106, 58–64.

Errington, F. P., Powell, E. O. and Thompson, N. (1965) Growth characteristics of some Gram-negative bacteria. J. Gen. Microbiol. 39, 109–123.

Fan, D. P. (1970) Autolysin(s) of *Bacillus subtilis* as dechaining enzyme. J. Bacteriol. 103, 494–499.

704

Fan, D. P. and Beckman, B. E. (1973) Structural difference between walls from hemispherical caps and partial septa of *Bacillus subtilis*. J. Bacteriol. 114, 790–797.

Fan, D. P., Beckman, B. E. and Beckman, M. M. (1974) Cell wall turnover at the hemispherical caps of *Bacillus subtilis*. J. Bacteriol. 117, 1330–1334.

Fan, D. P. and Beckman, M. M. (1972) New centrifugation technique for isolating enzymes from large cell structures: Isolation and characterization of two *Bacillus subtilis* autolysins. J. Bacteriol. 109, 1258–1265.

Fan, D. P., Pelvit, M. C. and Cunningham, W. P. (1972) Structural difference between walls from ends and sides of the rod-shaped bacterium *Bacillus subtilis*. J. Bacteriol. 109, 1266–1272.

Fantes, P. A., Grant, W. D., Pritchard, R. H., Sudbery, P. E. and Wheals, A. E. (1975) The regulation of cell size and the control of mitosis. J. Theoret. Biol. 50, 213–244.

Fiedler, F. and Glaser, L. (1973) Assembly of bacterial cell walls. Biochim. Biophys. Acta 300, 467–485.

Fiedler, F. and Glaser, L. (1974a) The synthesis of polyribitol phosphate. I. Purification of polyribitol phosphate polymerase and lipoteichoic acid carrier. J. Biol. Chem. 249, 2684–2689.

Fiedler, F. and Glaser, L. (1974b) The synthesis of polyribitol phosphate. II. On the mechanism of polyribitol phosphate polymerase. J. Biol. Chem. 249, 2690–2695.

Fiedler, F. and Glaser, L. (1974c) The attachment of poly(ribitol phosphate) to lipoteichoic acid carrier. Carbohydr. Res. 37, 37–46.

Fielding, P. and Fox, C. F. (1970) Evidence for stable attachment of DNA to membrane at the replication origin of *Escherichia coli*. Biochem. Biophys. Res. Commun. 41, 157–162.

Fitz-James, P. and Hancock, R. (1965) The initial structural lesion of penicillin action in *Bacillus megaterium*. J. Cell. Biol. 26, 657–667.

Fordham, W. D. and Gilvarg, C. (1974) Kinetics of crosslinking of peptidoglycan in *Bacillus megaterium*. J. Biol. Chem. 249, 2478–2482.

Forsberg, C. W. and Rogers, H. J. (1971) Autolytic enzymes in growth of bacteria. Nature 229, 272–273.

Forsberg, C. W. and Ward, J. B. (1972) N-acetylmuramyl-L-alanine amidase of *Bacillus licheniformis* and its L-form. J. Bacteriol. 110, 878–887.

Forsberg, C. W., Wyrick, P. B., Ward, J. B. and Rogers, H. J. (1973) Effect of phosphate limitation on the morphology and wall composition of *Bacillus licheniformis* and its phosphoglucomutase-deficient mutants. J. Bacteriol. 113, 969–984.

Frehel, C., Beaufils, A. M. and Ryter, A. (1971) Etude au microscope electronique de la croissance de la paroi chez *B. subtilis* et *B. megaterium*. Ann. Inst. Pasteur 121, 139–148.

Ganesan, A. T. and Lederberg, J. (1965) A cell-membrane bound fraction of bacterial DNA. Biochem. Biophys. Res. Commun. 18, 824–835.

Garrett, A. J. (1965) Rapid extraction of phosphorus during periodate oxidation of isolated cell walls of *Staphylococcus aureus* (Oxford). Biochem. J. 95, 6c–8c.

Ghuysen, J.-M. (1968) Use of bacteriolytic enzymes in determination of wall structure and their role in cell metabolism. Bacteriol. Rev. 32, 425–464.

Ghuysen, J.-M. (1977) This volume.

Ghuysen, J.-M. and Shockman, G. D. (1973) Biosynthesis of peptidoglycan. In: Bacterial Membranes and Walls (Leive, L., ed.), pp. 37–130, Marcel Dekker, New York.

Ghuysen, J.-M., Bricas, E., Leyh-Bouille, M., Lache, M. and Shockman, G. D. (1967) The peptide N-(L-alanyl-D-isoglutaminyl)-N-(D-isoasparaginyl)-L-lysyl-D-alanine and the disaccharide N-acetylglucosaminyl-β-1, 4-N-acetylmuramic acid in cell wall peptidoglycan of *Streptococcus faecalis* stain ATCC 9790. Biochemistry 6, 2607–2619.

Giesbrecht, P. (1972) Zur morphogenese der zellwand von Staphylokokken. Mikroskopie 28, 323–342.

Gilpin, R. W., Narrod, S., Wong, W., Young, F. E. and Chatterjee, A. N. (1974) Autolysis in *Staphylococcus aureus:* Preferential release of old cell walls. J. Bacteriol. 119, 672–676.

Glaser, L. (1973) Bacterial cell surface polysaccharides. Ann. Rev. Biochem. 42, 91–112.

Goodell, E. W. and Schwarz, U. (1974) Cell envelope composition of *Escherichia coli* K12: A comparison of the cell poles and the lateral wall. Eur. J. Biochem. 47, 567–572.

Green, E. W. and Schaechter, M. (1972) The mode of segregation of the bacterial cell membrane. Proc. Nat. Acad. Sci. U.S.A. 69, 2312–2316.

Greenwood, D. and O'Grady, F. (1973a) FL 1060: A new beta-lactam antibiotic with novel properties. J. Clin. Pathol. 26, 1–6.

Greenwood, D. and O'Grady, F. (1973b) The two sites of penicillin action in *Escherichia coli*. J. Infect. Dis. 128, 791–794.

Grover, N. B., Woldringh, C. L., Zaritsky, A. and Rosenberger, R. F. (1977) Elongation of rod-shaped bacteria. Biophys. J. in press.

Gudas, L. J. and Pardee, A. B. (1974) Deoxyribonucleic acid synthesis during the division cycle of *Escherichia coli*: A comparison of strains B/r, K-12, 15 and 15T⁻ under conditions of slow growth. J. Bacteriol. 117, 1216–1223.

Gudas, L. J., James, R. and Pardee, A. B. (1976) Evidence for the involvement of an outer membrane protein in DNA initiation. J. Biol. Chem. 251, 3470–3479.

Hakenbeck, R. and Messer, W. (1974) Activity of murein hydrolases and membrane synthesis in synchronized *Escherichia coli* B/r. Ann. Microbiol. Inst. Pasteur. 125 B, 163–166.

Hakenbeck, R., Goodell, E. W. and Schwarz, U. (1974) Compartmentalization of murein hydrolases in the envelope of *Escherichia coli*. FEBS Lett. 40, 261–264.

Hancock, I. and Baddiley, J. (1976) *In vitro* synthesis of the unit that links teichoic acid to peptidoglycan. J. Bacteriol. 125, 880–886.

Hartmann, R., Bock-Hennig, S. B. and Schwarz, U. (1974) Murein hydrolases in the envelope of *Escherichia coli*. Eur. J. Biochem. 41, 203–208.

Hartwell, L. H. (1974) *Saccharomyces cerevisiae* cell cycle. Bacteriol. Rev. 38, 164–198.

Harvey, R. J., Marr, A. G. and Painter, P. R. (1967) Kinetics of growth of individual cells of *Escherichia coli* and *Azotobacter agilis*. J. Bacteriol. 93, 605–617.

Hayashi, H., Araki, Y. and Ito, E. (1973) Occurrence of glucosamine residues with free amino groups in cell wall peptidoglycan from *Bacilli* as a factor responsible for resistance to lysozyme. J. Bacteriol. 113, 592–598.

Hebeler, B. H. and Young, F. E. (1976) Chemical composition and turnover of peptidoglycan in *Neisseria gonorrhoeae*. J. Bacteriol. 126, 1180–1185.

Helmstetter, C. E. (1967) Rate of DNA synthesis during the division cycle of *Escherichia coli* B/r. J. Mol. Biol. 24, 417–427.

Helmstetter, C. E. (1969) Methods for studying the microbial division cycle, in Methods in Microbiology (Norris, J. R. and Robbins, D. W., eds.), pp. 327–363, Academic Press, London and New York.

Helmstetter, C. E. (1974a) Initiation of chromosome replication in *Escherichia coli*. I. Requirements for RNA and protein synthesis and different growth rates. J. Mol. Biol. 84, 1–19.

Helmstetter, C. E. (1974b) Initiation of chromosome replication in *Escherichia coli*. II. Analysis of the control mechanism. J. Mol. Biol. 84, 21–36.

Helmstetter, C. E. and Cooper, S. (1968) DNA synthesis during the division cycle of rapidly growing *Escherichia coli* B/r. J. Mol. Biol. 31, 507–518.

Helmstetter, C. E. and Cummings, D. J. (1963) Bacterial synchronization by selection of cells at division. Proc. Nat. Acad. Sci. U.S.A. 50, 767–774.

Helmstetter, C. E. and Cummings, D. J. (1964) An improved method for the selection of bacterial cells at division. Biochim. Biophys. Acta 82, 608–610.

Helmstetter, C. E. and Pierucci, O. (1976) DNA synthesis during the division cycle of three substrains of *Escherichia coli* B/r. J. Mol. Biol. 102, 477–486.

Helmstetter, C. E., Cooper, S., Pierucci, O. and Revelas, E. (1968) On the bacterial life sequence. Cold Spring Harbor Symp. Quant. Biol. 33, 809–822.

Henning, U. (1975) Determination of cell shape in bacteria. Ann. Rev. Microbiol. 29, 45–60.

Henning, U. and Haller, I. (1975) Mutants of *Escherichia coli* K12 lacking all 'major' proteins of the outer cell envelope membrane. FEBS Lett. 55, 161–164.

Herbert, D. (1961) The chemical composition of micro-organisms as a function of their environment. Symp. Soc. Gen. Microbiol. 11, 391–416.

Herbold, D. R. and Glaser, L. (1975a) *Bacillus subtilis* N-acetylmuramic acid L-alanine amidase. J.

Biol. Chem. 250, 1676–1682.

Herbold, D. R. and Glaser, L. (1975b) Interaction of N-acetylmuramic acid L-alanine amidase with cell wall polymers. J. Biol. Chem. 250, 7231–7238.

Higgins, M. L. (1976) The three-dimensional reconstruction of whole cells of *Streptococcus faecalis* from thin sections of cells. J. Bacteriol. 127, 1337–1345.

Higgins, M. L. and Daneo-Moore, L. (1972) Morphokinetic reaction of cells of *Streptococcus faecalis* (ATCC 9790) to specific inhibition of macromolecular synthesis: Dependence of mesosome growth on deoxyribonucleic acid synthesis. J. Bacteriol. 109, 1221–1231.

Higgins, M. L. and Shockman, G. D. (1970a) Model for cell wall growth of *Streptococcus faecalis*. J. Bacteriol. 101, 643–648.

Higgins, M. L. and Shockman, G. D. (1970b) Early changes in the ultrastructure of *Streptococcus faecalis* after amino acid starvation. J. Bacteriol. 103, 244–254.

Higgins, M. L. and Shockman, G. D. (1971) Procaryotic cell division with respect to wall and membranes. CRC Crit. Rev. Microbiol. 1, 29–72.

Higgins, M. L. and Shockman, G. D. (1976) A study of a cycle of cell wall assembly in *Streptococcus faecalis* by three-dimensional reconstructions of thin sections of cells. J. Bacteriol. 127, 1346–1358.

Higgins, M. L., Coyette, J. and Shockman, G. D. (1973) Sites of cellular autolysis in *Lactobacillus acidophilus*. J. Bacteriol. 116, 1375–1382.

Higgins, M. L., Daneo-Moore, L., Boothby, D. and Shockman, G. D. (1974) Effect of inhibition of deoxyribonucleic acid and protein synthesis on the direction of cell wall growth in *Streptococcus faecalis* (ATCC 9790). J. Bacteriol. 118, 681–692.

Higgins, M. L., Pooley, H. M. and Shockman, G. D. (1970) Site of initiation of cellular autolysis in *Streptococcus faecalis* as seen by electron microscopy. J. Bacteriol. 103, 504–512.

Higgins, M. L., Pooley, H. M. and Shockman, G. D. (1971) Reinitiation of cell wall growth after threonine starvation of *Streptococcus faecalis*. J. Bacteriol. 105, 1175–1183.

Higgins, M. L., Tsien, H. C. and Daneo-Moore, L. (1976) Organization of mesosomes in fixed and unfixed cells. J. Bacteriol. 127, 1519–1523.

Highton, P. J. and Hobbs, D. G. (1971) Penicillin and cell wall synthesis: A study of *Bacillus licheniformis* by electron microscopy. J. Bacteriol. 106, 646–658.

Highton, P. J. and Hobbs, D. G. (1972) Penicillin and cell wall synthesis: A study of *Bacillus cereus* by electron microscopy. J. Bacteriol. 109, 1181–1190.

Hinks, R. P., Daneo-Moore, L. and Shockman, G. D. (1974) Cellular autolytic activity in synchronized populations of *Streptococcus faecalis* ATCC 9790. Abstr. Ann. Meet. Am. Soc. Microbiol. p. 25.

Hinks, R. P., Shockman, G. D. and Daneo-Moore, L. (1976) Cellular autolysis, peptidoglycan synthesis and the cell cycle in *Streptococcus faecium* ATCC 9790. Abstr. Ann. Meet. Am. Soc. Microbiol., p. 137.

Hirashima, A., Childs, G. and Inouye, M. (1973) Differential inhibitory effects of antibiotics on the biosynthesis of envelope proteins of *Escherichia coli*. J. Mol. Biol. 79, 373–389.

Hirota, Y., Ryter, A. and Jacob, F. (1968) Thermosensitive mutants of *Escherichia coli* affected in the processes of DNA synthesis and cellular division. Cold Spring Harbor Symp. Quant. Biol. 33, 677–693.

Hoffmann, B., Messer, W. and Schwarz, U. (1972) Regulation of polar cap formation in the life cycle of *Escherichia coli*. J. Supramol. Struct. 1, 29–37.

Höltje, J.-V. and Tomasz, A. (1974) Teichoic acid phosphorylcholine esterase—A novel enzyme activity in pneumococcus. J. Biol. Chem. 249, 7032–7034.

Höltje, J.-V. and Tomasz, A. (1975a) Lipoteichoic acid: A specific inhibitor of autolysin activity in pneumococcus. Proc. Nat. Acad. Sci. U.S.A. 72, 1690–1694.

Höltje, J.-V. and Tomasz, A. (1975b) Specific recognition of choline residues in the cell wall teichoic acid by the N-acetylmuramyl-L-alanine amidase of pneumococcus. J. Biol. Chem. 250, 6072–6076.

Höltje, J.-V., Mirelman, D., Sharon, N. and Schwarz, U. (1975) Novel type of murein transglycosylase in *Escherichia coli*. J. Bacteriol. 124, 1067–1076.

Hughes, R. C. and Stokes, E. (1971) Cell wall growth in *Bacillus licheniformis* followed by immunofluorescence with mucopeptide-specific antiserum. J. Bacteriol. 106, 694–696.

Hughes, R. C. and Tanner, P. J. (1968) The action of dilute alkali on some bacterial cell walls. Biochem. Biophys. Res. Commun. 33, 22–28.

Hughes, R. C., Tanner, P. J. and Stokes, E. (1970) Cell wall thickening in *Bacillus subtilis*. Biochem. J. 120, 159–170.

Inouye, M. and Pardee, A. B. (1970) Requirement of polyamines for bacterial division. J. Bacteriol. 101, 770–776.

Ivarie, R. D. and Pène, J. J. (1970) Association of the *Bacillus subtilis* chromosome with the cell membrane: Resolution of free and bound deoxyribonucleic acid on renografin gradients. J. Bacteriol. 104, 839–850.

Jacob, F., Brenner, S. and Cuzin, F. (1963) On the regulation of DNA replication in bacteria. Cold Spring Harbor Symp. Quant. Biol. 28, 329–348.

James, R. (1975) Identification of an outer membrane protein of *Escherichia coli*, with a role in the coordination of deoxyribonucleic acid replication and cell elongation. J. Bacteriol. 124, 918–929.

James, R. and Gudas, L. J. (1976) Cell cycle-specific incorporation of lipoprotein into the outer membrane of *Escherichia coli*. J. Bacteriol. 125, 374–375.

James, R., Haga, J. Y. and Pardee, A. B. (1975) Inhibition of an early event in the cell division cycle of *Escherichia coli* by FL 1060, an amidinopenicillanic acid. J. Bacteriol. 122, 1283–1292.

Johnson, K. G. and Campbell, J. N. (1972) Effect of growth conditions on peptidoglycan structure and susceptibility to lytic enzymes in cell walls of *Micrococcus sodonensis*. Biochemistry 11, 277–286.

Johnson, K. G. and McDonald I. J. (1974) Peptidoglycan structure in cell walls of parental and filamentous *Streptococcus cremoris* HP. Can. J. Microbiol. 20, 905–913.

Jones, N. C. and Donachie, W. D. (1973) Chromosome replication, transcription and control of cell division in *Escherichia coli*. Nature New Biol. 243, 100–103.

Jones, N. C. and Donachie, W. D. (1974) Protein synthesis and the release of the replicated chromosome from the cell membrane. Nature New Biol. 251, 252–253.

Joseph, R. and Shockman, G. D. (1974) Autolytic formation of protoplasts (autoplasts) of *Streptococcus faecalis* 9790: Release of cell wall, autolysin, and formation of stable autoplasts. J. Bacteriol. 118, 735–746.

Joseph, R. and Shockman, G. D. (1975) Synthesis and excretion of glycerol teichoic acid during growth of two streptococcal species. Infect. Immun. 12, 333–338.

Joseph, R. and Shockman, G. D. (1976) Autolytic formation of protoplasts (autoplasts) of *Streptococcus faecalis* 9790: Location of active and latent autolysin. J. Bacteriol. 127, 1482–1493.

Kamiryo, T. and Strominger, J. L. (1974) Penicillin-resistant temperature-sensitive mutants of *Escherichia coli* which synthesize hypo- or hypercross-linked peptidoglycan. J. Bacteriol. 117, 568–577.

Katz, W. and Martin, H. H. (1970) Peptide crosslinkage in cell wall murein of *Proteus mirabilis* and its penicillin-induced unstable L-form. Biochem. Biophys. Res. Commun. 39, 744–749.

Keglevic, D., Ladesic, B., Hadzija, O., Tomasic, J., Valinger, Z. and Pokorny, M. (1974) Isolation and study of the composition of a peptidoglycan complex excreted by the biotin-requiring mutant of *Brevibacterium divaricatum* NRRL-2311 in the presence of penicillin. Eur. J. Biochem. 42, 389–400.

Kennedy, L. D. and Shaw, D. R. D. (1968) Direction of polyglycerol phosphate chain growth in *Bacillus subtilis*. Biochem. Biophys. Res. Commun. 32, 861–865.

Kjeldgaard, N. O., Maaløe, O. and Schaechter, M. (1958) The transition between different physiological states during balanced growth of *Salmonella typhimurium*. J. Gen. Microbiol. 19, 607–616.

Knox, K. W. and Hall, E. A. (1965) The linkage between the polysaccharide and mucopeptide components of the cell wall of *Lactobacillus casei*. Biochem. J. 96, 302–309.

Koch, A. L. (1966) On evidence supporting a deterministic process of bacterial growth. J. Gen. Microbiol. 43, 1–5.

Koch, A. L. (1970) Overall controls on the biosynthesis of ribosomes in growing bacteria. J. Theoret. Biol. 28, 203–231.

Koch, A. L. (1971) The adaptive responses of *Escherichia coli* to a feast and famine existence. Adv. Microb. Physiol. 6, 147–217.

708

Koch, A. L. and Blumberg, G. (1976) The distribution of bacteria in the velocity gradient centrifuge. Biophys. J. 16, 389–405.

Koch, A. L. and Deppe, C. S. (1971) *In vivo* assay of protein synthesizing capacity of *Escherichia coli* from slowly growing chemostat cultures. J. Mol. Biol. 55, 549–562.

Koch, A. L. and Schaechter, M. (1962) A model for statistics of the cell division process. J. Gen. Microbiol. 29, 435–454.

Krulwich, T. A., Ensign, J. C., Tipper, D. J. and Strominger, J. L. (1967) Sphere-rod morphogenesis in *Arthrobacter crystallopoietes*. J. Bacteriol. 94, 741–750.

Krulwich, T. A. and Ensign, J. C. (1968) Activity of an autolytic N-acetylmuramidase during sphere-rod morphogenesis in *Arthrobacter crystallopoietes*. J. Bacteriol. 96, 857–859.

Kubitschek H. E. (1968a) Linear cell growth in *Escherichia coli*. Biophys. J. 8, 792–804.

Kubitschek, H. E. (1968b) Constancy of uptake during the cell cycle in *Escherichia coli*. Biophys. J. 8, 1401–1412.

Kubitschek, H. E. (1974) Constancy of the ratio of DNA to cell volume in steady-state cultures of *Escherichia coli* B/r. Biophys. J. 14, 119–123.

Kubitschek, H. E. and Freedman, M. L. (1971) Chromosome replication and the division cycle of *Escherichia coli* B/r. J. Bacteriol. 107, 95–99.

Lacks, S. (1970) Mutants of *Diplococcus pneumoniae* that lack deoxyribonucleases and other activities possibly pertinent to genetic transformation. J. Bacteriol. 101, 373–383.

Lancy, P., Jr. (1976) Investigations into the mechanisms of control of macromolecular biosynthesis *in vivo* in *Streptococcus faecalis* at different growth rates under regimens of nutritional limitation or increased osmotic pressure. Ph.D. Dissertation, Temple University School of Medicine.

Lancy, P., Jr., Rosenzweig, M., Carson, D. and Daneo-Moore, L. (1977) A possible constant efficiency of cellular proteins in the synthesis of surface constituents in glutamate limited cultures of *Streptococcus faecalis* ATCC 9790. In preparation.

Landman, O. E., Ryter, A. and Frehel, C. (1968) Gelatin-induced reversion of protoplasts of *Bacillus subtilis* to the bacillary form: Electronmicroscopic and physical study. J. Bacteriol. 96, 2154–2170.

Lapidus, I. R. (1971) Analysis of a model for multiseptation in bacteria. J. Bacteriol. 108, 607–608.

Lark, C. (1966) Regulation of deoxyribonucleic acid synthesis in *Escherichia coli*: Dependence on growth rates. Biochim. Biophys. Acta 119, 517–525.

Lark, K. G. and Lark, C. (1960) Changes during the division cycle in bacterial cell wall synthesis, volume, and ability to concentrate free amino acids. Biochim. Biophys. Acta 43, 520–530.

Lark, K. G. and Maaløe, O. (1954) The induction of cellular and nuclear division in *Salmonella typhimurium* by means of temperature shifts. Biochim. Biophys. Acta 15, 345–356.

Lederberg, J. (1956) Bacterial protoplasts induced by penicillin. Proc. Nat. Acad. Sci. U.S.A. 42, 574–577.

Leibowitz, P. J. and Schaechter, M. (1975) the attachment of the bacterial chromosome to the cell membrane. International Review of Cytology 41, 1–26.

Lin, E. C. C., Hirota, Y. and Jacob, F. (1971) On the process of cellular division in *Escherichia coli*. J. Bacteriol. 108, 375–385.

Linnet, P. E. and Tipper, D. J. (1974) Cell wall polymers of *Bacillus sphaericus*: Activities of enzymes involved in peptidoglycan precursor synthesis during sporulation. J. Bacteriol. 120, 342–354.

Lipmann, F. (1968) The relation between the direction and mechanism of polymerization. In: Essays in Biochemistry (Campbell, P. N. and Groville, G. D. eds.), pp. 1–23, Academic Press, London.

Lomnitzer, R. and Ron, E. (1972) Synchronization of cell division in *Escherichia coli* by elevated temperatures: a reinterpretation. J. Bacteriol. 109, 1316–1318.

Lorian, V. (1975) Some effects of subinhibitory concentrations of penicillin on the structure and division of staphylococci. Antimicrob. Agents Chemother. 7, 864–870.

Lorian, V. and Atkinson, B. (1975) Abnormal forms of bacteria produced by antibiotics. Am. J. Clin. Pathol. 64, 678–688.

Lorian, V. and Atkinson, B. (1976) Effects of subinhibitory concentrations of antibiotics on cross walls of cocci. Antimicrob. Agents Chemother. 9, 1043–1055.

Lund, F. and Tybring, L. (1972) 6β-amidinopenicillanic acids—a new group of antibiotics. Nature New Biol. 236, 135–137.

Lynch, J. L. and Neuhaus, F. (1966) On the mechanism of action of the antibiotic 0-carbamyl-D-serine in *Streptococcus faecalis*. J. Bacteriol. 91, 449–460.

Maaløe, O. and Hanawalt, P. (1961) Thymine Deficiency and the normal DNA replication cycle. I. J. Mol. Biol. 3, 144–155.

Maaløe, O. and Kjeldgaard, N. (1966) Control of Macromolecular Synthesis: A Study of DNA, RNA, and Protein Synthesis in Bacteria. W. A. Benjamin, New York.

Mann, N. and Carr, N. G. (1974) Control of macromolecular composition and cell division in the blue-green alga *Anacystis nidulans*. J. Gen. Microbiol. 83, 399–405.

Markham, J. L., Knox, K. W., Wicken, A. J. and Hewett, M. (1975) Formation of extracellular lipoteichoic acid by oral streptococci and lactobacilli. Infect. Immun. 12, 378–386.

Marunouchi, T. and Messer, W. (1973) Replication of a specific terminal chromosome segment in *Escherichia coli* which is required for cell division. J. Mol. Biol. 78, 211–228.

Maruyama, Y. and Yanagita, T. (1956) Physical methods for obtaining synchronous culture of *Escherichia coli*. J. Bacteriol. 71, 542–546.

Masters, M. and Pardee, A. B. (1965) Sequence of enzyme synthesis and gene replication during the cell cycle of *Bacillus subtilis*. Proc. Nat. Acad. Sci. U.S.A. 54, 64–70.

Masters, M., Kuempel, P. L. and Pardee, A. B. (1964) Enzyme synthesis in synchronous cultures of bacteria. Biochem. Biophys. Res. Commun. 15, 38–42.

Matney, T. S. and Suit, J. C. (1966) Synchronously dividing bacterial cultures. I. Synchrony following depletion and resupplementation of a required amino acid in *Escherichia coli*. J. Bacteriol. 92, 960–966.

Matsuhashi, S., Kamiryo, T., Blumberg, P., Linnett, P., Willoughby, E. and Strominger, J. (1974) Mechanism of action and development of resistance to a new amidino penicillin. J. Bacteriol. 117, 578–587.

Matsuhashi M., Takagaki, Y., Maruyama, I. N., Tamaki, S., Nishimura, Y., Suzuki, S., Ogino, U. and Hirota, Y. (1977) Proc. Nat. Acad. Sci. U.S.A., in press.

Mauck, J. and Glaser, L. (1970) Turnover of the cell wall of *Bacillus subtilis* W-23 during logarithmic growth. Biochem. Biophys. Res. Commun. 39, 699–706.

Mauck, J. and Glaser, L. (1972) On the mode of *in vivo* assembly of the cell wall of *Bacillus subtilis*. J. Biol. Chem. 247, 1180–1187.

Mauck, J., Chan, L. and Glaser, L. (1971) Turnover of the cell wall of Gram-positive bacteria. J. Biol. Chem. 246, 1820–1827.

Mauck, J., Chan, L., Glaser, L. and Williamson, J. (1972) Mode of cell wall growth of *Bacillus megaterium*. J. Bacteriol. 109, 373–378.

Meacock, P. A. and Pritchard, R. H. (1975) Relationship between chromosome replication and cell division in a thymineless mutant of *Escherichia coli* B/r. J. Bacteriol. 122, 931–942.

Mendelson, N. H. (1976) Helical growth of *Bacillus subtilis:* A new model of cell growth. Proc. Nat. Acad. Sci. U.S.A. 73, 1740–1744.

Mendelson, N. H. and Gross, J. D. (1967) Characterization of a temperature-sensitive mutant of *Bacillus subtilis* defective in deoxyribonucleic acid replication. J. Bacteriol. 94, 1603–1608.

Meyer, M., de Jong, M. A., Woldringh, C. L. and Nanninga, N. (1975) Physiological significance of folded genomes from amino acid-starved *Escherichia coli* K12 DG75 (abstract) Lunteren Lect. Mol. Gen. Sect. VII, p. 3.

Mindich, L. (1973) Synthesis and assembly of bacterial membranes. In: Bacterial Membranes and Walls (Leive, L., ed.), pp. 2–27. Marcel Dekker, New York.

Mirelman, D., Bracha, R. and Sharon, N. (1972) Role of the penicillin-sensitive transpeptidation reaction in attachment of newly synthesized peptidoglycan to cell walls of *Micrococcus luteus*. Proc. Nat. Acad. Sci. U.S.A. 69, 3355–3359.

Mirelman, D., Bracha, R. and Sharon, N. (1974a) Inhibition by penicillin of the incorporation and cross-linking of L-lysine in intact cells of *Micrococcus luteus*. FEBS Lett. 39, 105–110.

Mirelman, D., Bracha, R. and Sharon, N. (1974b) Penicillin-induced secretion of a soluble, uncross-linked peptidoglycan by *Micrococcus luteus* cells. Biochemistry 13, 5045–5053.

Mirelman, D., Shaw, D. R. D. and Park, J. T. (1971) Nature and origins of phosphorus compounds in isolated cell walls of *Staphylococcus aureus*. J. Bacteriol. 107, 239–244.

710

Mirelman, D., Yashouv-Gan, Y. and Schwarz, U. (1976) Peptidoglycan biosynthesis in a thermosensitive division mutant of *Escherichia coli*. Biochemistry 15, 1781–1790.

Mitchell, P. and Moyle, J. (1957) Autolytic release and osmotic properties of "protoplasts" from *Staphylococcus aureus*. J. Gen. Microbiol. 16, 184–194.

Mitchison, J. M. (1961) The growth of single cells. III. *Streptococcus faecalis*. Exp. Cell Res. 22, 208–225.

Mitchison, J. M. and Vincent, W. S. (1965) Preparation of synchronous cell cultures by sedimentation. Nature 205, 987–989.

Mitchison, J. M. (1971) The Biology of the Cell Cycle. Cambridge University Press, Cambridge, England.

Mosser, J. L. and Tomasz, A. (1970) Choline-containing teichoic acid as a structural component of pneumococcal cell wall and its role in sensitivity to lysis by an autolytic enzyme. J. Biol. Chem. 245, 287–298.

Murray, R. G. E., Francombe, W. H. and Mayall, B. H. (1959) The effect of penicillin on the structure of staphylococcal cell walls. Can. J. Microbiol. 5, 641–648.

Mychajlonka, M. and Shockman, G. D. (1976) Conservation of cell wall peptidoglycan in *Streptococcus mutans*. Abstr. Ann. Meet. Am. Soc. Microbiol., p. 137.

Oldmixon, E. H., Dezélée, P., Ziskin, M. and Shockman, G. D. (1976) A mechanism of forming peptide cross-links in the cell wall peptidoglycan of *Streptococcus faecalis*. Eur. J. Biochem. 68, 271–280.

Onken, A. and Messer, W. (1973) Cell division in *Escherichia coli*. Septation during synchronous growth. Mol. Gen. Genet. 127, 349–358.

Owen, P. and Salton, M. R. J. (1975) A succinylated mannan in the membrane system of *Micrococcus lysodeikticus*. Biochem. Biophys. Res. Commun. 63, 875–880.

Painter, P. R. and Marr, A. G. (1968) Mathematics of microbial populations. Ann. Rev. Microbiol. 22, 519–544.

Pardee, A. B. and Rozengurt, E. (1975) Role of the surface in production of new cells. In: MTP International Review of Science; Biochemistry of Cell Walls and Membranes, Biochemistry Series One, vol. 2, (Fox, C. F., ed.), pp. 155–185. Butterworths, London.

Park, J. T. and Burman, L. (1973) A new penicillin with a unique mode of action. Biochem. Biophys. Res. Commun. 51, 863–868.

Park, J. T., Shaw, D. R. D., Chatterjee, A., Mirelman, D. and Wu, T. (1974) Mutants of staphylococci with altered cell walls. Ann. N.Y. Acad. Sci. 236, 54–62.

Pato, M. L. (1975) Alterations of the rate of movement of deoxyribonucleic acid replication forks. J. Bacteriol. 123, 272–277.

Paulton, R. J. L. (1970) Analysis of the multiseptate potential of *Bacillus subtilis*. J. Bacteriol. 104, 762–767.

Pavlik, J. G. and Rogers, H. J. (1973) Selective extraction of polymers from cell walls of Gram-positive bacteria. Biochem. J. 131, 619–621.

Perkins, H. R. (1965) The action of hot formamide on bacterial cell walls. Biochem. J. 95, 876–882.

Perret, C. J. (1958) The effect of growth-rate on the anatomy of *Escherichia coli*. J. Gen. Microbiol. 18, 7–8.

Pierucci, O. and Zuchowski, C. (1973) Non-random segregation of DNA strands in *Escherichia coli* B/r. J. Mol. Biol. 80, 477–503.

Pitel, D. W. and Gilvarg, C. (1970) Mucopeptide metabolism during growth and sporulation in *Bacillus megaterium*. J. Biol. Chem. 245, 6711–6717.

Pless, D. D., Schmit, A. S. and Lennarz, W. J. (1975) The characterization of mannan of *Micrococcus lysodeikticus* as an acidic lipopolysacch aride. J. Biol. Chem. 250, 1319–1327.

Pollock, J., Nguyen-Disteche, M., Ghuysen, J.-M., Coyette, J., Linder, R., Salton, M., Kim, K., Perkins, H. and Reynolds, P. (1974) Fractionation of the DD-carboxypeptidase-transpeptidase activities solubilized from membranes of *Escherichia coli* K12, Strain 44. Eur. J. Biochem. 41, 439–446.

Pollock, J. and Sharon, N. (1970) Studies on the acceptor specificity of the lysozyme-catalyzed transglycosylation reaction. Biochemistry 9, 3913–3925.

Pooley, H. M. (1976a) Turnover and spreading of old wall during surface growth of *Bacillus subtilis*. J. Bacteriol. 125, 1127–1138.

Polley, H. M. (1976b) A layered distribution according to age, within the cell wall of *Bacillus subtilis*. J. Bacteriol. 125, 1139–1147.

Pooley, H. M. and Shockman, G. D. (1969) Relationship between the latent form and the active form of the autolytic enzyme of *Streptococcus faecalis*. J. Bacteriol. 100, 617–624.

Pooley, H. M. and Shockman, G. D. (1970) Relationship between the location of autolysin, cell wall synthesis and the development of resistance to cellular autolysis in *Streptococcus faecalis* after inhibition of protein synthesis. J. Bacteriol. 103, 457–466.

Pooley, H. M., Porres-Juan, J. M. and Shockman, G. D. (1970) Dissociation of an autolytic enzyme-cell wall complex by treatment with unusually high concentrations of salt. Biochem. Biophys. Res. Commun. 38, 1134–1140.

Powell, E. O. (1956) Growth rate and generation time of bacteria, with special reference to continuous culture. J. Gen. Microbiol. 15, 492–511.

Powell, E. O. and Errington, F. P. (1963) The size of bacteria, as measured with the Dyson image-splitting eyepiece. J. Roy. Microsc. Soc. 82, 39–49.

Previc, E. P. (1970) Biochemical determination of bacterial morphology and the geometry of cell division. J. Theoret. Biol. 27, 471–497.

Previc, E. P. and Lowell, N. (1975) Peptidoglycan compositions of a new strain of *Arthrobacter crystallopoietes* during sphere-rod morphogenesis. Biochim. Biophys. Acta 411, 377–385.

Pritchard, R. H. (1968) Control of DNA synthesis in bacteria. Heredity 23, 472.

Pritchard, R. H. (1974) Review lecture on the growth and form of a bacterial cell, in Phil. Trans. Roy. Soc. Lond. (Ser. B.) 267, 305–336.

Pritchard, R. H., Barth, P. T. and Collins, J. (1969) Control of DNA synthesis in bacteria. Symp. Soc. Gen. Microbiol. 19, 263–297.

Pritchard, R. H. and Zaritsky, A. (1970) Effect of thymine concentration on the replication velocity of DNA in a thymineless mutant of *Escherichia coli*. Nature 226, 126–131.

Rogers, H. J. (1965) The outer layers of bacteria: The biosynthesis of structure. Symp. Soc. Gen. Microbiol. 15, 186–219.

Rogers, H. J. (1967) Killing of staphylococci by penicillins. Nature 213, 31–33.

Rogers, H. J. (1970) Bacterial growth and the cell envelope. Bact. Rev. 34, 194–214.

Rogers, H. J. (1977) Envelope growth and synthesis in rod mutants and protoplasts of bacilli. Microbiology, pp. 25–34.

Rogers, H. J. and Forsberg, C. W. (1971) Role of autolysins in the killing of bacteria by some bactericidal antibiotics. J. Bacteriol. 108, 1235–1243.

Rogers, H. J., McConnell, M. and Hughes, R. C. (1971) The chemistry of the cell walls of *rod* mutants of *Bacillus subtilis*. J. Gen. Microbiol. 66, 297–308.

Rogers, H. J., Thurman, P. F., and Buxton, R. S. (1976) Magnesium and anion requirements of *rod B* mutants of *Bacillus subtilis*. J. Bacteriol. 125, 556–564.

Ron, E. Z., Rozenhak, S. and Grossman, N. (1975) Synchronization of cell division in *Escherichia coli* by amino acid starvation: Strain specificity. J. Bacteriol. 123, 374–376.

Rosenthal, R. and Shockman, G. D. (1975a) Characterization of the presumed peptide cross-links in the soluble peptidoglycan fragments synthesized by proptoplasts of *Streptococcus faecalis*. J. Bacteriol. 124, 410–418.

Rosenthal, R. and Shockman, G. D. (1975b) Synthesis of peptidoglycan in the form of soluble glycan chains by growing protoplasts (autoplasts) of *Streptococcus faecalis*. J. Bacteriol. 124, 419–423.

Rosenthal, R. S., Jungkind, D., Daneo-Moore, L. and Shockman, G. D. (1975) Evidence for the synthesis of soluble peptidoglycan fragments by protoplasts of *Streptococcus faecalis*. J. Bacteriol. 124, 398–409.

Rothfield, L. and Pearlman-Kothencz, M. (1969) Synthesis and assembly of bacterial membrane components. A lipopolysaccharide-phospholipid-protein complex excreted by living bacteria. J. Mol. Biol. 44, 477–492.

Rudner, R., Rejman, E. and Chargaff, E. (1965) Genetic implications of periodic pulsations of the rate of synthesis and the composition of rapidly labeled bacteria RNA. Proc. Nat. Acad. Sci. U.S.A. 54, 904–911.

Ryder, O. and Smith, D. (1974) Isolation of membrane-associated folded chromosomes from *Escherichia coli*: Effect of protein synthesis inhibition. J. Bacteriol. 120, 1356–1363.

712

Ryter, A., Hirota, Y. and Schwarz, U. (1973) Process of cellular division in *Escherichia coli*. Growth pattern of *E. coli* murein. J. Mol. Biol. 78, 185–195.

Salton, M. R. J. (1964) The Bacterial Cell Wall. Elsevier, Amsterdam.

Sargent, M. G. (1973) Membrane synthesis in synchronous cultures of *Bacillus subtilis* 168. J. Bacteriol. 116, 397–409.

Sargent, M. G. (1974) Nuclear segregation in *Bacillus subtilis*. Nature 250, 252–254.

Sargent, M. G. (1975a) Control of cell length in *Bacillus subtilis*. J. Bacteriol. 123, 7–19.

Sargent, M. G. (1975b) Anucleate cell production and surface extension in a temperature-sensitive chromosome initiation mutant of *Bacillus subtilis*. J. Bacteriol. 123, 1218–1234.

Sargent, M. G. (1975c) Control of membrane protein synthesis in *Bacillus subtilis*. Biochim. Biophys. Acta 406, 564–574.

Sayare, M., Daneo-Moore, L. and Shockman, G. D. (1972) Influence of macromolecular biosynthesis on cellular autolysis in *Streptococcus faecalis*. J. Bacteriol. 112, 337–344.

Schaechter, M., Maaløe, O. and Kjeldgaard, N. O. (1958) Dependency on medium and temperature of cell size and chemical composition during balanced growth of *Salmonella typhimurium*. J. Gen. Microbiol. 19, 592–606.

Schaechter, M., Williamson, J., Hood, Jr., J. R. and Koch, A. (1962) Growth, cell and nuclear divisions in some bacteria. J. Gen. Microbiol. 29, 421–434.

Scheefers-Borchel U. and Vielmetter, W. (1975) Properties of folded chromosomes during the synchronous division cycle of *Escherichia coli*. Lunteren Lect. Mol. Gen. Sect. VII, 2–3.

Scherrer, R. and Gerhardt, P. (1964) Molecular sieving by cell membranes of *Bacillus megaterium*. Nature 204, 649–650.

Scherrer, R. and Gerhardt, P. (1971) Molecular sieving by the *Bacillus megaterium* cell wall and protoplast. J. Bacteriol. 107, 718–735.

Schleifer, K. H. (1969) Substrate dependent modifications of the amino acid sequence of the murein of staphylococci. J. Gen. Microbiol. 57, xiv.

Schleifer, K. H. and Kandler, O. (1972) Peptidoglycan types of bacterial cell walls and their taxonomic implications. Bacterial. Rev. 36, 407–477.

Schwarz, U., Asmus, A. and Frank, H. (1969) Autolytic enzymes and cell division of *Escherichia coli*. J. Mol. Biol. 41, 419–429.

Schwarz, U. and Leutgeb, W. (1971) Morphogenetic aspects of murein structure and biosynthesis. J. Bacteriol. 106, 588–595.

Sebastian, J., Carter, B. L. A. and Halvorson, H. O. (1971) Use of yeast populations fractioned by zonal centrifugation to study the cell cycle. J. Bacteriol. 108, 1045–1050.

Sedgwick, E. G. and Paulton, R. J. L. (1974) Dimension control in bacteria. Can. J. Microbiol. 20, 231–236.

Sharon, N. and Seifter, S. (1964) A transglycosylation reaction catalyzed by lysozyme. J. Biol. Chem. 239, 2398–2399.

Shockman, G. D. (1959a) Bacterial cell wall synthesis: The effect of threonine depletion. J. Biol. Chem. 234, 2340–2342.

Shockman, G. D. (1959b) The reversal of cycloserine inhibition by D-alanine. Proc. Soc. Exp. Biol. Med. 101, 693–695.

Shockman, G. D. (1965) Symposium on the fine structure and replication of bacteria and their parts. IV. Unbalanced cell-wall synthesis: autolysis and cell-wall thickening. Bacteriol. Rev. 29, 345–358.

Shockman, G. D. and Cheney, M. C. (1969) Autolytic enzyme system of *Streptococcus faecalis*. V. Nature of the autolysin-cell wall complex and its relationship to properties of the autolytic enzyme of *Streptococcus faecalis*. J. Bacteriol. 98, 1199–1207.

Shockman, G. D. and Martin, J. T. (1968) Autolytic enzyme system of *Streptococcus faecalis*. IV. Electron microscopic observations of autolysin and lysozyme action. J. Bacteriol. 96, 1803–1810.

Shockman, G. D., Conover, M. J., Kolb, J. J., Phillips, P. M., Riley, L. S. and Toennies, G. (1961a) A study of lysis of *Streptococcus faecalis*. J. Bacteriol. 81, 36–43.

Shockman, G. D., Conover, M. J., Kolb, J. J., Riley, L. S. and Toennies, G. (1961b) Nutritional requirements for bacterial cell wall synthesis. J. Bacteriol. 81, 44–50.

Shockman, G. D., Daneo-Moore, L. and Higgins, M. L. (1974) Problems of cell wall and membrane growth, enlargement and division. Ann. N.Y. Acad. Sci. 235, 161–197.

Shockman, G. D., Kolb, J. J. and Toennies, G. (1958) Relations between bacterial cell wall synthesis, growth phase and autolysis. J. Biol. Chem. 230, 961–977.

Shockman, G. D., Pooley, H. M. and Thompson, J. S. (1967a) The autolytic enzyme system of *Streptococcus faecalis.* III. The localization of the autolysin at the sites of cell wall synthesis. J. Bacteriol. 94, 1525–1530.

Shockman, G. D., Thompson, J. S. and Conover, M. J. (1965) Replacement of lysine by hydroxylysine and its effects on cell lysis in *Streptococcus faecalis.* J. Bacteriol. 90, 575–588.

Shockman, G. D., Thompson, J. S. and Conover, M. J. (1967b) The autolytic enzyme system of *Streptococcus faecalis.* II. Partial characterization of the autolysin and its substrate. Biochemistry 6, 1054–1065.

Singer, H. J., Wise, Jr., E. and Park, J. T. (1972) Properties and purification of N-acetylmuramyl-L-alanine amidase from *Staphylococcus aureus* H. J. Bacteriol. 112, 932–939.

Slater, M. and Schaechter, M. (1974) Control of cell division in bacteria. Bact. Rev. 38, 199–221.

Smith, D. W. and Hanawalt, P. C. (1967) Properties of the growing point region in the bacterial chromosome. Biochim. Biophys. Acta 149, 519–531.

Smith, H. S. and Pardee, A. B. (1970) Accumulation of a protein required for division during the cell cycle of *Escherichia coli.* J. Bacteriol. 101, 901–909.

Snyder, R. W. and Young, F. E. (1969) Association between the chromosome and the cytoplasmic membrane in *Bacillus subtilis.* Biochem. Biophys. Res. Commun. 35, 354–362.

Spratt, B. G. (1975) Distinct penicillin binding proteins involved in the division, elongation, and shape of *Escherichia coli* K-12. Proc. Nat. Acad. Sci. U.S.A. 72, 2999–3003.

Spratt, B. G. and Pardee, A. B. (1975) Penicillin binding proteins and cell shape in *Escherichia coli.* Nature 254, 516–517.

Spratt, B. G. and Rowbury, R. J. (1970) A mutant in the initiation of DNA synthesis in *Salmonella typhimurium.* J. Gen. Microbiol. 64, 127–138.

Spratt, B. G. and Rowbury, R. J. (1971) Cell division in a mutant of *Salmonella typhimurium* which is temperature-sensitive for DNA synthesis. J. Gen. Microbiol. 65, 305–314.

Stonehill, E. H. and Hutchison, D. J. (1966) Chromosomal mapping by means of mutational induction in synchronous populations of *Streptococcus faecalis.* J. Bacteriol. 92, 136–143.

Streips, U. N., Doyle, R. J., Brown, W. C. and Sueoka, N. (1976) Transformation in *Bacillus subtilis* using cell wall-associated DNA. Abstr. Ann. Meet. Am. Soc. Microbiol., p. 97.

Sud, I. J. and Schaechter, M. (1964) Dependence of the content of cell envelopes on the growth rate of *Bacillus megaterium.* J. Bacteriol. 88, 1612–1617.

Sueoka, N. and Yoshikawa, H. (1965) The chromosome of *Bacillus subtilis* I. Theory of marker frequency analysis. Genetics 52, 747–757.

Sueoka, N. and Quinn, W. G. (1968) Membrane attachment of the chromosome replication origin in *Bacillus subtilis.* Cold Spring Harbor Symp. Quant. Biol. 33, 695–705.

Suganuma, A. (1962) Some observations on the fine structure of *Staphylococcus aureus.* J. Infect. Dis. 111, 8–16.

Tamura, T., Imae, Y. and Strominger, J. (1976) Purification to homogeneity and properties of two D-alanine carboxypeptidases I from *Escherichia coli.* J. Biol. Chem. 251, 411–423.

Tauro, P. and Halvorson, H. (1966) Effect of gene position on the timing of enzyme synthesis in synchronous cultures of yeast. J. Bacteriol. 92, 652–661.

Taylor, A., Das, B. C. and Van Heijenoort, J. (1975) Bacterial-cell-wall peptidoglycan fragments produced by phage λ or Vi endolysin and containing 1,6-anhydro-N-acetylmuramic acid. Eur. J. Biochem. 53, 47–54.

Tempest, D. W. and Ellwood, D. C. (1969) The influence of growth conditions on the composition of some cell wall components of *Aerobacter aerogenes.* Biotechnol. Bioeng. 11, 775–783.

Tempest, D. W., Dicks, J. W. and Ellwood, D. C. (1968) Influence of growth condition on the concentration of potassium in *Bacillus subtilis* var. *niger* and its possible relationship to cellular ribonucleic acid, teichoic acid and teichuronic acid. Biochem. J. 106, 237–243.

Tipper, D. J. and Strominger, J. L. (1965) Mechanism of action of penicillins: A proposal based on their structural similarity to acyl-D-alanyl-D-alanine. Proc. Nat. Acad. Sci. U.S.A. 54, 1133–1141.

714

Toennies, G. and Shockman, G. D. (1958) Growth chemistry of *Streptococcus faecalis*. Proc. 4th Int. Congr. Biochem., Vienna. 13, 365–394.

Toennies, G., Bakay, B. and Shockman, G. D. (1959) Bacterial composition and growth phase. J. Biol. Chem. 234, 3269–3275.

Toennies, G., Shockman, G. D. and Kolb, J. J. (1963) Differential effects of amino acid deficiencies on bacterial cytochemistry. Biochemistry 2, 294–296.

Tomasz, A. (1968) Biological consequences of the replacement of choline by ethanolamine in the cell wall of pneumococcus: Chain formation, loss of transformability, and loss of autolysis. Proc. Nat. Acad. Sci. U.S.A. 59, 86–93.

Tomasz, A. and Waks, S. (1975) Mechanism of action of penicillin: Triggering of the pneumococcal autolytic enzyme by inhibitors of cell wall synthesis. Proc. Nat. Acad. Sci. U.S.A. 72, 4162–4166.

Tomasz, A., Albino, A. and Zanati E. (1970) Multiple antibiotic resistance in a bacterium with suppressed autolytic system. Nature 227, 138–140.

Tomasz, A., Westphal, M., Briles, E. and Fletcher, P. (1975) On the physiological functions of teichoic acids. J. Supramol. Struct. 3, 1–16.

Tynecka, Z. and Ward, J. B. (1975) The inhibition of cross-linking by benzylpenicillin and cephaloridine *in vivo* accompanied by the formation of soluble peptidoglycan. Biochem. J. 146, 253–267.

Vambutas, V. and Salton, M. R. J. (1970) Incorporation of [^{14}C]glycine into *Micrococcus lysodeikticus* membrane protein and effects of protein synthesis inhibitors. Biochim. Biophys. Acta 203, 83–93.

Van Heijenoort, J., Parquet, C., Flouret, B. and Van Heijenoort, Y. (1975) Envelope-bound N-acetylmuramyl-L-alanine amidase of *Escherichia coli* K12. Eur. J. Biochem. 58, 611–619.

Van Tubergen, R. P. and Setlow, R. B. (1961) Quantitative radioautographic studies on exponentially growing cultures of *Escherichia coli*. The distribution of parental DNA, RNA, protein, and cell wall among progeny cells. Biophys. J. 1, 589–625.

Wadstrom, T. and Vesterberg, O. (1971) Studies on endo-β-N-acetyl-glucosaminidase, staphylolytic peptidase, and N-acetylmuramyl-L-alanine amidase in lysostaphin and from *Staphylococcus aureus*. Acta Path. Microbiol. Scand. Section B 79, 248–264.

Ward, C. B. and Glaser, D. A. (1971) Correlation between rate of cell growth and rate of DNA synthesis in *Escherichia coli* B/r. Proc. Nat. Acad. Sci. U.S.A. 68, 1061–1064.

Ward, J. B. (1973) The chain length of the glycans in bacterial cell walls. Biochem. J. 133, 395–398.

Ward, J. B. and Perkins, H. R. (1974) The direction of glycan synthesis in a bacterial peptidoglycan. Biochem. J. 135, 721–728.

Ward, J. B. and Perkins, H. R. (1971) Peptidoglycan biosynthesis by preparations from *Bacillus licheniformis*: Cross-linking of newly synthesized chains to preformed cell wall. Biochem. J. 139, 781–784.

Warth, A. D. and Strominger, J. L. (1971) Structure of the peptidoglycan from vegetative cell walls of *Bacillus subtilis*. Biochemistry 10, 4349–4358.

Watson, M. J. and Baddiley, J. (1974) The action of nitrous acid on C-teichoic acid (C-substance) from the walls of *Diplococcus pneumoniae*. Biochem. J. 137, 399–404.

Weidel, W. and Pelzer, H. (1964) Bagshaped macromolecules—a new outlook on bacterial cell walls. Adv. Enzymol. 26, 193–232.

White, D., Dworkin, M. and Tipper, D. J. (1968) Peptidoglycan of *Myxococcus xanthus*: Structure and relation to morphogenesis. J. Bacteriol. 95, 2186–2197.

Whitney, J. G. and Grula, E. A. (1964) Incorporation of D-serine into the cell wall mucopeptide of *Micrococcus lysodeikticus*. Biochem. Biophys. Res. Commun. 14, 375–381.

Wicken, A. J. and Knox, K. W. (1975) Lipoteichoic acids: A new class of bacterial antigen. Science 187, 1161–1167.

Wickus, G. G. and Strominger, J. L. (1972) Penicillin-sensitive transpeptidation during peptidoglycan biosynthesis in cell-free preparation from *Bacillus megaterium*. I. Incorporation of free diaminopimelic acid into peptidoglycan. J. Biol. Chem. 247, 5297–5306.

Wise, Jr., E. M. and Park, J. T. (1965) Penicillin: Its basic site of action as an inhibitor of a peptide cross-linking reaction in cell wall mucopeptide synthesis. Proc. Nat. Acad. Sci. U.S.A. 54, 75–81.

Woldringh, C. L. (1976) Morphological analysis of nuclear separation and cell division during the life cycle of *Escherichia coli* B/r. J. Bacteriol. 125, 248–257.

Wong, W., Young, F. E. and Chatterjee, A. N. (1974) Regulation of bacterial cell walls: Turnover of cell wall in *Staphylococcus aureus*. J. Bacteriol. 120, 837–843.

Worcel, A. and Burgi, E. (1974) Properties of a membrane-attached form of the folded chromosome of *Escherichia coli*. J. Mol. Biol. 82, 91–105.

Wu, P. C. and Pardee, A. B. (1973) Cell division of *Escherichia coli* Control by membrane organization. J. Bacteriol. 114, 603–611.

Wyke, A. W. and Ward, J. B. (1975) The synthesis of covalently-linked teichoic acid and peptidoglycan by cell-free preparations of *Bacillus licheniformis*. Biochem. Biophys. Res. Commun. 65, 877–885.

Yamada, M., Hirose, A. and Matsuhashi, M. (1975) Association of lack of cell wall teichuronic acid with formation of cell packets of *Micrococcus lysodeikticus (luteus)* mutants. J. Bacteriol. 123, 678–686.

Yamaguchi, K. and Yoshikawa, H. (1975) Association of the replication terminus of the *Bacillus subtilis* chromosome to the cell membrane. J. Bacteriol. 124, 1030–1033.

Yamaguchi, K., Murakami, S. and Yoshikawa, H. (1971) Chromosome-membrane association in *Bacillus subtilis*. I. DNA release from membrane fraction. Biochem. Biophys. Res. Commun. 44, 1559–1565.

Yoshikawa, H. and Sueoka, N. (1963) Sequential replication of *Bacillus subtilis* chromosome, I. Comparison of marker frequencies in exponential and stationary growth phases. Proc. Nat. Acad. Sci. U.S.A. 49, 559–566.

Young, F. E. (1965) Variation in the chemical composition of the cell walls of *Bacillus subtilis* during growth in different media. Nature 207, 104–105.

Young, F. E. (1966) Autolytic enzyme associated with cell walls of *Bacillus subtilis*. J. Biol. Chem. 241, 3462–3467.

Zaritsky, A. (1975) On dimensional determination of rod-shaped bacteria. J. Theoret. Biol. 54, 243–248.

Zaritsky, A. and Pritchard, R. H. (1973) Changes in cell size and shape associated with changes in the replication time of the chromosome of *Escherichia coli*. J. Bacteriol. 114, 824–837.

Zusman, D. R., Inouye, M. and Pardee, A. B. (1972) Cell division in *Escherichia coli:* Evidence for regulation of septation by effector molecules. J. Mol. Biol. 69, 119–136.

The synthesis and assembly
of plant cell walls:
possible control mechanisms

10

D. H. NORTHCOTE

1. Introduction

Most of the polysaccharide synthesis sites occur within the cell and a large proportion of the high-molecular-weight insoluble material of the cell wall is finally exported across the plasmalemma so that it is deposited external to the organized cytoplasm (Northcote, 1974a). The extent of the wall development must be constantly monitored by the enclosed cell and its composition and texture changed so that it adapts to the varying needs of the individual cell protoplast and the more complex requirements of the growth and function of a coordinated organ or tissue (Northcote, 1972). Thus, it is necessary to consider not only the mechanism of synthesis but also the control processes that influence the material synthesized so that it is changed or stopped as the cell differentiates. The control must involve the passage of information from the outer wall of the protoplast back to the synthetic sites within the cytoplasm. Particular polysaccharide synthetic systems are switched on or off, the ratios of the amounts of various polysaccharides synthesized are altered, the synthesis of lignin is induced, and material is deposited at particular sites within or on the wall. Eventually the control mechanisms cause a cessation of wall growth even though the cell may still be alive. This chapter will attempt to indicate some experimental approaches that describe the kind of information transmitted and the sites and mechanisms that are operated on to alter the synthesis of the wall.

2. Sequence of cell wall formation

The order of synthesis of cell wall constituents can be ascertained in several types of investigations, and the sequence indicates the order in which the various polysaccharide or lignin synthetic systems become active or inactive. Thus, some indication can be made about possible metabolic reactions on which the switching mechanism acts.

G. Poste and G.L. Nicolson (eds.)   *The Synthesis, Assembly and Turnover of Cell Surface Components*
© *Elsevier/North-Holland Biomedical Press, 1977.*

718

2.1. Cell plate and wall formation at telophase.

The mitotic cycle of cultures of bean and sycamore cells can be partially synchronized so that the culture has at least a mitotic index of 15 to 20%. Ultrastructural studies of these cells illustrate the events of mitosis and cytokinesis (Roberts
and Northcote, 1970). A proportion of the dividing cells undergo nuclear division when the mother nucleus lies immediately adjacent to the mother cell wall
on one side of the cell. Hence, at telophase, the cell plate is formed at this side of
the mother cell and grows across the vacuole of the cell to the opposite side. The
cell plate is formed by the fusion of numerous membrane-bound vesicles that are
aligned in the plane of the plate midway between the telophase nuclei. Some of
these vesicles are derived from the Golgi bodies found on each side of the plate
in the strand of cytoplasm, the phragmosome. Observation of the formation of
this cell plate and its gradual development into a recognizable wall yields two
experimental advantages. It concentrates attention on those organelles concerned with cell plate and wall formation since these are present within the
phragmosome, and it demonstrates a dynamic picture of cell plate and wall
formation from an immature region at the forming edge where the vesicles are
fusing progressively back to a mature cell wall at the side from which the development started. These studies also show that at the developing edge the
material contained within the vesicles is electron-transparent and no fibrillar
material is present. Even further back, at regions where the vesicles have fused
to give a continuous membrane-bound layer, that is equivalent to the new plasma
membranes of the daughter cells, very little fibrillar material can be detected.
Fibrils are present in large amounts only at the more mature regions. At the site
where it was developed initially and where it is attached to the mother cell wall, it
develops the same fibrillar texture as that of the mature wall. This developmental sequence suggests that the plate is first formed of nonfibrillar pectins and
hemicellulose, and into this matrix fibrils, probably composed of cellulose, are
woven at a later stage in development.

2.2. Cell wall formation in isolated protoplasts

The formation of a new cell wall can also be investigated as it develops around
isolated protoplasts prepared from intact cells. These protoplasts are prepared
by dissolving and weakening the walls of the cell by polysaccharidases so that the
protoplasts are released into an isotonic incubation medium containing the enzymes (Cocking, 1972). The enzymes may also degrade any oligosaccharides
exposed as glycoproteins within or on the cell membrane surface. The protoplasts can be filtered, washed, and incubated in an appropriate medium to reform
a new cell wall (Nagata and Takebe, 1971; Prat, 1973; Bright and Northcote,
1974; Burgess and Fleming, 1974; Shepard and Totten, 1975; Willison, 1976).
During the initial period of cell wall regeneration of soybean protoplasts in
liquid medium, polysaccharides are excreted into the media and this continues
for at least 20 hours during which the protoplasts are still sensitive to osmotic

shock and have no rigid wall (Hanke and Northcote, 1974). The material excreted into the media is pectin and is characteristic of a type in which the
polygalacturonic acid chain is substituted by neutral blocks composed of
arabinose and galactose (Barrett and Northcote, 1965). This pectin changes
during the 20-hour incubation period, but little material is retained at the cell
surface at this time. After 40 hours, glucans are found that can be seen in the
electron microscope to be microfibrils (Roland, 1973; Burgess and Fleming,
1974; Willison and Cocking, 1975). The meshwork of microfibrils that develops
seems able to entrap the continued production of matrix material by the
protoplast since from 40 hours onward the cells develop a wall that converts the
protoplast into a normal tissue culture cell (Hanke and Northcote, 1974). The
enzymic conditions used to prepare the protoplasts modify the cell a great deal,
so that the sequence of events occurring during the reformation of the wall must
depend on the recovery of the cells from shock and particularly on the regeneration and reorientation of the possibly degraded and exposed plasmalemma.
Nevertheless, the results again suggest that the sequence of wall formation is that
of synthesis and transport of matrix polysaccharides first (pectins and possibly
hemicelluloses) followed by cellulose synthesis, laid down as microfibrils at the
cell surface and woven into the matrix.

2.3. Cell wall formation during plasmolysis

To avoid some of the experimental difficulties involved in using protoplasts, and
to obtain additional information about the control mechanisms of cell wall formation, another type of experimental system has been employed. This involves
examining the regeneration of a new cell wall around a protoplast produced by
plasmolysis and therefore contained within the original cell wall.

Benbadis (1972) has shown that a wall can be regenerated within the old wall
of plasmolyzed tobacco mesophyll cells. By incubating the normal and freshly
plasmolyzed tissue with radioactive glucose, it is possible to compare the pattern
of polysaccharide synthesis in the two types of tissue (Boffey and Northcote,
1975). Plasmolysis generally decreases the incorporation of sugars into all the
polysaccharide fractions of the wall, but effects the incorporation of arabinose
into pectic material in a different way from its affect on other sugars (Table 1).
In freshly plasmolyzed tissue, the incorporation of arabinose into pectin was
stimulated whereas the incorporation into the other sugars and uronic acid was
decreased. The cell's response to the treatment that caused it to form a new cell
wall was thus again the synthesis of the pectic material and particularly the
arabinan components of this cell wall fraction.

3. Modifications of pectin synthesis

In many different tissues the composition of the wall pectins changes during
development and differentiation of the constituent cells (Thornber and North-

Table 1
Incorporation of Sugars into Wall Polysaccharides of Plasmolyzed Tobacco Leaf Cells[a]

	10^{-3} × Radioactivity per sample (cpm)											
	Unplasmolyzed			Plasmolyzed (0.7 M sorbitol)				Plasmolyzed (0.35 M KCl)			
	Sample			Sample			Ratio Plasmolyzed/	Sample			Ratio Plasmolyzed/
	1	2	Average	1	2	Average	Unplasmolyzed	1	2	Average	Unplasmolyzed
Sugar											
Uronic acids	15	9	12	4	5	5	0.4	5	3	4	0.3
Galactose + glucose	13	22	18	10	11	11	0.6	5	6	6	0.3
Arabinose	1.9	2.7	2.3	4.8	5.8	5.3	2.3	2.5	2.0	2.3	1.0
Ribose + rhamnose	4.7	4.9	4.8	2.2	1.9	2.1	0.4	1.7	1.1	1.4	0.3
Pectin before hydrolysis	38	40	39	25	25	25	0.6	16	12	14	0.4

[a] Levels of radioactivity in the pectins extracted from leaf discs incubated with D-U-^{14}C-glucose for 5 hours after 2 hours in nonradioactive media. Discs (approx. 0.9 cm diam.) were floated on media (5 ml) without any carbon source, with or without sorbitol or KCl, and were kept in the dark at 25°C before and during incubation with about 17 μCi of D-U-^{14}C-glucose per 10 discs, after which they were washed with media containing nonradioactive glucose (1%, w/v) for 2 hours. Pectin was extracted from each sample of 10 discs.

cote, 1961a,b; Stoddart and Northcote, 1967; Bowles and Northcote, 1972; Wright and Northcote, 1974). These changes are related to the alterations in the physical properties of the wall that occur as the functions of the cell alter during its development (Rees, 1969; Northcote, 1972). The pectin properties are considerably modified by the presence of large blocks of neutral sugars (arabinans, galactans, arabinogalactans) joined to the polygalacturonorhamnan backbone, and the cells must therefore be able to exert separate controls over the synthesis of these two constituent parts of their pectin. The type of pectin synthesized can also be modified by the application of a growth hormone such as 2,4-dichlorophenoxyacetic acid to a callus tissue, and in this instance the pattern of synthesis of the arabinan is influenced again (Rubery and Northcote, 1970).

Thus, the type of pectin synthesized can be influenced during normal growth and by a variety of experimental procedures that includes application of hormones, plasmolysis, and cell wall removal. The enzymes necessary for synthesis of the individual polysaccharides of the pectin and the transglycosylases by which they become linked, especially that transferring arabinogalactan portions onto the main galacturonorhamnan chain, are therefore of considerable importance (Stoddart and Northcote, 1967; Rubery and Northcote, 1970). These enzymic reactions not only control the texture of the wall but are targets for the controlling factors of its development (Dalessandro and Northcote, 1977a,b,c,d).

3.1. Possible control signals for changes in pectin synthesis

The process of plasmolysis involves several interdependent events. One consequence is the decrease in total surface area of the plasmalemma, and this must cause the general decrease in incorporation of radioactivity from glucose into the sugars of the polysaccharide that are formed after plasmolysis. However, this cannot explain the increase in arabinose incorporation. Loss of contact between much of the cell wall and the plasmalemma must cause significant changes in the environment of the membrane. For example, the ionic atmosphere at the outer surface of the membrane changes and this could result in disruption and alteration of those syntheses occurring at the cell surface. The removal of the acidic pectin molecules of the cell wall from the membrane surface could also alter the charge distribution across the membrane and the distribution and orientation of the membrane constituents, so that vesicle fusion from the Golgi bodies is changed and control of membrane flow from the endomembrane system could be influenced. Prat (1972) has shown that plasmolysis causes the disappearance of dictyosomes from onion root cells, and this could result in a control of those polysaccharides synthesized and transported by the Golgi bodies.

Another result of plasmolysis is the reduction of cell-cell interaction that must occur when an organized tissue is used, especially on the breakage of the plasmodesmata which link adjacent cells. The differentiation and coordinated response of the individual cells in an organized tissue probably involve the exchange of materials such as growth factors between adjacent cells, so that a relationship is established between the position of the cell in the tissue and its function (Northcote, 1969a). This function of the cell, in turn, depends on the nature and composition of its cell wall and the state of its development. The effect of plasmolysis might therefore be that of interference with these very complex cell-cell interactions that control the differentiation of the whole tissue and will be made apparent by the alteration in cell wall synthesis.

There are, however, several experiments indicating that the response to plasmolysis and the accompanying alteration of cell wall synthesis is due primarily to the withdrawal of the membrane from the immediate vicinity of the wall and the consequent change in the environment of the membrane (Boffey and Northcote, 1976). The effects of plasmolysis on the polysaccharide synthesis pattern of separated cells compared with those of intact leaf discs are very similar. The effects of plasmolysis on the incorporation of radioactivity are partially reversed when the plasmolyzed tissue is placed in a solution that causes deplasmolysis. During plasmolysis of the cells in a leaf disc, the alteration in the pattern of pectin synthesis is brought about immediately when the cells are placed in solutions just above the isoosmotic point as the plasmalemma begins to be withdrawn from the wall, and the plasmodesmata are not broken at this stage. This immediate effect is much greater than the subsequent responses to decreasing the plasmalemma surface as the osmotic pressure of the plasmolyzing solution is progressively increased above the isotonic point (Fig. 1).

4. Assembly of polysaccharide complexes within the cytoplasm

A problem with important consequences not only for the structure of the wall but for the mechanisms of synthesis is the extent of polysaccharide and protein material assembly before it is exported to the outside of the cell and the nature of the assembly and type of linkage that occurs between the various constituents in the cell wall after the material has been packed into the wall. Certain situations exist with some organisms where biological complexes that resemble a miniature cell wall or a building brick are formed. These complexes are almost completely assembled before they are exported to the outside of the cell.

One example of such a structure is the external scales of the haptophycean algae (Manton, 1966, 1967b; Manton and Leadbeater, 1974). These scales are produced continually by the algae, serve to form an exoskeletal covering of the cell, and can be seen to be assembled within the Golgi cisternae and vesicles of the endomembrane system of the cell (Manton, 1967a,b). They are composed of protein and polysaccharide and are complex, sculptured structures with a definite external pattern and form that varies for the different species (Green and Jennings, 1967; Brown et al., 1969, 1973; Allen and Northcote, 1975). Woven within the scales in a definite manner are microfibrils, composed mainly of β1-4 linked glucans, so that they resemble the α-cellulose fraction of higher plants (Brown et al., 1973; Allen and Northcote, 1975; Herth et al., 1975). In *Chrysochromulina chiton*, the matrix of the scales and the protein contain at least

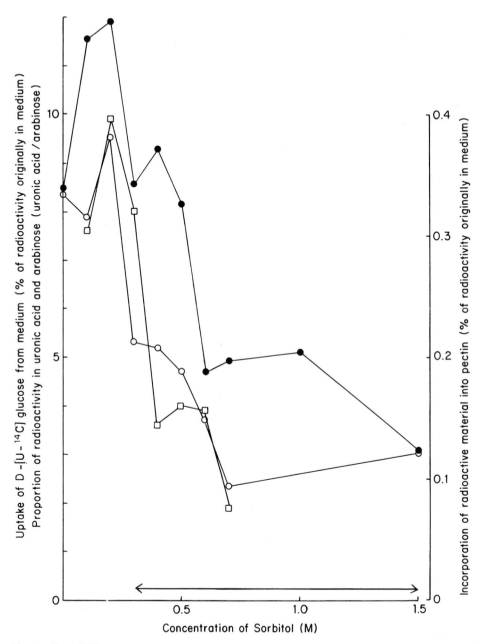

Fig. 1. Effect of different concentrations of sorbitol on D-U-¹⁴C-glucose incorporation into pectin by leaf discs of tobacco (*Nicotiana tabacum* cv. Xanthi). Leaf discs were floated on media containing approximately 20 μCi D-U-¹⁴C-glucose per sample for 5 hours after 1.5 hours on nonradioactive media. (●—●) Uptake of radioactive glucose; (○—○) incorporation into the pectic fraction of the walls; (□—□) ratio of incorporation into uronic acid: arabinose of the pectin fraction. The arrow represents the concentrations of sorbitol at which plasmolysis started and continued (Reproduced with permission from Boffey and Northcote, 1976).

three polysaccharides, two of which are acidic (Allen and Northcote, 1975). This whole complex of associated microfibrils and matrix material, assembled in a definite manner, is formed within the Golgi body and its vesicles before transfer to the outside of the cell.

Some mucilages and slimes are also complex polysaccharide aggregates that are assembled before export. Grant et al. (1969) suggested that mustard seed mucilage consisted of units composed of a cellulosic central polymer surrounded by pectinlike material. The unit was thus relatively stiff and fiberlike, but since it was surrounded by hydrophilic molecules it was soluble. The root-cap slime of corn seems to be a similar complex (Jones and Morré, 1967; Harris and North-cote, 1970; Wright and Northcote, 1974; Barlow, 1975; Wright and Northcote, 1975). The slime contains a glucose-rich polymer that can be isolated from the complex either by electrophoresis or degradation by acid hydrolysis (Wright and Northcote, 1974, 1976). This glucan is not hydrolysed by α-amylase and is almost completely degraded by periodate oxidation, it contains $\beta1\rightarrow4$ links and is insoluble in hot strong alkali. It is, however, associated with other sugars, particularly galacturonic acid, and these are covalently linked to the glucan chain (Wright and Northcote, 1975). The slime and the glucan appear fibrillar when precipitated from solution or freeze-dried, and these preparations are birefringent. When preparations of the glucan are shadowed with platinum/carbon and examined in the electron microscope, short microfibrillar structures can be seen (Wright and Northcote, 1976).

Some of the constituents of the slime are acidic polysaccharides and resemble pectin (Wright and Northcote, 1974). The slime, like the mucilage, is composed of coated glucan units. The synthesis of the glucan component probably occurs as part of the slime complex assembly within the dictyosome stacks and vesicles, since the slime in the vesicles is already fibrillar at this stage and similar in appearance to the packets of material that can be seen outside the plasmalemma after export (Mollenhauer and Whaley, 1963; Rougier, 1971). In addition, Bowles and Northcote (1972) have shown that membrane fractions rich in dictyosomes isolated from corn roots that had been incuated with D-U-^{14}C-glucose contain small but appreciable amounts of covalently bound radioactive glucose. This polymeric glucose was present within the membrane system of the cytoplasm before its export across the plasma membrane. Synthesis of the glucan within the dictyosome occurs in the presence of other polysaccharide synthetases (transglycosylases) so that substitution of the glucan chain by other sugars can occur. It can be shown that the glucan chains prepared from the slime by trans-elimination and isolation from a Biogel column carry neutral sugars (particularly fucose) and uronic acids linked to it covalently. The conformation of the central core of β-glucan chains makes the slime fibrillar while the surrounding sheath of hydrophilic molecules renders the whole complex soluble.

5. Polysaccharide associations within the cell wall

The interconnections of polysaccharides within the wall of higher plant cells is a problem that has been discussed constantly since 1937 (Norman, 1937). The difficulty is not that there are interconnections but of what type these are, and in particular whether covalent bonds occur between the major constituents in addition to the obviously extensive hydrogen bonding that must occur. This issue has received a new topicality because of the detailed studies of Albersheim and his colleagues (Bauer et al., 1973; Keegstra et al., 1973; Talmadge et al., 1973). Their work has attempted to show that the matrix material of the growing primary cell wall of cells such as those of tissue cultures, can be regarded as one large molecule, and that each polysaccharide chain of what could be regarded as a distinct polysaccharide, such as a xyloglucan, arabinanogalactan, or polygalacturonorhamnan is joined by the reducing end of the main chain to become a side branch of another polysaccharide or protein. Thus some molecules will act as bridges, having another chain attached to them as a side branch while they are similarly attached, via their reducing end, to become the side branch of another polymer. No such bridge molecules having both the connected molecules or parts of these molecules still attached to the bridge have ever been isolated. The method used to identify them has been to try and establish the linkages from the reducing end of each polysaccharide chain to that of the other molecule. The difficulty is that the ratio of the number of these bridging linkages to the number of constituent linkages in the main chain is very small. Nevertheless, it has been claimed that the linkages are there and pairs of them have been assigned to particular chains so that bridges can be postulated. The linkages were identified from a gas-liquid-chromatographic analysis of methylated cell walls and cell wall fractions in which each constituent was assumed to be fully methylated and a resolution of nearly all the various methylated derivatives of all the various sugars was obtained. Information about the possible interconnections was also obtained during the preparation of fractions from the wall by the use of purified hydrolytic polysaccharidases. The gas liquid chromatogram was coupled with a mass spectrograph and a quantitative identification of the numerous methylated derivatives of the various sugars was made. A model was proposed indicating the covalent linkages between all the constituents of the matrix wall material from the results of a study of sycamore suspension culture cells and later from work with bean suspension culture cells (Wilder and Albersheim, 1973). It was suggested that similar structures occurred in all primary cell walls. Other studies, using cell walls taken from intact plant tissues, have not confirmed this (Monro et al., 1975; Selvendran, 1975; Selvendran et al., 1975). The walls of specialized cells, such as those of barley aleurone, also do not resemble those of sycamore suspension cultures and they have different polysaccharides and interrelationships among their constituents (McNeil et al., 1975).

If these extensive covalent complexes are present, a problem arises about the sites and time of their formation, especially as in the early growing stage the texture of the wall is continually changing and the matrix is not rigid but more a viscous fluid in which the microfibrils can move to some extent. However, regardless of the presence of these covalent bonds, the main cohesion between the constituents, especially the polysaccharides, must be the hydrogen bonding that allows a flexibility to the structure of the wall which depends essentially on its most variable feature, the water content (Northcote, 1972). Many of the differ-

726

ent types and structures of the polysaccharides that form the matrix of the wall alter the hydrophilic or hydrophobic nature of the matrix and contribute to the properties of the wall by their relationship to water molecules. At later stages in development, the space occupied by the water in the wall becomes progressively filled with lignin. This preserves the tensile strength of the microfibril and makes a rigid matrix phase (Northcote, 1972).

6. The sites of polysaccharide synthesis

The membrane system of the endoplasmic reticulum, the Golgi bodies, and the plasmalemma is responsible for the transport and synthesis of cell wall material and other exportable polysaccharide substances from the plant cell (Northcote, 1970, 1972, 1974b; Dauwalder et al., 1972; Whaley, 1975). This can be clearly seen by radioautographic studies on rapidly growing tissue in which radioactive glucose is observed to be incorporated into the material of the Golgi cisternae and vesicles, and subsequently can be chased by incubation with nonradioactive glucose into the wall outside the plasmalemma. Similar results show that the Golgi vesicles give rise to some of the materials in secondary thickenings and, since the radioactive material detected can be isolated and analyzed, it has been shown to be wall polysaccharides (Northcote and Pickett-Heaps, 1966; Wooding, 1968; Northcote and Wooding, 1966, 1968).

Direct observations of the production of scales in the haptophycean algae and the formation of slime by roots (Hereward and Northcote, 1972), indicate assembly and modification of the material within the membrane system so that it resembles the material that is finally exported (section 4). Thus, the Golgi apparatus is involved in the synthesis and modification of the material during transport (Bowles and Northcote, 1976). Further direct evidence for the function of the Golgi apparatus in cell wall formation is its role in the construction of the cell plate at telophase which has already been described (Whaley and Mollenhauer, 1963; and section 2.1).

Although the radioautographic evidence clearly shows the function of the Golgi apparatus in cell wall assembly, no indication of the role of the endoplasmic reticulum is given in this type of investigation. However, many direct observations on tissue where cell wall synthesis is known to occur have shown characteristic organization and distribution patterns of the endoplasmic reticulum, suggesting that it has a distinct role in wall formation (Northcote, 1968). For example, the endoplasmic reticulum is found in layers near the developing cell plate and may be closely applied to the new plasmalemma surface in some locations (Roberts and Northcote, 1970). It has also been observed that the endoplasmic reticulum is essential for the synthesis of callose in the sieve plate of phloem (Northcote and Wooding, 1966, 1968) and is distributed in a definite pattern in the secondary thickening of xylem vessels of rapidly growing wheat roots (Pickett-Heaps and Northcote, 1966).

The function of the endoplasmic reticulum during sieve plate formation is particularly significant because the pores seem to be formed at sites demarcated by the distribution of the endoplasmic reticulum along the developing sieve plate. At these sites callose is deposited, and is subsequently removed during formation of the pore. Callose can be deposited much later at these same sites in the mature phloem sieve tubes, within the wall of the fully formed sieve plate at the pores (Northcote and Wooding, 1966). At this later time no organized endoplasmic reticulum or Golgi system is present within the sieve tube. Thus it is reasonable to suppose that the full complement of enzymes necessary for the synthesis of callose is either present within the wall or at the plasmalemma, and that their presence at these sites is associated with the earlier endoplasmic reticulum distribution during the initial stages of sieve plate formation.

More direct evidence for the function of the endoplasmic reticulum and the Golgi apparatus in cell wall formation is obtained by isolating the various parts of the membrane system separately from broken cells (Harris and Northcote, 1971; Bowles and Northcote, 1972, 1974, 1976). The enriched membrane fractions are isolated from corn and pea roots at definite stages of differentiation when polysaccharides of known composition are formed by the cells (Harris and Northcote, 1970). The polysaccharides that were synthesized prior to the isolation of the membranes are made radioactive by incubating the tissue with radioactive glucose. These experiments show that the endoplasmic reticulum and the Golgi apparatus contain polysaccharide material (hemicellulose and pectic substances) characteristic of the type of cell wall or other substance exported by the cell during isolation of the membranes. The only polysaccharide not found within the membrane system that is deposited in the wall in large amounts is cellulose. In the experiments the membranes were isolated either as intact Golgi bodies and closed membrane-bound sacs of rough endoplasmic reticulum or as closed smooth vesicles, so that the contents remained within the isolated membrane system. If cellulose were synthesized to a great extent at the plasmalemma surface, it would not be isolated within a membrane fraction even though a rapid synthesis of cellulose into the wall were occurring.

The direct evidence provided by the isolation of the membranes from higher plants thus indicates that both the endoplasmic reticulum and the Golgi apparatus are involved in the synthesis of pectic substances, hemicellulose, and root-cap slime but that these membrane fractions are not active in cellulose synthesis, which is carried out at the plasmalemma surface. However, the potential enzymic activity for cellulose synthesis, like the active enzymes responsible for the synthesis of hemicellulose and pectin, could have been present within the endoplasmic reticulum-Golgi apparatus system since the plasmalemma is derived in part by a direct contribution of membrane from the cytoplasmic part of the membrane system. The difference is that the enzymes responsible for cellulose synthesis only become fully active when the membrane is incorporated at the cell surface (Northcote 1972; Shore et al., 1975). That enzymic synthesis of glucan chains can sometimes occur within the Golgi apparatus and its vesicles before incorporation at the cell surface supports this idea; the synthesis of scales in the algae, root-cap slime in corn has been mentioned earlier (section 4). In addition, a xyloglucan has been isolated from the growth medium of sycamore suspension cultures (Aspinall et al., 1969) and is probably present in the walls of these cells and of higher plants (Keegstra et al., 1972). The polysaccharide consists of a central core of a $\beta 1 \to 4$ glucan with side branches of xylose linked $1 \to 6$; there are also small amounts of galactose and fucose in the side chains. The xyloglucan is probably a matrix polysaccharide and thus is synthesized like the slime within the endoplasmic reticulum-Golgi apparatus membrane system of the cell so that the glucose chain is synthesized in the presence of other transglycosylases.

Whether the endoplasmic reticulum contributes directly to the wall, or whether all the material in the endoplasmic reticulum has to be modified by passage through the Golgi apparatus is not yet established. Transfer of material to the Golgi apparatus may be necessary for several reasons. Additional synthesis may have to take place by reactions that occur only in this part of the system (Bowles and Northcote, 1976); the material may have to be assembled in a definite order before export, or it may have to be concentrated and covered with the different modified membrane of the Golgi vesicle. The endoplasmic reticulum can be very close to the cell membrane and yet not fuse with it. On the other hand, the Golgi vesicles do fuse with the plasmalemma, and this is probably due in part to the chemical similarity of the membranes at the dispersing face of the Golgi apparatus to that of the plasmalemma. The similarities of the membrane of the exported vesicles of the Golgi apparatus and the plasmalemma and their differences from the membranes of the forming face of the Golgi body and the endoplasmic reticulum has been indicated by microscopic observation and chemical analysis (Grove et al., 1968; Keenan and Morré, 1970; Hereward and Northcote, 1972; Hodson and Brenchley, 1976). The polymeric material containing the sugars is different in the two membrane systems and is consistent with the idea that although synthesis is begun in the endoplasmic reticulum the process is completed in the Golgi apparatus. In the membrane systems isolated from corn root cells, there was a greater proportion of high molecular weight pectins and xylans in the Golgi-rich fraction than in the endoplasmic reticulum-enriched fraction, and some of the sugar material more firmly bound to the membrane had a much lower molecular weight in the endoplasmic reticulum fraction than in the Golgi-rich fraction (Bowles and Northcote, 1976).

Although the mechanism of vesicle fusion and membrane extension is unclear, it seems likely that in addition to a similarity in chemical composition the sites of fusion may be marked by particular substructures at the plasmalemma and on the vesicle (Palade and Bruns, 1968; Northcote, 1969b; Lagunoff, 1973; Satir et al., 1973; Gratzl and Dahl, 1976). Also, a definite ionic environment may be necessary (Gratzl and Dahl, 1976; Morris and Northcote, 1977). This environment may in part be controlled by the polysaccharides present in the wall (see section 3.1).

7. Intermediates of polysaccharide synthesis

The donar substances of polysaccharide synthesis are the nucleoside diphosphate sugars. These compounds transfer the glycosyl radicals to form glycosidic bonds on other compounds, the acceptor substances, and the next stage in the synthesis is a transglycosylation step from the glycosylated acceptor to a growing polysaccharide chain or protein, so that eventually the completed polysaccharide or glycoprotein is formed (Northcote, 1969a). The initial acceptor substance can vary with the type of material that is eventually synthesized.

Compounds that have been identified as intermediates and initial acceptor substances during polysaccharide and glycoprotein synthesis are monosaccharides (Edelman and Jefford, 1968; Palmer et al., 1968), oligosaccharide, and polysaccharide chains (Helting and Rodén, 1969a, b; Ryman and Whelan, 1971), cyclitols (Kemp and Loughman, 1974), polyprenyl phosphates (Hemming, 1974), and proteins (Barengo et al., 1975; Krisman and Barengo, 1975; Tandecarz et al., 1975). The last two acceptors are of special interest since they are noncarbohydrate and could take part in the structure of the membrane system at which the polysaccharides are synthesized (section 6).

7.1. Polyprenyl intermediates

The polyprenyl phosphates were first recognized as acceptors of glycosyl radicals from the nucleoside diphosphate sugars during polysaccharide synthesis of the outer layers of bacterial membranes (Robbins et al., 1967; Nikaido and Hassid, 1971; Lennarz and Scher, 1972; Strominger et al., 1972). More recently, dolichol monophosphate and diphosphate sugars have been identified as intermediates during the biosynthesis of the oligosaccharide chains of certain glycoproteins found in animal cells (Behrens et al., 1973; Leloir et al., 1973; Parodi et al., 1973; Hsu et al., 1974; Levy et al., 1974; and chapter 2 in this volume by Cook), and possibly in plants (Alam and Hemming, 1973; Forsee and Elbein, 1975; Roberts and Pollard, 1975). They have also been identified as the primary acceptors during synthesis of the oligomannose units attached to serine/threonine and, via N-acetyl glucosamine, to aspartamide on the mannan protein of the yeast cell wall (Sentandreu and Northcote, 1968; Sentandreu and Lampen, 1971; Tanner et al., 1972; Babczinsky and Tanner, 1973; Lehle and Tanner, 1974; 1975). Evidence exists that they could act as intermediates during synthesis of some of the matrix polysaccharide in the cell walls of higher plants (Kauss, 1969; Villemez and Clark, 1969; Hinman and Villemez, 1975). In other instances, although polyprenyl phosphates and acidic lipids have been detected in particular tissues and have been shown to be acceptors of sugars, it is not always clear whether they are intermediates in polysaccharide or glycoprotein formation (Storm and Hassid, 1972; Alam and Hemming, 1973; Forsee and Elbein, 1973; Lezica et al., 1975).

The requirement for a lipid intermediate as a precursor for cellulose synthesis is still unknown. Colvin (1961) presented evidence that a lipidlike compound was a carrier in the biosynthesis of bacterial cellulose (Khan and Colvin, 1961) and possibly in plant cellulose. The lipid was not identified at that time, but more recently it has been partially purified and identified as an acidic lipid material (Colvin and Leppard, 1971), probably a mixture that includes polyprenyl phosphate galactose and diphosphate glucose and cellobiose (Garcia et al., 1974). Compounds with some of the chromatographic characteristics of polyprenyl phosphate sugars have been isolated from membrane preparations of pea roots after these were incubated with $UDP^{14}C$-D-glucose. These compounds contained oligosaccharides having $1{\rightarrow}3\beta$ and $1{\rightarrow}4\beta$ linkages. The membrane fraction also incorporated glucose into high molecular weight polysaccharides containing these linkages (Brett and Northcote, 1975). Cotton fibers have also been found to incorporate glucose from UDP-glucose into glucolipids, and these compounds were found in greater amounts during formation of the secondary rather than the primary wall (Delmer et al., 1974). Acceptor lipids may vary so that the different sugar lipids formed from nucleoside diphosphate sugars are used to form different polysaccharides. Thus a control of the synthetase activity could operate at the transglycosylase from the lipo-sugar to the polysaccharide or at the formation of the lipo-sugar from a particular nucleoside diphosphate sugar donor.

7.2. Protein intermediates

The participation of protein as an intermediate acceptor during polysaccharide synthesis was suggested for some cell wall polymers (Villemez, 1970). Recently this idea has been supported by much more substantial evidence during investigations on the synthesis of storage substances such as starch and glycogen. An acceptor substance must be present for the synthetases to transfer the glucose from UDPG or ADPG to form glycogen or starch chains. The enzymes can use the polysaccharides as acceptors and increase the chain lengths of the branches, which can then be employed to form a more extensive branching system by transglycosylations under the control of branching enzymes (Ryman and Whelan, 1971; Borovsky et al., 1976). Problems arise if a de novo synthesis of starch or glycogen is considered when no primer polysaccharide material is available as an acceptor.

Glycogen and starch are built up in close association with membrane systems. Degradation of glycogen by phosphorylase and its synthesis by the synthetase is under close and specific metabolic control, whereby the level of the glycogen in the cells is poised by the activities of the synthesis and breakdown reactions. The synthesis of glycogen also depends on the glycogen level in the cell and these factors for the control imply that the polysaccharide and enzymes for its synthesis and breakdown are present as a complex in close association, probably on a membrane system (Warson and Drachmans, 1972).

Recently, some parts of this complex of special relevance to the synthetic pathway have been investigated for both glycogen formation in *Escherichia coli* (Barengo et al., 1975) and liver (Krisman and Barengo, 1975), and starch in potato tissue (Tandecarz et al., 1975), *Oriza sativa* grains (Pisigan and Rosario, 1976), and algal tissue (Frederick, 1971). In these systems a transfer of glucose from UDPG is made to a protein acceptor. (With a nonsedimentable preparation from potato, and with the preparation from the algal tissue, glucose-1-P was the donor.) The glucoprotein found has unbranched $\alpha 1 \rightarrow 4$ glucosidic chains that can accept more glycosyl residues by transglycosylation systems using more UDPG or, in the case of starch, ADPG and UDPG (Tandecarz et al., 1975).

Protein intermediates have also been suggested for β-glucan synthetase systems, although the evidence so far is not as convincing as that for the α-glucans (Franz, 1975).

8. Control reactions during cell wall synthesis

The composition of the cell wall changes during growth and differentiation: the pectic substances are synthesized only during the period of primary growth, lignin is deposited only at secondary thickening, and callose is deposited during the formation of phloem sieve tubes. It is possible therefore, regardless of the signals that cause these changes in metabolic activity, to have some indication of the processes on which these signals act. The changes are effected by switching the activities of the various enzymes involved on or off. These enzymes are the epimerases (Dalessandro and Northcote, 1977a,b,c,d) the polysaccharide synthetases (Northcote, 1963), phenylalanine ammonia lyase (Rubery and Northcote, 1968), other enzymes involved in methylated polyphenol formation, and the oxidases for lignin synthesis (Towers, 1974). It has been possible to measure the activities of some of these enzymes either during a controlled stimulation of differentiation of a callus tissue (Haddon and Northcote, 1975; 1976a,b,c) or during the differentiation of xylem and phloem tissue in a stem in vivo (Dalessandro and Northcote, 1977a,b).

One of the obvious points of control would seem to be at the epimerase reactions shown in Fig. 2. These epimerases interconvert the precursor nucleotide sugars UDP-glucose $\rightleftharpoons$ UDP-galactose, UDP-glucuronic acid $\rightleftharpoons$ UDP-galacturonic acid, UDP-xylose $\rightleftharpoons$ UDP-arabinose and, as can be seen from Fig. 2, these should be active and functioning when polymers containing galactose, galacturonic acid, and arabinose (pectic substances) are laid down but not active during secondary thickening when no pectin is being formed. When the activities of these enzymes are measured in the cambium, differentiating xylem, and differentiated xylem of sycamore stem there is a small diminution of activity between the cambial and the differentiated xylem cell, but the enzymes do not become inactive in the differentiated cell and have the same order of activity as that present in the cambial tissue where only primary growth is occurring. However, the activities of the UDP-glucose dehydrogenase and the UDP-glucuronic acid decarboxylase increase markedly during the differentiation, which reflects the great increase in the deposition of the hemicellulose in the secondarily thickened walls (Dalessandro and Northcote, 1977a,b). Thus, although the activities of the various enzymes used for the interconversions of the nucleoside diphosphate sugars do allow a flux of carbohydrate from UDP-glucose into any particular precursor that is being used, the control for the type of polysaccharide that is formed is not exerted by direct action on the activities of these enzymes. The

732

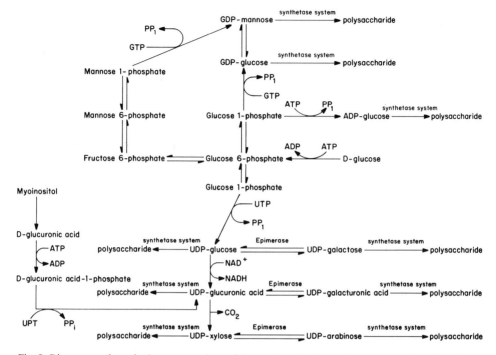

Fig. 2. Diagram to show the interconversions of the nucleoside diphosphate sugars, the initial donor molecules in polysaccharide synthesis. The synthetase systems may use intermediate acceptor substances. Heteropolysaccharides are formed from different donors and the sugars are eventually incorporated into single polysaccharides by the transglycosylase reactions.

control is operating at another point, probably at the polysaccharide synthetase stage, and some evidence for this has been obtained using callus tissue.

Callus tissue can be induced to differentiate by the application of varying ratios of auxin to kinetin (Wetmore and Rier, 1963; Jeffs and Northcote, 1966; 1967). This differentiation takes the form of cell nodules containing xylem and phloem. Roots and shoots may also be formed (Wright and Northcote, 1972). The differentiation may be measured by counting the number of differentiating cells in a cytological examination of the tissue. It may also be measured chemically by estimating lignin and by measuring the ratio of the amounts of arabinose to xylose in the wall (Jeffs and Northcote, 1966). This ratio indicates the relative amount of pectin in the wall and hence the state of cell development.

When the callus tissue is placed on the induction medium after transfer from a maintenance medium, differentiation occurs over a period of about 21 days. During this time various enzyme activities are induced (Haddon and Northcote, 1975). At least two enzymes concerned with lignin synthesis, phenylalanine ammonia lyase and caffeate-O-methyl transferase, are induced in a coordinated manner (Haddon and Northcote, 1976b). Callose synthetase activity also rises during vascular differentiation and this rise begins simultaneously with the rise of phenylalanine ammonia lyase activity, but the maximum value is reached about 3 days later and the rise is then more gradual. The switching mechanism used in these experiments was the action of auxin and kinetin. Abscissic acid represses the induction while gibberellic acid modified the time course and maximum activity of the response of phenylalanine ammonia lyase. The control for induction of the formation of the $\beta 1 \to 3$ glucan of the phloem was exerted by the induction of callose synthetase activity.

9. Summary

Experimentally, the changes in type of wall deposition can be modified in two ways, either by removal of the wall from around the protoplast by the formation of naked protoplasts and by plasmolysis, or by the application of growth factors to the growing tissues. Experiments with protoplasts and plasmolysis indicate that the metabolism of wall polysaccharides changes as the environment of the plasmalemma alters with the withdrawal of the wall. The results are consistent with the idea that control points exist for wall metabolism and deposition at the membrane surface that could possibly be influenced by the ionic environment of the membrane which is partially controlled by the type of polysaccharide present in the wall. The ionic atmosphere of the membrane could also be effected by growth factors, especially if these altered the permeability of the membrane. It is possible, however, that although an initial response is at the cell surface, subsequent effectors of the response, which may be the growth factors, operate at the synthetic sites, especially the polysaccharide synthetases, of the membrane system such as the Golgi bodies within the cell cytoplasm.

References

Alam, S. S. and Hemming, F. W. (1973) Polyprenol phosphates and mannosyl transferases in *Phaseolus aureus*. Phytochemistry 12, 1641–1649.
Allen, D. M. and Northcote, D. H. (1975) The scales of *Chrysochromulina chiton*. Protoplasma 83, 389–412.
Babczinsky, P. and Tanner, W. (1973) Involvement of dolicholmonophosphate in the formation of specific mannosyl-linkages in yeast glycoproteins. Biochem. Biophys. Res. Comm. 54, 1119–1124.
Barengo, R., Flawia, M. and Krisman, C. R. (1975) The initiation of glycogen biosynthesis in *Escherichia coli*. FEBS. Lett. 53, 274–278.
Barlow, P. W. (1975) The root cap. In: The development and function of roots (Torrey, J. G. and Clarkson, D. T. eds.), pp. 21–54. Academic Press, London.
Barrett, A. J. and Northcote, D. H. (1965) Apple fruit pectic substances. Biochem. J. 94, 617–627.
Bauer, W. D., Talmadge, K. W., Keegstra, K. and Albersheim, P. (1973) The structure of plant cell walls. II. The hemicellulose of the walls of suspension-cultured sycamore cells. Plant Physiol. 51, 174–187.

Behrens, N. H., Carminatti, H., Staneloni, R. J., Leloir, L. F. and Cantarella, A. I. (1973) Formation of lipid bound oligosaccharides containing mannose: Their role in glycoprotein synthesis. Proc. Nat. Acad. Sci. U.S.A. 70, 3390–3394.

Benbadis, A. (1972) Evolution ultrastructurale de protoplastes de mesophylle de tabac (*Nicotiana tabacum* L. CV Wis 38) culturés *in vitro*. C.R. Acad. Sci. Paris Ser. D274, 2492–2495.

Boffey, S. A. and Northcote, D. H. (1975) Pectin synthesis during the wall regeneration of plasmolysed tobacco leaf cells. Biochem. J. 150, 433–440.

Boffey, S. A. and Northcote, D. H. (1976) Unpublished results.

Borovsky, D., Smith, E. E. and Whelan, W. J. (1976) On the mechanism of amylose branching by potato Q-enzyme. Eur. J. Biochem. 62, 307–312.

Bowles, D. J. and Northcote, D. H. (1972) The sites of synthesis and transport of extracellular polysaccharides in the root tissues of maize. Biochem. J. 130, 1133–1145.

Bowles, D. J. and Northcote, D. H. (1974) The amounts and rates of export of polysaccharides found within the membrane system of maize root cells. Biochem. J. 142, 139–144.

Bowles, D. J. and Northcote, D. H. (1976) The size and distribution of polysaccharides during their synthesis within the membrane system of maize root cells. Planta (Berl) 128, 101–106.

Brett, C. T. and Northcote, D. H. (1975) The formation of oligoglucans linked to lipid during synthesis of β-glucan by characterized membrane fractions isolated from peas. Biochem. J. 148. 107–117.

Bright, S. W. J. and Northcote, D. H. (1974) Protoplast regeneration from normal and bromodeoxyuridine-resistant sycamore cells. J. Cell Sci. 16, 445–463.

Brown, R. M., Franke, W. W., Kleinig, H., Falk, H. and Little P. (1969) Cellulosic wall component produced by the Golgi apparatus of *Pleurochrysis scherffelii*. Science 166, 894–896.

Brown, R. M., Herth, W., Franke, W. W. and Romanovicz, D. (1973) The role of the Golgi apparatus in the biosynthesis and secretion of a cellulosic glycoprotein in *Pleurochrysis:* A model system for the synthesis of structural polysaccharides. In: Biogenesis of Plant Cell Wall Polysaccharides. (F. Loewus ed.), pp. 207–257. Academic Press, New York.

Burgess, J. and Fleming, E. N. (1974) Ultrastructural observations of cell wall regeneration around isolated protoplasts. J. Cell Sci. 14, 439–449.

Cocking, E. C. (1972) Plant cell protoplasts—isolation and development. Ann. Rev. Plant Physiol. 23, 29–50.

Colvin, J. R. (1961) Synthesis of cellulose from the ethanol-soluble precursors in green plants. Can. J. Biochem. Physiol. 39, 1921–1926.

Colvin, J. R. and Leppard, G. G. (1971) Purification of the precursor of bacterial cellulose. J. Polymer Sci (c) 36, 417–424.

Dalessandro, G. and Northcote, D. H. (1977a) Changes in enzymic activity of nucleoside diphosphate sugar interconversions during differentiation of cambium to xylem in sycamore and popular. Biochem. J. 162, 267–279.

Dalessandro, G. and Northcote, D. H. (1977b) Changes in enzymic activity of nucleoside diphosphate sugar interconversions during differentiation of cambium to xylem in pine and fir. Biochem. J. 162, 281–288.

Dalessandro, G. and Northcote, D. H. (1977c) Possible control sites of polysaccharide synthesis during cell growth and wall expansion of pea seedlings *(pisum sativuum L.)* Planta 134, 39–44.

Dalessandro, G. and Northcote, D. H. (1977d) Changes in enzymic activities of UDP-D-glucuronate decarboxylase and UDP-D-xylose 4-epimerase during cell division and xylem differentiation in cultured explants of Jerusalem artichoke. Phytochem. 16, 853–859.

Dauwalder, M., Whaley, W. G. and Kephart, J. E. (1972) Functional aspects of the Golgi apparatus. Sub-Cell. Biochem. 1, 225–275.

Delmer, D. P., Beasley, C. A. and Ordin, L. (1974) Utilization of nucleoside diphosphate glucoses in developing cotton fibres. Plant Physiol. 53, 149–153.

Edelman, J. and Jefford, T. G. (1968) The mechanism of fructosan metabolism in higher plants as exemplified in *Helianthus tuberosus*. New Phytol. 67, 517–531.

Forsee, W. T. and Elbein, A. D. (1973) Biosynthesis of mannosyl- and glucosyl-phosphoryl polyprenols in cotton fibres. J. Biol. Chem. 248, 2858–2867.

Forsee, W. T. and Elbein, A. D. (1975) Glycoprotein biosynthesis in plants. Demonstration of lipid linked oligosaccharides of mannose and N-acetylglucosamine. J. Biol. Chem. 250, 9283–9293.

Franz, G. (1975) Cellulose synthesis. The dependence of plasmalemma bound glucan synthetases on glycoproteins which can act as acceptor molecules. Abst. 8th Cellulose Conf. The Cellulose Research Inst. Syracuse, U.S.A.

Fredrick, J. F. (1971) De novo synthesis of polyglucans by a phorphorylase isoenzyme in algae. Physiol. Pantarum 25, 32–34.

Garcia, R. C., Recondo, E. and Dankert, M. (1974) Polysaccharide biosynthesis in Acetobacter xylinum. Enzymatic synthesis of lipid diphosphate and monophosphate sugars. Eur. J. Biochem. 43, 93–105.

Grant, G. T., McNab, C., Rees, D. A. and Skerrett, R. J. (1969) Seed mucilages as examples of polysaccharide denaturation. Chem. Commun. 805–806.

Gratzl, M. and Dahl, D. (1976) Ca++-induced fusion of Golgi derived secretory vesicles isolated from rat liver. F.E.B.S. Lett. 62, 142–145.

Green, J. C. and Jennings, D. H. (1967) A physical and chemical investigation of the scales produced by the Golgi apparatus within and found on the surface of the cells of Chrysochromulina chiton Parke et Manton. J. Exp. Bot. 18, 359–370.

Grove, S. N., Braeker, C. E. and Morré, D. J. (1968) Cytomembrane differentiation in the endoplasmic reticulum—golgi apparatus vesicle complex. Science 161, 171–173.

Haddon, L. E. and Northcote, D. H. (1975) Quantitative measurement of the course of bean callus differentiation. J. Cell Sci. 17, 11–26.

Haddon, L. E. and Northcote, D. H. (1976a) The influence of gibberellic acid and abscissic acid on cell and tissue differentiation of bean callus. J. Cell Sci. 20, 47–55.

Haddon, L. E. and Northcote, D. H. (1976b) Correlation of the induction of various enzymes concerned with phenylpropanoid and lignin synthesis during differentiation of bean callus. Planta 128, 255–262.

Haddon, L. E. and Northcote, D. H. (1976c) The effect of growth conditions and tissue of origin on the ploidy and morphogenetic potential of tissue cultures of bean (Phaseolus vulgaris L). J. Exp. Bot. 27, 1031–1051.

Hanke, D. E. and Northcote, D. H. (1974) Cell wall formation by soybean callus protoplast. J. Cell Sci. 14, 29–50.

Harris P. J. and Northcote, D. H. (1970) Patterns of polysaccharide biosynthesis in differentiating cells of maize root tips. Biochem. J. 120, 479–491.

Harris, P. J. and Northcote, D. H. (1971) Polysaccharide formation in plant Golgi bodies. Biochim. Biophys. Acta 237, 56–64.

Helting, T. and Rodén, L. (1969a) Biosynthesis of chondroitin sulfate. I. Galactosyl transfer in the formation of the carbohydrate-protein linkage region. J. Biol. Chem. 244, 2790–2798.

Helting, T. and Rodén, L. (1969b) Biosynthesis of chondroitin sulfate. II. Glucuronosyl transfer in the formation of the carbohydrate-protein linkage region. J. Biol. Chem. 244, 2799–2805.

Hemming, F. W. (1974) Lipids in glycan biosynthesis. In: M.T.P. International Review of Science—Biochemistry of Lipids (Goodwin, T. E., ed.), vol. 4, pp 39–97. University Park Press, Baltimore.

Hereward, F. V. and Northcote, D. H. (1972) A simple freeze-substitution method for the study of ultrastructure of plant tissues. Exp. Cell Res. 70, 73–80.

Herth, W., Kuppel, A., Franke, W. W. and Brown, R. M. (1975) The ultrastructure of the scale cellulose from Pleurochrysis scherffelii under various experimental conditions. Cytobiologie 10, 268–283.

Hinman, M. B. and Villemez, C. L. (1975) Glucomannan biosynthesis catalysed by Pisum sativum enzymes. Plant Physiol. 56, 608–612.

Hodson, S. and Brenchley, G. (1976) Similarities of the Golgi apparatus membrane and the plasma membrane in rat liver cells. J. Cell Sci. 20, 167–182.

Hsu, A-F., Baynes, J. W. and Heath, E. C. (1974) The role of dolichol-oligosaccharides in glycoprotein biosynthesis. Proc. Nat. Acad. Sci. U.S.A. 71, 2391–2395.

Jeffs, R. A. and Northcote, D. H. (1966) Experimental induction of vascular tissue in an undifferentiated plant callus. Biochem. J. 101, 146–152.

736

Jeffs, R. A. and Northcote, D. H. (1967) The influence of indol-3-yl acetic acid and sugar on the pattern of induced differentiation in plant tissue culture. J. Cell Sci. 2, 77–88.

Jones, D. D. and Morré, D. J. (1967) Golgi apparatus mediated polysaccharide secretion by outer root cap cells of Zea mays. II. Isolation and characteristics of the secretory product. Z. Pflanzenphysiol. 56, 166–169.

Kauss, H. (1969) A plant mannosyl-lipid acting in reversible transfer of mannose. FEBS. Lett. 5, 81–84.

Keegstra, K., Talmadge, K. W., Bauer, W. D. and Albersheim, P. (1973) The structure of plant cell walls. III. A model of the walls of suspension-cultured sycamore cells based on the interconnections of the macromolecular components. Plant Physiol. 51, 188–196.

Kemp, J. and Loughman, B. C. (1974) Cyclitol glucosides and their role in the synthesis of a glucan from UDPGlc in Phaseolus aureus. Characterization of some cyclitol glucosides and their synthesis. Biochem. J. 142, 153–159.

Khan, A. V. and Colvin, J. R. (1961) Synthesis of bacterial cellulose from labelled precursor. Science 133, 2014–2015.

Krisman, C. R. and Barengo, R. (1975) A precursor of glycogen biosynthesis: α1-4 glucan-protein. Eur. J. Biochem, 117–123.

Lagunoff, D. (1973) Membrane fusion during mast cell secretion. J. Cell Biol. 57, 252–259.

Lehle, L. and Tanner, W. (1974) Membrane-bound mannosyl transferase in yeast glycoprotein biosynthesis. Biochim. Biophys. Acta 350, 225–235.

Lehle, L. and Tanner, W. (1975) Formation of lipid bound oligosaccharides in yeast. Biochim. Biophys. Acta 399, 364–374.

Leloir, L. F., Staneloni, R. J., Carminatti, H. and Behrens, N. H. (1973) The biosynthesis of a N,N′diacetylchitobiose containing lipid by liver microsomes. A probable dolichol pyrophosphate derivative. Biochem. Biophys. Res. Commun. 52, 1285–1292.

Lennarz, W. J. and Scher, M. G. (1972) Metabolism and function of polyisoprenoid sugar intermediates in membrane-associated reactions. Biochim. Biophys. Acta 265, 417–441.

Levy, J. A., Carminatti, A., Cantarella, A., Behrens, N., Leloir, L. F. and Tábora, E. (1974) Mannose transfer to lipid linked di-N-acetylchitobiose. Biochem. Biophys. Res. Commun. 60, 118–125.

Lezica, R. P., Brett, C. T., Martinez, P. R. and Dankert, M. A. (1975) A glucose acceptor in plants with the properties of an α-saturated polyprenyl-monophosphate. Biochem. Biophys. Res. Commun. 66, 980–987.

McNeil, M., Albersheim, P., Taiz, L. and Jones, R. L. (1975) The structure of plant cell walls. VII. Barley aleurone cells. Plant Physiol. 55, 64–68.

Manton, I. (1966) Obervations on scale production in Prymnesium parvum. J. Cell Sci. 1, 375–380.

Manton, I. (1967a) Further observations on the fine structure of Chrysochromulina chiton with special reference to the haptonema, "peculiar" Golgi structure and scale production. J. Cell Sci. 2, 265–272.

Manton, I. (1967b) Further observations on scale formation in Chrysochromulina chiton. J. Cell Sci. 2, 411–418.

Manton, I. and Leadbeater, B. S. C. (1974) Fine-structural observations on six species of Chrysochromulina from wild Danish marine nanoplankton, including a description of C. campanulifera sp. nov. and a preliminary summary of the nanoplankton as a whole. Kongl. Danske Vidensk. Selsk. Biol. Shrift. 20, 3–36.

Mollenhauer, H. H. and Whaley, W. G. (1963) An observation on the functioning of the Golgi apparatus. J. Cell Biol. 17, 222–225.

Monro, J. A., Bailey, R. W. and Penny, D. (1975) Hemicellulose fractions and associated protein of lupin hypocotyl cell walls. Phytochem. 15, 175–181.

Morris, M. R. and Northcote, D. H. (1977) Influence of cations at the plasma membrane in controlling polysaccharide secretion from sycamore suspension cells. Biochem. J., in press.

Nagata, T. and Takabe, I. (1971) Plating of isolated tobacco mesophyll protoplasts on agar medium. Planta 99, 12–20.

Nikaido, H. and Hassid, W-Z. (1971) Biosynthesis of saccharides from glycopyranosyl esters of nucleoside pyrophosphates ("sugar nucleotides") Advanc. carb. chem. biochem. 26, 351–483.

Norman, A. G. (1937) The Biochemistry of Cellulose the Polyuronides Lignin etc. The Clarendon Press, Oxford.

Northcote, D. H. (1963) Changes in the cell walls of plants during differentiation. Symp. Soc. Exp. Biol. 17, 157–174.

Northcote, D. H. (1968) The organisation of the endoplasmic reticulum, the Golgi bodies and microtubules during cell division and subsequent growth. In: Plant cell organelles (Pridham, J. B., ed.), pp. 179–197. Academic Press, London.

Northcote, D. H. (1969a) The synthesis and metabolic control of polysaccharides and lignin during differentiation of plant cells. Essays Biochem. 5, 89–137.

Northcote, D. H. (1969b) Fine structure of cytoplasm in relation to synthesis and secretion in plant cells. Proc. R. Soc. Lond. Ser. B. 173, 21–30.

Northcote, D. H. (1970) The Golgi apparatus. Endeavour 30, 26–33.

Northcote, D. H. (1972) Chemistry of the plant cell wall. Ann. Rev. Plant Physiol. 23, 113–132.

Northcote, D. H. (1974a) Sites of synthesis of the polysaccharide of the cell wall. In: Plant Carbohydrate Chemistry (Pridham, J. B., ed.), pp. 165–181. Academic Press, London.

Northcote, D. H. (1974b) Complex envelope system. Membrane systems of plant cells. Phil. Trans. R. Soc. Lond. Ser. B. 268, 119–128.

Northcote, D. H. and Pickett-Heaps, J. D. (1966) A function of the Golgi apparatus in polysaccharide synthesis and transport in the root-cap cells of wheat. Biochem. J. 98, 159–167.

Northcote, D. H. and Wooding, F. B. P. (1966) Development of sieve tubes in Acer pseudoplatanus. Proc. R. Soc. Lond. Ser. B 163, 524–537.

Northcote, D. H. and Wooding, F. B. P. (1968) The structure and function of phloem tissue. Sci. Prog. Oxf. 56, 35–58.

Palade, G. E. and Bruns, R. R. (1968) Structural modulations of plasmalemmal vesicles. J. Cell Biol. 37, 633–644.

Parodi, A. J., Staneloni, R., Cantarella, A. I., Leloir, L. F., Behrens, N. H., Carminatti, H. and Levy, J. A. (1973) Further studies on a glycolipid formed from dolichyl-D-glucosyl monophosphate. Carbohydrate Res. 26, 393–400.

Pickett-Heaps, J. D. and Northcote, D. H. (1966) Relationship of cellular organelles to the formation and development of the plant cell wall. J. Exp. Bot. 17, 20–26.

Pisigan, R. A. and Rosario, E. J. (1976) Polysaccharide biosynthesis. Isoenzymes of soluble starch synthetase from Oriza sativa grains. Phytochem. 15, 71–73.

Prat, R. (1972) Plant protoplasts. Effect of the isolation procedure on cell structure. J. Microscopie (Paris) 14, 85–114.

Prat, R. (1973) Contribution a l'étude des protoplasts végétoux: du protoplast isolé et régeneration de sa paroi. J. Microscopie 18, 65–86.

Rees, D. A. (1969) Structure, conformation and mechanism in the formation of polysaccharide gels and networks. Biochem. 24, 267–332.

Robbins, P. W., Bray, D., Dankert, M. and Wright, A. (1967) Direction of chain growth in polysaccharide synthesis. Science 158, 1536–1542.

Roberts, K. and Northcote, D. H. (1970) The structure of sycamore callus cells during division in a partially synchronised suspension culture. J. Cell Sci. 6, 299–321.

Roberts, R. M. and Pollard, W. E. (1975) The incorporation of D-glucosamine into glycolipids and glycoproteins of membrane preparations from Phaseolus. Plant Physiol. 55, 431–436.

Roland, J-C. (1973) The relationship between the plasmalemma and plant cell wall. Int. Rev. Cytol. 36, 45–92.

Rougier, M. (1971) Etude cytochimique de la secrétion des polysaccharides végétaux a l'aide d'un material de choix: Les cellules de la coiffe de Zea mays. J. Microscopie 10, 67–82.

Rubery, P. H. and Northcote, D. H. (1968) Site of phenylalanine ammonia lyase activity and the synthesis of lignin during xylem differentiation. Nature 219, 1230–1234.

Rubery, P. H. and Northcote, D. H. (1970) The effect of auxin (2,4 dichlorophenoxyacetic acid) on the synthesis of cell wall polysaccharides in cultured sycamore cells. Biochim. Biophys. Acta 222, 95–108.

Ryman, B. E. and Whelan, W. J. (1971) New aspects of glycogen metabolism. Adv. Enzymol. 34, 285–443.

Satir, B., Schooley, C. and Satir, P. (1973) Membrane fusion in a model system. Mucocyst secretion in Tetrahymena. J. Cell Biol. 56, 153–176.

Selvendran, R. P. (1975) Cell wall glycoproteins and polysaccharides of parenchyma of *Phaseolus coccineus*. Phytochem. 14, 2175–2180.

Selvendran, R. P., Davies, A. M. C. and Tidder, E. (1975) Cell wall glycoproteins and polysaccharides of mature runner beans. Phytochem. 14, 2169–2174.

Sentandreu, R. and Lampen, J. O. (1971) Participation of a lipid intermediate in the biosynthesis of *Saccharomyces cerevisiae* LK 2G 12 mannan. FEBS. Lett. 14, 109–113.

Sentandreu, R. and Northcote, D. H. (1968) The structure of a glycopeptide isolated from the yeast cell wall. Biochem. J. 109, 419–432.

Shepard, J. F. and Totten, R. E. (1975) Isolation and regeneration of tobacco mesophyll cell protoplasts under low osmotic condtions. Plant Physiol. 55, 689–694.

Shore, G., Raymond Y. and Maclaehan, G. A. (1975) The site of cellulose synthesis. Cell surface and intracellular β1-4 glucan (cellulose) synthetase activities in relation to the stage and direction of cell growth. Plant Physiol. 56, 34–38.

Stoddart, R. W. and Northcote, D. H. (1967) Metabolic relationships of the isolated fractions of the pectic substances of actively growing sycamore cells. Biochem. J. 102, 194–204.

Storm, D. L. and Hassid, W. Z. (1972) The role of a D-mannosyl lipid as an intermediate in the synthesis of polysaccharide in *Phaseolus aureus* seedlings. Plant Physiol. 50, 473–476.

Strominger, J. L., Higashi, Y., Sandermann, H., Stone, K. J. and Willoughby, E. (1972) The role of polyisoprenyl alcohols in the biosynthesis of the peptidoglycan of bacterial cell walls and other complex polysaccharides. In: Biochemistry of the Glycosidic Linkage. An Integrated View (Piras, R. and Pontis, H. G., eds.), pp. 135–154. Academic Press, New York.

Talmadge, K. W., Keegstra, K., Bauer, W. D. and Albersheim, P. (1973) The structure of plant cell walls. I. The macromolecular components of the walls of suspension-cultured sycamore cells with a detailed analysis of the pectic polysaccharides. Plant Physiol. 51, 158–173.

Tandecarz, J., Lavintman, N. and Cardini, C. E. (1975) Biosynthesis of starch. Formation of a glucoproteic acceptor by a potato nonsedimentable preparation. Biochim. Biophys. Acta 345–355.

Tanner, W., Jung, P. and Linden, J. C. (1972) The role of a lipid intermediate in mannose-polymer biosynthesis in yeast. In: Biochemistry of the Glycosidic Linkage. An Integrated View. (Piras, R. and Pontis, H. G., eds.), pp. 227–231. Academic Press, New York.

Thornber, J. P. and Northcote, D. H. (1961a) Changes in the chemical composition of a cambial cell during its differentiation into xylem and phloem tissue in trees. 1. Main components. Biochem. J. 81, 449–455.

Thornber, J. P. and Northcote, D. H. (1961b) Changes in the chemical composition of a cambial cell during its differentiation into xylem and phloem tissue in trees. 2. Carbohydrate constituents of each main component. Biochem. J. 81, 455–464.

Towers, G. H. N. (1974) Enzymological aspects of flavonoid and lignin biosynthesis and degradation in plants. In: MTP International Review of Science. Plant Biochemistry (Northcote, D. H., ed.), Biochemistry 11, pp. 247–276. Butterworths, London.

Villemez, C. L. (1970) Characterization of intermediates in plant cell wall biosynthesis. Biochem. Biophys. Res. Commun. 40, 636–641.

Villemez, C. L. and Clark, A. F. (1969) A particle bound intermediate in the biosynthesis of plant cell wall polysaccharides. Biochem. Biophys. Res. Commun. 36, 57–63.

Warson, J-C. and Drochmans, P. (1972) Role of the sarcoplasmic reticulum in glycogen metabolism. Binding of phosphorylase, phosphorylase kinase and primer complexes to the sarcovesicles of rabbit skeletal muscle. J. Cell Biol. 54, 206–224.

Wetmore, R. and Rier, J. P. (1963) Experimental induction of vascular tissue in callus of angiosperms. Am. J. Bot. 50, 418–430.

Whaley, W. G. (1975) The Golgi Apparatus. Cell Biology Monographs, vol. 2. Springer-Verlag, Vienna.

Whaley, W. G. and Mollenhauer, H. H. (1963) The Golgi apparatus and cell plate formation—a postulate. J. Cell Biol. 17, 216–221.

Wilder, B. M. and Albersheim, P. (1973) The structure of the plant cell wall. IV. A structural comparison of the wall hemicellulose of cell suspension cultures of sycamore (*Acer pseudoplatanus*) and of red kidney bean *(Phaseolus vulgaris)* Plant Physiol. 51, 889–893.

Willison, J. H. M. (1976) An examination of the relationship between freeze-fractured plasmalemma and cell-wall microfibrils. Protoplasma 88, 187–200.

Willison, J. H. M. and Cocking, E. C. (1975) Microfibril synthesis at the surface of isolated tobacco mesophyll protoplasts, a freeze-etch study. Protoplasma 84, 147–159.

Wooding, F. B. P. (1968) Radioautographic and chemical studies of incorporation into sycamore vascular tissue walls. J. Cell Sci. 3, 71–80.

Wright, K. and Northcote, D. H. (1972) Root differentiation in sycamore callus. J. Cell Sci. 11, 319–337.

Wright, K. and Northcote, D. H. (1974) The relationship of root-cap slimes to pectins. Biochem. J. 139, 525–534.

Wright, K. and Northcote, D. H. (1975) An acidic oligosaccharide from maize slime. Phytochemistry 14, 1793–1798.

Wright, K. and Northcote, D. H. (1976) Identification of β1-4 glucan chains as part of a fraction of slime synthesised within the dictyosomes of maize root caps. Protoplasma. In press.

Envelopes of lipid-containing viruses as models for membrane assembly

11

Leevi KÄÄRIÄINEN and Ossi RENKONEN

1. Introduction

Cellular membranes consist of a lipid bilayer in which proteins are embedded by hydrophobic interactions (i.e., integral membrane proteins) or attached by polar interactions (i.e. peripheral proteins) (Singer and Nicolson, 1972; Wallach, 1972, 1975; Singer, 1974; Quinn, 1976; Nicolson et al., 1977). Many of the proteins can diffuse laterally in the plane of the membrane due to its fluidity (Singer and Nicolson, 1972; Capaldi, 1974). Some proteins span the lipid bilayer and are in contact with proteins on the other side (Bretscher, 1971, 1975). This transmembrane interaction between proteins limits their freedom to diffuse laterally (Bretscher and Raff, 1975). The membrane proteins exposed on the outside of the cell are often glycoproteins with complicated oligosaccharide chains (Gahmberg, 1976; Hughes, 1976;) and they discharge important functions in cell-cell interactions (Wallach, 1975; Hughes, 1976) and act as receptors for biologically active proteins such as hormones (Hughes, 1976).

The envelope of lipid-containing animal viruses shows many similarities to cellular membranes. Small angle x-ray scattering studies have shown that the lipids of Sindbis virus are arranged in a bilayer (Harrison et al., 1971). The outer envelope proteins are glycoproteins (Table 2) embedded in the lipids through hydrophobic interactions (Mudd, 1974; Uterman and Simons, 1974) Schloemer and Wagner, 1975). These glycoproteins can also span the lipid bilayer to make contact with the underlying nucleoprotein core (Garoff and Simons, 1974). In addition to glycoproteins, many viruses have another nonglycosylated membrane protein on the inner side of the bilayer (Choppin and Compans, 1975; Compans and Choppin, 1975; Wagner, 1975). Electron microscopic studies of infected cells suggest that the envelope of many animal viruses is derived from the plasma membrane, which has been modified by insertion of virus-specific projections (Acheson and Tamm, 1967; Compans and Dimmock, 1969; Lenard and Compans, 1974).

The similarities between viral envelopes and cellular membranes have raised

G. Poste and G.L. Nicolson (eds.)   The Synthesis, Assembly and Turnover of Cell Surface Components
© Elsevier/North-Holland Biomedical Press, 1977.

742

the hope that lipid-containing animal viruses can be used as models to study the biogenesis of the plasma membrane (Pfefferkorn and Hunter, 1963b). Viruses offer many obvious advantages for the study of membrane biogenesis: (1) their protein composition is much simpler than that of the plasma membrane (Simons et al., 1974); (2) most of the enveloped viruses are released from infected cells into the surrounding medium without concomitant lysis of the cell, which makes their purification relatively easy; (3) the replication of enveloped viruses can be synchronized by simultaneous infection of large amounts of cells, which enables study of translation, glycosylation, and transport of membrane proteins at a defined time during virus replication; (4) a considerable amount of virus-specific protein is synthesized during the replication period, thus facilitating the use of biochemical and immunological methods; and (5) in many cases the identification of the viral products is further facilitated by the effective shut off of host cell macromolecular synthesis (Metz, 1975).

In the present review the structure and assembly of viral envelopes will be discussed with particular reference to results obtained with Semliki Forest virus, which we have used as a membrane model for several years (Renkonen et al., 1974; Simons et al., 1974; Kääriäinen et al., 1975). Other aspects of the use of enveloped viruses as membrane models have been covered in a number of excellent recent reviews (Joklik and Zweerink, 1971; Choppin et al., 1972, 1973; Blough and Tiffany, 1973; Lenard and Compans, 1974; Casjens and King, 1975; Choppin and Compans, 1975; Compans and Choppin, 1975; Garoff and Simons, 1975; Simons et al., 1975).

2. General features of enveloped animal viruses

The properties of the main groups of enveloped animal viruses are listed in Tables 1 and 2. The complexity of the virion in specific taxonomic groups increases in relation to the accelerating size of the genome. This is reflected in the number of polypeptides serving as structural components, both in the nucleocapsid and the envelope. Thus pox- and herpesviruses are complicated models for the study of membrane structure and assembly. In vaccinia virus evidence exists that both the lipids and the carbohydrates are specified by the virus genome (Dales and Mosbach, 1968; Garon and Moss, 1971; Moss et al., 1971, 1973; Moss, 1974), making this virus a unique model for membrane biogenesis (Grimley et al., 1970; Weintraub and Dales, 1974; Dales et al., 1974). The carbohydrates of vaccinia virus consist of only a few glucosamine residues, which are associated with glycoproteins not exposed on the virion surface (Holowzak, 1970; Sarov and Joklik, 1972; Moss, 1974).

Herpesviruses acquire their envelope from the nuclear membrane and continue to mature in the cytoplasm (Roizman and Furlong 1974). Since only small amounts of virus are released into the medium, they are usually isolated from the infected cells. This tends to yield virus preparations contaminated with cellular material and virus-specific nonvirion material (reviews, Roizman, 1969; Kap-

Table 1
Basic Properties of Enveloped Animal Viruses

Virus group[a]	Shape	Size in Nanometers	Symmetry of Nucleocapsid	Nucleic Acid[b]	Total Molecular Weight of Nucleic Acid
Togaviruses (1)					
Alphavirus	Spherical	600–700	Cubical	ss linear RNA	4–4.5 × 10^6
Flavivirus	Spherical	500	Cubical?	ss linear RNA	3– 5 × 10^6
Bunyaviruses (2)	Spherical	900–1200	Helical?	ss segmented RNA	about 4 × 10^6
Myxoviruses (3)	Spherical	1000	Helical	ss segmented RNA	4 × 10^6
Paramyxoviruses (4)	Spherical	1200	Helical	ss linear RNA	5 × 10^6
Rhabdoviruses (5)	Bullet-shaped	700x 1750	Helical	ss linear RNA	4 × 10^6
Oncornaviruses (6)	Spherical	1200	Helical?	ss linear RNA	3.5 × 10^6
Arenaviruses (7)	Spherical	600–1200	Helical?	ss linear RNA	
Coronaviruses (8)	Spherical	800–1200	Helical?	?	?
Herpesviruses (9)	Spherical	1200	Cubical	ds linear DNA	100 × 10^6
Poxviruses (10)	Brick-shaped	2200-2200x 2800	Complex	ds linear DNA	160 × 10^6

[a] (1) Pfefferkorn and Shapiro, 1974; Strauss and Strauss, 1976. (2) Pettersson et al., 1971; Pettersson and Kääriäinen, 1973; Horzinek, 1975; Pettersson, 1975; Porterfield et al., 1976, (3) Compans and Choppin, 1975 (4) Choppin and Compans, 1975 (5) Wagner, 1975. (6) Bolognesi, 1974; Bader, 1975. (7) Lehman-Grube, 1973. (8) Bradburne and Tyrrell, 1971, Jackson and Muldoon, 1973; McIntosh, 1974. (9) Roizman and Furlong, 1974. (10) Moss, 1974.
[b] ss = single stranded; ds = double-stranded.

Table 2
Structural Proteins of Enveloped Animal Viruses

Virus group	Number of Different Polypeptides in				References
	Total virion	Nucleocapsid	Envelope membrane		
			Glycoproteins	Other proteins	
Togaviruses					
Alpha	3–4	1	2–3	—	Garoff and Simons, 1975
Flavi	3	2	1	—	Strauss and Strauss, 1976
Bunyaviruses	3	1	2		Porterfield et al., 1976
Myxoviruses	8	3	4	1	Compans and Choppin, 1975
Paramyxoviruses	5	2	2	1	Choppin and Compans, 1975
Rhabdoviruses	5	2	2	1	Wagner, 1975
Oncornaviruses	6–7	4	2–3	?	Bolognesi, 1974; Bader, 1975
Coronaviruses	6–7	?	4	?	McIntosh, 1974
Arenaviruses	4	1–2	2		Rawls et al., 1973
Herpesviruses	28	13	13	2	Perdue et al., 1974; Roizman and Furlong, 1974
Poxviruses	30	17	5^a	8^b	Sarov and Joklik, 1972; Moss, 1974

[a] nonglycosylated surface proteins.
[b] Two of these are glycoproteins.

lan, 1973; and Roizman and Furlong, 1974). Considering these difficulties, together with the complicated polypeptide composition (Table 2), it is understandable that herpesviruses may be too complex to serve as useful membrane models.

The large group of bunyaviruses, comprising about 100 members (Porterfield et al., 1976), as well as the Corona- and Arena-viruses have not been studied in sufficient detail to be discussed as membrane models (Jackson and Muldoon, 1973; Lehman-Grube, 1973).

The oncornaviruses have been omitted from this chapter mainly because the extensive literature on these agents, including their structure and assembly, has been reviewed in detail by Baltimore (1974), Bolognesi (1974), Eisenman et al., (1974), and Bader (1975). Further review would be redundant. Thus this review will deal only with the togaviruses, of which the alphaviruses are best known, and the myxo- paramyxo-, and rhabdoviruses.

Alpha-, paramyxo-, and rhabdoviruses multiply exclusively in the cytoplasm, and apparently no nuclear functions are necessary for their replication (Follet et al., 1974, 1975). They may be regarded as carriers of messenger RNA for foreign membrane proteins that are introduced to the cell by virus infection. The myxo- and flaviviruses require some nuclear functions early in infection (Follet et al., 1974; Compans and Choppin, 1975; Kos et al., 1975; Minor and Dimmock, 1975). For example, part of the RNA synthesis in myxovirus-infected cells occurs in the nucleus (Krug, 1972; Armstrong and Barry, 1975; Krug and Etkind, 1975). Despite differences in the mode of transcription and translation that will be discussed later, all these virus groups share many characteristics as models for membrane biogenesis.

3. Structure of viral envelopes

3.1. Alphaviruses

3.1.1. General properties

The alphaviruses, formerly known as group A arboviruses (Casals and Clarke, 1965), consist of RNA, protein, lipids, and protein-bound carbohydrate (Table 3). The properties of the two best known members, the Semliki Forest virus (SF virus), and the Sindbis virus will be discussed as representative examples of the alphavirus group. Negatively stained, purified virus preparations show spherical particles with a mean diameter of about 65 nm. The particle surfaces is comprised of spikes or projections about 7 nm in length (Fig. 1). Thin sections prepared from pelleted virus or virus-infected cell cultures show that the virus consists of an inner core, the nucleocapsid, which is surrounded by an envelope membrane containing the spikes (Fig. 1).

The virus envelope can be solubilized with detergents such as Nonidet P40, Triton X-100, or sodium deoxycholate (Strauss et al., 1968; Kääriäinen et al., 1969; Helenius and Söderlund, 1973). Various solubilization phases of the viral

Table 3
Chemical Composition of Some Enveloped Viruses

Virus	Percentage per Weight				References
	RNA	Protein	Lipid	Carbohydrate	
Alphaviruses					
SF virus	6.3	56.6	30.8	6.3	Laine et al., 1973
Sindbis	5.6	61	27	6.5	Pfefferkorn and Hunter, 1963a
Myxoviruses					
Influenza A	0.8–1.0	70	20	5–8	Ada and Perry, 1954; Frommhagen et al., 1959; Blough et al., 1967.
Paramyxoviruses					
SV5	0.9	73	20	6.1	Klenk and Choppin 1969a
Rhabdoviruses					
VSV	3	64	20	13	McSharry and Wagner 1971
Oncornaviruses					
RTV	2	62	30	6	Quigley et al., 1971

envelope have been analyzed in detail by Helenius and Söderlund (1973). Complete solubilization of the envelope is obtained at a Triton X-100 to lipid ratio of 20 to 1. Under these conditions the envelope proteins sediment at 4 S together with lipids and Triton X-100 micelles (Simons et al., 1973a). Lipid- and detergent-free envelope proteins can be isolated by sucrose gradient centrifugation in a detergent-free medium, where the envelope proteins aggregate to form octamers that sediment at about 29 S and retain their hemagglutinating activity (Helenius et al., 1976). The nucleocapsid, with a sedimentation value of about 150 S, can be isolated from the solubilized envelope components by sucrose gradient centrifugation. The nucleocapsid is a spherical particle about 39 nm in diameter that consists of about 240 copies of a lysine-rich polypeptide and one single-stranded RNA molecule with a molecular weight of 4 to 4.5×10^6 daltons (Strauss et al., 1968; Kääriäinen et al., 1969; Acheson and Tamm, 1970; Simons and Kääräinen, 1970; Levin and Friedman, 1971; Arif and Faulkner, 1972; Simmons and Strauss, 1972a; Laine et al., 1973). Since the unique properties of the SF virus nucleocapsid have been reviewed recently by Söderlund et al. (1975) they will not be discussed further. The general characteristics of SF virus are summarized in Table 4.

Table 4
General Properties of Semliki Forest Virus

Component	Diameter (Å)	Sedimentation Value	Density in CsCl (g/cm³)	Molecular Weight ($\times 10^6$)	Triangulation Number
Whole virion[a]	690	280 S	1.24	65–70	T=4
Nucleocapsid[b]	390	150 S	1.43	12–13	T=4 ?

[a] Kääriäinen et al., 1969; von Bonsdorff, 1973; Laine et al., 1973.
[b] Söderlund et al., 1975.

747

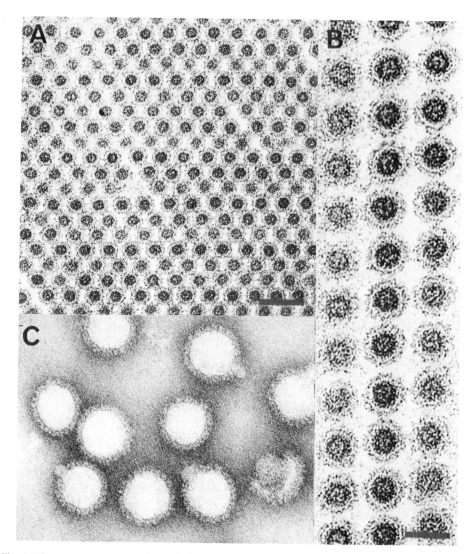

Fig. 1. Electron micrographs of purified Semliki Forest virus. (a) and (b) Thin sections of pelleted virus. (c) Negatively stained virus. Bars: a = 100 nm; b and c = 50 nm.

3.1.2. Envelope proteins

Three different envelope proteins with molecular weights of 49,000 (E-1), 52,000 (E-2), and 10,000 (E-3) daltons (Table 5) can be resolved by discontinuous polyacrylamide gel electrophoresis from the envelope fraction of SF virus (Garoff et al., 1974). These proteins have been purified in large amounts by chromatography in hydroxylapatite columns in the presence of sodium dodecyl sulfate.

The amino acid and carbohydrate composition of SF virus glycoproteins have

Table 5
Properties of Envelope Proteins of Some Lipid-Containing Viruses

Virus group	Designation of Protein	Molecular Weight ($\times 10^{-3}$)	Carbohydrate Composition	Function
Alphaviruses (SF virus)[a]	E-1	49	GlcNAc, Man, Gal, Fuc, NANA	Hemagglutinin
	E-2	52	GlcNAc, Man, Gal	Infectivity?
	E-3	10	GlcNAc, Gal, Fuc, NANA	?
(Sindbis)[b]	E-1	50–53	GlcNAc, Man, Gal, Fuc, NANA	Hemagglutinin
	E-2	50–53	GlcNAc, Man, Gal, Fuc, NANA	Infectivity
Myxoviruses[c]	HA	75	GlcNAc, Man, Gal, Fuc	Hemagglutinin
	HA$_1$	50		
	HA$_2$	25		
(Influenza A)	NA	55		Neuraminidase
	M	26	No carbohydrates	Matrix protein
Paramyxoviruses[d] (SV 5, Sendai, NDV)	HN	67–74	GlcNAc, Man, Gal, Fuc, NANA	Hemagglutinin and neuraminidase
	F	53–56	GlcNAc, Man, Gal, Fuc, NANA	Infectivity, cell fusion, and hemolysin
	M	38–41	No carbohydrates	Matrix protein
Rhabdoviruses[e] (VSV, rabies)	G	62–69	GlcNAc, Man, Gal, Fuc, NANA (2 type A chains)	Infectivity and hemagglutinin
	M	26–29	No carbohydrates	Matrix protein
Oncornaviruses[f]	gp1	70–115	GlcNAc, Man, Gal, Fuc, NANA	Infectivity and type-specific antigen
	gp2	37–50	GlcNAc, Man, Gal, Fuc, NANA	Group-specific antigen?

[a] Garoff et al., 1974; Helenius et al., 1976; Mattila et al., 1976.
[b] Burge and Strauss, 1970; Strauss et al., 1970; Schlesinger et al., 1972; Dalrymple et al., 1976.
[c] Ada and Gottschalk, 1956; Frommhagen et al., 1959; Compans and Choppin, 1975.
[d] Klenk et al., 1970; Moore and Burke, 1974; Choppin and Compans, 1975; Scheid and Choppin, 1975.
[e] Halonen et al., 1968; Sokol et al., 1971; Etchison and Holland; 1974a,b; Obijeski et al., 1974; Wagner, 1975.
[f] Bolognesi, 1974; Bader, 1975.

been determined from purified fractions by Garoff et al. (1974) The amino acid composition does not reveal any special hydrophobicity in any of the SF virus glycoproteins. However, the amphiphilic nature of the glycoproteins (E-1 and E-2) is reflected in their ability to bind detergents such as Triton X-100 and sodium deoxycholate, agents that bind to the hydrophobic regions of integral membrane proteins (Uterman and Simons, 1974; Becker et al., 1975; Helenius and Simons, 1975; Helenius et al., 1976; and for a review see Simons et al., 1976). Helenius et al. (1976) have shown that solubilization of SF virus membrane with sodium deoxycholate at concentrations above 2.2 mM induces the dissociation of E-1 and E-2 from each other. The E-1 protein bears the hemagglutinating activity. Separation of Sindbis virus E-1 and E-2 by isoelectric focusing in the presence of Triton X-100 has been reported recently by Dalrymple et al. (1976). E-1 had hemagglutinating activity and also cross-reacted with related alphavirus antisera, whereas E-2 was virus-specific, giving rise only to neutralizing antiserum against Sindbis virus.

The sugar chain composition of both Sindbis and SF virus envelope proteins have been investigated by chromatography in Biogel P 6 after extensive pronase digestion (Table 6). Using this technique, two different sizes of carbohydrate chains can be resolved (Sefton and Keegstra, 1974; Keegstra et al., 1975; Mattila et al., 1975). The sugar composition of each peak has been determined by isotope labeling of the different sugars (Burge and Strauss, 1970; Strauss et al., 1970). The Sindbis virus proteins E-1 and E-2 seem to contain two different chains: an A-type chain (Spiro, 1973) that contains N-acetylglucosamine, mannose, galatose, fucose, and sialic acid and one B-type chain containing N-acetyl, glucosamine and mannose (Table 6) (Sefton and Keegstra, 1974).

The carbohydrate chain composition of SF virus differs from Sindbis virus in that there is apparently one A-type chain in both E-1 and E-3, whereas E-2 has 2 to 3 B-type chains and one rather large X-type chain that may contain mainly galactose (Mattila et al., 1975; Renkonen et al., unpublished observations). The A-type chains, isolated from the whole SF virus after extensive pronase treatment, are characterized by successive degradations with pure exohydrolases. The reduction in molecular weight and the amount of released monosaccharides were determined by gel filtration (Pesonen and Renkonen, 1976). Sequential degradation was carried out with neuraminidase followed by L-α-fucosidase, D-β-galactosidase, D-β-N-acetyl-glucosaminidase, D-α-mannosidase, D-β-mannosidase and D-β-N-acetyl-glucosaminidase. The provisional sequence of the chains was deduced and, as shown in Figure 2, is a typical N-glycosidic A-type oligosaccharide chain recognized, for example, in the serum glycoproteins (Spiro, 1973, Hughes, 1975). The sialic acid residues are linked to the galactose residues, but the fucose is not. The fucose may be bound to one of the proximal N-acetyl-D-glucosamines.

Sefton (1976) has shown that the carbohydrate chain composition of Sindbis virus grown in normal and transformed chick cells in BHK hamster cells are basically the same. Vesicular stomatitis virus G-protein from the same cells contained only A-type chains whereas Sindbis virus showed both A- and B-type

Table 6
Carbohydrate Chain Composition of Semliki Forest (SF) and Sindbis Virus Glycoproteins

Carbohydrate	Monosaccharides per Mole of Glycopeptide					Molecular Weight of Glycopeptide	Number of Chains per Glycoprotein
	Sialic acid	Galactose	Fucose	Mannose	N-acetyl glucosamine		
SF virus:[a]							
Type A (E-1 and E-3)	3.4	3.1	0.7	3.7	6–7	3,700	1
Type B (E-2)	—	—	—	present	present	2,000	2–3
Type X (E-2)	—	present	—	—	— ?	3,000	1 ?
Sindbis virus:[b]							
Type A (E-1 and E-2)	1	5	1	2	7	3,000	1
Type B (E-1 and E-2)	—	—	—	3	6	1,740	1

[a] Mattila et al. (1975); Pesonen and Renkonen (1976).
[b] Strauss et al. (1970).

chains. Avian tumor viruses contained a large A-type oligosaccharide. These results strongly suggest that the oligosaccharide chain is "virus-specific," that is, the primary and perhaps the secondary structure of the polypeptide specify the type and, to some extent, the length of the oligosaccharide chain (Spiro, 1973; Hughes, 1976). The heterogeneity seen in the amount of terminal sialic acid residues may be host-dependent (Burge and Huang, 1970; Grimes and Burge, 1971; Keegstra et al., 1975). Removal of sialic acid residues from SF virus affected neither the hemagglutination nor the infectousness of the virus, whereas treatment with other sugar hydrolases lowered both activities (Kennedy, 1974).

Sindbis virus grown in *Aedes albopictus* mosquito cells does not contain sialic acid but is fully infectious and retains its hemagglutinating properties (Stollar et al., 1976). The recent results of Schlesinger et al. (1976) suggest that infectivity and hemagglutinating activities are preserved even if the terminal glucosamine residues and galactose residues (Fig. 2) are missing from the A-type chain. Thus, the glucosamine-mannose core may be sufficient for the biological activities of the virus.

3.1.3. Structure of the envelope

The glycoproteins of both Sindbis and SF virus are located on the surface of the virus, as shown by two different experimental approaches. The envelope proteins can first be removed by treating the intact virus with proteolytic enzymes such as bromelain or thermolysin (Compans, 1971; Uterman and Simons, 1974). The treated virus no longer has surface projections. Second, treatment of SF virus with ^{35}S-formyl methionylsulfate methylphosphate labels only the glycoproteins but not the capsid protein (Gahmberg et al., 1972). Similarly, only the glycoproteins of Sindbis virus are labeled by the lactoperioxidase iodination technique (Sefton et al., 1973).

The glycoproteins of Sindbis virus are arranged in the viral envelope in a highly regular formation. The structural units are clustered in a T = 4 icosahedral surface lattice (von Bonsdorff and Harrison, 1975). From the chemical composition of the closely related SF virus (Laine et al., 1973) this would

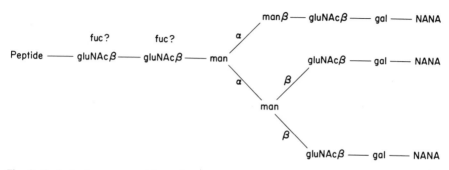

Fig. 2. Carbohydrate composition of type A oligosaccharide chain of SF virus. (Reproduced with permission from Luukkonen et al. [1976a]).

imply that there are 80 structural subunits, each consisting of three pairs of E-1—E-2 (and E-3 in SF virus).

Recently, Luukkonen et al. (1976a) have used the enzyme galactose oxidase together with ^{3}H-sodium borohydride to label the SF virus glycoproteins. Two envelope proteins E-1 and E-3 with similar A-types oligosaccharide chains were labeled with striking inequality in the intact virus, most of the label being associated with E-3 (Fig. 3a). This result suggests that the oligosaccharides of E-3 are located more "externally" than those of E-1. Interestingly, the labeling of E-1 was increased about fourfold in virus that was swollen with a small amount of Triton X-100 (Fig. 3b). At this concentration triton binds to the viral envelope without disrupting it (Helenius and Söderlund, 1973; Simons et al., 1973). It may be that the galactose residues of E-1 are covered by E-3 in the intact virus, and that the cover is removed when the virus is slightly swollen exposing the galactose residues to the enzyme. The galactose-containing oligosaccharides of E-2 were only slightly labeled in both intact and detergent-swollen virus. This poor labeling of E-2 is difficult to understand because the extent of the labeling did not increase significantly even in the completely solubilized virus preparation. This result may reflect specific structural features of the X-type oligosaccharide rather than the location of the E-2 protein in the detergent protein complexes or on the surface of the virus particle.

Alphavirus lipids comprise 25 to 30% of the dry weight of the particle (Table 3) and consist mainly of phospholipids and cholesterol. The lipid composition of SF virus closely resembles that of the host cell plasma membrane and clearly differs from that of the endoplasmic reticulum and whole cells (Table 7) and from mitochondrial and nuclear membranes (Renkonen et al., 1974). This finding is compatible with morphological studies showing that the viral nucleocapsid buds through the plasma membrane (Acheson and Tamm, 1967). The cholesterol-to-phospholipid molar ratio in the virus is, however, greater than that in the plasma membrane. This similarity in lipid composition between plasma membrane and virus is also reflected in the composition of glycolipids, which are minor components of the virus (Renkonen et al., 1971; Hirschberg and Robbins, 1974).

Small angle x-ray scattering studies carried out with purified Sindbis virus (Harrison et al., 1971) and SF virus (Harrison and Kääriäinen, unpublished observations) have shown that the lipids in both viruses are arranged in a bilayer with the polar groups in layers at radial distances of 210 Å and 258 Å for Sindbis and 202 Å and 250 Å for SF virus. Chemical analysis shows about 16,000 to 17,000 phospholipid-cholesterol pairs in the SF virus envelope (Laine et al., 1973), of which about 60% should be in the outer leaflet of the bilayer (Harrison et al., 1971).

As described above, the envelope proteins of SF virus can be digested with thermolysin to obtain spikeless particles with a marked tendency to aggregate (Uterman and Simons, 1974). Analysis of these particles labeled with radioactive amino acids reveals an intact capsid protein and a small 5,000 dalton fragment of envelope protein. This hydrophobic fragment can be extracted with the usual lipid solvents. Amino acid analysis has shown that it consists mainly of hyd-

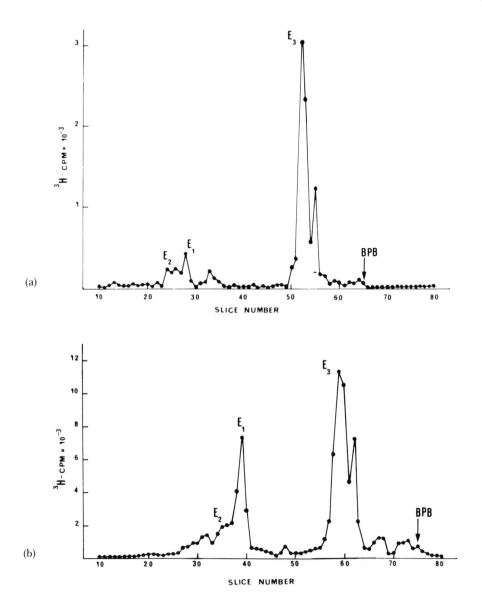

Fig. 3. Gel electrophoresis of Semliki Forest virus treated with neuraminidase and galactose oxidase followed by reduction with NaB³H₄: (a) intact virus; (b) Triton X-100 disrupted virus, which is indistinguishable from the triton-swollen virus (see text) (Luukkonen et al., submitted for publication).

rophobic amino acids. Comparision of the oligopeptides of this hydrophobic piece with those derived from E-1 and E-2 revealed that peptides from both envelope proteins were present in the fragment, strongly suggesting that both major envelope proteins are anchored to the lipid bilayer by a hydrophobic polypeptide fragment (Uterman and Simons, 1974; see also review by Simons et

Table 7
Phospholipid Composition of Semliki Forest Virus Grown in Hamster BHK 21 and *Aedes Albopictus* Cells Compared to the Phospholipid Composition of the Host Cell and Host Cell Plasma Membrane (BHK 21)

Lipid	Percentage of Distribution of Lipids					
	BHK 21 cells[a]				*Aedes albopictus* cells[b]	
	Whole cells	Endoplasmic reticulum	Plasma membrane	Virus	Whole cells	Virus
Phosphatidic acid	0.6	1.3	3.8	1.29	1.1	1.1
Cardiolipin	3.3	2.0	0.4	0.22	2.7	0
Phosphatidylethanolamine	23.0	14.0	18	23.2	47.6	62.4
Phosphatidylcholine	50.0	62.0	44	42.4	26.2	14.4
Phosphatidylserine	6.6	3.6	8.9	13.4	6.8	10.3
Phosphatidylinositol	5.7	4.3	2.3	1.6	6.7	1.55
Lysophosphatidylcholine	2.9	6.9	6.4	0.47	trace	0
Ceramide phosphorylcholine (sphingomyelin)	6.9	10.0	18	16.0	0.66	0.71
Ceramide phosphorylethanol-amine	0	0	0	0	8.5	9.5
Molar ratio Cholesterol/Phospholipids	0.28	0.29	0.70	0.99	0.07	0.22

[a] Renkonen et al., 1971, 1972a,b.
[b] Luukkonen et al., 1974, 1976b; Renkonen et al., 1974.

al., 1976). Sefton and Gaffney (1974) have shown that even after proteolytic digestion of the envelope protein the lipid bilayer in Sindbis virus envelope is in a less fluid state than in artificial bilayers made from lipids extracted from the virus with chloroform-methanol. These results suggest that the order imposed on Sindbis virus lipid may be due to the presence of a lipophilic polypeptide fragment in the membrane. It is tempting to speculate that the interaction between the hydrophobic fragment and the membrane lipids might be specific. If so, the envelope proteins would select particular lipids as suggested by David (1971). However, the similarity between SF virus and BHK plasma membrane lipids suggested that this might not be the case. The possibility remained that specificity lay in the interaction between fatty acid chains rather than in the polar groups. Each phospholipid class consists of numerous combinations of fatty acids with different lengths and degrees of saturation. Therefore the fatty acid distribution of the different lipid classes was analyzed from both the virus and the BHK cell plasma membrane (Laine et al., 1972), but no sign of specificity could be demonstrated.

Phospholipid analysis of SF virus grown in *Aedes albopictus* (mosquito) cells in vitro was done recently by Luukkonen et al. (1976b). The lipid composition of mosquito cells is remarkably different from that of BHK cells (Luukkonen et al., 1973). Almost two-thirds of the lipids of SF virus grown in these cells are phosphatidylethanolamine (Table 7). There was no sphingomyelin (ceramide phosphorylcholine). This was replaced by another sphingolipid, ceramide phosphorylethanolamine (Luukkonen et al., 1976b). Obviously, the possibility that some minor component such as phosphatidylserine would be selected specifically by the envelope proteins cannot be excluded (Semmel et al., 1975).

Other relevant information has been obtained by Garoff and Simons (1974), who used dimethyl suberimidate to cross-link SF virus proteins. This bifunctional reagent was shown to link almost all envelope proteins with the nucleocapsid. Most of the lipids could be removed and spherical particles consisting of nucleocapsid and envelope proteins could be isolated. Since this bifunctional reagent can react with amino groups maximally 11 Å apart, which is less than one-fourth the thickness of the lipid bilayer, these results strongly suggest that the envelope proteins actually span the membrane and are in close proximity to the nucleocapsid protein.

3.2. Myxo-, paramyxo-, and rhabdoviruses

The chemical composition of myxo-, paramyxo- and rhabdoviruses is given in Table 3. These viruses contain RNA, protein, protein-bound carbohydrate, and lipids in similar proportions.

3.2.1. Biological functions of the envelope proteins

The influenza A virus envelope has two easily recognizable biological activities, the capacity to agglutinate red blood cells and neuraminidase activity (Table 5). The former is associated with the envelope glycopolypeptides HA_1 and HA_2, which are joined by S-S-bridges (Laver, 1971; Skehel and Waterfield, 1975; Bucher et al., 1976). The active hemagglutinin consists of two such pairs and has a molecular weight of about 150,000 (Laver and Valentine, 1969; Schulze, 1972, 1973). In addition to HA_1 and HA_2, their immediate precursor, the HA protein, which has a molecular weight of about 75,000 to 80,000, is found in variable amounts in the virion depending on growth conditions and the host cell (Compans et al., 1970; Lazarowitz et al., 1971, 1973a,b; Skehel and Schild, 1971; and review by Compans and Choppin, 1975). In some cells posttranslational cleavage of HA_1 and HA_2 does not seem to occur at all, and the excreted virus remains uninfectious unless treated with proteolytic enzymes. The uncleaved HA protein has hemagglutinating activity (Klenk et al., 1975a; Lazarowitz and Choppin, 1975; Choppin et al., 1975). The other major glycoprotein in the influenza virus envelope is the neuraminidase (Table 5), which represents about one-fifth of the surface proteins (Schulze, 1972). The active enzyme is probably a tetramer with a molecular weight of about 250,000 (Bucher and Kilbourne, 1972; Lazdins et al., 1972; Wrigley et al., 1973; Kendal and Kiley, 1975).

The third major protein component of the myxovirus envelope is the so-called matrix or M-protein (Table 5). This protein does not contain carbohydrates and is not exposed outside the lipid bilayer, as shown by its resistance to treatment with proteolytic enzymes (Kendal et al., 1969; Compans et al., 1970; Schulze, 1970), and it cannot be labeled with reagents that react with surface glycoproteins (Stanley and Haslam, 1971; Rifkin et al., 1972). Several lines of indirect evidence strongly suggest that the M-protein is located beneath the lipid bilayer, probably mediating the contact between lipids and the inner ribonucleoprotein (Compans and Dimmock, 1969; Stanley and Haslam, 1971; Rifkin et al., 1972; Schulze, 1972, 1973; Griffith, 1975).

The paramyxovirus envelope consists of two major glycoproteins, HN and F (Table 5) (Klenk et al., 1970). Hemagglutinating and neuraminidase activities are associated with the HN protein (Scheid et al., 1972), whereas the F protein is responsible for the well-known cell-fusing and hemolyzing activities and also for the infectivity of the paramyxoviruses (Scheid and Choppin, 1974a,b, 1975; Poste and Waterson, 1975). Sendai virus grown in bovine kidney cells (MDBK cells) does not contain F protein but instead contains a larger protein, F_0, with an apparent molecular weight of 65,000. This virus is uninfectious and lacks both cell-fusing and hemolyzing activities. However, brief treatment with trypsin restores the biological activities of the virus with concomitant cleavage of F_0 to F, the latter having a molecular weight of 53,000 (Homma and Ohuchi, 1973; Scheid and Choppin, 1974a, 1976). In normal virus infection the cleavage of F_0 to F is carried out by host cell proteolytic enzymes (Scheid and Choppin, 1976). The cleavage product remains attached to the F protein, presumably by S-S-bridges (Moore and Burke, 1974; Nagai et al., 1976). The extent of the F_0 cleavage may be especially important for cell-fusing activity, since virions containing a considerable proportion of F_0 cannot fuse cells and are also non-infectious (Poste, 1975; Famulari and Fleissner, 1976a; Nagai et al., 1976).

As in the case of the myxoviruses, the paramyxovirus envelope also contains a matrix (M) protein (Table 5) that is not glycosylated (Klenk et al., 1970) and is again probably located beneath the lipid bilayer (reviews, Lenard and Compans, 1974; Choppin and Compans, 1975).

Only one type of glycoprotein, the G protein, is associated with the rhabdovirus envelope (Wagner et al., 1969, 1970, 1972; Burge and Huang, 1970; Sokol et al., 1971; Obijeski et al., 1974). This protein is responsible for both the hemagglutinating activity and infectivity of the virus (Bishop et al., 1975; Schloemer and Wagner, 1975b). The G protein contains about 11% carbohydrate in two A-type polysaccharide chains (Etchison and Holland, 1974) (Table 5). The sialic acid residues in the polysaccharides may be necessary for the biological activities of the virus (Schloemer and Wagner, 1974, 1975b,c; Schlesinger et al., 1976). A matrix protein is also present that is not glycosylated; it is located beneath the lipid bilayer (Wagner, 1975; Wagner et al., 1975), as in the myxo- and paramyxoviruses.

3.2.2. The amphiphilic nature of the envelope proteins
The mode of attachment of myxo- and paramyxovirus glycoproteins to the lipids is unknown at present (Choppin and Compans, 1975; Compans and Choppin, 1975; Lenard et al., 1975). The hemagglutinin of influenza virus is probably attached by a small fragment of HA_2 polypeptide, as suggested by the lower

molecular weight of the protease-released HA_2 polypeptide (Brand and Skehel, 1972). A hydrophobic fragment with a molecular weight of about 5,200 has been isolated recently from thermolysin-treated vesicular stomatitis virus (Mudd, 1974; Schloemer and Wagner, 1975a), suggesting that the amphiphilic nature of the G protein is responsible for its stability.

Considerable circumstancial evidence exists for the amphiphilic nature of myxo- and paramyxovirus glycoproteins. Like the G protein of vesicular stomatitis virus (Schloemer and Wagner, 1975a) and SF virus envelope proteins (Helenius and Simons, 1972; Helenius and von Bonsdorff, 1976), the HN and F envelope glycoproteins of SV 5 can be solubilized in biologically active form by Triton X-100 in low salt medium (Scheid et al., 1972). When the detergent is removed the HN and F proteins aggregate, forming fairly regular rosettes. Similar structures have been described for the isolated hemagglutinin of influenza virus (Laver and Valentine, 1969). All these properties are consistent with the idea that the glycoproteins of myxo-, paramyxo-, and rhabdoviruses are indeed amphiphilic integral membrane proteins (Helenius and Simons, 1975; also see reviews by Choppin and Compans, 1975; Compans and Choppin, 1975; and Wagner, 1975). The role of sulfate groups found recently in the envelope glycoproteins of these viruses remains unclear (Compans and Pinter, 1975; Pinter and Compans, 1975).

Recent experiments suggest that at least one of the viral integral proteins is a trans-membrane component. Experiments with formaldehyde-fixed vesicular stomatitis virus have shown that the G protein can be cross-linked to the M protein, allowing the removal of envelope lipids (Brown et al., 1974, 1975) and suggesting that the G protein might actually span the lipid bilayer.

The properties of the M protein are clearly different from those of the glycoproteins. Unlike the glycoproteins, however, it cannot be extracted with Triton X-100, and this is probably due to strong interaction with the ribonucleoprotein (Cartwright et al., 1970; György et al., 1971; Scheid et al., 1972; Scheid and Choppin, 1973). If the salt concentration is raised from 0.5 to 1 M KCl the M protein of SV 5 and Newcastle disease virus can be extracted from the virion, leaving the ribonucleoprotein intact. The solubilized M protein will, however, precipitate immediately when the salt concentration is reduced, even in the presence of Triton X-100 (Scheid and Choppin, 1973). The action of the high salt may be to dissociate the protein-protein interactions between the M protein and the ribonucleoprotein. To date, the amphilic nature of the M protein has not been demonstrated unequivocally (Gregoriades and Hirst, 1975).

3.2.3. Envelope lipids

The envelope of myxo-, paramyxo- and rhabdoviruses consists of phospholipids, cholesterol, and glycolipids (Kates et al., 1961, 1962; Blough and Lawson, 1968; Klenk and Choppin, 1969b; Blough and Merlie, 1970; McSharry and Wagner, 1971). The lipid composition of these viruses closely reflects that of their host cell plasma membrane, as demonstrated elegantly for paramyxoviruses (Klenk and Choppin, 1970a,b; see also reviews by Lenard and Compans, 1974; Choppin and

Compans, 1975; Compans and Choppin, 1975). The envelope lipids are arranged in a bilayer, as shown by electron spin resonance (Landsberger et al., 1971, 1973), fluorescent probe studies (Lenard et al., 1975), and small angle x-ray scattering. Cholesterol appears to be evenly distributed between the two leaflets of the bilayer (Lenard and Rothman, 1976).

3.3. Summary

The SF virus envelope shows several features characteristic of cellular membranes: (1) the lipids of the virus are organized in a bilayer; (2) the surface proteins are glycoproteins susceptible to proteolytic enzymes and to reagents which specifically label surface proteins; (3) two of the proteins, E-1 and E-2, are embedded in the lipid membrane by a hydrophobic segment that spans the whole bilayer to make contact with the nucleocapsid protein; and (4) the protein-lipid interaction in the membrane is not specific and allows the lipid composition of the membrane to vary greatly.

The envelope proteins of rhabdo-, paramyxo- and myxoviruses consist of one, two, and three glycoproteins, respectively and are located on the virion surface. In addition to these glycoproteins all of these viruses have a matrix protein (M protein) located on the inner surface of the lipid bilayer. The glycoproteins show amphipathic features and are probably associated with lipids by a hydrophobic fragment. The M protein is not released by neutral detergents unless high ionic strength is used, and it precipitates in the presence of detergent if the ionic strength is lowered.

4. Synthesis of envelope proteins

4.1. Synthesis of alphavirus proteins

The growth cycle of most alphaviruses is relatively short, about 8 to 10 hours. After adsorption of the virus there is a latent period of about 2 hours followed by an exponential increase in virus synthesis for 3 to 4 hours, and then a linear growth phase of 2 to 3 hours. During the exponential phase the release of virus takes about 1 minute after it has become infectious. At the end of the cycle each cell has produced and released several hundred or even several thousand, infectious units of virus into the medium. The fraction of virus-producing cells increases toward the end of the growth cycle, as does the amount of virus produced per cell (Dulbecco and Vogt, 1954; Rubin et al., 1955; Pfefferkorn and Shapiro, 1974; Kääriäinen et al., 1975; Strauss and Strauss, 1976). Since the particle-to-infectivity ratio varies from 5 to 20, the number of virus particles released into the medium can be as high as 20,000 per cell (Tuomi et al., 1975), representing about 7 to 10% of the phospholipid mass of the host cell plasma membrane. It should be noted, however, that the amount of viral protein and RNA synthesized in the infected cells is about ten times higher than that excreted as virus (Tuomi et al., 1975).

The coding capacity of the alphavirus genome is close to 450,000 daltons of protein (Levin and Friedman, 1971; Arif and Faulkner, 1972; Simmons and Strauss, 1972a). The four structural proteins C, E-1, E-2, and E-3 account for about 130,000 daltons, whereas most of the remainder is reserved for nonstructural proteins.

The 42 S RNA genome is of positive polarity, that is, it is infectious in protein-free form (Sonnabend et al., 1967) and can direct protein synthesis in vitro (Simmons and Strauss, 1974b; Smith et al., 1974; Glanville et al., 1976a). In infected cells another major messenger RNA, the 26 S RNA (mol. wt. 1.6×10^6) is synthesized (Sonnabend et al., 1967; Kääriäinen and Gomatos, 1969; Söderlund et al., 1973; Simmons and Strauss, 1974a; Tuomi et al., 1975); this is identical to about one-third of the 42 S RNA from the 3' terminal end (Simmons and Strauss, 1972b; S. I. T. Kennedy, personal communication). Translation of these two RNAs in cell-free protein synthesizing systems has shown that 26 S RNA is the messenger for the structural proteins, while the 42 S RNA is responsible for the translation of nonstructural proteins. (Simmons and Strauss, 1974b; Wengler et al., 1974; Clegg and Kennedy, 1975b; Glanville et al., 1976a,b).

Host cell protein synthesis declines rapidly in infected cells about 3 hours postinfection and is replaced by the synthesis of viral structural proteins. These are the capsid protein, envelope protein E-1, and a 62,000 dalton protein (p-62 or PE_2), which is the immediate precursor of envelope proteins E-2 and E-3 (Schlesinger and Schlesinger, 1973; Simons et al., 1973b). Several independent lines of evidence suggest that the virus structural proteins are translated as a polyprotein with a molecular weight of about 130,000: (1) only one ^{35}S-formyl-methionyl-tRNA labeled peptide can be obtained from the in vitro trypsin- (or pronase) digested product translated from the 26 S RNA in vitro (Burke, 1975; Glegg and Kennedy, 1975a, Glanville et al., 1976b); (2) temperature-sensitive mutants have been isolated, that accumulate a 130,000 dalton protein containing the amino acid sequences of the structural proteins (Schlesinger and Schlesinger, 1973; Keränen and Kääriäinen, 1974, 1975; Lachmi et al., 1975); (3) the structural proteins are translated sequentially in the order of capsid-p-62-E-1 in infected cells, where initiation of protein synthesis has been synchronized by high salt treatment (Clegg, 1975; Lachmi and Kääriäinen, 1976); and (4) tryptic peptide mapping of cleavage products containing capsid and p-62 (=p-86) and p-62 plus E-1 (=p-97) have confirmed the existence of polyprotein precursors (Lachmi et al., 1975).

Once synthesized, the polyprotein is quickly cleaved. Cleavage of the capsid protein from the polyprotein takes place rapidly after the translation of this protein has been completed (Clegg, 1975; Söderlund, 1976), whereas the cleavage between p-62 and E-1 occurs when the translation of E-1 is completed (Söderlund, 1976). Cleavage of capsid protein also takes place in cell-free protein synthesizing systems, even under conditions when the whole polyprotein is not translated (Cancedda and Schlesinger, 1974; Clegg and Kennedy, 1975b; Glanville et al., 1976a,b). In contrast to infected cells, the envelope proteins are not cleaved in vitro even when the complete polyprotein is translated (Simmons and Strauss, 1974b), which leads to the accumulation of capsid and the 97,000 dalton envelope precursor protein.

The synthesis of nonstructural proteins can barely be detected later than 5 hours postinfection, indicating an efficient translational control (Kääriäinen et al., 1976). Using a temperature-sensitive mutant of SF virus, ts-1, which is not able to shut off the synthesis of nonstructural proteins, we have recently established the mode of translation of these proteins (Keränen and Kääriäinen, 1974, 1975; Lachmi and Kääriäinen, 1976). Four nonstructural proteins with apparent molecular weights of 70,000 (ns-70), 86,000 (ns-86), 72,000 (ns-72), and 60,000 (ns-60) are synthesized sequentially after synchronizing the initiation of protein synthesis. Two short-lived precursors of 155,000 (A) and 135,000 (B) daltons can also be detected in infected cells. These results strongly suggest that the nonstructural proteins are synthesized as a polyprotein, which should have a molecular weight of approximately 300,000. Support for this view has been obtained from in vitro translation of 42 S RNA in the presence of ^{35}S-formyl-methionyl tRNA (Glanville et al., 1976b). Only one tryptic (or pronase) peptide labeled with ^{35}S could be isolated from the 42 S RNA product, and this differed from that detected in the 26 S RNA-directed product. The relationship between SF virus protein synthesis, genome size and messenger RNA is presented in Table 8 and Fig. 4.

4.2. Synthesis of myxovirus proteins

The genome of myxoviruses consists of at least seven different RNA fragments, ranging in size from 3.4 to 9.8 × 10^5 daltons (Table 8). The fragments are of negative polarity and must first be transcribed into positive strands in the host cell. This is accomplished primarily by virion-associated RNA polymerase (Chow and Simpson, 1971). Soon after infection the full genetic capacity of the virus is represented by poly-(A) containing messenger RNAs isolated in polysomes (Etkind and Krug, 1975), and these are translated to yield all of the virus-specific proteins found (Etkind and Krug, 1974). Polypeptides representing the entire coding capacity of the influenza virus genome have been identified in the infected cells (Lazarowitz et al., 1971; Skehel, 1972; Krug and Etkind, 1973). The only polypeptide that appears to be translated as a large precursor is the HA

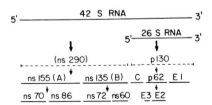

Fig. 4. Schematic presentation of the mode of translation of Semliki Forest virus 42 S RNA-coded proteins in infected cells. ns290 is hypothetical and has not yet been detected.

Table 8
Relationship Between Genome, Messenger RNA, and the Proteins Coded by Some Enveloped Animal Viruses

Virus group	Genome		Messenger RNA		Proteins		Mode of translation
	S-value (mol. wt.)	Polarity	S-value	Mol. Wt. × 10^-6	Code	Mol. Wt. × 10^-6	
Alphaviruses[a]	42 S (4-4.5 × 10^6)	Positive	42 S	4-4.5	ns-70	70	Nonstructural polyprotein
					ns-86	86	
					ns-72	72	
					ns-60	60	
SF virus			26 S	1.6	C	33	Structural polyprotein
					E-3	10*	
					E-2	52*	
					E-1	49*	
Myxoviruses[b]	17-20 S fragmented (total mol. wt. 4×10^6)	Negative	17-20 S	0.98	P₁	94	Final product
Influenza A				0.93	P₂	81	Final product
				0.82	HA	75*	Precursor
				0.70	NP	60	Final product
				0.58	NA	55*	Final product
				0.39	M	26	Final product
				0.34	NS	25	Final product
Rhabdoviruses[c]	42 S (4.4×10^6)	Negative	28 S	2.4	L	190	Final product
			12-16 S	0.7	G	69*	Final product
				0.55	N	50	Final product
				0.28	NS	45	Final product
				0.28	M	29	Final product
Paramyxoviruses[d]	50 S (5 × 10^6)	Negative	33 S	2.0	L	200	Final product
			18-20 S		P₂	79	Final product
					HN	67*	Final product
					F₀	60*	Precursor
					NP	56	Final product
					47K	47	Final product
					M	41	Final product

[a] Pfefferkorn and Shapiro, 1974; Kääriäinen et al., 1975; Lachmi and Kääriäinen, 1976; Strauss and Strauss, 1976.
[b] Bishop et al., 1971; Skehel, 1971; Compans and Choppin, 1975.
[c] Morrison et al., 1974; Knipe et al., 1975; Wagner, 1975.
[d] Collins and Bratt, 1973; Kolakofsky et al., 1974, 1975; Hightower et al., 1975; Lamb et al., 1975.
*glycoprotein.

protein. It is later cleaved to HA $_1$ and HA $_2$ (Compans and Choppin, 1975; and section 3.2.1.). The hemagglutinin (HA), neuraminidase (NA) and matrix protein (M) are found in infected cells in association with membrane structures, whereas the nucleocapsid (NP) protein is free in the cytoplasm (Compans, 1973a; Hay, 1974). The free and membrane-bound polysomes from fowl plague virus-infected cells contain the same classes of mRNA, suggesting a transient association of messenger RNA with the membranes (Glass et al., 1975).

All the virus-specific proteins can be detected at various times of infection but a considerable amount of regulation in their translation has been shown to exist. For example, there is a progressive increase in the rate of synthesis of the M protein later in the infection (Skehel, 1972; Meier-Ewert and Compans, 1974; Hay and Skehel, 1975; Klenk et al., 1975b).

4.3. Synthesis of rhabdovirus proteins

The vesicular stomatitis virus, 42 S RNA, is transcribed in infected cells by a virion-associated RNA transcriptase into several messenger RNA molecules, sedimenting at 28 S and 12 to 16 S (Baltimore et al., 1970; Huang, 1975). The 28 S RNA is responsible for the synthesis of the virion-associated transcriptase, the L protein (Table 8), as demonstrated by Morrison et al. (1974). The 12 to 16 S mRNAs have been separated by polyacrylamide gel electrophoresis into five different species, all of which have been translated in cell-free protein synthesizing systems (Knipe et al., 1975). Thus, it has been possible to show that each of the virus-coded proteins has its own messenger (Both et al., 1975a,b; Morrison et al., 1975). In a recent study by Morrison and Lodish (1975), membrane-bound and free polysomes from infected cells were analyzed for their virus-specific messenger RNAs. The messenger for the vesicular stomatitis virus G protein was found to be exclusively membrane-bound, whereas the mRNAs for the N and NS proteins were associated only with free polysomes. Messenger RNA for the M protein was both membrane-bound and free (see also Grubman et al., 1974, 1975).

4.4. Synthesis of paramyxovirus proteins

Recent findings indicate that there may be a striking similarity in the transcription of the paramyxovirus and rhabdovirus genomes. About 40% of the negative strand of Sendai virus and Newcastle disease virus 50 S genome RNA is transcribed into a 33 S mRNA and the rest is transcribed into several species sedimenting at about 18 S (Collins and Bratt, 1973; Bratt et al., 1975; Roux and Kolakovski, 1975) (Table 8). The largest 33 S RNA presumably codes for the 200,000 dalton L protein (Hightower and Bratt, 1974, 1975). Probably, the other proteins found in infected cells all have their own messengers since no posttranslational cleavage seems to occur in paramyxovirus-infected cells (Hightower et al., 1975; Zaides et al., 1975; Lamb et al., 1976) except for the cleavage of the F $_0$ protein into the active F fusion protein (review, Scheid and Choppin, 1975; and section 3.2.1.).

4.5. Summary

Translation of alphavirus-specific proteins occurs via two mRNAs, the 42 S RNA genome and the 26 S RNA, which is a partial copy of the genome found only in the infected cells. The structural proteins are synthesized as a 130,000 dalton polyprotein from the 26 S RNA in the following order: C, E-3, E-2, and E-1. The capsid protein is cleaved in the polysomes, whereas cleavage of E-1 takes place when the translation of the polyprotein has been completed. The four nonstructural proteins are translated from the 42 S RNA as a 290,000 dalton polyprotein from which the individual nonstructural proteins are rapidly cleaved.

Myxoviruses have a divided genome consisting of at least seven unique RNA pieces of negative polarity. These RNAs are transcribed by a virion-associated transcriptase to yield the corresponding positive strands, which are translated into viral structural proteins and one or two nonstructural proteins in the cytoplasm. The viral hemagglutinins (both glycoproteins) are the only proteins translated as a polyprotein; all the other proteins are translated in their final size.

The continuous, single-stranded linear genome of paramyxo- and rhabdoviruses is transcribed by a virion-associated transcriptase to yield one large and several small mRNAs. The large mRNA is the template for the virion transcriptase, whereas the small RNAs are messengers for individual proteins. The cell fusion glycoprotein F of paramyxoviruses is probably the only protein formed from a larger precursor (F_0). In vesicular stomatitis virus-infected cells the surface glycoprotein G is translated exclusively on membrane-bound polyribosomes, whereas the nonglycosylated "membrane protein" M is translated only partly, if at all on these polyribosomes.

5. Glycosylation of viral glycoproteins

5.1. Stepwise glycosylation of secretory and membrane glycoproteins

Our current understanding of glycosylation is based mainly on studies done on the secretory cells of pancreas and liver that have recently been reviewed by Molnar (1975) and Palade (1975) and by Morré (Chapter 1) and Cook (Chapter 2) in this volume.

The proteins to be secreted are synthesized on membrane-bound polysomes on the cytoplasmic side of the rough endoplasmic reticulum. Blobel and Dobberstein (1975a,b) have recently shown that the synthesis of immunoglobulin light chain probably starts on free ribosomes which quickly attach to the endoplasmic reticulum membrane. Blobel and Sabatini (1971) have advanced a "signal hypothesis" for translation of all secretory proteins, that must be on the cisternal side of the endoplasmic reticulum. Their hypothesis states that each messenger for secretory protein has in its 5′ terminal end a specific signal sequence for about 10 to 40 amino acids. The N-terminus of the nascent secretory protein attaches to the endoplasmic reticulum membrane via specific bonds to proteins in the membrane. These proteins then assemble to create a tunnel through which the growing polypeptide chain can be excreted into the cisternal space of the endoplasmic reticulum. Specific interaction between the membrane proteins and the large ribosomal subunit are also postulated. The "signal piece" of the immunoglobulin light chain with a length of about 40 amino acids is soon cleaved from the growing polypeptide chain by membrane-associated proteases. Blobel and Dobberstein (1975b) have shown convincingly that the product synthesized in association with the membranes is protected from added proteases as it is secreted across the membrane. Glycosylation of the secreted liver glycoproteins may begin before completion of the protein translation (Molnar, 1975). Addition of the first N-acetylglucosamine does not occur on the cytoplasmic side of the membrane (Molnar, 1975; Hughes, 1976). Instead, the growing polypeptide chain protruding into the cisternal space of the rough endoplasmic reticulum is glycosylated immediately when the available recognition sequences for glycosyl transferases reach the enzyme. The addition of the proximal glucosamine and mannose residues is rapid (Molnar, 1975; Weitzman and Scharff, 1976), and it has been suggested that the oligosaccharide core consisting of glucosamine and mannose could be donated as a preformed unit by a dolichol-oligosaccharide intermediate (Richards and Hamming, 1972; Hsu et al., 1974; Tkacz and Lampen, 1975; Hughes, 1976). Addition of galactose, fucose, and sialic acid occurs subsequently in the Golgi complex in a stepwise manner (Schachter et al., 1970; Whaley et al., 1972; Wagner et al., 1973; Autuori et al., 1975a,b; Elhammer et al., 1975; Molnar, 1975; Palade, 1975).

The same basic events of stepwise glycosylation have been suggested for membrane glycoproteins by Bosmann (1969) and by several other investigators (Whaley et al., 1972; Autuori et al., 1975a,b; Bretscher and Raff, 1975; Elhammer et al., 1975; Hughes, 1976). The only major difference would be that membrane glycoproteins, which are amphiphilic (Singer and Nicolson, 1972; Bretscher, 1975), remain attached to the membranes of the endoplasmic reticulum on their way to the Golgi complex from the rough endoplasmic reticulum (Hirano et al., 1972; Bretscher and Raff, 1975). The secretory proteins are taken into small vacuoles, the membrane of which is apparently derived from the Golgi membranes (Palade, 1959; Whaley et al., 1972; Palade, 1975; and Morré, chapter 1 this volume), and these migrate to the plasma membrane where they fuse with the membrane and release their contents to the outside of the cell. Several modifications in this process, such as storage vacuoles and fusion with lysosomes have been described by Caro and Palade (1964) and are reviewed by Palade (1975), Quinn (1976) and Hughes (1976).

Similar transport of membrane glycoproteins from the Golgi to the plasma membrane has been proposed by Hirano et al. (1972). The vesicles, in which the glycoproteins are attached to the membrane with their carbohydrate residues oriented toward the inside of the vesicle, fuse with the plasma membrane to bring the membrane glycoproteins to the outer (noncytoplasmic) surface of the plasma membrane. On the other hand, recent reports of Autuori et al. (1975a) and Elhammer et al. (1975) have demonstrated that some glycoproteins of the endoplasmic reticulum in rat liver cells receive their terminal sialic acid residues in the Golgi complex and are released from the Golgi membranes in nonvesicle form, possibly associated with lipids. These lipoglycoprotein "complexes" are then incorporated directly into the endoplasmic reticulum membrane. These authors have interpreted their results as suggesting that the endoplasmic reticulum glycoproteins are probably located on the cytoplasmic side of the membranes during their transport from the rough endoplasmic reticulum to the Golgi complex. This also implies that glycosylation occurs on the cytoplasmic side of the endoplasmic reticulum membranes utilizing sugar transferases different from those for the secreted glycoproteins, as suggested previously by Eylar (1965). This mode of processing membrane proteins is basically different from the "secretion type" model. It is, however, amenable to investigation using viruses containing only one or a few glycoproteins.

5.2. Alphaviruses as models for glycosylation of membrane proteins

5.2.1. Correlation between cleavage and glycosylation of the structural polyprotein

As discussed earlier, alphavirus glycoproteins are translated as a 130,000 dalton polyprotein in which the nonglycosylated capsid protein is N-terminal. If we suppose that the "secretory model" for processing the viral glycoproteins is correct, several interesting questions arise. Is the signal sequence here also at the 5' terminus of the 26 S RNA, that is, at the N-terminus of the capsid protein? The fact that nucleocapsid is found in the cytoplasmic side of the endoplasmic reticulum membranes (Acheson and Tamm, 1967; Erlandsson et al., 1967; Grimley et al., 1968) would indicate that the "signal sequence" cannot be at the N-terminus of the structural polyprotein. It could be part of the N-terminal end of the 97,000 dalton envelope precursor protein, which would then be a modification of the hypothesis. One would expect that when the cleavage of the capsid protein does not occur the 97,000 dalton protein is not excreted into the cisternal space and remains unglycosylated. The glycosylation of the 130,000 dalton precursor protein has been studied by Sefton and Burge (1973), using the ts-5 temperature-sensitive mutant of Sindbis virus which induces the synthesis of this precursor protein at 41°C. Labeling was carried out with glucosamine, mannose, galactose, and fucose and none of these were incorporated into the 130,000 dalton protein. Sefton and Burge also studied the glycosylation of the 97,000 dalton envelope precursor protein, which accumulates in BHK cells infected with wild-type Sindbis virus (Strauss et al., 1969) and has been designated as B protein. Again, no evidence of glycosylation of this protein was obtained. Recently, Duda and Schlesinger (1975) have suggested that the B protein is rapidly degraded and cannot be used for production of the envelope proteins. Possibly, the sites for the attachment of the signal sequence are rate-limiting in the transport of the glycoproteins. If the ribosome cannot secrete its product into the cisternal space it could be extruded into the cytoplasm and degraded.

Work with glycosylation inhibitors has revealed some interesting correlations between glycosylation and cleavage of SF virus-induced glycoproteins. 2-deoxyglucose, which is perhaps incorporated instead of mannose, is a potent inhibitor of SF virus multiplication (Kaluza et al., 1972, 1973; Scholtissek and Kaluza, 1975). Protein synthesis in infected cells is not inhibited in the presence of 2-deoxyglucose, but glycosylation of SF virus envelope proteins is severely impaired (Kaluza, 1975). Under these conditions, "new proteins" with apparent molecular weights of 94,000, 55,000, and 45,000 appear. The authors assume these to be the nonglycosylated precursors (or "apoproteins") of the 97,000 (p-97), 62,000 (p-62), and 50,000 (E-1 and possibly E-2) dalton proteins, respectively (Kaluza, 1975; Scholtissek and Kaluza, 1975). This would be analogous to results obtained with fowl plague virus (Klenk et al., 1972b). Since these proteins incorporate small amounts of radioactive glucosamine, mannose, and even fucose, they are probably located in the cisternal space of the endoplasmic reticulum if we accept the secretory model of membrane glycoprotein processing. The accumulation of the 94,000 dalton "apoprotein" may reflect the importance of glycosylation for the cleavage of the envelope proteins, as suggested by Schwarz and Klenk (1974) for fowl plague glycoproteins. The apoproteins cannot be converted to the real glycoprotein even if chased in the presence of mannose, indicating that deoxyglucose incorporation in the proximal end of the carbohydrate chain inhibits the further elongation of the chain (Kaluza, 1975). Fucose could be incorporated into the apoproteins, suggesting that it can attach to the proximal N-acetyl-glucosamine residues. It would therefore be interesting to compare the electrophoretic mobilities and the tryptic peptides of the 94,000 dalton apoprotein with the in vitro-produced p-97 and also with the nonglycosylated B protein from Sindbis virus-infected BHK cells (Sefton and Burge, 1973; Simmons and Strauss, 1974b).

5.2.2. Glycosylation of the precursor of E-2 and E-3, the p-62 (PE$_2$)

The precursor-product relationship between the 62,000 dalton glycoprotein p-62 (NSP 68, NVP 62, PE$_2$) and the envelope proteins E-2 and E-3 has been established convincingly by tryptic peptide mapping (see p. 000). In SF virus-infected cells the p-62 glycoprotein, when fully glycosylated, should contain one A-type oligosaccharide chain of E-3, 2 to 3 B-type chains of E-2 and the galactose-rich X-chain of E-2 (p. 759). The carbohydrate composition of p-62 from SF virus-infected cells has been analyzed recently after 7 hours labeling with tritiated sugars (Stenvall et al., to be published). The p-62 was subjected to thorough digestion with pronase followed by gel filtration (Mattila et al., 1976). Preliminary results indicate that the E-2 X-chain has attained its full size in p-62. All glucosamine and mannose label appears in a sharp peak with a molecular weight of about 2,000. This glycopeptide apparently does not contain galactose or fucose label, indicating that the E-3 A-chain is not complete in the p-62. The

2,000 dalton glycopeptide is of the same size as the E-2 B chains. Also, the behavior of this material on treatment with mannosidases suggests that E-2 B-oligosaccharides may be present; it is also possible that a part of the material may represent the glucosamine-mannose core of the E-3 A-type chain. If the 2,000 dalton glycopeptide actually contains the A-chain it implies that cleavage of p-62 occurs before the protein reaches the galactose, sialic acid, and fucose transferases located in the Golgi complex (Whaley et al., 1972; Wagner et al., 1973). The presence of galactose in the E-2 X-chain attached to the p-62 raises the interesting possibility that there may be two types of transferases for galactose: one would be responsible for the addition of galactose to the E-2 X-chain and the other for the terminal galactose residues of the E-3 A-chain. Another possible explanation for this result is that there is a steric hindrance that inhibits the completion of the A-type chain in the precursor molecule (Renkonen et al., 1977).

Sefton and Burge (1973) have investigated the labeling ratios of amino acids to different sugars in Sindbis virus-infected cells to determine what happens to the sugars when PE_2 is converted to E-2. It should be noted that since E-3 has not been reported in Sindbis virus-infected cells the results with this virus are not directly comparable to those reported above for SF virus. Another difference is the carbohydrate composition of E-1 and E-2 in the two viruses. Sindbis has apparently one A- and one B-type oligosaccharide chain in both glycoproteins, as discussed earlier. The conversion of PE_2 to E-2 seems to involve removal of mannose and addition of galactose and fucose, compatible with the idea that PE_2 has not reached the Golgi complex. The decrease in the relative amount of mannose may indicate the removal of B-unit or a mannose-rich core of the A-type chain. Another possibility is that the specific mannose radioactivity of p-62 is higher than that of E-2 because the former is the precursor of the latter.

Glycosylation inhibition of Sindbis virus glycoproteins with 20 mM glucosamine hinders the conversion of PE_2 to E-2, suggesting that proper glycosylation is a prerequisite for the cleavage (Duda and Schlesinger, 1975). Addition of canavanine 3 hours after infection with SF virus also results in an impairment of glycosylation of p-62 and E-1, and the conversion of p-62 to E-3 is inhibited completely (Ranki, 1972; Ranki et al., 1972). Since partial glycosylation of p-62 occurs even when the protein contains canavanine, the protein is probably secreted into the cisternal space of the endoplasmic reticulum.

Our inability to find apoproteins of p-62 or E-1 such as those described by Kaluza (1975) and Scholtissek and Kaluza (1975), even after short pulses, suggests that the glycosylation of the newly synthesized proteins must be extremely rapid. The possibility that the whole core of A-type oligosaccharide would be added as a unit, perhaps after the addition of the first glucosamine residue, could explain the rapidity of the glycosylation.

We have attempted to isolate lipid soluble, glucosamine-labeled derivatives from SF virus-infected cells, but were unable to identify any labeled polyisoprenyl sugar pyrophosphates in our system (Somerharju et al., unpublished).

768

5.3. Glycosylation of envelope proteins of other viruses

Glycosylation of influenza virus envelope proteins has been investigated by fractionation of infected cells after labeling with different radioactive sugars (Compans, 1973b; Hay, 1974; Hay and Skehel, 1975) as well as by inhibiting glycosylation using glucosamine or 2-deoxyglucose (Klenk et al., 1972b; Schwarz and Klenk, 1974; Scholtissek et al., 1975a). As discussed earlier, the carbohydrate chain composition of the influenza virus hemagglutinin polypeptides HA_1, HA_2, and the neuraminidase are unknown so far. Glucosamine, mannose, galactose, and fucose (but not sialic acid) are found in the influenza virions (Table 5), indicating that A-type oligosaccharide chains may be present (Fig. 3).

Support for the secretory model of membrane glycoprotein processing is obtained from cell fractionation studies. The HA protein, which is the precursor of HA_1 and HA_2 has been found regularly in the rough endoplasmic reticulum (Compans, 1973; Hay, 1974; Schwarz and Klenk, 1974). The HA protein can be labeled at this site with glucosamine but not with fucose, whereas in the smooth endoplasmic reticulum it is also labeled with fucose. The relative amounts of fucose and galactose in the HA protein are smaller than in the cleavage products (Schwarz and Klenk, 1974) found in the plasma membrane (Hay, 1974).

In the presence of glucosamine or 2-deoxyglucose, a poorly glycosylated protein with a molecular weight of about 64,000 can be detected in the fowl plague virus-infected cells (Schwarz and Klenk, 1974). This is assumed to be the non-glycosylated precursor or apoprotein of the HA protein and has been designated as HA_0. Apparently the HA_0 can give rise to two poorly glycosylated polypeptides HA_{01} and HA_{02} (Klenk et al., 1974; Klenk et al., 1975b), and cleavage seems to occur in the smooth endoplasmic reticulum. The conversion of HA_0 to HA_{01} and HA_{02}, as well as the conversion of HA to HA_1 and HA_2, is blocked by inhibitors of serine proteases (Schwarz and Klenk, 1974). Again, the results indicate the rapid glycosylation of the hemagglutinin.

Glycosylation of the vesicular stomatitis G protein, which probably contains two A-type oligosaccharide chains, may also be a stepwise process in which galactose, fucose, and sialic acid are added as terminal sugars (Lafay, 1974; Atkinson et al., 1976). Vesicular stomatitis and Newcastle disease viruses respond differently to 2-deoxyglucose present in glucose-containing medium compared with the alpha- and myxoviruses (Scholtissek et al., 1975a,b). Under conditions were SF and fowl plague virus growth is inhibited, multiplication of the rhabdo- and paramyxoviruses is almost unaffected. It should be mentioned Newcastle disease virus is extremely sensitive to the drug tunicamycin, which appears to be an inhibitor of the polyisoprenyl sugar pyrophosphate-mediated glycosylation (Tkacz and Lampen, 1975).

5.4. Summary

The synthesis of the glycoproteins of enveloped viruses offers a defined model for the study of glycosylation of membrane proteins. However, before full advantage can be taken of these models, the structure of the oligosaccharide chains will have to be determined. The studies done to date suggest that glycosylation, or at least part of it, is a stepwise process. Proximal sugars are rapidly attached in the rough endoplasmic reticulum, and the distal sugars added later in the smooth membranes. Some degree of glycosylation may be necessary for the proper cleavage of those glycoproteins made from a precursor. Work with drugs that impair glycosylation suggests that the transport of the protein into the cisternal space of the endoplasmic reticulum where they are accessible to glycosyltransferases takes place even if glycosylation does not occur.

6. Assembly of enveloped viruses

The assembly of enveloped viruses involves several steps: (1) the insertion of the evelope glycoproteins into the membrane which, in this review, is the host cell plasma membrane; (2) attachment of the inner membrane protein, the matrix (or M protein), to the cytoplasmic side of the plasma membrane; (3) assembly and transport of the viral nucleocapsid to the plasma membrane; (4) association of the nucleocapsid with the viral membrane components; and (5) budding of the nucleocapsid from the plasma (or cytoplasmic) membrane.

6.1. Alphaviruses

6.1.1. Appearance of envelope proteins
on the cell surface
The distribution of Sindbis virus envelope proteins on the cell surface has been studied elegantly by Birdwell and Strauss (1974a; Birdwell et al., 1973). They used the surface replica technique to localize envelope proteins, based on the distribution of hemocyanin-conjugated anti-virus antibodies. Cells fixed after the antibody treatment showed a clear aggregation of envelope proteins or "patch formation," whereas a "random" dispersed distribution of hemocyanin molecules was observed in cells fixed prior to antibody treatment. Two hours after infection, before virus release had started, envelope proteins were randomly distributed. Thereafter the amount of antigen increased rapidly but the distribution remained even. Labeling the cells at 37°C without prior fixation yielded clusters, whereas labeling at 4°C gave an even distribution similar to that in cells fixed before antibody labeling. The authors interpreted these results as indicating that the envelope proteins can diffuse laterally in the plane of the plasma membrane.

6.1.2. Kinetics of labeling of alphavirus proteins
in the released virus
Scheele and Pfefferkorn (1969) have studied the radiolabeling of proteins in released Sindbis virus. Infected cells were pulsed for 30 minutes and released virus harvested after different chase periods. About 90% of the radioactivity

incorporated into the viral proteins was never released from the cells. The protein label released in mature virus reached a maximum incorporation about 60 minutes after the end of the pulse, and declined rapidly thereafter. The capsid-to-envelope protein ratio in the earliest virions collected was higher than that in the virus released later, indicating that capsid protein was assembled and released more rapidly than the simultaneously synthesized envelope proteins. This result suggests that different pathways are involved in the transport of the nucleocapsid and envelope proteins to the plasma membrane. At the time of this study, the two envelope proteins of Sindbis virus could not be separated, so it was not determined whether E-1 and E-2 were transported together. The answer to this question came in a later study by Schlesinger and Schlesinger (1972), who showed that in the released virus both E-1 and E-2 are labeled with similar kinetics. This may mean that the envelope proteins are processed and transported together from the moment they are released from the ribosomes.

Whether p-62 (PE_2) is cleaved at the plasma membrane or earlier has not been definitely established. Iodination of infected cells in the presence of lactoperoxidase labels only E-1 but not E-2 or PE_2 (Sefton et al., 1973), whereas both E-1 and E-2 are almost equally accessible to iodination in the purified virus. This may mean that neither of these proteins is present in the plasma membrane before they are incorporated into the virion. Jones et al. (1974), using a temperature-sensitive mutant of Sindbis virus, have demonstrated the presence of PE_2 in cell fractions showing enrichment of 5′ nucleotidase activity, which is considered a plasma membrane marker. In our opinion it would be difficult to understand how E-2 could be transported to the plasma membrane independently of E-1, because as described above both proteins appear with the same kinetics in the released virus. One possible explanation for the lack of iodination of E-2 or PE_2 is to assume that their tyrosine residues are inaccessible to the lactoperoxidase. It would be interesting to repeat the experiments of Birdwell and Strauss (1974a) using specific antisera against E-1 and E-2 and perhaps also against E-3 of SF virus in conjunction with hemocyanin-conjugated anti-IgG.

The importance of the cleavage of p-62 (PE_2) for the maturation of alphaviruses has been stressed by many investigators (Jones et al., 1974; Duda and Schlesinger, 1975; Keränen and Kääriäinen, 1975; also see review by Strauss and Strauss, 1976). The released virus does not contain any significant amounts of p-62. Clearly this differs from the situation of the myxo- and paramyxoviruses, where uncleaved hemagglutinin or fusion protein is released in the form of noninfectious virus (p. 756).

6.1.3. The origin of alphavirus lipids

As discussed above, the phospho- and glycolipids of SF and Sindbis virus closely resemble those of the host cell plasma membrane (p. 754). Pfefferkorn and Hunter (1963b) demonstrated that a part of Sindbis virus phospholipids are synthesized before infection, whereas all the proteins and RNA are made during infection (Pfefferkorn and Clifford, 1964). The same is true of the lipids of myxo- (Kates et al., 1962), paramyxo- (Klenk and Choppin, 1969b), and rhabdoviruses (Schlesinger et al., 1973). When the cells are labeled with radioactive phosphorus, glycerol, or fatty acids during infection, radioactive lipids are incorporated into the released virions (Renkonen et al., 1974; and see review by Blough and Tiffany, 1973). The distribution of radioactive lipids in the virus reflects the different mode of synthesis of the individual phospholiopids as well as their mode of transport to the membrane from which the virus buds (Renkonen et al., 1972, 1974). Even in small virus samples the mass of individual phospholipids can be conveniently determined using equilibrium labeling with ^{32}P, as described by Renkonen et al. (1972c, 1974). By growing BHK cells in the presence of ^{32}P for two passages, a known constant specific activity of all phosphorus-containing components (Tuomi et al., 1975), including phospholipids, can be obtained (Renkonen et al., 1972c). When these cells were labeled with radioactive glycerol or serine at different times during infection with SF virus, it was possible to determine the specific activities of these precursors in different phospholipids in both the cells and the released virus (Renkonen et al., 1974). The presence of glycerol label in phosphatidylcholine, as well as the presence of serine label in the sphingomyelin, represents de novo synthesis of these compounds.

An interesting result was obtained when the sphingosine of sphingomyelin and phosphatidylserine was labeled with ^{14}C-serine 2 to 4 hours after infection and then chased with cold serine. The specific activities of labeled serine in sphingomyelin and phosphatidylserine in released virus (collected 10 hours post-infection) were higher than in the cells at any time after the pulse. One explanation for this finding could be that these lipids were transported with the envelope proteins from the endoplasmic reticulum, where they are synthesized (Wallach, 1975; Quinn, 1976), to the plasma membrane. This would imply some selective interaction between these lipids and the envelope proteins. Such "lipid flow" from the endoplasmic reticulum through Golgi complex to the plasma membrane has been suggested previously (Whaley et al., 1972; Hughes, 1976; Quinn, 1976).

6.1.4. Assembly of SF virus nucleocapsid
The kinetics of synthesis of the SF virus nucleocapsid has been studied by Söderlund (1973), who followed the incorporation of ^{35}S-methionine into the cytoplasmic nucleocapsid, after a short pulse and different chase periods. The newly labeled capsid protein reaches its maximum in the 140 S nucleocapsid afer a 5- to 7-minute chase. At this time about 70% of the capsid protein synthesized during the pulse is found in the nucleocapsids. The capsid protein is not found in the soluble protein fraction but is associated with fairly large, sedimenting structures (Söderlund, 1973; Söderlund and Kääriäinen, 1974).

We have found recently that the capsid protein attaches very rapidly to the large 60 S ribosomal subunit (Ulmanen et al., 1976). This binding of capsid protein also occurs in a cell-free protein-synthesizing system derived from wheat germ, and programmed with 26 S RNA, the messenger for viral structural proteins (Glanville and Ulmanen, 1976). Preliminary pulse-chase experiments carried out in SF virus-infected HeLa cells suggest that the ribosome-bound capsid protein is transferred to the cytoplasmic 140 S nucleocapsid (Kääriäinen et al., 1975). Details of how the 60 S capsid protein complex donates the capsid protein to 42 S RNA are not known at present. Assembly of the nucleocapsid takes place in the cytoplasm, as shown in numerous electron microscopic studies (Acheson and Tamm, 1967; Erlandson et al., 1967; Grimley et al., 1968; Lascano et al., 1969; Grimley and Friedman, 1970). About half of the nucleocapsid is free in the cytoplasm and the rest is associated with membranes (Bose and Brundige, 1972; Friedman et al., 1972; Ranki, 1972).

6.1.5. Budding of alphavirus through the plasma membrane

Since the alphavirus nucleocapsid in the virion is below the inner leaflet of the lipid bilayer, Harrison et al., (1971) suggested that the capsid protein be regarded as a membrane protein. If so, it should display an affinity for all membranes. Since a characteristic property of the intrinsic membrane proteins is the ability of their lipophilic part to bind detergents (Helenius and Simons, 1974), the ability of SF virus nucleocapsid to bind Triton X-100 was tested. The nucleocapsid behaved like the usual water-soluble proteins (Helenius and Söderlund, 1973), suggesting that the binding of nucleocapsid to the plasma membrane is not based on hydrophobic interactions. This also explains why substantial amounts of nucleocapsid remain free in the cytoplasm.

The finding of Garoff and Simons (1974) that nucleocapsid can be cross-linked to the envelope protein in the intact virus (p. 755) offered another possible explanation for the association of the nucleocapsid to the plasma membrane, namely that nucleocapsid could react with that part of the envelope protein which penetrates to the cytoplasmic side of the plasma membrane. Thus binding of nucleocapsid becomes specific only in those areas of the plasma membrane where viral glycoproteins are inserted. Interaction with only a few envelope proteins may be sufficient to keep the nucleocapsid attached to the plasma membrane. The bonds created between the nucleocapsid and envelope proteins would also be expected to inhibit the lateral diffusion of the proteins. All the envelope proteins coming into contact with the nucleocapsid would be similarly immobilized, leading to patch formation. If the binding specificity is high enough, host cell membrane proteins would remain free to diffuse away and would also be squeezed away by the envelope proteins (Simons et al., 1975).

Pfefferkorn and his colleagues (Waite and Pfefferkorn, 1970a; Brown et al., 1972; Waite et al., 1972) have reported an interesting phenomenon that helps us to understand the assembly of alphaviruses. When Sindbis virus is grown in a slightly hypotonic medium, virus release is inhibited. The synthesis of viral protein and RNA are unaffected and the assembly of the nucleocapsid and transport of envelope proteins to the plasma membrane also occurs. When the cells are transferred to normal medium a burst of virus release takes place within seconds, and about 2 minutes later half of the total virus yield has been released. This initial phase is followed by a slower release of virus that ends about 20 minutes later. At this time the cumulative yield of the whole virus cycle has been released to the medium.

Electron microscopy of thin-sectioned infected cells kept in the hypotonic medium revealed that the surface of the cells was covered with loosely attached particles that may be responsible for the rapid release of virus during the first 2 minutes after restoration of isotonicity. There was also a large amount of nucleocapsid arranged below the plasma membrane, which may account for the virus population which was released more slowly, but within 20 minutes. When the inhibited cultures were studied, using the freeze-etching technique, no signs of viral budding could be seen. When the hypotonic medium was changed to the normal ionic strength for 1½ minutes before freeze-etching, various phases of virus budding could be observed. Less frequently, the same phases were found in infected cells maintained in the normal medium continuously. Intramembranous 6 to 10 nm particles were visible, especially on the inner surface of the inner leaflet, during the early stages of maturation. The different budding stages are presented in the Fig. 5, which is a modification of the model presented by Brown et al., (1972). The nucleocapsid probably aligns under the plasma membrane by specific interaction between capsid and envelope proteins. At this stage (1) the intramembranous particles are clearly visible and this is also the stage where budding seems to stop when hypotonic medium is present. In the next stage, (2), increasing interactions between the capsid and envelope proteins cause bulging of the membrane and surface projections appear (Acheson and Tamm, 1967). The inner leaflet then detaches from the cell and closes around the nucleocapsid (stages 3–5), while the outer leaflet remains continuous with the cell plasma membrane and intramembranous particles no longer seen. The final event is detachment of the outer leaflet and closure of the lipid bilayer.

Stages 2 to 6 do not require energy since cyanide, sodium fluoride, sodium azide and iodoacetic acid do not inhibit the release of virus from cells restored to normal ionic strength (Waite and Pfefferkorn, 1970a). Thus final maturation

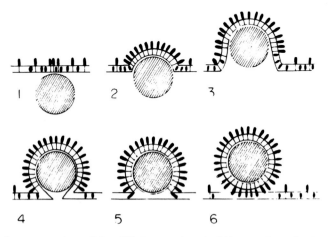

Fig. 5. Schematic presentation of Semliki Forest virus budding at the cell plasma membrane (modified from Brown et al., 1972). For description of stages 1 to 6 see text. The small particles within the plasma membrane represent host cell proteins and the clubs with tails are viral glycoproteins.

may be a self-assembly process (Casjens and King, 1975; Kaper, 1975), as suggested by von Bonsdorff and Harrison (1975). The essential requirement seems to be the free lateral diffusion of envelope proteins. If this is made impossible by "freezing" the lipids at 0°C or "patching" with divalent antibodies or plant lectins, budding does not occur (Waite and Pfefferkorn, 1970a; Birdwell and Strauss, 1973, 1974a; Rott et al., 1975; Finkelstein and McWilliams, 1976).

6.2. Assembly of other viruses

The assembly of myxo-, paramyxo- and rhabdoviruses is basically different from that of the alphaviruses since the former have a matrix (M) protein that is inserted between the nucleocapsid and the plasma membrane. Whether the M protein should be regarded as an integral or a peripheral membrane protein is not yet clear, but it is a membrane protein destined for the cytoplasmic side of the plasma membrane similar to those studied by Lodish and Small (1975) in reticulocytes.

6.2.1. Transport of proteins to the plasma membrane
The intracellular distribution and transport of influenza virus envelope proteins have been studied using the pulse-chase technique combined with cell fractionation (Compans, 1973b; Hay, 1974; Meier-Ewert and Compans, 1974; Compans and Caliguiri, 1975; Gregoriades and Hirst, 1975; Hay and Skehel, 1975). Synthesis of the major envelope glycoproteins (HA and NA) occurs on the rough endoplasmic reticulum from which they are transported to the smooth endoplasmic reticulum and presumably to the Golgi complex. The intracellular transport of the M protein is less clearly defined by pulse-chase experiments. This protein is found in the rough and smooth endoplasmic reticulum as well as in the plasma membrane very soon after its synthesis, and does not seem to migrate like the glycoproteins (Hay, 1974). Absorption of fragmented plasma membrane preparations from infected cells with red blood cells revealed that the M protein was constantly associated with the hemagglutinins, whereas the neuraminidase (NA) was more randomly distributed (Hay, 1974; Hay and Skehel, 1975). This indicates a closer association between HA and M protein than for NA and the M protein. One obvious explanation for this finding would be a specific interaction between HA and M protein similar to that between the alphavirus nucleocapsid and envelope proteins. The more random distribution of neuraminidase may mean that such an interaction does not occur between these two proteins.

Recent studies carried out with Newcastle disease virus (Nagai et al., 1976) have shown that the processing of paramyxovirus envelope proteins is very similar to that of influenza virus. The envelope glycoproteins F_0 and HN are transported from the rough endoplasmic reticulum to the plasma membrane, where the F_0 is cleaved to yield the active fusion protein F. In the smooth endoplasmic reticulum the HN protein acquires its ability to adsorb red blood cells. The M protein is associated with the rough and smooth endoplasmic reticulum as well as with the plasma membrane. The M protein reaches the plasma membrane somewhat earlier and becomes attached to the membrane fraction, which already contains the virus glycoproteins, as demonstrated by adsorption of plasma membrane fragments onto red blood cells.

The appearance of vesicular stomatitis virus glycoprotein G, matrix protein M, and NS protein in the plasma membrane has been reported recently by Atkinson et al. (1976). The glycosylated G protein appears in the plasma membrane after a lag period of about 20 minutes, whereas the M and NS proteins showed only a short lag period of less than 5 minutes. Only about one-third of the labeled G and M proteins reached the plasma membrane. Part of the incompletely glycosylated G protein, which could be detected after very short labeling period, was also found associated with the plasma membrane fraction for unknown reasons. This may partly explain the different results reported by David (1973), where an extremely rapid and equal labeling of vesicular stomatitis virus G and M proteins was found in the plasma membrane fraction. After a 30-second pulse with radioactive amino acids, the maximum specific activity of M and G proteins in the plasma membrane fraction was reached within 2 minutes.

Temperature-sensitive mutants defective in the processing of G protein (group V) and of M protein (group III) have been isolated from vesicular stomatitis virus (Lafay, 1974; Pringle and Wunner, 1975). Some mutants of group V synthesize G protein, which is not transported from the rough endoplasmic reticulum at the restrictive temperature (Lafay, 1974). This defect is reversible since G protein labeled at the restrictive temperature can be detected in the released virus when the temperature is lowered. Similar experiments with group III mutants revealed that M protein synthesized at the higher temperature was not chased into the virion when infected cultures were transferred to the permissive temperature. The G protein, which was synthesized at the high temperature but could not be released because assembly was blocked due to the defect in M protein, was successfully chased into virions when cultures were shifted downward to the permissive temperature (Lafay, 1974).

If we assume that the M protein is a peripheral membrane protein such as spectrin in the erythrocytes (Singer and Nicolson, 1972; Gahmberg, 1977; Quinn, 1976), the binding to the membrane could then be mediated by specific protein-protein interactions with the G protein. Temperature-sensitive mutations in M and G protein would then be expected in which this interaction cannot occur. In such cases the M protein might be "lost" in the cytoplasm in a soluble form.

6.2.2. Kinetics of labeling of proteins in released virus

The order in which the viral structural proteins are labeled when released virus from pulse-labeled cultures has also been investigated, and this offers an alternative strategy for studying the transport of different proteins to the site of assembly.

The kinetics of incorporation of the different influenza virus structural proteins into the released protein differs greatly (Hay, 1974; Hay and Skehel, 1975).

The M protein was incorporated most rapidly, whereas hemagglutinin showed a delay of about 20 minutes, presumably due to stepwise glycosylation. Interestingly, the neuraminidase was incorporated more slowly than the hemagglutinin. The lack of interaction between the NA and M proteins could perhaps explain this phenomenon. If the former were more randomly scattered over the plasma membrane, it would have a lower probability of being incorporated into the virion. The freedom of the neuraminidase to diffuse laterally might be a factor in release of the virus because it would allow the enzyme to digest sialic acid residues over a wide area and thereby inhibit aggregation and readsorption of newly released virions back to their host cell (Palese et al., 1974). Incorporation of the major nucleocapsid NP protein also follows different kinetics from those of the membrane proteins. There is a delay of about 45 minutes between the synthesis and release of NP protein, and the same is true of the replicase proteins P_1 and P_2. This delay probably reflects the kinetics of nucleocapsid assembly (Hay, 1974; Compans and Choppin, 1975; Hay and Skehel, 1975).

M protein is rapidly synthesized in other viruses as well. In released Sendai virus, the maximum level of radioactivity in the M protein is attained faster than the other viral proteins (Famulari and Fleissner, 1976b). These investigators also showed that the nucleocapsid is assembled into the released virus more slowly than the glycoproteins. The kinetics of labeling of vesicular stomatitis virus proteins has shown again that the M protein is the most rapidly incorporated, followed by the nucleocapsid protein N and the glycoprotein G (Kang and Prevec, 1971). This result is somewhat different from those obtained with influenza virus, in which the NP protein was incorporated more slowly than the glycoproteins. Atkinson et al. (1976) showed that the release of the G protein into the medium began soon after a 15-minute period of labeling with ^{3}H-fucose and ^{14}C-glucosamine. Since the pulse-labeling kinetics for the plasma membrane and the released virus were very similar, these authors suggest that the G protein is rapidly released after reaching the plasma membrane.

6.2.3. Maturation of other viruses

The assembly of the myxovirus nucleocapsid is not well understood at present (Compans and Choppin, 1975). In pulse-chase experiments with radioactive amino acids maximal labeling of cytoplasmic ribonucleoprotein (RNP) particles is obtained in 5 minutes (Krug, 1971, 1972). Part of the NP protein is attached to the viral RNA in the nucleus-forming nucleoplasmic RNPs which do not seem to be precursors of the virions (Krug and Etkind, 1975). Details of paramyxo- and rhabdovirus nucleocapsid assembly are also poorly understood (Blair and Robinson, 1970; and see reviews by Choppin and Compans, 1975; and Wagner, 1975). Presumably the N protein is released from the ribosomes into a soluble pool, and when the pool is large enough the protein is assembled with the RNA synthesized 1 to 2 hours earlier (Famulari and Fleissner, 1976b).

The mechanism(s) of association of nucleocapsid with the viral envelope proteins remains unresolved. It is generally assumed that the M protein plays a key role in this process (Lenard and Compans, 1974; Choppin and Compans, 1975;

Compans and Caliguiri, 1975; Compans and Choppin, 1975; Wagner, 1975). Since direct interaction between the viral glycoproteins and M protein has not been shown either in the purified virus or in infected cells, discussion of the assembly process remains at a speculative level (p. 774). The association of influenza virus hemagglutinins and M protein, as well as the Sendai virus HN protein and M protein in hemadsorbed fragments of plasma membrane (Hay, 1974; Nagai et al., 1976), suggests that the M protein and glycoproteins may interact. Assuming that such interaction exists, we are led to the following conclusions. The glycoproteins, which are transported to the plasma membrane by a different pathway, are not necessarily brought into contact with the M protein immediately. These glycoproteins would be free to diffuse laterally, like many cellular membrane proteins (Singer and Nicolson, 1972; Bretscher and Raff, 1975; Quinn, 1976), but when attached to the M protein they would be less mobile. Freedom for lateral diffusion is probably a prerequisite for successful maturation since plant lectins and divalent (but not monovalent) antibodies were shown to inhibit virus production (Becht et al., 1971, 1972; Poste et al., 1974; Lampert et al., 1975; Rott et al., 1975; Finkelstein and McWilliams, 1976). The concentration of glycoproteins into "virus-specific patches" (Compans and Caliguiri, 1975) or domains could occur in at least three different ways: (1) glycoproteins could interact to create clusters; (2) M proteins could also interact to bring the glycoproteins in patches; or (3) the nucleocapsid could interact with the M proteins assemble them as proposed for alphavirus maturation. Perhaps these alternatives could best be investigated by using temperature-sensitive mutants with defects in glycoproteins, M protein, and nucleocapsid assembly.

The recent electron microsopic study by Dubois-Dalcq and Reese (1975) with a measles variant (a paramyxovirus) has provided some support for the third alternative. Early in infection virus envelope glycoproteins at the cell surface, detected by antibodies coupled to horseradish peroxidase, were evenly scattered over the cell surface. Later, virus-specific strands were seen when nucleoprotein was aligned under the plasma membrane. The strands became wider and convoulted to form circular areas from which the nucleocapsid begun to bud. Since the strands followed the nucleocapsid, it is probable that interaction between the nucleocapsid in its extended form determines the orientation of the virus glycoproteins on the cell surface. Obviously, this interaction could occur without the M protein. The simplest explanation, however, would be that the primary interaction between the M protein and the nucleocapsid strand determined the surface structure. Freeze-fracturing also revealed that the intramembranous particles characteristic of the host cell membrane were extruded from areas where the viral proteins were attached. Formation of the circular bulges from which the final budding occurs may be mediated by interaction between matrix proteins. Evidence of the involvement of M protein in the assembly of Sendai virus (another paramyxovirus) has been obtained from the study of Portner et al. (1975) using a temperature-sensitive mutant that had a defect in the HN protein and was released into the medium at the restrictive temperature. The released virus did not contain the HN protein, indicating that it is not essential for budding of the virus. Apparently, interaction between the F and M proteins is sufficient for the formation of a virus-specific "patch" in the plasma membrane.

Selection of nucleoprotein strands in influenza virus-infected cells to yield particles in which the different RNAs are present in equimolar ratios remains a problem. The possibility that the polymerase proteins P_1 and P_2 take part in this selection should not be neglected. If we assume that in influenza virus-infected cells the hemagglutinins, but not the neuraminidase, are in contact with the M protein, the small quantities of neuraminidase in the virions (Schulze, 1972) would be explained. As indicated by Skehel and Hay (1975), the NA protein becomes the major viral protein of the plasma membrane late in the infection. If we assume direct interaction between NA and, say, the polymerase proteins, this might explain why neuraminidase is not excluded from the forming virions similar to the host proteins (Holland and Kiehn, 1970; Hay, 1974).

The surface glycoproteins are organized to form visible projections only when the nucleoprotein is under such an area of plasma membrane, (Compans et al., 1966; David et al., 1968; Bächi et al., 1969; Compans and Dimmock, 1969; Compans et al., 1970; Birdwell and Strauss, 1974b; Strauss and Strauss, 1976). In these studies the budding process on the plasma membrane has been visualized and is similar to that described for the alphaviruses, except that there seems to be a distinct dense layer of proteins under the membrane lipids, most probably the M proteins.

The specificity of the envelope formation has been studied by phenotypic mixing experiments between different enveloped viruses (Choppin and Compans, 1970; McSharry et al., 1971; Zavada, 1972; Huang et al., 1973). From these experiments it may be concluded that the possible interaction between envelope glycoprotein(s) and matrix protein is less specific than the interaction between nucleoprotein and matrix protein. For example, vesicular stomatitis-SV5 hybrid particles had M and N proteins of vesicular stomatitis virus but the envelope glycoproteins were from both parents (McSharry et al., 1971). The hybrids were shaped like typical rhabdoviruses, thus excluding the possibility that the shape of the virus is determined by the envelope glycoproteins.

6.3. Summary

Alpha-, myxo-, paramyxo-, and rhabdoviruses have been studied mainly under conditions in which the final assembly of the virions occurs in the plasma membrane. The glycoproteins are presumably transported in relatively small units from the smooth membranes to the plasma membrane. In the myxo-, paramyxo-, and rhabdoviruses, another membrane protein, the M protein, which is not glycosylated, is rapidly transported to the cytoplasmic side of the plasma membrane. Interaction between the M protein and the viral glycoproteins has yet to be demonstrated, but such an interaction would offer the simplest explanation for the concentration of the virus envelope glycoproteins into "patches" in the plasma membrane. The assembly of the viral nucleocapsid occurs in the cytoplasm and is not membrane-associated. In the alphaviruses the hydrophilic nucleocapsid proteins interact with the portion of the envelope glycoproteins that spans the lipid bilayer. Thus there is a specific interaction between nucleocapsid and glycoproteins that prevents the lateral diffusion of the glycoproteins in the plasma membrane and concentrates them in patches, whereas the host proteins diffuse away. In the myxo-, paramyxo, and rhabdoviruses the process may be similar. The hydrophilic nucleocapsid interacts with the membrane-bound M protein which, in turn, should be in contact with the surface glycoproteins. Again, concentration of the glycoproteins could be achieved by interaction with the nucleocapsid. The budding process, which does not require generation of energy, is probably a self-assembly process, whereby the increasing interaction between the nucleocapsid and glycoproteins (with or without the mediation of M protein) results in the final formation of spherical or bullet-shaped particles whose lipids are derived from the plasma membrane of the host cell.

7. General discussion

In this review we have discussed the assembly of enveloped animal viruses. We have restricted our discussion to the four simplest virus groups, the alpha-, myxo-, paramyxo-, and rhabdoviruses, all of which are RNA-containing viruses. The literature on the structure and multiplication of these virus group members has been cited selectively in this chapter, mainly because several excellent and comprehensive recent reviews are available (Pfefferkorn and Shapiro, 1974; Choppin and Compans, 1975; Compans and Choppin, 1975; Wagner, 1975; Strauss and Strauss, 1976).

The use of enveloped viruses as models for assembly of cellular membranes is relevant only if the structure of the viral envelopes is similar to that of cellular membranes. In section 3 we have summarized the evidence, which strongly supports the idea that viral membranes are actually simplified versions of host membranes. In the following discussion we have made a basic assumption that the viral membranes are synthesized along the same pathway as the cellular membranes. The validity of this assumption cannot be verified because the pathway of membrane synthesis is largely unknown in eukaryotic cells (Bretscher and Raff, 1975; Palade, 1975; Wallach, 1975; Atkinson, 1975; Hughes, 1976). It is based on the hope that investigation of viral membrane synthesis will reveal this normal pathway. It should be remembered, however, that during viral infection massive amounts of new messenger RNAs for viral membrane proteins are synthesized, leading to the overwhelming production of membrane proteins. This overloading of the machinery responsible for the synthesis of cellular membranes could lead to several "virus-specific artifacts." These artifacts may greatly obscure the detection of the "real pathway," but should be explicable as consequences of normal pathway overloading. Despite these inherent difficulties we will discuss the synthesis and transport of the viral glycoproteins and the nonglycosylated matrix protein as models for synthesis of host cell membranes.

7.1. Glycoproteins

The pathway of membrane glycoprotein synthesis in normal cells has not been fully elucidated to date (Autuori et al., 1975 a,b; Bretscher and Raff, 1975; Morrison and Lodish, 1975; Palade, 1975; Wallach, 1975; Quinn, 1976; Hughes, 1976; and chapter 1 by Mooré and chapter 2 by Cook, this volume). The basic question of whether the glycosylation of membrane proteins occurs inside the cisternal space of the endoplasmic reticulum as in secreted glycoproteins has not been answered.

If we assume that viral glycoproteins are glycosylated by a similar process to the excreted glycoproteins, they should be synthesized on membrane-bound polysomes (Borgese et al., 1974) and excreted into the cisternal space of the rough endoplasmic reticulum, as in the immunoglobulin light chain (Blobel and Dobberstein, 1975a,b). The fact that the messenger RNA for the vesicular stomatitis virus G protein is completely membrane-bound (Morrison and Lodish, 1975) suggests that this glycoprotein resembles the secretory glycoproteins. The cell fractionation studies done especially with myxo- and paramyxovirus-infected cells have shown that viral glycoproteins are actually associated with the rough and smooth membranes of the endoplasmic reticulum and are glycosylated stepwise during their transfer from the rough to the smooth endoplasmic reticulum and probably to the Golgi complex (section 5).

Due to their amphipathic nature, the transport of membrane glycoproteins should be different from that of the water-soluble glycoproteins since they are probably associated with the membranes by their lipophilic parts, as they are in the viral membrane. Thus their transport into various membranes should occur by lateral diffusion. Lateral movement of envelope proteins on the surface of infected cells has been demonstrated (Birdwell and Strauss, 1974b), and similar translational mobility may well occur in intracellular membranes. Assuming that terminal galactose, sialic acid, and fucose residues are added to the oligosaccharide chains only in the Golgi complex, the viral glycoproteins must spend some time there. In the transport of secretory proteins there seems to be a step requiring energy that may possibly be required for vesicle-mediated transport from the smooth endoplasmic reticulum to the Golgi complex (Palade, 1975). This kind of transport step has not been investigated in virus-infected cells. If the viral glycoproteins are transported to the Golgi complex for final glycosylation, they must be transported from there to the plasma membrane, where they have been shown to appear as early as 2 hours after infection.

Secretory proteins are excreted from the Golgi in vacuoles that fuse with the plasma membrane, emptying their contents on the outside of cell (Bennett et al., 1974, Palade, 1975). The membrane glycoproteins attached to the Golgi membrane by their lipophilic part and extending to the interior space of the Golgi could also be secreted within vesicles by a similar mechanism, as suggested by

Hirano et al. (1972). The fusion of these membrane-protein-containing vesicles with the plasma membrane would bring the glycoproteins to the outside aspect of the cell surface (Fig. 7). In artificially assembled vacuoles the glycoproteins are located both inside and outside the membrane (Hosaka, 1975). This mode of transport would also indicate that lipids from the Golgi complex are transported along with the glycoproteins.

Is there any support for the secretory model of membrane protein transport from studies done with enveloped viruses? As discussed earlier, the envelope proteins of SF virus span the lipid bilayer, are in contact with the nucleocapsid in the virus and probably also at the plasma membrane before final budding occurs (Simons et al., 1975). If we assume that the envelope proteins are transported from the rough to the smooth membranes of the endoplasmic reticulum to the Golgi complex in association with membranes, one would expect that under appropriate conditions the nucleocapsid would recognize the spanning part of the envelope protein also in these membranes. This interaction would be expected to occur relatively late in infection, when large amounts of both nucleocapsid and envelope proteins have been synthesized. Several ultrastructural studies have shown that nucleocapsid is bound to the membranes of the smooth endoplasmic reticulum and the Golgi complex (Acheson and Tamm, 1967; Erlandsson et al., 1967; Grimley et al., 1968, 1972; Lascano et al., 1969; Grimley and Friedman, 1970; and Fig. 6). In mouse brains infected with SF virus, attachment of nucleocapsid to the rough endoplasmic reticulum was shown to occur (Grimley and Friedman, 1970). The attachment of the nucleocapsid to the membranes is probably via a specific contact with the E-1 glycoprotein, since temperature-sensitive mutants of Sindbis virus in complementation group D (Burge and Pfefferkorn, 1966, 1968), which have a temperature-sensitive hemagglutinin (Yin, 1968), do not show attachment of nucleocapsid to cellular or plasma membranes (Brown and Smith, 1975; see also Tan, 1970). Thus the specific attachment of nucleocapsid to the rough and smooth endoplasmic reticulum and Golgi membranes would support the suggestion that the alphavirus glycoproteins are transported inside the cisternal space. Is there any evidence to support the presence of "transport vacuoles" in alphavirus-infected cells? If we accept nucleocapsid binding as a criterion for the presence of envelope glycoproteins, the answer is yes. There are typical cytoplasmic vacuoles lined with nucleocapsids, which Grimley et al. (1968) have designated as CPV II, found by all electronmicroscopists who have studied the morphogenesis of alphaviruses (Acheson and Tamm, 1967; Erlandsson et al., 1967; and see review by Pfefferkorn and Shapiro, 1974) (Figs. 6,7).

These vacuoles could be transport vacuoles between the Golgi and plasma membrane. Apparently, since nucleocapsids are bound to the whole surface area they cannot fuse with the plasma membrane and thus accumulate in the cytoplasm, representing a dead end. This interpretation would also explain why more than 90% of the labeled structural proteins are never released into the medium (Scheele and Pfefferkorn, 1969). The rapid release of the rest of the viral proteins would reflect the number of vacuoles that escaped attachment of nuc-

782

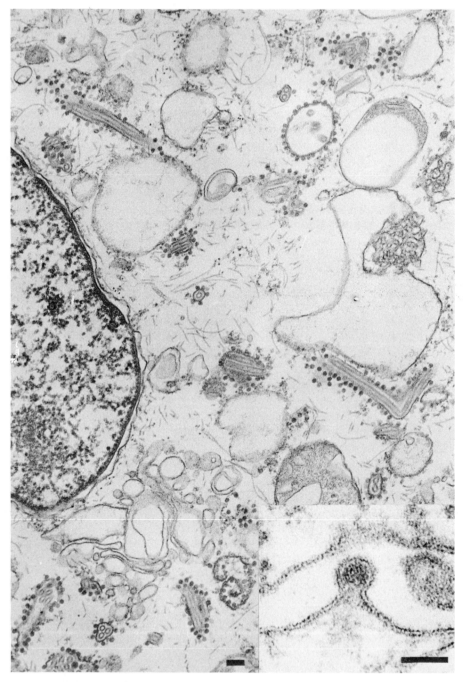

Fig. 6. Thin section of Semliki Forest virus-infected BHK21 cells 8 hours post infection. Bar=100 nm. Inset: budding profile; bar=50 nm.

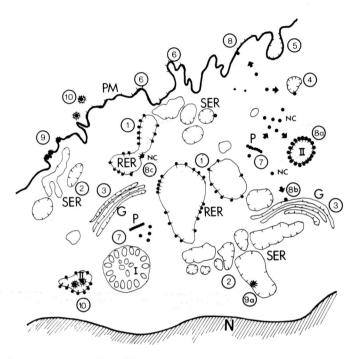

Fig. 7. Schematic representation of Semliki Forest virus assembly in infected cells.

Synthesis of envelope precursor protein p97 occurs on membrane-bound polysomes in the rough endoplasmic reticulum (RER), where the nascent protein is "secreted" into the cisternal space but remains attached to the membrane by a hydrophobic tail (1). The envelope proteins are cleaved to yield E-1 and p62 and are glycosylated primarily in RER. Then they are transported to the smooth endoplasmic reticulum (SER) by lateral diffusion (2). The final glycosylation occurs in the Golgi complex (G) (3), from which the glycoproteins are transported in "transport vacuoles" (4) to the plasma membrane (PM). At this point the vacuoles fuse with the plasma membrane (5). The glycoproteins are free to move laterally within the plasma membrane (6). The assembly of the nucleocapsid occurs in the free cytoplasm (7). The nucleocapsid recognizes the spanning glycoprotein in the plasma membrane (8). This recognition can occur in "transport vacuoles," creating virus specific cytoplasmic vacuoles lined with nucleocapsids (8a). It can also take place in the Golgi (8b) and in SER and RER (8c). Interaction between nucleocapsid and increasing amounts of spanning glycoproteins organizes the latter into virus-specific patches and budding through the plasma membrane occurs (9) (see Fig. 4). Occasionally budding may also occur at RER and SER (9a). Mature virions are released outside the cells (10). Conditions resembling those at the plasma membrane are sometimes created in the cytoplasmic membranes, leading to formation of intracellular virions (II). I refers to cytoplasmic vacuole I, the site of viral RNA synthesis (Grimley et al., 1968).

leocapsid during transport. One would thus expect more successful transport of the vacuoles to the plasma membrane, and also release as infectious virus, early during the infection than later on. This exact result was reported by Scheele and Pfefferkorn in 1969, who found that the specific radioactivity expressed per plaque-forming unit decreased by about two-thirds from 4 to 9 hours postinfection.

If the transport vacuole is not too densely covered with nucleocapsids, virus could well bud into the vacuole leading to intracellular formation of virus. This type of intracellular budding occurs in SF virus-infected chick embryo cells (Grimley et al., 1968) and also in SF virus-infected mouse brains (Grimley and Friedman, 1970). At late stages of infection in mouse brain cells, virus budding also takes place in the rough endoplasmic reticulum and Golgi membranes. Intracellular budding of this type, together with the cytoplasmic vacuoles surrounded by nucleocapsid, is presumably an example of those "virus-specific artifacts" that are bound to occur when excessive amounts of envelope protein and nucleocapsid are formed in the later stages of infection. Thus the real transport vacuoles should be identified early in the alphavirus infection.

Depending on the size of the transport vacuoles, which may be a property of the host cell, we could assume that considerable amounts of virus could bud into cytoplasmic vacuoles. Zee et al. (1970) have demonstrated well the different modes of virus maturation by using six different host cells for growing vesicular stomatitis virus. In L and Vero cells most of the virus matured at the plasma membrane, whereas in pig kidney cells almost all the virus matured into large cytoplasmic vacuoles. In all these cases the viral G protein was the same, suggesting that the size of the vacuoles is a property of the cell.

The flaviviruses mature into large cytoplasmic vacuoles, from which they are transported to the extracellular space similarly to the secretory proteins (McGee and Gosztonyi, 1967; Murphy et al., 1968; Oyanagi et al., 1969; Matsumura et al., 1971; Demsey et al., 1974; Stohlman et al., 1975). The same mode of intracellular maturation has been described when Sindbis and SF virus are grown in *Aedes albopictus* (mosquito) cells (Raghow et al., 1973; Gliedman et al., 1975).

An interesting modification of the assembly process is found in the maturation of Uukuniemi virus (a bunyavirus) at the membranes of the Golgi complex (von Bonsdorff et al. 1970). However, the virus has a phospholipid composition that is indistinguishable from the BHK cell plasma membrane and SF virus grown in the BHK cells (Renkonen et al., 1972c), despite appearing to bud from intracellular membranes.

Common features of the flaviviruses, mosquito cell-grown alphaviruses, and Uukuniemi virus are their relatively slow growth and inability to inhibit host protein synthesis, which may interfere with the transport vacuole formation.

If the "transport vacuole hypothesis" is correct, they should also carry lipids from the Golgi to the plasma membrane. Lipid analyses of isolated Golgi membranes have been carried out in relatively few cases (Keenan et al., 1970; Quinn, 1976.) In rat liver, the phospholipid composition of the Golgi is similar to that of the plasma membrane and differs from the endoplasmic reticulum. Despite extensive virus formation in SF virus-infected BHK cells, there was no difference between plasma membrane lipids from uninfected and infected cells (Renkonen et al., 1971). Thus if lipids are transported, they must be similar in phospholipid composition to those already in the plasma membrane. The only positive proof of lipid transfer during infection has been obtained by analyzing the newly synthesized phospholipids in the released virus, as described earlier.

Phosphatidylserine and sphingomyelin had the highest specific activities in the released virus. We could explain this only by assuming that newly synthesized lipids in the smooth endoplasmic reticulum become associated with the envelope proteins, and migrate with them to the plasma membrane and, ultimately, into the released virus. This would be compatible with the idea of membrane flow presented by Siekevitz et al (1967), Dallner et al., (1966), and Hughes, (1976).

The "secretory model" of membrane glycoprotein migration is by no means verified by the above evidence, though this hypothesis does at least explain many of the observations that are otherwise difficult to account for. More direct evidence, such as the isolation of Golgi-derived vesicles containing the viral glycoproteins, together presumably with some host cell proteins, is needed to substantiate the hypothesis. These vesicles should have similar lipid composition to that of the plasma membrane.

7.2. Matrix proteins

Enveloped viruses also offer opportunities for studying the nonglycosylated matrix protein, which remains on the cytoplasmic side of the plasma membrane. As discussed previously, the key question of whether M protein is a peripheral membrane protein or an amphiphilic integral membrane protein cannot be answered at present. In fractionation experiments on infected cells, M protein is always found in association with membranes, as expected for an amphiphilic protein. Yet the vesicular stomatitis virus-induced messenger for the M protein is found in the free polysome fraction (Morrison and Lodish, 1975). These results can be explained in two ways: (1) the amphiphilic M protein has a lipophilic part near the C-terminal end (Bretscher, 1975) and associates with the membranes only after this part of the protein has been translated. This would cause a situation where the M protein attaches to the nearest membrane by protein-lipid interaction, and would explain why the M protein is found in all membranes and cannot be chased, for example, from the rough endoplasmic reticulum to the plasma membrane even though most of the glycoproteins can be chased. The rapid and linear appearance of the M protein in the released virus could be explained by assuming that those M proteins that are made in close proximity to the plasma membrane are rapidly attached to it, recognized by the nucleocapsid, and excreted immediately. The rate-limiting role of the M protein in the synthesis of the virus would then reflect the probability with which the M protein happens to be "precipitated" onto the plasma membrane; and (2) the peripheral water soluble M protein is synthesized on free polysomes such as the cytoplasmic components of mitochondrial ATPase (Tzagoloff, 1973) and then diffuses to sites where the viral glycoproteins span the lipid bilayer. The M protein, by attaching to the spanning glycoprotein, thus becomes a peripheral membrane protein through protein-protein interaction.

Lodish and Small (1975) have shown that membrane proteins located on the inner side of the plasma membrane in reticulocytes, are made on membrane-free polysomes. They suggest that all proteins destined for the inner cytoplasmic side of the plasma membrane are made on free polysomes. If the first model suggested above for the M protein is true for the cellular proteins, a considerable amount of wasted proteins would be needed to ensure the proper membrane structure. The model based on protein-protein interaction would thus be much more economical. The use of temperature-sensitive mutants with a G protein that fails to span the bilayer should result in the appearance of soluble M protein, analogous to the Sindbis virus mutant in which the E-1 probably does not span the membrane thus leading to failure in nucleocapsid attachment.

In this final section studies we have summarized of the biogenesis of cellular membranes, particularly the plasma membrane, using viruses as models. We have also pointed out areas in which further research is needed. We hope that these ideas, whether right or wrong, will stimulate further research in the field.

Acknowledgments

We are most grateful to Dr. Carl-Henrik von Bonsdorff who has kindly supplied the electron micrographs. Thanks are also due to Dr. Niall Glanville for his constructive criticism and grammatical correction of the manuscript. Our personal research work has been supported by grants from the Finnish Academy and the Sigrid Juselius Foundation.

References

Acheson, N. H. and Tamm, I. (1967) Replication of Semliki Forest virus: An electron microscopic study. Virology 32, 128–143.
Acheson, N. H. and Tamm, I. (1970) Structural proteins of Semliki Forest virus and its nucleocapsid. Virology 41, 321–329.
Ada, G. L. and Gottschalk, A. (1956) The component sugars of the influenza-virus particle. Biochem. J. 62, 686–689.
Ada, G. L. and Perry, B. T. (1954) The nucleic acid content of influenza virus. Aust. J. Exp. Biol. Med. Sci. 32 453–468.
Arif, B. M. and Faulkner, P. (1972) Genome of Sindbis virus J. Virol. 9, 102–109.
Armstrong, S. J. and Barry, R. D. (1975) The detection of virus-induced RNA synthesis in the nuclei of cells infected with influenza viruses. In: Negative Strand Viruses (Mahy, B. W. J., and Barry, R. D., eds.) vol. 1 pp. 491–499. Academic Press, London.
Atkinson, P. H. (1975) Synthesis and assembly of HeLa cell plasma membrane glycoproteins and proteins. J. Biol. Chem. 250, 2123–2134.
Atkinson, P. H., Moyer, S. A. and Summers, D. F. (1976) Assembly of vesicular stomatitis virus glycoprotein and matrix protein into HeLa cell plasma membranes. J. Mol. Biol. 102, 613–631.
Autuori, F., Svensson, H. and Dallner, G. (1975a) Biogenesis of microsomal membrane glycoproteins in rat liver. I. Presence of glycoprotein in microsomes and cytosol. J. Cell. Biol. 67, 687–699.
Autuori, F., Svensson, H. and Dallner, G. (1975b) Biogenesis of microsomal membrane glycoprotein

in rat liver. II. Purification of soluble glycoproteins and their incorporation into microsomal membranes. J. Cell. Biol. 67, 700–714.

Bächi, T., Gerhard, W., Lindenmann, J. and Kühlethaler, K. (1969) Morphogenesis of influenza A virus in Ehrlich ascites tumor cells as revealed by thin-sectioning and freeze-etching. J. Virol. 4, 769–776.

Bader, J. P. (1975) Reproduction of RNA tumor viruses, In: Comprehensive Virology (Fraenkel-Conrat, H., and Wagner, R. R., eds.) vol. 4, pp. 253–332. Plenum Press, New York.

Baltimore, D. (1974) Tumor viruses: 1974. Cold Spring Harb. Symp. Quant. Biol. 39, 1187–1200.

Baltimore, D., Huang, A. S. and Stampfer, M. (1970) Ribonucleic acid synthesis of VSV. II. An RNA polymerase in the virion. Proc. Natl. Acad. Sci. U.S.A. 66, 572–576.

Becht, H., Hämmerling, U. and Rott, R. (1971) Undisturbed release of influenza virus in the presence of univalent antineuraminidase antibodies. Virology 46, 337–343.

Becht, H., Rott, H. and Klenk, H.-D. (1972) Effect of concanavalin A on cells infected with enveloped RNA viruses. J. Gen. Virol. 14, 1–8.

Becker, R., Helenius, A. and Simons, K. (1975) Solubilization of the Semliki Forest virus membrane with sodium dodecyl sulfate. Biochemistry 14, 1835–1841.

Bennett, G., Leblond, C. P. and Haddad, A. (1974) Migration of glycoprotein from the Golgi apparatus to the surface of various cell types as shown by radioautography after labeled fucose injection into rats. J. Cell. Biol. 60, 258–284.

Birdwell, C. R. and Strauss, J. H. (1973) Agglutination of Sindbis virus and of cells infected with Sindbis virus by plant lectins. J. Virol. 11, 502–507.

Birdwell, C. R. and Strauss, J. H. (1974a) Replication of Sindbis virus. IV. Electron microscope study of the insertion of viral glycoproteins into the surface of infected chick cells. J. Virol. 14, 366–374.

Birdwell, C. R. and Strauss, J. H. (1974b) Maturation of vesicular stomatitis virus: Electron microscopy of surface replicas of infected cells. Virology 59, 587–590.

Birdwell, C. R., Strauss, E. G. and Strauss, J. H. (1973) Replication of Sindbis virus. III. An electron microscopic study of virus maturation using the surface replica technique. Virology 56, 429–438.

Bishop, D. H. L., Obijeski, J. F. and Simpson, R. W. (1971) Transcription of the influenza ribonucleic acid genome by a virion polymerase II. Nature of the in vitro polymerase product. J. Virol. 8, 74–80.

Bishop, D. H. L., Roy, P., Bean, W. J. and Simpson, R. W. (1972) Transcription of the influenza ribonucleic acid genome by a virion polymerase. III. Completeness of the transcription process. J. Virol. 10, 689–697.

Bishop, D. H. L., Repik, O., Obijeski, J. F., Moore, N. F. and Wagner, R. R. (1975) Restitution of infectivity to spikeless vesicular stomatitis virus by solubilized viral components. J. Virol. 16, 75–84.

Blair, C. D. and Robinson, W. S. (1970) Replication of Sendai virus II. Steps in virus assembly. J. Virol. 5, 639–650.

Blobel, G. and Sabatini, D. D. (1971) Ribosome-membrane interaction in eukaryotic cells. In: Biomembranes (Manson, L.A., ed.) vol. 2, pp. 193–195. Plenum Press, New York.

Blobel, G. and Dobberstein, B. (1975a) Transfer of proteins across membranes. I. Presence of proteolytically processed and unprocessed nascent immunoglobulin light chains on membrane-bound ribosomes of murine myeloma. J. Cell. Biol. 67, 835–851.

Blobel, G. and Dobberstein, B. (1975b) Transfer of proteins across membranes II. Reconstitution of functional rough microsomes from heterologous components. J. Cell Biol. 67, 852–862.

Blough, H. A. and Lawson, D. E. M. (1968) The lipids of paramyxoviruses: A comparative study of Sendai and Newcastle disease virus. Virology 36, 286–292.

Blough H. A. and Merlie, J. P. (1970) The lipids of incomplete influenza virus. Virology 40, 685–692.

Blough, H. A. and Tiffany, J. (1973) Lipids in viruses. Adv. Lipid Res. 11, 267–339.

Blough, H. A., Weinstein, D. B., Lawson, D. E. M. and Kodicek, E. (1967) The effect of vitamin A on myxoviruses. II. Alterations in the lipids of influenza virus. Virology 33, 459–466.

Bolognesi, D. (1974) Structural components of RNA tumor viruses. Advan Virus Res. 19, 315–359.

von Bonsdorff, C.-H. (1973) The structure of Semliki Forest virus. Commentationes Biologicae 74, 1–53. Societes Scientarium Fennica Helsinki, Finland.

von Bonsdorff, C.-H. and Harrison, S. C. (1975) Sindbis virus glycoproteins form a regular icosahedral surface lattice. J. Virol. 16, 141–145.

von Bonsdorff, C.-H., Saikku, P. and Oker-Blom, N. (1970) Electron microscope study on the development of Uukuniemi virus. Acta Virol. 14, 109–114.

Borgese, N., Mok, W., Kreibich, G. and Sabatini, D. D. (1974) Ribosomal-membrane interaction: In vitro binding of ribosomes to microsomal membranes. J. Mol. Biol. 88, 559–580.

Bose, H. R. and Brundige, M. A. (1972) Selective association of Sindbis virion proteins with different membrane fractions of infected cells. J. Virol. 9, 785–791.

Bosmann, H. B., Hagopian, A. and Eylar, E. H. (1969) Cellular membranes: The biosynthesis of glycoprotein and glycolipid in HeLa cell membrane. Arch. Biochem. Biophys. 130, 573–583.

Both, G. W., Moyer, S. A. and Banerjee, A. K. (1975a) Translation and identification of the viral mRNA species isolated from subcellular fractions of vesicular stomatitis virus-infected cells. J. Virol. 15, 1012–1019.

Both, G. W., Moyer, S. A. and Banerjee, A.K. (1975b) Translation and identification of the mRNA species synthesized in vitro by the virion-associated RNA polymerase of vesicular stomatitis virus. Proc. Natl. Acad. Sci. U.S.A. 72, 274–278.

Bradburne, A. F. and Tyrrell, D. A. J. (1971) Coronaviruses of man. Progr. Med. Virol. 13, 373–403.

Brand, C. M. and Skehel, J. J. (1972) Crystalline antigen from influenza virus envelope. Nature New Biol. 238, 145–147.

Bratt, M. A., Collins, B. S., Hightower, L. E., Kaplan, J., Tsipis, J. E. and Weiss, S. R. (1975) Transcription and translation of Newcastle disease virus. In: Negative Strand Viruses. (Mahy, B. W. J., and Barry, R. D., eds.), Vol. 1, pp. 387–408, Academic Press, London.

Bretscher, M. S. (1971) A major protein which spans the human membrane erythrocyte membrane. J. Mol. Biol. 59, 351–357.

Bretscher, M. S. (1975) C-terminal region of the major erythrocyte sialoglycoprotein is on the cytoplasmic side of the membrane. J. Mol. Biol. 98, 831–833.

Bretscher, M. S. and Raff, M. C. (1975) Mammalian plasma membranes. Nature 258, 43–49.

Brown, D.T. and Smith, J. F. (1975) Morphology of BHK-21 cells infected with Sindbis virus temperature-sensitive mutants in complementation groups D and E. J. Virol. 15, 1262–1266.

Brown, D. T., Waite, M. R. F. and Pfeffercorn, E. R. (1972) Morphology and morphogenesis of Sindbis virus as seen with freeze-etching techniques. J. Virol. 10, 524–536.

Brown, F. (1975) Lipid and protein organization in vesicular stomatitis virus. In: Negative Strand Viruses (Mahy, B. W. J., and Barry, R. D., eds.), vol. 1, pp. 21–24. Academic Press, London.

Brown, F., Smale, C. J. and Horzinek, M. C. (1974) Lipid and protein organization in vesicular stomatitis and Sindbis viruses. J. Gen. Virol. 22, 455–458.

Bucher, D. J. and Kilbourne, E. D. (1972) A 2 (N2) neuraminidase of the X-7 influenza virus recombinant: Determination of molecular size and subunit composition of the active unit. J. Virol. 10, 60–66.

Bucher, D. J., Li, S. S.-L., Kehoe, J. M. and Kilbourne, E. D. (1976) Chromatographic isolation of the hemagglutinin polypeptides from influenza virus vaccine and determination of their aminoterminal sequences. Proc. Natl. Acad. Sci. U.S.A. 73, 238–242.

Burge, B. W. and Huang, A. S. (1970) Comparison of membrane protein glycopeptides of Sindbis virus and vesicular stomatitis virus. J. Virol. 6, 176–182.

Burge, B. W. and Pfefferkorn, E. R. (1966) Complementation between temperature-sensitive mutants of Sindbis virus. Virology 30, 214–223.

Burge, B. W. and Pfefferkorn, E. R. (1968) Functional defects of temperature-sensitive mutants of Sindbis virus. J. Mol. Biol. 35, 193–205.

Burge, B. W. and Strauss, J. H. (1970) Glycopeptides of the membrane glycoprotein of Sindbis virus. J. Mol. Biol. 47, 449–466.

Burke, D. C. (1975) Processing of alphavirus-specific proteins in infected cells. Med. Biol. 53, 352–356.

Cancedda, R. and Schlesinger, M. J. (1974) Formation of Sindbis virus capsid protein in mammalian cell-free extracts programmed with viral messenger RNA. Proc. Natl. Acad. Sci. U.S.A. 71, 1843–1947.

Capaldi, R. A. (1974) A dynamic model of cell membranes. Scient. Amer. 230, 27–33.

Caro, L. G. and Palade, G. E. (1964) Protein synthesis, storage, and discharge in the pancreatic exocrine cell. An autoradiographic study. J. Cell Biol. 20, 473–495.

Cartwright, B., Talbot, P. and Brown, F. (1970) The proteins of biologically active sub-units of vesicular stomatitis virus. J. Gen. Virol. 7, 267–272.

Casals, J. and Clarke, D. H. (1965) Arboviruses; Group A. In: Viral and Ricketsial Infections of Man (Horsfall, F. L., and Tamm, I., eds.), pp. 583–605. Pitman Medical Publishing Co., London.

Casjens, S. and King, J. (1975) Virus assembly. Ann. Rev. Biochem. 44, 555–611.

Chen, W. W., Lennarz, W. J., Tarentino, A. L. and Maley, F. (1975) A lipid-linked oligosaccharide intermediate in glycoprotein synthesis in oviduct. J. Biol. Chem. 250, 7006–7013.

Choppin, P. W. and Compans, R. W. (1970) Phenotypic mixing of envelope proteins of the parainfluenza virus SV5 and vesicular stomatitis virus. J. Virol. 5, 609–616.

Choppin, P.W. and Compans, R. W. (1975) Reproduction of paramyxoviruses. In: Comprehensive Virology. (Frankel-Conrat, H., and Wagner, R. R., eds.) Vol 4, pp. 95–178. Plenum Press, New York.

Choppin, P. W., Compans, R. W., Scheid, A., McSharry, J. J. and Lazarowitz, S. G. (1972) Structure and assembly of viral membranes. In: Membrane Research (Fox, C. F., ed.), pp. 163–179. Academic Press, New York.

Choppin, P. W., Sheid, A., Lazarowitz, S. G., McSharry, J. J. and Compans, R. W. (1973) Proteins of viral membranes. In: Advances in Biosciences. (Raspe, G., ed.), vol. 11, pp. 83–108, Pergamon Press, Vieweg.

Choppin, P. W., Lazarowitz, S. G. and Goldberg, A. R. (1975) Studies on proteolytic cleavage and glycosylation of the hemagglutinin of influenza A and B viruses, In: Negative Strand Viruses (Mahy, B. W. J., and Barry, R. D., eds.), Vol. 1, pp. 105–119. Academic Press, London.

Chow, N.-L. and Simpson, R. W. (1971) RNA-dependent RNA polymerase activity associated with virions and subviral particles of myxoviruses. Proc. Natl. Acad. Sci. U.S.A. 68, 752–756.

Clegg, J. C. S. (1975) Sequential translation of capsid and membrane protein genes in alphaviruses. Nature 254, 454–455.

Clegg, J. C. S. and Kennedy, S. I. T. (1975a) Initiation of synthesis of the structural proteins of Semliki Forest virus. J. Mol. Biol. 97, 40 –411.

Clegg, I. and Kennedy, I. (1975b) Translation of Semliki Forest virus intracellular 26-S RNA: Characterization of the products synthesised in vitro. Eur. J. Biochem. 53, 175–183.

Collins, B. S. and Bratt, M. A. (1973) Separation of the messenger RNAs of Newcastle disease virus by gel electrophoresis. Proc. Natl. Acad. Sci. U.S.A. 70, 2544–2548.

Compans, R.W. (1971) Location of the glycoprotein in the membrane of Sindbis virus. Nature New Biol. 229, 114–116.

Compans, R. W. (1973a) Influenza virus proteins. II. Association with components of the cytoplasm. Virology 51, 56–70.

Compans, R. W. (1973b) Distinct carbohydrate components of influenza virus glycoproteins in smooth and rough cytoplasmic membranes. Virology 55, 541–545.

Compans, R. W. and Caliguiri, L. A. (1975) Replication of influenza virus: Studies on synthesis of membrane components and on a purified viral RNA polymerase. In: Negative Strand Viruses. (Mahy, B. W. J., and Barry, R. D., eds.) Vol 2, pp. 573–594, Academic Press, London

Compans, R. W. and Choppin, P. W. (1975) Reproduction of myxoviruses. In: Comprehensive Virology (Fraenkel-Conrat, H., and Wagner, R. R., eds.) Vol. 4, pp. 179–252, Plenum Press, New York.

Compans, R. W. and Dimmock, N. J. (1969) An electron microscopic study of single-cycle infection of chick embryo fibroblasts by influenza virus. Virology 39, 499–515.

Compans, R. and Pinter, A. (1975) Incorporation of sulfate into influenza virus glycoproteins. Virology 66, 151–160.

Compans, R. W., Holmes, K. V., Dales, S. and Choppin, P. W. (1966) An electron microscopic study of moderate and virulent virus-cell interactions of the parainfluenza virus SV5. Virology 30, 411–426.

Compans, R. W., Klenk, H.-D., Caliguiri, L. A. and Choppin, P. W. (1970a) Influenza virus proteins. I. Analysis of polypeptides of the virion and identification of spike glycoproteins. Virology 42, 880–889.

Compans, R. W., Dimmock, N. J. and Meier-Ewert, H. (1970b) An electron microscopic study of the influenza virus-infected cell. In: The Biology of Large RNA Viruses (Barry, R. D., and Mahy, B. W. J., eds.), pp. 87–108. Academic Press, New York.

Dales, S. and Mosbach, E. H. (1968) Vaccinia as a model for membrane biogenesis. Virology 35, 564–583.

Dallner, G., Siekewitz, P. and Palade, G. E. (1966) Biogenesis of endoplasmic reticulum membranes. II. Synthesis of constitutive microsomal enzymes in developing rat hepatocytes. J. Cell Biol. 30, 97–117.

Dalrymple, J. M., Schlesinger, S. and Russell, P. K. (1976) Antigenic characterization of two Sindbis envelope glycoproteins separated by isoelectric focusing. Virology 69, 93–103.

David-West, T. S. and Labzoffsky, N. A. (1968) Electron microscopic studies on the development of vesicular stomatitis virus. Arch. Ges. Virusforsch. 23, 105–125.

David, A. E. (1971) Lipid composition of Sindbis virus. Virology 46, 711–720.

David, A. E. (1973) Assembly of the vesicular stomatitis virus envelope: Incorporation of viral polypeptides into the host cell plasma membrane. J. Mol. Biol. 76, 135–148.

Demsey, A., Steere, R. L., Brandt, W. E. and Veltri, B. J. (1974) Morphology and development of Dengue-2 virus employing freeze-fracture and thin-section techniques. J. Ultrastruct. Res. 46, 103–116.

Dubois-Dalcq, M. and Reese, T. S. (1975) Structural changes in the membrane of Vero cells infected with a paramyxovirus. J. Cell Biol. 67, 551–565.

Duda, E. and Schlesinger, M. J. (1975) Alterations in Sindbis viral envelope proteins by treating BHK cells with glucosamine. J. Virol. 15, 416–419.

Dulbecco, R. and Vogt, M. (1954) One-step growth curve of western equine encephalomyelitis virus on chicken embryo cells grown in vitro and analysis of virus yeilds from single cells. J. Exp. Med. 99, 183–199.

Eisenman, R., Vogt, V. M. Diggelmann, H. (1974) Synthesis of avian RNA tumor virus structural proteins. Cold Spring Harbor Symp. Quant. Biol. 39, 1067–1075.

Elhammer, A., Svensson, H., Autuori, F., and Dallner, G. (1975) Biogenesis of microsomal membrane glycoproteins in rat liver. III. Release of glycoproteins from the Golgi fraction and their transfer to microsomal membranes. J. Cell Biol. 67, 715–724.

Erlandson, R. A., Babcock, V. I., Southam, C. M., Bailey, R. B. and Shipkey, F. H. (1967) Semliki Forest virus in HEp-2 cell cultures. J. Virol. 1, 996–1009.

Etchison, J. R. and Holland, J. J. (1974a) Carbohydrate composition of the membrane glycoprotein of vesicular stomatitis virus. Virology 60, 217–229.

Etchison, J. R. and Holland, J. J. (1974a) Carbohydrate composition of the membrane glycoprotein of vesicular stomatitis virus grown in four mammalian cell lines. Proc. Natl. Acad. Sci. U.S.A. 71, 4011–4014.

Etkind, P. R. and Krug, R. M. (1974) Influenza viral mRNA. Virology 62, 38–45.

Etkind, P. R. and Krug, R. M. (1975) Purification of influenza viral complementary RNA: Its genetic content and activity in wheat germ cell-free extracts. J. Virol. 16, 1464–1475.

Eylar, E. H. (1965) On the biological role of glycoproteins. J. Theoret. Biol. 10, 89–113.

Famulari, N. G., and Fleissner, E. (1976a) High-titer replication of nondefective Sendai virus in MDBK cells. J. Virol. 17, 597–604.

Famulari, N. G. and Fleissner, E. (1976b) Kinetics of utilization of Sendai virus RNA and protein in the process of virion assembly. J. Virol. 17, 605–613.

Finkelstein, M. S. and McWilliams, M. (1976) Effects of plant lectins on virus growth in nonlymphoid cells. Virology 69, 570–586.

Follett, E. A. C., Pringle, C. R., Wunner, W. H. and Skehel, J. J. (1974) Virus replication in enucleate cells: vesicular stomatitis and influenza virus. J. Virol. 13, 394–399.

Follett, E. A. C., Pringle, C. R. and Pennington, T. H. (1975) Virus development in enucleate cells:

Echovirus, poliovirus, pseudorabies virus, reovirus, respiratory syncytial virus and Semliki Forest virus. J. Gen. Virol. 26, 183–196.

Fox, C. F. (1972) The structure of cell membranes. Scient. Amer. 226, 31–38.

Friedman, R. M., Levin, J. G., Grimley, P. M. and Berezesky, I. K. (1972) Membrane-associated replication complex in arbovirus infection. J. Virol. 10, 504–515.

Frommhagen, L. H., Knight, C. A. and Freeman, N. K. (1959) The ribonucleic acid, lipid, and polysaccharide constituents of influenza virus preparations. Virology 8, 176–197.

Gahmberg, C. G. (1977) Cell surface proteins. Changes during cell growth and malignant transformation. In: Dynamic Aspects of Cell Surface Organization (Poste, G., and Nicolson, G. L., eds.), Cell Surface Reviews, Vol. 3. Elsevier/North-Holland, Amsterdam.

Gahmberg, C. G., Simons, K., Renkonen, O. and Kääriäinen, L. (1972) Exposure of proteins and lipids in the Semliki Forest virus membrane. Virology 50, 259–262.

Garoff, H. (1974) Cross-linking of the spike glycoproteins in Semliki Forest virus with dimethylsuberimidate. Virology 62, 385–392.

Garoff, H. and Simons, K. (1974) Location of the spike glycoproteins in the Semliki Forest virus membrane. Proc. Natl. Acad. Sci. U.S.A. 71, 3988–3992.

Garoff, H. and Simons, K. (1975) Viral envelope composition: proteins. In: Cell Membranes and Viral Envelopes (Blough, H. A., and Tiffany, J., eds.), Academic Press, New York.

Garoff, H., Simons, K. and Renkonen, O. (1974) Isolation and characterization of the membrane proteins of Semliki Forest virus. Virology 61, 493–504.

Garon, C. F. and Moss, B. (1971) Glycoprotein synthesis in cells infected with vaccinia virus. II. A glycoprotein component of the virion. Virology 46, 233–246.

Glanville, N. and Ulmanen, I. (1976) Biological activity of in vitro synthesised protein: Binding of Semliki Forest virus capsid protein to the large ribosomal subunit. Biochem. Biophys. Res. Commun., 71, 393–399.

Glanville, N., Morser, J., Uomala, P. and Kääriäinen, L. (1976a) Simultaneous translation of structural and nonstructural proteins from Semliki Forest virus RNA in two eukaryotic systems in vitro. Eur. J. Biochem. 64, 167–175.

Glanville, N., Ranki, M., Morser, J., Kääriäinen, L. and Smith, A. E. (1976b) Initiation of translation directed by Semliki Forest virus 42S and 26S RNAs in vitro. Proc. Nat. Acad. Sci. U.S.A., in press.

Glass, S. E., McGeoch, D. and Barry, R. D. (1975) Characterization of the mRNA of influenza virus. J. Virol. 16, 1435–1443.

Gliedman, J. B., Smith, J. F. and Brown, D. T. (1975) Morphogenesis of Sindbis virus in cultured Aedes albopictus cells. J. Virol. 16, 913–926.

Gregoriades, A. and Hirst, G. K. (1975) The membrane protein of influenza virus and its distribution within the infected cell. In: Negative Strand Viruses (Mahy, B. W. J., and Barry, R. D., eds.), Vol. 2, pp. 595–609, Academic Press, London.

Griffith, I. P. (1975) The fine structure of influenza virus, In: Negative Strand Viruses (Mahy, B. W. J., and Barry, R. D., eds.), Vol. 1, pp. 121–132. Academic Press, London.

Grimes, W. J. and Burge, B. W. (1971) Modification of Sindbis virus glycoprotein by host-specific glycosyl transferases. J. Virol. 7, 309–313.

Grimley, P. M. and Friedman, R. M. (1970) Development of Semliki Forest virus in mouse brain: An electron microscopic study, Exp. Mol. Pathol. 12, 1–13.

Grimley, P.M., Berezesky, I.K. and Friedman, R.M. (1968) Cytoplasmic structures associated with an arbovirus infection: loci of viral ribonucleic acid synthesis. J. Virol. 2, 1326–1338.

Grimley, P. M., Rosenblum, E. N., Mims, S. J. and Moss, B. (1970) Interruption by rifampicin of an early stage in vaccinia virus morphogenesis: Accumulation of membranes which are precursors of virus envelopes. J. Virol. 6, 519–533.

Grimley, P. M., Levin, J. G., Berezesky, I. K. and Friedman, R. M. (1972) Specific membranous structures associated with the replication of group A arboviruses. J. Virol. 10, 492–503.

Grubman, M. J., Ehrenfeld, E. and Summers, D. F. (1974) In vitro synthesis of proteins by membrane-bound polyribosomes from vesicular stomatitis virus-infected HeLa cells. J. Virol. 14, 560–571.

Grubman, M. J., Moyer, S. A., Banerjee, A. K. and Ehrenfeld, E. (1975) Sub-cellular localization of vesicular stomatitis virus messenger RNAs. Biochem. Biophys. Res. Commun. 62, 531–538.

György, E., Sheehau, M. C. I., and Sokol, F. (1971) Release of envelope glycoprotein from rabies virions by a nonionic detergent. J. Virol. 8, 649–655.

Halonen, P. E., Murphy, F. A., Fields, B. N. and Reese, D. R. (1968) Hemagglutinin of rabies and some other bullet-shaped viruses. Proc. Soc. Exp. Biol. Med. 127, 1037–1042.

Harrison, S. C., Jumblatt, A. D. and Darnell, J. E. (1971) Lipid and protein organization in Sindbis virus. J. Mol. Biol. 60, 523–528.

Hay, A. J. (1974) Studies on the formation of the influenza virus envelope. Virology 60, 394–418.

Hay, A. J. and Skehel, J. J. (1975) Studies on the synthesis of influenza virus proteins. In: Negative Strand Viruses, (Mahy, B. W. J., and Barry, R. D., eds.), Vol 2. pp. 635–655. Academic Press, London.

Helenius, A. and Simons, K. (1972) The binding of detergents to lipophilic and hydrophilic proteins. J. Biol. Chem. 247, 3656–3661.

Helenius, A. and Simons, K. (1975) Solubilization of membranes by detergents. Biochim. Biophys. Acta. 415, 29–79.

Helenius, A. and Söderlund, H. (1973) Stepwise dissociation of the Semliki Forest virus membrane with Triton X-100. Biochim. Biophys. Acta. 307, 287–300.

Helenius, A. and von Bonsdorff, C.-H. (1976) Semliki Forest virus membrane proteins: Preparation and characterization of spike complexes soluble in detergent-free medium. Biochim. Biophys. Acta., 436, 895–899.

Helenius, A., Fries, E., Garoff, H. and Simons, K. (1976) Solubilization of Semliki Forest virus membrane proteins with sodium deoxycholate. Biochim. Biophys. Acta., 436, 319–334.

Hightower, L. E. and Bratt, M. A. (1974) Protein synthesis in Newcastle disease virus-infected chicken embryo cells. J. Virol. 13, 788–800.

Hightower, L. E. and Bratt, M. A. (1975) Protein metabolism during steady state of Newcastle disease virus infection I. Kinetics of amino acid and protein accumulation. J. Virol. 15, 696–706.

Hightower, L. E., Morrison, T. G. and Bratt, M. A. (1975) Relationships among the polypeptides of Newcastle disease virus. J. Virol. 16, 1599–1607.

Hirano, H., Parkerhouse, B., Nicolson, G. L., Lennox, E. S. and Singer, S. J. (1972) Distribution of saccharide residues on membrane fragments from a myeloma-cell homogenate: its implications for membrane biogenesis. Proc. Nat. Acad. Sci. U.S.A. 69, 2945–2949.

Hirschberg, C. B. and Robbins, P. W. (1974) The glycolipids and phospholipids of Sindbis virus and their relation to the lipids of the host cell plasma membrane. Virology 61, 602–608.

Holland, J. J. and Kiehn, E. D. (1970) Influenza virus effects on cell membrane proteins. Science 167, 202–205.

Holowczak, J. A. (1970) Glycopeptides of vaccinia virus. I. Preliminary characterization and hexosamine content. Virology 42, 87–99.

Homma, M. and Ohuchi, M. (1973) Trypsin action on the growth of Sendai virus in tissue culture cells. III. Structural difference of Sendai viruses grown in eggs and tissue culture cells. J. Virol. 12, 1457–1465.

Horzinek, M. C. (1975) The structure of Togaviruses and Bunyaviruses. Med. Biol. 53, 406–411.

Hosaka, Y. (1975) Artificial assembly of active envelope particles of HVJ (Sendai virus). In: Negative Strand Viruses (Mahy, B. W. J., and Barry, R. D., eds.), Vol. 2, 885–903. Academic Press, London.

Hsu, A.-F., Baynes, J. W. and Heath, E. C. (1974) The role of a dolichol-oligosaccharide as an intermediate in glycoprotein biosynthesis. Proc. Nat. Acad. Sci. U.S.A. 71, 2391–2395.

Huang, A. S. (1975) Ribonucleic acid synthesis of vesicular stomatitis virus. In: Negative Strand Viruses (Mahy, B. W. J., and Barry, R. D. eds.) Vol. 1, 353–359, Academic Press, New York.

Huang, A. S., Baltimore, D. and Bratt, M. A. (1971) Ribonucleic acid polymerase in virions of Newcastle disease virus: Comparison with the vesicular stomatitis virus polymerase. J. Virol. 7. 389–394.

Huang, A. S., Besmer, P., Chu, L. and Baltimore, D. (1973) Growth of pseudotypes of vesicular stomatitis virus with N-tropic murine leukemia virus coats in cells resistant to N-tropic viruses. J. Virol. 12, 659–662.

Hughes, R. C. (1976) In: Membrane Glycoproteins. Butterworths, London.

Jackson, G. G. and Muldoon, R. L. (1973) Viruses causing common respiratoy infections in man. III. Respiratory syncytial viruses and Corona viruses. J. Inf. Dis. 128, 674–702.

Jensik, S. C. and Silver, S. (1976) Polypeptides of mumps virus. J. Virol. 17, 363–373.

Joklik, W. K. and Zweerink, H. J. (1971) the morphogenesis of animal viruses. Ann. Rev. Genetics 5, 297–259.

Jones, K. J., Waite, M. R. F. and Bose, H. R. (1974) Cleavage of a viral envelope precursor during the morphogenesis of Sindbis virus. J. Virol. 13, 809–817.

Kääriäinen, L. and Gomatos, P. J. (1969) A kinetic analysis of the synthesis of BHK 21 cells of RNAs specific for Semliki Forest virus. J. Gen. Virol. 5, 251–265.

Kääriäinen, L., Simons, K. and von Bonsdorff, C.-H. (1969) Studies in subviral components of Semlike Forest virus. Ann. Med. Exp. Biol. Fenn. 47, 235–248.

Kääriäinen, L., Keränen, S., Lachmi, B., Söderlund, H., Tuomi, K., and Ulmanen, I. (1975) Replication of Semliki Forest virus. Med. Biol. 53, 342–352.

Kääriäinen, L., Lachmi, B. and Glanville, N. (1976) Translational control in Semliki Forest virus infected cells. Ann. Microbiol. (Inst. Pasteur) 127A, 197–203.

Kaluza, G. (1975) Effect of impaired glycosylation on the biosynthesis of Semliki Forest virus glycoproteins. J. Virol. 16, 602–612.

Kaluza, G., Scholtissek, C and Rott, R. (1972) Inhibition of the multiplication of enveloped RNA-viruses by glucosamine and 2-deoxy-D-glucose. J. Gen. Virol. 14, 251–259.

Kaluza, G., Schmidt, M. F. G. and Scholtissek, C. (1973) Effect of 2-deoxy-D-glucose on the multiplication of Semliki Forest virus and the reversal of the block by mannose. Virology 54, 179–189.

Kang, C. Y. and Prevec, L. (1971) Proteins of vesicular stomatitis virus. III. Intracellular synthesis and extracellular appearance of virus-specific proteins. Virology 46, 678–690.

Kaper, J. M. (1975) In: The Chemical Basis of Virus Structure. Dissciation and Reassembly. Elsevier/North-Holland, Amsterdam.

Kaplan, A. S. (1973) Ed. The Herpesviruses. Academic Press, New York, 739.

Kates, M., Allison, A. C., Tyrrell, D. A. J. and James, A. T. (1961) Lipids of influenza virus and their relation to those of the host cell. Biochim. Biophys. Acta 52, 455–466.

Kates, A., Allison, A.C., Tyrrell D. A. J. and James, A. T. (1962) Origin of lipids in influenza virus. Cold Spring Harb. Symp. Quant. Biol. 27, 293–301.

Keegstra, L., Sefton, B. and Burke, D. (1975) Sindbis virus glycoproteins: Effect of the host cell on the oligosacchrides. J. Virol. 16, 613–620.

Keenan, T. W. and Morré, D. J. (1970) Phospholipid class and fatty acid composition of Golgi apparatus isolated from rat liver and comparison with other cell fractions. Biochemistry 9, 19–25.

Kendal, A.P. and Kiley, M. P. (1975) Structural comparisons of influenza A neuraminidases. In: Negative Strand Viruses (Mahy, B. W. J., and Barry, R. D., eds.), Vol. 1, 145–159. Academic Press, New York.

Kendal, A. P., Apostolev, K. and Belyavin, G. (1969) The effect of protease treatment on the morphology of influenza A, B, and C viruses. J. Gen. Virol. 5, 141–143.

Kennedy, S. I. T. (1974) The effect of enzymes on structural and biological properties of Semlike Forest virus. J. Gen. Virol. 23, 129–143.

Kennedy, S. I. T. and Burke, D. C. (1972) Studies on the structural proteins of Semliki Forest virus. J. Gen. Virol. 14, 87–98.

Keränen, S. and Kääriäinen, L. (1974) Isolation and basic characterization of temperature-sensitive mutants from Semliki Forest virus. Acta Pathol. Microbiol. Scand. 82B, 810–820.

Keränen, S. and Kääriäinen, L. (1975) Proteins synthesised by Semliki Forest virus and its 16 temperature-sensitive mutants. J. Virol. 16, 388–396.

Klenk, H.-D. and Choppin, P. W. (1969a) Chemical composition of the parainfluenza virus SV5. Virology 37, 155–157.

Klenk, H.-D. and Choppin, P. W. (1969b) Lipids of plasma membranes of monkey kidney and hamster kidney cells and of parainfluenza virions grown in these cells. Virology 38, 255–268.

Klenk, H.-D. and Choppin, P. W. (1970a) Glycosphingolipids of plasma membranes of cultured cells and an enveloped virus (SV5) grown in these cells. Proc. Nat. Acad. Sci. U.S.A. 66, 57–64.

Klenk, H.-D. and Choppin, P. W. (1970b) Plasma membrane lipids and parainfluenza virus assembly. Virology 40, 393–947.

Klenk, H.-D., Caliguiri, L. A. and Choppin, P. W. (1970) The proteins of the parainfluenza virus SV5. II. The carbohydrate content and glycoproteins of the virion. Virology 42, 473–481.

Klenk, H.-D., Scholtissek, C. and Rott, R. (1972) Inhibition of glycoprotein biosynthesis of influenza virus by D-glucosamine and 2-deoxy-D-glucose. Virology 49, 723–734.

Klenk, H.-D., Wöllert, W., Rott, R. and Scholtissek, C. (1974) Association of influenza virus proteins with cytoplasmic fractions. Virology 57, 28–41.

Klenk, H.-D., Rott, R., Orlich, M. and Blödorn, J. (1975a) Activation of influenza A viruses by trypsin treatment. Virology 68, 426–439.

Klenk, H.-D., Wöllert, W., Rott, R. and Scholtissek, C. (1975b) The biosynthesis of the influenza virus hemagglutinin, In: Negative Strand Viruses (Mahy, B. W. J., and Barry, R. D., eds.), Vol. 2, pp. 621–634. Academic Press, London.

Knipe, D., Rose, J. K. and Lodish, H. (1975) Translation of individual species of vesicular stomatitis viral mRNA. J. Virol. 15, 1004–1011.

Kolakofsky, D., de la Tour, E. B. and Delius, H. (1974) Molecular weight determination of Sendai and Newcastle disease virus RNA. J. Virol. 13, 261–268.

Kolakofsky, D., de la Tour, E. B. and Delius, H. (1975) Molecular weight determination of parainfluenza virus RNA. In: Negative Strand Viruses (Mahy, B. W. J. and Barry, R. D., eds.), Vol. 1, 243–257, Academic Press, London.

Kos, K. A., Osborne, B. A. and Goldsby, R. A. (1975) Inhibition of group B arbovirus antigen production and replication in cells enucleated with cytochalasin B. J. Virol. 15, 913–917.

Krug, R. M. (1971) Influenza viral RNPs newly synthesised during the latent period of viral growth in MDCK cells. Virology 44, 125–136.

Krug, R. M. (1972) Cytoplasmic and nucleoplasmic viral RNPs in influenza virus infected MDCK cells. Virology 50, 103–113.

Krug, R. M. and Etkind, P. R. (1973) Cytoplasmic and nuclear virus-specific proteins in influenza virus-infected MDCK cells. Virology 56, 334–348.

Krug, R. M. and Etkind, P. R. (1975) Influenza virus-specific products in the nucleus and cytoplasm of infected cells. In: Negative Strand Viruses (Mahy, B. W. J. and Barry, R. D., eds.), Vol. 2, pp. 555–572. Academic Press, London.

Kurth, R. and Bauer, H. (1975) Avian RNA tumor viruses. A model for studying tumor associated cell surface alterations. Biochim. Biophys. Acta 417, 1–23.

Lachmi, B. and Kääriäinen, L. (1976) Sequential translation of nonstructural proteins in cells infected with a Semliki Forest virus mutant. Proc. Nat. Acad. Sci. U.S.A. 73.

Lachmi, B., Glanville, N., Keränen, S. and Kääriäinen, L. (1975) Tryptic peptide analysis of nonstructural and structural precursor proteins from Semliki Forest virus mutant-infected cells. J. Virol. 16, 1615–1629.

Lafay, F. (1974) Envelope proteins of vesicular stomatitis virus: Effect of temperature-sensitive mutations in complementation groups III and V. J. Virol. 14, 1220–1228.

Laine, R., Kettunen, M.-L., Gahmberg, C. G., Kääriäinen L. and Renkonen, O. (1972) Fatty chains of different lipid classes of Semliki Forest virus and host cell membranes. J. Virol. 10, 433–438.

Laine, R., Söderlund, H. and Renkonen, O. (1973) Chemical composition of Semliki Forest virus. Intervirology 1, 110–118.

Lamb, R. A. and Mahy, B. W. J. (1975) The polypeptides and RNA of Sendai virus. In: Negative Strand Viruses (Mahy, B. W. J. and Barry, R. D., eds.), Vol. 1, pp. 65–87, Academic Press, London.

Lamb, R. A., Mahy, B. W. J. and Choppin, P. W. (1976) The synthesis of Sendai virus polypeptides in infected cells. Virology 69, 116–131.

Lampert, P. W., Joseph, B. S. and Oldstone, M. B. A. (1975) Antibody-induced capping of measles virus antigens on plasma membrane studied by electron microscopy. J. Virol. 15, 1248–1255.

Landsberger, F. R., Lenard, J., Paxton, J. and Compans, R. W. (1971) Spin-label electron spin resonance study of the lipid-containing membrane of influenza virus. Proc. Nat. Acad. Sci. U.S.A. 68, 2579–2583.

Landsberger, F. R., Compans, R. W., Choppin, P. W. and Lenard, J. (1973) Organization of the lipid phase in viral membranes. Effects of independent variation of the lipid and the protein composition. Biochemistry 12, 4498–4502.

Lascano, E. F., Berria, M. I. and Oro, J. G. B. (1969) Morphogenesis of Aura virus. J. Virol. 4, 271–282.

Laver, W. G. (1971) Separation of two polypeptide chains from the hemagglutinin subunit of influenza virus. Virology 45, 275–288.

Laver, W. G. and Valentine, R. C. (1969) Morphology of the isolated hemagglutinin and neuraminidase subunits of influenza virus. Virology 38, 105–119.

Lazarowitz, S. G. and Choppin, P. W. (1975) Enhancement of the infectivity of influenza A and B viruses by proteolytic cleavage of the hemagglutinin polypeptide. Virology 68, 440–454.

Lazarowitz, S. G., Compans, R. W. and Choppin, P. W. (1971) Influenza virus structural and nonstructural proteins in infected cells and their plasma membranes. Virology 46, 830–843.

Lazarowitz, S. G., Goldberg, A. R. and Choppin, P. W. (1973a) Proteolytic cleavage by plasmin of the HA polypeptide of influenza virus: Host cell activation of serum plasminogen. Virology 56, 172–180.

Lazarowitz, S. G., Compans, R. W. and Choppin, P. W. (1973b) Proteolytic cleavage of the hemagglutinin polypeptide of influenza virus. Function of the uncleaved polypeptide HA. Virology 52, 199–212.

Lazdins, I., Haslam, E. A. and White, D. O. (1972) The polypeptides of influenza virus. VI. Composition of the neuraminidase. Virology 49, 758–765.

Lehman-Grube, F. (1973) Editor Lymphocytic choriomeningitis virus and other Arenaviruses, 339 pp. Springer-Verlag, Berlin.

Lenard, J. and Compans, R. W. (1974) The membrane structure of lipid-containing viruses. Biochem. Biophys. Acta 344, 51–94.

Lenard, J. and Rothman, J. E. (1976) Transbilayer distribution and movement of cholesterol and phospholipid in the membrane of influenza virus. Proc. Nat. Acad. Sci. U.S.A. 73, 391–395.

Lenard, J., Landsberger, F. R., Wong, C. Y., Choppin, P. W. and Compans, R. W. (1975) Organization of lipid and protein in viral membranes: Spin label and fluorescent probe studies. In: Negative Strand Viruses (Mahy, B. W. J. and Barry, R. D., eds.), Vol. 2, pp. 823–833, Academic Press, New York.

Levin, J. G. and Friedman, R. M. (1971) Analysis of arbovirus nucleic acid forms by polyacrylamide gel electrophoresis. J. Virol. 7, 504–514.

Lodish, H. F. (1973) Biosynthesis of reticulocyte membrane proteins by membrane-free polyribosomes. Proc. Nat. Acad. Sci. U.S.A. 70, 1526–1530.

Lodish, H. F. and Small, B. (1975) Membrane proteins synthesized by rabbit reticulocytes. J. Cell. Biol. 65, 51–64.

Luukkonen, A., Brummer-Korvenkontio, M. and Renkonen, O. (1973) Lipids of cultured mosquito cells (Aedes albopictus) comparison with cultured mammalian fibroblasts (BHK 21 cells). Biochim. Biophys. Acta 326, 256–261.

Luukkonen, A., Gahmberg, C. G. and Renkonen, O. (1976a) Surface labeling of Semliki Forest virus glycoproteins using galactose oxidase exposure of E3-glycoprotein. Virology 76, 55–57.

Luukkonen, A., Kääriäinen, L. and Renkonen, O. (1976b) Phospholipid composition of Semliki Forest virus grown in Aedes albopictus cells. Biochim. Biophys. Acta 450, 109–120.

Matsumura, T., Stollar, V. and Schlesinger, R. W. (1971) Studies on the nature of dengue viruses. V. Structure and development of dengue virus in Vero cells. Virology 46, 344–355.

Mattila, K., Luukkonen, A. and Renkonen, O. (1976) Protein-bound oligosaccharides of Semliki Forest virus. Biochim. Biophys. Acta 419, 435–444.

McGee-Russel, S. M. and Gosztonyi, G. (1967) Assembly of Semliki Forest virus in brain. Nature 214, 1204–1206.

McIntosh, K. (1974) Coronaviruses: A comparative review. Current Topics. Microbiol. Immunol. 63, 85–129.

McSharry, J. J. and Wagner, R. R. (1971) Lipid composition of purified vesicular stomatitis viruses. J. Virol. 7, 59–70.

McSharry, J. J., Compans, R. W. and Choppin, P. W. (1971) Proteins of vesicular stomatitis virus and of phenotypically mixed vesicular stomatitis virus-simian virus 5 virions. J. Virol. 8, 722–729.

Meier-Ewert, H. and Compans, R. W. (1974) Time course of synthesis and assembly of influenza virus products. J. Virol. 14, 1083–1091.

Metz, D. H. (1975) Discrimination between viral and cellular synthesis. In: Control Processes in Virus Multiplication (Burke, D. C. and Russell, W. C., eds.), pp. 323–353, Cambridge University Press, Cambridge.

Minor, P. D. and Dimmock, N. J. (1975) Inhibition of synthesis of influenza virus protein: Evidence for two host cell dependent events during multiplication. Virology 67, 114–123.

Molnar, J. (1975) A proposed pathway of plasma glycoprotein synthesis. Mol. Cell. Biochem. 6, 3–14.

Moore, N. F. and Burke, D. C. (1974) Characterization of the structural proteins of different strains of Newcastle disease virus. J. Gen. Virol. 25, 275–289.

Morrison, T. G. and Lodish, H. F. (1975) Site of synthesis of membrane and nonmembrane proteins of vesicular stomatitis virus. J. Biol. Chem. 250, 6955–6962.

Morrison, T. G., Stampfer, M., Baltimore, D. and Lodish, H. F. (1974) Translation of vesicular stomatitis virus messenger RNA by extracts from mammalian and plant cells. J. Virol. 13, 62–72.

Morrison, T. G., Stampfer, M., Lodish, H. F. and Baltimore, D. (1975) In vitro translation of vesicular stomatitis virus messenger RNAs and the existence of a 40 S "plus" strand. In: Negative Strand Viruses (Mahy, B. W. J. and Barry, R. D., eds.), Vol. 1, pp. 293–300, Academic Press, New York.

Moss, B. (1974) Reproduction of poxviruses. In: Comprehensive Virology (Fraenkel-Conrat, H. and Wagner, R. R., eds.), Vol. 3, pp. 405–474. Plenum Press, New York.

Moss, B., Rosenblum, E. N. and Garon, C. F. (1971) Glycoprotein synthesis in cells infected with vaccinia virus. I. Non-virion glycoproteins. Virology 46, 221–232.

Moss, B., Rosenblum, E. N. and Garon, C. F. (1973) Glycoprotein synthesis in cells infected with vaccinia virus. III. Purification and biosynthesis of the virion glycoprotein. Virology 55, 143–156.

Mudd, J. A. (1974) Glycoprotein fragment associated with vesicular stomatitis virus after proteolytic digestion. Virology 62, 573–577.

Murphy, F. A., Harrison, A. K., Gary, G. W., Whitfield, S. G. and Forrester, F. T. (1968) St. Louis encephalitis virus infection of mice. Electron microscopic studies of central nervous system. Lab. Invest. 19, 652–662.

Mussgay, M., Enzmann, P.-J., Horzinek, M. C. and Weiland, E. (1975) Growth cycle of arboviruses in vertebrate and arthropod cells. Progr. Med. Virol. 19, 257–323.

Nagai, Y., Ogura, H. and Klenk, H.'D. (1976) Studies on the assembly of the envelope of Newcastle disease virus. Virology 69, 523–538.

Nicolson, G. L., Poste, G. and Ji, T. H. (1977) The dynamics of cell membrane organization. In: Dynamic Aspects of Cell Surface Organization (Poste, G. and Nicolson, G. L., eds.), pp. 1–73, Elsevier/North-Holland, Amsterdam.

Obijeski, J. F., Marchenko, A. T., Bishop, D. H. L., Cann, B. W. and Murphy, F. A. (1974) Comparative electrophoretic analysis of the virus proteins of four rhabdoviruses. J. Gen. Virol. 22, 21–33.

Oyanagi, S., Ikuta, F. and Ross, E. R. (1969) Electron microscopic observations in mice infected with Japanese encephalitis. Acta Neuropath. 13, 169–181.

Palade, G. (1959) In: Subcellular Particles (Hayeshi, T., ed.), pp. 64–83. Ronald, New York.

Palade, G. (1975) Intracellular aspects of the process of protein synthesis. Science 189, 347–358.

Palese, P., Tobita, K., Ueda, M. and Compans, R. W. (1974) Characterization of temperature sensitive influenza virus mutants defective in neuraminidase. Virology 61, 394–410.

Perdue, M. L., Kemp, M. C., Randall, C. C. and O'Callahan, D. J. (1974) Studies of the molecular anatomy of the L-M cell strain of equine herpes virus type 1: Proteins of the nucleocapsid and intact virion. Virology 59, 201–216.

Pesonen, M. and Renkonen, O. (1976) Sequence and anomeric configuration of monosaccharides in type A glycopeptides of Semliki Forest virus. Biochim. Biophys. Acta 455, 510–525.

Pettersson, R. (1975) The structure of Uukuniemi virus a proposed member of the Bunyaviruses. Med. Biol. 53, 418–424.

Pettersson, R. and Kääriäinen, L. (1973) The ribonucleic acids of Uukuniemi virus, a non-cubical tick-borne arbovirus. Virology 56, 608–619.

Pettersson, R., Kääriäinen, L., von Bonsdorff, C.-H. and Oker-Blom, N. (1971) Structural components of Uukuniemi-virus, a non-cubical tick-borne arbovirus. Virology 46, 712–729.

Pfefferkorn, E. R. and Hunter, H. S. (1963a) Purification and partial chemical analyses of Sindbis virus. Virology 20, 433–445.

Pfefferkorn, E. R. and Hunter, H. S. (1963b) The source of the ribonucleic acid and phospholipid of Sindbis virus. Virology 20, 446–456.

Pfefferkorn, E. R. and Clifford, R. L. (1964) The origin of the protein of Sindbis virus. Virology 23, 217–223.

Pfefferkorn, E. R. and Shapiro, D. (1974) Reproduction of Togaviruses. In: Comprehensive Virology, (Fraenkel-Conrat, H. and Wagner, R. R., eds.), Vol. 2, pp. 171–230. Plenum Press, New York.

Pinter, A. and Compans, R. W. (1975) Sulfated components of enveloped virus. J. Virol. 16, 859–866.

Porter, A., Scroggs, R. A., Marx, P. A. and Kingsbury, D. W. (1975) A ts-mutant of Sendai with an altered hemagglutinin-neuraminidase polypeptide: consequences for virus assembly and cytopathology. Virology 67, 179–187.

Porterfield, J. S., Casals, J., Chumakov, M. P., Gaidamovich, S. Ya., Hannoun, C., Holmes, I. H., Horzinek, M. C., Mussgay, M., Oker-Blom, N. and Russell, P. K. (1975/76) Bunyaviruses and Bunyaviridae. Intervirology 6, 13–24.

Poste, G. (1975) Interaction of concanavalin A with the surface of virus-infected cells. In: Concanavalin A (Chowdhury, T. K. and Weiss, A. K., eds.), pp. 117–152. Plenum Press, New York.

Poste, G. and Waterson, A. P. (1975) Cell fusion by Newcastle disease virus. In: Negative Strand Viruses (Mahy, B. W. J. and Barry, R. D., eds.), Vol. 2, pp. 905–922. Academic Press, New York.

Poste, G., Alexander, D. J., Reeve, P. and Hewlett, G. (1974) Modification of Newcastle disease virus release and cytopathogenicity in cells treated with plant lectins. J. Gen. Virol. 23, 255–270.

Pringle, C. R. and Wunner, W. H. (1975) A comparative study of the structure and function of the vesicular stomatitis virus genome. In: Negative Strand Viruses (Mahy, B. W. J. and Barry, R. D., eds.), Vol. 2, pp. 707–723. Academic Press, London.

Quigley, J. P. and Rifkin, D. B. (1971) Phospholipid composition of Rous sarcoma virus, host cell membranes and other enveloped RNA viruses. Virology 46, 106–116.

Quinn, P. J. (1976) The Molecular Biology of Cell Membranes, Macmillan, London, 229 pp.

Raghow, R. S., Davye, M. W. and Dalgarno, L. (1973) The growth of Semliki Forest virus in cultured mosquito cells: Ultra structural observations. Arch. Ges. Virosforch. 43, 165–168.

Ranki, M. (1972a) Semliki Forest virus replication and canavanine. Thesis. University of Helsinki, Finland.

Ranki, M. (1972b) Nucleocapsid and envelope protein of Semliki Forest virus as affected by canavanine. J. Gen. Virol. 15, 59–67.

Ranki, M., Kääriäinen, L. and Renkonen, O. (1972) Semliki Forest virus glycoproteins and canavanine. Acta. path. micróbiol. Scand. 80B, 760–768.

Rawls, W. E., Ramos, B. A. and Carter, M. F. (1973) Biophysical and biochemical studies of Pichincle virus. In: Lymphocytic Choriomeningitis Virus and Other Arenaviruses (Lehman-Grube, F., ed.), pp. 257–272, Springer-Verlag, Berlin.

Renkonen, O., Kääriäinen, L., Simons, K., and Gahmberg, C. G. (1971) The lipid class composition of Semliki Forest virus and of plasma membranes of the host cells. Virology 46, 318–326.

Renkonen, O., Gahmberg, C. G., Simons, K. and Kääriäinen, L. (1972a) The lipids of the plasma membranes and endoplasmic reticulum from cultured baby hamster kidney cells (BHK21) Biochim. Biophys. Acta 255, 66–78.

Renkonen, O., Kääriäinen, L., Gahmberg, C. G. and Simons, K. (1972b) Lipids of Semliki Forest virus and host cell membranes. In: Current Trends in the Biochemistry of Lipids (Ganguly, J. and Smellie, R. M. S., eds.), pp. 407–422, Academic Press, New York.

Renkonen, O., Kääriäinen, L., Pettersson, R. and Oker-Blom, N. (1972c) The phospholipid composition of Uukuniemi virus, a non-cubical tick-borne arbovirus. Virology 50, 899–901.

Renkonen, O., Luukkonen, A., Brotherus, J. and Kääriäinen, L. (1974) Composition and turnover of membrane lipids in Semliki Forest virus and in host cells. In: Control of Proliferation in Animal Cells (Clarkson, B. and Baserga, R., eds.), pp. 495–504, Cold Spring Harbor Laboratory, New York.

Renkonen, O., Pesonen, M. and Mattila, K. (1977) Oligosaccharides of the membrane glycoproteins of Semliki Forest virus. Nobel Symposium 34 (Abrahamsson, S. and Pascher, I., eds.),pp. 407–426, Plenum Press, New York.

Richards, J. B., and Hamming, F. W. (1972) The transfer of mannose from guanosine diphosphate mannose to dolichol phosphate and protein by pig liver endoplasmic reticulum. Biochem. J. 130, 77–93.

Rifkin, D. B., Compans, R. W. and Reich. E. (1972) A specific labeling procedure for proteins on the outer surface of membranes. J. Biol. Chem. 247, 6432–6437.

Roizman, B. (1969) The herpesviruses—a biochemical definition of the group. Curr. Topics Microbiol. Immunol. 49, 1–79.

Roizman, B. and Furlong, D. (1974) The replication of herpesviruses. In: Comprehensive Virology (Fraenkel-Conrat, H. and Wagner, R. R., eds.), Vol. 3, pp. 229–403, Plenum Press, New York.

Rolleston, F. S. (1974) Membrane bound and free ribosomes. Sub-cell biochemistry 3, 91–117.

Rott, R., Becht, H., Hammer, G., Klenk, H.-D. and Scholtissek, C. (1975) Changes in the surface of the host cell after infection with envelope viruses. In: Negative Strand Viruses (Mahy, B. W. J., and Barry, R. D., eds.), Vol. 2, pp. 843–857. Academic Press, London.

Roux, L. and Kolakofsky, D. (1975) Isolation of RNA transcripts from the entire Sendai viral genome. J. Virol. 16, 1426–1439.

Rubin, H., Baluda, M. and Hotchin, J. E. (1955) The maturation of western equine encephalomyelitis virus and its release from chick embryo cells in suspension. J. Exp. Med. 101, 205–212.

Sarov, I. and Joklik, W. K. (1972) Studies on the nature and location of the capsid polypeptides of vaccinia virions. Virology 50, 579–592.

Schachter, H. I., Jabbal, I., Hudgin, R. L. and Pinteric, L. (1970) Intracellular localization of host sugar nucleotide glycoprotein glycosyl transferases in a Golgi-rich fraction. J. Biol. Chem. 245, 1090–1100.

Scheele, C. M. and Pfefferkorn, E. R. (1969) Kinetics of incorporation of structural proteins into Sindbis virions. J. Virol. 3, 369–375.

Scheid, A. and Choppin, P. W. (1973) Isolation and purification of the envelope protein of Newcastle disease virus. J. Virol. 11, 263–271.

Scheid, A., and Choppin, P. W. (1974a) Identification of biological activities of paramyxovirus glycoproteins. Activation of cell fusion, hemolysis and infectivity by proteolytic cleavage of an inactive precursor protein of Sendai virus. Virology 57, 475–490.

Scheid, A., and Choppin, P. W. (1974b) The hemagglutinating and neuraminidase protein of a paramyxovirus: Interaction with neuraminic acid in affinity chromatography. Virology 62, 125–133.

Scheid, A., and Choppin, P. W. (1975) Isolation of paramyxovirus glycoproteins and identification of their biological properties. In: Negative Strand Viruses (Mahy, B. W. J., and Barry, R. D., eds.) Vol. 1, pp. 177–192. Academic Press, London.

Scheid, A., and Choppin, P. W. (1976) Protease activation mutants of Sendai virus. Virology 69, 265–277.

Scheid, A., Caliguiri, L. A., Compans, R. W. and Choppin, P. W. (1972) Isolation of paramyxovirus glycoproteins. Association of both hemagglutinating and neuraminidase activities with the larger SV5 glycoprotein. Virology 50, 640–652.

Schlesinger, S. and Schlesinger, M. J. (1972) Formation of Sindbis virus proteins: Identification of a precursor for one of the envelope proteins. J. Virol. 10, 925–932.

Schlesinger, M. J. and Schlesinger, S. (1973) Large-molecular-weight precursors of Sindbis virus proteins. J. Virol. 11, 1013–1016.

Schlesinger, H. R., Wells, H. J. and Hummeler, K. (1973) Comparison of the lipids of intracellular and extracellular rabies viruses. J. Virol. 12, 1028–1030.

Schlesinger, M. J., Schlesinger, S. and Burge, B. W. (1972) Identification of a second glycoprotein in Sindbis virus. Virology 47, 539–541.

Schlesinger, S., Gottlieb, C., Feil, P., Gelb, N. and Kornfield, S. (1976) Growth of enveloped RNA viruses in a line of chinese hamster ovary cells with deficient N-acetyl-glucosaminyltransferase activity. J. Virol. 17, 239–246.

Schloemer, R. H. and Wagner, R. R. (1974) Sialoglycoprotein of vesicular stomatitis virus: role of the neuraminic acid in infection. J. Virol. 14, 270–281.

Schloemer, R. H. and Wagner, R. R. (1975a) Association of vesicular stomatitis glycoprotein with virion membrane: characterization of the lipophilic tail fragment. J. Virol. 16, 237–249.

Schloemer, R. H. and Wagner, R. R. (1975b) Cellular adsorption function of the sialoglycoprotein of vesicular stomatitis virus and its neuraminic acid. J. Virol. 15, 882–893.

Schloemer, R. H. and Wagner, R. R. (1975c) Mosquito cells infected with vesicular stomatitis virus yield unsialylated virions of low infectivity. J. Virol. 15, 1029–1032.

Scholtissek, C. and Kaluza, G. (1975) Interference with the glycosylation of Semliki Forest virus proteins. Med. Biol. 53, 357–364.

Scholtissek, C., Kaluza, G., Schmidt, M. and Rott, R. (1975a) Influence of sugar derivatives on glycoprotein synthesis of enveloped viruses. In: Negative Strand Viruses (Mahy, B. W. J. and Barry, R. D., eds.), Vol. 2, pp. 669–683. Academic Press, London.

Scholtissek, C., Rott, R. and Klenk, H. D. (1975b) Two different mechanisms of the inhibition of the multiplication of enveloped viruses by glycosamine. Virology 65, 191–200.

Schulze, I. T. (1970) The structure of influenza virus. I. The polypeptides of the virion. Virology 42, 890–904.

Schulze, I. T. (1972) The structure of influenza virus. II. A model based on the morphology and composition of subviral particles. Virology 47, 181–196.

Schulze, I. T. (1973) Structure of the influenza virion. Adv. Virus Res. 18, 1–55.

Schwartz, R. T. and Klenk, H.-D. (1974) Inhibition of glycosylation of the influenza virus hemag-glutinen. J. Virol. 14, 1023–1034.

Sefton, B. M. (1976) Virus-dependent glycosylation. J. Virol. 17, 85–93.

Sefton, B. M. and Burge, B. W. (1973) Biosynthesis of the Sindbis virus carbohydrates. J. Virol. 12, 1366–1374.

Sefton, B. M. and Keegstra, K. (1974) Glycoproteins of Sindbis virus: Preliminary characterization of the oligosaccharides. J. Virol. 14, 522–530.

Sefton, B. M. and Gaffney, B. J. (1974) Effect of the viral proteins on the fluidity of the membrane lipids in Sindbis virus. J. Mol. Biol. 90, 343–358.

Sefton, B. M., Wickus, G. G. and Burge, B. W. (1973) Enzymatic iodination of Sindbis virus proteins. J. Virol. 11, 730–735.

Semmel, M., Israel, A. and Auderbert, F. (1975) Phospholipids in surface membranes of Newcastle disease virus infected cells. In: Negative Strand Viruses (Mahy, B. W. J. and Barry, R. D., eds.), Vol. 2, pp. 875–883. Academic Press, London.

Siekevitz, P., Palade, G. E., Dallner, G., Ohad, I. and Omura, I. (1967) The biogenesis of intracellular membranes, In: Organizational Biosynthesis (Vogel, H. J., Lampen, J. O. and Bryson, V., eds.), pp. 331–362. Academic Press, New York.

Simmons, D. T. and Strauss, J. H. (1972a) Replication of Sindbis virus. I. Relative size and genetic content of 26 S and 49 S RNA. J. Mol. Biol. 71, 599–614.

Simmons, D. T. and Strauss, J. H. (1972b) Replication of Sindbis virus: Multiple forms of double-stranded RNA isolated from infected cells. J. Mol. Biol. 71, 615–631.

Simmons, D. T. and Strauss, J. H. (1974a) Replication of Sindbis virus: V. Polyribosomes and mRNA in infected cells. J. Virol. 14, 552–559.

Simmons, D. T. and Strauss, J. H. (1974b) Translation of Sindbis virus 26 S RNA and 49 S RNA in lysates of rabbit reticulocytes. J. Mol. Biol. 86, 397–409.

Simons, K. and Kääriäinen, L. (1970) Characterization of the Semliki Forest virus core and envelope protein. Biochem. Biophys. Res. Commun. 38, 981–988.

Simons, K., Helenius, A. and Garoff, H. (1973a) Solubilization of the membrane proteins from Semliki Forest virus with Triton X-100. J. Mol. Biol. 80, 119–133.

Simons, K., Keränen, S. and Kääriäinen, L. (1973b) Identification of a precursor for one of the Semliki Forest virus membrane proteins. FEBS Lett. 29, 87–91.

Simons, K., Kääriäinen, L., Renkonen, O., Ghamberg, C. G., Garoff, H., Helenius, A., Keränen, S., Laine, R., Ranki, M., Söderlund, H. and Utermann, G. (1973c) Semliki Forest virus envelope as a simple membrane odel. In: Membrane Mediated Information (Kent, P. W., ed.), Vol. 2, pp. 81–99. Medical and Technical Publishing Co. Lancaster, England.

Simons, K., Garoff, H., Helenius, A., Kääriäinen, L. and Renkonen, O. (1975) Structure and assembly of virus membranes. In: Perspectives in Membrane Biology (Estrado, O. and Gitler, G., eds.). Academic Press, New York.

Simons, K., Garoff, H. and Helenius, A. (1977) The glycoproteins of the Semliki Forest virus membrane. In: Membrane Techniques (Capaldi, R. A., ed.), Vol. 1, Marcel Dekker, New York, in press.

Singer, S. J. (1974) The molecular organization of membranes. Ann. Rev. Biochem. 43, 805–833.

Singer, S. J. and Nicolson, G. L. (1972) The fluid mosaic model of the structure of cell membrane. Science 175, 720–731.

Skehel, J. J. (1971) Estimation of the molecular weight of the influenza virus genome. J. Gen. Virol. 11, 103–109.

Skehel, J. J. (1972) Polypeptide synthesis in influenza virus infected cells. Virology 49, 23–36.

Skehel, J. J. and Waterfield, M. D. (1975) Studies on the primary structure of influenza hemagglutinin. Proc. Nat. Acad. Sci. U.S.A. 72, 93–97.

Smith, A. E., Wheeler, T., Glanville, N. and Kääriäinen, L. (1974) Translation of Semliki-Forest-virus 42-S RNA in a mouse cell-free system to give virus-coat proteins. Eur. J. Biochem. 49, 101–110.

Söderlund, H. (1973) Kinetics of formation of the Semliki Forest virus nucleocapsid. Intervirology 1, 354–361.

Söderlund, H. (1976) The post-translational processing of Semliki Forest virus structural polypeptides in puromycin treated cells. FEBS Lett. 63, 56–58.

Söderlund, H. and Kääriäinen, L. (1974) Association of capsid protein with Semliki Forest virus messenger RNAs. Acta Pathol. Microbiol. Scand. 82B, 33–40.

Söderlund, H., Glanville, N. and Kääriäinen, L. (1973/74) Polysomal RNAs in Semliki Forest virus-infected cells. Intervirology 2, 110–113.

Söderlund, H., Kääriäinen, L. and von Bonsdorff, C.-H. (1975) Properties of Semliki Forest virus nucleocapsid. Med. Biol. 53, 412–417.

Sokol, F., Stancek, D. and Koprowski, H. (1971) Structural proteins of rabies virus. J. Virol. 7, 241–242.

Sonnabend, J. A., Martin, E. M. and Mecs, E. (1967) Viral specific RNAs in infected cells. Nature 213, 365–367.

Spiro, R. G. (1973) Glycoproteins. Adv. Protein Chem. 27, 349–467.

Stanley, P. and Haslam, E. A. (1971) The polypeptides of influenza virus. V. Localization of polypeptides in the virion by iodination techniques. Virology 46, 764–773.

Stohlman, S. A., Wisseman, C. L., Eylar, O. R. and Silverman, D. J. (1975) Dengue virus-induced modification of host cell membranes. J. Virol. 16, 1017–1026.

Stollar, V., Stollar, B. D., Koo, R., Harrap, K. A. and Schlesinger, R. W. (1976) Sialic acid contents of Sindbis virus from vertebrate and mosquito cells. Equivalence of biological and immunological viral properties. Virology 69, 104–115.

Strauss, J. H., Burge, B. W., Pfefferkorn, E. R. and Darnell, J. E. (1968) Identification of the membrane protein and "core" protein of Sindbis virus. Proc. Nat. Acad. Sci. U.S.A. 59, 533–537.

Strauss, J. H., Burge, B. W. and Darnell, J. E. (1969) Sindbis virus infection of chick and hamster cells: Synthesis of virus-specific proteins. Virology 37, 367–376.

Strauss, J. H., Burge, B. W. and Darnell, J. E. (1970) Carbohydrate content of the membrane protein of Sindbis virus. J. Mol. Biol. 47, 347–448.

Strauss, J. H. and Strauss, E. G. (1976) Togaviruses. In: The Molecular Biology of Animal Viruses (Nayak, D. P., ed.). Marcel Dekker, New York. In press.

Tan, K. B., Sambrook, J. F. and Bellet, A. J. P. (1969) Semliki Forest virus temperature-sensitive mutants: Isolation and characterization. Virology 38, 427–439.

Tan, K. B. (1970) Electron microscopy of cells infected with Semliki Forest virus temperature-sensitive mutants: Correlation of ultrastructural and physiological observations. J. Virol. 5, 632–638.

Tkacz, J. S. and Lampen, O. (1975) Tunicamycin inhibition of polyisopronyl N-acetylglucosaminyl pyrophosphate formation in calf-liver microsomes. Biochem. Biophys. Res. Commun. 65, 248–257.

Tsai, K. H. and Lenard, J. (1975) Assembly of influenza virus membrane bilayer demonstrable with phospholipase C. Nature 253, 554–555.

Tuomi, K., Kääriäinen, L. and Söderlund, H. (1975) Quantitation of Semliki Forest virus RNAs in infected cells using ^{32}P-equilibrium labeling. Nucleic Acid. Res. 2, 555–565.

Tzagoloff, A., Rubin, M. S. and Dierra, M. F. (1973) Biosynthesis of mitochondrial enzymes. Biochim. Biophys. Acta 301, 71–104.

Ulmanen, I., Söderlund, H. and Kääriäinen, L. (1976) Semliki Forest virus capsid protein associates with the 60 S ribosomal subunit in infected cells. J. Virol. 20, 203–210.

Utermann, G. and Simons, K. (1974) Studies on the amphipathic nature of the membrane proteins in Semliki Forest virus. J. Mol. Biol. 85, 569–587.

Wagner, R. R. (1975) Reproduction of rhabdoviruses. In: Comprehensive Virology (Fraenkel-Conrat, H. and Wagner, R. R., eds.) Vol. 4, pp. 1–93. Plenum Press. New York.

Wagner, R. R., Schnaitman, T. C., Snyder, R. M. and Schnaitman, C. A. (1969) Protein composition of the structural components of vesicular stomatitis virus. J. Virol. 3, 611–618.

Wagner, R. R., Snyder, R. M. and Yamazaki, S. (1970) Proteins of vesicular stomatitis virus: Kinetics and cellular sites synthesis. J. Virol. 5, 548–558.

Wagner, R. R., Prevec, L., Brown, F., Summers, D. F., Sokol, F. and MacLeod, R. (1972) Classification of rhabdovirus proteins: a proposal. J. Virol. 10, 1228–1230.

Wagner, R. R., Pettersson, E. and Dallner, G. (1973) Association of the two glycosyl transferase activities of glycoprotein synthesis with low equilibrium density smooth microsomes. J. Cell. Sci. 12, 603–615.

Wagner, R. R., Emerson, S. V., Imblum, R. L. and Kelley, I. M. (1975) Structure function relationship of the proteins of vesicular stomatitis virus. In: Negative Strand Viruses (Mahy, B. W. J. and Barry, R. D., eds.), Vol. 1, pp. 1–19, Academic Press, London.

Waite, M. R. F. and Pfefferkorn, E. R. (1970) Inhibition of Sindbis virus production by media of low ionic strength: Intracellular events and requirements for reversal. J. Virol. 5, 60–71.

Waite, M. R. F., Brown, D. T. and Pfefferkorn, E. R. (1972) Inhibition of Sindbis virus release by media of low ionic strength: an electron microscope study. J. Virol. 10, 537–544.

Wallach, D. F. H. (1972) The dispositions of proteins in the plasma membranes of animal cells. Biochim. Biophys. Acta 265, 61–69.

Wallach, D. F. H. (1975) Membrane Molecular Biology of Neoplastic Cells. 525 pp. Elsevier/North-Holland, Amsterdam.

Weintraub, S. and Dales, S. (1974) Biogenesis of poxviruses: Genetically controlled modifications of structural and functional components of the plasma membrane. Virology 60, 96–127.

Weitzman, S. and Scharff, M.D. (1976) Mouse myeloma mutants blocked in the assembly, glycosylation and secretion of immunoglobulin. J. Mol. Biol. 102, 237–252.

Wengler, G. and Wengler, G. (1976) Localization of the 26 S RNA sequence on the viral genome type 42 S RNA isolated from SFV infected cells. Virology 73, 190–199.

Wengler, G., Beato, M. and Hackemank, B.-A. (1974) Translation of 26 S virus-specific RNA from Semliki Forest virus-infected cells in vitro. Virology 61, 120–128.

Whaley, W. G., Dauwalder, M. and Kephart, J. E. (1972) Golgi Apparatus: Influence on cell surfaces. Science 175, 596–599.

Wrigley, N. G., Skehel, J. J., Charlwood, P. A. and Brand, C. M. (1973) The size and shape of influenza virus neuraminidase. Virology 51, 525–529.

Yin, F. H. (1969) Temperature-sensitive behavior of the hemagglutinin in a temperature-sensitive mutant virion of Sindbis virus. J. Virol. 4, 547–548.

Zaides, V. M., Selimova, L. M., Zhirnov, O. P. and Bukrinskaya, A. G. (1975) Protein synthesis in Sendai-virus-infected cells. J. Gen. Virol. 27, 319–327.

Závada, J. (1972) Pseudotypes of VSV with the coat of murine leukemia and of avian myeloblastosis in virus. J. Gen. Virol. 15, 183–191.

Zee, Y. C., Hackett, A. J. and Talens, L. (1970) Vesicular stomatitis virus maturation sites in six different host cells. J. Gen. Virol. 7, 95–102.

In vitro and in vivo assembly of bacteriophage PM2: a model for protein-lipid interactions

12

Richard M. FRANKLIN

1. Introduction

The physicochemical interactions that lead to the formation of complex biological structures are now amenable to experimental analysis using simple biological systems. When such interactions are understood in simple systems it will be possible to look for similar interactions in more complex systems. In any case, the most suitable systems are those that can be purified easily in large quantities and in which there is some readily measurable biological activity. Some bacterial viruses fulfill these criteria admirably and it has been possible to reconstitute very simple RNA bacteriophages completely, and some of the very complex coliphages have been partially reconstituted (Casjens and King, 1975). Many details of protein-protein interactions have also been revealed in studies on the reconstitution of both rod-shaped and spherical plant viruses (Casjens and King, 1975). Thus it was reasonable to utilize a lipid-containing bacterial virus for studies on the protein-lipid and protein-protein interactions that occur during the formation of a biological membrane. Until recently bacteriophage PM2, which grows on the marine pseudomonad *Pseudomonas* BAL-31, was the only lipid-containing bacteriophage that had been studied from the viewpoint of molecular biology. Now a second lipid-containing bacteriophage, ϕ6, that grows on a phytopathogenic pseudomonad has also been studied (Vidaver et al., 1973). This virus will be discussed and compared with PM2.

2. Chemical composition of bacteriophage PM2

The basic facts concerning the composition of bacteriophage PM2 are given in Table 1. There are four proteins in PM2, designated simply as proteins I to IV. The properties of these proteins will be discussed later. The nucleic acid is a circular double-stranded DNA, which is superhelical and has a molecular weight of 6.3 to 6.4×10^6 (Camerini-Otero and Franklin, 1975). Although the composi-

G. Poste and G.L. Nicolson (eds.)   *The Synthesis, Assembly and Turnover of Cell Surface Components*
© *Elsevier/North-Holland Biomedical Press, 1977.*

Table 1
Chemical Composition of Bacteriophage PM2

Protein	72% by weight
Nucleic Acid	14.3% by weight
Lipid	12.6–14.0% by weight
	Phospholipid 92–93% of Total Lipid
	Neutral Lipid 7–8% of Total Lipid

tion is very similar to that of the host cell, both having a G-C content of 42 to 43%, this likeness seems to be fortuitous since there are no regions of homology between the viral and host cell DNA (Franklin et al., 1969).

Most of the lipid is phospholipid and the composition of the phospholipids, as a percentage of the total lipid fraction in the virion grown under standard conditions (broth or synthetic medium at 25°C) in wild-type *Pseudomonas* BAL-31 is 59 to 64% phosphatidylglycerol (PG), 28 to 39% phosphatidylethanolamine (PE), and less than 1% acylphosphatidylglycerol (APG). Although this variation in composition may depend on growth conditions, the phospholipid composition is very different from that of the host cell, which has 20 to 22% PG, 76 to 78% PE, and 0.3 to 0.6% APG (Braunstein and Franklin, 1971; Tsukagoshi et al., 1975a). Braunstein and Franklin (1971) suggested that the unknown phospholipid, compound X, was a derivative of phosphatidylglycerol, since alkaline hydrolysis of compound X yielded glycerolphosphorylglycerol. A tentative identification of compound X as acylphosphatidylglycerol was made by Diedrich and Cota-Robles (1974) by comparing the thin-layer chromatographic behavior of compound X with the published values of a *Salmonella typhimurium* acylphosphatidylglycerol (Olsen and Ballou, 1971). Identification by comparison of R_f values can be rather dangerous, however. Compound X also has an R_f similar to an enzymatically synthesized cardiolipin, using an *Eschericia coli* membrane preparation as a source of cardiolipin synthetase and *Pseudomonas* BAL-31 PG as substrate (Tsukagoshi et al., 1975b). No cardiolipin synthetase activity could be detected, however, in *Pseudomonas* BAL-31 extracts (Tsukagoshi et al., 1975b). Positive identification of compound X was achieved recently by a combination of direct chemical analysis and nuclear magnetic resonance spectroscopy (Tsukagoshi et al., 1976b). The third acyl group is located on the primary hydroxyl group of the second glycerol moiety. Unfortunately the biological significance of this minor phospholipid component is unclear, as will be discussed further on.

3. Structure of bacteriophage PM2

Bacteriophage PM2 appears to be isometric and there are suggestions of icosahedral symmetry in the electron micrographs of the virion. The particles appear hexagonal in projection and evidence of two- and three-fold symmetry axes can be seen (Silbert et al., 1969). Unfortunately, it has been impossible to visualize individual morphological units. The average diameter of the particles negatively stained with phosphotungstate is 614 Å. Short, splayed spikelike structures project from the vertices of the particle. In some negatively stained particles it is possible to distinguish an inner hexagonal form separated from the outer shell by a stain impenetrable region. From thin sections of virus stained with uranyl acetate and lead citrate, it appears that there may be a bilayer arrangement of the phospholipids. The densely staining central region has been interpreted as being the DNA. The structure suggested from ultrastructural studies is therefore a bilayer with an outer shell of protein and an inner nucleocapsid. This structure was confirmed by low angle x-ray diffraction studies (Harrison et al., 1971) in which a structure was obtained to a resolution of about 25 Å (Fig. 1). It was possible to demonstrate an outer shell and an inner shell of protein and a phospholipid bilayer between these two protein shells. An estimate was made of the absolute electron density in the particle and, although this is certainly subject to considerable error, the density in the middle of the bilayer of 0.30 electron/Å^3 was certainly higher than that expected when only hydrocarbon chains were present in the middle of the bilayer. From the chemical composition discussed above and from the volume occupied by the bilayer only about 50% of this region can be occupied by phospholipid (Camerini-Otero and Franklin, 1972). By implication, the remaining volume is probably occupied by protein and current evidence supports this hypothesis (Franklin et al., 1976).

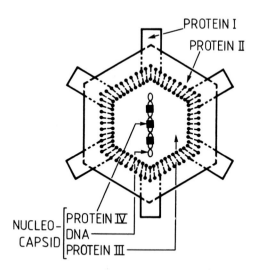

Fig. 1. Model of bacteriophage PM2. Protein I forms the spikes projecting from the vertices of the icosahedrally-shaped particle, protein II the outer protein shell, and protein III the inner (nucleocapsid) shell. Boundaries of the outer and inner protein shells are not shown since both regions probably penetrate into the bilayer. Protein IV is closely associated with viral DNA and may also interact with protein III and the bilayer. The outer lamella of the bilayer is predominantly phosphatidylglycerol and the inner lamella is mainly phosphatidylethanolamine.

4. Localization of the PM2 viral proteins

Several properties of the four proteins of PM2 are described in Table 2. The molecular weights calculated from the amino acid composition of the individual proteins are compared with the approximate molecular weights obtained by gel electrophoresis (PAGE) in the presence of SDS (Schäfer et al., 1974; Hinnen et al., 1976). Molecular weights obtained by PAGE and those obtained from chemical analyses are in good agreement except in the case of protein IV. Basing our considerations of molecular weight on the number of cysteine residues in the protein, it is clear that the molecular weight of protein IV is 6480 with one cysteine residue rather than 12960 with two cysteine residues. Anomalies in polypeptide molecular weight determinations by PAGE are not unusual, however, in very small polypeptides such as protein IV.

Proteins I, III, and IV are slightly acidic with isoelectric points near neutrality (Table 4; Schäfer et al., 1974). Protein II, however, is very basic. In 6M urea plus 10 mM $CaCl_2$, the isoelectric point is 12.3; if 10 mM EDTA is substituted for $CaCl_2$, the pI drops to 12.1. After treatment with 6M guanidine hydrochloride plus 10 mM EDTA followed by dialysis against 6M urea plus 10 mM EDTA, the isoelectric point is 9 (Schäfer et al., 1974). Thus protein III is a basic protein in which the degree of basicity is controlled by Ca^{2+}. Probably protein II is not totally denatured in 6 M urea, and the only way to remove the Ca^{2+} completely seems to be to denature the protein in guanidine hydrochloride in the presence of EDTA. The isoelectric point of protein II calculated from the amino acid composition is 8.4 (Hinnen et al., 1976), which is in reasonable agreement with the experimentally determined pI. If all of the carboxyl groups, a total of 10 or 11, are complexed with Ca^{2+}, the isoelectric point will be raised to about 10.5. This also agrees with the experimental findings. An attempt was made to demonstrate the binding of calcium directly to protein II using equilibrium dialysis and radioactive Ca^{2+}. Unfortunately it was necessary to use a very high urea concentration (8 M) to dissolve sufficient quantities of protein II for this experiment, and in 8 M urea there was little or no binding of Ca^{2+} to protein II (Hinnen et al., 1976). This may be due to a conformational dependence of the binding of Ca^{2+} to protein II or to a very low binding constant. The dissociation constant would have to be below 10^{-3} M to detect binding

Table 2
Some Properties of Bacteriophage PM2 Proteins

Protein	Molecular weight[a]	Molecular weight[b]	Cysteine	Absences
I	43,000	43,640	2	—
II	26,000–27,000	27,310	2	—
III	12,500	13,250	1	Tyr, His
IV	4,700	6,480	1	His

[a] Determined by PAGE.
[b] Calculated from the amino acid composition.

under the conditions used. Since the dissociation constants of several calcium-binding proteins are of the order of magnitude of 10^{-5} M (Kretsinger, 1974), it would seem that the conformational effects are largely responsible for the lack of binding.

Protein I forms the spikes, as shown by correlating the loss of protein I following bromelain digestion with disappearance of the spikes (Hinnen et al., 1974). Spikeless particles cannot absorb to the host cell, implying a role for the spikes in attachment of the virion to the bacterium.

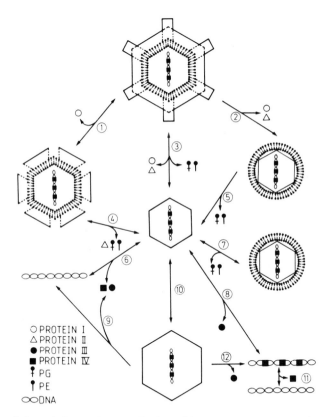

Fig. 2. Scheme of degradation and reconstitution of bacteriophage PM2. Solvent conditons involved in the individual steps are tabulated below. The nature of the structures involved can be derived by reference to Fig. 1 and to the text. NTC is the standard buffer used in our work on PM2 and has the following composition: 1 M NaCl; 20 mM Tris. HCl, pH 7.2 at 20°C; 0.01 M CaCl₂: **1–2:** NTC plus 1 M urea. Pathway (1) gives a yield of 95%, and pathway (2) a yield of 5%; **3–5:** 4–6 M urea. Protein II is completely removed from the nucleocapsid particle only in 6 M urea; **6,9:** 8 M urea; 1% β-alanine, acetic acid to pH 4.5; **7:**4 M urea, NTC plus phospholipids; dialyzed to 1 M urea, NTC; **8,12:** 8.5 to 9 M urea, NTC, 20°C; a yield of 20–25%, based on the initial amount of DNA; **11:** 8.5 M urea, NTC plus a further addition of NaCl to 3 M final concentration, 20°C; **10:**7.2 M urea, NTC dialyzed to 7.3 M urea, NTC.

Protein II forms the outer protein shell. The external location of protein II was first suspected from calculations of the volume occupied by this protein (Harrison et al., 1971). Under conditions in which the virion bilayer is intact, only protein II is labeled with the diazonium salt of sulfanilic acid, a bulky charged reagent that couples to tyrosine and histidine and does not penetrate the viral bilayer under normal conditions (Hinnen et al., 1974).

The external location of proteins I and II has also been confirmed by stepwise degradation of the virion in the presence of urea (Hinnen et al., 1974). Urea will dissociate protein-protein complexes, as well as unfold proteins at higher concentration, by virtue of the hydrophobic effect (Kauzmann, 1959; Nozaki and Tanford, 1963; Tanford, 1964). The hydrophobic effect is the energy gained by the reduction in the number of nonpolar contacts with water that occur when a protein folds. This energy gain compensates for the loss of chain configurational energy. Stabilization and specific configuration of either individual polypeptides or oligomers then occurs by point-to-point interactions via hydrogen bonds, electrostatic forces, and Van der Waals' forces (Kauzmann, 1959; Ramachandran and Sasisekharan, 1968; Chothia, 1975; Chothia and Janin, 1975). Urea provides an environment favorable to the solubilization of nonpolar amino acid moieties. The mechanistic basis of this solubilization effect is not well understood (Nozaki and Tanford, 1963). In the presence of 1 M urea, PM2 dissociates in two ways. The major particle is a lipid-containing nucleocapsid. This structure contains proteins III and IV, the viral DNA, the lipids, and small and variable amounts of protein II (Hinnen et al., 1974). The minor particle is a spikeless virion with only protein I missing. Protein I and most of protein II (derived from the major particle) are solubilized. In 4.5 M urea the only stable structure is the nucleocapsid, which contains proteins III and IV and the DNA (Hinnen et al., 1974). Proteins I and II and the lipid are solubilized but some of the lipid is probably complexed with protein II, as we shall see later. From volume occupancy considerations, protein III must form the nucleocapsid shell (Franklin, 1974).

Localization of protein IV was achieved using the method of Boni and Budowsky (1973) to identify proteins that are in close proximity to nucleic acids (Hinnen et al., 1974). In this method the C_5-C_6 double bond of cytosine is saturated by reaction with sodium bisulfite ($Na_2S_2O_5$), a strong nucleophilic reagent. This in turn increases the reactivity of the C_4 electrophilic center, allowing substitution of the exocyclic amino group by weak nucleophilic reagents, such as a nearby protein amino group, under very mild conditions. Thus activation of cytosine in DNA or RNA results in a covalent cross-linking to any adjacent protein. The protein-DNA (or RNA) complex can be isolated by appropriate means, such as phenol extraction followed by ultracentrifugation in a sucrose gradient. The protein can then be cleaved from the nucleic acid by a nucleophilic reagent such as o-methylhydroxylamine or its equimolar mixture with bisulfite. Bacteriophage PM2 was treated with 1M $Na_2S_2O_5$ at pH 7.4 and the virus was then dissociated by heating to 55°C for 5 minutes in the presence of 1%

SDS plus 1% 2-mercaptoethanol.[1] The dissociated virus was sedimented on a sucrose gradient and the DNA-containing fraction treated with 1M $Na_2S_2O_5$ plus 1M o-methylhydroxylamine to cleave any associated protein. The cleaved protein was identified as protein IV by gel electrophorsis.

The association of protein IV with viral DNA is of interest in relation to the polynucleotide-dependent polynucleotide-pyrophosphorylase activity of this protein (Schäfer and Franklin, 1975a). This enzyme catalyzes polynucleotide polymerization but is not template-specific, and the product did not bind covalently to the 3'-termini of the primer nucleic acids or polynucleotides. Both ribonucleoside and deoxyribonucleoside triphosphates were converted into an acid-insoluble product. We do not know the possible role of this enzyme in the biosynthesis of PM2, particularly since the isolated DNA is itself infectious (van der Schans et al., 1971). Protein IV is a rather small polypeptide and probably the enzymatic activity is present only in a dimer of it (Schäfer and Franklin, 1975a).

Protein IV, as well as protein III, have the solubility properties ascribed to proteolipids (Camerini-Otero et al., 1972; Schäfer et al., 1974; Franklin et al., 1976). Furthermore, protein IV must lie adjacent to protein III and the bilayer as well as to the viral DNA, since after cross-linking with glutaraldehyde or dimethyl suberimidate, phosphatidylethanolamine became associated with the protein III plus IV fraction obtained after viral lipid extraction and proteins III and IV disappeared completely from the standard PAGE (Schäfer et al., 1975). Some of the large molecular weight polypeptides that appeared in the PAGE pattern may possibly be cross-linked complexes of proteins II and III. Protein IV is acidic with an isoelectric point of 5.5 (Schäfer et al., 1974). This raises the question of how can the proteolipid and acidic properties of this protein be reconciled to its association with DNA? A partial amino acid sequence has been determined for protein IV (Hinnen et al., 1976). The first 27 amino acids from the N-terminus are polar with a clustering of basic amino acids. This is followed by a hydrophobic segment of at least 16 amino acids, which has yet to be sequenced. The C-terminal sequence is polar in nature. The N-terminal basic amino acids probably interact with the DNA, whereas the hydrophobic central region may form a loop that interacts either with protein III or/and the phospholipid bilayer. These sequence studies also explain why protein IV has limited solubility in water despite its high hydrophobicity index and low polarity (Table 4).

Besides the polynucleotide-dependent polynucleotide-pyrophosphorylase activity of protein IV, two further virion-associated enzymatic activities have been reported: (1) an endolysin activity, measured by the hydrolysis of ^{3}H-murein

---

[1]In the study by Hinnen et al. (1974), the virus was treated in both the presence and absence of 0.05% Triton X-100. In the presence of Triton X-100, however, the virus dissociated under the incubation conditions given (but not in the standard buffer in which the virus was stored), and therefore Triton X-100 was not used in the main experiments.

prepared from *E. coli* W945T3282 (a mutant requiring several amino acids, which is diaminopimelate decarboxylase negative), grown in the presence of ^{3}H-diaminopimelate, is found in the PM2-infected host cell but not in the uninfected cell (Tsukagoshi et al., 1977). This activity is also found in highly purified PM2 virus preparations. There is an increased activity in virus that has been disrupted by a freeze-thaw cycle. The enzymatic specific activity is highest in the protein III fraction, and we can make a definitive assignment of the endolysin activity to protein III; and (2) an endonuclease activity that converts PM2 superhelical DNA into a linear form has also been reported to be associated with the virion (Laval, 1974). The activity is the same in the presence or absence of nonionic detergents and could well be a host cell contaminant in partially purified virus preparations. We have confirmed this supposition while testing for endonuclease activity at different stages of viral purification (Franklin et al., 1976). After the second CsCl gradient centrifugation (Hinnen et al., 1974) endonuclease activity could still be found in the banded virus. Following the final purification step, however, no endonuclease activity remained associated with the virion fraction. In this step the virus is sedimented in a sucrose gradient in the presence of 3 M NaCl and all remaining traces of host cell components, particularly membrane fragments, are separated from the virus fraction (Hinnen et al., 1974).

5. Protein-lipid interactions in bacteriophage PM2

The basicity of protein II, which forms the outer protein shell, and the hydrophobicity of protein III, which forms the inner protein shell, lead us to postulate that strong electrostatic forces between protein II and the acidic phospholipid PG play a dominant role in both the stabilization of the virus bilayer and in the selection of PG as the predominant phospholipid (Schäfer et al., 1974). At the same time we consider that protein III must undergo a hydrophobic interaction with the bilayer (Schäfer et al., 1974). At pH 7, the isoelectric point of the intact virion, protein II should be positively charged and thus interact electrostatically with the negatively charged PG.

Several experiments support our hypothesis concerning protein-lipid interactions in bacteriophage PM2:

1. We have already mentioned the dissociation of PM2 in increasing urea concentrations. The ready solubility of protein I in 1 M urea suggests it is the protein most weakly bound to the virion. There is no evidence that this protein, the spike protein, penetrates the bilayer. Protein I is the only water-soluble protein and has the lowest hydrophobicity index and the highest polarity (Table 4; Hinnen et al., 1976). This would fit well with the position of the spikes on the viral surface, the necessity of the spikes to interact with the bacterial surface in an aqueous milieu, and the weak

bonding and lack of interaction with the viral bilayer mentioned earlier.

When protein II is solubilized in 4 M urea, some phospholipid remains associated with it (R. Schäfer and R. M. Franklin, unpublished observations) and this must be PG, according to the two-phase distribution experiments mentioned below in item 3. In unpublished experiments, PM2 was labeled in vivo with ^{32}P -phosphate and upon dissociation in 4 M urea, the protein II was found to carry ^{32}P -phospholipid label with it during electrophoresis on Cellogel in the presence of 4 M urea. This observation favors the existence of strong electrostatic interactions between protein II and PG. The hydrophobicity of protein II is similar to that of a number of globular proteins (Bigelow, 1967). The polarity would correlate with the electrostatic interaction with PG but does not relate well to the solubility properties of this protein. It is insoluble in aqueous solutions but can be solubilized in 8 M urea or in 1% SDS. Part of the difficulty in solubilizing this protein may be due, however, to residual PG associated with it.

The nucleocapsid remaining in 4 M urea must have a hydrophobic surface since it aggregates when the urea concentration decreases to 1 M or less. When the urea concentration increases a stepwise change occurs in the nucleocapsid where the normal-sized nucleocapsid of 400Å diameter undergoes a 40% swelling to reach a diameter of about 560Å (Table 3; and Schäfer et al., 1977a). This change occurs between 7.2 and 8.0 M urea and is accompanied by a large increase in solvation that must be internal since the Stokes diameter is comparable to the diameter determined from electron microscopy. We interpret this sudden change in diameter as being due to a weakening of the hydrophobic interactions between the nucleocapsid protein subunits, probably leading to a cooperative change. Protein III has a hydrophobicity index similar to that of protein II, but a low polarity (Table 4). This would favor hydrophobic interaction with lipid or other hydrophobic proteins.

2. In a mixed liposomal vesicle of PG and PE, the thermodynamically fa-

Table 3
Properties of the PM2 Nucleocapsid

	Nucleocapsid (7.2 M urea, NTC)	Swollen Nucleocapsid (8.0 M urea, NTC)
Diameter (EM) in Å	456	544
Diameter (neutron) in Å[a]	470 ± 10	—
Diffusion Constant (cm²/sec)	5.88×10^{-8}	3.93×10^{-8}
Stokes Diameter (Å)	458	550
Particle Weight[b]	13.6×10^6	13.6×10^6
ϕ_2 (cm³/gm)	0.800	0.800
"Dry" Diameter (Å)	326	326
Solvation (gm/gm)	1.42	3.04

[a] Schneider et al., manuscript in preparation (1977).
[b] From sedimentation-diffusion.

812

vored distribution would place PG in the outer lamella (Israelachvili, 1973) but we cannot predict a priori the distribution in a natural protein-lipid bilayer. Electrostatic interactions between protein II and the bilayer would require PG to be localized in the outer lamella, and this has been demonstrated directly using chemical labeling methods (Schäfer et al., 1974).

3. A direct demonstration of electrostatic interactions between protein II and PG was made by studying the distribution of PG and PE in a two-phase system, the phospholipids being in chloroform and the proteins in 1 M urea (Schäfer and Franklin, 1975b). In these binding experiments, proteins I and II formed one soluble fraction isolated in 4.5 M urea. Proteins III and IV were isolated as a second soluble fraction in 8 M urea. In the mixture of proteins I and II (case 1), protein II was present in ten-fold higher concentration than protein I, and thus we can assume that the observed effects were due chiefly to the former protein. In the mixture of proteins III and IV (case 2), however, there was only twice as much III as IV, and therefore the observed effects could well be due to a mixture of the two proteins. In case 1, PG was distributed preferentially in the urea phase in the presence or absence of up to 0.5 M NaCl and also in the absence of exogenous Ca^{2+} at pH 7. However, at pH 11.5, in the complete absence of Ca^{2+}, the distribution was nonpreferential. This pH was above the isoelectric point of protein II under the given condition. In case 2, the distribution in the urea phase was nonpreferential in 0.5 M NaCl or 0.01 M NaCl (both with added Ca^{2+}).

Table 4
Comparison of PM2 Proteins With Those of Bovine Myelin

Protein	pI	$\langle H\Phi\rangle$	R1[a]	R2[a]	R3[a]	R4[a]	Polarity
BACTERIOPHAGE PM2							
I	6.2	983	1.67	1.89	1.25	1.41	49.4
II	12.3	1139	1.33	1.47	1.00	1.11	44.9
III	5.8	1019	1.46	1.46	0.97	0.97	40.1
IV	5.5	1241	1.10	1.21	0.71	0.79	36.2
CNS MYELIN *(white matter)*							
Basic	10.3	907	2.87	3.29	2.07	2.38	51.6
Folch-Lees	—	1180	1.01	1.14	0.59	0.66	36.4
Wolfgram	—	1090	1.62	1.77	1.17	1.28	48.3
PNS MYELIN *(sciatic)*							
Basic	—	1067	1.82	1.93	1.26	1.33	50.9
Folch-Lees	—	1409	0.62	0.68	0.36	0.40	29.6
Wolfgram	—	1063	1.31	1.45	0.91	1.01	43.0

[a] R1 = hydrophilic/hydrophobic; R2 = hydrophilic/apolar; R3 = total charged/hydrophobic; R4 = total charged/apolar.

4. A comparison of the electron-spin resonance parameters of PM2 with those of egg-yolk lecithin provides some evidence for protein-lipid interactions without specifying the relative contributions of the different types of interactions (Scandella et al., 1974). The high polarity, determined from the isotropic hyperfine splitting constant, might be due to extension of some of the viral protein into the hydrocarbon region of the bilayer and would therefore provide some support for protein-lipid hydrophobic interactions.

5. Cross-linking experiments, some of which have already been mentioned, provide further evidence for protein-lipid interactions. Glutaraldehyde, dimethyl suberimidate, and tolylene-2,4-diisocyanate all reacted with proteins III and IV, and during the course of this reaction PE became associated with the protein III-IV fraction suggesting an intimate contact between PE and protein III and, perhaps, protein IV (Schäfer et al., 1975). These cross-linking reagents would not be expected to react with PG, and therefore no statement can be made concerning a possible contact between PG and proteins III and IV.

After mild glutaraldehyde treatment, under the conditions leading to cross-linking, there is an increase in the order parameter calculated from ESR spectra physically equivalent to a temperature decrease of 5°C in the native membrane (Schäfer et al., 1975). These effects are more pronounced in the outer regions of the bilayer where the strongest protein-lipid interactions might be expected to occur.

There are two further remarkable features of these cross-linking experiments. First, none of the reagents mentioned above cross-linked protein II to itself or to protein I, even though there are 15 lysines in protein II and lysine is one of the major reactive sites for glutaraldehyde (Korn et al., 1972) and probably also for dimethyl suberimidate and tolylene-2,4-diisocyanate. All of these reagents react with primary amino groups such as that found in the lysyl moiety (Habeb and Miramoto, 1968; Davies and Stark, 1970; Wold, 1972). One might argue that the lysines of protein II are so far apart in neighboring polypeptides that cross-linking could not occur. Considering the number of lysines in protein II, this seems a less likely interpretation than lack of availability of the lysines due to electrostatic interaction with PG. Secondly, dimethyl adipimidate, which is closely related to dimethyl suberimidate, will react both with erythrocyte membrane proteins and with hemoglobin within the intact erythrocyte (Niehaus and Wold, 1970). Considering the charge of these imidoesters, however, it is difficult to imagine that they could pass through a pure phospholipid bilayer. Actually, the erythrocyte membrane is known to contain at least two glycoproteins that span the membrane (Bretscher and Raff, 1975). Evidence also exists than one or both of the major spike glycoproteins (E_1, E_2) of Semliki forest virus span the viral membrane, and in this case dimethyl suberimidate can cross-link the glycoproteins to the nucleocapsid protein located on the inner side of the membrane

(Garoff and Simons, 1974; and chapter 11 by Kääriänen and Renkonen, this volume). The reaction of dimethyl suberimidate with PM2 proteins III and IV would suggest, therefore, that at least protein III must penetrate the bilayer and possibly form protein channels to the outer protein shell. A similar argument can be made on the basis of the glutaraldehyde reaction with proteins III and IV since glutaraldehyde is also relatively polar. Tolylene-2,4-diisocyanate, on the other hand, is relatively apolar. The proposed presence of protein III and possibly protein II (see below) in the bilayer might be the basis for the high electron density there (Harrison et al., 1971) and the disparity between lipid volume and bilayer volume (Camerini-Otero and Franklin, 1972).

6. Although ^{35}S-sulfanilic acid diazonium salt (DSA) will only label protein II when the intact virus is treated under mild conditions, protein III is labeled in concentrations of LiCl greater than 0.75 M. Li$^+$ has a very high field strength resulting from its very small ionic radius (Stein, 1962) and correspondingly has the highest hydration of the alkali metals (Glueckauf, 1955). Therefore Li$^+$ will interact chiefly with the PG of the outer lamella, weakening the interaction between protein II and PG and allowing the diazonium salt to interact with protein III. Because of the size and charge (zwitterionic) of the diazonium salt it is unlikely that it can penetrate very deeply into the bilayer under any condition, providing a further indication for channels of at least protein III in the bilayer.

7. Alterations in the fatty acid composition of PM2 can result in changes in phospholipid composition, presumably through alterations in the balance of electrostatic and hydrophobic protein-lipid interactions (Tsukagoshi et al., 1975a). Using a mutant strain of *Pseudomonas* BAL-31, which has a deficiency in the fatty acid oxidation pathway (mutant UFA), it is possible to alter the fatty acid composition of the host cell to almost 90% of a given unsaturated fatty acid. Virus grown in such cells will have the fatty acid composition of the modified host cell. In this way we have obtained virus with almost 90% *cis* C$_{16:1}$ or *trans* C$_{16:1}$ fatty acid. In both cases the infectious virus is produced with approximately the same efficiency when the yield is compared with that from wild-type cells, but the phospholipid compositions are different. In *cis* 16:1 there is about 53% PG in the virions; on the other hand, in *trans* 16:1, about 42% PG is present. The yield of virus in cells containing *cis* 16:1 is about 2×10^{12} PFU/mg of viral protein and about 8×10^{11} PFU/mg in cells containing *trans* 16:1. This suggests that both preparations are relatively homogeneous biologically and chemically.

Certainly the packing of the acyl chains will be different in *cis* and *trans* unsaturated fatty acids and will vary from virus grown in wild-type host cells, where there is about 50% *cis* 16:1 fatty acid. Packing of fatty acids must be very critical in the PM2 bilayer since the apparent icosahedral shape of the entire particle and also of the nucleocapsid means that the corners must be relatively sharp.

6. Comparison of protein-lipid interactions in bacteriophage PM2 and in myelin

One of the major protein species of both central nervous system (CNS) and peripheral nervous system (PNS) myelin is the basic protein that comprises about 30% of the myelin proteins (Eylar, 1974; and chapter 5 by Rumsby and Crang, this volume). The isoelectric point of the basic myelin protein from human white matter (Al protein) was 10.8 as determined by extrapolation of mobilities on cellulose acetate sheets (Cellogel) to zero mobility (Schäfer and Franklin, 1975c). We believe that this species of protein interacts electrostatically with acidic lipids in the myelin membrane. Actually, myelin contains more phosphatidylserine (negatively charged) and sphingomyelin (zwitterion) and less phosphatidylserine (zwitterion) than most other mammalian plasma membranes (Rouser et al., 1968). In addition, a considerable amount of cerebroside sulfate is present in myelin and this is also an acidic lipid (Eylar, 1973). The basic protein of myelin is known to form complexes with the acidic phospholipids triphosphoinositide, phosphatidic acid, phosphatidylserine, and phosphatidylinositol, as can be demonstrated directly using a two-phase system (Palmer and Dawson, 1969). Binding of triphosphoinositide is suppressed at ionic strengths greater than 0.1 M as might be expected for electrostatic interactions of this type. A further direct demonstration of the interaction of myelin basic protein with acidic lipids is the formation of a lamellar phase with such lipids, as shown by x-ray diffraction experiments on the phase (Mateu et al., 1973). The repeat distance is 154 or 175Å, depending on whether the basic protein is isolated from the CNS or PNS myelin, respectively. This distance is similar to the 150 to 160Å repeat of CNS myelin and the 170 to 180Å repeat of PNS myelin (Blaurock and Worthington, 1969).

Although isolated basic protein is completely hydrolyzed in 30 minutes by α-chymotrypsin, trypsin, bromelain, and thermolysin, it is stable in the presence of these enzymes when in intact myelin or when complexed with sulfolipids that had been prepared from myelin (Schäfer and Franklin, 1975c). Similar experiments have also been carried out with PM2 protein II but only with thermolysin, since it was necessary to work with the isolated protein in 2 M urea and of the four proteolytic enzymes, only thermolysin was active in 2 M urea (Schäfer and Franklin, 1975c). Whereas isolated protein II was rapidly degraded by thermolysin, protein II in the intact virion as well as protein II complexed with PG were about 99% resistant to thermolysin. The model of the basic protein of myelin proposed by Eylar (1974) predicts electrostatic interactions with the acidic lipids of the bilayer and hydrophobic sequences penetrating into the bilayer. A similar working model for the relationship of protein II to the PM2 bilayer might be useful. It would imply that a large proportion of the PM2 polypeptide be buried in the bilayer, considering particularly the analogous case of Semliki forest virus, where most of the two major envelope proteins can be readily digested with proteolytic enzymes, leaving a hydrophobic peptide embedded in the bilayer (Utermann and Simons, 1974).

Several properties of bovine CNS and PNS myelin calculated from the amino acid analyses of Eng and coworkers (1968) are compared with those of the PM2 proteins in Table 4. The calculated parameters are those defined by Barrantes (1973), the polarity defined by Capaldi and Vanderkooi (1972), and the average hydrophobicity by Bigelow (1967). The basic proteins have the lowest hydrophobicities and the highest polarities of the myelin proteins and these compare with PM2 protein I. The solubility properties of these proteins are also similar, the major difference being the isoelectric point. The isoelectric point for basic protein from bovine CNS myelin presented in Table 4 was also determined by Cellogel electrophoresis (R. Schäfer, M. Thonney, and R. M. Franklin, unpublished observations). The Folch-Lees proteolipids, which form the major fraction of myelin protein (Barrantes et al., 1972; Folch, 1974), are most similar to protein IV with respect to hydrophobicity, polarity, and the Barrantes ratios. The hydrophobicity of protein III is lower and the polarity higher than those of the proteolipids. Unlike the myelin proteolipid, PM2 proteins III and IV do not have fatty acids covalently linked to them (Hinnen et al., 1976). The Wolfgram proteins have parameters most similar to those of PM2 protein I. These comparisons emphasize the limited usefulness of the various parameters derived from amino acid compositions. Certainly, the limited sequence data we have for proteins II and IV have been much more helpful in correlating chemical information with function.

7. In vitro assembly of bacteriophage PM2

The stepwise degradation of PM2 in increasing concentrations of urea suggested a method for reconstitution of the virus from its subunits (Schäfer and Franklin, 1975b). In the first step, proteins III and IV and the viral DNA, which are not associated in 8 M urea at pH 4.5, were dialyzed to 4.5 M urea at neutral pH with the ensuing formation of a nucleocapsid, identical to the nucleocapsid formed by degradation of the virion. All properties investigated were the same for these two nucleocapsids, including densities in CsCl, chemical compositions, A_{260}/A_{280} ratios, and specific infectivity for spheroplasts of the DNA isolated from the particles by phenol extraction. The ultrastructure of the two nucleocapsids was also the same. Reconstitution took place in 4.5 M urea, 0.5 M NaCl, 10 mM $CaCl_2$, 20 mM β-mercaptoethanol, and 25 mM Tris buffer (pH 7.5 at 4°C). Ca^{2+} could not be replaced by other divalent cations, including Mg^{2+} or Mn^{2+}. Recently, Sr^{2+} and Ba^{2+} were also tested and were also ineffective as substitutes for Ca^{2+} in the reconstitution of nucleocapsid (Schäfer et al., 1977b). To date, however, these cations have not been tested in the reconstitution of virus from nucleocapsid. The yield of nucleocapsid was low (2.3%) but could be improved over tenfold to 33% by the addition of bovine serum albumin to the reconstitution mixture. This apparently diminished the nonspecific aggregation of proteins III and IV. Recently, we have been able to reconstitute nucleocapsid from the individual purified polypeptides III and IV rather than from the mixture of

817

proteins III and IV. In this case there was an excellent yield of nucelocapsid (35%) and no bovine serum albumin was necessary (Schäfer et al., 1977b). This demonstrates unequivocally that nucleocapsid contains only proteins III and IV and that these proteins, isolated by SDS gel filtration, can be renatured to their native configuration. Under the conditions used for such reconstitution, protein III alone does not form a nucleocapsid shell but rather a nonspecific aggregate. It may be that the DNA-protein IV complex forms a center of nucleation that is necessary for the formation of the nucleocapsid. Some preliminary evidence exists for formation of a DNA-protein IV complex as the first stage of *in vitro* assembly (Table 5; and Schäfer and Franklin, 1975a; Schäfer et al., 1977b). Furthermore, the first step in which protein III is involved appears to be the formation of a dimer of this protein (Table 5; and Schäfer et al., 1977b). We may consider reconstitution of the nucleocapsid as a process driven by the difference in free energy of solution of the polypeptides in high and low urea concentrations. (Nozaki and Tanford, 1963). Entropy differences resulting from changes in water structure compared with enthalpy differences arising from protein-protein interactions would give rise to the negative free energy difference needed to force assembly (Lauffer, 1975).

Despite improvements in yield, the amount of reconstituted nucleocapsid recovered has always been very small. Since we had demonstrated the identity of the reconstituted nucelocapsid (NC) with that from the virion, the latter was used for the further steps in reconstitution (Schäfer and Franklin, 1975b). The next step was formation of a nucleocapsid with lipid (NCL), formed by dialyzing a mixture of nucleocapsid and phospholipids from the 4.5 M urea solution to a solution containing 1 M urea, 1 M NaCl, 0.01 M CaCl$_2$, and 25 mM Tris buffer (pH 7.5 at 4°C). NCL was obtained in high yield with a lipid content 90% of that expected on the basis of the lipid content of the virion. The lipid composition, however, was that in the incubation mixture, not that found normally in the virion. This is understandable on the basis of the postulated hydrophobic interactions between protein III and the bilayer. Furthermore, NCL particles could not be used for reconstitution of the infectious particle because protein II

Table 5
Some Intermediates in Bacteriophage PM2 Assembly

1)	Protein IV-DNA complex	Formed in 20 mM Tris, pH 7.5; 1 mM EDTA	
2)	Protein III *dimer* molecular weight:	(8 M urea, 1% β-alanine, pH 4.5) 24,000 27,300	gel filtration analytical ultracentrifugation
3)	Protein II *monomer* molecular weight:	(4 M urea, NTC, pH 7.2) 28,000 28,500	gel filtration analytical ultracentrifugation
4)	Protein II-PG complex		

did not bind to such particles. To form infectious particles it was necessary to start with NC plus a mixture of phospholipids and proteins I and II in 4.5 M urea and dialyze this mixture to a urea-free buffer (0.5 M NaCl, 0.01 M CaCl $_2$, 10 mM β-mercaptoethanol, 25 mM Tris buffer, pH 7.5 at 4°C). Infectious virus was obtained in the presence of 30 to 50 mg/ml of bovine serum albumin, which prevented nonspecific aggregation. Since the ratio of physical to infectious reconstituted particles was about 3×10^{-7}, as compared to 0.6 for our standard virus preparation, we must describe the particles seen in the electron microscope as viruslike. The yield of infectious particles depended on the PG/PE ratio in the reconstitution mixture. The yield was higher if this ratio was similar to that in the virion than when the ratio was similar to that in the host cell. In both cases, however, the PG/PE ratio in the virion was similar to that in native virus, probably due to the selective interaction of PG with protein II combined with hydrophobic interaction of the phospholipids with the nucleocapsid surface.

With good yields of reconstituted nucleocapsid now available, using the individually purified proteins III and IV as starting material, it has been possible to reconstitute infectious particles using reconstituted NC instead of NC obtained by urea degradation of the virion (Schäfer et al., 1977b). Thus it has been possible to achieve a complete reconstitution of PM2. A further dissociation of the steps in reconstitution has been accomplished by adding protein II plus the phospholipids to the nucleocapsid in 4.0 M urea and dialyzing to 1 M urea. Protein II is present as a monomer in the 4 M urea used in the starting conditions (Table 4; and Schäfer et al., 1977b) and probably forms a complex with PG prior to formation of the outer protein shell, as discussed in detail earlier. The resulting noninfectious particle becomes infectious when protein I is added and this incubation mixture is dialyzed to 1M NaCl with Tris buffer and 10mM CaCl $_2$.

In further reconstitution experiments the role of acyl phosphatidylglycerol (APG) and the effect of *cis* versus *trans* fatty acids was investigated (Tsukagoshi et al., 1977). NC derived from virus by urea treatment was mixed with proteins I and II, and various mixtures of lipids that were isolated from *Pseudomonas* BAL-31, strain UFA (unsaturated fatty acid auxotroph), which had been grown in the presence of ^{32}P-phosphate and either *cis* 16:1 or *trans* 16:1 fatty acid. In the case of phospholipids containing approximately 90% *cis* 16:1 fatty acids, the yield of reconstituted infectious virus was the same in the presence or absence of 0.5% APG. If, however, PG was replaced by APG, then no infectious particles were formed. In the comparison of *cis* versus *trans* fatty acids, viruslike particles that could be isolated by centrifugation were formed in both cases but only those formed with phospholipids containing *cis* fatty acids were infectious (Table 6). In the experiment shown in Table 6, the input percentage of APG was high in the case of *trans* 16:1 to simulate the conditions found in vivo (Tsukagoshi et al., 1975a). The ratio of PG to PE seemed to be controlled by protein II in both the *cis* and *trans* cases, as in the normal case described above. Also, the amount of APG in the particles was higher than in the input mixture and this might be due to electrostatic interaction between protein II and APG. This situation is different from the in vivo case, where the particles formed in the presence of *trans* 16:1 fatty acid are infectious but with a PG/PE ratio different from that found in virus grown in wild-type cells or in the presence of *cis* 16:1 fatty acid (see above). The lack of infectivity in the case of reconstitution with *trans* 16:1 particles would indicate that there may be problems in packing the *trans* fatty acids into the bilayer. These packing problems may be partially compensated for during assembly in vivo. They do not seem to be compensated for in vitro either below or above the temperature at which a phase transition occurs in *Pseudomonas* BAL-31 membranes containing *trans* 16:1 fatty acids (Tsukagoshi et al., 1976a) since no infectious particles were obtained below or above the phase transition temperature. Despite possible packing problems there was no preferential uptake of *cis* fatty acids during reconstitution in the presence of mixtures of *cis* and *trans* fatty acids, distinguished by ^{32}P or ^{33}P labeling.

Further information was obtained on the noninfectious particles containing *trans* fatty acids using particles reconstituted with either *cis* or *trans* 16:1 fatty acids and with radioactively labeled viral polypeptides (Tsukagoshi et al., 1977). Both the *cis* and *trans* particles contain similar proportions of all four viral polypeptides. Furthermore, the radioactivity associated with both *cis* and *trans* particles adsorbs to the host pseudomonad and this adsorption is prevented by prior treatment of the particles with bromelain, which destroys the spikes and prevents adsorption of the normal virion (section 4). Therefore the noninfectious *trans* particles adsorb to the host cell, and the lack of infectivity must be due to blockage of some step between adsorption and initiation of infection presum-

Table 6
Reconstitution with *cis* and *trans* fatty acids

Fatty acid	Phospholipid	Input[a]	Particle[b]	Infectivity
cis 16:1	PE	69.1	40.6	2.5×10^3 PFU
	PG	29.8	57.0	
	APG	1.1	2.4	
cis 16:1	PE	36.7	41.5	2.6×10^3 PFU
	PG	62.3	56.2	
	APG	1.0	2.3	
trans 16:1	PE	61.8	35.0	0
	PG	30.3	55.0	
	APG	7.9	10.0	
trans 16:1	PE	30.5	36.2	0
	PG	61.2	52.0	
	APG	8.3	11.8	

[a] Input phospholipid composition (%).
[b] Phospholipid composition (%) in the isolated reconstituted particles. Reconstitution and isolation of particles was carried out as described by Schäfer and Franklin (1975b).

ably occurring after release of the viral DNA from the particles. One could imagine that this release step might be associated with fusion of the virion with the host cell plasma membrane and that such a fusion is prevented by stereochemical effects in the case of the bilayer containing *trans* fatty acids. Unfortunately, too little is known concerning initiation of PM2 infection so that only these speculative comments can be offered at present.

8. In vivo assembly of bacteriophage PM2

In view of the difference in phospholipid composition between PM2 and its host cell BAL-31, it was considered important to investigate phospholipid metabolism in the infected cell. In initial experiments a trend toward an increase in PG synthesis and a decrease in PE synthesis was observed (Braunstein and Franklin, 1971). This observation was confirmed and extended in later experiments (Tsukagoshi and Franklin, 1974). PE present prior to infection is degraded and the rate of PE synthesis is reduced. No degradation of PG is present prior to infection and its rate of synthesis increases. One-third of the lipids found in the completed virion are derived from the preexisting host cell lipids and the remainder are synthesized after infection. The enzymatic basis for these changes has been established (Tsukagoshi et al., 1975b). Phospholipase A activity against PE increases about twofold during infection, whereas activity against PG remains constant. Whether two different enzymes are involved and whether there is a viral-specific phospholipase A have not been investigated. There is also an increase in PG synthetase activity during the first 30 minutes after infection. PS synthetase and decarboxylase remain constant during this time; then the decarboxylase activity decreases, effectively blocking PE synthesis despite the rise in PS synthetase activity between 30 and 45 minutes after infection.

We call these metabolic changes *active* control processes as opposed to the *passive* control process of electrostatic interaction between protein II and PG. The reconstitution experiments that demonstrated an increased efficiency of reconstitution of the infectious virus in the presence of the viral PG/PE ratio as compared to reconstitution in the presence of the host cell PG/PE ratio leads us to believe that the active process, increasing the PG content and decreasing the PE content of the cell, leads to a more efficient in vivo assembly.

Intracellular assembly does not seem to occur in conjunction with the plasma membrane of the host bacterium (Dahlberg and Franklin, 1970). One might imagine that a phospholipid exchange protein (Wirtz and Zilversmit, 1968; Harvey et al., 1973; and review by Wirtz, 1974) plays a role in the transport of the lipids to an assembly site. Such a protein seems to be a factor in vaccinia virus assembly (Stern and Dales, 1974) but a similar protein has not yet been found in bacteria.

In the assembly of PM2, Ca^{2+} is required both for the formation of nucleocapsid in vitro (Schäfer and Franklin, 1975b) and for the assembly of the virion in vivo (Snipes et al., 1974). In the former case, Ca^{2+} may be involved in conformational changes of protein III that might be occurring during nucleocapsid formation. In the latter case Ca^{2+} must be associated with protein II and plays a crucial role in determining the isoelectric point of this protein (Schäfer et al., 1974; Schäfer and Franklin, 1975b). It could also interact with the phospholipids by analogy with the effects of Ca^{2+} on mixed phosphatidylserine-phosphatidylcholine phases that result in the formation of solid phosphatidylserine patches in a fluid phosphatidylcholine matrix (Ito et al., 1975).

It is clear from this discussion that the events during assembly of PM2 in vivo are still obscure. The difficulties in handling the nucleocapsid, which aggregates even in low concentrations of urea, make it almost impossible to demonstrate such a structure in infected cells. Although we foresee an even deeper understanding of the assembly process in vitro that is amenable to physicochemical study and chemical manipulation, the prospects for rapid progress in understanding assembly in vivo are not as good. Certainly a set of stable PM2 mutants would be most useful for analysis of in vivo assembly, and several laboratories are presently working on this aspect of the problem.

9. Bacteriophage φ6

Bacteriophage φ6 is the lipid-containing bacteriophage that infects the phytopathogenic pseudomonad, *Pseudomonas phaseolicola* (Vidaver et al., 1973). Evidence for the presence of lipids in this virus includes its partial inactivation by toluene, ethyl ether, and 0.05% sodium deoxycholate, and its complete inactivation by 0.5% sodium deoxycholate and chloroform. Chemical analyses reported in this and subsequent papers must be considered with caution, however, since the criteria for purity used by the authors mentioned above did not include immunological studies designed to show that the purified virus was actually free of host cell proteins or membrane components. This criterion is important since the virus has a buoyant density in CsCl of 1.27 gm/ml, similar to that of PM2 (Camerini-Otero and Franklin, 1975) and also to that of bacterial membranes (Franklin et al., 1971). The criteria for purity of φ6 used by Vidaver and coworkers are as follows: (1) both in sucrose gradient velocity sedimentation and CsCl isopycnic sedimentation the peaks of infectivity and ultraviolet absorbing material (λ = 260nm) are located at the same place and the PFU/OD$_{260}$ ratio was approximately uniform throughout the peak region; (2) the chemical composition of the virus peak isolated from a sucrose gradient was the same as that isolated from a CsCl gradient; and (3) only double-stranded RNA characteristic of that found in the virion was found in the peak region, according to a statement in the discussion of the paper of Vidaver and coworkers (1973). This is not quite true, however (see below). The purified virus contains 25% lipid, 62% protein, and 13% RNA. The fatty acid composition of the virus was very similar, if not identical, to that of the host cell with three major fatty acids $-C_{16:0}$, $C_{16:1}$, and $C_{18:1}$.

The virus appears to be polyhedral in shape, 60 to 70 nm in diameter, and has a particle weight of approximately 80×10^6, based on estimates made from the RNA content and RNA molecular weight (van Etten et al., 1974). Vidaver and coworkers (1973) report an amorphous, sacklike tail, but in the published electron micrographs the tail appears more like an extrusion of some material from a partially damaged particle. In some of the micrographs spikelike projections from the vertices of the particles appear to be present but these are not mentioned by Vidaver and coworkers. They do claim, however, that the particle is surrounded by an amorphous membranelike structure, and Sands (1973) has emphasized this "unusual" feature as a structural difference between $\phi 6$ and PM2, as far as the structure of the membrane is concerned. It is questionable whether such a difference exists, particularly since more recent electron micrographs of the virion do not indicate such pleomorphism (Sinclair et al., 1975). Before further ultrastructural (electron microscopy) and structural (x-ray diffraction) studies are made, I would defer any attempts to propose a structure for $\phi 6$. We know, for example, that PM2 in the presence of nonionic detergents is far more labile when stained with phosphotungstic acid than with uranyl acetate (Hinnen et al., 1974) and we can readily obtain images of damaged PM2 similar to the images of $\phi 6$ published by Vidaver and coworkers (1973) or by Ellis and Schlegel (1974). It will be necessary to examine $\phi 6$ stained with a variety of reagents with and without fixation and also sectioned viral pellets and then to correlate such data with an electron density profile as has been done for bacteriophage PM2 (Harrison et al., 1971).

Bacteriophage $\phi 6$ is more complex tham PM2, containing a total of 10 structural proteins with molecular weights varying from 93,000 to less than 6,000 daltons (Sinclair et al., 1975). Proteins P3 (84,000 daltons) P9 (8,700 daltons), and P10 (< 6000 daltons) were completely extracted from the virion with 1% Triton X-100; protein P6 (21,000 daltons) was only partially extracted.

The phospholipid composition of $\phi 6$ was compared with that of its host cell by Sands (1973). The phospholipids were labeled with ^{32}P and the virus purified according to Vidaver et al. (1973). In a sucrose-gradient velocity centrifugation there was a peak of ^{32}P located at the same position as the virus infectivity, but no additional criteria for purity were given and therefore the criticism made earlier concerning viral purity also applies to this study. The host cell had 15% "cardiolipin," and 29% PG, whereas the virus had 8% "cardiolipin," 35% PE, and 57% PG. The "cardiolipin" was identified by thin-layer chromatographic comparison with commercially available cardiolipin but the source of this standard was not given. This is not a very convincing way to identify cardiolipin, since cardiolipins from different sources do not have the same chromatographic properties. It will be necessary to identify cardiolipin (or whatever the compound is) by other methods such as comparison with cardiolipins synthesized from host cell phospholipid precursors and by analysis of the products of deacylation (Braunstein and Franklin, 1971; Diedrich and Cota-Robles, 1974). Nevertheless the results of Sands are interesting, and suggest that there is a preponderance of the acidic phospholipid, PG, in the virion, the ratio of PG/PE being almost the reverse of that in the host cell, recalling the situation with PM2 (Braunstein and Franklin, 1971). It will be interesting to see if there is a basic protein in $\phi6$, as has been found in PM2 (Schäfer et al., 1974). There is a slight shift in the composition of ^{32}P-labeled phospholipid after infection, the tendency being toward an increase in PG (Sands et al., 1974). Further studies of the phospholipid metabolism after $\phi6$ infection should be very interesting.

There is a temperature-sensitive step in the synthesis of $\phi6$ since the virus does not grow above 30°C, whereas the host cell will grow up to 34°C (Sands et al., 1974). When the temperature is shifted from a permissive (25°C) to a non-permissive (31°C) temperature at different times after infection, inhibition of virus growth occurs up to 50 miutes postinfection. In the reverse experiment, virus appears to have a normal time course up to 50 minutes postinfection and then with a delay when the shift occurs at later times. These experiments suggest that the temperature-sensitive step occurs late in infection. This step does not appear to be related to lysis, since infected cells incubated at 31°C for 90 minutes and then lysed artificially do not release infectious virus. Electron-spin resonance experiments using a hydrocarbon spin label, oxazolidine, suggested that there might be some change in the molecular properties of both the host cell membrane and the viral membrane at about 30°C. Similar experiments with liposomes prepared from host cell phospholipids did not result in such changes and thus might be due to protein-lipid interactions in the case of virus or host cell. The shift in labeled phospholipids toward PG that occurs after infection at 25°C also occurs at 31°C (Sands et al., 1974). Therefore the block does not seem to be in phospholipid metabolism. Whether the change in membrane properties at 30°C can be correlated with the late block in virus growth above 30°C is a matter for future investigation.

Bacteriophage $\phi6$ has a segmented double helical RNA (Semancik et al., 1973). This RNA has an unusually high G-C content (58%) with a T_m of 91°C and a buoyant density in Cs$_2$SO$_4$ of 1.605 g cm^{-3}. This RNA is segmented with segments of molecular weight 2.2, 2.8 and 4.5 $\times$ 10^{-6}. Although the three segments have similar melting temperatures and base compositions, hybridization experiments demonstrated that the three segments do not have any common sequences (van Etten et al., 1974). In the presence of high salt (2 $\times$ SSC), 2.5 to 3.1% of the radioactivity of ^{32}P-labeled ds $\phi6$ RNA was digested by a combination of pancreatic RNase plus T1 RNase without altering the sedimentation properties of the bulk of the label. Thus $\phi6$ RNA may have a short single-stranded region at one or possibly both ends although there are alternative explanations, such as the presence of a contaminating ss RNA or the presence of a small percentage of incomplete RNA molecules among the majority of complete ds molecules. RNA polymerase activity is associated with the virion, and it has been suggested that the enzyme may have to complete short single-stranded regions before the RNA can initiate a new cycle of infectivity (van Etten et al., 1973).

824

Acknowledgment

Some of the work described in this review was supported by grant number 3.8530.72 SR from the *Fonds National Suisse de la Recherche Scientifique.*

References

Barrantes, F.J. (1973) A comparative study of several membrane proteins from the nervous system. Biochem. Biophys. Res. Commun. 54: 395–402.

Barrantes, F.J., LaTorre, J.L., de Carlin, M.C.L., and de Robertis, E. (1972) Studies on proteolipid proteins from cerebral cortex. I. Preparation and some properties. Biochim. Biophys. Acta 263: 368–381.

Bigelow, C.C. (1967) On the average hydrophobicity of proteins and the relation between it and protein structure. J. Theoret. Biol. 16: 187–211.

Blaurock, A.E., and Worthington, C.R. (1969) Low-angle x-ray diffraction patterns from a variety of myelinated nerves. Biochim. Biophys. Acta 173: 419–426.

Boni, I.Y. and Budowsky, E.I. (1973) Transformation of non-covalent interactions in nucleoproteins into covalent bonds induced by nucleophilic reagents. I. The preparation and properties of the products of bisulfite ion-catalyzed reaction of amino acids and peptides with cytosine derivatives. J. Biochem. (Tokyo) 73: 821–830.

Braunstein, S.N. and Franklin, R.M. (1971) Structure and synthesis of a lipid-containing bacteriophage. V. Phospholipids of the host BAL-31 and of the bacteriophage PM2. Virology 43: 685–695.

Bretscher, M.S. and Raff, M.C. (1975) Mammalian plasma membranes. Nature 258: 43–49.

Camerini-Otero, R.D. and Franklin, R.M. (1972) Structure and synthesis of a lipid-containing bacteriophage. XII. The fatty acids and lipid content of bacteriophage PM2. Virology 49: 385–393.

Camerini-Otero, R.D. and Franklin, R.M. (1975) Structure and synthesis of a lipid containing bacteriophage. XVII. The molecular weight and other physical properties of bacteriophage PM2. Eur. J. Biochem. 53: 343–348.

Camerini-Otero, R.D., Datta, A., and Franklin, R.M. (1972) Structure and synthesis of a lipid-containing bacteriophage. XI. Studies on the structural glycoprotein of the virus particle. Virology 49: 522–536.

Capaldi, R.D. and Vanderkooi, G. (1972) The low polarity of many membrane proteins. Proc. Nat. Acad. Sci., U.S.A., 69: 930–932.

Casjens, S. and King, J. (1975) Virus Assembly. Ann. Rev. Biochem. 44: 555–611.

Chothia, C. (1975) Structural invariants in protein folding. Nature 254: 304–308.

Chothia, C. and Janin, J. (1975) Principles of protein-protein recognition. Nature 256: 705–708.

Davies, G.E. and Stark, G.R. (1970) Use of dimethyl suberimidate, a cross-linking reagent, in studying the subunit structure of oligomeric proteins. Proc. Nat. Acad. Sci., U.S.A. 66: 651–656.

Diedrich, D.L. and Cota-Robles, E.H. (1974) Heterogeneity in lipid composition of the outer membrane and cytoplasmic membrane of *Pseudomonas* BAL-31. J. Bacteriol. 119: 1006–1018.

Ellis, L.F. and Schlegel, R.A. (1974) Electron microscopy of *Pseudomonas* φ6 bacteriophage. J. Virol 14: 1547–1551.

Eng, L.F., Chao, F.-C., Gerstl, B., Pratt, D., and Tavaststjerna, M.G. (1968) The maturation of human white matter mylein. Fractionation of the myelin membrane proteins. Biochemistry 7: 4455–4465.

Eylar, E.K. (1973) Myelin-Specific Proteins. In: Proteins of the Nervous System (Schneider, D.J. et al., ed.). pp. 27–44. Raven Press, New York.

Folch-Pi, J. (1973) Proteolipids. In: Proteins of the Nervous System (Schneider, D.J., et al., ed.), pp. 45–66. Raven Press, New York.

Franklin, R.M., Salditt, M., and Silbert, D.A. (1969) Structure and synthesis of a lipid-containing bacteriophage. I. Growth of bacteriophage PM2 and alterations in nucleic acid metabolism in the infected cell. Virology 38: 627–640.

Franklin, R.M., Datta, A., Dahlberg, J.E., and Braunstein, S.N. (1971) The cell membranes of a marine pseudomonad, *Pseudomonas* BAL-31; physical, chemical, and biochemical properties. Biochim. Biophys. Acta 233: 521–537.

Franklin, R.M., Hinnen, R., Schäfer, R., and Tsukagoshi, N. (1976) Structure and assembly of lipid-containing viruses, with special reference to bacteriophage PM2 as one type of model system. Proc. Roy. Acad. Sci. B276:63–80.

Garoff, H. and Simons, K. (1974) Location of the spike glycoproteins in Semliki forest virus membrane. Proc. Nat. Acad. Sci., U.S.A. 71: 3988–3992.

Glueckauf, E. (1955) The influence of ionic hydration on activity coefficients in concentrated electrolyte solutions. Trans. Faraday Soc. 51: 1235–1244.

Habeb, A.F.S.A. and Hiramoto, R. (1968) Reaction of proteins with glutaraldehyde. Arch. Biochem. Biophys. 126: 16–26.

Harrison, S.C., Caspar, D.L.D., Camerini-Otero, R.D., and Franklin, R.M. (1971) Lipid and protein arrangement in bacteriophage PM2. Nature New Biol. 229: 197–201.

Harvey, M.S., Wirtz, K.W.A., Kamp, H.H., Zeegers, B.J.M., and van Deenen, L.L.M. (1973) A study on phospholipid exchange proteins present in the soluble fractions of beef liver and brain. Biochim. Biophys. Acta 323: 234–239.

Hinnen, R., Schäfer, R., and Franklin, R.M. (1974) Structure and synthesis of a lipid-containing bacteriophage. XIV. Preparation of virus and localization of the structural proteins. Eur. J. Biochem. 50: 1–14.

Hinnen, R., Chassin, R., Schäfer, R., Franklin, R. M., Hitz, H., and Schäfer, D. (1976) Structure and synthesis of a lipid-containing bacteriophage. XXIII. Purification, chemical composition, and partial sequences of the structural proteins. Eur. J. Biochem. 68:139–152.

Israelachvili, J.N. (1973) Theoretical considerations on the asymmetric distribution of charged phospholipid molecules on the inner and outer layers of curved bilayer membranes. Biochim. Biophys. Acta. 323: 659–663.

Ito, T., Ohnishi, S., Ishinaga, M., and Kito, M. (1975) Synthesis of a new phosphatidylserine spin-label and calcium-induced lateral phase separation in phosphatidylserine-phosphatidylcholine membranes. Biochemistry 14: 3064–3069.

Kauzmann, W. (1959) Some Factors in the Interpretation of Protein Denaturation, In: Advances in Protein Chemistry, (Anfinsen, C.B. Jr., Anson, M.L., Bailey, K., and Edsall, J.T., eds.) vol. XIV, pp. 1–63. Academic Press, New York.

Korn, A.H., Feairheller, S.H., and Filachione, E.M. (1972) Glutaraldehyde: Nature of the reagent. J. Mol. Biol. 65: 525–529.

Kretsinger, R.H. (1974) Calcium Binding Proteins and Natural Membranes. In: Perspectives in Membrane Biology (Estrada, S.O. and Gitler, C., eds.) pp. 229–262. Academic Press, New York.

Lauffer, M.A. (1975) Entropy-Driven Processes in Biology. Springer-Verlag, Berlin.

Laval, F. (1974) Endonuclease activity associated with purified PM2 bacteriophages. Proc. Nat. Acad. Sci., U.S.A. 71: 4965–4969.

Mateu, L., Luzzati, V., London, Y., Gould, R.M., Vossberg, F.G.A., and Olive, J. (1973) X-ray diffraction and electron microscope study of the interactions of myelin components. The structure of a lamellar phase with a 150 to 180 Å repeat distance containing basic proteins and acidic lipids. J. Mol. Biol. 75: 697–709.

Niehaus, W.G. Jr. and Wold, F. (1970) Cross-linking of erythrocyte membranes with dimethyl adipimate. Biochim. Biophys. Acta. 196: 170–175.

Nozaki, Y. and Tanford, C. (1963) The solubility of amino acids and related compounds in aqueous urea solutions. J. Biol. Chem. 238: 4074–4081.

Olsen, R.W. and Ballou, C.E. (1971) Acyl phosphatidylglycerol. A new phospholipid from *Salmonella typhimurium*. J. Biol. Chem. 246: 3305–3313.

826

Palmer, F.B. and Dawson, R.M.C. (1969) Complex-formation between triphosphoinositide and experimental allergic encephalitogenic protein. Biochem. J. 111: 637–646.

Ramachandran, G.N. and Sasisekharan, V. (1968) Conformation of polypeptides and proteins. In: Advances in Protein Chemistry (Anfinsen, C.B., Jr., Anson, M.L., Edsall J.T., and Richards, F.M., eds.), vol XXIII, pp. 283–437. Academic Press, New York.

Rouser, G., Nelson, G.J., Fleischer, S., and Simon, G. (1968) Lipid composition of animal cell membranes, organelles and organs. In: Biological Membranes. Physical Fact and Function. (Chapman, D. ed.), pp. 5–69. Academic Press, London.

Sands, J.A. (1973) The phospholipid composition of bacteriophage ϕ6. Biochem. Biophys. Res. Commun. 55: 111–116.

Sands, J.A., Cupp, J., Keith, A., and Snipes, W. (1974) Temperature sensitivity of the assembly process of the enveloped bacteriophage ϕ6. Biochim. Biophys. Acta 373: 277–285.

Scandella, C., Schindler, H., Franklin, R.M., and Seelig, J. (1974) Structure and synthesis of a lipid-containing bacteriophage. XVI. Acyl-chain motion in the PM2 virus membrane. Eur. J. Biochem. 50: 29–32.

Schäfer, R. and Franklin, R.M. (1975a) Structure and synthesis of a lipid-containing bacteriophage. XX. A polynucleotide-dependent polynucleotide-pyrophosphorylase activity in bacteriophage PM2. Eur. J. Biochem. 58: 81–85.

Schäfer, R. and Franklin, R.M. (1975b) Structure and synthesis of a lipid-containing bacteriophage. XIX. Reconstitution of bacteriophage PM2 in vitro. J. Mol. Biol. 97: 21–34.

Schäfer, R. and Franklin, R.M. (1975c) Resistance of the basic membrane proteins of myelin and bacteriophage PM2 to proteolytic enzymes. FEBS Lett. 58: 265–268.

Schäfer, R., Hinnen, R., and Franklin, R.M. (1974) Structure and synthesis of a lipid-containing bacteriophage. XV. Properties of the structural proteins and distribution of the phospholipid. Eur. J. Biochem. 50: 15–27.

Schäfer, R., Huber, U., Franklin, R.M., and Seelig, J. (1975) Structure and synthesis of a lipid-containing bacteriophage. XXI. Chemical modifications of bacteriophage PM2 and the resulting alterations in acyl-chain motion in the PM2 membrane. Eur. J. Biochem. 58: 291–296.

Schäfer, R., Marcoli, R., Lustig, A. and Franklin, R.M. (1977a) Manuscript in preparation.

Schäfer, R., Hinnen, R., and Franklin, R.M. (1977b) Manuscript in preparation.

Semancik, J.S., Vidaver, A.K., and van Etten, J.L. (1973) Characterization of a segmented double-helical RNA from bacteriophage ϕ6. J. Mol. Biol. 78: 617–625.

Silbert, J.A., Salditt, M., and Franklin, R.M. (1969) Structure and synthesis of a lipid-containing bacteriophage. III. Purification of bacteriophage PM2 and some structural studies on the virion. Virology 39: 666–681.

Sinclair, J.F., Tzagoloff, A., Levine, D., and Mindich, L. (1974) Proteins of bacteriophage ϕ6. J. Virol. 16: 686–695.

Snipes, W., Cupp, J., Sands, J.A., Keith, A., and Davis, A. (1974) Calcium requirement for assembly of the lipid-containing bacteriophage PM2. Biochim. Biophys. Acta. 339: 311–322.

Stein, W.D. (1962) Behaviour of Molecules in Solution. In Comprehensive Biochemistry. (Florkin, M. and Stolz, E.H., eds.), vol. 2, pp. 219–282. Elsevier/North-Holland, Amsterdam.

Stern, W. and Dales, S. (1974) Biogenesis of vaccinia: concerning the origin of the envelope phospholipids. Virology 62: 293–306.

Tanford, C. (1964) Isothermal unfolding of globular proteins in aqueous urea solutions. J. Amer. Chem. Soc. 86: 2050–2059.

Tsukagoshi, N. and Franklin, R.M. (1974) Structure and synthesis of a lipid-containing bacteriophage. XIII. Studies on the origin of the viral phospholipids. Virology 59: 408–417.

Tsukagoshi, N., Petersen, M.H., and Franklin, R.M. (1975a) Structure and synthesis of a lipid-containing bacteriophage. XVIII. Modification of the lipid composition in bacteriophage PM2. Virology 66: 206–216.

Tsukagoshi, N., Petersen, M.H., and Franklin, R.M. (1975b) Structure and synthesis of a lipid-containing bacteriophage. XXII. Characterization of some enzymes of glycerophosphatide metabolism of Pseudomonas BAL-31 and alterations in their activity after infection with bacteriophage PM2. Eur. J. Biochem. 60: 603–613.

Tsukagoshi, N., Petersen, M.H., Huber, U., Franklin, R.M., and Seelig, J. (1976a) Phase transitions in the membrane of a marine bacterium, *Pseudomonas* BAL-31. Eur. J. Biochem. 62: 257–262.

Tsukagoshi, N., Kania, M.N., and Franklin, R.M. (1976b) Identification of acyl phosphatidylglycerol as a minor phospholipid of *Pseudomonas* BAL-31. Biochim. Biophys. Acta 450:131–136.

Tsukagoshi, N., Schäfer, R., and Franklin, R.M. (1976c) Structure and synthesis of a lipid-containing bacteriophage. XXIV. Effects of lipids containing *cis* or *trans* fatty acids on the reconstitution of bacteriophage PM2. Eur. J. Biochem. 73:469–476.

Tsukagoshi, N., Shäfer, R. and Franklin, R. M. (1977) Structure and synthesis of a lipid-containing bacteriophage. XXV. An endolysis activity associated with bacteriophage PM2. Eur. J. Biochem., in press.

Utermann, G. and Simons, K. (1974) Studies on the amphipathic nature of the membrane proteins in Semliki forest virus. J. Mol. Biol. 85: 569–587.

Van der Schans, G.P., Weyermans, J.P., and Bleichrodt, J.F. (1971) Infection of spheroplasts of *Pseudomonas* with DNA of bacteriophage PM2. Mol. Gen. Genetics. 110: 263–271.

Van Etten, J.L., Vidaver, J.K., Koski, R.K., and Semancik, J.S. (1973) RNA polymerase activity associated with bacteriophage ϕ6. J. Virol. 12: 461–471.

Van Etten, J.L., Vidaver, A.K., Koski, R.K., and Burnett, J.P. (1974) Base composition and hybridization studies of the three double-stranded RNA segments of bacteriophage ϕ6. J. Virol. 13: 1254–1262.

Vidaver, A.K., Koski, R.K., and van Etten, J.L. (1973) Bacteriophage ϕ6: a lipid-containing virus of *Pseudomonas phaseolicola*. J. Virol. 11: 799–805.

Wirtz, K.W.A. (1974) Transfer of phospholipids between membranes. Biochim. Biophys. Acta 344: 95–117.

Wirtz, K.W.A. and Zilversmit, D.B. (1968) Exchange of phospholipids betweeen liver mitochondria and microsomes *in vitro*. J. Biol. Chem. 243: 3596–3602.

Wold, F. (1972) Bifunctional Reagents. In: Methods of Enzymology (Hirs, C.H.W. and Timasheff, S.N., eds.), vol. 25, pp. 623–651. Academic Press, New York.

Local differentiation of the cell surface of Ciliates: their determination, effects, and genetics*

13

T. M. SONNEBORN

The Ciliate cell surface is marked by known local differentiations of every order from the molecular to that of structures conspicuous in optical microscopy at low magnifications. The vast literature on these differentiations will not be reviewed comprehensively here. Instead, I shall discuss some of the main findings in a few Ciliates that have been studied more extensively, namely Stentor, Tetrahymena, Paramecium and Oxytricha. Frankel (1974) and Sonneborn (1975a) include accounts of some of the important results on other Ciliates, especially those of Jerka-Dziadosz on Urostyla, Kaczanowska on Chilodonella, Suzuki on Blepharisma, and Tucker on Nassula. This chapter will begin with the most conspicuous surface differentiations and end with ultrastructural and molecular differentiations, pointing at each level to anything that is known or indicated about their determination, effects, and genetics.

1. The oral apparatus and the site of the oral primordium

In the four Ciliates discussed here, as in most Ciliates, all or part of the food intake system is the most conspicuous local differentiation of the cell surface. Actually, much of this system is often inside the cell as an internal extension of the cell cortex; nevertheless the surface parts leading to the internal parts are highly conspicuous. The food intake system, or a major part of it, usually arises in a delimited area of the cell cortex called the primordium site. There it develops in preparation for cell division and passes to the *opisthe*, the posterior product of cell division. In the four Ciliates discussed here, a primordium also arises during oral regeneration or replacement, and in Oxytricha during excystment. The site where a primordium will develop (except perhaps in the cyst)

---

*Contribution #1042 from the Department of Zoology, Indiana University.

G. Poste and G.L. Nicolson (eds.)   The Synthesis, Assembly and Turnover of Cell Surface Components can be specified in relation to visible landmarks of the cell surface. We shall consider whether these correlates are determinative, whether the primordium site or what develops from it has more far-reaching effects, and how the site and its productions are inherited.

1.1 Stentor

Stentor is the ciliate par excellence for microsurgical experiments that have been performed in abundance and have yielded highly important results (Tartar 1961, chapter 10 and earlier; Uhlig, 1960; de Terra, 1974, 75), including one of the first and fullest analyses of the determination of the primordium site. The site is located posterior to the existing oral apparatus in the so-called stripe-contrast region of the cell cortex (Fig. 1A). This region is where the two ends of a stripe-width gradient meet, the stripes being longitudinal pigmented bands varying in width in a graded series around the cell. The primordium site is in the area of the diagonally oriented narrowest stripes, close to where they abut on the longitudinal widest stripes; it extends across a number of the narrowest stripes and the rows of cilia (i.e., kineties) that alternate with them. The primoridum develops into the whole oral apparatus, including the conspicuous peristome (Fig. 1A).

Tartar (1961 and earlier) demonstrated by microsurgery that a primordium site is established wherever an area of stripe contrast is created, regardless of its position, orientation (Fig. 1B), or magnitude. His findings were confirmed and elaborated by Uhlig (1960). When an extra primordium site is constructed to have the same polarity and orientation as the original one and to be diametrically opposite the original one, a doublet cell is created which, under favorable conditions of culture, reproduces true to the doublet type. Thus the primordium site or something closely associated with it directs its own reproduction.

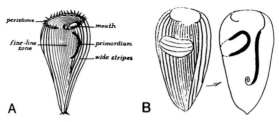

A   B

Fig. 1. Diagrams of Stentor to illustrate the position of the oral primordium. (A) An early oral primordium arising in the oral primordium site, or the narrowest stripe region close to where it abuts on the widest stripe region. The pigmented stripes are represented as clear areas bounded by lines that represent ciliary rows (kineties). (B) A patch of wide stripes grafted in a narrow stripe region at right angles to the narrow stripes. This results in a supernumerary oral primordium in the narrow stripe region near where it surrounds the graft. An oral primordium also arises at the normal wide-narrow stripe juncture. (Reproduced with permission from Tartar [1956] (A) and Tartar [1961] (B)).

Although a stripe-contrast zone regularly determines a primordium site, Tartar (1961) has occasionally observed primordia to arise in areas where there was no perceptible stripe contrast. He therefore concluded that determination is not due to the visible stripe contrast itself but to some as yet unidentified invisible factors that are usually associated with stripe-contrast areas. As in other cases that will be mentioned later, microscopically visible correlates of determinative events may not be causal but serve merely as signposts of the presence of more elusive, presumably molecular, causative factors.

1.2. Tetrahymena

A primordium can arise in at least three different positions in Tetrahymena. The usual site is somewhat anterior to the cell equator immediately to the cell's left of the first postoral kinety (Fig. 2A). This is where primordia normally arise in preparation for cell division; growth before cell division then shifts them to a position just posterior to the equator. The oral apparatus near the anterior end of the parent cell is retained by the *proter,* the anterior product of transverse cell division; the newly developed one becomes the oral apparatus of the opisthe. Frequently, a second site (Fig. 2B) operates at cell division in certain stocks of *T. cosmopolitanis* and *T. canadensis.* This site is latitudinally the same as the one just described, but it is shifted one or a few kineties away from, usually to the left of,

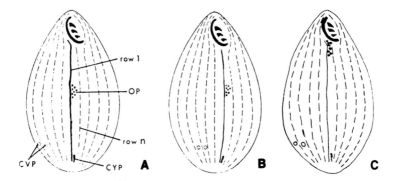

Fig. 2. Diagrams of the ventral (oral) surface of Tetrahymena to illustrate positions of the oral primordium. The four very heavy, short lines near the anterior end represent the three membranelles and the undulating membrane of the oral apparatus. The positions in which an oral primordium (OP) arises are marked by a group of dots in the space to the cell's left (viewer's right) of a solid longitudinal line representing a kinety (a ciliary row). Other kineties are shown by broken lines. The kineties that terminate anteriorly at the oral apparatus are called postorals; the postoral kinety furthest to the cell's right is the first postoral kinety or row or kinety 1. (A) Normal position of oral primordium origin in preparation for cell division. (B) One of the varied positions (one row further to the viewer's right than in Fig. 2A) where the oral primordium can arise in an abnormal stock of *T. thermophila* (Nanney, 1967). Note the corresponding shift in position of cytoproct (CYP) and contractile vacuole pores (CVP). (C) Position of oral primordium origin during process of oral replacement; in the absence of cell divisions, the old oral apparatus is replaced by a new one. (Parts A and B reproduced with permission from Nanney, 1967.)

the first postoral kinety (Nanney, 1967). Thus, the primordium site is not determined by any particular kinety. A third site is, like the usual one at cell division, immediately to the left of the first postoral kinety, but far forward and close to the existing oral apparatus instead of equatorial (Fig. 3C). This site operates in the absence of cell division during the process of oral replacement (Frankel and Williams, 1973) that occurs under certain nutritive and other conditions.

The occurrence of both longitudinal and latitudinal variations in the position of the primordium site raises the question of whether there is but a single large primordium "field" any part of which can give rise to a primordium (different parts responding to different signals), or whether these signals activate one or another of two or three or more discrete primordium sites. What determines the whole primordium field (if it is unitary) or the separate ones (if they are separate) is not really known.

The development of primordia follows essentially the same course, wherever it is located (Frankel and Williams, 1973). Starting as a loose field of nonciliated basal bodies, the field becomes more compact and the basal bodies develop certain appendages; then they become ordered into three membranelles and an undulating membrane. Eventually each membranelle comes to consist of three parallel rows of ciliated basal bodies with appendages. The undulating membrane then contains two parallel rows of basal bodies only one of which is ciliated; it later becomes associated with a ribbed wall and deep fibers.

As Nanney (1966a) has stressed, other features of the cortex have definite cytogeometric relations with the meridian of the primordium site (Fig. 2A). The cytoproct (cell anus) is a local cortical differentiation near the posterior end of the same meridian. Slightly anterior to the latitude of the cytoproct and about 90° of circumference to the cell's right of the primordium meridian are the pores of the contractile vacuoles, another surface differentiation. When the primordium site shifts position laterally, so do the cytoproct and contractile vacuole pores (Fig. 2B); their geometric relations are retained (Nanney, 1967). Thus there is either a common cause for all three positional determinations, or one of them—presumably the primordium site—determines the others.

Apparently the inheritance of these cortical relations is due at least in part to a nonnuclear basis. As in Stentor, doublet cells reproduce true to type. Thus, each oral segment of cortex with its primordium site is reproduced. The hereditary difference between cells with one or with two oral segments exists within a single clone, that is, presumably among cells of identical genotype. According to Nanney, clones of doublets of T. thermophila always become singlets eventually. This differs from what happens in Colpidium campylum, a close relative of Tetrahymena, in which very stable clones of doublets have long been known (Sonneborn, 1932).

Loss of the doublet condition in Tetrahymena follows gradual reduction of the originally nearly double number of kineties in the course of successive cell generations. As the number of kineties approaches the normal upper limit for singlets, cells appear with only one oral apparatus, but they still possess two cytoprocts and two sets of contractile vacuole pores. Later, the supernumerary pores disappear, and last of all the extra cytoproct. Nanney (unpublished observations) points out that losses proceed from anterior to posterior structures.

1.3. Paramecium

More is known about the *genetics* of the primordium site and the oral apparatus in *P. tetraurelia* than in any other ciliate. As in Stentor and Tetrahymena, the position of the site is uniquely defined by its relation to visible structures. It lies between the innermost ciliary row of the right wall of the vestibule, a subequatorial depression of the cortex, and the wall of the peristomal cavity (Fig. 3A). During the morphostatic stage of the cell cycle, the primordium site is devoid of ciliated basal bodies but has many nonciliated basal bodies in the subsurface cortex (Jones, 1976). In the morphogenetic stage preceding cell division, these basal bodies rise to the cell surface (Fig. 3B), become ciliated, and organize into the 12 ciliated rows of the peristomal wall. This oral anlage passes at cell division to the opisthe where its development into a mature oral apparatus is completed (Jones, 1976). Meanwhile, the vestibule elongates and is cleaved by the fission furrow; its posterior part passes to the opisthe along with the developing oral structures.

Whether determination of the primordium site is due to interaction between

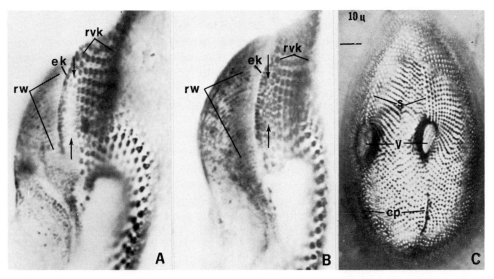

Fig. 3. *Paramecium tetraurelia*, silver preparations. (A) The site (arrows) where the oral primordium will appear during prefission morphogenesis. (B) An early stage of formation of the oral primordium (arrows). (C) Cell with two adjacent ventral cortices and one dorsal cortex; a typical descendant of a normal cell that picked a piece of the vestibular area from its mate (see text). *Key:* cp, cytoprocts; ek, endoral kinety at boundary between right vestibular and peristomal walls; rvk, right vestibular kineties; rw, ribbed wall of peristome; s, anterior sutures; v, vestibules. (Parts A and B courtesy of Warren Jones; Part C reproduced with permission from Sonneborn, 1963).

the right vestibular rows and the nearby peristomal wall or whether it is due to invisible factors associated with these marker structures is as uncertain as in the case of the stripe-contrast zone in Stentor. However, something associated with the right vestibular wall and immediately adjacent structures is clearly determinative, and the determiner is both reproduced by a nonnuclear mechanism and capable, directly or indirectly, of organizing the major features of the ventral cortex and cell surface, as will now be set forth.

When conjugants fail to separate after fertilization, as sometimes occurs spontaneously and can be induced by exposure to immobilizing antiserum, they may fuse to create homopolar doublets. These in turn produce clones of doublets that are remarkably stable through fissions and fertilizations. The doublets have a continuous endoplasm without partitions, the whole enclosed by a single, continuous cortical sheet containing two complete sets of cortical structures with corresponding components of the sets 180° apart. At their origin the doublets have two macronuclei, but they soon come to have only one.

Although the mode of origin of doublets in Paramecium and other Ciliates strongly indicates that they do not have any relevant genomic difference from singlets, the full genetic proof of the basis for the hereditary difference between singlets and doublets has been given only in *P. tetraurelia* (Sonneborn, 1963). Mendelian analysis of crosses between doublets and singlets, with marker genes, excluded the possibility of a genotypic difference. Using an endoplasmic marker (κ) in either parent, endoplasmic transfer was demonstrated in both directions (from singlet to doublet and vice versa) and the results excluded an endoplasmic basis for the singlet-doublet difference. Nuclear differentiation, that is, stable differences in genic activity, and differences in nuclear size were likewise excluded by exploiting special genetic tricks. These exclusions seemed to leave only a difference in the cortex itself as the basis for the hereditary difference between singlets and doublets. The critical cortical area was then demonstrated to be the area of the primordium site itself—singlets having one such area, doublets two. In essence, the demonstration consisted of adding the critical area to singlets and destroying or otherwise removing one of the two areas from doublets.

Addition of the critical area was achieved (Sonneborn, 1963) by "cortical picking." When separation of conjugants is delayed following exposure to immobilizing antiserum, very rarely it occurs unequally, one cell being the recipient of a piece of the vestibular area that belonged to its mate. This natural graft takes, develops a supernumerary vestibule, primordium site, and oral apparatus to the right of the host's comparable structures, and a second typical ventral cortex, including cytoproct and the typical asymmetric kinety pattern, develops around it (Fig. 3C). The cell with these two complete and immediately adjacent ventral surfaces has only one dorsal surface. The new cell type is faithfully reproduced for a considerable number of cell generations. This important result demonstrates that a small area of the cortex contains what is required to determine a primordium site and has the capacity to induce, directly or indirectly, the hereditary development of the highly structured pattern of the whole ventral cortex.

Loss of one primordium site and associated area in a doublet was induced by Hanson (1955, 1962) using a fine beam of ultraviolet irradiation. When the beam was directed to the primordium site, that is, the right wall of the vestibule, opisthes were formed that lacked one oral apparatus but were otherwise doublets. The progeny of these opisthes by cell divisions never reacquired the second oral apparatus. Once gone, it was lost forever. However, Hanson and Ungerleider (1973) report one such incomplete opisthe, some of whose immediate progeny regained the missing oral apparatus after autogamy if they underwent autogamy within four cell generations after the loss. The basis for this exception remains obscure.

Sonneborn (1963, 1975a) reported the accompaniments and consequences of the occasional spontaneous loss of one oral area from a doublet. Approaching loss is signaled by a shift in position of the contractile vacuole pores, which come to lie much closer to the oral meridian. Also, the oral apparatus shifts to a more anterior position, and the size and anteroposterior position of the cytoproct vary. After loss of one oral apparatus, the residual ventral pattern of the cortex persists for about 20 cell generations and is then gradually converted to a dorsal pattern. The contractile vacuole pores shift closer and closer to the meridian of the cytoproct and eventually disappear. The anterior ends of the ventral kineties shift forward toward, and then reach, the anterior polar region. Finally the cytoproct disappears. This completes the transformation of a doublet lineage into singlets. Thus the structures that are induced to form in definite relative positions by the presence of a primordium and oral apparatus have limited hereditary stability when the primordium site is lost; they undergo eventual regression and loss in the absence of the inducer.

The sequence of events in the doublet to singlet series resembles that mentioned earlier in Tetrahymena. However, the neatness of the anteroposterior progression in Tetrahymena is blurred in Paramecium by the behavior of the contractile vacuole pores. One is more anterior than the vestibule, the other is as far posterior as the cytoproct; yet both shift positions before the oral apparatus is lost and both persist for many generations after the oral apparatus has disappeared. If there is a gradient of loss in Paramecium it is not strictly anteroposterior, but it may radiate from the oral area. Since the oral area is anterior in Tetrahymena but subequatorial in Paramecium, the two alternatives are identical in the former but distinct in the latter.

1.4. Oxytricha

Oxytricha is of special interest mainly because it undergoes encystment. During encystment it loses all of its visible surface differentiations except the cell membrane; nevertheless, all of the surface differentiations are redeveloped during excystment (Grimes, 1973a,b; Hammersmith, 1976).

The surface of Oxytricha (Fig. 4) is patterned very differently from that of the three organisms discussed so far. Kineties, longitudinal rows of single, well-separated cilia (the so-called dorsal bristles), occur only on the dorsal (aboral)

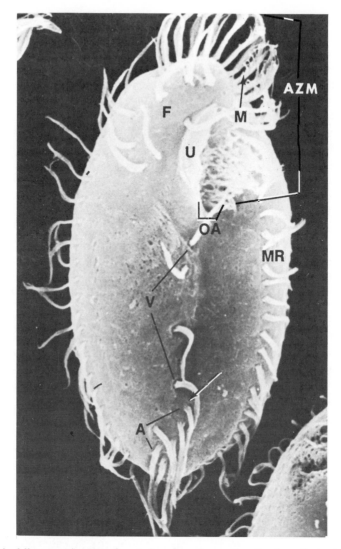

Fig. 4. *Oxytricha fallax*, scanning EM of ventral surface. Arrow, the site where the oral primordium will appear during prefission morphogenesis; the same primordium also gives rise to certain cirri (see text). A, group of six anal cirri. AZM, adoral zone of membranelles; F, group of eight frontal cirri. M, a membranelle of the AZM. MR, marginal row of cirri. OA, oral apparatus. U, undulating membrane. V, group of four ventral cirri. (Courtesy of G. Grimes.)

surface. The organelles of the ventral (oral) surface of this flattened cell are composed of cilia packed into two main kinds of organelle: cirri, composed of several short rows and columns of tightly packed cilia; and membranelles composed of four parallel rows, three long and one short, of tightly packed cilia. Cirri are disposed in regular numbers and positions on the ventral surface and also are lined up in two rows, one along the right and the other along the left margin of the cell. Many membranelles are lined up in parallel array to form an arc extending from a point anterior to the center of the ventral surface (where the cytostome or mouth is) to the left and forward around the anterior end of the cell. This arc is called the adoral zone of membranelles (AZM).

The primordium site lies considerably posterior to the AZM, just to the left of the most anterior cirrus of the anal group (Fig. 4, arrow). In preparation for cell division, many ciliary basal bodies appear at the primordium site and move forward in unordered array to a position just behind and to the left of the AZM. There, the group breaks up into a subgroup that will form all of the cirri of the opisithe except the marginals and another subgroup that will form the AZM of the opisthe. The proter retains the parental AZM. All cirri (except the marginals) disintegrate and disappear, being replaced by new ones, those for the proter developing from groups of new basal bodies that arise in the anterior ventral region and migrate to the proper final positions. The marginal rows of cirri proliferate from a region in their middle. Then the cell divides.

During encystment all ciliary organelles disintegrate and disappear, as do all other recognizable structures of the cortex except the cell membranes. The cell has two plasma membranes and lays down the cyst wall between them. Grimes (1973a) examined almost complete sets of thin sections of cysts and was unable to find even a single basal body or any other structure of the cortex except the two cell membranes. During excystment, a primordium appears and gives rise to more than it does at cell division, in fact to the whole ventral ciliature including the two marginal rows of cirri.

Homopolar doublets arise both by fusion of conjugants and by faulty cell division and are perpetuated through cell divisions and conjugations. Again, since doublets and singlets of the same clone reproduce true to type, primordium sites are reproduced at cell divisions. The highly dedifferentiated cyst remembers whether it was formerly a singlet or doublet. Encysted singlets excyst as singlets, encysted doublets excyst as doublets, even when the doublet cyst is smaller than some singlet cysts (Grimes, 1973b). Hammersmith (1976) has carried this further by obtaining irregular multiple monsters with various numbers of AZM in varied positions relative to one another. Serial sections of encysted monsters were examined by electronmicroscopy, with results that agreed with those of Grimes: except for the plasma membranes, no recognizable cortical structures were present. Observing the number and relative positions of the AZM in the same cells before and after excystment, Hammersmith found that with minor exceptions the same number of AZM reappeared after excystment and in similar relative positions.

These results on doublets and multiple monsters show clearly that the primordium site or its determiner is a local differentiation that is reproduced at cell divisions and persists through the global dedifferentiation occurring at encystment. Moreover, when removed by amputation, a primordium site is irreparably lost (Grimes, 1973b). Other parts of the cell cortex can be regenerated if the primordium site is still present. When one of the two primordium sites of a precystic doublet is removed, a singlet emerges at excystment.

Recently, Grimes (1976) irradiated one primordium site of a doublet with a laser beam and found, as expected, that subsequent generations from the first opisthe lacked one primordium site and all the ventral ciliature to which it normally gives rise at cell divisions, namely AZM, oral apparatus, and all cirri of the ventral surface except the marginal rows which normally reproduce themselves. These two rows persisted and continued to reproduce in the absence of the corresponding primordium site. Recall that at excystment, the primordium gives rise to the marginal rows as well as to the ventral cirri, AZM, and oral apparatus. Thus, certain products of the primordium at excystment—the marginal rows—have the capacity to reproduce themselves at fissions even when the primordium that initially gave rise to them has been destroyed. This is not to say that the marginal rows can arise at excystment in the absence of the oral primordium. Grimes has not yet reported the fate at excystment of the marginal rows of cirri that persist during fissions after loss of the primordium site. A priori, they would obviously not be expected to reappear at excystment.

The fact that all cortical differentiations discernible at the EM level, except the plasma membranes, disappear in the cyst while the primordium site persists (or is redetermined) at excystment suggests that the primordium site may be a differentiation of the plasma membrane(s). Thus the definite positional relations of the primordium site to visible cortical structures in the free-living cell seem to lack determinative significance. These cortical structures appear to be merely signposts, like the stripe-contrast zone in Stentor. The basic nature and molecular determinants of the primordium site in Oxytricha, as in other Ciliates, remain elusive problems.

2. The repeating unit of cortical structure

The cortex of some ciliates, such as Tetrahymena and Paramecium, is largely composed of more or less longitudinal rows of unit territories, each of which has the same or nearly identical structural plan. Most of the relevant work on them has been done on *P. tetraurelia* and *T. thermophila*.

2.1. Paramecium tetraurelia

For present purposes, it is not necessary to go into full detail about the electronmicroscopic structure of the cortical unit, but only to note several salient features showing that the units are asymmetrical with definite markers of anterior, posterior, right and left. First, I shall describe these features for units that possess only one basal body and cilium, then for those that possess two. These are the only two kinds of outer surface cortical units (Fig. 5).

All cortical units are bounded by ridges, the ciliary basal bodies being located at the bottom of the enclosed valleys. The basal body and cilium are slightly to the cell's right of the cortical unit center. A kinetodesmal fiber extends from the anterior right margin of the basal body, near its base, upward to the right ridge and anteriorly in the ridges for the length of three or more units. The parasomal sac, an invagination of the cell surface, lies slightly to the right of and anterior to the basal body. When two basal bodies are present in a cortical unit, they are equally off-center to the right, one anterior to the other. A kinetodesmal fiber arises only from the posterior one. The parasomal sac lies to the right of both basal bodies and between them.

The units are lined up in longitudinal rows (kineties). The number of kineties per cell, exclusive of vestibule and peristome, varies from about 65 to 77; the total number of units is about 4,000. Successive units in a row are contiguous, sharing bounding ridges. In the middle of the transverse ridge between units of a row, the tip of a trichocyst is inserted. Adjacent rows of units are also contiguous, their units also sharing bounding ridges. The whole surface, the units and their cilia, is covered by a continuous sheet of plasma membrane.

At cell division, the fission furrow cuts across all of the outer surface rows so that each row of the parent cell is represented in both daughter cells. Meanwhile, the number of units per row doubles. This is accomplished by a definite pattern of unit reproduction (Sonneborn and Chen-Shan, unpublished observations). Some units, the most anterior and most posterior ones in each row, do not reproduce. Most units on the dorsal (aboral) surface give rise to two units. Those

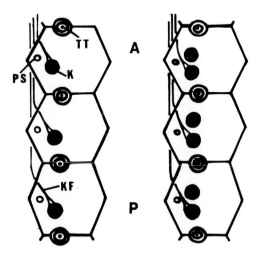

Fig. 5. *Paramecium tetraurelia:* diagrams of the two kinds of units of cortical structure, one with one cilium and the other with two, as viewed from outside the cell. (Viewer's left is cell's right.) A, anterior; K, kinetosome (ciliary basal body). KF, kinetodesmal fiber. P, posterior; PS, parasomal sac. TT, tip of trichocyst. (Reproduced with permission from Sonneborn, 1975a.)

nearest the cell equator give rise to more than two units. This zone of super-reproduction is narrow on the aboral surface, but much more extensive on the ventral (oral) surface.

The mode of unit reproduction (Fig. 6) has been described by Dippell (1964, 1968). The unit elongates and a new basal body develops from a plaque located immediately anterior to the inner end of an existing basal body. As with centrioles, the axis of the new basal body is at first perpendicular to the axis of the adjacent mature one. Later, the new basal body tilts up toward and moves to the cell surface; there it inserts into the plasma membrane and a cilium grows from it. Meanwhile, its "appendages"—the kinetodesmal fiber and the parasomal sac—develop around it. Then one or more ridges (into which trichocysts insert) grow in transversely to subdivide the "parent" unit into "daughter" units. Some units divide directly into three or four and some probably undergo more than one round of reproduction at the same cell division.

The amount of unit reproduction and the kinds of units produced are correlated with and, directly or indirectly, determined by the distance and direction (vector) from the oral area. Correlation is shown by the fact that the distribution of one- and two-cilia units accords with a standard pattern (Fig. 7) that is reproduced at cell division. Determination is shown by the facts that the same ventral pattern is developed around "grafted" supernumerary oral areas (Fig. 3C and p. 834) and that this pattern reverts to the simple dorsal pattern when the supernumerary oral apparatus is lost (p. 835). Moreover, essentially normal distribution of the two kinds of units develops in inverted rows (Fig. 8), which I now will discuss for other reasons.

Inverted rows can be produced in any one of several ways, most of them due to growth of rows from one cell into another when the two cells are joined side by side in heteropolar array, either in conjugants (Beisson and Sonneborn, 1965) or in dividing cells (Sonneborn, unpublished observations) that fail to disjoin. The units in such inverted rows (Fig. 8) have their cilia and basal bodies, parasomal sacs, and kinetodesmal fibers to the cell's left of unit center, instead of to its right; and their kinetodesmal fibers extend toward the posterior instead of the anterior end of the cell. At cell division inverted units reproduce the inverted orientation (Dippell, 1964; Beisson and Sonneborn, 1965). In other words, the position of origin and direction of migration of new basal bodies remains unchanged with respect to the structures within the unit. The "appendages" of the new basal bodies are positioned with reference to other structures in the unit exactly as in normally oriented rows. Hence, the positional information for new developments within a unit is determined within the unit and is independent of the unit's orientation with respect to the rest of the cell.

Consequently, once a unit becomes inverted (i.e., 180° off normal), its progeny units thereafter are all inverted. A whole inverted row is perpetuated in descendants (Fig. 9) because its units reproduce in inverted orientation and the growing row is transversely bisected at each fission. Lines of descent characterized by cells having one, a few, or more inverted rows have been obtained (Beisson and Sonneborn, 1965). Some were followed for more than 800 cell generations and some were crossed to normals. The inverted rows persisted in the one mate

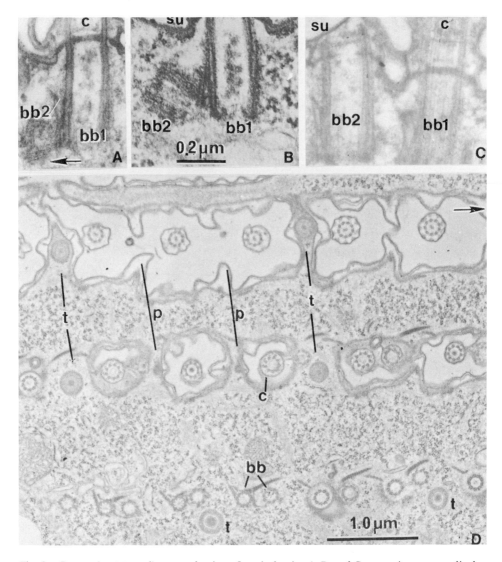

Fig. 6. *Paramecium tetraurelia:* reproduction of cortical units. A, B, and C are sections perpendicular to cell surface; D is a subtangential section cutting deeper into cell toward bottom of photo. Arrows point to cell's anterior. (A) young basal body (bb2) developing anterior to proximal end of, and at right angles to, a mature, ciliated (C) basal body (bb1); (B) later stage, bb2 more elongated and tilting toward cell surface (su). (C) still later stage, bb2 fully elongated, perpendicular to cell surface, and about to attach to plasma membrane. (D) greatly elongated reproducing cortical units in three adjacent rows. Trichocysts (t) in transverse ridges marking the anterior and posterior borders of the units. Top row: the unit between two trichocysts contains three cilia shown in cross-section; the partitions (p) are growing across the unit, subdividing it into three parts each of which will become a complete cortical unit. Middle row: the trichocyst-bounded unit is cut near the bases of the cilia, a level at which partitioning into three units has been completed. Bottom row: each of the three subdivisions of the original trichocyst-bounded unit contains two basal bodies (bb); as the other two rows show, one basal body of each of the three nascent units has not yet become ciliated. (Photos courtesy of R. V. Dippell; from Sonneborn, 1970.)

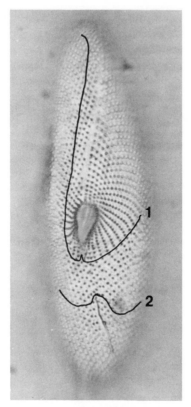

Fig. 7. *Paramecium tetraurelia:* silver preparation indicating distribution of the two kinds of cortical units of the ventral surface. All units anterior to solid line 1 contain two cilia; those posterior to solid line 2 contain one cilium. The region between lines 1 and 2 contains both kinds of units.

Fig. 8. *Paramecium tetraurelia:* diagram of an inverted row (2) of cortical units between two normally oriented rows (1, 3). The wide space (between basal bodies of rows 1 and 2) and narrow space (between those of rows 2 and 3) on the two sides of an inverted row (or group of rows) are created by the eccentricity of the position of parts of the units: basal bodies (bb), parasomal sacs (s) and kinetodesmal fibers (f) are all off-center to the same side of a unit, to the viewer's left (cell's right, R) in normal rows, viewer's right in inverted rows. Fiber extends anteriorly (A) in normal rows, posteriorly in inverted rows. (Courtesy of R. V. Dippell; from Sonneborn, 1970.)

Fig. 9. *Paramecium tetraurelia:* silver preparation of a prefission cell bearing two inverted rows (i1, i2) separated by three normally oriented rows. In the latter, the triangular silver deposits point to the viewer's left; in the inverted rows, to the viewer's right. Each triangle represents three adjacent silver dots, one at each apex; two mark the bases of cilia and one the parasomal sac. The opposite orientation of the triangles in the two kinds of rows is due to the position of the sac on opposites side of the ciliary bases (see Fig. 8). That the cell is preparing to divide is shown by the presence of reproducing cortical units (arrows) and of a developing oral primordium (OP) seen as a wide, dark subsurface area in the (cell's) right vestibular wall (cf. Fig. 3). The arrowed units are recognized as reproducing because the silver deposit in them is elongated due to the presence in the unit of more than two cilia and basal bodies. Since the cell is preparing to divide, and division cuts transversely across all rows near the cell's equator, the figure shows that inverted rows, like normal rows, are inherited at cell division. (Photo courtesy of R. V. Dippell.)

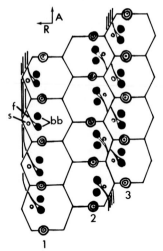

Fig. 8.

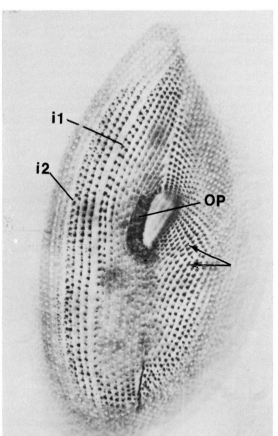

Fig. 9.

and its cytoplasmic progeny. Inverted rows, like normal rows, can be lost or duplicated when row numbers change as they frequently do (see below), but they never give rise to a normally oriented row except by the same kinds of processes that yield inverted rows from normally oriented rows (see above).

There are a number of physiological and developmental consequences of inverted rows that are also inherited as an indirect consequence of the inheritance of the inversions themselves. They indicate the importance for the cell of cytoplasmically inherited variations of surface structure. Beisson and Sonneborn (1965) observed that the swimming of cells was different when they had a patch of several inverted rows. They referred to this as "twisty" swimming. Normal cells swim in a straight path (Fig. 10A) or a helix of very small diameter and large pitch. The helical path of twisty cells has a larger diameter and smaller pitch, deviation from the normal path increasing with the number of rows in the inverted patch (Fig. 10B). With still more rows in the inverted patch, the path becomes circular, the diameter of the circle decreasing as the patch becomes larger (Fig. 10B,C). With about 20 rows or more in the patch, the cells simply rotate in one spot about their midpoint (Fig. 10C). Unable to locomote, they lie on the bottom of the container where (in shallow cultures) they feed, grow, and reproduce. Since they cannot swim up to the surface, they asphyxiate and die in deep culture.

From the beginning, we inferred that the abnormalities were probably due to the fact that cilia in inverted rows beat in the opposite direction from normally oriented rows. This was demonstrated by placing the cells in a suspension of carmine particles and observing that the particles outside but near the cell are swept in opposite directions on different parts of the cell surface. Complete proof was presented by Tamm et al. (1975) by scanning electronmicroscopic pictures of instantaneously fixed cells that show the reversed beat of individual cilia in exactly the position where the inverted patch is known to be. The pictures further showed reversal of the metachronal waves. Thus direction of ciliary beat is determined by something built into the structure of the cortex and is also inherited cytoplasmically.

For several years my coworkers (Dippell, Schneller) and I have been studying other indirect effects of inversions in a clone with a large inverted patch on the left side of the cells (Sonneborn, 1975a). Fig. 11 shows some of these effects. The anterior end of a normally oriented, left ventral row stops at the anterior suture (the midventral space between right and left rows); instead, the anterior end of an inverted row bends forward parallel to the suture (Fig. 11A). At the border where an inverted patch begins, the opposite orientation of the units in the last normal row and the first inverted row creates a wide space between these adjacent rows. Into this space a branch of the cytoproct extends when the wide space abuts on the normal cytoproct (Fig. 11B). At the fission plane, inverted rows fail to cleave or do so imperfectly (Fig. 11C). Where the internal path of food vacuole migration meets the inverted patch, the further motion of food vacuoles is blocked.

These effects of inverted ciliary rows tell something about the roles of the cell

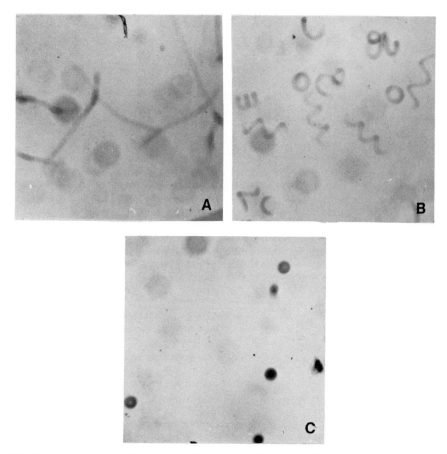

Fig. 10. *Paramecium tetraurelia:* two second exposures showing paths of movement. (A) Normal cells. (B) Cells with a patch of 6 to 10 inverted rows. The tighter helices are paths of cells with more inverted rows; circles are paths of cells with most inverted rows. (C) Cells with a patch of 15 to 20 inverted rows. Paths are very small circles or dark spots due to rotation of the cell about a point within the cell.

surface and cortex and about interactions between parts of the surface of the same cell. First, the two ends of a cortical unit of structure interact differently with the anterior suture and with the plane of cell division. Second, in the cytoproct area the space between two normally oriented rows reacts differently from the larger space between a normal and an inverted row. Third, the movement of external organelles (cilia) and of internal organelles (food vacuoles) is correlated with (and determined by) the orientation of cortical structures.

From the earlier discussion about the perpetuation of rows by transection at cell division, one would expect the number of rows to be a stable hereditary character, but this is not the case in Paramecium or Tetrahymena. The number of surface rows in a normal singlet clone can vary from about 65 to 77 during log

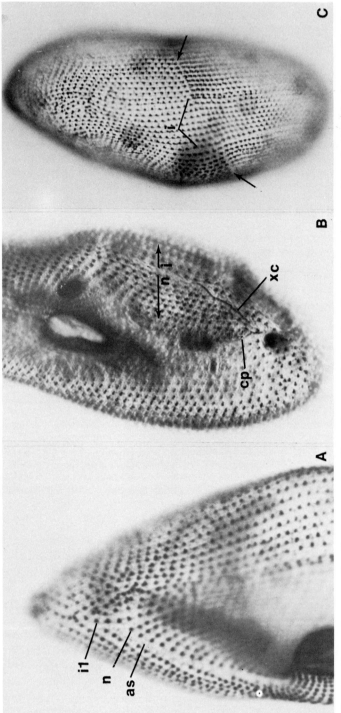

Fig. 11. *Paramecium tetraurelia*: silver preparations illustrating some indirect hereditary effects of inverted kinetics. (A) The first row (i1) of an inverted patch extends anteriorly in the anterior suture (as) instead of stopping at the suture as normally oriented rows (n) do. (B) An extra cytoproct (xc) close to the normal cytoproct (cp) in the wide space (cf. Fig. 8) between normally oriented rows (n) and inverted rows (i). (C) A patch of nine inverted rows (i) in a dividing cell. The division space (arrows) does not appear in the inverted patch. Note the wide and narrow inter-row spaces on the two orders of the inverted patch. (Reproduced with permission from Sonneborn, 1975a).

growth in *P. tetraurelia* (Sonneborn and Chen-Shan, unpublished observations) and from about 14 to 34 in certain clones of *T. thermophila* (Nanney, 1966b). The ways in which rows are gained and lost is imperfectly known in both species, but changes in cultural conditions are an important factor in triggering these changes.

More interesting are Nanney's studies of what happens when cells of the same clone but with different row numbers are put under the same uniform cultural conditions. The frequency of row number change varies with the amount of deviation from a certain number of rows that is typical of the clone under those conditions. This characteristic number shows the lowest frequency of change; it is the "stability center." The row number changes in cells that deviate from this most stable row number are predominantly toward it. The stability center for a given set of cultural conditions is probably under genic control, as shown in Euplotes by Heckmann and Frankel (1968). Nevertheless, both in Tetrahymena and in Euplotes, differences in row number that are close to the stability center tend strongly to be maintained by cytoplasmic heredity even after crosses.

Not surprisingly, our account of cell surface genetics has come at last, in the concept of the stability center, to an example of genic effects on the cell surface. In the next section, when we turn to the molecular differentiations of the cell surface, interplay between genic and cytoplasmic inheritance becomes prominent.

3. Molecular differentiations of the cell surface

3.1. Surface antigens

Chemically the best known molecules of the cell surface in ciliates are the immobilization or i-antigens of Paramecium, especially *P. primaurelia, P. biaurelia,* and *P. tetraurelia* (reviewed by Sommerville, 1970; Finger, 1974; Sonneborn, 1975b). They are on the plasma membrane and are exposed to the environment. In the presence of homologous antiserum, cilia of the same cell are bound together, immobilizing the cell. The antigens are proteins with molecular weights of 240,000 to 320,000. They comprise about 30% of the protein content of isolated cilia, but are also located on the cell surface. Earlier workers reported them to be composed of up to 9 polypeptides of two or three kinds. Recent workers (Reisner et al., 1969; Hansma, 1975) claim that the whole antigen is a single polypeptide. The i-antigens are very rich in disulfide bonds. The amino acid composition is known for some of them.

Ordinarily a cell has only one species of i-antigens on its surface at any one time; this defines the i-serotype to which the cell belongs. However, the cells of even a completely homozygous clone can differ in serotype, that is, different cells have different i-antigens on their surface. The different i-antigens producible by a single homozygous stock are coded by genes at unlinked loci. A dozen or more different i-antigens are known in single, intensively studied homozygous stocks of *P. tetraurelia*. Allelic differences between stocks are also known to occur. In *P. tetraurelia,* heterologous reactions usually occur between allelic i-antigens; but in *P. primaurelia* allelic cross-reactions are often weak or nonexistent. Antigens coded by allelic genes, regardless of cross-reactions, are designated variations of the same serotype; antigens coded by nonallelic genes are designated different serotypes.

Various external conditions, such as exposure to homologous antiserum, change of temperature and cultivation in different media, can bring about serotype transformations, often with a high degree of directivity, (i.e., directing which of the alternative serotypes will replace the original one). Remarkably, disappearance of one i-antigen is regularly accompanied by appearance of another. The regularity of replacement and the invariable presence of i-antigens indicate that they have an important function, but this funtion is unknown.

Although the i-antigens are coded by genes, the genetics of serotypes also includes an extranuclear aspect. Two or more serotypes of *P. tetraurelia* can persist hereditarily through asexual and sexual reproduction in genetically identical and completely homozygous cell lineages under identical cultural conditions (Sonneborn, 1948). Generally, regardless of which i-antigen is on the surface of a cell, the cell progeny tend strongly to have the same i-antigen on their surfaces unless the conditions induce transformation. One of the transforming conditions is exposure to homologous antiserum. Molecular interaction at the cell surface between antibody and antigen leads to a nuclear event: the gene coding for the interacting antigen is repressed and a previously repressed gene coding for another antigen is derepressed. Finally, it should be emphasized that a degree of nuclear differentiation is also involved: different genes remain hereditarily active in different cells of the same genotype, as is believed to be true in the case of cells in different tissues or organs of the same multicellular organism. In the serotypes of *P. tetraurelia,* however, this autocatalytic or positive feedback aspect of genic activity has limited stability since it can be terminated by transforming conditions.

The system of i-serotypes varies somewhat from species to species of Paramecium, as can be learned from the reviews cited above. A comparable and well-studied system with additional points of interest exists in *Tetrahymena thermophila* (reviews, Nanney, 1968; Sonneborn, 1975c).

3.2. Mating type substances and mating types

Important in the present context are the mating type substances, sometimes called gamones, of ciliates (reviews, Hiwatashi, 1969; Miyake, 1974, and further reported on by Braun and Miyake, 1975; Miyake and Bleyman, 1976). These are substances that lead cells to interact specifically with each other and unite for fertilization. Cells of different mating types produce different gamones. To date, the only ciliate gamones characterized chemically are the two gamones of one species of Blepharisma. One is a glycoprotein of 20,000 molecular weight; the other is calcium-3 (2'-formylamino-5'-hydroxybenzoyl) lactate. These gamones are secreted into the medium, the first by cells of one mating type and the second by cells of the other mating type. Each gamone is believed to bind to a specific receptor molecule on the cell surface of the other mating type. Interaction between gamones and receptors triggers the events that lead Blepharisma to conjugate.

Less is known chemically, but more in other respects, about the mating type substances of Paramecium. They are firmly bound to the cell surface. All efforts have failed to detect them in the medium or to isolate them in solution from cell fractions. They are still active, however, on cells killed in certain ways (e.g., by formalin, picric acid, or ammonium sulfate), on isolated cilia, and on small membrane-bound vesicles isolated from cilia (Kitamura and Hiwatashi, 1976). Indirect evidence suggests that one or both may be proteins or essentially associated with proteins.

In Paramecium, the test for the presence of mating type substance on the plasma membrane of one cell is the capacity of the cilia of that cell to adhere specifically to the cilia of mating reactive cells of the other mating type. It should be recognized in what follows that this test does not exclude the possibility of presence of the substance in an unavailable position (e.g., deeper in the membrane or in the cell) or in a masked form. With this limitation in mind, it can be said that the gamones are not present on cells that are feeding and growing, but only on those that are entering stationary phase or have been in that phase for a limited time. In some stocks and species they are not present until after the passage of a considerable number of cell generations (the period of sexual immaturity) after conjugation. In some, they are present on the surface only in a certain phase of a circadian rhythm. They are present in all species of Paramecium on the cilia of the ventral (oral) surface but only when cells are cultured within a range of temperatures, different for different species, that is usually narrower than the range of temperature for growth or viability. In *Paramecium bursaria,* which has a circadian rhythm of mating reactivity, Cohen (1964) demonstrated that at the onset of the reactive phase of the rhythm only the posterior ventral cilia were reactive, that reactivity then spread anteriorly and laterally on the ventral surface, and that at the end of the reactive phase loss of reactivity occurred first near the anterior end of the cell and spread posteriorly. Thus, unlike the i-antigens which are always present over the whole cell surface and all cilia, the mating type substances are localized differentiations of the plasma membrane and are present there only under certain external and internal conditions.

The remainder of this account of mating types will deal only with *Paramecium tetraurelia.* However, the systems in most other species of Paramecium and Tetrahymena are either the same or show some of the same phenomena. *P. tetraurelia* has two mating types, commonly designated O and E, and each is hereditary according to a system to be set forth below. When reactive cultures of the two types are brought together, the cells are not attracted to each other, but cells of different mating type adhere to each other by the gamone mechanism on their ventral cilia when these cilia happen to contact each other by chance in the course of random swimming. The cells remain together in ciliary contact for 1 to 2 hours, and this plasma membrane interaction between two cells of different mating type initiates the following sequence of events of conjugation: loss of a band of ventral cilia and a wider band of trichocysts beginning near the anterior end of the cells and spreading posteriorly; adhesion between cells, first near the anterior end (holdfast region) and then further posteriorly to the oral area and slightly beyond; complete loss of gamone activity; meiosis; loss and replacement of the oral apparatus; reciprocal cross-fertilization between the two mates; and final separation of mates. The whole process at 27°C takes 5.5 to 6.0 hours. Beisson and Capdeville (1966) found that the whole complex series of events of conjugation can occur when continuously exposed to inhibitory concentrations of puromycin, beginning when the cells are mating reactive and continuing until after mates separate.

The genetics of mating types in *P. tetraurelia* includes three components: genes, stable nuclear differentiation, and a gene-dependent cytoplasmic differentiator of nuclei (reviewed by Nanney, 1968; Butzel, 1974; Sonneborn, 1975b). Recessive genic mutations at any one of three loci restrict homozygotes for them to mating type O (Byrne, 1973). Wild-type lineages of identical and completely homozygous genotype can be either mating type O or mating type E. In this case, the unit of inheritance is the *caryonide,* a lineage in which the macronuclei of all the cells are descended from a common ancestral macronucleus. Each fertilized cell gives rise to two caryonides. Two macronuclei develop from different products of division of the synkaryon, the fertilization nucleus; they segregate at the first cell division and divide at all subsequent divisions.

Only at the time of their origin and development do macronuclei become irreversibly determined for control of mating type, some for type O, others for type E. All descendants of a determined macronucleus are alike: determination is stable and hereditary; inheritance is caryonidal. The two developing new macronuclei in a fertilized cell tend strongly to be determined for the same mating type. The simplest interpretation is that both are determined for type O if a cytoplasmic differentiator of nuclei is absent or present in low concentration, and both for type E if the differentiator is present in high concentration. At intermediate, near-threshold concentrations, one macronucleus may be determined for type O, one for type E, or one or both may be determined for selfing. Selfing caryonides contain both O and E cells. The cytoplasmic differentiator is held not only to determine macronuclei for type E but also to determine macronuclei for production of high concentrations of the differentiator itself. Thus the operation of this system usually results in the two sister caryonides having the same mating type as that of their cytoplasmic (i.e., "maternal") parent. Mating type therefore tends to remain unchanged at cell division, autogamy (self-fertilization), and conjugation (unless there is considerable cytoplasmic exchange). The determination and differentiation of macronuclei may be viewed as based on stable hereditary states of genic repression (type O, low concentration of differentiator or none) or derepression (type E, high concentration of differentiator).

3.3. Trichocysts, the cell membrane, and nuclei

Trichocysts are carrot-shaped organelles (Fig. 12A) that arise and develop deep in the cell, enter the cortex, and become inserted into the plasma membrane (Fig. 12B) in the cross-ridges bounding the units of cortical structure. There are thus about 4,000 trichocysts per cell. Their function is obscure. Plattner et al. (1973) and Beisson et al. (1976) have described regular configurations of groups of granules, believed to be proteins, in the plasma membrane at the sites of trichocyst attachment (Fig. 12C). The central rosette of granules is present only when trichocysts are attached and the form of the outer aggregation of granules changes under the same condition. On stimulation of the cell by various mechanical, physical, or chemical agents (e.g., picric acid), the trichocyst is discharged from the cell in the form of a long thread (Fig. 12D) composed of a single protein (Steers et al., 1969). The capacity to discharge normally attached trichocysts varies with the number of granules in the central rosette (Beisson et al., 1976). Discharged trichocysts are replaced by others from the intracellular pool.

Recessive gene mutations at more than 15 loci in stock 51 of *P. tetraurelia* affect the development, structure, migration, insertion, or discharge of trichocysts (listed in Sonneborn, 1975b). Two classes of mutants are of special interest: those in which trichocysts insert into the plasma membrane but cannot discharge from intact cells, and those which fail to insert. The former class has structurally normal trichocysts that are capable of discharge in cell squashes, not in intact cells. The block is in the response system to external stimuli, perhaps to a decreased number of rosette granules in the cell membrane at the point where trichocysts are inserted.

We (Sonneborn and Schneller, in preparation) have discovered that the genetics of the first class of mutants is strikingly parallel to the genetics of mating types described above. Several recessive gene mutations exist, any one of which blocks the discharge response to external stimuli, like the recessive mutants that prevent development of mating type E. The dominant wild type (Fig. 13) can be either capable of trichocyst discharge (D) or incapable, that is, non-discharge (N), just as wild type can be either mating type O or E. As with the two alternative mating types, the two trichocyst alternatives (N vs. D) are based on macronuclear differentiation, determination occurring at the time of origin and development of new macronuclei; these nuclear differentiations are stable and hereditary, heredity being caryonidal; the two new macronuclei in the same cell tend to be determined alike; and mixed caryonides, like selfers in the mating type system, sometimes occur. In the N versus D alternative we have found a strong influence of the medium during the sensitive period: for example, depleted medium for 1 day at this stage yields virtually 100% determination as N among the sexual progeny of N parents, while fresh culture medium yields virtually 100% determination as D. (Whether the medium effect also applies to mating types has not yet been investigated.) The interest of the N versus D alternative, as of the mating types, is that cell surface character alternatives (different gamones in the case of mating types and possibly the number of intramembrane granules in the

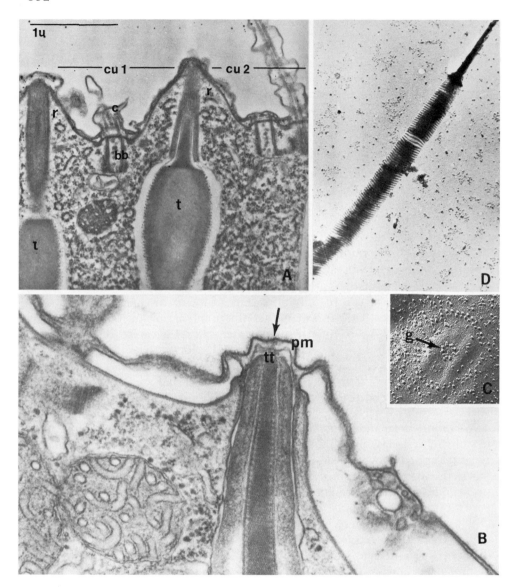

Fig. 12. *Paramecium tetraurelia:* electron micrographs of trichocysts and their relation to the plasma membrane. (A) Trichocyst (t) in transverse ridge (r) bounding contiguous cortical units (cu1, cu2) of same row; bb, basal body; c, cilium. (B) Detail showing connection (arrow) between tip of trichocyst (tt) and plasma membrane (pm). (C) Freeze-fracture of plasma membrane showing groups of granules (g) marking site where trichocyst tip is inserted (X 90,000). (D) Discharged trichocyst. (Parts A, B, and D courtesy of R. V. Dippell; Part C courtesy of B. Satir.)

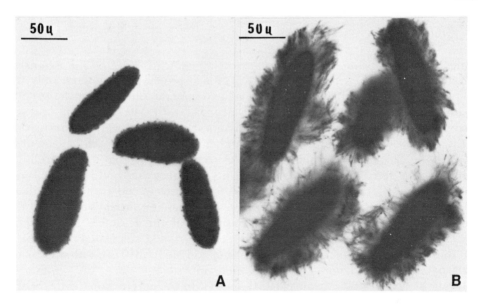

Fig. 13. *Paramecium tetraurelia,* wild-type stock d4-113 (a derivative of standard stock 51) in picric acid to show the two alternative hereditary phenotypes. (A) The non-discharge (N) phenotype; only cilia seen on surface. (B) The discharge (D) phenotype; thick mass of discharged trichocysts masks the cilia.

rosettes at the point of trichocyst insertion in the case of N vs. D) can be due to stable nuclear differentiations among cells of identical genotype, differentiations established at a limited developmental stage and thereafter irreversible through subsequent cell reproduction. Moreover, the discovery of the medium effect seems to open the door to identification of the determinative agent, how it operates on the nuclei, and the nature of the nuclear differentiation.

The other class of trichocyst mutants, those that fail to insert into the cell membrane, have been investigated by Ruiz et al. (1976). There is a remarkable correlation: macronuclei in homozygotes for these mutant genes fail to move to a definite site near the cell cortex in preparation for cell division and show other abnormalities in nuclear behavior, whereas other trichocyst mutants that insert but are blocked in discharge do not show these abnormalities in nuclear movement and localization. The authors present the intriguing hypothesis that failure of trichocysts to attach to the plasma membrane is the cause of the nuclear abnormalities. As indicated earlier, it is known that insertion of trichocysts alters the array of granules in the plasma membrane. The authors therefore suggest that these or other alterations in the plasma membrane (or cortex), due to trichocyst insertion, function also in the guidance of nuclear movements and localizations during cell division (and postzygotic developments). In general agreement with the existence of relations between the cell cortex and nuclei are the elegant studies of de Terra (1974, 1975) on the cortical control of nuclear behavior in another Ciliate, Stentor.

4. Epilogue

The preceding examples are only an introduction to some of the extensive studies on the cell surface of Ciliates. The Ciliate cell performs not only the "housekeeping" functions common to cells in general, but also functions comparable to those usually associated with neurons, with cells of the immune system, with gametes, with secretory and phagocytic cells, and with differentiating cells and nuclei in the development of higher organisms. The cell lineage of at least some Ciliates also exhibits life cycles comparable to that of multicellular animals with periods of immaturity, maturity, and senescence. All of these functions and changes have been and are being extensively and fruitfully studied. Among those not dealt with at all in this review, mention should be made of the mutational dissection of the electrophysiology of the excitable Paramecium plasma membrane (Kung, 1975; Kung et al., 1975) and correlated biochemical studies of the mutant membranes (Browning and Nelson, 1976), with their obvious parallels to and implications for neurophysiology. Also, the reviews of Hiwatashi (1969) and Miyake (1974) on cell-to-cell interactions provide the background for an area of cell surface studies that seem to be on the threshold of important advances.

From the ensemble of research mentioned in this chapter (and others not mentioned here at all), the reviewer senses that these initially independent areas of Ciliate research are moving toward convergence and, perhaps, synthesis. The cell membrane and adjacent cortex appear to be a common matrix in which the varied and separately studied phenomena have remarkable parallels that suggest close and systematically regulated interrelations. This matrix has, moreover, a high degree of genetic autonomy; it has visible as well as molecular polarity, asymmetry, and positional and temporal differentiation and specialization; and it has functional—and at least in some instances structural—connections with internal organelles such as trichocysts, food vacuoles, and nuclei, and with the "external" ciliary organelles. If a synthetic view of the properties and functions of the cell surface is to be achieved in any type of cell, the range already partially explored in Ciliates such as Paramecium and Tetrahymena suggests that these are the cells of choice for further study.

Acknowledgments

The work of the author and his associates on Paramecium and Oxytricha was supported by grant #GM 15410-09 from the U. S. Public Health Service. The author is greatly indebted to Prof. Ruth V. Dippell and Mr. Warren R. Jones for preparation of the figures and to Professors D. L. Nanney and J. Frankel for helpful comments.

References

Beisson, J. and Capdeville, Y. (1966). Sur la nature possible des étages de differentiation conduisant à l'autogamy ·hez *Paramecium aurelia*. C. R. Acad. Sci. Paris, Ser. D. 233, 1258–1261.

Beisson, J., Lefort-Tran, M., Pouphile, M., Rossignol, M. and Satir, B. (1976) Genetic analysis of membrane differentiation in Paramecium. Freeze-fracture study of the trichocyst cycle in wild type and mutant strains. Cell Biol. 69, 126–143.

Beisson, J. and Sonneborn, T.M. (1965) Cytoplasmic inheritance of the organization of the cell cortex in *Paramecium aurelia*. Proc. Nat. Acad. Sci., U.S.A., 53, 275–282.

Braun, V. and Miyake, A. (1975) Composition of blepharmone, a conjugation-inducing glycoprotein of the Ciliate Blepharisma. FEBS Lett. 53, 131–134.

Browning, J.L. and Nelson, D.L. (1976) Amphipathic amines affect membrane excitability in Paramecium: role for bilayer couple. Proc. Nat. Acad. Sci., U.S.A., 73, 452–456.

Butzel, H.M. Jr. (1974) Mating type determination and development in *Paramecium aurelia*. In: Paramecium: A Current Survey (van Wagtendonk, W.J., ed.), pp. 91–130. Elsevier, New York.

Byrne, B.C. (1973) Mutational analysis of mating type inheritance in syngen 4 of *Paramecium aurelia*. Genetics 74, 63–80.

Cohen, L. (1964) Diurnal intracellular differentiation in *Paramecium bursaria*. Exp. Cell Res. 36, 398–406.

Dippell, R.V. (1964) Perpetuation of cortical structure and pattern in *P. aurelia*. Excerpta M∩dica 77, 16–17.

Dippell, R.V. (1968) The development of basal bodies in Paramecium. Proc. Nat. Acad. Sci., U.S.A., 61, 461–468.

Finger, I. (1974) Surface antigens of *Paramecium aurelia*. In: Paramecium: A Current Survey (van Wagtendonk, W.J., ed.), pp. 131–164. Elsevier, New York.

Frankel, J. (1974) Positional information in unicellular organisms. J. Theor. Biol. 47, 439–481.

Frankel, J. and Williams, N.E. (1973) Cortical development in Tetrahymena. In: Biology of Tetrahymena (Elliott, A.M., ed.), pp. 375–409. Halsted Press, New York.

Grimes, G.W. (1973a) Differentiation during encystment and excystment in *Oxytricha fallax*. J. Protozool. 20, 92–104.

Grimes, G.W. (1973b) An analysis of the determinative difference between singlets and doublets of *Oxytricha fallax*. Genet. Res. Camb. 21, 57–66.

Grimes, G. W. (1976) Laser microbeam induction of incomplete doublets of *Oxytricha fallex*. Genet. Res. Camb. 27, 213–226.

Hammersmith, R.L. (1976) The redevelopment of doublets and monster cells of *Oxytricha fallax* after cystment. J. Cell Sci. 22, 563–573.

Hansma, H.G. (1975) The immobilization antigen of *Paramecium aurelia* is a single polypeptide chain. J. Protozool. 22, 257–259.

Hanson, E.D. (1955) Inheritance and regeneration of cytoplasmic damage in *Paramecium aurelia*. Proc. Nat. Acad. Sci., U.S.A. 41, 783–786.

Hanson, E.D. (1962) Morphogenesis and regeneration of oral structures in *Paramecium aurelia*. J. Exp. Zool. 150, 45–68.

Hanson, E.D. and Ungerleider, R.M. (1973) The formation of the feeding organelle in *Paramecium aurelia*. J. Exp. Zool. 185, 175–188.

Heckmann, K. and Frankel, J. (1968) Genic control of cortical pattern in Euplotes. J. Exp. Zool. 168, 11–38.

Hiwatashi, K. (1969) Paramecium. In: Fertilization (Metz, C.B. and Monroy, A., eds.), vol. 2: 255–293. Academic Press, New York.

Jones, W.R. (1976) Oral morphogenesis during asexual reproduction in *Paramecium tetraurelia*. Genet. Res. Camb. 27, 187–204.

Kitamura, A. and Hiwatashi, K. (1976) Mating-reactive membrane vesicles from cilia of *Paramecium caudatum*. J. Cell Biol. 69, 736–740.

Kung, C. (1975) Genetic dissection of the excitable membrane of Paramecium. Genetics 79, 423–431.

Kung, C., Chang, S.Y., Satow, Y., Van Houten, J. and Hansma, H. (1975) Genetic dissection of behavior in Paramecium. Science 188, 898–904.

Miyake, A. (1974) Cell interaction in conjugation of Ciliates. Current Topics Microbiol. Immunol. 64, 49–77.

Miyake, A. and Bleyman, L.K. (1976) Gamones and mating types in the genus Blepharisma and their possible taxonomic application. Genet. Res. Camb. 27, 267–275.

Nanney, D.L. (1966a) Cortical integration in Tetrahymena: an exercise in cytogeometry. J. Exp. Zool. 161, 307–318.

Nanney, D.L. (1966b) Corticotype transmission in Tetrahymena. Genetics 54, 955–968.

Nanney, D.L. (1967) Cortical slippage in Tetrahymena. J. Exp. Zool. 166, 163–169.

Nanney, D.L. (1968) Ciliate genetics: patterns and programs of gene action. Ann. Rev. Genet. 2, 121–140.

Plattner, H., Miller, F. and Bachmann, L. (1973) Membrane specializations in the form of regular membrane to membrane attachment sites in Paramecium. A correlated freeze-etching and ultrathin-sectioning analysis. J. Cell Sci. 13, 687–719.

Reisner, A.H., Rowe, J. and Macindoe, H.M. (1969) The largest known monomeric globular proteins. Biochim. Biophys. Acta 188, 196–206.

Ruiz, F., Adoutte, A., Rossignol, M. and Beisson, J. (1976) Genetic analysis of morphogenetic processes in Paramecium. I. A mutation affecting trichocyst formation and nuclear division. Genet. Res. Camb. 27, 109–122.

Sommerville, J. (1970) Serotype expression in Paramecium. Adv. Microbial. Physiol. 4, 131–178.

Sonneborn, T.M. (1932) Experimental production of chains and its genetic consequences in the Ciliate Protozoan *Colpidium campylum*. Biol. Bull. 63, 187–211.

Sonneborn, T.M. (1948) The determination of hereditary antigenic differences in genically identical Paramecium cells. Proc. Nat. Acad. Sci., U.S.A., 34, 413–418.

Sonneborn, T.M. (1963) Does preformed cell structure play an essential role in cell heredity? In: The Nature of Biological Diversity (Allen, J.M., ed.), pp. 165–221. McGraw-Hill, New York.

Sonneborn, T.M. (1970) Gene action in development. Proc. Roy. Soc. London B176, 347–366.

Sonneborn, T.M. (1975a) Positional information and nearest neighbor interactions in relation to spatial patterns in Ciliates. Ann. Biol. 14, 565–584.

Sonneborn, T.M. (1975b) *Paramecium aurelia*. In: Handbook of Genetics (King, R.C., ed.), vol. 2, pp. 469–594. Plenum Press, New York.

Sonneborn, T.M. (1975c) Tetrahymena. In: Handbook of Genetics (King, R.C., ed.), vol. 2, pp. 433–467. Plenum Press, New York.

Steers, E., Jr., Beisson, J. and Marchesi, V.T. (1969) A structural protein extracted from the trichocyst of *Paramecium aurelia*. Exp. Cell Res. 57, 392–396.

Tamm, S.L., Sonneborn, T.M. and Dippell, R.V. (1975) The role of cortical orientation in the control of the direction of ciliary beat in Paramecium. J. Cell Biol. 64, 98–112.

Tartar, V. (1956) Grafting experiments concerning primordium formation in *Stentor coeruleus*. J. Exp. Zool., 131, 75–122.

Tartar, V. (1961) The Biology of Stentor. Pergamon Press, New York.

Terra, N. de (1974) Cortical control of cell division. Science 184, 530–537.

Terra, N. de (1975) Evidence for cell surface control of macronuclear DNA synthesis in Stentor. Nature 258, 300–303.

Uhlig, G. (1960) Entwicklungsphysiologische Untersuchungen zur Morphogenese von *Stentor coeruleus*. Arch. Prostistenk. 105, 1–109.

Subject Index

Absorption, optical (*see* Circular dichroism)
Acetylcholine (*see* Nerve Cell; Synapse; Vesicles, synaptic)
Acylation, membrane (*see* Lipids, membrane)
Adherence (*see* Wall, bacterial)
Adhesion (*see* Peptidoglycan, bacterial)
Adoral zone (*see* Ciliates)
Adrenalin (*see* Hormones; Secretion; Synapse)
Agglutination, cell (*see* Lectins)
Agranular reticulum (*see* Nerve cell; Transport, axonal)
Algae (*see* Wall, plant)
Alkaloids (*see* Colchicine)
Alpha virus (*see* Membrane, virus; Virus, animal)
Anesthetics (*see* Drug effects; Fluidity, membrane)
Antibiotics
— actinomycin D, 148, 617
— ampicillin, 544, 545, 640, 642–3, 645–9
— cephalexin, 644, 648, 678
— cephalothin, 640, 644
— cephchloridine, 641, 644, 647
— chloramphenicol, 617, 640, 684
— division, synchronization, 664–5
— effects, electron microscopy, 642–3
— β-lactam, 546, 548, 549, 639, 640, 644, 647
— novobiocin, 552
— oxicillin, 641, 644
— penicillin, 535, 536, 539–40, 542, 543, 546, 550, 552, 559, 600, 611, 629, 631, 639–41, 644, 646, 649, 678
— penicillin-binding proteins, 546, 647
— peptidoglycan, bacterial (*see* Peptidoglycan, bacterial)
— phagocytosis (*see* Phagocytosis)
— puromycin, 94, 96, 100, 148, 563, 564
— rifampicin, 684
— ristocetin, 543
— shape, bacterial, 639–47, 649, 678
— vanomycin, 543, 629
— wall, bacterial, 535–6, 539–49, 552, 559, 600, 611, 639–48

Antibody (*see* Antigens, cell surface; Immunoglobulin)
Antigens, cell surface
— ABH blood group, 105, 422, 429–31, 435
— bacterial, 505, 510–4, 516, 529, 566, 569–71
— basic protein, myelin, 326–7
— ciliates, 847–8
— endocytosis, antibody-induced, 149
— envelope, bacterial (*see* Envelope, bacterial)
— Forssman, 430–1, 494, 550–1, 653
— glycolipid, 325–7, 422
— glycoprotein, 423
— H–2, 149, 154, 429–31
— H blood group, 105, 429
— HL–A, 145, 428–31
— immunoglobulin, surface (*see* Immunoglobulins)
— lipopolysaccharide, 510–6
— lipoprotein, 505
— MN blood group, 58–90, 422
— phospholipids, 326
— radioimmune assay, 326, 327
— surface-Ig, 149
— teichoic acids, 487, 492, 494
— TL, 145
— virus antigen, 429
Assembly, membrane
— autolysins, 630–2, 649–53
— bacteria, 532–3
— bacteriophage, 816–21
— biosynthesis (*see* Biosynthesis)
— division, cell, 403–13, 415–7, 421, 442–52
— lipopolysaccharides, 570–2
— membrane, plasma, 62–3, 149–51, 166, 175–6, 266, 281, 332, 440–51, 463, 742–86
— peptidoglycan, 558–62, 600, 653
— teichoic acids, 556–61, 600
— virus, 752, 769–86
— wall, bacterial, 167, 532–52, 556–61, 570–2, 600, 622–8, 630–2, 639–46, 687–99
— wall, plant, 717–9, 726–31
Asymmetry (*see* Labeling, surface; Lipids,

868

Transport, axonal; Virus, cell membrane modification)

Labeling, surface
— acetic anhydride, 151
— asymmetry, membrane, 451
— carbodiimide, 327
— cell cycle, 425–7, 451
— DIDS, 328
— fluorescent probes, 310, 325
— formylmethionyl sulfate methyl phosphate, 751
— galactose oxidase, 329, 425–27, 751–3
— lactoperoxidase-catalyzed iodination, 138, 146, 152, 155, 156, 166, 329, 451, 751
— metabolic labeling, 98, 102, 126–219, 140–6, 150–1
— phenylglyoxal, 425–7
— sulfanilic acid, 814
— TNBS, 328
Lactoperoxidase iodination (see Labeling, surface)
Lateral compressibility (see Phase transition, membrane)
Lateral mobility (see Fixation)
Lateral phase separation (see Lipids, membrane; Phospholipids, membrane)
Lectins
— agglutination, 423–4, 432–4
— binding studies, 423–4, 432–3
— capping, 149
— cell cyde effects, 416, 423–4, 432–4
— concanavalin A (Con A), 26, 117, 129, 149, 297, 374–5, 416, 423–34, 432–4
— endocytosis, induced, 149, 166, 224
— fusion, effects on, 166, 374–5
— lectin-induced rearrangements, 423–4, 432–4
— mitosis, 416, 423
— phytohemagglutinin (PHA), 89
— receptors, membrane (see Receptor, membrane)
— receptors, myelin (see Myelin)
— ricin (Ricinus communis), 129, 224
— transmembrane perturbations, 275
— tumor cell (see Tumor cell)
— wall, bacterial cell, 492–3
— wheat germ agglutinin (WGA), 90, 149, 423, 432–4
Lipase (see Enzyme effects)
Lipid A (see Lipopolysaccharides, bacterial; Liposome)
Lipid intermediates (see Glycoproteins,

membrane; Glycolipids, membrane; Glycosyltransferase; Peptidoglycan, bacterial; Teichoic acids; Wall, bacterial)
Lipid modification (see Flow, membrane)
Lipids, membrane
— acylation, 370–2
— asymmetric distribution, 337, 385
— bacteriophage, 804, 811–4
— biosynthesis, 6–11, 62, 106–20, 166, 370–2, 422
— cardiolipin, 385, 487, 557, 754, 822
— carrier proteins, 7, 372
— cell cycle, 422, 426–7, 435–42
— cerebroside, 253, 255, 287, 302–4, 309–10
— cholesterol, 7, 41, 61, 173, 253–4, 387, 290–1, 752, 754
— composition, 249–54, 286–96, 299–302, 754
— diffusion, lateral, 252–3, 255, 298–9
— dolicol intermediates, 27, 106–14, 115–22, 729
— enzymes, 6–11, 106–20, 176
— exchange, 11, 209, 372
— fluidity, membrane (see Fluidity, membrane)
— freeze-fracture electron microscopy, 278
— glycolipids (see Glycolipids, membrane)
— head group orientation, 255–8
— hydration, 254–5
— isoprenoids, 26, 27, 106–14, 526, 529–31
— lateral phase separation, 252–4
— lipid A (see Lipopolysaccharides)
— lipid-dense droplets, 422
— lipopolysaccharide (see Lipopolysaccharides)
— liposome (see Liposome)
— local synthesis, 166, 370–2
— lysosome, 61
— membrane, plasma (see Membrane, plasma)
— myelin, 250–5
— peroxidation, 377
— phagocytosis, changes during, 366, 369–72
— phase transition properties, 251–4, 297–9
— phosphoglycerides, 288–9
— phospholipids (see Phospholipids, membrane)
— plasmalogens, 287, 301–2
— polyprenyl intermediates, 729–30
— properties, 250–3, 255–8, 268
— protein interactions, 307–13
— sphingolipids, 41, 42, 61, 173, 250, 257–8, 287, 290, 300, 302–4, 307–10
— sphingomyelin, 41, 42, 61, 173, 287, 771
— sterols, 11, 287
— sulfatides, 287, 303–4
— undecaprenol, 27
— unsaturation, 251

872